Library of Congress Cataloging-in-Publication Data

Singh, V. P. (Vijay P.)
Hydrologic systems.

Bibliography: v. 1, p.
Includes index.
Contents: v. 1. Rainfall-runoff modeling.
1. Hydrology—Mathematical models. I. Title.
GB656.2.M33S56 1988 551.48′072′4 87-13937
ISBN 0-13-448051-1 (v. 1)

Editorial/production supervision
and interior design: *Carolyn Fellows*
Cover design: *Edsal Enterprises*
Manufacturing buyers: *Rhett Conklin* and *Gordon Osbourne*

Printed in the United States of America
10 9 8 7 6 5 4 3 2 1

ISBN 0-13-448051-1 025

Prentice-Hall International (UK) Limited, *London*
Prentice-Hall of Australia Pty. Limited, *Sydney*
Prentice-Hall Canada Inc., *Toronto*
Prentice-Hall Hispanoamericana, S.A., *Mexico*
Prentice-Hall of India Private Limited, *New Delhi*
Prentice-Hall of Japan, Inc., *Tokyo*
Simon & Schuster Asia Pte. Ltd., *Singapore*
Editora Prentice-Hall do Brasil, Ltda., *Rio de Janeiro*

To my parents

Contents

Preface

Since the development of unit hydrograph theory, applications of systems and control-theoretic concepts to analysis and synthesis of hydrologic systems have mushroomed phenomenally. The underlying motivation has partly been the relative simplicity and more immediate need to obtain practical solutions to hydrologic problems. Fragmented accounts of some of the approaches based on these concepts have been presented in many textbooks on hydrology. A comprehensive account did not, however, appear until the authoritative work of J. C. I. Dooge described in Technical Bulletin No. 1468, "Linear Theory of Hydrologic Systems," published in 1973 by the U.S. Department of Agriculture. Indeed, no one can write on the subject matter of this book without becoming indebted to Dooge for this treatise. The subject of hydrologic systems has expanded greatly in recent years both in variety of techniques and diversity of applications. The motivation for writing this book stemmed from the desire to provide, under one cover, a comprehensive account of some aspects of this rapidly expanding field. It should, however, be emphasized that this book is not intended to present an encyclopedic treatment of hydrologic systems but to present pertinent models as well as what is perceived to be required for developing new and more comprehensive models for analysis and synthesis of hydrologic systems.

This book is an outgrowth of lecture notes prepared for courses on hydrologic modeling taught to geoscience students at New Mexico Institute of Mining and Technology and to civil engineering students at George Washington University, Mississippi State University, and Louisiana State University. The criticisms and discussions offered by the students to some extent influenced the style of presentation in the book. Considerable effort has been made to discuss concepts and illustrate them with examples, even at the expense of increasing size and having repetition. This style may appear somewhat less rigorous to a hydrologist, but it is deemed desirable for the person not knowing the subject. Experience indicates that the contents of this book may occupy two one-semester courses either at an advanced undergraduate level or at the beginning graduate level. Depending upon the particular requirements of a course, one can choose material from different chapters of the book. It is hoped that the material covered would also be of some interest to those engaged in practice of agricultural engineering, civil engineering, earth sciences, environmental sciences, forest and range sciences, hydrology, or water resource technology. Extensive literature cited at the end of the book may be of value to those contemplating to pursue further the topics covered.

The subject matter of this book is divided into three parts. The first part, encompassing seven chapters, presents preliminaries perceived to be background information essential for development of hydrologic models. This part constitutes the foundation upon which rests the development of subsequent parts. The philosophy of systems approach is introduced in Chapter 1, and definitions and concepts are presented in Chapter 2. Types of hydrologic problems and the models to be developed to address them are reported in Chapter 3. Elementary functions and their hydrologic responses are discussed in Chapter 4, and applications of these functions to development of superposition integrals are shown in Chapter 5. Presented in Chapter 6 are the methods of moments and cumulants commonly used for estimating parameters of hydrologic models derived by applying the elementary functions or otherwise. Chapter 7 discusses seven additional methods of parameter estimation, some of which are of recent origin.

This book concentrates on rainfall-runoff systems. A system is composed of three parts: (1) input, (2) system function, and (3) output. Mean areal rainfall and geomorphologic characteristics usually constitute input data to a rainfall-runoff system and are discussed in the companion Volume II. The system function is derived by considering various aspects of rainfall-runoff relationship that constitute the subject matter of the second and third parts. The watershed runoff is introduced in Chapter 8 and its time characteristics, in Chapter 9. Different types of mathematical models for an effective rainfall–direct runoff relationship have been presented in hydrologic literature. Both continuous-time and discrete-time linear time-invariant models are discussed in Chapters 10 and 11, respectively. Differential equations and the characteristics of rainfall-runoff systems that they exhibit are treated in Chapter 12. Also shown there is the relationship with conceptual models. The continuous-time and discrete-time linear time-invariant forms of these models are discussed in Chapters 13 and 14 respectively. Linear time-varying models are introduced in Chapter 15. The discussion is extended to nonlinear conceptual models in Chapter 16 and to nonlinear empirical models in Chapter 17. Both linear and nonlinear models applicable to ungaged watersheds are presented in Chapter 18.

The concluding third part deals with flood routing. The methods of reservoir flood routing are discussed in Chapter 19. Linear and nonlinear as well as empirical and conceptual models are included. Chapter 20 presents the methods of channel flood routing. Again, linear and nonlinear, empirical and conceptual, and time-invariant and time-variant models are considered.

Vijay P. Singh
Baton Rouge, Louisiana

Acknowledgments

My interest in the subject matter of this book has been inspired by friends and colleagues in education, research, and practice of engineering and water resource technology as well as those whom I have seldom met but have come to know through their research contributions. It is a pleasure to acknowledge the assistance of the many scientists, engineers, and hydrologists who have so generously allowed their work to be included here. I have tried to make my acknowledgments as specific as possible in the body of the text. Any omission on my part has been entirely inadvertent, and I offer my apologies in advance. I would benefit and be obliged if readers would transmit to me any errors, omissions, or criticisms.

Several sections of this book are based on the work supported in part through the years by the National Science Foundation and the Office of Water Research and Technology of the U.S. Department of Interior. This support is gratefully acknowledged.

H. Aminian, K. Arora, L. M. Chacon, P. K. Chowdhury, P. F. Krstanovic, D. Jain, S. K. Jain, C. A. Quiroga, S. Raudales, and K. Singh assisted in the solution of some example problems as well as proofreading of some chapters. Mr. F. X. Yu, Augustfirst Agricultural College, The People's Republic of China, reviewed the entire manuscript and made many suggestions. Among those reviewing parts of the manuscript were Dr. N. R. Bhaskar, Department of Civil Engineering, University of Louisville, and K. Arora, S. K. Jain, P. F. Krstanovic, C. A. Quiroga, and Dr. J. F. Cruise, all of the Department of Civil Engineering, Louisiana State University. I, of course, accept the full responsibility for any omissions, shortcomings, or mistakes that may remain. S. Sartwell did the typing with skill and care. N. Duffy and P. K. Chowdhury did part of the drafting. Dr. R. K. Seals, Chairman of the Department of Civil Engineering, Louisiana State University, provided encouragement and support whenever needed. My wife, Anita, and children, Vinay and Arti, allowed me to spend many hours away from them; my parents allowed me to study and work in the United States. I owe my gratitude to them all for their sacrifices, patience, endurance, love, and understanding. Without their support this book would not have been completed.

The search for truth is in one way hard and in another way easy. For it is evident that no one can master it fully nor miss it wholly. But each adds a little to our knowledge of Nature, and from all the facts assembled there arises a certain grandeur.

Aristotle

PART I
PRELIMINARIES

CHAPTER ONE
Introduction Hydrologic Investigation

Many approaches to the study of hydrologic problems have evolved over the years as evident from proliferating hydrologic books and periodicals (see Appendix A). These approaches are so diverse and hybrid that it is not always possible to distinguish one type of approach from another (National Research Council, 1982). Nonetheless, these can be divided broadly into two groups (Amorocho and Hart, 1964): (1) physical science approach and (2) systems approach. The former is also referred to as a basic, pure, causal, dynamic, or theoretical approach; and the latter, as an operational, applied, empirical, black-box, or parametric approach. These are, in fact, most appropriately called basic research and applied research. Aggregations of studies involving the former can be called physical hydrology; those involving the latter, systems hydrology. These groups reflect the philosophy of research workers in hydrology as elucidated by Amorocho and Hart (1964):

> The first group espouses the pursuit of scientific research into the basic operation of each component of the hydrologic cycle in order to gain a full understanding of their mechanisms and interactions. Although the immediate motivation of an individual researcher may not transcend the narrow confines of a set of special phenomena, it is implicit that a full synthesis of the hydrologic cycle may eventually be sought. This concept of a full synthesis is held to be the only rational approach to hydrology. The second group is motivated by the need to establish workable relationships between measured parameters in the hydrologic cycle to be used in solving pressing practical technological problems. These people generally hold that the vast complexity of the systems involved in these studies, and the inadequacy of the knowledge now available and the knowledge likely to exist in the foreseeable future, make the possibility of a full synthesis so remote in certain cases that it must be discarded for practical purposes. Under these premises, a logical approach would consist of measuring those observed variables in the hydrologic cycle, which appear pertinent to the problem, and then attempting to establishing explicit algebraic relationships between them. It is hoped that these relationships hold true within the range of conditions normally encountered in practice.

1.1 PHYSICAL APPROACH

In the physical approach the primary motivation is the study of physical phenomena and their understanding; the eventual practical application of this knowledge for engineering and other purposes is recognized but not explicitly sought. We strive for explanations of how things happen and why they behave as they do. A physical approach to determine output from a given system would normally require specification of (a) system input, (b) system structure (geometry), and (c) physical laws, together with initial and boundary conditions. To illustrate, briefly consider some examples.

EXAMPLE 1.1 Flood Routing

Routing a flow through an open channel extending from $x = x_0$ to $x = x_1$, as shown in Fig. 1.1, would require specification of (1) inflow hydrograph $Q(x_0, t)$ to be routed, (2) channel geometry, (3) governing laws, including the equations of continuity and momentum as well as knowledge of the friction factor, and (4) initial and boundary conditions. With that knowledge, it is to be hoped that the problem

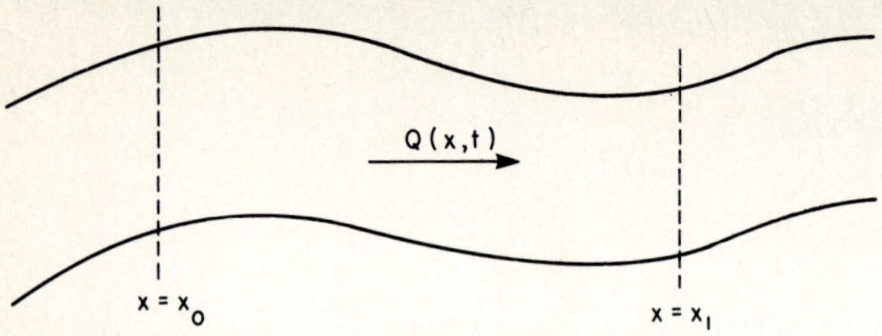

FIGURE 1.1(a) Channel reach of length x_1 to x_0.

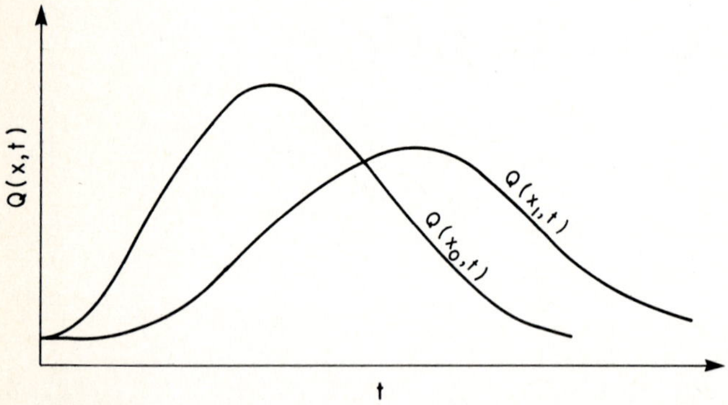

FIGURE 1.1(b) Inflow hydrograph $Q(x_0, t)$ at $x = x_0$ and routed outflow hydrograph $Q(x_1, t)$ at $x = x_1$ for the channel reach.

of routing can be solved to yield the outflow hydrograph $Q(x_1, t)$ using standard mathematical tools. The work by Fread (1982) is an example.

EXAMPLE 1.2 Rainfall-Runoff Relation

The problem is to determine direct runoff due to a rainfall event on a given watershed (see Fig. 1.2). To establish the relationship between rainfall and runoff would require specification of (1) space-time distribution of rainfall, (2) space-time distribution of infiltration, (3) watershed physiographic characteristics, (4) physical laws, including equations of continuity and momentum as well as friction law, (5) antecedent soil moisture conditions, and (6) initial and boundary conditions. Here evaporation, interception, and other abstractions are considered vanishingly small. The problem of determining direct runoff can then be attempted. The work by Woolhiser (1982b) can be cited as an example.

EXAMPLE 1.3 Infiltration

If we consider the problem of determining infiltration due to rainfall, then we need to specify (1) antecedent soil moisture condition, (2) rainfall characteristics, (3) physical laws governing flow through porous media including continuity equation and Darcy's law, (4) characteristics of porous media, and (5) initial and boundary conditions. With these specifications we can hope to estimate infiltration. The work by Morel-Seytoux (1973) is an example.

EXAMPLE 1.4 Watershed Sediment Yield

If we want to determine sediment yield (wash load) due to rainfall from agricultural watersheds, then we need to specify (1) rainfall characteristics, (2) infiltration characteristics, (3) physiographic characteristics, (4) antecedent soil moisture conditions, (5) physical laws, including laws of conservation of mass and momentum of water,

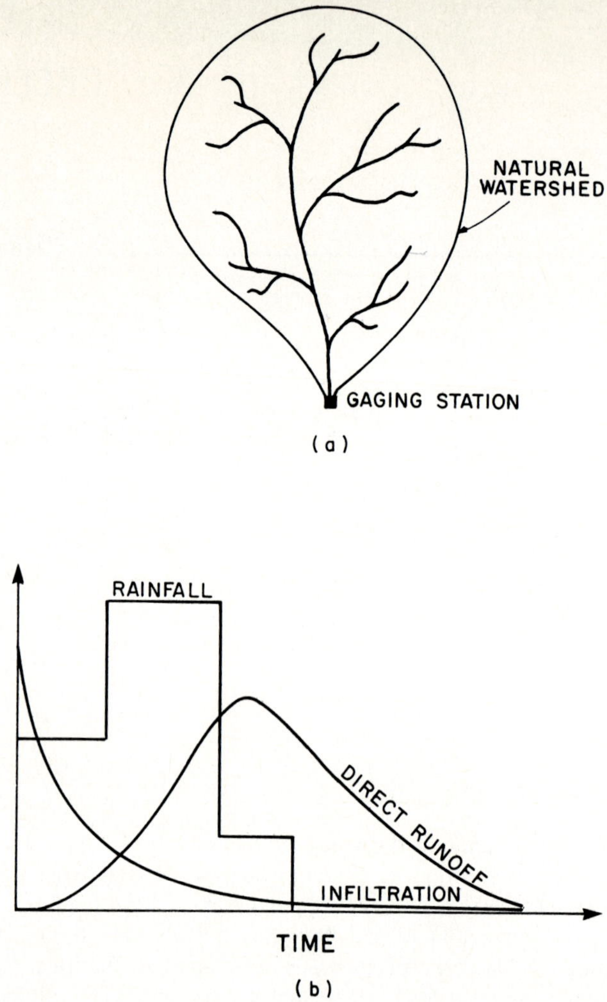

FIGURE 1.2. Rainfall-runoff relationship for a natural watershed.

law of conservation of mass for sediment, and a relationship between sediment concentration and runoff, and (6) initial and boundary conditions. The work by Singh and Regl (1983) can be cited as an example.

EXAMPLE 1.5 Evaporation from a Water Body

If the problem is one of determining evaporation from a water body, then we need to know (1) hydrometeorological conditions, (2) governing physical laws, including laws of conservation of energy and mass, and (3) initial and boundary conditions. The work by Brutsaert (1982) is an example.

EXAMPLE 1.6 Flood-Frequency Analysis

To perform a frequency analysis requires knowledge of (a) flood-precipitation relationship based on atmospheric dynamics, and (b) probability distribution of precipitation based on physical laws. Then the probability distribution of floods can be derived. The work by Eagleson (1972) typifies this mode of investigation.

Thus a model of a hydrological process based on physical approach must have, according to Klemes (1972), the following characteristics:

1. A hydrological model must be independent of any intended application. In other words, it must be possi-

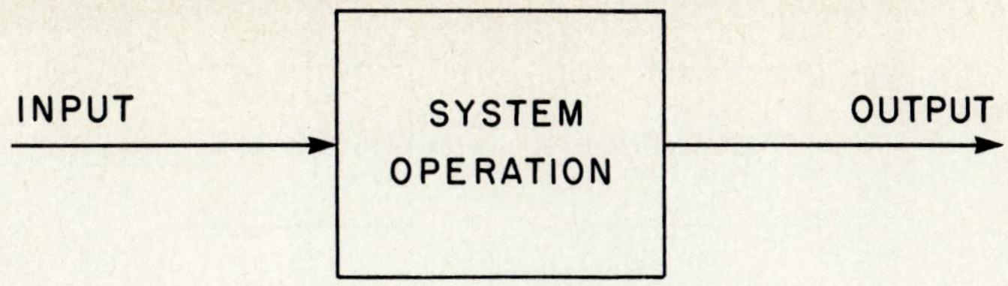

FIGURE 1.3. Systems representation for input-output relationship.

ble to construct a plausible model of a hydrological process without knowing in advance for what it will eventually be used. In nature a given process behaves as it does regardless of any eventual use. If a model is to describe the process, it must be based on information derived from the process itself, not from its use. *Use* is a sociological category describing the relation between the process and people but not the process itself. By incorporating it into the model, we are in fact introducing an ideology into it.

2. A hydrological model must incorporate—or at least be compatible with—the known physical structure and mechanism of the whole modeled process. If the model is aimed at and designed for an explanation or description of only one particular aspect of the process, it is very likely to end up on a dead end. As an example, it is easy to show that by a purely pragmatic approach based on operational criteria, one can arrive at such diverse models for various aspects of the same process that it may be impossible to enclose them in a general model. It is quite conceivable that in the same watershed, a precipitation series is well fitted by a white noise, the corresponding series, by a fractional noise, and the transforming effect of the watershed, by a linear reservoir. Considered separately, the three models can represent excellent tools for the intended purposes. The difficulty is that they cannot be combined into one model of a rainfall-runoff relationship, since a linear reservoir fed with white noise can never yield fractional noise as output. This indicates that the recognition of the need for generalization in applied hydrological modeling is incompatible with a purely pragmatic approach and requires inclusion of pure-science criteria into the model development.
3. All characteristics of the observed hydrological process must be within the confidence limits derived for them from the model. Obviously the more parameters fall "out of range," the less credibility the model commands.

These criteria may seem too academic to be relevant in the context of applied hydrological modeling. This is not so. They should indeed be implied in operational decisions and criteria governing model operation.

1.2 SYSTEMS APPROACH

Here we investigate a system for the explicit purpose of establishing an input-output relationship that can be used for reconstructing past events or prediction of future events. In this approach we are concerned with the system operation, not the nature of the system itself (its components, their connection with one another, and so on) or the physical laws governing its operation. Thus the intent is to circumvent the difficulties arising from the complexity of the input, the complexity of the system structure, and the complexity of the governing physics—that is, the complexity of the physical approach. The physical laws and the nature of the system are combined into a single concept of system operation. It is this concept that constitutes a so-called black-box, as shown in Fig. 1.3. The system operation is the link between input and output. If the nature of the system or the physical laws are changed, the system operation also changes. Thus the systems approach is an overall one and is not concerned with details that may or may not be important and that, in any event, may not be known. The philosophy of the systems approach can best be illustrated by reconsidering Examples 1.1 through 1.6.

EXAMPLE 1.1a Flood Routing

In this approach the entire channel reach is considered as a unit. The inflow hydrograph is considered as the input into the system and the outflow hydrograph, the output. In place of the St. Venant equations, we use a spatially lumped form of continuity equation and an assumed relation between storage and discharge. These two equations are combined, and the resulting equation is solved for routing the inflow hydrograph to produce the outflow hydrograph. The Muskingum method is a classic example (Singh and McCann, 1980b). We are not interested in knowing the discharge hydrograph at every point of the channel. The parameters in the storage-discharge relation are determined using records of previously observed hydrographs. We are obviously not concerned with the channel geometry and its spatial variation. Nor are we concerned with how the flow regime changes in the space of the channel and in time. We do not require knowledge of the friction factor and its variability. Instead we assume that the complex channel geometry and the complex hydraulics of flow are lumped together and described for that channel by the storage-discharge relation and the simplified continuity equation. Thus the essential feature of systems approach centers on the three elements: (1) input, (2) system function, and (3) output.

EXAMPLE 1.2a Rainfall-Runoff Relation

To approach the rainfall-runoff problem from a systems viewpoint, the watershed is considered as a system. It is then represented by a configuration of storage elements, often referred to as *reservoirs* or *channels*. These may be linear or nonlinear (see Fig. 1.4). Rainfall and infiltration are lumped together, and effective rainfall is computed; this forms the input into the system. In place of the St. Venant equations, we use a spatially lumped continuity equation and a storage-discharge relation. These two equations are combined, and the resulting equation is solved for each reservoir and channel. The resulting output is the hydrograph of direct runoff. The parameters in this approach are determined using historical records. The models of Nash (1958) and Dooge (1959) are typical examples.

In this representation we are not interested in the knowledge

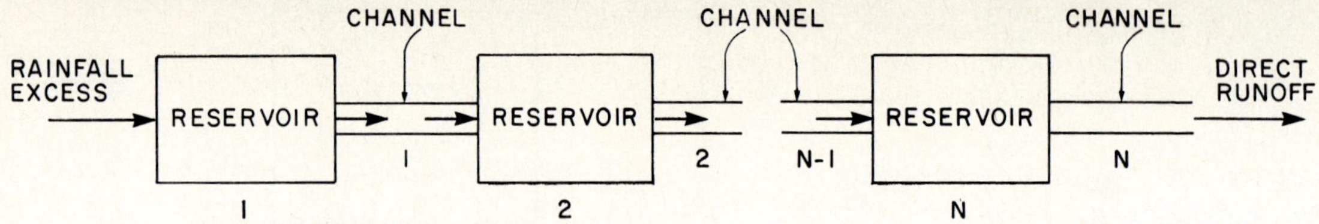

FIGURE 1.4. A systems approach for determining direct runoff from rainfall.

of the runoff hydrograph at every point in the watershed. We are obviously not concerned with geometric and topographic details and their spatial variability. Nor do we require knowledge of the friction factor and its variability. Although infiltration and runoff occur simultaneously due to a given rainfall episode, we account for infiltration through effective rainfall. We do not worry about hydraulics of direct runoff and its space-time variability. Thus our emphasis is more on the solution of the problem and less on the method of arriving at it.

EXAMPLE 1.3a Infiltration

The problem of rainfall infiltration is substantially simplified by using a systems approach. The entire soil column is considered as a system. The rate of water infiltrating into the soil is assumed to follow a simple function, exponential or power. The parameters appearing in the function are determined using observed data. The Horton equation (Horton, 1940) and the Kostyakov equation (Kostyakov, 1932) are two of the examples. It is clear that in this approach we do not consider capillary, viscous, or gravitational forces affecting entry of water into the soil. We do not explicitly consider the effect of air movement on infiltration. We do not explicitly worry about space-time variability of hydraulic conductivity or, for that matter, other physical properties of the porous media. We do not care if Darcy's law is satisfied or not. We do not know if the parameters have physical significance. What we care about, however, is a simple method that can determine infiltration with reasonable accuracy.

EXAMPLE 1.4a Watershed Sediment Yield

A systems representation of watershed sediment yield due to rainfall consists of considering the watershed as a system. The system operation involves construction of (1) a rainfall-runoff relationship and (2) a runoff-sediment relationship. A universal soil-loss equation (Wischmeir and Smith, 1960) and a unit sediment graph (Rendon-Herrero, 1978; Williams, 1978; Singh, Baniukiewicz, and Chen, 1982; Singh and Chen, 1982) are two of the examples. Obviously, as noted in the previous examples, a systems approach to estimation of sediment yield is a simplified one. In this approach we are not concerned with the hydraulics of sediment movement or the dynamic forces that govern sediment discharge. Watershed physiographic characteristics are not explicitly accounted for. The rainfall-runoff relationship is not physically based and neither is the runoff-sediment one. The parameters appearing in this approach are estimated using historical records.

EXAMPLE 1.5a Evaporation from a Water Body

In a systems representation the evaporation is represented as a function of the factors that principally affect it. Temperature is considered as one of the principal factors. Equations proposed by Thornthwaite (1931), Blaney and Morin (1942), and Penman (1948) are some of the examples. The parameters in these equations are estimated using observed data. In another systems representation, evaporation may be estimated from water budget or energy budget. Whatever the systems approach, the physical laws of turbulence and diffusion are not explicitly considered. The hydrometeorological factors affecting evaporation are considered empirically.

EXAMPLE 1.6a Flood-Frequency Analysis

The systems approach to flood-frequency analysis is based on historical records of floods and statistical analysis of these records. This analysis tries to fit an appropriate form of probability distribution to the historical record. The best-fit distribution then serves as a model for frequency analysis. The Gumbel distribution (Gumbel, 1941), and lognormal distribution (Hazen, 1930) are two of the examples. It is thus seen that this approach is strictly based on observed data. There is no consideration of the dynamics of flood-precipitation relationship. We do not explicitly consider the effect of changing watershed characteristics on the probability distribution. We do not consider the distribution of precipitation at all.

We may, however, mistakenly believe that the nature of the system or the physical laws are completely ignored in the systems approach. This obviously is not the case. Suppose we do not have any record of inflow and outflow hydrographs for the channel under consideration in Example 1.1, and we want to synthesize the system. Then the parameters are determined by correlation with channel characteristics. Thus the effect of structure on system operation is taken into account. The hydraulics of open-channel flow is considered through the storage-discharge relation. If this were assumed to be linear and time-invariant, the entire operation of routing would be reduced to the derivation of an impulse response for the channel. This would imply that the system is linear and time-invariant. Thus, the principle of superposition would be applicable.

1.3 JUSTIFICATION FOR A SYSTEMS APPROACH

The need for a systems approach arises in many instances, primarily from the complexities inherent in the physical approach and a belief that the system behavior may be approximated by major physical factors, thus permitting the neglect of minor features and exact spatial distribution of physical parameters. To illustrate, let us consider the routing problem mentioned previously. The problem of routing in a natural situation is not as simple as it might appear to be. The difficulty arises from complexities in specification of inflow hydrograph, channel geometry, dynamics of flow, and initial and boundary conditions. An explicit solution for unsteady, nonuniform flow in any realistic situation is not tractable. Even a numerical solution is not easy in such a case. Whether the method of characteristics, an explicit finite difference scheme, an implicit finite difference scheme, the finite element method, or the boundary element method is used, difficulties of one kind or another arise in the numerical solution of the problem. Such difficulties are typical of a physical approach.

Frequency analysis is another illustrative example. Frequency analysis involves (1) discovery of the form of the distribution of the quantity with which we are dealing, and (2) evaluation of the parameters of that distribution. It is very difficult to perform frequency analysis purely through the physical laws. Consider flood as the quantity of interest. Floods are caused by precipitation and snowmelt. Let us assume that the relationship between floods and precipitation is known. Then the flood frequency can be expressed only through the frequency of precipitation. If this relationship is not simple, it may be very difficult to relate frequencies. To be specific, if the mean daily discharge Q from a river basin is related to the mean daily precipitation P as

$$Q = f(P) \tag{1}$$

where f is simple and represents a one-to-one correspondence, then the frequency of Q is easily expressible in terms of the frequency of P. If Q is also related to duration D and maximum intensity I of precipitation by

$$Q = f(P, D, I) \tag{2}$$

then the frequency of Q is no longer easy to determine.

On the other hand, we observe that major floods may occur under similar climatic conditions on neighboring watersheds. In such situations one watershed can act as a model of the other and hence their differences are partly irrelevant. Thus in many situations the immediate need to solve a practical problem requires the use of a systems approach.

Furthermore, the use of the systems approach stems from a particular mode of thinking toward hydrology. Perhaps Dooge (1972) summed it up best:

> It is important to remember that in the context of water resources development a model is something to be used rather than something to be believed, that individual models must always be regarded as tools which are designed to be useful for a particular purpose rather than as dogmas which support an ideology.

1.4 COMPLEMENTARITY OF PHYSICAL AND SYSTEMS APPROACHES

It is clear from the previous discussion that in a physical approach the emphasis is on the topics of study, whereas in a systems approach the emphasis is on the method used. Nevertheless, this seeming dichotomy in the modes of approach is more fictitious than real, as argued by Klemes (1972):

> It is easy to recognize the difference in their aims. What is not always easy to see is that these two directions of inquiry are complementary rather than exclusive, that it makes little sense to argue which of the two approaches is "better," and to judge the faculties of one by the criteria of the other. To improve the understanding between what often seems to be two hostile camps, it is desirable to talk to the pure scientists about the challenges of applied research and to applied scientists about the virtues of basic research.

Indeed, for continuing progress of hydrology it is essential that these two modes of approaches are properly interpreted, emphasized, and pursued. This certainly seems to be the prevalent attitude in pure as well as applied sciences, as noted by Rosenblueth and Wiener (1945):

> The intention and result of a scientific inquiry is to obtain an understanding and control of some part of the universe. This statement implies a dualistic attitude on the part of the scientists. Indeed science does and should proceed from this dualistic basis. But even though the scientist behaves dualistically, his dualism is operational and does not necessarily imply strict dualistic metaphysics.

Further, it can be argued that the difference between the physical approach and the systems approach, although significant, is more of degree or emphasis than of kind. All hydrologic problems begin as closed-box or black-box problems; that is, only a few of the significant variables are recognized. Scientific progress consists in the progressive opening of these boxes. The successive addition of variables leads gradually to more elaborate theoretical models and then to a hierarchy in these models, from the relatively simple, highly abstract ones to the more complex, more elegant theoretical models. For an excellent discussion of physical and systems approaches, we refer to Klemes (1978, 1982, 1983). We revisit these approaches in the next chapter.

1.5 CHOOSING BETWEEN APPROACHES

The choice of a particular method of hydrologic investigation depends on the requirements of (1) the user (accuracy, simplicity, amount of desired information), (2) the problem, (3) the approach (availability of data, computational facility), and (4) the economics. Some of these requirements may not be independent of each other. It is conceivable that a given approach will seldom satisfy all of these requirements, and consequently one approach will seldom be uniformly better than the other under all circumstances (Singh and Woolhiser, 1976; Singh, 1979).

A choice among approaches depends on their systematic evaluations, which, in turn, entails (1) construction of an objective function, (2) use of goodness-of-fit criterion, (3) sensitivity analysis, (4) error analysis, and (5) comparison and ranking (Dawdy, 1969). These considerations are important from the standpoint of theoretical research as well as practical applications and are discussed at some length later in the text.

EXERCISES

1.1. It is required to perform high-flow forecasting for a given river basin. Discuss a physical approach. Then develop a systems approach.

1.2. Consider a small town located near a stream. The waste effluent generated by the town is discharged into the stream. Discuss physical and systems approaches to model water quality of the stream downstream to the point of waste discharge.

1.3. Discuss physical and systems approaches to determining transpiration from plants of a given species.

1.4. Maximizing crop production involves modeling the soil-water-fertilizer-plant-atmospheric system. Discuss briefly physical and systems approaches to this problem.

1.5. Consider that industrial chemical wastes are disposed through deep well-injection. Discuss physical and systems approaches for determining the migration of these wastes in space and time.

1.6. A knowledge of soil moisture is required for irrigation scheduling. Discuss physical and systems approaches to determine it.

1.7. Consider an urban area that is subject to frequent flooding. It is required to design a drainage system to alleviate the problem of flooding. Discuss the physical and systems approaches to determine its design.

1.8. Consider a watershed subject to rapid land use changes, including urbanization, forest clearing and cutting, and highway construction. Discuss the physical and systems approaches to quantify the effect of these changes on the watershed hydrology.

1.9. A forest watershed has considerable snow cover. Due to warming weather, the snow begins to melt. Discuss the physical and systems approaches to determining snowmelt runoff from this watershed.

1.10. A government agency is considering construction of a reservoir near a metropolitan area for alleviating flood hazards and supplying water for municipal and industrial uses. Discuss systems and physical approaches to its design.

1.11. Discuss data requirements of physical and systems approaches.

1.12. Which approach is better suited for ungaged watersheds and why?

1.13. Can an approach be always classified as a physical or a systems approach? If not, then why not?

1.14. A chemical spill occurs in a river. Which approach should be used to assess its environmental impact?

1.15. What approach should be used for precipitation forecasting?

CHAPTER TWO

Systems Approach: Definitions and Concepts

2.1 SYSTEMS TERMINOLOGY

Systems engineering contains terminology that must be properly understood before applying systems tools and techniques to modeling in hydrology.

2.1.1 System Definition

There are many definitions of a system, perhaps as many as there are books on the subject of systems analysis and synthesis (Truxal, 1955; Beer, 1959; Ackoff, 1962; Zadeh and Desoer, 1963; Schwartz and Friedland, 1965; de Neufville and Stafford, 1971). Intuitively, a system can be thought of as a collection of interacting components subject to various inputs and producing various outputs. Chow (1964) defined a system as an aggregate or assemblage of parts, being either objects or concepts, united by some form of regular interaction or interdependence. The definition of a system by Dooge (1967), however, is more comprehensive and meaningful:

> Any structure, device, scheme or procedure,
> Real or Abstract,
> That interrelates in a given time reference,
> An input, cause, or stimulus,
> Of matter, energy, or information
> And
> An output, effect, or response
> Of information, energy, or matter.

This definition is instructive and brings out some important characteristics of the system:

1. A system can consist of more than one component; it normally does.
2. These components are separate; they may be interdependent.
3. These components are put together following some sort of scheme. That is, a system is an ordered arrangement.
4. A system interrelates input and output, cause and effect, or stimulus and response.
5. The performing of this interrelationship is the main function of a system; the emphasis is on the system's function rather than on its structure.
6. This interrelationship is performed in a specified time frame.
7. A system does not require that input and output be alike or have the same nature.
8. Reference to an input is not confined to consideration of a single input. The input can be a vector or can be completely distributed in time and space.
9. A system can be real or abstract.

To define mathematically the system in a systems-theoretic sense (V. Vemuri and N. Vemuri, 1970), a system S is a set of ordered pairs of inputs and outputs, simply called signals,

$$S = \{x_i(t), y_i(t)\}, \qquad i = 1, 2, 3, \ldots \tag{2.1}$$

where the braces denote the set of ordered pairs (x_i, y_i), $i = 1, 2, \ldots$, with an underlying functional relation characterizing the mapping of $x_i(t)$ on $y_i(t)$. Essentially a system S is composed of three parts:

1. The input $x(t)$
2. The output $y(t)$
3. The system operation or function H

2.1.2 Input

The members of the domain of S, that is, the set $x_i(t)$, are the inputs to the system function H.

2.1.3 Output

The members of the range of S, that is, the set $y_i(t)$, are the outputs of the system function H.

2.1.4 System Function

The system function H of a system S contains a mathematical characterization of the process that relates the inputs and outputs. A system function may be characterized by a differential operator, an integral operator, an integro-differential operator, or a functional operator, and it completely describes the input-output relation.

It is often convenient to represent a system schematically by means of a box, as shown in Fig. 2.1. Consider a simple system receiving the input x, having the system function H operating on x, and producing the output y. The relation between the input x and the output y is simply written as

$$y = Hx \tag{2.2}$$

which should read: y is the response of H to x. The symbol H serves to identify a system as well as its operator. When H precedes an input as in equation (2.2), it is an operator that operates on the input x to produce the output y.

2.1.5 Function

A *function* $f(\cdot)$ is a single-valued set of ordered pairs (x_i, y_i), $i = 1, 2, 3, \ldots$, that is,

$$f(\cdot) = \{(x_1, y_1), (x_2, y_2), \ldots, (x_n, y_n)\} \tag{2.3}$$

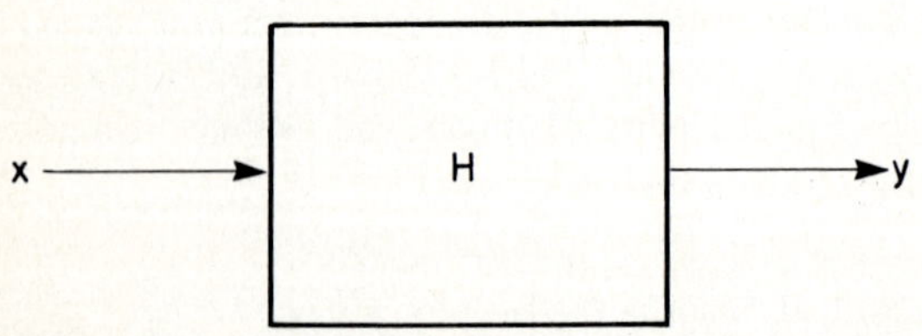

FIGURE 2.1 Schematic representation of a system.

A signal can then be regarded either as a function of time or as a set of ordered pairs.

2.1.6 Functional

A *functional* $F[\cdot]$ is a single-valued set of ordered pairs of scalar-valued functions of one real variable $f_k(\cdot)$, and scalars z_k, that is,

$$F[\cdot] = \{(f_k, z_k)\} \tag{2.4}$$

A function defines a mapping from one point to another point, whereas a functional defines a mapping from a function to a point. A simple example of a functional is the pair $(g(w), \int_a^b g(w)\, dw)$, where a mapping from the function $g(w)$ to the point whose value is defined by the definite integral $\int_a^b g(w)\, dw$ is implied. If we let $z = \int_a^b g(w)\, dw$, a functional can be written in an alternative form:

$$z = F[g(w)]\Big|_{w=a}^{w=b}$$

2.1.7 Variable

A *variable* is a characteristic of a system that may be measured and that may assume different numerical values at different times. Daily rainfall, runoff, evaporation, infiltration, temperature, pressure and soil moisture are some of the examples. The terms *variate* and *variables* are synonymous.

2.1.8 Parameter

A *parameter* is a quantity characterizing a system. It may or may not remain constant in time. Hydraulic conductivity, transmissivity, storativity, Manning's roughness factor, storage delay time, time of concentration, and coefficients in the Muskingum method of flood routing are some examples.

2.1.9 State

The *state* of a system is quantified in terms of system variables. Any change in any variables of the system produces a change of state. If all the state variables are known, the state of the system is completely known. For example, if we know all the components of the hydrologic cycle at a specified time, the cycle is completely determined at that time. Frequently, our objective in engineering is to ascertain the state of a system at some future time from the knowledge of its present or initial state. For example, we may wish to determine tomorrow's river flow from today's river conditions. Mathematically, this problem is formulated in terms of differential equations. If $y(t)$ represents the state of a system at any time t, a differential equation can be utilized; this equation may have the form

$$\frac{dy(t)}{dt} = f(y); \qquad y(t = 0) = y_0 \tag{2.5}$$

to obtain the value of $y(t)$ at all future times and, therefore, at any particular time. In most engineering problems, the basic equation is a nonlinear partial differential equation or a functional equation more complicated in form than the ordinary differential equation just shown.

Besides a mathematical approach, there can be other ways to determine the state of a system. Not all approaches, however, are applicable to determining the state of a given system. For example, some systems are determined historically. Their past states determine their present states. The states of other systems are described by certain laws of chance or probability. In hydrology most systems are part deterministic, part stochastic.

2.1.10 Heuristic

A *heuristic* is a rule of thumb, strategy, trick, simplification, or any other kind of device that drastically reduces the effort in search of solutions to problems. It lacks mathematical rigor. A useful heuristic offers solutions that are sometimes acceptable.

2.1.11 Model

A *model* is a simplified representation of a complex system. It simulates some but not all the characteristics of the system. It involves similarity without identity. A mathematical model represents the system by a set of equations expressing relationships between system variables and parameters; for example,

$$f\left[x(t), y(t); \frac{\partial x}{\partial t}, \frac{\partial y}{\partial t}; \frac{\partial^2 x}{\partial t^2}, \frac{\partial^2 y}{\partial t^2}; \ldots; a_1, a_2, a_3, \ldots\right] = 0 \tag{2.6}$$

where $f[\cdot]$ is a function of specified form and a_i, $i = 1, 2, \ldots$, are measured or estimated parameters.

2.1.12 Algorithm

An *algorithm* is a procedure that ensures the desired solution, given enough time.

2.2 SYSTEM REPRESENTATION

A system can be composed of a number of subsystems, each of which can have a distinct input-output linkage. A system or a subsystem can also be divided into components, each of which is an input-output element. It is conventional to represent a system by rectangular blocks. Such a representation is often referred to as the *black-box representation*. The simplest representation of a system is shown in Fig. 2.2(a), where there is a single input and single output. A more realistic representation occurs when the black box has a number of inputs and outputs, as shown in Fig. 2.2(b). A frequently encountered situation in atmospheric, geologic, hydrologic and environmental systems, as shown in Fig. 2.2(c), is one in which a series of elements each has its own input-output relationship. Fig. 2.2(d) shows a single element receiving input and discharging output in different forms (y, I, and E) and in different directions. Although highly idealized, this is the situation frequently experienced in hydrologic practice. A generalized and more realistic representation (Fig. 2.2(e)) consists of a series of elements each receiving input x, allowing infiltration I and evaporation E, and producing output y. In a similar manner, a system can be represented in a number of ways, depending upon the degree of simplification desired.

To illustrate, a hydrologic system, which pertains to water in its natural state, can be defined as any system in accordance with the preceding definitions. The hydrologic cycle can be regarded as a hydrologic system. Various components of this system might include precipitation, interception, evaporation, transpiration, infiltration, detention storage or retention storage, surface runoff, interflow, and groundwater flow. Each component is a subsystem if it satisfies the characteristics of a system set out in its definition. Thus, the various components of the hydrologic system can be regarded as hydrologic subsystems. Whether a particular component is to be treated as a system or subsystem depends on the objective of the inquiry.

There are many illustrations representing the hydrologic cycle; some are said to be qualitative (Horton, 1931), some descriptive (Ackerman, Coleman, and Ogrosky, 1955) and some quantitative (Lettau, 1954). As Dooge (1967) pointed out, such representations involve a high degree of abstraction from the actual hydrologic cycle despite touches of verisimilitude. All the land surface, ocean surface, and atmosphere have been considered single elements, and the complexities of local variation have been ignored. It is more instructive to represent the hydrologic cycle as a system (Fig. 2.3). Although austere in form and even lacking in the color of modern art, this figure reflects a lesser degree of abstraction than earlier, more picturesque representations. Fig. 2.4 shows another representation of the hydrologic cycle developed by Kulandaiswamy (1964). This representation is similar to the systems representations used by electrical and control engineers.

A complete study of the entire hydrologic cycle is too vast to be included in hydrology. It obviously requires an interdisciplinary integrated scientific effort and has to draw heavily upon several branches of science, as elucidated by Meinzer (1942) and Chow (1964). Recognizing the scope of hydrology, the system of Fig. 2.3 is simplified for watershed runoff in Fig. 2.5. The lines of water transport representing precipitation, evaporation, transpiration and runoff are either inputs or outputs to this new system. Clearly, precipitation is an input and runoff is an output. Whether evaporation and transpiration are inputs or outputs is not always clear. A somewhat more detailed representation of watershed runoff is shown in Fig. 2.6.

Even the system of Fig. 2.5 is complex; it is composed

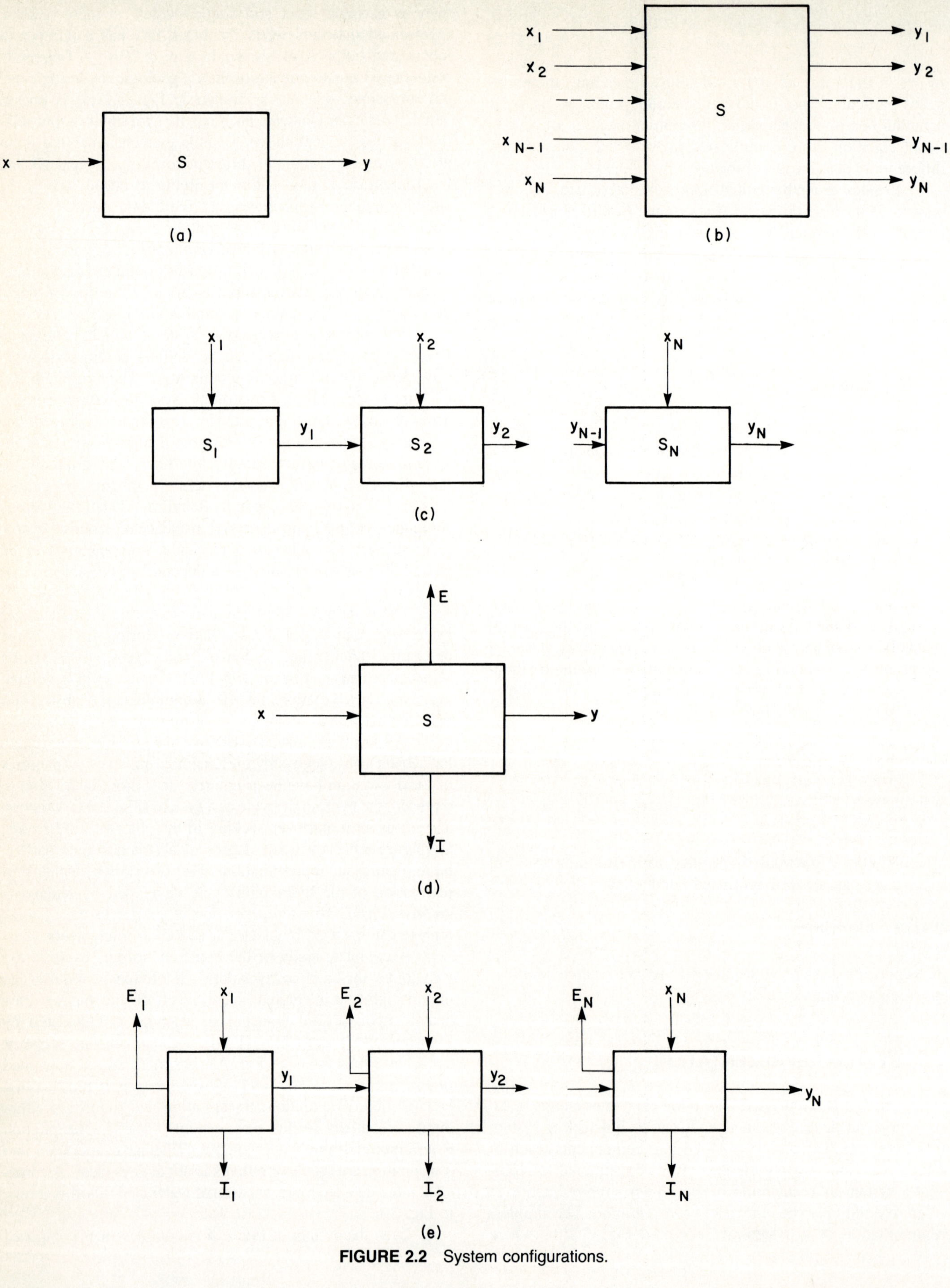

FIGURE 2.2 System configurations.

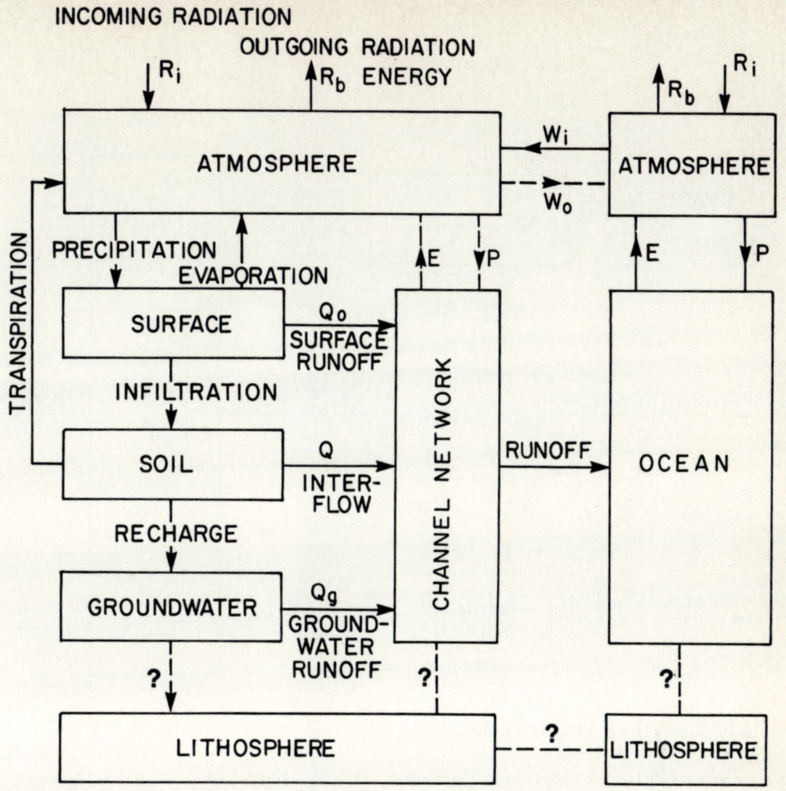

FIGURE 2.3 A systems representation of the hydrologic cycle (after Dooge, 1973).

of several subsystems. Some of these subsystems are the surface runoff system, the soil moisture system, the groundwater system, and the streamflow system. For illustration, let us consider runoff and soil moisture systems here.

EXAMPLE 2.1 Runoff System

In a runoff system rainfall is divided into (1) rainfall excess, (2) infiltration, (3) interception, (4) evaporation and transpiration (evapotranspiration), and (5) depression and detention storage. The rainfall excess produces surface runoff (Fig. 2.6). Infiltration replenishes soil moisture storage, which is extracted by evapotranspiration. Interception is usually consumed by evaporation. Depression storage contributes to both infiltration and evapotranspiration. During a given short-duration rainfall episode interception, evapotranspiration and depression storage normally constitute a small portion of the rainfall volume. Hence these are often either ignored or lumped together and taken as a small fraction of the volume. It is the infiltration that is usually significant and must be estimated as accurately as possible. Any infiltrated water in excess of soil moisture storage recharges groundwater storage, which eventually emerges as baseflow. Similarly, runoff is divided into (1) surface runoff, (2) interflow, and (3) groundwater flow. Because it is virtually impossible to separate these three components accurately, the distinction between them is vague. It is more practical to divide runoff into direct runoff and base flow. This leads to a simplified direct runoff system, as shown in Fig. 2.7. There are many plausible representations of the watershed runoff system. For description of an idealized runoff cycle, see Hoyt (1942a, 1942b).

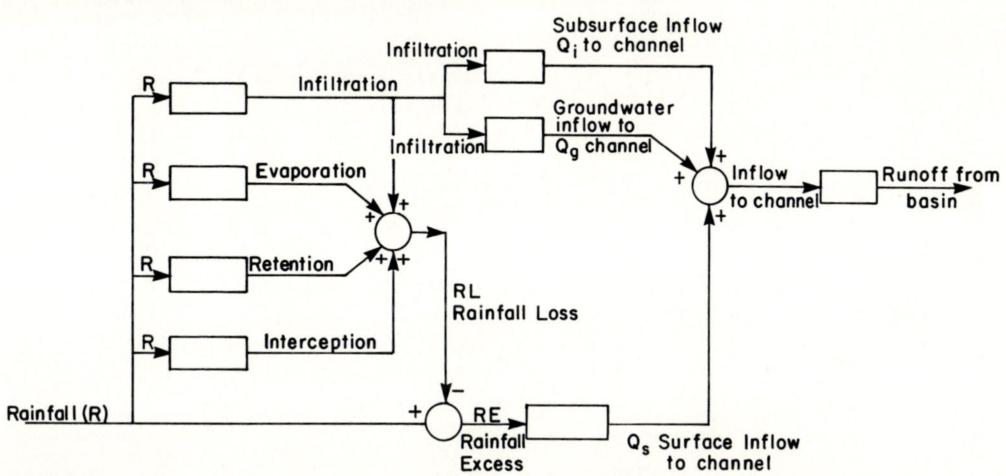

FIGURE 2.4 Another systems representation of the hydrologic cycle (after Kulandaiswamy, 1964).

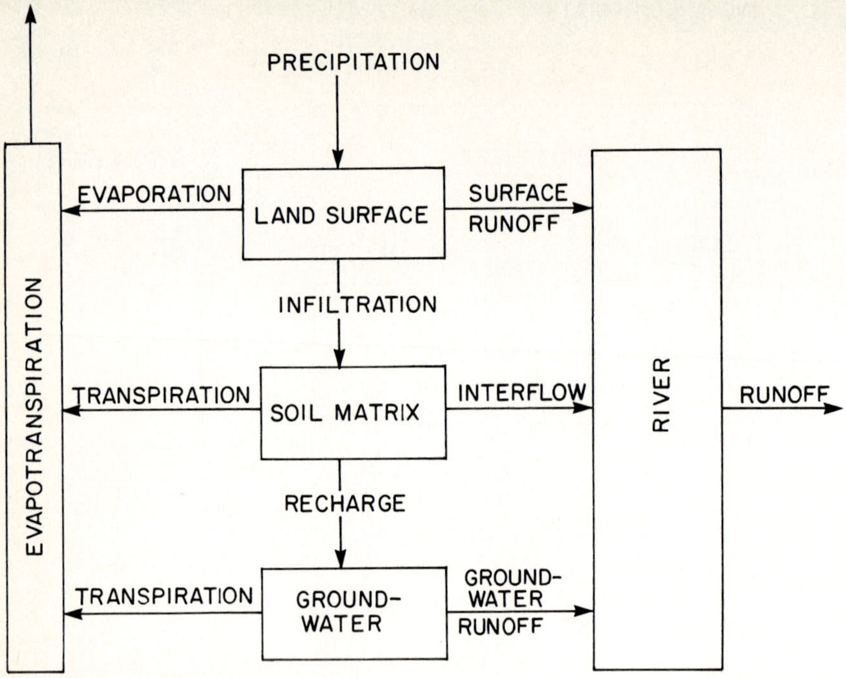

FIGURE 2.5 A systems representation of watershed runoff.

EXAMPLE 2.2 Soil Moisture System

In a soil moisture system (Singh and Dickinson, 1975), extraction of soil moisture from upper horizons of the soil profile occurs through both evaporation and transpiration. The rate of extraction from various portions of the profile is not the same and depends on soil genesis and development of plant root system. This extraction occurs until the soil becomes almost dry. When it rains, depending upon rainfall characteristics, a portion of the rainfall infiltrates into the ground and replenishes soil moisture storage. Infiltrated water in excess of soil moisture storage leads to groundwater recharge. The remainder of rainfall fills detention and depression storage over the ground and also appears as surface runoff. A systems representation of the soil moisture system is illustrated in Fig. 2.8.

2.3 CLASSIFICATION OF SYSTEMS

There are many ways to classify systems and their models. Each classification is based upon a particular set of system characteristics. The same system can therefore fit into different

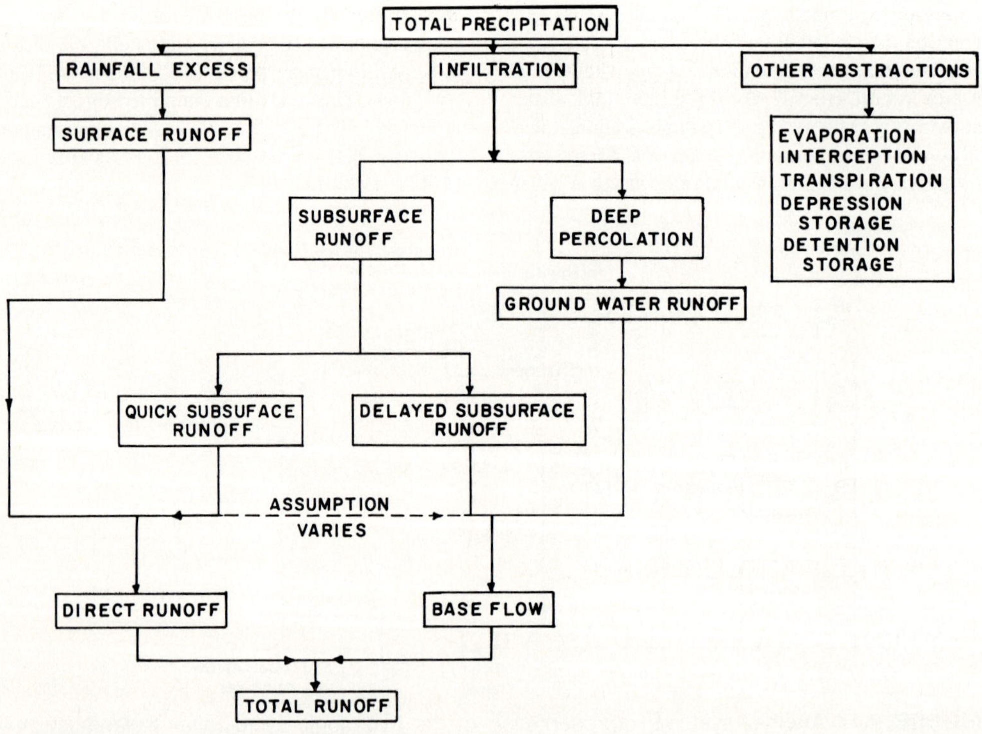

FIGURE 2.6 A systems representation of watershed runoff (adapted from Chow, 1964).

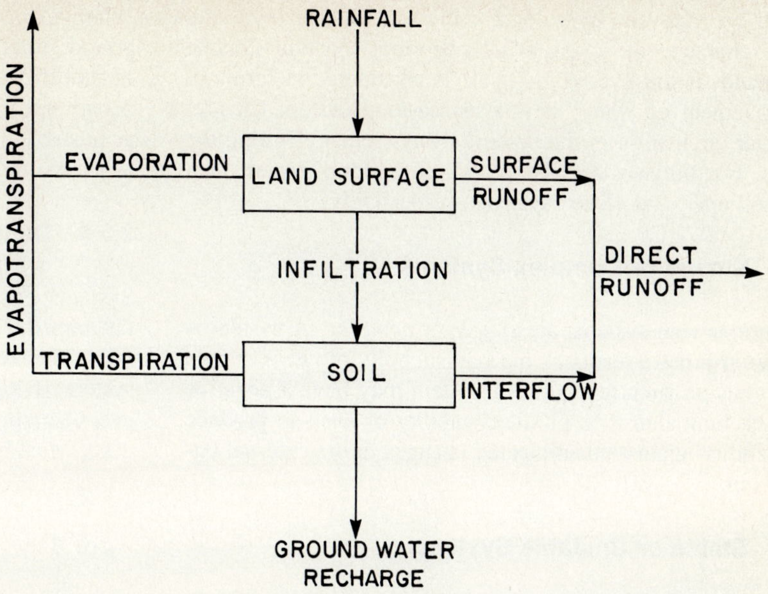

FIGURE 2.7 A simplified direct runoff system.

categories. The following characteristics are particularly noted for classification of systems:

1. Abstractness
2. Naturality
3. Openness
4. Simplicity
5. Stability
6. Dampness
7. Causality
8. Memory
9. Linearity
10. Time-invariance
11. Continuousness
12. Lumpedness
13. Determinism

2.3.1 Abstract or Physical Systems

A system can be *abstract* or *physical*. Engineering and hydrological systems are physical. A physical system is a system in the real world. A sequential system is a physical system involving input, output, and some working medium (matter, energy, or information) that passes through the system. A dynamic system is a physical system receiving quantitative input and acting accordingly under specified conditions to produce quantitative output. A hydrologic system is normally a physical, sequential, dynamic system.

2.3.2 Natural or Devised Systems

Sometimes a distinction is made between natural and devised systems. The input, output, and other state variables for natural systems are measurable but not controllable. For devised systems they are both measurable and controllable. Most hydrologic systems are natural systems. A laboratory watershed is, however, an example of a devised system. Lysimeters and rainfall simulators are other examples of devised systems.

2.3.3 Open-Loop or Closed-Loop Systems

Systems can be *open-loop* or *closed-loop*. When the system operation is independent of the output produced, the system is an open-loop system. The operation of a closed-loop system depends on the output with a feedback action. Most hydrologic systems are open-loop because their output does not significantly affect their operation. On the other hand, the groundwater system of an island and the global hydrological system are certainly closed-loop systems because their output greatly affects their balance.

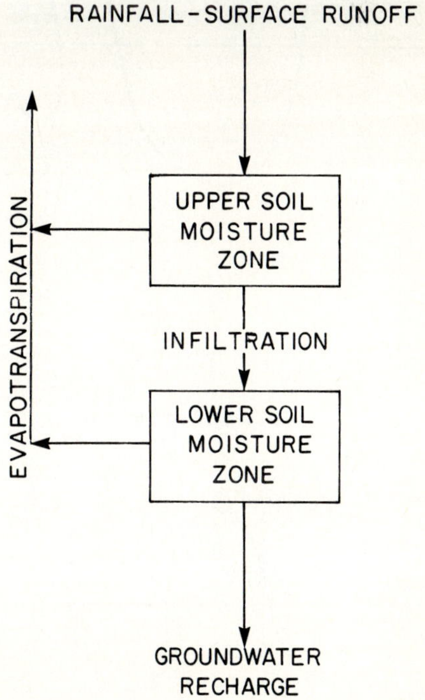

FIGURE 2.8 A simplified soil moisture system.

If we view the hydrologic cycle as a system involving energy changes, the cycle is an open-loop system. On the other hand, if the hydrologic cycle is considered in terms of the movement of water, it is a closed-loop system because the water circulating in the system always remains within the system. For brevity, open-loop and closed-loop systems can be called open and closed systems, respectively.

2.3.4 Simple or Complex Systems

In a similar vein systems are *simple* or *complex*. A system is complex if it has a feedback mechanism built into it. A simple system has no such feedback. A system may have a negative feedback built into it to produce stability or even to produce ultrastability against unanticipated changes in the external environment.

2.3.5 Stable or Unstable Systems

Systems can be classified as stable or unstable. In a stable system if the input is bounded, the output is also bounded. Virtually all hydrologic systems are stable or extremely stable. When a finite period rainfall is input to a watershed, a finite period runoff hydrograph is its output, although the period of runoff occurrence will be larger than the period of rainfall occurrence. This is an example of a stable system. A system is unstable if a finite input produces an unbounded output, as shown in Fig. 2.9.

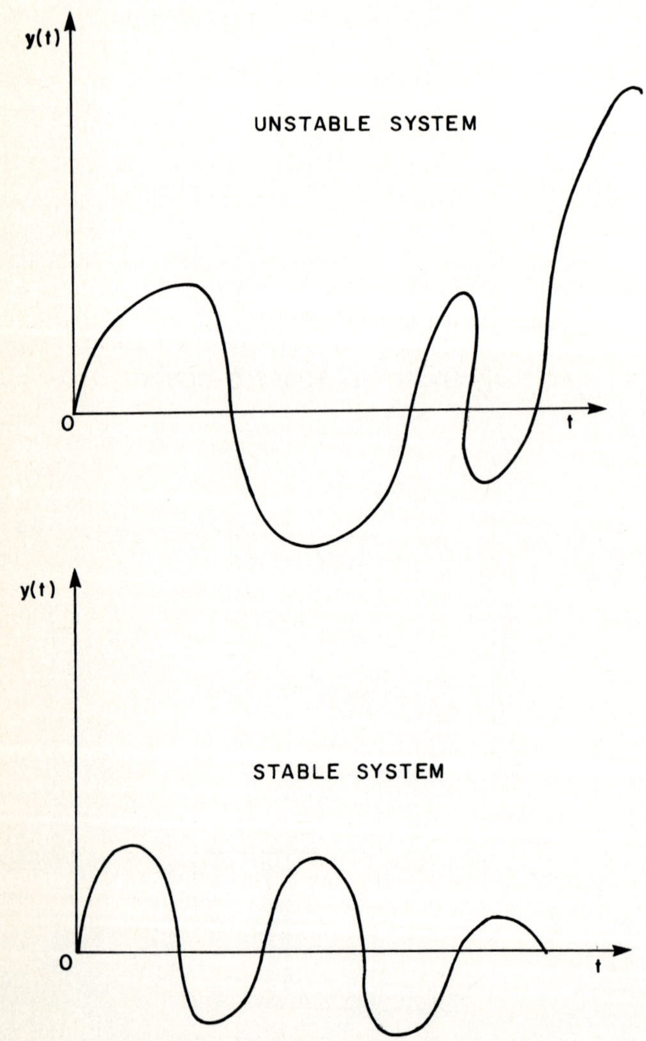

FIGURE 2.9 Stable and unstable systems.

2.3.6 Damped or Undamped Systems

Systems can be *damped* or *undamped*. A stable system is undamped if, when disturbed, it produces an indefinite oscillation. The system is damped if its oscillation dies out. The system is highly damped if the oscillation dies out without ever changing sign. These characteristics are illustrated in Fig. 2.10. Most hydrologic systems are damped or highly damped.

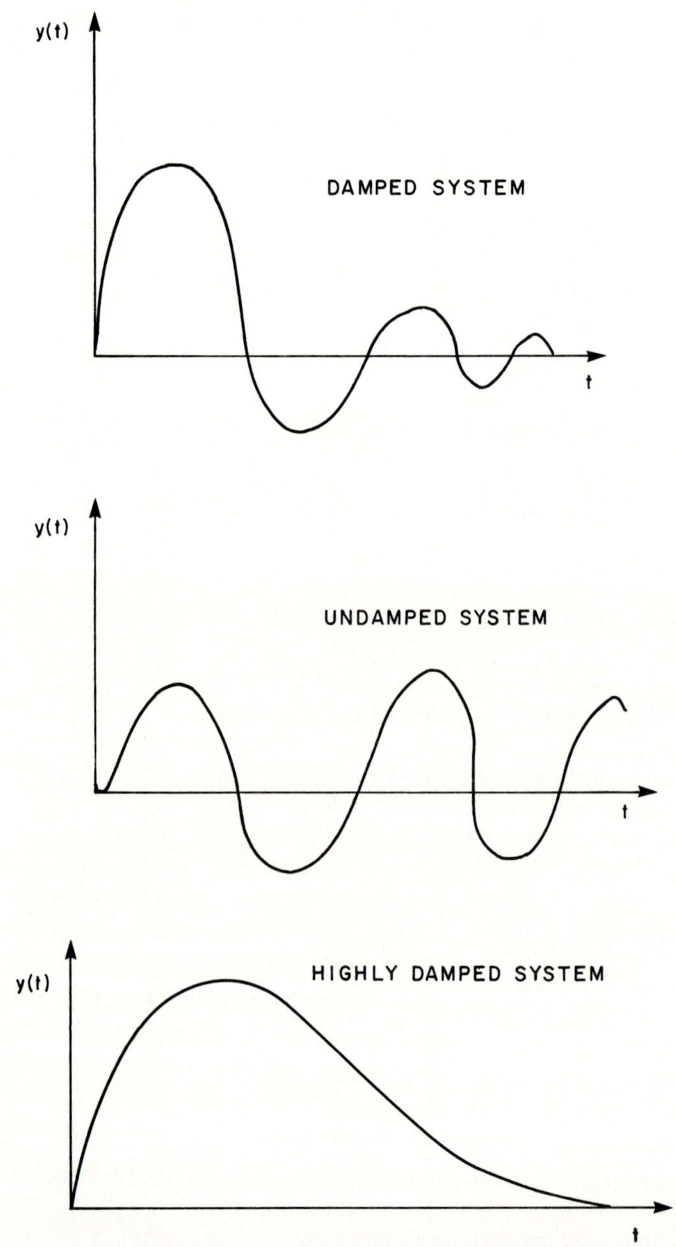

FIGURE 2.10 Damped and undamped systems.

2.3.7 Continuous-Time or Discrete-Time Variables and Systems

If the input to and the output from the system can change at any point in time, the system is called a *continuous-time* system, or a continuous system for short. To make continuous change evident, we write the input and the output as functions of continuous-time variable t, that is, $x = x(t)$, $y = y(t)$. A continuous system is described by a differential equation. The term *continuous-time* does not imply that all inputs and outputs are mathematically continuous functions. They are functions of a continuous variable t.

A *discrete-time* system, or discrete system, is a system whose inputs and outputs change at discrete instants in time. Between these instants the input and output may change, may not be defined, or may be constant. A discrete system is described by a difference equation.

In a quantized system the variables (inputs and outputs) may assume only a countable number of values, but the change from one value to another may occur at any time, as shown in Fig. 2.11. A quantized variable is also defined at discrete instants of time, but its value remains unchanged between two successive instants.

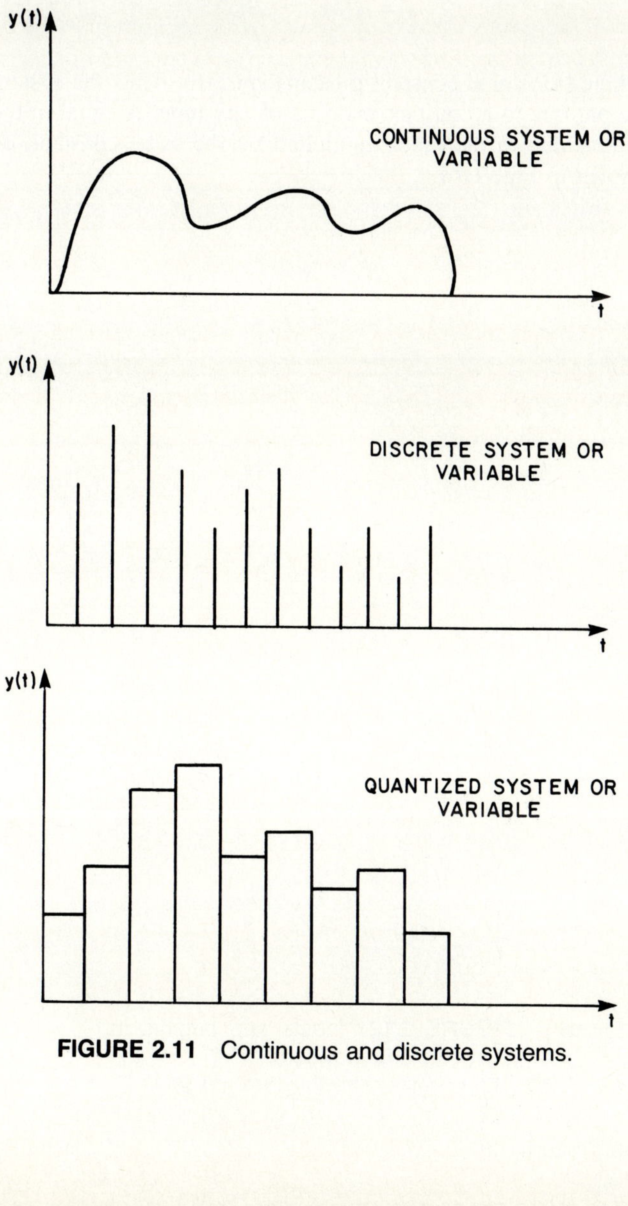

FIGURE 2.11 Continuous and discrete systems.

To emphasize the discrete nature of the system, we write the input and output as functions of discrete variable t_i, $i = 1, 2, \ldots$; that is, $x = x(t_i)$, or $x = x(i)$ and $y = y(t_i)$, or $y = y(i)$.

In hydrology, rainfall is defined as a quantized variable, whereas streamflow, infiltration, and evaporation are some of the continuous variables. In hydrologic practice many variables are sampled at discrete times and are assumed to vary linearly between sampling times. Furthermore, often it is the output that constitutes the basis to classify hydrologic systems in the above manner; that is, if the output of a hydrologic system is continuous, discrete, or quantized, the system is, respectively, continuous, discrete, or quantized.

2.3.8 Causal or Noncausal Systems

Systems can be characterized as *causal* or *noncausal*. In a causal system the output cannot precede its corresponding input. In other words, its response due to an input does not depend on future values of input. All hydrologic systems are causal. Rainfall must always occur before runoff. In a noncausal system the output does not have to follow the input. Some systems learn from their past histories and improve their performances. Such systems are called *adaptive* systems. Causal systems are also referred to as nonanticipatory or physical systems, whereas noncausal systems are referred to as anticipative systems.

2.3.9 Memory or No-Memory Systems

System *memory* represents the length of time in the past over which the input affects the present system behavior. A system can have zero memory, finite memory, or infinite memory. Zero-memory systems are also referred to as instantaneous systems, and memory systems are called dynamic systems.

A system is instantaneous if its output at any time t depends at most on the input values at the same time but not on the past or future values of the input. Otherwise the system is a finite memory system. In other words, the output at any time t depends not only on the input at t but also on at least some of the past input values (and, in a noncausal system, on some future input values also). If the system output at time t is completely specified by the input in the interval $[t - T, t]$, the system has a memory of length T. If $T = 0$, the system is instantaneous, or has zero memory. If T is finite or infinite, the system is dynamic.

EXAMPLE 2.3

For illustration, consider the rational method that mathematically relates discharge $Q(t)$ to rainfall intensity $P(t)$ as

$$Q(t) = \frac{C}{T} \int_{t-T}^{t} P(t)\, dt \tag{2.7}$$

where C is a rational coefficient and T is a period (greater than zero) characteristic of the watershed, called the time of concentration. This system has finite memory. On the other hand, if $t - T$ is assumed to be zero, the system has an infinite memory. If equation (2.7) were to be expressed as

$$Q(t) = C\,P(t) \tag{2.8}$$

this would imply a zero-memory system. In hydrology all three kinds of systems are believed to exist.

2.3.10 Time-Invariant or Time-Variant Systems

Another concept used to classify systems is *time-invariance*. A system is time-invariant if its input-output relationship does not change with time. The form of the output depends only on the form of the input and not on the time at which the input is applied. More precisely, a system having the input-output relation

$$y(t) = Hx(t) \tag{2.9}$$

is time-invariant if and only if

$$Hx(t \pm T) = y(t \pm T) \tag{2.10}$$

for any $x(t)$ and any T. The system must be at rest prior to the application of $x(t)$. Systems that do not have this property are called *time-variant* systems. Other labels, such as fixed and stationary, are also used to denote time-invariant systems, whereas labels such as nonstationary, time-varying, and time-variable signify time-variant systems. Graphically, these systems are illustrated in Figs. 2.12 and 2.13. Most hydrologic systems are time-variant due to variations in solar activity during the day and seasonal variations during the year. For simplicity, they are assumed to be time-invariant. Time-invariant systems are characterized by differential (or difference) equations with constant coefficients. If the coefficients are time-dependent, these differential (or difference) equations characterize time-variant systems.

2.3.11 Linear or Nonlinear Systems

Classification or hydrologic systems is often based on the concept of linearity. Systems can be *linear* or *nonlinear*. There is more than one way to interpret linearity in hydrology.

2.3.11.1 Linearity in Systems Theory

In systems theory a system is linear if it satisfies (1) the property of proportionality or homogeneity and (2) the property of superposition or additivity. Otherwise, the system is nonlinear.

Proportionality means that the system input and output have the same scale ratio, that is,

$$y = H(ax) = aH(x) \tag{2.11}$$

where a is some constant greater than zero. Thus the system output due to a constant multiple of any input is equal to the output due to the input multiplied by the same constant, as shown in Fig. 2.14.

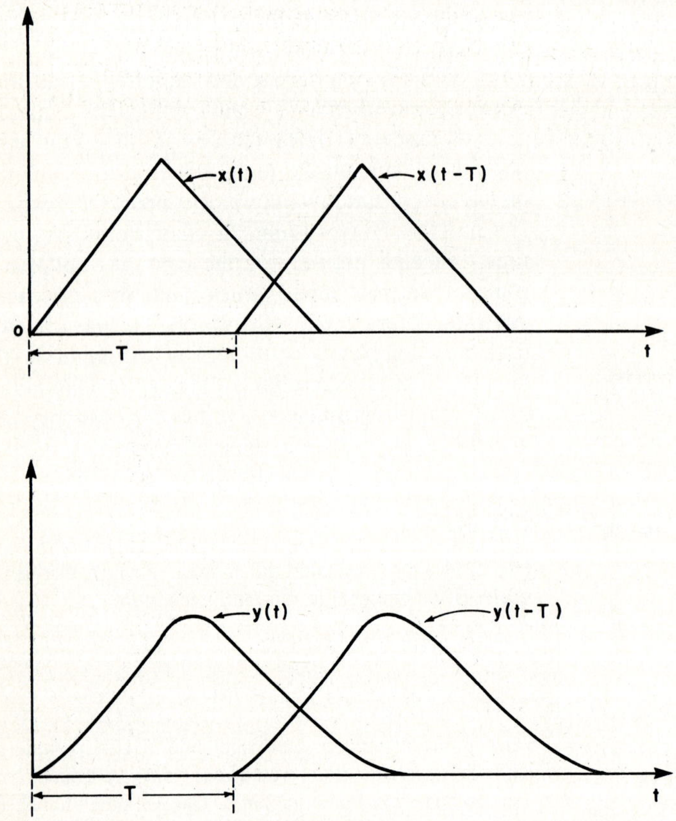

FIGURE 2.12 Inputs and outputs in a time-invariant system.

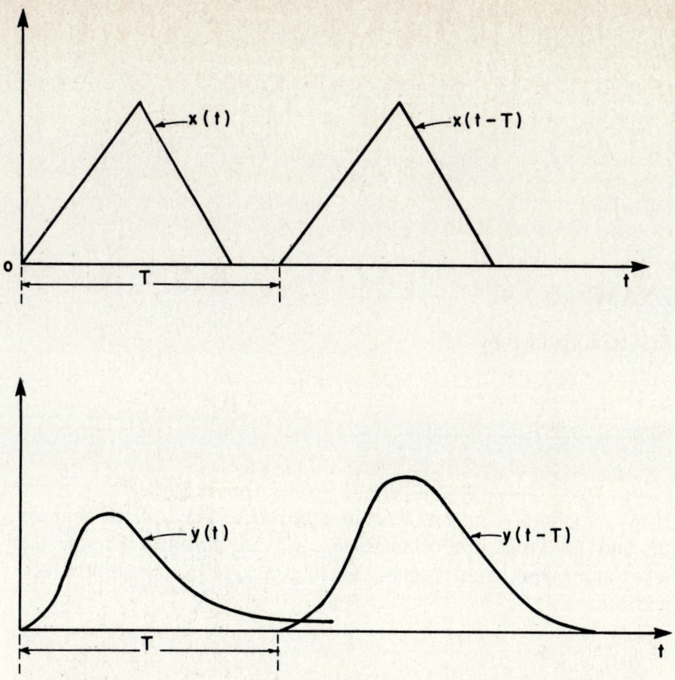

FIGURE 2.13 Inputs and outputs in a time-invariant system.

Superposition can be defined as follows: If

$$y_1 = H(x_1)$$

and

$$y_2 = H(x_2)$$

then

$$\begin{aligned} y_1 + y_2 &= H(x_1) + H(x_2) \\ &= H(x_1 + x_2) = H(x_1) + H(x_2) \end{aligned} \tag{2.12}$$

It is important to note that the system be at rest at the time of application of x_1 and x_2. Put simply, this asserts that the system output due to a sum of two inputs is equal to the sum of the two outputs, as shown in Fig. 2.15. Therefore,

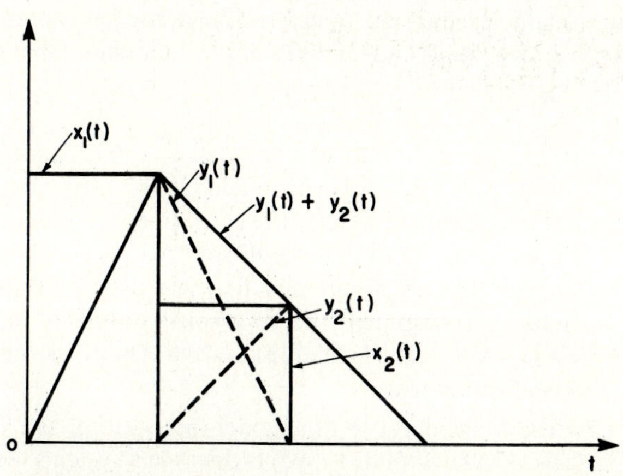

FIGURE 2.15 Inputs and outputs in a linear system following the principle of superposition.

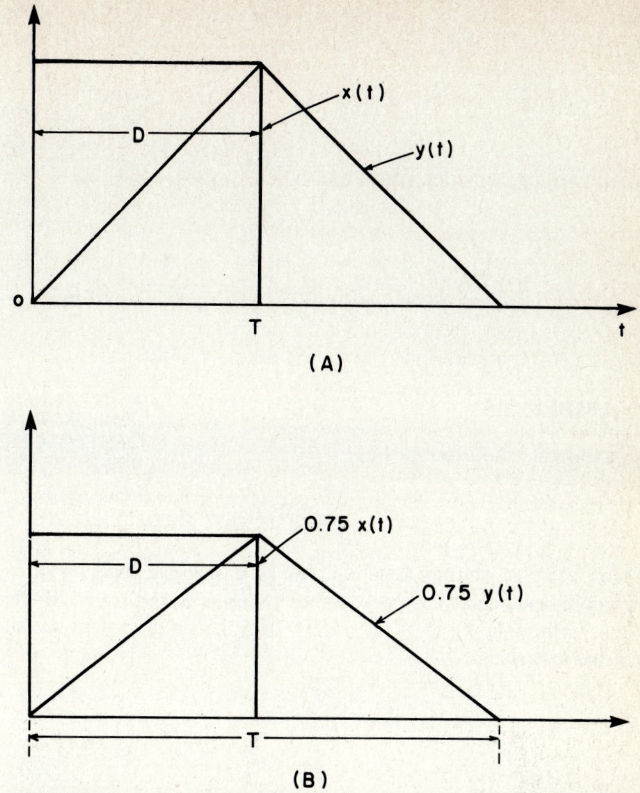

FIGURE 2.14 Inputs and outputs in a linear system following the principle of proportionality.

the output due to the combined effect of a number of separate inputs is equal to the sum of separate outputs due to their corresponding individual inputs. Thus a complex input can be broken down into simple elements, and the output can be obtained by summing the outputs due to these simple elements. This is the reason that the output from one complex input can be used as a basis for computing the output from another complex input of a totally different pattern. In hydrologic parlance, superposition means that the time distribution of system output from a given input is independent of concurrent outputs due to antecedent inputs.

A linear system possesses both the property of proportionality and the property of superposition. It can be easily shown that any system possessing these properties is linear. Hence, a system is linear if and only if it obeys both the properties of proportionality and superposition; that is,

$$H(ax_1 + bx_2) = aH(x_1) + bH(x_2) \tag{2.13}$$

where a, b are any constants greater than zero. By induction it can be shown that

$$H\left(\sum_{i=1}^{N} a_i x_i\right) = \sum_{i=1}^{N} a_i H(x_i) \tag{2.14}$$

provided N is finite. We shall assume that this property is valid for $N \to \infty$ and also for integrals; that is,

$$H\left(\sum_{i=1}^{\infty} a_i x_i\right) = \sum_{i=1}^{\infty} a_i H(x_i) \tag{2.15}$$

and

$$H\left(\int_a^b a(s)x(t, s)\, ds\right) = \int_a^b a(s)H(x(t, s))\, ds \tag{2.16}$$

To illustrate, consider some examples.

EXAMPLE 2.4

Any system characterized by linear differential equations or difference equations is linear.

EXAMPLE 2.5

Let the input x be expressed as

$$x(t) = \sum_{j=0}^{M} a_j W_j(t)$$

where W_j is an arbitrary function, say an orthogonal function, a_j is the coefficient, and M is the number of terms in the series. If the output corresponding to $W_j(t)$ is $Z_j(t)$, then the linear system output $y(t)$ due to $x(t)$ is

$$y(t) = \sum_{j=0}^{M} a_j Z_j(t)$$

EXAMPLE 2.6

The system defined by

$$y(t) = Hx(t) = x^2(t)$$

is not linear because of the following statement:

$$H(x_1 + x_2) = (x_1 + x_2)^2 \neq Hx_1 + Hx_2 = x_l^2 + x_2^2$$

Thus it is seen that a system characterized by

$$y(t) = Hx(t) = x^n(t), \qquad n \neq 1$$

is not linear.

EXAMPLE 2.7

The system defined by

$$y(t) = Hx(t) = C$$

in which C = constant, is not linear because

$$H(x_1 + x_2) = C \neq Hx_1 + Hx_2 = C + C = 2C$$

EXAMPLE 2.8

The system characterized by

$$y = H\begin{bmatrix} x_1 \\ x_2 \\ \cdot \\ \cdot \\ \cdot \\ x_N \end{bmatrix} = \prod_{i=1}^{N} x_i$$

is not linear because

$$y = Ha\begin{bmatrix} x_1 \\ x_2 \\ \cdot \\ \cdot \\ \cdot \\ x_N \end{bmatrix} = H\begin{bmatrix} ax_1 \\ ax_2 \\ \cdot \\ \cdot \\ \cdot \\ ax_N \end{bmatrix} = \prod_{i=1}^{N} ax_i = a^N \prod_{i=1}^{N} x_i$$

$$\neq a\prod_{i=1}^{N} x_i$$

EXAMPLE 2.9

A system defined by

$$y = ax + b$$

where a and b are constants, is not linear since

$$H(x_1 + x_2) = ax_1 + ax_2 + b \neq Hx_1 + Hx_2 = a(x_1 + x_2) + 2b$$

However, this system can be considered to be linear if the changes Δx and Δy about a benchmark (x_0, y_0) are considered to be the input and output, respectively. Then $x = x_0 + \Delta x$, $y = y_0 + \Delta y$, and

$$\Delta y = a\Delta x$$

2.3.11.2 Linearity in Statistical Regression

Clarke (1971, 1973) showed that linearity in the systems theory sense is not necessarily equivalent to linearity in the statistical sense. In other words, linearity has an alternative meaning. The system is linear in the statistical regression sense if it is linear in the parameters to be estimated, and it is in this sense that it is used by mathematical modelers in the fields other than hydrology (Ross, 1970, 1972). If input $x(t)$ and output $y(t)$ are related by

$$y = a + bx + cx^2 \tag{2.17}$$

then this system is linear in the statistical regression sense but nonlinear in the systems theory sense. The converse is true for

$$y = a + \frac{x}{a} \tag{2.18}$$

Similar examples that are linear in one sense but nonlinear in another can be found in hydrologic literature. We give two examples cited by Clarke (1971):

1. The Amorocho model (Amorocho, 1963; Amorocho and Hart, 1965; Amorocho and Orlob, 1961; Amorocho and Brandstetter, 1971) is linear in statistical regression sense but nonlinear in systems theory sense. This is the premise that was used by Bidwell (1971) to estimate the functional ordinates in the Amorocho model. We discuss the Amorocho model later in the text.
2. The Mandelbrot-Wallis model (Mandelbrot and Wallis, 1969a, 1969b, 1969c) is linear in a systems theory sense but nonlinear in a statistical regression sense.

Most hydrologists (O'Donnell, 1960, 1966; Nash,1958, 1959; Dooge, 1959, 1973, 1977) use linearity in the systems theory sense. Nevertheless, the statistical interpretation should be distinguished to avoid ambiguity.

Most hydrologic systems are nonlinear. They are assumed to be linear for simplicity. Some hydrologic systems, however, behave almost linearly in certain ranges of their operation. For example, recent evidence suggests that in a certain range of system operation, the watershed runoff response is linear. Pilgrim (1966) showed that the time until peak of labeled water tended to approach an asymptotic value at high discharge. This is a significant result because if runoff response were indeed linear, the travel time of a given watershed would be constant. Pilgrim (1966) also showed that at low discharges, the runoff response was highly nonlinear. The above result explains in part the apparent success of the unit hydrograph method for flood estimation in some cases. The question frequently arises, How good is linear approximation of a nonlinear system? Or, How can the linearity of a hydrologic system be measured? To address this question briefly, the following is in order.

2.3.12 Measures of Linearity of Hydrologic Systems

Several measures to determine linearity of hydrologic systems have been suggested. We discuss four of them here: two applicable to rainfall-runoff system and two of a general kind. It should be emphasized that a system may behave linearly in one respect but nonlinearly in another. For example, a watershed may be linear with regard to its flood-producing characteristics but nonlinear with respect to infiltration. A hydrologic system is seldom linear across the board.

2.3.12.1 The Amorocho Measure

It is well known that the output $y(t)$ from a time-variable, linear system can be expressed as the convolution of a kernel function $h(t)$ with the system input $x(t)$:

$$y(t) = \int_0^{\infty} h(t, \tau)x(\tau)\, d\tau \tag{2.19}$$

Amorocho (1963, 1967, 1973) generalized equation (2.19) to represent a general hydrologic system with input $x(t)$ and output $y(t)$ by a functional series

$$y(t) = \sum_{n=1}^{\infty} \int_{-\infty}^{\infty} \cdots \int_{-\infty}^{\infty} h_n(t; \tau_1, \tau_2, \ldots, \tau_n) \prod_{i=1}^{n} x(\tau_i)\, d\tau_1 \cdot d\tau_2 \cdot \ldots, d\tau_n \tag{2.20}$$

Equation (2.20) represents a nonlinear system.

Amorocho (1963) suggested absolute linearity and unit linearity to measure the linearity of hydrologic systems. Absolute linearity or simply linearity $\lambda(t)$ is defined as the ratio between the linear term of the functional series and the summation of the full series. Mathematically this can be expressed as

$$\lambda(t) = \frac{\int_{-\infty}^{\infty} h_1(t; \tau)x(\tau)\, d\tau}{\sum_{n=1}^{\infty} \int_{-\infty}^{\infty} \cdots \int_{-\infty}^{\infty} h_n(t; \tau_1, \tau_2, \ldots, \tau_n) \prod_{i=1}^{n} x(\tau_i)\, d\tau_1 \ldots d\tau_n} \tag{2.21}$$

For a system to be linear, $\lambda(t)$ must equal one. Since the value of linearity at a specified time depends on the particular form of the input sequence, $\lambda(t)$ does not characterize the system uniquely.

To eliminate the effects of arbitrary input variations in the characterization of a system response, unit linearity was suggested. This can be defined as the linearity function of a system when it is responding to a unit step input and can be expressed as

$$\lambda(t) = \frac{m_1(t)}{\sum_{n=1}^{\infty} m_n(t)} \tag{2.22}$$

where

$$m_n(t) = \int_0^t \cdots \int_0^t h_n(t; \tau_1, \tau_2, \ldots, \tau_n) d\tau_1 \ldots d\tau_n$$

and t is a particular value of time. The numerical values of the function for particular values of t now depend only on the units of measurement defining the unit step input and on the characteristics of the system. Under the general assumptions that the denominator in equation (2.22) always converges to positive numbers and that the numerator always has positive values, it follows that $\lambda(t) \le 1$ for all values of t. A linear system is then defined by $\lambda(t) = 1$ for all t. $\lambda(t)$ measures how linear a hydrologic system is.

2.3.12.2 The Rogers Measure

In a series of papers (Rogers, 1980, 1982; Rogers and Zia, 1982), Rogers attempted to establish a relationship between peak discharge Q_P and runoff volume V as

$$\text{Log}\, Q_P = b + m\, \text{Log}\, V \tag{2.23}$$

in which b is the intercept and m the slope of the line best fitting the data. This equation was called *standardized peak discharge distribution* because baseflow was not separated from the hydrograph. Drainage basins exhibiting linearity are, according to Rogers, indicated by a slope m that equals 1. A slope of 1 preserves the linear relation that the peak discharge is directly proportional to runoff volume, which is inherent in the unit hydrograph theory.

Most watersheds are not truly linear. Thus the slope m in equation (2.23) is often less than 1. Such watersheds are nonlinear, wherein the peak discharge is a power function of runoff volume (Rogers, 1982):

$$Q_p = a + V^m \tag{2.24}$$

where a is the arithmetic intercept. The value of m quantifies the degree of nonlinearity. Its numerical value can vary from slightly less than 1 to −0.83.

2.3.12.3 Correlation Function

The linearity of a hydrologic system can be measured by correlation analysis in a time domain. If $x(t)$ and $y(t)$ are stationary stochastic processes, the cross-correlation function $\rho_{xy}(k)$ can be defined as

$$\rho_{xy}(k) = \frac{G_{xy}(k)}{[G_{xx}(0)\, G_{yy}(0)]^{0.5}} \tag{2.25}$$

where k is the time lag, $G_{xy}(k)$ is the cross-covariance function of $x(t)$ and $y(t)$, and $G_{xx}(0)$ and $G_{yy}(0)$ are the variances of $x(t)$ and $y(t)$, respectively. $\rho_{xy}(k)$ measures the linearity of a system in the time domain and may vary between 0 and 1.

2.3.12.4 Coherence Function

Kisiel (1969) proposed the coherence function $C^2_{xy}(f)$ as a measure of linearity:

$$C^2_{xy}(f) = \frac{|G_{xy}(f)|^2}{(G_{xx}(f)\, G_{yy}(f))}, \qquad 0 \le C^2_{xy}(f) \le 1 \tag{2.26}$$

where $G_{xy}(f)$ is the cross-spectral density function (Fourier transform) of the cross-correlation function $\rho_{xy}(\cdot)$, $G_{xx}(f)$ is the input spectral density, and $G_{yy}(f)$ is the output spectral density. This function is analogous to the correlation coefficient squared (equation (2.25)) and defines a linear association as a function of frequency. For a linear system the coherence function should be unity at all frequencies. This function gives a dynamic view of the cross correlation of the time series $x(t)$ in relation to $y(t)$; that is, a linear predictability of $y(t)$ from $x(t)$. If the coherence function is 0, $x(t)$ and $y(t)$ are completely unrelated. A coherence between 0 and 1 implies one or more of three possible situations: (1) Extraneous noise is present in the measurement; (2) the system relating $y(t)$ and $x(t)$ is not linear; and (3) $y(t)$ is an output due to an input $x(t)$ as well as other inputs. Studies of monthly rainfall-runoff data clearly indicate a strong measure of linear association at periods of 6 to 12 months.

It would appear that the measure of linearity proposed by Amorocho (1963) is incomplete. Deviation from linearity may be due to errors in (1) identifying time variability of the unit impulse response (Chiu, 1968; Chiu and Bittler, 1969), (2) defining system interrelationships (Jenkins and Watts, 1968), (3) space-time sampling, (4) ascertaining effect of initial conditions in distributed input functions, and (5) measurement errors (Kisiel, 1969).

2.3.13 Deterministic or Stochastic Systems

A deterministic system uniquely defines its output for specified input and initial and boundary conditions. This output is defined with certainty. That is, the same input gives rise to the same output. The input to a deterministic system may be either deterministic or stochastic.

On the other hand, a system is stochastic if its behavior is governed by laws of probability. That is to say, there is a certain element of chance associated with obtaining an output for a specified input. Let us quote Kisiel (1969):

> In the stochastic approach uncertainty by way of probability laws is woven on to the fabric of hydrodynamic and phenomenological relations which define mean value behavior of a system with zero mean-square error.

This definition is in accord with Chow (1964):

> If the chance of occurrence of the variables involved in such a process is ignored and the model is considered to follow a definite law of certainty, but not any law of probability, the process and its model are described as deterministic. On the other hand if the chance of occurrence of the variables is taken into consideration and the concept of probability is introduced in formulating the model, the process and its model are described as stochastic or probabilistic.

Chow (1970) further classified stochastic systems into pure probabilistic systems and nonpure probabilistic systems. In a pure probabilistic system the uncertain influence may vary in accordance with some unknown causes following a probabilistic distribution. If the uncertain influence is superimposed as a random variation over a deterministic system, the system is nonpure probabilistic. If the probability distribution of the uncertain influence varies with time, the stochastic system is nonstationary; otherwise, it is stationary. An example of a quasi-stochastic system is to treat rainfall and runoff over a watershed as stochastic and the watershed itself as deterministic.

The definitions of stochastic and deterministic systems are not as obvious as they might appear at first sight. The following example by Clarke (1973) illustrates how these terms are sometimes used differently by different investigators.

EXAMPLE 2.10

Consider the models (with $y_t = y(t)$):

$$y_t = a + b \cos \frac{2\pi t}{12} + \epsilon_t \tag{2.27}$$

where ϵ_t is a random variable having zero mean and constant variance σ^2_ϵ, and ϵ_{t-k} uncorrelated for all $k \neq 0$; and

$$\frac{d^2y(t)}{dt^2} = -\frac{4\pi^2}{144} by \tag{2.28}$$

where neither y nor b is a random variable. Parameters of equation (2.27) can be easily estimated by the least squares method, and the estimated output $\hat{y}_{t+k}$ at time $t + k$ is then

$$\hat{y}_{t+k} = a + b\cos\left[2\pi\frac{(t+k)}{12}\right] \quad (2.29)$$

where circumflexes denote estimated quantities. This has the same form as an integral of equation (2.28); some workers would therefore regard equation (2.29) as a deterministic model, or a model giving a deterministic forecast, because no residual term ϵ_{t+k} appears there. However, the random variable ϵ_t has been replaced by its expected value of 0, and y_{t+k} remains a random variable since a and b, calculated from data values containing random components, are random variables. This shows that the absence of an explicit term ϵ_t in an equation such as equation (2.29) is not a sufficient condition for it to represent a deterministic model. Most hydrologic systems are part deterministic and part stochastic; the degree of stochasticity varies from one system to another.

2.3.14 Lumped or Distributed Systems

Systems are often classified as *lumped* or *distributed* according to spatial characteristics of system variables and parameters. A lumped variable or parameter ignores spatial variability. On the other hand, if variation in one or more dimensions is taken into account, the system variable, or parameter, is distributed. Thus a system is lumped (Eagleson, 1967; Nash, 1969; Holtan, 1970) if its variables and parameters are lumped; it is distributed if its variables and parameters are distributed. Lumped systems are described by ordinary differential equations, and distributed sytems are described by partial differential equations. Difficulties arising from ''lumping in time''—that is, from treating daily rainfall and discharge from a watershed as average values occurring throughout that time interval—have been noted by Murray (1970).

A lumped system is often treated as a black box. The system function is determined by input-output data. No consideration is given to the internal structure of the system. A unit hydrograph representation is a classical example of a lumped rainfall-runoff system. The unit hydrograph is essentially derived by using historical rainfall-runoff data. Many unit hydrograph representations of lumped rainfall-runoff systems have been developed in hydrology. These are discussed later in the text. Because such representations do not explain the mechanics of system operation, they are sometimes alleged to be an exercise in curve fitting.

Becker (1973) distinguished two classes of distributed systems that account for spatial variability in input variables or system parameters: (a) the probability distributed model and (b) the geometrically distributed model.

A probability distributed model describes spatial variability without reference to the geometrical configuration of the points in the network at which an input variable such as rainfall is measured or estimated. Becker points out that the Stanford watershed model (Crawford and Linsley, 1962, 1964, 1966) is of this type, since infiltration is assumed to have a rectangular probability distribution over the watershed.

A geometrically distributed model expresses spatial variability in terms of the orientation of the network points, their positions and areal distribution. Nash and Sutcliffe (1969) state that a geometrically-distributed model treats a catchment not as a random assembly of different parts but a system whose parts are related to each other by their common geomorphological history.

EXAMPLE 2.11

If a network of N raingages records rainfall $x_1(t)$, $x_2(t)$, . . . , $x_N(t)$ in the time interval $(t-1, t)$, then the spatially lumped model combines the N readings into a single input variable $\bar{x}(t)$. A probability distributed model considers the rainfalls $x_1(t)$, $x_2(t)$ . . . to be values selected at random from a probability distribution. A geometrically distributed model describes the variability of the N readings by statistical relations such as those discussed by Whittle (1962, 1964), Matern (1960), and Bartlett (1966).

EXERCISES

2.1 Consider a hydrologic system with which you are familiar. Specify its input, output, system function, and parameters.

2.2 Describe a part of the hydrologic cycle with which you are familiar using the nomenclature of the systems approach. Show the relationship of this part of the hydrologic cycle to other parts by means of a sketch. Subdivide your system into components and describe it with a sketch.

2.3 Consider the part of the hydrologic cycle in 2.2. What are the basic physical laws describing its operation? Would you say that they are well known? Can you think of a useful material model as well as a symbolic model to help understand the system?

2.4. Give at least two examples, preferably from hydrology, of each type of system discussed in Section 2.3.

2.5. Consider a groundwater system in which stream and aquifer interact. Give two simple mathematical formulations representing the system such that one is linear in a systems theory sense and the other is linear in a statistical regression sense.

2.6. Select a watershed for which at least 30 rainfall-runoff events are available. Apply the correlation function and the Rogers measure to measure linearity of this watershed.

2.7. Many hydrologic models are nonlinear but can be reduced to linear models through appropriate transformations. Give at least two examples of nonlinear models that can be transformed to be linear using (a) scale transformation and (b) logarithmic transformation.

2.8. Give at least one example of each type of system. Select hydrologic examples as far as possible.

2.9. Why are most hydrologic systems highly damped and stable?

2.10. Many hydrologic systems exhibit different characteristics at different space-time scales. Give two examples illustrating this behavior.

CHAPTER THREE

Problems and Models in Hydrology

The concepts of systems approach presented in the preceding chapter can form the basis for classification of problems and models in hydrology. Classification of models assists in recognizing the features they have in common and the respects in which they differ. Two benefits accrue as a result: (1) consistency of parameter estimation with model assumptions is evaluated and (2) various types of models are brought together as complementary rather than competing alternatives.

3.1 CLASSIFICATION OF PROBLEMS IN HYDROLOGY

Given the scope of hydrology, the hydrologic problems can be broadly classified into two groups: (1) scientific or technological problems and (2) management or behavioral problems. In this text we shall deal with some aspects of scientific problems.

Although science and technology are vital to the continued development of natural resources to serve people's needs, the social structure within which technology is applied is equally vital. Often hydrologic systems have to operate within a constraining behavioral framework. It is therefore appropriate that the study of scientific aspects of hydrologic systems be complemented by the study of behavioral aspects.

3.1.1 Technological Problems

We introduce these problems in the next section in some detail and try to develop techniques based on the systems approach to solve some of these problems. The study, control, and operation of hydrologic systems continue to remain the central technological problems. These problems are compounded by lack of sufficient data, uncertainty in measurements, and their stochastic behavior. Most hydrologic problems are nonlinear, part deterministic, part stochastic, distributed, time-variant, and dynamic in nature. They experience inputs that cannot be controlled. Their study requires conceptualization in both space and time domains. In many cases, this conceptualization forms the basis for development of a particular methodology.

3.1.2 Management Problems

The obvious natural habits of water often ignore the jurisdictional boundaries created by men and give rise to the need for collective action by both public and private organizations. In fact, scientific and technological capabilities to handle water—water management needs—are almost powerless unless translated by effective and adequate institutional arrangements into significant social values. Modernization of water law, political institutions related to water resources development and management, and assessment of existing water resources policies are some of the examples of management problems.

Many hydrological problems can be studied effectively by resorting to the systems approach. The extent to which the systems approach can be profitably employed becomes evident by consideration of an example relating to the task of developing a generalized planning system to predict and study the future water requirements of a community. Various facets of such an effort are diagrammatically shown in Fig. 3.1. It is clear from the figure that the social structure within which science and technology are applied is as vital as the technology itself.

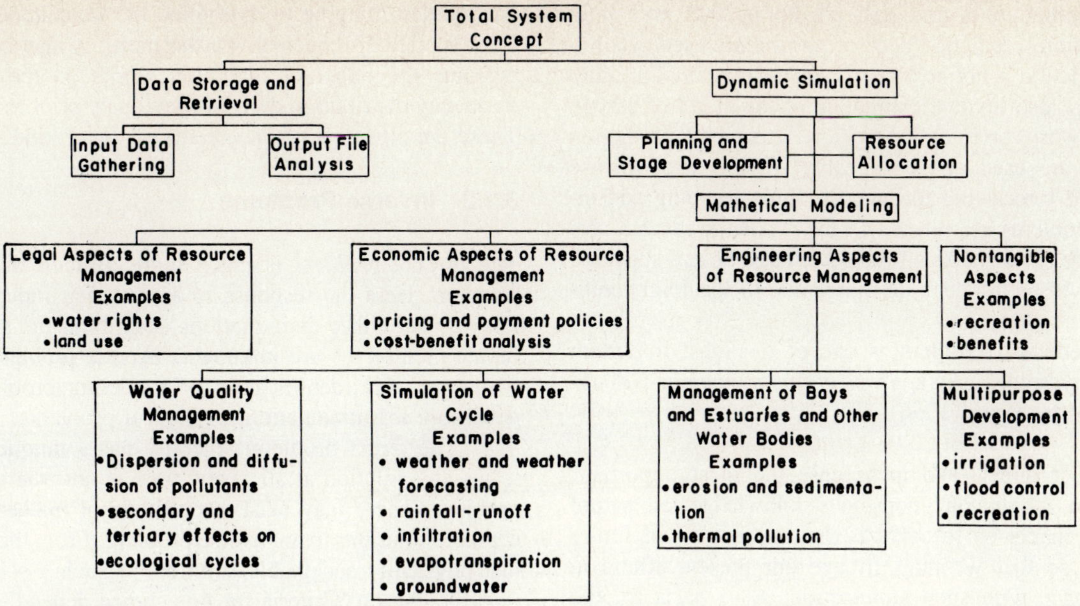

FIGURE 3.1 Nature of hydrologic problems from a systems point of view in the context of resource planning and management (after V. Vemuri and N. Vemuri, 1970).

3.2 PROBLEMS ASSOCIATED WITH THE SYSTEMS APPROACH

The problems associated with hydrologic systems can be broadly classified, as shown in Table 3.1, into three groups: (1) direct (forward) problems, (2) inverse (indirect) problems, and (3) simulation problems. Of these, direct problems are in the classical tradition.

3.2.1 Direct Problems

Direct problems are characterized by a complete specification of the system, such as in terms of a mathematical equation. The objective is to study or predict the system response for a specified input. Mathematically, a direct problem consists in the solution of a differential equation, integral, integro-differential, or some such functional equation (Lee, 1964).

Direct problems can be of four types: (1) analysis problems, (2) estimation problems, (3) prediction problems, and (4) frequency problems. Many problems in hydrology and water resources are direct in kind.

TABLE 3.1

Problems associated with hydrologic systems

Problem	Input	System Function	Output
Direct			
1. Analysis	×	×	?
2. Estimation	×	×	?
3. Prediction	×	×	?
4. Frequency	×	×	?
Indirect			
1. Identification	×	?	×
2. Detection	?	×	×
Simulation			
1. Identification	×	?	×
2. Analysis	×	×	?

× denotes *known* and ? denotes *to be determined.*

3.2.1.1 Analysis Problems

The system and the input along with initial and boundary conditions are specified, and the system response is to be determined. As an example, given the instantaneous unit hydrograph for a watershed, an analysis problem may be to determine storm runoff due to a specified rainfall excess pattern on the watershed. Another example may be to determine the time of concentration for known watershed characteristics.

3.2.1.2 Estimation Problems

The estimation problem consists in determining past, present, and future states of a system in the presence of noise. These lead to smoothing, filtering, and prediction problems, respectively (Liebelt, 1967). For given measurements of the observed state $x(t)$ up to time T, the estimation problem is to determine $\bar{x}(t|T)$ that is a best estimate of $x(t)$ in some sense. If $t < T$, the time series $x(t)$ is smoothed to estimate the past state of the system. For $t = T$, the time series $x(t)$ is filtered to estimate the present state. If $t > T$, the time series $x(t)$ is extrapolated to estimate the future states. The prediction problem is of particular significance in hydrology and is discussed separately.

3.2.1.3 Prediction Problems

The term *prediction* is not used consistently in hydrology. In estimation theory it explicitly means estimation of future states of a system at specified future times and locations. Weather, flood, and drought forecasting fall into this category. In other instances, it simply means specifying the magnitude of an event with an average probability of occurrence in the future.

Here the assumption is that past conditions will prevail in the future. Often, prediction and forecasting are used synonymously. Prediction is not necessarily the estimation of a future event but may also involve estimation of the effect of certain management works on river flow regime. Frequently prediction and analysis are used interchangeably. In this text we use prediction and forecasting to have the same meaning and include such problems as: When will the discharge reach a certain magnitude? How large will the discharge be at a specified time? What will be the peak discharge due to specified conditions?

The prediction problem is one of the most important problems in hydrology. An analysis of this problem is perhaps best described by Hertz (1956):

> The most direct, and in a sense the most important, problem which our conscious knowledge of nature should enable us to solve is the anticipation of future events, so that we may arrange our present affairs in accordance with such anticipation. As a basis for the solution of this problem, we always make use of our knowledge of events which have already occurred, obtained by chance observation or by prearranged experiment. In endeavoring then to draw inferences as to the future from the past, we always adopt the following process. We form for ourselves images or symbols of external objects; and the form which we give them is such that the necessary consequences of the images in thought are always the images of the necessary consequents in nature of the things pictured. In order that this requirement may be satisfied, there must be a certain conformity between nature and our thought. Experience teaches us that the requirement can be satisfied, and hence that such a conformity does in fact exist. When from our accumulated previous experience we have once succeeded in deducing images of the desired nature, we can then in short time develop by means of them, as by means of models, the consequences which in the external world only arise in a comparatively long time, or as the result of our own interposition. We are then enabled to be in advance of the facts, and to decide as to present affairs in accordance with the insight so obtained. The images which we here speak of are our conception of things. With the things themselves they are in conformity in *one* important respect, namely, in satisfying the above mentioned requirement. For our purpose it is not necessary that they should be in conformity with the things in any other respect whatever. As a matter of fact, we do not know, nor have we any means of knowing, whether our conceptions of things are in conformity with them in any other than this *one* fundamental respect.

3.2.1.4 Frequency Problems

A *frequency problem* may involve the determination of frequency of a specified magnitude of a hydrologic event. The form of the frequency distribution from which the desired frequency is to be determined is known. On the other hand, the problem may be to determine the magnitude of an event for a specified frequency. Furthermore, it may be required to evaluate the effect of hydraulic works on the form of the frequency distribution derived previously prior to these works. These are all frequency problems of direct kind.

3.2.2 Inverse Problems

The *inverse problem* is much more difficult than the direct problem. Here the response to a particular input or inputs is known, but either the equations describing the system or the inputs themselves are unknown. Inverse problems can be of two types: (1) identification or modeling problems and (2) detection, instrumentation, or control problems.

The direct problem generally has a unique solution, if it has any solution at all. This same is not true of an inverse problem. There may exist a multitude of mathematically acceptable solutions to an inverse problem. From these solutions, one physically acceptable solution, if such a solution exists, must be chosen. Criteria for this choice depend on the nature of the inverse problem.

3.2.2.1 Identification or Modeling Problems

The identification problem is one of the most difficult problems in hydrology (V. Vemuri and N. Vemuri, 1970) and generally arises from our need to control the behavior of a physically existing system. An identification problem consists of determining a mathematical characterization of the system, given a finite set of observations on input and output.

There may exist more than one mathematical solution to the identification problem. Therefore, criteria must be specified to help choose one out of these many solutions. The criterion to be employed here is different from that in the design problem and must be based on physical plausibility, not physical realizability of the system.

The identification problem is somewhat different from the design problem or the synthesis problem. While the validity of a particular design can always be checked either by studying the model or the prototype, it is almost practically impossible to verify the validity of an identified model. There are many reasons for this predicament. The input is generally not controllable. The period of observation is always limited, and consequently the validity of the model is limited to the range of variation of input data used in identifying the model. Obviously a model cannot contain more information than is available in the data from which it is derived (Brillouin, 1962).

A typical example of an identification problem is to develop a rainfall-runoff model for a watershed. Observations on rainfall and runoff for a watershed under study are given. The objective is to determine the system function. Derivation of a unit hydrograph using rainfall-runoff records is one way to specify the system function. This specification, however, implies certain assumptions regarding the nature of the system.

The identification problem entails estimation of system parameters. Therefore, a method must be developed to estimate the system parameters, which may vary as a function of the environment. The best estimate of these parameters should be determined. Because "best" implies a criterion of fit or

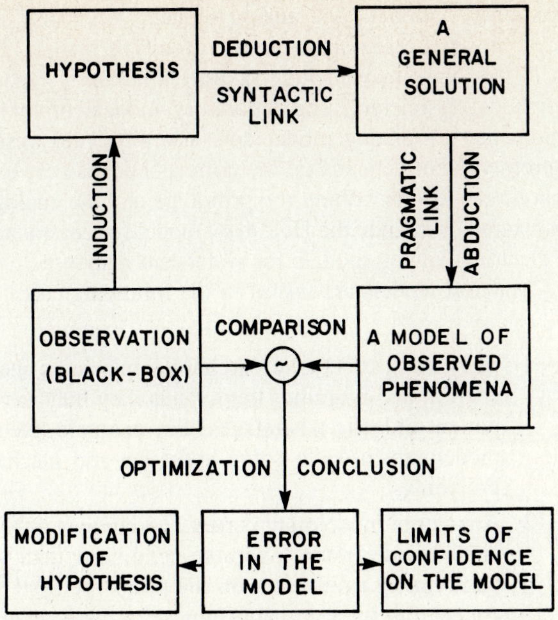

FIGURE 3.2 A block diagram of the problem-solving procedure (after Kisiel, 1969).

measure that is not absolute, the resulting estimate is not unique. Normally the quadratic error criterion, or the least squares criterion, is employed because of its convex properties, but it can be argued that nature has no such criterion (Guillemin, 1954). This essentially becomes a problem of optimization, wherein some prior knowledge of the system would be useful.

There are many ways to characterize the system. Whatever the method of system characterization, the computation process involved in solving the identification problem consists of three parts (Fig. 3.2):

1. Determination of the form of the model (that is, selection of the model dynamic equation) and isolation of unknown parameters.
2. Selection of a criterion by which the goodness of fit of the model responses to the actual system responses can be evaluated.
3. Selection of an algorithm for adjustment of parameters in such a way that the differences between the model and system responses, as measured by the second criterion, are minimized.

3.2.2.2 Detection, Instrumentation, or Control Problems

This problem is characterized by determination of an unknown input, given the output and the system function. Many situations arise in hydrology when the input is not known and the problem becomes one of detection. For example, given a streamflow record at a downstream point, it is desired to reconstruct the streamflow time series at an upstream point where no record exists. Another example is when watershed response is given but the corresponding effective rainfall is to be determined. Still another situation is to identify the parameters of a groundwater basin using explosives at certain places and study the disturbances caused by these detonations on the transient elevation of the water table elsewhere in the basin. The point is to study the system sensitivity (V. Vemuri et al., 1969) to an external disturbance and determine the system parameters using sensitivity analysis. The immediate instrumentation is to find the optimum nature of the externally introduced disturbance.

The problem of input detection is inherent in instrumentation. The instrumentation problem can also be formulated as a control problem. There our interest is to determine the characteristics of an external signal that drives a system from a given initial state to a desired terminal state. The essential elements of a control problem are (1) a mathematical model of the system to be controlled, (2) a description of the system output, (3) a set of admissible inputs or controls, and (4) a performance or cost function that is a measure of the effectiveness of a given control criterion. The task of determining a pumping-recharge policy toward an optimum groundwater basin management is an example of a control problem.

3.2.3 Design, Synthesis, or Simulation Problems

In a design problem the nature of the expected input and the nature of the required system response are specified. A system having this input-response relationship has to be physically realized or designed. Thus the design problem involves design of a nonexistent system that is normally required to satisfy certain desirable social goals according to some criterion such as cost, benefit, safety, or life span. As an example, it is possible to have many design alternatives for a surge tank that satisfy some stipulated performance requirements. To choose among many possibilities, it is essential to specify design specifications or criteria for choosing. Perhaps it is necessary to keep the cost low, meet the peak load demands, not allow the hydraulic head to fall below a prescribed level, maximize the safety, or increase the life span. In such a situation, more than one design may satisfy a specified criterion or no design may exist that satisfies all the specifications. Thus, an optimum design is optimum only for a particular criterion. Once the criterion is changed, the design may no longer be optimum.

The development of a computer-aided methodology for automatic design of wastewater collection and treatment systems is a design problem. If information about topography, flows, pipe, costs and excavation are specified, a minimum-cost trunk sewer can be designed. This problem can be solved using standard mathematical techniques. In the same vein, irrigation scheduling by computer is a typical design problem. Given information on crops, climate, water availability, soils and labor, an automated method of efficient irrigation scheduling leading to maximum crop production can be developed. In many parts of the world this method of irrigation scheduling has been in use for the past several years.

We do not design hydrologic systems in exactly the same sense as do the structural or mechanical engineers. We do, however, attempt to simulate complex hydrologic systems using simpler models. This essentially is a synthesis problem, wherein we wish to design a system converting a known input to a known output within specified limits of precision. This is a two-step procedure. First, a model is built. Second, its operation is tested by analysis. Thus identification, prediction, and analysis problems jointly constitute a simulation problem. That is why simulation problems are the most difficult of all. There are many simulation models in hydrology. Some notable examples include the Stanford Watershed Model (Crawford and Linsley, 1962, 1964, 1966), the USGS Flood Hydrograph Model (Dawdy, Litchy and Bergmann, 1970), the USDA Watershed Hydrology Model (Holtan and Lopez, 1973), and the National Weather Service Model (National Weather Service, 1972).

3.3 MODELING IN HYDROLOGY

Many discussions regarding modeling have appeared in the scientific literature, but the rationale for model building was perhaps best expressed by Rosenblueth and Wiener (1945):

> No substantial part of the universe is so simple that it can be grasped and controlled without abstraction. Abstraction consists in replacing the parts of the universe under consideration by a model of similar but simpler structure. Models, formal or intellectual on the one hand, or material on the other, are thus a central necessity of scientific procedure.

Most hydrologic systems are extremely complex, and we cannot hope to understand them in all detail. Therefore, abstraction is necessary if we are to understand or control some aspects of their behavior. Abstraction consists in replacing the system under consideration with a model of similar but simpler structure. The basic purpose of a model is to simulate and predict the operation of the system that is unduly complex and the effect of changes on this operation. The use of hydrologic models for prediction purposes arises largely because of the inadequacy of hydrologic data (Dooge, 1972). Even were the amount of data presently collected to be enormously expanded, it would still not give sufficient information about important hydrologic variables because the records would not cover for every set of interests either the extreme events or the behavior over a long period of time of the hydrological variables. Apart from these deficiencies, the increasing effect of human activity on the elements of hydrologic cycle will tend more and more to render hydrological data of limited use for the direct prediction of corresponding behavior in the future.

Hydrologic models can be variously classified (Dawdy, 1969a; Dawdy and Kalinin, 1969; DeCoursey, 1971; Snyder, 1971; Snyder and Stall, 1965; Woolhiser, 1971, 1973, 1975, 1982; Miller and Woolhiser, 1975). We distinguish them as (1) material and (2) symbolic or formal.

3.3.1 Material Models

A *material model* is the representation of the real system by another system, which has similar properties but is much easier to work with. Material models can be classified as iconic, or "look-alike," models and analog models. An iconic model bears a physical resemblance to the prototype system. Examples in this class may include rainfall simulators, laboratory watersheds, lysimeters, and laboratory models of hydraulic phenomena. An analog model does not physically resemble the prototype but depends on the correspondence between the symbolic models describing the prototype and the analog system. Examples include the Heleshaw model for coastal aquifer and electrical analog models for watershed response.

Material models are useful in the following cases:

1. They may assist the researcher in replacing a phenomenon in an unfamiliar field. Thus they may have important didactic advantages. For example, electrical models are used to solve hydraulic and mechanical problems.
2. A material model may permit experiments to be conducted under more favorable conditions than would be normally available with the prototype system. Of course, this translation presumes that there are reasonable grounds for supposing similarity between the two situations; it thus presupposes the possession of an adequate formal model with a structure similar to that of the two material systems. The formal model need not be thoroughly comprehended; the material model then serves to supplement the formal one.

Because the material model must share the most important properties of the complex prototype system, it should not be grossly oversimplified or too small to exhibit important scale-dependent behavior. At times the relation between a material model and its prototype may involve nothing more than a change of scale in space or time. Biological experiments are a good example of change in spatial scale, whereas genetic experiments show change in temporal scale.

In hydrology the physical size of the system is important. Also important is the nature of input. For example, experiments on natural watersheds are very time consuming and the experimenter has no control over the input. Although scale models are useful in very special cases and do not offer a viable tool, some hydrologic systems can be modeled utilizing a change in scale (Chery, 1966; Grace and Eagleson, 1966).

A material model that does not involve a change in scale may still be valuable because experiments can be carried out more conveniently or can be repeated at will (Woolhiser and Schulz, 1973). The experimental watershed systems at the University of Illinois (Chow, 1967), Kyoto University (Yano and Ishihara, 1968), and Colorado State University (Holland, 1969; and Woolhiser and Schultz, 1973) can be considered to be of prototype scale.

Even with prototype scale models the question of size is still important. Obviously a large model is more expensive and difficult to operate than a small one. The important consideration is that the material model must duplicate the most important characteristics of the complex system. So a large model is the one that exhibits all important scale-dependent behavior.

TABLE 3.2

Characteristics of language of models (after Lee, Adams, and Gaines, 1968)

Language of Models	*Characteristics*				
	Descriptive Capability	*Ambiguity*	*Manipulation Capability*	*Implementation Capability*	*Primary Function*
English text many words, voluminous	Good	Very ambiguous	None	Limited	Descriptive explanations and directions
Drawings and block diagrams (concise)	Good	Not ambiguous (not always)	None	Good	Design, assembly, and construction
Logical flow charts and decision tables (concise)	Fair	Not ambiguous	None	Good	Computer programming
Curves, tables, nomographs (concise)	Fair	Not ambiguous	Good	None	Expressing simple relationships between a few variables
Mathematics (concise)	Poor with respect to layperson	Not ambiguous unless assumptions are inconsistent	Excellent	Good	Problem solution and optimization—many variables

3.3.2 Formal Models

A *formal model* is a symbolic expression in logical terms of an idealized, relatively simple situation sharing the structural properties of the original system. Symbolic models can be variously expressed (Lee, Adams, and Gaines, 1968), as shown in Table 3.2. In this text we are concerned with symbolic models of mathematical nature.

A mathematical model expresses the system behavior by a set of equations, perhaps together with logical statements expressing relationships between variables and parameters. Equation (2.6) is an example of a mathematical model.

When the system to be modeled is complex, it may be adequate for many purposes to use relatively simple form $f^*(\cdot)$ for the system function $f(\cdot)$ of equation (2.6) and to express the lack of fit by means of a residual error. Then we can write the system function of equation (2.6) as (replacing partial derivatives by their finite difference approximations):

$$f^*(y_t, x_t; y_{t-1}, x_{t-1}; y_{t-2}, x_{t-2}; \ldots a_1, a_2; \ldots) + \epsilon_t = 0 \quad (3.1)$$

where ϵ_t is the residual at time t. Choosing a function $f^*(\cdot)$ sufficiently simple for the problem at hand is the art of model building.

Mathematical models (Dooge, 1977) are sometimes divided into (1) theoretical or physical models, (2) conceptual models, and (3) empirical models, as shown in Fig. 3.3. The latter two types of models are popular in systems hydrology. This division depends upon whether or not the form of a particular model is suggested by consideration of physical processes transforming input into output.

The division of mathematical models is rather arbitrary because one person's empiricism may be another person's theory. To illustrate, Darcy's law, like Newton's laws of motion, is a matter of observation and is hence empirical by strict definition—a point recognized by Mandelbrot (1970) and Nelder (1972). Furthermore, models formulated without explicit consideration of physical processes may have parameters that may have physical significance. On the other hand, models based on physical considerations may contain empirical components. Thus Diskin (1970) gave quasi-physical interpretations for the parameters of the linear regression equation

$$y_t = a + bx_t + \epsilon_t \quad (3.2)$$

where x_t and y_t are annual rainfall and runoff from a watershed. Although the differential equations describing turbulent open-channel flow are derived from considerations of conservation of mass and momentum, they require an estimate of empirical roughness coefficient, including Manning's n or Chezy's C (Eagleson, 1970).

This classification may obscure the fact that there are many models that do not satisfy all the general laws or describe the empirical circumstances accurately but still could not be called purely empirical. This characterization depends on the problem under study. For example, the St. Venant equations are overly simplified or are of little use for studying turbulence characteristics of flow. Similarly, the boundary layer concept is semiempirical from the standpoint of molecular theory of liquid.

Nevertheless, these models can be defined in an approximate sense. An empirical model is not based on physical laws. It merely presents the facts, that is, it is a representation of the data. If the conditions change, it has no predictive capability. Therefore, all empirical models have some chance of being fortuitous and in principle should not be used outside the range of data from which they were derived. Examples of empirical models may include the rational method, the models of Amorocho (1963) and Bidwell (1971), and unit hydrograph models based on harmonic analysis (O'Donnell, 1960), the least squares method (Snyder, 1955), and Laguerre polynomials (Dooge, 1965). Frequency models based on the concept of entropy (Sonuga, 1972; Singh, Rajagopal, and Singh, 1986), frequency factor (Chow, 1964), and distribution fitting also fall into this category.

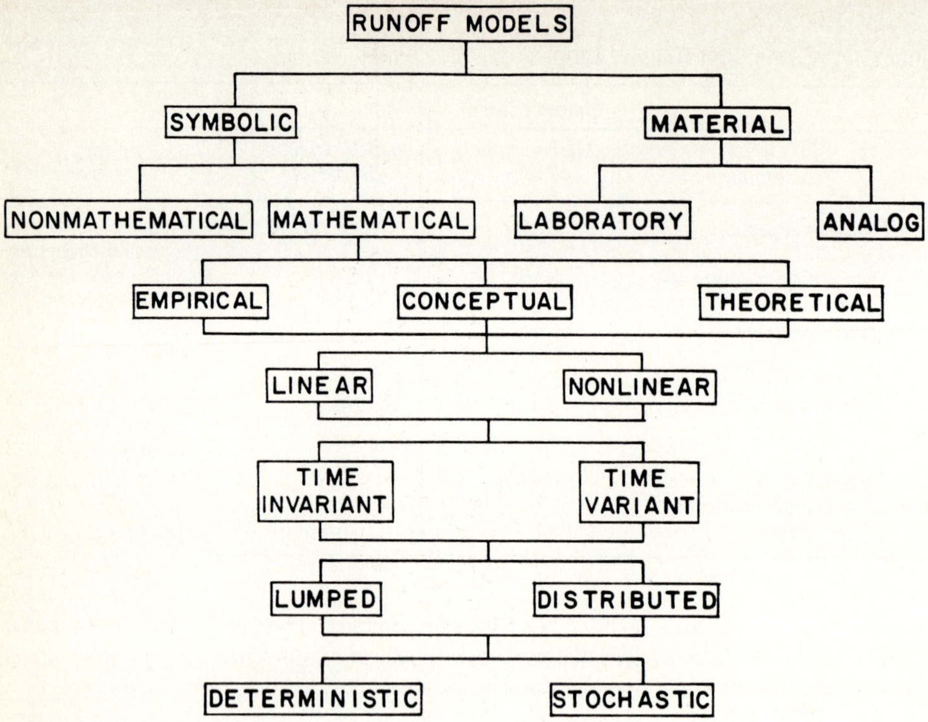

FIGURE 3.3 A classification of runoff models.

Theoretical models presumably are the consequences of the most important laws governing the phenomena. A theoretical model has a logical structure similar to the real-world system and may be helpful under changed circumstances. All theoretical models simplify and, therefore, are more or less wrong. It must be emphasized that the only perfect model of a physical system is the system itself. There is no such thing as a perfect mathematical model because an abstract quantity cannot perfectly represent a physical entity.

Examples of theoretical models may include watershed runoff models based on St. Venant equations, infiltration models based on two-phase flow theory of porous media (Morel-Seytoux, 1978), evaporation models based on theories of turbulence and diffusion (Brutsaert and Mawdsley, 1976), and groundwater models based on fundamental transport equations (Freeze, 1971).

Conceptual models are intermediate between theoretical and empirical models, although they can be used broadly to embrace both of these types of models. Generally, conceptual models consider physical laws but in highly simplified form. Examples of conceptual models may include rainfall-runoff models based on the spatially lumped form of the continuity equation and the storage-discharge relationship. Models of Nash (1958) and Dooge (1959) are conceptual models. On the other hand, conceptual models may also include those based on highly simplified versions of St. Venant equations. For example, models derived from linear diffusion analogy and linearized versions of St. Venant equations (Harley, 1967; O'Meara, 1968) may be considered conceptual models.

All three types of mathematical models are useful but in somewhat different circumstances. Each has its own effectiveness, depending upon the objective of study, the degree of complexity of the problem, and the degree of accuracy desired. For example, theoretical models aid in understanding a process and generally yield information in greater detail in both time and space. They contain parameters that, in principle, have physical significance and can be estimated by independent measurements.

Empirical models do not aid in physical understanding. They contain parameters that may have little direct physical significance and can be estimated only by using concurrent measurements on input and output. In many situations, empirical models can yield accurate answers and can, therefore, serve a useful purpose in decision-making.

Conceptual models provide useful results efficiently and economically for some problems. They contain parameters, some of which may have direct physical significance and can, therefore, be estimated by using concurrent observations on input and output.

A theoretical model generally yields more information than an empirical one. However, there is no conflict between the two; they represent different levels of approximation of reality. The former can supplement the latter by indicating its limits of validity or by pointing to external factors that may modify it. An excellent essay on empirical and physical models in hydrology was recently presented by Klemes (1982). This essay provides in detail a philosophical perspective on (1) hydrologic modeling in general, (2) merits and demerits of empirical modeling, (3) why hydrological models work, and (4) drawbacks, dangers, and potential benefits of causal modeling.

3.4 SENSITIVITY ANALYSIS

Sensitivity analysis studies the changes in the optimal solution with the changes in the system components, the input, or the parameters including those of the objective function. It also includes the rate of change in one factor with respect to change in another factor. Generally, sensitivity analysis is divided into two parts: (1) parametric sensitivity and (2) component sensitivity.

3.4.1 Parameter Sensitivity

Consider a solution S to be a function of parameters p_i, $i = 1, 2, \ldots, n$:

$$S = f(p_1, p_2, p_3, \ldots, p_n) \tag{3.3}$$

The change in S resulting from a change in p_i can be expressed explicitly by considering a Taylor series expansion,

$$f(p_i + \Delta p_i, p_j, j \neq i) = S + \frac{\partial S}{\partial p_i}\Delta p_i + \frac{1}{2}\frac{\partial^2 S}{\partial p_i^2}\Delta p_i^2 + \cdots \tag{3.4}$$

If the nonlinear terms are small in comparison with linear terms, equation (3.4) reduces to

$$f(p_i + \Delta p_i, p_j, j \neq i) = S + \frac{\partial S}{\partial p_i}\Delta p_i \tag{3.5}$$

Therefore,

$$\Delta S = f(p_i + \Delta p_i, p_j, j \neq i) - S = \frac{\partial S}{\partial p_i}\Delta p_i \tag{3.6}$$

Equation (3.6) can be considered as a linearized sensitivity equation. It measures the change in S that results from the change in p_i. This equation can be extended to the case where more than one parameter is changed simultaneously. The general definition of sensitivity s can thus be derived by inserting equation (3.3) in equation (3.6):

$$s = \frac{\partial S}{\partial p_i} = \frac{[f(p_i + \Delta p_i, p_j, j \neq i) - f(p_i, p_2, \ldots p_i, \ldots p_n)]}{\Delta p_i} \tag{3.7}$$

Thus, s can be determined either by direct differentiation or perturbation methods.

The importance of parameter sensitivity for developing, evaluating, and using hydrologic models has long been recognized (Dawdy and O'Donnell, 1965; Dawdy, 1969b; DeCoursey and Snyder, 1969; Ibbit and O'Donnell, 1970; Green, 1970; Vemuri et al., 1979). Indeed, most optimization procedures for model calibration entail sensitivity analysis in one way or another. The steepest-descent method (Cauchy, 1847) the pattern-search method (Green, 1970), the DeCoursey-Snyder method (DeCoursey and Snyder, 1969), and the Rosenbrock method (Rosenbrock, 1960; Dawdy and O'Donnell, 1965) are some of the optimization methods frequently used in hydrology. The relative sensitivity of a parameter has been used as a measure of the importance of that parameter to the physical process. However, the mathematical rationale for such a correspondence has yet to be presented.

Parameter sensitivity can be used to determine an appropriate optimization algorithm to obtain optimal model parameters. For example, the greater the sensitivity of the model response to a parameter, in general, the closer and the sooner that parameter will be optimized. Insensitive parameters often do not reach their true values. This suggests that only those parameters that are sufficiently sensitive should be included in optimization. A criterion that would indicate the potential of a parameter to converge to the true value would be needed. Such a convergence criterion could be used to identify which parameters should be included in optimization and subsequently related to physical characteristics.

3.4.2 Component Sensitivity

Component sensitivity measures the effect of variation in the input function I on the output function S (McCuen, 1973a, 1973b; Singh and Buapeng, 1977).

$$s = \frac{\partial S}{\partial I} = \frac{\Delta S}{\Delta I} \tag{3.8}$$

Component sensitivity can be used to examine if some components of a model can be dropped without seriously jeopardizing its predictive value. It can be used to identify the relative importance of individual subsystems within the system. Thus, it is an important tool in planning, design, and management of water resource systems (James, Bower, and Metalas, 1969; Texas Water Department Board, 1971). To illustrate, sensitivity of input parameters such as total water demand and power costs provide the means for selecting a water management plan other than the minimum cost plan if the risk for that plan is considered too high.

3.5 ERROR ANALYSIS

Error analysis deals with the effects of errors resulting from various sources on model output. Four sources of error can be identified (von Neumann and Goldstein, 1947): (1) Mathematical models are approximate idealized representations of physical systems, (2) model parameters are estimated from experimental or historical data and are thus subject to data and estimation errors, (3) model solutions entail truncation error because of the need to use transcendental operations and implicit definitions, and (4) computing procedures introduce roundoff errors in their performance of elementary operations. Of these the first two are of major significance.

Errors resulting from these sources in model results are of two types: (1) random, and (2) systematic. Random errors occur when the model shows no tendency to overestimate or underestimate for a number of successive time intervals. On the other hand, systematic errors occur when the error tends to persist over a series of time intervals without changing its

TABLE 3.3

Qualitative comparison of errors in hydrologic modeling with analogous errors in standard statistical analysis (after Dawdy, 1969b).

Sources of Error	*Size of Error*	*Statistical Analogous*
Data	a	Measurement and sampling error
Comparison of measured to simulated during period used for fitting	$a - b$	Standard error of estimate
Comparison of measured to simulated during period not used for fitting	$a + c$	Standard error of prediction

sign. Random errors in the data will undoubtedly produce random errors in the output. Systematic errors in the data will be reflected in incorrect parameter values (Aitken, 1973). The importance of one type of error relative to the other depends on the problem at hand.

Errors in data are reflected in the optimized model parameters. If perfect input data were routed through a perfect model, the output record would agree perfectly with an error-free output record. If errors are introduced into the input or output record or both, the output is not exactly reproduced by a perfect model. If an optimization procedure is used, the parameters deviate from their true values in order to minimize the deviations between the observed and predicted values as specified in the objective function. The optimal set of parameters is now in error, and the fitted values of the objective function are different from its true value.

This process is analogous to statistical least squares analysis. The optimized parameters deviate from their population values because of random errors in the data. The standard error of estimate is a measure of error of reproduction in the fitted data. The standard error of prediction, however, is somewhat greater than the standard error of estimate, for it includes both the measure of lack of fit of the data to calibrate the model and the measure of error in the optimized parameters. These relationships are shown in Table 3.3. If the assumptions of regression theory hold, the error of prediction can be computed from the standard error, the deviations of the independent variables can be computed from their mean, and the error in the coefficients can be determined from the independent variables. These assumptions seldom hold, however, and there is no theory by which to compute the error of prediction.

Nonlinearity of hydrologic processes precludes any theoretical description of the mechanism by which errors in data are transferred to model parameters and combined with input errors in the test period to produce errors in the simulated output. Such an analysis will give insight into the calibration process, will show how errors in data are transmitted to the model, and will give a measure of the effect of data errors on the accuracy of predictions. Data errors have different results depending upon the type of data used. All errors transmit part of their error to model parameters and then to model results but each one does so differently. Table 3.3 sketches the effect of errors in data on the model calibration process.

Transmittal of errors in input to output will depend on model structure. One model may be more sensitive to these errors than the other, as shown in Fig. 3.4 (Singh and Woolhiser, 1976). In evaluating a model and then comparing models, transmittal of errors must be carefully studied. Equations for error analysis are similar to those for sensitivity analysis. Indeed, clues may be obtained if modifications in a particular

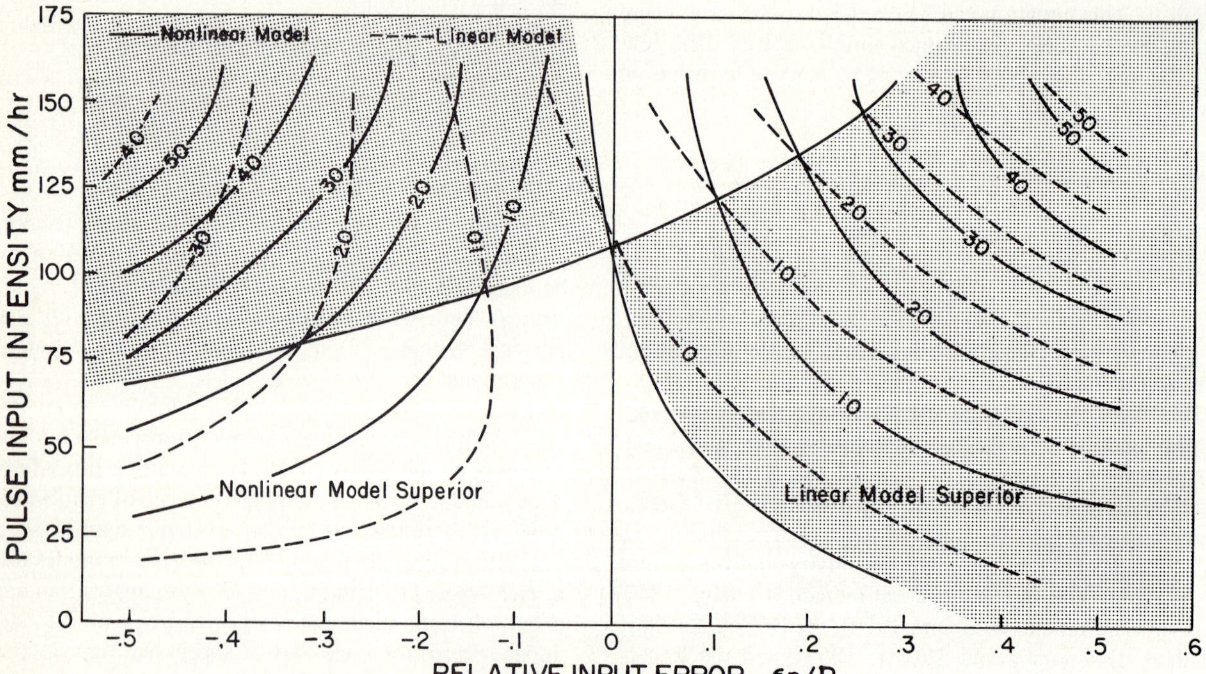

FIGURE 3.4 Output error surfaces for n = 1.5 (mm/h) (after Singh and Woolhiser, 1976).

model are necessary or if one model is superior to the other. In hydrology, error analysis has not been effectively used as a tool in model development, calibration, and testing.

3.6 METHODOLOGY FOR USING HYDROLOGIC MODELS

Dooge (1972) outlined a rational methodology for the use of hydrologic models for prediction purposes. This methodology consists of a number of steps. These, with slight modifications, are as follows:

1. Define the problem.
2. Specify the objective.
3. Study the data availability.
4. Determine the computing facilities available.
5. Specify the economic and social constraints.
6. Choose a particular class of hydrologic models.
7. Select a particular type of model from the given class.
8. Calibrate the model (that is, optimize the parameters).
9. Evaluate the performance of the model.
10. Use the model for prediction purposes.
11. Embed the model in a more general model.

In a systematic application of the methodology, steps 6–9 could be iterated until a satisfactory model was obtained. Step 9 is, however, crucial in this entire operation. Only when a model has been objectively calibrated and evaluated can it be applied to a specific problem with assurance that the best use is being made of the data and that something is known about the order of magnitude of the accuracy of prediction. There exists a multitude of hydrologic models. However, a rational methodology for their choice is yet to be developed. Consequently, the user does not know which model to choose and why.

EXERCISES

3.1. Give at least two examples from hydrology of each type of problem associated with systems approach.

3.2. Give at least two examples from hydrology of each type of model discussed in Section 3.2.

3.3. Consider a simple model as in equation (2.24). Numerically show its component as well as its parameter sensitivity.

3.4. A flood-control study is being called for before determining whether and what kinds of flood control works should be undertaken to protect a given metropolitan area from flooding. Briefly outline both the management and technological problems that should be addressed.

3.5. What kinds of problems can be envisioned in interbasin transfer of water?

3.6. Outline the problems to be addressed in settling disputes on sharing water of an international river.

3.7. Summarize advantages and disadvantages of each type of hydrologic model.

3.8. What role does sensitivity analysis play in parameter optimization?

3.9. What types of hydrologic problems are more amenable to laboratory experimentation?

3.10. Outline possible sources of error in a hydrologic model. Comment on their relative significance. Which sources of errors are better understood?

CHAPTER FOUR

Elementary Functions and their Hydrologic Responses

The elementary functions play a very important role in analysis and synthesis of hydrologic systems. We can resolve any arbitrary system input or output into these elementary functions. We can conveniently study the response of a hydrologic system to the elementary functions and combine their responses to obtain the response to the applied input. It is, therefore, necessary to understand these elementary functions and some of their properties. We study these functions briefly for both continuous-time and discrete-time systems. For further details the reader is referred to such standard texts on linear systems as Eykhoff (1974), Gupta (1971), Schwarz and Friedland (1965), and Zadeh and Desoer (1963), among others.

4.1 ELEMENTARY FUNCTIONS FOR CONTINUOUS-TIME SYSTEMS

The elementary functions, also called singularity functions, can be defined as

$$u_{i+1}(t) = \int_{-\infty}^{t} u_i(\tau)d\tau, \qquad i = \ldots, -2, -1, 0, 1, 2, \ldots \tag{4.1}$$

$$u_1(t) = \begin{cases} 0 & t = 0 \\ 1 & t \geq 0 \end{cases} \tag{4.2}$$

The functions $u_2(t)$, $u_1(t)$, $u_0(t)$ and $u_{-1}(t)$ are generally referred to as the unit ramp, unit step, unit impulse, and unit doublet, respectively. The unit impulse and unit step are most frequently used in hydrology and deserve special emphasis.

4.1.1 Unit Ramp

The *unit ramp* $u_2(t)$ is shown in Fig. 4.1 and can be defined as:

$$u_2(t) = \begin{cases} 0 & t < 0 \\ t & t \geq 0 \end{cases} \tag{4.3}$$

4.2.1 Unit Step

The *unit step* $u_1(t)$, as shown in Fig. 4.2, can be defined as

$$u_1(t) = \begin{cases} 0 & t < 0 \\ 1 & t \geq 0 \end{cases} \tag{4.4}$$

This has several important hydrologic implications. As will be seen, it is an integral of the unit impulse. It defines the input for the *S*-hydrograph (abbreviation for summation-hydrograph). It can be used to define the unit pulse which forms input for the unit hydrograph.

4.1.3 Unit Pulse

The *unit pulse* of width τ and height $1/\tau$, as shown in Fig. 4.3, can be defined as

$$u_\tau(t) = \begin{cases} \dfrac{1}{\tau} & 0 \leq t < \tau \\ 0 & t \geq \tau \end{cases} \tag{4.5}$$

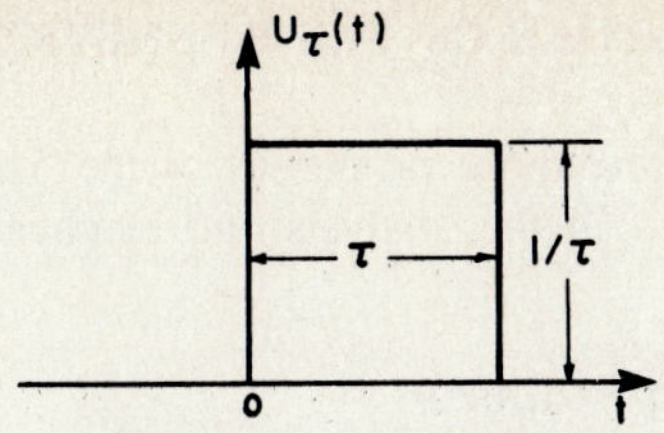

FIGURE 4.1 The unit ramp.

The unit pulse is a rectangular pulse whose area is unity, and it can be defined by using the unit step. To illustrate, the unit pulses of duration τ of Fig. 4.5(a)–(c) can be defined respectively as

$$\text{(a)} \quad u_\tau = \frac{1}{\tau}\left[f\left(t + \frac{\tau}{2}\right) - f\left(t - \frac{\tau}{2}\right)\right] \tag{4.6}$$

$$\text{(b)} \quad u_\tau = \frac{1}{\tau}\left[f(t) - f(t - \tau)\right] \tag{4.7}$$

$$\text{(c)} \quad u_\tau = \frac{1}{\tau}\left[f(t + \tau) - f(t)\right] \tag{4.8}$$

The unit pulse is frequently used in hydrology and can be used to define the unit impulse in limit. For example, if we allow $\tau \rightarrow 0$, equations (4.5)–(4.8) will represent the unit impulse.

4.1.4 Unit Impulse

The *unit impulse* $u_0(t)$, the Dirac delta function, or delta function (δ-function) $\delta(t)$ for brevity, can be defined as

$$u_0(t) = \delta(t) = 0, \qquad t \neq 0 \tag{4.9}$$

and

$$\int_{-\infty}^{t} \delta(w)\, dw = 1, \qquad t \neq 0 \tag{4.10}$$

From equation (4.4) it is evident that

$$u_1(t) = \int_{-\infty}^{t} \delta(w)\, dw = 1, \qquad t \neq 0 \tag{4.11}$$

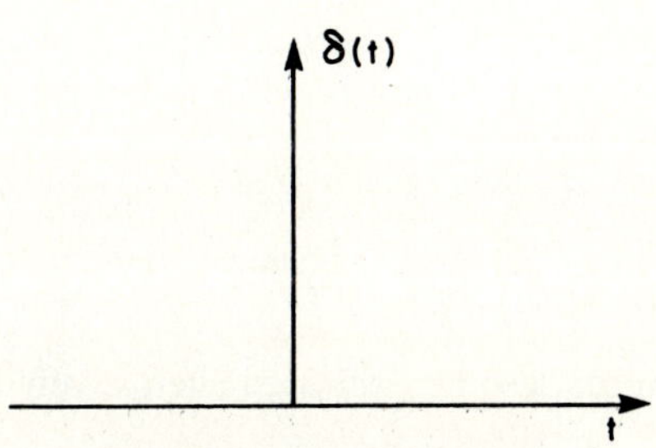

FIGURE 4.2 The unit step.

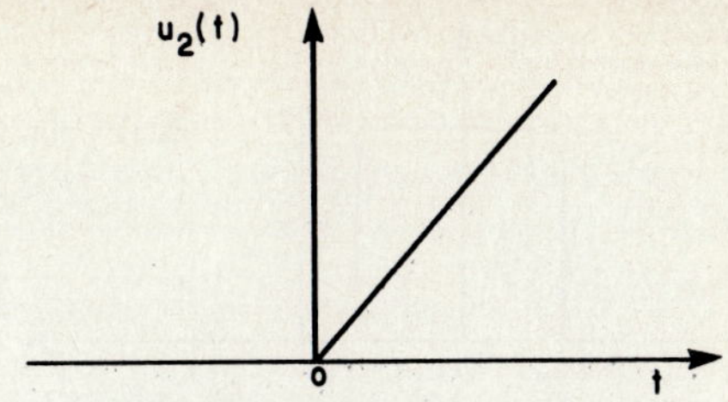

FIGURE 4.3 The unit pulse.

The δ-function is sketched in Fig. 4.4. Equation (4.9) states that the δ-function possesses undefined magnitude at the time of occurrence and is zero elsewhere. Further, the area under this function is unity. Indeed, such a function is physically difficult to imagine. However, in a limiting sense we can think of it as a pulse having a very high magnitude and infinitesimally small duration such that its area is unity. The pulse can be of any arbitrary shape. In fact, there are many functions that satisfy equation (4.10) in limit and can serve to represent the δ-function. All these various representations can be considered equivalent; some are shown in Fig. 4.5.

4.1.5 Unit Doublet and Higher-Order Delta Functions

Higher-order δ-functions are defined as derivatives of $\delta(t)$ and can be expressed as

$$\delta^{(n)}(t) = \frac{d^n(\delta(t))}{dt^n} \tag{4.12}$$

When $n = 1$, equation (4.12) defines the unit doublet, that is,

$$u_{-1}(t) = \delta^{(1)}(t) \tag{4.13}$$

It can be considered as the derivative of functions that define the δ-function in limit. Similarly, higher order δ-functions define the unit triplet, unit quadruplet, and so on, although these are hard to imagine physically.

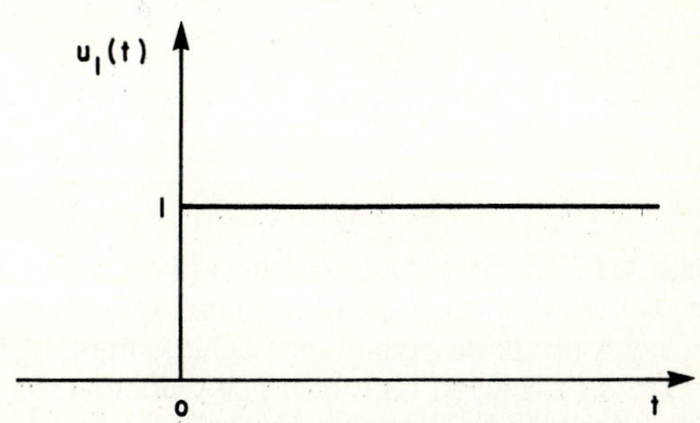

FIGURE 4.4 The δ-function.

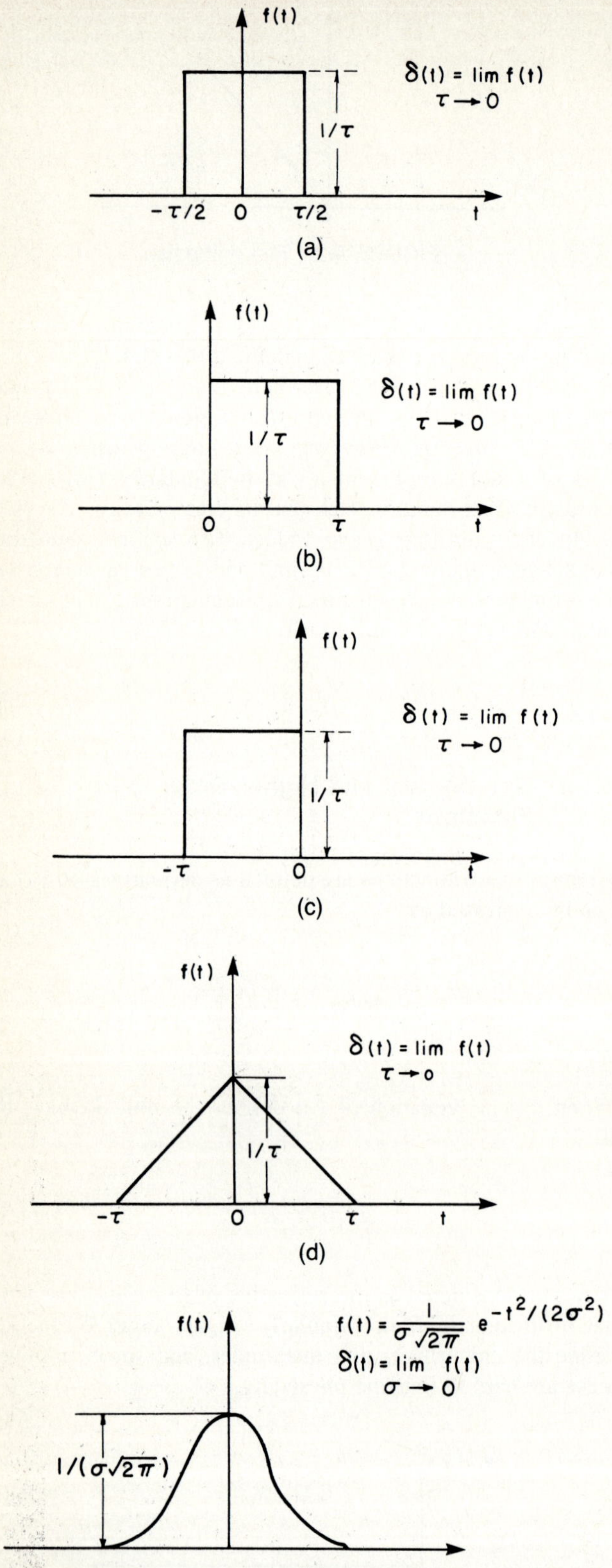

FIGURE 4.5 Representations of the δ-function. (a) δ-function as a limiting form of an even pulse. (b) δ-function as a limiting form of a pulse approaching from right. (c) δ-function as a limiting form of a pulse approaching from left. (d) δ-function as a limiting form of an even triangular pulse. (e) δ-function as a limiting form of Gaussian probability function.

4.2 PROPERTIES OF THE δ-FUNCTION

We briefly define those properties of the δ-function that are frequently used in the analysis and synthesis of hydrologic systems.

4.2.1 Sifting Property

The *sifting* property states that the δ-function assumes the value of the function at the time the δ-function is applied, provided that function is continuous, has continuous derivatives of all orders and is zero outside a finite interval. This property can be mathematically expressed as

$$\int_{-\infty}^{\infty} f(t)\,\delta(t-a)\,dt = f(a) \tag{4.14}$$

in which a is constant. This can be easily shown using integration by parts. It is, however, more intuitive to derive it by decomposing the limits of integration. Rewriting equation (4.14) as

$$\begin{aligned} F &= \int_{-\infty}^{\infty} f(t)\delta(t-a)\,dt \\ &= \int_{-\infty}^{a-\epsilon} f(t)\delta(t-a)\,dt + \int_{a-\epsilon}^{a+\epsilon} f(t)\delta(t-a)\,dt \\ &\quad + \int_{a+\epsilon}^{\infty} f(t)\delta(t-a)\,dt \end{aligned}$$

Referring to Fig. 4.6, since the product of two functions within the integral sign vanishes everywhere except at $t = a$, the first and third integrals vanish for any positive value of the time increment ϵ. This yields

$$F = \int_{a-\epsilon}^{a+\epsilon} f(t)\delta(t-a)\,dt$$

As $\epsilon \to 0$, $f(t) \to f(a)$. Hence,

$$F = f(a)\int_{a-\epsilon}^{a+\epsilon} \delta(t-a)\,dt = f(a)$$

or

$$\int_{-\infty}^{\infty} f(t)\delta(t-a)\,dt = f(a)$$

This property can be generalized for any value of a, that is, a is variable or any value of t. Let the variable a be denoted by s. Then

$$f(s) = \int_{-\infty}^{\infty} f(t)\delta(t-s)\,dt$$

This relation can also be expressed alternatively as

$$f(s) = \int_{-\infty}^{\infty} f(t)\delta(s-t)\,dt, \qquad \text{for all } s$$

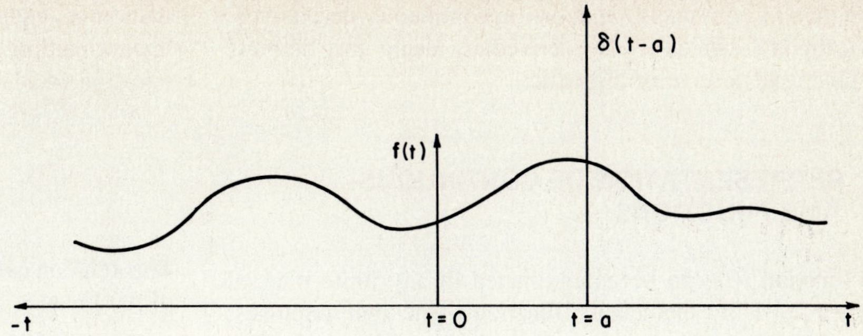

FIGURE 4.6 Graph showing the sifting property of the Dirac delta function.

This is seen by noting that $\delta(t - s) = \delta(s - t)$ not only at $t = s$, where the δ-function occurs, but also everywhere where both functions vanish. This property will be utilized in resolving any function $f(t)$ into a continuum of δ-functions.

This is the most important property of the δ-function and is frequently used as its definition. This property forms the basis of the concept of impulse response, to be discussed later. To illustrate this property, we consider an example.

EXAMPLE 4.1

Consider $f(t) = e^t$. Then

$$\int_{-\infty}^{\infty} e^t \delta(t - a)\, dt = e^a$$

If $a = 0$, then

$$\int_{-\infty}^{\infty} e^t \delta(t)\, dt = e^0 = 1$$

4.2.2 Relation to the Unit Step Function

From equation (4.11) it is seen that the δ-function is the derivative of the unit step function:

$$\delta(t) = \frac{d}{dt}(u_1(t)) \tag{4.15}$$

This property has, as is shown later in the chapter, a very important hydrologic implication.

4.2.3 Even and Odd Property

The *even and odd* property is defined as

$$\delta^{(n)}(t) = (-1)^n \delta^{(n)}(-t) \tag{4.16}$$

That is, odd-order δ-functions are odd functions and even-order δ-functions are even functions.

4.2.4 Sifting Higher-Order δ-Functions

If we combine properties expressed by equations (4.14) and (4.16), then

$$\int_{-\infty}^{\infty} \delta^{(n)}(t) f(t)\, dt = (-1)^n f^{(n)}(0) \tag{4.17}$$

This property can be proved using integration by parts and noting that $f(t)$ is zero outside a finite interval.

4.2.5 δ-Function of Another Function

The δ-function of another function can be expressed as

$$\delta[f(t)] = \frac{\delta(t - a)}{|f^{(1)}(a)|} \tag{4.18}$$

where $f(t)$ is a monotonic function such that $f(a) = 0$, and $f^{(1)}(\cdot)$ is first derivative of $f(\cdot)$. Note that if $f(t)$ vanishes for $t = t_i$, $i = 1, 2, 3, \ldots, m$, then

$$\delta[f(t)] = \sum_{i=1}^{m} \frac{\delta(t - t_i)}{|f^{(1)}(t_i)|} \tag{4.19}$$

where $f(t)$ must be monotonic near each of the vanishing points.

EXAMPLE 4.2

Consider the function $f(t) = \sin t$, $t_i = i\pi$, $i = 0, \pm 1, \pm 2, \ldots$. We want to represent $\delta(\sin t)$.

Then

$$f^{(1)}(t) = \cos t$$

and applying equation (4.19),

$$\delta[\sin t] = \sum_{i=-\infty}^{\infty} \frac{\delta(t - i\pi)}{|\cos i\pi|} = \sum_{i=-\infty}^{\infty} \delta(t - i\pi)$$

EXAMPLE 4.3

Consider $f(t) = at$, $t \geq 0$. We want to represent $\delta(f(t))$. Then

$$f^{(1)}(t) = au_1(t)$$

Therefore,

$$\delta[f(t)] = \frac{\delta(t)}{|au_1(t)|}$$

4.2.6 Multiplication of a Function with Higher-Order δ-Functions

The product of a function with higher-order δ-functions can be expressed as

$$\delta^{(n)}(t) f(t) = \sum_{i=0}^{n} (-1)^i \binom{n}{i} f^i(0)\, \delta^{n-i}(t) \tag{4.20}$$

where $f(t)$ is continuous and has n continuous derivatives. This property can easily be derived by using the Leibnitz rule for the derivative of a product.

4.3 REPRESENTATION OF CONTINUOUS-TIME FUNCTIONS

Any function $f(t)$ can be approximated in any finite interval $-T \leq t \leq T$ by elementary functions—the unit impulses, unit pulses, and unit steps. In hydrologic applications such approximations are frequently used.

4.3.1 Resolution into Unit Pulses

Consider a function $f(t)$ as shown in Fig. 4.7. This function can be resolved into a finite number of the unit pulses each of width D, occurring at $t = jD$, $j = 0, 1, 2, \ldots, N$; $ND = T$. Since the volume of the unit pulse is 1 and the height is $1/D$, the pulse occurring at $t = jD$ must be multiplied by $f(jD)$ to make its amplitude equal to the value of the function approximated. Therefore,

$$f(t) = \sum_{j=0}^{N} f(jD)\, u_D(t - jD)D, \qquad 0 \leq t \leq T \tag{4.21}$$

where

$$u_D(t - jD) = \begin{cases} \dfrac{1}{D}, & jD < t < (j+1)D \\ 0, & \text{otherwise} \end{cases} \tag{4.22}$$

Equation (4.22) represents a rectangular pulse of duration D and unit volume.

4.3.2 Resolution into Unit Impulses

Consider the function $f(t)$, as shown in Fig. 4.7, to be resolved into unit impulses. As D becomes smaller and N becomes larger, the approximation of $f(t)$ becomes better. In the limiting case, $D \rightarrow 0$ and $N \rightarrow \infty$, so that ND remains constant ($ND = T$). Then the argument jD of f can assume all values in the interval $(0, T)$, and can be replaced by a continuous variable, say s. The pulses would turn into impulses and the summation in equation (4.21) would turn into an integral; the approximation would be exact. Thus as $D \rightarrow 0$ and $N \rightarrow \infty$, equation (4.21) would become

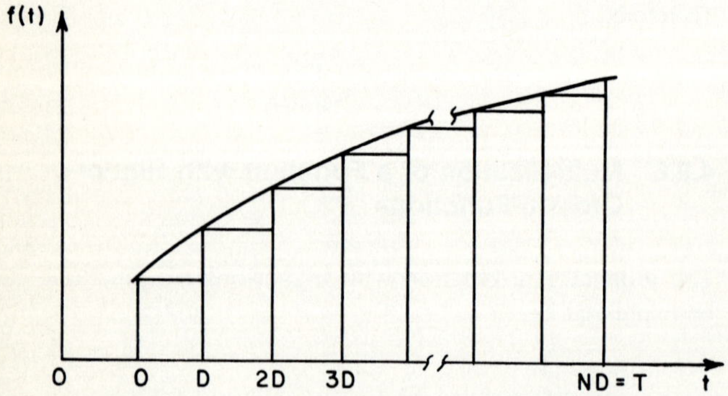

FIGURE 4.7 Approximation of $f(t)$ by unit pulses.

$$f(t) = \int_0^T f(s)\, \delta(t - s)\, ds \tag{4.23}$$

This relation can be extended to the entire domain of definition of $f(t)$ $(-\infty < t < \infty)$,

$$f(t) = \int_{-\infty}^{\infty} f(s)\, \delta(t - s)\, ds \tag{4.24}$$

This equation is the same as equation (4.14), expressing the sifting property of the δ-function. This equation is not valid at a point of discontinuity, but it would specify $f(t)$ to the left as well as to the right of the discontinuity. Thus any $f(t)$ can be resolved into a continuum of unit impulses.

4.3.3 Resolution into Unit Steps

Consider a function $f(t)$ for $t_0 \leq t \leq t_0 + T$, as shown in Fig. 4.8. This function can be resolved into a finite number of unit steps. The resolution can be expressed by

$$f(t) = f(t_0)\, u_1(t - t_0) + \sum_{j=1}^{N} [f(t_0 + k\Delta s) - f(t_0 + (k-1)\,\Delta s)]\; u_1(t - t_0 - k\Delta s), \qquad t \geq t_0 \tag{4.25}$$

Dividing and multiplying the terms within the summation by Δs gives

$$f(t) = f(t_0)u_1(t - t_0) + \sum_{j=1}^{N} \frac{[f(t_0 + k\Delta s) - f(t_0 + (k-1)\Delta s)]}{\Delta s}\; u_1(t - t_0 - k\Delta s)\Delta s \tag{4.26}$$

Evidently, as $\Delta s \rightarrow 0$ and $N \rightarrow \infty$ with $N\Delta s = T$, the approximation becomes exact and the summation becomes an integral. Recall, however, that

$$\lim_{\Delta s \to 0} \frac{f(t_0 + k\Delta s) - f(t_0 + (k-1)\,\Delta s)}{\Delta s} = \lim_{\Delta s \to 0} \frac{\Delta f}{\Delta s} = \frac{df}{ds}$$

is the derivative of $f(t)$ at the point $t = t_0 + s$. Therefore,

$$f(t) = f(t_0)\, u_1(t - t_0) + \int_0^T \dot{f}(t_0 + s)\, u_1(t - t_0 - s)\, ds \tag{4.27}$$

Let $w = t_0 + s$ and $T \rightarrow \infty$. Then

$$f(t) = f(t_0)\, u_1\,(t - t_0) + \int_{t_0}^{\infty} \dot{f}(w)\, u_1(t - w)\, dw, \quad t \geq t_0 \tag{4.28}$$

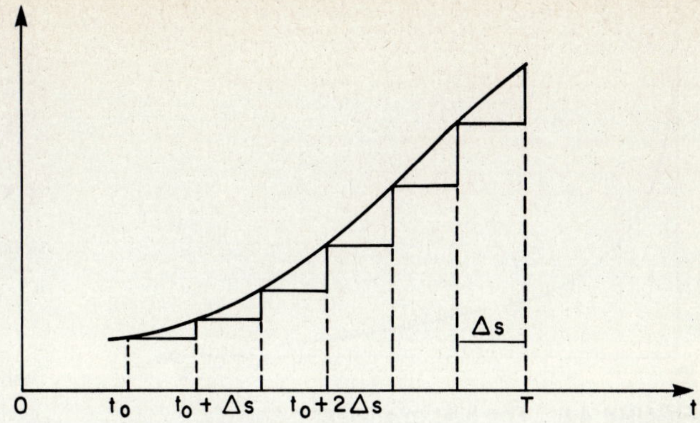

FIGURE 4.8 Approximation of $f(t)$ by unit steps.

Note that the function $f(t)$ is assumed continuous at $t = t_0$. If $f(t)$ is not continuous, then an impulse resulting from differentiating at a point of discontinuity must be included in equation (4.28). In most cases, however, $t_0 = 0$ and $f(0) = 0$; equation (4.28) can simply be written as

$$f(t) = \int_0^\infty \dot{f}(w)\, u_1(t - w)\, dw, \qquad t > 0 \tag{4.29}$$

where $\dot{f}(s)$ includes any impulse which may occur at the origin. Equation (4.29) can be easily derived from equation (4.24) using integration by parts.

4.4 HYDROLOGIC RESPONSE OF ELEMENTARY FUNCTIONS

The preceding section shows that system inputs and outputs can be resolved into elementary functions. The reason for performing such resolutions is that the response of a linear system to an arbitrary input can be determined from the response to the elementary input.

4.4.1 Unit Impulse Response

The unit impulse response $h(t)$, UIR for short, is the output of a system due to the unit impulse or the δ-function. The UIR is a function of time and is taken with zero initial conditions on the system. For a linear system the UIR gives as complete a description of the system behavior as is needed. In surface water hydrology the instantaneous unit hydrograph (IUH) is the UIR of a given watershed. The UIR is variously known as the characteristic response, weighting function, kernel function, Green's function, or influence function. Mathematically,

$$h(t) = H\delta(t) \tag{4.30}$$

in which H is the system function. The UIR of a linear system depends on the time of application of the unit impulse. Thus we can define

$$h(t, \tau) = H\delta(t - \tau) \tag{4.31}$$

Here τ denotes the time at which the unit impulse is applied.

If the system is causal, the UIR is subject to the restriction

$$h(t, \tau) = 0, \qquad t < \tau \tag{4.32}$$

If the system is time-invariant, the UIR becomes

$$h(t, \tau) = H\delta(t - \tau) = h(t - \tau) \tag{4.33}$$

The UIR can be obtained by solving the differential equation(s) relating the system input and output, with input reduced to the unit impulse.

EXAMPLE 4.4

Consider the differential equation frequently utilized in hydrology

$$k\frac{dh(t)}{dt} + h(t) = \delta(t), \qquad t > 0$$

where k is a parameter that is constant in time. The initial condition is $h(0) = 1/k$. Then the UIR can be obtained in two ways. For $t > 0$, the δ-function vanishes. Therefore, the differential equation becomes homogeneous with right side disappearing and can be solved directly. The other approach is to consider the differential equation as given. Then, with the integrating factor $\exp(t/k)$, the differential equation can be immediately solved. Thus

$$h(t) = \frac{e^{-t/k}}{k}, \qquad t > 0$$

and $h(t) = 0$ for $t < 0$. Therefore, we can write

$$h(t) = \frac{e^{-t/k}}{k} u_1(t) \qquad \text{for all } t$$

This is the IUH of a linear reservoir whose lag time is k, as is seen later in the text, and is shown in Fig. 4.9.

4.4.2 Unit Step Response

The unit step response $U(t)$, or USR, is the output of a system, initially at rest, due to the unit step. Mathematically,

$$U(t) = Hu_1(t) \tag{4.34}$$

In surface water hydrology the S-hydrograph (or summation-hydrograph) is the USR of a given watershed.

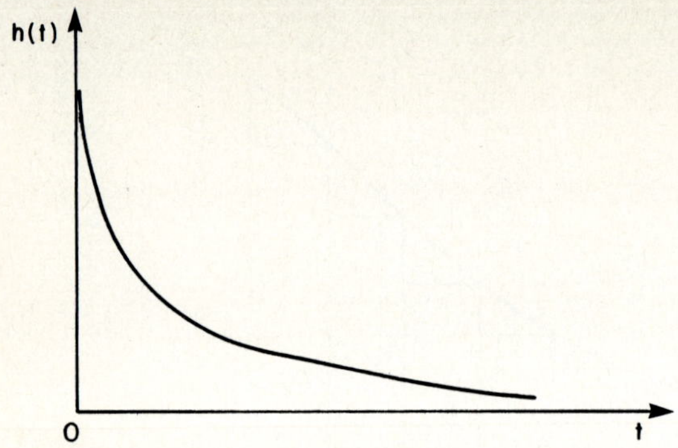

FIGURE 4.9 The IUH of a linear reservoir.

If the unit step is applied at time τ, then

$$U(t, \tau) = Hu_1(t - \tau)$$

If the system is causal, then

$$U(t, \tau) = 0 \qquad \text{for } t < \tau \tag{4.35}$$

If the system is causal and time-invariant, then

$$U(t - \tau) = Hu_1(t - \tau) \tag{4.36}$$

EXAMPLE 4.5

Consider the differential equation

$$\frac{dy(t)}{dt} + y(t) = u_1(t), \qquad t > 0$$

with the initial condition $y(0) = 0$. Then the solution is

$$y(t) = U(t) = Hu_1(t) = \begin{cases} 0, & t < 0 \\ (1 - e^{-t}), & t \geq 0 \end{cases}$$

The USR is graphed in Fig. 4.10.

4.4.3 Unit Pulse Response

The unit pulse response $h_D(t)$, or UPR, is the output of a system initially at rest due to the unit pulse of duration D time units. Mathematically,

$$h_D(t) = Hu_D(t) \tag{4.37}$$

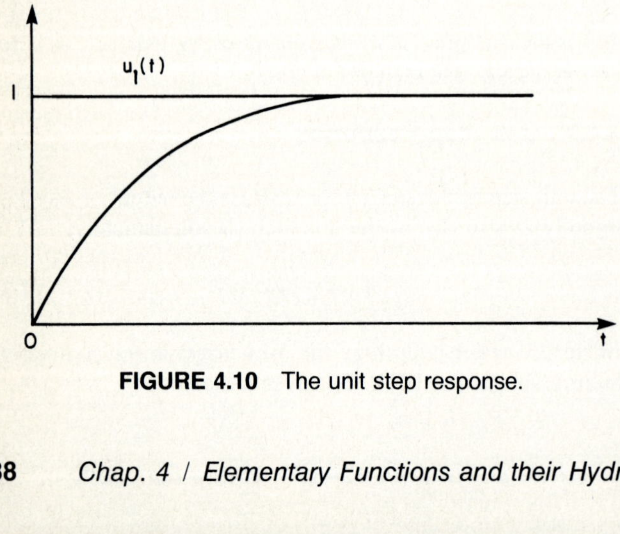

FIGURE 4.10 The unit step response.

In hydrology the unit hydrograph is the UPR of a given watershed. The UPR can be determined by solving the differential equation(s) relating the system input and output, with input specified as a unit pulse. Consider the following example.

EXAMPLE 4.6

Consider the differential equation,

$$\frac{dy(t)}{dt} + y(t) = u_D(t), \qquad t > 0$$

with $y(0) = 0$. Note that $u_D(t)$ can be expressed as

$$u_D(t) = \frac{1}{D}[u_1(t) - u_1(t - D)]$$

and

$$U(t) = Hu_1(t) = \begin{cases} 0, & t \leq 0 \\ 1 - e^{-t}, & t < 0 \end{cases}$$

Then, using the principles of proportionality and superposition,

$$\begin{aligned} y(t) = Hu_D(t) = h_D(t) &= H\left\{\frac{1}{D}[u_1(t) - u_1(t - D)]\right\} \\ &= \frac{1}{D}Hu_1(t) - \frac{1}{D}Hu_1(t - D) \\ &= \frac{1}{D}(1 - e^{-t}) - \frac{1}{D}(1 - e^{-(t-D)}) \\ &= \frac{1}{D}(1 - e^{-D})\, e^{-(t-D)} \end{aligned}$$

Therefore, the solution is

$$y(t) = h_D(t) = \begin{cases} 0, & t \leq 0 \\ \dfrac{1 - e^{-t}}{D}, & 0 < t \leq D \\ \dfrac{(1 - e^{-D})\, e^{-(t-D)}}{D}, & t > D \end{cases}$$

This is the D-time period unit hydrograph of a linear reservoir having a unit lag time; see Fig. 4.11.

This solution can also be obtained by solving the differential equation directly first with $u_D(t)$ and then without it. This is the approach more frequently used in hydrology. That is,

$$y(t) = \frac{1}{D}(1 - e^{-t}), \qquad 0 \leq t \leq D$$

When $u_D(t)$ is 0 for $t \geq D$,

$$\frac{dy(t)}{dt} + y(t) = 0$$

with the condition $y(D) = y_0$. Then

$$y(t) = y_0 e^{-(t-D)}, \qquad t \geq D$$

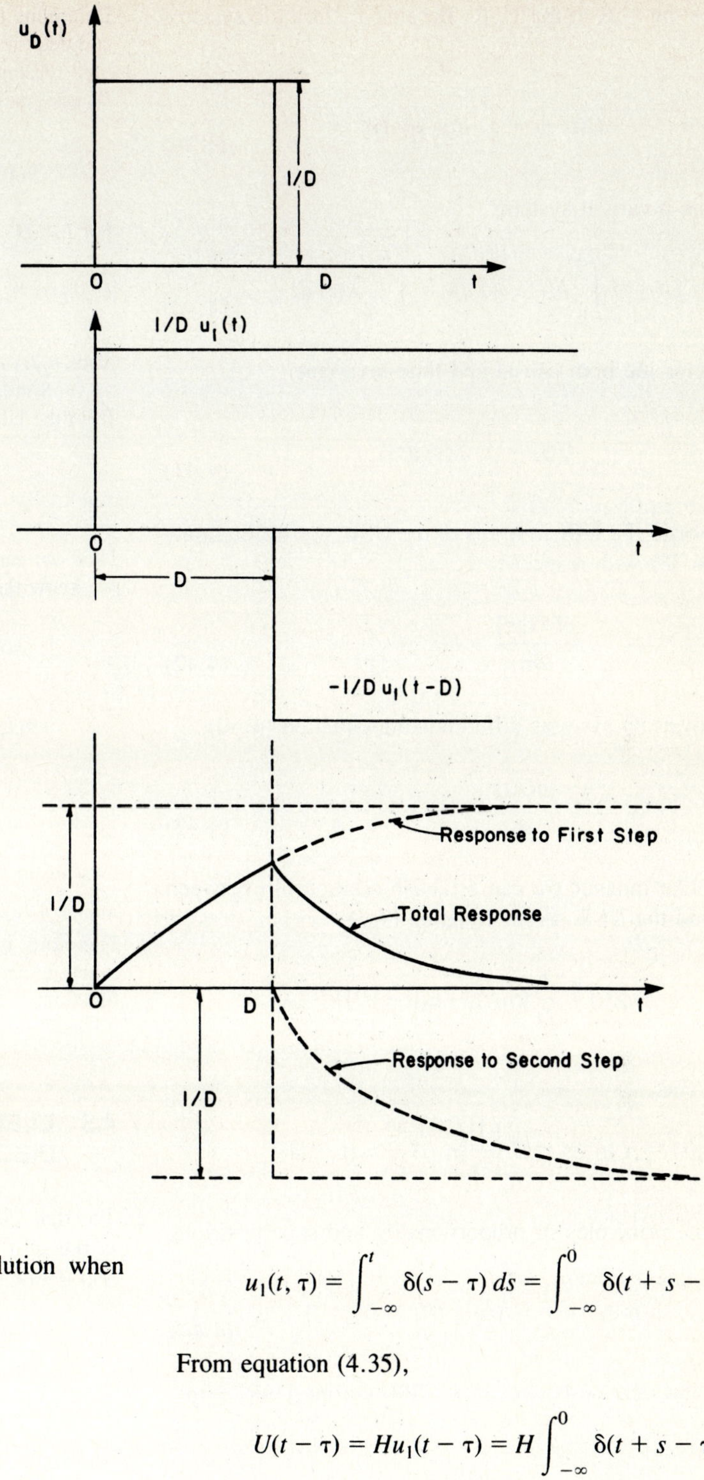

FIGURE 4.11 The UPR of a linear reservoir.

y_0 is given by substituting D for t into the solution when $u_D(t)$ is not zero,

$$y_0 = \frac{1}{D}(1 - e^{-D})$$

Therefore,

$$y(t) = \frac{1}{D}(1 - e^{-D})\, e^{-(t-D)}$$

which is the same as obtained before using the unit step response.

4.4.4 Relationship Among Responses of Elementary Functions

The UIR, UPR, and USR are related to each other in the same manner as their corresponding elementary functions. From equation (4.11) we can write

$$u_1(t, \tau) = \int_{-\infty}^{t} \delta(s - \tau)\, ds = \int_{-\infty}^{0} \delta(t + s - \tau)\, ds$$

From equation (4.35),

$$U(t - \tau) = Hu_1(t - \tau) = H \int_{-\infty}^{0} \delta(t + s - \tau)\, ds$$

Because the operator H satisfies the property of additivity,

$$U(t, \tau) = \int_{-\infty}^{0} H\, \delta(t + s - \tau)\, ds = \int_{-\infty}^{0} h(t, s - \tau)\, ds$$

Let $w = \tau - s$. Then

$$U(t, \tau) = \int_{\tau}^{\infty} h(t, w)\, dw \tag{4.38}$$

This relates the UIR to the USR. Because hydrologic systems are causal,

$$U(t, \tau) = \int_{\tau}^{t} h(t, w)\, dw \tag{4.39}$$

For the time-invariant systems,

$$U(t) = \int_{0}^{\infty} h(t - w)\, dw = \int_{-\infty}^{t} h(z)\, dz \tag{4.40}$$

If the systems are both causal and time-invariant,

$$U(t) = \int_{0}^{t} h(s)\, ds \tag{4.41}$$

To obtain the UIR in terms of the USR, we differentiate equation (4.38) with respect to τ,

$$\frac{\partial U(t, \tau)}{\partial \tau} = h(t, \tau) \tag{4.42}$$

For time-invariant systems differentiating equation (4.40),

$$\frac{dU(t)}{dt} = h(t) \tag{4.43}$$

In a like manner we can establish relationship between the UPR and the USR. From equation (4.7),

$$u_D(t) = \frac{1}{D}[u_1(t) - u_1(t - D)]$$

Therefore,

$$h_D(t) = H\, u_D(t) = H\left\{\frac{1}{D}[u_1(t) - u_1(t - D)]\right\}$$

Applying the principles of proportionality and superposition,

$$h_D(t) = \frac{1}{D}[U(t) - U(t - D)] \tag{4.44}$$

This gives the relation between the UPR and the USR. Similarly,

$$h_D(t) = \frac{1}{D}\int_{t-D}^{t} h(t)\, dt \tag{4.45}$$

which gives the relation between the UPR and the UIR.

EXAMPLE 4.7

Consider the Example 4.4. Its UIR, as determined previously is,

$$h(t) = \frac{1}{k} e^{-t/k}, \qquad t > 0$$

Determine the UPR and USR. Then assume that the USR is known and determine the UIR and UPR.

Using equation (4.45), $h_D(t)$ can be determined. For $0 \leq t \leq D$, integration is to be performed from 0 to t,

$$h_D(t) = \frac{1}{D}\int_{0}^{t} \frac{e^{-s/k}}{k}\, ds = \frac{1}{D}(1 - e^{-t/k}), \qquad 0 < t \leq D$$

For $t \geq D$,

$$h_D(t) = \frac{1}{D}\int_{t-d}^{t} \frac{1}{k} e^{-s/k}\, ds = \frac{1}{D}(1 - e^{-D/k})\, e^{-(t-D)/k}, \qquad t \geq D$$

Thus $h_D(t)$ is completely determined from the UIR.

Similarly, by using equation (4.41) the USR can be determined from the UIR,

$$U(t) = \int_{0}^{t} \frac{1}{k} e^{-t/k}\, dt = (1 - e^{-t/k}), \qquad t > 0$$

Now we can reverse our order of computation. Let us assume that we know the USR and wish to determine the UIR and UPR. Then

$$h(t) = \frac{d}{dt}(1 - e^{-t/k}) = \frac{1}{k} e^{-t/k}, \qquad t > 0$$

$$h_D(t) = \frac{1}{D}(1 - e^{-t/k}), \qquad 0 < t \leq D$$

$$= \frac{1}{D}[(1 - e^{-t/k}) - (1 - e^{-(t-D)/k})], \qquad t \geq D$$

$$= \frac{1}{D}(1 - e^{-D/k})\, e^{-(t-D)/k}, \qquad t \geq D$$

Therefore, it can be seen that the system response to one elementary function can be determined from its response to another elementary function.

4.5 ELEMENTARY FUNCTIONS FOR DISCRETE-TIME SYSTEMS

The discrete-time counterpart of the unit impulse or δ-function is the unit function, often referred to as the Kronecker delta and denoted by δ_{nk}, which can be defined as

$$u_0(n - k) = \delta(n - k) = \begin{cases} 0, & n \neq k \\ 1, & n = k \end{cases} \tag{4.46}$$

The unit function is illustrated in Fig. 4.12.

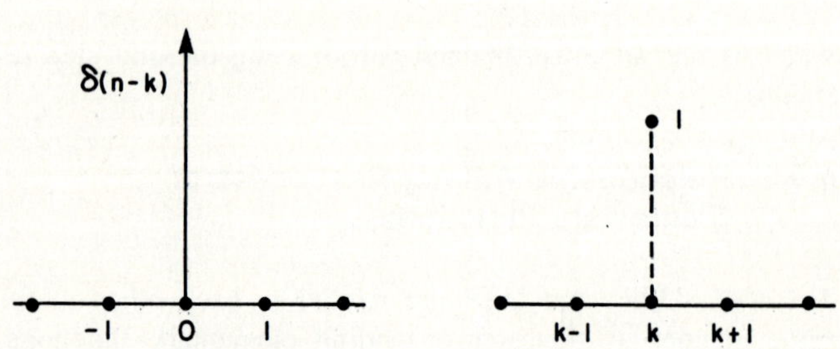

FIGURE 4.12 Unit function.

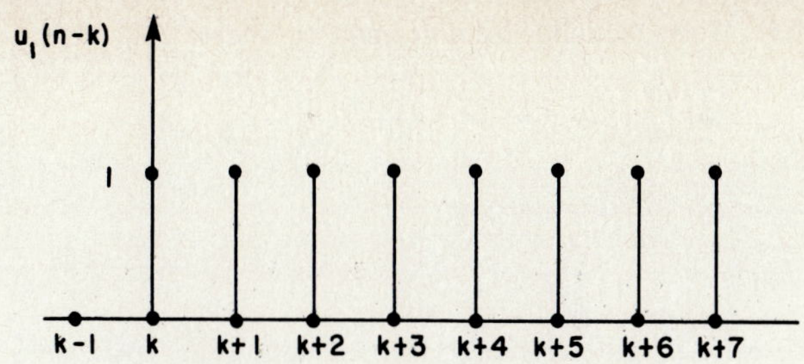

FIGURE 4.13 Discrete-time unit step.

The discrete-time unit step can be defined as

$$u_1(n-k) = \begin{cases} 0, & n < k \\ 1, & n \geq k \end{cases} \tag{4.47}$$

This is illustrated in Fig. 4.13 and is not as frequently used as the unit function.

4.6 REPRESENTATION OF DISCRETE-TIME FUNCTIONS

The representation of a discrete-time function $f(n)$ by unit functions is straightforward:

$$\begin{aligned} f(n) &= f(-k)\delta(n+k) + f(-k+1)\delta(n+k-1) + \cdots \\ &\quad + f(-1)\delta(n+1) + f(0)\delta(n) + f(1)\delta(n-1) + \cdots \\ &= \sum_{k=-\infty}^{\infty} f(k)\delta(n-k) \end{aligned} \tag{4.48}$$

It should be noted that equation (4.48) is exact and is the discrete-time counterpart of equation (4.24).

In a similar manner, a discrete-time function $f(n)$ can be resolved into discrete-time unit steps. The discrete-time counterpart of equation (4.27) can be written as

$$\begin{aligned} f(n) &= f(n_0)\, u_1(n-n_0) \\ &\quad + \sum_{k=n_0+1}^{\infty} [f(k) - f(k-1)]\, u_1(n-k), \qquad n \geq n_0 \end{aligned} \tag{4.49}$$

4.5.2 Unit Function Response

The unit function response of a discrete-time system can be defined as

$$h(n, k) = H\delta(n-k) \tag{4.50}$$

The left side of equation (4.50) gives the response at nth instant to a unit function applied at kth instant. For causal systems

$$h(n, k) = 0, \qquad n < k \tag{4.51}$$

For time-invariant systems

$$h(n-k) = H\delta(n-k) \tag{4.52}$$

Equations (4.50) and (4.52) can be expressed in matrix form. For example,

$$[h(n, k)] = \begin{bmatrix} h(0,0) & 0 & 0 & \cdots \\ h(1,0) & h(1,1) & 0 & \cdots \\ h(2,0) & h(2,1) & h(2,2) & \cdots \\ h(3,0) & h(3,1) & h(3,2) & \cdots \\ \vdots & \vdots & \vdots & \vdots \end{bmatrix} \tag{4.53}$$

Similarly,

$$[h(n-k)] = \begin{bmatrix} h(0) & 0 & 0 & 0 & \cdots \\ h(1) & h(0) & 0 & 0 & \cdots \\ h(2) & h(1) & h(0) & 0 & \cdots \\ h(3) & h(2) & h(1) & 0 & \cdots \\ \vdots & \vdots & \vdots & \vdots \end{bmatrix} \tag{4.54}$$

These matrices are often called the *transmission matrices*. It is appropriate to mention that the kth column of the transmission matrix comprises the sequence of numbers obtained in response of the system to the unit function applied at the kth instant.

EXERCISES

4.1. Show the following:

(a) $\delta(t-s) f(t) = \delta(t-s) f(s)$

(b) $f(t)\, \delta^{(1)}(t-s) = f(s)\, \delta^{(1)}(t-s) - f^{(1)}(s)(t-s)$

4.2. Show that the following functions can be considered as δ-functions in the limit as $a \to \infty$:

(a) $f(t) = \dfrac{a}{\pi(1 + a^2t^2)}$

(b) $f(t) = \dfrac{a \exp(-a^2t^2)}{\sqrt{\pi}}$

(c) $f(t) = \dfrac{\sin at}{\pi t}$

4.3. Show that the following functions can be considered as a unit step $u_1(t)$ in the limit as $a \to \infty$:

(a) $2^{-\exp(-at)}$

(b) $\dfrac{1}{\pi}\displaystyle\int_{-\infty}^{at} \frac{\sin u}{u}\, du$

(c) $\dfrac{1}{\sqrt{\pi}}\displaystyle\int_{-at}^{\infty} \exp(-u^2)\, du$

4.4. Show that

$$f(t) = \int_{-\infty}^{\infty} \delta(t-s) f(s)\, ds$$

4.5. Show that

$$f(t) = \int_0^{\infty} \dot{f}(s)\, u_1(t-s)\, ds, \qquad f(t) = 0 \quad \text{for } t < 0$$

4.6. Prove equation (4.49).

4.7. The impulse response of a time-invariant linear system is given by

$$h(t) = [\exp(-t) - \exp(-2t)]\, u_1(t)$$

Find the unit step response of the system.

4.8. The response of a time-varying system at time t to a unit step applied at time s is

$$y(t, s) = s[1 - \exp(-t)]\, u_1(t-s)$$

where u_1 is the unit step function. Find the unit impulse response.

4.9. Solve the following differential equations:

(a) $\dfrac{dy}{dt} + 2y = \delta^{(2)}(t)$

(b) $t\dfrac{dy}{dt} + y = \delta^{(1)}(t)$

(c) $\exp(-t)\dfrac{dy}{dt} - y = \delta^{(1)}(t)$

4.10. A linear time-variable system is described by the differential equation

$$\frac{dy}{dt} + \frac{y}{t} = x(t)$$

(a) Determine the unit step response of the system.
(b) Find the unit impulse response of the system.

CHAPTER FIVE

Superposition Integrals

The analysis and synthesis of linear hydrologic systems are conveniently performed by using superposition integrals, particularly convolution. The usefulness of elementary functions and their responses discussed in the preceding chapter are perhaps best illustrated through application of these integrals. In this chapter we briefly discuss superposition integrals and their applications for both continuous-time and discrete-time systems. For further details see Dooge (1973), Gupta (1971), Schwarz and Friedland (1965), and Zadeh and Desoer (1963), among others.

5.1 CONTINUOUS-TIME SYSTEMS

5.1.1 Derivation of the Superposition Integral Using Impulse Response

The superposition integral can be easily derived by using the sifting property of the δ-function and the principles of proportionality and superposition. It was shown previously that an arbitrary input $x(t)$ can be resolved into a continuum of the unit impulses. This resolution can be expressed as

$$x(t) = \int_{-\infty}^{\infty} x(\tau)\delta(t-\tau)\, d\tau$$

The system output $y(t)$ can be expressed as

$$y(t) = Hx(t) = H\int_{-\infty}^{\infty} x(\tau)\delta(t-\tau)\, d\tau$$

where H is the system function. Invoking the principle of superposition,

$$y(t) = \int_{-\infty}^{\infty} H[x(\tau)\delta(t-\tau)]\, d\tau$$

Involving the principle of proportionality,

$$y(t) = \int_{-\infty}^{\infty} x(\tau)H\delta(t-\tau)\, d\tau$$

The unit impulse response is

$$h(t, \tau) = H\delta(t-\tau)$$

Therefore,

$$y(t) = \int_{-\infty}^{\infty} x(\tau)h(t, \tau)\, d\tau \tag{5.1}$$

Equation (5.1) is the superposition integral. This integral can be thought of as the summation of the responses to the input impulses occurring at time τ and having the volume $x(\tau)\, d\tau$, with τ varying over the entire range $(-\infty, \infty)$. It expresses the relationship between input, output, and impulse response. In hydrological terms it means that the response of a linear system for any input can be determined by equation (5.1), given the impulse response of the system.

5.1.2 Forms of the Superposition Integral in Equation (5.1)

The forms of superposition integral reflect special properties of the system or the input. To illustrate, if the system is causal, the integration may be terminated at $\tau = t$ because $h(t, \tau) = 0$ for $\tau > t$. Therefore,

$$y(t) = \int_{-\infty}^{t} x(\tau)h(t, \tau)\, d\tau \tag{5.2}$$

If the system is time-invariant, $h(t, \tau) = h(t - \tau)$ and

$$y(t) = \int_{-\infty}^{\infty} x(\tau)h(t - \tau)\, d\tau \tag{5.3}$$

Alternatively,

$$y(t) = \int_{-\infty}^{\infty} x(t - w)h(w)\, dw \tag{5.4}$$

Equation (5.4) is obtained from equation (5.3) with the change of variable $w = t - \tau$.

Customarily, the system input starts at $t = 0$; that is, $x(t) = 0$ for $t < 0$. Therefore,

$$y(t) = \int_{0}^{\infty} x(\tau)h(t, \tau)\, d\tau \tag{5.5}$$

With this condition, if the system is causal and time-invariant, then

$$y(t) = \int_{0}^{t} x(\tau)h(t - \tau)\, d\tau \tag{5.6}$$

Alternatively,

$$y(t) = \int_{0}^{t} x(t - w)h(w)\, dw \tag{5.7}$$

Equation (5.6) is frequently written as

$$y(t) = x(t) * h(t) \quad \text{or} \quad x * h \tag{5.8}$$

Alternatively,

$$y(t) = h(t) * x(t) \quad \text{or} \quad h * x \tag{5.9}$$

If the system has a finite memory L, equations (5.6)–(5.7) become

$$y(t) = \int_{t-L}^{t} x(\tau)h(t - \tau)\, d\tau \tag{5.10}$$

and

$$y(t) = \int_{0}^{L} x(t - w)h(w)\, dw \tag{5.11}$$

If the length of input is M and the length of output is N, then

$$N = L + M \tag{5.12}$$

In hydrologic terms, M may denote the duration of rainfall excess, L is the base length of IUH, and N is the duration of surface runoff. Equations (5.3)–(5.4), (5.6)–(5.7) and (5.10)–(5.11) are convolution integrals. The convolution integral is also called the Duhamel integral. The process of obtaining it is shown in Fig. 5.1 and is simply called convolution. In Table 5.1 we summarize the limits of integration in equations (5.3)–(5.4) for various system properties.

5.1.3 More on Superposition Integral

The system properties assumed for obtaining the various forms of superposition integral can be easily derived. To this end we employ the convolution integral of equation (5.6).

5.1.3.1 Time-Invariance

Time-invariance implies that $h(t - \tau)$ does not depend on τ, the time when the input is applied, but only on the time difference $(t - \tau)$. In other words, if $y_1(t)$ is the output due to an input $x_1(t)$, then $y_2(t) = y_1(t - a)$ is the output due to the input $x_2(t) = x_1(t - a)$, where a is a constant greater than 0. This can be seen as follows:

$$y_2(t) = \int_{0}^{t} x_1(\tau - a)h(t - \tau)\, d\tau$$

Since $x(\tau - a) = 0$ for $\tau < a$, the lower limit of integration is $\tau = a$,

$$y_2(t) = \int_{a}^{t} x_1(\tau - a)h(t - \tau)\, d\tau$$

Let $\tau - a = s$. Then

$$y_2(t) = \int_{0}^{t-a} x_1(s)h(t - a - s)\, ds$$

But

$$y_1(t - a) = \int_{0}^{t-a} x_1(\tau)h(t - a - \tau)\, d\tau$$

Since both these integrals are identical, $y_2(t) = y_1(t - a)$ for any a.

5.2.3.2 Proportionality

If

$$y_1(t) = \int_{0}^{t} x(\tau)h(t - \tau)\, d\tau$$

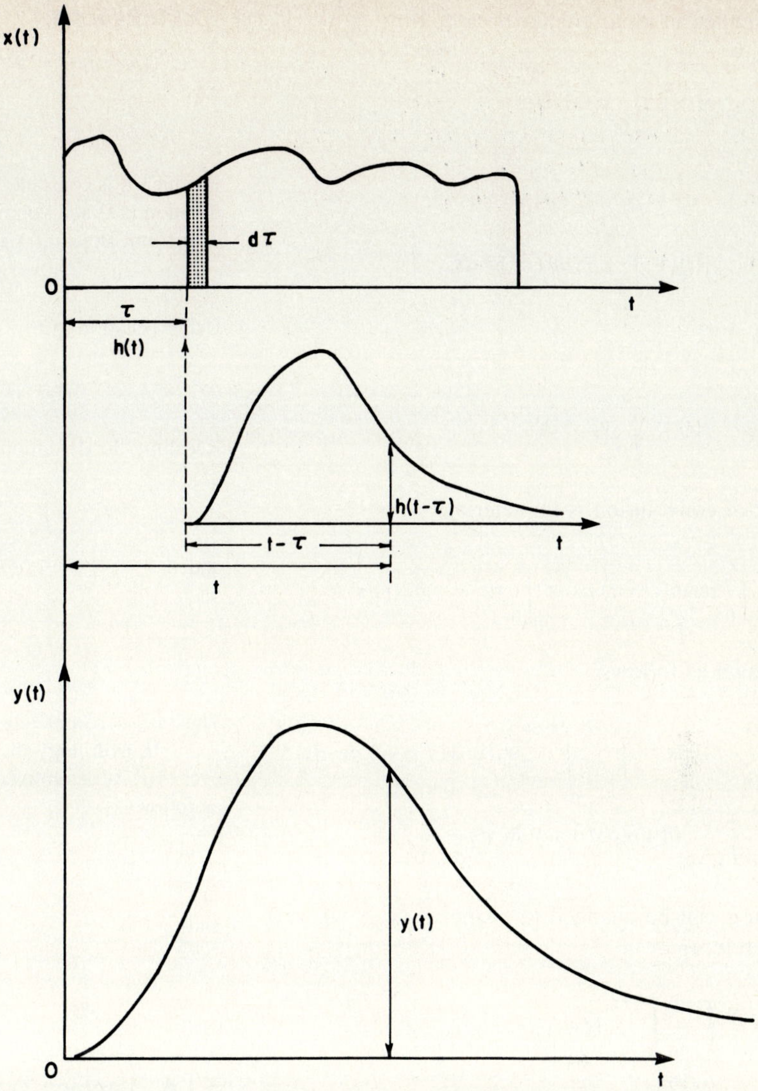

FIGURE 5.1 Convolution operation of a linear time-invariant system.

and $x_2(t) = ax_1(t)$, where a is a constant greater than 0, then

$$y_2(t) = \int_0^t x_2(\tau)h(t-\tau)\,d\tau = \int_0^t ax_1(\tau)h(t-\tau)\,d\tau$$
$$= ay_1(t)$$

TABLE 5.1

Limits of integration in equations (5.3)–(5.4) for various system properties

System Properties	*Equation (5.3) Limits of Integration*		*Equation (5.4) Limits of Integration*	
	Lower	*Upper*	*Lower*	*Upper*
General	$-\infty$	∞	$-\infty$	∞
Causal	$-\infty$	t	0	∞
	0	t	0	t
Memory				
(a) Finite L	$t-L$	t	0	L
(b) Infinite	0	t	0	t

5.1.3.3 Superposition

If $x(t) = x_1(t) + x_2(t)$, then

$$y(t) = \int_0^t x_1(\tau)h(t-\tau)\,d\tau + \int_0^t x_2(\tau)h(t-\tau)\,d\tau$$
$$= y_1(t) + y_2(t)$$

5.1.3.4 Order of Convolution

For a linear time-invariant system the order of convolution is immaterial, that is,

$$y(t) = \int_0^t x(\tau)h(t-\tau)\,d\tau = \int_0^t h(\tau)x(t-\tau)\,d\tau$$

This can be shown by changing the variable of integration. Let $s = t - \tau$. Then $\tau = t - s$ and $d\tau = -ds$. Therefore,

$$y(t) = \int_0^t x(\tau)h(t-\tau)\,d\tau = \int_0^t x(t-s)h(s)\,ds$$
$$= \int_0^t x(t-\tau)h(\tau)\,d\tau$$

This can be extended to a more general case. If

$$x(t) = \int_0^t x_1(s)\,ds$$

and

$$y_1(t) = \int_0^t x_1(s)h(t-s)\,ds$$

then

$$y(t) = \int_0^t y_1(s)\,ds$$

Since the order of convolution is immaterial,

$$y(t) = \int_0^t x(s)h(t-\tau)\,ds = \int_0^t x(t-s)h(s)\,ds$$

This can be shown as follows:

$$y(t) = \int_{s=0}^{s=t} y_1(s)\,ds = \int_{s=0}^{s=t}\int_{w=0}^{w=s} x_1(w)h(s-w)\,dw\,ds$$

$$= \int_{s=0}^{s=t}\int_{w=0}^{w=s} h(w)x_1(s-w)\,dw\,ds$$

The upper limit s can be changed to t since $x_1(s-w)=0$ for $w>s$. Therefore,

$$y(t) = \int_{s=0}^{s=t}\int_{w=0}^{w=t} h(w)x_1(s-w)\,dw\,ds$$

$$= \int_{w=0}^{w=t} h(w)\int_{s=0}^{t} x_1(s-w)\,ds\,dw$$

Let $z = s - w$. Then $dz = ds$. Hence

$$y(t) = \int_{w=0}^{w=t} h(w)\left[\int_{z=-w}^{z=0} x_1(z)\,dz + \int_{z=0}^{z=t-w} x_1(z)\,dz\right]dw$$

The first integral within the brackets vanishes because the system is initially at rest. The second integral can be written as

$$\int_{z=0}^{z=t-w} x_1(z)\,dz = x(t-w)$$

by definition of $x(t)$. Therefore,

$$y(t) = \int_{w=0}^{w=t} h(w)x(t-w)\,dw = \int_{w=0}^{w=t} x(w)h(t-w)\,dw$$

The unit step response function is a direct consequence of this property.

EXAMPLE 5.1

Given $h(t) = e^{-t}$, determine $y(t)$ for $x(t)$ defined as

$$x(t) = \begin{cases} A, & 0 \le t \le D \\ 0, & t \ge D \end{cases}$$

in which A is constant. The function $x(t)$ is a rectangular pulse whose duration is D and volume AD. Therefore, we can define $x(t)$ in terms of the unit steps $u_1(\cdot)$ as

$$x(t) = A[u_1(t) - u_1(t-D)]$$

Using equation (5.6),

$$y(t) = \int_0^t A[u_1(\tau) - u_1(\tau - D)]\,e^{-(t-\tau)}\,d\tau$$

Recalling the definition of the unit step,

$$y(t) = A\int_0^t e^{-(t-\tau)}\,d\tau = A(1-e^{-t}), \qquad 0 \le t \le D$$

and

$$y(t) = A\int_0^t e^{-(t-\tau)}\,d\tau - A\int_D^t e^{-(t-\tau)}\,d\tau, \qquad t \ge D$$

$$= A(1-e^{-D})\,e^{-(t-D)}, \qquad t \ge D$$

Thus $y(t)$ is completely specified.

In hydrology the unit step is not very frequently used. However, $y(t)$ is determined more frequently using convolution directly as follows:

$$y(t) = \int_0^t A\,e^{-(t-\tau)}\,d\tau = A(1-e^{-t}), \qquad 0 \le t \le D$$

and

$$y(t) = \int_0^D A\,e^{-(t-\tau)}\,d\tau = A(1-e^{-D})\,e^{-(t-D)}, \qquad t \ge D$$

5.1.4 Laplace Transform of Convolution Integral

The Laplace transform of the convolution integral is of fundamental importance in the analysis and synthesis of linear hydrologic systems. The *Laplace transform* (unilaterial) *of a function* $f(t)$ can be defined as

$$L\{f(t)\} = f(s) = \int_0^\infty f(t)\,e^{-st}\,dt \tag{5.13}$$

This is a transform in the s-domain. Taking the Laplace transform of the convolution integral,

$$L[y(t)] = L\left[\int_0^\infty x(\tau)h(t-\tau)\,d\tau\right]$$

or

$$y(s) = \int_0^\infty e^{-st}\left[\int_0^\infty x(\tau)h(t-\tau)\,d\tau\right]dt$$

This can be written as

$$y(s) = \int_0^\infty x(\tau)\, e^{-s\tau} \left[\int_0^\infty e^{-s(t-\tau)} h(t - \tau)\, d(t - \tau) \right] d\tau$$

Thus

$$y(s) = x(s)\, h(s) \tag{5.14}$$

Equation (5.14) shows that the Laplace transform of the convolution of two time functions is the product of their Laplace transforms. Therefore, if any two of $x(t)$, $h(t)$, and $y(t)$ are known and their Laplace transforms are available, the Laplace transform of the unknown third can be determined. By inverting its Laplace transform, the unknown as a function of t can then be determined.

EXAMPLE 5.2

Determine $y(t)$ if $x(t) = A\, u_1(t)$ and $h(t) = e^{-t}$, using the Laplace transform.

We determine $x(s) = A/s$ and $h(s) = 1/(1 + s)$. Then $y(s) = 1/s(1 + s)$, which we can write as

$$y(s) = \left(\frac{1}{s} - \frac{1}{1+s}\right) A$$

Inverting the Laplace transform,

$$y(t) = A(1 - e^{-t})$$

This is the answer we want.

EXAMPLE 5.3

Using the Laplace transform determine $y(t)$ for $h(t)$ and $x(t)$ specified in Example 5.1, we determine

$$x(s) = A\left(\frac{1}{s} - \frac{e^{-Ds}}{s}\right)$$

$$h(s) = \frac{1}{1+s}$$

Therefore,

$$y(s) = A\left(\frac{1}{s} - \frac{e^{-DS}}{s}\right)\frac{1}{1+s} = A\left(\frac{1}{s} - \frac{1}{1+s}\right) - A\, e^{-Ds}\left(\frac{1}{s} - \frac{1}{1+s}\right)$$

Inverting the Laplace transform,

$$y(t) = A(1 - e^{-t}) - A[1 - e^{-(t-D)}]\, u_1(t - D)$$

Therefore,

$$y(t) = A(1 - e^{-t}), \qquad 0 \le t \le D$$

$$= A(1 - e^{-t}) - A[1 - e^{-(t-D)}], \qquad t \ge D$$

or

$$y(t) = A(1 - e^{-D})\, e^{-(t-D)}, \qquad t \ge D$$

This is the answer we want.

5.1.5 Superposition Integral Using the Step Response

Equation (4.28) defines resolution of a function $x(t)$ into unit steps. Rewriting it,

$$x(t) = x(t_0)\, u_1(t - t_0) + \int_{t_0}^\infty \dot{x}(s)\, u_1(t - s)\, ds, \qquad t \ge t_0$$

Let us assume that $x(t) = 0$ for $t < t_0$. Then the system output $y(t)$ is

$$y(t) = Hx(t) = H\left[x(t_0)\, u_1(t - t_0) + \int_{t_0}^\infty \dot{x}(s)\, u_1(t - s)\, ds\right]$$

Recall the unit step response,

$$U(t, \tau) = H\, u_1(t - \tau)$$

Invoking the principles of proportionality and superposition,

$$y(t) = x(t_0) U(t, t_0) + \int_{t_0}^\infty \dot{x}(s) U(t, s)\, ds, \qquad t \ge t_0 \tag{5.15}$$

Equation (5.15) is based on the assumption that $x = 0$, $t < t_0$. The first term on the right side can be omitted by adjusting the lower limit of the integral from t_0^+ to t_0^-. If $\lim_{t \to -\infty} x(t) = 0$, equation (5.15) can be written for $t \to \infty$ as

$$y(t) = \int_{-\infty}^\infty \dot{x}(s) U(t, s)\, ds \tag{5.16}$$

Equation (5.16) is the superposition integral in terms of step response.

5.1.6 Forms of the Superposition Integral in Equation (5.16)

The superposition integral is written in various forms, reflecting special characteristics of the system or the input. To illustrate, if the system is causal,

$$y(t) = \int_{-\infty}^t \dot{x}(s) U(t, s)\, ds \tag{5.17}$$

If the system is time-invariant,

$$y(t) = \int_{-\infty}^\infty \dot{x}(s) U(t - s)\, ds \tag{5.18}$$

Alternatively,

$$y(t) = \int_{-\infty}^\infty \dot{x}(t - w) U(w)\, dw \tag{5.19}$$

Equation (5.19) is easily obtained from equation (5.18) using the change of variable $w = t - s$.

If the system is causal and time-invariant and $x(t)$ starts at $t = 0$,

$$y(t) = \int_0^t \dot{x}(s) U(t - s)\, ds \tag{5.20}$$

Alternatively,

$$y(t) = \int_0^t \dot{x}(t - w) U(w)\, dw \tag{5.21}$$

Again,

$$y(t) = \dot{x}(t) * U(t) = U(t) * \dot{x}(t) \tag{5.22}$$

Equations (5.18)–(5.21) are called convolution integrals.

EXAMPLE 5.4

Consider the function $x(t)$ as specified in the Example 5.1. Its $U(t)$ is as shown previously. Determine $y(t)$ using the convolution integral of equation (5.20).

The function $U(t)$ is

$$U(t) = (1 - e^{-t})$$

We can express the function $x(t)$ as

$$x(t) = A[u_1(t) - u_1(t - D)]$$

Then

$$\dot{x}(t) = A[\delta(t) - \delta(t - D)]$$

Using equation (5.20) for $0 \le t \le D$,

$$\begin{aligned} y(t) &= \int_0^t A\delta(s)\,(1 - e^{-(t-s)})\,ds \\ &= A(1 - e^{-t}), \qquad 0 \le t \le D \end{aligned}$$

For $t \ge D$ we obtain

$$\begin{aligned} y(t) &= \int_0^t A[\delta(s) - \delta(s - D)](1 - e^{-(t-s)})\,ds \\ &= \int_0^t A\delta(s)\,(1 - e^{-(t-s)})\,ds - \int_0^t A\delta(s - D)(1 - e^{-(t-s)})\,ds \\ &= A(1 - e^{-t}) - A(1 - e^{-(t-D)}) \\ &= A(1 - e^{-t})e^{-(t-D)}, \qquad t \ge D \end{aligned}$$

This is the same answer that we have obtained previously in several different ways.

5.1.7 Graphical Evaluation of the Convolution Integral

The graphical interpretation is helpful in understanding the process of convolution. It aids in evaluation of the convolution integral when one or both functions in the integrand are continuous. The graphical interpretation is particularly important in determining the limits of integration for evaluating convolution.

We want to evaluate graphically

$$y(t) = \int_{-\infty}^{\infty} h(t - \tau)x(\tau)\,d\tau \tag{5.23}$$

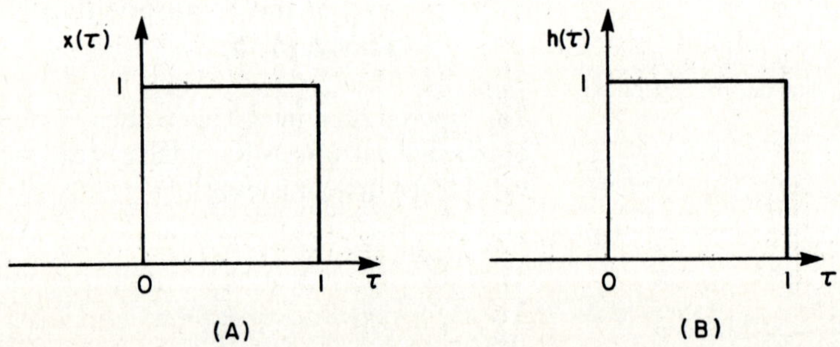

FIGURE 5.2 (a) A function $x(\tau)$. (b) A function $h(\tau)$.

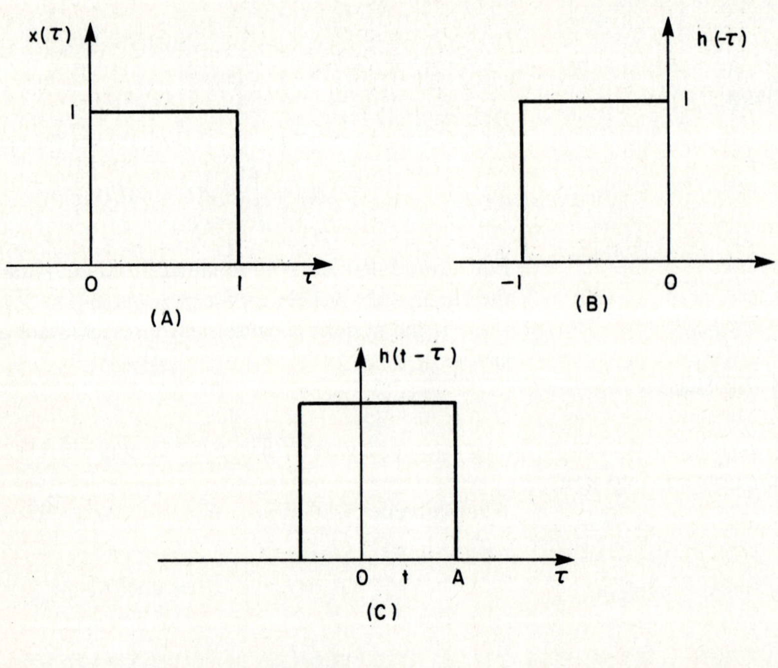

FIGURE 5.3 (a) A function $x(\tau)$. (b) A function $h(-\tau)$. (c) A function $h(t - \tau)$.

5.2 DISCRETE-TIME SYSTEMS

5.2.1 Derivation of Superposition Summation Using Unit Functions

The superposition summation for discrete-time systems can be derived in a like manner by using the principles of proportionality and superposition. From the previous chapter, a discrete-time input $x(n)$ can be resolved into unit functions as

$$x(n) = \sum_{k=-\infty}^{\infty} x(k)\delta(n-k)$$

The system output, $y(n)$, can be written as

$$y(n) = Hx(n) = H \sum_{k=-\infty}^{\infty} x(k)\delta(n-k)$$

Using the principle of superposition,

$$y(n) = \sum_{k=-\infty}^{\infty} H[x(k)\delta(n-k)]$$

Using the principle of proportionality,

$$y(n) = \sum_{k=-\infty}^{\infty} x(k)H\delta(n-k)$$

Using the unit function response,

$$h(n, k) = H\delta(n-k)$$

we obtain the superposition summation,

$$y(n) = \sum_{k=-\infty}^{\infty} x(k)h(n, k) \tag{5.25}$$

Equation (5.25) is the discrete counterpart of the superposition integral in equation (5.1).

5.2.2 Forms of Superposition Summation in Equation (5.25)

The forms of superposition summation reflect special properties of the system, the input or the output. To illustrate, if the system is causal, the summation may be terminated at n because $h(n, k) = 0$ for $k > n$. Therefore,

$$y(n) = \sum_{k=-\infty}^{\infty} x(k)h(n, k) \tag{5.26}$$

If the system is time-invariant, $h(n, k) = h(n-k)$ and

$$y(n) = \sum_{k=-\infty}^{\infty} x(k)h(n-k) \tag{5.27}$$

Alternatively, noting that k plays the same role as a variable of integration and making a change of variable $m = n - k$,

$$y(n) = \sum_{k=-\infty}^{\infty} x(n-m)h(m) \tag{5.28}$$

Customarily, the system input starts at $n = 0$; that is, $x(n) = 0$, $n < 0$. Therefore,

$$y(n) = \sum_{k=0}^{\infty} x(k)h(n, k) \tag{5.29}$$

Further, if the system is both causal and time-invariant,

$$y(n) = \sum_{k=0}^{n} x(k)h(n-k) \tag{5.30}$$

Alternatively,

$$y(n) = \sum_{m=0}^{n} x(n-m)h(m) \tag{5.31}$$

Equations (5.30) and (5.31) are called convolution summations—discrete counterparts of the convolution integral. Frequently, equation (5.30) is written as

$$y(n) = x(n) * h(n) \tag{5.32}$$

Similarly, equation (5.31) is written as

$$y(n) = h(n) * x(n) \tag{5.33}$$

Equations (5.30) and (5.31) are also written as

$$y_n = \sum_{k=0}^{n} x_k h_{n-k} \tag{5.34}$$

and

$$y_n = \sum_{k=0}^{n} x_{n-k} h_k \tag{5.35}$$

Equations (5.34) and (5.35) are equivalent,

$$y_n = \sum_{k=0}^{n} x_k h_{n-k} = \sum_{k=0}^{n} h_k x_{n-k}, \qquad n = 0, 1, 2, 3, \ldots$$

and constitute a set of algebraic equations. It should be noted that x represents the volume of input. Equations (5.34) and (5.35) can be written in matrix form. If the system memory is of length L and the input of length M, then using equation (5.12) we can write equations (5.34) and (5.35) as

$$[y]_{N,1} = [x]_{N,L}\,[h]_{L,1} \tag{5.36}$$

and

$$[y]_{N,1} = [h]_{N,M}\,[x]_{M,1} \qquad (5.37)$$

The matrix $[h]$ is often called as the *transmission matrix*. If x_k, $k = 0, 1, \ldots, M$, has $M + 1$ elements and h_k, $k = 0, 1, 2, \ldots, L$, has $(L + 1)$ elements, then y_k, $k = 0, 1, 2, \ldots, (L + M)$, has $L + M + 1$ elements and can be obtained as

$$\begin{bmatrix} y_0 \\ y_1 \\ y_2 \\ \cdot \\ \cdot \\ \cdot \\ y_L \\ \cdot \\ \cdot \\ \cdot \\ y_M \\ \cdot \\ \cdot \\ \cdot \\ y_{L+M} \end{bmatrix} = \begin{bmatrix} x_0 & 0 & 0 & \cdots & & 0 \\ x_1 & x_0 & 0 & \cdots & & 0 \\ x_2 & x_1 & x_0 & \cdots & & 0 \\ \cdot & \cdot & x_1 & \cdot & & \cdot \\ \cdot & \cdot & \cdot & & \cdot & \cdot \\ \cdot & \cdot & \cdot & & \cdot & \cdot \\ x_L & x_{L-1} & x_{L-2} & \cdots & x_1 & x_0 \\ \cdot & \cdot & & \cdot & & \cdot \\ \cdot & \cdot & & & \cdot & \cdot \\ \cdot & \cdot & & & \cdot & \cdot \\ x_M & x_{M-1} & x_{M-2} & \cdots & & x_{M-L} \\ \cdot & \cdot & & \cdot & & \cdot \\ \cdot & \cdot & & & \cdot & \cdot \\ \cdot & \cdot & & & \cdot & \cdot \\ 0 & 0 & 0 & \cdots & & x_M \end{bmatrix} \begin{bmatrix} h_0 \\ h_1 \\ h_2 \\ \cdot \\ \cdot \\ \cdot \\ h_L \end{bmatrix}$$

if $M > L$. On the other hand, if $M < L$ then we obtain

$$\begin{bmatrix} y_0 \\ y_1 \\ y_2 \\ \cdot \\ \cdot \\ \cdot \\ y_M \\ \cdot \\ \cdot \\ \cdot \\ \cdot \\ y_L \\ \cdot \\ \cdot \\ \cdot \\ y_{L+M} \end{bmatrix} = \begin{bmatrix} x_0 & 0 & 0 & \cdots & & 0 \\ x_1 & x_0 & 0 & \cdots & & 0 \\ x_2 & x_1 & x_0 & \cdots & & 0 \\ \cdot & \cdot & \cdot & & \cdot & \cdot \\ \cdot & \cdot & \cdot & & \cdot & \cdot \\ \cdot & \cdot & \cdot & & \cdot & \cdot \\ x_M & x_{M-1} & 0 & \cdots & & 0 \\ 0 & x_M & x_{M-1} & \cdots & & 0 \\ \cdot & \cdot & \cdot & & \cdot & \cdot \\ \cdot & \cdot & & & \cdot & \cdot \\ \cdot & \cdot & & & \cdot & \cdot \\ 0 & 0 & 0 & \cdots & x_1 & x_0 \\ \cdot & \cdot & \cdot & \cdot & & \\ \cdot & \cdot & \cdot & & \cdot & \\ \cdot & \cdot & \cdot & & \cdot & \\ 0 & 0 & 0 & \cdots & & x_M \end{bmatrix} \begin{bmatrix} h_0 \\ h_1 \\ h_2 \\ \cdot \\ \cdot \\ \cdot \\ h_M \\ \cdot \\ \cdot \\ \cdot \\ h_L \end{bmatrix}$$

The matrix X is a Toeplitz band matrix.

If $x_k = 0$ for all $k > M$ and $h_k = 0$ for all $k > L$, then for a system initially at rest ($x_k = h_k = y_k = 0$ for $k = 0$) we can write

$$\begin{aligned} y_n &= \sum_{k=0}^{n} x_k h_{n-k} \\ &= \sum_{k=0}^{n} h_k x_{n-k}, && \text{for } 0 \le n \le M \\ &= \sum_{k=0}^{M} x_k h_{n-k} \\ &= \sum_{k=n-M}^{n} h_k x_{n-k}, && \text{for } (M + L) \le k \le L \\ &= \sum_{k=n-L}^{M} x_k h_{n-k} \\ &= \sum_{k=n-M}^{L} h_k x_{n-k}, && \text{for } (L + 1) \le n \le (L + M) \end{aligned}$$

for $L > M$ and

$$\begin{aligned} y_n &= \sum_{k=0}^{n} x_k h_{n-k} \\ &= \sum_{k=0}^{n} h_k x_{n-k}, && \text{for } 0 \le n \le L \\ &= \sum_{k=n-L}^{M} x_k h_{n-k} \\ &= \sum_{k=0}^{L} h_k x_{n-k}, && \text{for } (L + 1) \le n \le M \\ &= \sum_{k=n-L}^{M} x_k h_{n-k} \\ &= \sum_{k=n-M}^{L} h_k x_{n-k}, && \text{for } (M + 1) \le n \le (L + M) \end{aligned}$$

for $L < M$.

EXAMPLE 5.7

Take $L = 4$ and $M = 2$. Write the matrix and algebraic equations of convolution summation. Since $L + M = 6$,

$$\begin{aligned} y_0 &= x_0 h_0 \\ y_1 &= x_1 h_0 + x_0 h_1 \\ y_2 &= x_2 h_0 + x_1 h_1 + x_0 h_2 \\ y_3 &= x_2 h_1 + x_1 h_2 + x_0 h_3 \\ y_4 &= x_2 h_2 + x_1 h_3 + x_0 h_4 \\ y_5 &= x_2 h_3 + x_1 h_4 + x_0 h_5 \\ y_6 &= x_2 h_4 + x_1 h_5 + x_0 h_6 \end{aligned}$$

In matrix form,

$$\begin{bmatrix} y_0 \\ y_1 \\ y_2 \\ y_3 \\ y_4 \\ y_5 \\ y_6 \end{bmatrix} = \begin{bmatrix} x_0 & 0 & 0 & 0 & 0 \\ x_1 & x_0 & 0 & 0 & 0 \\ x_2 & x_1 & x_0 & 0 & 0 \\ 0 & x_2 & x_1 & x_0 & 0 \\ 0 & 0 & x_2 & x_1 & x_0 \\ 0 & 0 & 0 & x_2 & x_1 \\ 0 & 0 & 0 & 0 & x_2 \end{bmatrix} \begin{bmatrix} h_0 \\ h_1 \\ h_2 \\ h_3 \\ h_4 \\ h_5 \end{bmatrix}$$

EXAMPLE 5.8

Take $L = 2$ and $M = 4$. Write the matrix and algebraic equations of convolution summation. Again, $L + M = 6$. Therefore,

$$\begin{aligned} y_0 &= x_0 h_0 \\ y_1 &= x_1 h_0 + x_0 h_1 \end{aligned}$$

$$y_2 = x_2h_0 + x_1h_1 + x_0h_2$$
$$y_3 = x_3h_0 + x_2h_1 + x_1h_2$$
$$y_4 = x_4h_0 + x_3h_1 + x_2h_2$$
$$y_5 = x_4h_1 + x_3h_2$$
$$y_6 = x_5h_1 + x_4h_2 + x_3h_3$$

In matrix form,

$$\begin{bmatrix} y_0 \\ y_1 \\ y_2 \\ y_3 \\ y_4 \\ y_5 \\ y_6 \end{bmatrix} = \begin{bmatrix} x_0 & 0 & 0 \\ x_1 & x_0 & 0 \\ x_2 & x_1 & x_0 \\ x_3 & x_2 & x_1 \\ x_4 & x_3 & x_2 \\ 0 & x_4 & x_3 \\ 0 & 0 & x_4 \end{bmatrix} \begin{bmatrix} h_0 \\ h_1 \\ h_2 \end{bmatrix}$$

EXAMPLE 5.9

Consider the discrete-time input and the UIR as shown in Fig. 5.8. Determine the discrete-time output.

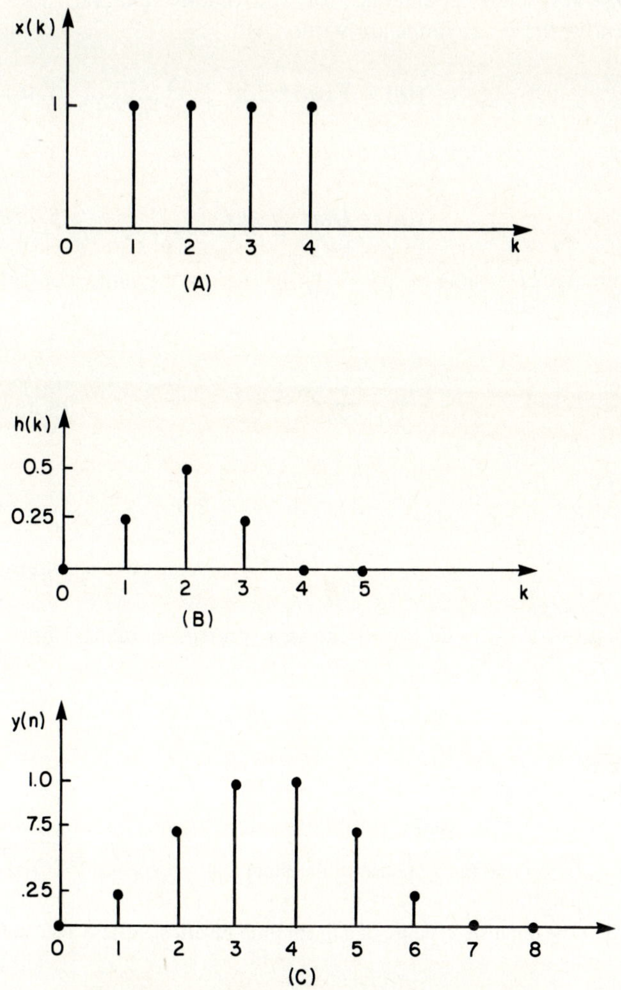

FIGURE 5.8 Example 5.9: (a) Discrete-time input. (b) Discrete-time UIR. (c) Discrete-time output.

We first consider equation (5.34).

$$y_0 = x_0h_0 = 0$$
$$y_1 = x_0h_1 + x_1h_0 = 1 \times .25 + 1 \times 0 = 0.25$$
$$y_2 = x_0h_2 + x_1h_1 + x_2h_0$$
$$= 1 \times .5 + 1 \times .25 + 1 \times 0 = 0.75$$
$$y_3 = x_0h_3 + x_1h_2 + x_2h_1 + x_3h_0$$
$$= 1 \times .25 + 1 \times .5 + 1 \times .25 + 1 \times 0 = 1$$
$$y_4 = x_0h_4 + x_1h_3 + x_2h_2 + x_3h_1 + x_4h_0$$
$$= 1 \times 0 + 1 \times .25 + 1 \times .5 + 1 \times .25 + 0 \times 0 = 1$$
$$y_5 = x_0h_5 + x_1h_4 + x_2h_3 + x_3h_2 + x_4h_1 + x_5h_0$$
$$= 1 \times 0 + 1 \times 0 + 1 \times .25 + 1 \times .5 + 0 \times .25 + 0 \times 0 = 0.75$$
$$y_6 = x_0h_6 + x_1h_5 + x_2h_4 + x_3h_3 + x_4h_2 + x_5h_1 + x_6h_0$$
$$=$$
$$1 \times 0 + 1 \times 0 + 1 \times 0 + 1 \times .25 + 0 \times .5 + 0 \times .25 + 0 \times 0$$
$$= 0.25$$
$$y_7 = x_0h_7 + x_1h_6 + x_2h_5 + x_3h_4 + x_4h_3 + x_5h_2 + x_6h_1 + x_7h_0$$
$$= 1 \times 0 + 1 \times 0 + 1 \times 0 + 1 \times 0 + 0 \times .25 + 0 \times .5 + 0 \times .5$$
$$+ 0 \times 0 = 0$$

The output $y(n)$ is graphed in Fig. 5.8(c).

If we consider equation (5.34), we immediately notice that the calculation is identical. Further, we can also perform the same calculation using equation (5.36) or (5.37). That is,

$$\begin{bmatrix} y_0 \\ y_1 \\ y_2 \\ y_3 \\ y_4 \\ y_5 \\ y_6 \\ y_7 \end{bmatrix} = \begin{bmatrix} x_0 & 0 & 0 & 0 & 0 \\ x_1 & x_0 & 0 & 0 & 0 \\ x_2 & x_1 & x_0 & 0 & 0 \\ x_3 & x_2 & x_1 & x_0 & 0 \\ x_4 & x_3 & x_2 & x_1 & x_0 \\ 0 & x_4 & x_3 & x_2 & x_1 \\ 0 & 0 & x_4 & x_3 & x_2 \\ 0 & 0 & 0 & x_4 & x_3 \end{bmatrix} \begin{bmatrix} h_0 \\ h_1 \\ h_2 \\ h_3 \\ h_4 \end{bmatrix}$$

Substituting the values of $x(k)$ and $h(k)$,

$$\begin{bmatrix} y_0 \\ y_1 \\ y_2 \\ y_3 \\ y_4 \\ y_5 \\ y_6 \\ y_7 \end{bmatrix} = \begin{bmatrix} 1 & 0 & 0 & 0 & 0 \\ 1 & 1 & 0 & 0 & 0 \\ 1 & 1 & 1 & 0 & 0 \\ 1 & 1 & 1 & 1 & 1 \\ 0 & 1 & 1 & 1 & 1 \\ 0 & 0 & 1 & 1 & 1 \\ 0 & 0 & 0 & 1 & 1 \\ 0 & 0 & 0 & 0 & 1 \end{bmatrix} \begin{bmatrix} 0 \\ 0.25 \\ 0.5 \\ 0.25 \\ 0 \end{bmatrix}$$

Thus we can solve for $y(n)$ and get the same answer as before.

5.2.3 Derivation of Superposition Summation Using the Step Response

From the previous chapter we know that a discrete-time input $x(n)$ can be resolved into unit steps as

$$x(n) = x(n_0)\, u_1(n - n_0) + \sum_{k=n_0+1}^{\infty} [x(k) - x(k-1)]\, u_1(n - k)$$

The system output $y(n)$ can be expressed as

$$y(n) = Hx(n)$$

$$= H\left[x(n_0)u_1(n - n_0) + \sum_{k=n_0+1}^{\infty} \{x(k) - x(k-1)\}\, u_1(n-k)\right]$$

Using the principles of proportionality and superposition,

$$y(n) = x(n_0)\, H[u_1(n - n_0)] + \sum_{k=n_0+1}^{\infty} [x(k) - x(k-1)]\, H[u_1(n-k)]$$

$$= x(n_0)\, U(n, n_0) + \sum_{k=n_0+1}^{\infty} [x(k) - x(k-1)]\, U(n, k) \qquad (5.38)$$

where $U(\cdot, \cdot)$ is the unit step response. Equation (5.38) is based on the assumption that $x = 0$, $n < n_0$. If $\ln_{k\to\infty} x(k) = 0$, then equation (5.38) can be written as

$$y(n) = \sum_{k=-\infty}^{\infty} [x(k) - x(k-1)]U(n, k) \qquad (5.39)$$

Equation (5.39) is the superposition summation—a discrete counterpart of the superposition integral of equation (5.16).

5.2.4 Forms of Superposition Summation in Equation (5.39)

If the system is causal, the summation may be terminated at n because $U(n, k) = 0$ for $k > n$. Therefore,

$$y(n) = \sum_{k=-\infty}^{n} [x(k) - x(k-1)]U(n, k) \qquad (5.40)$$

If the system is time-invariant, $U(n, k) = U(n - k)$ and

$$y(n) = \sum_{k=-\infty}^{\infty} [x(k) - x(k-1)]U(n-k) \qquad (5.41)$$

Alternatively, noting that k plays the same role as a variable of integration and making a change of variable $m = n - k$,

$$y(n) = \sum_{k=-\infty}^{\infty} [x(n-m) - x(n-m-1)]U(m) \qquad (5.42)$$

If $x(k)$ starts at $k = 0$ with $x(k) = 0$, $k < 0$, then

$$y(n) = \sum_{k=0}^{\infty} [x(k) - x(k-1)]U(n, k) \qquad (5.43)$$

Further, if the system is both causal and time-invariant,

$$y(n) = \sum_{k=0}^{n} [x(k) - x(k-1)]U(n-k) \qquad (5.44)$$

Alternatively,

$$y(n) = \sum_{m=0}^{n} [x(n-m) - x(n-m-1)]U(m) \qquad (5.45)$$

Equations (5.44) and (5.45) are called *convolution summation*—a discrete counterpart of convolution integral. Frequently, these equations are written as

$$y(n) = \nabla x(n) * U(n) \qquad (5.46)$$

and

$$y(n) = U(n) * \nabla x(n) \qquad (5.47)$$

where $\nabla x(k) = x(k) - x(k-1)$. These can be more conveniently written as

$$y_n = \sum_{k=0}^{n} \nabla x_k U_{n-k} \qquad (5.48)$$

and

$$y_n = \sum_{k=0}^{n} \nabla x_{n-k} U_k \qquad (5.49)$$

Equations (5.48) and (5.49) can be expressed in matrix form.

EXERCISES

5.1 The impulse response of a linear time-invariant system is given by

$$h(t) = \exp(-(t-1))\, u_1(t-1)$$

Determine the response to the input $x(t) = \exp(-t)\, u_1(t)$ by using each of equations (5.6), (5.7), (5.20), and (5.21). Graph each response.

5.2 The impulse response of a linear time-invariant system is given by

$$h(t) = [\exp(-t) - \exp(-2t)]\, u_1(t)$$

Determine the response to the input $x(t) = \exp(-t)u_1(t)$, and plot it.

5.3 The impulse response of a linear time-invariant system is given by

$$h(t) = \begin{cases} \sin t, & 0 \le t \le \pi \\ 0, & \text{otherwise} \end{cases}$$

Find the response to the input

$$x(t) = \begin{cases} \dfrac{t}{\pi}, & 0 \le t \le \pi \\ 0, & \text{otherwise} \end{cases}$$

and graph it.

5.4 The impulse response of a linear time-invariant system is given by

$$h(t) = \exp(-t)\cos t$$

Evaluate the response to a step input applied at $t = 0$ and plot it.

5.5 Use the graphical technique to solve Exercise 5.4.

5.6 The impulse response of a linear time-invariant system is given by

$$h(t) = \begin{cases} \dfrac{t}{\pi}, & 0 \le t \le \pi \\ 0, & \text{otherwise} \end{cases}$$

Find the response to the input

$$x(t) = \begin{cases} \cos t, & 0 \le t \le \pi \\ 0, & \text{otherwise} \end{cases}$$

5.7 Use the graphical technique to solve Exercise 5.6.

5.8 The impulse response to a linear time-invariant system is given by

$$h(t) = \begin{cases} \exp(-t), & 0 \le t \le \tau \\ 0, & \text{otherwise} \end{cases}$$

Evaluate the response to the input

$$x(t) = \begin{cases} t, & 0 \le t \le 1 \\ 0, & \text{otherwise} \end{cases}$$

5.9 Use the graphical method to solve Exercise 5.8.

5.10 A two-dimensional inpulse function $\delta(x, y)$ may be defined in a manner analogous to that used for $\delta(t)$:

$$\delta(x, y) = 0 \qquad \text{if } x \ne 0 \quad \text{or} \quad y \ne 0$$

$$\int_{-\infty}^{\infty}\int_{-\infty}^{\infty} \delta(s, w)\, ds\, dw = 1$$

If the response of a two-dimensional system to $\delta(x, y)$ is $h(x, y)$, what is its response to an input $f(x, y)$?

5.11 For a linear time-invariant system the impulse response is assumed to be represented by a rectangular pulse extending over a finite period of time. If the input is assumed to be represented by the same rectangular pulse, derive the system response and graph it.

5.12 Solve Exercise 5.11 by the graphical method.

CHAPTER SIX

Parameter Estimation

There are several methods for estimating parameters of hydrologic models. The methods of moments (Nash, 1959; Dooge, 1973) and cumulants (Dooge, 1973; Harley, 1967; O'Meara, 1968; Van de Nes and Hendriks, 1971) are discussed here. Some of the other methods are discussed in the next chapter.

6.1 METHOD OF MOMENTS FOR CONTINUOUS SYSTEMS

This method is frequently utilized to estimate parameters of the linear hydrologic models (Nash, 1959; Diskin, 1967; Diskin and Boneh, 1968; Dooge, 1973). We therefore discuss it in detail.

6.1.1 Definition and Notation

Let x be a continuous variable and $f(x)$ its function satisfying some necessary conditions. Then the rth *moment* M_r^a of $f(x)$ about an arbitrary point a can be defined as

$$M_r^a(f) = M_r^a = \int_{-\infty}^{\infty} (x-a)^r f(x)\, dx \tag{6.1a}$$

This is the definition used normally in engineering. In statistics, however, this definition is

$$M_r^a(f) = \frac{\int_{-\infty}^{\infty} (x-a)^r f(x)\, dx}{\int_{-\infty}^{\infty} f(x)\, dx} \tag{6.1b}$$

As the denominator in equation (6.1b) defines the area under the curve, which is usually unity or made to unity by normalization, the two definitions are numerically the same. In this text we use the definition of equation (6.1a) with $f(x)$ normalized beforehand. The variable x may or may not be a random variable.

Analogous to equation (6.1), the *absolute moment* W_r^a of order r about point a can be defined as

$$W_r^a = \int_{-\infty}^{\infty} |x-a|^r f(x)\, dx \tag{6.1c}$$

Clearly, if r is even then the absolute moment is equal to the ordinary moment. Furthermore, if the range of the function is positive, then the absolute moments about any point to the left of the start of the function are equal to the ordinary moments of corresponding order.

Here, $r \geq 0$ is the order of the moment. This notation will be employed throughout this chapter. To be explicit, M denotes the moment, the subscript denotes the order of the moment, the superscript denotes the point about which to take the moment, and the quantity within the parentheses denotes the function, in normalized form, whose moment is taken. It is, of course, assumed here that the integral in equation (6.1) converges. There are some functions that possess moments of lower order, and some do not possess any except the moment of zero order. However, if a moment of higher order exists, moments of all lower orders must exist.

If $r = 0$, then

$$M_0^a = \int_{-\infty}^{\infty} (x-a)^0 f(x)\, dx = \int_{-\infty}^{\infty} f(x)\, dx = 1$$

Thus, the zero-order moment is the area under the curve defined by $f(x)$ subject to $-\infty < x < \infty$.

If $r = 1$, then

$$M_1^a = \int_{-\infty}^{\infty} (x - a)^1 f(x)\, dx = \mu - a$$

where μ is the *centroid* of the area or mean. Thus the first moment is the weighted mean about the point a. If $a = 0$, the first moment gives the mean. When $a = \mu$, then the rth moment about the mean is

$$M_r^\mu = \int_{-\infty}^{\infty} (x - \mu)^r f(x)\, dx \tag{6.2}$$

In particular, if $r = 1$ then

$$M_1^\mu = \int_{-\infty}^{\infty} (x - \mu) f(x)\, dx = \mu - \mu = 0$$

Henceforth we drop the superscript if the moment is taken about 0. The descriptive properties of the moments with respect to a specific function can be summarized as

M_0 = area

M_1 = lag or mean

M_2^μ = variance, a measure of dispersion of the function about the mean

M_3^μ = measure of skewness of the function

M_4^μ = *kurtosis*, a measure of the peakedness of the function

EXAMPLE 6.1

Consider a function $f(x) = a \exp(-ax)$, $x > 0$, $a > 0$. Determine M_r^μ and M_r.

$$M_r^\mu = a \int_0^{\infty} (x - \mu)^r \exp(-ax)\, dx$$

Therefore,

$$M_0^\mu = -\exp(-ax)|_0^\infty = 1$$

$$M_1^\mu = a \int_0^{\infty} x \exp(-ax)\, dx - a\mu \int_0^{\infty} \exp(-ax)\, dx$$

$$= \mu - \mu = 0$$

and so on.

Likewise,

$$M_r = a \int_0^{\infty} x^r \exp(-ax)\, dx$$

Therefore,

$$M_0 = -\exp(-ax)\,|_0^\infty = 1$$

$$M_1 = a\left[-\frac{x}{a}\exp(-ax) - \frac{1}{a^2}\exp(-ax)|_0^\infty\right] = \frac{1}{a}$$

and so on.

6.1.2 Moment-Generating Function (MGF)

The moments of a function can be determined either directly by using their definition in equation (6.1) or by using the generating functions. This latter approach is more viable and more frequently used and has other advantages. One generating function is the *moment-generating function* (MGF), denoted by $G(\theta)$. The MGF of $f(x)$ can be defined as

$$G(\theta) = E[e^{\theta x}] \tag{6.3}$$

where *E is the expectation operator and* θ is the transform variable of MGF. We can write equation (6.3) as

$$G(\theta) = \int_{-\infty}^{\infty} e^{\theta x} f(x)\, dx \tag{6.4}$$

If $G(\theta)$ exists, then it is continuously differentiable in some neighborhood of the origin. If $G(\theta)$ in equation (6.4) is differentiated r times with respect to θ,

$$\frac{d^r}{d\theta^r} G(\theta) = \int_{-\infty}^{\infty} x^r e^{\theta x} f(x)\, dx$$

If $\theta = 0$, we obtain

$$\frac{d^r}{d\theta^r} G(\theta)\,|_{\theta=0} = \int_{-\infty}^{\infty} x^r f(x)\, dx = E[x^r]$$
$$= M_r \tag{6.5}$$

The symbol on the left-hand side is interpreted to mean the rth derivative of $G(\theta)$ evaluated at $\theta = 0$. Thus, the moments of a function can be obtained by differentiation of its MGF. Furthermore, if we expand $G(\theta)$ about $\theta = 0$ by Taylor series,

$$G(\theta) = \sum_{r=0}^{\infty} \left[\frac{d^r G(\theta)}{d\theta^r}\,|_{\theta=0}\right] \frac{\theta^r}{r!}$$

This can be written as

$$G(\theta) = \sum_{r=0}^{\infty} M_r \frac{\theta^r}{r!} \tag{6.6}$$

Equation (6.6) shows that M_r's are nothing but coefficients of θ^r in the Taylor series expansion of the MGF. This same result can be obtained by expanding the exponential term in equation (6.3),

$$e^{\theta x} = \sum_{r=0}^{\infty} \frac{(\theta x)^r}{r!}$$

On taking expectation of both sides, we can write

$$G(\theta) = E\left[\sum_{r=0}^{\infty} \frac{(\theta x)^r}{r!}\right]$$

$$= \sum_{r=0}^{\infty} \frac{\theta^r}{r!} E[x^r]$$

$$= \sum_{r=0}^{\infty} M_r \frac{\theta^r}{r!} \tag{6.7}$$

Equation (6.7) shows that M_r's are the coefficients of θ^r in the exponential expansion.

EXAMPLE 6.2

Determine the MGF of the function in Example 6.1, and then determine the moments.

$$G(\theta) = a\int_0^{\infty} \exp(\theta x)\exp(-ax)\, dx = a\int_0^{\infty} \exp[-(a-\theta)x]\, dx$$

$$= \frac{a}{a-\theta}$$

$$M_r = \frac{d^r G(\theta)}{d\theta^r}\bigg|_{\theta=0} = \frac{d^r}{d\theta^r}\left(\frac{a}{a-\theta}\right)\bigg|_{\theta=0} = r!\frac{a}{(a-\theta)^{r+1}}\bigg|_{\theta=0} = r!\frac{1}{a^r}$$

$$M_0 = 1$$

$$M_1 = \frac{1}{a}$$

$$M_2 = \frac{2}{a^2}$$

and so on.

6.1.3 Characteristic Function (CF)

Unfortunately the MGFs do not always exist. Often it is better to work with the *characteristic function* (CF), which always exists. A characteristic function can be defined as

$$C(\theta) = E[e^{i\theta x}]$$

or

$$C(\theta) = \int_{-\infty}^{\infty} e^{i\theta x} f(x)\, dx \tag{6.8}$$

where $C(\theta)$ can be viewed as an integral transform of $f(x)$. The integral of $C(\theta)$ converges absolutely and uniformly for the functions of our interest in linear theory of hydrologic systems.

If $C(\theta)$ is differentiated r times with respect to θ,

$$\frac{d^r}{d\theta^r} C(\theta) = (i)^r \int_{-\infty}^{\infty} e^{i\theta x} x^r f(x)\, dx$$

$$\frac{d^r}{d\theta^r} C(\theta)\bigg|_{\theta=0} = (i)^r \int_{-\infty}^{\infty} x^r f(x)\, dx = (i)^r E[x^r]$$

$$= (i)^r M_r$$

or

$$M_r = (i)^{-r} \frac{d^r}{d\theta^r} C(\theta)\bigg|_{\theta=0} \tag{6.9}$$

Thus, the moments of a function can be obtained by differentiation of the corresponding characteristic function. Hence, if $C(\theta)$ is known explicitly, the moments about the origin can be obtained.

Moreover, if we expand $C(\theta)$ by Taylor series about the origin,

$$C(\theta) = \sum_{r=0}^{\infty} \frac{(\theta)^r}{r!}\left[\frac{d^r}{d\theta^r} C(\theta)\bigg|_{\theta=0}\right]$$

$$= \sum_{r=0}^{\infty} \frac{(i\theta)^r}{r!} M_r \tag{6.10}$$

We find that the moments about the origin are the coefficients of θ_r in the expansion of $C(\theta)$. This same result is obtained by expanding the exponential term in equation (6.8),

$$C(\theta) = E\left[\sum_{r=0}^{\infty} \frac{(i\theta x)^r}{r!}\right]$$

$$= \sum_{r=0}^{\infty} \frac{(i\theta)^r}{r!} E[x^r] = \sum_{r=0}^{\infty} \frac{(i\theta)^r}{r!} M_r$$

It is interesting to contrast $C(\theta)$ with the Fourier transform, which can be defined as

$$F(w) = \int_{-\infty}^{\infty} f(x)\, e^{-iwx}\, dx$$

where

$$f(x) = \frac{1}{2\pi}\int_{-\infty}^{\infty} F(w)\, e^{iwx}\, dw$$

Likewise,

$$f(x) = \frac{1}{2\pi}\int_{-\infty}^{\infty} C(\theta)\, e^{-i\theta x}\, d\theta$$

EXAMPLE 6.3

Determine the CF of the function in Example 6.1, and then determine the moments.

$$C(\theta) = a\int_0^{\infty} \exp(i\theta x)\exp(-ax)\, dx$$

$$= a\int_0^{\infty} \exp[-(a-i\theta)x]\, dx$$

$$= \frac{a}{a-i\theta}$$

$$\frac{d^r}{d\theta^r} C(\theta)\bigg|_{\theta=0} = r!(i)^r \frac{a}{(a-i\theta)^{r+1}}\bigg|_{\theta=0} = (i)^r M_r$$

Therefore,

$$C(\theta)\,|_{\theta=0} = M_0 = 1$$

$$\left.\frac{d^rC(\theta)}{d\theta^r}\right|_{\theta=0} = r!\left(\frac{i}{a}\right)^r$$

$$M_1 = \frac{1}{a}$$

$$M_2 = \frac{2}{a^2}$$

and so on.

6.1.4 Laplace Transform

For certain functions the integral represented by $G(\theta)$ or $F(w)$ may not exist. It is then necessary to use the *Laplace transform*. For functions that are zero for $x < 0$, the ordinary Laplace transform can be used, in which we have

$$F(s) = L[f(x)] - \int_0^{\infty} f(x)\,e^{-sx}\,dx$$

For functions that have values for $x < 0$, we must use the bilateral Laplace transform, given by

$$F(s) = L[f(x)] = \int_{-\infty}^{\infty} f(x)e^{-sx}\,dx$$

Differentiating the function $F(s)$ r times with respect to s and evaluating the derivatives at $s = 0$,

$$\left.\frac{d^r}{ds^r}F(s)\right|_{s=0} = \int_0^{\infty} (-1)^r x^r f(x)\,dx$$

$$= (-1)^r M_r \tag{6.11}$$

Thus we see that once the Laplace transform of a function is known, its moments can be obtained by differentiation.

EXAMPLE 6.4

Determine the moments of the function in Example 6.1 by using the Laplace transform.

$$F(s) = a\int_0^{\infty} \exp[-(a+s)x]\,dx$$

$$= \frac{a}{a+s}$$

$$\frac{d^r}{ds^r}F(s) = (-1)^r\frac{r!a}{(a+s)^{r+1}}$$

$$M_r = (-1)^{-r}\left.\frac{d^r}{ds^r}F(s)\right|_{s=0} = \frac{r!}{a^r}$$

Therefore,

$$M_0 = 1$$

$$M_1 = \frac{1}{a}$$

$$M_2 = \frac{2}{a^2}$$

and so on.

6.1.5 Fourier Transform

The moments of a function $f(x)$ can also be obtained by employing the Fourier transform expressed as

$$F(w) = T[f(x)] = \int_{-\infty}^{\infty} f(x)e^{-iwx}\,dx$$

in which T denotes the Fourier transform. Differentiating the function $F(w)$ r times with respect to w and evaluating the derivatives at $w = 0$,

$$\left.\frac{d^r}{dw^r}F(w)\right|_{w=0} = \int_{-\infty}^{\infty} (-ix)^r f(x)\,dx$$

$$= (-i)^r M_r \tag{6.12}$$

Thus if the Fourier transform of a function is known, its moments about origin can be obtained by differentiation.

EXAMPLE 6.5

Determine the moments of the function in Example 6.1 by using the Fourier transform.

$$F(w) = a\int_0^{\infty} e^{-(a+iw)x}\,dx$$

$$= \frac{a}{a+iw}$$

Therefore,

$$\frac{d^r}{dw^r}F(w) = (-i)^r\frac{r!a}{(a+iw)^{r+1}}$$

$$M_r = \frac{r!}{a^r}$$

Thus,

$$M_0 = 1$$

$$M_1 = \frac{1}{a}$$

$$M_2 = \frac{2}{a^2}$$

and so on.

6.1.6 Change of Reference Points for Moments

Let a and b be two constants and let $c = b - a$. Let us denote the rth moments of a function about a and b, respectively, by M_r^a and M_r^b. Expanding binomially,

$$(x-a)^r = (x-b+b-a)^r = (x-b+c)^r$$

$$= \sum_{j=0}^{r}\binom{r}{j}(x-b)^{r-j}c^j \tag{6.13}$$

Then we write

$$M_r^a = \int_{-\infty}^{\infty} (x-a)^r f(x)\, dx$$

Substituting equation (6.13),

$$M_r^a = \int_{-\infty}^{\infty} \sum_{j=0}^{r} \binom{r}{j} (x-b)^{r-j} c^j f(x)\, dx$$

$$= \sum_{j=0}^{r} \binom{r}{j} c^j \int_{-\infty}^{\infty} (x-b)^{r-j} f(x)\, dx$$

$$= \sum_{j=0}^{r} \binom{r}{j} c^j M_{r-j}^b \quad (6.14a)$$

This gives the rth moment of a function about a in terms of its rth moment and lower moments about b. Writing the suffixes as power indices (without, of course, interpreting them as such except for the purpose of expansion) or the symbolic mnemonic form of the above relationship,

$$M_r^a = [M^b + c]^r, \qquad \text{for all } r \quad (6.14b)$$

in which $[M^b]^r$ is interpreted as M_r^b for all r. If we specialize by taking a as the origin and b the centroid or the first moment μ,

$$M_r = \sum_{j=0}^{r} \binom{r}{j} (\mu)^j M_{r-j}^{\mu} \quad (6.15a)$$

The symbolic mnemonic form of this relationship is

$$M_r = [M^{\mu} + \mu]^r, \qquad \text{for all } r \quad (6.15b)$$

In particular this leads to

$$\begin{aligned}
M_0 &= M_0^{\mu} = 1 \\
M_1 &= M_1^{\mu} + \mu M_0^{\mu} = \mu \\
M_2 &= M_2^{\mu} + \mu^2 \\
M_3 &= M_3^{\mu} + 3\mu M_2^{\mu} + \mu^3 \\
M_4 &= M_4^{\mu} + 4\mu M_3^{\mu} + 6 M_2^{\mu} \mu^2 + \mu^4 \quad (6.15c)
\end{aligned}$$

and so on. Equation (6.15b) can be manipulated to express moments about the centroid in terms of moments about the origin. In particular, this yields

$$\begin{aligned}
M_0^{\mu} &= 1 \\
M_1^{\mu} &= 0 \\
M_2^{\mu} &= M_2 - \mu^2 \\
M_3^{\mu} &= M_3 - 3M_2\mu + 2\mu^3 \\
M_4^{\mu} &= M_4 - 4M_3\mu + 6\mu^2 M_2 - 3\mu^4 \quad (6.15d)
\end{aligned}$$

Equation (6.15d) can also be derived directly. Writing the rth moment about point b,

$$M_r^b = \int_{-\infty}^{\infty} (x-b)^r f(x)\, dx$$

Expanding biominally,

$$(x-b)^r = (x-a-b+a)^r = (x-a-c)^r$$

$$= \sum_{j=0}^{r} \binom{r}{j} (x-a)^{r-j} (-c)^j$$

Inserting in the general expression for M_r^b,

$$M_r^b = \int_{-\infty}^{\infty} \sum_{j=0}^{r} \binom{r}{j} (x-a)^{r-j} (-c)^j f(x)\, dx$$

$$= \sum_{j=0}^{r} \binom{r}{j} (-c)^j M_{r-j}^a \quad (6.15e)$$

The symbolic mnemonic form of this relationship is

$$M_r^b = [M^a - c]^r, \qquad \text{for all } r \quad (6.15f)$$

In particular, if $a = 0$ and $b = \mu$, then

$$M_r^{\mu} = \sum_{j=0}^{r} (-1)^j \binom{r}{j} \mu^j M_{r-j} \quad (6.15g)$$

Likewise,

$$M_r^{\mu} = [M - \mu]^r, \qquad \text{for all } r \quad (6.15h)$$

Again, the coefficients of the expansions in equation (6.15h) are those of the binomial expansion. The powers of terms containing the quantity μ are real powers; the powers of terms containing M are not real powers but only indices of the moments concerned.

6.1.7 Invariance Property

The moments have an invariance property, which states that when the variate values are multiplied by a constant, the rth moment M_r is multiplied by a^r. This is evident at once from its definition.

6.1.8 Theorem of Moments

Moments represent parameters that are useful in the analysis and synthesis of hydrologic systems. Nash (1959) derived a theorem relating the moments of input, output, and impulse response of linear systems about their corresponding centroids. Diskin and Boneh (1968) extended this theorem to the moments of these functions taken about their arbitrary points. We first derive the Nash theorem of moments and then consider its extension. The derivations can be performed either by using the binomial expansion or the Laplace transform.

6.1.8.1 Use of the Binomial Expansion

The convolution operation for a linear time-invariant system is shown in Fig. 6.1. For convenience it is assumed that the integrals of input, output, and impulse response functions are all normalized to unity; this assumption, however, is not essential for the validity of the conclusions. Let a, b, and c be the centers of the areas of output, input, and impulse response functions, respectively. The center of area of output will always be later than the centroid of area of input by c. This implies the relation between the first moments about the origin, that is,

$$a = b + c$$

or

$$M_1^0(y) = M_1^0(x) + M_1^0(h) \tag{6.16}$$

Writing the convolution integral for the output at γ,

$$y(\gamma) = \int_0^\infty x(\tau)h(\epsilon)\, d\tau \tag{6.17}$$

where $\gamma = \epsilon + \tau$. The rth moment of y about the centroid a can be expressed as

$$\begin{aligned} M_r^a(y) &= \int_0^\infty y(\gamma)\gamma^r\, d\gamma \\ &= \int_0^\infty y(\gamma)(\epsilon + \tau)^r\, d\epsilon \\ &= \int_0^\infty \int_0^\infty x(\tau)h(\epsilon)(\epsilon + \tau)^r\, d\tau\, d\epsilon \end{aligned}$$

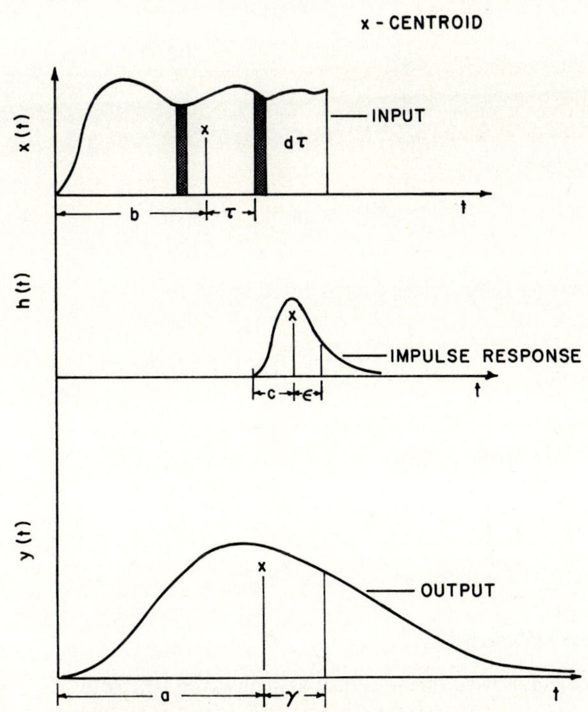

FIGURE 6.1 Convolution operation of a linear time-invariant system.

We can expand $(\epsilon + \tau)^r$ by the binomial series,

$$(\epsilon + \tau)^r = \sum_{j=0}^{r} \binom{r}{j} \tau^j \epsilon^{r-j}$$

Therefore,

$$M_r^a(y) = \int_0^\infty \int_0^\infty x(\tau)h(\epsilon)\left[\sum_{j=0}^{r} \binom{r}{j} \tau^j \epsilon^{r-j}\right] d\epsilon\, d\tau$$

Recall that

$$M_r^b(x) = \int_0^\infty x(\tau)\tau^r\, d\tau$$

$$M_r^c(h) = \int_0^\infty h(\epsilon)\epsilon^r\, d\epsilon$$

Then

$$\begin{aligned} M_r^a(y) &= \sum_{j=0}^{r} \binom{r}{j} \int_0^\infty x(\tau)\tau^j\, d\tau \int_0^\infty h(\epsilon)\epsilon^{r-j}\, d\epsilon \\ &= \sum_{j=0}^{r} \binom{r}{j} M_j^b(x) M_{r-j}^c(h) \end{aligned} \tag{6.18}$$

Writing the symbolic mnemonic form,

$$M_r^a(y) = [M^b(x) + M^c(h)]^r \tag{6.19}$$

This is the Nash theorem of moments.

Thus, the theorem of moments gives the rth moment of y in terms of all the moments of rth order and lower of x and those of h. It is observed that if the moments are evaluated successively for $r = 1, 2, 3, \ldots$, there will be just one unknown in each equation, which is the highest moment of h in the equation concerned. It is thus possible to obtain from the moments of the known $x(t)$ and $y(t)$ the unknown moments of $h(t)$. We can write the first four moments as

$$\begin{aligned} M_1^\mu(y) &= M_1^\mu(x) = M_1^\mu(h) = 0 \\ M_2^\mu(y) &= M_2^\mu(x) + M_2^\mu(h) \\ M_3^\mu(y) &= M_3^\mu(x) + M_3^\mu(h) \\ M_4^\mu(y) &= M_4^\mu(x) + M_4^\mu(h) + 6M_2^\mu(x)\, M_2^\mu(h) \end{aligned}$$

These equations enable us to evaluate the moments of h from those of x and y.

6.1.8.2 Use of Laplace Transform

The convolution integral can be written in an alternative form as

$$y(t) = \int_0^t x(\tau)h(t - \tau)\, d\tau$$

If we take the Laplace transform of this equation, we obtain

$$Y(s) = X(s)H(s) \tag{6.20}$$

where s is the Laplacian variable. Also, Y, X, and H are Laplace transforms of y, x and h, respectively. Adopting the notation $D = d/ds$, $X(s) = D^0X(s)$, and $Y(s) = D^0y(s)$ and differentiating equation (6.20) r times using the Leibnitz rule for rth derivative of a product,

$$\begin{aligned} D^r(Y) = D^r(XH) &= (D^rX)H \\ &\quad + \binom{r}{1}(D^{r-1}X)(DH) + \cdots + X(D^rH) \\ &= \sum_{j=0}^{r}\binom{r}{j}(D^{r-j}X)(D^jH) \end{aligned} \tag{6.21a}$$

or in symbolic form,

$$D^rY(s) = (DX(s) + DH(s))^r, \qquad r \geq 1 \tag{6.21b}$$

where it is implied that the terms in the expansion are derivatives, not power terms, and the first and last terms, which include the derivatives of zero order, are

$$X(s)\frac{d^r}{ds^r}H(s) \quad \text{and} \quad H(s)\frac{d^r}{ds^r}X(s)$$

From equation (6.11) it is clear that

$$\frac{d^r}{ds^r}Y(s)\big|_{s=0} = (-1)^r M_r(y) \tag{6.22}$$

$$\frac{d^r}{ds^r}X(s)\big|_{s=0} = (-1)^r M_r(x) \tag{6.23}$$

$$\frac{d^r}{ds^r}H(s)\big|_{s=0} = (-1)^r M_r(h) \tag{6.24}$$

Inserting equations (6.22)–(6.24) in equation (6.21),

$$M_r(y) = \sum_{j=0}^{r}\binom{r}{j}M_{r-j}(x)M_j(h)$$

Writing a few terms explicitly,

$$M_1(y) = M_0(x)\,M_1(h) + M_1(x)\,M_0(h)$$

$$M_2(y) = M_0(x)\,M_2(h) + 2M_1(x)\,M_1(h) + M_2(x)\,M_0(h)$$

$$\begin{aligned} M_3(y) = M_0(x)\,M_3(h) &+ 3M_1(x)\,M_2(h) \\ &+ 3M_2(x)\,M_1(h) + M_3(x)\,M_0(h) \end{aligned}$$

and so on.

The general form may thus be written in the form of a binomial expansion,

$$M_r(y) = (M(x) + M(h))^r \tag{6.25}$$

The terms in equation (6.25) are moments, not power terms, and the first and last terms include moments of zero order. These are the moments about the origin rather than the centroid of the areas. Restriction is placed on the units of the input and output. If we desire moments about the centroid, it is easy to obtain them by using the relationships developed in equations (6.14)–(6.15).

6.1.8.3 Use of the Fourier Transform

If we take the Fourier transform of the convolution integral,

$$F[y(t)] = F\left[\int_0^{\infty} x(\tau)\,h(t-\tau)\,d\tau\right]$$

then it can be shown that

$$Y(w) = X(w)\,H(w) \tag{6.26}$$

in which Y, X, and H are Fourier transforms of y, x and h, respectively, and w is Fourier variable. Differentiating equation (6.26) r times using the Leibnitz rule,

$$D^r(Y) = D^r(XH) = \sum_{j=0}^{r}\binom{r}{j}(D^{r-j}X)(D^jH)$$

or following the convention of equation (6.21),

$$D^r(Y) = [DX(w) + DH(w)]^r, \qquad r \geq 1 \tag{6.27}$$

with $D \equiv d/dw$, $X(w) = D^0\,X(w)$ and $H(w) = D^0\,H(w)$. From equation (6.12) it is seen that

$$\frac{d^r}{dw^r}Y(w)\big|_{w=0} = (-i)^r M_r(y)$$

$$\frac{d^r}{dw^r}X(w)\big|_{w=0} = (-i)^r M_r(x)$$

$$\frac{d^r}{dw^r}H(w)\big|_{w=0} = (-i)^r M_r(h)$$

Inserting these relationships in equation (6.27),

$$M_r(y) = \sum_{j=0}^{r}\binom{r}{j}M_{r-j}(x)M_j(h)$$

or in symbolic form,

$$M_r(y) = [M(x) + M(h)]^r$$

which is the same as equation (6.25).

EXAMPLE 6.6

Consider a linear time-invariant system whose UIR is

$$h(t) = \exp(-t)$$

When it is subject to the input

$$x(t) = \begin{cases} x_0, & 0 \le t \le D \\ 0, & \text{otherwise} \end{cases}$$

the output produced is

$$y(t) = \begin{cases} x_0[1 - \exp(-t)], & 0 \le t \le D \\ y_0\exp[-(t - D)], & t \ge D \end{cases}$$

in which

$$y_0 = x_0[1 - \exp(-D)].$$

Determine the moments of $y(t)$ by the theorem of moments as well as by computing them independently.

Solution: The function $h(t)$ is normalized. Its rth moment about origin is obtained as

$$M_r(h) = r!$$

The function $x(t)$ is normalized by dividing by x_0D. Its rth moment about origin is

$$M_r(x) = \frac{D^r}{r + 1}$$

Therefore, the rth moment of $y(t)$ is obtained by the theorem of moments as

$$M_r(y) = \sum_{j=0}^{r} \binom{r}{j} \frac{D^{r-j}}{r - j + 1} j!$$

$$M_1(y) = 1 + \frac{D}{2}$$

$$M_2(y) = \frac{D^2}{3} + D + 2$$

$$M_3(y) = \frac{D^3}{4} + D^2 + 3D + 6$$

and so on.

Now we compute the moments of $y(t)$ independently and check if they match those just computed. First we compute the area under this curve, A_y,

$$A_y = \int_0^D X_0[1 - \exp(-t)]dt + \int_D^\infty y_0 \exp[-(t - D)]\, dt$$

$$= X_0D$$

The normalized function $y(t)$ is

$$y(t) = \begin{cases} \dfrac{1}{D}[1 - \exp(-t)], & 0 \le t \le D \\ y_1 \exp[-(t - D)], & t \ge D \end{cases}$$

where

$$y_1 = \frac{1}{D}[1 - \exp(-D)]$$

Therefore,

$$M_1(y) = \int_0^D t\frac{1}{D}[1 - \exp(-t)]\, dt + \int_D^\infty ty_1\exp[-(t - D)]\, dt$$

$$= \frac{1}{D}\left[\frac{t^2}{2} - te^{-t} + e^{-t}\right]\Bigg|_0^D + y_1[-te^{-(t-D)} + e^{-(t-D)}]\Bigg|_D^\infty$$

$$= \frac{D}{2} + 1$$

$$M_2(y) = \int_0^D \frac{t^2}{D}[1 - \exp(-t)]\, dt + y_1 \int_D^\infty t^2 \exp[-(t - D)]\, dt$$

$$= \frac{1}{D}\left[\frac{t^3}{3} + t^2e^{-t} + 2te^{-t} + 2e^{-t}\right]\Bigg|_0^D$$

$$+ y_1[-t^2e^{-(t-D)} - 2te^{-(t-D)} - 2e^{-(t-D)}]\Bigg|_D^\infty$$

$$= \frac{D^2}{3} + D + 2$$

$$M_3(y) = \int_0^D \frac{t^3}{D}[1 - \exp(-t)]\, dt + y_1 \int_D^\infty t^3 \exp[-(t - D)]\, dt$$

$$= \frac{1}{D}\left[\frac{t^4}{4} + t^3e^{-t} - 3t^2e^{-t} + 6(te^{-t} - e^{-t})\right]\Bigg|_0^D$$

$$+ y_1[-t^3e^{-(t-D)} + 3t^2e^{-(t-D)} - 6te^{-(t-D)} + 6e^{-(t-D)}]\Bigg|_D^\infty$$

$$= \frac{D^3}{4} + D^2 + 3D + 6$$

and so on. These moments are the same as before.

6.1.8.4 Moments about an Arbitrary Point

The theorem of moments also holds if the moments are taken about an arbitrary, fixed point. Let this point be a. Then

$$M_r^a(y) = \int_0^\infty (t - a)^r y(t)\, dt \tag{6.28}$$

Substituting equation (6.17),

$$M_r^a(y) = \int_0^\infty \int_0^t (t - a)^r x(\tau)h(t - \tau)\, d\tau\, dt$$

Changing the order of integration,

$$M_r^a(y) = \int_0^\infty \int_\tau^\infty x(\tau)(t - a)^r h(t - \tau)\, dt\, d\tau$$

or

$$M_r^a(y) = \int_0^\infty x(\tau) \int_\tau^\infty (t - a)^r h(t - \tau)\, dt\, d\tau$$

The double integration is evaluated in the (t, τ) plane, as shown in Fig. 6.2. We can evaluate the inner integral by taking $u = t - \tau$,

$$\int_\tau^\infty (t - a)^r h(t - \tau)\, dt = \int_0^\infty [\tau + (u - a)]^r h(u)\, du$$

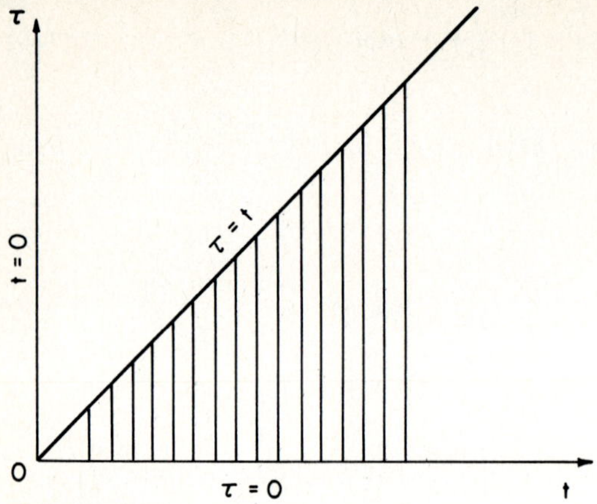

FIGURE 6.2(a) Region in the (t, τ) plane encompassed by convolution integral.

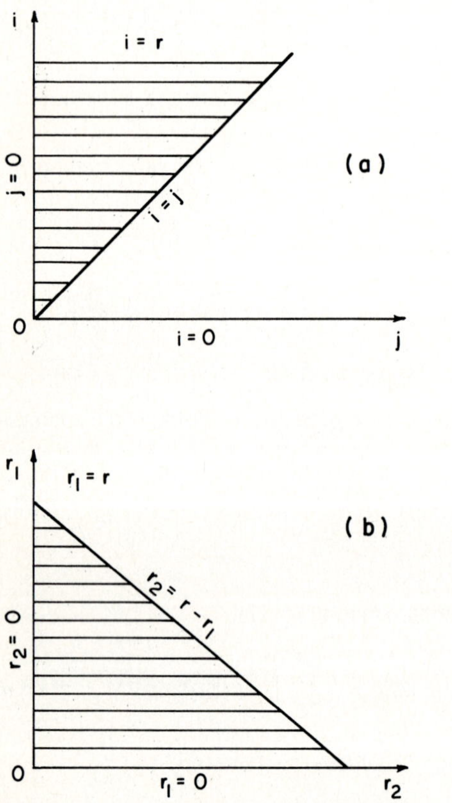

FIGURE 6.2(b) Regions in the (j, i) and (r_2, r_1) planes encompassed by double summations.

Expanding the bracketed term in the binomial series,

$$\begin{aligned}\int_{\tau}^{\infty} (t-a)^r h(t-\tau)\,dt &= \int_0^{\infty} \sum_{j=0}^{r} \binom{r}{j} (u-a)^j \tau^{r-j} h(u)\,du \\ &= \sum_{j=0}^{r} \binom{r}{j} \tau^{r-j} \int_0^{\infty} (u-a)^j h(u)\,du \\ &= \sum_{j=0}^{r} \binom{r}{j} \tau^{r-j} M_j^a(h)\end{aligned}$$

Therefore,

$$\begin{aligned}M_r^a(y) &= \int_0^{\infty} x(\tau) \sum_{j=0}^{r} \binom{r}{j} \tau^{r-j} M_j^a(h)\,d\tau \\ &= \sum_{j=0}^{r} \binom{r}{j} M_j^a(h) \int_0^{\infty} \tau^{r-j} x(\tau)\,d\tau \\ &= \sum_{j=0}^{r} \binom{r}{j} M_j^a(h) M_{r-j}(x)\end{aligned} \tag{6.29}$$

This can be written in compact symbolic form as

$$M_r^a(y) = [M^a(h) + M(x)]^r \tag{6.30a}$$

In equation (6.29) the expansion is performed according to the binomial theorem. This says that if the moments of y and h are known about the point a, then the moments of x are obtained about the origin. Moreover, if equation (6.17) is expressed in another equivalent form (equation (4.4)), equation (6.29) takes the form

$$M_r^a(y) = [M(h) + M^a(x)]^r \tag{6.30b}$$

This says that if the moments of x and y are known about a, then the moments of h are obtained about the origin. If $a = 0$, it is seen that equations (6.29) and (6.30) are identical to equation (6.25).

6.1.9 Generalized Theorem of Moments

For generalization of the theorem of moments (Nash, 1959), we closely follow the work of Diskin and Boneh (1968). Let there be three arbitrary points a, b, and c about which moments of $y(t)$, $x(t)$, and $h(t)$, respectively, are taken. The functions $x(t) = 0$, $y(t) = 0$, $t \le 0$. Let us define

$$M_r^a(y) = \int_0^{\infty} (t-a)^r y(t)\,dt$$

$$M_r^b(x) = \int_0^{\infty} (t-b)^r x(t)\,dt$$

$$M_r^c(h) = \int_0^{\infty} (t-c)^r h(t)\,dt$$

We derive the relationship between $M_r^a(y)$, $M_r^b(x)$ and $M_r^c(h)$ using, again, the binomial expansion as well as the Laplace transform.

6.1.9.1 Use of the Binomial Expansion

We first derive the relationships between $M_r^a(h)$ and $M_r^c(h)$ as well as between $M_r^a(x)$ and $M_r^b(x)$. Let us write

$$\begin{aligned}M_r^a(h) &= \int_0^{\infty} (t-a)^r h(t)\,dt \\ &= \int_0^{\infty} [(t-c) + (c-a)]^r h(t)\,dt\end{aligned}$$

This can be written as

$$M_r^a(h) = \int_0^\infty \sum_{j=0}^{r} \binom{r}{j}(t-c)^j(c-a)^{r-j}h(t)\,dt$$
$$= \sum_{j=0}^{r} \binom{r}{j}(c-a)^{r-j}\int_0^\infty (t-c)^j h(t)\,dt$$
$$= \sum_{j=0}^{r} \binom{r}{j}(c-a)^{r-j}M_j^c(h)$$

Symbolically,

$$M_r^a(h) = [(c-a) + M^c(h)]^r \qquad (6.31)$$

By analogy,

$$M_r^a(x) = [(b-a) + M^b(x)]^r \qquad (6.32)$$

In equations (6.31) and (6.32) the expansions are carried out binomially where the powers of $(c - a)$ and $(b - a)$ are real but represent orders for moments. Equation (6.31) or (6.32) is a general expression relating the moments of all orders of any function about any two arbitrary points.

We can specialize in equation (6.32) by taking $a = 0$,

$$M_r(x) = [b + M^b(x)]^r \qquad (6.33)$$

Substituting equations (6.31) and (6.33) into equation (6.29),

$$M_r^a(y) = \sum_{j=0}^{r} \binom{r}{j}\left\{\sum_{i=0}^{j}\binom{j}{i}(c-a)^{j-i}M_i^c(h)\right\}$$
$$\left\{\sum_{k=0}^{r-j}\binom{r-j}{k}b^{r-j-k}M_k^b(x)\right\} \qquad (6.34)$$

Rewriting the binomial coefficients of equation (6.34),

$$\binom{r}{j}\binom{j}{i}\binom{r-j}{k} = \frac{r!}{j!(r-j)!}\,\frac{j!}{i!(j-i)!}\,\frac{(r-j)!}{k!(r-j-k)!}$$
$$= \frac{r!}{i!(j-i)!k!(r-j-k)!}$$

Therefore,

$$M_r^a(y) = \sum_{j=0}^{r}\sum_{i=0}^{j}\sum_{k=0}^{r-j}\frac{r!}{i!(j-i)!k!(r-j-k)!}$$
$$(c-a)^{j-i}\,b^{r-j-k}M_i^c(h)\,M_k^b(x) \qquad (6.35)$$

Equation (6.35) is similar to the binomial expansion for the sum of four variables. Let A_1, A_2, A_3 and A_4 be these variables. Then

$$(A_1 + A_2 + A_3 + A_4)^r = \sum_{r_1=0}^{r}\sum_{r_2=0}^{r-r_1}\sum_{r_3=0}^{r-r_1-r_2}$$
$$\frac{r!}{r_1!r_2!r_3!r_4!}A_1^{r_1}A_2^{r_2}A_3^{r_3}A_4^{r_4} \qquad (6.36)$$

where $r_4 = r - r_1 - r_2 - r_3$.

In order to take advantage of equation (6.36) for simplification of equation (6.35), let

$$r_1 = j - i; \qquad r_2 = i; \qquad r_3 = k; \qquad r_4 = r - j - k$$

This yields $r_1 + r_2 + r_3 + r_4 = r$. In equation (6.35) the limits of the summation on k can be changed in terms of r_3: from $[k, r - j]$ to $[r_3, r - r_1 - r_2]$. The limits on i can be changed from $[i, j]$ to $[r_2, r - r_1]$. The limits on j can be changed from $[j, r]$ to $[r_1, r]$. These last two limits can be obtained by considering the mapping of (j, i) plane onto (r_1, r_2) plane and noting that the line $i = j$ is transformed into the line $r_1 = 0$, and line $i = 0$ into the line $r_2 = 0$ and the line $i = r$ into the line $r_1 = r - r_2$ or $r_2 = r - r_1$, as shown in Fig. 6.2. Thus equation (6.35) can be written as

$$M_r^a = \sum_{r_1=0}^{r}\sum_{r_2=0}^{r-r_1-r_2}\frac{r!}{r_1!r_2!r_3\,r_4!}(c-a)^{r_1}b^{r_4}M_{r_2}^c(h)\,M_{r_4}^b(x) \qquad (6.37)$$

Equation (6.37) can be written symbolically as

$$M_r^a(y) = [M^b(x) + M^c(h) + (c-a) + b]^r \qquad (6.38a)$$

or

$$M_r^a(y) = [M^b(x) + M^c(h) + S]^r \qquad (6.38b)$$

where $S = c - a + b$, a linear combination of the times about which moments are taken. Diskin and Boneh (1968) named S as characteristic time. Equation (6.38) can be obtained more directly as follows. Writing the binomial expansion for the sum of four variables as

$$(A_1 + A_2 + A_3 + A_4)^r = \sum_{j=0}^{r}\sum_{i=0}^{j}\sum_{k=0}^{r-j}\frac{r!}{i!\,(j-1)!\,k!\,(r-j-k)!}$$
$$A_3^{i-1}A_1^{r-j-k}A_4^{i}A_2^{k}$$

A comparison with equation (6.35) immediately leads to equation (6.38).

Equations (6.36)–(6.38) are alternative expressions for the general theorem of moments, of which the Nash theorem of moments is a special case.

6.1.9.2 Use of the Laplace Transform

Let the Laplace transforms of the functions of $y(t)$, $x(t)$, and $h(t)$ be denoted, respectively, by $Y(s)$, $X(s)$, and $H(s)$, where these functions have their origin at $t = 0$. Let these functions be shifted to right by a, b and c, respectively:

$$y_1(t) = y(t-a)\,u_1(t-a) \tag{6.39a}$$

$$x_1(t) = x(t-b)\,u_1(t-b) \tag{6.39b}$$

$$h_1(t) = h(t-c)\,u_1(t-c) \tag{6.39c}$$

where $u_1(t-a)$ is the unit step function at $t = a$. The Laplace transforms of $y_1(t)$, $x_1(t)$, and $h_1(t)$ can be written, using the shifting theorem, as

$$Y_1(s) = Y(s)\exp(-as) \tag{6.40a}$$

$$X_1(s) = X(s)\exp(-bs) \tag{6.40b}$$

$$H_1(s) = H(s)\exp(-cs) \tag{6.40c}$$

The moments of $y_1(t)$ about its origin are equal to the moments of $y(t)$ about the point $t = -a$, that is,

$$M_r(y_1) = M_r^{-a}(y)$$

The moments of $y_1(t)$ can be obtained by differentiating its Laplace transform and evaluating the differentiation at $s = 0$. Taking the rth derivative of equation (6.40a) by using the Leibnitz rule,

$$D^r[Y_1(s)] = D^r[Y(s)\exp(-as)]$$

$$= \sum_{j=0}^{r} \binom{r}{j} [D^{r-j}\, Y(s)] D^j(\exp(-as))]$$

Evaluating at $s = 0$,

$$M_r(y_1) = [M(y) + a]^r \tag{6.41}$$

or

$$M_r^{-a}(y) = [M(y) + a]^r \tag{6.42}$$

If we convolute $x_1(t)$ and $h_1(t)$, then

$$y_2(t) = \int_0^t x_1(\tau)\, h_1(t-\tau)\, d\tau \tag{6.43}$$

Taking its Laplace transform,

$$Y_2(s) = X_1(s)H_1(s) = X(s)H(s)\exp[-(b+c)s] \tag{6.44}$$

Eliminating $Y_2(s)$, $X(s)$, $H(s)$, and $Y(s)$ by using equations (6.39), (6.40), and (6.44),

$$Y_1(s) = X_1(s)\, H_1(s)\exp[(b + c - a)s]$$

or

$$Y_1(s) = X_1(s)\, H_1(s)\exp(sS) \tag{6.45}$$

in which $S = b + c - a$ as before.

The relationship between moments of $y_1(t)$, $x_1(t)$, and $h_1(t)$ can be obtained, as before, by differentiating equation (6.45) with respect to s and evaluating the derivatives at $s = 0$. This yields

$$M_r(y_1) = [M(x_1) + M(h_1) - S]^r \tag{6.46}$$

Therefore,

$$M_r^{-a}(y) = [M^{-b}(x) + M^{-c}(h) - S]^r \tag{6.47}$$

If a, b, and c are assumed to be negative, $a = -a_1$, $b = -b_1$, $c = -c_1$, then $S = -S_1$ and equation (6.47) is identical to equation (6.38).

6.1.10 Inversion of Moments

In analysis and synthesis of linear hydrologic systems the moments can be used as parameters to represent the response function. A question arises: Can the response function be derived from a knowledge of the values of the moments? Let us recall that

$$M_r = (-1)^r \left[\frac{d^r}{ds^r} F(s)\right]\Bigg|_{s=0}$$

If the moments about the origin are known for a well-behaved function, the Laplace transform can be expressed in terms of these moments by means of Taylor series,

$$F(s) = \sum_{r=0}^{\infty} (-s)^r \frac{M_r}{r!}$$

Therefore, knowledge of the Laplace transform gives certain information about the behavior of the original function. If, however, we wish explicitly to know the complete function exactly, it is necessary to invert the Laplace transform numerically. This can perhaps be best done by using orthogonal functions.

6.1.11 Dimensionless Moments

It is often convenient to use dimensionless moments, which are independent of one another, in model calibration. Nash (1959) used dimensionless moments about the mean defined as

$$m_2 = \frac{M_2^{\mu}}{(M_1)^2\,(M_0)} \tag{6.48}$$

$$m_3 = \frac{M_3^{\mu}}{(M_1)^3\,(M_0)} \tag{6.49}$$

$$m_4 = \frac{M_4^{\mu}}{(M_1)^4\,(M_0)} \tag{6.50}$$

Here m_2, m_3 and m_4 are generally called shape factors. The objective of dividing by the first moment is to remove the time-scale effect from higher moments and thus make them dimensionless. These shape factors can be used to compare hydrologic models by constructing an m_3 versus m_2 diagram, as suggested by Nash (1959) and done by Harley (1967), O'Meara (1968), and Dooge (1980). We use this approach in several chapters in the text.

6.2 METHOD OF MOMENTS FOR DISCRETE SYSTEMS

6.2.1 Definition

If the function is discrete, represented as f_j, $j = -\infty, \ldots, -1, 0, 1, 2, \ldots, \infty$, then its rth moment about the origin or any other arbitrary point can be defined in an analogous manner. For convenience let the arbitrary point be origin. Then the rth moment is defined as

$$M_r = \sum_{m=-\infty}^{\infty} m^r f_m \tag{6.51}$$

It is assumed here that f_m is normalized, that is,

$$\sum_{m=-\infty}^{\infty} f_m = 1 \tag{6.52}$$

Otherwise,

$$M_r = \frac{\sum_{m=-\infty}^{\infty} m^r f_m}{\sum_{m=-\infty}^{\infty} f_m} \tag{6.53}$$

It is thus seen that equations (6.51) and (6.53) are analogous to equation (6.1).

6.2.2 MGF

Again, the moments of a discrete function can be determined either directly from their definition or by using generating functions. If x is discrete and takes the value j with probability p_j, then the MGF is

$$G(\Theta) = \sum_j e^{\Theta j} p_j = E[e^{\Theta j}] = P[e^{\Theta}] \tag{6.54}$$

in which P is probability-generating function. The rth moment can be determined by differentiating $G(\Theta)$ r times with respect to Θ and then equating to zero,

$$\left.\frac{d^r G(\Theta)}{d\Theta^r}\right|_{\Theta=0} = M_r \tag{6.55}$$

Therefore, equations (6.6) and (6.7) also hold for discrete functions.

For discrete functions the Z-transform can be used as an MGF in the same way as the Laplace transform is used for continuous functions. The Z-transform of the function f_j is defined for the bilateral case as

$$Z(f_j) = F(z) = \sum_{j=-\infty}^{\infty} f_j z^{-j} \tag{6.56}$$

and for the unilateral case as

$$Z(f_j) = F(z) = \sum_{j=0}^{\infty} f_j z^{-j} \tag{6.57}$$

in which $F(z)$ signifies the Z-transform of the function f_j. In most hydrologic cases, $f_j = 0$, $j < 0$ so the bilateral transform reduces to the unilateral case. The moments of f_j about the origin can be obtained from the Z-transform in the following manner:

$$M_0 = \sum_{j=-\infty}^{\infty} f_j = F(z)|_{z=1} \tag{6.58a}$$

$$M_1 = \sum_{j=-\infty}^{\infty} j f_j = -z\frac{d}{dz}(F(z)|_{z=1} \tag{6.58b}$$

$$M_2 = \sum_{j=-\infty}^{\infty} j^2 f_j = -z\frac{d}{dz}\left[-z\frac{d}{dz}F(z)\right]\Bigg|_{z=1} \tag{6.58c}$$

$$M_3 = \sum_{j=-\infty}^{\infty} j^3 f_j = \left(-z\frac{d}{dz}\left\{-z\frac{d}{dz}\left[-z\frac{d}{dz}F(z)\right]\right\}\right)\Bigg|_{z=1} \tag{6.58d}$$

and so on. The moments of discrete functions can be obtained analogously about any reference points. Consequently, the relationships of equations (6.14) and (6.15) hold. Likewise, the theorem of moments, derived for continuous functions, is also valid for discrete functions. It can be easily seen by noting the Z-transform of the convolution summation,

$$y_j = \sum_{i=0}^{j} x_i h_{j-i} = \sum_{i=0}^{j} h_i x_{j-i}, \qquad j = 0, 1, 2, \ldots \tag{6.59}$$

yields

$$Z[y_j] = Z\left[\sum_{i=0}^{j} x_j h_{j-i}\right] = Z\left[\sum_{i=0}^{j} h_i x_{j-i}\right]$$
$$Y(z) = X(z)H(z) = H(z)X(z) \tag{6.60a}$$

which is identical to equation (6.20). Taking log to the base e,

$$\ln Y(z) = \ln X(z) + \ln H(z) \tag{6.60b}$$

6.2.3 Inversion of the Z-Transform

The inverse Z-transform can be defined as

$$Z^{-1}[F(z)] = f_j = \frac{1}{2\pi i}\int F(z)\, z^{j-1}\, dz$$
$$= \sum_{\text{poles}} \text{residues of } [F(z)\, z^{j-1}] \quad \text{at the poles of } F(z) \tag{6.61}$$

For a multiple pole of order m at $z = w$,

$$\text{The residue} = \left[\frac{1}{(m-1)!}\frac{d^{m-1}}{dz^{m-1}}\{(z-w)^m F(z)\, z^{j-1}\}\right]\bigg|_{z=w}$$

EXAMPLE 6.7

Determine the moments of the following function about the origin and about the centroid:

$$f_j = \left(\frac{1}{1+a}\right)\left(\frac{a}{1+a}\right)^j, \qquad j = 0, 1, 2, \ldots$$

This function is the unit impulse of a discrete linear reservoir (of delay time $a > 0$).

Solution: The Z-transform of f_j is

$$F(z) = \frac{z}{(1+az) - a}$$

$$M_0 = \sum_{j=0}^{\infty} f_j = F(z)|_{z=1} = 1$$

$$M_1 = \left[-z\frac{d}{dz}F(z)\right]\bigg|_{z=1} = a = \mu$$

$$M_2 = \left\{-z\frac{d}{dz}\left[-z\frac{d}{dz}F(z)\right]\right\}\bigg|_{z=1} = a(2a+1)$$

Similarly,

$$M_3 = a(1 + 6a + 6a^2)$$

$$M_4 = a(1 + 2a)(1 + 12a + 12a^2)$$

and so on.

The moments of f_j about the mean are:

$$M_3^{\mu} = a(1+a)$$

$$M_3^{\mu} = a(1+a)(1+2a)$$

$$M_4^{\mu} = a(1+a)(1+9a+9a^2)$$

and so on.

6.2.4 Factorial Moments

Bree (1977) pointed out that factorial moments rather than ordinary moments lead, in general, to a more straightforward method of parameter estimation for discrete systems. The rth factorial moment $M_{[r]}$ of f_j about origin can be defined (Kendall and Stuart, 1958) as

$$M_{[r]}(f_j) = \sum_{j=-\infty}^{\infty} j^{[r]} f_j, \qquad r = 0, 1, 2, \ldots \tag{6.62}$$

where the quantity $j^{[r]}$ is defined as

$$j^{[r]} = j(j-1)(j-2)\cdots(j-r+1) = \frac{j!}{(j-r)!} \tag{6.63}$$

The factorial moments can also be defined about any arbitrary point a,

$$M_{[r]}^{a}(f_j) = \sum_{j=-\infty}^{\infty} (j-a)^{[r]} f_j, \qquad r = 0, 1, 2, \ldots \tag{6.64}$$

By direct expansion,

$$M_{[1]}^{a} = M_1$$

$$M_{[2]}^{a} = M_2 - wM_1^{a}$$

$$M_{[3]}^{a} = M_3^{a} - 3wM_2^{a} + 2w^2M_1^{a}$$

$$M_{[4]}^{a} = M_4^{a} - 6wM_3^{a} + 11w^2M_2^{a} - 6w^3M_1^{a}$$

and conversely,

$$M_2^{a} = M_{[2]}^{a} + wM_{[1]}^{a}$$

$$M_3^{a} = M_{[3]}^{a} + 3wM_{[2]}^{a} + w^2M_{[1]}^{a}$$

$$M_4^{a} = M_{[4]}^{a} + 6wM_{[3]}^{a} + 7w^2M_{[2]}^{a} + w^3M_{[1]}^{a}$$

in which w is the data interval (for discretization), which is assumed to be unity in equation (6.64). The factorial moments also obey equations (6.14) and (6.15); that is,

$$M_{[r]}^{a}(f_j) = \sum_{j=0}^{r}\binom{r}{j} M_{[r-j]}^{b}\, c^{[j]} \tag{6.65}$$

which can be written symbolically as

$$M_{[r]}^{a} = [M_{[\;]}^{b} + c]^{[r]}$$

6.2.4.1 Factorial MGF (FMGF)

If the frequency-generating function $P(x)$ is

$$P(x) = \sum_{j=0}^{\infty} x^j p_j$$

then on replacing x by $(1 + x)$, we get

$$P(1+x) = \sum_{j=0}^{\infty} (1+x)^j p_j = \sum_{i=0}^{\infty} \frac{x^i}{i!} M_{[i]} \tag{6.66}$$

Therefore, $P(1 + x)$ is a *factorial MGF* (FMGF) in which the factorial moment of order i is the coefficient of $x^i/i!$.

The factorial moments can also be derived from the Z-transform. Let B be the backward shift order. Then

$$g\left[\frac{1}{(1-B)}\right] = F(z)|_{z=(1-B)^{-1}} = \sum_{j=0}^{\infty} f_j z^{-j} = \sum_{j=0}^{\infty} f_j(1-B)^j$$

$$= \sum_{j=0}^{\infty} f_j \sum_{i=0}^{j}\binom{j}{i}(-B)^i$$

$$= \sum_{i=0}^{\infty}\frac{(-B)^i}{i!}\sum_{j=0}^{\infty} f_j j^{[i]}$$

$$= \sum_{i=0}^{\infty} M_{[i]}\frac{(-B)^i}{i!} \tag{6.67}$$

It can also be noted that

$$\frac{d^r}{dB^r} g(B)|_{B=1} = M_{[r]} \tag{6.68}$$

where $g(B) = F(z)|_{z=B-1}$.

6.2.4.2 Factorial Moments Theorem

From the Z-transform of the convolution summation in equation (6.61a), we can obtain

$$Y\left(\frac{1}{1-B}\right) = X\left(\frac{1}{1-B}\right) H\left(\frac{1}{1-B}\right) \tag{6.69}$$

By using equation (6.67),

$$Y\left(\frac{1}{1-B}\right) = \sum_{r=0}^{\infty} M_{[r]}(y) \left[\frac{(-B)^r}{r!}\right] \tag{6.70a}$$

$$X\left(\frac{1}{1-B}\right) = \sum_{r=0}^{\infty} M_{[r]}(x) \left[\frac{(-B)^r}{r!}\right] \tag{6.70b}$$

$$H\left(\frac{1}{1-B}\right) = \sum_{r=0}^{\infty} M_{[r]}(h) \left[\frac{(-B)^r}{r!}\right] \tag{6.70c}$$

Inserting equation (6.70) in equation (6.69) gives

$$M_{[r]}(y) = [M_{[\ ]}(x) + M_{[\ ]}(h)]^r, \qquad \text{for } r = 1, 2, \ldots \tag{6.71}$$

The factorial moments theorem is valid for the moments about origin as well as the moments about the centroid.

EXAMPLE 6.8

Compute factorial moments of the function in Example 6.7. Applying the definition of the generating function,

$$M_{[r]} = r!\, a^r$$

6.3 METHOD OF CUMULANTS

Cumulants are another set of descriptors for estimating parameters of hydrologic models. These are, from a theoretical standpoint, more useful than moments.

6.3.1 Definition and Notation

We can define the cumulants using either the MGF or CF. The cumulant generating function CGF, denoted by $K(\Theta)$, is the logarithm of MGF or CF,

$$K(\Theta) = \ln G(\Theta) \tag{6.72}$$

or

$$K(\Theta) = \ln C(\Theta) \tag{6.73}$$

Thus the cumulants are defined by

$$K(\Theta) = \sum_{r=1}^{\infty} K_r \frac{\Theta^r}{r!} \tag{6.74}$$

or

$$K(\Theta) = \sum_{r=1}^{\infty} K_r \frac{(i\Theta)^r}{r!} \tag{6.75}$$

where K_r is the rth cumulant. Note that whereas M_r is the coefficient of $(i\Theta)^r/r!$ in $C(\Theta)$, K_r is the coefficient of $(i\Theta)^r/r!$ in $\ln C(\Theta)$, if an expression in power series exists. Thus we can determine the cumulants by differentiation of either $\ln G(\Theta)$ or $\ln C(\Theta)$,

$$K_r = \left[\frac{d^r}{d\Theta^r} \ln G(\Theta)\right]\bigg|_{\Theta=0} \tag{6.76a}$$

$$K_r = \left[\frac{d^r}{d\Theta^r} \ln C(\Theta)\right]\bigg|_{\Theta=0} \tag{6.76b}$$

Similarly, if the Fourier transform or the Laplace transform of the function is known, K_r can be determined as

$$K_r = (-1)^r \left[\frac{d^r}{dw^r} \ln F(w)\right]\bigg|_{w=0} \tag{6.76c}$$

or

$$K_r = (-1)^r \left[\frac{d^r}{ds^r} \ln F(s)\right]\bigg|_{s=0} \tag{6.76d}$$

EXAMPLE 6.9

Determine the cumulants of the function in Example 6.1. Its $G(\Theta)$ as shown in Example 6.2 is

$$G(\Theta) = \frac{a}{a-\Theta}$$

Therefore,

$$K(\Theta) = \ln G(\Theta)$$

$$K_r = \frac{(r-1)!}{(a-\Theta)^r}\bigg|_{\Theta=0} = \frac{(r-1)!}{a^r}$$

Specifically,

$$K_0 = 0$$

$$K_1 = \frac{1}{a}$$

$$K_2 = \frac{1}{a^2}$$

$$K_3 = \frac{2}{a^3}$$

and so on.

6.3.2 Relationship between Cumulants and Moments

By using the generating functions, we can derive the relationships between cumulants and moments. When $\Theta = 0$,

$$G(\Theta) = 1$$

$$\ln[G(\Theta)] = 0$$

Therefore,

$$K_0 = 0$$

For $r = 1$,

$$\begin{aligned} K_1 &= \frac{d}{d\Theta} \ln G(\Theta)|_{\Theta=0} \\ &= \frac{1}{G(\Theta)} \frac{d}{d\Theta} G(\Theta)|_{\Theta=0} \\ &= \frac{d}{d\Theta} G(\Theta)|_{\Theta=0} = M_1 = \mu \end{aligned} \tag{6.77a}$$

For $r = 2$,

$$\begin{aligned} K_2 &= \frac{d^2}{d\Theta^2} \ln(G(\Theta)|_{\Theta=0} \\ &= \left[\frac{1}{G(\Theta)} \frac{d^2G(\Theta)}{d\Theta^2} - \frac{1}{G(\Theta)^2} \frac{dG(\Theta)}{d\Theta}\right]\Bigg|_{\Theta=0} \\ &= M_2 - (M_1)^2 \end{aligned} \tag{6.77b}$$

which, by relation of moments about the origin to moments about the centroid, is equal to

$$K_2 = M_2^{\mu} \tag{6.77c}$$

By a further differentiation, we can show that

$$K_3 = M_3 - 3M_1M_2 + 2M_1^3 \tag{6.77d}$$

which by definition is M_3. Thus

$$K_3 = M_3^{\mu} \tag{6.77e}$$

Similarly, by successive differentiation we can show that

$$K_4 = M_4 - 4M_3M_1 - 3(M_2)^2 + 12M_2(M_1)^2 - 6(M_1)^4 \tag{6.77f}$$

In terms of moments about the centroid,

$$K_4 = M_4^{\mu} - 3(M_2^{\mu})^2 \tag{6.77g}$$

Similarly,

$$K_5 = M_5^{\mu} - 10M_3^{\mu}M_2^{\mu} \tag{6.77h}$$

and so on.

Inversely,

$$M_1 = K_1 \tag{6.78a}$$

$$M_2 = K_2 + K_1^2 \tag{6.78b}$$

$$M_3 = K_3 + (2K_2 + K_1^2)K_1 \tag{6.78c}$$

$$M_4 = K_4 + 4K_3K_1 + 3K_2^2 + 6K_2K_1^2 + K_1^4 \tag{6.78d}$$

6.3.3 Invariance Property

If a frequency function is multiplied by a constant, then the rth cumulant will be multiplied by this constant to the power r. This is evident from its definition. The cumulants have another property of great practical importance. If the origin is shifted, the first cumulant is changed by the amount of the shift, but all other cumulants remain unchanged. The cumulants are thus invariant under a change of origin. Thus a linear transformation

$$\xi = ax + b$$

leaves the cumulants unaltered as far as the constant b is concerned and multiplies K_r by a^r. The exception to this property is the first cumulant, which is equal to the mean.

EXAMPLE 6.10

Show that the cumulants of all orders except the first order remain unchanged with the position of the reference point.

If the origin is changed from a to b in equation (6.8), where $c = b - a$ as before, then the effect on $C(\Theta)$ is one of multiplying it by $\exp(-i\Theta x)$, for $\int \exp(i\Theta x)f(x)\,dx$ becomes $\int \exp(i\Theta(x - c))f(x)\,dx$. Therefore, the effect on $\ln C(\Theta)$ is merely to add the term $-i\Theta C$, and consequently the coefficients in $\ln C(\Theta)$ are unchanged, except the first, which is reduced by C.

6.3.4 Theorem of Cumulants

A relatively simple theorem can be derived for the relationship between the cumulants of the input, response function, and the output (O'Connor and Nash, 1968). Taking the logarithm of equation (6.20),

$$\ln(Y(s)) = \ln(H(s)) + \ln(X(s)) \tag{6.79}$$

By differentiating both sides of equation (6.80) r times and taking the derivatives at $s = 0$, we obtain

$$K_r(y) = K_r(h) + K_r(x), \qquad r = 1, 2, \ldots \tag{6.80}$$

Thus, for a linear time-invariant system, the rth cumulant of the output is equal to the sum of the rth cumulants of the input and the response function. Except for the first cumulant, all others are independent of the point of reference. The simple form of the Nash theorem of moments for the second and third moments is seen by noting that the second and third moments about the centroid are equal to the second and third cumulants, respectively.

A very significant consequence of this theorem for linear

time-invariant systems is as follows. If a system is represented by a cascade of linear elements, then the rth cumulant of the IUH of the cascade is equal to the sum of the rth cumulants of the IUHs of individual elements. This can be shown by the successive application of the theorem of cumulants to each element of the cascade. That is,

$$K_r(h) = \sum_{i=1}^{N} K_r(h_i) \tag{6.81}$$

in which h_i is the IUH of the ith element and N is number of elements in the cascade.

EXAMPLE 6.11

Consider Example 6.6. Determine the cumulants of $y(t)$ independently as well as by the theorem of cumulants.

The function $h(t)$ is normalized. Its $G(\Theta)$ is

$$G(\Theta) = \int_0^\infty e^{\Theta t} e^{-t}\, dt = \int_0^\infty e^{-t(1-\Theta)}\, dt = \frac{1}{1-\Theta}$$

$$\ln G(\Theta) = -\ln(1-\Theta)$$

Therefore,

$$K_r(h) = \left[\frac{d^r}{d\Theta^r} \ln G(\Theta)\right]\bigg|_{\Theta=0}$$

$$K_1(h) = 1$$

$$K_2(h) = 1$$

and so on.

The function $x(t)$ is normalized by dividing by $x_0 D$. The natural log of the MGF is

$$\ln G(\Theta) = \ln\left(\int_0^D \frac{e^{\Theta t}}{D}\, dt\right)$$

Therefore,

$$K_1(x) = \frac{D}{2}$$

$$K_2(x) = \frac{D^2}{12}$$

and so on.

The function $y(t)$ is normalized by dividing by $x_0 D$. It has

$$\ln G(\Theta) = \ln\left[\int_0^D \frac{e^{\Theta t}}{D}(1-e^{-t})dt + \int_d^\infty \frac{y_0}{x_0 D} e^{\Theta t} e^{-(t-D)}\, dt\right]$$

Therefore,

$$K_1(y) = 1 + \frac{D}{2}$$

$$K_2(y) = 1 + \frac{D^2}{12}$$

and so on.

By applying the theorem of moments we find the cumulants of y as obtained directly.

6.3.5 Dimensionless Cumulants

Like dimensionless moments, dimensionless cumulants have been found useful in model calibration. These can be defined as

$$s_2 = \frac{K_2}{(K_1)^2} = m_2 \tag{6.83a}$$

$$s_3 = \frac{K_3}{(K_1)^3} = m_3 \tag{6.83b}$$

$$s_4 = \frac{K_4}{(K_1)^4} = m_4 - 3m_2^2 \tag{6.83c}$$

These dimensionless cumulants may also be termed as shape factors and can be used to compare hydrologic models.

6.3.6 Factorial Cumulants

A *factorial cumulant* $K_{[r]}$ can be defined as the coefficient of $\Theta^r/r!$ in the expansion of the logarithm of the FMGF. The factorial cumulant generating function (FCGF) is therefore defined as either $\ln P(x)$ or $\ln y(B)$,

$$\ln g(B) = \ln F\left(\frac{1}{1-B}\right) = \sum_{r=1}^{\infty} K_{[r]}\left[\frac{(-B)}{r!}\right] \tag{6.83a}$$

in which $K_{[r]}$ is the rth factorial cumulant. It may be noted that

$$K_{[r]} = \frac{d^r}{dz^r} \ln F\left(\frac{1}{z}\right)\bigg|_{z=1} \tag{6.83b}$$

The relations between cumulants and factorial cumulants are formally similar to those between moments and factorial moments. Analogous to equation (6.65), with unit interval,

$$K_{[1]} = K_1$$
$$K_{[2]} = K_2 - K_1$$
$$K_{[3]} = K_3 - 3K_2 + 2K_1$$
$$K_{[4]} = K_4 - 6K_3 + 11K_2 - 6K_1 \tag{6.84a}$$

and so on. Conversely,

$$K_2 = K_{[2]} + K_{[1]}$$
$$K_3 = K_{[3]} + 3K_{[2]} + K_{[1]}$$
$$K_4 = K_{[4]} + 6K_{[3]} + 7K_{[2]} + K_{[1]} \tag{6.84b}$$

and so on.

The theorem of factorial cumulants can also be derived in the same way as the theorem of ordinary cumulants. Formally, these theorems are similar in appearance. Thus,

$$K_{[r]}(y) = K_{[r]}(x) + K_{[r]}(h), \qquad r = 1, 2, 3, \ldots \tag{6.85}$$

EXAMPLE 6.11

Compute factorial cumulants of the function in Example 6.7. By applying equation (6.84a),

$$K_{[r]} = (r-1)!K^r, \qquad r = 1, 2, 3, \ldots$$

EXERCISES

6.1. Consider the normal distribution with mean $\bar{x}$ and variance S_x^2,

$$f(x) = \frac{1}{S_x\sqrt{2\pi}} \exp\left[-\frac{(x-\bar{x})^2}{2S_x^2}\right],$$

$$-\infty < \bar{x} < \infty \quad \text{and} \quad S_x^2 > 0$$

Determine the first four moments about the origin as well as about the centroid.

6.2. Determine the MGF of the normal distribution in Exercise 6.1. Determine the first four moments about origin by using the MGF.

6.3. Determine the characteristic function of the normal distribution in Exercise 6.1. Then determine the first four moments about origin.

6.4. By using the moments about origin as well as about centroid as obtained in Exercise 6.1, show that equation (6.15) is valid.

6.5. Let the UIR of a linear time-invariant system be denoted by

$$h(X) = \frac{1}{a\Gamma(b)}\left(\frac{x}{a}\right)^{b-1} \exp\left(\frac{-x}{a}\right), \qquad a > 0, b > 0$$

If the input to the system is denoted by a unit pulse of duration D, then determine the first three moments of the system output by using the theorem of moments.

6.6. Determine the system output in Exercise 6.5 either numerically or analytically. Graph it.

6.7. Determine the first four cumulants of the distribution in Exercise 6.1. Verify equations (6.79a)–(6.79d) using these quantities.

6.8. The discrete UIR of a cascade of equal linear reservoirs is

$$h_j = \frac{\Gamma(j+b-1)}{\Gamma(b)\,\Gamma(j+1)}\left(\frac{1}{1+a}\right)^b \left(\frac{a}{1+a}\right)^j$$

This is discrete equivalent of the UIR in Exercise 6.5. Determine the MGF and the first four moments.

6.9. Determine the CGF and the first four cumulants for the UIR in Exercise 6.8.

6.10. Show by using the UIR in Exercise 6.8 that the theorem of moments is valid.

6.11. Show by using the UIR in Exercise 6.8 that the theorem of cumulants is valid.

6.12. Determine the FMGF and the first four factorial moments for the UIR in Exercise 6.8.

6.13. Determine the FCGF and the first four factorial moments for the UIR in Exercise 6.8.

CHAPTER SEVEN

More on Parameter Estimation

Of the various methods for estimating parameters of hydrologic models, the methods of moments and cumulants were discussed in the preceding chapter. Some of the other popular methods used in hydrology include (1) the graphical method (Yevjevich, 1972); (2) the method of incomplete means (Houghton, 1978); (3) the method of probability weighted moments (Greenwood et al., 1979); (4) the method of mixed moments (Rao, 1980, 1983; Shrader et al., 1981); (5) maximum likelihood estimation (Douglas, Clarke, and Newton, 1976; Sorooshian, Gupta, and Fulton, 1983; Phien and Jivajirajah, 1984); (6) the least squares method (Jones, 1971; Snyder, 1972; Bree, 1978a, 1978b); (7) constrained parameter estimation (Natale and Todini, 1974; Todini and Wallis, 1977); (8) Bayesian estimation (Natale and Todini, 1974; Kuczera, 1983a, 1983b); (9) optimization methods (Himmelblau, 1972; Rosenbrock, 1960; Palmer, 1969; DeCoursey and Snyder, 1969; Sorooshian, 1980; Sorooshian and Gupta, 1983; Sehitoglu, 1983); (10) regression and correlation (Yevjevich, 1972; Clarke, 1973); and (11) the method of principle of maximum entropy (Tribus, 1969).

7.1 METHOD OF INCOMPLETE MEANS

Houghton (1978) developed a procedure called the *method of incomplete means* to estimate parameters of the Wakeby distribution used for analysis of hydrologic extremes. The method uses no moments higher than the first and yields stable estimates with little bias. It uses a combination of means calculated over only part of the range. To explain this method, we follow Houghton (1978). Consider a sample of N ranked observations $x_1, x_2, x_3, \ldots, x_N$, having a probability density function $f(x)$. Calculate the mean $\bar{x}$,

$$\bar{x} = \int_{-\infty}^{\infty} x f(x)\, dx \tag{7.1}$$

Let $y = F(x)$, where $F(x)$ is the cumulative density function. Then $x = F^{-1}(y)$, and $dy = f(x)\, dx$. Therefore,

$$\bar{x} = \int_0^1 F^{-1}(y)\, dy \tag{7.2}$$

For illustration let us consider a special case of the Wakeby distribution, defined as

$$x = c(1 - F)^{-d} + e \tag{7.3}$$

where c, d, and e are parameters and

$$F(x) = \int_0^x f(s)\, ds \tag{7.4}$$

Substituting equation (7.3) in equation (7.2),

$$\bar{x} = \int_0^1 [c(1 - F)^{-d} + e]\, dF = e + \frac{1}{1 - d} \tag{7.5}$$

$\bar{x}$ will fall between two adjacent observations in the sample and divide it effectively into two disjointed sets. Calculate

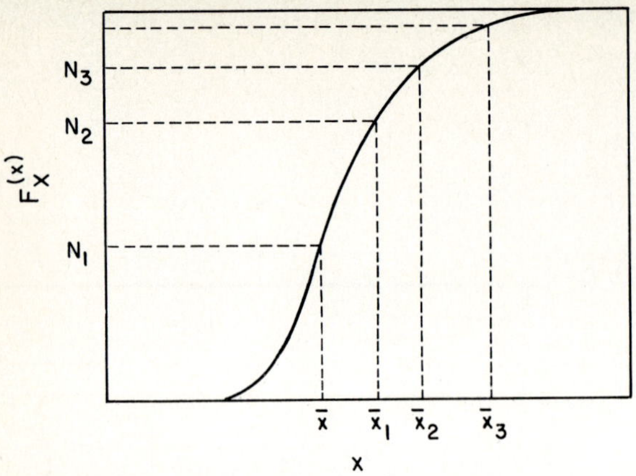

FIGURE 7.1 Incomplete means computation (after Houghton, 1978).

the mean of the upper set, say $\bar{x}_1$, as shown in Fig. 7.1. Calculate the mean of all observations above $\bar{x}_1$ and call that incomplete mean $\bar{x}_2$, and so on. To compute these means, define $\bar{x}_{(a,b)}$ to be the mean of the interval (x_a, x_b) in x space or of the interval (a, b) in F space. Therefore,

$$\bar{x}_{(a,b)} = \frac{1}{b-a}\int_a^b F^{-1}(y)\,dy \tag{7.6}$$

The endpoints are determined by functions of N_i. For example,

$$\bar{x}_2 = \frac{1}{1-(N_2/N)}\int_{(N_2/N)}^1 [c(1-F)^{-d} + e]\,dF$$
$$= \frac{c}{1-d}\left[1-\frac{N_2}{N}\right]^{-d+1} + e\left(1-\frac{N_2}{N}\right) \tag{7.7}$$

Likewise,

$$\bar{x}_i = e\left(1-\frac{N_2}{N}\right) + \frac{c}{1-d}\left[1-\frac{N_i}{N}\right]^{-d+1}, \qquad i = 1, 2, 3, \ldots \tag{7.8}$$

The incomplete means $\bar{x}_i$ and the corresponding N_i can be related to the parameters of the Wakeby distribution in a straightforward manner. This is due to its inverse definition. The first two incomplete means are used to calculate c and e. If d is specified, then the relationship is linear in c and e. Houghton applied this procedure to small samples ($N = 20$). He assumed a value of d and estimated c and e by linear regression analysis. For 48 flood records (Houghton, 1978), d was observed to be in the neighborhood of 0.2. He noted that the estimate of T-year flood ($T \leq 100$) was not very sensitive to d, at least for a small neighborhood around it.

The method of incomplete means can be applied to any function that is expressible in inverse form. Examples are exponential and extreme-value type 1 distributions. The relation between incomplete means and parameters of the function can then be determined. However, if the function is not easily expressible in inverse form, then this relation would not be as simple. A case in point is the lognormal distribution or log-Pearson type 3 distribution.

EXAMPLE 7.1

Consider the extreme value type 1 probability distribution,

$$f(x) = a\exp\{-a(x-b) - \exp[-a(x-b)]\}, \qquad a > 0,\ -\infty < b < x$$

Determine the parameters a and b by the method of incomplete means.

$$F(x) = \int_0^x f(x)\,ds = \int_0^x a\exp\{-a(s-b) - \exp[-a(s-b)]\}\,ds$$
$$= \exp\{-\exp[-a(x-b)]\}$$

Taking the log of $F(x)$,

$$\ln F(x) = -\exp[-a(x-b)]$$

Taking the log again,

$$\ln[-\ln F[(x)] = -a(x-b)$$

Therefore,

$$x = b - \frac{1}{a}\ln[-\ln F(x)]$$

This is the inverse form of the extreme-value type 1 distribution.

For a sample of N ranked observations $x_1, x_2, x_3, \ldots, x_N$, the endpoints are determined by functions N_i. For example,

$$\bar{x}_i = \frac{1}{(1-(N_i/N))}\int_{N_i/N}^1 \left[b - \frac{1}{a}\ln(-\ln F)\right]dF$$

The first term within the integrand can be written as

$$\int_{N_i/N}^1 b\,dF = bF|_{N_i/N}^{i} = b\left[1-\frac{N_i}{N}\right]$$

The second term within the integrand can be expressed as

$$\int_{N_i/N}^1 \ln\{-\ln F\}\,dF = \int_{N_i/N}^1 \ln(\ln F^{-1})\,dF$$

Let $u = \ln F^{-1}$. Then $F = \exp(-u)$ and $dF = -\exp(-u)\,du$. Hence,

$$\int_{\ln N/N_i}^1 \ln u[-\exp(-u)]\,du = \int_0^{\ln N/N_i} \ln u\exp(-u)\,du$$

Expressing the exponential term as a series,

$$= \int_0^{\ln N_i/N}\left[1 - u + \frac{u^2}{2!} - \frac{u^3}{3!} + \cdots\right]\ln u\,du$$
$$= \int_0^{\ln N_i/N}\left[\ln u - u\ln u + \frac{u^2}{2!}\ln u - \frac{u^3}{3!}\ln u + \cdots\right]du$$

We perform integration term by term,

$$\int \ln u\,du = u\ln u - \int \frac{1}{u}u\,du = u\ln u - u$$

$$\int u\ln u\,du = \ln u\frac{u^2}{2} - \int \frac{1}{u}\frac{u^2}{2}\,du = \frac{u^2}{2}\ln u - \frac{u^2}{4}$$

$$\int u^2\ln u\,du = \ln u\frac{u^3}{3} - \int \frac{1}{u}\frac{u^3}{3}\,du = \frac{u^3}{3}\ln u - \frac{u^3}{9}$$

$$\int u^3\ln u\,du = \ln u\frac{u^4}{4} - \int \frac{1}{u}\frac{u^4}{4}\,du = \frac{u^4}{4}\ln u - \frac{u^4}{6}$$

and so on. Inserting these in the main integrand,

$$\int_0^{\ln N/N_i}\left[\ln u - u\ln u + \frac{u^2}{2!}\ln u - \frac{u^3}{3!}\ln u + \cdots\right]du$$

$$= \left[u\ln u - u - \frac{u^2}{2}\ln u + \frac{u^2}{4} + \frac{u^3}{2!3}\ln u \right.$$

$$\left. - \frac{u^3}{2!9} - \frac{u^4}{3!4}\ln u + \frac{u^4}{3!6}\cdots\right]\Bigg|_0^{\ln N/N_i}$$

$$= \left[u\ln u - \frac{u^2}{2}\ln u + \frac{u^3}{2!3} - \frac{u^4}{3!4} + \cdots \right.$$

$$\left. - u + \frac{u^2}{4} - \frac{u^3}{2!9} + \frac{u^4}{3!16} - \cdots\right]\Bigg|_0^{\ln N/N_i}$$

$$= \left[-\ln u\left\{-u + \frac{u^2}{2!} - \frac{u^3}{3!} + \frac{u^4}{4!} - \cdots\right\}\right.$$

$$\left. - u + \frac{u^2}{2!2} - \frac{u^3}{3!3} + \frac{u^4}{4!4} - \cdots\right]\Bigg|_0^{\ln N/N_i}$$

$$= \left[-\ln u\sum_{j=1}^{\infty}\frac{(-u)^j}{j!} - \sum_{j=1}^{\infty}\frac{(-r)^j}{j!j}\right]\Bigg|_0^{\ln N/N_i}$$

Therefore,

$$\bar{x}_i = \frac{1}{(1-(N_i/N))}\left[b\left(1-\frac{N_i}{N}\right)\right.$$

$$\left. + \frac{1}{a}\left\{\ln\left(\ln\frac{N}{N_i}\right)\sum_{j=1}^{\infty}\frac{(-1)^j}{j!}\ln\left(\frac{N}{N_i}\right) + \sum_{j=1}^{\infty}\frac{(-1)^j}{j!j}\ln\left(\frac{N}{N_i}\right)\right]$$

On simplification,

$$\bar{x}_i = b + \frac{1}{a}\left[\ln\left(\ln\frac{N}{N_i}\right)\sum_{j=1}^{\infty}\frac{(-1)^j}{j!}\ln\left(\frac{N}{N_i}\right)\right.$$

$$\left. + \sum_{j=1}^{\infty}\frac{(-1)^j}{j!j}\ln\left(\frac{N}{N_i}\right)\right], \qquad i = 1, 2, \ldots$$

The parameters a and b can be estimated by computing $\bar{x}_1$ and $\bar{x}_2$ for N/N_1 and N/N_2.

7.2 METHOD OF PROBABILITY WEIGHTED MOMENTS (PWM)

Greenwood et al. (1979) introduced the method of probability weighted moments (PWM) and showed its usefulness in deriving explicit expressions for the parameters of the distributions whose inverse forms $x = x(F)$ can be explicitly defined. They derived relations between the parameters and the PWMs for the generalized lambda, Wakeby, Weibull, Gumbel, logistic, and kappa distributions. Let a probability distribution function be denoted by $F = F(x) = P[X \leq x]$. The PWMs can be defined as

$$M_{i,j,k} = E[x^i F^j (1-F)^k] = \int_0^1 [x(F)]^i F^j (1-F)^k \, dF \tag{7.9}$$

where $M_{i,j,k}$ is the probability weighted moment of order (i, j, k), E is the expectation operator, and i, j, and k are real numbers. If $j = k = 0$ and i is a nonnegative integer, then M_i,0,0 represents the conventional moment about the origin of order i. If $M_{i,0,0}$ exists and x is a continuous function of F, then $M_{i,j,k}$ exists for all nonnegative real numbers j and k.

For nonnegative integers j, k,

$$M_{i,0,k} = \sum_{j=0}^{k}\binom{k}{j}(-1)^j M_{i,j,0} \tag{7.10}$$

$$M_{i,j,0} = \sum_{k=0}^{j}\binom{j}{k}(-1)^k M_{i,0,k} \tag{7.11}$$

If $M_{i,0,k}$ exists and x is a continuous function of F, then $M_{i,j,0}$ exists. When the inverse $x = x(F)$ of the distribution $F = F(x)$ cannot be analytically defined, it may in general be difficult to derive $M_{i,j,k}$ analytically.

When i, j, and k are nonnegative integers, the probability weighted moment of order (i, j, k), $M_{i,j,k}$, is proportional to $E[x^i_{j+1,j+k+1}]$, the ith moment about the origin of the $(j+1)$st order statistic for a sample of size $n = k + j + 1$. Symbolically,

$$M_{i,j,k} \propto E[x^i_{j+1,j+k+1}] \tag{7.12}$$

Specifically,

$$E[x^i_{j+1,j+k+1}] = \frac{M_{i,j,k}}{B[j+1, k+1]} \tag{7.13}$$

where $B[\cdot,\cdot]$ denotes the beta function. If $j = 0$,

$$E[x^i_{i,k+1}] = (k+1)M_{i,0,k} \tag{7.14}$$

Here $(k + 1)M_{i,0,k}$ represents the ith moment about the origin of the first-order statistic for a sample of size $k + 1$. Likewise, if $k = 0$,

$$E[x^i_{j+1,j+1}] = (j+1)M_{i,j,0} \tag{7.15}$$

where $(j + 1)M_{i,j,0}$ represents the ith moment about the origin of the $(j + 1)$th order statistic for a sample of size $j + 1$.

The expected value of the range of x in a sample of size $n = k + 1 = j + 1$ can be written as

$$E[x_{n,n} - x_{1,n}] = n(M_{1,n-1,0} - M_{1,0,n-1}) \tag{7.16}$$

The PWMs can be derived for a distribution, which can be expressed as $x = x(F)$. These can then be related to the distribution parameters. The resulting relations may be of simpler structure than those between the conventional moments and the parameters. The simpler structure may be due in part to x being taken to only the first power.

EXAMPLE 7.2

Consider the distribution in Example 7.1. Determine its parameters by the method of probability weighted moments.

As before,

$$F = \exp[-\exp\{-a(x-b)\}]$$

and

$$x = b - \frac{1}{a}\ln[-\ln F]$$

Since it has two parameters, we need to determine two PWMs. When $i = l, j, k = 0$, then by definition

$$M_{l,j,0} = \int_0^1 [x(F)]\, F^j (1 - F)^0\, dF$$

$$= \int_0^1 \left[b - \frac{1}{a}\ln(-\ln F)\right] F^j\, dF$$

$$= b\frac{F^{j+1}}{j+1}\Big|_0^1 - \frac{1}{a}\int_0^1 \ln(-\ln F)\, F^j\, dF$$

$$= \frac{b}{1+j} + \frac{1}{a}\left[\frac{\ln(1+j) + \epsilon}{1+j}\right]$$

where ϵ is Euler's number + 0.5772. Since

$$M_{i,j,0} = \sum_{k=0}^{j} \binom{j}{j} (-1)^k M_{i,0,k}$$

we have

$$M_{1,1,0} = \sum_{k=0}^{1} \binom{1}{k} (-1)^k M_{1,0,k}$$

$$= M_{1,0,0} - M_{1,0,1}$$

Therefore,

$$M_{1,0,0} = b + \frac{\epsilon}{a}$$

$$M_{1,1,0} = \frac{b}{2} + \frac{1}{a}\left[\frac{\ln 2 + \epsilon}{2}\right]$$

where $M_{1,0,0}$ is the ordinary moment of first order about the origin.

$$M_{1,0,1} = \int_0^1 \left[b - \frac{1}{a}\ln(-\ln F)\right](1 - F)\, dF$$

$$= \int_0^1 b\, dF = \int_0^1 \frac{1}{a}\ln(-\ln F)\, dF$$

$$- \int_0^1 bF\, dF + \int_0^1 \frac{1}{a} F \ln(-\ln F)\, dF$$

$$= b - \frac{b}{2} - \int_0^1 \frac{1}{a}\ln(-\ln F)\, dF + \int_0^1 \frac{1}{a} F\ln(-\ln F)\, dF$$

$$= \frac{b}{2} + \frac{\epsilon}{a} - \frac{1}{a}\left[\frac{\ln 2 + \epsilon}{2}\right]$$

$$= \frac{b}{2} + \frac{\epsilon}{a} - \frac{\epsilon}{2a} - \frac{\ln 2}{2a}$$

$$= \frac{b}{2} + \frac{\epsilon}{2a} - \frac{\ln 2}{2a}$$

Therefore,

$$M_{1,0,0} - 2M_{1,0,1} = \frac{\ln 2}{a}$$

or

$$\frac{M_{1,0,0} - 2M_{1,0,1}}{\ln 2} = \frac{1}{a}$$

and

$$b = M_{1,0,0} - \frac{\epsilon}{a}$$

7.3 METHOD OF MIXED MOMENTS (MIXM)

Rao (1980, 1983) proposed a *method of mixed moments* (MIXM) for fitting log-Pearson type (LPT) III distribution. The MIXM method is applicable to any log probability distribution. As the name suggests, the MIXM method is based on mixing the moments of real and logarithmically transformed data. Thus only the first two moments (mean and variance) of the data are used. For example, if it is desired to fit the LPT III distribution, then its parameters can be estimated in two ways: (1) the first method uses mean ($\bar{x}$) and variance S_x^2 of real data and mean of logarithmically transformed values ($\bar{y}$), and (2) the second method uses the mean of the real data $\bar{x}$, mean $\bar{y}$, and variance S_y^2 of logarithmically transformed data. Rao (1980) showed by Monte Carlo experimentation that the first method possessed superior statistical properties as compared to the second method. We consider an example to illustrate this method.

EXAMPLE 7.3

Let the log-Pearson type (LPT) III distribution have the parameters a, b, and c and let $y = \ln x$. The density function (Pearson type III distribution) of y is

$$f(y) = \frac{|a|}{\Gamma(b)} [a(y - c)]^{b-1} \exp[-a(y - c)]$$

The density function of the LPT III distribution is

$$f(x) = \frac{|a|}{\Gamma(b)} \frac{\exp(ax)}{x^{1+a}} [a(\ln x - c)]^{b-1}$$

For the LPT III distribution to be positively skewed, $a > 0$, $c \le y < \infty$, and $\exp(c) \le x < \infty$. Determine the parameters a, b, and c by the MIXM method.

Let $\bar{y}$, S_y^2, and C_{sy} be the mean, variance, and the coefficient of skewness of the Pearson type (PT) III distribution. Let $\bar{x}$ and S_x^2 be the mean and variance of the PT III distribution. Following Rao (1983),

$$\bar{y} = c + ba^{-1}$$

$$S_y^2 = ba^{-2}$$

$$C_{sx} = \frac{a}{|a|} 2b^{-0.5}$$

$$b = \frac{\ln\left(\dfrac{S_x^2 + \bar{x}^2}{\bar{x}^2}\right)}{\ln\left(\dfrac{(1 - 1/a)^2}{(1 - 2/a)}\right)}$$

$$c = \ln \bar{x} + b \ln\left(1 - \frac{1}{a}\right)$$

Rearranging the expression for b to solve for the parameter a explicitly from a known value of b,

$$a = 1 \pm \left(1 + \frac{1}{Z}\right)^{0.5}$$

where

$$Z = \left[\frac{x^{-2} + S_x^2}{x^{-2}}\right]^{b-1} - 1$$

The parameter a has the same sign as C_{sy}, which is related to b. Therefore, an initial estimate of C_{sy} will provide an initial estimate of b and the algebraic sign of a. Final estimates of the parameters can be obtained as follows. Compute Z and a and then c. Compute $\bar{y}$ from the values of a, b, and c obtained before. Repeat the calculations until computed $\bar{y}$ equals observed $\bar{y}$ by adjusting b. For additional discussion we refer to Rao (1983).

7.4 METHOD OF MAXIMUM LIKELIHOOD ESTIMATION (MLE)

Let $f(x; a_1, a_2, \ldots, a_m)$ be a probability density function (pdf) of x with parameters a_i, $i = 1, 2, \ldots, m$, to be estimated. Let there be a random sample of $x_1, x_2, \ldots, x_n$ from this density. Then their joint pdf is $f(x_1, x_2, x_3, \ldots, x_n; a_1, a_2, \ldots, a_m)$. Since the sample is random, the joint pdf can be written as

$$f(x_1, x_2, x_3, \ldots, x_n; a_1, a_2, \ldots, a_m) = \prod_{i=1}^{n} f(x_i; a_1, a_2, \ldots, a_m)$$

Interpreted conceptually, the probability of obtaining a given value of x, say x_1, is proportional to $f(x_1; a_1, a_2, \ldots, a_m)$. Likewise, the probability of obtaining the random sample $x_1, x_2, \ldots, x_n$ from the population of x is proportional to the product of their individual probability densities or their joint pdf. This joint pdf is also called the *likelihood function* or ML function, denoted by L,

$$L = \prod_{i=1}^{n} f(x_i; a_1, a_2, \ldots, a_m) \tag{7.17}$$

The parameters a_i, $i = 1, 2, \ldots, m$, are unknown.

The method of maximum likelihood estimates the unknown parameters by maximizing L, that is, maximizing the likelihood that the sample under consideration is the one that would be obtained if n random observations were selected from $f(x; a_1, a_2, \ldots, a_m)$. The values of the parameters so obtained are known as the *maximum likelihood estimation* (MLE) estimators. Since $\ln L$ attains its maximum for specified values of a_i, $i = 1, 2, \ldots, m$, as does L, the ML function can also be expressed as

$$\ln L = L^* = \ln \prod_{i=1}^{n} f(x_i; a_1, a_2, \ldots, a_m) = \prod_{i=1}^{n} \ln f(x_i; a_1, a_2, \ldots, a_m) \tag{7.18}$$

It is sometimes easier to find the maximum of the logarithm of the maximum likelihood.

The procedure for estimating the parameters or determining the point where the MLE function achieves its maximum involves differentiating L or $\ln L$ partially with respect to each parameter and equating it to zero. This results in m equations,

$$\frac{\partial L(a_1, a_2, \ldots, a_m)}{\partial a_1} = 0$$
$$\frac{\partial L(a_1, a_2, \ldots, a_m)}{\partial a_2} = 0$$
$$\vdots$$
$$\frac{\partial L(a_1, a_2, \ldots, a_m)}{\partial a_m} = 0 \tag{7.19}$$

These m equations in m unknowns can then be solved for the m unknown parameters. A number of workers have used this method in hydrology (Douglas, Clarke, and Newton, 1976; Sorooshian, Gupta, and Fulton, 1983; Phien and Jivajirajah, 1984).

EXAMPLE 7.4

Consider the function $f(x) = a \exp(-ax)$, $x > 0$, $a > 0$. Determine its parameter by the MLE method.

$$L = a^n \exp[-a \Sigma x_i]$$

Taking its logarithm,

$$\ln L = L^* = n \ln a - a \Sigma x_i$$

Differentiating with respect to a and equating to 0,

$$a = \frac{1}{n} \Sigma x_i$$

7.5 METHOD OF PRINCIPLE OF MAXIMUM ENTROPY (POME)

Entropy was first mathematically represented by Shannon (1498) and has since been called the Shannon entropy functional, SEF for short, denoted by $I[f]$ or $I[x]$. It is a numerical measure of uncertainty associated with $f(x)$ in describing the variable x and is defined as

$$I[f] = I[x] = -K \int_0^\infty f(x) \ln[f(x)]\, dx, \qquad 0 \le x \le \infty \tag{7.20}$$

where $k > 0$ is an arbitrary constant or scale factor depending upon the choice of measurements and can be absorbed into the base of the logarithm. Equation (7.20) can simply be written as

$$I[f] = -\int_0^\infty f(x) \ln[f(x)]\, dx \tag{7.21}$$

We may think of $I[f]$ as the mean value of $-\ln[f(x)]$. Actually, $-I$ measures strength and $+I$ measures the weakness of $f(x)$. The SEF allows choosing the $f(x)$ that minimizes the uncertainty. Note that $f(x)$ satisfies the normality condition

$$\int_0^\infty f(s)\, ds = 1 \tag{7.22}$$

For a discrete system,

$$I[f] = -\sum f_j \ln[f_j]; \sum_{j=0}^{\infty} f_j = 1 \tag{7.23}$$

Jaynes (1961, 1982) formulated the *principle of maximum entropy* (POME), according to which the least biased $f(x)$ is obtained by maximizing the entropy subject to the given information. Mathematically, it can be stated as follows: Given n linearly independent constraints C_i in the form

$$C_i = \int_0^\infty g_i(x) f(x)\, dx, \qquad i = 1, 2, \ldots, m \tag{7.24}$$

where $g_i(x)$ are some functions whose averages over $f(x)$ are specified, then the maximum of I subject to equation (7.24) is given by

$$f(x) = \exp\left[-a_0 - \sum_{i=1}^{m} a_i g_i(x)\right] \tag{7.25}$$

where a_i, $i = 1, 2, \ldots, m$, are the Lagrange multipliers and can be determined from equations (7.24) and (7.25) along with equation (7.22). This can be done as follows.

According to POME, we maximize equation (7.21) or (7.23) subject to equation (7.24). $I[f]$ can be maximized by the method of Lagrange multipliers. This introduces parameters $(a_0 - 1)$, a_1, a_2, $\ldots$, a_m, chosen such that variations in a functional of $f(x)$ vanish. Therefore,

$$I[f] = a_0 + \sum_{i=1}^{m} a_i C_i \tag{7.26}$$

To estimate parameters of a given function involves two steps: (1) specification of constraints based on experimental data required to derive the function and (2) derivation of relationships between parameters and constraints. Consequently, this method may not be suitable for any arbitrary function but is, however, applicable to most distribution functions commonly employed in hydrology.

Let a function be of the form

$$f(x) = \alpha x^k \exp\left[-\sum_{i=1}^{m} a_i g_i(x)\right] \tag{7.27}$$

where $g_i(x)$ are known functions and α and k are constants. By applying POME we explicitly construct

$$I[f] = -\ln \alpha - kE[\ln x] + \sum_{i=1}^{m} a_i E[g_i(x)] \tag{7.28}$$

We may then seek to maximize $I[f]$ subject to the constraints $E[\ln x]$ and $E[g_i(x)]$, which can be evaluated numerically by means of empirical data, to obtain the function of equation (7.27).

To maximize $I[f]$ we insert equation (7.25) in equation (7.22),

$$\int_0^\infty \exp\left[-a_0 - \sum_{i=1}^{m} a_i g_i(x)\right] dx = 1 \tag{7.29}$$

and get

$$a_0 = \ln \int_0^\infty \exp\left[-\sum_{i=1}^{m} a_i g_i(x)\right] dx \tag{7.30}$$

The Lagrange multipliers are related to the given constraints by

$$-\frac{\partial a_0}{\partial a_i} = C_i \tag{7.31}$$

It can also be shown that

$$\frac{\partial^2 a_0}{\partial a_i^2} = \text{var}\,[g_i(x)]; \quad \frac{\partial^2 a_0}{\partial a_i \partial a_j} = \text{cov}[g_i(x), g_j(x)] \tag{7.32}$$

With the Lagrange multipliers estimated from equations (7.31) and (7.32), the function given by equation (7.25) is uniquely defined. It is implicit that parameters in the function are uniquely related to the Lagrange multipliers. The procedure is illustrated by an example.

EXAMPLE 7.4

Determine the parameters $a > 0$ and $b > 0$ of the function

$$f(x) = \frac{1}{a\Gamma b}\left(\frac{x}{a}\right)^{b-1} \exp\left[-\frac{x}{a}\right], \qquad 0 < x$$

This function is a two-parameter gamma distribution and is used frequently in hydrology. First, we need to deduce the constraints required for derivation of $f(x)$. To this end,

$$I[f] = -\int_0^\infty f(x) \ln f(x)\, dx = [\ln a\Gamma b + (b-1)\ln a] \int_0^\infty f(x)\, dx - (b-1)\int_0^\infty \ln x\, f(x)\, dx + \frac{1}{a}\int_0^\infty x\, f(x)\, dx$$

Thus, the constraints can be specified as

$$\int_0^\infty f(x)\, dx = 1$$

$$\int_0^\infty x f(x)\, dx = E[x]$$

$$\int_0^\infty \ln x f(x)\, dx = E[\ln x]$$

The least biased $f(x)$ consistent with the above constraints is determined by POME and takes the form

$$f(x) = \exp[-a_0 - a_1 x - a_2 \ln x]$$

where a_0, a_1, and a_2 are Lagrange multipliers. From the condition of normality,

$$\int_0^\infty f(x)\, dx = \int_0^\infty \exp[-a_0 - a_1 x - a_2 \ln x]\, dx = 1$$

or

$$\exp(a_0) = \int_0^\infty \exp[-a_1x - a_2\ln x]\, dx$$

$$= \int_0^\infty x^{-a_2}\exp(-a_1x)\, dx$$

Let $y = a_1x$. Then $dy = a_1\, dx$, and

$$\exp(a_0) = \frac{1}{a_1^{1-a_2}}\int_0^\infty y^{-a_2}\exp(-y)\, dy$$

Using the definition of the gamma function (Abramowitz and Stegun, 1965),

$$\exp(a_0) = \frac{1}{a_1^{1-a_2}}\Gamma(1 - a_2)$$

or

$$a_0 = (a_2 - 1)\ln a_1 + \ln \Gamma(1 - a_2)$$

Differentiating the integral equation for a_0 with respect to a_1 and a_2, respectively,

$$\frac{\partial a_0}{\partial a_1} = -\frac{\int_0^\infty x\exp[-a_1x - a_2\ln x]\, dx}{\int_0^\infty \exp[-a_1x - a_2\ln x]\, dx}$$

$$= -\int_0^\infty x\exp[-a_0 - a_1x - a_2\ln x] = -\int_0^\infty xf(x)\, dx = -E[x]$$

$$\frac{\partial a_0}{\partial a_2} = -\frac{\int_0^\infty \ln x\exp[-a_1x - a_2\ln x]\, dx}{\int_0^\infty \exp[-a_1x - a_2\ln x]\, dx}$$

$$= -\int_0^\infty \ln x\exp[-a_0 - a_1x - a_2\ln x]\, dx$$

$$= -\int_0^\infty \ln x f(x)\, dx = -E[\ln x]$$

Also from the algebraic expression for a_0,

$$\frac{\partial a_0}{\partial a_1} = \frac{a_2 - 1}{a_1}$$

$$\frac{\partial a_0}{\partial a_2} = \ln a_1 + \frac{\partial}{\partial a_2}\ln \Gamma(1 - a_2)$$

$$= \ln a_1 + \Psi(1 - a_2)$$

where $\Psi[\cdot]$ is the psi function. If $k = 1 - a_2$, then $\partial k/\partial a_2 = -1$. Hence

$$\frac{\partial a_0}{\partial a_2} = \ln a_1 - \Psi(k)$$

Equating the expressions for $\partial a_0/\partial a_1$,

$$\frac{a_2 - 1}{a_1} = -E[x] \quad \text{or} \quad \frac{k}{a_1} = E[x]$$

and doing likewise for $\partial a_0/\partial a_2$,

$$\ln a_1 - \Psi(k) = -E[\ln x]$$

or

$$\ln a_1 = \Psi(k) - E[\ln x]$$

Inserting $a_1 = k/E[x]$,

$$\Psi(k) - E[\ln x] = \ln k - \ln E[x]$$

This leads to

$$E[\ln x] - \ln E[x] = \Psi(k) - \ln k$$

This can be used to determine k, which when substituted in the other equation yields a_1. On substituting expressions for a_0, a_1, and a_2 in $f(x) = \exp[-a_0 - a_1x - a_2 \ln x]$, we find that $a_1 = 1/a$ and $a_2 = 1 - b$. Thus, the parameters a and b of the gamma distribution are specified in terms of constraints,

$$ab = E[x]$$

$$\ln(b) - \Psi(b) = \ln(E[x]) - E[\ln x]$$

7.6 METHOD OF LEAST SQUARES (MOLS)

Let there be a function $y = f(x; a_1, a_2, \ldots, a_m)$, where a_i, $i = 1, 2, \ldots, m$, are parameters to be estimated. The method of least squares (MOLS) involves estimating the parameters by minimizing the sum of squares of all deviations between observed and computed values of y. Mathematically, this sum S can be expressed as

$$S = \sum_{i=1}^{n} d_i^2 = \sum_{i=1}^{n} [y_0(i) - y_c(i)]^2$$

$$= \sum_{i=1}^{n} [y_0(i) - f(x_i; a_1, a_2, \ldots, a_m)]^2 \tag{7.33}$$

where $y_0(i)$ is the ith observed value of y, $y_c(i)$ the ith computed value of y, and $n > m$ is the number of observations. The minimum of S in equation (7.33) can be obtained by differentiating S partially with respect to each parameter and equating to 0:

$$\frac{\partial \sum_{i=1}^{n} [y_0(i) - f(x_i; a_1, a_2, \ldots, a_m)]^2}{\partial a_1} = 0$$

$$\frac{\partial \sum_{i=1}^{n} [y_0(i) - f(x_i; a_1, a_2, \ldots, a_m)]^2}{\partial a_2} = 0$$

$$\vdots$$

$$\frac{\partial \sum_{i=1}^{n} [y_0(i) - f(x_i; a_1, a_2, \ldots, a_m)]^2}{\partial a_m} = 0 \tag{7.34}$$

Thus m equations, usually called normal equations, are obtained for the estimation of m parameters.

EXAMPLE 7.5

Let a straight line be expressed as $y = a + bx$. Determine the parameters a and b by the MOLS for a sample of n values of x and y.

$$S = \sum_{i=1}^{n} [y_0(i) - a - bx_i]^2$$

Differentiating S partially with respect to a and b and equating each time to 0,

$$\frac{\partial S}{\partial a} = -\Sigma y_0 + na + 2b \Sigma x_i = 0$$

$$\frac{\partial S}{\partial b} = -\Sigma y_0 x_i + a \Sigma x_i + 2b \Sigma x_i^2 = 0$$

The normal equations then are

$$na + b \Sigma x_i = \Sigma y_0$$

$$a \Sigma x_i + b \Sigma x_i^2 = \Sigma x_i y_0$$

In these equations all the summation quantities are known from the sample. The parameters a and b can therefore be determined by solving these equations.

EXAMPLE 7.6

Let there be an expression $y = a + bx + cx^2$. Determine the parameters a, b, and c by the MOLS for a sample of n values of x and y.

We follow the same procedure. Constructing S as

$$S = \sum_{i=1}^{n} [Y_0(i) - a - bx_i - cx_i^2]^2$$

Differentiating S partially with respect to a, b, and c are equating each time to zero results in three normal equations:

$$an + b \Sigma x_i + c \Sigma x_i^2 = \Sigma y_0$$

$$a \Sigma x_i + b \Sigma x_i^2 + c \Sigma x_i^3 = \Sigma x_i y_0$$

$$a \Sigma x_i + b \Sigma x_i^3 + c \Sigma x_i^4 = \Sigma x_i^2 y_0$$

These equations can be solved for a, b, and c, since all summation terms are known from the sample.

EXAMPLE 7.7

Let a general linear model be expressed as $y = a_0 + \sum_{i=1}^{m} a_i x_i$. Estimate the parameters a_i, $i = 0, 1, 2, \ldots, m$, from a sample of n values of $y, x_1, x_2, \ldots, x_m, n > m$.

This expression is similar in form to the linear equation in Example 7.5. Therefore, the normal equations of that example can be extended to the general case. Following the same procedure, the normal equations can be written as

$$na_0 + a_1 \Sigma x_1 + a_2 \Sigma x_2 + a_3 \Sigma x_3 + \cdots + a_m \Sigma x_m = \Sigma y_0$$

$$a_0 \Sigma x_1 + a_1 \Sigma x_1^2 + a_2 \Sigma x_1 x_2 + a_3 \Sigma x_1 x_3 + \cdots + a_m \Sigma x_1 x_m = \Sigma x_1 y_0$$

$$a_0 \Sigma x_2 + a_1 \Sigma x_1 x_2 + a_2 \Sigma x_2^2 + a_3 \Sigma x_2 x_3 + \cdots + a_m \Sigma x_2 x_m = \Sigma x_2 y_0$$

$$\vdots$$

$$a_0 \Sigma x_m + a_1 \Sigma x_1 x_m + a_2 \Sigma x_2 x_m + a_3 \Sigma x_3 x_m + \cdots + a_m \Sigma x_m^2 = \Sigma x_m y_0$$

Thus, there are as many equations as there are unknowns. These equations can therefore be solved to obtain the parameters.

7.6.1 Matrix Representation of the MOLS

Many of the computations performed in fitting multiple variable models to observed data can be expressed more efficiently using matrix notation. The normal equations in the LS method can often be reduced to the form

$$\mathbf{XA} = \mathbf{Y} \tag{7.35}$$

where X is a matrix of $(n + 1, m + 1)$ dimensions, $\mathbf{A}$ is a vector of the parameters having $(m + 1, 1)$ dimensions, and $\mathbf{Y}$ is a vector having $(n + 1, 1)$ dimensions. In this section, uppercase letters signify matrices or vectors. The superscript T signifies the transpose. The deviations between observed and computed values of y can be written as

$$\mathbf{D} = \mathbf{Y} - \mathbf{XA}$$

where $\mathbf{D}$ is a vector of deviations d_i, $i = 1, 2, \ldots, n$.

The sum of squares of deviations can be obtained by using the inner product, which is found by multiplying $\mathbf{D}$ by its transpose $\mathbf{D}^T$,

$$\begin{aligned} \Sigma d_i^2 &= \mathbf{D}^T\mathbf{D} \\ &= [\mathbf{Y}^T - \mathbf{A}^T X^T][\mathbf{Y} - \mathbf{XA}] \\ &= \mathbf{Y}^T\mathbf{Y} - \mathbf{Y}^T X\mathbf{A} - \mathbf{A}^T X^T \mathbf{Y} + \mathbf{A}^T X^T X\mathbf{A} \end{aligned} \tag{7.36}$$

Since $\mathbf{A}$ and and $\mathbf{Y}$ are column vectors, their transposes are row vectors. Thus, the second and third terms on the right side of equation (7.36) are scalar in form. Since a scalar and its transpose are the same,

$$d_i^2 = \mathbf{Y}^T Y - 2\mathbf{A}^T X^T \mathbf{Y} + \mathbf{A}^T X^T X\mathbf{A}$$

Differentiating with respect to $\mathbf{A}$ and equating to zero,

$$\mathbf{X}^T X\mathbf{A} = X^T\mathbf{Y}$$

The Matrix $[X^T\mathbf{X}]$ and the vector $[X^T\mathbf{Y}]$ are generated from observed values of the variates. The parameters can then be determined as

$$\mathbf{A} = [X^T\mathbf{X}]^{-1} X^T \mathbf{Y} \tag{7.37}$$

Equation (7.37) expresses a least squares solution in compact form.

7.7 METHOD OF NONLINEAR LEAST SQUARES (MONLS)

Many hydrologic processes cannot be adequately represented by linear models. We then resort to nonlinear models. The LS method, discussed previously in the context of linear models, can be extended to nonlinear models. The NLS method is a numerical procedure satisfying the objective embodied in the LS method. Snyder (1972, 1980) has used it extensively and has given an excellent discussion on it. We follow his work here. To motivate the discussion of the NLS method, let us consider estimating parameters a, b, and c in

$$z = ax^b y^c \tag{7.38}$$

Here b and c are shape parameters, which cannot be estimated by the LS method. By taking a logarithmic transform, we can write equation (7.38) in a linear form,

$$\ln z = \ln a + b \ln x + c \ln y \tag{7.39}$$

To apply the LS method to equation (7.39), the objective must be stated in terms of ln z. This is not the same thing as stating the objective in terms of z. This change in objective may produce pronounced distortion by giving higher weight to smaller values of z and lower weight to larger values of z.

Let us modify equation (7.38) as

$$z = a(x - d)^b(y - e)^c \tag{7.40}$$

where d and e are thresholds of x and y and can also be called shift parameters. It is clear that equation (7.40) is not amenable to linearization through logarithmic transformation. Even if $b = c = 1$, it cannot be cast in a linear form, as evident from

$$z = axy - ady - aex + ade \tag{7.41}$$

Equation (7.41) contains a linear combination of four terms, but the products of parameters are nonlinear. Therefore, the parameters cannot be obtained by the LS method. We might argue that a trial-and-error procedure could be used by changing values of the parameters until we found the values giving the minimum residual sum of squares. This is not, however, viable for any significant number of model parameters. If a model contains m parameters and we attempt k levels of each parameter, then the number of parameter combinations is k^m. As seen from Table 7.1, the number of trials needed to evaluate all combinations in order to find the best parameter set becomes too large for this method to be of any practical value. Clearly, another method is needed to produce optimal parameter values with relatively few iterations.

The *method of nonlinear least squares* (MONLS) is based on Taylor series expansion of the function whose parameters are to be determined. Before presenting the method for

TABLE 7.1

Number of parameter combinations, k^m

Number of Parameters	Number of Levels		
	3	4	5
5	2.4×10^2	1.0×10^3	3.1×10^3
10	5.9×10^4	1.0×10^6	9.8×10^6
15	1.4×10^6	1.1×10^9	3.1×10^{10}
20	3.5×10^9	1.1×10^{12}	9.6×10^{13}

a function having m parameters, we first consider a one-parameter nonlinear function,

$$y = f(x; a) \tag{7.42}$$

where an optimum value of a is desired for a set of values of x and y. For a different value of a, say $a + h$, where h is a small increment, the value of the function can be given by

$$y_1 = f(x; a + h) \tag{7.43}$$

A Taylor series expansion of $f(x; a + h)$ yields

$$f(x; a + h) = f(x; a) + hf'(x; a) + \frac{h^2}{2!}f''(x; a) + \frac{h^3}{3!}f'''(x; a) + R_k \tag{7.44}$$

or

$$y_1 = y + hf'(x; a) + \frac{h^2}{2!}f''(x; a) + \frac{h^3}{3!}f'''(x; a) + R_k$$

where the residual term

$$R_k = \frac{h^k}{k!}f^{(k)}(x; a_1), \qquad a \le a_1 \le a + h \tag{7.45}$$

Two points must be made. First, to find an optimal value of a, we change a to $a + h$ and compute the function $f(x; a + h)$ and then keep incrementing the parameter and computing the function until the desired a is obtained. Second, in practice a very accurate expansion of the function by Taylor series in equation (7.44) is not desired. In this manner we can find a leading to the smallest residual sum of squares.

We now extend this discussion to the case where the function takes a more general form,

$$y = f(x_1, x_2, \ldots, x_m; a_1, a_2, \ldots, a_k) \tag{7.46}$$

Note that x_i, $i = 1, 2, \ldots, m$, are fixed and a_i, $i = 1, 2, \ldots, k$, are considered to vary. Let h_i, $i = 1, 2, \ldots, k$, denote increments of the parameters a_i, $i = 1, 2, \ldots, k$. Then expanding the function in equation (7.46) by Taylor series,

$$\begin{aligned} y_1 &= f(x_1, x_2, \ldots, x_m; a_1 + h_1, a_2 + h_2, \ldots, a_k + h_k) \\ &= \sum_{j=0}^{N-1} \frac{1}{j!}\left(h_1\frac{\partial}{\partial a_1} + h_2\frac{\partial}{\partial a_2} + \cdots + h_k\frac{\partial}{\partial a_k}\right)^j f(x_1, x_2, \ldots, x_m; \\ &\qquad a_1, a_2, \ldots, a_k) + R_N \end{aligned} \tag{7.47}$$

where $(h_1\, \partial/\partial a_1 + h_2\, \partial/\partial a_2 + \cdots + h_k\, \partial/\partial a_k)^j$ is a repeating differential operator on the function $f(x_1, x_2, \ldots, x_m; a_1, a_2, \ldots, a_k)$ and

$$R_N = \frac{1}{N!}\left(h_1 \frac{\partial}{\partial a_1} + h_2 \frac{\partial}{\partial a_2} + \cdots + h_k \frac{\partial}{\partial a_k}\right)^N f(x_1, x_2, \ldots, x_m; a_1 + ch_1, a_2 + ch_2, \ldots, a_k + ch_k), \qquad 0 \le c \le 1 \tag{7.48}$$

If we limit the Taylor series expansion to the first differential,

$$f(x_1, x_2, \ldots, x_m; a_1 + h_1, a_2 + h_2, \ldots, a_k + h_k) = f(x_1, x_2, \ldots, x_m; a_1, a_2, \ldots, a_k) + \sum_{j=1}^{k} h_j \frac{\partial f}{\partial a_j} \tag{7.49}$$

As an example consider a function having four parameters. Then

$$y_1 = y + \sum_{j=1}^{4} h_j \frac{\partial f}{\partial a_j} \tag{7.50}$$

This gives

$$d = y_1 - y = \sum_{j=1}^{4} h_j \frac{\partial f}{\partial a_j} \tag{7.51}$$

Equation (7.14) gives the difference between observed values of f and those estimated with specific values of the parameters. Thus, by changing each of the parameters with appropriate increment h, we change the functional value from y to y_1. Interpreted differently, we can adjust the parameters by eliminating the error $y_1 - y$, or d.

The MONLS is essentially a linear adjustment process by which, using the LS solutions repeatedly, parameters of nonlinear equations are estimated to a desired precision. If we compare equation (7.51) with a general case, then it is clear that they are similar in linear structure with d replacing y, the partial derivatives replacing x_i's, and h_i's replacing a_i's. Therefore, this allows us to compute h_i's as regression coefficients. The normal equations are formed on $\partial f/\partial a_i$ and d. Following evaluation of h_i's, new values of the parameters become

$$a_{i,j+1} = a_{i,j} + h_{i,j}, \qquad j = 1, 2, \ldots \tag{7.52}$$

where j signifies the iteration. This process is repeated with these new values of the parameters. In other words, we compute the differences and partial derivatives and then the h_i's. This produces further improvement in the parameters. We repeat the process until the changes in the h_i's become very small. This implies that a_i's can no longer be improved and $\Sigma\, d_i^2$ is a minimum. Hence, the NLS method is completed.

To implement this method, however, requires computation of the partial derivatives of the function. This usually means that the function must be of relatively simple form, which can be differentiated. Many hydrologic models are not simple equations. For example, a model using rainfall would require an explicit function representing rainfall hyetograph. Such a function for rainfall, however, does not exist. What is normally given is rain falling in discrete increments of time. A procedure is therefore needed to accommodate such functions. One procedure (Snyder, 1980) is the method of partial finite divided differences, where the divided differences replace the partial derivatives in equation (7.51).

Let y_c denote the computed value of y. The relation between y_c and the parameter a can be expressed as $y_c = f(a)$, as shown in Fig. 7.2. The partial derivative $\partial f/\partial a$ evaluated at a_0 is the tangent of this relation at (a_0, y_c). To obtain the derivative by divided difference, we change a slightly by adding Δa_0. A new value of y_{c1} can be computed corresponding to $a_0 + \Delta a_0$. The slope of the line between the two points $(a_0 + \Delta a_0, y_{c1})$ and (a_0, y_c) is simply given as $(y_{c1} - y_c)/\Delta a_0$. For a small value of Δa_0, $\partial f/\partial a$ and $(y_{c1} - y_c)/\Delta a_0$ will not differ significantly and the former can therefore be replaced by the latter. Likewise, the partial derivatives corresponding to other parameters can be evaluated. Thus, equation (7.51) can be expressed as

$$d = y - y_c = h_1 \frac{y_{a1} - y_c}{\Delta a_1} + h_2 \frac{y_{a2} - y_c}{\Delta a_2} + h_3 \frac{y_{a3} - y_c}{\Delta a_3} + h_4 \frac{y_{a4} - y_c}{\Delta a_4} + \cdots \tag{7.53}$$

where y_{ai}, $i = 1, 2, \ldots$, are computed values of y corresponding to the appropriate parameters.

There may, however, arise myriad difficulties in seeking an NLS solution. While starting at specified parameter values, the method may converge to a local minimum, which is not the global minimum sought. However, this can be checked by starting the method at different initial parameter values to see if the same minimum is obtained. The convergence procedure may not be monotonic; that is, the ith iteration may sometimes be inferior to the $(i - 1)$th iteration. This can be corrected by minimizing roundoff errors. Sometimes two different sets of parameter values give the same minimum value for the sum of squares of deviations. Which solution is selected should be based on hydrologic considerations.

7.8 OPTIMIZATION METHODS

There exists a plethora of optimization methods (Rosenbrock, 1960; Palmer, 1969; DeCoursey and Snyder, 1969; Himmelblau, 1972; Sorooshian, 1980; Sorooshian and Gupta, 1983; Sehitoglu, 1983), which can be used to estimate model parameters. An excellent discussion of those methods used frequently

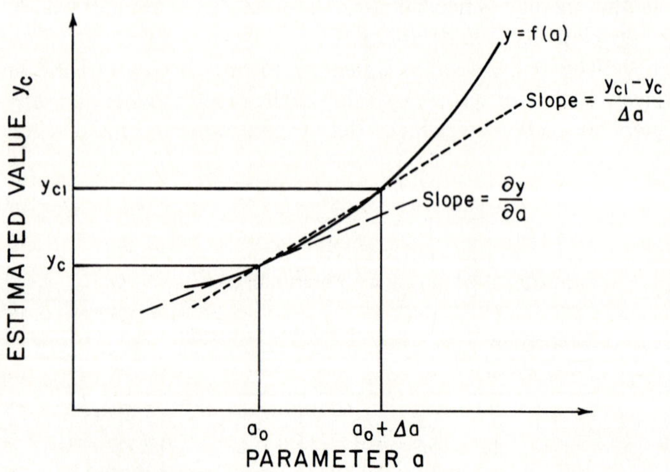

FIGURE 7.2 Definition of slope by divided finite difference.

to fit conceptual hydrologic models has been presented by Morel-Seytoux (1982). We do not dwell on these methods here. All optimization methods require specification of an objective function or error criterion.

7.8.1 Objective Functions

The concept of automatically determining optimum model parameters requires the objective function to be compatible with the intended use. It is, however, difficult to define an error criterion, which, upon minimization, will produce optimum parameter values without undesirable bias. A comparison of models also requires that the same objective function is used in determining parameters of each of the models. A number of objective functions have been used or suggested in hydrology (Diskin and Simon, 1977). Some of them are given here.

7.8.1.1 Sum of Squares of Deviations

The objective function based on the sum of squares of deviations may be defined as

$$F = \sum_{i=0}^{N} [y_0(i) - y_e(i)]^2 \tag{7.54}$$

where F is the index of disagreement or error, $y_0(i)$ is the ith observed value of the variate, y, $y_e(i)$ is the ith estimated value of y, and N is the number of observations. When divided by N, this F is analogous to the residual variance of a regression analysis. The minimization of F is designed to match the entire curve of y even though greater weight is assigned to its higher values. In the process of matching the entire curve, it may happen that too great a weight is placed on small values because of their larger numbers. This may result in poor curve peak matching. Another weakness of this objective function is that it is strongly affected by poor time synchronization between the measured independent and dependent variables.

The value of F may be computed for each event if the fitting for an individual event is desired. If the model parameters are to be optimized over a set of events, then F needs to be computed over the entire set. In that case F will be a very poor index of measuring disagreement because the event having the least weight may be the most important of all. This is a rather severe limitation for flood studies.

7.8.1.2 Sum of Squares of Peak Deviations

This can be expressed as

$$F = \sum_{j=1}^{N} [y_{p0}(j) - y_{pe}(j)]^2 \tag{7.55}$$

where $y_{p0}(j)$ is jth observed peak value of the variate y, $y_{pe}(j)$ is the jth estimated peak value of y, and N is the number of peak values in the optimization set. This is particularly suitable in flood studies and seems to have some attractive features. Obviously, among peaks greater weight is palced on higher peaks. Implicitly, it assumes that from a risk viewpoint, loss increases parabolically with the peak.

If F is divided by the number of events, the mean squared error results. This shows, on the average, how much error occurs as the optimization is performed over a set of events. Because it requires only the peak value from each event, it is efficient computationally. However, its use is not recommended where peak is not an important consideration, as in the case of low flow studies.

7.8.1.3 Sum of Squares of Logarithmic Deviations of Peak and Volume

The objective function comprises two components:

$$F_1 = \sum_{j=1}^{M} [\ln y_{po}(j) - \ln y_{pe}(j)]^2 \tag{7.56}$$

$$F_2 = \sum_{j=1}^{M} [\ln V_0(j) - \ln V_e(j)]^2 \tag{7.57}$$

Then

$$F = aF_1 + (1 - a)F_2 \tag{7.58}$$

where $V_0(j)$ is jth observed volume of the variate y, $V_e(j)$ is the jth estimated volume of y, and a is an arbitrarily chosen weighting factor ($0 \le a \le 1$).

It is evident that this objective function gives weight to both the volume and shape characteristics of y. The error between observed and estimated values is computed in terms of logarithmic deviations rather than natural units to prevent the parameters from being biased to fit only large-magnitude events. To make the objective function even more flexible, an arbitrary constant is introduced, as shown in equation (7.58). The constant can be chosen in accordance with the emphasis that is intended to be placed on a particular error component. For example, if the peak is emphasized, a should be greater than .5.

7.8.1.4 Sum of Squares of Logarithmic Deviations of Normalizing Time

This is defined as

$$F = \sum_{j=1}^{N} [\ln T_0(j) - \ln T_e(j)]^2 \tag{7.59}$$

where $T_0(j)$ is the observed normalizing time for the jth event and $T_e(j)$ is the estimated normalizing time for the jth event. The normalizing time can be defined in many ways. It is, however, not usable in all model optimization problems. This may be a useful criterion where dimensionless solutions are obtained. If the curve-fitting is good using this objective function, then the model structure would appear to be adequate.

7.8.1.5 Sum of Relative Differences Raised to the Power *a*

The sum of relative differences raised to the power a can be written as

$$F = \sum_{i=0}^{N} \left[\frac{y_0(i) - y_e(i)}{y_0(i)^a}\right]^a, \qquad a \ge 1 \tag{7.60}$$

For $a = 1$, this objective function places equal emphasis on all values of y. This is, in fact, a normalized verison of equation (7.55). For $a > 1$, the effect will be to fit the smaller values better. This may be more appropriate for fitting drought models.

All the foregoing objective functions share some common properties. (1) For a certain class of problems (linear models), the contours of F become quadratic surfaces, which are amenable to very powerful optimizing techniques. (2) For linear models, F has statistical significance, since $F/(N - 1)$ (where N = number of data points) is the variance and can be used to place confidence limits upon the optimized parameter values (assuming there is no dependence between data points). (3) For flood forecasting, greater emphasis is placed on matching the flood peak of a hydrograph. Peak ordinates are often an order of magnitude greater than the average ordinates and, therefore, near equal absolute errors mean proportional errors in the large ordinates.

7.8.1.6 Sum of Absolute Differences

The sum of absolute differences can be expressed as

$$F = \sum_{i=0}^{N} |y_0(i) - y_e(i)| \tag{7.61}$$

This objective function is stable in the sense that reduction of F guarantees a reduction in the area between the two curves. It places less emphasis on large ordinates of y.

7.8.1.7 Sum of Squares of Deviations Raised to the Power a

The sum of squares of deviations raised to the power a can be expressed as

$$F = \sum_{i=0}^{N} [y_0(i) - y_e(i)]^a, \qquad a = 1, 2, \ldots \tag{7.62}$$

By increasing a, the emphasis in fitting is increasingly placed on the larger values.

7.8.1.8 Absolute Differences of Peak and its Time

The absolute differences of peak and its time are expressed as

$$F_1 = \left| \frac{y_{po}(j) - y_{pe}(j)}{y_{po}(j)} \right| \tag{7.63}$$

$$F_2 = \left| \frac{t_{po}(j) - t_{pe}(j)}{t_{pe}(j)} \right| \tag{7.64}$$

$$F = aF_1 + (1 - a)F_2 \tag{7.65}$$

where $t_{po}(j)$ is observed peak time for the jth event and $t_{pe}(j)$ is estimated peak time for the jth event.

If equal emphasis is placed on both F_1 and F_2, then a should be .5. This function can be used in fitting only one point on the curve for each event, for example, the hydrograph peak and its time.

7.8.1.9 Absolute Difference of Peak

The absolute difference of peak is

$$F = \left| \frac{y_{po}(i) - y_{pe}(i)}{y_{po}(i)} \right| \tag{7.66}$$

This is frequently utilized in fitting only the peak.

7.9 PROBLEMS OF PARAMETER ESTIMATES

We estimate parameters of a distribution function from sample values. There are, of course, myriad ways by which to obtain parameter estimates. A question then arises: How should we best use the data to form estimates? This immediately raises another question: What do we mean by the best estimates? Also, are these estimates unique?

Each estimate of a parameter is a function of sample values, which are observations of a random variable. Thus, the parameter estimate itself is a random variable having its own sampling distribution. An estimate obtained from a given set of values can be regarded as an observed value of the random variable. Thus, the goodness of an estimate can be judged from its distribution. Excellent discussions on some of these issues have been reported by Kuczera (1982, 1983a, 1983b), Sorooshian and Gupta (1983), and Gupta and Sorooshian (1983), among others. A number of statistical properties are available that can be used to address the above questions.

7.9.1 Unbiased Estimates

Let the parameter be a and its estimate a_c. The estimate a_c is called an *unbiased* estimate of a if $E(a_c) = a$. In general, an estimate will have a certain bias $b(a)$ depending on a so that

$$E(a_c) = a + b(a) \tag{7.67}$$

Obviously, $b(a) = 0$ for an unbiased estimate. It should, however, be noted that an individual a_c is not equal to or even close to a even if $b(a) = 0$. It simply implies that the average of many independent estimates of a will be equal to a.

7.9.2 Consistency

Let the sample size be n. Then a_c is called a *consistent* estimate of a if it converges to a with probability 1 as n tends to infinity. Because many unbiased estimates have variances of the type

$$\mathrm{Var}(a_c) \cong \frac{C}{(n)^{.5}} \tag{7.68}$$

the condition of consistency is satisfied in most cases. Here C is constant. What is, however, desirable in practice is to have $\mathrm{Var}(a_c)$ as small as possible. This would imply that the probability density function of a_c would be more concentrated about a.

7.9.3 Efficiency

An estimate a_c of a is said to be *efficient* if it is unbiased and its variance is at least as small as that of any other unbiased estimate of a. If there are two estimates of a, say a_1 and a_2, then the relative efficiency of a_1 with respect to a_2 is

$$e = \frac{E[a_1 - a]^2}{E[a_2 - a]^2} \leq 1$$

if $E[a_2 - a]^2 > E[a_1 - a]^2$. An efficient estimate has $e = 1$. If an efficient estimate exists, it can be obtained by the MLE or entropy method.

7.9.4 Sufficiency

An estimate a_c of a is said to be *sufficient* if it uses all of the information relevant to a that is contained in the sample. More precisely, let a_1 and a_2 be two independent estimates of a. Then a_1 is considered a sufficient estimate if the joint probability distribution of a_1 and a_2 has the property

$$f(a_1, a_2) = f(a_1)f(a_2|a_1) = f(a_1)K(x_1, x_2, \ldots, x_n) \quad (7.69)$$

in which $f(a_1)$ is the distribution of a_1, $f(a_2|a_1)$ is the conditional distribution of a_2 given a_1, and $K(x_1, x_2, \ldots, x_n)$ is not a function of a but only of x_i's. If this equation holds, then a_2 does not produce any new information about a that is not already contained in a_1. In this case, a_1 is a sufficient estimate.

EXERCISES

7.1. Determine the parameter of an exponential distribution

$$f(x) = a\exp(-ax), \qquad a > 0$$

by the method of probability weighted moments.

7.2. Let the UIR of a linear time-invariant system be denoted by

$$h(x) = \frac{1}{a\Gamma(b)}\left(\frac{x}{a}\right)^{b-1}\exp\left(\frac{-x}{a}\right), \qquad a > 0, b > 0$$

Determine its parameters by the method of incomplete means.

7.3. Consider a three-parameter lognormal distribution and determine its parameters by the method of mixed moments.

7.4. Determine the parameter of the UIR in Exercise 7.2 by the MLE method.

7.5. Determine parameters of a Pearson type III distribution by using the principle of maximum entropy.

7.6. Use the least squares method to determine the parameters of the UIR in Exercise 7.2. The observed data on the UIR are generated from the following Gamma distribution:

$$h(t) = \frac{1}{k\Gamma(n)}\left(\frac{t}{k}\right)^{n-1}\exp(-t/k), \qquad k > 0, t > 0$$

in which $h(t)$ is the UIR, t is time in seconds, $n = 3$, and $k = 100$ seconds.

7.7. Table 7.2 lists instantaneous annual peak discharges for a river. It is desired to fit the LPT III distribution to the annual peak discharges. Estimate the parameters by the methods of moments, cumulants, maximum likelihood estimation, mixed moments, least squares, and principle of maximum entropy.

7.8. Catalog strengths and weaknesses of different methods of parameter estimation. Can you suggest guidelines for choosing these methods?

TABLE 7.2

Peak discharges for the Kentucky River near Salvisa, Kentucky (after McCabe, 1962)

Year	*Discharge (cfs)*	*Year*	*Discharge (cfs)*	*Year*	*Discharge (cfs)*
1895	47,300	1917	111,000	1939	84,300
1896	54,400	1918	71,700	1940	45,000
1897	87,200	1919	96,100	1941	28,400
1898	65,700	1920	92,500	1942	46,000
1899	91,500	1921	34,100	1943	80,400
1900	53,500	1922	69,000	1944	55,000
1901	67,800	1923	73,400	1945	72,900
1902	70,000	1924	99,100	1946	71,200
1903	66,900	1925	79,200	1947	46,800
1904	34,700	1926	62,600	1948	84,100
1905	58,000	1927	93,700	1949	61,300
1906	47,000	1928	68,700	1950	87,100
1907	66,300	1929	80,100	1951	70,500
1908	80,900	1930	32,300	1952	77,700
1909	80,000	1931	43,100	1953	77,200
1910	52,300	1932	77,000	1954	20,600
1911	58,000	1933	53,600	1955	85,000
1912	67,200	1934	70,800	1956	82,900
1913	115,000	1935	89,400	1957	88,700
1914	46,100	1936	62,600	1958	60,200
1915	52,400	1937	112,000	1959	40,300
1916	94,300	1938	44,000	1960	50,500

PART II

EFFECTIVE RAINFALL–DIRECT RUNOFF RELATIONSHIP

CHAPTER EIGHT

Watershed Runoff Modeling: Preliminary Remarks

The phenomenon of watershed runoff is complex. Our understanding of physical principles and the mathematical formulations governing it is limited. First, we give a short account following the Soil Conservation Service (1971), from a systems viewpoint, of this phenomenon. At the end of the chapter a brief discussion of some aspects of its physics is provided.

8.1 RUNOFF COMPONENTS

The watershed runoff is composed of three components, which may, at a given time, occur separately or simultaneously with varying magnitudes. These, as illustrated in Fig. 8.1, are (1) *surface runoff*, (2) *subsurface runoff*, or interflow, and (3) *baseflow*, or groundwater runoff. As the name suggests, the surface runoff travels over the ground surface and through channels to the watershed outlet. As shown in the figure, the travel is from point 1 to point 2. The surface runoff may be composed of (1) overland flow and (2) channel flow. The portion of surface runoff that flows over the land surface to stream channels is called *overland flow*.

Surface runoff usually occurs when the rainfall intensity exceeds the initial demands of interception, infiltration and surface storage. It varies during the storm and may cease either during its occurrence or shortly after its cessation. As surface runoff flows toward the watershed outlet, a portion of it is infiltrated into the soil or channel bed; the infiltration taking place in a channel is often referred to as *transmission loss*. This situation is particularly common in arid, semiarid, or subhumid climate areas. Indeed, the transmission losses may sometimes be so large that they entirely eliminate surface runoff from appearing at the outlet. As shown in the figure, the travel distance from point 3 to point 4 depends upon (1) the amount of surface runoff, (2) soil moisture characteristics, and (3) watershed physiography.

The subsurface runoff represents that portion of infiltrated rainfall that moves laterally through the upper soil horizons until it reaches the stream channel. Its other names are subsurface flow, interflow, subsurface storm flow, and storm seepage. As seen in Fig. 8.1, the water infiltrated at point 5 eventually returns to the surface at point 6, continuing as surface runoff to point 7. This moves more slowly than surface runoff and may join the surface runoff during or after the storm. The proportion of streamflow that occurs as subsurface runoff depends on geologic characteristics of the watershed and space-time properties of rainfall. It is commonly found in humid climates and in watersheds with high infiltration capacities and moderate to steep slopes.

The base flow is that portion of infiltrated water that reaches the water table and then discharges into the stream. As shown in Fig. 8.1, the water entering at point 8 goes directly to the water table and then eventually enters a stream at point 9. This type of flow moves much more slowly and has little effect on flood peaks in small watersheds. Depending upon permeability of the soil, the baseflow response widely varies from one groundwater body to another.

These components of runoff may not appear regularly on a watershed. For example, in arid areas runoff on small watersheds is nearly always surface runoff. In humid areas, subsurface runoff is predominant. However, a long succession of rainfall storms may produce subsurface runoff and even base flow in arid areas. Therefore, the type of runoff occurring on a watershed is determined by a combination of climatic and physiographic factors in conjunction with space-time characteristics of rainfall.

Although these components of runoff originate from dif-

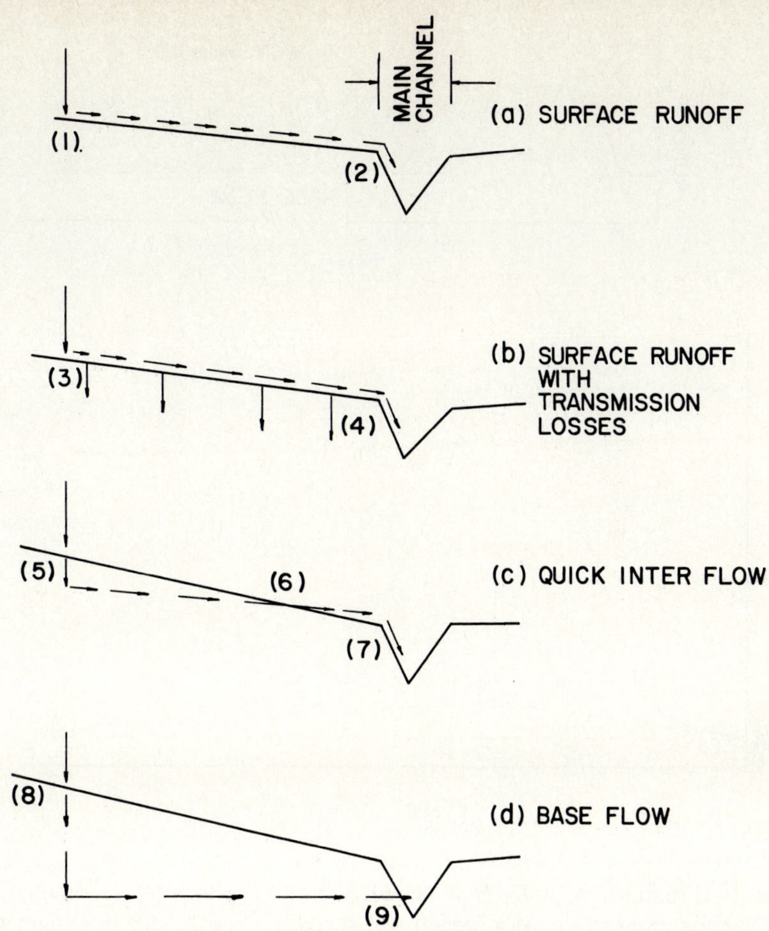

FIGURE 8.1 Runoff components (adapted from Soil Conservation Service, 1971).

ferent sources or have different paths of travel, the distinction between them remains essentially arbitrary. In the course of their travels, these components may have mixed paths and change forms. For example, interflow may surface during part of its way and appear as surface runoff. The converse is equally likely. Similarly, depending upon the geologic formations, the base flow and interflow may share some of the features. Therefore, in practice it may be very difficult to separate these components accurately. That may partly explain the customary division of runoff into (1) direct, or storm, runoff, and (2) baseflow. The direct runoff is considered to consist of surface runoff and so-called quick interflow, whereas baseflow consists of delayed interflow and groundwater runoff. In hydrologic literature direct runoff, storm runoff, and surface runoff appear to be freely used interchangeably. This appears to be true of groundwater runoff, baseflow, and base runoff. It may be noted that this latter division of runoff into components is based on the time of travel rather than the path of travel.

8.2 SEPARATION OF RUNOFF COMPONENTS

In a typical streamflow hydrograph, as shown in Fig. 8.2, two portions appear easily distinguishable. There is a lower portion of the hydrograph, which occurs fairly regularly and is subject to only a gradual change throughout the year. This portion is frequently referred to as base flow. On the other hand, there is a rapidly fluctuating portion, which represents the immediate response to rainfall. This portion is referred to as direct runoff. The addition of these two portions is the streamflow.

As noted previously, the separation of direct runoff and baseflow is made in an entirely arbitrary manner. Nevertheless, the separation is made to avoid the very long tails that must otherwise be associated with the part of the hydrograph due to each rainfall storm. There are several methods of hydrograph separation encountered in hydrologic literature (Barnes, 1939, 1959; Hursh and Brater, 1941; Laurenson, 1961; Kunkle, 1962; Knisel, 1963; Levi and Valdes, 1964; Toebes and Strang, 1964; Nash, 1966; Hall, 1968; Kulandaiswamy and Seetharaman, 1969; Toebes et al., 1969; James and Thompson, 1970; Rogers, 1981, 1982; K. P. Singh and Stall, 1971; Mitchell, 1972; Klassen and Pilgrim, 1975; Gilman, 1977a, 1977b; Jones and McGilchrist, 1978; Tjomsland, Ruud and Nordseth, 1978; Anderson and Burt, 1980). Two very simple methods are shown in Fig. 8.3. In one method, the baseflow is represented by *ABC*, where *AB* and *CB* are drawn tangentially to the beginning of rise and the end of recession portions, respectively. In the other method, the base flow represented by *ADE* is simply assumed constant. Likewise, baseflow can be separated in many different arbitrary ways.

The separation of runoff components for each rainfall storm can perhaps be best done by availing of their recession characteristics (Toebes and Strang, 1964; Nash, 1966). The

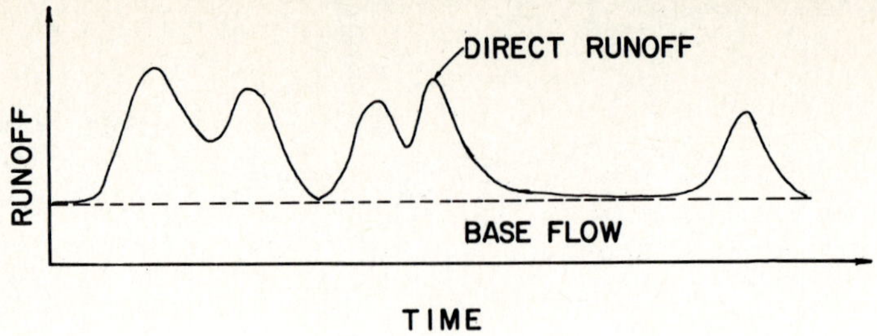

FIGURE 8.2 A typical streamflow hydrograph.

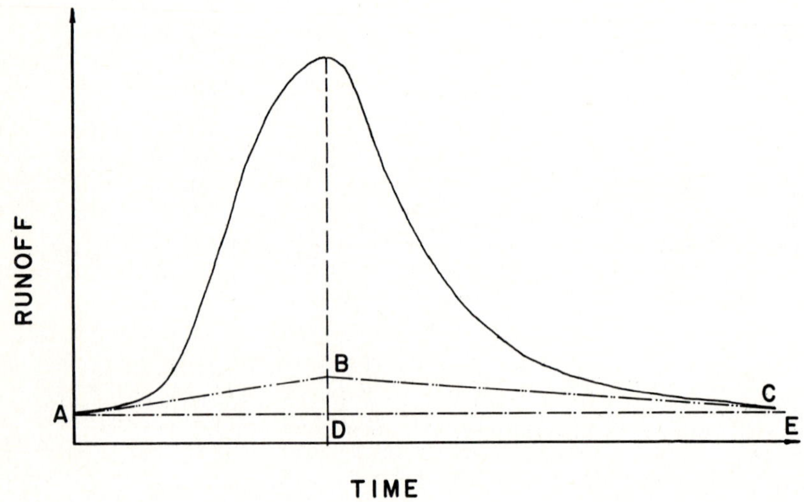

FIGURE 8.3 Two simple procedures for hydrograph separation.

recession represents withdrawal of water from storage within the watershed. Therefore, during recession, runoff will depend upon the storage remaining. If the storage-runoff relationship is assumed to be linear,

$$S = kQ \tag{8.1}$$

where S is storage, Q is discharge, and k is a constant having the dimension of time. Coupling it with

$$\frac{dS}{dt} = -Q, \qquad Q(t_0) = Q_0 \tag{8.2}$$

yields

$$Q = Q_0 e^{-(t-t_0)/k} \tag{8.3}$$

where Q_0 is the runoff at $t = t_0$ at which the recession began. Fig. 8.4 sketches equation (8.3), which shows k to be equal to the time difference between any Q and Q/e. Equation (8.3) assumes that Q goes to zero asymptotically and roughly approximates the recessions of direct runoff and base flow. This plots as a straight line on a semilogarithmic paper and renders an easy determination of k and Q_0.

If equation (8.3) is taken to represent the recession of the three runoff components, then runoff Q at any time can be given by the sum of the values of the three components at that time, that is,

$$Q = Q_1 e^{-t/k_1} + Q_2 e^{-t/k_2} + Q_3 e^{-t/k_3} \tag{8.4}$$

where Q_1, Q_2, and Q_3 are the values of the three runoff components at $t = 0$, and k_1, k_2, and k_3 are their respective time constants. Equation (8.4) is sketched in Fig. 8.5 and has six unknowns. One way to determine these unknowns might be to write equation (8.4) for six different points on the actual recession and then solve the six equations for six unknowns. There is, of course, no restriction placed on the choice of these points. This would mean that we could choose any six points along a recession and determine three exponential components whose sum would pass through these six points, thus decomposing the runoff recession into three exponential components.

However, as pointed out by Nash (1966), it can be shown that if there exists even a small error in the position of one or more of the six points, the resulting errors in the k's of

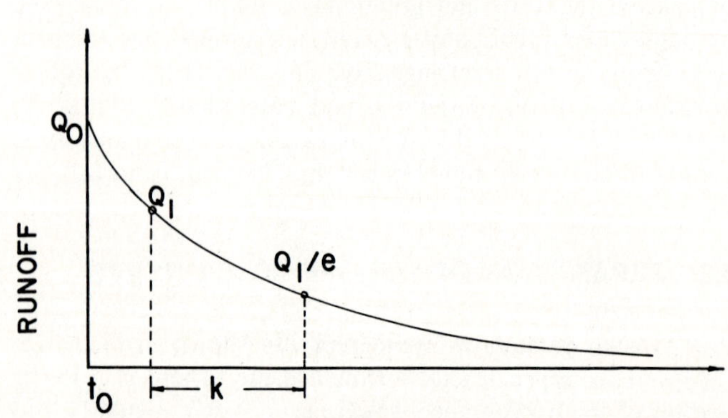

FIGURE 8.4 Exponentially declining recession.

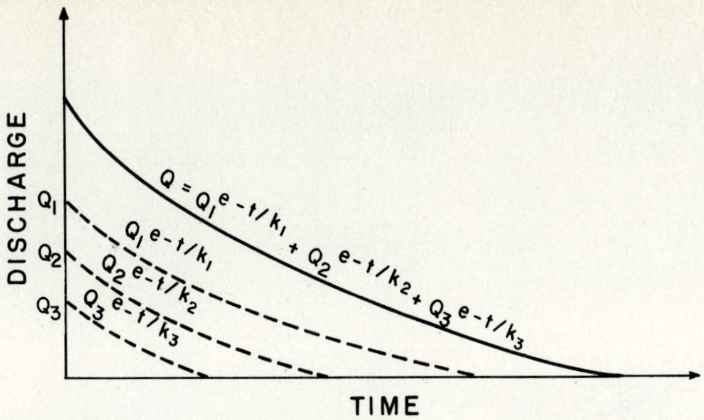

FIGURE 8.5 Exponentially declining recessions of runoff components.

the components may be large. Since there can be innumerably different ways in which a given runoff recession can be closely approximated by three components, it is very difficult to decide which are the true components. If the recession is approximated by two components, the parameters may be determined more easily and perhaps more accurately.

Another way of looking at the recession (or depletion) is that it can be described by a characteristic depletion equation (Laurenson, 1961; Knisel, 1963; James and Thompson, 1970; Klassen and Pilgrim, 1974; Anderson and Burt, 1980)

$$Q_t = Q_{t-1} k_r \tag{8.5a}$$

where Q_t is runoff at any time t, Q_{t-1} is runoff at one time unit earlier $t - 1$, and k_r is a dimensionless recession constant $\epsilon(0, 1)$ that depends on the time unit. Equation (8.5a) can be written as

$$Q_t = Q_0 k_r^t \tag{8.5b}$$

where Q_0 is runoff at $t = 0$. The time unit is frequently taken as 1 day, but any other unit can be chosen appropriately.

Equations (8.5a) and (8.3) are similar, and their k and k_r may be related to one another. To establish this relationship, it may be preferable to express equation (8.5a) in discrete-time form as

$$Q_{j\Delta t} = k_r Q_{(j-1)\Delta t} = k_r^j Q_0, \qquad j = 0, 1, 2, \ldots \tag{8.5c}$$

where $t = j\Delta t$. On comparing equation (8.5c) with equation (8.3),

$$k = \frac{-\Delta t}{\ln k_r} \tag{8.6}$$

This shows an explicit dependence of k_r on Δt.

Substituting equation (8.5b) into equation (8.2) and integrating with the condition $S = 0$, $t \rightarrow \infty$,

$$S_t = \frac{-Q_t}{\ln k_r} \tag{8.7}$$

Equation (8.7) gives the storage S_t remaining in the watershed at time t.

Equation (8.5b) plots as a straight line on a semilogarithmic paper. When the runoff recession is plotted, it is usually not a straight line but a curve with gradually decreasing slope or increasing k_r. This is partly because the water is derived from three types of storage, including stream channels, soil surface, and groundwater, which have different lag characteristics (as shown in Fig. 8.1).

If equation (8.5b) is taken to represent the three-component recession, then

$$Q = Q_1 k_{r1}^t + Q_2 k_{r2}^t + Q_3 k_{r3}^t \tag{8.8}$$

where k_{r1}, k_{r2}, and k_{r3} are the recession constants of the three runoff components, which are Q_1, Q_2, and Q_3, respectively, at $t = 0$. As discussed in the case of equation (8.4), the case of equation (8.8) has six unknowns, which can be estimated in like manner but will experience similar problems.

Barnes (1940) suggested that the recession could be approximated by three straight lines on a semilogarithmic paper. The slope of the last portion of the recession should represent k_r for groundwater. By projecting this slope backward in time, determining the difference between the total runoff and groundwater runoff, and repeating this procedure, the recession consisting of surface runoff and interflow can be obtained. This process can be repeated to obtain k_r for interflow and surface runoff. This is shown in Fig. 8.6.

Another method of hydrograph separation was presented by Rogers and Zia (1982). This is based on the relation

$$Q_T = Q_0 + Q_p^{.6} \tag{8.9}$$

in which Q_0 is the base flow in cubic feet per second prior to hydrograph rise marking the beginning of the hydrograph, Q_p is the peak discharge of hydrograph in cubic feet per second, and Q_T is the discharge in cubic feet per second marking the termination of the direct runoff. Implicit here is that $Q_p > 1.0$. By using equation (8.9), the time base of direct runoff can be computed. During this period, the baseflow can be separated by any standard procedure. For example, the

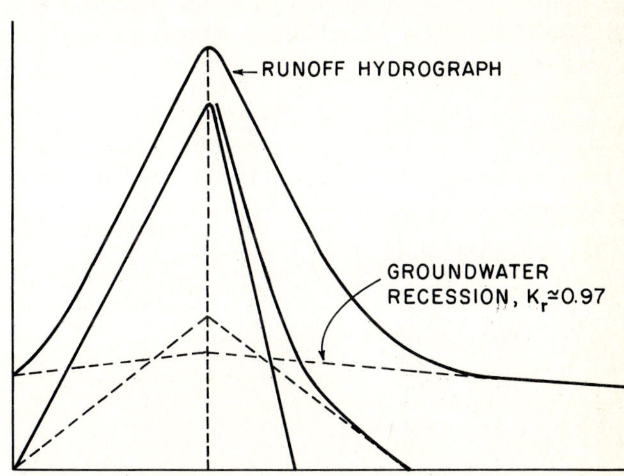

FIGURE 8.6 Recession analysis using semilog plot.

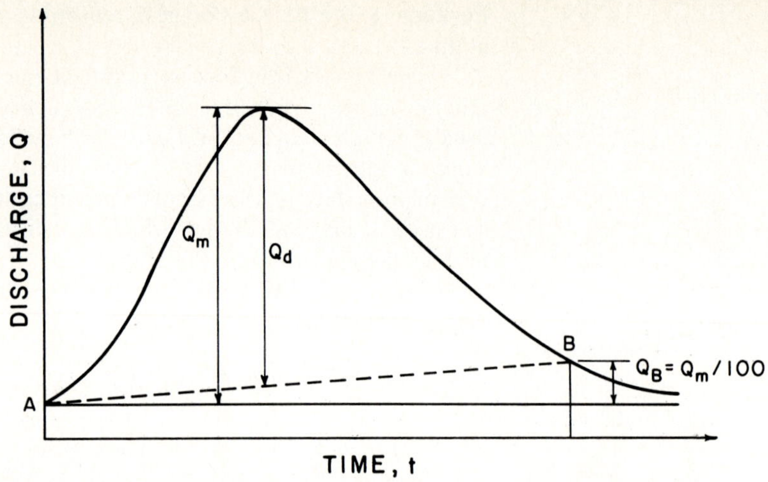

FIGURE 8.7 Baseflow separation (after Rao and Delleur, 1974).

baseflow can be represented by a linear rise from Q_0 to a point under Q_p and then a linear decline to Q_T.

Rao and Delleur (1974) used another method of baseflow separation in quantifying the effects of urbanization on watershed runoff. Referring to Fig. 8.7, let A and B represent the points at which the direct runoff begins and ends, respectively. The point A can be easily located, especially in isolated storms. By drawing a horizontal line through point A, the value of the maximum discharge Q_m can be determined. Point B is then located on the recession limb of the hydrograph such that the discharge Q_B at B in excess of that at A is one-hundredth of Q_m. The line joining points A and B represents the baseflow separation line. The direct runoff is then obtained by subtracting the base flow from the total runoff. The peak of the direct runoff is Q_d.

EXAMPLE 8.1

The streamflow of the Amite River at Denham Springs, Louisiana, was observed during April 20 to May 3, 1979, as given in Table 8.1. Perform the hydrograph separation to determine surface runoff, interflow and groundwater runoff, and their depletion constants k_{r1}, k_{r2}, and k_{r3}. Compute the volumes of surface runoff and interflow.

The streamflow hydrograph was plotted on a semilog paper, as shown in Fig. 8.8. A straight line was fitted to the groundwater recession. This was projected back to the time of peak flow. The baseflow rise was represented by a straight line joining the initial flow and the time to peak flow. This baseflow hydrograph was subtracted from the streamflow hydrograph, yielding a hydrograph of surface runoff and interflow. This procedure was repeated for separating interflow. The hydrographs of interflow and surface runoff are plotted in Figs. 8.9 and 8.10 and tabulated in Table 8.1. The volumes of surface runoff and interflow as well as their depletion constants were computed:

Volume of surface runoff $= 5.25 \times 10^8 \text{ m}^3$

Volume of interflow $= 1.77 \times 10^7 \text{ m}^3$

Surface runoff: $k_{r1} = 0.5$

Interflow: $k_{r2} = 0.75$

Groundwater: $k_{r3} = 0.97$

EXAMPLE 8.2

Plot the streamflow recession on semilog paper for the data in Example 8.1. Compute the recession constants k_1, k_2, and k_3 of surface runoff, interflow, and baseflow respectively.

The total streamflow recession is plotted on a semilog paper, as shown in Fig. 8.11. It is seen that the recession is comprised of three parts each fitted by a straight line representing surface runoff, interflow, and groundwater recessions. Since each part can be represented by equation (8.3), the recession constant is given by the slope of the corresponding line:

$$k_1 = \frac{36}{\ln(800/260)} = \frac{36}{1.12} = 32.03 \text{ hours for surface runoff}$$

$$k_2 = \frac{18}{\ln(120/84)} = \frac{18}{.356} = 50.47 \text{ hours for interflow}$$

$$k_3 = \frac{42}{\ln(60/48)} = \frac{42}{.223} = 188.2 \text{ hours for groundwater}$$

EXAMPLE 8.3

Consider the streamflow data of Example 8.1. Perform the hydrograph separation into direct runoff and baseflow and plot. The baseflow separation is shown in Fig. 8.12 on p. 94.

8.3 EFFECTIVE RAINFALL

The effective rainfall is that portion of rainfall that contributes to direct runoff. Similarly, rainfall excess constitutes the portion of rainfall that contributes to surface runoff. The difference between effective rainfall and rainfall excess is that the former includes the latter plus abstractions. However, the terms *effective rainfall* and *rainfall excess* are freely interchanged in hydrologic usage. Thus, a rainfall storm is considered to be composed of two portions: one that contributes to runoff and another that contributes to abstractions, including interception,

TABLE 8.1

Streamflow hydrograph of the Amite River basin at Denham Springs, Louisiana, during April 21 to May 3, 1979, and hydrograph separation

			Three-Part Separation			
Date	Time	Total Flow m^3/s	Groundwater Flow m^3/s	Interflow + Surface Runoff m^3/s	Interflow m^3/s	Surface Runoff m^3/s
4/20	6 a	33				
	12 p	33				
	6 p	32				
	12 a	32				
4/21	6 a	32				
	12 p	127	35	92	6.2	86
	6 p	263	39	224	7.6	216
	12 a	425	42	383	9	374
4/22	6 a	555	47	508	11	497
	12 p	759	52	707	13	694
	6 p	1034	56	978	16	962
	12 a	1294	62	1232	19	1213
4/23	6 a	1478	68	1410	23	1387
	12 p	1617	75	1542	28	1514
	6 p	1727	82	1645	33	1612
	12 a	1798	90	1708	40	1668
4/24	6 a	1886	100	1786	50	1736
	12 p	1940	110	1830	60	1770
	6 p	1903	107	1796	54	1742
	12 a	1778	104	1674	50	1624
4/25	6 a	1597	102	1495	45	1450
	12 p	1419	100	1319	41	1278
	6 p	1240	98	1142	37	1105
	12 a	1059	95	964	34	930
4/26	6 a					
	12 p	719	91	628	28	600
	6 p	589	89	500	26	474
	12 a	464	86	378	23	355
4/27	6 a	374	85	289	21	268
	12 p	281	83	198	19	179
	6 p	207	80	127	18	109
	12 a	159	79	80	16	64
4/28	6 a	145	77	68	14	54
	12 p	124	76	48	13	35
	6 p	106	74	32	12	20
	12 a	93	72	21	11	10
4/29	6 a	85	70	15	10	5
	12 p	78	69	9	9	0
	6 p	75	67	8	8	0
	12 a	74	66	8		
4/30	6 a	71	64	7		
	12 p	67	63	4		
	6 p	64	61	3		
	12 a	61	60	1		
5/1	6 a	59				
	12 p	57				
	6 p	56				
	12 a	54				
5/2	6 a	53				
	12 p	52				
	6 p	51				
	12 a	50				
5/3	6 a	49				
	12 p	48				
	6 p	47				
	12 a	46				

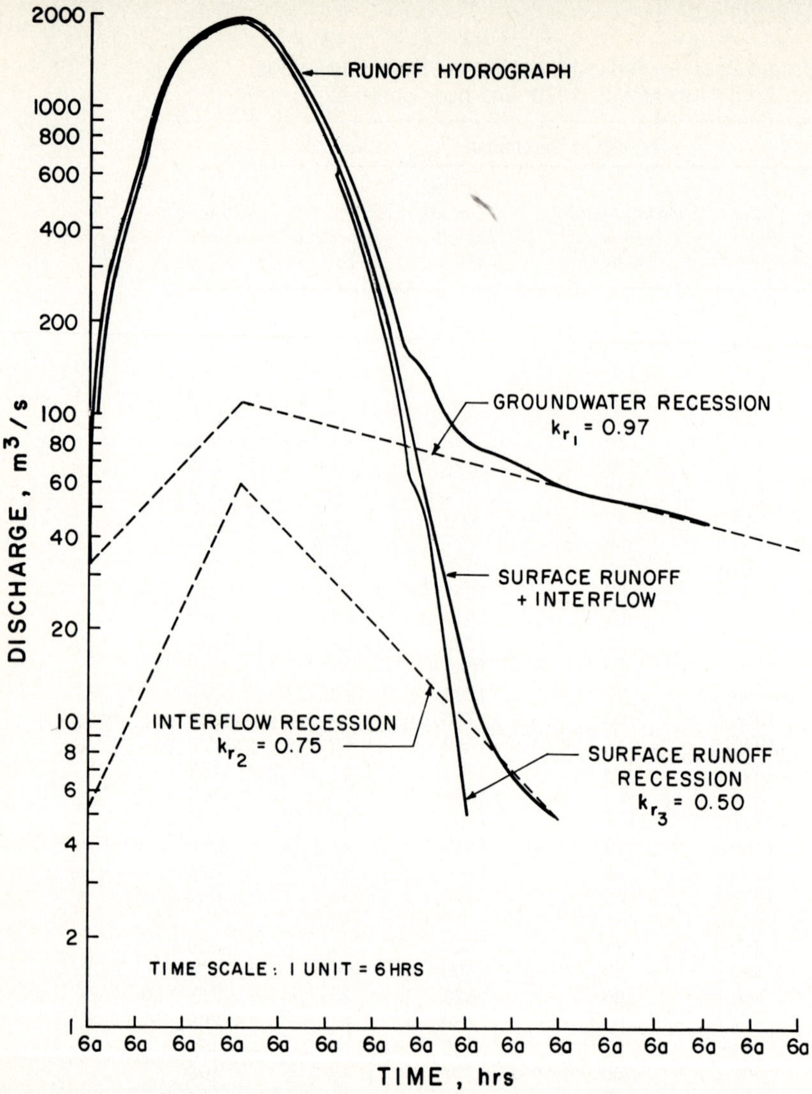

FIGURE 8.8 Example 8.1: Separation of streamflow hydrograph of the Amite River basin at Denham Springs, Louisiana, during April 21 to May 3, 1979.

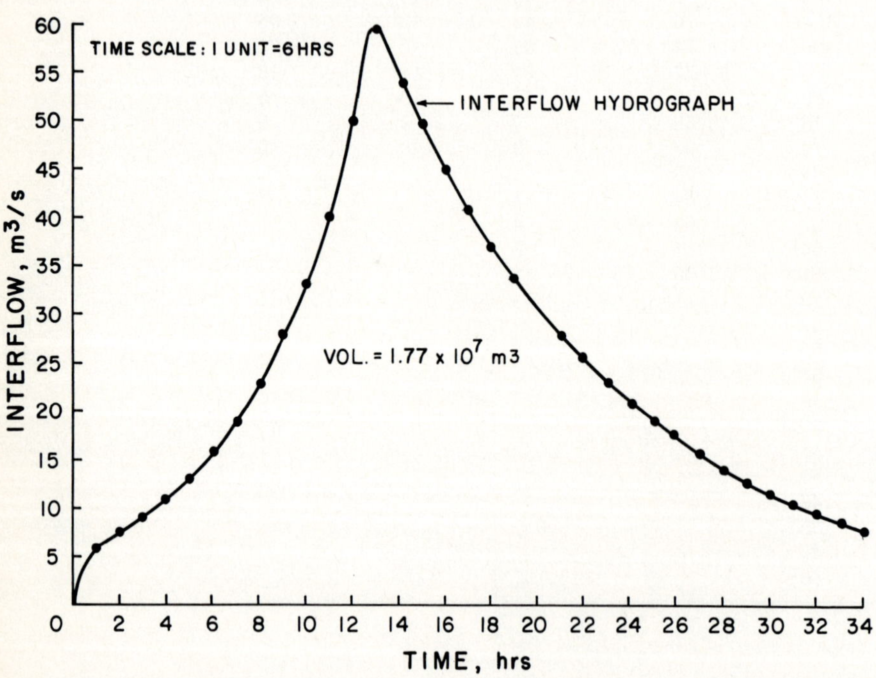

FIGURE 8.9 Example 8.1: Interflow hydrograph of the Amite River basin at Denham Springs, Louisiana, during April 21 to May 3, 1979.

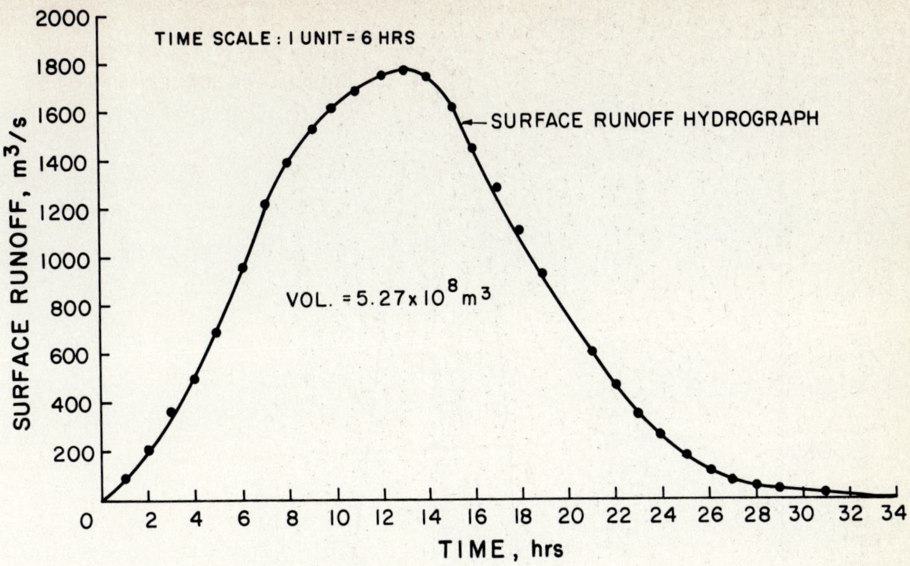

FIGURE 8.10 Example 8.1: Surface runoff hydrograph of the Amite River basin at Denham Springs, Louisiana, during April 21 to May 3, 1979.

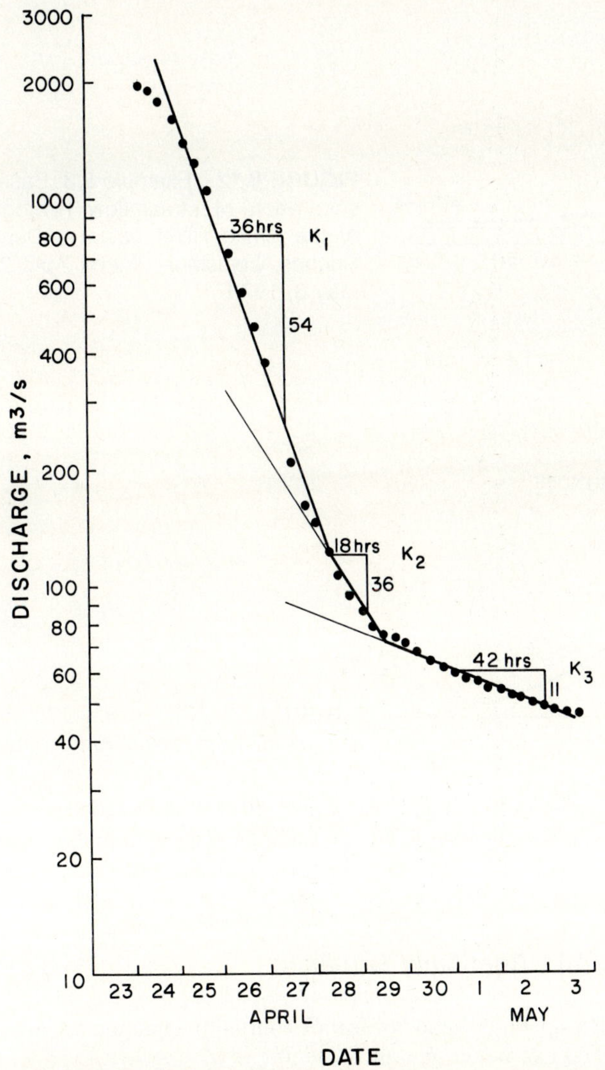

FIGURE 8.11 Example 8.2: Recession of streamflow of the Amite River basin at Denham Springs, Louisiana, during April 21 to May 3, 1979.

evaporation, transpiration, depression and detention storage, and infiltration. Because the effective rainfall entirely becomes direct runoff, its volume must equal the volume of direct runoff.

For a given rainfall-runoff event, the volume of direct runoff is determined by separating the baseflow. This gives the volume of effective rainfall and, in turn, the volume of rainfall that is used up by abstractions. The distribution of this volume in time (determination of loss function) and then its subtraction from rainfall intensity yields effective rainfall. This is illustrated in Fig. 8.13. However, due to extreme space-time heterogeneity in antecedent conditions, extrapolation of this loss function for any other rainfall event is extremely difficult. The loss function can be determined in one of several ways. An accurate determination of this loss function is of fundamental importance in watershed runoff modeling (V. P. Singh, 1976b; Lowing and Mein, 1981).

Let us pause for a moment. If, for simplicity, we neglect all abstractions other than infiltration, we immediately see that during and after a rainfall episode, runoff and infiltration occur simultaneously. Indeed, infiltration continues to occur as long as surface runoff occurs or there is water over the ground. Looking at durations of rainfall and runoff, it is usually true that the duration of surface runoff is much larger than the duration of rainfall. But infiltration is allowed to occur only during the period of rainfall. The question is, What happens to the infiltration during the period of surface runoff in excess of that of rainfall? One might argue that infiltration may be very small after the cessation of rainfall and hence be neglected. This may not be true. Besides, from a physical standpoint, in order to compensate for infiltration accounted for after the period of rainfall, infiltration has to be allowed to occur at a rate higher than that at which it actually occurs. This is inevitable to ensure volume continuity of rainfall excess and surface runoff. Therefore, the concept of rainfall excess or effective rainfall is an artificial one. The same thing can be said about direct runoff. In nature, rainfall occurs as a continuum and so does runoff.

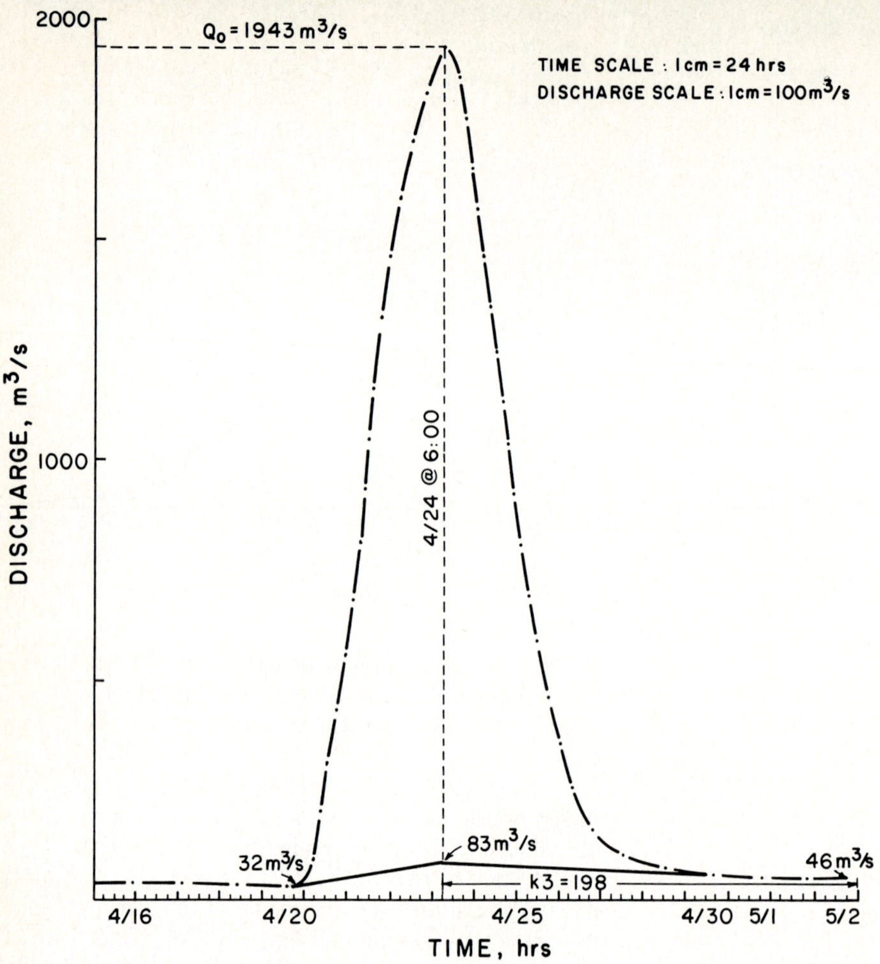

FIGURE 8.12 Example 8.3: Baseflow separation of streamflow hydrograph of the Amite River basin at Denham Springs, Louisiana, during April 21 to May 3, 1979.

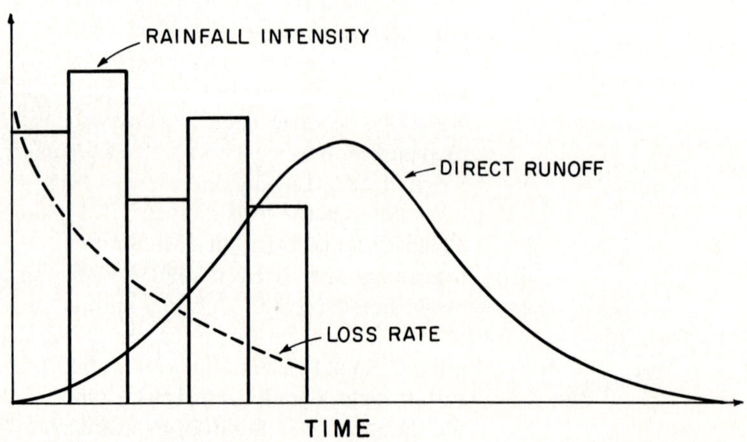

FIGURE 8.13 Determination of effective rainfall from rainfall-runoff data.

8.4 RAINFALL-RUNOFF RELATION

The relation between direct runoff and rainfall may be considered to have three aspects:

1. The relation between the volume of rainfall in a given storm and the resulting volume of direct runoff.
2. The relation between the time distribution or hydrograph of direct runoff and the time distribution or hyetograph of the effective rainfall.
3. The relation between rainfall frequency and discharge frequency.

In the succeeding chapters we discuss the second aspect in considerable detail. The first will be dealt with in the companion volume. Here our objective is to assemble the rudiments required in development of the rainfall-runoff relationship.

8.4.1 Continuity Equation

The spatially lumped form of continuity equation for a watershed can be written as

$$I = Q + \frac{dS}{dt} \tag{8.10}$$

where I is the rate of inflow, Q is the rate of outflow, and S is the storage. These quantities are all functions of time t. Appendix C provides its derivation. Equation (8.10) is also referred to as an equation of water budget or volume balance. Although the watershed is truly a distributed system and must be represented by a partial differential equation(s), equation (8.10) is one of the basic equations used in representing hydrologic systems. This equation has two unknowns, namely, Q and S. Since Q is the variable for which the differential equation is to be solved, it must be retained. Then we must have another equation that can be used to eliminate S from equation (8.10).

8.4.2 Storage Characteristics

The functional relationships for storage used or suggested so far have been special cases or variants of the general storage equation (Kulandaiswamy, 1964)

$$S = \sum_{i=0}^{M} a_i(I, Q) \frac{d^iQ}{dt^i} + \sum_{j=0}^{M} b_j(I, Q) \frac{d^jI}{dt^j} \tag{8.11}$$

which is derived in Appendix D. Here a_i and b_j are coefficients and depend on both I and Q. Let us first consider two simplest storage-discharge relationships,

$$S = kQ \tag{8.12}$$

$$S = kQ^n \tag{8.13}$$

If we graph S versus Q for a watershed, the plot is normally a loop, as shown in Fig. 8.14. Equation (8.12) is a straight-line approximation of the loop and equation (8.13) a curvilinear approximation. Both are, however, single-valued, whereas the actual relationship between Q and S is double-valued. A plot of Q versus time and S versus time for actual rainfall storms will show that for watersheds that are not too small, the peak of storage, which more or less occurs at or near the end of the effective rainfall, precedes the peak of surface runoff by a significant amount of time. However, equations (8.12) and (8.13) (with $n > 0$) yield that S and Q peak concurrently. Further, substituting equations (8.12) and (8.13) into equation (8.10) gives, respectively,

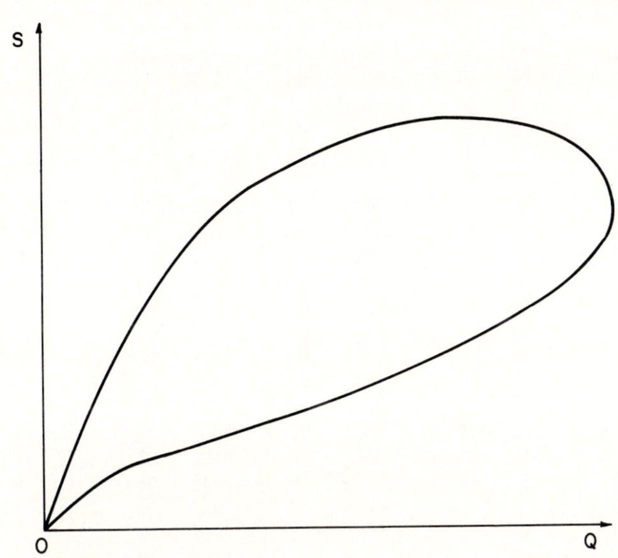

FIGURE 8.14 A typical storage-discharge relationship.

$$I = Q + k\frac{dQ}{dt} \tag{8.14}$$

$$I = Q + knQ^{n-1}\frac{dQ}{dt} \tag{8.15}$$

The peak of Q occurs mostly after I has ceased to exist. If $I = 0$, equations (8.14) and (8.15) become

$$-Q = k\frac{dQ}{dt} \tag{8.16}$$

$$-Q = knQ^{n-1}\frac{dQ}{dt} \tag{8.17}$$

When $Q = Q_p$, $dQ/dt = 0$, where Q_p is the peak of Q. Then both equations (8.16) and (8.17) gives $Q = 0$ at $Q = Q_p$, a patently absurd result. This indicates that only the recession side of the runoff hydrograph, preferably after the inflection point, can be approximated by equations (8.14) and (8.15). These may not be appropriate for the rising limb of the hydrograph. Besides Q, there are obviously other variables on which S depends.

Storage in a channel reach consists of prism storage and wedge storage. Under steady-state conditions only prism storage exists. The wedge storage is associated with unsteady flow. The wedge storage is positive for the rising hydrograph and negative for the falling hydrograph. A plot of Q versus S for steady-flow conditions in a channel is a single curve, and the loop is caused by the superposition of the wedge storage over the prism storage. From the nature of the loop it is apparent that the unsteady effects of flow are considerable and must be taken into consideration. Even in channel reaches it has been found that the storage is a function not only of inflow but of outflow also. It is all the more so when the storage on the entire watershed is considered. It then appears reasonable that storage on a watershed should be considered as a function of inflow, outflow, and their time derivatives.

EXAMPLE 8.4

The inflow-outflow data observed on the reach from Burrendong to Wellington of the Macquarie River in Australia for a major flood that occurred in March, 1926, are given in Table 8.2. Compute the storage. Plot storage against outflow. Attempt to represent the curve by $S = kQ$, and compute k. Do likewise for $S = kQ^n$ and determine k and n.

By employing equation (8.10), inflow, outflow, and storage are computed as given in Table 8.3. Storage and average outflow values are plotted giving storage-outflow loop. Figures 8.15 and 8.16 show fitting of this loop by equations (8.12) and (8.13), respectively.

TABLE 8.2

The inflow-outflow data observed in March, 1926, on the reach from Burrendong to Wellington of the Macquarie River, New South Wales, Australia (after The Institution of Engineers, Australia, 1977)

Date	*Time*	*I* *(10^3 m^3/s)*	*Q* *(10^3 m^3/s)*
24	12	.18	.12
	15	.32	.20
	18	.70	.40
	21	1.70	.80
	24	3.22	1.36
25	03	4.56	2.00
	06	5.66	2.64
	09	4.40	3.28
	12	3.12	3.80
	15	3.94	3.94
	18	2.16	3.62
	21	1.80	3.06
	24	1.46	2.52
26	03	1.18	2.14
	06	.98	1.76
	09	.80	1.44
	12	.66	1.18
	18	.48	.74
	24	.34	.50
27	12	.32	.30
	24	.14	.18
28	24	.06	.09

8.4.3 General Storage Equation

Equation (8.11) represents a general storage equation derived by Kulandaiswamy (1964) and Chow and Kulandaiswamy (1971). However, its application to determining storage on natural watersheds presents many formidable problems. The principal difficulty lies in determination of a_i's and b_j's, which are functions of both Q and I. If the nonlinearity of the watershed behavior is not large, a remarkable simplification is possible. Instead of considering these coefficients to be continuous functions of I and Q, these can be considered to be functions of a constant $\bar{Q}$ and a constant $\bar{I}$, which may be obtained for a given effective rainfall and direct runoff hydrograph. For example, $\bar{Q}$ and $\bar{I}$ can be taken at the equilibrium condition. Then equation (8.11) becomes

$$S = \sum_{i=0}^{M} a_i(\bar{I}, \bar{Q}) \frac{d^iQ}{dt^i} + \sum_{j=0}^{N} b_j(\bar{I}, \bar{Q}) \frac{d^jI}{dt^j} \tag{8.18}$$

Equation (8.18) can provide a meaningful criterion to verify whether the system is linear or not. If a_i's and b_j's are independent of $\bar{Q}$ and $\bar{I}$, the system is linear; otherwise it is nonlinear. For a particular storm, $\bar{Q}$ and $\bar{I}$ are constants, and consequently the coefficients are constants.

In a watershed system, $\bar{Q} = \bar{I}$ at equilibrium and so $F(\bar{Q}, \bar{I})$, given in Appendix D by equation (D.15), can be reduced to a function of one variable, say $\bar{Q}$. Therefore, a

TABLE 8.3

Inflow, outflow, and storage in March, 1926, for the Macquarie River, New South Wales, Australia

Time Hours	*Inflow 1000 m^3/s*	*Average Inflow 1000 m^3/s*	*Outflow 1000 m^3/s*	*Average Outflow 1000 m^3/s*	*Change in Storage 1000 m^3*	*Storage 1000 m^3*
.000E 00	.180E 00	.000E 00	.120E 00	.000E 00	.000E 00	.000E 00
.300E 01	.320E 00	.250E 00	.200E 00	.160E 00	.900E −01	.972E 03
.600E 01	.700E 00	.510E 00	.400E 00	.300E 00	.210E 00	.324E 04
.900E 01	.170E 01	.120E 01	.800E 00	.600E 00	.600E 00	.972E 04
.120E 02	.322E 01	.246E 01	.136E 01	.108E 01	.138E 01	.246E 05
.150E 02	.456E 01	.389E 01	.200E 01	.168E 01	.221E 01	.485E 05
.180E 02	.566E 01	.511E 01	.264E 01	.232E 01	.279E 01	.786E 05
.210E 02	.440E 01	.503E 01	.328E 01	.296E 01	.207E 01	.101E 06
.240E 02	.312E 01	.376E 01	.380E 01	.354E 01	.220E 00	.103E 06
.270E 02	.394E 01	.353E 01	.394E 01	.387E 01	−.340E 00	.997E 05
.300E 02	.216E 01	.305E 01	.362E 01	.378E 01	−.730E 00	.918E 05
.330E 02	.180E 01	.198E 01	.306E 01	.334E 01	−.136E 01	.771E 05
.360E 02	.146E 01	.163E 01	.252E 01	.279E 01	−.116E 01	.646E 05
.390E 02	.118E 01	.132E 01	.214E 01	.233E 01	−.101E 01	.537E 05
.420E 02	.980E 00	.108E 01	.176E 01	.195E 01	−.870E 00	.433E 05
.450E 02	.800E 00	.890E 00	.144E 01	.160E 01	−.710E 00	.366E 05
.480E 02	.660E 00	.730E 00	.118E 01	.131E 01	−.580E 00	.303E 05
.540E 02	.480E 00	.570E 00	.740E 00	.960E 00	−.390E 00	.219E 05
.600E 02	.340E 00	.410E 00	.500E 00	.620E 00	−.210E 00	.174E 05
.720E 02	.320E 00	.330E 00	.300E 00	.400E 00	−.700E −01	.144E 05
.840E 02	.140E 00	.230E 00	.180E 00	.240E 00	−.100E −01	.139E 05
.108E 03	.600E −01	.100E 00	.900E −01	.135E 00	−.350E −01	.109E 05

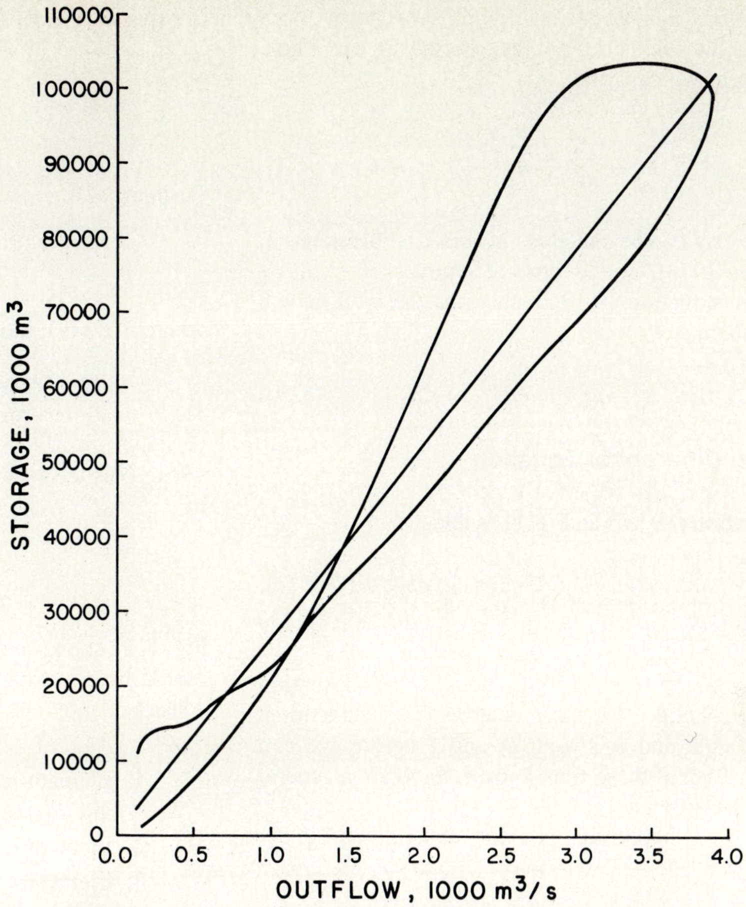

FIGURE 8.15 Storage-outflow curve for March, 1926, reresented by $S = kQ$, k = 26,324 seconds, for the Macquarie River, New South Wales, Australia.

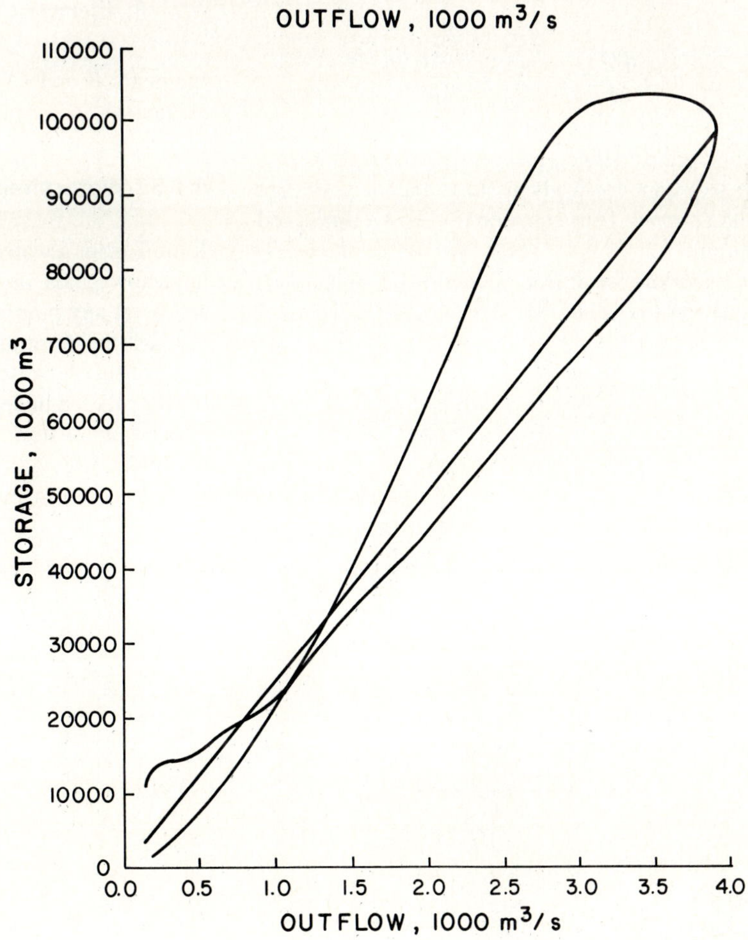

FIGURE 8.16 Storage-outflow curve for March, 1926, represented by $S = kQ^n$, k = 24,996 seconds and n = 1.016, for the Macquarie River, New South Wales, Australia.

plot of a_i's and b_j's against $\bar{Q}$ for a number of storms on a watershed will show whether or not the system is linear. Thus, for a linear system we can write

$$S = \sum_{i=0}^{M} a_i \frac{d^iQ}{dt^i} + \sum_{j=0}^{N} b_j \frac{d^jI}{dt^j} \tag{8.19}$$

Equation (8.19) can also be called the generalized Muskingum method (Dooge, 1973). If the time derivatives of Q and I are insignificant, equation (8.19) reduces to the well-known Muskingum equation,

$$S = a_0Q + b_0I \tag{8.20}$$

8.4.4 System Differential Equation

Combining equations (8.10) and (8.11) yields

$$\sum_{i=0}^{M+1} \left(a_{i-1} + \frac{da_i}{dt}\right) \frac{d^iQ}{dt^i} = - \sum_{j=0}^{N+1} \left(b_{j-1} + \frac{db_j}{dt}\right) \frac{d^jI}{dt^j} \tag{8.21}$$

where $a_{-1} = 1$, $a_{M+1} = 0$, $b_{-1} = -1$, and $b_{N+1} = 0$. Equation (8.21) is quite general. We must determine the functional relationships of a_i's and b_j's with Q and I before we can hope to solve it. Even if these were known, to obtain a general solution would be difficult.

If the linearity hypothesis is valid, equation (8.21) reduces to

$$\sum_{i=0}^{M+1} a_{i-1} \frac{d^iQ}{dt^i} = - \sum_{j=0}^{N+1} b_{j-1} \frac{d\,I}{dt^j} \tag{8.22}$$

where a_{-1} and b_{-1} are as defined previously. The coefficients a_i's and b_j's are constant for a given storm but may change from one storm to another. Thus equation (8.22) is a linearized differential equation.

Using the notation for linear operators ($D \equiv (d/dt)$), we can write equation (8.22) as

$$Q(t) = - \frac{\left(\sum_{j=1}^{N} b_{j-1} D^j\right) + 1}{\left(\sum_{i=1}^{M} a_{i-1} D^i\right) + 1} I(t) \tag{8.23}$$

Writing in a more compact form,

$$Q(t) = \frac{P(D)}{R(D)} I(t) \tag{8.24}$$

Further,

$$Q(t) = F(D)\, I(t) \tag{8.25}$$

where $F(D)$ is a rational function given by $P(D)/R(D)$. Hence,

$$Q(t) = HI(t)$$

where H is the system function performed on $I(t)$ and is given by

$$H = F(D) \tag{8.26}$$

Fig. 8.17 shows a block diagram of a watershed system represented by equation (8.26). The system can be considered to consist of three elements:

1. Element A, which transforms quantities $I(t)$ and $Q(t)$ into $S(t)$.
2. Element B, which transforms $S(t)$ into dS/dt.
3. Element C, which merely adds $I(t)$ and $-dS/dt$.

Elements B and C are linear. Nonlinearity, if any, is in the element A.

8.4.5 Watershed Response

Runoff from a watershed consists of two phases: (1) the overland flow phase or the land phase and (2) the channel phase. Inflow at any point in the watershed travels a certain distance to reach the point of measurement. During this travel it undergoes changes caused by overland and channel storage. The transformation undergone by the inflow is due to (1) translation effect and (2) storage effect, consisting of time lag and shape modification.

Let $f_1(t)$ in Fig. 8.18 be the inflow rate and $f_2(t)$ be the

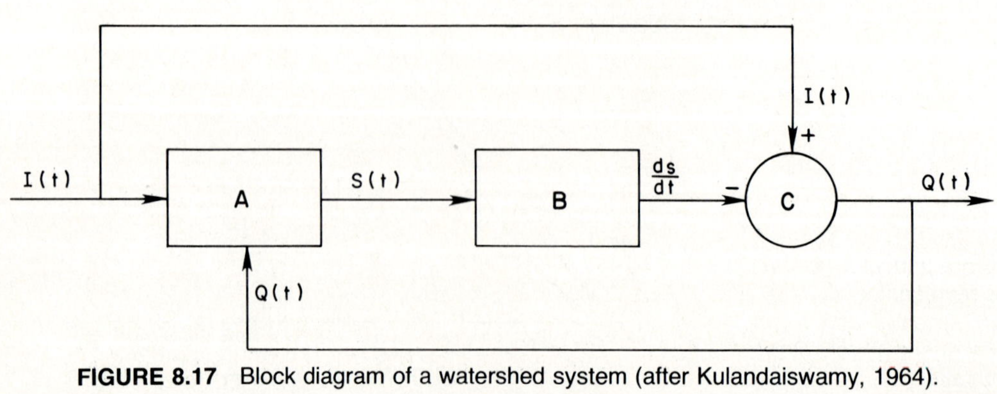

FIGURE 8.17 Block diagram of a watershed system (after Kulandaiswamy, 1964).

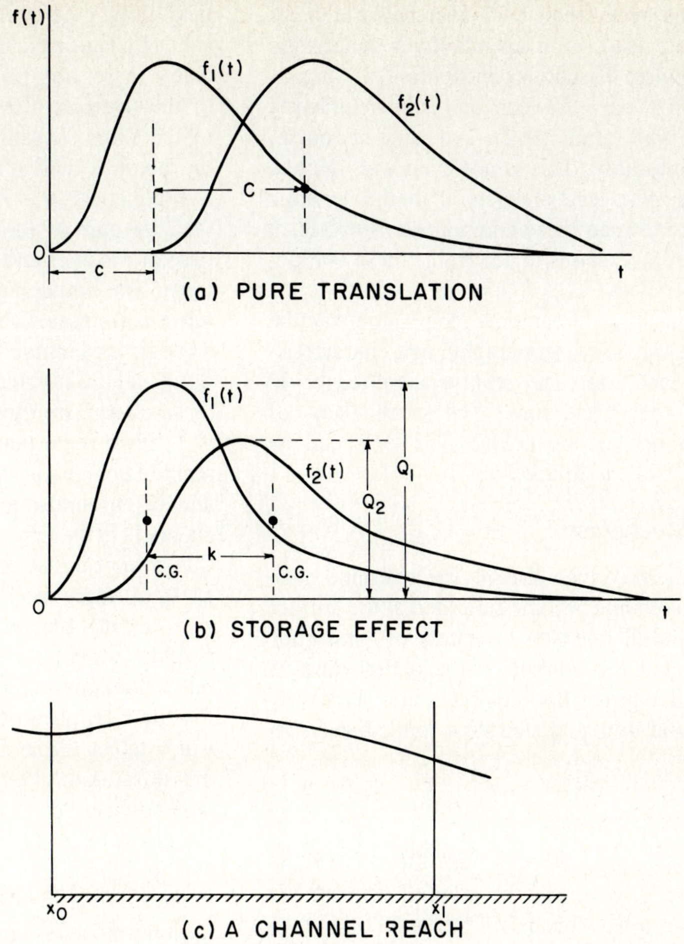

FIGURE 8.18 (a) Pure translation. (b) Storage effect. (c) A channel reach.

outflow from a system; $f_2(t)$ is an exact reproduction of $f_1(t)$ shifted by a time interval C.

$$f_2(t) = f_1(t - C), \qquad t \geq C$$
$$f_2(t) = 0, \qquad t < C \tag{8.27}$$

This is an example of pure translation, or we can write

$$f_1(t) = f_2(t + C), \qquad t \geq 0 \tag{8.28}$$

Fig. 8.18 also gives an example of storage effect. The inflow is given by $f_1(t)$ and the outflow is given by $f_2(t)$. The center of area of inflow is shifted by k hours, which is a measure of lag time. There is at present no quantitative measure of "shape modification," but the reduction of peak from Q_1 to Q_2 may be taken as a measure of the change in shape.

8.4.6 Streamflow Continuum

Streamflow evolves as a continuum (V. P. Singh, Prasad, and Ubertini, 1983). Its components occur concurrently, but their relative magnitudes vary with time. If we consider that there is a sudden burst of rainfall, then surface runoff predominates during the rising part of streamflow hydrograph, interflow during the early part of its recession, and baseflow during the delayed part of its recession. The mechanisms and, therefore, the governing equations of these components are different but influenced by dynamic interactions prevailing between them. Surface runoff interacts with the saturated flow occurring in the immediate vicinity of the ground and is governed by partial differential equations of hyperbolic type. Interflow occurs as a result of saturated flow occurring below the ground between the overland flow length and water table. It interacts with surface runoff and the unsaturated flow surrounding its domain. Its governing partial differential equations are of parabolic type. Baseflow is also governed by partial differential equations of elliptic type and is influenced by recharge as a result of infiltration. The governing partial differential equations for unsaturated flow and infiltration are of a parabolic type.

Treatment of streamflow as a continuum and taking into account dynamic interactions between its components are based on the concepts embodied in Horton's infiltration theory of runoff (Horton, 1933). According to this theory, rainfall is absorbed for intensities not exceeding infiltration capacity, and for excess rainfall there is a constant rate of absorption as long as the infiltration capacity is unchanged. Thus, infiltration divides rainfall into two parts. One part travels over the surface, giving rise to surface runoff, and the other part infiltrates into the ground, resulting eventually in interflow and baseflow. These components have been treated at various levels of mathematical sophistication but in virtual isolation of one another as if one component has nothing to do with the other. Surface flow has been studied for over half a century (Woolhiser, 1982). As a result, it independently is understood reasonably well. Likewise, baseflow contribution to streamflow has been studied for nearly 30 years and is independently understood reasonably well (Hall, 1982). The same of course, cannot be said about interflow. This is not even well defined and is least understood. Also least understood are the dynamic interactions prevailing between these components. Further, it remains yet to establish procedures to determine these components on ungaged watersheds.

The factors controlling streamflow generation can be summarized as climate, geology, topography, soil characteristics, vegetation, and land use. The relative significance of each factor varies in space and time. The mechanisms of streamflow generation verified by field experiments can be summarized (Freeze, 1974) as follows.

8.4.6.1 Hortonian Mechanism

The *Hortonian mechanism* gives rise to the overland flow from partial areas, where soils become saturated at the surface from above due to rainfall intensity in excess of infiltration capacity (Horton, 1933). The control on the partial areas is the distribution of soil types. This illustrates interaction between overland flow and thin saturated flow immediately beneath the ground.

8.4.6.2 Dunne-Black Mechanism

The *Dunne-Black mechanism* (Dunne and Black, 1970a, 1970b), overland flow is generated by rainfall on variable source areas adjacent to stream channels. Overland flow occurs when soils become saturated at the surface from below by rising water tables. The control on the variable source areas is constituted by the topographic and hydrogeological configuration of the hillslope. This illustrates dynamic interactions between the three streamflow components.

8.4.6.3 Subsurface Flow Mechanism

The subsurface flow (interflow and groundwater) delivers water directly to a permanent stream or an expanding intermittent channel network in a variable source area. The control on the subsurface flow system is hydrogeology. The recharge to groundwater is controlled by the overlying unsaturated zone. This illustrates dynamic interactions between unsaturated and saturated flows.

Of these mechanisms, the Hortonian mechanism has perhaps been most widely used. This mechanism and the Dunne-Black mechanism can occur concurrently and merge into one another in a given watershed. The predominance of one mechanism over the other depends on the controls existing at a given point in time and space. An understanding of these mechanisms and the controls for their existence is essential for modeling streamflow.

8.4.6.4 A Simple Case Illustrating Dynamic Interactions

We consider a plane discharging into a channel. The plane is porous and initially dry. Its soil properties are spatially homogeneous. If it is subjected to a uniform rainfall, the overland flow units will begin forming with the ponding time $t = t_p(L)$, starting at the incised channel boundary. For monotonically increasing rainfall rate, the upper boundary $x = x_p(t)$ of the overland flow will migrate upward toward the divide of the slope. If rainfall is of sufficient duration, say $t = T_1$, the point x_p will reach the end $x \doteq 0$ at time, say $t = T_0$. Subsequent to $t = T_1$, the end $x = x_p(t)$ will recede.

A part of the volume of water infiltrated through the base of the overland flow remains in the soil mass in saturated form. The remaining part of this water is absorbed by unsaturated soil mass. The three modes of transport are coupled through the continuity of flux between surface flow, saturated flow, and unsaturated flow domains. The boundaries separating these phases are dynamic and are unknown.

Two important cases can be distinguished: (1) The element is ephemeral in nature (absence of nearby groundwater); and (2) groundwater or baseflow contributes to the incised channel. In the first case the sequence of growth and redistribution of soil moisture is depicted in Fig. 8.19(a)–(e). For $t > t_p(L)$, all three modes of transport exist.

In the second case the presence of water table gives rise to a richer but complicated growth of saturated and unsaturated domains. This is depicted in Fig. 8.20(a)–(f). Again, for $t \leq t_p(L)$, unsaturated flow occurs in soil mass above the water table. Groundwater recharge may occur as a result of this unsaturated flow. The significance of this contribution is not known at this time, but it is hoped that a qualitative estimate can easily be made. For $t_p(L) < t < T_0$, overland flow develops and its length grows as seen from Fig. 8.20(c) and (d). At the same time interflow may emerge. As the time progresses, the saturated flow may contribute to overland flow as shown in Fig. 20(d)–(f). Thus the total contribution to streamflow from overland flow in this case consists of the Hortonian overland flow plus the Dunne-Black flow over the domain $x_r \leq x \leq L$, where x_r is the locus of the Dunne-Black flow. The determination of $x_r(t)$ needs special attention.

8.4.7 Source Area Response

It is commonly assumed that surface runoff occurs over the entire watershed area. This is largely based on the assumption that rainfall intensity exceeds the infiltration capacity over the whole watershed. However, in many geographic regions surface runoff is rarely observed. Further, many watersheds yield well-defined hydrographs from storms whose rainfall intensity is less than the infiltration capacity of the soil-cover complex.

Several possible explanations now exist for the sources of water generating a storm hydrograph: (1) Overland flow occurs over the entire area or a portion thereof. (2) Overland flow occurs on an area saturated by downslope groundwater flow as well as infiltrating water. (3) Subsurface flow occurs due to rapid infiltration and transmission through unsaturated and saturated zones. Experimental evidence indicates that one or all of these mechanisms may be effective on a given watershed during a rainfall event or from one event to another.

The mechanisms (1) and (2) can be dynamic in nature although partial areas are thought of as fixed in location (Arteaga and Rantz, 1973; Lane et al., 1978; Pearce and Mc-

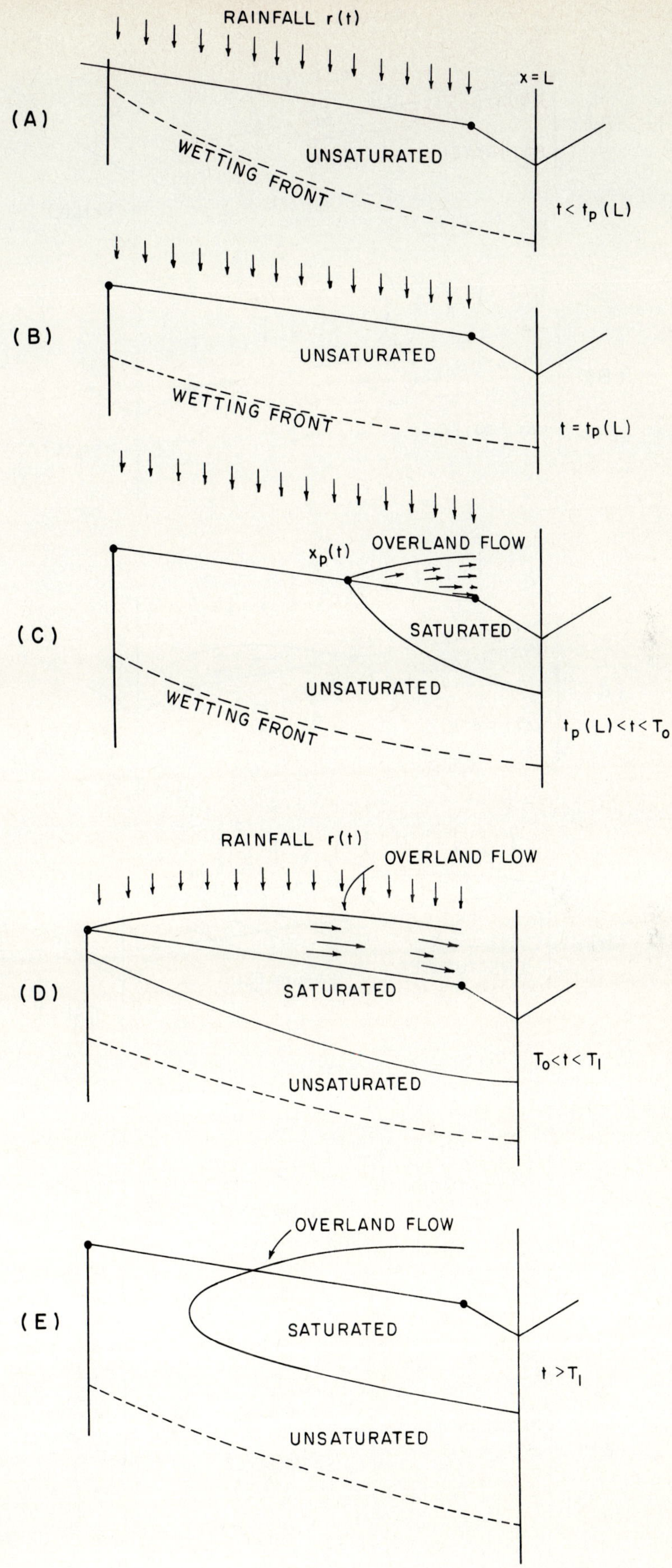

FIGURE 8.19 Evolution of streamflow due to rainfall in an ephemeral drainage basin.

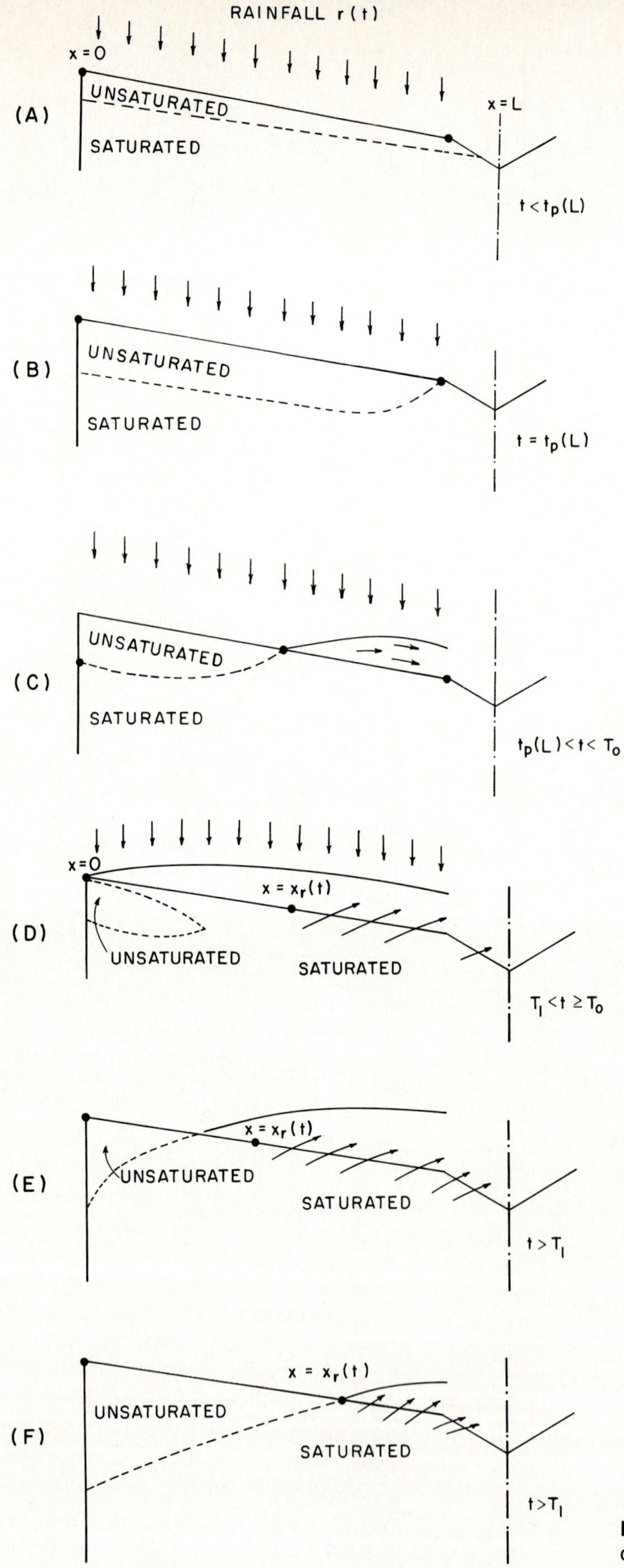

FIGURE 8.20 Evolution of streamflow due to rainfall in a drainage basin.

Kerchar, 1979). Saturated areas grow during a storm and shrink during interstorm periods (Bissell and Ragan, 1969; Lee and Delleur, 1976; Beven, Gilman, and Newson, 1979; Beven and Kirkby, 1979). Therefore, the initial size of contributing area depends on the time history of rainfall during the prestorm period and the growth depends on the time history of rainfall during the storm.

Therefore, the term *partial area response*, or *source area response*, is used to designate the response of a watershed when only a portion of the total drainage area is contributing runoff at a point of interest (Engman and Rogowski, 1974; Lane and Wallace, 1976; Weisman, 1978). As pointed out by Lane et al. (1978), if the watershed is assumed to contribute uniformly over the entire area, whereas actually less than the entire area is contributing, then the assumed specific runoff rates, expressed as runoff per unit area, are in error by the factor of the proportion of area contributing. The errors in estimating surface runoff rates carry over into estimated values of, say, erosion rates.

Excellent surveys of literature pertaining to the source-area concept as well as subsurface streamflow have recently been reported by Freeze (1974), Dunne, Moore, and Taylor (1975), Hewlett and Troendle (1975), Yair and Lavee (1974, 1976), and Troendle (1979). The source-area concepts evolved from the work done by Hewlett (1961), Betson (1964), Amerman (1965), Ragan (1968), Betson and Marius (1969), Dunne (1970), Dunne and Black (1970a, 1970b), and Patten (1975), among others. There exists an argument in the literature between source-area concept and subsurface storm flow contribution (Hewlett, 1961; Hewlett and Hibbert, 1967; Hewlett and Nutter, 1970; Whipkey, 1965), although Dunne and Black (1970a, 1970b) and Freeze (1972a, 1972b) have lent greater credibility to the source-area concept. In particular, Freeze (1972a, 1972b) defined limitations on the significant occurrence of subsurface streamflow.

EXERCISES

8.1 Consider the streamflow hydrograph as given in Table 8.4. Perform the hydrograph separation by four different methods. How different are the results of these methods?

TABLE 8.4

Streamflow hydrograph of the Albert River observed in April, 1972, at Bromfleet, Australia (after the Institution of Engineers, Australia, 1977)

Date	*Time (h)*	*Recorded Hydrograph (m^3/s)*
2	1800	2.12
	1900	2.69
	2000	4.53
	2100	12.72
	2200	25.13
	2300	34.67
	2400	40.37
3	100	50.31
	200	70.90
	300	121.02
	400	212.27
	500	310.74
	600	367.70
	700	351.98
	800	310.74
	900	270.54
	1000	233.82
	1100	199.46
	1200	173.42
	1300	156.00
	1400	147.59
	1500	136.68
	1600	128.61
	1700	19.83
	1800	113.51

TABLE 8.4. *(continued)*

Date	*Time (h)*	*Recorded Hydrograph (m^3/s)*
	1900	102.55
	2000	93.76
	2100	87.79
	2200	77.62
	2300	71.10
	2400	67.42
4	200	53.82
	400	46.06
	600	38.81
	1200	27.96
	1800	21.81
	2400	17.56
5	1200	13.23
	2400	10.36
6	1200	8.61
7	1200	6.94

8.2 Which method of hydrograph separation would you recommend, and why?

8.3 It is reasonable to state that some methods of hydrograph separation are more accurate than others? If yes, which are those methods? If no, what makes them so?

8.4 Consider the inflow-outflow data of Example 8.5. Does the storage in the reach depend on both inflow and outflow or on outflow alone? Support your answer with evidence.

8.5 Assume that the storage in the river reach of Example 8.5 depends on both inflow and outflow in proportion of one to four. Plot storage versus the weighted flow. Can this be approximated by a straight line?

8.6 Equation (8.12) or equation (8.13) may be a reasonable approximation for representing the storage-discharge relation during recession. What kind of storage-discharge relation would you suggest during hydrograph rise? Can you substantiate your answer with data?

8.7 Why do surface runoff, interflow, and baseflow have different recession constants? What are their approximate ranges of values?

8.8 Consider a forested watershed subjected to rainfall. Describe streamflow generation for different rainfall intensities.

8.9 Plot equation (8.6) on a rectangular graph paper and discuss how k_r varies with Δt. What are its practical implications?

8.10 From a practical point of view, how important is the loop behavior of the storage-discharge relationship? How much error is likely to be caused by its linear or curvilinear representation?

CHAPTER NINE

Response Time Characteristics

There are several measures of watershed response time characteristics, as shown in Fig. 9.1. The time parameters most frequently used in hydrology are the time of concentration, lag time, time base, time to equilibrium, time to peak, time of travel, residence time, infiltration opportunity time, and time to ponding.

The time characteristics constitute an important part of hydrologic modeling and design (Schulz and Lopez, 1974). For example, the design of urban drainage systems using the rational formula requires an estimate of the time of concentration to determine the critical rainfall intensity (Rossmiller, 1980; Ben-Zvi, 1984). The Snyder method of unit hydrograph synthesis requires an estimate of the lag time (Snyder, 1938). The kinematic wave flood routing uses the time to equilibrium, which is closely associated with the time of concentration (Overton and Meadows, 1976; Singh, 1976b; Singh and Agiralioglu, 1980). The Muskingum method of flood routing in channels makes use of the time of travel (Singh and McCann, 1979, 1980a, 1980b; Singh and Chowdhury, 1980). The SCR-TR-20 computer model (Soil Conservation Service, 1969), which is used for spillway design and other hydrologic analyses, requires an estimate of the time of concentration. The Clark method of hydrograph synthesis (Clark, 1945) also requires the time of concentration. The design of surface irrigation systems makes use of the infiltration opportunity time (Singh and Ram, 1983). The infiltration-runoff analyses require an estimate of the time to ponding (Smith, 1972; Prasad and Romkens, 1983). Flood-frequency analyses use the time to peak (Rodriguez-Iturbe, Sanabria and Caamano, 1983). Wycoff and Singh (1976) provide a method for estimating the required volume of stormwater detention. Their method uses as input both the time to peak and the time base of a hydrograph. In summary, the bulk of hydrologic analyses require the value of one time parameter or another as input, with the time of concentration and the lag time being most frequently used.

The accuracy of the design flood estimate is directly related to the accuracy with which an estimate of watershed response time is made. Meynink (1978) showed that varying the time of concentration from one half to twice the estimated value changed the peak discharge from 1.64 to .48 times the estimated value for a typical 5-square-kilometer watershed on the Darling Downs in Australia (Table 9.1). The range of half to twice the estimated time of concentration is probably consistent with the error associated with most empirical estimates of response times on ungaged watersheds. For a specified recurrence interval, a first approximation of the design flood estimate is inversely proportional to the response time. For example, if the response time is half the real value, then the flood will be twice the real value, and vice versa. Indeed, little is known about the accuracy of estimates of response times on ungaged watersheds. Empirical expressions are usually used to estimate response time characteristics. We present some of these expressions in the ensuing discussion.

9.1 SYMBOLISM OF RESPONSE TIME CHARACTERISTICS

Referring to Fig. 9.1., we catalog the symbols designating various watershed response time characteristics. Similar but less comprehensive catalogs have been presented by Rao and

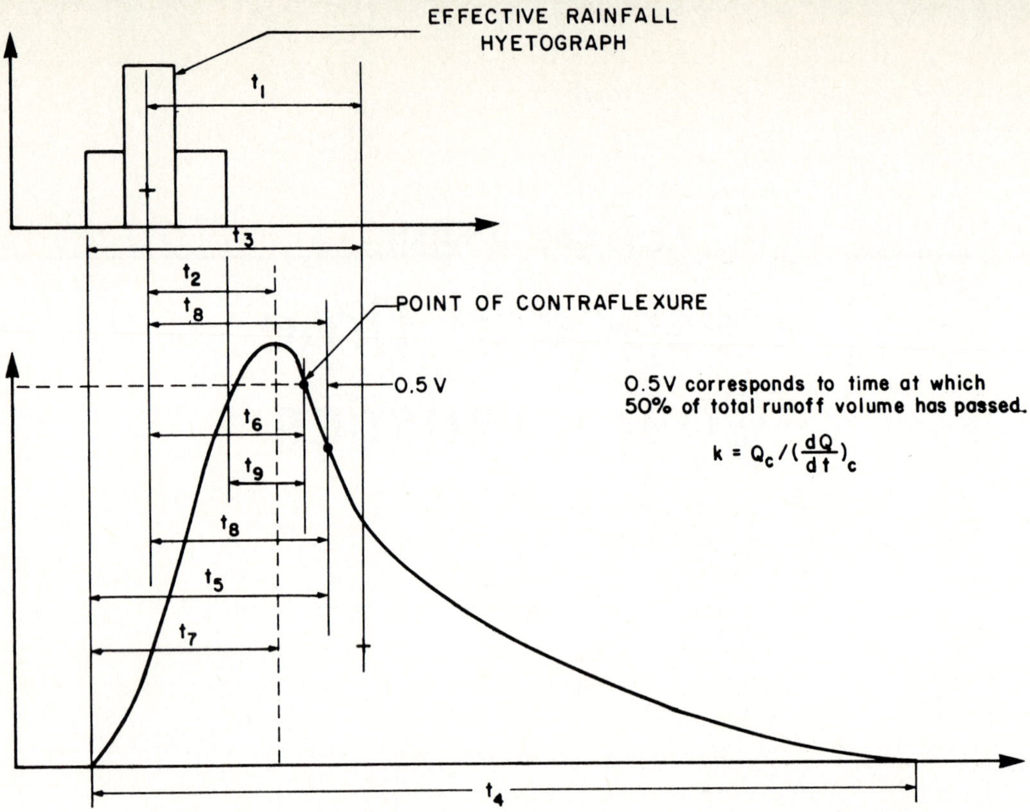

FIGURE 9.1 Definition sketch for various measures of time characteristics.

TABLE 9.1

Effect of time of concentration on peak discharge (after Meynink, 1978)

Time of Concentration (hs)	*Rainfall Intensity (mm/h)*	*Runoff Coefficient*	*Design Peak Discharge (m^3/s)*
1	54	.71	21.3
2	37	.63	13.0
4	24	.51	6.2

Delleur (1974), Schulz and Lopez (1974), and Meynink (1978). (See Box 9–1 for our catalog.)

9.2 FACTORS AFFECTING TIME CHARACTERISTICS

The factors affecting response time characteristics can be classified as (1) physiographic factors, including areal extent, form, slope, surface topographic characteristics, vegetation, and land use; (2) rainfall characteristics, including intensity in time and space, rainfall duration, and direction of the storm; and (3) other factors, such as antecedent soil moisture condition, infiltration characteristics, wind velocity, and weather condition. A time characteristic is generally not a constant for a given watershed. It is constant only by assumption. Following McCuen, Wong, and Rawls (1984), these factors can be classified into five groups: slope, watershed size, flow resistance, channel, and flow. These are either for the overland flow, pipe flow, or channel flow portion of the watershed. Thus, the methods of estimating time characteristics can be distinguished by both their input data requirements and the dominant flow regime, as shown in Table 9.2.

If overland flow dominates on a watershed, then the input data should represent this feature. The length and slope of overland flow are used frequently to characterize it. Measures of flow resistance may include Manning's roughness coefficient n_m, the runoff coefficient of the rational method C, the SCS curve number C_N, the percentage of imperviousness C_u, or the land cover type. A rainfall intensity, such as 2-year, 2-hour rainfall intensity, is often used as input to include the effect of flow on the time of travel.

The channel characteristics should be included in the input data if the channel flow makes a significant contribution to the total travel time of runoff. The length L_c and slope S_c of the main channel are used most frequently to incorporate the effect of slope and size. Manning's or Chezy's roughness coefficient is the most widely used index of flow resistance. Espey and Winslow (1968) used a coefficient C_f to indicate the degree of channelization, which can also be considered as an indicator of flow resistance. The rainfall intensity, volume of direct runoff, or hydraulic radius R corresponding to full flow can be used to represent the flow characteristics.

In many small watersheds, the pipe flow plays a significant role in governing runoff characteristics. The input data

BOX 9–1

Symbol	Explanation	References
t_1	*Lag time*: time interval between the centroid of effective rainfall and the centroid of direct runoff.	Horner and Flynt (1936); Mitchell (1948)
t_2	*Lag time*: time interval between the centroid of effective rainfall and peak of direct runoff.	Snyder (1938); Taylor and Schwarz (1952); Eagleson (1962)
t_3	*Lag time*: time interval between the beginning of effective rainfall and the centroid of direct runoff.	Wilson (1972); Schulz and Lopez (1974); U.S. Army Corps of Engineers (1949)
t_4	*Lag time*: time interval between the beginning of direct runoff and the end of direct runoff or the time base of direct runoff hydrograph.	Snyder (1938); Meynink (1978)
t_5	*Lag time*: time interval between the beginning of effective rainfall and the time when 50 percent of direct runoff has passed the gaging station.	Rao and Delleur (1974)
t_6	*Lag time*: time interval between the centroid of effective rainfall and the point of contraflexure on direct runoff recession.	Rao and Delleur (1974)
t_7	*Lag time* or *time to peak*: time interval between the beginning of effective rainfall and the peak of direct runoff.	Linsley, Kohler and Paulhus (1958)
t_8	*Lag time*: time interval between the centroid of effective rainfall and the time when 50 percent of direct runoff has passed the gaging station.	Wilson (1972); U.S. Bureau of Reclamation (1965)
t_9	*Time of concentration*: time interval between the end of effective rainfall and the point of contraflexure on direct runoff recession.	Viessman, et al. (1978); McCuen, Wong and Rawls (1984)
t_{10}	Time interval between the beginning of rainfall (not effective rainfall) and the peak of runoff (not direct runoff).	Lopez (1973)
t_{11}	Time interval between the centroid of effective rainfall and the peak of unit hydrograph.	Lopez (1973)
t_{12}	Time interval between the beginning of effective rainfall and the peak of unit hydrograph.	Lopez (1973)
t_{13}	Time interval required for the hydrograph to rise from low flow to the maximum stage (might be equivalent to t_7).	Ramser (1927); Kirpich (1940); Gray (1961); Wu (1969a)
t_{14}	*Equilibrium time*: time interval required for the runoff rate to become equal to the supply rate (rainfall intensity).	Izzard (1946); Wei and Larson (1971)
t_{15}	*Lag time*: time interval between 50 percent of effective rainfall and 50 percent of direct runoff.	Overton (1971); Singh (1975); Agiralioglu and Singh (1981); Singh and Agiralioglu (1982)
t_{16}	*Time to ponding*: time interval between the start of rainfall and the start of direct runoff.	Smith (1972)
t_{17}	*Infiltration opportunity time*: the time period for which water is available for infiltration at a given point in space.	Singh and Sherman (1983)

then should reflect this feature. Flow resistance can be estimated by Manning's, Chezy's, or Darcy-Weisbach's friction factor. Slope and size can be represented in a variety of ways. The flow characteristics can be reflected by a maximum flow rate Q_p, the largest or smallest pipe diameter, or mean hydraulic radius.

In a real-world situation, more than one flow regime may exist. For example, in an urbanized watershed all three flow regimes may exist: overland, channel, and pipe flows. Consequently, a method of estimating a time characteristic must include factors representing these flow regimes. In other words, the input data must reflect these features.

TABLE 9.2

Criteria for classifying time parameters and variables commonly used to represent the input data (adapted from McCuen, Wong, and Rawls, 1984)

	Input Data				
Flow Regime	*Flow Resistance*	*Watershed Size*	*Slope*	*Channel*	*Flow*
Overland	n, C, CN, C_u	L, A	S	—	I
Channel	c_z, n_m, C_f	L_c, $L_{10\text{–}85}$, L_{ca}	S_c, $S_{10\text{–}85}$	R	I, Q
Pipe	n	L	S	R	Q_p

A = drainage area
c_z = Chezy's roughness coefficient
C = runoff coefficient of the rational method
C_f = channelization factor
CN = runoff curve number
C_u = percentage of imperviousness
I = effective rainfall intensity for a duration equal to the time of concentration
L = length of overland flow
L_c = length of the main channel
L_{ca} = length to center of area
$L_{10\text{–}85}$ = length of the main channel between points 10 and 85 percent of the total length from the watershed outlet
n_m = Manning's roughness coefficient
Q = flow rate
Q_P = maximum flow rate
R = hydraulic radius
S = slope
S_c = slope of main channel

9.3 TIME OF CONCENTRATION, T_c

The time of concentration T_c is the most frequently used time parameter. Like other parameters it is usually defined in terms of either the physical characteristics of a watershed or the effective rainfall hyetograph and direct runoff hydrograph. There are two commonly accepted definitions of T_c.

First, T_c is the time required for water to travel from the most remote portion of the watershed to its outlet or design point. The methods of estimation based on this definition use watershed characteristics and sometimes a precipitation index such as the 2-year, 2-hour rainfall intensity. A number of empirical equations based on this definition are given in the ensuing discussion.

The second definition of T_c is based on a rainfall hyetograph and the resulting runoff hydrograph. The effective rainfall hyetograph and direct runoff hydrograph are computed from these observations. The time of concentration is the time between the center of mass of effective rainfall and the inflection point on the recession of direct runoff hydrograph. Alternatively, T_c is often considered as the time from the end of effective rainfall to the point of inflection on direct runoff hydrograph recession, as shown in Fig. 9.1 (Horton, 1939).

Neither definition yields a true value or even a reproducible value of T_c due to the difficulty of uniquely defining and then measuring the factors affecting it.

9.4 METHODS OF ESTIMATING T_c

There exist a multitude of methods for estimating T_c. Most of these methods can be classified by the input data required and the flow regime considered by them. These methods and their input data are shown in Table 9.2. Some of these methods are designed primarily for overland flow, some primarily for channel flow, and some for both channel and overland flows. Some of these methods are presented briefly here.

Most of the equations for estimating T_c can be expressed as a function of a length measure L_p, a slope measure S_p, and a coefficient C_p, which is a measure of surficial watershed characteristics and which may or may not be a constant,

$$T_c = C_p L_p^a S_p^b \tag{9.1}$$

in which a and b are exponents that may vary from one watershed to the other. These equations have been derived either empirically or by using equations of open channel hydraulics.

9.4.1 Hydraulically Derived Equations

Various forms of equations of hydraulics can be applied to the estimate T_c provided the watershed geometry can be represented in a simple manner and the friction characteristics are known (Singh, 1976b; Chen and Evans, 1977; Singh and Agiralioglu, 1981; Hirano, 1980, Gregory, 1982). For example, Henderson and Wooding (1964) derived the time of concentration for overland flow on a plane using kinematic wave theory, which can be stated as

$$T_c = I^{(1-m)/m}\left(\frac{L}{\alpha}\right)^{1/m} \tag{9.2}$$

where m and α are parameters in a kinematic depth-discharge relationship. If we use the Chezy relation, then

$$\alpha = c_z S^{0.5} \quad \text{and} \quad m = 1.5 \tag{9.3}$$

If the Manning relation is used, then

$$\alpha = \frac{1}{n_m} S^{0.5} \quad \text{and} \quad m = \frac{5}{3} \tag{9.4}$$

where c_z is the Chezy friction coefficient, n_m is the Manning friction coefficient, L is the length of the plane, and S is the ground slope. The overland flow lengths ranged from 50 to

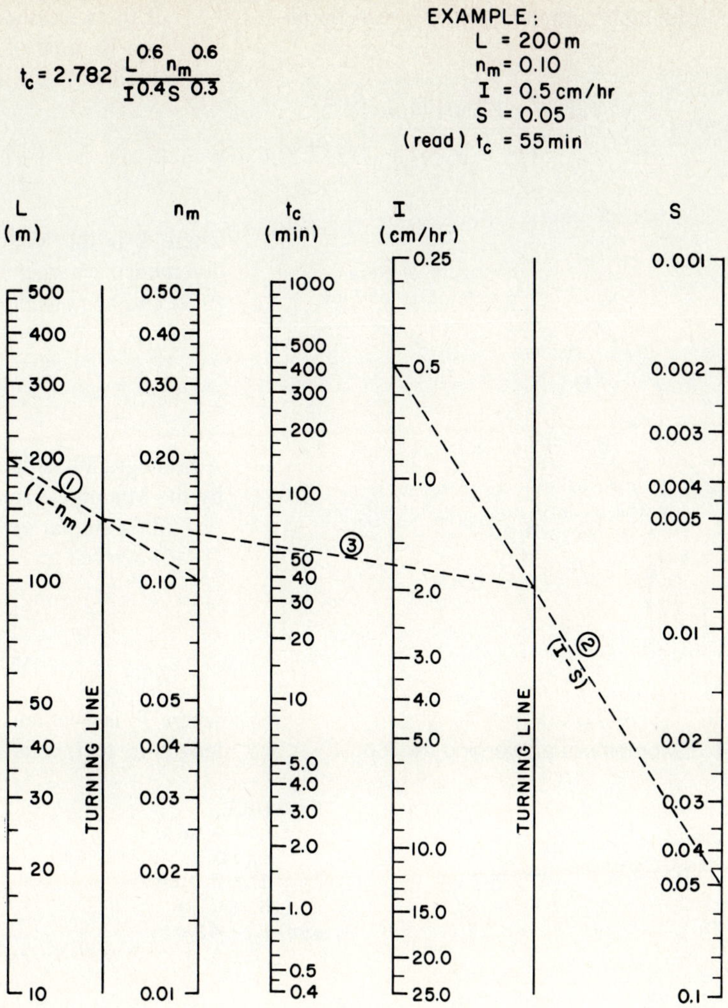

FIGURE 9.2 Nomograph for time of concentration of a plane using Manning's roughness coefficient (after Agiralioglu and Singh, 1981).

100 feet. Ragan and Duru (1972) developed a nomograph to compute T_c using equation (9.2) in conjunction with equation (9.4). Agiralioglu and Singh (1981) developed nomographs for equation (9.2) with α estimated by Manning's, Chezy's, and Darcy-Weisbach's roughness coefficients. Fig. 9.2 is a nomograph with the Manning coefficient.

Singh (1976) derived T_c using kinematic wave theory for plane and converging geometries for different conditions imposed on effective rainfall and friction parameters. For a converging geometry, as shown in Fig. 9.3, the relation for T_c can be stated as

$$T_c = \frac{1}{m} I^{(1-m)/m} \left(\frac{L}{2\alpha}\right)^{1/m} [B(a;b) - B_w(a;b)] \tag{9.5}$$

where $a = 1 - (.5/m)$, $b = 1/m$, $w = r^2$, r is the convergence parameter, and L the length of the section. The quantity B without a subscript denotes a complete beta function and can be expressed in terms of the gamma function as

$$B(a, b) = \frac{\Gamma(a)\Gamma(b)}{\Gamma(a + b)} \tag{9.6}$$

However, B with a subscript denotes an incomplete beta function and has the following mathematical connotation:

$$B_w(a, b) = \int_a^w x^{a-1} (1 - x)^{b-1}\, dx, \qquad a > 0, b > 0, x \in (0, 1)$$

$$= \frac{w^a(1 - w)^b}{a} \left\{1 + \sum_{j=0}^{\infty} \frac{B(a + 1; j + 1)}{B(a + b; j + 1)} w^{j+1}\right\} \tag{9.7}$$

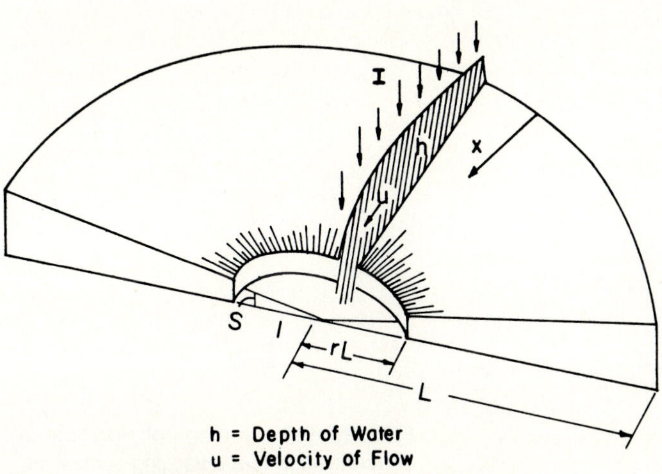

FIGURE 9.3 Geometry of a converging section.

Equation (9.5) is considerably simplified when r can be assumed to be small,

$$T_c = \frac{1}{m} I^{(1-m)/m} \left(\frac{L}{2\alpha}\right)^{1/m} \left[\frac{\Gamma(1 - 1/2m)\Gamma(1/m)}{\Gamma(1 + 1/2m)}\right] \tag{9.8}$$

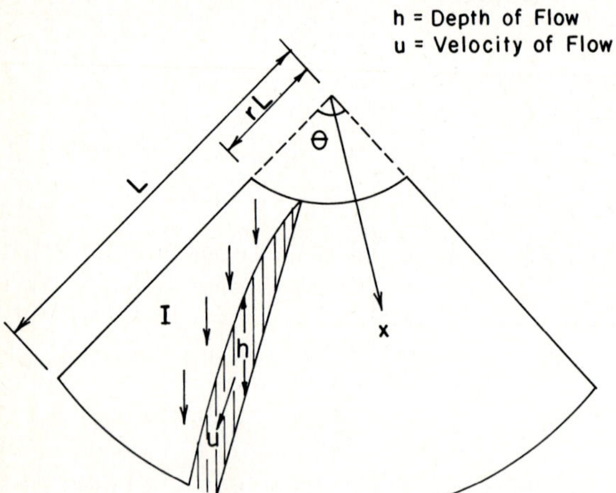

FIGURE 9.4 Geometry of a diverging section.

If the watershed is linearly diverging, as shown in Fig. 9.4, then its time of concentration, as derived by Singh and Agiralioglu (1980), is given by

$$T_c = \left(\frac{T}{2}\right)^{(1-m)/m} \left(\frac{1}{\alpha}\right)^{1/m} \int_{rL}^{L} \left(\frac{x^2 - r^2L^2}{x}\right)^{(1-m)/m} dx \tag{9.9}$$

where L is the length of the diverging section and r is the divergence parameter. If r is assumed to be small, equation (9.9) can be considerably simplified:

$$T_c = m \left(\frac{I}{2}\right)^{(1-m)/m} \left(\frac{1}{\alpha}\right)^{1/m} L^{1/m} \tag{9.10}$$

A nomographic solution of equation (9.9), with α represented by the Manning's roughness coefficient, is shown in Fig. 9.5.

In a similar vein, using hydraulic considerations, Izzard (1946) derived an expression for T_c as

$$T_c = \frac{1}{60}\left(\frac{.007I + c}{S^{1/3}}\right) L \left(\frac{IL}{43{,}200}\right)^{-2/3} \tag{9.11}$$

where T_c is overland flow time in minutes, L is the watershed length in feet, S is the slope of overland flow path in feet

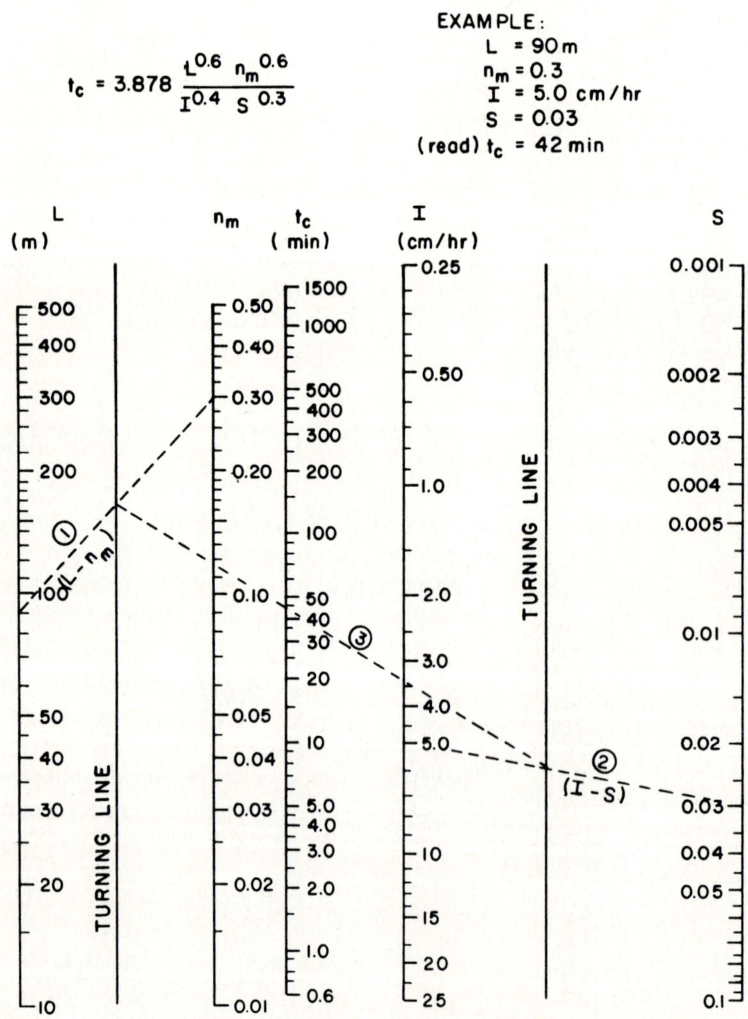

FIGURE 9.5 Nomograph for time of concentration of a diverging section using Manning's roughness coefficient (after Agiralioglu and Singh, 1981).

TABLE 9.3

Values of *c* for Izzard's formula

Surface	*Value*
Smooth asphalt surface	.007
Concrete pavement	.012
Tar and gravel pavement	.017
Closely slipped soil	.046
Dense bluegrass turf	.060

per foot, I is the effective rainfall in inches per hour, and c is the surface roughness coefficient. The values of c are given in Table 9.3. Equation (9.11) is valid for laminar overland flow for which $IL < 500$ and was derived for developed surfaces. This equation should be used cautiously for slopes steeper than .04. Equations similar to equation (9.11) were also derived by Butler (1977a, 1977b, 1980). Gill (1977a) argued in a discussion of the paper by Butler (1977a) that the assumption of considering slope to be invariant with distance was unjustified. However, in a reply Butler (1977b) showed by using experimental data that replacing friction slope with bed slope caused an error of less than 1 percent and was thus reasonable. Butler extended his derivations to incorporate both laminar and turbulent flows taking place on the watershed plane. It may be noted that equation (9.11) is similar to (9.2). Indeed the former can be derived from the latter. A derivation of equation (9.11) is given in Appendix E.

9.4.2 Empirically Derived Equations

Since T_c is most extensively used in the rational method, for which the basis is empirical, a simple relation for T_c may be perhaps more appropriate. The Bransby-Williams formula (Williams, 1922) is one such relation, which can be written for T_c in hours as

$$T_c = \frac{LA^{.4}}{DS^{.2}} \tag{9.12}$$

where L is the longest distance in miles from the edge of the drainage basin to its outfall, A is the area of the basin in square miles, D is the diameter in miles of a circular basin whose area is A, and S is the average fall in feet per 100 feet of the ground measured along the most direct route to the outfall. The drainage area can be up to 50 square miles.

Another simple relation is the Kirpich formula (Kirpich, 1940), which he calibrated for small watersheds in Pennsylvania and Tennessee. His formula for the Tennessee watersheds (ranging in size from 1 to 112 acres; slope varying from 3 to 10 percent) is

$$T_c = 13 \times 10^{-5} \frac{L^{.77}}{S^{.385}} \tag{9.13}$$

and for the Pennsylvania watersheds is

$$T_c = 21.67 \times 10^{-6} \frac{L^{.77}}{S^{.5}} \tag{9.14}$$

where L is the length of the watershed in feet, measured along the channel from the outlet and in a direct line from the upper end of the channel to the farthest point on the watershed, and S is the dimensionless ratio, the elevation difference between the outfall and the farthest point. Thus S is approximately the average slope of the basin. If the slope is nonuniformly changing along L, a weighted S may be desirable. T_c should be multiplied by .4 for watersheds where the overland flow path is either concrete or asphalt and by .2 for watersheds where the channel is concrete-lined. This is classified as a channel-flow method. However, given the relatively small drainage areas, it should reflect significant portions of overland-flow travel time.

Based on an analysis of characteristics of 19 watersheds having areas between 25 and 1624 square miles in the Scotie and Sandusky River watersheds, Johnstone and Cross (1949) developed the following relationship for T_c:

$$T_c = \frac{4.7}{r^2}\left(\frac{L}{S}\right)^{0.5} \tag{9.15}$$

where T_c is measured in hours, L (length of the principal stream in the watershed) is in miles, S (average slope of the mainstream) is in feet per mile, and r is a branching factor based on the stream pattern. They found a good approximation of equation (9.15) by neglecting r:

$$T_c = 5\left(\frac{L}{S}\right)^{.5} \tag{9.16}$$

A similar relation for T_c was developed by Eaton (1954), who performed a correlation study for seven Tasmanian rivers having watershed areas from 48 to 322 square miles. His equation is

$$T_c = 1.35\left(\frac{AL}{r}\right)^{.37} \tag{9.17}$$

where the notations have the same meaning as before.

Kerby (1959), following the work of Hathaway (1945), proposed an empirical relation and developed a nomograph for determining the time of concentration for overland flow. His relation can be expressed as

$$T_c = \left(\frac{2LN}{3S^{.5}}\right)^{1/2.14} \tag{9.18}$$

where T_c is the time of concentration for overland flow in minutes, L is the length in feet of overland flow defined as the distance from the extremity of the drainage area in a direction parallel to the slope until a defined channel is reached, S is the difference in elevation between the extreme edge and the end of overland flow length divided by L, and N is the average surface retardance coefficient of overland flow. This is classified as a surface-flow method. Watersheds ranging in areas less than 10 acres, slopes less than 1 percent, and N less than .8 were used. Table 9.4 gives values of N for some surfaces. Kerby (1959) noted that L would be within 1200

TABLE 9.4

Values of *N* for Kerby's formula

Type of Surface	*N*
Smooth impervious surface	.02
Smooth, bare packed soil	.10
Poor grass, cultivated row crops, or moderately rough bare surfaces	.20
Deciduous timberland	.60
Pasture or average grass	.40
Conifer timberland, deciduous timberland with deep forest litter, or dense grass	.80

feet. Similar relations and nomographs have also been developed by Mockus (1961), Seelye (1968), and the Denver Regional Council of Governments (1969), as shown in Fig. 9.6.

The Federal Aviation Agency (1970) developed an equation for T_c from airfield drainage data:

$$T_c = 0.03(1.1 - C)L^{.5}S^{.333} \tag{9.19}$$

in which C is the runoff coefficient of the rational method, L is the maximum length of overland flow in feet, and S is the slope in percent of the longest overland flow path. This is a surface-flow method.

An equation of T_c developed by the Soil Conservation Service (1975) is

$$T_c = 10^{-5} \times 27.78 \sum_i (LKS^{-.5})_i \tag{9.20}$$

in which L is the length of overland flow in feet, K is an empirical coefficient indicative of land cover, and S is the slope in percent. T_c is regarded as the ratio of the length and velocity. A watershed is divided into a number of flow segments. The graph, shown in Fig. 9.7 and provided in TR-55 (Soil Conservation Service, 1975) for the SCS velocity method, is used to compute the velocity of flow through the segment. A travel time is obtained for each flow segment. The time of concentration is set equal to the sum of the segment travel times. This is basically an overland-flow method but can be extended to include channel flow.

McCuen, Wong, and Rawls (1983) calibrated the following equations for urban areas:

$$T_c = .01462L_f^{.5552}I_2^{-.7164}S_{fm}^{-.207} \tag{9.21}$$

and

$$T_c = .0469L_f^{.445}I_2^{-.7231}C_f^{.5517}S_{fm}^{-.226} \tag{9.22}$$

where L_f is the total length of flow path in feet, I_2 is 2-year, and T_c-hour rainfall intensity in inches per hour, S_{fm} is the channel slope in feet per mile, and C_f is the Espey channelization coefficient (Espey, Morgan, and Masch, 1965). These equations include both overland flow and channel flow regimes. Equation (9.22) includes C_f, which is a flow-resistance variable. Equation (9.21) should be used except when the channel is undergoing a hydraulic change or is significantly different from the average channel for an area; in such cases, an estimate of C_f should be made, and equation (9.22) should be used to estimate T_c.

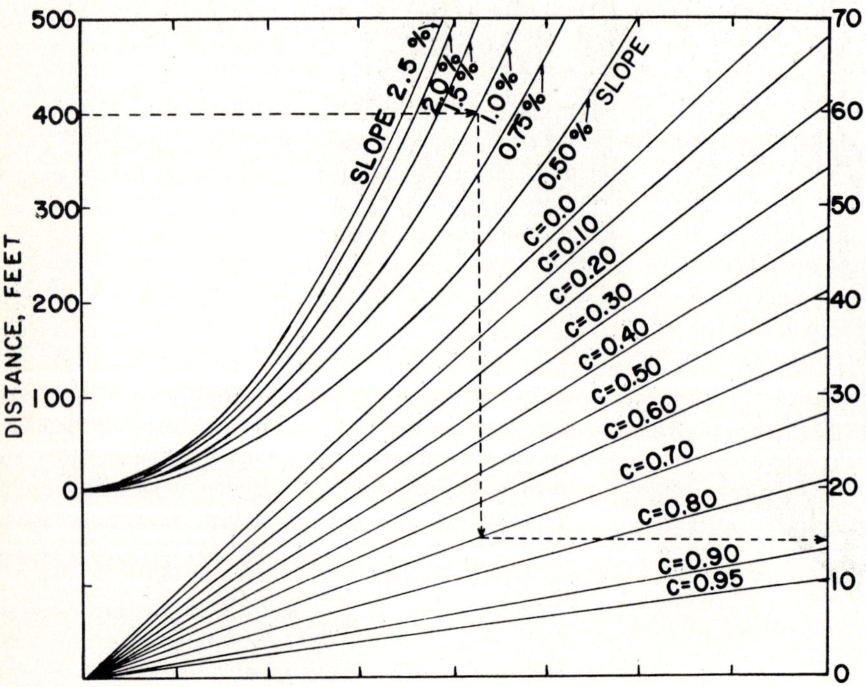

FIGURE 9.6 Overland flow time curves (after Denver Regional Council of Governments, 1969).

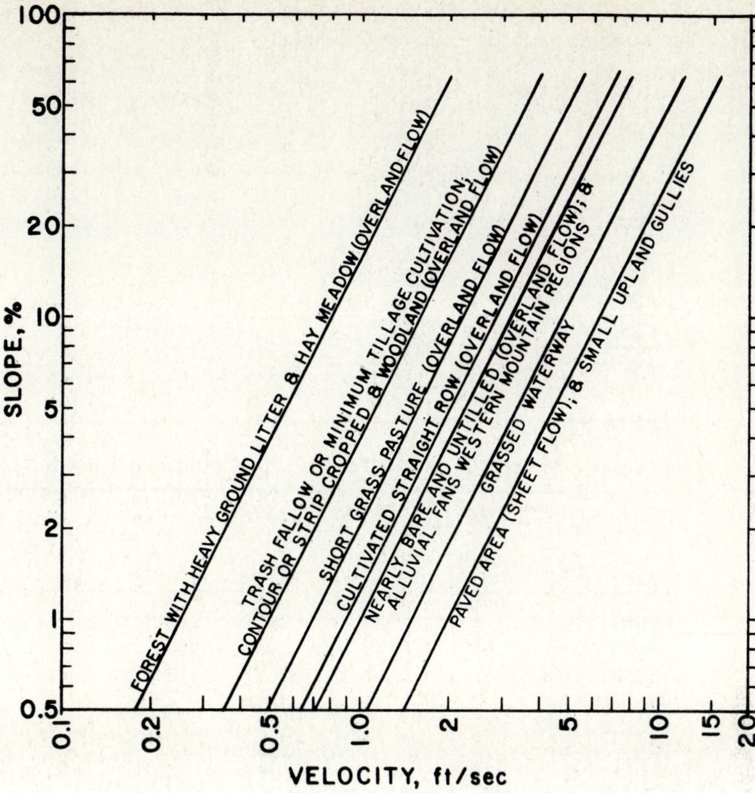

FIGURE 9.7 Velocities for estimating overland flow time (after Soil Conservation Service, 1972).

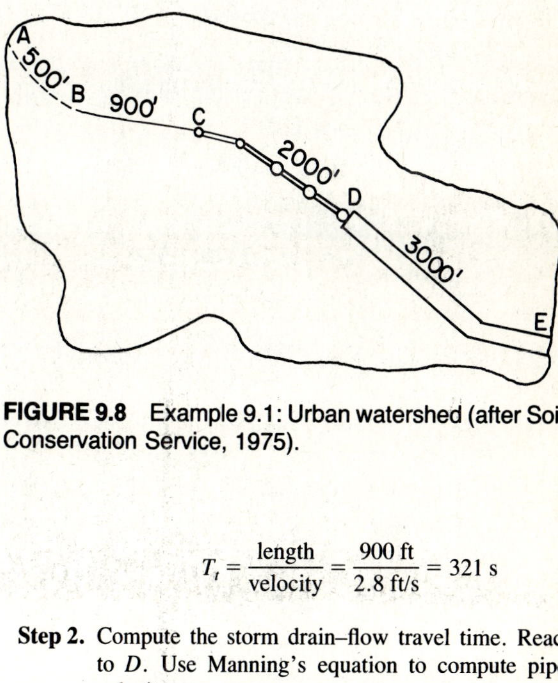

FIGURE 9.8 Example 9.1: Urban watershed (after Soil Conservation Service, 1975).

EXAMPLE 9.1

An urbanized watershed is shown in Fig. 9.8. Three types of flow conditions exist from the most remote point of the watershed to the outlet. This example is taken from Soil Conservation Service (1975). Compute the travel time (T_t) and time of concentration (T_c) based on the following data:

Reach	Description of Flow	Slope (%)	Length (ft)
A to *B*	Overland (forest)	7	500
B to *C*	Overland (shallow gutter)	2	900
C to *D*	Storm drain with manhole covers, inlets, and so on (n_m = .015; diameter 3 feet)	1.5	2000
D to *E*	Open channel, gunite trapezoidal (b = 5; d = 3; z = 1.1; n_m = .019)	.5	3000

Step 1. Compute the overland flow travel time. Reach *A* to *B* (forest cover). From Fig. 9.7 for a slope of 7 percent, read v = 0.7 ft/s.

$$T_t = \frac{\text{length}}{\text{velocity}} = \frac{500 \text{ ft}}{.7 \text{ ft/s}} = 714 \text{ s}$$

Reach *B* to *C* (street gutter). From Fig. 9.7 for a slope of 2 percent, read v = 2.8 ft/s.

$$T_t = \frac{\text{length}}{\text{velocity}} = \frac{900 \text{ ft}}{2.8 \text{ ft/s}} = 321 \text{ s}$$

Step 2. Compute the storm drain–flow travel time. Reach *C* to *D*. Use Manning's equation to compute pipefull velocity.

$$v = \frac{1.49}{n_m}\left(\frac{D}{4}\right)^{2/3} S^{1/2}$$

$$v = \frac{1.49}{.015}\left(\frac{3}{4}\right)^{2/3} (.015)^{1/2} = 10 \text{ ft/s}$$

$$T_t = \frac{\text{length}}{\text{velocity}} = \frac{2000 \text{ ft}}{10 \text{ ft/s}} = 200 \text{ s}$$

Step 3. Compute the open-channel-flow travel time. Reach D to E. Use Manning's equation to compute bankfull velocity.

$$v = \frac{1.49}{n_m} R^{2/3} S^{1/2}$$

$n = .019$ for gunite channel

$S = .005$

$$v = \frac{1.49}{.019}(1.78)^{2/3}(.005)^{1/2} = 8.2 \text{ ft/s}$$

$$T_t = \frac{\text{length}}{\text{velocity}} = \frac{3000 \text{ ft}}{8.2 \text{ ft/s}} = 366 \text{ s}$$

Step 4. Summary

Reach	Description of flow	Length (ft)	Velocity (ft/s)	Travel time(s)
A to B	Overland	500	.7	714
B to C	Overland	900	2.8	321
C to D	Storm drain	2000	10.0	200
D to E	Open channel	3000	8.2	366
			Total	1601

Thus

$$T_c = \frac{1601 \text{ s}}{(3600 \text{ s/h})} = .44 \text{ h}$$

TABLE 9.5

Equations of lag time in hours according to definition 1 (t_1 from Fig. 9.1)

Author	*Equation*	*Equation Number for Referencing*
Clark (1945)	$t_1 = KL/\sqrt{S_c}, \quad k \in [0.8, 2.2]$	(9.24)
Linsley (1945)	$t_1 = KL\sqrt{A/S_c}$	(9.25)
Mitchell (1948)	$t_1 = KA^{.6}, \quad K = 1.05$	(9.26)
Johnstone and Cross (1949)	$t_1 = K + 90\frac{W}{S}, \quad K = 1.5$	(9.27)
Eaton (1954)	$t_1 = K(WA/(Lr))^{1/3}, \quad K = 1.2, \quad r \in [1, 2]$	(9.28)
O'Kelly (1955)	$t_1 = KS_0^d$	(9.29)
Dooge (1955)	$t_1 = KA^{.25}/\sqrt{S_0}, \quad K \in [10, 13]$	(9.30)
Hoyt and Langbein (1955)	$t_1 = KA^{.4}, \quad K \in [1, 3]$	(9.31)
Nash (1960)	$t_1 = KA^{.3} S_0^{-.3}, \quad K = 27.6$	(9.32)
	$t_1 = KL^{0.3} S_c^{-.33}, \quad K = 20$	(9.33)
Carter (1961)	$t_1 = K(L/\sqrt{S_a})^{.6}, \quad K = 1.7$	(9.34)
Morgan and Johnson (1962)	$t_1 = Kt_2^{.81}, \quad K = 2.8$	(9.35)
	$T_1 = KA^{.6}, \quad K = 105$	
Wu (1963)	$t_1 = KA^{.94} L^{-1.47} S_c^{-1.47}, \quad K = 780$	(9.36)
Bell (1967	$t_1 = KA^{.33}, \quad K \in [0.5, 3]$	(9.37)
Kennedy and Watt (1967)	$t_1 = KL^{.66} S^{-.33} S_b^{1.21}, \quad K = 6.71$	(9.38)
Cordery (1968)	$t_1 = K\left[\frac{W}{\sqrt{S_0}} + \frac{L_{nm}}{2\sqrt{S_c}}\right]$	(9.39)
Wu (1969)	$t_1 = KA^{0.23}$	(9.40)
Bell and Omkar (1969)	$t_1 = KA^{0.33}$	(9.41)
Rao and Delleur (1974)	$t_1 = .78\,A^{.496} L^{.073} S^{-.075} (1 + I_A)^{-1.289}$	(9.42)
	$t_1 = .78A^{.542} S^{-.081} (1 + I_A)^{-1.21}$	(9.43)
	$t_1 = .803A^{.512} (1 + I_A)^{-1.433}$	(9.44)
Boyd (1978)	$t_1 = 1.066\,(2\mu - 1)^{.404}$	(9.45)
	$t_1 = 2.51\,A^{.38}$	(9.46)

A = watershed area in square miles
A_m = area in square kilometers
D_r = duration of the effective rainfall in hours
I_A = percentage impervious area
I_d = depth of the effective rainfall in millimeters
K = constant
L = mainstream length in miles
L_r = watershed lag ratio
n_m = manning roughness coefficient
r = branching factor
S = mean basin slope in feet per mile
S_a = mainstream slope calculated by equal area method in feet per mile
S_b = A dimensionless factor representing the ratio of area A_b of lakes, marshes, and ponds in the upper two-thirds of the watershed to the watershed area, $S_b = 1 + 20(A_b/A)$
S_c = mean slope of the mainstream in feet per mile
S_d = measure of overland slope, a dimensionless ratio
S_0 = average overland slope in parts per 10,000
$(2\mu - 1)$ = total number of interior and exterior links in the channel network
W = average width of watershed in miles

9.5 CHOOSING BETWEEN EQUATIONS

McCuen, Wong, and Rawls (1984) evaluated 11 equations for estimating T_c from 48 urban watersheds located in most regions of the United States. These equations are (9.2), (9.13), (9.14), (9.18), (9.19), (9.20), (9.21), and (9.22). The watersheds had areas of less than 4000 acres, with average impervious area of 29.1 percent and the mean time of concentration 1.49 hours, having a range from 0.21 to 6.14 hours. These equations resulted in errors in the estimated value of T_c that were greater than 30 minutes for more than 50 percent of the watersheds. The maximum errors for these equations ranged from 1.48 to 10.29 hours. The standard deviation of the errors ranged from .37 to 2.27 hours, with only 1 of the 11 equations having a value less than 1 hour. If such errors are translated into errors in design discharges, then the design errors can be significant indeed. Given the wide variation in the estimated T_c values and the very large errors, which equation should be used? For watersheds having nonhomogeneous land uses and significant channel flow, a mixed method is certainly preferable. A velocity method is most likely to provide an accurate estimate for such watersheds. Equations (9.2) and (9.20) represent such a method. Similarity in watershed characteristics between the watershed where a design is required and the watersheds used to calibrate an equation is perhaps the most important criterion for selecting an equation. The other considerations are availability of data and physical characteristics of the watershed. The equation(s) selected for computing T_c must reflect pertinent flow and physiographic characteristics.

9.6 LAG TIME

There are many definitions of lag time, as evident from the foregoing discussion. Horner and Flynt (1936) remarked, "The term, 'lag' . . . has reference only to the difference in phase between salient features of the rainfall and runoff curves, . . . Values, subject to correction, are taken from the time difference between salient features . . . , or between the centers of mass of the rainfall and runoff rate curves." This time difference between salient features appears to have been interpreted in many different ways. We briefly outline some of the more frequently utilized interpretaions.

9.6.1 Definition 1

Perhaps the most popular definition of lag time is the time difference between the centroid of effective rainfall and that of direct runoff. If $h(t)$ is the IUH, then its first moment defines the lage time,

$$T_l = \int_0^\infty t\, h(t)\, dt \tag{9.23}$$

Further, if a single linear reservoir is used to define the IUH, then its storage coefficient k specifies the lag time; that is, $T_l = k$. The lag time defined in this way appears to be the most stable measure (Schulz and Lopez, 1974; Espey, Morgan, and Masch, 1965). Many attempts have been made to estimate T_l empirically in terms of watershed characteristics. The resulting empirical equations are given in Table 9.5.

9.6.2 Definition 2

Another frequently employed definition of lag time is the time difference between the centroid of the effective rainfall and the peak of direct runoff. Several empirical equations have been derived to relate lag time T_p to watershed characteristics. Some of these equations are listed in Table 9.6. The lag time defined in this way is more variable than the one of definition 1 above.

9.6.3 Definition 3

This definition is due to Overton (1970), who defined the lag time as the lapse between the time of occurrence of 50 percent input volume and 50 percent output volume. For the effective rainfall resulting in equilibrium hydrograph, Overton derived this lag time T_l to be the ratio of storage at the equilibrium condition V_e to the intensity of effective rainfall I,

$$T_l = \frac{V_e}{I} \tag{9.55}$$

TABLE 9.6

Equations of lag time in hours according to Definition 2 (t_2 from Fig. 9.1)

Author	*Equation*	*Equation Number for Referencing*
Snyder (1938)	$t_2 = C_t(LL_c)^{.3}$, $\quad C_t \in [1.8, 2.2]$	(9.47)
Linsley (1943)	$t_2 = C_t(LL_c)^{.3}$, $\quad C_t \in [0.3, 1.2]$	(9.48)
Taylor and Schwarz (1952)	$t_2 = C\exp(m_1 D)$, $\quad C = \frac{.6}{\sqrt{S}},$ $\quad m_1 = 0.212(LL_c)^{-.36}$	(9.49)
Mockus (1957)	$t_2 = \frac{K L_w^{.8}(M+1)^{1.67}}{S_0^{.5}}$, $\quad K = \frac{1}{1900}$	(9.50)
Hickok, Keppel, and Rafferty (1959)	$t_2 = C\left(\frac{A^{.3}}{S_0 D^{.5}}\right)^{.61}$	(9.51)
	$t_2 = C\left[\frac{(L_c + W)^{.5}}{S_0 D^{.5}}\right]^{.65}$	(9.52)
Eagleson (1962)	$t_2 = \frac{L_a n_m}{1.5} R_a^{2/3} S_a^{-1/2}$	(9.53)
Soil Conservation Service (1975)	$t_2 = \frac{L_w^{.8}(1000/CN - 9)^{.7}}{1900\, S_0^{.5}}$	(9.54)

CN = runoff curve number
D = drainage density
L = length of the main channel in miles
L_a = mean travel distance in feet equal to that portion of the sewer which flows full
L_c = distance in miles up the main stream to the center of area
L_w = hydraulic length of watershed in feet
M = soil cover parameter
n_m = Manning's roughness factor
R_a = weighted hydraulic radius in feet of the main sewer flowing full
S = average channel slope
S_0 = average watershed slope in percent
S_a = weighted slope of the main sewer in feet per foot
W = watershed width

Using kinematic wave equations, T_l was derived by Overton (1971) for a plane, by Singh (1975) for a converging surface, and by Singh and Agiralioglu (1980) for a diverging surface. For a plane surface, T_l can be expressed as

$$T_l = \frac{m}{m+1}\left[\frac{LI^{1-m}}{\alpha}\right]^{1/m} \tag{9.56}$$

where m is an exponent and α is the resistance parameter in kinematic depth-discharge relationship. A nomographic solution is presented in Fig. 9.9 (Agiralioglu and Singh, 1981).

For a converging surface,

$$T_l = \left\{\left[\frac{LI}{2\alpha}\right]^{1/m} \frac{1}{2\,I(1-r)}\left[\frac{\Gamma(1+1/m)\,\Gamma(1-1/2m)}{\Gamma(2+1/2m)}\right.\right. - \frac{r^{2(m+1)/m}(1-r^2)^{1-1/2m}}{(1+1/m)} \left.\left.\left[1+\sum_{j=0}^{\infty}\frac{B(2+1/m;\, j+1)}{B(2+1/2m;\, j+1)}r^{2(j+1)}\right]\right]\right\} \tag{9.57}$$

in which $B(\cdot\,,\,\cdot)$ represents the beta function. If r is only partly accounted for,

$$T_l = \left[\frac{LI}{2\alpha}\right]^{1/m}\frac{\Gamma(1+1/m)\,\Gamma(1-1/2m)}{2I(1-r)\,\Gamma(2+1/2m)} \tag{9.58}$$

where I is the time-invariant intensity of effective rainfall and α and m are, as before, kinematic wave parameters in the depth-discharge relationship

For a diverging surface,

$$T_l = \frac{1}{L(1-r)} I^{(1-m)/m}\left(\frac{1}{2\alpha}\right)^{1/m}\int_{rL}^{L}\left(\frac{x^2-r^2L^2}{x}\right)^{1/m} dx \tag{9.59}$$

Using the binomial theorem,

$$T_l = \frac{I^{m^*-1}}{L(1-r)}\left(\frac{1}{2\alpha}\right)^{m^*}\sum_{j=0}^{\infty}\binom{m^*}{j} (-1)^j (rL)^{2j}\left[\frac{L^{m^*-2j+1}-(rL)^{m^*-2j+1}}{m^*-2j+1}\right] \tag{9.60}$$

where $m^* = 1/m$. If r is negligibly small,

$$T_l = \left(\frac{L}{2\alpha}\right)^{m^*}\left(\frac{m}{m+1}\right) I^{m^*-1} \tag{9.61}$$

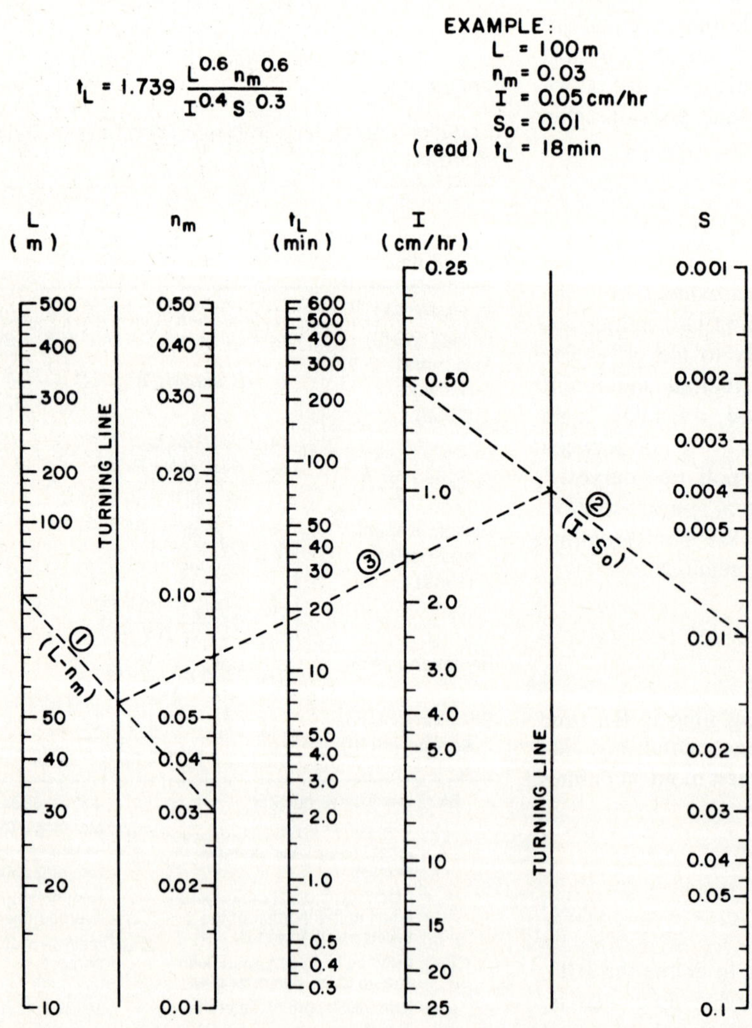

FIGURE 9.9 Nomograph for lag time of a plane using Manning's roughness coefficient (after Agiralioglu and Singh, 1981). *Note:* $t_L = T_l$

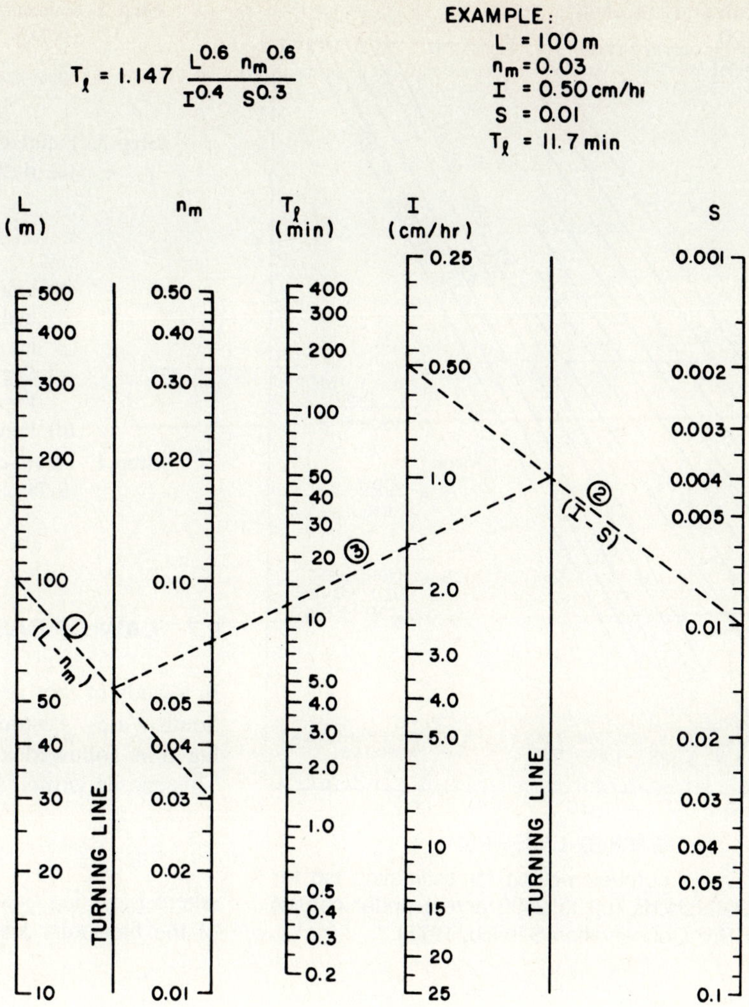

FIGURE 9.10 Nomograph for lag time of a diverging section using Manning's roughness coefficient (after Agiralioglu and Singh, 1981).

Agiralioglu and Singh (1981) presented a nomograph based on equation (9.61), as shown in Fig. 9.10.

It is evident that area, channel length, channel slope, and overland slope are the most frequently utilized geomorphologic characteristics for estimation of watershed lag time. Other characteristics, utilized much less frequently, include drainage density, stream order, roughness coefficient, and watershed width.

Soil Conservation Service (1975) presented a nomograph for equation (9.54), as shown in Fig. 9.11. This is a curve number method applicable to small urban areas less than 2,000 acres. This yields adequate lag time for an area which is completely paved, such as a parking lot. However, it overestimates for composite land-use areas where streets, gutters, or sewers provide a more efficient flow pattern than lawns, forests, or other pervious areas. This method was modified by Soil Conversation Service to incorporate the effect of changes in the main channel and the watershed surface. For example, if the main channel is partly lined or made hydraulically more efficient, then its travel time will be reduced, as shown in Fig. 9.12. Likewise, by increasing impervious portion of its area, the watershed lag can be modified as shown in Fig. 9.13. We illustrate this method by an example.

The U.S. Army Corps of Engineers (1949) presented curves for computing lag (defined as t_3 in Fig. 9.1) from watershed characteristics. Three types of watersheds were distinguished: (1) mountainous drainage areas, (2) foothill drainage areas, and (3) valley drainage areas. By including slope S of the main channel in Snyder's definition of lag (t_2), one curve was obtained for each type of watershed, as shown in Fig. 9.14.

EXAMPLE 9.2

A watershed of 1000 acres has a present-condition curve number of 75, average watershed slope of 4 percent, and hydraulic length of 13,200 feet. Urban development is expected to modify about 70 percent of the hydraulic length, increase the impervious area to 40 percent, and increase the runoff curve number to 80. This example is taken from Soil Conservation Service (1975). Compute lag time and time of concentration, using the curve number method, for the present and future conditions.

Step 1. Present-condition lag from equation (9.54) or Fig. 9.1 with CN = 75:

$$\text{Lag} = \frac{(13{,}200)^{.8}\,(3.33 + 1)^{.7}}{1900(4)^{0.5}} = 1.45 \text{ h}$$

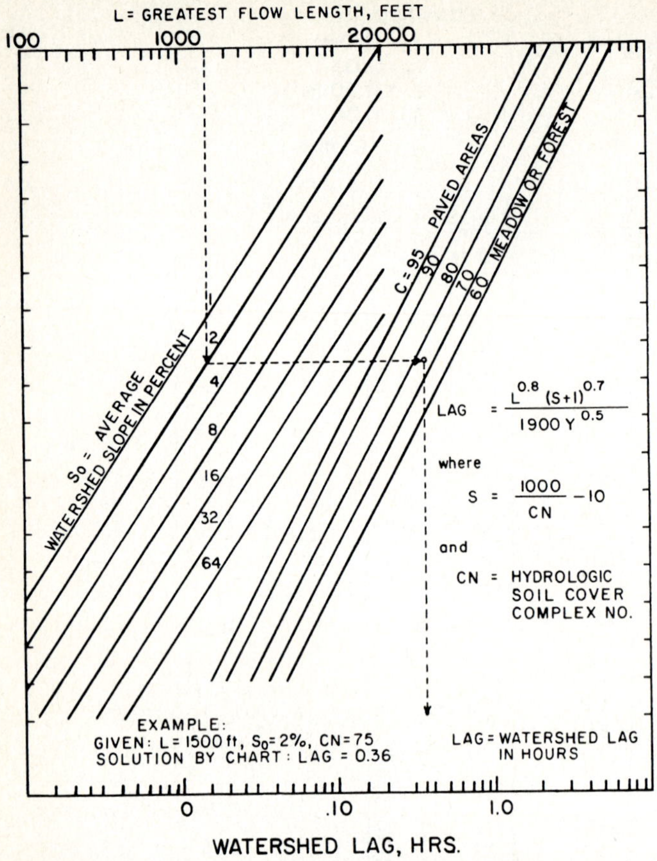

FIGURE 9.11 Curve number method for estimating lag for homogeneous watersheds (up to 2000 acres) under natural conditions (after Soil Conservation Service, 1975).

Step 2. Present-condition time of concentration from equation (9.74):

$$T_c = 1.67(1.45) = 2.42 \text{ h}$$

Step 3. Future-condition lag:

(a) Basic future-condition lag with $C_N = 80$:

$$\text{Lag} = \frac{(13{,}200)^{.8}\,(2.5 + 1)^{.7}}{1900(4)^{.5}} = 1.25 \text{ h}$$

(b) Lag factor for modification of 70 percent of the hydraulic length from Fig. 9.11: hydraulic-length lag factor = .59.

(c) Lag factor for 40 percent impervious area from Fig. 9.12: impervious-area lag factor = .76.

(d) Future-condition lag = 1.25(.59)(.76) = 0.56 h.

Step 4. Future-condition time of concentration from equation (9.74):

$$T_c = 1.67(.56) = 0.94 \text{ h}$$

9.7 LAW OF BASIN LAG

In a study of four nested drainage basins near Sydney in New South Wales, Australia, Boyd (1978) discovered that the basin lag time followed a law similar to the law of basin areas. This can be written as

$$L_w = L_1 R_l^{w-1} \qquad (9.62)$$

where L_w is lag time of the basin of order w, L_1 is lag time of the first-order basin, and R_l is basin lag ratio. He found

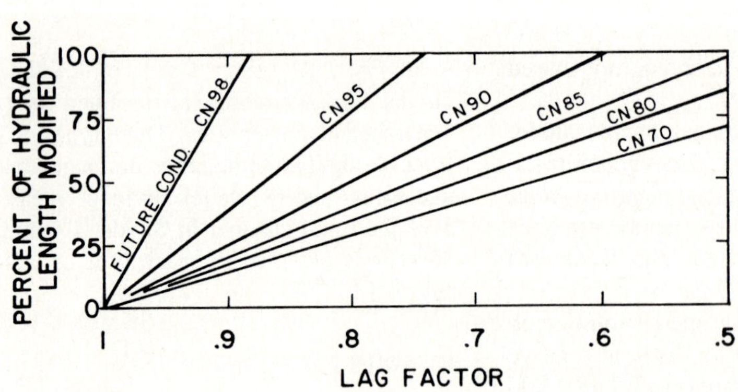

FIGURE 9.12 Factors for adjusting lag from equation (9.54) or Fig. 9.11 when the main channel has been hydraulically improved (after Soil Conservation Service, 1975).

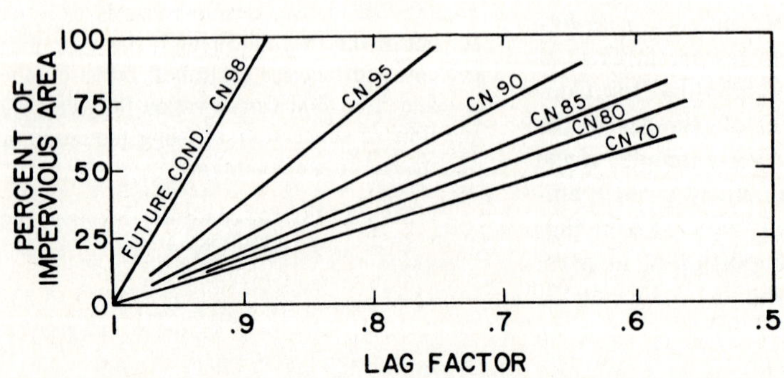

FIGURE 9.13 Factors for adjusting lag from equation (9.54) or Fig. 9.11 when impervious areas occur in the watershed (after Soil Conservation Service, 1975).

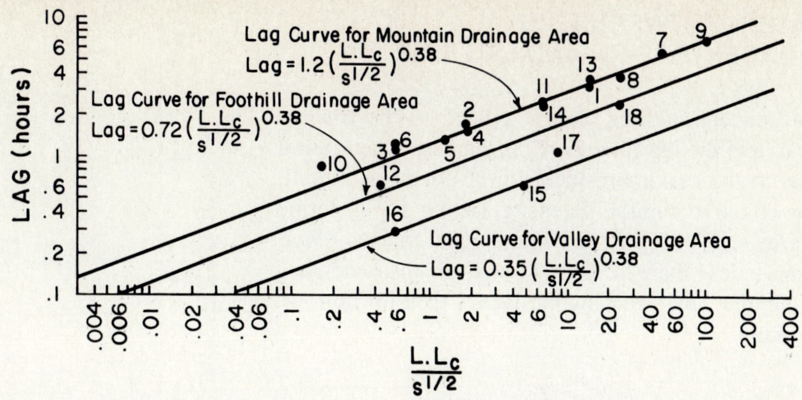

FIGURE 9.14 Relationship between lag (defined as t_3) and basin characteristics (after U.S. Army corps of Engineers, 1949).

R_l to be 1.737 for his data. Here L is the same as T_l. Equation (9.62) can be used conveniently in hydrologic synthesis of ungaged basins.

9.8 VARIATION IN LAG TIME

The lag time, defined according to definition 1, is more stable and, therefore, more suitable in hydrologic modeling. If the watershed is assumed to be linear and time-invariant, the time parameter is constant and independent of storm properties, whereas most of the other measures particularly, the duration of effective rainfall (Meynink, 1978), depend on storm characteristics.

The lag time is frequently found to vary with watershed and rainfall characteristics. The watershed characteristics can be of two types: (1) morphological, such as area, main stream length, overland slope, main stream slope, and drainage density, and (2) nonmorphological, such as vegetation and land use. The morphological characteristics may be considered constant within the time frame of hydrologic design. However, vegetation and land-use practices may change within a season and from one season to another. Further, these are influenced by climatic conditions.

Recognizing the variability of lag time, several investigators have attempted to relate it to the characteristics of effective rainfall and discharge (Laurenson, 1962; Askew, 1968, 1970a, 1970b; Meynink, 1978; Overton, 1970, 1971; Rao and Delleur, 1974; Singh, 1975; Singh and Agiralioglu, 1980). Some of the variable lag time models are given in Table 9.7. Of all the studies on Lag time, Meynink (1978) conducted perhaps the most comprehensive investigation. The variability of lag time is also supported by kinematic wave theory.

TABLE 9.7

Equations of variable lag time in hours (as defined in Fig. 9.1)

Author	*Equation*	*Equation Number for Referencing*
Laurenson (1962, 1964)	$t_1 = a\,Q_m^b, \quad a = 64,\ b = -.27$	(9.63)
Askew (1968, 1970a, 1970b)	$t_1 = KA^{.57}\,Q_m^{-.23}, \quad K = 8.28$	(9.64)
	$t_1 = KA^{.54}\,S_d^{-.16}\,Q_m^{-.23}, \quad K = 2.91$	(9.65)
	$t_1 = KL^{.8}\,S_d^{-.33}\,Q_m^{-.23}, \quad K = 2.91$	(9.66)
Rao and Delleur (1974)	$t_1 = .831A^{.458}(1 + I_A)^{-1.66}\,P_E^{.267}\,D_r^{.371}$	(9.67)
	$t_2 = .731L^{.943}(1 + I_A)^{-4.303}\,P_E^{-2.114}\,D_r^{.238}$	(9.68)
Meynink (1978)	$t_1 = KA_m^{.3405}\,Q_p^{-.3619}, \quad K = 2.55$	(9.69)
	$t_1 = KA_m^{1/3}\,Q_p^{-2/15}\left(\frac{D_r}{I_d}\right)^{1/5}, \quad K = 3.12$	(9.70)
Boyd (1978)	$t_1 = 2.12\,A_m^{.57}\,Q_s^{-.23}$	(9.71)

A = watershed area in square miles
A_m = watershed area in square kilometers
D_r = duration of the effective rainfall in hours
I_A = percentage impervious area
I_d = depth of the effective rainfall in millimeters
I_e = depth of effective rainfall in inches
L = length of the main stream in miles
Q_m = average discharge rate associated with the surface runoff hydrograph in cubic feet per second
Q_p = peak discharge in millimeters per hour
Q_s = peak discharge in cubic meters per second
S_d = measure of overland slope, a dimensionless ratio

9.9 RELATION BETWEEN TIME OF CONCENTRATION AND LAG TIME

The time of concentration is closely related to the time of equilibrium T_e and the lag time T_l. T_e is the time when runoff peak is equal to the maximum intensity of effective rainfall. For uniform effective rainfall intensity lasting for a period equal to or greater than T_c, T_c and T_e are equal to each other. If the duration is less than T_c, then $T_c > T_e$. If the watershed geometry is simple, say a rectangular plane, then for uniform effective rainfall,

$$T_c \geq T_e \geq T_l \tag{9.72}$$

For natural watersheds of larger size and complex drainage pattern, the water originating on the most remote part may arrive at the outlet too late to contribute to the peak runoff. Therefore, T_c is generally greater than T_l. It is physically plausible that on natural watersheds, T_e will be equal to or greater than T_c. Therefore,

$$T_e \geq T_c \geq T_l \tag{9.73}$$

It is evident that these time characteristics are not independent of each other for a watershed. The Soil Conservation Service (1975) suggested that

$$T_c = 1.67t_2 \tag{9.74}$$

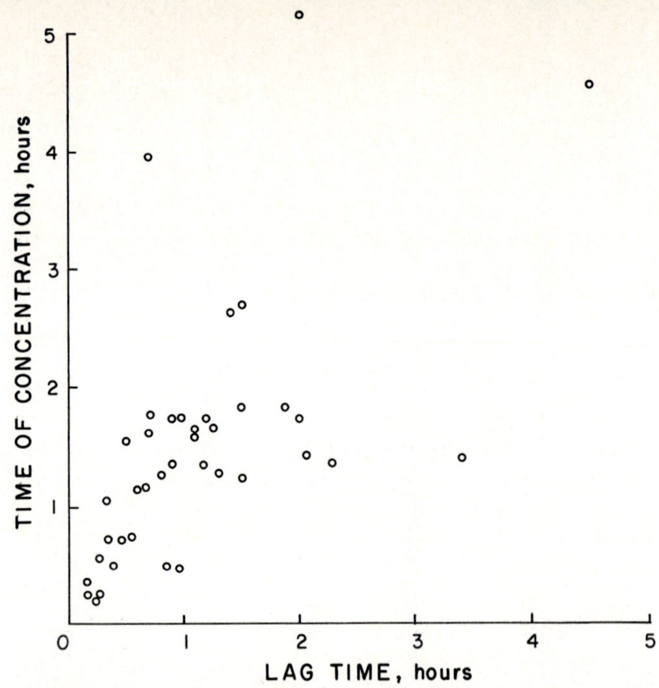

FIGURE 9.15 Measured times of concentration versus mean lag times from rainfall-runoff data (after McCuen, Wong, and Rawls, 1983).

TABLE 9.8

Other measures of time characteristics (as defined in Fig. 9.1)

Author	*Equation*	*Equation Number for Referencing*
Williams (1922)	$t_8 = \dfrac{KL}{A^{.1} S^{.2}}$	(9.77)
Chow (1962)	$t_8 = K\left(\dfrac{L_a}{\sqrt{S_a}}\right)^{.64}, \quad K = .00054$	(9.78)
Espey, Morgan, and Masch (1965)	$t_8 = KL^{.12} S^{-.52}$	(9.79)
Bruce and Clark (1966)	$t_8 = K\left[\dfrac{LV}{S^{0.5}}\right]^{.467}$	(9.80)
Schulz (1969)	$t_8 = K\left[\dfrac{LL_c}{\sqrt{S}}\right]^{.162}, \quad K = 1.9$	(9.81)
Black (1972)	$t_8 = KA^{.29}$	(9.82)
Chu and Lytle (1972)	$t_4 = \dfrac{KL}{\sqrt{S}}$	(9.83)

A = watershed area in square miles
L = length of main stream length in miles
L_a = length of main channel in feet
L_c = length to centroid of area in miles
S = overall slope of main stream
S_a = slope of main stream calculated by equal area method in feet per foot

Overton and Meadows (1976) indicated

$$T_c = 1.6t_2 \tag{9.75}$$

However, it can be shown by using the Soil Conservation Service (SCS) triangular unit hydrograph with 37.5 percent of the volume under the rising limb that $t_1 = 1.176t_2$. By inserting equation (9.74),

$$T_c = 1.417t_1 \tag{9.76}$$

McCuen, Wong, and Rawls (1984) examined the validity of equations (9.74) and (9.76) by using hyetograph and hydrograph data from 39 urban watersheds. They computed the mean values of T_c and t_1 for the entire sample of watersheds. These were 1.51 hours and 1.12 hours, respectively. The ratio of 1.51/1.12 is 1.35, which agrees well with the value of 1.417. Thus, it appears that on the average, computed T_c values agree with T_c values from rainfall-runoff data. However, values on a given watershed show considerable scatter, as seen in Fig. 9.15.

9.10 OTHER TIME CHARACTERISTICS

In addition to lag time, several other measures of the time characteristics of runoff hydrograph have been used in hydrograph synthesis. These are defined in Fig. 9.1. Some of these have been related to watershed characteristics, as shown by equations listed in Table 9.8.

EXERCISES

9.1 Data from eight urban and five rural watersheds in Indiana and Texas are given in Table 9.9. Compute the time of concentration for these watersheds by as many equations as require the data available in the table. Compare these values and comment on their respective equations.

9.2 Develop guidelines for choosing among methods of computing the time of concentration.

9.3 Compute the lag time for the watersheds in Exercise 9.1 by as many equations as you can. Compare the results and comment on their respective equations.

9.4 Develop guidelines for choosing among methods of computing the lag time.

9.5 Choose a rural watershed and an urban watershed from Table 9.9. Compute the lag time according to different definitions. Compare the values and comment.

9.6 Which definition of the lag time would you use and why? Which definition is the most meaningful of all?

9.7 Select a third-order watershed. Compute the lag time for this watershed using any five empirical formulas as well as using the law of lag time. Comment on the results.

9.8 Select a third-order watershed. Compute the lag time for this watershed using any five empirical formulas as well as using the law of lag time. Comment on the results.

9.9 For the watershed in Exercise 9.7, check the validity of the law of lag time.

9.10 How valid is the relation between lag time and time of concentration?

TABLE 9.9

Physiographic characteristics of the basins (after Rao and Delleur, 1974)

Number	Name	Area *A* (mi²)	Length of Stream *L* (mi)	Mean Basin Slope *S* (ft/mi)	Percentage Impervious Area I_A (%)	No. of Storms Used for Analysis	Average Time Lag (h)
1	Ross Ade (upper)[(a)]	.0455	.6133	112.0	38.0	21	.21
2	Ross Ade (lower)[(a)]	.6125	2.1765	112.0	37.4	10	.61
3	Purdue Swine Farm (upper)[(a)]	.2776	.6439	3.9	21.3	4	.30
4	Purdue Swine Farm (lower)[(a)]	.4562	1.067	3.40	13.3	2	.40
5	Pleasant Run (Arlington)[(b)]	7.580	3.822	14.26	10.5	13	3.57
6	Pleasant Run (Brookville)[(b)]	10.10	5.644	12.67	15.5	12	3.72
7	Little Eagle Creek[(b)]	19.31	11.10	23.23	2.1	8	9.34
8	Lawrence Creek[(b)]	2.86	1.705	32.21	0	12	2.05
9	Bear Creek[(c)]	7.00	3.864	39.07	0	5	5.44
10	Bean Blossom Creek[(c)]	14.60	6.44	32.74	0	7	6.56
11	Waller Creek (38th Street)[(d)]	2.31	4.371	47.0	27.0	17	1.44
12	Waller Creek (23rd Street)[(d)]	4.13	5.23	47.0	37.0	12	1.58
13	Wilbarger Creek[(d)]	4.61	3.17	45.94	0	8	2.89

[(a)] West Lafayette, Indiana [(b)] Indianapolis, Indiana [(c)] Indiana; [(d)] Austin, Texas

CHAPTER TEN

Linear Black-Box Models for Continuous-Time Systems

The rational method, the time-area methods, and the unit hydrograph method are perhaps the most popular black-box models of the rainfall-runoff relationship.

10.1 THE RATIONAL METHOD

The origin of the rational method is somewhat obscure. In the United States, it is often referred to as the Kuichling method (Kuichling, 1889) and in Great Britain, as the Lloyd-Davies method (Lloyd-Davies, 1906). However, Mulvany (1850) was, according to Dooge (1957), perhaps the first to give a clear exposition of this method, as follows:

> The first matter of importance to be ascertained in the case of a small or mountainy catchment is the time which a flood requires to attain its maximum height, during the continuance of a uniform rate of fall of rain. This may be assumed to be the time necessary for the rain which falls on the most remote portion of the catchment to travel to the outlet, for it appears to me that the discharge must be greatest when supply from every portion of the catchment arrives simultaneously at the point of discharge supposing, as above premised, the rate of supply to remain constant, and this length of time being ascertained, we may assume that the discharge will be the greatest possible under the circumstances of a fall of rain occurring, of the maximum uniform rate of fall for that time.

He went on to discuss the factors affecting the time of concentration as follows:

> This question of time, as regards any catchment, must depend chiefly on the extent, form and rate of inclination of its surface; and therefore one great object for investigation is the relation between these causes and their effect; so that, having ascertained the extent, form and average inclination of any catchment, we may be able to determine in the first place, the duration of constant rain required to produce a maximum discharge, and consequently to fix upon the maximum rate of rainfall applicable to the case. It is evident that, as a space of time is reduced, the rate of maximum rate of rain is increased.

Mulvany (1850) assumed a constant rate of rainfall and was concerned with the maximum rate of runoff.

Independently, Kuichling (1889) discussed the need to incorporate the relation between rainfall intensity, drainage area and time of concentration as follows:

> The writer was long since impressed with the fact that during showers the volume of water discharged at the mouths of several large outlet sewers in the City of Rochester, N.Y., appeared to increase and diminish directly with the intensity of the rain at different stages, but that a certain length of time was required in each case after the termination of a brief and heavy downpour before corresponding flood showed itself at the outfall; these floods, moreover, seemed to last about as long as the said showers themselves, and the conclusion was therefore reached that there must be some definite relation between these fluctuations of discharge and the intensity of the rain, also between the magnitude of the drainage area and the time required for the floods to

appear and subside. . . . The conclusion is accordingly irresistible that the rates of rainfall adopted in computing the dimensions of a main sewer must correspond to the time required for the concentration of the drainage waters from the whole tributary area when small, or from so much thereof as will produce an absolute maximum discharge when the area is very large.

The rational method presents the concept of time of concentration and its relation to maximum runoff. This method can be expressed as

$$Q = CPA \tag{10.1}$$

where Q is the peak discharge (volumetric rate), C is the runoff coefficient (dimensionless) depending upon the drainage area characteristics, A is the area of the drainage basin, and P is the average intensity of rainfall of specified frequency for a duration equal to the time of concentration T_c of the drainage area. If A is specified in acres, and P in inches per hour, then Q is in cubic feet per second. This takes advantage of the fact that one acre-inch per hour is very nearly equal to one cubic foot per second.

It should be noted that the storm duration and time of concentration are not necessarily equal. Normally, the storm duration is considerably longer than the time of concentration. Furthermore, the time of concentration (the period for which rainfall intensity is averaged) can occur during any segment of the storm duration at the beginning, before, during, or after the middle portion, or near the end.

Equation (10.1) implies that the peak discharge occurs when the entire area is contributing at the mouth, which will occur at a period T_c after the rainfall starts. Therefore, rainfall occurring after the time of concentration will not contribute to a greater peak discharge but will only prolong the discharge period. The peak discharge will equal the average intensity of effective rainfall I during T_c.

Kuichling (1889) suggested that C approached a constant value for a given drainage basin as the magnitude of the rain storm increased. The rational method can be used to determine the frequency of discharges from the frequency of rainfall (Nash, 1958). Indeed, a consequence of the rational method is that the frequency of occurrence of any given peak discharge is equal to the frequency of the required amount of rainfall on the watershed in a period equal to the time of concentration.

10.1.1 Assumptions

The assumptions made in the rational method can be stated as follows:

1. 1 cubic foot per second = 1 acre-inch per hour

$$\frac{1 \text{ acre-in.}}{\text{h}} = \frac{1 \text{ acre-in.}}{\text{h}} \times \frac{43{,}560 \text{ ft}^2}{\text{acre}} \times \frac{\text{h}}{3600 \text{ s}} \times \frac{\text{ft}}{12 \text{ in.}}$$
$$= 1.0083 \text{ cfs} \cong 1 \text{ cfs}$$

2. The watershed area is the area draining or tributary to the point of design.
3. The rainfall intensity does not vary during the time period equal to the time of concentration.
4. The time of concentration is the time required for water to run off from the hydraulically most remote point in the watershed to the point of design.
5. The peak discharge at any point is a direct function of the average rainfall intensity during the time of concentration to that point, since A and C are constant.
6. The frequency of peak discharge is the same as the frequency of the average rainfall intensity, since A and C are constant.
7. The runoff frequency curve is parallel to the rainfall frequency curve. This is a consequence of constant C for all recurrence intervals.
8. The rationale behind the rational formula has been expressed in various ways. Three versions are given here.
   (a) The rational method estimates peak discharge on small watersheds, usually less than 5 square miles. It is based on the premise that for rainfall storms of uniform intensity in space and time, the maximum rate of runoff occurs when the entire watershed area is contributing to the flow at the outlet. This rate of runoff is a fraction of the rainfall intensity.
   (b) For a rainfall storm of uniform intensity in space and time occurring over a critical period of time, the runoff reaches its maximum rate when all parts of the watershed contribute to the flow at the outlet or point of design.
   (c) The rain producing the greatest peak runoff for any particular area will be the rain of the highest intensity expected in that area with a duration equal to the time of concentration for that area or the time required for the most distant drop of rain water in the area to find its way overland to the point where it enters the sewer. This last point suggests that the water bypassing an inlet or inlets can change the area contributing to the flow at the outlet or point of design.

10.1.2 Mathematical Representation

If it is assumed that the time of concentration T_c is constant for a given watershed and that the runoff peak is directly proportional to the average intensity of effective rainfall for T_c, then the rational method represents a lumped linear time-invariant system. Write equation (10.1) as

$$Q = \frac{A}{T_c} \int_{t-T_c}^{t} I(\tau)\, d\tau \tag{10.2}$$

where I is the intensity of effective rainfall. On comparing equation (10.2) with equation (4.10), it is immediately clear that the UIR of the rational method is

$$h(t) = \frac{A}{T_c} \tag{10.3}$$

Note that $h(t) = 0$ outside of $0 \le t \le T_c$. Equation (10.3) was derived by Nash (1958) and is graphed in Fig. 10.1. Further, the unit pulse response (UPR) and unit step response (USR) of the rational method can be determined using equation (10.3). Let D be the duration for which the UPR is desired. Then

$$h_D = \frac{1}{D}[U(t) - U(t - D)]$$

Using equation (10.3),

$$U(t) = \int_0^t \frac{A}{T_c}\,dx = \frac{A}{T_c}T$$

Therefore,

$$U(t) = \begin{cases} \dfrac{A}{T_c}t, & 0 \le t \le T_c \\ A, & t \ge T_c \end{cases} \tag{10.4}$$

Equation (10.4) is graphed in Fig. 10.2. Using equation (10.4)

$$h_D(t) = \begin{cases} \dfrac{A}{DT_c}t, & 0 \le t \le D \\ \dfrac{A}{T_c}, & D \le t \le T_c \\ \dfrac{A}{D}\left[1 - \dfrac{1}{T_c}(t - D)\right], & T_c \le t \le T_c + D \end{cases} \tag{10.5}$$

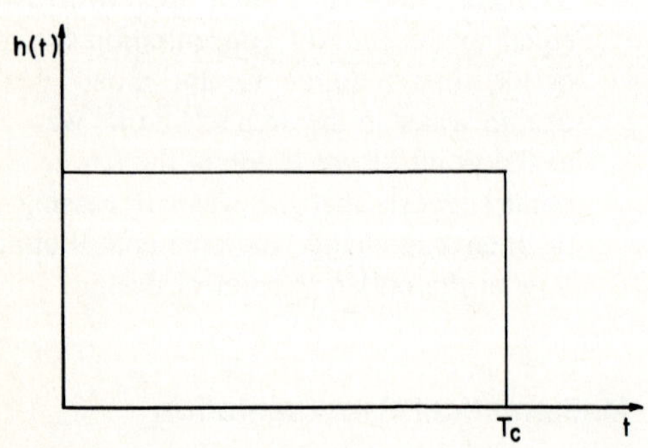

FIGURE 10.1 The UIR of the rational method.

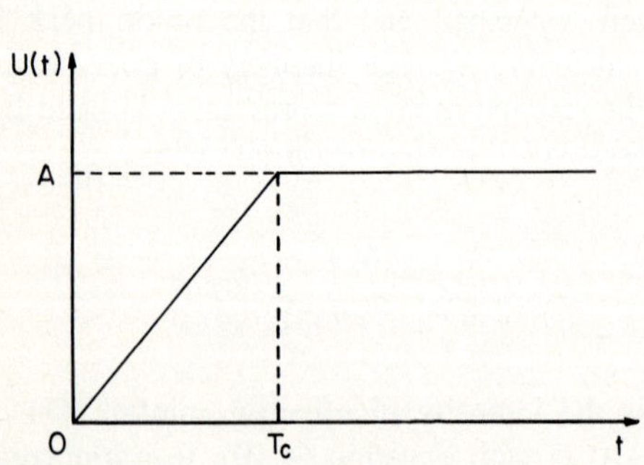

FIGURE 10.2 The USR of the rational method.

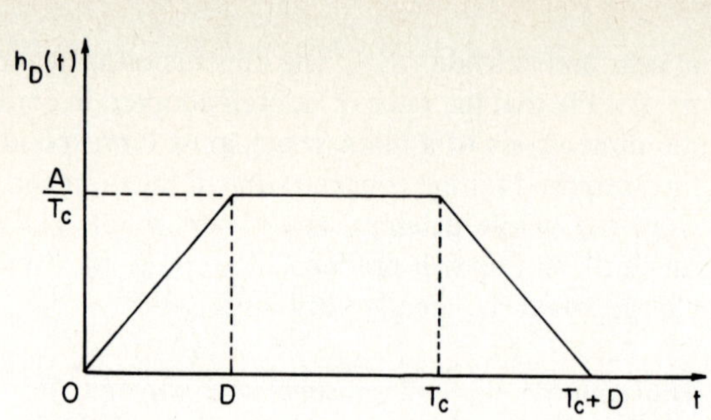

FIGURE 10.3 The UPR of the rational method.

Fig. 10.3 is a pictorial representation of $h_D(t)$. Here D is assumed to be less than T_c. If, however, D is equal to T_c, then

$$h_{T_c}(t) = \begin{cases} \dfrac{At}{T_c^2}, & 0 \le t \le T_c \\ \dfrac{A}{T_c}\left[1 - \dfrac{1}{T_c}(t - T_c)\right], & T \le t \le 2T_c \end{cases} \tag{10.6}$$

Equation (10.6) is graphed in Fig. 10.4.

10.1.3 Physical Limitations

Although limitations of the rational method are implicit in the oversimplifying assumptions that it employs, the consequence of these assumptions is most vividly brought to light by equation (10.3). It is physically inconceivable that the response of a drainage basin to the unit impulse of effective rainfall will be a unit pulse of duration T_c and height $1/T_c$.

Sometimes the rational method gives a greater discharge, for a given frequency, for part of a drainage basin than for the whole. To illustrate, consider a drainage area as shown in Fig. 10.5. Let A_1 be the area of a portion of the drainage basin between A and B, and let T_1 be its time of concentration. Let A_2 be the area of the entire drainage basin, and let T_2 be its time of concentration. Using the rational method, the discharge from the area between A and B is given by $Q_1 = CP_1 A_1$, where P_1 is the average intensity of rainfall of given frequency lasting for T_1. Similarly, the discharge for the entire area will be given by $Q_2 = CP_2 A_2$, where P_2 is the average intensity of rainfall of given frequency lasting over a period

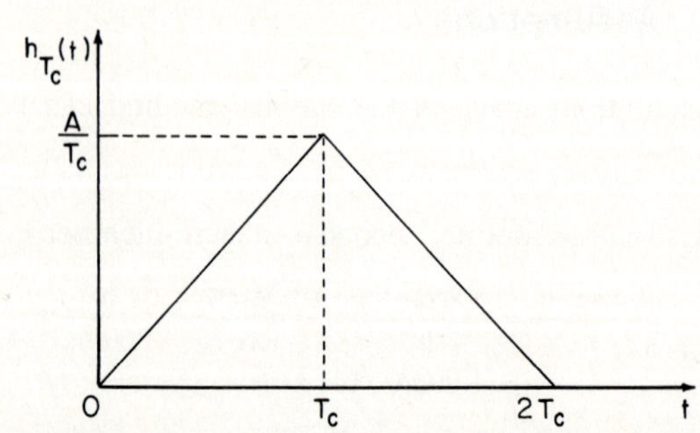

FIGURE 10.4 The UPR of the rational method for a duration equal to T_c.

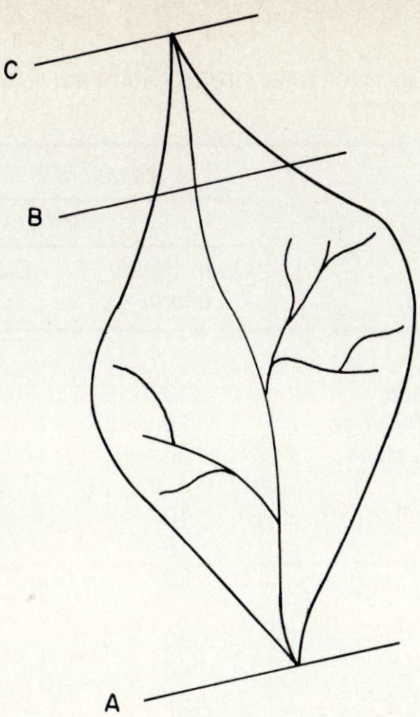

FIGURE 10.5 A hypothetical drainage area.

T_2. Since T_2 is much greater than T_1, P_2 is much smaller than P_1. Therefore, it is plausible that the ratio of P_1/P_2 is much greater than the ratio A_2/A_1, and if this is true, then Q_1 will be greater than Q_2. This inconsistency of the rational method stems from the fundamental assumption that Q equals the average intensity of effective rainfall I during T_c.

Reid (1927) and Norris (1946) put forward graphical methods for proportioning the subareas contributing to the discharge at the outlet. Although these methods corrected, to some extent, the effect of the stated inconsistency, they made no fundamental change in the rational method and, hence, in no way removed the cause of the error. These graphical methods are also referred to as the *tangent methods*.

The limitations of the rational method can be summarized as follows:

1. The runoff rate reaches a maximum value at a time equal to the time of concentration; that is, the time of peak discharge equals the time of concentration.
2. As originally conceived, the method yields a peak discharge—a single point on the flood hydrograph. This limitation is, however, removed by following the mathematical representation developed in the preceding section.
3. The method does not accommodate spatially variable rainfall.
4. Since rainfall intensity is related to T_c, a higher rainfall intensity over a portion of the total watershed can produce a higher peak discharge than a lower intensity over the entire watershed.
5. An average rainfall intensity used in the method may have no time sequence relation to the actual rainfall pattern during the storm. The intensity-duration-frequency (I-D-F) curve employed in the method is not a time-sequence curve of rainfall.
6. A point should be noted about the I-D-F curves. The maxima of several durations from a given storm as used in the method are not necessarily in their original sequential order; and the resulting tabulation of maxima ordered according to size or duration may possess little resemblance to the original storm pattern. In many cases, the rainfall intensities on the same frequency curve for various durations are not from the same storm.
7. The method involves a considerable degree of subjective judgment in its application. Consequently, there exist variations in its interpretation. This may mar reproducibility of results.
8. If a system is designed using the rational method, then it is difficult to logically analyze modifications, such as provisions for relief, rearrangement of sewers, or reusing the method.

10.1.4 A Comment on Its Application

Although the rational method has recently come under increasing attack for its lack of physical realism, it continues to be used frequently in design of urban drainage (McPherson, 1969). We quote Papadakis and Preul (1972):

> Large expenditures are continuously made for the design and construction of storm water systems for urban runoff. Yet rather crude analytical methods are often used for such design. A survey of most municipal engineering departments would likely show that the rational formula is most commonly used.

This is not only true of the United States but also of Australia, the United Kingdom, and India, to name but a few (UNESCO, 1977). In a comparative evaluation of several methods for the design of storm sewers, Burke and Gray (1979) concluded that the rational method is not always adequate for design purposes. They suggested that the use of the rational method should be discouraged.

Dooge (1973) pointed out that the errors in the rational method resulting from the assumption of uniform rainfall intensity and the neglect of storage effects are opposite in sign. Therefore, the method may yield reasonable results. This inference was also reached by Schaake, Geyer, and Knapp (1967) in an experimental examination of the method. They remarked:

> Such analyses as are possible suggest that, although the method can be wide of the mark in any given case, it may be on the average giving reasonable results.

Few would dispute that the assumptions made in the rational method are unduly restrictive. Nevertheless, these might nearly hold for airport runways, parking lots, and paved areas with gutters and sewers of fixed geometric and hydraulic characteristics. This might partly explain its popularity in design of drainage systems for these areas.

10.1.5 Parameter Estimation

The rational method has two parameters, namely, the proportionality constant or runoff coefficient C and the time of concentration T_c. Determination of T_c is discussed in detail in Chapter 9. Here we discuss determination of C.

10.1.5.1 Runoff Coefficient C

The runoff coefficient C is the ratio of the runoff peak to the average rainfall intensity, and it has no dimension. It represents the fraction of rainfall that appears as surface area from the drainage area. Ramser, Norton, and Ellison (1929) listed six major factors influencing C: (1) nature of the soil, (2) surface slope, (3) surface cover, (4) rainfall intensity, (5) degree of saturation of the watershed at the beginning of rainfall, and (6) surface storage. To this list we can also add: (1) percentage of impervious area, (2) duration of rainfall, (3) shape of drainage area, (4) design frequency, (5) interception, and (6) roof drainage—its connection to a storm sewer, driveway, or pervious surface. In short, topographic, infiltration, and rainfall characteristics are the principal determinants upon which C depends. Although C is usually assumed constant for a specified watershed, it does vary from one rainfall event to another (Ramser, 1927; Gregory and Arnold, 1932). Indeed, C does not even remain constant during the storm duration (Horner, 1910; Horner and Jens, 1941). If T_c occurs at the beginning of the storm, then the antecedent moisture conditions become important. If T_c occurs near the end of a long, heavy rainfall, the ground will be nearly saturated and all the depressions filled. In these two cases, the value of C will be different. Horn and Schwab (1963) gave an excellent discussion on the evaluation of the rational runoff coefficient for small agricultural watersheds (see Fig. 10.6).

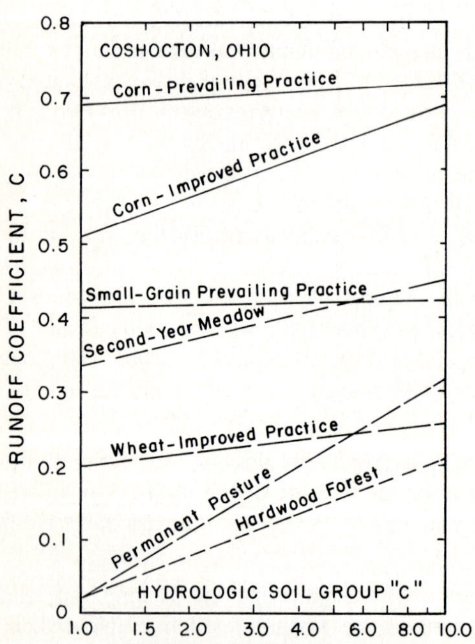

FIGURE 10.6 Runoff coefficients for agricultural lands (adapted from Horn and Schwab, 1963).

TABLE 10.1

Runoff coefficient for rural areas (Schwab, Frevert, Edminster, and Barnes, 1971)

Topography and Vegetation	*Values of C in Q = CPA* — *Soil Texture*: *Open Sandy Loam*	*Clay and Loam*	*Tight Clay*
Woodland			
Flat 0.5% slope	.10	.30	.40
Rolling 5–10% slope	.25	.35	.50
Hilly 10–30% slope	.30	.50	.60
Pasture			
Flat	.10	.30	.40
Rolling	.16	.36	.55
Hilly	.22	.42	.60
Cultivated			
Flat	.30	.50	.60
Rolling	.40	.60	.70
Hilly	.52	.72	.82

There exists a considerable amount of literature providing values of C under a wide variety of conditions. Horner and Flynt (1936), American Concrete Pipe Association (1970), and American Society of Civil Engineers (1969) have given values for urban areas; Chow (1964), for cultivated, pasture, and woodlands; Gray (1970), for agricultural watersheds; and American Concrete Pipe Association (1958), for culvert design, highway surface, airport drainage, and storm sewer design. For example, the values of C for rural areas are given in Table 10.1. Table 10.2 presents the values of C for particular types of urban areas, and Table 10.3 presents the values that can be used to compute a weighted C based on the actual percentage of lawns, streets, roofs, and so on (American Society of Civil Engineers, 1976).

In a comprehensive study, Miller (1968) developed a systematic procedure for evaluation of C from small natural

TABLE 10.2

Runoff coefficient for urban areas for storms of 5–10 year frequencies (American Society of Civil Engineers, 1976)

Description of Area	*Runoff Coefficients*
Business	
Downtown	.70 to .95
Neighborhood	.50 to .70
Residential	
Single-family	.30 to .50
Multiunits, detached	.40 to .60
Multiunits, attached	.60 to .75
Residential (suburban)	.25 to .40
Apartment	.50 to .70
Industrial	
Light	.50 to .80
Heavy	.60 to .90
Parks, cemeteries	.10 to .25
Playgrounds	.20 to .35
Railroad yard	.20 to .35
Unimproved	.10 to .30

TABLE 10.3

Values for determination of a composite runoff coefficient for urban areas for storms of 5–10 year frequencies (American Society of Civil Engineers, 1976)

Character of Surface	*Runoff Coefficients*
Pavement	
Asphaltic and concrete	.70 to .95
Brick	.70 to .85
Roofs	.75 to .95
Lawns, sandy soil	
Flat, 2%	.05 to .10
Average, 2–7%	.10 to .15
Steep, 7%	.15 to .20
Lawns, heavy soil	
Flat, 2%	.13 to .17
Average, 2–7%	.18 to .22
Steep, 7%	.25 to .35

drainage areas. He divided C into two components: one relating to overland flow, designated by C_0, and another relating to channel flow, designated by C_s. His procedure was for the determination of C_0. A set of curves was developed, based on 50-year event at Lexington, Kentucky, to determine C_0 from five measurable watershed characteristics, including soil depth, permeability, overland slope, impervious area, and soil exposure (Clarke, 1968). These curves are shown in Fig. 10.7. Further, variation of C_0 was studied with regard to mean annual rainfall, rainfall intensity, and frequency, as shown in Figs. 10.8 and 10.9.

10.1.5.2 Optimization of T_c and C

Nash (1967) pointed out that the parameters T_c and C of the rational method could be estimated by optimization. By examining a record of rainfall and runoff for a watershed, the optimum values of C and T_c can be determined to obtain the best agreement between computed and observed runoff peaks according to a specified error criterion. A measure of agreement can be defined in a least squares sense. Using different parts of the record, we can obtain a number of estimates of these parameters and can even judge the significance of these estimates. These parameters can be obtained in a like manner for a number of representative watersheds and can then be related to watershed characteristics. The relations thus developed can be used to obtain the parameters for ungaged watersheds.

10.1.6 Procedure for Computation of Peak Discharge

Although the determination of Q from equation (10.1) is obvious, from a design standpoint a brief comment may be appropriate. The procedure may entail the following steps:

1. Determine C for specified topographic, soil, cover, and land-use characteristics. If the watershed consists of several distinct zones, a weighted C may be necessary. Tables 10.1 through 10.3 may appropriately be used.
2. Determine T_c for specified topographic characteristics using any of the equations discussed in Chapter 9.
3. Determine the rainfall intensity P for a duration equal to T_c using rainfall I-D-F curves developed for the watershed in question. Fig. 10.10 gives a typical illustration of such curves. These curves give P for various return periods or average recurrence frequencies.
4. For specified A, and P, and C determined as specified earlier, compute Q from equation (10.1).

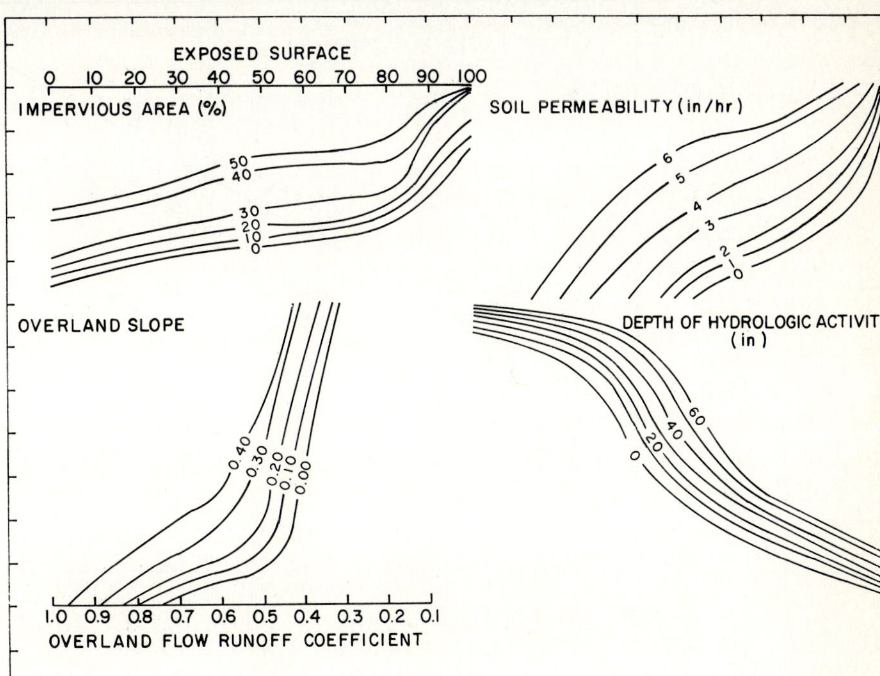

FIGURE 10.7 Curves for estimating overland flow runoff coefficients from watershed characteristics (after Miller, 1968).

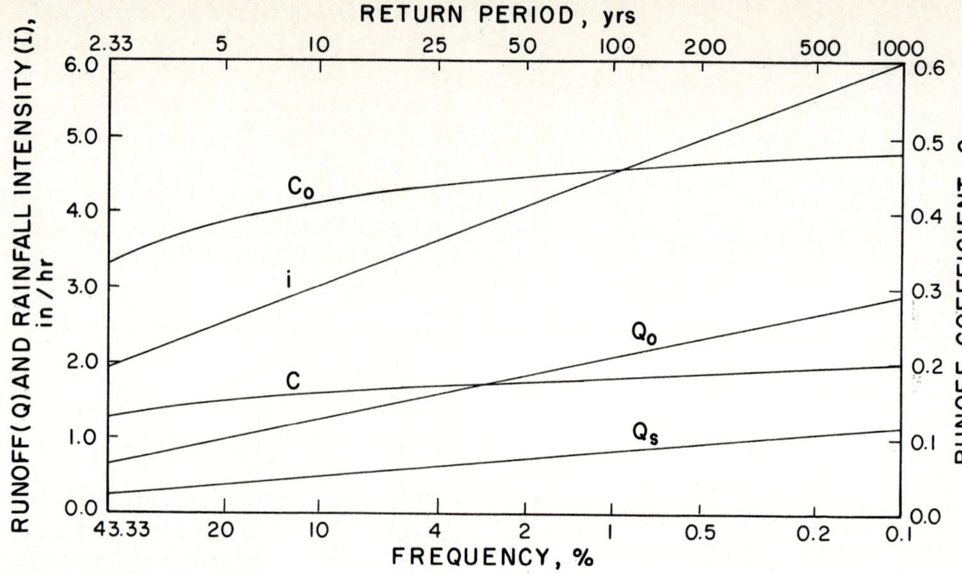

FIGURE 10.8 Rainfall, runoff, and runoff coefficients for Cave Creek Watershed at Cairo, Illinois (after Miller, 1968).

In design, the selection of a return period is necessary to execute Step 3. In practice, this selection depends upon the particular circumstances of the design structure.

EXAMPLE 10.1

Compute the peak discharge at the mouth of a watershed located near Toronto, Ontario, having the following characteristics: pasture land, fairly flat with average slope of about 1 percent, tight clay soils, drainage area, 40 ha, distance from the farthest point to the outlet, 2000 m, and difference in elevation from the farthest point to the outlet, 10 m.

Solution:

The value of runoff coefficient from Table 10.1 is $C = 0.4$. The watershed gradient is $\frac{10}{2000} \times 100 = .5\%$.

For the length of flow equal to 2000 m and a gradient of 0.5%, the time of concentration from equation (9.13) is 20 min. From Fig. 9.10, rainfall intensity corresponding to $T_c = 20$ min is:

1. For a 10-yr return period, $P = 8.5$ cm/h.
2. For a 20-yr return period, $P = 9.75$ cm/h.

Therefore,

1. For a 10-yr flood, $Q = 0.4 \times 8.5 = 3.4$ cm/h.
2. For a 25-year flood, $Q = 0.4 \times 9.75 = 3.9$ cm/h.

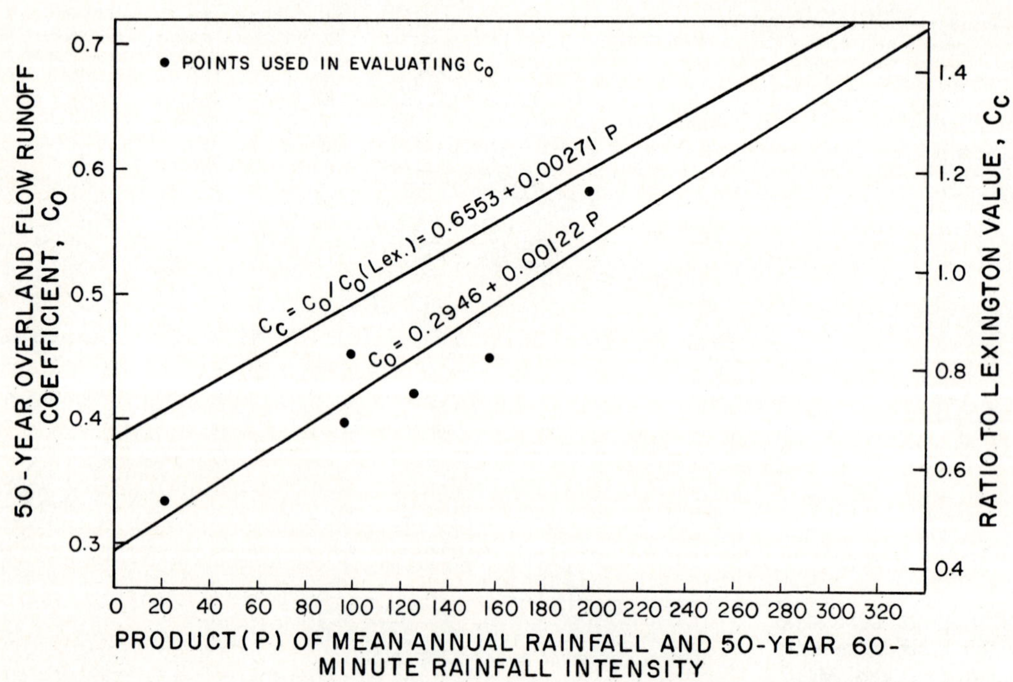

FIGURE 10.9 Variation of overland flow runoff coefficient with rainfall regime (after Miller, 1968).

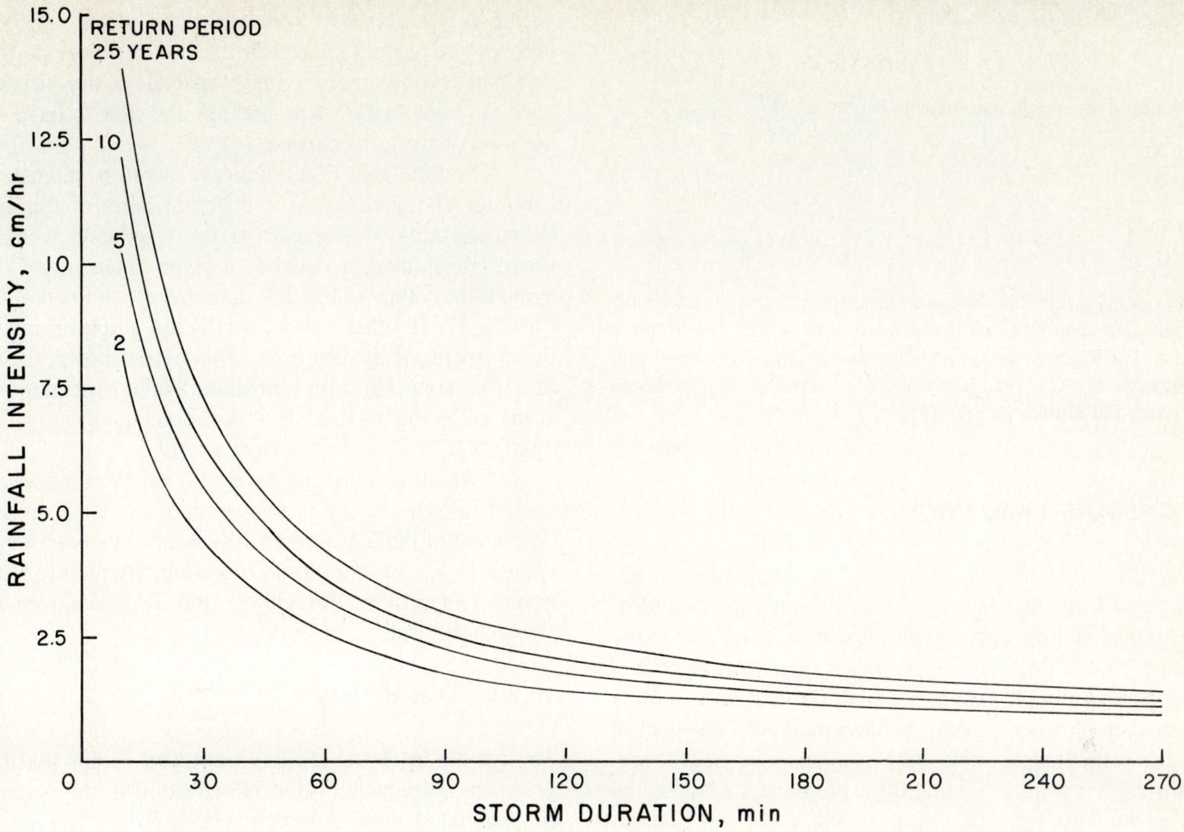

FIGURE 10.10 Illustrative rainfall I-D-F curves, drawn for Toronto, Ontario, Canada.

EXAMPLE 10.2

Determine the peak discharge and direct runoff hydrograph by using the rational method for Ward Creek at Capital Heights Avenue in Baton Rouge, Louisiana. The drainage area is 4.10 mi^2. The main channel length and slope are 16,210 ft and 7.2 ft/mi, respectively. Compare the peak discharge values computed by B. L. Neely (1976) in "Floods in Louisiana, Magnitude and Frequency," page 207, published by Louisiana Department of Highways, Baton Rouge, Louisiana.

Solution: The time of concentration from equation (9.13) is

$$T_c = \frac{.0078 \times 16{,}210^{0.77}}{.0014^{-0.385}} \cong 172 \text{ min} \cong 3 \text{ hours}$$

The frequencies selected are 5, 10, 15, 25, and 50 years. The rainfall depth V corresponding to each of these frequencies* is:

$$V_5 = 3.6 \text{ in.} \qquad P = \frac{3.5}{3} = 1.17 \text{ in./h}$$

$$V_{10} = 4.0 \text{ in.} \qquad P = \frac{4}{3} = 1.33 \text{ in./h}$$

$$V_{15} = 4.3 \text{ in.} \qquad P = \frac{4.3}{3} = 1.43 \text{ in./h}$$

$$V_{25} = 4.8 \text{ in.} \qquad P = \frac{4.8}{3} = 1.6 \text{ in./h}$$

$$V_{50} = 5.2 \text{ in.} \qquad P = \frac{5.2}{3} = 1.73 \text{ in./h}$$

*From D. Hershfield. Rainfall frequency atlas of the United States. *U.S. Weather Bureau Technical Paper* 40 (1961).

For the Ward Creek at Capital Heights Avenue, the drainage area includes neighborhoods and light industries, where C can vary from .5 to .7. If $C = .5$, then

$$Q_5 = .5 \times 1.17 \times 2624 = 1540 \text{ cfs}$$

$$Q_{10} = .5 \times 1.33 \times 2624 = 1748 \text{ cfs}$$

$$Q_{15} = .5 \times 1.43 \times 2624 = 1880 \text{ cfs}$$

$$Q_{25} = .5 \times 1.6 \times 2624 = 2100 \text{ cfs}$$

$$Q_{50} = .5 \times 1.73 \times 2624 = 2270 \text{ cfs}$$

Likewise we can carry out these calculations for $C = .6$ and $C = .7$. These values of Q can be compared with those computed by Neely (1976) who used LPT III distribution to analyze peak flow. Results of this comparison are shown in the following table. From this it appears that $C = .5$.

Peak Discharge cfs	Rational Method			*Values of Neely (1976)*
	C = .5	C = .6	C = .7	
Q_5	1540	1840	2150	1580
Q_{10}	1740	2090	2440	1800
Q_{15}	1880	2250	2630	—
Q_{25}	2100	2520	2940	2060
Q_{50}	2270	2720	3180	2250

The discharge hydrographs for each effective rainfall can be computed by using equation (10.5). The hydrographs will simply be triangular in shape, that is,

$$T_c = D = 3 \text{ h}$$

$$I = .5 \times 1.17 = .585 \text{ in./h}$$

The discharge hydrograph can then be expressed as

$$Q(t) = \frac{CAV}{9}t = \frac{CAI}{3}t$$

$$= \frac{CAV}{3}\left[1 - \frac{1}{3}(t-3)\right] = CIA\left[1 - \frac{1}{3}t - 3)\right], \qquad 3 \leq t \leq 6$$

It is thus readily seen that the peak discharge of the direct runoff hydrograph corresponding to a given value of C will be as computed before. The hydrograph peak will occur at $t = 3$ h. It will rise and recede linearly. For $C = .6$, the hydrograph corresponding to values of I are plotted as shown in Fig. 10.11.

10.2 TIME-AREA METHODS

The time-area methods were developed in recognition of the importance of time distribution of rainfall on runoff in hydrologic design of storage and regulation works. There are many time-area methods that have appeared in the literature; many of them differ only in the method of presentation. Dooge (1973) gave an excellent survey of these methods. The central idea in these methods is of a time contour or an isochrone. An isochrone is a contour joining those points in the watershed that are separated from the outlet by the same travel time. The isochrones cannot cross one another, cannot close, and can only originate or terminate on the watershed boundary (Dooge, 1959). In general, the isochrones have discontinuities at ridge lines because there the travel time depends on which of the two alternative flow paths the water takes. The isochrones are closely spaced in the flatter, lower portion of the watershed while more widely spaced in the steeper, upper portion. Ross (1921) was perhaps the first to have suggested the idea of using isochrones.

The time-area (TA) diagram therefore indicates the distribution of travel times of different parts of the watershed. When the area of that part of the watershed whose time of travel is less than or equal to a given value, say τ, is plotted against the value τ, the TA diagram is obtained as shown in Fig. 10.12. In other words, this is the graph of area enclosed by an isochrone against time. This figure shows, for any value of τ, the area that will contribute to the maximum discharge at the outlet due to rainfall of duration equal to τ. It is obvious that $0 \leq \tau \leq T_c$ and $0 \leq A(\tau) \leq A(T_c)$.

Often it is more convenient to use the time-area-concentration (TAC) curve, as shown in Fig. 10.12, which is the derivative of the TA diagram. Evidently, its base length corresponds to T_c. In hydrologic literature, frequently the TA diagram, TAC curve, TA curve, and TAC diagram are interchangeably used.

10.2.1 Travel Time

The definition of travel time employed in this text is the one given by Laurenson (1964). Each point in the watershed can be associated with a particular time period representing the time difference between the occurrence of an element of the effective rainfall at that point and the realization of the effect of this element at the outlet. This time period is the travel time of that point. This definition can be made more specific

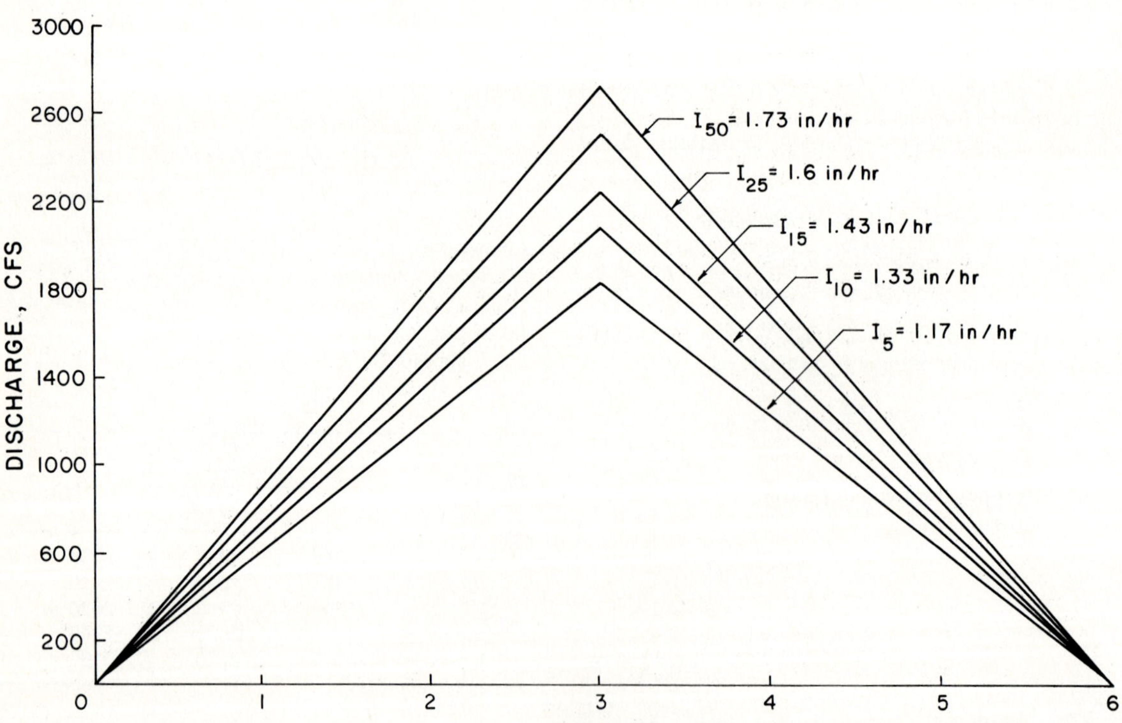

FIGURE 10.11 Example 10.2: Direct runoff hydrographs computed by the rational method ($C = .6$) for various rainfall intensities.

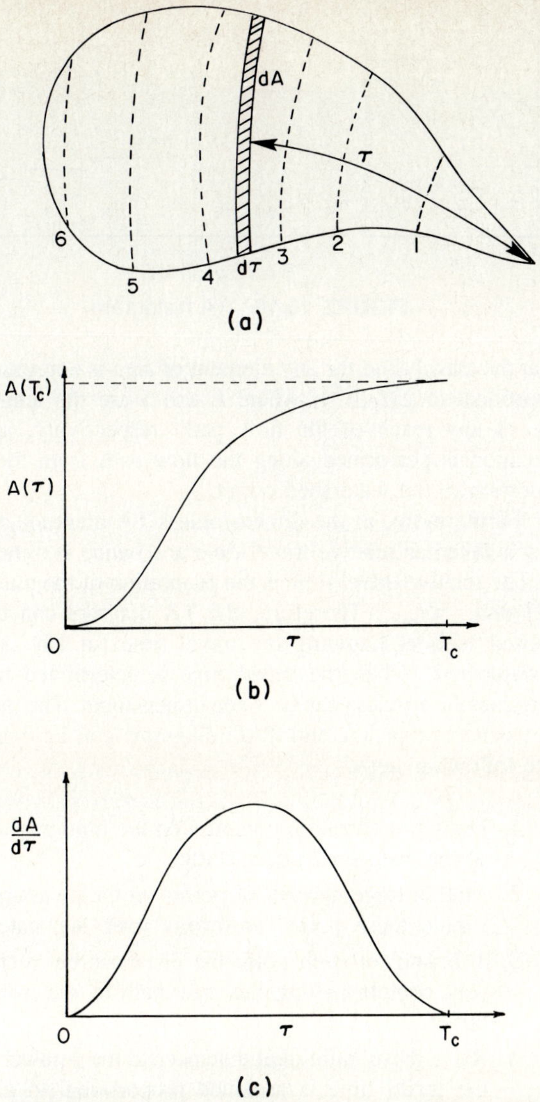

FIGURE 10.12 (a) Isochrones of travel time. (b) TA diagram. (c) TAC curve.

by considering the time of travel of a point to be equal to the time difference between the occurrence of the elemental effective rainfall at the point and the center of mass of the resulting runoff at the outlet. Therefore, this implies travel of an effect rather than a drop of water. This effect is transmitted by both wave movement and translation of water. This travel time can also be called *storage delay* time.

This definition is at variance with the one that regards it as the time it would take a drop of water to travel from the point to the outlet. As emphasized by Laurenson (1964), it is unrealistic to talk in terms of drops of water. On the ground and in storms, water does not exist as a collection of independent drops but as an amorphous mass.

10.2.2 Construction of TA and TAC Diagrams

We can construct TA and TAC diagrams in many different ways. Three of these methods are discussed here.

10.2.2.1 Method 1: Use of Channel Profile

The construction of TA and TAC diagrams for a given watershed is outlined in the following steps.

1. Compute the time of concentration T_c for the watershed in question. This can be done using one of the formulas presented in Chapter 9. If observations on rainfall, infiltration, and runoff are available, then T_c can be taken to be the time difference between the end of the effective rainfall and the point of inflection on the recession limb of the runoff hydrograph as shown in Fig. 10.13.
2. Select an appropriate time interval and divide T_c by it. This selection should be such that enough points are obtained on the TA diagram. For example, if T_c is 10 hours, the time interval may be chosen to be 1 hour. This gives 10 points to construct the TA diagram.
3. Construct isochrones on the watershed map at a spacing equal to the time interval chosen in Step 2, as shown in Fig. 10.14. Here τ_i, $i = 1, 2, \ldots, 10$, denote isochrones on the figure. Drawing of isochrones is tedious. If a detailed topographic survey of the watershed is available, the isochrones can be determined by allowing for the time of overland flow to a channel and then computing the time of travel in the channel using a standard open-channel-flow formula.

   However, a more practical way of constructing the isochrones is by plotting a profile of the main channel and using the contour lines on the topographic map. The profile of the main channel is shown in Fig. 10.15. This profile is segmented into as many parts as the number of time intervals just selected. By superimposing the T_c scale over the channel distance scale (abscissa of Fig. 10.15), the elevations of the intersections of the isochrones with the main channel can be determined. Similar calculations can be undertaken for the major tributaries. The isochrones can then be drawn by joining the points of intersections having the same time of travel.

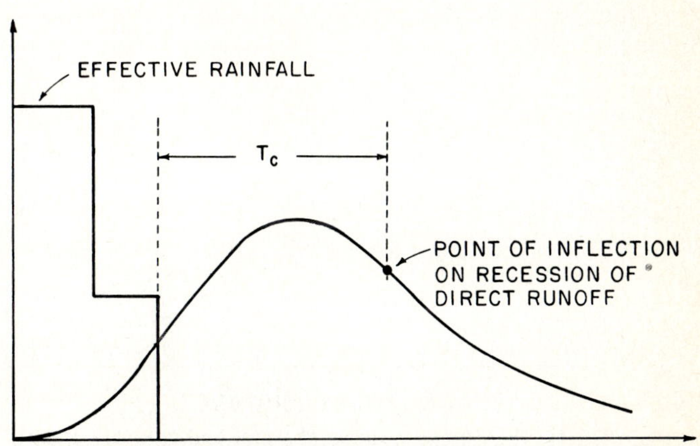

FIGURE 10.13 Determination of time of concentration.

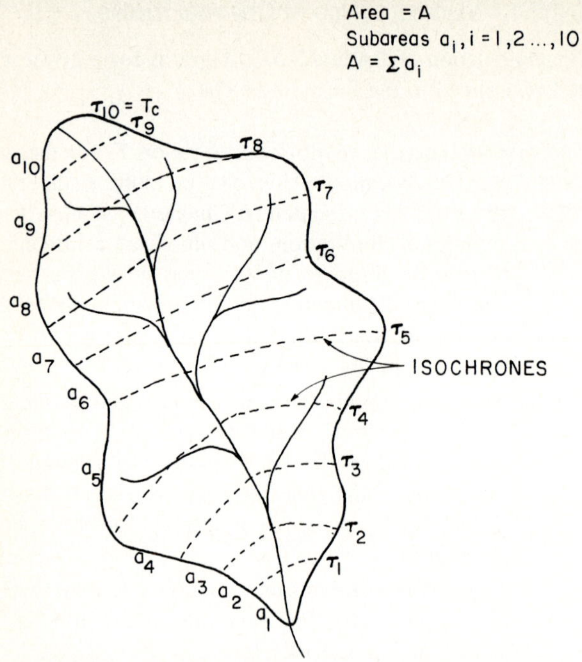

FIGURE 10.14 Isochrones in a hypothetical watershed.

4. Compute the area between the isochrones. This can be easily done by planimetering. These areas are denoted by a_i, $i = 1, 2, \ldots, 10$, in Fig. 10.14.
5. Plot these areas a_i, $i = 1, 2, \ldots, 10$, against τ_i, $i = 1, 2, \ldots,$ as shown in Fig. 10.16. This figure gives the TA histogram. Its continuous-time approximation is the TAC diagram.
6. Plot the cumulative area, Σa_i, $i = 1, 2, \ldots, 10$, against τ_i, as shown in Fig. 10.17. This figure gives the TA diagram.

10.2.2.2 Method 2: Dimensionless TA Diagram

This method of constructing TA and TAC diagrams is due to Laurenson (1964). The fundamental assumption made here

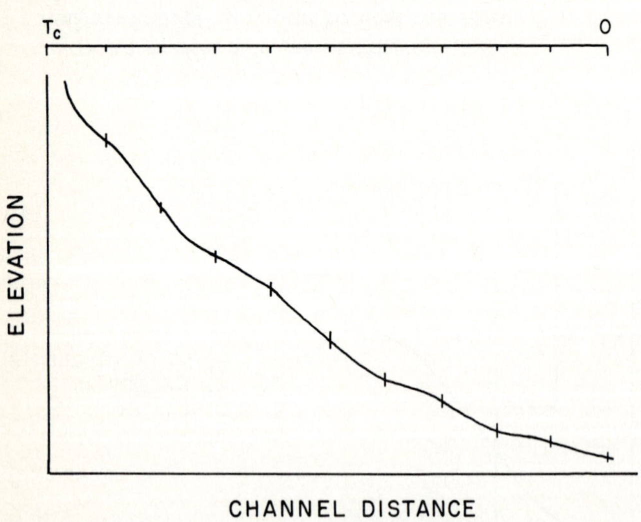

FIGURE 10.15 Profile of main channel and time of concentration.

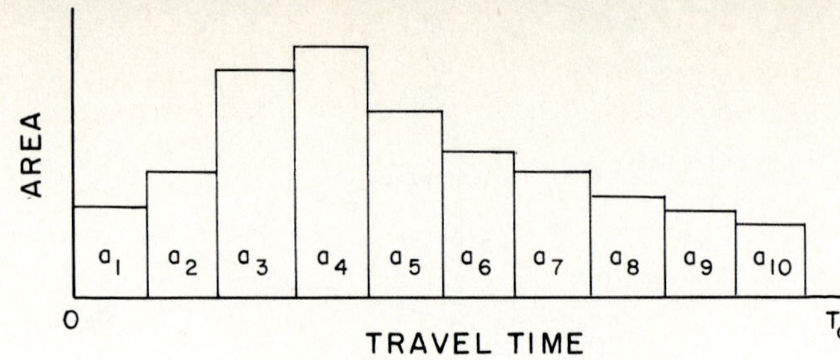

FIGURE 10.16 TA histogram.

is that the travel time for any element of area is approximately proportional to $\Sigma(L/S^{1/2})$, where L and S are the length and slope of any reach of the flow path, respectively, and the summation is performed along the flow path from the point in question to the watershed outlet.

Furthermore, in the dimensionless TA diagram, the abscissa is taken as relative travel time and hence is dimensionless. The relative travel time is the proportion of the maximum travel time, $\tau/\tau_{\max}$. Therefore, the TA diagram can be first proposed without knowing the travel time for any point in the watershed. When the travel time is determined for any point, then the abscissa can be made dimensional. The mechanics of constructing TA and TAC diagrams can be explained in the following steps:

1. Draw the elevation contours on the topographic map of the watershed under study.
2. Mark a large number of points on the contours. Distribute these points uniformly over the watershed.
3. Tabulate for each point the distances between adjacent contours along the flow path to the watershed outlet.
4. Raise these individual distances to the $\frac{3}{2}$ power, since the travel time is assumed proportional to $L/S^{.5}$ or $L^{3/2}/E^{1/2}$, where E is the contour interval. Since E is constant, the travel time is proportional to $L^{3/2}$.

   The watershed outlet may not be on a contour, and a correction would have to be made for the lowest channel segment. This correction amounts to multi-

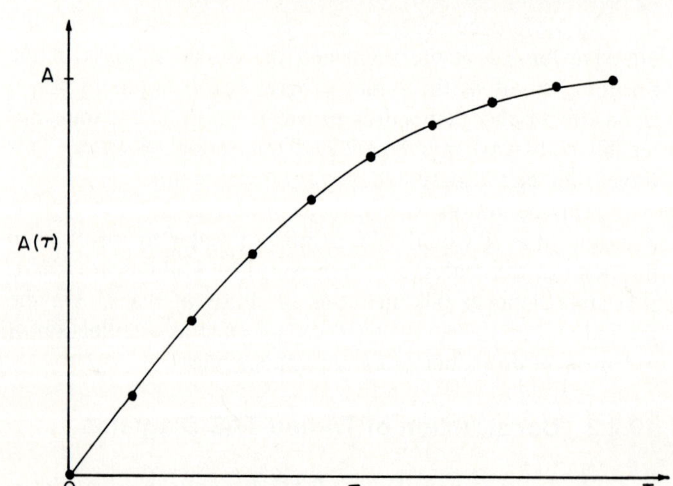

FIGURE 10.17 TA diagram.

plying its length by $(E/E_1)^{1/2}$, where E_1 is the elevation drop of this length.

5. Sum the lengths raised to the $\frac{3}{2}$ power for each point.
6. Reduce the sums obtained in Step 5 proportionately to give a delay time of unity for the extreme upstream portion of the watershed.
7. Mark the relative delay time obtained in Step 6 on the points in the map. Draw the isochrones of relative travel times.
8. The remainder of the steps for construction of TA and TAC diagrams are the same as in the previous method.

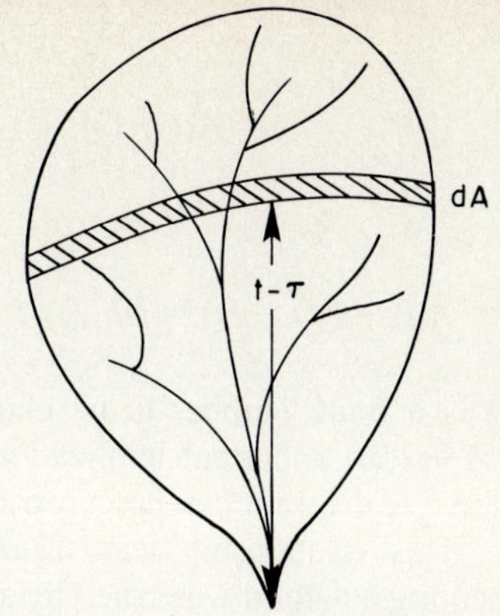

FIGURE 10.18 Watershed area.

10.2.2.3 Method 3: Average Velocity Method

The average velocity method was developed by Soil Conservation Service (1972) and is based on computing the length of flow and the average velocity of flow. The procedure of constructing TA or TAC diagrams using this method is explained in the following steps:

1. Mark a large number of points, preferably uniformly distributed along elevation contours, on the topographic map of the watershed.
2. Measure the distance of flow from these points to the contributing channels. These are overland flow distances.
3. Compute the average velocity of flow corresponding to each of these distances by the SCS method discussed in Chapter 9.
4. Divide the distance of flow by the corresponding velocity and obtain the travel time.
5. From a point on the channel to the watershed outlet, the travel time can be computed using an open channel hydraulics formula.
6. For each point, sum these two times of travel. This gives the time of travel for that point. The remainder of the steps are identical.

The overland flow time is normally a significant portion of the total travel time in smaller watersheds but a much smaller portion in larger watersheds. The computation of overland flow time should be restricted to watersheds smaller than 2000 acres and to subwatershed portions of larger areas above and beyond the point where hydraulic measurements are impractical to make.

10.2.3 Mathematical Representation

From the previous discussion, it is clear that the system represented by the time area methods is linear. Therefore, we may want to determine its UIR in the usual manner. First, consider an elemental area dA within a watershed, as shown in Fig. 10.18. Let $(t - \tau)$ be its time of concentration. Then the effective rainfall at time τ falling on this area will contribute to the discharge at the outlet at time t. Let this contribution be $dQ(t)$. Therefore,

$$dQ(t) = I(\tau)dA(t - \tau) \tag{10.7}$$

Dividing and multiplying the right side by $d\tau$,

$$dQ(t) = \frac{dA(t - \tau)}{d\tau} I(\tau)\, d\tau \tag{10.8}$$

Upon integration and using the initial condition $Q(0) = 0$,

$$Q(t) = \int_0^t \frac{dA(t - \tau)}{d\tau} I(\tau)\, d\tau \tag{10.9}$$

On comparing equation (10.9) with equation (5.10), it is evident that the UIR of the time area methods is

$$h(t - \tau) = \frac{dA(t - \tau)}{d\tau} \tag{10.10}$$

or simply,

$$h(t) = \frac{dA(t)}{d\tau} \tag{10.11}$$

Therefore, the UIR of the TA methods is the TA concentration curve. This is the fundamental assumption in all TA methods. Equation (10.11) was also shown by Nash (1958).

Similarly, the UPR and USR can be determined from the TA methods. The D-time unit UPR, $h_D(t)$, can be written as

$$h_D(t) = \frac{1}{D}[U(t) - U(t - D)] \tag{10.12}$$

Then

$$U(t) = \int_0^t h(x)\, dx = \int_0^t \frac{dA}{d\tau} d\tau = A(t) \tag{10.13}$$

Similarly,

$$U(t - D) = A(t - D) \tag{10.14}$$

Therefore,

$$A(t) - A(t - D) = Dh_D(t) \tag{10.15}$$

The quantity D is a finite number to be chosen arbitrarily. Equation (10.15) has an important implication in hydrology. This says that the TA diagram can be constructed from a D-unit-time UPR. This relationship was utilized by Mathur (1974) in representing a natural watershed by a series of linear channels.

10.2.4 Assumptions and Limitations

The fundamental assumption in the TA methods is one of translation. They allow for the delay experienced by water in reaching the watershed outlet. They also allow for spatially nonuniform effective rainfall. However, they do not allow for the storage effects, which are primarily responsible for producing attenuation in peak flow and for increasing the time base of the hydrograph. It is therefore not surprising that the TA methods tend to overestimate the peak discharge. This is evident from the results reported by Mathur (1974). These methods may yield acceptable results for small watersheds, where storage effects are minimal.

10.2.5 Procedure for Application of TA Methods

If a watershed is divided into a number of strips by drawing isochrones and if the time of travel from the strip to the outlet is applied to the effective rainfall occurring on each strip, the resulting outflow from the whole watershed at its outlet will specify the hydrograph of surface runoff. In other words, the rate of runoff from any strip is equal to the intensity of effective rainfall on that strip, and this runoff will arrive at the watershed outlet by the time of concentration of the strip. The time of concentration of each strip is assumed to be independent of the rate of runoff from the strip and from the surrounding strips. Nash (1958) perhaps explained it best by an example:

> The runoff at any time t is equal to the area enclosed by the 1-hour contour multiplied by the mean intensity of effective rainfall of the hour previous to t plus the additional area enclosed by the 2-hour contour multiplied by the mean intensity of effective rainfall during the period from 1 to 2 hours before t and so on.

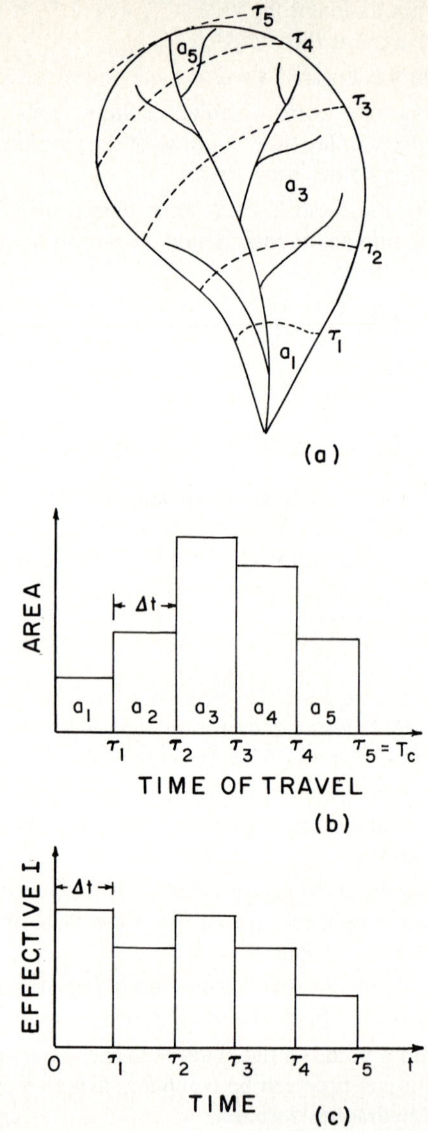

FIGURE 10.19 (a) Watershed decomposition by isochrones. (b) TA histogram. (c) Effective rainfall hyetograph.

For purposes of applying the TA method, let us consider the watershed shown in Fig. 10.19(a). Determine its T_c, and divide it into a number of time intervals of equal length. Draw the isochrones to divide the watershed into the same number of strips as the number of intervals. Determine the area of these strips and plot the time area histogram as shown in Fig. 10.19(b).

Consider the hyetograph of effective rainfall for which runoff hydrograph is desired. Replot this hyetograph such that the time interval is the same as above. This is shown in Fig. 10.19(c). Applying the time area concept, the runoff hydrograph can be mathematically expressed as

$$Q_i = \sum_{j=1}^{i} a_j I_{i-j} \tag{10.16}$$

where a_i is the area enclosed by the ith isochrone and I_i is the intensity of effective rainfall in the ith interval. Equation (10.16) is a discrete convolution of I and a is as discussed in Chapter 4.

EXAMPLE 10.3

Consider an effective rainfall hyetograph as

Time (min)	*Intensity (cm/h)*
0	5
5	4
10	5
15	4
20	2
25	0

and the time area histogram for a specified watershed as

Time (min)	*Area (ha)*
5	1
10	2
15	4
20	3
25	2

Compute the runoff hydrograph due to the above effective rainfall.

Solution: The watershed area is 12 ha. We compute the runoff hydrograph ordinates using equation (10.16).

$Q_0 = 0$

$Q_1 = a_1 I_0 = 1 \times 5 = 5$

$Q_2 = a_1 I_1 + a_2 I_0 = 1 \times 4 + 2 \times 5 = 14$

$Q_3 = a_1 I_2 + a_2 I_1 + a_3 I_0 = 1 \times 5 + 2 \times 4 + 4 \times 5 = 33$

$Q_4 = a_1 I_3 + a_2 I_2 + a_3 I_1 + a_4 I_0 = 1 \times 4 + 2 \times 5 + 4 \times 4 + 3 \times 5 = 45$

$Q_5 = a_1 I_4 + a_2 I_3 + a_3 I_2 + a_4 I_1 + a_5 I_0 = 1 \times 2 + 2 \times 4 + 4 \times 4 + 3 \times 4 + 2 \times 5 = 52$

$Q_6 = a_2 I_4 + a_3 I_3 + a_4 I_2 + a_5 I_1 = 2 \times 2 + 4 \times 4 + 3 \times 5 + 2 \times 4 = 43$

$Q_7 = a_3 I_4 + a_4 I_3 + a_5 I_2 = 4 \times 2 + 3 \times 5 + 2 \times 4 = 31$

$Q_8 = a_4 I_4 + a_5 I_3 = 3 \times 2 + 2 \times 4 = 14$

$Q_9 = a_5 I_4 = 2 \times 4 = 8$

$Q_{10} = 0$

Dividing Q_i, $i = 1, 2, \ldots, 10$, by the watershed area in order to get runoff in centimeters per hour,

Time (min)	*Runoff (cm/h)*
0	0
5	.42
10	1.17
15	2.75
20	3.75
25	4.33
30	3.58
35	2.58
40	1.17
45	.67
50	0

EXAMPLE 10.4

Determine the streamflow hydrograph due to a rainfall storm of April 22, 1979, by the TA diagram method for the Comite River basin near Olive Branch, Louisiana. Construct the TA diagram by the channel profile method. The rainfall hyetograph is given in Fig. 10.20. The drainage area is shown in Fig. 10.21. Other pertinent basin characteristics are shown in Table 10.4.

Solution: First the TA diagram is constructed by using the channel profile method. The time of concentration T_c can be computed by equation (9.13) as

$$T_c = 0.0078 \times (128{,}830)^{.77} (0.0015)^{-.385}$$

$$\cong 820 \text{ min} \cong 13.7 \text{ h}$$

$$\cong 14 \text{ h}$$

The time interval to be selected is 1 h. Thus 14 points will be available to construct the TA diagram. Fig. 10.22 shows a profile

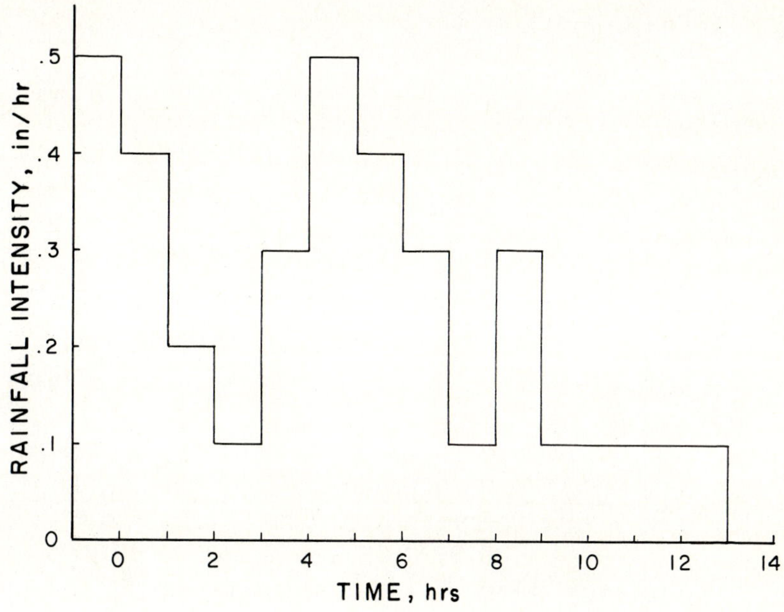

FIGURE 10.20 Example 10.4: Rainfall hyetograph observed on April 22, 1979 (total rainfall = 3.4 in.), at Clinton, Louisiana.

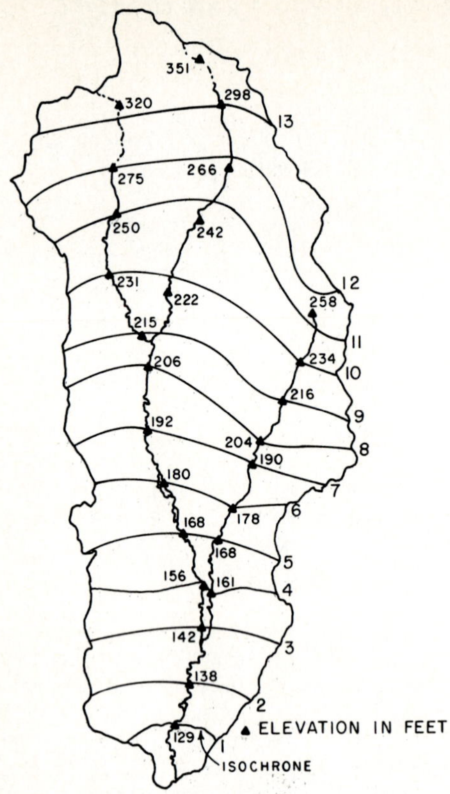

FIGURE 10.21 Example 10.4: The Comite River basin near Olive Branch, Louisiana.

of the main channel segmented into 14 parts. By superimposing the T_c scale over the abscissa, the elevations of the intersections of the isochrones with the main channel are determined. The isochrones are then drawn as shown in Fig. 10.22. The same is done with the major tributaries in the basin. Two major tributaries, the Little Comite River and the Pretty Creek, are considered. Their channel profiles are plotted as shown in Figs. 10.23 and 10.24. The time of concentration for the Little Comite River at its mouth is computed. Its characteristics are L = 9.23 mi = 48,734 ft, and S = .002. Therefore, T_c = 4.64 h ≅ 5 h. By selecting a 1-h time interval, 5 elevation-distance-T_c points are computed:

TABLE 10.4

Physical characteristics of drainage basin of the Comite River

	Profile Data of Longest Watercourse:		*Total*
Point	*Distance Upstream from Gage*	*Streambed Elevation*	*Fall = 235 ft* *Length = 24.4 mi* *Slope = 9.6 ft/mi*
0	0 mi	120 ft	10%–85%
10%	2.4	132	Fall = 148 ft
50%	12.2	189	Length = 18.3 mi
85%	20.7	280	Slope = 8.1 ft/mi
100%	24.4	355	

Station number and name: 7–3775 Comite River Near Olive Branch
Drainage area = 145 mi^2 Average land elevation = 206 ft
Wooded area = 61.1% Ponds, swamps, and lakes = 0.8%
Cleared area = 38.9% Basin mean length = 12.8 mi

T_c (*h*)	*Elevation* (*ft*)
1	215
2	231
3	250
4	275
5	320

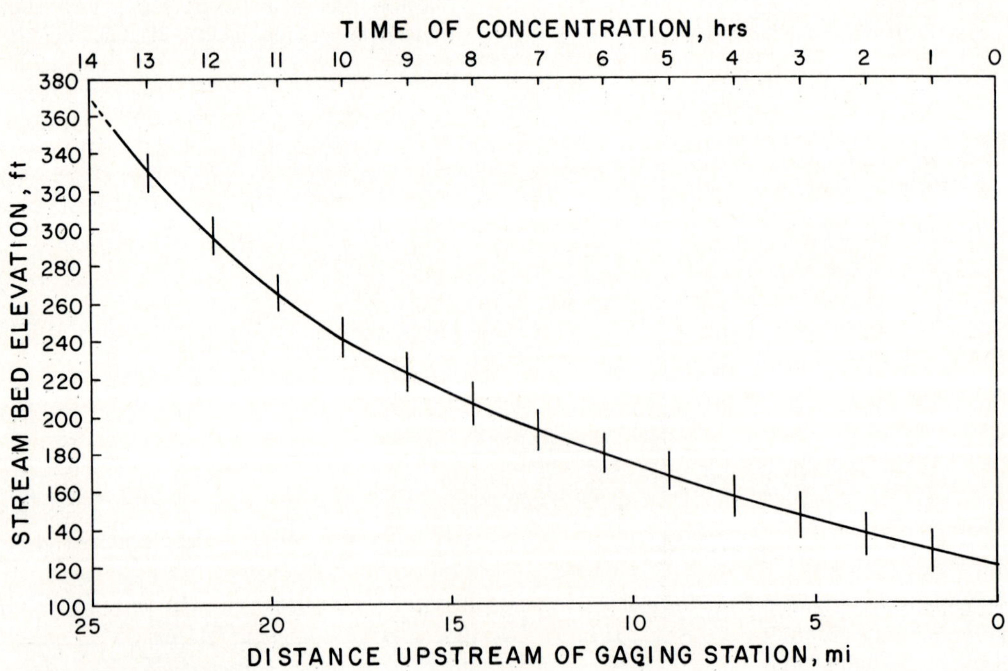

FIGURE 10.22 Example 10.4: Channel bed profile of the Comite River with superimposition of time of concentration.

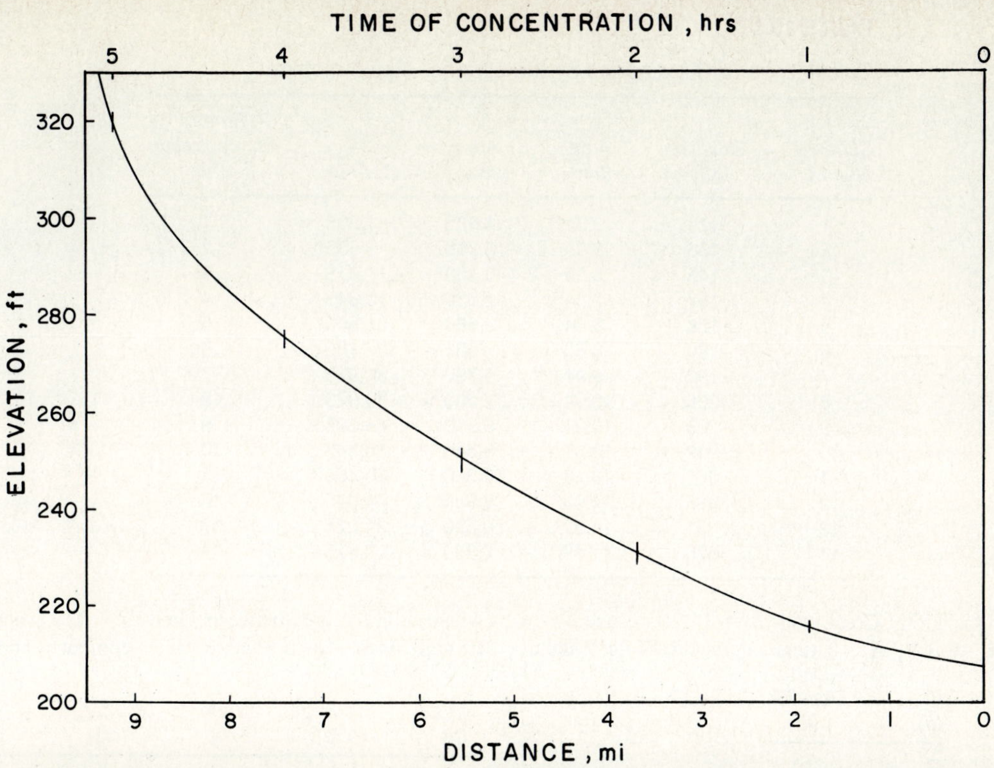

FIGURE 10.23 Example 10.4: Channel bed profile of the Little Comite River with superimposition of time of concentration.

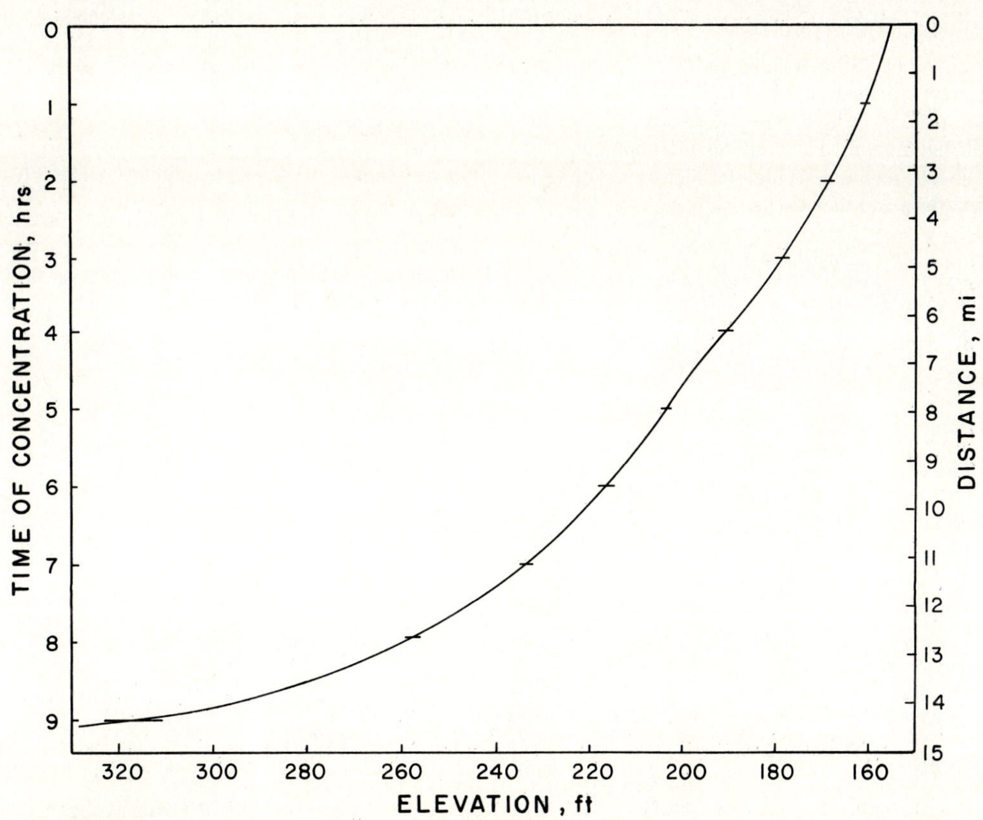

FIGURE 10.24 Example 10.4: Channel bed profile of the Pretty Creek with superimposition of time of concentration.

TABLE 10.5

Calculations for TA diagram by the channel profile method

Subarea Number	Elevation (ft)	Area (mi^2)	Area (acres)	Σ Area	Time of Concentration (h)
1	129	2.82	1,805	1,805	1
2	138	5.83	3,730	5,535	2
3	148	8.56	5,480	11,015	3
4	157	7.54	4,830	15,845	4
5	168	8.84	5,660	21,505	5
6	180	8.88	5,680	27,185	6
7	192	9.04	5,790	32,975	7
8	206	10.94	7,000	39,975	8
9	223	10.21	6,530	46,505	9
10	242	13.81	8,840	55,345	10
11	266	16.13	10,320	65,665	11
12	296	14.03	8,980	74,645	12
13	331	14.97	9,580	84,225	13
14	370	11.49	7,350	9,575	14

Likewise, for the Pretty Creek with L = 14.26 mi and S = .0016, T_c = 8.82 h $\cong$ 9 h. A 1-h time interval leads to 9 data points:

T_c (h)	Elevation (ft)
1	161
2	168
3	178
4	190
5	204
6	216
7	234
8	258
9	320

The areas between isochrones are then computed as in Table 10.5. The resulting TA diagram is shown in Fig. 10.25.

Note that an estimate of abstractions or losses is required for computation of runoff hydrograph. For purposes of simplicity, it is assumed that these abstractions are negligible. Equation (10.9) is first written in discrete form and then used:

$$Q_i = \sum_{j=0}^{i} a_j I_{i-j} \Delta t$$

where a_j is the area between j and $(j - 1)$st isochrones, which are Δt units apart. The discharge values computed are:

Time (h)	Discharge (cfs)	Time (h)	Discharge (cfs)	Time (h)	Discharge (cfs)
0	0	10	18,036	20	14,300
1	903	11	20,840	21	9,130
2	2,587	12	22,685	22	8,550
3	4,593	13	23,926	23	5,980
4	6,250	14	24,460	24	3,030
5	6,773	15	22,200	25	2,570
6	8,640	16	20,500	26	1,690
7	11,013	17	19,800	27	735
8	13,741	18	18,900	28	0
9	15,396	19	18,000		

EXAMPLE 10.5

Solve Example 10.4 with the dimensionless TA diagram method.

Solution: A number of points are marked on the contours, as shown in Fig. 10.26. Distances of these points and their corresponding values of T_c are computed. A time interval of 1 h is selected. The

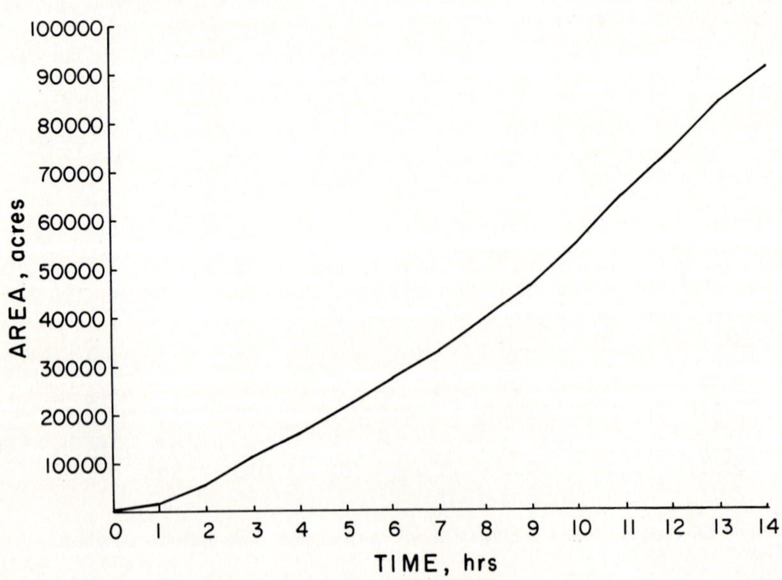

FIGURE 10.25 Example 10.4: TA diagram for the Comite River near Olive Branch, Louisiana.

resulting TA diagram computations are given in Table 10.6 and are graphed in Fig. 10.27. The discharge computations are as follows:

Time (h)	*Discharge (cfs)*	*Time (h)*	*Discharge (cfs)*	*Time (h)*	*Discharge (cfs)*
0	0	9	18,583	19	14,103
1	951	10	20,099	20	9,681
2	2,498	11	19,968	21	7,342
3	3,767	12	23,504	22	5,914
4	4,976	13	26,983	23	3,418
5	8,210	14	25,573	24	2,897
6	9,213	15	20,171	25	1,290
7	11,307	16	18,661	26	202
8	14,770	17	19,500	27	0
		18	18,104		

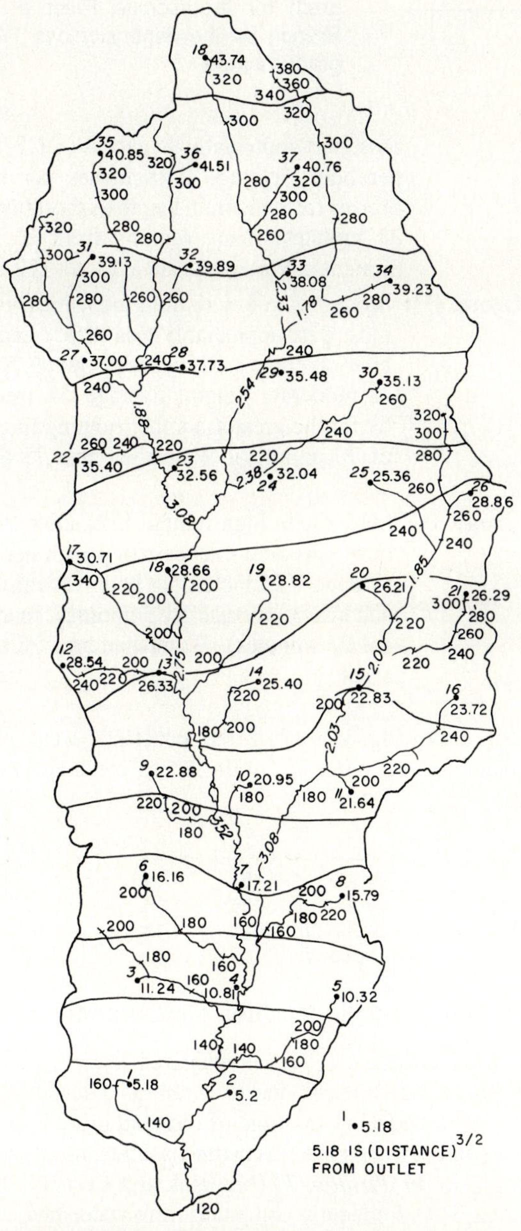

FIGURE 10.26 Example 10.5: The isochrones for the Comite River basin near Olive Branch, Louisiana.

TABLE 10.6

Calculations for the dimensionless TA diagram method

Subarea	*Area (mi^2)*	*Area (acres)*	*Σ Area (acres)*	*T_c (h)*
1	2.97	1,901	1,901	1
2	5.43	3,475	5,376	2
3	6.24	3,994	9,370	3
4	7.79	4,986	14,355	4
5	14.06	8,998	23,354	5
6	6.95	4,448	27,802	6
7	11.04	7,066	34,867	7
8	14.69	9,402	44,269	8
9	15.44	9,882	54,151	9
10	11.30	7,232	61,383	10
11	9.65	6,176	67,559	11
12	17.29	11,066	78,624	12
13	17.00	10,880	89,504	13
14	3.15	2,016	91,520	14

10.3 MODIFICATIONS OF THE RATIONAL METHOD

Several modifications of the rational method have long been suggested. We briefly discuss some of them.

10.3.1 TA Modification

This modification aims at determining a weighted C for a given watershed. By drawing isochrones the watershed is divided into a number of zones. Each zone is assigned an appropriate value of C, depending upon its imperviousness, which is reduced progressively with increasing distance from the outlet. By combining zonal areas with their corresponding values of C, a weighted C is obtained. The total discharge is computed by summing the discharges from the various zones due to the design pattern of rainfall distribution.

10.3.2 The Gregory-Arnold Modification

Gregory and Arnold (1932) modified the rational method by taking into account such factors as shape and slope of the watershed, pattern of the stream system, and characteristics of channel flow. Their modification can be expressed as

$$P = \frac{KT^a}{D^b} \tag{10.17}$$

where K is a coefficient depending on geographic location, T is the recurrence interval in years for a rainfall intensity P in inches per hour for a duration of D in minutes to be equalled or exceeded, and a and b are exponents depending upon geographic location. D must be equal to the time of concentration T_c, which can be expressed as

$$T_c = \frac{F^{1/b}}{(CAK)^s T^{as}} \tag{10.18}$$

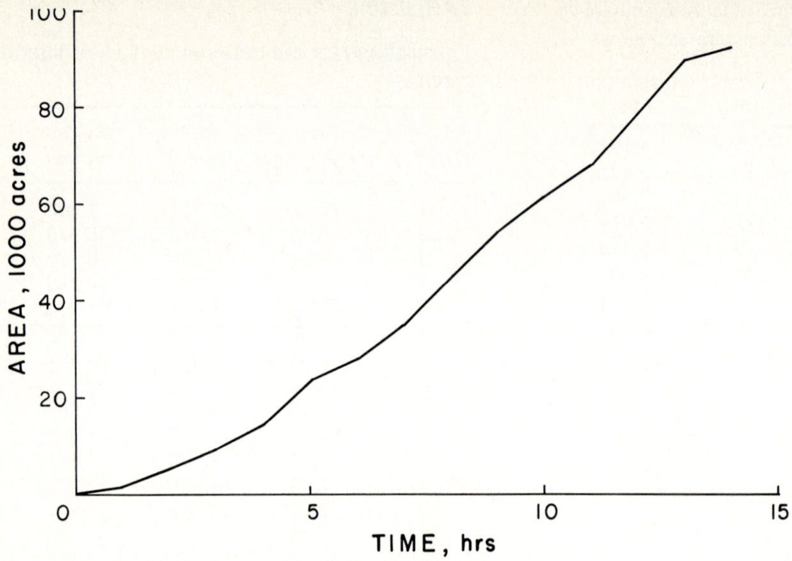

FIGURE 10.27 Example 10.5: TA diagram for the Comite River at Olive Branch by the dimensionless TA diagram method.

where $s = 1/(4 - b)$, A is watershed area, F is the watershed factor, equal to $1/(GBE)$, with G as shape factor, B channel hydraulic factor, and E slope factor.

10.3.3 The Bernard Modification

Bernard (1938) modified the rational coefficient in the Gregory-Arnold modification to

$$C = C_m \left(\frac{T}{100}\right)^x \tag{10.19}$$

where C_m is a limiting runoff coefficient corresponding to a recurrence interval of 100 years, and x is constant. In his modification, he also prepared charts for values of C_m and x in equations (10.19), b, and K in equation (10.17) for the humid central and eastern half of the United States.

10.3.4 The Roe and Snyder Modification

In a study of the rational method for small agricultural areas in Ohio, Roe and Snyder (1943) proposed a modification of the form:

$$Q = CI^a A \tag{10.20}$$

where C and a are constants. Based on a statistical analysis of 119 storms, they found that C and a would depend on rainfall and watershed characteristics and could, therefore, be specified a priori. They classified the watersheds into four groups, as follows.

Group I: Low rainfall intensities (for the most part appreciably less than .5 inch per hour, average .27 inch per hour). Long storms (average 114.5 minutes, the great majority in excess of 1 hour). Watershed area 75 to 373 acres.

Group II: Moderate to high rainfall intensities (for the most part appreciably in excess of 1.5 inches per hour, average 2.28 inches per hour). Storms from medium length to short (average 38 minutes, many running from 20 to 30 minutes). Watershed area 21 to 112 acres.

Group III: Moderate to low rainfall intensities (for the most part appreciably less than 1 inch per hour, average 1.01 inches per hour). Storms of moderate length (average 51 minutes, by far the greater number running in excess of 30 minutes). Watershed area 75 to 373 acres.

Group IV: Moderately high rainfall intensities (for the most part well in excess of 1 inch per hour, average 1.25 inches per hour). Storms short and sharp (average 33.5 minutes, many under 25 minutes). Watershed area 75 to 373 acres.

The following values of the constants were recommended for these groups (if I is in inches per hour and A is in acres):

	C	a
Group I	.6	1.14
Group II	.36	1.14
Group III	.22	1.14
Group IV	.07	1.14

10.3.5 The Bondelid-McCuen Modification

Bondelid and McCuen (1979) emphasized that the peak discharge may actually occur due to a rainfall intensity lasting for a period not equal to the time of concentration. The storm period leading to the peak discharge is often referred to as the *critical storm duration* T_s (Meynink and Cordery, 1979). In practice, T_c is frequently computed from watershed characteristics alone. Thus, the assumption that T_s equals T_c would imply that T_s is also a function of watershed characteristics only. This assumption may lead to a nonoptimum design.

This modification, therefore, refines the rational method for those cases where T_s is less than T_c. Two types of watersheds are distinguished according to land use: (1) homogeneous and (2) nonhomogeneous. First, consider the homogeneous case.

By drawing the isochrones a given watershed can be decomposed into subareas of equal travel time. If the area in the rational method is taken to be the effective area A_e—the area actually contributing to runoff during the storm—then A_e can be considered as a function of rainfall duration D. Obviously, A_e will be the area enclosed by an isochrone of duration equal to the storm duration. A plot of A_e versus D can be generated by constructing isochrones for several values of D. This curve would be analogous to the TA curve. An analytical expression can also be determined for $A_e(D)$ using curve-fitting techniques. Thus expressing the rational method as

$$Q = CP(D)A_e(D) \tag{10.21}$$

where C is constant for homogeneous land use and varies spatially for nonhomogeneous land uses. From equation (10.21) it is clear that the peak discharge will occur when

$$A_e \frac{dP}{dD} + P \frac{dA_e}{dD} = 0 \tag{10.22}$$

Therefore, T_s values are those nonnegative values of D that satisfy equation (10.22). If analytical expressions for $P(D)$ and $A_e(D)$ are known, T_s can be explicitly determined. If these are not known analytically, a graphical solution is easy to obtain. Rearranging equation (10.22),

$$-\frac{dP/dD}{P} = \frac{dA_e/dD}{A_e} \tag{10.23}$$

On plotting the left and right sides of equation (10.23), it is evident that T_s will occur at their intersection.

If the watershed is nonhomogeneous with regard to its land use, then

$$Q = KP \tag{10.24}$$

where $K = CA$. K can be made a function of D alone by determining A_e and the weighted C for each isochrone. Thus $K(D)$ is obtained either graphically or analytically. T_s can then be determined by

$$K \frac{dP}{dD} + P \frac{dK}{dD} = 0 \tag{10.25}$$

Again, an analytical solution is possible if $P(D)$ and $K(D)$ are analytically known. If not, a graphical solution is easily obtained in the manner described above:

$$-\frac{dP/dD}{P} = \frac{dK/dD}{K} \tag{10.26}$$

A comparison of equations (10.23) and (10.26) shows that the rainfall curves remain unaltered, but the watershed curve is expressed in terms of K rather than A_e.

10.4 THE UNIT HYDROGRAPH METHOD

10.4.1 The Unit Hydrograph Theory

In 1932, Sherman developed the unit hydrograph concept for determining the hydrograph of direct runoff from the hyetograph of effective rainfall, which has since been widely used in applied hydrology. The unit hydrograph (UH), or simply unit graph, of a watershed can be defined as the hydrograph of direct runoff resulting from one unit (1 inch or 1 centimeter) of effective rainfall occurring uniformly over the watershed at a uniform rate during a unit period of time. This unit period is not necessarily equal to unity; it can be any finite number. This unit period of effective rainfall is the period for which the unit hydrograph is determined. As soon as this period changes, so will the unit hydrograph for a specified watershed. Thus, there can be as many unit hydrographs for a watershed as there are periods of effective rainfall.

The basic postulates of the unit hydrograph concept are perhaps best described by Johnstone and Cross (1949):

1. For a given drainage basin, the duration of surface runoff is essentially constant for all uniform-intensity storms of the same length, regardless of differences in the total volume of surface runoff.
2. For a given drainage basin, if two uniform-intensity storms of the same length produce different total volumes of surface runoff, then the rates of surface runoff at corresponding times t, after beginning of two storms, are in the same proportion to each other as the total volumes of surface runoff.
3. The time distribution of surface runoff from a given storm period is independent of concurrent runoff from antecedent storm periods.

The definition of the unit hydrograph together with the above postulates constitute what is now called *unit hydrograph theory*.

10.4.2 Mathematical Representation for Continuous-Time Systems

The mathematical representation of unit hydrograph theory is facilitated by proper interpretation of its assumptions. It can be seen that the postulates 1 and 2 together make up the principle of proportionality. In other words, if the duration of input is fixed but its volume changes, then the duration of outflow does not change but its ordinates do in proportion to the volume of input. This is illustrated in Fig. 10.28. The third postulate is the principle of superposition, which allows the decomposition of a complex input into elementary functions and then superimposes the outputs of these functions on one another to obtain the total output. This is illustrated

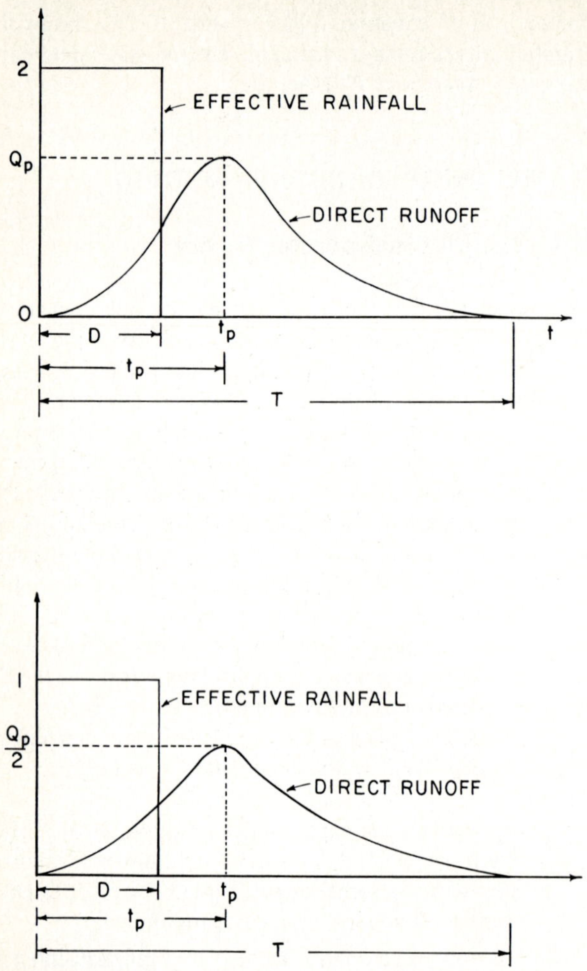

FIGURE 10.28 Application of the principle of proportionality in hydrograph calculation.

in Fig. 10.29. The effective rainfall hyetograph can be broken down into three rectangular pulses. The direct runoff hydrograph due to each pulse is determined. The hydrographs of direct runoff due to these pulses are added to obtain the direct runoff hydrograph due to the complex input.

Thus, unit hydrograph theory assumes that the system is linear and time-invariant. That is, the direct runoff is derived from the effective rainfall by a linear operation. To avoid ambiguity in interpretation of the UH, it must be remembered that it corresponds to a particular duration of effective rainfall and has a unit volume. We often use 1-hour UH, 6-hour UH, or 1-day UH. Here 1-hour, 6-hour, and 1-day are not the duration of the UH, but they are the duration of effective rainfall. Because effective rainfall is assumed to occur uniformly during this period, its duration defines its intensity. This duration is referred to as *unit storm duration* and may change from one application to another. Thus, for a given watershed there can be a large number of UHs, each corresponding to a specific duration of the effective rainfall.

The difficulty arising from the dependence of the UH on the duration D of the effective rainfall is circumvented by letting D be diminished indefinitely. The UH so obtained is called the *instantaneous unit hydrograph* (IUH). Thus, the IUH is a hypothetical UH due to the effective rainfall whose duration tends to zero as a limit but whose volume remains unity (say 1 centimeter). It is then evident that the IUH is independent of the duration of effective rainfall.

If, on the other hand, the duration of effective rainfall is indefinitely long and its intensity is one unit per unit of time (say 1 centimeter per hour), then the hydrograph so obtained is termed the summation hydrograph, S-hydrograph, or simply SH. This hydrograph assumes a deformed S-shape and its ordinates ultimately approach the rate of effective rainfall in the limit or at the time of equilibrium. If the definitions of UH, IUH, and SH are closely scrutinized, the connections between them become apparent.

From Chapters 4 and 5, it is seen that the UH is equivalent to the UPR, the IUH is equivalent to the UIR, and the SH is equivalent to the USR. Denoting the effective rainfall by $I(t)$, the direct runoff by $Q(t)$, and the IUH by $h(t)$, equation (5.6) can be recast as

$$Q(t) = \int_0^t h(t-\tau)I(\tau)\,d\tau, \qquad h(\tau) = 0 \quad \text{for} \quad \tau < 0 \tag{10.27a}$$

or

$$Q(t) = \int_0^t h(\tau)I(t-\tau)\,d\tau \tag{10.27b}$$

with

$$h(t) \geq 0 \qquad \text{for any } t \geq 0$$

$$\lim_{t\to\infty} h(t) = 0$$

$$\int_0^\infty h(s)\,ds = 1$$

Experimental evidence shows that

$$\int_0^\infty (t-\bar{t}\,)h(t)\,dt > 0 \tag{10.27c}$$

where

$$\bar{t} = \int_0^\infty t\,h(t)\,dt \tag{10.27d}$$

Equation (10.27c) states that the IUH must be skewed to the right; that is, its peak should be to the left of the centroid with a long tail extending to the right. If the UH is denoted by $h(D, t)$, equation (3.45) can be rewritten as

$$h(D, t) = \frac{1}{D}\int_{t-D}^t h(s)\,ds \tag{10.28}$$

Similarly, if $U(t)$ denotes the SH, then

$$U(t) = \int_0^t h(s)\,ds \tag{10.29}$$

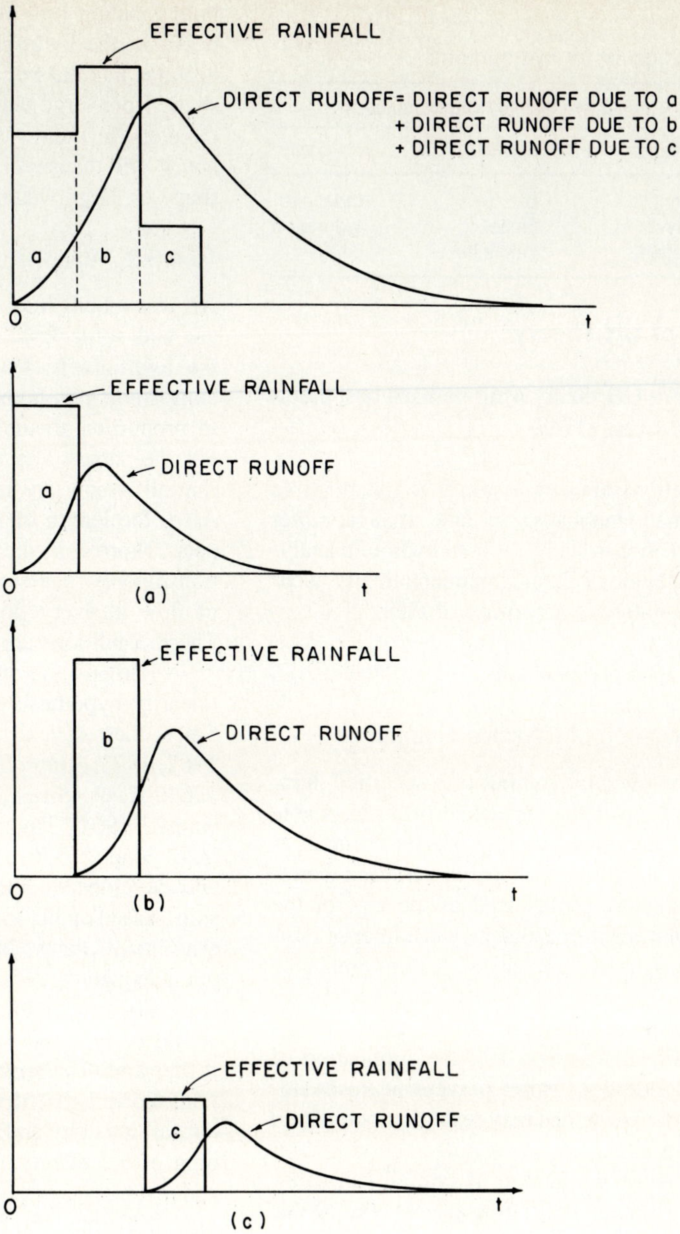

FIGURE 10.29 Application of the principle of superposition in hydrograph calculation.

and

$$Q(t) = \int_0^t U(t-\tau)\,\dot{I}(\tau)\,d\tau, \qquad \dot{I}(t) = \frac{dI(t)}{dt} \tag{10.30}$$

The UH and SH are related by

$$h(D, t) = \frac{1}{D}[U(t) - U(t - D)] \tag{10.31}$$

Similarly,

$$h(t) = \frac{dU(t)}{dt} \tag{10.32}$$

These relationships (equations (10.27)–(10.32)) are derived in Chapters 4 and 5 and are not rederived here.

It may be didactic to summarize that the three hydrographs—IUH, UH, and SH—are ascribed to three distinct characteristics of effective rainfall—volume, duration and intensity—as shown in Table 10.7. Further, it may be noted that the time to the peak of $h(D, t)$ can be determined by differentiating equation (10.31) and equating it to zero:

$$h(t) - h(t - D) = 0 \tag{10.33}$$

This says that the peak of $h(D, t)$ occurs at a time when the ordinate of the IUH is equal to the ordinate at D time units earlier. The SH attains its maximum value at a time equal to D hours less than the time base of the initial $h(D, t)$.

TABLE 10.7

Effective rainfall characteristics for hydrographs

Hydrograph	*Effective Rainfall Characteristics*		
	Volume	*Duration*	*Intensity*
IUH	Unity	0	Indefinite
UH	Unity	Finite	1/duration
SH	Indefinite	Indefinite	Unity

10.4.3 Limitations of UH Theory

The propositions made in UH theory were perhaps best elucidated by Johnstone and Cross (1949):

> All these propositions are empirical. It is not possible to prove them mathematically. In fact, it is a rather simple matter to demonstrate by rational hydraulic analysis that not a single one of them is mathematically accurate. Fortunately, nature is not aware of them.

Let us briefly examine these propositions.

10.4.3.1 Space Invariance of Effective Rainfall

It is seldom true that the effective rainfall of a specified duration will occur uniformly over the watershed of a reasonable size. Therefore, spatial invariance of the effective rainfall is only an assumption, not a reality. The spatial variation generally becomes more and more pronounced as the size of the watershed increases. Rainfall storms that produce intense rainfall usually do not extend over large areas. A fair semblance of a uniform spatial distribution of rainfall may seldom be obtained on watersheds exceeding 500 square kilometers—an upper limit conventionally observed in the application of the unit hydrograph. In addition, storm movement upstream, downstream, or across the watershed may result in nonuniform areal distribution of rainfall.

The nonuniform areal distribution of rainfall can cause variation in hydrograph shape (Dickinson and Ayers, 1965). For example, if high-intensity rainfall occurs over the area near the mouth, then the resulting hydrograph usually experiences a rapid rise, sharp peak, and rapid recession. On the other hand, if this same rainfall occurs over the upstream portion of the watershed, then the resulting hydrograph usually experiences a slow rise, lower and broader peak, and slow recession. However, errors in the unit hydrograph due to nonuniformity in areal distribution of rainfall can be minimized by decomposing the watershed into subwatersheds, each being small to ensure uniform areal distribution, deriving the UH for each subwatershed, and then appropriately combining these UHs to develop the UH for the entire watershed.

10.4.3.2 Time-Invariance of Effective Rainfall

The effective rainfall usually does not occur uniformly, even for as short a duration as 5 minutes. However, the effect of temporal variation of rainfall intensity on the UH shape depends principally on the watershed size. For example, rainfall bursts lasting only a few minutes may produce well-defined peaks in the hydrograph from small watersheds (say, a few hectares in area) but may cause little change in the hydrograph shape from large watersheds (say, several hundred square kilometers in area). Only those changes that last for several hours will cause distinguishable change in the hydrograph shape on large watersheds (Dickinson and Ayers, 1965).

10.4.3.3 Validity of Linearity Hypothesis

All watersheds in nature are nonlinear; some are more nonlinear and some less. They are linear only by assumption. If the hydrographs from the storms of the same duration are compared, it is commonly found that their ordinates are not in proportion to their volumes and that their time bases are not the same. The peaks of the unit hydrographs for small rainfall events are usually lower than those for larger ones. Also, the length of the recession depends on the hydrograph peak. Moreover, the watershed linearity requires a linear relationship between storage and discharge and that the velocity of flow at every point must be constant for all discharges. These conditions are too stringent and are almost never valid.

Perhaps the first conclusive evidence invalidating the linearity hypothesis appeared in a study by Minshall (1960). Since then several investigators (Amorocho, 1961, 1963, 1967, 1973; Amorocho and Orlob, 1961; Amorocho and Hart, 1964; K. P. Singh, 1962, 1964; Diskin, 1964; Kulandaiswamy, 1964; Dooge, 1967; Pilgrim, 1976, 1977, 1980; V. P. Singh, 1976, 1979) have probed the question of linearity and developed nonlinear models for rainfall-runoff relationship. Based on his analysis of small watersheds in the midwestern United States, Minshall (1960) found that one UH could not adequately define the shape of the hydrograph derived from a storm of unit duration. Consequently, different UHs would be required to represent the watershed runoff response if the rainfall intensity varied (see Fig. 10.30). His investigation showed that there may be a good relationship between rainfall intensity and the UH peak and the time from the beginning of the effective rainfall to runoff peak (see Figs. 10.31 and 10.32).

Nevertheless, many investigators support the view that, at least as a practical tool, the unit hydrograph hypothesis is useful (Lowing and Mein, 1982). This hypothesis is particularly adequate in the range of floods experienced on natural watersheds. Besides, this hypothesis is amenable to linear mathematics whose methods are simple and best understood. Therefore, the UH will continue to enjoy its popularity until workable nonlinear methods are developed that are accurate without being unduly complex and costly.

10.4.4 Dimensional Considerations

The dimensions of the UH can be expressed in two ways. If the UH is derived from a direct runoff hydrograph due to the effective rainfall of specified duration whose depth is not unity, then the dimension of the UH ordinates will be equal to the dimension of the runoff divided by the dimension of the depth of effective rainfall. However, runoff can be expressed either in terms of discharge having volumetric rate units or specific

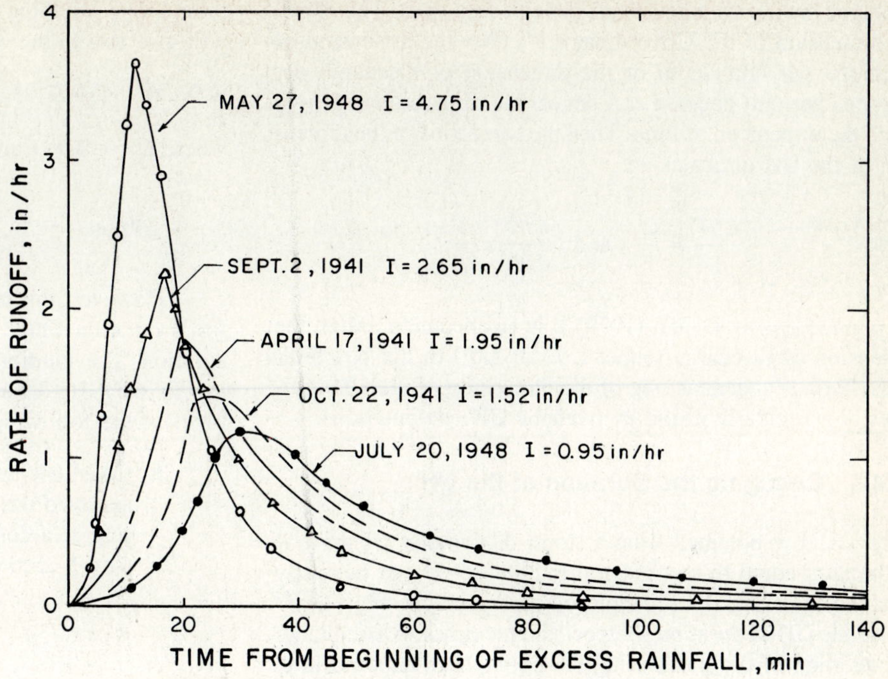

FIGURE 10.30 Unit hydrographs of watershed *W*-1 for different intensities (after Minshall, 1960).

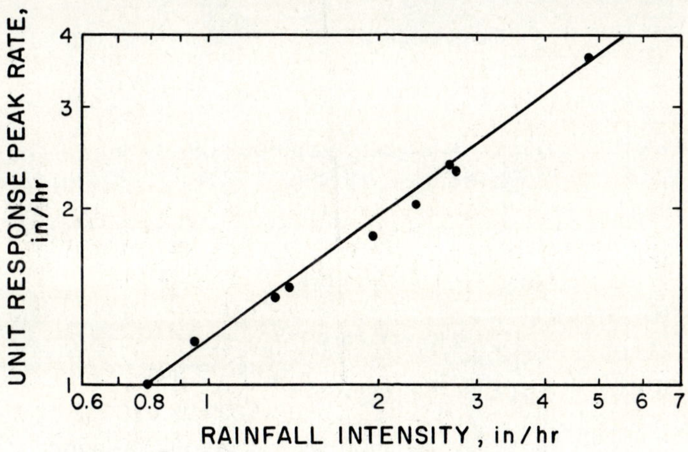

FIGURE 10.31 Relationship between UH peak rate and rainfall intensity (after Minshall, 1960).

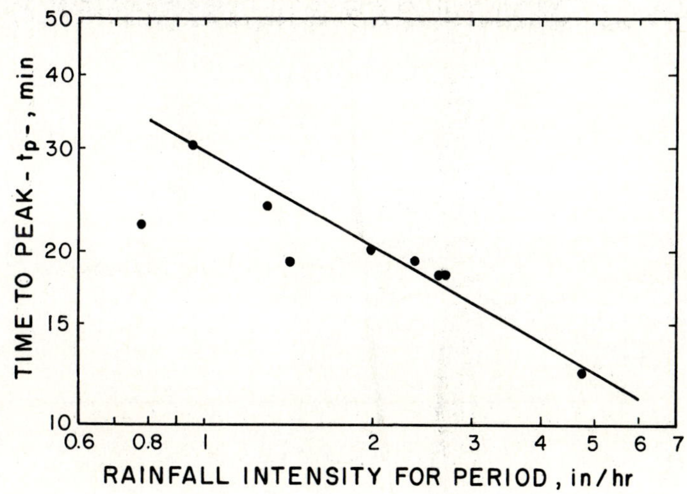

FIGURE 10.32 Relationship between time to UH peak rate and rainfall intensity (after Minshall, 1960).

discharge having volumetric rate per unit area units. Therefore, the dimension of the UH ordinates is either the dimension of discharge per unit depth or the dimension of discharge per unit area per unit depth. Let L denote the dimension of length and T the dimension of time. Then the two forms of the dimension of the UH ordinates are

$$\frac{L^3/T}{L} = \frac{L^2}{T} \quad \text{and} \quad \frac{L^3/T}{L^2 L} = \frac{1}{T}$$

As pointed out by Diskin (1979) it is erroneous to assign the dimension of discharge (either L^3/T or L/T) to the UH ordinates. Proper understanding of the dimension of the UH ordinates is particularly useful in synthetic UH computations.

10.4.5 Changing the Duration of the UH

Once a UH is obtained from a storm of duration D, the UH of duration equal to any integral multiple of D can be easily obtained by using the principle of superposition. Fig. 10.33 shows the UH of duration D repeated three times. A simultaneous addition of ordinates using the principle of superposition will result in runoff hydrograph having three units of volume as a result of the storm having the duration of three units of time ($3D$). Dividing the ordinates of this hydrograph by 3 will give rise to the $3D$ UH, that is,

$$h(D_*, t) = \tfrac{1}{3}\left[h(D, t) + h(D, t - D) + h(D, t - 2D)\right] \quad (10.34a)$$

where $D_* = 3D$. Equation (10.34a) can be generalized as

$$h(D_*, t) = \frac{1}{n}\sum_{j=1}^{n} h(D, t - jD), \qquad D_* = nD \quad (10.34b)$$

However, this method will not work if the UH is required for a duration other than an integral multiple of the given duration. The solution in such cases can be obtained by resorting to the SH. This is referred to as the SH method and involves the following steps (see Fig. 10.34):

1. Construct the SH using the given UH. This SH is assumed due to a continuous effective rainfall occurring at a constant rate of 1 unit depth per unit time (say 1 centimeter per hour). This follows:

$$U(t) = D[h(D,t) + h(D,t - D) + h(D, t - 2D) + \cdots] \quad (10.35a)$$

$$= D\sum_{j=1}^{t} h(D, t - jD) \quad (10.35b)$$

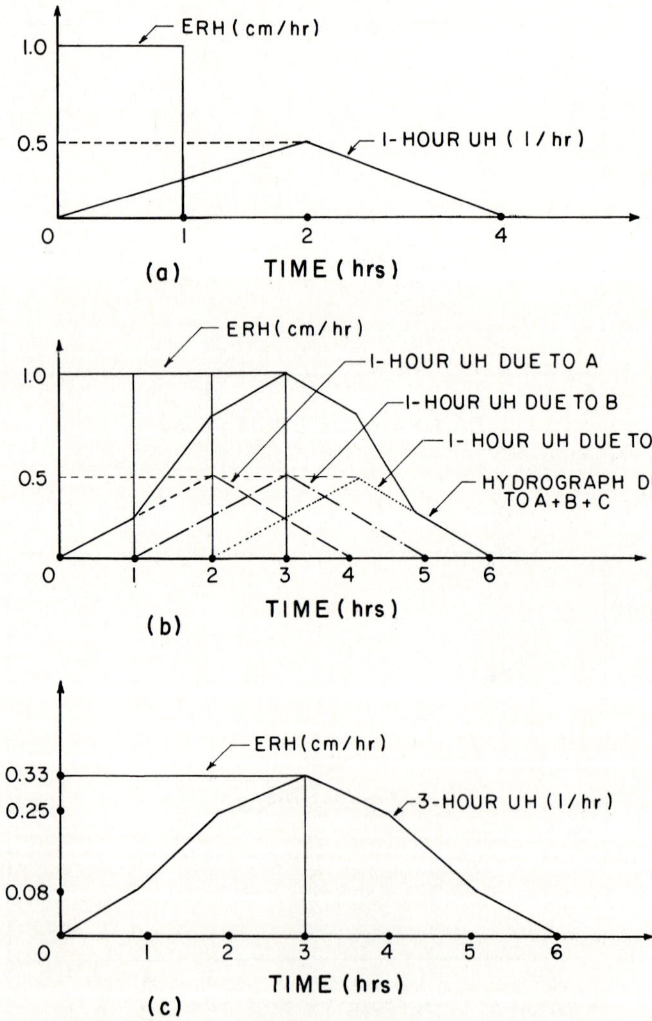

FIGURE 10.33 Derivation of a 3-hour UH from a 1-hour UH.

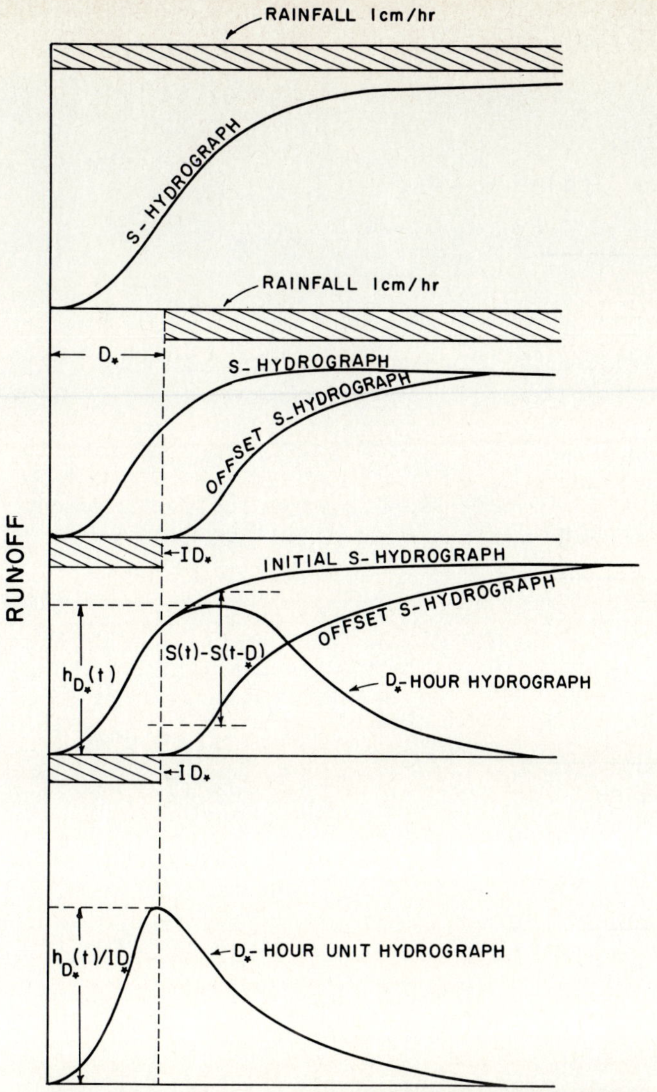

FIGURE 10.34 Derivation of the UH from the S-hydrograph.

2. Advance or offset the position of the SH by a period equal to the desired duration of D_* hours. Call this SH an offset SH, expressed by $U(t - D_*)$.
3. Compute the difference between the ordinates of these two SHs, that is, $U(t) - U(t - D_*)$.
4. Divide the difference in the ordinates by D_*. This gives the desired UH of duration D_*,

$$h(D_*, t) = \frac{U(t) - U(t - D_*)}{D_*} \tag{10.36}$$

A word of caution is appropriate to mention here. If possible, the derivation of a short period UH from one of longer duration should be avoided. Referring to equation (10.36), it is seen that the process of deriving a greater-duration UH from a shorter one entails averaging of ordinates. Therefore, small errors in the ordinates of the short-duration UH are ironed out in calculation of the ordinates of the UH of greater duration. On the other hand, any small errors in the greater duration may lead to disproportionately large errors in calculation of the ordinates of a short-duration UH. This would indicate that the watershed does not respond in an exactly linear manner.

Frequently, the SH constructed from a given UH becomes wobbly when it approaches unity. If the UH of a desired duration is calculated from this SH, its ordinates may be rapidly fluctuating, sometimes even becoming negative. This may be because the UH with which we started is not the true UH. Further, the SH, strictly speaking, may not be continuously known. This would also correspond to the duration of the UH from which it was derived. Therefore, errors in the UH will result in oscillations in the SH. This may again indicate nonlinearity of the watershed. Measurement and data errors may also cause these oscillations.

EXAMPLE 10.6

A rainfall-runoff event is selected to derive a 2-h UH for a watershed. The corresponding effective rainfall hyetograph and direct runoff hydrograph have been computed as

$$I(t) = \begin{cases} 5 \text{ cm/h}, & 0 \le t \le 2 \\ 0, & t \ge 2 \end{cases}$$

$$Q(t) = \begin{cases} 2.5\,t & 0 \le t \le 2 \\ 10 - 2.5\,t, & 2 \le t \le 4 \end{cases}$$

where t is measured in hours. Determine the 2-h unit hydrograph. Determine the S-hydrograph for the watershed. Compute 3-h and 4-h unit hydrographs.

Solution: Since $I(t)$ satisfies the assumptions required by unit hydrograph theory, the 2-h UH can be obtained by simply dividing $Q(t)$ by its volume, which is 10 cm. Therefore,

$$h(2, t) = \begin{cases} 0.25\,t, & 0 \le t \le 2 \\ 1 - 0.25, & 2 \le t \le 4 \end{cases}$$

To determine $U(t)$, we first need to determine $Q(t)$ as a result of $I(t)$, whose intensity is 1 cm/h. The duration of $I(t)$ is, of course, 2 h. This is given as

$$Q(2, t) = \begin{cases} 0.5\,t, & 0 \le t \le 2 \\ 2 - 0.5\,t, & 2 \le t \le 4 \end{cases}$$

Then the S-hydrograph is given as

$$U(t) = \begin{cases} 0.5\,t, & 0 \le t \le 2 \\ 1, & t \ge 2 \end{cases}$$

By using $S(t)$ we can compute the 3-hour UH. To this end, we compute the direct runoff due to 3-h effective rainfall, which can be expressed as

$$Q(3, t) = \begin{cases} 0.5t, & 0 \le t \le 2 \\ 1, & 2 \le t \le 3 \\ 2.5 - 0.5\,t, & 3 \le t \le 5 \end{cases}$$

Therefore, the 3-h UH is

$$h(3, t) = \begin{cases} \dfrac{t}{6}, & 0 \le t \le 2 \\ \dfrac{1}{6}, & 2 \le t \le 3 \\ \left(\dfrac{5}{6}\right) - \left(\dfrac{t}{6}\right), & 3 \le t \le 5 \end{cases}$$

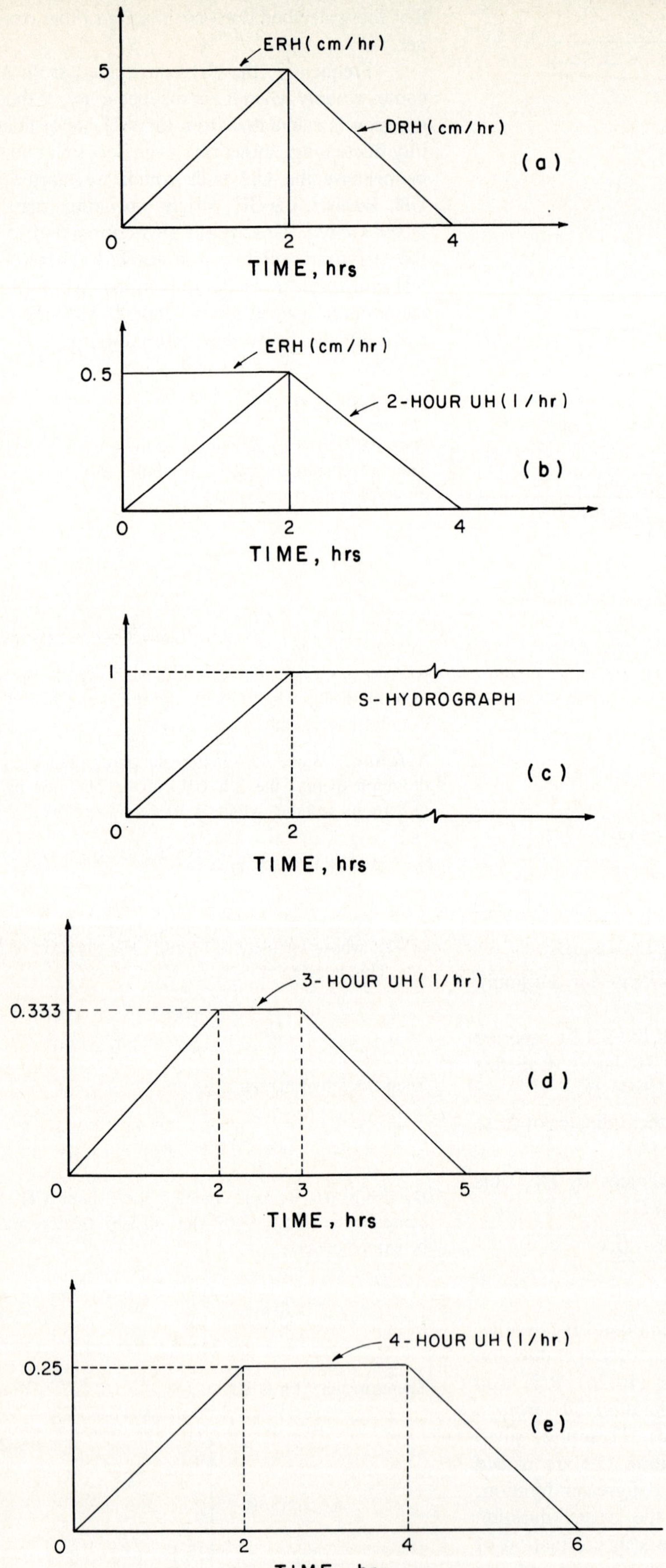

FIGURE 10.35 Example 10.6: (a) Effective rainfall and direct runoff, (b) 2-hour UH. (c) S-hydrograph. (d) 3-hour UH. (e) 4-hour UH.

Likewise, we can compute the 4-h UH from the SH or directly from the 2-hour UH. In any event this is given by

$$h(4, t) = \begin{cases} .125t, & 0 \le t \le 2 \\ .25, & 2 \le t \le 4 \\ .75 - .125t, & 4 \le t \le 6 \end{cases}$$

These hydrographs are sketched in Fig. 10.35.

10.4.6 Distribution Graph

Bernard (1935) expressed the unit hydrograph concept in terms of the distribution graph (DG). A DG is a UH of direct runoff volume, as shown in Fig. 10.36, and 100 percent of the area is under the curve. According to UH theory, if the time base of the UH is divided into any given number of equal time intervals, then the percentage of the total volume of flow occurring during a given time interval will approximately be the same irrespective of the amount of total runoff. The DG will normally be more stable than the UH for a given watershed. Its shape may vary from one watershed to another, reflecting the differences in their runoff characteristics. It may therefore be preferable to the UH for comparing hydrograph characteristics from areas of different sizes.

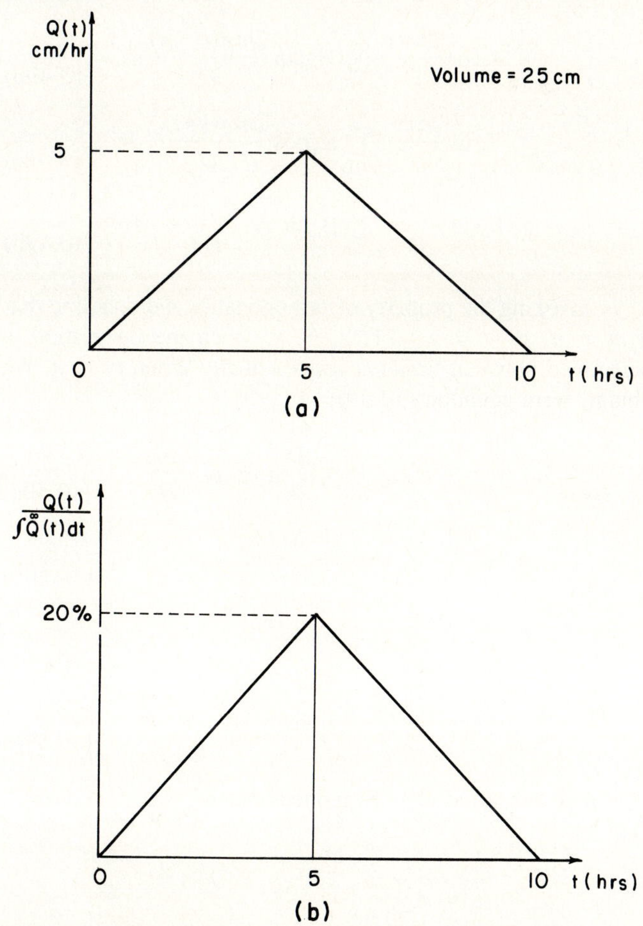

FIGURE 10.36 (a) A direct runoff hydrograph. (b) Distribution graph.

10.4.7 The Uses of the UH

The UH of a specified duration has many uses in environmental and water resource management. It is not necessary to provide any exhaustive catalog of these uses here but we do mention a couple of illustrative uses.

The UH can be used to determine the watershed response due to a given rainfall event if an estimate can be made of the abstractions not contributing to runoff. This has immediate application in flood forecasting and warning. With the knowledge of the time distribution of runoff response, both the flood peak and its time of occurrence are known. For the maximum possible rainfall, maximum possible flood can be determined. Thus, the UH can serve as an indicator of the flood-producing characteristics of a given watershed.

Another important application is in determination of the effect of flood-protection works on watershed response. This can be done by comparing the UH of a stated duration derived prior to the protection works with the UH of the same duration derived after completion of the works (see Fig. 10.37). This comparison will help in proper management of water resources.

Still another application is in determining soil erosion from upland watersheds. If sediment concentration for a given watershed is known, the unit sediment graph for a specified duration can be constructed (Rendon-Herrero, 1979; Williams, 1979; V. P. Singh, Baniukiewicz, and Chen, 1982). This, in turn, can be utilized to estimate erosion due to specified rainfall on the watershed.

10.4.8 Derivation of the UH or IUH for Continuous-Time Systems

There are several methods used to derive the UH, which are also applicable for deriving the IUH. When the duration of the UH is very short, then the UH and IUH are practically indistinguishable from each other. In practice the question arises, How short should this duration be? The answer to this question depends on the shape of the particular unit hydrograph. We can determine it approximately as follows. Assume that the calculated UH of duration D is the IUH. Use equation (10.28) to calculate the corresponding UH of duration D. If this UH does not significantly differ from the original UH, the assumption is justified. If the difference is significant,

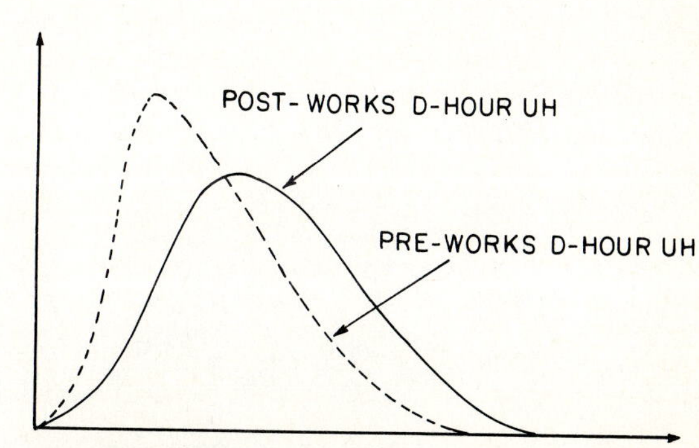

FIGURE 10.37 Comparison of pre- and post-works UHs of a stated duration.

the dissimilarity may suggest the probable shape of the IUH, which may be sketched and tested as before. However, we must bear in mind that the watersheds are usually nonlinear and time-variant. Some of these differences are due to these attributes. Therefore, if the difference is within a few percent, further corrections may not be warranted. Some of the analytical methods of derivation are discussed here.

10.4.8.1 Harmonic Analysis

Harmonic analysis was first applied by O'Donnell (1960) and has since been used by some other investigators (Phillippe and Wiggert, 1969; Hall, 1977; V. P. Singh and Baniukiewicz, 1981; V. P. Singh, Baniukiewicz, and Ram, 1982). It assumes that the effective rainfall, the IUH, and the resultant runoff hydrograph can each be represented by the sum of a harmonic series, each series having the same fundamental time period equal to or greater than the time of storm runoff. The coefficients of the nth harmonics of the three series are found to be related to each other. Therefore, the harmonic coefficients of the IUH can be derived from the harmonic coefficients of effective rainfall and runoff hydrograph.

If the effective rainfall is assumed to be periodic with a period D for the system of memory L, equation (4.10) can be modified to

$$Q(t \pm kD) = \int_{t-L}^{t} I(\tau \pm kD)\, h(t-\tau)\, d\tau, \qquad k = 1, 2, \ldots \tag{10.37}$$

and equation (4.11), to

$$Q(t \pm kD) = \int_{0}^{t} I(t-\tau \pm D)\, h(\tau)\, d\tau, \qquad k = 1, 2, \ldots \tag{10.38}$$

where

$$I(t \pm kD) = 0, \qquad M < t < D \tag{10.39}$$

$$D \geq N = L + M \tag{10.40}$$

Here M denotes the length of effective rainfall and N is the length of runoff hydrograph. If equation (4.10) holds, Q returns to 0 during each period D and equation (10.37) can be replaced by

$$Q(t) = \int_{t-L}^{t} I(\tau)\, h(t-\tau)\, d\tau \tag{10.41}$$

Equation (10.41) holds for an isolated storm event.

Expressing $I(t)$, $h(t)$, and $Q(t)$ by the infinite Fourier expansion,

$$I(t) = \sum_{n=0}^{\infty}\left(a_n\cos\frac{2n\pi t}{D} + b_n\sin\frac{2n\pi t}{D}\right) \tag{10.42}$$

$$h(t) = \sum_{m=0}^{\infty}\left(\alpha_m\cos\frac{2m\pi t}{D} + \beta_m\sin\frac{2m\pi t}{D}\right) \tag{10.43}$$

$$Q(t) = \sum_{r=0}^{\infty}\left(A_r\cos\frac{2r\pi t}{D} + B_r\sin\frac{2r\pi t}{D}\right) \tag{10.44}$$

It may be noted that $I(t)$ and $h(t)$ will be 0 over parts of the range 0 to D. These equations remain valid, however.

It may be recalled that

$$a_n = \frac{1}{2\pi}\int_0^{2\pi} I(t)\cos(nt)\, dt, \qquad n \geq 1 \tag{10.45}$$

$$a_0 = \frac{1}{2\pi}\int_0^{2\pi} I(t)\, dt \tag{10.46}$$

$$b_n = \frac{1}{2\pi}\int_0^{2\pi} I(t)\sin(nt)\, dt, \qquad n \geq 0 \tag{10.47}$$

We can write similar equations for α_m, β_m and A_r, B_r.

Substituting equations (10.42) and (10.43) into equation (10.41),

$$Q(t) = \int_{t-L}^{t}\left[\sum_{n=0}^{\infty}\left(a_n\cos\frac{2n\pi\tau}{D} + b_n\sin\frac{2n\pi\tau}{D}\right)\right] \times \left[\sum_{m=0}^{\infty}\left(\alpha_m\cos\frac{2m\pi(t-\tau)}{D} + \beta_m\sin\frac{2m\pi(t-\tau)}{D}\right)\right] d\tau \tag{10.48}$$

Let us define

$$p_1 = \int_{t-L}^{t}\sum_{n=0}^{\infty} a_n\cos\frac{2n\pi\tau}{D}\sum_{m=0}^{\infty}\alpha_m\cos\frac{2m\pi(t-\tau)}{D}\, d\tau \tag{10.49a}$$

$$p_2 = \int_{t-L}^{t}\sum_{n=0}^{\infty} a_n\cos\frac{2n\pi\tau}{D}\sum_{m=0}^{\infty}\beta_m\sin\frac{2m\pi(t-\tau)}{D}\, d\tau \tag{10.49b}$$

$$p_3 = \int_{t-L}^{t}\sum_{n=0}^{\infty} b_n\sin\frac{2n\pi\tau}{D}\sum_{m=0}^{\infty}\alpha_m\cos\frac{2m\pi(t-\tau)}{D}\, d\tau \tag{10.49c}$$

$$p_4 = \int_{t-L}^{t}\sum_{n=0}^{\infty} b_n\sin\frac{2n\pi\tau}{D}\sum_{m=0}^{\infty}\beta_m\sin\frac{2m\pi(t-\tau)}{D}\, d\tau \tag{10.49b}$$

Invoking the property of orthogonality and recalling that if $m \neq n$, $p_i = 0$, $i = 1, 2, 3, 4$, when the integration is performed between limits D spaced apart. When $m = n$, we obtain, from equation (10.49),

$$p_1 = D\, a_0\alpha_0 + \sum_{n=1}^{\infty}\frac{D}{2}a_n\alpha_n\cos\frac{2\pi nt}{D} \tag{10.50a}$$

$$p_2 = \frac{D}{2}\sum_{n=1}^{\infty} a_n\beta_n\sin\frac{2n\pi t}{D} \tag{10.50b}$$

$$p_3 = \frac{D}{2}\sum_{n=1}^{\infty} b_n\alpha_n\sin\frac{2n\pi t}{D} \tag{10.50c}$$

$$p_4 = -\frac{D}{2}\sum_{n=1}^{\infty} b_n\beta_n\cos\frac{2n\pi t}{D} \tag{10.50d}$$

Therefore, equation (10.48) reduces to

$$Q(t) = D\, a_0\alpha_0 + \sum_{n=1}^{\infty}\frac{D}{2}(a_n\alpha_n - b_n\beta_n)\cos\frac{2n\pi t}{D} + \sum_{n=0}^{\infty}\frac{D}{2}(a_n\beta_n + b_n\alpha_n)\sin\frac{2n\pi t}{D} \tag{10.51}$$

On comparing equation (10.51) with equation (10.44),

$$A_n = \frac{D}{2}(a_n\alpha_n - b_n\beta_n), \qquad n \geq 1; A_0 = a_0\alpha_0 \tag{10.52}$$

$$B_n = \frac{D}{2}(a_n\beta_n + b_n\alpha_n) \tag{10.53}$$

Equations (10.51) and (10.53) define the linkage between the harmonic coefficients of I, h, and Q. Solving for the harmonic coefficients α_n and β_n of the IUH,

$$\alpha_n = \frac{2}{D}\frac{a_nA_n + b_nB_n}{a_n^2 + b_n^2} \qquad \text{for } n \geq 1; \alpha_0 = \frac{1}{D}\frac{A_0}{a_0} \tag{10.54}$$

$$\beta_n = \frac{2}{D}\frac{a_nB_n + b_nA_n}{a_n^2 + b_n^2} \tag{10.55}$$

Thus, it is seen that the harmonic coefficients of the IUH can be determined from the harmonic analysis of effective rainfall and direct runoff hydrograph. Hence, the IUH can be derived from its harmonic components.

The derivation of equations (10.54) and (10.55) can be accomplished in an alternative manner (Dooge, 1973) by expressing $I(t)$, $h(t)$, and $Q(t)$ as

$$I(t) = \sum_{n=0}^{\infty} \gamma_n e^{in2\pi t/D} \tag{10.56}$$

$$h(t) = \sum_{m=0}^{\infty} \rho_m e^{im2\pi t/D} \tag{10.57}$$

$$Q(t) = \sum_{r=0}^{\infty} \xi_r e^{ir2\pi t/D} \tag{10.58}$$

Recall that

$$\xi_r = \frac{1}{D}\int_0^D Q(t)e^{-ir2\pi t/D}\, dt \tag{10.59}$$

Replacing $Q(t)$ in equation (10.59) by equation (10.41) and then substituting equations (10.56) and (10.57) for $I(t)$ and $h(t)$ and simplifying,

$$\xi_r = D\gamma_r\rho_r \tag{10.60}$$

This is the linkage equation between the Fourier coefficients of $I(t)$, $h(t)$, and $Q(t)$. These coefficients can be expressed in terms of cosine and sine coefficients of equations (10.56) through (10.58) as

$$\xi_k = \frac{1}{2}(A_k - iB_k) \tag{10.61}$$

$$\gamma_k = \frac{1}{2}(a_k - ib_k) \tag{10.62}$$

$$\rho_k = \frac{1}{2}(\alpha_k - i\beta_k) \tag{10.63}$$

Substituting equations (10.61) through (10.63) into equation (10.60),

$$A_k = \frac{D}{2}(\alpha_k a_k - b_k\beta_k) \tag{10.64}$$

$$B_k = \frac{D}{2}(a_k\beta_k + b_k\alpha_k) \tag{10.65}$$

These equations are the same as equations (10.54) and (10.55) and can be used to obtain α_k and β_k as given in equation (10.43).

The above analysis utilizes infinite harmonic series. Such a series sums exactly to the value of the function it represents at every one of the infinite number of points in the range of its application. Its application to a set of data consisting of only a finite number of points is discussed later in the chapter.

10.4.8.2 Fourier Transform Method

The operation of this method is analogous to the method of harmonic analysis. Taking the Fourier transforms $X(w)$, $H(w)$, and $Y(w)$, respectively, of the effective rainfall, the IUH, and the runoff hydrograph,

$$F[I(t)] = X(w) = \int_{-\infty}^{\infty} I(t)\, e^{-iwt}\, dt \tag{10.66}$$

$$F[h(t)] = H(w) = \int_{-\infty}^{\infty} h(t)\, e^{-iwt}\, dt \tag{10.67}$$

$$F[Q(t)] = Y(w) = \int_{-\infty}^{\infty} Q(t)\, e^{-iwt}\, dt \tag{10.68}$$

Substituting equation (10.27) into equation (10.68),

$$Y(w) = \int_{-\infty}^{\infty} e^{-iwt}\int_{-\infty}^{\infty} I(\tau)\, h(t-\tau)\, d\tau\, dt \tag{10.69}$$

Upon simplification,

$$Y(w) = X(w)H(w) \tag{10.70}$$

Equation (10.70) is the linkage between Fourier transforms of $I(t)$, $h(t)$, and $Q(t)$ and shows that convolution in the time domain is equivalent to multiplication in the transform domain. For purposes of calculation, the real and imaginary parts of the transforms (denoted by subscripts R and I, respectively) can be separated, for example,

$$X(w) = X_R(w) + iX_I(w) \tag{10.71}$$

$$Y(w) = Y_R(w) + iY_I(w) \tag{10.72}$$

$$H(w) = H_R(w) + iH_I(w) \tag{10.73}$$

Substituting equations (10.71) through (10.73) into equation (10.70) and separating the real and imaginary parts gives

$$Y_R(w) = H_R(w)X_R(w) - H_I(w)X_I(w) \quad (10.74)$$

$$Y_I(w) = H_R(w)X_I(w) + H_I(w)X_R(w) \quad (10.75)$$

Solving for $H_R(w)$ and $H_I(w)$,

$$H_R(w) = \frac{X_R(w)Y_R(w) + X_I(w)Y_I(w)}{[X_R(w)]^2 + [X_I(w)]^2} \quad (10.76)$$

$$H_I(w) = \frac{X_R(w)Y_I(w) + X_I(w)Y_R(w)}{[X_R(w)]^2 + [X_I(w)]^2} \quad (10.77)$$

Equations (10.76) and (10.77) specify the IUH in the frequency domain. It is necessary to invert $H(w)$ to obtain $h(t)$. Therefore,

$$h(t) = \frac{1}{2\pi}\int_{-\infty}^{\infty} [H_R(w)\cos wt - H_I(w)\sin wt]\, dw \quad (10.78)$$

This can be written, noting that $H_R(-w) = H_R(w)$, $H_I(-w) = -H_I(w)$, as

$$h(t) = \frac{1}{\pi}\int_{0}^{\infty} [H_R(w)\cos wt - H_I(w)\sin wt]\, dw \quad (10.79)$$

This can be further reduced to

$$h(t) = \frac{2}{\pi}\int_{0}^{\infty} H_R(w)\cos wt\, dw \quad (10.80)$$

$$h(t) = \frac{2}{\pi}\int_{0}^{\infty} H_I(w)\sin wt\, dw \quad (10,81)$$

The Fourier transform method relaxes the restriction due to isolated inputs and finite memories. However, it is difficult to invert the transform particularly when it is known only numerically. Nevertheless, this method has been used in modeling hydrologic systems (Levi and Valdes, 1964; Blank and Delleur, 1968; Blank, Delleur, and Giorgini, 1971; Delleur and Rao, 1971a, 1971b; Hall, 1977). We discuss the problem of inversion in the next section.

10.4.8.3 Laplace Transform Method

Following the procedure of Fourier transform, let us define the Laplace transforms

$$L[I(t)] = X(s) = \int_0^{\infty} I(t)e^{-st}\, dt \quad (10.82)$$

$$L[h(t)] = H(s) = \int_0^{\infty} h(t)e^{-st}\, dt \quad (10.83)$$

$$L[Q(t)] = Y(s) = \int_0^{\infty} Q(t)e^{-st}\, dt \quad (10.84)$$

Substituting equation (10.27) into equation (10.84),

$$Y(s) = \int_0^{\infty} e^{-st}\int_0^{\infty} I(t)\, h(t-\tau)\, d\tau\, dt \quad (10.85)$$

On simplifying,

$$Y(s) = X(s)H(s) \quad (10.86)$$

This is the linkage between the Laplace transforms of $I(t)$, $h(t)$, and $Q(t)$. Thus if $X(s)$ and $Y(s)$ are known, then $H(s)$ can be determined and inverted to obtain $h(t)$. A distinct disadvantage of the Laplace transform method is that its numerical inversion is painful. Bellman, Kalaba, and Lockett (1966) have suggested that an efficient way of performing this inversion is to expand the Laplace transform in terms of series of orthogonal polynomials and then invert the series term by term. Of all the transforms, the Laplace transform is most frequently utilized in hydrologic modeling (Diskin, 1964, 1967; Blank and Delleur, 1968; Blank, Delleur, and Giorgini, 1971; Delleur and Rao, 1971a, 1971b; Dooge, 1973).

10.4.8.4 Laplace-Gamma Method

The Laplace-gamma method consists in considering the Laplacian variable s in equations (10.82)–(10.84) to be complex, as suggested by Chambers (1965). By doing so, the complex inversion of the Laplace transform can be circumvented. This approach was pursued by Blank and Delleur (1968), Blank, Delleur, and Giorgini (1971), and Rao and Delleur (1971). Following their work, we rewrite equation (10.83) as

$$H(s) = \int_0^{\infty} h(t)e^{-st}\, ds \quad (10.87)$$

where $s = x + iy$. Here x is taken to be a real constant to the right of the singularities of $H(s)$. Then $ds = i\, dy$, so that

$$\begin{aligned} H(x+iy) = H(y) &= \int_0^{\infty} e^{-(x+iy)t}\, h(t)\, dt \\ &= \int_0^{\infty} e^{-xt}\, h(t)(\cos yt - i\sin yt)\, dt \\ &= H_R(y) + iH_I(y) \end{aligned} \quad (10.88)$$

The subscripts R and I denote the real and imaginary parts, respectively. We have

$$H_R(y) = \int_0^{\infty} h(t)\, e^{-xt}\cos yt\, dt \quad (10.89)$$

$$H_I(y) = -\int_0^{\infty} e^{-xt}\, h(t)\sin yt\, dt \quad (10.90)$$

Therefore, the term $H(y)$ is the Fourier transform of $h(t)e^{-xt}$. The inverse transform can be expressed as

$$e^{-xt}h(t) = \frac{1}{2\pi}\int_{-\infty}^{\infty} H(y)\, e^{iyt}\, dy, \qquad t > 0 \quad (10.91)$$

and because $h(t) \equiv 0$, $t < 0$,

$$0 = \frac{1}{2\pi}\int_{-\infty}^{\infty} H(y)\, e^{-iyt}\, dy, \qquad t < 0 \tag{10.92}$$

Summing equations (10.91) and (10.92),

$$e^{-xt}h(t) = \frac{1}{\pi}\int_{-\infty}^{\infty} H(y)\cos yt\, dt \tag{10.93}$$

Since $h(t)$ is real, $H_R(-y) = H_R(y)$ and $H_I(-y) = -H_I(y)$. In other words, $H_R(y)$ is an even function of y and $H_I(y)$ an odd function of y. Then equation (10.93) can be written as

$$h(t) = \frac{2e^{xt}}{\pi}\int_{0}^{\infty} H_R(y)\cos yt\, dt \tag{10.94}$$

Here $H_R(y)$ and $H_I(y)$ can be computed using equation (10.86), that is,

$$H(y) = H_R(y) + iH_I(y) = \frac{Y_R(y) + iY_I(y)}{X_R(y) + iX_I(y)}$$

Upon simplification,

$$H_R(y) = \frac{Y_R(y)X_R(y) + Y_IX_I(y)}{[X_R(y)]^2 + [X_I(y)]^2} \tag{10.95}$$

$$H_I(y) = \frac{X_R(y)Y_I(y) - Y_R(y)X_I(y)}{(X_R(y))^2 + (X_I(y))^2} \tag{10.96}$$

These are the linkage equations for the Laplace-gamma transforms of $h(t)$ in terms of those of $I(t)$ and $Q(t)$. It may be noted that when $X = 0$, the Laplace-gamma transform reduces to the Fourier transform. The choice of X is dictated by system stability. Since hydrologic systems are stable, X can be taken as a positive number.

We now consider an example to show the use of the Laplace-gamma method and to compare it with the Laplace transform method.

EXAMPLE 10.7

This example is from Blank and Delleur (1968). Let $I(t) = te^{-bt}$ and $Q(t) = t^2 e^{-at}$, where both a and b are constants greater than 0. Determine the IUH.

Solution:

$$Y_R(y) = \int_0^{\infty} e^{-(x+a)t}\, t^2\cos yt\, dt$$

$$Y_I(y) = -\int_0^{\infty} e^{-(x+a)t}t^2\sin yt\, dt$$

$$X_R(y) = \int_0^{\infty} e^{-(x+b)t}t\cos yt\, dt$$

$$X_I(y) = -\int_0^{\infty} e^{-(x+a)t}t\sin yt\, dt$$

Using the following identities (Dwight, 1966),

$$\int_0^{\infty} t\, e^{-ct}\sin mt\, dt = \frac{2\,cm}{(c^2 + m^2)^2}$$

$$\int_0^{\infty} t\, e^{-ct}\cos mt\, dt = \frac{c^2m^2}{(c^2 + m^2)^2}$$

$$\int_0^{\infty} t^2 e^{-ct}\cos mt\, dt = \frac{2c(c^2 - 3m^2)}{(c^2 + m^2)^3}$$

$$\int_0^{\infty} t^2 e^{-ct}\sin mt\, dt = \frac{2m(3c^2 - m^2)}{(c^2 + m^2)^3}$$

We obtain

$$Y_R(y) = \frac{2(x + a)^3 - 6(x + a)y^2}{[(x + a)^2 + y^2]^3}$$

$$Y_I(y) = \frac{2y^3 - 6(x + a)^2y}{[(x + a)^2 + y^2]^3}$$

$$X_R(y) = \frac{(x + b)^2 - y^2}{[(x + b)^2 + y^2]^2}$$

$$X_I(y) = -\frac{2(x + b)y}{[(x + b)^2 + y^2]^2}$$

and from equation (10.95),

$$H_R(y) = \frac{2(x + a)^3(x + b)^2}{[(x + a)^2 + y^2]^3} + \frac{12(x + a)^2(x + b) - 6(x + b)^2(x + a) - 2(x + a)^3}{[(x + a)^2 + y^2]^3}y^2 + \frac{6(x + a) - 4(x + b)}{[(x + a)^2 + y^2]^3}y^4$$

We can substitute $H_R(y)$ into equation (10.94) and obtain $h(t)$. This will involve evaluation of the following integrals:

$$\int_0^{\infty}\frac{\cos yt}{[(x + a)^2 + y^2]^3}dy, \quad \int_0^{\infty}\frac{y^2\cos yt}{[(x + a)^2 + y^2]^3}dy, \quad \text{and} \quad \int_0^{\infty}\frac{y^4\cos t}{[(x + a)^2 + y^2]^3}dy$$

If we let $\delta = x + a$, then the integrals

$$\int_0^{\infty}\frac{e^{izt}}{[\delta^2 + z^2]^3}dz, \quad \int_0^{\infty}\frac{z^2e^{izt}}{[\delta^2 + z^2]^3}dz, \quad \text{and} \quad \int_0^{\infty}\frac{z^4e^{izt}}{[\delta^2 + z^2]^3}dz$$

will be computed around the semicircle, as shown in Fig. 10.38. Each integral has a pole of order three at $z = i\delta$. To illustrate, consider

$$\int\frac{e^{izt}}{(\delta^2 + z^2)^3}dz = \int_{-\infty}^{\infty}\frac{\cos yt}{(\delta^2 + y^2)^3}dy + i\int_{-\infty}^{\infty}\frac{\sin yt}{(y^2 + \delta^2)^3}dy + \int_{\Gamma}\frac{e^{izt}}{(z^2 + \delta^2)^3}dz = \frac{2\pi i}{2!}\lim_{z\to i\delta}\frac{d^2}{dz^2}\left\{\left[(z - i\delta)^3\frac{e^{izt}}{(z^2 + \delta^2)^3}\right]\right\}$$

Since

$$\frac{d^2}{dz^2}\left[\frac{e^{izt}}{(z + i\delta)^3}\right] = \frac{-t^2(z + i\delta)\, e^{izt} - 2it\, e^{izt}}{(z + i\delta)^4} - \frac{4[it(z + i\delta)\, e^{izt} - 3e^{izt}}{(z + i\delta)^5}$$

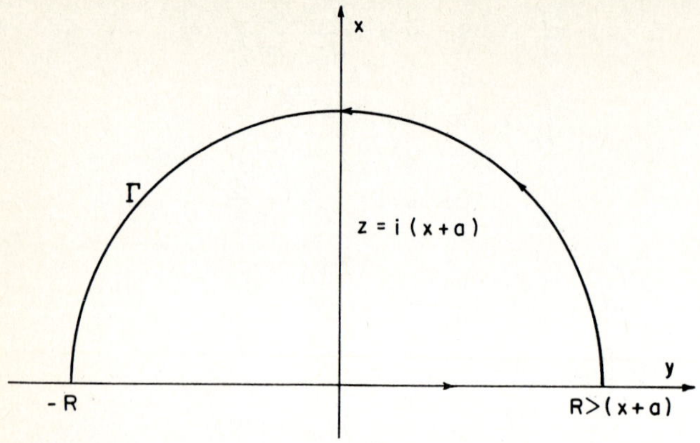

FIGURE 10.38 Contour integration for $\exp(izt)/[x+a)^2+z^2]^3$.

we obtain,

$$J_1 = \int_0^\infty \frac{\cos yt}{[(x+a)^2+y^2]^3}\,dy = \frac{t^2(x+a)^2 + 3t(x+a) + 3}{8(x+a)^3}\, e^{-at}\frac{\pi}{2}e^{-xt}$$

In a similar manner,

$$J_2 = \int_0^\infty \frac{y^2\cos yt}{[(x+a)^2+y^2]^3}\,dy$$

$$= \frac{4t(x+a)^3 - 4t^2(x+a)^4 + 4(x+a)^2}{32(x+a)^3}\, e^{-at}\frac{\pi}{2}e^{-xt}$$

and

$$J_3 = \int_0^\infty \frac{y^4\cos yt}{[(x+a)^2+y^2]^3}\,dy$$

$$= \frac{4t^2(x+a)^6 - 20t(x+a)^5 + 12(x+a)^2}{32(x+a)^3}\, e^{-at}\frac{\pi}{2}e^{-xt}$$

Therefore,

$$h(t) = \frac{2e^{xt}}{\pi}\{2(x+a)^3(x+b)^2 J_1 + [12(x+a)^2(x+b)$$

$$- 6(x+b)^2(x+a) - 2(x+a)^3]J_2 + [6(x+a) - 4(x+b)]J_3\}$$

Substituting for J_1, J_2, and J_3,

$$h(t) = [2 + 4(b-a)t + (a-b)^2t^2]\,e^{-at}$$

To facilitate a comparison, we now use the regular Laplace transform. We know that

$$L\left[\frac{t^{n-1}e^{-ct}}{(n-1)!}\right] = \frac{1}{(s+c)^n}$$

Then

$$Y(s) = \frac{2}{(s+a)^3}, \qquad X(s) = \frac{1}{(s+b)^3}, \quad \text{and} \quad H(s) = \frac{2(s+b)^3}{(s+a)^3}$$

Expanding the latter expression into partial fractions,

$$H(s) = \frac{2}{s+a} + \frac{4(b-a)}{(s+a)^2} + \frac{2(b-a)^2}{(s+a)^3}$$

Taking the inverse Laplace transform,

$$h(t) = [2 + 4(b-a)t + (a-b)t^2]\,e^{-at}$$

This is the same result that was obtained by using the Laplace-gamma method.

From this example, it appears that although the closed-form calculations for the Laplace-gamma method are lengthier than for the Laplace transform method, it has two important advantages for numerical computations when the functions $I(t)$ and $Q(t)$ are known pointwise only, as is normally the case in hydrology: (1) It does not require representation of $I(t)$ and $Q(t)$ by polynomials of arbitrary degrees, as would be the case with the Heaviside expansion in inversion of the Laplace transform. (2) It eliminates the need for integration in the complex plane in the inversion procedures and replaces it with an integral on the real line that can be computed numerically. Therefore, the Laplace-gamma method can be applied to discrete data more conveniently.

10.4.8.5 Laguerre Analysis

Dooge (1965) was the first to have proposed and used Laguerre functions for modeling hydrologic systems. The method has been discussed by O'Donnell (1966) and again by Dooge (1973). As Dooge (1973) pointed out, this method shares with the Fourier series the advantage of orthogonality and the property of covering the range from zero to infinity. As will be seen from the ensuing discussion, a disadvantage of this method is that the linkage equation is more complicated, which may render the determination of Laguerre coefficients of the IUH numerically less stable.

Laguerre functions can be defined as

$$f_n(t) = e^{-t/2}\left[\sum_{k=0}^{n}(-1)^k\binom{n}{k}\frac{t^k}{k!}\right] \tag{10.97}$$

It may be noted that these functions are orthogonal over the entire positive range 0 to ∞ of their variable t and are made up of gamma distributions—a distribution widely used to represent hydrologic systems. Equation (10.97) is based on the definition of the Laguerre polynomials $L_n(t)$, which are given by the expression in the brackets. These polynomials are orthonormal with respect to the weighting factor e^{-t} within the range 0 to $+\infty$. Each Laguerre function consists of a sum of weighted powers of t multiplied by an exponential. For example,

$$f_0(t) = e^{-t/2} \tag{10.98}$$

$$f_1(t) = (1-t)\,e^{-t/2} \tag{10.99}$$

$$f_2(t) = (1 - 2t + \tfrac{1}{2}t^2)e^{-t/2} \tag{10.100}$$

$$f_3(t) = (1 - 3t + \tfrac{3}{2}t^2 - \tfrac{1}{6}t^3)e^{-t/2} \tag{10.101}$$

and so on.

Following Dooge (1965, 1973), we can express $I(t)$, $h(t)$, and $Q(t)$ in terms of Laguerre functions,

$$I(t) = \sum_{m=0}^{\infty} a_n f_m(t) \tag{10.102}$$

$$h(t) = \sum_{n=0}^{\infty} a_n f_n(t) \tag{10.103}$$

$$Q(t) = \sum_{p=0}^{\infty} c_p f_p(t) \tag{10.104}$$

Due to the orthonormality property, we can write

$$a_m = \int_0^\infty I(t) f_m(t)\, dt \tag{10.105}$$

$$b_n = \int_0^\infty h(t) f_n(t)\, dt \tag{10.106}$$

$$c_p = \int_0^\infty Q(t) f_p(t)\, dt \tag{10.107}$$

To develop the linkage between c_p, b_n, and a_m, we follow the derivation proposed by Diskin (1977), which is based on the use of Laplace transform. Taking the Laplace transform of equations (10.102) and (10.103),

$$X(s) = \sum_{m=0}^{\infty} a_m F_m(s) \tag{10.108}$$

$$H(s) = \sum_{n=0}^{\infty} b_n F_n(s) \tag{10.109}$$

$$Y(s) = \sum_{p=0}^{\infty} c_p F_p(s) \tag{10.110}$$

Substituting equations (10.108) and (10.109) into equation (10.86),

$$\sum_{p=0}^{\infty} c_p F_p(s) = \left[\sum_{m=0}^{\infty} a_m F_m(s)\right]\left[\sum_{n=0}^{\infty} b_n F_n(s)\right] \tag{10.111}$$

which can be written as

$$\sum_{p=0}^{\infty} c_p F_p(s) = \sum_{m=0}^{\infty}\sum_{n=0}^{\infty} a_m b_n F_m(s)\, F_n(s) \tag{10.112}$$

If the product $F_m(s)\, F_n(s)$ can be expressed in the terms that correspond to the terms contained in the expression for $F_p(s)$, then equation (10.112) can be used to obtain the linkage equation.

The Laplace transform of a Laguerre function is obtained by taking the transform of a function defined as a product of a power term and an exponential. For example, if

$$z(t) = t^k e^{-wt} \tag{10.113}$$

then its Laplace transform is

$$Z(s) = \frac{k!}{(s + w)^{k+1}} \tag{10.114}$$

In a like manner, the Laplace transform of equation (10.97) is

$$F_n(s) = \sum_{k=0}^{\infty} \binom{n}{k} \frac{(-1)^k}{(s + \frac{1}{2})^{k+1}} \tag{10.115}$$

which can also be written as

$$F_n(s) = \frac{1}{(s + \frac{1}{2})^{n+1}} \sum_{k=0}^{\infty} \binom{n}{k} (-1)^k \left(s + \frac{1}{2}\right)^{n-k} \tag{10.116}$$

Recognizing the summation term as the binomial expansion for $[(-1) + (s + \frac{1}{2})]^n = (s - \frac{1}{2})^n$, equation (10.116) can be reduced to

$$F_n(s) = \frac{(s - \frac{1}{2})^n}{(s + \frac{1}{2})^{n+1}} \tag{10.117}$$

To develop the linkage relation, the convolution of two Laguerre functions should be evaluated. Let $g(t)$ represent the result of convoluting two Laguerre functions,

$$g(t) = \int_0^t f_m(\tau) f_n(t - \tau)\, d\tau \tag{10.118}$$

Then the Laplace transform of this equation is

$$G(s) = F_m(s) F_n(s) \tag{10.119}$$

Using equation (10.117) we can write equation (10.119) as

$$G(s) = \frac{(s - \frac{1}{2})^m}{(s + \frac{1}{2})^{m+1}} \cdot \frac{(s - \frac{1}{2})^n}{(s + \frac{1}{2})^{n+1}} = \frac{(s - \frac{1}{2})^{m+n}}{(s + \frac{1}{2})^{m+n+2}} \tag{10.120}$$

This can also be written as

$$G(s) = \frac{(s - \frac{1}{2})^{m+n}}{(s + \frac{1}{2})^{m+n+1}} - \frac{(s - \frac{1}{2})^{m+n+1}}{(s + \frac{1}{2})^{m+n+2}} \tag{10.121}$$

Substituting equations (10.117) and (10.121) into equation (10.112) gives

$$\sum_{p=0}^{\infty} c_p \frac{(s - \frac{1}{2})^p}{(s + \frac{1}{2})^{p+1}} = \sum_{m=0}^{\infty}\sum_{n=0}^{\infty} a_m b_n \left[\frac{(s - \frac{1}{2})^{m+n}}{(s + \frac{1}{2})^{m+n+1}} - \frac{(s - \frac{1}{2})^{m+n+1}}{(s + \frac{1}{2})^{m+n+2}}\right] \tag{10.122}$$

On comparing the terms of equal powers on both sides,

$$c_0 = a_0 b_0 \tag{10.123}$$

$$c_1 = a_0 b_1 + a_1 b_0 - a_0 b_0 \tag{10.124}$$

$$c_2 = a_0 b_2 + a_1 b_1 - (a_0 b_1 + a_1 b_0) + a_2 b_0 \tag{10.125}$$

$$\vdots$$

$$c_j = \sum_{i=0}^{j} a_i b_{j-i} - \sum_{i=0}^{j-1} a_i b_{j-i-1} \tag{10.126}$$

If we sum up the coefficients c_j for increasing values of j,

$$c_0 + c_1 = a_0 b_1 + a_1 b_0$$

$$c_0 + c_1 + c_2 = a_0 b_2 + a_1 b_1 + a_2 b_0$$

$$\vdots$$

$$\sum_{j=0}^{k} c_j = \sum_{i=0}^{k} a_i b_{k-i} = \sum_{i=0}^{k} b_i a_{k-i} \qquad (10.127)$$

Equations (10.126) and (10.127) establish the linkage between a_m, b_n, and c_p. Equation (10.127) can be used to evaluate the coefficients b_j in an ascending order. Separating the term that involves b_k, equation (10.127) can be written as

$$\sum_{j=0}^{k} c_j = a_0 b_k + \sum_{i=0}^{k-1} b_i a_{k-i} \qquad (10.128)$$

or

$$\sum_{j=0}^{k} c_j = a_0 b_k + \sum_{i=1}^{k} a_i b_{k-i} \qquad (10.129)$$

Therefore, we can solve equations (10.128) and (10.129) for any b_k if b_i, $i = 0, 1, 2, \ldots, k-1$ have previously been evaluated.

10.4.8.6 Gamma Distribution Method

Dooge (1973) suggested the use of gamma distribution to derive the IUH from the data of effective rainfall and direct runoff. Expressing $I(t)$, $h(t)$, and $Q(t)$ in terms of the gamma distribution,

$$I(t) = \sum_{i=0}^{\infty} A_i \frac{e^{-t/2}(t/2)^i}{2(i!)} \qquad (10.130)$$

$$h(t) = \sum_{j=0}^{\infty} B_j \frac{e^{-t/2}(t/2)^k}{2(j!)} \qquad (10.131)$$

$$Q(t) = \sum_{k=0}^{\infty} C_k \frac{e^{-t/2}(t/2)^k}{2(k!)} \qquad (10.132)$$

Substituting equations (10.130)–(10.131) into equation (10.27),

$$Q(t) = \int_0^t \sum_{i=0}^{\infty} A_i \frac{e^{-\tau/2}(\tau/2)^i}{2(i!)} \sum_{j=0}^{\infty} B_j \frac{e^{-(t-\tau)/2}((t-\tau)/2)^j}{2(j!)} d\tau$$

$$= \sum_{i=0}^{\infty} A_i \sum_{j=0}^{\infty} B_j \int_0^t \frac{e^{-\tau/2}(\tau/2)^i}{2(i!)} \frac{e^{-(t-\tau)/2}((t-\tau)/2)^j}{2(j!)} d\tau$$

$$= \sum_{i=0}^{\infty} A_i \sum_{j=0}^{\infty} B_j \frac{e^{-t/2} t^{i+j+1}}{2(i!)(j!)2} \beta(i+1, j+1)$$

where $\beta(i + 1, j + 1)$ is a beta function. Expressing the beta function in terms of the factorials,

$$Q(t) = \sum_{i=0}^{\infty} A_i \sum_{j=0}^{\infty} B_j \frac{e^{-t/2}(t/2)^{i+j+1}}{2(i+j+1)!} \qquad (10.133)$$

Replacing $Q(t)$ by equation (10.132),

$$\sum_{k=0}^{\infty} C_k \frac{e^{-t/2}(t/2)^k}{2(k)!} = \sum_{i=0}^{\infty} A_i \sum_{j=0}^{\infty} B_j \frac{e^{-t/2}(t/2)^{i+j+1}}{2(i+j+1)!}$$

This leads to

$$\sum_{k=0}^{\infty} C_k \frac{(t/2)^k}{k!} = \sum_{i=0}^{\infty} A_i \sum_{j=0}^{\infty} B_j \frac{(t/2)^{i+j+1}}{(i+j+1)!}$$

On comparing the terms on the two sides of this equation,

$$C_k = 2 \sum_{i=0}^{k-1} A_i B_{k-i-1} \qquad (10.134)$$

This is the linkage relation between the gamma distribution coefficients of $I(t)$, $h(t)$, and $Q(t)$. We need to express B_j in terms of A_i and C_k. Therefore,

$$A_0 B_k = C_{k+1} - \sum_{i=1}^{k} A_i B_{k-i} \qquad (10.135)$$

If the value of B_0 is zero, we can use

$$A_1 B_{k-1} = C_{k+1} - \sum_{i=2}^{k} A_i B_{k-i} \qquad (10.136)$$

Since the gamma distributions are not orthogonal to one another, it may not be possible to obtain their coefficients A_i and C_k directly, but they can be expressed in terms of the corresponding Laguerre coefficients, which can be obtained from equations (10.102) and (10.104), respectively. The relationship between them is

$$A_i = (-2)^{i+1} \sum_{m=i}^{\infty} \binom{m}{i} a_m \qquad (10.137)$$

$$C_k = (-2)^{k+1} \sum_{n=k}^{\infty} \binom{n}{k} c_n \qquad (10.138)$$

10.4.8.7 Power Spectrum Analysis

The power spectrum analysis method was perhaps first used by Bayazit (1966) for derivation of the IUH. He noted that one of the main advantages of using power spectrum analysis is the possibility of handling the errors due to imperfect measurements and finite length of the record. Power spectrum analysis is essentially the study in the frequency domain of a phenomenon originally recorded in a time domain.

Let us first define the correlation function. The autocorrelation of a times series, say $x(t)$, is the covariance of the lagged products of two values of the time series,

$$\rho(\tau) = \lim_{D\to\infty} \frac{1}{D} \int_{-D/2}^{D/2} x(t)x(t+\tau)\, dt \qquad (10.139)$$

where D is the period with which the event repeats itself.

It may be noted that $\rho(0)$ is the variance of $x(t)$. Similarly, the cross-correlation between two time series $x(t)$ and $y(t)$ can be defined as

$$\gamma(\tau) = \lim_{D\to\infty} \frac{1}{D} \int_{-D/2}^{D/2} x(t)y(t+\tau)\, dt \qquad (10.140)$$

The power spectrum of $x(t)$ is the Fourier cosine transform of its autocorrelation,

$$G(w) = \int_{-\infty}^{\infty} \rho(\tau)\cos(2\pi\, w\tau)\, d\tau \qquad (10.141)$$

It is clear from equation (10.141) that $G(w)$ represents $\rho(\tau)$ in the frequency domain. By inverse transform,

$$\rho(\tau) = \int_{-\infty}^{\infty} G(w)\cos(2\pi\, w\tau) dw \qquad (10.142)$$

It can be seen that $G(w)$ represents the contribution to the variance from frequencies between w and $w + dw$.

The cospectrum $R(w)$ and quadspectrum $Z(w)$ are similarly defined as the Fourier cosine and sine transforms of the cross-correlation function, respectively:

$$R(w) = \int_{-\infty}^{\infty} \gamma(\tau)\cos(2\pi\, w\tau)\, d\tau \qquad (10.143)$$

$$Z(w) = \int_{-\infty}^{\infty} \gamma(\tau)\sin(2\pi w\tau)\, d\tau \qquad (10.144)$$

If n denotes the wave number, then $w = n/D$ and equations (10.141), (10.143), and (10.144) can be written as

$$G(n) = \int_{-\infty}^{\infty} \rho(\tau)\cos n\frac{2\pi\tau}{D}\, d\tau \qquad (10.145)$$

$$R(n) = \int_{-\infty}^{\infty} \gamma(\tau)\cos n\frac{2\pi\tau}{D}\, d\tau \qquad (10.146)$$

$$Z(n) = \int_{-\infty}^{\infty} \gamma(\tau)\sin n\frac{2\pi\tau}{D}\, d\tau \qquad (10.147)$$

It will now be shown that α_n and β_n in equations (10.54)–(10.55) can be interpreted in terms of power spectra of the correlations between $I(t)$ and $Q(t)$.

Writing $\rho(\tau)$ of $I(t)$,

$$\rho(\tau) = \lim_{D\to\infty} \frac{1}{D} \int_{-D/2}^{D/2} I(t)\, I(t+\tau)\, dt \qquad (10.148)$$

where $I(t)$ is expressed by equation (10.42). Rewriting it,

$$I(t) = \sum_{n=0}^{\infty} \left(a_n\cos n\frac{2\pi t}{D} + b_n\sin n\frac{2\pi t}{D}\right) \qquad (10.149)$$

and

$$I(t+\tau) = \sum_{m=0}^{\infty} \left(a_m\cos m\frac{2\pi s}{D} + b_m\sin m\frac{2\pi s}{D}\right),$$
$$s = t + \tau \qquad (10.150)$$

and then substituting in equation (10.148) to solve for $\rho(\tau)$,

$$\int_{-D/2}^{D/2} I(t)I(t+\tau)\, dt$$
$$= \int_{-D/2}^{D/2} \sum_{n=0}^{\infty} \left(a_n\cos n\frac{2\pi t}{D} + b_n\sin n\frac{2\pi t}{D}\right) \cdot \sum_{m=0}^{\infty} \left(a_m\cos m\frac{2\pi s}{D} + b_m\sin m\frac{2\pi s}{D}\right) dt \qquad (10.151)$$

Evaluating the integral

$$\int_{-D/2}^{D/2} I(t)\, I(t+\tau) dt$$
$$= \frac{D}{2}\left[2a_0^2 + \sum_{n=1}^{\infty} a_n^2\cos n\frac{2\pi\tau}{D} + \sum_{n=1}^{\infty} a_n b_n \sin n\frac{2\pi\tau}{D} - \sum_{n=1}^{\infty} a_n b_n \sin n\frac{2\pi\tau}{D} + \sum_{n=1}^{\infty} b_n^2 \cos n\frac{2\pi\tau}{D}\right]$$
$$= \frac{D}{2}\left[2a_0^2 + \sum_{n=1}^{\infty} (a_n^2 + b_n^2)\cos n\frac{2\pi\tau}{D}\right] \qquad (10.152)$$

Thus $\rho(\tau)$ becomes

$$\rho(\tau) = \lim_{D\to\infty} \frac{1}{2}\left[2a_0^2 + \sum_{n=1}^{\infty} (a_n^2 + b_n^2)\cos n\frac{2\pi\tau}{D}\right] \qquad (10.153)$$

In a manner analogous to equation (10.153), $\gamma(\tau)$ between $I(t)$ and $Q(t)$ can be defined using equation (10.42) and (10.44) as

$$\gamma(\tau) = \lim_{D\to\infty} \frac{1}{2}\left[2A_0a_0 + \sum_{n=1}^{\infty} (a_nA_n + b_nB_n)\cos n\frac{2\pi\tau}{D} + \sum_{n=1}^{\infty} (a_nB_n - A_nb_n)\sin n\frac{2\pi\tau}{D}\right] \qquad (10.154)$$

$G(w)$ for $I(t)$ can be derived by substituting equation (10.153) into equation (10.141),

$$G(n) = \int_{-D/2}^{D/2} \left\{\lim_{D\to\infty} \frac{1}{2}\left[2a_0^2 + \sum_{n=1}^{\infty} (a_n^2 + b_n^2)\cos n\frac{2\pi\tau}{D}\right]\right\} \cos n\frac{2\pi\tau}{D}\, d\tau \qquad (10.155)$$

On simplifying it,

$$G(n) = \begin{cases} \dfrac{a_n^2 + b_n^2}{4} D & \text{for } n \neq 0 \\ \dfrac{a_0^2}{2} D & \text{for } n = 0 \end{cases} \tag{10.156}$$

In an analogous manner, $R(n)$ and $Z(n)$ for $\gamma(\tau)$ between $I(t)$ and $Q(t)$ can be obtained,

$$R(n) = \begin{cases} \dfrac{a_n A_n + b_n B_n}{4} D & \text{for } n \neq 0 \\ \dfrac{a_0 A_0}{2} D & \text{for } n = 0 \end{cases} \tag{10.157}$$

$$A(n) = \frac{a_n B_n - A_n b_n}{4} D \tag{10.158}$$

Substituting equations (10.156)–(10.158) into equations (10.54)–(10.55),

$$\alpha_n = \frac{2}{D} \frac{R(n)}{G(n)} \text{ for } n \neq 0; \qquad \alpha_0 = \frac{1}{D} \frac{R(0)}{G(0)} \tag{10.159}$$

$$\beta_n = \frac{2}{D} \frac{Z(n)}{G(n)} \tag{10.160}$$

Equations (10.153)–(10.160) define the coefficients of harmonic components of $h(t)$ in terms of the correlations between $I(t)$ and $Q(t)$.

10.4.8.8 Wiener-Hopf Equation

Eagleson, Mejia, R. and March (1966) developed an approach based on the Wiener-Hopf equation (Wiener, 1949) for computation of optimum realizable unit hydrographs. If we let $r(t)$ define the difference between observed and computated $Q(t)$ as

$$r(t) = Q(t) - \int_0^\infty h(\tau)\, I(t-\tau)\, d\tau \tag{10.161}$$

then the optimum $h(t)$ is one that minimizes $r(t)$ according to some criterion. If the criterion is one of least squares, then

$$E = \lim_{D\to\infty} \frac{1}{D} \int_{-D/2}^{D/2} [r(t)]^2\, dt \tag{10.162}$$

where E is the integral square error or error criterion which we wish to minimize. Following Dooge (1973),

$$E = \lim_{D\to\infty} \frac{1}{D} \int_{-D/2}^{D/2} [Q(t) - \int_0^\infty h(\tau)\, I(t-\tau) d\tau]^2\, dt \tag{10.163}$$

The problem, therefore, reduces to determining $h(t)$ such that E in equation (10.163) is minimized. Simplifying it,

$$E = \lim_{D\to\infty} \frac{1}{D} \int_{-D/2}^{D/2} \Big[Q(t)Q(t) - 2Q(t) \int_0^\infty h(\tau)\, I(t-\tau)\, d\tau\, dt + \Big\{ \int_0^\infty h(\tau) I(t-\tau)\, d\tau \Big\}^2 \Big]\, dt \tag{10.164}$$

This can be further simplified by using the definitions of correlations as

$$E_1 = \lim_{D\to\infty} \frac{1}{D} \int_{-D/2}^{D/2} Q(t)\, Q(t)\, dt = \rho_{QQ}(0) \tag{10.165}$$

$$E_2 = \lim_{D\to\infty} \frac{1}{D} \int_{-D/2}^{D/2} 2Q(t) \int_0^\infty h(\tau)\, I(t-\tau)\, d\tau\, dt \tag{10.166}$$

Interchanging the integral sign,

$$E_2 = -2 \int_0^\infty h(\tau)\, \gamma_{IQ}(\tau)\, d\tau \tag{10.167}$$

Similarly,

$$E_3 = \lim_{D\to\infty} \frac{1}{D} \int_{-D/2}^{D/2} \int_0^\infty h(\tau_1)\, I(t-\tau_1)\, d\tau_1 \cdot \int_0^\infty h(\tau_2)\, I(t-\tau_2)\, d\tau_2\, dt = \int_0^\infty h(\tau_1)\, d\tau_1 \cdot \int_0^\infty h(\tau_2)\, \rho_{II}(\tau_1 - \tau_2)\, d\tau_2 \tag{10.168}$$

Therefore,

$$E = E_1 + E_2 + E_3 \tag{10.169}$$

Differentiating equation (10.169) with respect to $h(t)$ and equating to zero,

$$\gamma_{IQ}(\tau_1) - \int_0^\infty h(\tau_2)\, \rho_{II}(\tau_1 - \tau_2)\, d\tau_2 = 0 \tag{10.170}$$

Equation (10.170) is the Wiener-Hopf equation, whose solution yields the optimum $h(t)$ for the system for which I and Q are specified as continuous-time series. It is thus seen that the optimal $h(t)$ depends on the autocorrelation function $\rho_{II}(\tau)$ of the effective rainfall and the cross-correlation function $\gamma_{IQ}(\tau)$ between the effective rainfall and discharge.

10.4.8.9 Other Polynomial Approximations

A number of other polynomials that are orthogonal over a finite interval may be used for derivation of the IUH. Papazafiriou (1975) used the Chebyshev polynomials and the orthogonal polynomials introduced by Stiefel (1955) and analyzed by Forsythe (1957). The procedure presented by Papazafiriou can be outlined as follows. Let the set of orthogonal polynomials be $y(t) = [P_n(t), n = 1, 2, \ldots, N]$. Then the IUH can be approximated as

$$\rho(\tau) = \lim_{D\to\infty} \frac{1}{D} \int_{-D/2}^{D/2} x(t)x(t+\tau)\, dt \tag{10.139}$$

where D is the period with which the event repeats itself.

It may be noted that $\rho(0)$ is the variance of $x(t)$. Similarly, the cross-correlation between two time series $x(t)$ and $y(t)$ can be defined as

$$\gamma(\tau) = \lim_{D\to\infty} \frac{1}{D} \int_{-D/2}^{D/2} x(t)y(t+\tau)\, dt \tag{10.140}$$

The power spectrum of $x(t)$ is the Fourier cosine transform of its autocorrelation,

$$G(w) = \int_{-\infty}^{\infty} \rho(\tau)\cos(2\pi w\tau)\, d\tau \tag{10.141}$$

It is clear from equation (10.141) that $G(w)$ represents $\rho(\tau)$ in the frequency domain. By inverse transform,

$$\rho(\tau) = \int_{-\infty}^{\infty} G(w)\cos(2\pi w\tau) dw \tag{10.142}$$

It can be seen that $G(w)$ represents the contribution to the variance from frequencies between w and $w + dw$.

The cospectrum $R(w)$ and quadspectrum $Z(w)$ are similarly defined as the Fourier cosine and sine transforms of the cross-correlation function, respectively:

$$R(w) = \int_{-\infty}^{\infty} \gamma(\tau)\cos(2\pi w\tau)\, d\tau \tag{10.143}$$

$$Z(w) = \int_{-\infty}^{\infty} \gamma(\tau)\sin(2\pi w\tau)\, d\tau \tag{10.144}$$

If n denotes the wave number, then $w = n/D$ and equations (10.141), (10.143), and (10.144) can be written as

$$G(n) = \int_{-\infty}^{\infty} \rho(\tau)\cos n\frac{2\pi\tau}{D}\, d\tau \tag{10.145}$$

$$R(n) = \int_{-\infty}^{\infty} \gamma(\tau)\cos n\frac{2\pi\tau}{D}\, d\tau \tag{10.146}$$

$$Z(n) = \int_{-\infty}^{\infty} \gamma(\tau)\sin n\frac{2\pi\tau}{D}\, d\tau \tag{10.147}$$

It will now be shown that α_n and β_n in equations (10.54)–(10.55) can be interpreted in terms of power spectra of the correlations between $I(t)$ and $Q(t)$.

Writing $\rho(\tau)$ of $I(t)$,

$$\rho(\tau) = \lim_{D\to\infty} \frac{1}{D} \int_{-D/2}^{D/2} I(t)\, I(t+\tau)\, dt \tag{10.148}$$

where $I(t)$ is expressed by equation (10.42). Rewriting it,

$$I(t) = \sum_{n=0}^{\infty} \left(a_n \cos n\frac{2\pi t}{D} + b_n \sin n\frac{2\pi t}{D} \right) \tag{10.149}$$

and

$$I(t+\tau) = \sum_{m=0}^{\infty} \left(a_m \cos m\frac{2\pi s}{D} + b_m \sin m\frac{2\pi s}{D} \right), \quad s = t + \tau \tag{10.150}$$

and then substituting in equation (10.148) to solve for $\rho(\tau)$,

$$\begin{aligned}\int_{-D/2}^{D/2} I(t)I(t+\tau)\, dt &= \int_{-D/2}^{D/2} \sum_{n=0}^{\infty} \left(a_n \cos n\frac{2\pi t}{D} + b_n \sin n\frac{2\pi t}{D} \right) \cdot \\ &\quad \sum_{m=0}^{\infty} \left(a_m \cos m\frac{2\pi s}{D} + b_m \sin m\frac{2\pi s}{D} \right) dt\end{aligned} \tag{10.151}$$

Evaluating the integral

$$\begin{aligned}\int_{-D/2}^{D/2} I(t)\, I(t+\tau)dt &= \frac{D}{2}\Bigg[2a_0^2 + \sum_{n=1}^{\infty} a_n^2 \cos n\frac{2\pi\tau}{D} + \sum_{n=1}^{\infty} a_n b_n \sin n\frac{2\pi\tau}{D} \\ &\quad - \sum_{n=1}^{\infty} a_n b_n \sin n\frac{2\pi\tau}{D} + \sum_{n=1}^{\infty} b_n^2 \cos n\frac{2\pi\tau}{D} \Bigg] \\ &= \frac{D}{2}\left[2a_0^2 + \sum_{n=1}^{\infty} (a_n^2 + b_n^2) \cos n\frac{2\pi\tau}{D} \right]\end{aligned} \tag{10.152}$$

Thus $\rho(\tau)$ becomes

$$\rho(\tau) = \lim_{D\to\infty} \frac{1}{2}\left[2a_0^2 + \sum_{n=1}^{\infty} (a_n^2 + b_n^2) \cos n\frac{2\pi\tau}{D} \right] \tag{10.153}$$

In a manner analogous to equation (10.153), $\gamma(\tau)$ between $I(t)$ and $Q(t)$ can be defined using equation (10.42) and (10.44) as

$$\begin{aligned}\gamma(\tau) = \lim_{D\to\infty} \frac{1}{2}\Bigg[2A_0a_0 &+ \sum_{n=1}^{\infty} (a_nA_n + b_nB_n)\cos n\frac{2\pi\tau}{D} \\ &+ \sum_{n=1}^{\infty} (a_nB_n - A_nb_n)\sin n\frac{2\pi\tau}{D} \Bigg]\end{aligned} \tag{10.154}$$

$G(w)$ for $I(t)$ can be derived by substituting equation (10.153) into equation (10.141),

$$\begin{aligned}G(n) = \int_{-D/2}^{D/2} \Bigg\{ \lim_{D\to\infty} \frac{1}{2}\Bigg[2a_0^2 &+ \sum_{n=1}^{\infty} (a_n^2 + b_n^2) \cos n\frac{2\pi\tau}{D} \Bigg] \Bigg\} \cos n\frac{2\pi\tau}{D}\, d\tau\end{aligned} \tag{10.155}$$

On simplifying it,

$$G(n) = \begin{cases} \dfrac{a_n^2 + b_n^2}{4} D & \text{for } n \neq 0 \\ \dfrac{a_0^2}{2} D & \text{for } n = 0 \end{cases} \tag{10.156}$$

In an analogous manner, $R(n)$ and $Z(n)$ for $\gamma(\tau)$ between $I(t)$ and $Q(t)$ can be obtained,

$$R(n) = \begin{cases} \dfrac{a_n A_n + b_n B_n}{4} D & \text{for } n \neq 0 \\ \dfrac{a_0 A_0}{2} D & \text{for } n = 0 \end{cases} \tag{10.157}$$

$$A(n) = \frac{a_n B_n - A_n b_n}{4} D \tag{10.158}$$

Substituting equations (10.156)–(10.158) into equations (10.54)–(10.55),

$$\alpha_n = \frac{2}{D} \frac{R(n)}{G(n)} \text{ for } n \neq 0; \qquad \alpha_0 = \frac{1}{D} \frac{R(0)}{G(0)} \tag{10.159}$$

$$\beta_n = \frac{2}{D} \frac{Z(n)}{G(n)} \tag{10.160}$$

Equations (10.153)–(10.160) define the coefficients of harmonic components of $h(t)$ in terms of the correlations between $I(t)$ and $Q(t)$.

10.4.8.8 Wiener-Hopf Equation

Eagleson, Mejia, R. and March (1966) developed an approach based on the Wiener-Hopf equation (Wiener, 1949) for computation of optimum realizable unit hydrographs. If we let $r(t)$ define the difference between observed and computated $Q(t)$ as

$$r(t) = Q(t) - \int_0^{\infty} h(\tau)\, I(t-\tau)\, d\tau \tag{10.161}$$

then the optimum $h(t)$ is one that minimizes $r(t)$ according to some criterion. If the criterion is one of least squares, then

$$E = \lim_{D\to\infty} \frac{1}{D} \int_{-D/2}^{D/2} [r(t)]^2\, dt \tag{10.162}$$

where E is the integral square error or error criterion which we wish to minimize. Following Dooge (1973),

$$E = \lim_{D\to\infty} \frac{1}{D} \int_{-D/2}^{D/2} [Q(t) - \int_0^{\infty} h(\tau)\, I(t-\tau) d\tau]^2\, dt \tag{10.163}$$

The problem, therefore, reduces to determining $h(t)$ such that E in equation (10.163) is minimized. Simplifying it,

$$E = \lim_{D\to\infty} \frac{1}{D} \int_{-D/2}^{D/2} \left[Q(t)Q(t) - 2Q(t) \int_0^{\infty} h(\tau)\, I(t-\tau)\, d\tau\, dt + \left\{ \int_0^{\infty} h(\tau) I(t-\tau)\, d\tau \right\}^2 \right] dt \tag{10.164}$$

This can be further simplified by using the definitions of correlations as

$$E_1 = \lim_{D\to\infty} \frac{1}{D} \int_{-D/2}^{D/2} Q(t)\, Q(t)\, dt = \rho_{QQ}(0) \tag{10.165}$$

$$E_2 = \lim_{D\to\infty} \frac{1}{D} \int_{-D/2}^{D/2} 2Q(t) \int_0^{\infty} h(\tau)\, I(t-\tau)\, d\tau\, dt \tag{10.166}$$

Interchanging the integral sign,

$$E_2 = -2 \int_0^{\infty} h(\tau)\, \gamma_{IQ}(\tau)\, d\tau \tag{10.167}$$

Similarly,

$$E_3 = \lim_{D\to\infty} \frac{1}{D} \int_{-D/2}^{D/2} \int_0^{\infty} h(\tau_1)\, I(t-\tau_1)\, d\tau_1 \cdot \int_0^{\infty} h(\tau_2)\, I(t-\tau_2)\, d\tau_2\, dt = \int_0^{\infty} h(\tau_1)\, d\tau_1 \cdot \int_0^{\infty} h(\tau_2)\, \rho_{II}(\tau_1 - \tau_2)\, d\tau_2 \tag{10.168}$$

Therefore,

$$E = E_1 + E_2 + E_3 \tag{10.169}$$

Differentiating equation (10.169) with respect to $h(t)$ and equating to zero,

$$\gamma_{IQ}(\tau_1) - \int_0^{\infty} h(\tau_2)\, \rho_{II}(\tau_1 - \tau_2)\, d\tau_2 = 0 \tag{10.170}$$

Equation (10.170) is the Wiener-Hopf equation, whose solution yields the optimum $h(t)$ for the system for which I and Q are specified as continuous-time series. It is thus seen that the optimal $h(t)$ depends on the autocorrelation function $\rho_{II}(\tau)$ of the effective rainfall and the cross-correlation function $\gamma_{IQ}(\tau)$ between the effective rainfall and discharge.

10.4.8.9 Other Polynomial Approximations

A number of other polynomials that are orthogonal over a finite interval may be used for derivation of the IUH. Papazafiriou (1975) used the Chebyshev polynomials and the orthogonal polynomials introduced by Stiefel (1955) and analyzed by Forsythe (1957). The procedure presented by Papazafiriou can be outlined as follows. Let the set of orthogonal polynomials be $y(t) = [P_n(t), n = 1, 2, \ldots, N]$. Then the IUH can be approximated as

$$h(\tau) \sum_{i=1}^{N} a_i P_i(\tau) \tag{10.171}$$

where N is the order of the polynomial expansion of the IUH. Substituting equation (10.171) into equation (10.27b),

$$\begin{aligned} Q(t) &= \int_0^t I(t-\tau) \sum_{i=1}^{N} a_i P_i(\tau)\, d\tau \\ &= \sum_{i=1}^{N} a_i \int_0^t I(t-\tau)\, P_i(\tau)\, d\tau \end{aligned} \tag{10.172}$$

For convenience of notation, let

$$C_i(t) = \int_0^t I(t-\tau)\, P_i(\tau)\, d\tau \tag{10.173}$$

Then

$$Q(t) = \sum_{i=1}^{N} a_i\, C_i(t) \tag{10.174}$$

The quantities $C_i(t)$, $i = 1, 2, 3, \ldots, N$, can be computed from equation (10.173) by knowing effective rainfall and first N orthogonal polynomials. For purposes of computation this equation can be expressed in discrete form,

$$C_i(t) = \sum_{\tau=1}^{M} P_i(\tau)\, I(t-\tau)\, \Delta t \tag{10.175}$$

in which M is the length of the finite memory of the system. With $Q(t)$ known from observations and $C_i(t)$ computed as above, the orthogonal coefficients a_i can be computed by solving the system of equations,

$$\begin{aligned} Q(t_0) &= a_0 C_0(t_0) + a_1 C_1(t_0) + \cdots + a_N C_N(t_0) \\ Q(t_1) &= a_0 C_0(t_1) + a_1 C_1(t_1) + \cdots + a_N C_N(t_1) \\ Q(t_2) &= a_0 C_0(t_2) + a_2 C_2(t_2) + \cdots + a_N C_N(t_2) \\ &\;\;\vdots \\ Q(t_L) &= a_0 C_0(t_L) + a_1 C_1(t_L) + \cdots + a_N C_N(t_L) \end{aligned} \tag{10.176}$$

where t_L denotes the length of the input record. In compact matrix notation,

$$[Q_k] = [C_{k,i}]\,[a_i], \qquad k = 0, 1, 2, \ldots, L;\ i = 1, 2, \ldots, N \tag{10.177}$$

If $N = L$, the system will have the same number of unknowns as equations and an exact solution may be obtained. If $N < L$, the system is underdetermined and one is then free to use any appropriate method to find a solution. A multiple linear regression analysis may be appropriate here. Harder and Zand (1969), Brandstetter and Amorocho (1970), Bidwell (1971), Amorocho and Brandstetter (1971), and Zand and Harder (1973) used multiple regression analysis.

If the polynomials are the Chebyshev polynomials, then these can be generated by the following recurrence relation:

$$P_{n+1}(t) = 2tP_n(t) - P_{n-1}(t), \qquad P_0(t) = 1, \qquad P_1(t) = t \tag{10.178}$$

If $t_k = \cos(k\pi/n)$, the Chebyshev polynomials are orthogonal with respect to summation over the finite interval $k = 0, 1, 2, \ldots, N$.

$$.5\, P_m(t_0)\, P_n(t_0) + \sum_{k=1}^{N-1} P_m(t_k)\, P_n(t_k) + .5 P_m(t_N) P_n(t_N) = \begin{cases} 3 & \text{for } m \neq n \\ \dfrac{N}{2} & \text{for } m = n \neq 0 \text{ or } N \\ N & \text{for } m = n = 0 \text{ or } N \end{cases} \tag{10.179}$$

The polynomials S_n suggested by Stiefel (1955), orthogonal over a finite interval, can be generated by the recurrence relation,

$$S_{n+1}(t) = [(t - a_{n+1}) S_n(t) - b_n S_{n-1}(t)] w^2(t), \qquad S_0(t) = w^2(t) \tag{10.180}$$

where $w^2(t)$ is an arbitrary positive weight function, and

$$a_{n+1} = \frac{\sum_{i=1}^{N} t_i S_n^2(t_i) w^2(t_i)}{\sum_{i=1}^{N} S_n^2(t_i) w^2(t_i)} \tag{10.181}$$

$$b_n = \frac{\sum_{i=1}^{N} t_i S_n(t_i) S_{n-1}(t_i) w^2(t_i)}{\sum_{i=1}^{N} S_{n-1}^2(t_i) w^2(t_i)} \tag{10.182}$$

The orthogonal form of these polynomials can be obtained from the relation,

$$y_n(t_i) = \frac{S_n(t_i)}{\left[\sum_{i=1}^{N} S_n^2(t_i) w^2(t_i)\right]^{.5}} \tag{10.183}$$

The polynomials of equation (10.183) with weight function unity can be called *unit weighted orthonormal polynomials.*

For derivation of the IUH, the order of polynomial expansion N and the length of the system memory M must be estimated. These were estimated by Papzafiriou (1975) by trial and error. He noted that these polynomials produced physically realizable IUHs with minimal error and were superior to harmonic analysis.

10.4.9 Continuous Convolution with Hydrologic Data

Because hydrologic data are usually defined by discrete values sampled at equal time increments, the integral form of convolution equation, as in equation (10.27), is replaced by the convolution summation and the hydrologic system identified therefrom. Recent work by Chapman (1985) shows, however, that the integral form of convolution equation can be used for hydrologic data sampled at equal time increments if simple functional forms are assumed to describe the change in hydrologic variables between sampling points. Some examples of such functional forms are the step function, linear segment, or sliding polynomial (Snyder, 1961a, 1976). This allows for the sampling increments to be different for the effective rainfall and IUH, which contrasts with the usual restrictions in convolution summation, and the direct runoff to be defined continuously in time, not just at the sampling points of the effective rainfall and IUH.

EXERCISES

10.1. Determine the peak discharge required for the design of a highway crossing a stream whose watershed has the following characteristics: mainstream length = 2.5 km; drainage area = 8.5 km^2; mainstream slope = 5.5 m/km; and watershed type = medium soil, cropped.

10.2. Determine the direct runoff hydrograph by using the rational method. The watershed area is 5 km^2 and has a time of concentration of 2 h. The effective rainfall intensity is 5 cm/h and lasts uniformly for a period of 2 h.

10.3. Solve Exercise 10.2 if the duration of effective rainfall is 5 h.

10.4. Solve Exercise 10.2 if the duration of effective rainfall is 1 h.

10.5. Data for a rainfall-runoff event for the Eastern Creek at the Great Western Highway in Australia are available in Table 10.8. The drainage area is 24.86 km^2. Derive the instantaneous unit hydrograph for the Eastern Creek by harmonic analysis.

TABLE 10.8
Data for a rainfall-runoff event of May 20–21, 1959, observed for the Eastern Creek at the Great Western Highway in Australia

Date	*Time*	*Rainfall in the Preceding Hour (mm)*	*Recorded Hydrograph (m^3/s)*
5/20/59	1500	0.00	
	1600	2.28	
	1700	1.02	
	1800	4.06	
	1900	0.50	
	2000	3.30	
	2100	4.82	
	2200	6.86	0.00
	2300	1.93	0.04
	2400	0.50	0.16
5/21/59	100	4.32	0.70
	200	6.35	2.23
	300	2.28	5.60
	400	1.52	10.26
	500	0.76	11.25
	600		11.49
	700		10.25
	800		8.39
	900		6.80
	1000		5.45
	1100		4.20
	1200		3.37
	1300		2.80
	1400		2.40
	1500		2.10
	1600		1.88
	1700		1.70
	1800		1.63
	1900		1.45
	2000		1.37
	2100		1.28
	2200		1.20
	2300		1.05

10.6. Derive the IUH for the data in Exercise 10.5 by Laguerre analysis.

10.7. Derive the IUH for the data in Exercise 10.5 by the gamma distribution method.

10.8. Derive the IUH for the data in Exercise 10.5 by power spectrum analysis.

10.9. Derive the IUH for the data in Exercise 10.5 by using Chebyshev polynomials.

10.10. Compare the IUH's derived in Exercises 10.5 and 10.9. Which method is better if one is?

10.11. Which method would you recommend for deriving the IUH and why?

10.12. Consider the rainfall-runoff data given in Table 10.9 for an event observed at an experimental agricultural watershed

TABLE 10.9
Data for a rainfall-runoff event of April 24, 1957, observed at watershed *Y* (area 309 acres) near Riesel (Waco) Texas

Rainfall Data					
Hour	*Minute*	*Intensity (in./h)*	*Hour*	*Minute*	*Intensity (in./h)*
2	33	.50	3	01	2.31
2	39	1.50	3	08	1.50
2	43	2.52	3	14	.80
2	48	4.00	3	20	.30
2	51	2.31	3	32	.03
2	58	3.2	4	18	.00

TABLE 10.9. *(continued)*

		Runoff Data			
Hour	*Minute*	*Discharge (in./h)*	Hour	Minute	Discharge (in./h)
2	51	.0077	3	14	1.300
2	54	.0144	3	16	1.450
2	56	.0204	3	18	1.610
2	57	.0283	3	20	1.750
2	58	.0512	3	22	1.810
2	59	.098	3	25	1.770
3	00	.153	3	34	1.440
3	01	.212	3	36	1.350
3	02	.296	3	42	1.030
3	03	.398	3	43	.967
3	04	.455	3	45	.860
3	05	.500	3	48	.702
3	09	.661	3	55	.547
3	10	.743	4	15	.298
3	11	1.020	4	47	.191
			4	51	.183
			4	54	.163

Y located near Riesel (Waco), Texas. Derive the IUH by any two methods of your choice. Compare these methods. How do these compare for the rainfall-runoff data of Exercise 10.5?

10.13. Introduce some error in the effective rainfall computed in Exercise 10.12. Then derive the IUH by the same two methods and compare them.

10.14. Find the Laguerre coefficients for the following functions:

(a) $t^3 \exp(-t)/3!,\ 0 < t < \infty$

(b) $\frac{t}{10}\exp(8 - t) - \frac{t}{10},\ 0 < t < 8$

(c) $125 - 75\cos\left(\frac{\pi t}{48}\right),\ -\infty < t < \infty$

CHAPTER ELEVEN

Linear Black-Box Models for Discrete-Time Systems

Hydrologic data are usually available in discrete or quantized form, although some are given at such a high frequency that they can be treated, without appreciable error, as a continuous record. Nevertheless, from a practical standpoint it may be more appropriate to treat a discrete record as it is and develop techniques for system identification accordingly. Strictly speaking, for discrete data the UH developed will be a finite period UH and not an IUH. However, the period can be chosen so small that for all practical purposes the UH and the IUH are indistinguishable from one another. Therefore, we will imply in the discussion here that the same techniques can be applied to derive the IUH for discrete data.

11.1 MATHEMATICAL REPRESENTATION

The mathematical representation of the unit hydrograph follows naturally from a discussion of discrete-time systems in Chapters 4–5. To reemphasize, consider Fig. 11.1. Here the effective rainfall is given in blocks of duration D. Then the discrete-time or finite difference equivalent of equation (10.27a) is

$$Q(nD) = \sum_{k=1}^{n} I(kD)Dh[(n-k)D] \tag{11.1a}$$

where $I(k)$ is the mean intensity of effective rainfall in the kth time period. Or simply (with D equal to unity),

$$Q(n) = \sum_{k=1}^{n} I(k)h(n-k) = \sum_{k=1}^{n} I_k h_{n-k} \tag{11.1b}$$

Thus, given the unit hydrograph of duration D, the runoff hydrograph can be obtained from equation (11.1b) for any effective rainfall provided it is composed of blocks each of duration D. From a practical standpoint, it may often be convenient to carry out such calculations in tabular form, as pointed out by Nash (1966).

Let us define $I_1, I_2, \ldots, I_k$ to be the quantities of effective rainfall in intervals of D beginning at $t = 0$, that is, I_2 is the amount of effective rainfall in the interval $2D$; $h_1, h_2, \ldots, h_n$, are the ordinates of the UH of duration D at intervals D after the beginning of effective rainfall; and $Q_1, Q_2, \ldots, Q_n$, are the ordinates of direct runoff hydrograph per unit watershed area at intervals of D beginning at $t = 0$; that is, Q_2 is at $t = 2D$. The tabular arrangement is shown in Table 11.1. There, Q's are equal to the sum of the terms in the corresponding columns. For example,

$$Q_0 = 0$$
$$Q_1 = I_1h_1$$
$$Q_2 = I_2h_1 + I_1h_2$$
$$Q_3 = I_3h_1 + I_2h_2 + I_1h_3$$
$$Q_4 = I_4h_1 + I_3h_2 + I_2h_3 + I_1h_4$$
$$\vdots$$
$$Q_{k+n-2} = I_{k+n-2}h_1 + \cdots + I_1h_{k+n-2}$$
$$Q_{k+n-1} = \sum_{i=1}^{k+n-1} h_i I_{k+n-i} \tag{11.2a}$$

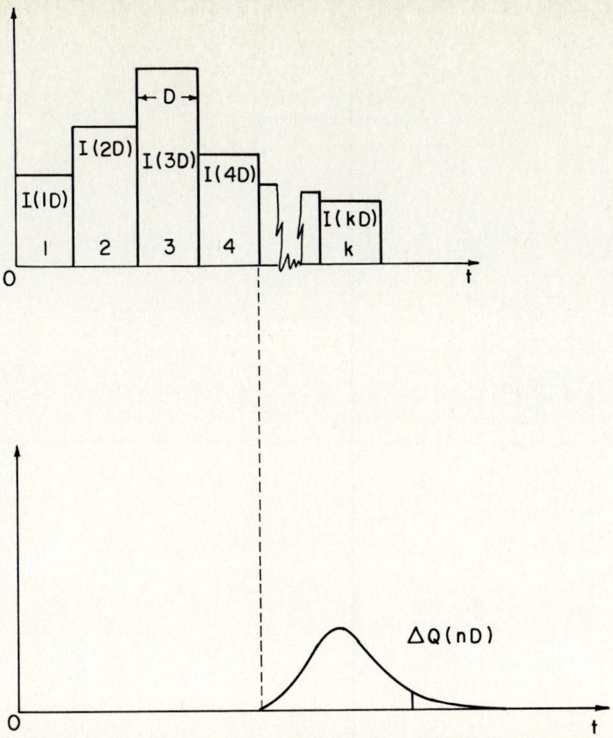

FIGURE 11.1 UH representation for discrete-time systems.

As shown in Chapter 4, equation (11.2a) can be conveniently written in matrix notation as

$$[Q]_{k+n-1,1} = [h]_{k+n-1,k+n-1}\,[I]_{k+n-1,1} \tag{11.2b}$$

where

$$[Q] = \begin{bmatrix} Q_1 \\ Q_2 \\ \cdot \\ \cdot \\ \cdot \\ Q_{k+n-2} \\ Q_{k+n-1} \end{bmatrix}; \quad [I] = \begin{bmatrix} I_1 \\ I_2 \\ \cdot \\ \cdot \\ \cdot \\ I_{k+n-2} \\ I_{k+n-1} \end{bmatrix}; \quad \text{and}$$

$$[h] = \begin{bmatrix} h_1 & 0 & 0 & \cdots & 0 \\ h_2 & h_1 & 0 & \cdots & 0 \\ h_3 & h_2 & h_1 & \cdots & 0 \\ \cdot & \cdot & \cdot & \cdot & \cdot \\ \cdot & \cdot & \cdot & \cdot & \cdot \\ \cdot & \cdot & \cdot & \cdot & \cdot \\ h_{k+n-1} & \cdot & \cdot & \cdots & h_1 \end{bmatrix}$$

Alternatively, equation (11.2a) can also be written as

$$[Q]_{k+n-1,1} = [I]_{k+n-1,k+n-1}\,[h]_{k+n-1} \tag{11.2c}$$

where

$$[h] = \begin{bmatrix} h_1 \\ h_2 \\ \cdot \\ \cdot \\ \cdot \\ h_{k+n-1} \end{bmatrix} \quad \text{and} \quad [I] = \begin{bmatrix} I_1 & 0 & 0 & \cdots & 0 \\ I_2 & I_1 & 0 & \cdots & 0 \\ I_3 & I_2 & I_1 & \cdots & 0 \\ \cdot & \cdot & \cdot & \cdot & 0 \\ \cdot & \cdot & \cdot & \cdot & \cdot \\ \cdot & \cdot & \cdot & \cdot & 0 \\ I_{k+n-1} & \cdot & \cdot & \cdots & I_1 \end{bmatrix}$$

If the rainfall and runoff records are continuous then they should be discretized first. A preliminary step to the discretization scheme is the selection of a sampling time interval D. The continuous functions of $I(t)$, $Q(t)$, and, hence, $h(t)$ are discretized to have the same time interval and can be represented by sets of discrete pulses at regular time interval. The discrete effective rainfall pulses are assigned to starting points of intervals and discrete UH and direct runoff are assigned to endpoints of the intervals. The value of a sampled function at each sampling point is given by the integral of the corresponding function over a time interval D. For example, kth pulse of $h(t)$ is given by

$$h_k = \frac{1}{D}\int_{t_k}^{t_{k+1}} h(t)\,dt \tag{11.3}$$

where $t_k = kD$ and $t_{k+1} = (k+1)D$. The positions of the pulses of three functions according to this discretization scheme are shown in Fig. 11.2. The number of pulses of $I(t)$, N_I, and of $Q(t)$, N_Q for a storm is given by the number of time intervals needed to encompass their respective time periods. The number of pulses of $h(t)$, N_h, can be expressed as

$$N_h = N_Q - N_I + 1 \tag{11.4a}$$

Another point regarding discretization is to ensure that the volumes (or depths) of discretized $Q(t)$ and $I(t)$ equal each other,

$$\sum_{j=1}^{N_I} I_j = \sum_{j=1}^{N_Q} Q_j \tag{11.4b}$$

Usually there is a small error in total depths due to discretization. This can be compensated by taking the ratio of the left side to the right side of equation (11.4b) and then multiplying Q_j, $j = 1, 2, \ldots, N_Q$ by this ratio.

Kavvas and Schulz (1972) investigated the effect of sampling interval on the UH oscillations. If the sampling interval is too small, the number of data points increases. It also brings higher frequencies and their unwanted contributions leading to oscillations. Taking too large an interval has an opposite effect and decreases resolution of the rising limb, crest segment, and early part of the recession limb of the UH. Thus, an optimum sampling interval giving the most data points without violating the conditions of a low-pass filter suppressing oscillations by attenuating high frequency components should be used.

TABLE 11.1

Computation of direct runoff hydrograph from unit hydrograph of duration D

h/t	0	D	2D	3D	4D	. . .	$(k-1)D$	kD	$(k+1)D$	$(k+2)D$	$(k+3)D$	$(k+4)D$	. . .	$(k+n-2)D$	$(k+n-1)D$
		I_1	I_2	I_3	I_4	. . .	I_{k-1}	I_k							
h_1		I_1h_1	I_2h_1	I_3h_1	I_4h_1	. . .	$I_{k-1}h_1$	I_kh_1							
h_2			I_1h_2	I_2h_2	I_3h_2	. . .	$I_{k-2}h_2$	$I_{k-1}h_2$	I_kh_2						
h_3				I_1h_3	I_2h_3	. . .	$I_{k-3}h_3$	$I_{k-2}h_3$	$I_{k-1}h_3$	I_kh_3					
h_4					I_1h_4	. . .	$I_{k-4}h_4$	$I_{k-3}h_4$	$I_{k-2}h_4$	$I_{k-1}h_4$	I_kh_4				
h_5						. . .	$I_{k-5}h_5$	$I_{k-4}h_5$	$I_{k-3}h_5$	$I_{k-2}h_5$	$I_{k-1}h_5$	I_kh_5			
.															
.															
.								I_1h_{n-1}	I_2h_{n-1}	I_3h_{n-1}	I_4h_{n-1}	I_5h_{n-1}	. . .	I_kh_{n-1}	
h_n								I_1h_n	I_2h_n	I_3h_n	I_4h_n	I_5h_n	. . .	$I_{k-1}h_n$	I_kh_n
	Q_0	Q_1	Q_2	Q_3	Q_4	. . .	Q_k	. . .	. . .	. . .	. . .	. . .	. . .	Q_{k+n-2}	Q_{k+n-1}

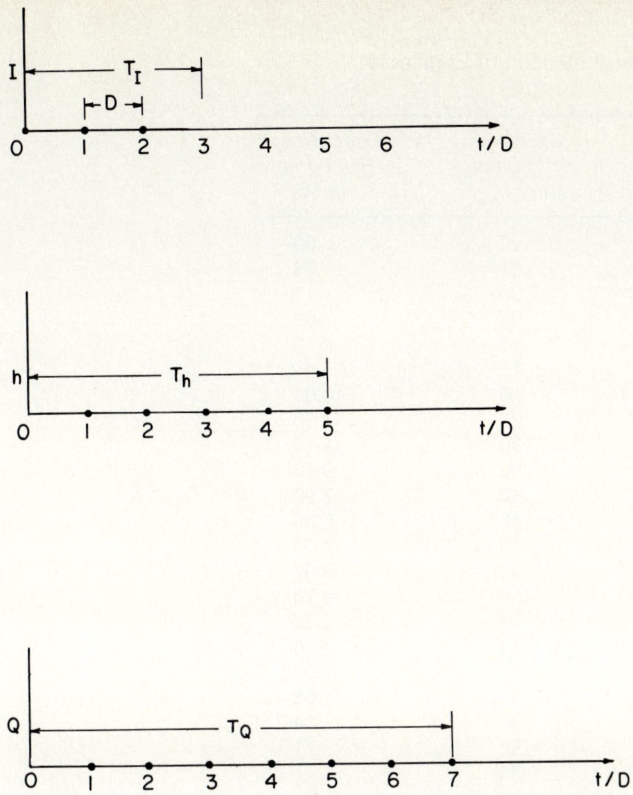

FIGURE 11.2 Location of pulses in discretization scheme.

11.2 DERIVATION OF A UH OR IUH FROM A SINGLE-STORM EVENT

11.2.1 Single-Event Method

The *single-event* method consists in choosing a single, isolated, short-duration, high-intensity, flood-producing storm that is reasonably uniform throughout its duration over the space of the watershed. The runoff hydrograph obtained from this storm should be separated into base flow and direct runoff. The hydrograph of direct runoff can then be reduced to the unit hydrograph for the duration equal to the duration of effective rainfall. This can be taken as a representative UH for that duration. This duration must be much smaller relative to the period during which flood rises.

The advantages of using a short-duration storm are manifold. (1) The variations in the time distribution of intensity of effective rainfall are much less. (2) The determination of effective rainfall is rendered much easier. (3) It is more reliable to derive the UH of a storm of longer duration from the one of a shorter duration than vice versa. However, such storms are not a common occurrence. Short-duration storms do not always produce floods. In such cases, we are forced to choose storms of longer duration. In any event, the UH so obtained must represent the true conditions during floods.

It may often be necessary to use complex storms for derivation of the UHs. If the storm pattern is such that it can be separated into isolated periods of rain, then the UHs can be derived for all the periods using hydrograph separation. These can then be combined in some manner to obtain a single average UH. However, if the pattern is such that this separation is not possible, then the following methods can be employed.

11.2.2 The Collins Method

The *Collins* method (Collins, 1939) is an iterative trial procedure and is described in detail by Gray (1970). It involves the following basic steps:

1. Determine the duration for which the UH is to be derived. This period should usually be equal to or less than one-fourth of the period of rise of the direct runoff.
2. Determine the time base of the UH which is equal to the time from the end of the effective rainfall to the end of the direct runoff plus the UH duration.
3. Assume a UH with the above qualifications. This can be done in many ways. The easiest way is to take the UH as a rectangular pulse over its time base. Another way is to assume it as a triangle, preferably with its rise over one-third and recession over two-thirds of the time base.
4. Apply the assumed UH to all effective rains except the largest. Compute the direct runoff hydrograph.
5. Subtract this computed hydrograph from the observed hydrograph. Adjust the residuals to the UH terms so that they have a unit volume. This yields the first trial UH.
6. Determine the initial UH for the second trial, and start a second round of adjustment by keeping, of course, unit volume. This can be done by averaging the initial and first trial UH's as

$$\bar{h} = \frac{h^0 + h^1 F}{1 + F} \tag{11.5}$$

where h^0 is ordinate of the initial UH, h^1 is the ordinate of the first trial UH, $\bar{h}$ is the ordinate of the initial UH for the second trial, and

$$F = \frac{\Sigma(Q - \Sigma I h^0)}{\Sigma I h^0}$$

7. Repeat this procedure until the differences between the UHs of successive trials are within acceptable limits. This method is illustrated by Example 11.1.

EXAMPLE 11.1

A rainfall-runoff record is given in Tables 11.2 and 11.3 for the Eastern Creek at Great Western Highway in Australia. Its drainage area is 24.86 km^2. Derive a UH for this multiperiod storm by the Collins method. This example is taken from the Institution of Engineers, Australia (1977).

TABLE 11.2

Derivation of direct runoff hydrograph (after Institution of Engineers, Australia, 1977)

Date	Time	*Recorded Hydrograph* (m^3/s)	*Baseflow Hydrograph* (m^3/s)	*Direct Runoff Hydrograph* (m^3/s)
5/20/59	2200	.00	.00	.00
	2300	.04	.01	.03
	2400	.16	.05	.11
5/21/59	100	.70	.08	.62
	200	2.23	.11	2.12
	300	5.60	.15	5.45
	400	10.26	.20	10.06
	500	11.25	.25	11.00
	600	11.49	.30	11.19
	700	10.25	.35	9.90
	800	8.39	.40	7.99
	900	6.80	.45	6.35
	1000	5.45	.49	4.96
	1100	4.20	.54	3.66
	1200	3.37	.59	2.78
	1300	2.80	.64	2.16
	1400	2.40	.70	1.70
	1500	2.10	.75	1.35
	1600	1.88	.80	1.08
	1700	1.70	.86	.84
	1800	1.63	.92	.71
	1900	1.45	.98	.47
	2000	1.37	1.00	.37
	2100	1.28	1.00	.28
	2200	1.20	1.00	.20
	2300	1.05	.96	.09

Solution

1. Separate the baseflow from the hydrograph, as in Table 11.2.
2. Calculate the amount of direct runoff in millimeters.

$$\begin{aligned}\text{Volume of runoff} &= \Sigma\ (\text{hourly ordinates}) \times 3600 \\ &= 85.47 \times 3600\ \text{m}^3 \\ &= 308 \times 10^3\ \text{m}^3\end{aligned}$$

Therefore,

$$\begin{aligned}\text{Depth of runoff} &= \frac{\text{volume}}{\text{watershed area}} \\ &= \frac{3.08 \times 10^5 \times 10^3}{24.86 \times 10^6} \\ &= 12.4\ \text{mm}\end{aligned}$$

3. Calculate the loss rate by trial and error, knowing that the total effective rainfall must equal 12.4 mm. This is demonstrated in Table 11.3. This gives an initial loss of 22.84 mm and a constant infiltration of 0.80 mm/h.
4. Calculate the volume of runoff for the UH.

$$\begin{aligned}\text{Watershed area} &= 24.86\ \text{km}^2 \\ &= 24.86 \times 10^6\ \text{m}^2\end{aligned}$$

For 1 mm of runoff,

$$\begin{aligned}\text{Volume} &= 24.86 \times 10^6 \times 10^{-3}\ \text{m}^3 \\ &= 24.86 \times 10^4\ \text{m}^3\end{aligned}$$

5. Assume an initial unit hydrograph of the given volume. For the first estimate, a rectangular unit hydrograph (that is, one of constant flow) is assumed. Choosing a duration of 20 h, the ordinate is 0.345 m^3/s, since $20 \times 3600 \times 3.45 = 2.486 \times 10^5\ m^3$.
6. Calculate first trial unit hydrograph, as shown in Table 11.4. To calculate the first trial unit hydrograph (Column 13 of Table 11.4), we work only on the time periods where maximum I has effect, that is, neglect calculation for the time units 0, 1, 2, and 24, 25, and 26 as the differences between Columns 10 and 11 in these time units are not caused by the maximum value of I. These differences approach zero as the unit hydrograph approaches its correct value.
7. Start a second round of adjustments by determining the initial unit hydrograph for the second trial and adjusting to unit volume. Do this by averaging the initial and first trial unit hydrographs using Equation (11.5). This is shown in Table 11.5. Therefore,

$$F = \frac{40.54}{43.85} = 0.925$$

$$\bar{h}_1 = \frac{h^0 + 0.925\,h^1}{1.925}$$

The process is continued until the differences between the final unit hydrographs of successive trials are acceptable. This is demonstrated in the calculations shown in Table 11.6.

TABLE 11.3

Calculation of effective rainfall (after Institution of Engineers, Australia, 1977)

Date	Time	*Rainfall in the Preceding Hour (mm)*	*Rainfall − Initial Loss (mm)*	*Rainfall − Initial Loss − Continuing Loss = Effective Rainfall (mm)*
5/20/59	1500	.00		
	1600	2.28		
	1700	1.02		
	1800	4.06		
	1900	.50		
	2000	3.30		
	2100	4.82		
	2200	6.86		
	2300	1.93	1.93	1.13
	2400	.50	.50	.00
5/21/59	100	4.32	4.32	3.52
	200	6.35	6.35	5.55
	300	2.28	2.28	1.48
	400	1.52	1.52	.72
	500	.76	.76	.00
			17.66	12.40

TABLE 11.4

Calculation of first trial unit hydrograph (after Institution of Engineers, Australia, 1977)

Time Unit	*Effective Rainfall* I *(mm)*	*Assumed Initial Unit Hydrograph* h^0	Ih^0						ΣIh^0	*Actual Hydrograph* (Q) *(m^3/s)*	$Q - \Sigma Ih^0$	$\frac{Q - \Sigma Ih^1}{I_{max}}$ (h^1)
(1)	(2)	(3)	(4)	(5)	(6)	(7)[a]	(8)	(9)	(10)	(11)	(12)	(13)[b]
0	1.13	.345								.00		
1	.00	.345	.389						.389	.03		
2	3.52	.345	.389	0					.389	.11		
3	5.55	.345	.389	0	1.214				1.603	.62		
4	1.48	.345	.389	0	1.214	—			1.603	2.12	.517	.093
5	.72	.345	.389	0	1.214	—	.511		2.114	5.45	3.336	.601
6	.00	.345	.390	0	1.214	—	.511	.248	2.362	10.06	7.698	1.387
7		.345	.389	0	1.214	—	.511	.248	2.362	11.00	8.638	1.556
8		.345	.389	0	1.214	—	.511	.248	2.362	11.19	8.828	1.591
9		.345	.389	0	1.214	—	.511	.248	2.362	9.90	7.538	1.358
10		.345	.389	0	1.214	—	.511	.248	2.362	7.99	5.628	1.014
11		.345	.389	0	1.214	—	.511	.248	2.362	6.35	3.988	.719
12		.345	.389	0	1.214	—	.511	.248	2.362	4.96	2.598	.468
13		.345	.389	0	1.214	—	.511	.248	2.362	3.66	1.298	.234
14		.345	.389	0	1.214	—	.511	.248	2.362	2.78	.418	.075
15		.345	.389	0	1.214	—	.511	.248	2.362	2.16	−.202	−.036
16		.345	.389	0	1.214	—	.511	.248	2.362	1.70	−.662	−.119
17		.345	.389	0	1.214	—	.511	.248	2.362	1.35	−.012	−.182
18		.345	.389	0	1.214	—	.511	.248	2.362	1.08	−1.282	−.231
19		.345	.389	0	1.214	—	.511	.248	2.362	.84	−1.552	−.280
20			.389	0	1.214	—	.511	.248	2.362	.71	−1.652	−.298
21				0	1.214	—	.511	.248	1.973	.47	−1.503	−.271
22					1.214	—	.511	.248	1.973	.37	−1.603	−.289
23						—	.511	.248	.759	.28	−.479	−.086
24							.511	.248	.759	.20		
25								.248	.248	.09		
26										.00		

[a] Do not include the highest effective rainfall period.

[b] Column 13 lists the first trial UH values (called h^1).

TABLE 11.5

Derivation of initial UH for second trial (after Institution of Engineers, Australia, 1977)

h_i	$0.925h^1$	$\bar{h}_0$	Initial Unit Hydrograph Ordinates for Second Trial[a]
.345	.086	.224	.218
.345	.556	.468	.456
.345	1.283	.846	.823
.345	1.439	.927	.902
.345	1.472	.944	.919
.345	1.256	.832	.810
.345	.938	.666	.648
.345	.665	.525	.511
.345	.433	.404	.394
.345	.216	.291	.284
.345	.069	.215	.210
.345	−.033	.162	.158
.345	−.110	.122	.119
.345	−.168	.092	.089
.345	−.214	.068	.066
.345	−.259	.045	.044
.345	−.276	.036	.035
.345	−.251	.049	.048
.345	−.267	.041	.040
.345	−.080	.138	.134
		7.095	

[a] Adjusted to the required unit hydrograph volume of 24.86×10^3 m³.

$$\text{Ordinate} = \frac{\bar{h}^0 \times 24.86 \times 10^4}{3600 \times 7.095}$$

11.2.3 Method of Forward Substitution

The set of simultaneous equations given in matrix form in equation (11.2b) can be reexpressed as

$$
\begin{aligned}
Q_1 &= I_1h_1 \\
Q_2 &= I_1h_2 + I_2h_1 \\
Q_3 &= I_1h_3 + I_2h_2 + I_3h_1 \\
&\vdots \\
Q_m &= I_mh_1 + I_{m-1}h_2 + \cdots + I_1h_m \\
&\vdots \\
Q_{m+n-2} &= I_mh_{n-1} + I_{m-1}h_n \\
Q_{m+n-1} &= I_mh_n
\end{aligned}
\tag{11.6}
$$

Therefore, if Q_i, $i = 1, 2, \ldots, m + n - 1$, and I_i, $i = 1, 2, \ldots, m$, are known, then h_i, $i = 1, 2, \ldots, n$, can be determined from equation (11.6) by forward substitution. For example, given I_1 and Q_1, h_1 can be computed. Then given I_1, I_2, Q_1, Q_2, and h_1, h_2 can be computed, and so on. Thus h_i, $i = 1, 2, \ldots, n$, can be completely specified by

$$h_i = \frac{Q_i - \sum_{j=2}^{i} I_j h_{i-j+1}}{I_1} \tag{11.7}$$

In practice, however, this method is not reliable because of the presence of errors in the data. Each computation depends on all preceding computations. It can be shown that if there are errors in the measurement of I or Q, unrealistic h would be obtained.

From equation (11.6) it is obvious that we can also undertake backward substitution with equal ease. That is, first compute h_n, then h_{n-1}, and so on, until h_1. Evidently this method of backward substitution has the same disadvantage as its counterpart, the method of forward substitution.

EXAMPLE 11.2

Take a truly linear system (Kavvas, 1972): $[h] = [5, 2]$, $[I] = [3, 4, 1]$, and $\Delta t =$. Then

$$
\begin{aligned}
Q_1 &= 5 \times 3 = 15 \\
Q_2 &= 5 \times 4 + 2 \times 3 = 26 \\
Q_3 &= 5 \times 1 + 2 \times 4 = 13 \\
Q_4 &= 4 \times 1 = 2 \\
Q_5 &= 0
\end{aligned}
$$

Therefore, $[Q] = [15, 26, 13, 2]$

Suppose an error was made in the observation of $[I]$ as $[I] = [3, 0, 1]$. Then using equation (11.7),

$$h_1 = \frac{Q_1}{I_1} = \frac{15}{3} = 5$$

$$h_2 = \frac{Q_2 - h_1I_2}{I_1} = \frac{26}{3} = 8.66$$

$$h_3 = \frac{Q_3 - I_2h_2 - I_3h_1}{I_1} = \frac{13 - 5}{3} = 2.66$$

$$h_4 = \frac{Q_4 - I_2h_3 - I_3h_2 - I_4h_1}{I_1} = \frac{2 - 8.66 - 1}{3} = \frac{6.66}{3} = -2.22$$

$$h_5 = \frac{Q_5 - I_2h_4 - I_3h_3}{I_1} = -\frac{2.66}{3}$$

$$h_6 = +\frac{I_3h_4}{I_1} = +\frac{2.22}{3}$$

Therefore, $[h] = [5, 8.66, 2.66, -2.22, -\frac{2.66}{3}, \frac{2.22}{3}, \ldots]$

It is apparent that negative and oscillating values of the kernel $[h]$ are obtained.

Let us now convolute observed $[I]$ having error with true, error-free $[h]$,

$$
\begin{aligned}
Q_1 &= 3 \times 5 = 15 \\
Q_2 &= 3 \times 2 + 0 \times 15 = 6 \\
Q_3 &= 1 \times 5 + 0 \times 2 = 5 \\
Q_4 &= 1 \times 2 = 2
\end{aligned}
$$

Therefore, $[\tilde{Q}] = [15, 6, 5, 2]$. If we compare this with $[Q]$ specified previously, we observe that the two discharges differ considerably.

However, if an error is introduced in the original set of values for the discharge, which are assumed to be error-free, as $[Q] = [15, 16, 13, 2]$, then taking the error-free values of $[I]$ and using equation (11.7),

TABLE 11.6

Subsequent trials of UH calculations (after Institution of Engineers, Australia, 1977)

										Second Trial										
Initial UH for second trial	.218	.456	.823	.902	.919	.810	.648	.511	.394	.284	.210	.158	.119	.089	.066	.044	.035	.048	.040	.134
Final UH for second trial	.072	.337	.860	.931	.969	.856	.675	.534	.420	.302	.229	.179	.141	.112	.089	.067	.041	.046	.004	.086
										Third Trial										
Final	.029	.283	.898	.940	.982	.869	.678	.535	.426	.302	.228	.179	.142	.115	.091	.075	.046	.053	.000	.066
										Fourth Trial[a]										
Final	.021	.247	.925	.941	.984	.875	.678	.534	.430	.302	.228	.180	.142	.115	.089	.079	.047	.088	.002	.056
										Fifth Trial[a]										
Final	.027	.221	.944	.940	.983	.877	.678	.533	.432	.302	.228	.180	.141	.115	.087	.080	.047	.059	.006	.050
										Sixth Trial[a]										
Final	.038	.201	.957	.939	.980	.879	.678	.531	.433	.302	.227	.181	.140	.116	.085	.081	.048	.058	.009	.046
										Seventh Trial[a]										
Final	.050	.185	.965	.939	.977	.879	.678	.529	.434	.302	.226	.182	.139	.117	.084	.081	.048	.057	.012	.043
										Eighth Trial[a]										
Final	.061	.172	.971	.939	.974	.879	.678	.528	.434	.302	.225	.183	.138	.118	.083	.081	.049	.055	.015	.042
Difference[b]	.011	−.013	.006	.000	−.003	.000	.000	−.001	.000	.000	−.001	.001	−.001	.001	−.001	.000	.001	−.002	.003	−.001

[a] Note that there is little advantage in proceeding past the third trial.

[b] Difference between final unit hydrographs from the seventh and eighth trials.

$$h_1 = \frac{Q_1}{I_1} = \frac{15}{3} = 5$$

$$h_2 = \frac{Q_2 - I_2 h_1}{I_1} = \frac{16 - 4 \times 5}{3} = -1.33$$

$$h_3 = \frac{Q_3 - I_2 h_2 - I_3 h_1}{I_1} = \frac{13 - 4(-1.33) - 1 \times 5}{3} = \frac{33.3}{3}$$

$$h_4 = \frac{Q_4 - I_2 h_3 - I_3 h_2 - I_4 h_1}{I_1} = 2 - 1.33 = 0.67$$

Therefore, $[\tilde{h}] = [5, -2, \frac{13.33}{3}, .67, \ldots]$. It is apparent, again, that $[h]$ now obtained is oscillating and has negative values.

Now take the error-free $[h]$ and error prone $[Q]$. Then, using forward substitution in equation (11.6),

$$I_1 = \frac{Q_1}{h_1} = \frac{15}{5} = 3$$

$$I_2 = \frac{Q_2 - I_1 h_2}{h_1} = \frac{16 - 3 \times 2}{5} = 2$$

$$I_3 = \frac{Q_3 - h_2 I_2 - h_3 I_1}{h_1} = \frac{13 - 2 \times 2}{5} = \frac{9}{5}$$

$$I_4 = \frac{Q_4 - h_2 I_3 - h_3 I_2}{h_1} = \frac{2 - 2(9/5)}{5} = -\frac{8}{25}$$

$$I_5 = \frac{[-h_2 I_4]}{h_1} = \frac{-2(-8/25)}{5} = \frac{16}{125}$$

$$\vdots$$

Therefore, $[\tilde{I}] = [3, 2, 9/5, -8/25, 16/125, \ldots]$. It is evident that $[\tilde{I}]$ has unrealistic values. This example demonstrates the need for a method that would given both stable $[h]$ and real and stable $[Q]$.

11.2.4 Method of Successive Approximation

The method of *successive approximation* assumes a UH initially and then constructs the direct runoff hydrograph using this UH. By comparing the observed and constructed hydrographs, an estimate of error is obtained. If the error is not within reasonable limits, the assumed UH is modified and the procedure repeated until a UH that leads to the hydrograph comparing favorably with the observed one is obtained. Mathematically, this method is analogous to linear iteration or the method of simultaneous displacements and to the Collins method discussed before. It is closely analogous to the method of linear iteration for solving nonlinear equations (Conte, 1965). This method was employed by Newton and Vinyard (1967) to determine the UH from floods on the computer.

11.2.5 The DeLaine Method

DeLaine (1970) proposed a method for deriving the UH without using rainfall data. The method requires two error-free runoff hydrographs. The method has since been used by Raghavendran and Reddy (1975) and can be formulated as follows.

If $I(t)$ is sampled at m points and $h(t)$ is sampled at n points one unit time apart, then $Q(t)$ will have $m + n - 1$ ordinates, as seen from equation (11.2). For convenience, let these be normalized so that

$$\sum_{i=1}^{m} I_i = 1 \tag{11.8}$$

$$\sum_{i=1}^{n} h_i = 1 \tag{11.9}$$

$$\sum_{i=1}^{m+n-1} Q_i = 1 \tag{11.10}$$

Rewriting the discrete convolution,

$$Q_j = \sum_{i=1}^{j} I_i h_{j-i+1} \tag{11.11}$$

As seen from the previous discussion, the discrete representation of convolution consists in a set of simultaneous algebraic equations. If h_i, $i = 1, 2, \ldots, n$, are the only unknowns, then the equations are linear; however, if both I_i, $i = 1, 2, \ldots, m$, and h_i, $i = 1, 2, \ldots, n$, are unknown, then the equations are nonlinear.

Equations (11.8)–(11.11) contain $(m + n)$ independent equations. If Q_i, $i = 1, 2, \ldots, m + n - 1$, are known, there are $m + n$ unknowns, namely, I_i, $i = 1, 2, \ldots, m$, and h_i, $i = 1, 2, \ldots, n$. If equations (11.8)–(11.11) are solved, there will be at least one set of real values that satisfy these equations, since the system is real and physical. However, the exact number of real and complex sets of solutions satisfying these nonlinear equations is not clear.

If the runoff hydrographs are chosen and equations (11.9)–(11.11) are solved for I_i, $i = 1, 2, \ldots, m$, and h_i, $i = 1, 2, \ldots, n$, separately for each hydrograph, then there will be one common set of h_i, $i = 1, 2, \ldots, n$, in these two sets of solutions. This is because the system is assumed invariant from one input to another. Therefore, the problem now reduces to determining the common h_i, $i = 1, 2, \ldots, n$.

One way to solve these equations is by expressing

$$I(k) = \sum_{i=1}^{m} I_i k^{i-1} \tag{11.12}$$

$$h(k) = \sum_{i=1}^{n} h_i k^{i-1} \tag{11.13}$$

$$Q(k) = \sum_{i=1}^{m+n-1} Q_i k^{i-1} \tag{11.14}$$

Substituting equation (11.11) into equation (11.14),

$$\begin{aligned} Q(k) &= \sum_{j=1}^{m+n-1} \left[\sum_{i=1}^{j} I_i h_{j-i+1} \right] k^{j-1} \\ &= I_1 h_1 + (I_1 h_2 + I_2 h_1)k + (I_1 h_3 + I_2 h_2 + I_3 h_1)k^2 \\ &\quad + \cdots + (I_m h_n) k^{m+n-2} \end{aligned} \tag{11.15}$$

TABLE 11.7

Direct runoff hydrographs from two independent storms

Hour	*Time (min)*	*Runoff I (cfs)*	*Runoff I (normalized)*	*Runoff II (cfs)*	*Runoff (normalized)*
0	0	.0	0	0	0
0	15	.15	.00075	.75	.0833
0	30	1.05	.0525	.95	.2166
0	45	2.8	.14	.23	.2555
1	0	4.15	.2075	2.05	.2277
1	15	4.05	.2025	1.35	.15
1	30	3.35	.1675	.5	.555
1	45	2.40	.12	.1	.0111
2	0	1.15	.0575	.0	.0
2	15	.6	.03		
2	30	.30	.015		
2	45	.0	.0		

If we take the product of equations (11.12)–(11.13), then we see that

$$Q(k) = I(k)\,h(k) \qquad (11.16)$$

Consequently, we see that the polynomial in equation (11.16), whose coefficients are the successive ordinates Q_i, $i = 1, 2, \ldots, m + n - 1$, equals the product of two polynomials given by equations (11.12)–(11.13) that have coefficients which are the successive ordinates I_i, $i = 1, 2, 3, \ldots, m$, and h_i, $i = 1, 2, 3, \ldots, n$. The factors of the polynomial in equation (11.14) are also the factors of equations (11.12)–(11.13).

To determine these factors equation (11.14) is factorized for each of the two sets of hydrograph data. The factors of equations (11.13) are common to both the polynomials of equation (11.14). Perhaps this can be best illustrated by an example.

EXAMPLE 11.3

This example is due to DeLaine (1970). Consider a watershed whose area is 49.6 acres. Table 11.7 provides two runoff hydrographs. Compute the 15-minute UH and the distribution of effective rainfall corresponding to the two hydrographs.

Solution: The two hydrographs are normalized so that equation (11.10) holds. The normalized values are given in Table 11.7. Let these (nontrivial values) be $Q_1, Q_2, \ldots, Q_{10}$, for the first storm and $Q_1, Q_2, \ldots, Q_7$ for the second storm. Inserting these in equation (11.15) and equating to zero,

$$0 = .0075 + .0525k + 0.14k^2 + .2075k^3 + \cdots + .015k^9$$

$$0 = .0833 + .2166k + 0.2555k^2 + .2277k^3 + \cdots + .0111k^6$$

We determine the roots of these two equations as given in Table 11.8. The roots common to both equations are then marked. These yield all the factors of equation (11.13), that is,

$$(k + 1.0)(k + 1.0)(k + 1.225\,i)(k - 1.225\,i) = 1.5 + 3.0\,k + 2.5\,k^2 + 2\,k^3 + 1.0\,k^4$$

The sum of ordinates of h from equation (11.9) must equal unity, so that coefficients of this expression are normalized to yield

$$.15 + .3k + .25k^2 + .2k^3 + .1k^4$$

Therefore, these coefficients according to equation (11.13) constitute the successive ordinates of the 15-min UH,

$$[h] = [.15, .3, .25, .2, .1]$$

Now we determine the effective rainfall.

$$\text{Total volume of effective rainfall for the first storm} = \text{total volume of runoff for this storm} = (.15 + 1.05 + 2.8 + \cdots + 0.3) \times 15 \times 60 \text{ ft}^3 = 18{,}000 \text{ ft}^3$$

$$\text{Total depth of effective rainfall} = \frac{\text{volume of effective rainfall}}{\text{area of watershed}} = \frac{18{,}000 \times 12}{49.6 \times 10 \times 66 \times 66} = .1 \text{ in.}$$

The remaining roots in Table 11.8 lead to factors of equation (11.12). The effective rainfall for the first storm is obtained from

$$(k + .307)(k + .524 + .425i)(k + .524 - .423i)(k - .677 + 1.391i)(k - .677 - 1.391i) = 1.0 + 5.0k + 7.0k + 4.0k^2 + .0k^4 + 3.0k^5$$

In order for the coefficients to sum to unity, these coefficients are multiplied by .005:

$$.005 + .025k + .035k^2 + .02k^3 + .0k^4 + .015k^5$$

TABLE 11.8

Roots of equation (11.15) for two runoff hydrographs

Roots of Equation for Runoff I	*Roots of Equation for Runoff II*
− .307	
−1.0	−1.0
−1.0	−1.0
−1.225*i*	−1.225*i*
1.225*i*	1.225*i*
−0.524 − 0.423*i*	−1.5 − 1.658*i*
−0.524 + 0.423*i*	−1.5 + 1.658*i*
+0.677 − 1.391*i*	
+0.677 + 1.391*i*	

Thus the distribution of effective rainfall for the first storm with 15-min time interval is

$$[I] = [.005,\ .025,\ .035,\ .02,\ .0,\ .015 \text{ in.}]$$

Similarly, the distribution of effective rainfall for the second storm is found as

$$[I] = [0.0175,\ 0.0105,\ 0.0035 \text{ in.}]$$

11.2.6 Harmonic Analysis

In actual practice, a harmonic series is fit to a set of data where only a finite amount of information is available. This means that the harmonic series is finite with a finite number of coefficients. For example, if there are m observations, we can find only m coefficients divided between the sine and cosine terms. The coefficients of a finite harmonic series are referred to as *harmonic coefficients*.

Let $Q(t)$ be defined by Q_k, $k = 0, 1, \ldots, m$, with $Q_k = Q_0$ at $t = 0, 2\pi/m, 2(2\pi/m), \ldots, (m-1)(2\pi/m), 2\pi$. Then at these points we can exactly fit a function of the form

$$Q(s) = \frac{A_0}{2} + \sum_{k=1}^{n} A_k \cos\left(k\frac{2\pi s}{m}\right) + \sum_{k=1}^{n} B_k \sin\left(k\frac{2\pi s}{m}\right) \quad (11.17)$$

where $m = 2n + 1$ is the number of discharge data points. Because there are m observations available, only m meaningful coefficients A_k and B_k can be derived from the data. Since the sine and cosine functions are orthogonal under summation, the coefficients can be written, following O'Donnell (1960), as

$$A_r = \frac{2}{m}\sum_{k=0}^{m-1} Q_k \cos\frac{2\pi rk}{m}, \qquad 1 \le r \le \frac{m}{2} \quad (11.18)$$

$$B_r = \frac{2}{m}\sum_{k=0}^{m-1} Q_k \sin\frac{2\pi rk}{m}, \qquad 1 \le r \le \frac{m}{2} \quad (14.19)$$

$$A_0 = \frac{1}{m}\sum_{k=0}^{m-r} Q_k \quad (11.20)$$

and if m is even,

$$A_{m/2} = \frac{1}{m}\sum_{k=0}^{m-r} Q_k \cos(\pi k) \quad (11.21)$$

Similarly,

$$h(s) = \frac{\alpha_0}{2} + \sum_{k=1}^{n} \alpha_k \cos\left(k\frac{2\pi s}{m}\right) + \sum_{k=1}^{n} \beta_k \sin\left(k\frac{2\pi s}{m}\right) \quad (11.22)$$

$$I(s) = \frac{a_0}{2} + \sum_{k=1}^{n} a_k \cos\left(k\frac{2\pi s}{m}\right) + \sum_{k=1}^{n} b_k \sin\left(k\frac{2\pi s}{m}\right) \quad (11.23)$$

As Dooge (1973) discussed, this formulation can be conveniently expressed in exponential form,

$$Q(s) = \sum_{k=-n}^{n} \xi_k e^{ik2\pi s/m} \quad (11.24)$$

$$\xi_k = \frac{1}{m}\sum_{s=0}^{m-1} Q(s) e^{-ik2\pi s/m} \quad (11.25)$$

$$I(s) = \sum_{k=-n}^{n} \gamma_k e^{ik2\pi s/m} \quad (11.26)$$

$$\gamma_k = \frac{1}{m}\sum_{s=0}^{m-1} I(s) e^{-ik2\pi s/m} \quad (11.27)$$

$$h(s) = \sum_{k=-n}^{n} \rho_k e^{ik2\pi s/m} \quad (11.28)$$

$$\rho_k = \frac{1}{m}\sum_{s=0}^{m-1} h(s) e^{-ik2\pi s/m} \quad (11.29)$$

The discrete functions representing the effective rainfall, runoff, and UH are assumed to be periodic with a period equal to mD. Therefore, equation (11.2) can be written as

$$Q(s) = \sum_{w=s-m+1}^{s} I(w)h(s-w) \quad (11.30)$$

Substituting equation (11.30) into equation (11.25), replacing $I(s)$ and $h(s)$ by equations (11.26) and (11.28), and using the orthogonality property,

$$\xi_k = m\gamma_k\rho_k \quad (11.31)$$

This linkage relation is analogous to equation (10.60). The linkage relation analogous to equations (10.61)–(10.63) for the coefficients of the trigonometric functions can be derived in a similar manner and will be of the form

$$\alpha_k = \frac{n}{2}\left(\frac{a_k A_k + b_k B_k}{a_k^2 + b_k^2}\right) \quad (11.32)$$

$$\beta_k = \frac{n}{2}\left(\frac{a_k B_k - b_k A_k}{a_k^2 + b_k^2}\right) \quad (11.33)$$

The harmonic analysis, or Fourier series method, is largely applied to periodic data. It can, however, be applied to an isolated input by assuming that the input and output are periodic with a period equal to or greater than the length of the output. A particular advantage of this method is that the transformation involved is easy to perform in practical applications.

11.2.6 Fourier Transform

Levi and Valdes (1964) applied the Fourier transform to the analysis of a complex discharge hydrograph of Tuxpan River at Quito, Guerrero, Mexico. Let $I(t)$ and $Q(t)$ be periodic with a period D. Let their values be known at $2m + 1$ equidistant points,

$$t_{-m} = 0,\ t_{-m+1},\ t_{-m+2},\ \ldots,\ t_0,\ t_1,\ \ldots\ t_{m-1},\ t_m = D$$

We wish to study them during the time interval (0, D). Let us transform the time interval (0, D) into ($-\pi$, π) using the relation

$$s = \pi\left(\frac{2t}{D} - 1\right) \tag{11.34}$$

It is clear that these functions will be periodic with 2π time interval and will be known at $2m + 1$ corresponding equidistant points,

$$s_{-m} = -\pi,\ s_{-m+1},\ s_{-m+2},\ \ldots,\ s_0,\ s_1,\ \ldots,\ s_m = \pi$$

Then it is known from Fourier series theory that $Q(s)$ and $I(s)$ can be expressed by equations (11.24) and (11.26), respectively:

$$Q(s) = \sum_{k=-m}^{m} \xi_k e^{iks} \tag{11.35}$$

$$I(s) = \sum_{k=-m}^{m} \gamma_k e^{iks} \tag{11.36}$$

If $Q(w)$ is the Fourier transform of $Q(t)$, $I(w)$ the Fourier transform of $I(t)$, and $h(w)$ the Fourier transform of $h(t)$, then $Q(w) = I(w)\ h(w)$. The inverse Fourier transform of $Q(w)$ is

$$Q(t) = \frac{1}{2\pi}\int_{-\infty}^{\infty} h(w)I(w)e^{itw}\,dw$$

and that of $I(w)$ is

$$I(t) = \frac{1}{2\pi}\int_{-\infty}^{\infty} I(w)e^{itw}\,dw$$

On comparing these with equations (11.35) and (11.36) and after some manipulation, $Q(s)$ can be expressed as

$$Q(s) = \sum_{k=-m}^{m} \gamma_k h(w_k)e^{iks} \tag{11.37}$$

where w_k is the Fourier transform variable w corresponding to s_k of s. Because equations (11.37) and (11.35) must be equivalent,

$$h(w_k) = \frac{\xi_k}{\gamma_k} \tag{11.38}$$

Using equation (11.38) it is possible to calculate the Fourier transform of $h(t)$ at equidistant points s_k, or t_k for that matter, for known values of ξ_k and γ_k. The coefficients ξ_k and γ_k can be determined following a well-known procedure (Lanczos, 1956), which permits adjustment of these coefficients according to the least squares method:

$$\xi_k = \frac{1}{2m}\sum_{j=-m+1}^{m-1} Q\left(j\frac{\pi}{m}\right)e^{-ikjs} + \frac{1}{4m}[Q(-\pi)e^{ik\pi} + Q(\pi)e^{-ik\pi}] \tag{11.39}$$

and

$$\gamma_k = \frac{1}{2m}\sum_{j=-m+1}^{m-1} I\left(j\frac{\pi}{m}\right)e^{-ikjs} + \frac{1}{4m}[I(-\pi)e^{ik\pi} + I(\pi)e^{-ik\pi}] \tag{11.40}$$

The problem now reduces to determining the inverse transform of $h(w_k)$,

$$h(w) = \int_0^{\infty} h(t)e^{-iwt}\,dt = \int_0^{\infty} h(t)\cos(wt)dt - i\int_0^{\infty} h(t)\sin wt\,dt$$

Since $h(t)$ is real, the real part of $h(w)$ is

$$R[h(w)] = \int_0^{\infty} h(t)\cos wt\,dt$$

and

$$h(t) = \frac{2}{\pi}\int_0^{\infty} R[h(w)]\cos wt\,dw$$

To carry out calculations, the graph of $R[h(w)]$ may be approximated by means of trapezes (Solodovnikov, 1969) fitting the graph of the real part of $H(w)$, $R[h(w)]$, as

$$h(t) = \frac{2}{\pi}\sum_k d_k w_k \frac{\sin w_k t + \sin \delta_k t}{w_k t\ \delta_k t} \tag{11.41}$$

The symbols are shown in Fig. 11.3. The sum is to be performed over all the trapezes through which $R[h(w)]$ is fitted.

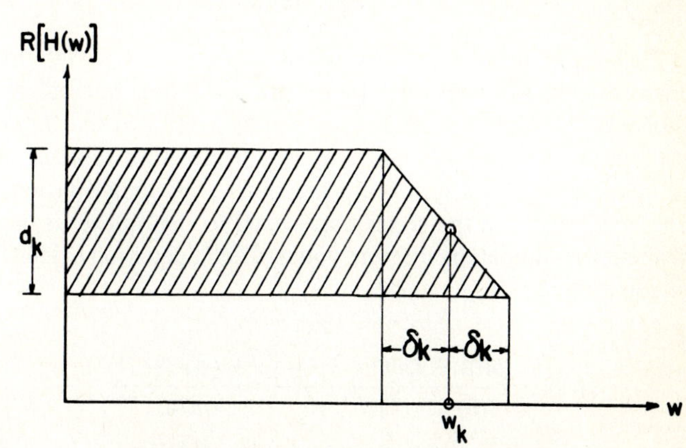

FIGURE 11.3 Trapeze and its symbols.

To summarize, the following steps may be noted: (1) Define the interval $(0, D)$ and transform the variable t to s by equation (11.34). (2) Evaluate the coefficients ξ_k and γ_k, $k = 1, 2, \ldots, (2m + 1)$, from known values of $Q(t)$ and $I(t)$. (3) Determine the corresponding values of $h(s)$ by equation (11.38). (4) Approximate the graph $R[h(w)]$ by trapezes and evaluate $h(t)$ by equation (11.41).

11.2.8 The Z-Transform

This transform was specifically developed for discrete-time systems. Let a function be known at discrete points T intervals apart. Then such a function can be defined by equation (4.48).

$$f(t) = \sum_{n=0}^{\infty} f(t)\delta(t - nT) \tag{11.42}$$

where n is an integer. Taking the Laplace transform of equation (11.42),

$$L[f(t)] = L\left[\sum_{n=0}^{\infty} f(t)\delta(t - nT)\right] = \sum_{n=0}^{\infty} f(nT)e^{-snT} \tag{11.43}$$

Let $z = e^{sT}$. Then the Z-transform can be defined as

$$F(z) = Z[f(nT)] = \sum_{n=0}^{\infty} f(nT)z^{-n} \tag{11.44}$$

where

$$f(nT) = \frac{1}{2\pi j}\oint F(z)z^{n-1}\,dz \tag{11.45}$$

Here z is a complex variable and the contour integral is evaluated on the unit circle. The properties of the Z-transform are similar to those of the Laplace transform. Therefore, the relationship between Z-transforms of $I(t)$, $h(t)$, and $Q(t)$, denoted by $X(z)$, $H(z)$ and $Y(z)$, respectively, is similar to equation (10.86),

$$Y(z) = X(z)H(z) \tag{11.46}$$

To obtain $h(t)$, we must obtain the inverse of $H(z)$. In practice it is not necessary to use the inversion integral of equation (11.45). It is seen from equation (11.46) that $H(z)$ is the ratio of two polynomials in z^{-1}. It therefore suffices to express the ratio in a series of powers of z^{-1} by long division. The coefficients in the series correspond to the values of the discrete function at the discrete points. Using equation (11.44) in equation (11.46),

$$H(z) = \frac{Q(0) + Q(T)z^{-1} + \cdots + Q(nT)z^{-n}}{I(0) + I(T)z^{-1} + \cdots + I(mT)z^{-m}} \tag{11.47}$$

By long division

$$H(z) = h(0) + h(1)z^{-1}h(2)z^{-2} + \cdots + h(n)z^{-n} + \cdots \tag{11.48}$$

It is seen that $h(t)$ is given by the coefficients in the series expansion.

11.2.9 Meixner Functions

Dooge (1966) is perhaps the first to have suggested the use of the *Meixner functions* for the analysis of linear direct runoff systems. However, no mathematical discussion on the use of these functions to derive the UH appeared prior to Dooge (1973). His discussion was based on the premise that Meixner polynomials are a discrete-time analog of Laguerre polynomials (Dooge, 1965). Other investigators (Laurenson and O'Donnell, 1969; Dooge and Garvey, 1970; Brandstetter and Amorocho, 1970; Amorocho and Brandstetter, 1971; V. P. Singh and Baniukiewicz, 1980; V. P. Singh, Baniukiewicz, and Ram, 1982; V. P. Singh and McCann, 1983) have since used Meixner functions. The method of their use is based on representing each of the three functions in equation (11.2) by a finite sum of weighted Meixner functions. A linkage equation is then developed to relate the Meixner coefficients of the UH to those of effective rainfall and direct runoff.

Meixner functions are special cases of the Meixner polynomials. A detailed account of these functions and their mathematical properties is given by Lee (1966). These functions $f_n(t)$ can be defined as

$$f_n(t) = (\tfrac{1}{2})^{(t+n+1)/2} \sum_{k=0}^{n} (-1)^k \binom{n}{k}\binom{t}{k}, \qquad n = 0, 1, 2, \ldots \tag{11.49}$$

where t is a discrete-time variable and $\binom{n}{k} = 0$ when $n < k$. The functions $\{f_n(t)\}$ are orthogonal with respect to the discrete inner product and complete in the space of all continuous functions. $I(t)$, $h(t)$, and $Q(t)$ of equation (11.2) can be represented in terms of Meixner functions as

$$I(t) = \sum_{m=0}^{\infty} a_m f_m(t) \tag{11.50}$$

$$h(t) = \sum_{n=0}^{\infty} \alpha_n f_n(t) \tag{11.51}$$

$$Q(t) = \sum_{r=0}^{\infty} A_r f_r(t) \tag{11.52}$$

The problem then reduces to expressing α_m in terms of a_m and A_m. Dooge (1973) presented a linkage relation without a proof. V. P. Singh and McCann (1983) derived this relation by using the Z-transform. The linkage relation can be expressed as

$$\alpha_n a_0 = \sum_{k=0}^{n} (\tfrac{1}{2})^{(n-k+1)/2} A_k - \sum_{k=0}^{n-1} \alpha_k a_{n-k} \tag{11.53}$$

A derivation of equation (11.53) is given in Appendix F.

11.2.10 Method of Least Squares

The *least squares method* is based on minimizing the sum of the squares of differences between observed hydrograph and computed hydrograph ordinates for any IUH. This method was first used by Snyder (1955, 1961b); this method has since been used by several investigators (Ramaseshan, Chaube, and Sharma, 1973; Natale and Todini, 1974; Mays and Coles, 1980; V. P. Singh and Baniukiewicz, 1980; V. P. Singh, Baniukiewicz, and Ram, 1982). Following Snyder's work, we can write equation (11.6) using matrix notation as

$$[Q] = [I][h] \tag{11.54}$$

that is,

$$\begin{bmatrix} Q_1 \\ Q_2 \\ \cdot \\ \cdot \\ \cdot \\ Q_p \end{bmatrix}_{p\times 1} = \begin{bmatrix} I_1 & 0 & 0 & 0 & & \cdot & \cdot & \cdot & 0 \\ I_2 & I_1 & 0 & 0 & & \cdot & \cdot & \cdot & 0 \\ \cdot & \cdot & \cdot & \cdot & & \cdot & & & \cdot \\ \cdot & \cdot & \cdot & \cdot & & \cdot & & & \cdot \\ \cdot & \cdot & \cdot & \cdot & & \cdot & & \cdot & \cdot \\ 0 & 0 & 0 & \cdots & I_{n-1} & \cdot & & \cdot & I_{m-1} \\ 0 & 0 & 0 & \cdots & & \cdot & \cdot & \cdot & I_m \end{bmatrix}_{p\times n} \begin{bmatrix} h_1 \\ h_2 \\ \cdot \\ \cdot \\ \cdot \\ h_n \end{bmatrix}_{n\times 1}$$

where $p = m + n - 1$. It is evident that there are $m + n - 1$ equations to solve for n unknowns. Therefore, there are $m - 1$ redundant equations. The least squares criterion can be used to reduce the number of equations to n and thus obtain a stable $[h]$.

Let r_i, $i = 1, 2, \ldots, p$, denote residuals or errors and be defined by

$$[r]_{p,1} = [Q]_{p,1} - [I]_{p,n}[h]_{n,1} = [Q] - [I]\,[h]$$

Then

$$r^2 = [[Q] - [I]\,[h]]^T [[Q] - [[I]\,[h]]$$

If the values of the effective rainfall are expressed as a set of coefficients a_{ij}, the system of equations (11.54) can be written as

$$\begin{aligned} a_{11}h_1 + a_{12}h_2 + \ldots + a_{1n}h_n &= Q_1 \\ a_{21}h_1 + a_{22}h_2 + \ldots + a_{2n}h_n &= Q_2 \\ a_{31}h_1 + a_{32}h_2 + \ldots + a_{3n}h_n &= Q_3 \\ &\vdots \\ a_{p1}h_1 + a_{p2}h_2 + \ldots + a_{pn}h_n &= Q_p \end{aligned}$$

Expressing the errors,

$$\begin{aligned} (a_{11}h_1 + a_{12}h_2 + \ldots + a_{1n}h_n - Q_1)^2 &= r_1^2 \\ (a_{21}h_1 + a_{22}h_2 + \ldots + a_{2n}h_n - Q_2)^2 &= r_2^2 \\ (a_{31}h_1 + a_{32}h_2 + \ldots + a_{3n}h_n - Q_3)^2 &= r_3^2 \\ &\vdots \\ (a_{p1}h_1 + a_{p2}h_2 + \ldots + a_{pn}h_n - Q_p)^2 &= r_p^2 \end{aligned}$$

The error function E can be expressed as

$$E = \sum_{i=1}^{p} r_i^2 = \sum_{i=1}^{p} (a_{i1}h_1 + a_{i2}h_2 + \ldots + a_{in}h_n - Q_i)^2 \tag{11.56}$$

To minimize E in equation (11.56), partial derivatives are taken with respect to h_i, $i = 1, 2, \ldots, n$, and are equated to zero. Then

$$\frac{\partial E}{\partial h_1} = \sum_{i=1}^{p} 2\,(a_{i1}h_1 + a_{i2}h_2 + \ldots + a_{in}h_n - Q_i)a_{i1} = 0$$

This yields

$$h_1 \sum_{i=1}^{p} a_{i1}a_{i1} + h_2 \sum_{i=1}^{p} a_{i1}a_{i2} + \ldots + h_n \sum_{i=1}^{p} a_{i1}a_{in} = \sum_{i=1}^{p} a_{i1}Q_i$$

Similarly, for other partial derivatives with respect to h_i,

$$\begin{aligned} h_1 \sum_{i=1}^{p} a_{i2}a_{i1} + h_2 \sum_{i=1}^{p} a_{i2}a_{i2} + \ldots + h_n \sum_{i=1}^{p} a_{i2}a_{in} &= \sum_{i=1}^{p} a_{i2}Q_i \\ &\vdots \\ h_1 \sum_{i=1}^{p} a_{in}a_{i1} + h_2 \sum_{i=1}^{p} a_{in}a_{i2} + \ldots + h_n \sum_{i=1}^{p} a_{in}a_{in} &= \sum_{i=1}^{p} a_{in}Q_i \end{aligned}$$

Writing original equations in matrix form,

$$\begin{bmatrix} a_{11} & a_{12} & \cdot & \cdot & \cdot & a_{1n} \\ a_{21} & a_{22} & \cdot & \cdot & \cdot & a_{2n} \\ \cdot & \cdot & \cdot & & & \cdot \\ \cdot & \cdot & & \cdot & & \cdot \\ \cdot & \cdot & & & \cdot & \cdot \\ a_{p1} & a_{p2} & \cdot & \cdot & \cdot & a_{pn} \end{bmatrix}_{p\times n} \begin{bmatrix} h_1 \\ h_2 \\ \cdot \\ \cdot \\ \cdot \\ h_n \end{bmatrix}_{p\times 1} \downarrow = \begin{bmatrix} Q_1 \\ Q_2 \\ \cdot \\ \cdot \\ \cdot \\ Q_p \end{bmatrix}_{n\times 1} \tag{11.57}$$

or

$$[A]\,[h] = [Q]$$

Writing the derived equations in matrix form,

$$\begin{bmatrix} \sum_{i=1}^{p} a_{i1}^2 & \sum_{i=1}^{p} a_{i1}a_{i2} & \cdots & \sum_{i=1}^{p} a_{i1}a_{in} \\ \sum_{i=1}^{p} a_{i2}a_{i1} & \sum_{i=1}^{p} a_{i2}^2 & \cdots & \sum_{i=1}^{p} a_{i2}a_{in} \\ \vdots & \vdots & & \vdots \\ \sum_{i=1}^{p} a_{in}a_{i1} & \sum_{i=1}^{p} a_{in}a_{i2} & \cdots & \sum_{i=1}^{p} a_{in}^2 \end{bmatrix}_{n\times n} \begin{bmatrix} h_1 \\ h_2 \\ \vdots \\ h_n \end{bmatrix}_{n\times 1} = \begin{bmatrix} \sum_{i=1}^{p} a_{i1}Q_i \\ \sum_{i=1}^{p} a_{i2}Q_i \\ \vdots \\ \sum_{i=1}^{p} a_{in}Q_i \end{bmatrix}_{n\times 1} \quad (11.58)$$

We can write

$$\begin{bmatrix} \sum_{i}^{p} a_{i1}Q_i \\ \sum_{i}^{p} a_{i2}Q_i \\ \vdots \\ \sum_{i}^{p} a_{in}Q_i \end{bmatrix}_{n\times 1} = \begin{bmatrix} a_{11} & a_{21} & \cdots & a_{p1} \\ a_{12} & a_{22} & \cdots & a_{p2} \\ \vdots & \vdots & & \vdots \\ a_{1n} & a_{2n} & \cdots & a_{pn} \end{bmatrix}_{n\times p} \begin{bmatrix} Q_1 \\ Q_2 \\ \vdots \\ Q_p \end{bmatrix}_{p\times 1}$$

which, in compact form, is

$$[AQ] = [A]^T [Q]$$

Taking the first matrix on the left side of equation (11.58), which is of the form:

$$[\quad] = [A]^T [A]$$

Therefore,

$$[A]^T [A] [h] = [A]^T [Q]$$
$$[h] = [[A]^T [A]]^{-1} [A]^T [Q]$$

Replacing $[A]$ by $[I]$,

$$[h] = [[I]^T [I]]^{-1} [I]^T [Q] \qquad (11.59)$$

The above derivation is much lengthier but is given so one can better appreciate what is involved in the least squares operation.

A more compact and elegant derivation is now given for equation (11.59). We want to get the sum of squares of the differences specified by equation (11.55). This is easily obtained by using the inner products, that is, by multiplying $[r]$ by its transpose $[r]^T$. Therefore,

$$E = \sum_{i=1}^{p} r_i^2 = [r]^T [r] = [Q - Ih]^T [Q - Ih] \qquad (11.60)$$

Differentiating E with respect to $[h]$ and equating it to 0,

$$\frac{dE}{d[h]} = -[I]^T[Q - Ih] - [I][Q - Ih]^T = 0$$
$$-[I]^T[Q] + [I]^T[I][h] - [I][Q]^T - [I][h]^T[I] = 0$$
$$2[I]^T[I][h] = 2[I]^T[Q]$$

Therefore,

$$[h] = [[I]^T [I]]^{-1} [I]^T [Q]$$

11.2.11 Linear Programming (LP) with Minimum Sum of Deviations

Linear programming (LP) was probably first applied by Eagleson, Mejia-R., and March (1966) to solve the Wiener-Hopf equations to obtain the optimum UH. It has since been em-

$$\begin{bmatrix} \sum_{i}^{p} a_{i1}^2 & \sum_{i}^{p} a_{i1}a_{i2} & \cdots & \sum_{i}^{p} a_{i1}a_{in} \\ \sum_{i}^{p} a_{i2}a_{i1} & \sum_{i}^{p} a_{i2}^2 & \cdots & \sum_{i}^{p} a_{i2}a_{in} \\ \vdots & \vdots & & \vdots \\ \sum_{i}^{p} a_{in}a_{i1} & \sum_{i}^{p} a_{in}a_{i2} & \cdots & \sum_{i}^{p} a_{in}^2 \end{bmatrix} = \begin{bmatrix} a_{11} & a_{21} & \cdots & a_{p1} \\ a_{12} & a_{22} & \cdots & a_{p2} \\ \vdots & \vdots & & \vdots \\ a_{1n} & a_{2n} & \cdots & a_{pn} \end{bmatrix} \begin{bmatrix} a_{11} & a_{12} & \cdots & a_{1n} \\ a_{21} & a_{22} & \cdots & a_{2n} \\ \vdots & \vdots & & \vdots \\ a_{p1} & a_{p2} & \cdots & a_{pn} \end{bmatrix}$$

ployed by Deininger (1969), Diskin and Boneh (1973, 1974), Natale and Todini (1976a, 1976b), Newman and deMarsily (1976), K. P. Singh (1976), Mays and Coles (1980), Dreiss (1984), and Unver and Mays (1984) for estimating the UH directly from convolution. The UH ordinates must be subject to the volume conservation constraint

$$\sum_{k=0}^{\infty} h_i = 1 \tag{11.61}$$

the nonnegativity constraint

$$h_i \geq 0, \qquad i = 0, 1, 2, \ldots \tag{11.62}$$

and the ordering constraint,

$$h_i \leq h_{i+1} \leq h_p, \qquad i = 1, 2, \ldots, p \tag{11.63}$$

$$h_p \geq h_i \geq h_{i+1}, \qquad i = p, p+1, \ldots \tag{11.64}$$

where h_p denotes the peak ordinate of the UH. The previous methods do not explicitly consider these constraints. The least squares formulation with inequality constraint loses its linearity and cannot be solved by LP techniques. Mays and Taur (1982) and Unver and Mays (1984) have presented a nonlinear programming model to solve the optimization problem. However, rather than minimizing the sum of the squares of the deviations, one could minimize the sum of the absolute deviations,

$$E = \sum_{i=1}^{N} |e_i| \tag{11.65}$$

where

$$e_i = Q_i - \sum_{j=1}^{i} x_{i-j+1} h_j, \qquad i = 1, 2, 3, \ldots, N \tag{11.66}$$

To make the objective function defined by equation (11.65) admissible for LP, we define the error, which could be positive or negative, to be the difference between two nonnegative variables u_i and v_i. Equation (11.66) can be written as

$$u_i - v_i = Q_i - \sum_{j=1}^{i} x_{i-j+1} h_j, \qquad i = 1, 2, \ldots, N \tag{11.67}$$

or

$$Q_i = \sum_{j=1}^{i} x_{i-j+1} h_j + u_i - v_i \tag{11.68}$$

The objective function of equation (11.65) can be expressed equivalently as minimizing

$$E = \sum_{i=1}^{N} (u_i - v_i) \tag{11.69}$$

Thus, the problem of minimizing the sum of absolute deviations is transformed into an LP problem (Dantzig, 1963), namely, to minimize the linear function in equation (11.69) subject to the deviation constraint of equation (11.66), volume conservation constraint in equation (11.61), inequality constraint in equations (11.63)–(11.64), and nonnegativity constraint in equation (11.62).

11.3 DERIVATION OF UH FROM MULTISTORM EVENTS

The preceding section dealt with the methods of deriving a finite-period UH from a single-storm event (complex rainfall-runoff record). Although it is assumed generally that a watershed possesses a unique UH (of a given duration), an analysis of several rainfall-runoff records often produces different UHs (V. P. Singh and Baniukiewicz, 1980). This lack of uniqueness is normally explained either by the presence of noise in effective rainfall or direct runoff or by the inadequacy of the assumption of linearity. Diskin and Boneh (1975) argue that it is predominantly the noise which causes the nonuniqueness. Consequently, a representative UH is found by averaging several UHs corresponding to different events, although many difficulties arise in doing so. For example, the base lengths of the UHs, the time to peak, and the peak may be all different. Therefore, the best way to obtain this average is not clear. Should base lengths be averaged with common location of the peak ordinates? Should the ordinates of the storms be weighted? We discuss some of the methods to derive the representative UH.

11.3.1 Method of Averaging

The method of averaging consists in selection of more than one storm satisfying the qualifications specified in the previous method. The unit hydrographs for these storms can be constructed in the usual manner. It is normally found that these UHs have peaks occurring at different times. Consequently, the representative UH can be obtained by first computing the average peak time and the average peak ordinate and then drawing a UH having these values and conforming to the shape of the other UHs. We are not taking the arithmetic average of the ordinates of the UHs to derive the representative UH, because this would result in a peak lower than the individual peaks. Figure 11.4 illustrates the method for two hypothetical storms.

11.3.2 Least Squares Method: General Remarks

When several (say N) effective rainfall-direct runoff events are considered simultaneously for derivation of a representative (or an average in some sense) UH, then equation (11.1) can be expressed for every event. However, the first step toward this end is the adjustment of the time bases of direct runoff hydrographs (DRH) of the various storms so that a constant

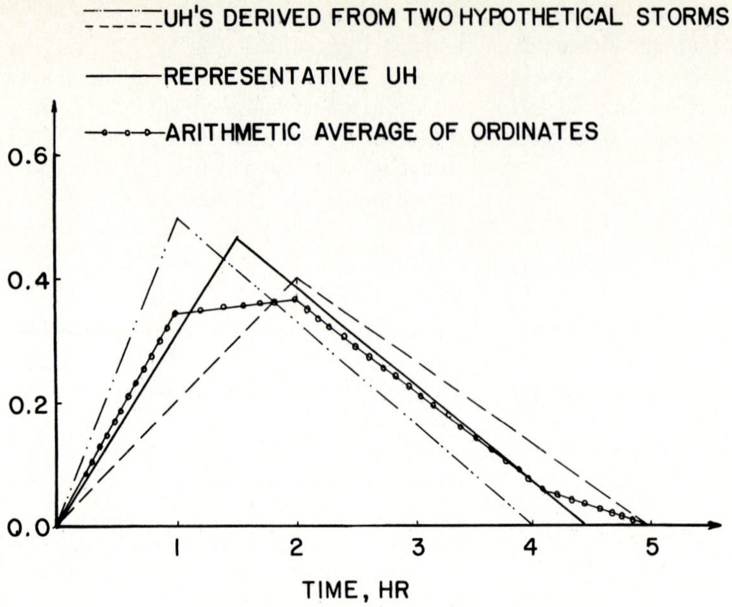

FIGURE 11.4 Derivation of an average UH from two hypothetical storms.

difference between the time bases of these hydrographs and those of effective rainfall hyetographs (ERH) is produced for all events. Let T_{Ij} and I_{Qj} denote the time bases for the jth event of ERH and DRH, respectively. To be physically meaningful, both these time bases must be finite. If T_h is the time base of the UH, then

$$T_h = T_{Qj} - T_{Ij}, \qquad j = 1, 2, 3, \ldots, N \qquad (11.70)$$

Experience as well as experimental evidence show that equation (11.70) is not satisfied for all events. Therefore, T_{Ij} and T_{Qj}, $j = 1, 2, \ldots, N$, should be adjusted to meet equation (11.70). There can be a multitude of reasons for nonconstancy of T_h. Diskin and Boneh (1975) ascribed it principally to hydrograph separation, especially locating the termination of direct runoff on the observed runoff hydrograph recession. By assuming that T_{Qj} is usually underestimated, they defined T_h as

$$T_h = \max_{1 \le j \le N} (T_{Qj} - T_{Ij}) \qquad (11.71)$$

With this definition of T_h, T_Q should be modified for those events not fulfilling equation (11.70) as

$$T^*_{Qj} = T_{Ij} + T_h \qquad (11.72)$$

Note that T_{Ij} is kept unaltered. The values of direct runoff in the extended range between T^*_{Qj} and T_{Qj} are all taken to be zero, that is,

$$Q_j(t) = 0, \qquad T_{Qj} \le t \le T^*_{Qj}, \qquad j = 1, 2, \ldots, N \qquad (11.73)$$

This method of adjustment of the time base is not the only one, but it is convenient and appears reasonable.

The scheme for discretizing the continuous-time functions $I(t)$ and $Q(t)$, and hence $h(t)$, is the same as adopted in the preceding section on single-event analysis. These functions are discretized to have the same sampling time interval D and can be represented by finite pulses obeying the relation

$$N_h = N_{Qj} - N_{Ij}, \qquad j = 1, 2, \ldots, N \qquad (11.74)$$

Equation (11.4b) can be extended as

$$\sum_{j=1}^{N_{Ij}} I_{ij} = \sum_{j=1}^{N_{Qi}} Q_{ij}, \qquad i = 1, 2, 3, \ldots, N \qquad (11.75)$$

Equation (11.1) can be expressed as

$$Q_{ij} = \sum_{k=J}^{L} h_k I_{i,j-k+1}, \qquad i = 1, 2, \ldots, N; \quad j = 1, 2, \ldots, N_{Qi} \qquad (11.76)$$

where the limits of the summation are $J = \max(1; j + 1 - N_{Ii})$ and $L = \min(j, N_h)$. Equation (11.76) is a set of linear equations whose number N_e is

$$N_e = \sum_{j=1}^{N} N_{Qj} \qquad (11.77)$$

Evidently this depends upon D and N. The number of unknowns, N_h, is prescribed by equation (11.74) and also depends on D. Since $N_e > N_h$ even for $N = 1$, equation (11.76) represents an overdetermined set. The unknowns h_j, $j = 1, 2, \ldots, N_h$, can be determined optimally by using a least squares objective function.

Writing equation (11.76) in matrix notation as

$$\mathbf{Y} = X \cdot \mathbf{H} \qquad (11.78)$$

where X is a rectangular matrix of effective rainfall having N_e rows and N_h columns, $\mathbf{H}$ is a column vector of the UH containing N_h elements, and $\mathbf{Y}$ is a column vector of DR having N_e elements. The matrix X is actually a triangular matrix having nonzero elements only in its lower half. One way of solving equation (11.78) is by using matrix inversion, which is an extension of the least squares method described in the previous section for a single event. This yields

$$\mathbf{H} = (X^T \cdot X)^{-1} (X^T \cdot \mathbf{Y}) \tag{11.79}$$

which leads to the least sum of squared deviation, e_{ij},

$$\sum_{i=1}^{N} \sum_{j=1}^{N_{Qi}} e_{ij}^2 \rightarrow \min \tag{11.80}$$

Although this solution can be carried out on a computer without any great difficulty, it is plagued by practical problems, as elucidated by Diskin and Boneh (1975), Bree (1978a, 1978b), and Kitanidis and Bras (1979). In real-world situations the square matrix $(X^T \cdot X)$ is often found to be ill-conditioned. The ill-conditioning can be measured by the *condition number* C_N of the matrix Z defined as $|Z| \cdot |Z^{-1}|$, where $|Z|$ is the norm of the matrix Z. A greater value of this number implies greater degree of ill-conditioning. Diskin and Boneh (1975) used the Euclidean norm (or spectral norm), for which C_N is equal to the ratio of $|a_{\max}|/|a_{\min}|$, where $a_{\max}$ and $a_{\min}$ are the maximum and minimum eigenvalues of the matrix. Roughly speaking, C_N represents the difference between the largest and smallest elements of the matrix. For several cases, Diskin and Boneh found this number to be on the order of 10^{10}, as compared with 10 to 10^2 for well-behaved matrices. Another problem arises from collinearity and stability estimation of the IUH. Kitanidis and Bras (1979) suggested using ridge-regression techniques for treating collinearity, and Kalman filtering for stability.

Another factor is related to the size of N_h, which is usually large (dictated by physical considerations), and D, which in turn is small. This, however, means that successive pulses, especially of effective rainfall and direct runoff, are not entirely independent. Bree (1978a, 1978b) has given an excellent treatment of this issue. This results in ill-conditioning. Another consideration is that nontrivial errors are introduced in computing $(X^T \cdot X)$, which magnify during inversion of this quantity.

11.3.3 Least Squares Solution: Steepest Descent Method

Diskin and Boneh (1975) imposed two constraints on the IUH. First, its ordinates must be nonnegative. This is specified by equation (11.62) and can further be written as

$$h_j \geq 0, \qquad j = 1, 2, \ldots, N_h \tag{11.81}$$

Second, the area enclosed by it must be unity. It follows directly from equation (11.2) and is expressed by equation (11.61) but can be recast as

$$\sum_{j=1}^{N_h} h_j = 1 \tag{11.82}$$

They found that the least squares solution with matrix inversion did not satisfy these constraints and did not yield a satisfactory IUH. Therefore, they employed the steepest descent method to obtain the least squares solution for determination of optimum IUH subject to equations (11.81)–(11.82). The objective function to be used is expressed by equation (11.80) but can be recast as a function of the vector of unknowns h,

$$S = g(h) \tag{11.83}$$

This method was proposed by Fletcher and Powell (1963). This method is a rapidly converging, efficient descent iterative scheme for obtaining the minimum value of a multivariable function in a finite number of iterations. A short account of this method is given here. For complete details see Fletcher and Powell (1963) or Diskin and Boneh (1975). The method requires initial estimates of the vector of variable h and the auxiliary matrix before starting the iterative search. An initial guess of the auxiliary matrix can be taken as a unit matrix. An initial estimate of the vector h can be obtained from another solution. Then the function S is computed for the initial guess of h. Also computed is the steepest direction on the surface given by S in the multidimensional space defined by the elements of h. This direction is determined from the elements of the gradient vector and those of an auxiliary matrix, which is an approximation to the inverse of the matrix of the second-order derivatives (the Hessian matrix) of the function S. The method searches along the direction of the steepest descent for the point at which S attains its minimum value. At this point, the new values of the elements of the gradient vector and the corrections to the current values of the elements of the auxiliary matrix are computed. These are then used to determine the direction of the steepest descent of S. The procedure continues until the objective function is satisfied. Diskin and Boneh have shown that as the solution progresses, the auxiliary matrix converges to the second-order derivative matrix, whose elements are

$$d_{ij} = \frac{\partial^2 S}{\partial h_i \, \partial h_j} \tag{11.84}$$

where h_i and h_j are the elements of the unknown vector h.

11.3.4 Least Squares Solution: The Householder Method

The Householder method (Reid, 1967; Mawdsley and Tagg, 1981) gives the least squares solution and overcomes the difficulties associated with the matrix inversion method. This method involves the orthogonal transformation of equation (11.78)

$$\mathbf{Z} = \mathbf{RH} \tag{11.85}$$

where $\mathbf{Z} = WY$, $\mathbf{R} = WX$, and W is an orthogonal matrix chosen such that

$$\mathbf{R} = \begin{bmatrix} \tilde{R} \\ \cdot \\ \cdot \\ \cdot \\ 0 \end{bmatrix} \quad (11.86)$$

in which $\tilde{R}$ is an $N_h \times N_h$ upper triangular matrix and 0 is an $(N_Z - N_h) \times N_h$ null matrix. If the vector $\mathbf{Z}$ is written as

$$\mathbf{Z} = \begin{bmatrix} \tilde{\mathbf{Z}} \\ \cdot \\ \cdot \\ \cdot \\ \mathbf{Z}^* \end{bmatrix} \quad (11.87)$$

where $\tilde{\mathbf{Z}}$ is an $N_h \times 1$ vector and $\mathbf{Z}^*$ is an $(N_Q - N_h) \times 1$ vector, then equation (11.85) can be expressed as

$$\begin{bmatrix} \tilde{\mathbf{Z}} \\ \cdot \\ \cdot \\ \cdot \\ \mathbf{Z}^* \end{bmatrix} = \begin{bmatrix} \tilde{\mathbf{R}} \\ \cdot \\ \cdot \\ \cdot \\ \mathbf{0} \end{bmatrix} \mathbf{H} \quad (11.88)$$

The least squares solution of equation (11.88) can be obtained from normal equations as

$$\begin{bmatrix} \tilde{\mathbf{R}}^T & \vdots & 0 \end{bmatrix} \begin{bmatrix} \tilde{\mathbf{Z}} \\ \cdot \\ \cdot \\ \cdot \\ \mathbf{Z}^* \end{bmatrix} = \begin{bmatrix} \tilde{\mathbf{R}}^T & \vdots & 0 \end{bmatrix} \begin{bmatrix} \tilde{\mathbf{R}} \\ \cdot \\ \cdot \\ \cdot \\ 0 \end{bmatrix} \mathbf{H} \quad (11.89)$$

which reduces to

$$[\tilde{R}^T \tilde{\mathbf{Z}}] = [\tilde{R}^T \tilde{R}]\, \mathbf{H} \quad (11.90)$$

or

$$\tilde{\mathbf{Z}} = \tilde{R}\mathbf{H}$$

where R is an upper triangular matrix. Therefore, equation (11.90) can be solved by back substitution. Clearly the choice of $\mathbf{W}$ is crucial; it is found by successive applications of eliminating orthogonal transformations (Wilkinson, 1965). Although this method involves more multiplication steps than Gaussian elimination, its numerical stability is guaranteed, which also implies existence of $\mathbf{W}$. This method has been successfully applied by Mawdsley and Tagg (1981) to two watersheds in Great Britain.

11.3.5 Linear Programming

The LP method, discussed in Section 11.2.10, can be extended to derive a composite UH for a watershed considering several rainfall-runoff events simultaneously. This was first done by Mays and Coles (1980). V. P. Singh and Baniukiewicz (1980) and V. P. Singh, Baniukiewicz, and Ram (1982) have since used this method. Additional discussion of this method is given by Diskin and Boneh (1980), Wormleaton (1980), and Meredith and Byrones (1981). The mathematical development presented here follows Mays and Coles (1980).

Let us write the convolution summation of equation (11.1) for a number of rainfall-runoff events as

$$Q_{k,n} = I_{k,n}h_1 + I_{k,n-1}h_2 + I_{k,n-2}h_3 + \cdots + I_{k,n-m+1}h_m$$
$$k = 1, 2, \ldots, K; \quad n = 1, 2, \ldots, N; \quad m = 1, 2, \ldots, M \quad (11.91)$$

where k refers to the observed hydrograph, K is the total number of observed hydrographs, n refers to the ordinate number of each observed hydrograph, N is the total number of ordinates of the kth observed hydrograph, m is the unit hydrograph ordinate, and M is the total number of unit hydrograph ordinates. The relationship between the number of observed runoff hydrograph ordinates, the number of unit hydrograph ordinates, and the number of periods of effective rainfall L_k for kth observed runoff hydrograph is

$$N_k = M + L_k - 1; \qquad k = 1, 2, \ldots, K \quad (11.92)$$

Obviously, the number of UH ordinates must be the same for each observed runoff hydrograph k. Also these ordinates must be spaced at equal Δt time intervals apart. Equation (11.91) can be written in a matrix form as

$$[Q] = [I]\,[h] \quad (11.93)$$

This equation can be solved for $[h]$. By definition it must satisfy

$$\Delta t \sum_{m=1}^{M} h_m = 1 \quad (11.94)$$

Further, the UH must take on positive values only,

$$h_m \geq 0, \qquad m = 1, 2, \ldots, M \quad (11.95)$$

The linear programming method can be used to determine $[h]$ for known $[Q]$ and $[I]$ subject to the above constraints. The objective is to derive a *UH* that minimizes the difference between the observed $[Q]$ and computed $[Q_*]$. This can be accomplished by minimizing the sum of the absolute deviations between $[Q]$ and $[Q_*]$. We can include nonnegative slack variables x_k and y_k to account for the deviations.

We can express

$$Q_{k,n} = Q_{*k,n} + x_{k,n} - y_{k,n}; \quad k = 1, 2, \ldots, K; \quad n = 1, 2, \ldots, N \quad (11.96)$$

where x is the deviation of computed Q_* below observed Q and y is the deviation of Q_* above Q. The objective is to minimize the sum of the derivations,

$$\text{Min } E = \sum_{k=1}^{K} \sum_{n=1}^{N} (x_{k,n} + y_{k,n}) \tag{11.97}$$

in which E is the objective function to be minimized subject to the constraint in equation (11.95) and

$$x_{k,n} \geq 0 \quad \text{and} \quad y_{k,n} \geq 0; \qquad k = 1, 2, \ldots, K \quad n = 1, 2, \ldots, N \tag{11.98}$$

We can modify equation (11.96) by replacing $Q_{k,n}$ by $Q_{*k,n}$ and then substituting $Q_{k,n} - x_{k,n} + y_{k,n}$ for $Q_{*k,n}$ therein,

$$Q_{k,n} = I_{k,n}h_1 + I_{k,n-1}h_2 + \ldots + I_{k,n-1+m}h_m + x_{k,n} - y_{k,n}; \quad k = 1, 2, \ldots, K; \quad n = 1, 2, \ldots, N \quad m = 1, 2, \ldots, M \tag{11.99}$$

We can specify additional constraints to insure monotonic decline in the UH ordinates beyond the peak by

$$h_m \geq h_{m+1}, \qquad m \geq p \tag{11.100}$$

Thus, the decision variables in the LP method are h_m, $x_{k,n}$, and $y_{k,n}$. The total number of decision variables is determined using the number of decision variables (DV).

$$DV = 2\left(\sum_{k=1}^{K} N_k\right) + M \tag{11.101}$$

The total number of constraints (CN) for the LP method is determined using

$$CN = \sum_{k=1}^{K} N_k + 1 \tag{11.101}$$

An algorithm for solution of the LP problem can be written using the simplex method (Dantzig, 1963). This method involves the solution of sets of linear simultaneous equations. The idea is to progress from one basic feasible solution of constraints to another in such a way as to continually change the objective function values until an optimal solution is reached.

11.4 A COMPARISON OF THE METHODS OF UH DERIVATION

For purposes of comparison, the following criteria can be specified: (1) The IUH ordinates are all positive. (2) the IUH shape is preserved. (3) Errors in the input data are not amplified during the IUH derivation. (4) The method is capable of admitting a number of events simultaneously for the IUH derivation. (5) Computationally, the method is simple, efficient and easily programmable on digital computer. Most methods do not satisfy these requirements. Each method has its strengths and weaknesses. For example, linear programming (Mays and Coles, 1980) can simultaneously utilize a number of rainfall-runoff events for derivation of a composite IUH, but other methods, with the exception of successive over-relaxation (Newton and Vinyard, 1970) cannot. Wormleaton (1980) reported recently that he had developed a least squares method to consider more than one rainfall-runoff event at a time. On the other hand, linear programming and constrained least squares can guarantee all nonnegative IUH ordinates. A detailed comparison of these methods is given by K. P. Singh (1976). He found the two methods to be comparable for two storms. The least squares methods showed gyrations about the UH recession obtained from the LP method for two other storms. Eagleson, Mejia-R., and March (1966) pointed out that time series and spectral analysis did not ensure realizable hydrographs, that is, hydrographs with only nonnegative ordinates. Boneh and Golan (1978, 1979) have given an excellent discussion on the question of negative and nonnegative IUH ordinates. Although a number of studies on various methods of UH derivation have been reported, none appear on an objective comparison of all these methods.

Laurenson and O'Donnell (1969) examined four methods including: (1) harmonic analysis, (2) Meixner polynomials, (3) least squares, and (4) Nash models. These methods were compared with respect to their relative strengths and weaknesses as well as their sensitivity to errors in rainfall-runoff data. They concluded that no one method was uniformly better than the other. In a separate study, Dooge and Garvey (1970) compared these same four methods. For error-free data, the least squares method was found to be the best, but it was also the worst for erroneous data. Of the remaining three methods, Meixner polynomials were the best in the both cases (with and without error), followed by the Nash model, which was then followed by harmonic analysis.

V. P. Singh and Baniukiewicz (1981) and V. P. Singh, Baniukiewicz, and Ram (1982) compared nine methods: (1) matrix method (MT), (2) forward substitution (FS), (3) successive over-relaxation (SOR), (4) least squares (LS), (5) harmonic analysis (HA), (6) Laguerre polynomials (LP), (7) Meixner polynomials (MP), (8) time series (TS) method, and (9) linear programming (LP). Eight rainfall-runoff events were used: four for derivation of the composite IUH and four for prediction of the direct runoff. The methods were evaluated and compared with one another in regard to (1) regeneration performance, (2) representative UH, (3) prediction performance, and (4) sensitivity to data errors. For each aspect, several statistical measures were used to evaluate and compare the methods: (1) correlation coefficient, (2) coefficient of efficiency, (3) integral square error, (4) absolute error to the mean, (5) relative error to the mean, (6) standard error of estimate, and (7) special correlation coefficient. The LP, HA, and MP methods were found to be better than MT, FS, and SOR methods. The LP, HA, and LS methods were found to be the best. The MP, TS, and LP methods were also found to be effective, but increased prudence was required in their applications. When data errors were large, the LP, LS, and LP methods were the best. When the errors were unusually high, the LP method was found to be superior.

EXERCISES

11.1 Data on effective rainfall and direct runoff are given for two rainfall events that occurred on watershed *C*2 near Riesel (Waco), Texas. Its drainage area is 2.43 km^2.

Rainfall Event of 6-10-41

Effective rainfall is:

Time mn	Intensity cm/h	Time mn	Intensity cm/h	Time mn	Intensity cm/h	Time mn	Intensity cm/h
.0	1.6618	4.0	.2416	6.0	2.4623	9.0	.5512
11.0	1.0173	15.0	3.0039	19.0	2.9127	24.0	4.6349
28.0	4.4087	31.0	9.5085	34.0	6.9971	42.0	3.9745
45.0	1.4458	48.0	2.9800	51.0	1.4654	54.0	.9673
58.0	.0000	59.0	.2179	63.0	.0000	66.0	.9933
68.0	.0000	71.0	.2429	75.0	.0000	79.0	.2548
81.0	.0000	85.0	.0000	115.0	.0000		

Observed DRH is:

Time mn	Intensity cm/h	Time mn	Intensity cm/h	Time mn	Intensity cm/h	Time mn	Intensity cm/h
.0	.0000	7.0	.0010	15.0	.0013	23.0	.0025
28.0	.0071	30.0	.0152	33.0	.0406	35.0	.0704
38.0	.1349	42.0	.2344	45.0	.3531	50.0	.5182
57.0	.7874	62.0	1.227	67.0	1.7628	73.0	2.1234
75.0	2.1920	81.0	2.2403	85.0	2.1920	90.0	2.1006
100.0	1.8491	115.0	1.4478	135.0	1.0490	157.0	.7264
187.0	.4191	207.0	.2946	237.0	.1600	262.0	.1031
293.0	.0640	337.0	.0386	415.0	.196	495.0	.0119

Rainfall Event of 6-23-59

Effective rainfall is:

Time mn	Intensity cm/h	Time mn	Intensity cm/h	Time mn	Intensity cm/h	Time mn	Intensity cm/h
.0	2.5579	4.0	.0000	11.0	6.9719	13.0	5.8759
17.0	10.1227	21.0	13.9754	24.0	6.3927	28.0	2.2400
32.0	6.4651	36.0	8.0126	38.0	.4203	44.0	.2884
53.0	.0000	83.0	.0000	93.0	.0000	103.0	.0000

Observed DRH is:

Time mn	Intensity cm/h	Time mn	Intensity cm/h	Time mn	Intensity cm/h	Time mn	Intensity cm/h
.0	.0000	8.0	.0030	24.0	.0211	28.0	.1184
31.0	.1740	33.0	.2075	37.0	.2167	43.0	.1989
51.0	.1671	57.0	.1405	70.0	.2441	74.0	.4318
77.0	.7010	83.0	1.0820	89.0	1.3868	99.0	1.5494
107.0	1.5875	121.0	1.5494	143.0	1.3868	157.0	1.2319
174.0	1.0312	191.0	.7874	209.0	.5791	229.0	.4318
294.0	.2075	359.0	.1031				

Determine the IUH for this watershed by the DeLaine method.

11.2 Determine the IUH by the method of successive approximation using data for each event in Question 11.1. Do the two IUHs differ significantly from one another? Compute an average IUH.

11.3 Determine the IUH by the method of least squares using data for each event separately in Question 11.1. Compute an average IUH and compare it with that obtained in Question 11.2.

11.4 Determine the IUH by linear programming using data for each event separately in Question 11.1. Compute an average IUH and compare it with that obtained in Question 11.3.

11.5 Determine the representative IUH by linear programming using data simultaneously for both events in Question 11.1 and another event given below:

Rainfall Event of 5-13-57

Effective rainfall is:

Time *mn*	*Intensity* *cm/h*	*Time* *mn*	*Intensity* *cm/h*	*Time* *mn*	*Intensity* *cm/h*	*Time* *mn*	*Intensity* *cm/h*
.0	1.5784	3.0	1.3273	9.0	5.9357	11.0	2.5858
15.0	6.8527	19.0	4.3600	22.0	3.8871	26.0	1.7889
31.0	3.0391	36.0	.6253	41.0	2.4726	45.0	.1147
61.0	.0000	77.0	.1996	88.0	.2402	101.0	.0000
115.0	.0000	135.0	.0000	165.0	.0000		

Observed DRH is:

Time *mn*	*Intensity* *cm/h*	*Time* *mn*	*Intensity* *cm/h*	*Time* *mn*	*Intensity* *cm/h*	*Time* *mn*	*Intensity* *cm/h*
.0	.0000	13.0	.0013	19.0	.0183	22.0	.1453
26.0	.2418	30.0	.3099	34.0	.3429	41.0	.3556
50.0	.3302	53.0	.3302	61.0	.4140	67.0	.6299
73.0	.9855	80.0	1.2802	89.0	1.4376	92.0	1.4376
108.0	1.2421	116.0	1.0922	131.0	.8230	145.0	.6401
163.0	.4877	175.0	.4216	233.0	.2367		

Compare it with the average IUH obtained in Question 11.4.

11.6 Determine the IUH by the method of least squares using data simultaneously for both events in Question 11.5. Compare it with the average IUH in Question 11.3.

11.7 Determine the IUH by the Collins method for each event in Question 11.5. Compute an average IUH and compare it with that obtained in Question 11.2.

11.8 Which method would you choose for determining the IUH for a watershed and why?

CHAPTER TWELVE

Differential Equations and Hydrologic System Characteristics

In Chapter 8 we derived a general differential equation for a lumped linear time-invariant effective rainfall-direct runoff system. In Chapters 10 and 11 we discussed the operation of such a system using the IUH without explicitly resorting to the use of the differential equation. In this chapter we discuss some of the characteristics of hydrologic systems which affect the forms of their governing differential equations.

12.1 A GENERAL DIFFERENTIAL EQUATION

Equation (8.11), derived by Chow and Kulandaiswamy (1971), is sufficiently general in the hydrologic context and can be recast as

$$\sum_{i=0}^{M} a_i \frac{d^i Q}{dt^i} = \sum_{j=0}^{N} b_j \frac{d^j I}{dt^j} \tag{12.1}$$

where I is effective rainfall, Q is direct runoff, a_i and b_i are coefficients, t is time, and M and N are positive and finite. Let $D \equiv d/dt$. Then we obtain

$$Q(t) = \frac{\sum_{j=0}^{N} b_j D^j}{\sum_{i=0}^{M} a_i D^i} I(t) \tag{12.2}$$

We can enumerate characteristics of hydrologic systems governed by equation (12.2)

12.2 SYSTEM CHARACTERISTICS WITH $M > N$

12.2.1 Continuity

If a system is stable and damped—a property satisfied by most hydrologic systems—then

$$\int_0^\infty Q\, dt = \int_0^\infty I\, dt \tag{12.3}$$

must hold. This condition is satisfied by equation (12.2) if and only if a_0 and $b_0 = 1$. Therefore,

$$Q(t) = \frac{1 + \sum_{j=1}^{N} b_j D^j}{1 + \sum_{i=1}^{M} a_i D^i} I(t) = \frac{f_1 D)}{f_2(D)} I(t) \tag{12.4}$$

in which f_1 (D) and f_2 (D) are polynomials in D. If we use the Laplace transform, then we simply replace D by the Laplace variable s,

$$Q(s) = \frac{1 + \sum_{j=1}^{N} b_j s^j}{1 + \sum_{i=1}^{M} a_i s^i} I(s) = \frac{f_1 (s)}{f_2 (s)} I(s) \tag{12.5}$$

If $I(t)$ is replaced by $\delta(t)$ then $Q(t)$ will be reduced to the IUH by $h(t)$. Therefore,

$$h(s) = \frac{1 + \sum_{j=1}^{N} b_j s^j}{1 + \sum_{i=1}^{M} a_i s^i} = \frac{f_1(s)}{f_2(s)} \tag{12.6}$$

It will be seen from the ensuing discussion that for stable damped system, $M > N$ and, therefore, the ratio of the two polynomials on the right side of equation (12.6) can be expressed as sum of a set of partial fractions. By inverting $h(s)$, the IUH can be obtained. Then the integral of the IUH can be shown to be 1, thereby proving equation (12.3).

EXAMPLE 12.1

Let $f_1(s) = 1$ and $f_2(s) = 1 + s$. Then

$$h(s) = \frac{1}{1 + s}$$

Inverting,

$$h(t) = e^{-t}$$

Therefore,

$$\int_0^\infty h(t)\,dt = \int_0^\infty e^{-t}\,dt = 1$$

This is equal to $\int_{-\infty}^{\infty} \delta(t)\,dt = 1$. Therefore, the system satisfies continuity.

12.2.2 Stability and Damping

To illustrate these properties, let us assume that the order of $f_2(s)$ is greater than that of $f_1(s)$ and then examine the implications of this assumption. This may be best done by citing an example. We distinguish two cases: (1) the roots are distinct, and (2) the roots are repeated.

12.2.2.1 Distinct Roots

Let $f_1(s) = 1 + 3s$ and $f_2(s) = 1 + 3s + 2s^2$. Then

$$h(s) = \frac{1 + 3s}{(1 + s)(1 + 2s)}$$

Using partial fractions,

$$h(s) = \frac{A}{1 + s} + \frac{B}{1 + 2s} = \frac{2}{1 + s} - \frac{1}{1 + 2s}$$

Upon inversion,

$$h(t) = 2e^{-t} - \tfrac{1}{2}e^{-t/2}$$

It is then seen that under the condition $M > N$, $h(t)$ can be written as a sum of exponentials. These exponential elements are due to factors of the denominator $f_2(s)$, and their associated coefficients are due to the numerator $f_1(s)$. Each exponential represents an operation; the two represent two operations in parallel, as shown in Fig. 12.1(a). Writing $h(t)$ as

$$h(t) = \frac{a_1}{k_1} e^{-t/k_1} + \frac{a_2}{k_2} e^{-t/k_2}$$

where k_1 and k_2 are the roots of $f_2(s)$ and can be found from

$$(1 + k_1 s)(1 + k_2 s) = f_2(s)$$

Obviously these roots can be positive, negative, real, imaginary, or complex depending upon the form of the denominator $f_2(s)$. Let us examine each case.

Case 1: Positive Roots

If k_1 and k_2 are positive, then $h(t)$ consists of two parallel elements, as shown in Fig. 12.1(a), which recede exponentially, as shown in Fig. 12.1(b). This means that the system is stable and highly damped.

Case 2: Negative Roots

If k_1 and k_2 are negative, then $h(t)$ consists of two parallel elements, as shown in Fig. 12.1(a), which increase exponentially, as shown in Fig. 12.1(c). This implies that the system is unstable and undamped. It may be noted that if one root is positive and the other negative, the system would still be unstable and undamped.

Case 3: Imaginary Roots

If k_1 is imaginary, then it will be associated with its conjugate k_2. Let $k_1 = ik$; then the factors of $f_2(s)$ are $(1 + iks)$ and $(1 - iks)$. Writing $h(s)$ as

$$h(s) = \frac{a}{(1 + iks)(1 - iks)} = \frac{a}{(1 + k^2 s^2)}$$

This represents the operation of two elements. On inversion,

$$h(t) = ak \sin(t/k)$$

The IUH is sinusoidal and therefore is stable but undamped, as shown in Fig. 12.1(d).

Case 4: Complex Roots

If k_1 is complex, then k_2 will be its conjugate. Let $k_1 = a + ib$ then $k_2 = a - ib$. Then the factors of $f_2(s)$ are $[1 + (a + ib)s]$ and $[1 + (a - ib)s]$, and $h(s)$ is

$$h(s) = \frac{c}{[1 + (a + ib)s]\,[1 + (a - ib)s]}$$

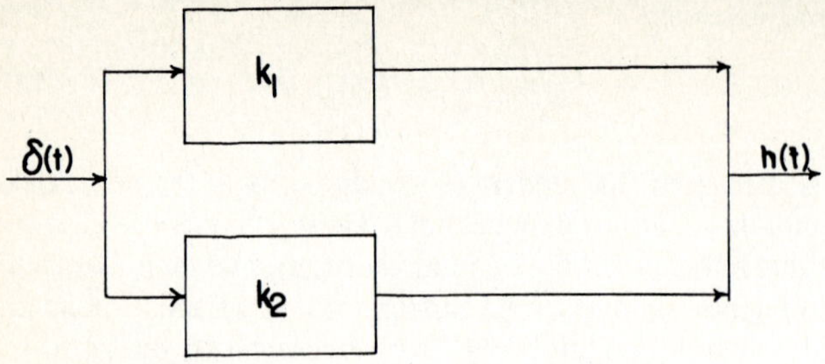

FIGURE 12.1(a) Two elements in parallel.

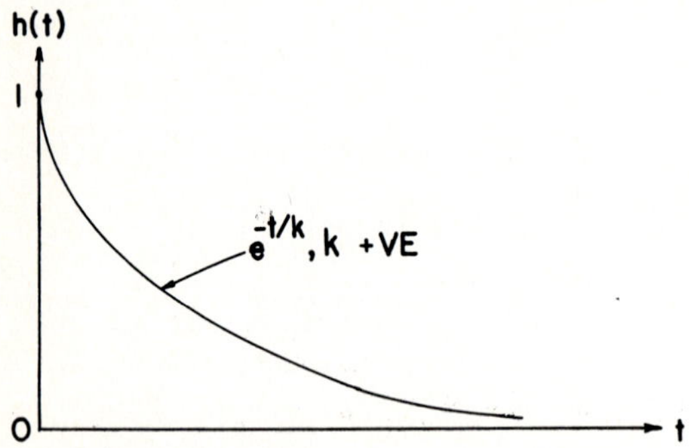

FIGURE 12.1(b) Exponential recession.

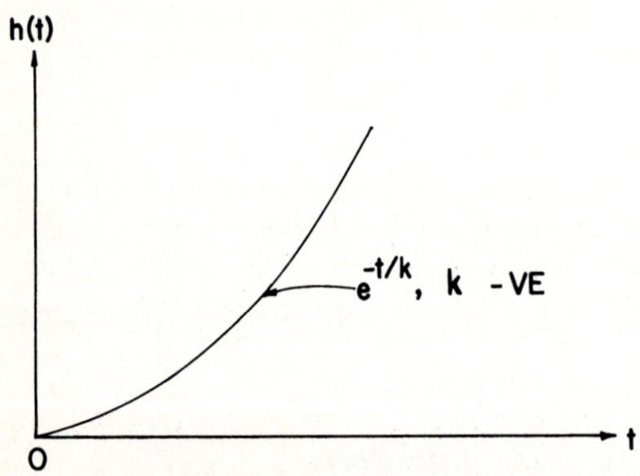

FIGURE 12.1(c) Exponential rise.

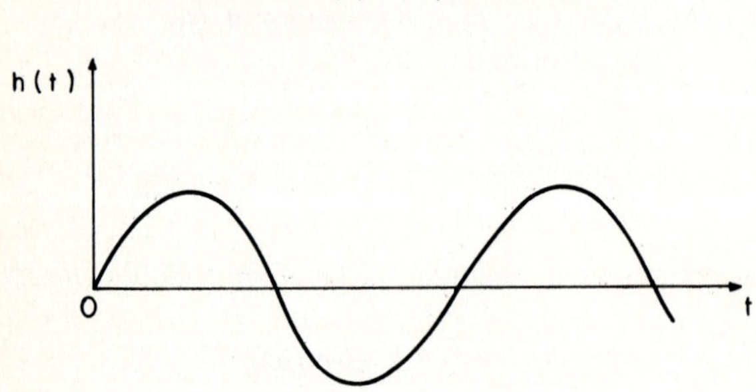

FIGURE 12.1(d) Sinusoidal IUH.

This represents the operation of two elements. We can write this in terms of partial fractions:

$$\frac{c}{[1+(a+ib)s]\,[1+(a-ib)s]}$$

$$= \frac{A}{1+(a+ib)s} + \frac{B}{1+(a-ib)s}$$

$$= c\left[\frac{(a+ib)}{(2ib)\,[1+(a+ib)s]} - \frac{(a-ib)}{(2ib)\,[1+(a-ib)s]}\right]$$

Upon inversion,

$$h(t) = \frac{c(a+ib)}{(2ib)\,(a+ib)} e^{-t/(a+ib)} - \frac{c(a-ib)}{2ib\,(a-ib)} e^{-t/(a-ib)}$$

On simplifying,

$$h(t) = \frac{c}{2ib} e^{-ta/(a^2+b^2)} [e^{-ibt/(a^2+b^2)} - e^{ibt/(a^2+b^2)}]$$

$$= \frac{c}{2ib} e^{-ta/(a^2+b^2)} \sin\left[\frac{bt}{(a^2+b^2)}\right]$$

It is evident that the stability and damping of $h(t)$ will depend upon a, the real part of the root. If a is positive, then the exponential term decays in time because $a^2 + b^2$ is always positive. The sinusoidal term oscillates. Therefore, the product of the two terms represents an oscillation with diminishing amplitude. However, if a is negative, then clearly the system is unstable and undamped. The oscillation grows unbounded.

It can therefore be concluded that if the real part of the root is positive, the system is stable and damped. If it is negative, the system is unstable and undamped. This conclusion can be extended to the systems having more than two elements. If any of the roots have negative real parts, the corresponding elements—and consequently the entire system—will be unstable. Therefore, the occurrence of positive real parts in all the roots is a necessary and sufficient condition for system stability.

12.2.2.2 Repeated Roots

If the roots in equation (12.6) repeat themselves, then additional terms must be accordingly provided for obtaining partial fractions. For example, if two terms of the denominator in equation (12.6) are equal, then we must have an additional term $A/(1 + ks)^2$ to get partial fractions. To illustrate, let

$$f_1(s) = 1 + 2s$$

$$f_2(s) = \frac{1}{(1+s)^2}$$

Then we can obtain the partial fractions as

$$\frac{A}{1+s} + \frac{B}{(1+s)^2} = \frac{(1+2s)}{(1+s)^2}$$

We obtain $A = 2$ and $B = -1$. Therefore.

$$h(s) = \frac{2}{1+s} - \frac{1}{(1+s)^2}$$

Upon inversion,

$$h(t) = 2\,e^{-t} - te^{-t}$$

Thus, if k is the root that occurs twice, then $1/(1 + ks)^2$ must be allowed to obtain partial fractions. Fig. 12.2 illustrates

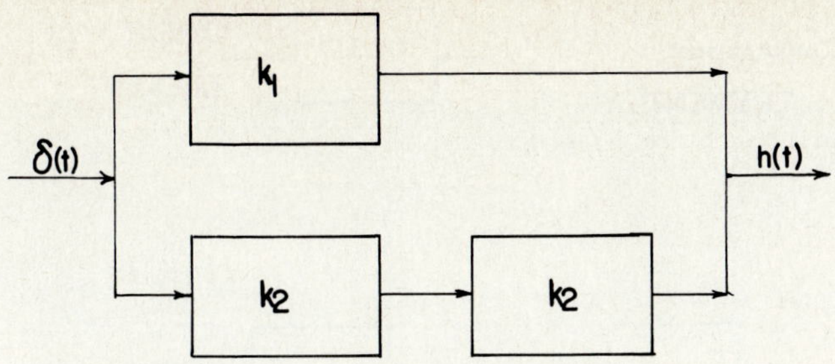

FIGURE 12.2 Elements arranged in series and parallel.

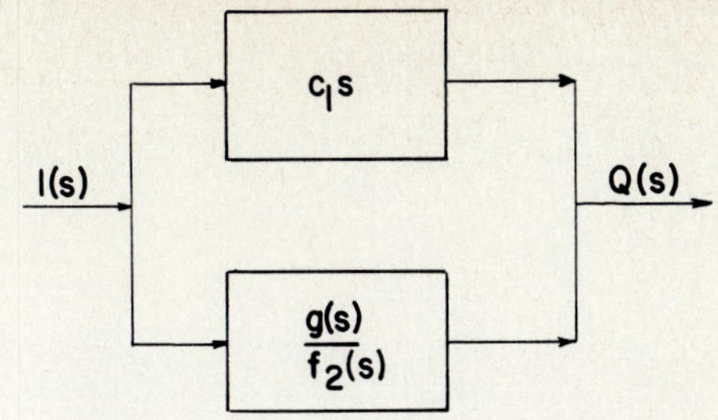

FIGURE 12.3 Two elements in parallel.

this case if $k_1 = k_2 = k_3$. Similarly if there are n roots that are equal, then we must allow

$$\frac{a_1}{(1 + ks)} + \frac{a_2}{(1 + ks)^2} + \cdots + \frac{a_n}{(1 + ks)^n}$$

for the partial fractions. Having obtained partial fractions, we can study if the system is stable and damped according to the criteria developed in the preceding section.

Alternatively, we can view the equality in the limit. That is, if two roots are equal, then the factors of the denominators can be written as $[1 + (k + \epsilon)s]$ and $[1 + (k - \epsilon)s]$, where ϵ is a small number and can be made as small as desired. Thus, $(k + \epsilon)$ and $(k - \epsilon)$ can be viewed as two different roots. Consequently, the relationship between the kind of system and the roots developed for distinct roots applies.

12.3 SYSTEM CHARACTERISTICS WITH $M < N$

If $f_1(s)$ is of higher order than $f_2(s)$, then equation (12.6) can be expressed as

$$c_1 s^r + c_2 s^{r-1} + \cdots + \frac{g(s)}{f_2(s)}$$

where $g(s)$ is not of higher order than $f_2(s)$. To illustrate, let $r = 1$. This means $f_1(s)$ is one degree higher than $f_2(s)$. Then

$$h(s) = \left[c_1 s + \frac{g(s)}{f_2(s)}\right]$$

On inverting,

$$h(t) = \delta'(t) - L^{-1}\left[\frac{g(s)}{f_2(s)}\right]$$

where $\delta'(t)$ is the first derivative of the unit impulse, or δ-function. This operation can be represented by two elements in parallel, as shown in Fig. 12.3. The derivative of $\delta(t)$, as discussed in Chapter 3, represents a positive impulse and a negative impulse occurring simultaneously at $t = 0$. In the IUH this can be illustrated as in Fig. 12.4. Hydrologically, this means that one part of the IUH will be proportional to the derivative of the unit impulse. From a computational stand-

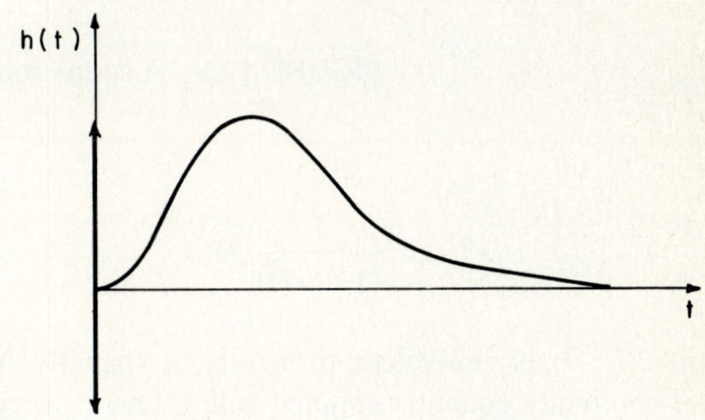

FIGURE 12.4 An IUH with $\delta'(t)$.

point, this gives rise to a somewhat undesirable situation because it is difficult to estimate reliably the derivative of a quantity available in discrete time. Furthermore, if the order of $f_1(s)$ is r times higher than that of $f_2(s)$ in equation (12.6), then computation of correspondingly higher derivatives of $\delta(t)$ will be even more difficult. It can therefore be inferred at least provisionally that if $f_1(s)$ is of higher order than $f_2(s)$, then numerical difficulties will be experienced in computation of the IUH. This condition can be characterized as ill-conditioning.

12.4 NETWORK REPRESENTATION

Let us consider a network representation of the system whose operation is described by equation (12.1). If we factor the numerator and the denominator in equation (12.2),

$$Q(t) = a \frac{\prod_{j=1}^{N} (1 + c_j D)}{\prod_{j=1}^{M} (1 + k_i D)} I(t) \tag{12.7}$$

where a is some constant more than 0 and the c's and k's are the roots of the numerator and denominator, respectively. It is then clear that there are two kinds of first-order elements in equation (12.7), one corresponding to the numerator and the other corresponding to the denominator. These elements can be defined as

$$Q(t) = (1 + cd)\, I(t) \tag{12.8}$$

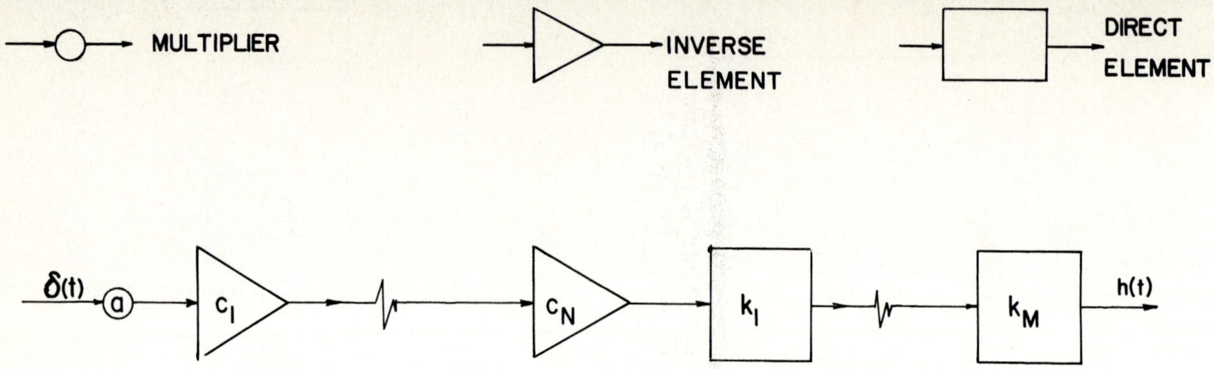

FIGURE 12.5 A series representation of the system of equation (12.7).

and

$$Q(t) = \frac{1}{(1 + kD)} I(t) \tag{12.9}$$

Equation (12.9) is equivalent to writing a spatially lumped form of continuity equation coupled with a linear storage-discharge relation expressed in equations (8.10) and (8.12). As will be seen in the next chapter, this defines what is commonly called as a linear reservoir, or a direct element. Therefore, equation (12.9) represents the operation of routing through a linear reservoir. Note that k is unrestricted at this point; in other words, it may be real, imaginary, or complex with its real part being position or negative.

On the other hand, equation (12.8) can be written as

$$I(t) = \frac{1}{(1 + cD)} Q(t) \tag{12.10}$$

which appears similar in form to equation (12.9) with $I(t)$ and $Q(t)$ exchanging places. No restriction is imposed on c. It is clear that equation (12.8) represents an element, called the *inverse element*, whose operation is the inverse of that of equation (12.9). Equation (12.8) differentiates $I(t)$, multiplies it by c, and then adds this to $I(t)$ to produce $Q(t)$. In equation (12.10), $I(t)$ is produced by multiplying the derivative of $Q(t)$ by c and adding this to $Q(t)$. Thus the two elements operate in opposite manner.

It is clear that the system whose operation is described by equation (12.7) can be represented by a series of these first-order elements and a single multiplying factor, as shown in Fig. 12.5, with $I(t)$ replaced by $\delta(t)$ and $Q(t)$ replaced by $h(t)$. Note that a single numerical multiplier is required and the order of the elements is immaterial in the series representation. This representation, however, is by no means unique. Alternative representations can be developed for this system depending upon the nature of numerator and denominator in equation (12.7).

An alternative representation for equation (12.7) can be obtained by using partial fractions for its denominator if we rule out the occurrence of multiple roots. That is,

$$h(t) = a \prod_{i=1}^{N} (1 + c_i D) \left[\sum_{j=1}^{M} \frac{r_j}{(1 + k_j D)} \right] \delta(t)$$

where r_j represents the jth coefficient of the jth partial fraction. This is shown in Fig. 12.6. Here $\delta(t) \rightarrow [^{r_1}_{r_2}$ indicates symbolically that $\delta(t)$ must be divided between parallel paths in proportion to their coefficients, which must, therefore, add to one. Similarly $] \rightarrow h(t)$ indicates that $h(t)$ is made up of the functions in the parallel paths.

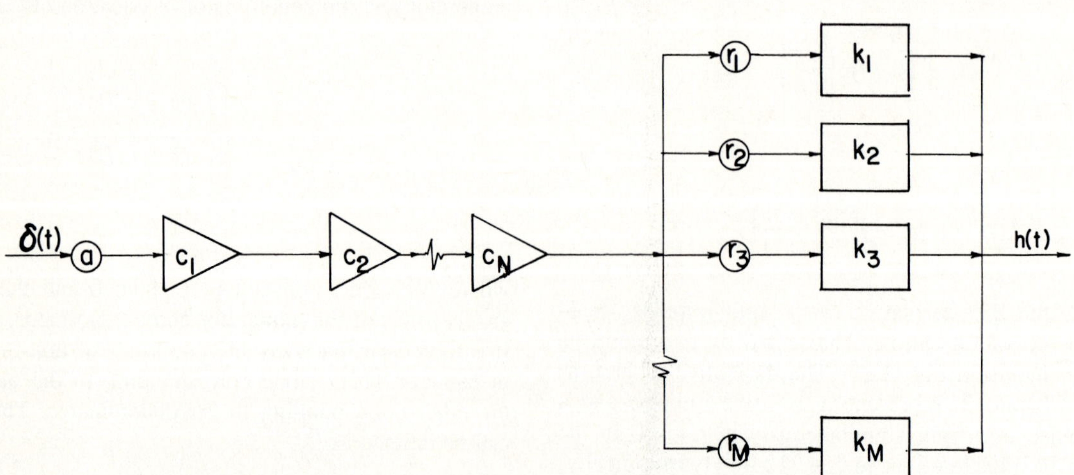

FIGURE 12.6 A series cum parallel representation of equation (12.7).

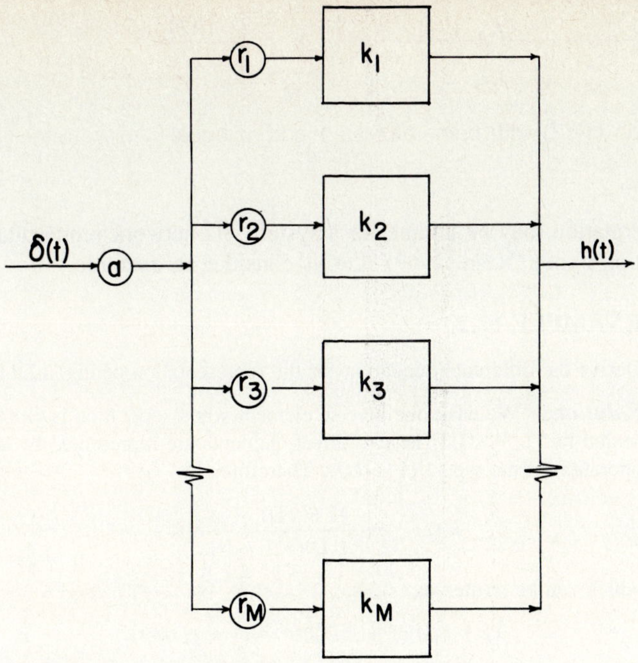

FIGURE 12.7 A parallel representation of equation (12.7).

12.4.1 Representation when $M > N$

The discussion in the foregoing section shows that equation (12.7) can be represented by a sum of partial fractions, again ruling out the occurrence of repeated roots, as

$$h(t) = a \sum_{j=1}^{M} \frac{r_j}{(1 + k_j D)} \delta(t) \qquad (12.11)$$

This indicates that the network representation will consist only of direct elements in parallel. These elements will obviously correspond to the factors of the denominator. The numerical coefficients r_j will be obtained by partial fractions. This representation is shown in Fig. 12.7.

12.4.2 Representation with Repeated Roots

The discussion in the previous section enumerated two possibilities to deal with the systems having repeated roots. One possibility is to consider the equal roots slightly different from one another and then use partial fractions. To illustrate, consider

$$h(t) = \frac{1}{(1 + k_1 D)(1 + k_1 D)} \delta(t)$$

We can make one k_1 different from the other by introducing into it a small error, for instance, and then write

$$h(t) = \frac{1}{(1 + k_1 D)(1 + k_2 D)} \delta(t)$$

Now using partial fractions,

$$h(t) = \left[\frac{A}{1 + k_1 D} + \frac{B}{1 + k_2 D}\right] \delta(t)$$

The coefficients A and B can be computed and will be

$$A = -\frac{k_1}{k_1 - k_2}, \qquad B = \frac{k_2}{k_1 - k_2}$$

It is evident, that $k_1 \rightarrow k_2$, $A \rightarrow -\infty$ and $B \rightarrow \infty$. This will, therefore, present numerical difficulties. However, the system can be represented by Fig. 12.7.

The other possibility is more practically viable and exact. If there are M equal roots in the denominator, then it is clear from the previous discussion that the system representation will consist of M parallels each consisting of M, $M - 1$, $M - 2$, . . . , 1 elements in series. Therefore, equation (12.7) can be represented by Fig. 12.8. Similarly, other representations can be developed depending upon the nature of the nu-

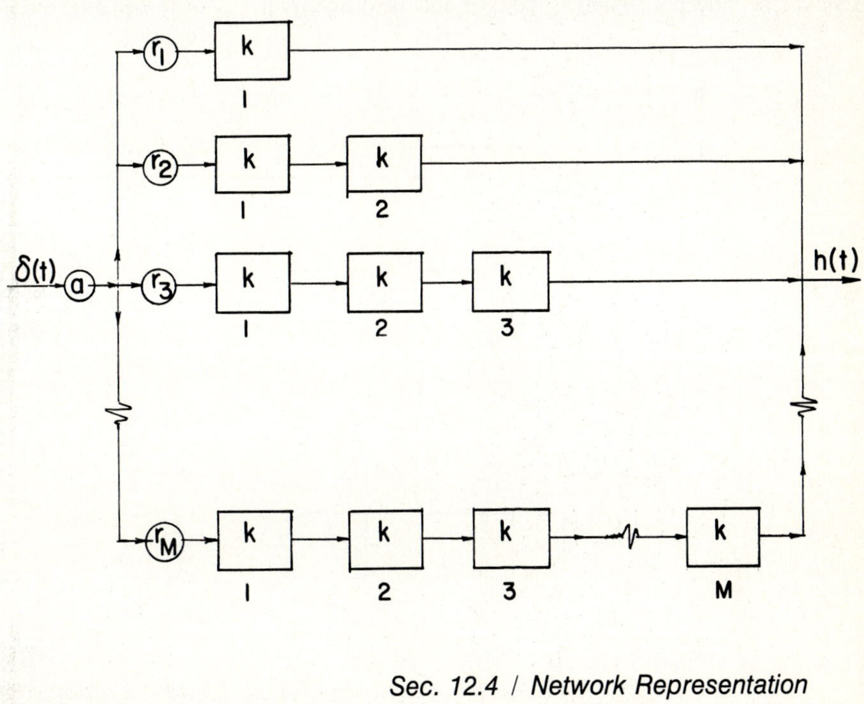

FIGURE 12.8 A network representation of equation (12.7) with M repeated roots.

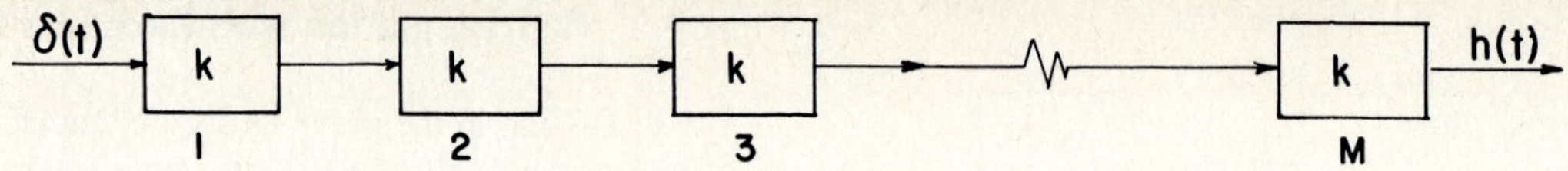

FIGURE 12.9 A series representation of equation (12.7) with numerator as 1 and m equal roots.

merator. For example, if the numerator in equation (12.7) is taken as 1, then Fig. 12.8 can be used to represent it, as can Fig. 12.9. Other representations (see Fig. 12.10 as an example) can be developed in a like manner.

12.4.3 Representations with $M < N$

If the numerator in equation (12.7) is of higher order than the denominator, then by factoring the numerator, sufficient factors can be separated out to reduce it to the order of the denominator. The factors separated out constitute the inverse elements, which can be arranged in series. The remainder can be represented by the direct elements arranged in parallel. To illustrate, if $R = N - M$, then equation (12.7) can be represented by Fig. 12.6. Other alternative representations can be developed, depending upon the nature of the roots of the denominator.

12.5 NETWORK AND DIFFERENTIAL EQUATION

We have discussed various kinds of network representations for equation (12.7). These representations are all composed of the elements, whether direct or inverse, of first order. These can be utilized to derive the IUH of the system. Reversing the order of analysis, it is equally plausible to derive a differential equation for a specified representation of these elements and then derive the IUH. Therefore, it can be concluded that a differential equation, network of first-order elements, and an IUH can represent one and the same systems and are therefore equivalent concepts. Whereas the IUH and the differential equation may be unique for a system, its network representation is not (Nash, 1969). Let us consider an example.

EXAMPLE 12.2

Derive the differential equation for the representation of Fig. 12.11.

Solution: We have one inverse element whose operation is represented by $(1 + cD)$. The two direct elements are represented by an operation of the type $1/(1 + kD)$. Therefore,

$$h(t) = \frac{(1 + cD)}{(1 + k_1D)(1 + k_2D)}\,\delta(t)$$

which can be written as

$$(1 + k_1D)(1 + k_2D)h(t) = (1 + cD)\delta(t)$$

Hence,

$$k_1 k_2 \frac{d^2h(t)}{dt^2} + (k_1 + k_2)\frac{dh(t)}{dt} + h(t) = \delta(t) + c\frac{d}{dt}\delta(t)$$

This is a linear second-order ordinary differential equation describing the operation of the system represented by Fig. 12.11.

12.6 HYDROLOGIC SYSTEM REPRESENTATION

In the previous discussion we did not impose specific restrictions on the coefficients in equation (12.7). If a hydrologic system, such as a watershed system, is represented by linear elements, then it may be appropriate to assume that all the network elements would be direct in kind and that all the roots would be positive real. This would mean that the numerator in equation (12.7) would be unity. Similarly, other assump-

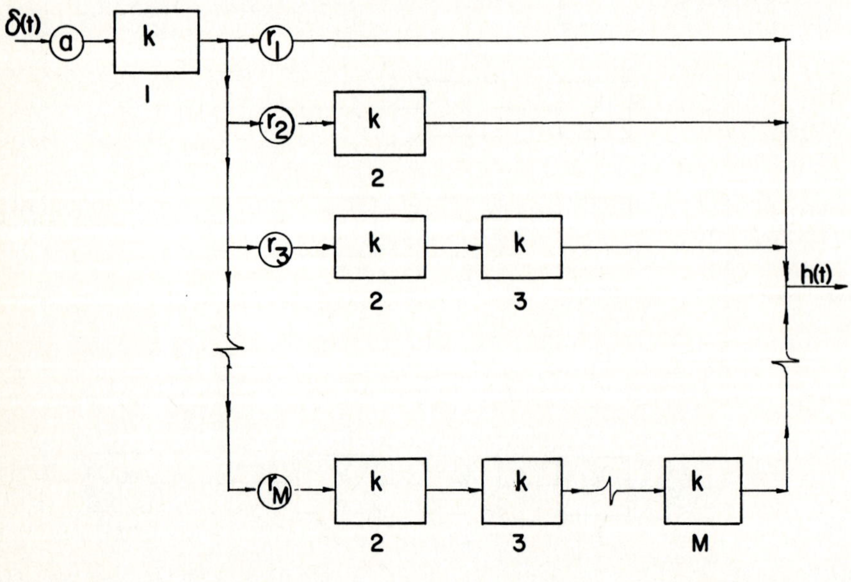

FIGURE 12.10 A network representation of equation (12.7).

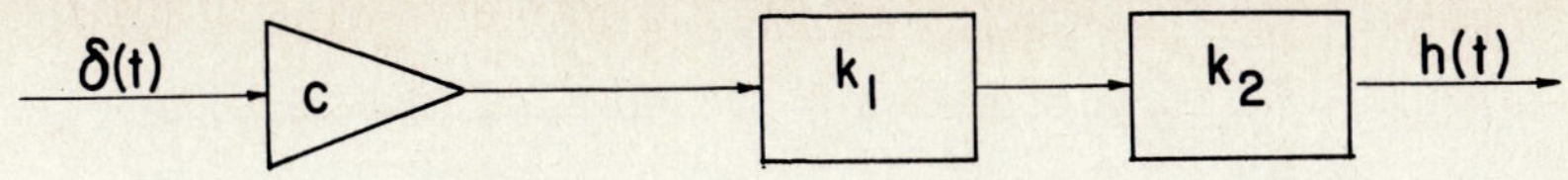

FIGURE 12.11 A series representation with one inverse element and two direct elements.

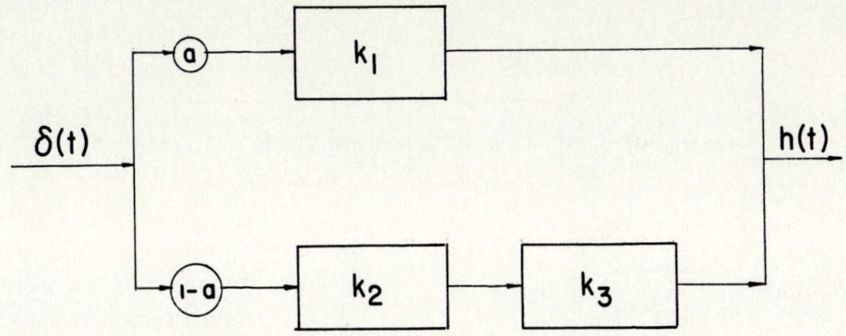

FIGURE 12.12 A network representation of direct elements.

tions can be made depending upon the nature of a particular hydrologic system under consideration.

If a network representation of a hydrologic system is developed using physical characteristics and the above assumption, then the derivation of the governing differential equation becomes simple. The study of the network and the corresponding differential equation might give useful insight into the system behavior. Let us consider a simple example and expand on this idea a little further. Consider a network representation of Fig. 12.12. The differential equation can be derived by writing

$$h(t) = \left[a\frac{1}{1 + k_1D} + (1 - a)\frac{1}{(1 + k_2D)(1 + k_3D)}\right]\delta(t)$$

Hence,

$$h(t) = \frac{a[(1 + k_2D)(1 + k_3D)] + (1 - a)(1 + k_1D)}{(1 + k_1D)(1 + k_2D)(1 + k_3D)}\delta(t)$$

$$[1 + (k_1 + k_2 + k_3)D + (k_1k_2 + k_2k_3 + k_1k_3)D^2 + k_1k_2k_3D^3]\,h(t)$$
$$= [1 + (ak_2 + ak_3 + (1 - a)k_1)D + ak_2k_3D^2]\,\delta(t)$$

Thus, the resulting differential equation is of the third order. This can be interpreted as equal to the number of elements in the arrangement provided that they are distinct. A question then arises, What happens when some of the elements are not distinct? That is, some are equal. To answer this question, let us assume that $k_1 = k_3$. Then we obtain

$$[1 + (k_1 + k_2)D + k_1k_2D^2]\,h(t) = (1 + ak_2D)\delta(t)$$

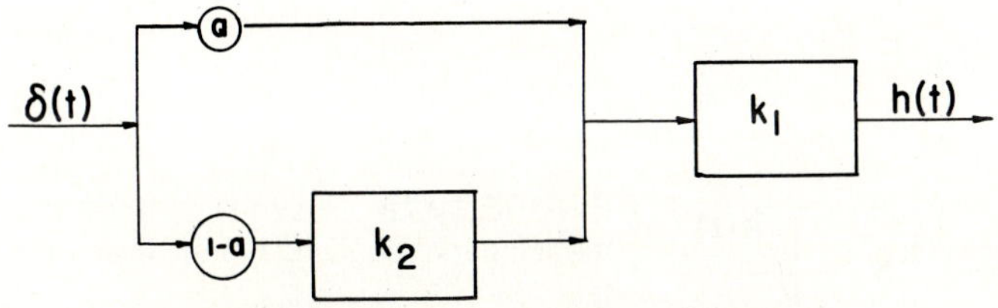

FIGURE 12.13 A network representation.

The resulting differential equation is of second order. The network has two distinct elements. Its representation is shown in Fig. 12.13–12.14.

What happens if $k_1 = k_2 = k_3 = k$? Then we can write

$$[1 + 2kD + k^2D^2]\,h(t) = (1 + akD)\delta(t)$$

from the representation, as shown in Fig. 12.15. The resulting differential equation is still of second order, although all the elements are equal. Therefore, the order of the differential equation is equal to the number of elements in the longest series provided the elements in other parallels are in common with its elements.

Although no inverse element is present in the network representations of Fig. 12.12–12.15, we do have a numerator in each case. The numerator is produced by the parallels. It can therefore be said that if there were no parallels, there would be no numerators. Further, if the number of elements in each parallel were the same and if all the elements were equal, then there would be no numerators.

Furthermore, referring to Fig. 12.12, the factors of the numerator may not be any of those of the denominator. The expression $1 + [a(k_2 + k_3) + (1 - a)k_1]D + ak_2k_3D^2$, when divided by the denominator, may not have all real positive roots. Therefore, it may be plausible to say something about the shape of the IUH from a network representation and then about the realism of the representation itself. One might ask, Should the elements be the same in each parallel? Should all the parallels be the same? Of course, this will depend upon the physical features of the system under consideration. For

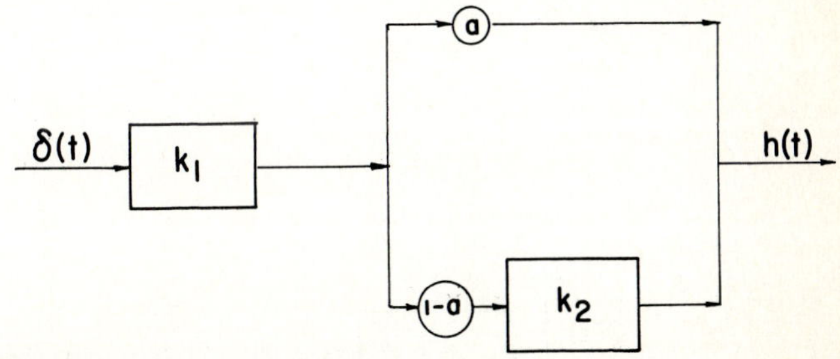

FIGURE 12.14 An alternative representation of the network of Fig. 12.13.

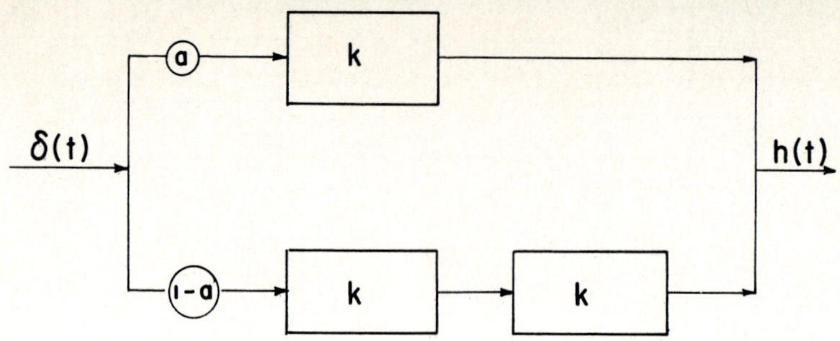

FIGURE 12.15 A network representation of two parallels having equal elements.

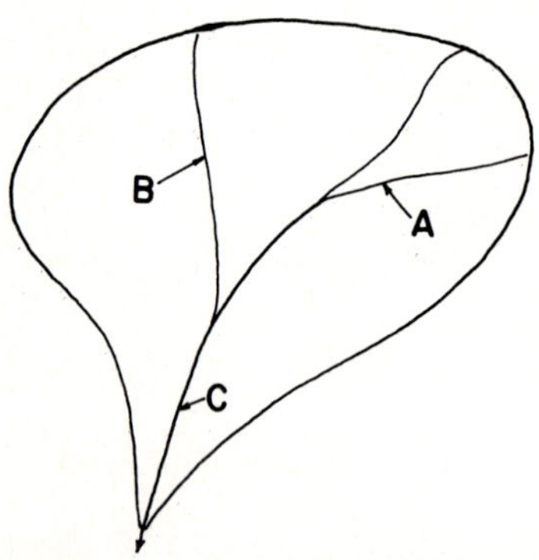

FIGURE 12.16 An example watershed.

example, if we were to consider the effective rainfall–direct runoff relation on a natural watershed, we could assume linearity and represent the relation by a number of cascades arranged in parallel, each cascade representing a certain portion of the watershed, as shown in Fig. 12.16. This watershed has two tributaries, *A* and *B*, and a main channel, *C*. Each stream can be represented by a series of linear direct elements, as shown in Fig. 12.17. Many representations of the watershed can be developed depending upon (1) distribution of the effective rainfall and (2) elements in the network. To illustrate this point, consider that the watershed is represented by a cascade of elements, each receiving effective rainfall in accor-

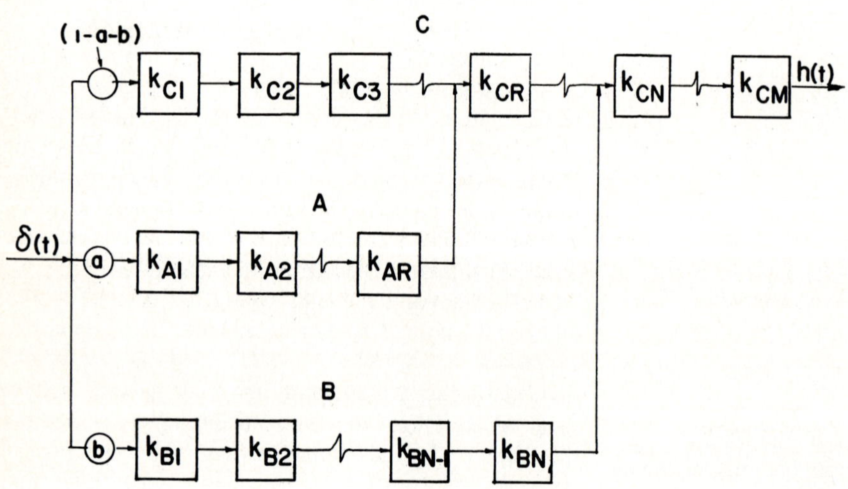

FIGURE 12.17 A network representation of the watershed in Fig. 12.16.

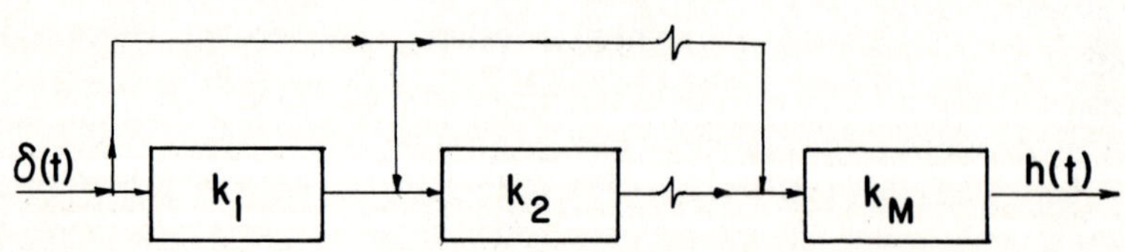

FIGURE 12.18 A series representation with multiple entry points for the effective rainfall.

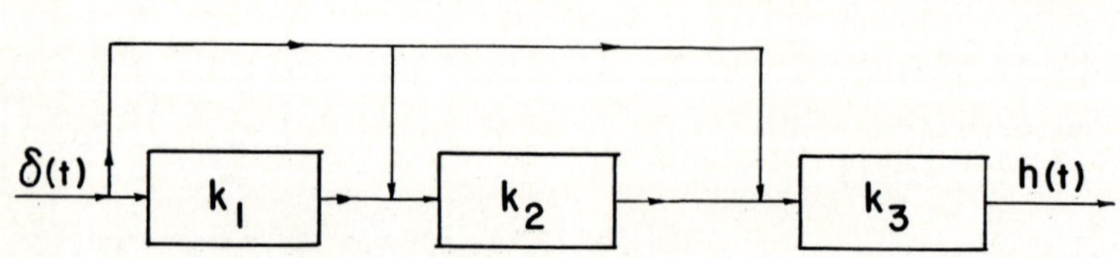

FIGURE 12.19 A series of three elements, each receiving the effective rainfall.

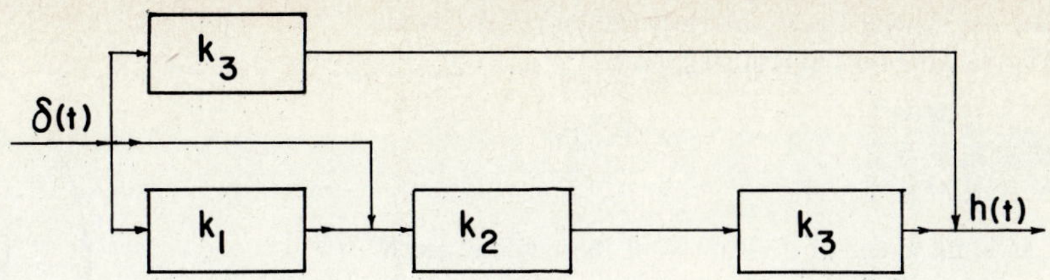

FIGURE 12.20 An alternative representation for Fig. 12.19.

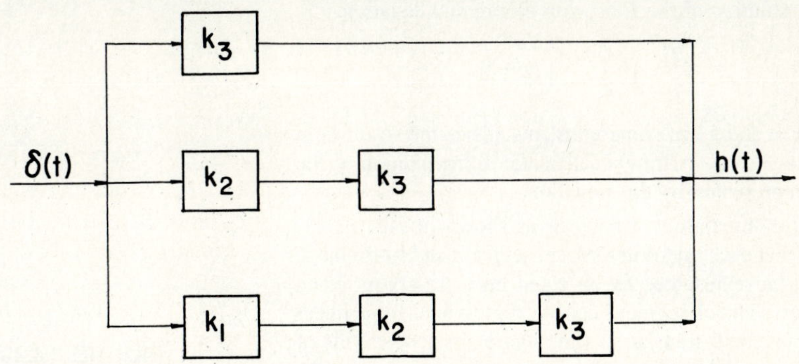

FIGURE 12.21 An equivalent representation of Fig. 12.19.

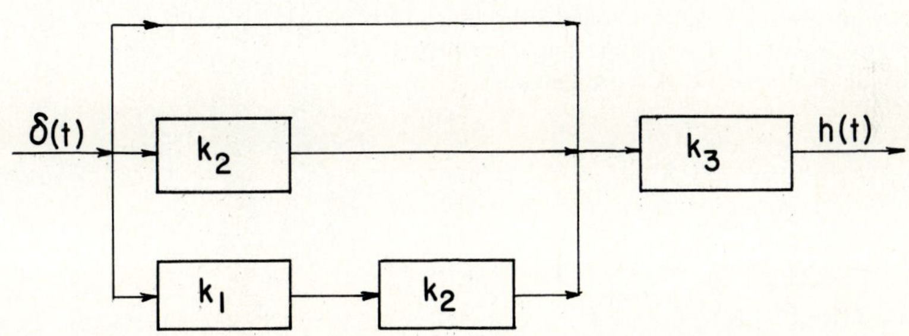

FIGURE 12.22 An equivalent representation of Fig. 12.19.

dance with its area of influence. For simplicity let us assume that the area represented by each element is the same. The network representation can then be shown by Fig. 12.18.

It is easily shown that the representation of Fig. 12.18 is not unique. Indeed several alternative possibilities exist for representation, at least as many as the number of entry points. To illustrate this, consider three direct elements in series with three entry points of the effective rainfall, as shown in Fig. 12.19. Alternative representations, which are supposedly equivalent, are sketched in Figs. 12.20–12.22.

EXERCISES

12.1. A linear hydrologic system is represented by two parallel cascades. Each cascade has two unequal elements. Assume effective rainfall is equally divided into the two cascades at the upstream. Derive the system differential equation.

12.2. If one element is the same in both cascades of Question 12.1, what will be the system differential equation?

12.3. Draw an alternative system representation for Question 12.2.

12.4. If each element of the two cascades receives the same effective rainfall, then derive the system differential equation.

12.5. Draw alternative system representations for Question 12.4.

12.6. The following equation is the UIR of a linear system:

$$h(t) = (a_1 t + a_2 t^2 + a_3 t^3)\exp(-t)$$

Draw two alternative arrangements of equal linear storage reservoirs of unit storage delay time that would have the same UIR as this system.

12.7. The input to and output from a linear time-invariant system are given as

$$I(t) = \frac{1}{2}\left(t + \frac{t^2}{2}\right)\exp\left(\frac{-t}{2}\right), \qquad 0 < t < \infty$$

and

$$Q(t) = \frac{1}{2}\left(\frac{t^5}{5!} + \frac{t^6}{6!}\right)\exp\left(\frac{-t}{2}\right), \qquad 0 < t < \infty$$

Find the unit impulse response of the system.

12.8. Consider the general linear storage equation (12.1). What would be the UIR for the following cases?

(a) $N = 1, M = 3$
(b) $N = 1, M = 2$
(c) $N = 1, M = 1$

where M is the order of derivatives of the outflow and N that of inflow. Provide a schematic representation of the system in each case.

12.9. Find the solution of the following differential equation:

$$\frac{dQ}{dt} + \left(\frac{1}{k} - \frac{n}{t}\right) Q = 0$$

in which n and k are some constants. Does the result hold for all values of n? Draw a schematic representation of the system represented by this equation.

12.10. Consider a three-order watershed as shown in Fig. 12.23. Assume that each one-order stream segment and its drainage area can be represented by an equal linear reservoir. Each two-order channel segment can be represented by a linear channel of equal translation time and a linear reservoir of equal storage delay time. Each three-order segment can be represented by a linear channel of equal translation time and a linear reservoir of equal storage delay time. Derive the differential equation representing the watershed. Draw at least five alternative network representations.

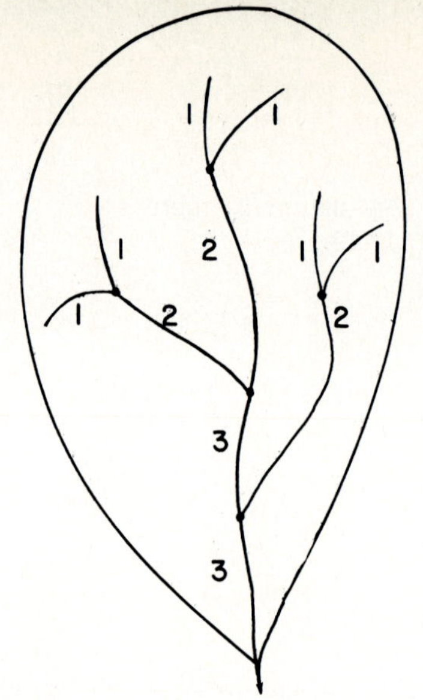

FIGURE 12.23 Question 12.10: A third-order watershed.

CHAPTER THIRTEEN

Linear Conceptual Models for Continuous-Time Systems

The approaches utilized to develop linear conceptual models of rainfall-runoff relationship can be classified into three groups. The first group employs a differential equation that supposedly governs the operation of a specified system (Kulandaiswamy, 1964; Chow, 1964; Shen, 1965; Chaudhry, 1976; Jackson, 1968; Chow and Kulandaiswamy, 1971, 1982; V. P. Singh and McCann, 1979; McCann and V. P. Singh, 1980, 1981; Te and Kay, 1983). The second group utilizes an arrangement of the so-called conceptual elements, including linear channels and linear reservoirs (Nash, 1957; Dooge, 1959, 1977; Chow, 1964; S. Bravo et al., 1970; Maddaus and Eagleson, 1969; Harley, 1967; O'Meara, 1968; V. P. Singh and McCann, 1980a). The third group makes some hypotheses about rainfall-runoff relationship more or less on intuitive grounds (Lienhard, 1964, 1972). We shall study these approaches and the models resulting from them in detail in this chapter.

In Chapter 8 we derived a general differential equation for a lumped linear time-invariant hydrologic system; we discussed its network representations in Chapter 12. In Chapters 10 and 11 we described the operation of such a system using the IUH without explicitly resorting to the use of the differential equation. Nash (1960) showed that these two groups of approaches may converge to one another under certain conditions. At this point it may be pertinent to introduce the widely used concepts of linear channel, linear reservoir, and linear differential equation and identify the mathematical operations that they represent.

From Chapter 4 we see that there are two ways to employ these concepts to determine direct runoff (DR). First, the IUH of the system is determined. It is then convoluted with the specified effective rainfall (ER). Second, the Laplace transform of the DR is obtained by multiplying the Laplace transform of the IUH by that of the ER, which is then inverted (Te and Kay, 1983). Both methods are popular in hydrology. We utilize both in this chapter.

13.1 LINEAR CHANNEL

Let us write the equation of continuity in one-dimensional form for a channel with no lateral inflow,

$$\frac{\partial Q}{\partial x} + \frac{\partial A}{\partial t} = 0 \tag{13.1}$$

and assume the properties of the channel to be

$$A = C(x)Q \tag{13.2}$$

where A is the cross-sectional area of flow, Q is the rate of outflow, and C is a constant at a section of the channel varying with distance. Combining equations (13.1) and (13.2),

$$\frac{\partial Q}{\partial x} + C(x)\frac{\partial Q}{\partial t} = 0 \tag{13.3}$$

Referring to Fig. 13.1, let inflow at x_0 be denoted by $Q(x_0, t)$ and let

$$Q(x_0, t) = f_0(t) \tag{13.4}$$

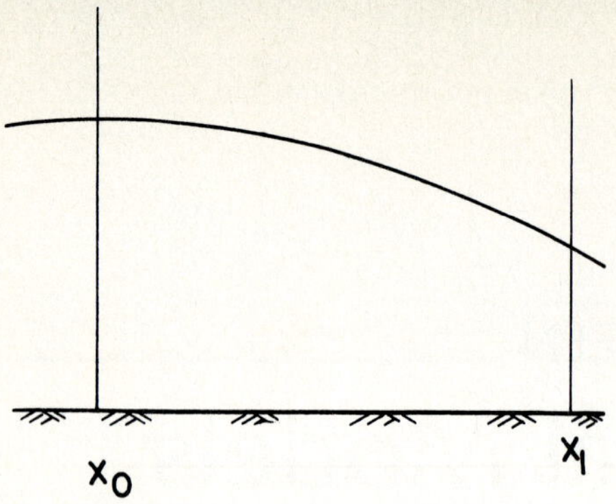

FIGURE 13.1 A channel reach.

Given $Q(x_0, t)$, the solution of equation (13.3) will give Q at any x. Let

$$Q(x_1, t) = f_1(t) \tag{13.5}$$

It can be shown by solving equation (13.3) that

$$Q(x_1, t) = f_1(t) = f_0(t - T) \tag{13.6}$$

where T is the time of travel from x_0 to x_1. The outflow at x_1 is merely the inflow at x_0 delayed by a time T. Equation (13.2) defines a linear channel, and from this equation it is seen that its effect on inflow is pure translation.

Let

$$D_x = \frac{\partial}{\partial x}$$

$$D_t = \frac{\partial}{\partial t}$$

Equation (13.3) can be written in the operator notation as

$$[D_x + C(x)D_t]Q = 0 \tag{13.7}$$

$$Q = \frac{1}{D_x + C(x)D_t}(0) \tag{13.8}$$

The operator $1/[D_x + C(x)D_t]$ represents the effect of a linear channel on inflow defined by equation (13.4); $C(x)$ is not a constant and, in its present form, the operator is not commutative. If C is assumed constant, the operator becomes $1/(D_x + CD_t)$, and its effect is to shift the function $Q(x_0, t)$ by a time interval equal to T where

$$T = C(x_1 - x_0) \tag{13.9}$$

Then

$$Q(x_1, t) = HQ(x_0, t) = Q(x_0, t - T), \qquad t \geq T \tag{13.10}$$

Here the operator H represents the effect on the inflow of a linear channel reach whose travel time is T. The translation time of a linear channel is also referred to as delay time, and the operator H is the delay operator. This operator is commutative.

The linear channel is analogous to the idea of a linear reservoir, to be discussed in the ensuing section, where the storage-outflow curve is a straight line. The channel unaffected by backwater has a definite rating curve at every point of the channel. For a linear channel, the rating curve at every point specifies a linear relationship between discharge and area. This implies that the velocity is constant for all discharges but may vary from point to point along the reach.

13.1.1 The IUH

We can now specify the IUH of the linear channel. If the inflow, represented by $\delta(t)$, is taking place only at the upstream end of the channel, then referring to Fig. 13.2,

$$h(t) = \delta(t - T) \tag{13.11}$$

where T is the translation time. If, however, the input is considered to occur uniformly along its length, then the IUH is

$$h(t) = \frac{1}{T}\left[u_1(t) - u_1(t - T)\right] \tag{13.12}$$

where u_1, as before, signifies the unit step function. The form of the IUH is shown in Fig. 13.2(c).

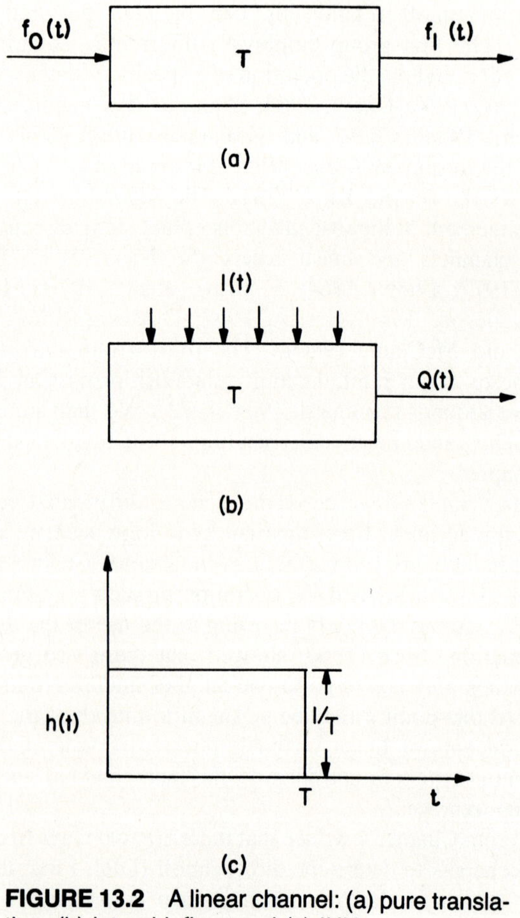

FIGURE 13.2 A linear channel: (a) pure translation, (b) lateral inflow, and (c) IUH.

13.1.2 Moments and Cumulants

The IUH in equation (13.12) has only one parameter T, which can be estimated by the method of moments (M) or cumulants (K). The first four moments of the IUH about its origin are:

$$M_1 = \frac{T}{2} \tag{13.13a}$$

$$M_2 = \frac{T^2}{3} \tag{13.13b}$$

$$M_3 = \frac{T^3}{4} \tag{13.13c}$$

$$M_4 = \frac{T^4}{5} \tag{13.13d}$$

Likewise, the first four cumulants are:

$$K_1 = \frac{T}{2} \tag{13.14a}$$

$$K_2 = \frac{T^2}{12} \tag{13.14b}$$

$$K_3 = 0 \tag{13.14c}$$

$$K_4 = -\frac{T^4}{120} \tag{13.14d}$$

13.1.3 The DRH

The DRH of a linear channel can be determined for any arbitrary ERH. Since an ERH is composed of a number of pulses, it can be expressed by using unit step functions $u_1(t)$,

$$I(t) = I_0\, u_1(t) + \sum_{j=1}^{m} (I_j - I_{j-1}) u_1(t - jD) + I_m\, u_1[t - (m+1)D] \tag{13.15}$$

in which D is the time interval, m is the number of pulses, and I_j, $j = 0, 1, 2, \ldots, m$, are ER intensities, as shown in Fig. 13.3. If we let

$$w_0 = I_0, \qquad w_j = I_j - I_{j-1}, \qquad j = 1, 2, \ldots, m$$

then

$$I(t) = \sum_{j=0}^{m} w_j u_1(t - jD) \tag{13.16}$$

The DRH of a linear channel with its IUH given by equation (13.11) is

$$Q(t) = \int_0^t I(\tau)\, \delta(t - T - \tau)\, d\tau = I(t - T)$$

$$= \sum_{j=0}^{m} w_j\, u_1(t - jD - T) \tag{13.17}$$

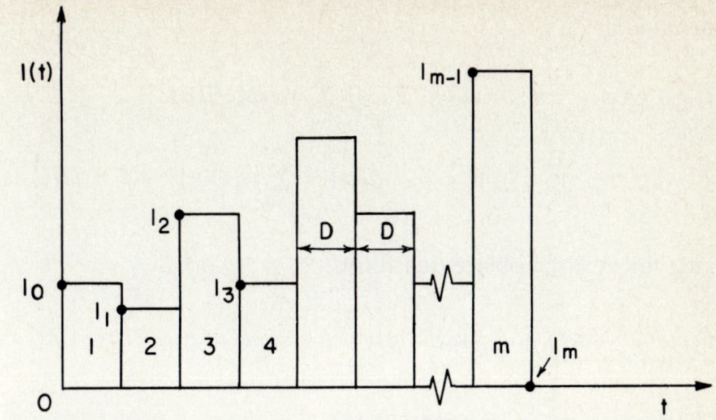

FIGURE 13.3 An ERH composed of m pulses, each of duration D time units.

On the other hand, if the IUH is prescribed by equation (13.12), then

$$Q(t) = \frac{1}{T}[u_1(t) - u_1(t - T)] \int_0^t I(\tau)\, d\tau$$

$$= \frac{1}{T}\int_{jD}^{t} [u_1(t) - u_1(t - T)] \sum_{j=0}^{m} w_j\, u_1(t - jD)\, d\tau$$

Since the ERH and IUH are of finite duration, the limits of the integral should be properly considered.

$$Q(t) = \frac{1}{T}\sum_{j=0}^{m} w_j [u_1(t - jD)(t - jD) - u_1(t - jD - T)(t - jD - t)] \tag{13.18}$$

EXAMPLE 13.1

Determine the DRH of a linear channel by using the Laplace transform for the ERH given by equation (13.16).

The Laplace transform of a unit step function is

$$L[u_1(t - D) = \frac{\exp(-Ds)}{s}$$

Taking the Laplace transform of equation (13.16),

$$L[I(t)] = I(s) = \sum_{j=0}^{m} \frac{w_j \exp(-jDs)}{s}$$

Likewise, from equation (13.11)

$$L[h(t)] = L(\delta(t - T)] = \exp(-Ts)$$

Since

$$Q(s) = L[Q(t)] = h(s)\, I(s) = \sum_{j=0}^{m} \frac{w_j \exp[-s(jD + T)]}{s}$$

Taking the inverse Laplace transform,

$$Q(t) = \sum_{j=0}^{m} w_j u_1(t - jD - T)$$

which is the same as equation (13.17).

If $h(t)$ is given by equation (13.12), then

$$L[h(t)] = \frac{1}{Ts}[1 - \exp(-Ts)]$$

Therefore,

$$Q(s) = \frac{1}{Ts^2}[1 - \exp(-Ts)] \sum_{j=0}^{m} w_j exp[-jDs]$$

$$= \frac{1}{Ts^2}\left\{\sum_{j=0}^{m} w_j \exp[-jDs] - \sum_{j=0}^{m} w_j \exp[-s(T+jd)]\right\}$$

Taking the inverse Laplace transform,

$$Q(t) = \frac{1}{T}\sum_{j=0}^{m} w_j[u_1(t-jD)(t-jD) - u_1(t-jD-T)(t-jD-T)]$$

which is the same as equation (13.18).

EXAMPLE 13.2

Consider a linear channel having a translation time of 6 h. An ERH is given by equation (13.15) with $I_0 = 5$ cm/h, $I_1 = 2$ cm/h, $I_2 = 1$ cm/hr, and $I_3 = 2$, $I_4 = 0$ cm/h, $D = 0.5$ h, and $m = 4$. Compute the DRH with the IUH given by equation (13.12).

The ERH can be expressed by equation (13.16), in which

$w_0 = 5$, $w_1 = -3$, $w_2 = -1$, $w_3 = 1$, and $w_4 = -2$.

Therefore,

$$I(t) = 5\,u_1(t) - 3\,u_1(t-0.5) - u_1(t-1) + u_1(t-1.5) - 2\,u_1(t-2)$$

Its Laplace transform is

$$I(s) = \frac{1}{s}[5 - 3\exp(-0.5s) - \exp(-s) + \exp(-1.5s) - 2\exp(-2s)]$$

We are given

$$h(t) = \tfrac{1}{6}[u_1(t) - u_1(t-6)]$$

Its Laplace transform is

$$h(s) = \frac{1}{6s}[1 - \exp(-6s)]$$

Thus the DRH is given by

$$Q(s) = \frac{1}{6s}[1 - \exp(-6s)][5 - 3\exp(-0.5s) - \exp(-s) + \exp(-1.5s) - 2\exp(-2s)]\frac{1}{s}$$

$$= \frac{1}{6s^2}[5 - 3\exp(-0.5s) - \exp(-s) + \exp(-1.5s) - 2\exp(-2s) - 5\exp(-6s) + 3\exp(-6.5s) + \exp(-7s) - \exp(-7.5s) + 2\exp(-8s)]$$

Inverting the Laplace transform,

$$Q(t) = \frac{1}{6}[5t - 3\,u_1(t-0.5)(t-0.5) - u_1(t-1)(t-1) + u_1(t-1.5)(t-1.5) - 2\,u_1(t-2)(t-2) - 5u_1(t-6)(t-6) + 3u_1(t-6.5)(t-6.5) + u_1(t-7)(t-7) - u_1(t-7.5)(t-7.5) + 2\,u_1(t-8)(t-8)$$

Doing the calculations gives the following:

Time (h)	Discharge (cm/h)	Time (h)	Discharge (cm/h)
0	0	4.5	.833
.5	.417	5.0	.833
1.0	.583	5.5	.833
1.5	.667	6.0	.833
2.0	.833	6.5	.417
2.5	.833	7.0	.25
3.0	.833	7.5	.167
3.5	.833	8.0	.00
4.0	.833		

13.2 LINEAR RESERVOIR

The continuity equation for a reservoir can be written as

$$I(t) - Q(t) = \frac{dS(t)}{dt} \tag{13.19}$$

where $I(t)$ is inflow rate, $Q(t)$ is the outflow rate, and S is storage. Assume that the discharge properties of the reservoir are such that

$$S = kQ \tag{13.20}$$

where k is the coefficient of proportionality or average residence time (Bolin and Rodhe, 1973). Combining equations (13.19)–(13.20),

$$I - Q = k\frac{dQ}{dt} \tag{13.21}$$

Utilizing the operator notation $D = d/dt$,

$$Q = \frac{1}{1+kD}I(t)$$

Mathematically, this is equivalent to

$$Q = e^{-t/k}\int e^{t/k}I\,dt \tag{13.22}$$

Equation (13.20) defines what is known as a linear reservoir and the operator $1/(1+kD)$ represents its effect on the inflow. Let

$$H = \frac{1}{1+kD}$$

The operator H represents the effect of a linear reservoir with proportionality coefficient k and is often referred to as the *storage operator*. This operator is commutative.

13.2.1 The IUH

The IUH of the linear reservoir obtained by replacing $I(t)$ by $\delta(t)$ in equation (13.22) is

$$h(t) = \frac{e^{-t/k}}{k} \tag{13.23}$$

It may be noted that the IUH can also be derived rather conveniently from the S-hydrograph $U(t)$ of the linear reservoir,

$$U(t) = 1 - e^{-t/k} \tag{13.24}$$

Differentiating with respect to t,

$$h(t) = \frac{dU(t)}{dt} = \frac{e^{-t/k}}{k}$$

Its appearance is shown in Fig. 13.4. If the inflow were a unit pulse of duration D, the UH $h(D; t)$, as shown in Fig. 13.5, and the direct runoff $Q(t)$ (ignoring the dimensions of $h(D; t)$ and $Q(t)$) would be

$$Q(t) = h(D;\ t) = I[1 - e^{-t/k}], \qquad t \leq D \tag{13.25a}$$

and

$$Q(t) = h(D;\ t) = Q_p\, e^{-(t-D)/k}, \qquad t \leq D \tag{13.25b}$$

where Q_p represents the hydrograph peak and is given by

$$Q_p = I[1 - e^{-D/k}] \tag{13.26}$$

Here $I = 1/D$. The hydrograph peak will occur at the end of the duration of unit pulse.

As $t \to \infty$, $Q(t) = I(t)$. This implies an equilibrium condition: outflow becoming equal to inflow. As $t \to 0$, $Q = 0$. As $I(t)$ terminates at $t = D$, the recession starts immediately. For an instantaneous inflow, which fills the reservoir of storage S in $t = 0$, $Q_p = S/k$. The equation of outflow is simply

$$Q(t) = \frac{S}{k} e^{-t/k} \tag{13.27}$$

For a unit inflow, $S = 1$ and $I(t) = \delta(t)$. Consequently $Q(t) \to h(t)$, the IUH that is the same as given by equation (13.23).

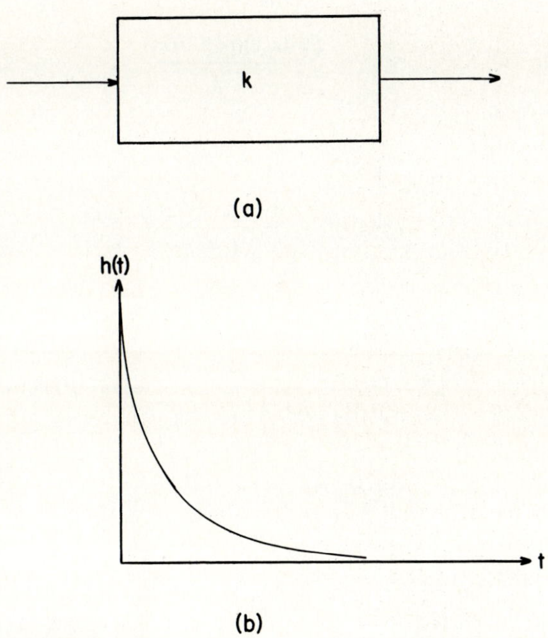

FIGURE 13.4 A linear reservoir: (a) lag time k, (b) IUH.

13.2.2 Properties of Impulse Response

An impulse form of the effective rainfall has a finite volume V or S. Equation (13.27) prescribes the response of the linear reservoir due to such an impulse. At $t = 0$, $Q(t) = Q_0$. Therefore,

$$Q_0 = \frac{S}{k}$$

The initial state Q_0 is directly proportional to the impulse effective rainfall volume S and inversely proportional to the parameter k. Equation (13.27) can then be written as

$$Q(t) = Q_0 \exp\left(\frac{-t}{k}\right)$$

The volume of the impulse response is

$$V = \int_0^\infty Q(t)\, dt = \int_0^\infty Q_0 \exp\left(\frac{-t}{k}\right) dt = Q_0 k$$

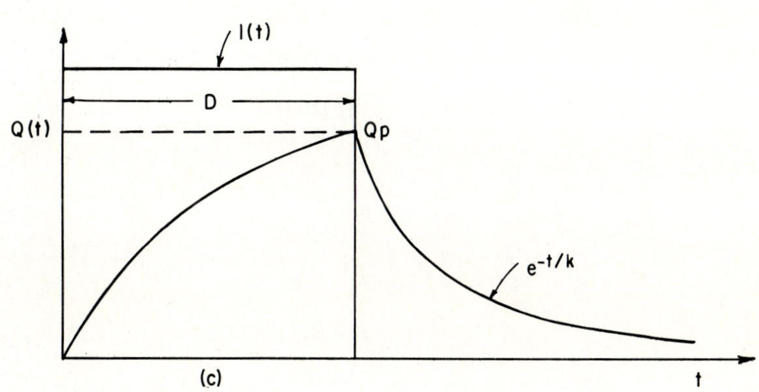

FIGURE 13.5 A linear reservoir: hydrograph due to a pulse of duration D hours duration.

We can determine the lag of this response. To this end, we compute the first moment about the Q-axis,

$$M_1^Q = \int_0^\infty Q(t) \cdot t \cdot dt = \int_0^\infty t \cdot Q_0 \exp\left(\frac{-t}{k}\right) dt = Q_0 k^2$$

Therefore, lag time T_l is

$$T_l = \frac{M_1^Q}{V} = \frac{Q_0 k^2}{Q_0 k} = k$$

Likewise, we can compute the mean discharge. We compute the first moment about the t-axis,

$$M_1' = \int_0^\infty \frac{[Q(t)]^2}{2} dt = \frac{Q_0^2}{2} \int_0^\infty \exp\left(\frac{-2t}{k}\right) dt = \frac{Q_0^2 k}{4}$$

Therefore, mean discharge Q_m is

$$Q_m = \frac{M_1'}{V} = \frac{Q_0^2 k}{4} \cdot \frac{1}{Q_0 k} = \frac{Q_0}{4}$$

13.2.3 Parameter Estimation

The parameter k has the dimension of time and is equal to the average lag time imposed on the inflow by the reservoir. This is referred to as *lag time* and equals the difference in time between the centers of areas of inflow and outflow. Evidently, for a particular event on a gaged watershed, k can be estimated from records of inflow and outflow provided certain qualifications are met (Pedersen, Peters, and Helweg, 1980). For example, for a rainfall-runoff event these qualifications are: (1) The event is relatively isolated in time. (2) The event is fairly uniformly distributed over the watershed. (3) The event has a single well-defined peak. The value of k is constant if the effective rainfall-direct runoff relationship is linear. However, its variation with rainfall and watershed characteristics has been reported by various investigators (Rao, Delleur, and Sarma, 1972; Sarma, Delleur, and Rao, 1969; Schulz and Lopez, 1974), as discussed in Chapter 9.

The moments of the IUH in equation (13.23) about its origin can be specified as

$$M_1 = k \quad (13.28a)$$

$$M_2 = 2k^2 \quad (13.28b)$$

$$M_3 = 3k^3 \quad (13.28c)$$

$$M_4 = 22k^4 \quad (13.28d)$$

Likewise, its cumulants are

$$K_1 = k \quad (13.29a)$$

$$K_2 = k^2 \quad (13.29b)$$

$$K_3 = 2k^3 \quad (13.29c)$$

$$K_4 = 6k^4 \quad (13.29d)$$

13.2.4 The DRH

The DRH of a linear reservoir for an arbitrary ERH can be determined either by direct convolution or by using Laplace transform.

$$Q(t) = \int_0^t \frac{1}{k} \exp\left[\frac{-(t-\tau)}{k}\right] I(\tau)\, d\tau$$

Inserting equation (13.16) for $I(t)$

$$Q(t) = \sum_{j=0}^{m} w_j u_1(t - jD) \frac{1}{k} \int_{jD}^{t} \exp\left[-\frac{(t-\tau)}{k}\right] d\tau$$

$$= \sum_{j=0}^{m} w_j u_1(t - jD) \left[1 - \exp\left(-\frac{(t - jD)}{k}\right)\right] \quad (13.30)$$

This same expression can also be obtained by using the Laplace transform. Note that

$$h(s) = \frac{1}{1 + ks}$$

$$I(s) = \sum_{j=0}^{m} \frac{w_j \exp(-jDs)}{s}$$

Therefore,

$$Q(s) = \sum_{j=0}^{m} \frac{w_j \exp(-jDs)}{s(1 + ks)}$$

Upon inversion,

$$Q(t) = \sum_{j=0}^{m} w_j u_1(t - jD) \left\{1 - \exp\left[\frac{-(t - jD)}{k}\right]\right\}$$

which is the same as equation (13.30).

EXAMPLE 13.3

Consider a linear reservoir with $k = 6$ h. Compute the DRH for the ERH of Example 13.2.

$$I(t) = 5u_1(t) - 3u_1(t - .5) - u_1(t - 1) + u_1(t - 1.5) - 2u_1(t - 2)$$

Its Laplace transform is

$$I(s) = \frac{1}{s}[5 - 3\exp(-.5s) - \exp(-s) + \exp(-1.5s) - 2\exp(-2s)]$$

The IUH is

$$h(t) = \frac{1}{k} \exp\left(\frac{-t}{k}\right)$$

Let $a = 1/k$. The Laplace transform of $h(t)$ is

$$h(s) = a\left[\frac{1}{s + a}\right]$$

Therefore,

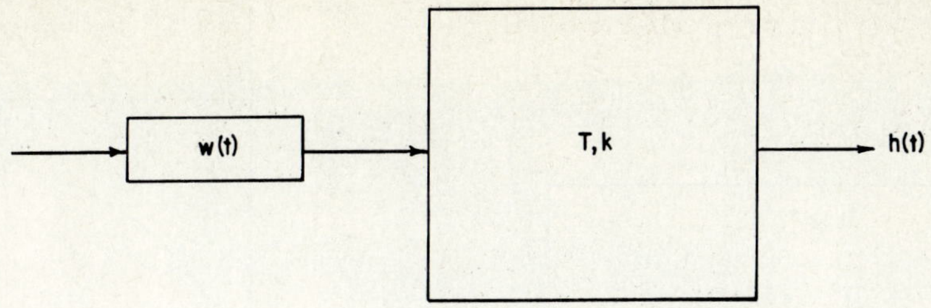

FIGURE 13.6 A model routing the effective rainfall through a TAC curve ($w(t)$) and an arrangement of linear channels (T) and linear reservoirs (k). This structure represents most of the conceptual models discussed in this chapter.

$$
\begin{aligned}
Q(s) &= h(s)I(s) \\
&= a\left(\frac{1}{s+a}\right)[5 - 3\exp(-.5s) - \exp(-s) \\
&\quad + \exp(-1.5s) - 2\exp(-2s)]\frac{1}{s} \\
&= \frac{1}{s(s+a)}[5 - 3\exp(-.5s) - \exp(-s) \\
&\quad + \exp(-1.5s) - 2\exp(-2s)]
\end{aligned}
$$

The quantity

$$\frac{1}{s(s+a)} = \frac{1}{sa} - \frac{1}{a(s+a)}$$

Therefore,

$$
\begin{aligned}
Q(s) = \Bigg[& 5\left(\frac{1}{s} - \frac{1}{s+a}\right) - 3\exp(-.5s)\left(\frac{1}{s} - \frac{1}{s+a}\right) \\
& - \exp(-s)\left(\frac{1}{s} - \frac{1}{s+a}\right) + \exp(-1.5s)\left(\frac{1}{s} - \frac{1}{s+a}\right) \\
& - 2\exp(-2s)\left(\frac{1}{s} - \frac{1}{s+a}\right)\Bigg]
\end{aligned}
$$

On inverting,

$$
\begin{aligned}
Q(t) = {} & 5\,u_1(t)\left[1 - \exp\left(\frac{-t}{k}\right)\right] - 3\,u_1(t-0.5) \\
& \cdot\left\{1 - \exp - \left[\frac{(t-0.5)}{k}\right]\right\} - u_1(t-1) \\
& \cdot\{1 - \exp[-(t-1)/k]\} + u_1(t-1.5) \\
& \cdot\{1 - \exp[-(t-1.5)/k]\} \\
& - 2u_1(t-2)\left\{1 - \exp\left[\frac{-(t-2)}{k}\right]\right\}
\end{aligned}
$$

Doing the calculations gives the following:

Time (h)	Discharge (cm/h)	Time (h)	Discharge (cm/h)
0	0	5.5	.38
.5	.4	6.0	.35
1.0	.53	6.5	.32
1.5	.57	7.0	.30
2.0	.68	7.5	.27
2.5	.63	8.0	.25
3.0	.58	9.0	.21
3.5	.53	12.0	.13
4.0	.49	16.0	.07
4.5	.45	20.0	.03
5.0	.41		

Linear channels and linear reservoirs, in conjunction with the TAC curve, have been variously combined to develop models of the effective rainfall-direct runoff relationship. Most of the conceptual models employ some arrangement of these components, as shown in Fig. 13.6. We discuss some of these models here.

13.3 THE ZOCH MODEL

The Zoch model (Zoch, 1934, 1936, 1937) consists in representing a watershed by a TAC curve and then routing it through a linear reservoir of equations (13.20)–(13.21), as shown in Fig. 13.7. Zoch assumed the effective rainfall to occur uniformly for a finite period of time. Four shapes of TAC curves $w(t)$ were considered: (1) rectangular, (2) triangular, (3) elliptical, and (4) arbitrary. These shapes are shown in Fig. 13.8. He derived analytical expressions for the unit hydrograph in each case. These analytical expressions can be derived more conveniently if we resort to the IUH concept. We consider the above shapes of $w(t)$ in the ensuing discussion.

13.3.1 Rectangular TAC Curve

If the effective rainfall is $\delta(t)$ then its TAC is as shown in Fig. 13.8a,

$$w(t) = \begin{cases} V_0, & 0 < t < T_c \\ 0, & t \geq T_c \end{cases} \tag{13.31}$$

where V_0 is constant, which equals A/T_c, with A being the watershed area. Then the IUH of the Zoch model is

$$h(t) = \int_0^t \frac{e^{-(t-\tau)/k}}{k} w(\tau)\, d\tau \tag{13.32}$$

which on a unit area basis yields

$$h(t) = \begin{cases} \dfrac{1}{T_c}(1 - e^{-t/k}), & 0 \leq t \leq T_c \\ \dfrac{1}{T_c}(1 - e^{-T_c/k})\, e^{-(t-T_c)/k}, & t \geq T_c \end{cases} \tag{13.33}$$

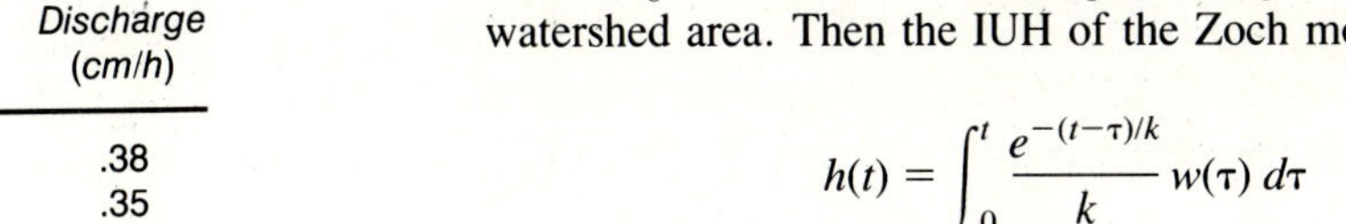

FIGURE 13.7 A model routing the effective rainfall through a TAC curve ($w(t)$) and a linear reservoir (k). This schematic represents the Zoch, Clark, and O'Kelly models.

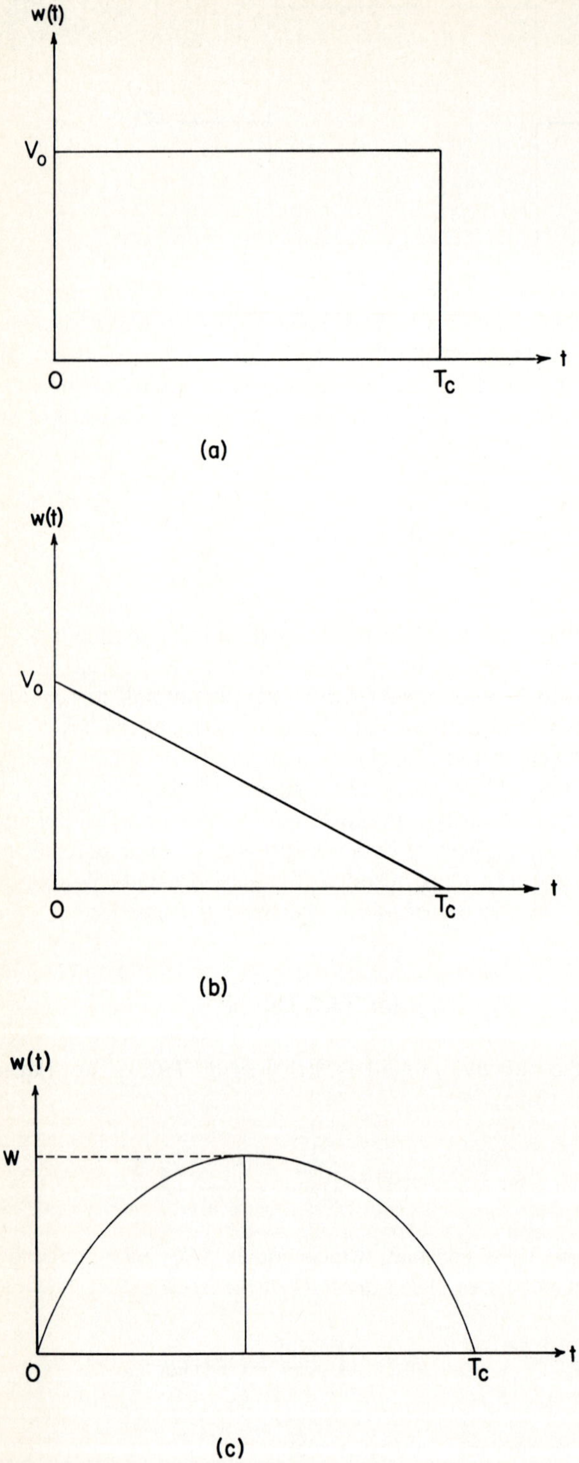

FIGURE 13.8 TAC curves for the Zoch model: (a) rectangular, (b) triangular, and (c) elliptical.

Obviously the maximum of $h(t)$ occurs at $t = T_c$, as shown in Fig. 13.9. The moments of the IUH about its origin can be expressed as

$$M_1 = T_c\left(.5 + \frac{k}{T_c}\right) \tag{13.34a}$$

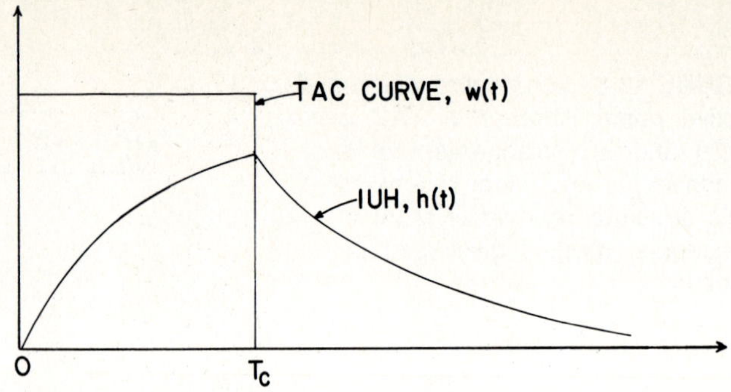

FIGURE 13.9 The IUH by the Zoch and Clark models for a rectangular watershed.

$$M_2 = T_c^2\left[\frac{1}{3} + \frac{k}{T_c} + 2\left(\frac{k}{T_c}\right)^2\right] \tag{13.34b}$$

$$M_3 = \frac{T_c^3}{4} + T_c^2 + 3\,k^2\,T_c + 6k^3 \tag{13.34c}$$

$$M_4 = \frac{T_c^4}{5} + kT_c^3 + 4k^2T_c^2 + 12k^3T_c + 24k^4 \tag{13.34d}$$

Similarly, its cumulants are given as

$$K_1 = T_c\left(.5 + \frac{k}{T_c}\right) \tag{13.35a}$$

$$K_2 = T_c^2\left[\frac{1}{12} + \left(\frac{k}{T_c}\right)^2\right] \tag{13.35b}$$

$$K_3 = T_c^3\left[2\left(\frac{k}{T_c}\right)^3\right] \tag{13.35c}$$

$$K_4 = T_c^4\left[-\frac{1}{120} + 6\left(\frac{k}{T_c}\right)^4\right] \tag{13.35d}$$

The DRH for this case can now be derived for an arbitrary $I(t)$. For purposes of simplicity, $I(t)$ is considered to be a rectangular pulse of duration D and height M, that is, $I(t) = M[u_1(t) - u_1(t - D)]$. Because the IUH has a discontinuity at $t = T_c$, three cases are distinguished: (1) $T_c = D$, the duration of $I(t)$, (2) $T_c < D$, and (3) $T_c > D$. For these cases we derive the DRH.

Case 1: $T_c = D$

$$\begin{aligned} Q(t) &= \int_0^t I(\tau)\,h(t-\tau)\,d\tau \\ &= \int_0^t \frac{M}{T_c}\left[1 - \exp\left\{\frac{-(t-\tau)}{k}\right\}\right]d\tau, \qquad 0 \le t \le T_c \\ &= \frac{M}{T_c}\left\{t - k\left[1 - \exp\left(\frac{-t}{k}\right)\right]\right\}, \qquad 0 \le t \le T_c \end{aligned} \tag{13.36a}$$

$$Q(t) = \int_{t-T_c}^{T_c} \frac{M}{T_c}\{1 - \exp[-(t-\tau)/k]\}\, d\tau$$
$$+ \int_{T_c}^{t} \frac{M}{T_c}\left[1 - \exp\left(\frac{-T_c}{k}\right)\right]$$
$$\exp\left\{\frac{-(t-\tau-T_c)}{k}\right\} d\tau, \qquad T_c \le T \le 2T_c$$
$$= \frac{M}{T_c}\left[\left\{(2T_c - t) + k\left[\exp\left(\frac{-T_c}{k}\right)\right] - \exp\left[\frac{-(t-T_c)}{k}\right]\right\}\right.$$
$$+ \left\{k\left[1 - \exp\left(\frac{-T_c}{k}\right)\right]\right.$$
$$\left.\left.\left(1 - \exp\left[-\frac{-(t-T_c)}{k}\right]\right)\right\}\right], \quad T_c < t \le 2T_c \tag{13.36b}$$

$$Q(t) = \int_{t-T_c}^{t} \frac{M}{T_c}\left[1 - \exp\left(\frac{-T_c}{k}\right)\right]\exp\left[\frac{-(t-T_c-\tau)}{k}\right] d\tau$$
$$= \frac{kM}{T_c}\left[1 - \exp\left(\frac{-T_c}{k}\right)\right]\exp\left(\frac{-T_c}{k}\right)\left\{\exp\left[\frac{-(t-T_c)}{k}\right]\right.$$
$$\left. - \exp\left(\frac{-t}{k}\right)\right\}, \qquad t > 2T_c \tag{13.36c}$$

Case 2: $T_c < D$

$$Q(t) = \frac{M}{T_c}\left\{t - k\left[1 - \exp\left(\frac{-t}{k}\right)\right]\right\}, \qquad 0 \le t \le T_c \tag{13.37a}$$

$$Q(t) = \int_0^{T_c} \frac{M}{T_c}\left[1 - \exp\left\{\frac{-(t-\tau)}{k}\right\}\right] d\tau$$
$$+ \int_{T_c}^{t} \frac{M}{T_c}\left[1 - \exp\left(\frac{-T_c}{k}\right)\right]\exp\left[\frac{-(t-\tau-T_c)}{k}\right] d\tau$$
$$= \frac{M}{T_c}\left(\left\{T_c - k\left[1 - \exp\left(\frac{-T_c}{k}\right)\right]\right\}\right.$$
$$+ \left\{k\left[1 - \exp\left(\frac{-T_c}{k}\right)\right]\right.$$
$$\left.\left.\left(1 - \exp\left[\frac{-(t-T_c)}{k}\right]\right)\right\}\right), \qquad T_c \le t \le D \tag{13.37b}$$

$$Q(t) = \int_{t-D}^{t} \frac{M}{T_c}\left\{1 - \exp\left[\frac{-(t-\tau)}{k}\right]\right\} d\tau$$
$$+ \int_{T_c}^{t} \frac{M}{T_c}\left[1 - \exp\left(\frac{-T_c}{k}\right)\right]\exp\left[\frac{-(t-T_c-\tau)}{k}\right] d\tau$$
$$= \frac{M}{T_c}\left[\left(D + k\left\{\exp(-t/k) - \exp\left[\frac{-(t-D)}{k}\right]\right\}\right)\right.$$
$$+ k\left[1 - \exp\left(\frac{-T_c}{k}\right)\right]\exp\left(\frac{T_c}{k}\right)\left[\exp\left(\frac{-T_c}{k}\right)\right.$$
$$\left.\left. - \exp\left(\frac{-t}{k}\right)\right]\right], \qquad D < t < T_c + D \tag{13.37c}$$

$$Q(t) = \int_{t-D}^{t} \frac{M}{T_c}\exp\left[\frac{-(t-\tau-T_c)}{k}\right]$$
$$[1 - \exp(-T_c/k)]\, d\tau, \qquad t > T_c + D$$
$$= \frac{kM}{T_c}\left[1 - \exp\left(\frac{-T_c}{k}\right)\right]\exp\left(\frac{T_c}{k}\right)\left[\exp\left(\frac{-(t-T_c}{k}\right)\right.$$
$$\left. - \exp\left(\frac{-t}{k}\right)\right], \qquad t > T_c + D \tag{13.37d}$$

Case 3: $T_c > D$

$$Q(t) = \int_0^t \frac{M}{T_c}\{1 - \exp[-(t-\tau)/k]\}\, d\tau$$
$$= \frac{M}{T_c}\{t - k[1 - \exp(-t/k)]\}, \qquad 0 \le t \le D \tag{13.38a}$$

$$Q(t) = \int_{t-D}^{t} \frac{M}{T_c}\left\{1 - \exp\left[\frac{-(t-\tau)}{k}\right]\right\} d\tau$$
$$= \frac{M}{T_c}\left(\mathrm{D} + \mathrm{k}\left\{\exp\left(\frac{-t}{k}\right)\right.\right.$$
$$\left.\left. - \exp\left[\frac{-(t-D)}{k}\right]\right\}\right), \qquad 0 \le t \le T_c \tag{13.38b}$$

$$Q(t) = \int_{t-D}^{t} \frac{M}{T_c}\left\{1 - \exp\left[\frac{-(t-\tau)}{k}\right]\right\} d\tau$$
$$+ \int_{T_c}^{t} \frac{M}{T_c}\left[1 - \exp\left(\frac{-T_c}{k}\right)\right]$$
$$\exp\left[\frac{-(t-\tau-T_c)}{k}\right) d\tau, \qquad T_c \le t \le T_c + D$$
$$= \frac{M}{T_c}\left[\left\{T_c + D - t + k\left(\exp\left(\frac{-T_c}{k}\right)\right.\right.\right.$$
$$\left.\left. - \exp\left[-\frac{(t-D)}{k}\right]\right)\right\}$$
$$+ k\left(\left[1 - \exp\left(\frac{-T_c}{k}\right)\right]\left\{1 - \exp\right.\right.$$
$$\left.\left.\left.\left[\frac{-(t-T_c)}{k}\right]\right\}\right)\right], \qquad T_c \le t \le T_c + D \tag{13.38c}$$

$$Q(t) = \int_{t-D}^{t} \frac{M}{T_c}\left[1 - \exp\left(\frac{-T_c}{k}\right)\right]\exp\left[\frac{-(t-\tau-t_c)}{k}\right] d\tau$$
$$= \frac{M}{T_c}\exp\left(\frac{T_c}{k}\right)\left[1 - \exp\left(\frac{-T_c}{k}\right)\right]\left\{\exp\left[\frac{-(t-D)}{k}\right]\right.$$
$$\left. - \exp\left(\frac{-t}{k}\right)\right\}, \qquad t \ge T_c + D \tag{13.38d}$$

These solutions can be extended to the case where $I(t)$ is expressed by equations (13.16).

EXAMPLE 13.4

Plot the IUH of the Zoch model for a rectangular watershed for various values of k/T_c.

For purposes of this plot $T_c = 4$ and $V_0 = 0.25$. Using equation (13.33), the IUH is plotted as shown in Fig. 13.10.

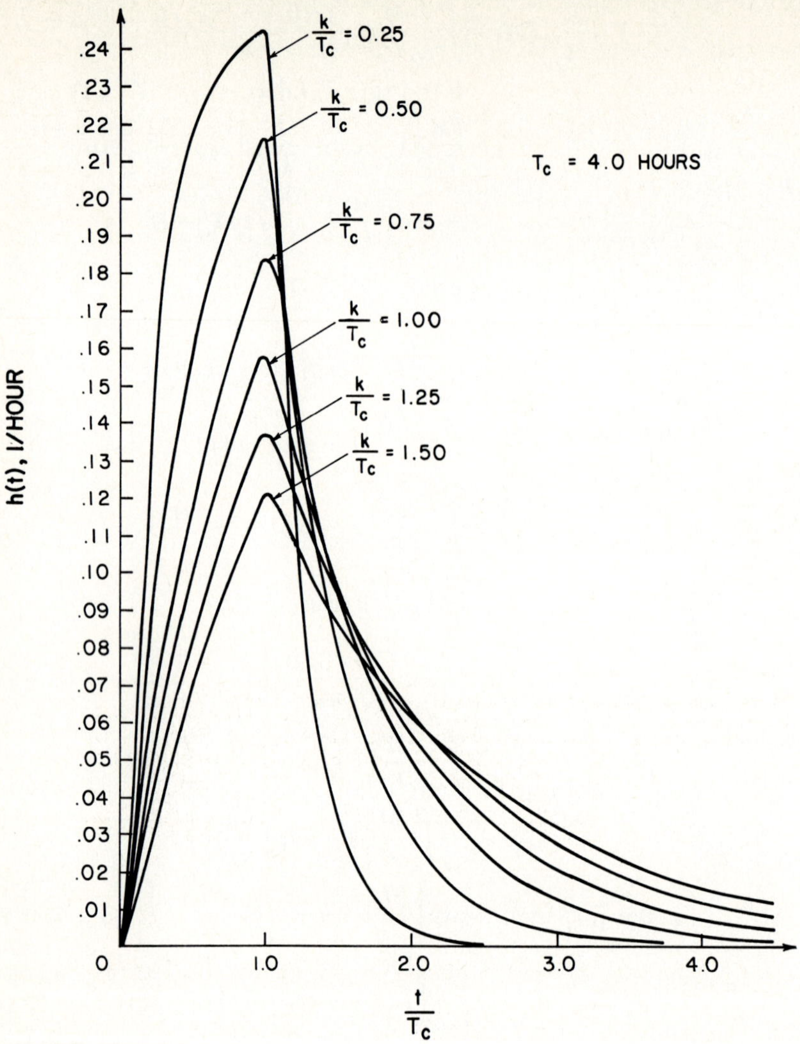

FIGURE 13.10 Example 13.4: The IUH of the Zoch model for a rectangular watershed for various values of k/T_c.

EXAMPLE 13.5

Plot the IUH of the Zoch model for a rectangular watershed as $h(t)T_c$ versus t/T_c for various values of k/T_c.

By taking $T_c = 4$ and $V_0 = 0.25$, the IUH is plotted as shown in Fig. 13.11.

EXAMPLE 13.6

Plot the IUH of the Zoch model for a rectangular watershed as $h(t)T_\ell$ versus t/T_ℓ for various values of k/T_c.

By taking $T_c = 4$, $V_0 = 0.25$, and $T_\ell = 2 + k$, the IUH is plotted as shown in Fig. 13.12.

EXAMPLE 13.7

Compute the DRH by the Zoch model with $w(t)$ represented by a rectangle of base 6 h. The watershed area is 1 km^2. The value of k is 3 h. Use the effective rainfall of Example 13.2.

The DRH is obtained following equations (13.36)–(13.38) and is tabulated in Table 13.1.

13.3.2 Triangular TAC Curve

If $w(t)$ is as represented in Fig. 13.8(b) with $V_0 = 2A/T_c$,

$$w(t) = \begin{cases} V_0\left(1 - \dfrac{t}{T_c}\right), & 0 \le t \le T_c \\ 0, & t \ge T_c \end{cases} \tag{13.39}$$

then on a unit area basis

$$h(t) = \begin{cases} \dfrac{2}{T_c}\left[\left(1 + \dfrac{k}{T_c}\right)(1 - e^{-t/k}) - \dfrac{t}{T_c}\right], & 0 \le t \le T_c \\ \dfrac{2}{T_c}\left[\left(1 + \dfrac{k}{T_c}\right)(e^{-(t-T_c)/k} - e^{-t/k}) \right. \\ \qquad \left. - e^{-(t-T_c)/k}\right], & t \ge T_c \end{cases} \tag{13.40}$$

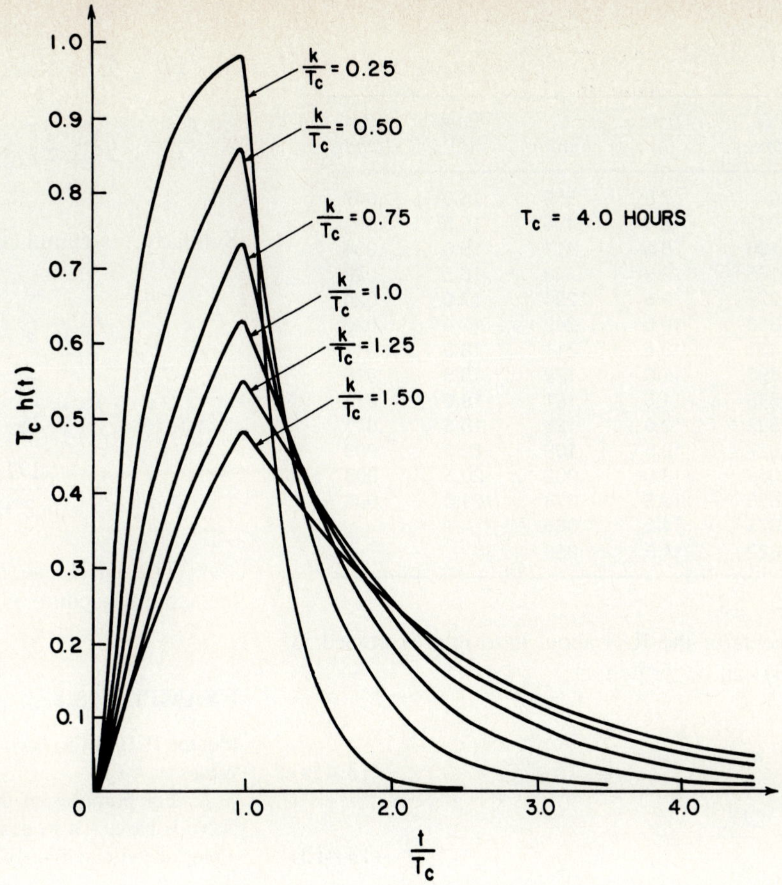

FIGURE 13.11 Example 13.5: The dimensionless IUH, $T_c h(t)$, of the Zoch model for a rectangular watershed for various values of k/T_c.

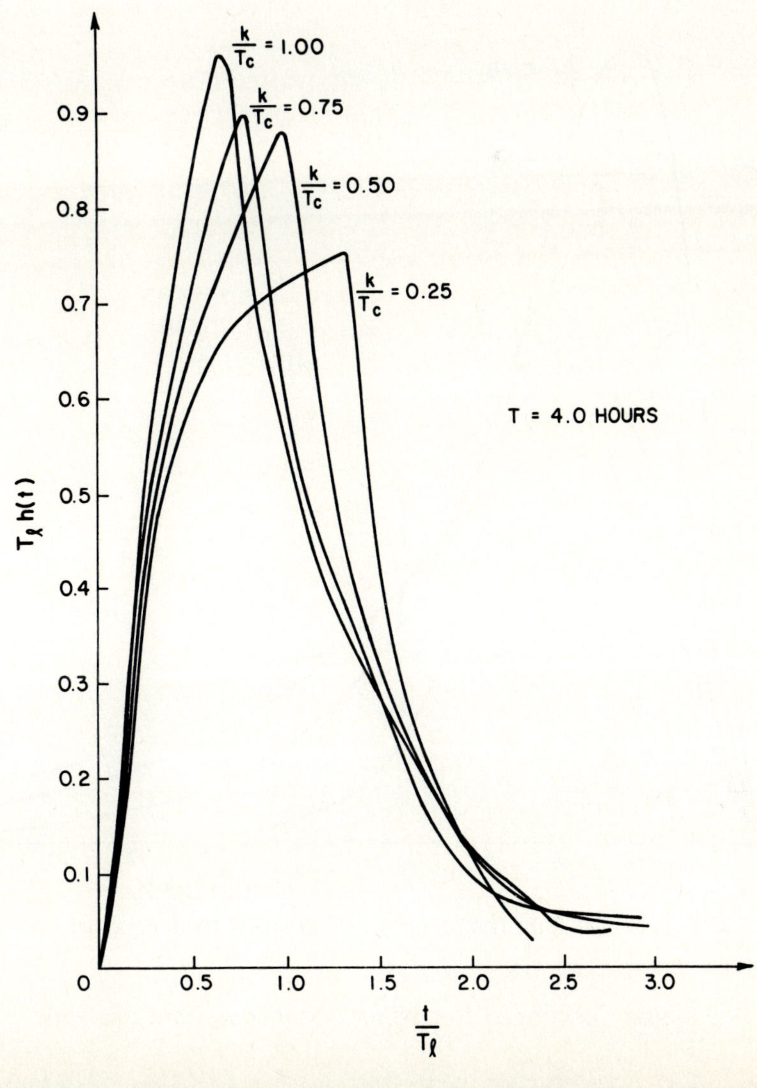

FIGURE 13.12 Example 13.6: The dimensionless IUH, $T_\ell h(t)$, of the Zoch model for a rectangular watershed for various values of k/T_c.

TABLE 13.1

Time (h)	Q (cm/h)	Time (h)	Q (cm/h)	Time (h)	Q (cm/h)
.0	.0	7.5	.559	15.0	.047
.5	.033	8.0	.485	15.5	.04
1.0	.105	8.5	.411	16.0	.034
1.5	.185	9.0	.348	16.5	.029
2.0	.272	9.5	.294	17.0	.024
2.5	.358	10.0	.249	17.5	.020
3.0	.431	10.5	.211	18.0	.017
3.5	.493	11.0	.179	18.5	.015
4.0	.545	11.5	.151	19.0	.012
4.5	.589	12.0	.128	19.5	.011
5.0	.627	12.5	.108	20.0	.009
5.5	.658	13.0	.092	20.5	.008
6.0	.685	13.5	.078	21.0	.006
6.5	.675	14.0	.066		
7.0	.622	14.5	.056		

The moments of the IUH about its origin expressed in equation (13.38) can be written as

$$M_1 = \frac{T_c}{3} + k \tag{13.41a}$$

$$M_2 = \frac{T_c^2}{6} + \frac{2kT_c}{3} + 2k^2 \tag{13.41b}$$

$$M_3 = \frac{T_c^3}{10} + \frac{kT_c^2}{3} + 2k^2T_c + 6k^3 \tag{13.41c}$$

$$M_4 = \frac{T_c^4}{15} + \frac{2}{5}kT_c^3 + 2k^2T_c^2 + 8k^3T_c + 24k^4 \tag{13.41d}$$

Similarly, its cumulants are

$$K_1 = \frac{T_c}{3} + k \tag{13.42a}$$

$$K_2 = \frac{T_c^2}{18} + k^2 \tag{13.42b}$$

$$K_3 = \frac{17T_c^3}{270} + \frac{23}{18}kT_c^2 + \frac{20k^2T_c}{9} + 4k^3 \tag{13.42c}$$

The DRH can now be derived for any $I(t)$ following the same procedure as before.

EXAMPLE 13.8

Plot the IUH of the Zoch model for a triangular watershed for various values of k/T_c.

For purposes of this plot, T_c = 4 h, and the triangular *TAC* curve is base = 4 h, peak = .5, occurring at 1 h. Fig. 13.13 shows a plot of the resulting IUH.

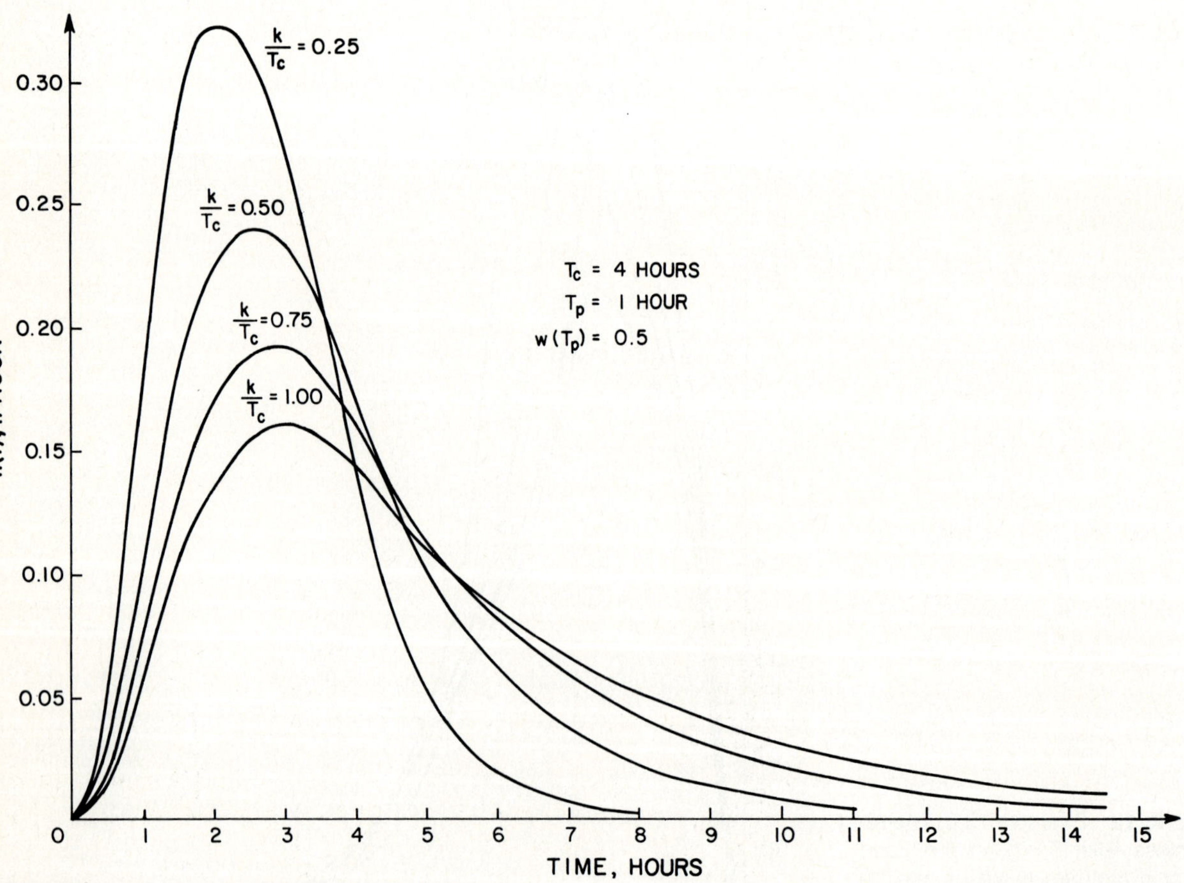

FIGURE 13.13 Example 13.8: The IUH of the Zoch model for a triangular watershed for various values of k/T_c.

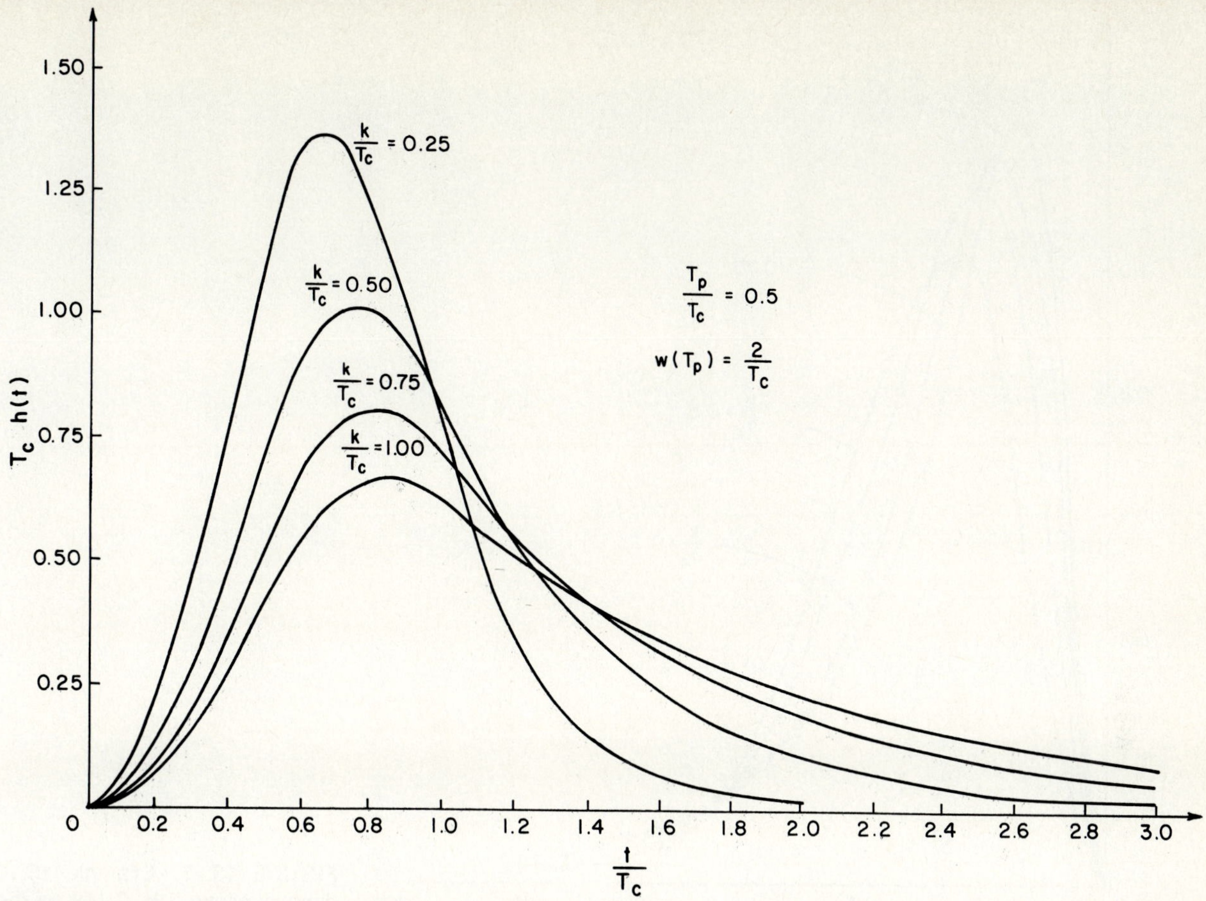

FIGURE 13.14 Example 13.9: The dimensionless IUH, $T_c h(t)$, of the Zoch model for a triangular watershed for various values of k/T_c.

EXAMPLE 13.9

Plot the IUH of the Zoch model for a triangular watershed for various values of k/T_c as $h(t)T_c$ versus t/T_c.

Using the data of Example 13.8, the IUH is plotted in Fig. 13.14.

EXAMPLE 13.10

Plot the IUH of the Zoch model for a triangular watershed for various values of k/T_c as $h(t)T_\ell$ versus t/T_ℓ.

Using the data of Example 13.8, the IUH is plotted in Fig. 13.15.

13.3.3 Elliptic TAC Curve

If $w(t)$ is as shown in Fig. 13.8(c),

$$w(t) = \begin{cases} W\left[1 - \left(\frac{2t}{T_c} - 1\right)^2\right]^{0.5}, & 0 \le t \le T_c \\ 0, & t \ge T_c \end{cases} \quad (13.43)$$

then the IUH follows:

$$h(t) = \int_0^t \frac{e^{-(t-\tau)/k}}{k} W\left[1 - \left(\frac{2\tau}{T_c} - 1\right)^2\right]^{.5} d\tau, \qquad 0 \le t \le T_c$$

Let $x = 2\tau/T_c - 1$. Then $d\tau = \frac{1}{2} T_c\, dx$. On substitution, with $z = 2t/T_c - 1$,

$$h(t) = \frac{W}{2k} T_c e^{-(t-.5T_c)/k} \int_{-1}^{z} e^{xT_c/(2k)} (1 - x^2)^{.5}\, dx, \qquad 0 \le t \le T_c \quad (13.44a)$$

Equation (13.44a) is hard to solve explicitly but can be easily solved numerically using standard techniques. However, when $t \ge T_c$, then

$$h(t) = \frac{W}{2k} T_c e^{-(t-.5T_c)/k} \int_{-1}^{1} e^{xT_c/(2k)} (1 - x^2)^{.5}\, dx$$
$$= \pi W e^{-(t-.5T_c)/k} J_1\left(\frac{T_c}{2k}\right), \qquad t \ge T_c \quad (13.44b)$$

where $J_1(T_c/(2k))$ is the modified Bessel function of the first kind of order one with argument $T_c/(2k)$.

On the other hand, $w(t)$ may be arbitrary subject to the following conditions: (1) It is single-valued except at the points of discontinuity. (2) It is finite everywhere; that is, $w(t)$ is bounded. (3) It is positive everywhere except at $t = T_c$, where it may have zero value. (4) It does not have more than a finite number of discontinuities on a finite range. These condi-

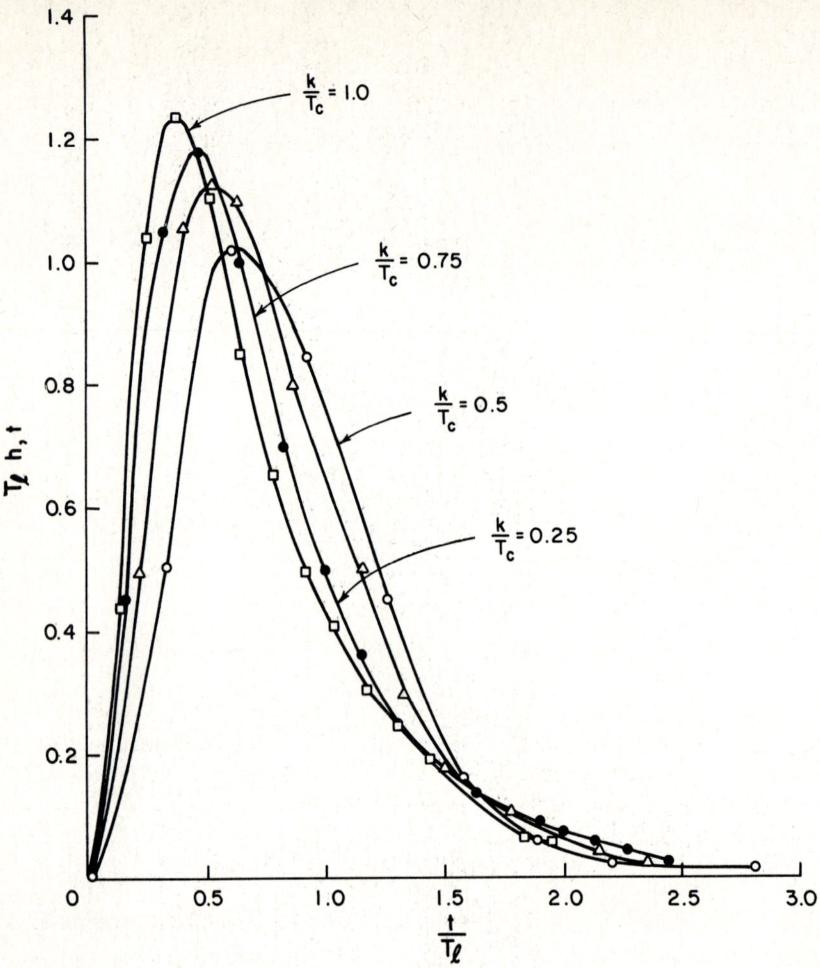

FIGURE 13.15 Example 13.10: The dimensionless IUH, $T_\ell h(t)$, of the Zoch model for a triangular watershed for various values of k/T_c.

tions are satisfied in runoff hydrology. For an arbitrary $w(t)$, equation (13.32) specifies $h(t)$, which can be easily computed numerically.

13.4 THE CLARK MODEL

Clark (1945) suggested that the UH for a watershed due to an instantaneous rainfall can be determined by routing its TAC curve through a single linear reservoir of equations (13.20)–(13.21). Physically, this is equivalent to the Zoch model, but the concept of UH is replaced by one of IUH. Its conceptual framework is shown in Fig. 13.7. Clark applied his method to a hypothetical watershed area of rectangular shape with uniform instantaneous effective rainfall, as shown in Fig. 13.9. If the effective rainfall lasts for a D-hour period, then the D-hour UH by the Clark model is given by convolving equation (13.33) with the D-hour effective rainfall hyetograph and has the appearance as shown in Fig. 13.16. To illustrate, consider $I(t)$ to be uniform in time as

$$I(t) = \begin{cases} \dfrac{1}{D}, & 0 < t < D \\ 0, & t \geq D \end{cases} \tag{13.45}$$

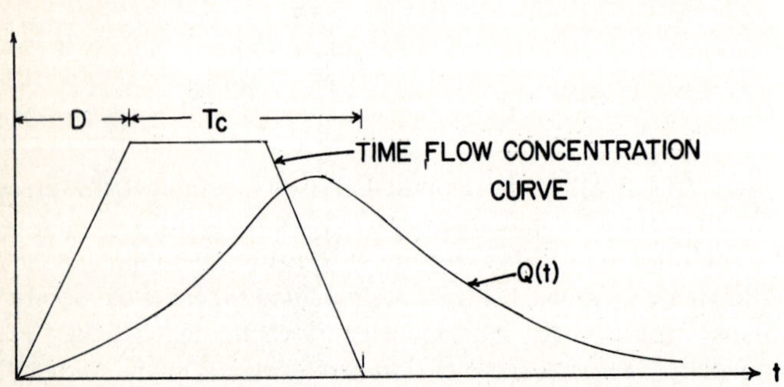

FIGURE 13.16 Direct runoff hydrograph by the Clark model for D-hour effective rainfall.

In convolution, two cases have to be distinguished: (1) $D \leq T_c$, and (2) $D \geq T_c$. Let us first consider case 1. If we disregard the units, then $h(D; t)$ and $Q(t)$ equal each other numerically.

$$h(D; t) = Q(t) = \int_0^t V_0(1 - e^{-(t-\tau)/k}) \frac{1}{D} d\tau, \qquad 0 \leq T \leq D$$

$$= \frac{V_0}{D}[t - k(1 - e^{-t/k})], \qquad 0 \leq t \leq D \tag{13.46a}$$

For $t \geq D$,

$$
\begin{aligned}
h(D; t) &= Q(t) \\
&= \frac{V_0}{D} \int_0^D (1 - e^{-(t-\tau)/k})\, d\tau, \qquad D \le t \le T_c \\
&= \frac{V_0}{D} [D - k(e^{-(t-D)/k} - e^{-t/k})], \\
& \quad D \le T \le T_c
\end{aligned} \tag{13.46b}
$$

Similarly,

$$
\begin{aligned}
h(D; t) &= Q(t) \\
&= \frac{V_0}{D} \int_0^D (1 - e^{-T_c/k}) e^{-(t-\tau-T_c)/k}\, d\tau, \qquad t \ge T_c \\
&= \frac{kV_0}{D} [1 - e^{-T_c/k})(e^{-(t-D-T_c)/k} - e^{-(t-T_c)/k}), \qquad t \ge T_c
\end{aligned} \tag{13.46c}
$$

In case 2,

$$
h(D; t) = Q(t) = \frac{V_0}{D} [t - k(1 - e^{-t/k})], \qquad 0 \le t \le T_c \tag{13.47a}
$$

$$
\begin{aligned}
h(D; t) &= Q(t) \\
&= \frac{V_0}{D} k \left[\frac{T_c}{k} + (e^{T_c/k} - 1) \right. \\
& \quad \left. (1 - e^{-(t-T_c)/k} - e^{-T_c/k}) \right], \qquad T_c \le t \le D
\end{aligned} \tag{13.47b}
$$

and

$$
\begin{aligned}
h(D; t) = Q(t) &= \frac{V_0}{D} k \left[\frac{T_c}{k} + (e^{T_c/k} - 1) \right. \\
& \quad \left. \{e^{-(t-D)/k} - e^{-(t-T_c)/k} - e^{-T_c/k}\} \right], \\
& \quad t \ge D
\end{aligned} \tag{13.47c}
$$

13.5 THE O'KELLY MODEL

O'Kelly (1955) defined the TAC curve by an isosceles triangle, as shown in Fig. 13.17 and routed it through a linear reservoir to produce the IUH for the watershed. Thus the O'Kelly model is equivalent to the Clark model except for the definition of TAC curve. Its schematic is shown in Fig. 13.7. To derive its IUH, let us consider a general triangle whose special case will be an isosceles triangle. That is,

$$
w(t) = \begin{cases} \dfrac{W}{t_1} t, & 0 \le t \le t_1 \\[2ex] -\dfrac{W}{T_c - t_1} t + \dfrac{WT_c}{T_c - t_1}, & t_1 \le t \le T_c \end{cases} \tag{13.48}
$$

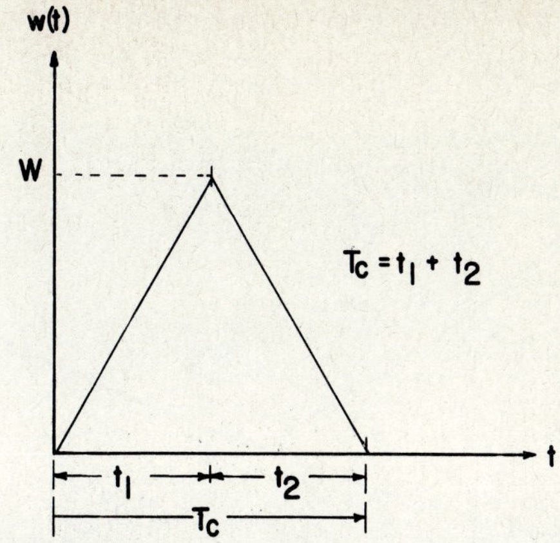

FIGURE 13.17 An isosceles triangle for the TAC curve of the O'Kelly model.

where $W = 2A/T_c$, with A as area of the triangle. Convoluting equation (13.23) with equation (13.49) on a unit area basis,

$$
h(t) = \frac{2}{T_c t_1} (t - k + ke^{-t/k}), \qquad 0 \le t \le t_1 \tag{13.49a}
$$

$$
\begin{aligned}
h(t) &= \frac{2}{T_c t_2} (T_c - t + k) + \frac{2k}{T_c t_1} \left[1 - \left(1 + \frac{t_1}{t_2} \right) e^{t_1/k} \right] e^{-t/k}, \\
& \quad t_1 \le t \le T_c
\end{aligned} \tag{13.49b}
$$

$$
\begin{aligned}
h(t) &= \frac{2k}{T_c} \left[\frac{1}{t_1} + \frac{e^{T_c/k}}{t_2} - \left(\frac{1}{t_1} + \frac{1}{t_2} \right) e^{t_1/k} \right] e^{-t/k}, \\
& \quad t \ge T_c
\end{aligned} \tag{13.49c}
$$

For an isosceles triangle, $t_1 = T_c/2$ and $W = (\sqrt{3}/2)T_c$.

The maximum of $h(t)$, called h_p, occurs when the storage is maximum in equation (13.19), that is, inflow = outflow. Therefore, differentiating equation (13.49b) with respect to t and equating it to zero,

$$
e^{t_p/k} + \frac{t_2}{t_1} = \left(1 + \frac{t_2}{t_1} \right) e^{t_1/k} \tag{13.50}
$$

where t_p is the time to h_p. Further, Dooge (1955) showed that

$$
\frac{h_p}{W} = \frac{T_c - t_p}{t_2} \tag{13.51}
$$

and

$$
\frac{W(t_1 + t_2)}{2} = \frac{WT_c}{2} = V = \text{volume} \tag{13.52}
$$

Dooge (1955) stated that for values of t_p/k less than unity, the values of h_p and t_p would not be materially affected by the ratio of t_2/t_1. For an isosceles triangle, $t_2/t_1 = 1$. Therefore,

$$
\frac{h_p}{W} = 2 \left(1 - \frac{t_p}{T_c} \right) \tag{13.53}
$$

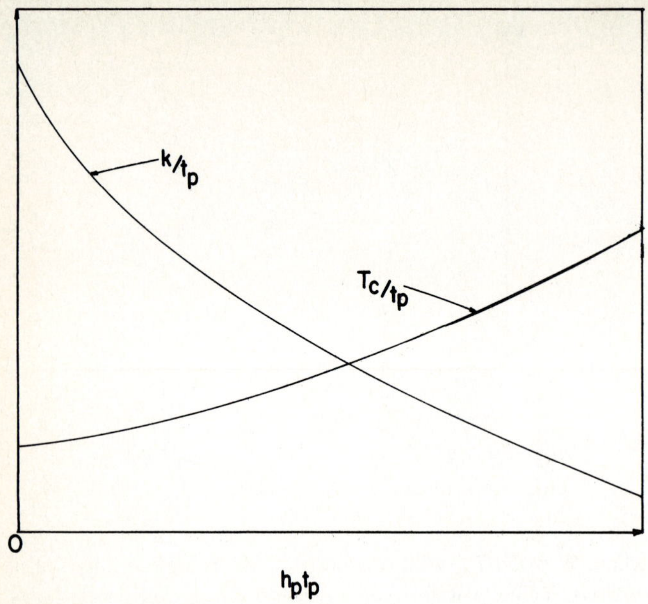

FIGURE 13.18 Graphical solution for the IUH peak and its time for the O'Kelly model.

and

$$e^{t_p/k} + 1 = 2\,e^{T_c/(2k)} \tag{13.54}$$

Therefore,

$$\frac{t_p}{k} = \ln(2\,e^{T_c/(2k)} - 1) \tag{13.55}$$

Combining equations (13.52) and (13.54),

$$\frac{h_p t_p}{4V} = \frac{t_p}{T_c}\left(1 - \frac{t_p}{T_c}\right) \tag{13.56}$$

TABLE 13.2

Variation in dimensionless peak of the IUH with k/T_c (after Dooge, 1959)

$\frac{k}{T_c}$	$h_p T_\ell/V$ Rectangle	$h_p T_\ell/V$ Triangle
.25	.74	1.03
.50	.87	1.02
1.0	.95	1.01
2.0	.98	1.00
3.0	.99	1.00
4.0	1.00	1.00

Equations (13.54)–(13.56) specify h_p and t_p for known k and T_c. Dooge (1955) presented a graphical solution by plotting $h_p t_p$ versus T_c/t_p and k/t_p, as shown in Fig. 13.18. The ratio k/T_c determines the shape of the IUH and is related to the watershed slope (G. B. Williams, 1952). Dooge (1959) showed that

$$h_p T_\ell = V \tag{13.57}$$

$$\frac{t_p}{T_\ell} = 1 - \left(\frac{k}{T_\ell}\right)^2 \tag{13.58}$$

where T_ℓ is the first moment of the IUH or its lag time,

$$T_\ell = \int_0^\infty t h(t)\, dt \tag{13.59}$$

The peak of the dimensionless IUH remains remarkably constant for various values of k/T_c. This is tabulated in Table 13.2.

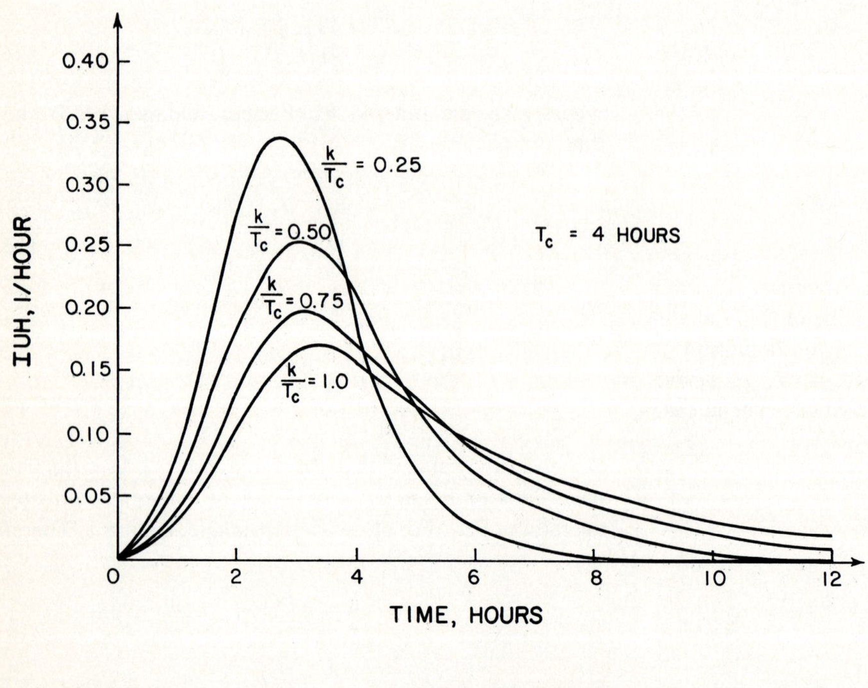

FIGURE 13.19 Example 13.11: The IUH of the O'Kelly model for various values of k/T_c.

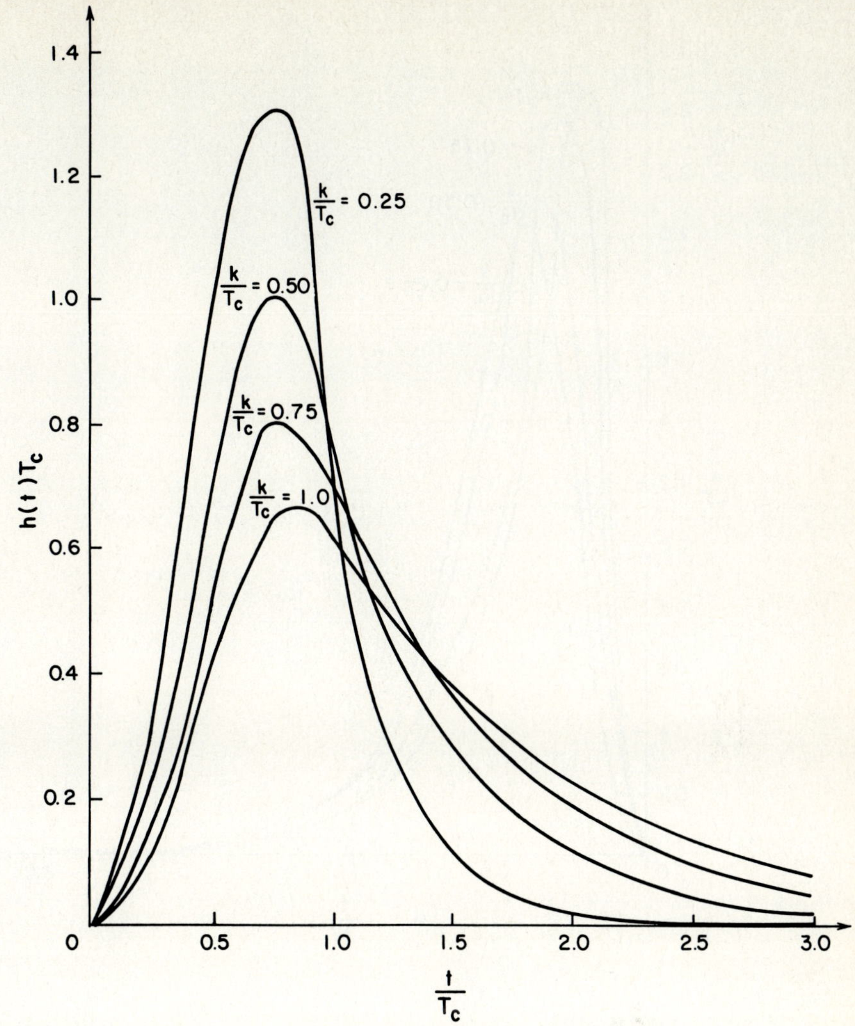

FIGURE 13.20 Example 13.12: The dimensionless IUH, $T_c h(t)$, of the O'Kelly model for various values of k/T_c.

EXAMPLE 13.11

Plot the IUH of the O'Kelly model for various values of k/T_c. For this example, $T_c = 4$ h. Fig. 13.19 shows the resulting IUH.

EXAMPLE 13.12

Plot the IUH of the O'Kelly model for various values of k/T_c as $h(t)T_c$ versus t/T_c. $T_c = 4$ h.

Using the data of Example 13.13, the IUH is plotted in Fig. 13.20.

EXAMPLE 13.13

Plot the IUH of the O'Kelly model for various values of k/T_c as $h(t)T_\ell$ versus t/T_ℓ. $T_c = 4$ h.

Using the data of Example 13.12, the IUH is plotted in Fig. 13.21. $T_\ell = [(\frac{7}{6}) + 2k + 2k^2 + (k^3/2)] - \exp(-2/k)\,[3k^2 + (k^3/2) + 4k]$

EXAMPLE 13.14

Compute the DRH by the O'Kelly model for the data of Example 13.2. The TAC curve has peak at midpoint of its base of 6 h.

From the expressions for the IUH,

$$h(t) = \begin{cases} \dfrac{1}{9}\left[t - 3 - 3\exp\left(\dfrac{-t}{3}\right)\right], & 0 \le t \le 3 \\ 1 - \dfrac{t}{9} - 1.48\exp\left(\dfrac{-t}{3}\right), & 3 \le t \le 6 \\ 0.984\exp\left(\dfrac{-t}{3}\right), & t \ge 6 \end{cases}$$

$$I(t) = 5u_1(t) - 3u_1(t - .5) - u_1(t - 1.0) + u_1(t - 1.5) - 2u_1(t - 2.0)$$

By performing convolution, the DRH is obtained, as shown in Fig. 13.22.

13.6 A LINEAR CHANNEL AND A LINEAR RESERVOIR IN SERIES

This method consists of a linear channel and a linear reservoir in series, as shown in Fig. 13.23. The IUH of this model is

$$h(t) = \frac{e^{-(t-T)/k}}{k} \tag{13.60}$$

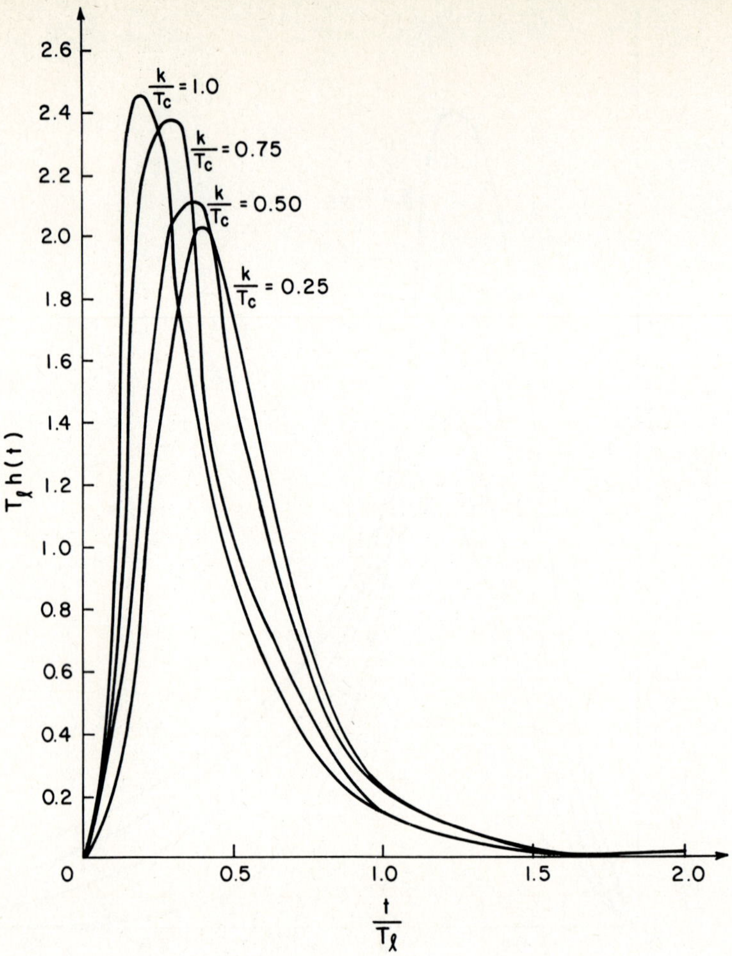

FIGURE 13.21 Example 13.13: The dimensionless IUH, $T_\ell h(t)$, of the O'Kelly model for various values of k/T_c.

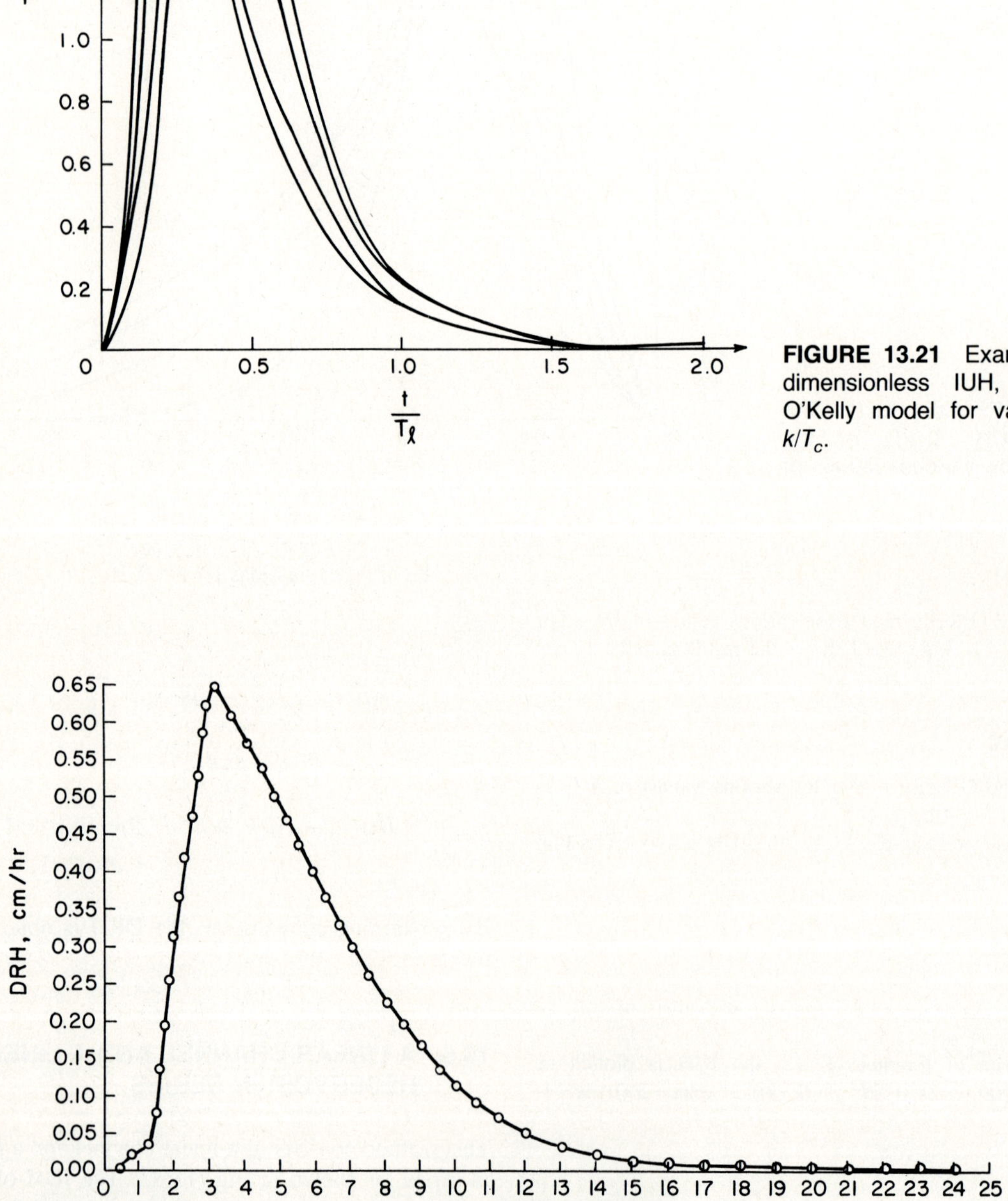

FIGURE 13.22 Example 13.14: The direct runoff hydrograph by the O'Kelly model for the ERH of Example 13.2.

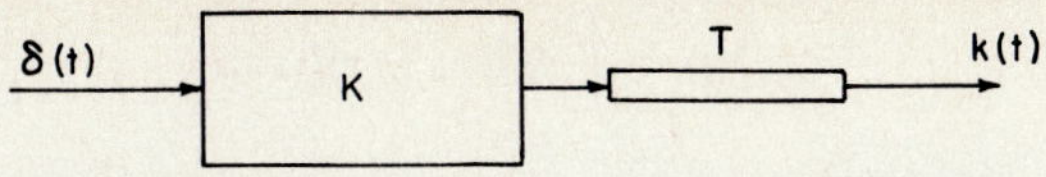

FIGURE 13.23 A linear reservoir and a linear channel in series.

where T denotes translation time of the channel. It may be noted that the order in which these two elements occur is immaterial, since the translation due to the linear channel merely involves the shifting of time scale. Further, the effective rainfall constitutes the input to the model. By convolving equation (13.60) with the effective rainfall, the direct runoff hydrograph can be easily computed. This model can indeed be considered as a special case of any of the Zoch, Clark, and O'Kelly models in that $w(t)$ is replaced by a constant amount of translation. As will be clear from the ensuing discussion, there is a one-to-one correspondence between linear channels and TAC curve.

The moments and cumulants of the IUH expressed by equation (13.60) can be expressed as

$$M_1 = k + T \tag{13.61a}$$

$$M_2 = 2k^2 + 2kT + T^2 \tag{13.61b}$$

$$M_3 = 2k^3 + (k + T)\,[k^2 - 2kT - 2T^2] \tag{13.61c}$$

$$M_4 = 17k^4 + 8k^3T + (k + T)^2\,[8k^2 + T^2 + 2kT) \tag{13.61d}$$

and

$$K_1 = k + T \tag{13.62a}$$

$$K_2 = k^2 \tag{13.62b}$$

$$K_3 = 2k^3 \tag{13.62c}$$

$$K_4 = 6k^4 \tag{13.62d}$$

The DRH due to any $I(t)$ can be obtained by convoluting equation (13.60) with equation (13.17),

$$Q(t) = \sum_{j=0}^{m} w_j u_1(t - jD) \int_{jD}^{t} \frac{1}{k} exp\left[\frac{-(t - T - \tau)}{k}\right] d\tau$$

$$= \sum_{j=0}^{m} w_j u_1(t - jD)\left[1 - \exp\left(\frac{-(t - T - jD)}{k}\right)\right] \tag{13.63}$$

EXAMPLE 13.15

For an arbitrary $I(t)$ derive the DRH for the above model by using the Laplace transform.

$$h(s) = L\left[\frac{\delta(t - T)}{1 + kD}\right] = \exp\frac{(-Ts)}{1 + ks}$$

Therefore,

$$Q(s) = \sum_{j=0}^{m} \frac{w_j \exp(-jDs)\exp(-Ts)}{s(1 + ks)}$$

Taking the inverse Laplace transform,

$$Q(t) = \sum_{j=0}^{m} w_j u_1(t - jD)\left\{1 - \exp\left[\frac{-(t - T - jD)}{k}\right]\right\}$$

which is the same as equation (13.63).

13.7 THE CASCADE OF LINEAR CHANNELS

Mathur (1972a, 1972b) proposed to represent a natural watershed conceptually by a set of linear channels arranged in series. This involves folding a two-dimensional area into a unidimensional translatory scheme. To illustrate, let us assume that travel times from various portions of the watershed to the outlet are known. In other words, its TAC diagram is known. Referring to Fig. 13.24, the travel time from portion (1) to the outlet is marked as T. For a uniform translation time ΔT, the tributary B can be decomposed into three reaches representing the areas ΔA, $\Delta A'$, and $\Delta A''$. Similarly, the main channel and other tributaries can be divided into several reaches each having ΔT translation time. Then subareas of tributary reaches can be superimposed on the subareas of the reaches of main channel having the same translation time. This is shown in Fig. 13.25. These superimposed areas can be construed as subwatersheds or mean delay time areas.

It is thus seen that a linear channel, as used here, functionally represents a specific portion of the watershed area. Its position is determined by the topographic characteristics. Therefore, the order in which these linear channels are arranged is also fixed. It is in this sense that the use of a linear channel made by Mathur (1972a) is different from that by Dooge (1959).

The inflow to a linear channel can come from its upstream channel and its subwatershed area. Distributed runoff from any of the intermediate subwatersheds is assumed to enter the upper bound of its corresponding linear channel, along with the outflow translated down to it from its previous reaches. Translated inflow contributed by previous reaches is defined as primary inflow, whereas distributed runoff from its corresponding subwatershed area reaching the upper bound as secondary inflow.

Mathematically, referring to Fig. 13.26, if we let,

$$dQ(t) = I(\tau)\, dA \tag{13.64}$$

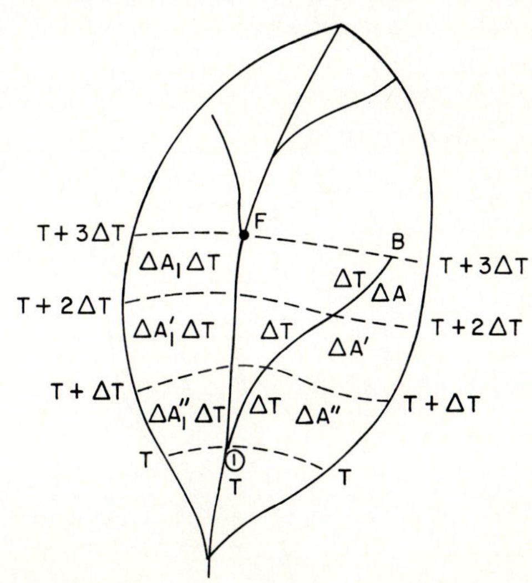

FIGURE 13.24 Delineation of subwatershed areas.

$$Q(t) = \int_0^t h(0, t - \tau) I(\tau)\, d\tau \tag{13.67}$$

Note that equations (13.66) and (13.67) refer to runoff from the same subwatershed for the same prevailing conditions. We can express $h(0, t)$ in terms of the S-hydrograph $U(t)$,

$$U(t) = \int_0^t h(0, s)\, ds \tag{13.68}$$

Then

$$U(t - \tau) = \int_0^{t-\tau} h(0, t - \tau)\, d\tau$$

and

$$U(t - T - \tau) = \int_0^{t-T-\tau} h(0, t - T - \tau)\, d\tau$$

Since the UH ordinates are proportional to differences in ordinates at two positions on $U(t)$ separated by the UH duration T,

$$h(T, t - \tau) = \frac{1}{T}[U(t - \tau) - U(t - T - \tau)]$$

We can choose T as small as we wish. Substituting $U(t)$ in terms of $h(0, t)$ from equation (13.68),

$$h(T, t - \tau) = \frac{1}{T}\left[\int_0^{t-\tau} h(0, t - \tau)\, d\tau - \int_0^{t-T-\tau} h(0, t - T - \tau)\, d\tau\right]$$

Substituting equation (13.66) to replace $h(0, t)$,

$$h(T, t - \tau) = \frac{1}{T}\left[\int_0^{t-\tau} \frac{dA}{d\tau}\, d\tau - \int_0^{t-T-\tau} \frac{dA}{d\tau}\, d\tau\right]$$

Therefore,

$$Th(T, t - \tau) = A_{t-\tau} - A_{t-T-\tau} \tag{13.69}$$

If $T = 1$ hour,

$$h(1, t) = A_t - A_{t-1} \tag{13.70}$$

Equation (13.70) shows that the subwatershed areas represented by linear channels are directly given by the ordinates of the UH.

It is seen that the TAC curve is a response of linear channels arranged in series and parallel as shown in Fig. 13.27. Each channel represents a particular portion of the watershed and has a particular area to drain. If storage effects are neglected, then the two-dimensional flow pattern in the watershed

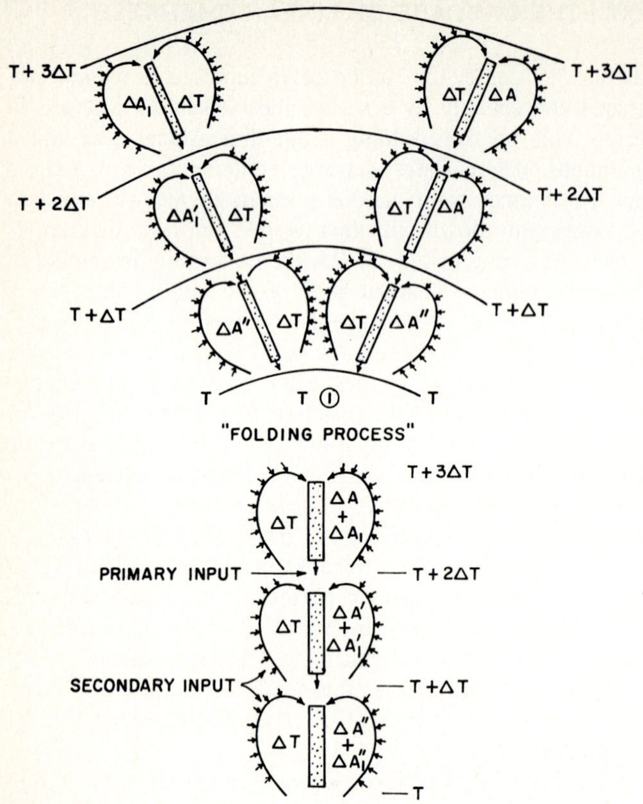

FIGURE 13.25 Equivalent linear channel cascade (after Mathur, 1972).

then

$$Q(t) = \int_0^t I(\tau) \frac{dA}{d\tau}\, d\tau \tag{13.65}$$

Equation (13.65) is the same as equation (13.66) and relates three parameters of the folding process: t refers to the total translation from the confluence section to the endpoint of the influence area for a tributary reach; $d\tau$ is the elementary translation adopted as a basic unit for each linear channel; and dA is the equivalent area represented by the channel reach. Therefore,

$$h(0, t - \tau) = \frac{dA}{d\tau} \tag{13.66}$$

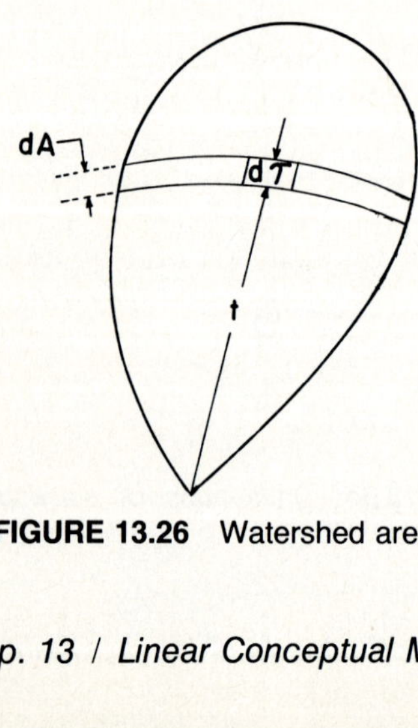

FIGURE 13.26 Watershed area.

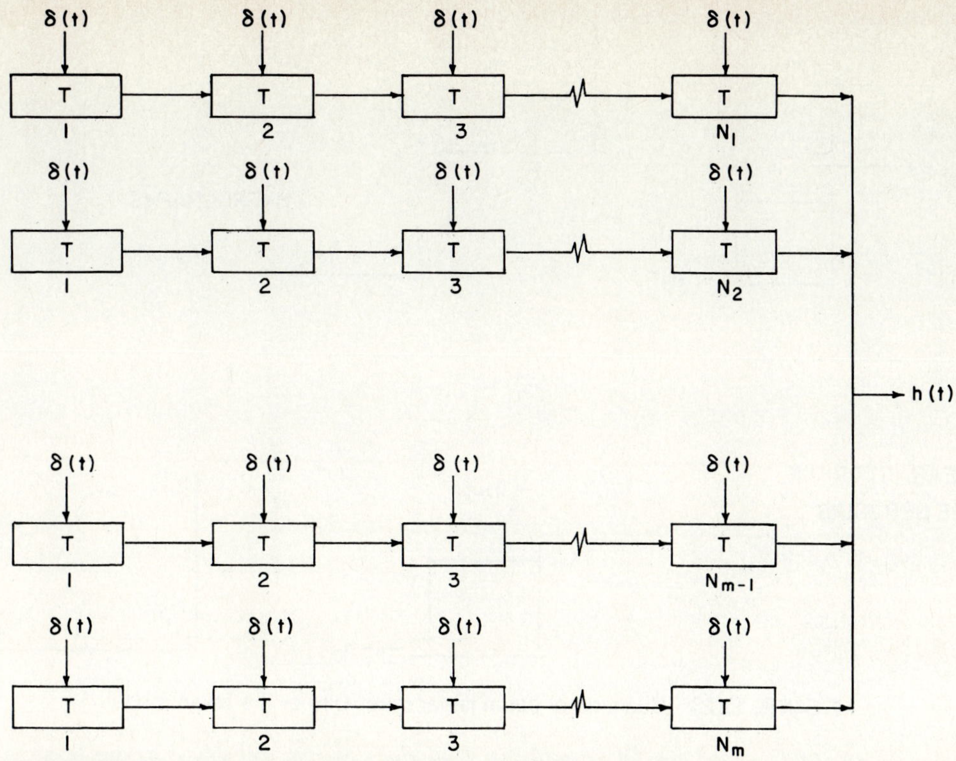

FIGURE 13.27 Cascade of linear channels arranged in parallel. Linear channels may or may not have the same translation time. Each channel represents a particular part of the drainage area.

can be simulated by such an arrangement of linear channels. Thus, any TAC curve can be represented by a particular arrangement of linear channels and vice versa.

13.8 THE NASH MODEL

In a series of papers, Nash (1957, 1958, 1959, 1960) developed a model, as shown in Fig. 13.28, based on a cascade of equal linear reservoirs for derivation of the IUH for a natural watershed. Kalinin and Milyukov (1958) independently derived an identical model for routing of flows in open channels. This is one of the most popular and frequently used models in applied hydrology (Bloomsburg, 1960; Palmer and Johnson, 1964; Shen, 1965; DeCoursey, 1966; Harley, 1967; O'Meara, 1968; Maddaus and Eagleson, 1969; Bravo-S. et al., 1970; Haan, 1970; Wu, 1963, 1969a; Schulz, 1969; Watt, 1971; Schulz, Pinkayan, and Komsartra, 1971. Wang and Wu, 1972; Gupta and Moin, 1974; Gupta, Thongchareon, and Moin, 1974; Serban, 1976; Shelburne and V. P. Singh, 1976; V. P. Singh, 1976; Croley, 1980; Aron and White, 1982). We derive it in three different ways.

13.8.1 Application of Linear Operators

When the instantaneous unit effective rainfall, specified by $\delta(t)$, is fed into the first reservoir then equation (13.21) becomes

$$h_1(t) = \frac{1}{(1 + kD)} \delta(t)$$

For the second reservoir, the inflow is constituted by the outflow from the first reservoir. Therefore,

$$h_2(t) = \frac{1}{(1 + kD)} h_1(t) = \frac{1}{(1 + kD)^2} \delta(t)$$

Likewise,

$$h_n(t) = h(t) = \frac{1}{(1 + kD)^n} \delta(t) \tag{13.71}$$

The solution of equation (13.71), as seen in the next section, is

$$h(t) = \sum_{i=1}^{n} c_i t^{i-1} e^{-t/k} \tag{13.72}$$

where the c_i's are constants to be evaluated using the initial conditions. For each reservoir,

$$h_i(0) = 0, \qquad i = 1, 2, \ldots, n-1 \tag{13.73}$$

Then it is found that

$$c_i = 0, \qquad i = 1, 2, \ldots, n-1 \tag{13.74}$$

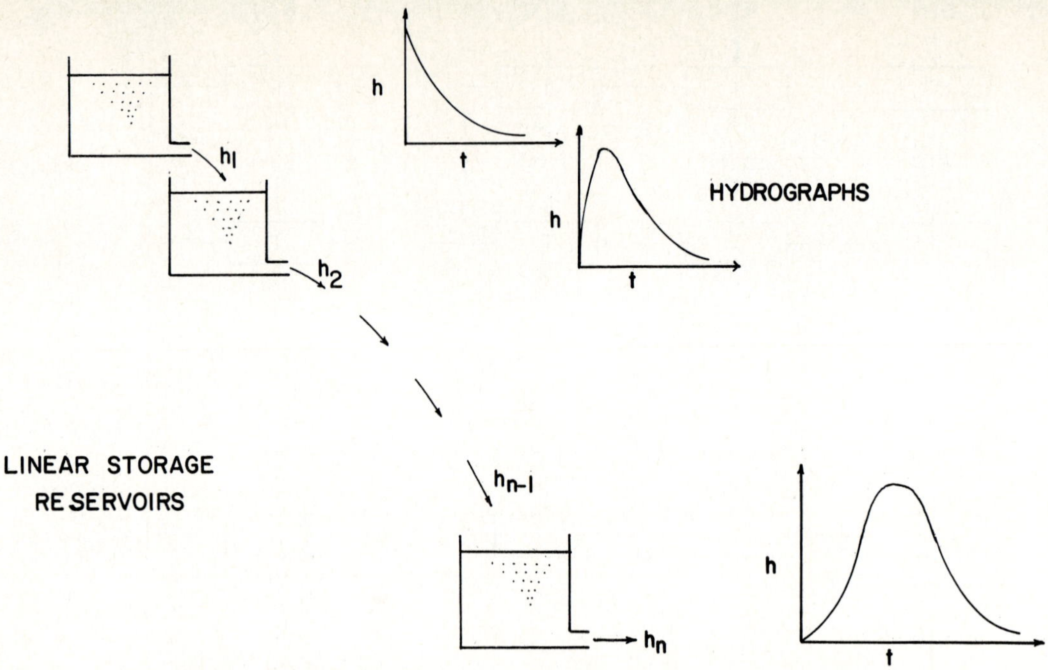

FIGURE 13.28 Cascade of equal linear reservoirs or the Nash model.

and

$$c_n = \frac{1}{(n-1)!k^n} \tag{13.75}$$

Therefore, the IUH of the Nash model is

$$h(t) = \frac{1}{(n-1)!}\left(\frac{t}{k}\right)^{n-1}\frac{e^{-t/k}}{k} = \frac{1}{\Gamma(n)}\left(\frac{t}{k}\right)^{n-1}\frac{e^{-t/k}}{k} \tag{13.76}$$

Note that equation (13.76) is a gamma distribution with parameters n and k, which is extensively tabulated in statistical books. The Nash model can also be derived by using either the convolution theorem or the Laplace transform.

13.8.2 Application of Convolution Theorem

When an instantaneous inflow takes place to the first reservoir, its level is raised by an amount sufficient to accommodate the increased storage and the discharge rises instantaneously from 0 to $1/k$. The IUH and the outflow of the first linear reservoir are given by

$$h_1(t) = \frac{e^{-t/k}}{k}$$

This outflow from the first reservoir constitutes the inflow to the second reservoir, whose outflow is

$$h_2(t) = \int_0^t \frac{1}{k}e^{-\tau/k}\frac{1}{k}e^{-(t-\tau)/k}\,d\tau$$
$$= \frac{t}{k^2}e^{-t/k}$$

Similarly, successively routing through n reservoirs or convoluting n times obtains

$$h_n(t) = h(t) = \frac{1}{k}\frac{1}{(n-1)!}\left(\frac{t}{k}\right)^{n-1}e^{-t/k}$$

which is the same as equation (13.76). This method is also called as the method of successive routing.

13.8.3 Application of Laplace Transform

For a linear storage reservoir we can write

$$h + k\frac{dh}{dt} = \delta(t)$$

Taking the Laplace transform,

$$L\left(h + k\frac{dh}{dt}\right) = L[\delta(t)]$$

$$h(s) + ks\,h(s) = 1$$

$$h(s) = \frac{1}{1+ks} = \frac{1}{k(1/k+s)}$$

The outflow from the first reservoir due to an instantaneous inflow is

$$h_1(s) = \frac{1}{k(1/k+s)}$$

For the second reservoir,

$$h_2 + k\frac{dh_2}{dt} = h_1$$

Then

$$L\left[h_2 + k\frac{dh_2}{dt}\right] = L[h_1]$$

$$h_2(s)\,(1 + ks) = \frac{1}{k(1/k + s)}$$

Thus

$$h_2(s) = \frac{1}{k^2(1/k + s)^2}$$

Similarly for the *n*th reservoir we can write

$$h_n(s) = \frac{1}{k^n(1/k + s)^n}$$

Taking inverse of the Laplace transform,

$$h_n(t) = \frac{1}{k}\left(\frac{t}{k}\right)^{n-1}\frac{e^{-t/k}}{\Gamma(n)}$$

which is the same as equation (13.76)

13.8.4 Some Characteristics of the Nash Model

Equation (13.76) shows that the outflow is expressed by a single term identical to the gamma probability density function. Hence, we can write

$$h(t) = \frac{1}{k}p\left(\frac{t}{k}; n - 1\right) \tag{13.77}$$

where $p(\cdots)$ is the Gamma probability density function, which is extensively tabulated in statistical books.

The time to peak t_p may be obtained by differentiating equation (13.76) with respect to t and putting $(dh/dt) = 0$:

$$t_p = k(n - 1) \tag{13.78}$$

$$h_p = \frac{1}{k(n-1)!}[(n-1)^{n-1}e^{-(n-1)}] \tag{13.79}$$

$$= \frac{1}{k}p(n - 1; n - 1) \tag{13.80}$$

Fig. 13.29 shows a plot of h_p versus n. It can be seen that as n increases, routing a storage of the same lag time has comparatively less peak-reducing effect.

Another important characteristic of storage is observed when a constant lag time T is assumed. We can determine T as

$$T = \frac{\int_0^\infty ht\,dt}{\int_0^\infty h\,dt} \tag{13.81}$$

where T denotes the time between the centers of the areas of inflow and outflow. This lag time is distributed over a number of reservoirs each of lag time k. That is,

$$T = nk \tag{13.82}$$

Each reservoir has a lag time $k = T/n$. While T remains constant, n is increased; that is, the total lag time is distributed over more and more reservoirs. The ultimate lag effect remains T irrespective of the value of n. We can then study the effect on the outflow hydrograph of varying n. Thus, the value of h_p for various n can be used as a measure of change in shape. Fig. 13.29 gives a plot of h_p versus n for $n \geq 2$. From this figure and equation (13.79), it can be seen that as n increases, the peak increases. That is, the peak-reduction effect reduces as T is distributed over more and more reservoirs, which means that the shape-modification effect decreases as the storage is distributed. Because the lag time remains the same, in the extreme case of distributed storage the storage effect tends to become one of translation only. In other words, as $n \to \infty$, the outflow is

$$h_n = \delta[(n - 1)K] \tag{13.83}$$

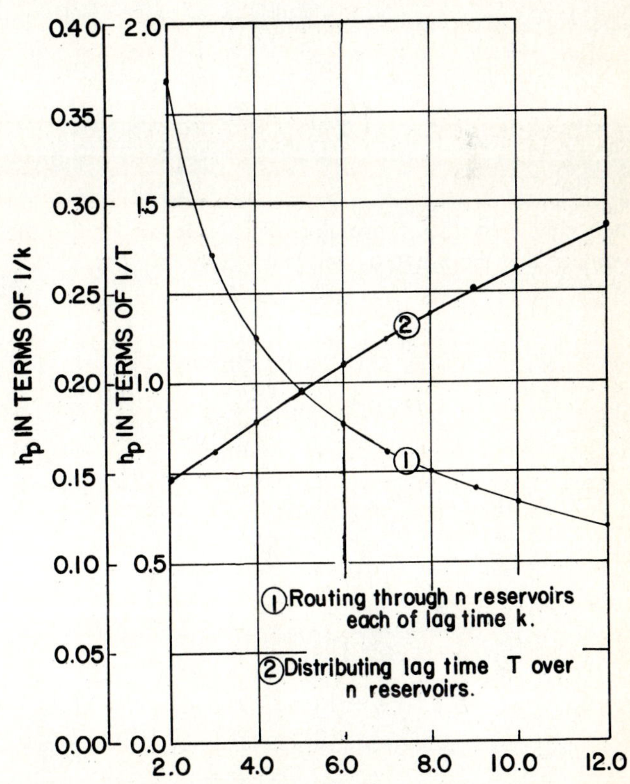

FIGURE 13.29 Routing $\delta(t)$ of the effective rainfall through a cascade of linear reservoirs.

This indicates that if a finite total delay time is divided among a large number of equal linear reservoirs, the effect on an instantaneous inflow is equivalent to translation by an amount equal to the total delay time. Thus, we can perceive the reservoir action as being related to concentrated storage and translatory action as being related to completely distributed storage. We can show this result mathematically. Routing an inflow $f_1(t)$ through reservoirs of lag time T/n to obtain the final outflow $f_2(t)$ can be written as

$$f_2(t) = \frac{1}{\left(1 + \frac{T}{n}D\right)^n} f_1(t) \tag{13.84}$$

Taking the limits,

$$\lim_{n\to\infty} \frac{1}{\left(1 + \frac{T}{n}D\right)^n} = e^{-TD} \tag{13.85}$$

$$\begin{aligned} f_2(t) &= e^{-TD} f_1(t) \\ &= \sum_{j=0}^{\infty} (-1)^j \frac{(TD)^j}{j!} f_1(t) \end{aligned} \tag{13.86}$$

The right side of equation (13.86) is equal to expansion of $f_1(t - T)$ in Taylor series about t. Therefore,

$$f_2(t) = f_1(t - T) \tag{13.87}$$

Hence,

$$\lim_{n\to\infty} \frac{1}{\left(1 + \frac{T}{n}D\right)^n} = H(T) \tag{13.88}$$

which equals the delay operator. The Nash model has the advantage that the S-hydrograph is given by

$$\begin{aligned} U(t) &= \int_0^t h(0, t)\, dt \\ &= \frac{1}{k}\int_0^t \left(\frac{t}{k}\right)^{n-1} \frac{e^{-t/k}}{(n-1)!}\, dt \\ &= \int_0^m \frac{m^{n-1} e^{-m}}{(n-1)!}\, dm, \qquad m = t/k \\ &= \frac{\int_0^m m^{n-1} e^{-m}\, dm}{(n-1)!} \\ &= G\left(\frac{t}{k}; n-1\right) \end{aligned}$$

where $G(\cdot)$ is the ratio of the incomplete to the complete gamma functions. Thus, the ordinates of the finite period UH can be found from the difference between two tabulated values:

$$\begin{aligned} h(D, t) &= \frac{U(t) - U(t-D)}{D} \\ &= \frac{1}{D}\left\{G\left(\frac{t}{k}; n-1\right) - G\left(\frac{t-D}{k}; n-1\right)\right\} \end{aligned}$$

13.8.5 Differential Equation of the Nash Model

From equation (13.71),

$$(1 + kD)^n Q(t) = I(t)$$

Expanding by the binomial series,

$$\left[\sum_{j=0}^{n} \binom{n}{j} (kD)^j\right] Q(t) = I(t)$$

Let $a_j = \binom{n}{j} k^j$. Then

$$\left(\sum_{j=0}^{n} a_j D^j\right) Q(t) = I(t)$$

Therefore,

$$a_n \frac{d^n Q}{dt^n} + a_{n-1} \frac{d^{n-1} Q}{dt^{n-1}} + \cdots + a_1 \frac{dQ}{dt} + a_0 Q = I(t) \tag{13.89}$$

where the coefficient a_i's are constant. This is the general differential equation of a stable highly damped system, which obeys the law of continuity and whose elements do not change in time.

13.8.6 Parameter Estimation

The Nash model has two parameters n and k, which can be estimated in a number of ways as discussed in Chapter 5. The moments and cumulants of the IUH expressed by equation (13.76) can be written as

$$M_1 = nk \tag{13.90a}$$

$$M_2 = nk^2(n + 1) \tag{13.90b}$$

$$M_3 = nk^3(2 + 3n - 2n^2) \tag{13.90c}$$

$$M_4 = nk^4(6 + 11n + 6n^2 - 2n^3) \tag{13.90d}$$

and

$$K_1 = M_1 \tag{13.91a}$$

$$K_2 = nk^2 \tag{13.91b}$$

$$K_3 = 2nk^3 \tag{13.91c}$$

$$K_4 = 6nk^4 \tag{13.91d}$$

In addition, some workers (Haan, 1970; DeCoursey, 1966; Croley,. 1980; Aron and White, 1982; Collins, 1983) have suggested graphical methods to estimate k and n. These are simple and can be programmed on small pocket calculators.

We briefly outline them here. Since k and n are estimated from the characteristics of the gamma function, it may be preferable to consider a direct runoff hydrograph, since this is what is normally available. Therefore, we can write

$$Q(t) = \frac{V}{k\Gamma(n)}\left(\frac{t}{k}\right)^{n-1} \exp[-t/k],$$

$$t > 0, \qquad \int_0^\infty Q(t)dt = V \quad (13.92)$$

Equation (13.92) possesses all the characteristics of the IUH or UH discussed in Chapter 11. Note that if V corresponds to the volume of direct runoff produced by a unit depth of effective rainfall applied uniformly in time and space, then $Q(t)$ will reduce to the UH for the duration of effective rainfall.

Referring to Fig. 13.30, Croley (1980) has argued that any parameter set that involves two equations in n and k may be employed to calculate n and k. The potential parametric sets can be divided into the methods based on point boundary conditions and those based on planar boundary conditions. In the former category are ($Q = Q_p$, $t = T_p$), ($Q = Q_i$, $t = t_i$), and ($Q = Q_a$, $t = t_a$); and in the latter category are ($Q = Q_p$, $t = t_i$), ($Q = Q_p$, $t = t_a$), ($t = t_p$, $t = t_i$), and ($t = t_p$, $t = t_a$). Here Q_p is the peak discharge, t_p is the time to peak discharge, t_i is the time to the inflection point on the recession limb, t_a is the time up to which volume V of direct runoff equals aV with $0 < a < 1$, Q_i is the discharge at $t = t_i$, and Q_a corresponds to the discharge at $t = t_a$. Five of these have been proposed by Croley (1980), and all have been investigated by V. P. Singh and Chowdhury (1985). There may be other parametric sets having practical or physical significance. Based on the preceding sets, four basic sets of conditions can be derived. The first set is

$$Q(t_p) = Q_p \quad (13.93)$$

$$\left.\frac{dQ(t)}{dt}\right|_{t=t_p} = 0 \quad (13.94)$$

The second set is

$$\left.\frac{dQ(t)}{dt}\right|_{t=t_p} = 0 \quad (13.94)$$

$$\left.\frac{d^2Q(t)}{dt^2}\right|_{t=t_i} = 0 \quad (13.95)$$

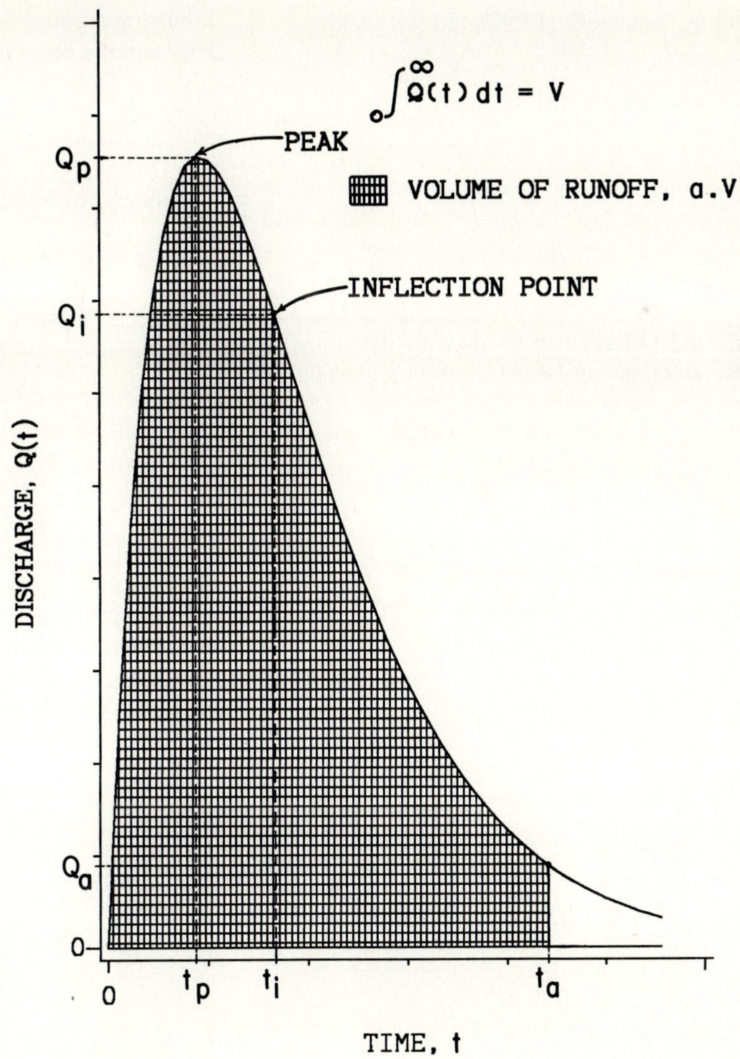

FIGURE 13.30 Direct runoff hydrograph represented by a gamma distribution.

The second condition states that the hydrograph slope stops decreasing and begins increasing at the inflection point. The third set follows:

$$\left.\frac{dQ(t)}{dt}\right|_{t=t_p} = 0 \tag{13.94}$$

$$\int_0^{t_a} Q(t)\, dt = aV \tag{13.96}$$

The fourth set is

$$Q(t_p) = Q_p \tag{13.93}$$

$$\left.\frac{d^2Q(t)}{dt^2}\right|_{t=t_i} = 0 \tag{13.95}$$

By employing these conditions, equations for n and k can be derived from equation (13.93); these are tractable with small programmable calculators. Croley (1980) cautioned against using nomographs or tables because small errors in t_p, Q_p, t_i and t_a can lead to large errors in the values of n and k. However, the findings of Haan (1970), Aron and White (1982), and Collins (1983) do not lend support to this conclusion. We now derive equations for n and k.

Given t_p and Q_p, equations (13.92)–(13.94) yield

$$\frac{t_p}{k} = n - 1, n > 1 \tag{13.96}$$

$$\frac{Q_p\, t_p}{V} = (n-1)^n \exp(1-n)/[k\Gamma(n)]$$
$$= \frac{[\exp(1)/(t_p/k)]^{-t_p/k}}{\Gamma(t_p/k)} \tag{13.97}$$

Equations (13.96) and (13.97) can be used to determine n and k. An iterative solution can be obtained on a calculator.

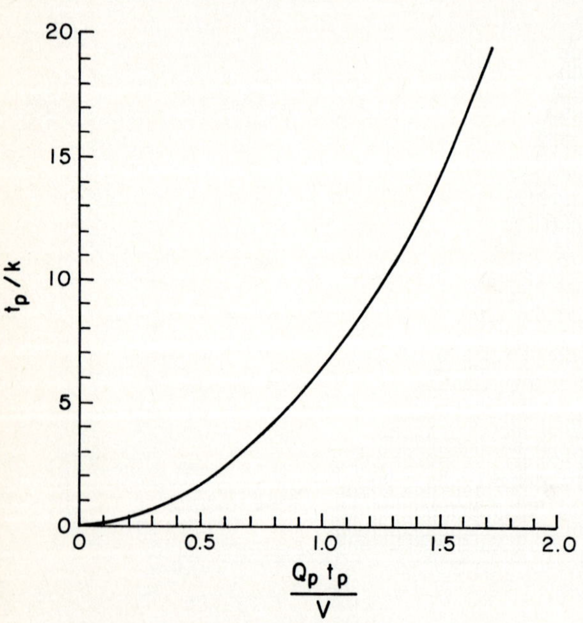

FIGURE 13.31 A graph relating h_pt_p to t_p/k.

Graphical solutions of equation (13.97) for t_p/k versus Q_pt_p/V have been given by Haan (1970) and Haan and Barfield (1978), as shown in Fig. 13.31. Aron and White (1982) expressed the right side of equation (13.97) as

$$G(n) = \frac{(n-1)^n \exp(1-n)}{\Gamma(n)} \tag{13.98}$$

which can be graphed as in Fig. 13.32 and tabulated as in Table 13.3.

If t_p and t_i are given, then equations (13.97) and (13.95) yield

$$(t_i/k)^2 - 2(n-1)\left(\frac{t_i}{k}\right) + (n-1)(n-2) = 0 \tag{13.99}$$

This is a quadratic function having two roots. The smaller and larger roots correspond to the inflection points preceding and succeeding the peak, respectively. The larger root is of interest here; it is given by

$$t_i = k[n - 1 + (n-1)^{.5}] \tag{13.100}$$

If we substitute equation (13.96) into equation (13.100), then a quadratic equation with two roots is obtained; one root is zero and the other is

$$n - 1 = \left(\frac{t_p}{t_i - t_p}\right)^2 \tag{13.101}$$

Equations (13.96) and (13.101) can be used to estimate n and k.

If t_p, t_a, and a are specified, then

$$\int_0^{t_a} \frac{V}{k\Gamma(n)} \left(\frac{t}{k}\right)^{n-1} \exp\left(\frac{-t}{k}\right) dt = aV \tag{13.102}$$

Let $x = t/k$. Then equation (13.92) becomes

$$\frac{1}{\Gamma(n)} \int_0^{t_a} x^{n-1} \exp(-x)\, dx = a \tag{13.103}$$

which is the incomplete gamma function. Equations (13.96) and (13.103) can be used to estimate n and k for specified values of t_p, t_a, and a.

If Q_p and t_i are given, then equations (13.96), (13.97), and (13.100) can be utilized. From equations (13.96) and (13.97),

$$Q_p = V(n-1)^{n-1} \frac{\exp[-(n-1)]}{[k\Gamma(n)]} \tag{13.104}$$

Eliminating k between equations (13.100) and (13.104),

$$\frac{Q_p\, t_i}{V} = \frac{\{[(n-1) + (n-1)^{0.5}]\,(n-1)^{n-1} \exp(1-n)\}}{\Gamma(n)} \tag{13.105}$$

which can be used to compute n.

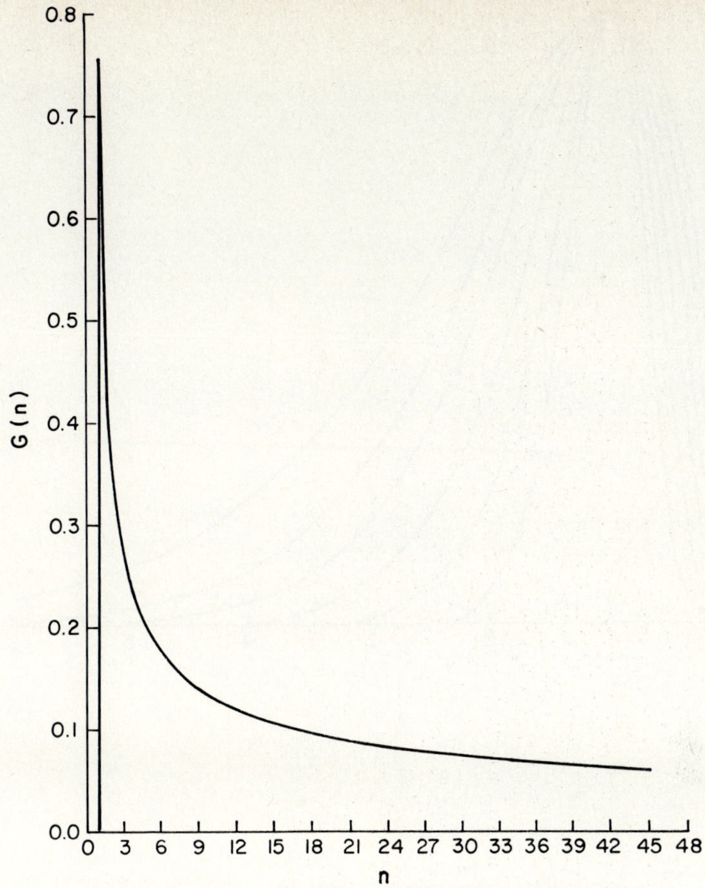

FIGURE 13.32 Function $G(n)$.

If $Q = Q_p$, $t = t_a$, and $V(t = t_a) = aV$ are given, then equations (13.105) and (13.103) can be used to compute n and k. If $Q = Q_i$ and $t = t_i$ are given, then equations (13.100) and (13.92) can be employed to compute n and k. Finally, if $Q = Q_a$ and $t = t_a$ are given, then equations (13.92) and (13.102) can be used to compute n and k.

V. P. Singh and Chowdhury (1985) compared 12 methods of computing n and k. Five of them were the methods of moments, cumulants, maximum likelihood estimation, least squares, and principle of maximum entropy; the remaining seven were the aforementioned methods based on point and planar boundary conditions. They concluded that the five statistical methods were reliable, easier, and efficient and should be preferred. Except for the method based on ($Q = Q_p$, $t = t_p$), all other methods based on boundary conditions were unsatisfactory and should be abandoned for purposes of achieving a mathematical fit.

TABLE 13.2

Values of $G(n)$

n	G(n)	n	G(n)
1.0	.368	4.0	.781
1.5	.463	5.0	.877
2.0	.541	6.0	.964
2.5	.610	7.0	1.043
3.0	.672	8.0	1.117

13.8.7 Dimensionless IUH

Dimensionless forms of the IUH derived from the Nash model have been used for synthesis of direct runoff (Haan, 1960; DeCoursey, 1966; Schultz, 1969; Schulz, Pinkayan, and Komsartra, 1971). Frequently the normalizing quantities have been taken as h_p and t_p or T_ℓ. The dimensionless IUH becomes

$$\frac{h(t)}{h_p} = \left\{\frac{t}{t_p}\exp\left[1 - \left(\frac{t}{t_p}\right)\right]\right\}^{n-1} \tag{13.106}$$

Equation (13.108) contains only one parameter n. Under appropriate circumstances it can be used for regionalization. By computing n and k for one watershed, the dimensionless IUH can be constructed from equation (13.106) and can be used to determine the IUH for other watersheds provided h_p and t_p are known. Soil Conservation Service (1972) used $(n - 1) = t_p/k \cong 4$ to plot dimensionless UH. Fig. 13.33 plots dimensionless IUHs for various values of t_p/k.

13.8.8 The DRH

The DRH for any $I(t)$ is now derived by using the Nash model. We use the Laplace transform here:

$$h(s) = \frac{1}{(1 + ks)^n} \tag{13.107}$$

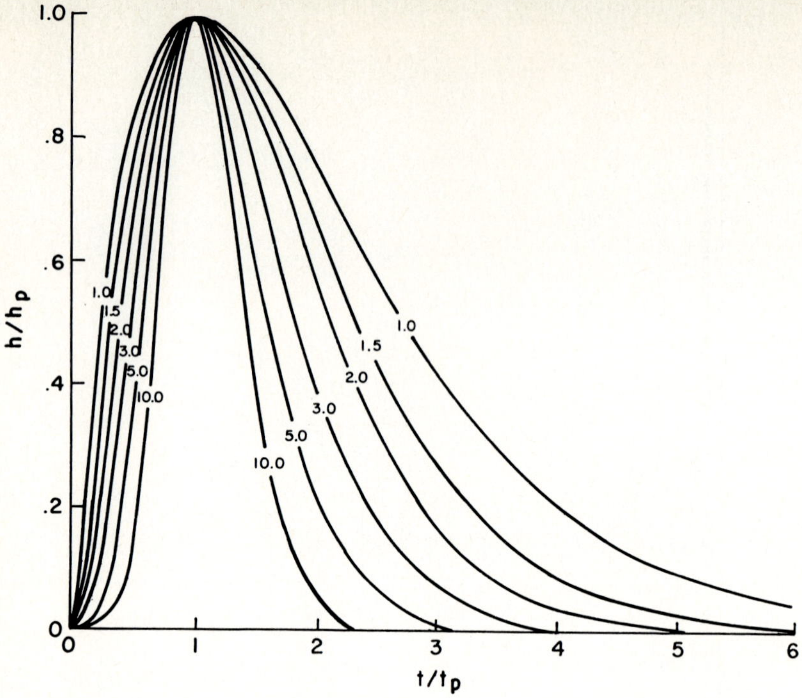

FIGURE 13.33 Dimensionless IUHs by the Nash model for various values of t_p/k.

Therefore,

$$Q(s) = \sum_{j=0}^{m} w_j \exp(-jDs) \frac{1}{s(1+ks)^n} \qquad (13.108)$$

On inverting,

$$Q(t) = \sum_{j=0}^{m} w_j u_1(t-jD) - \sum_{j=0}^{m} w_j u_1(t-jD) \sum_{i=0}^{n-1} \frac{1}{i!}\left(\frac{t-jD}{k}\right)^i \exp\left[\frac{-(t-jD)}{k}\right] \qquad (13.109)$$

EXAMPLE 13.16

Compute the DRH by the Nash model for the ERH in Example 13.2. Take $k = 2$ hours and $n = 2$, 3, and 4. Fig. 13.34 shows the direct runoff hydrographs computed by the Nash model for various values of n.

13.8.8 Inclusion of Translation

The Nash model can easily be extended to include translation T by inserting a linear channel, as shown in Fig. 13.35. The governing equation then becomes

$$h(t-T) = \frac{1}{(1+kD)^n} \delta(t-nT) \qquad (13.110)$$

Let $t^* = t - nT$. Then

$$h(t^*) = \frac{1}{(1+kD)^n} \delta(t^*) \qquad (13.111)$$

$$= \frac{1}{k\Gamma(n)} \left(\frac{t^*}{k}\right)^{n-1} \exp\left(\frac{-t^*}{k}\right) \qquad (13.112)$$

Its moments can be written as

$$M_1 = nT + nk \qquad (13.113a)$$

$$M_2 = (nT + nk)^2 + nk^2 \qquad (13.113b)$$

$$M_3 = 2nk^3 + (nT + nk)[3nk^2 - 2(nT + nk)^2] \qquad (13.113c)$$

Likewise, its cumulants can be specified as

$$K_1 = nK + nk \qquad (13.114a)$$

$$K_2 = nk^2 \qquad (13.114b)$$

$$K_3 = 2nk^3 \qquad (13.114c)$$

$$K_4 = 6nk^4 \qquad (13.114d)$$

It is seen that all save the first cumulant are identical with those of the two-parameter Nash model.

The DRH for any $I(t)$ by this model can be derived directly from equation (13.109) by introducing the translation time,

$$Q(t) = \sum_{j=0}^{m} w_j u_1(t-jD) - \sum_{j=0}^{m} w_j u_1(t-jD) \sum_{i=0}^{n-1} \frac{1}{i!}\left(\frac{t-nT-jD}{k}\right)^i \exp\left[-\frac{(t-nT-jD)}{k}\right] \qquad (13.115)$$

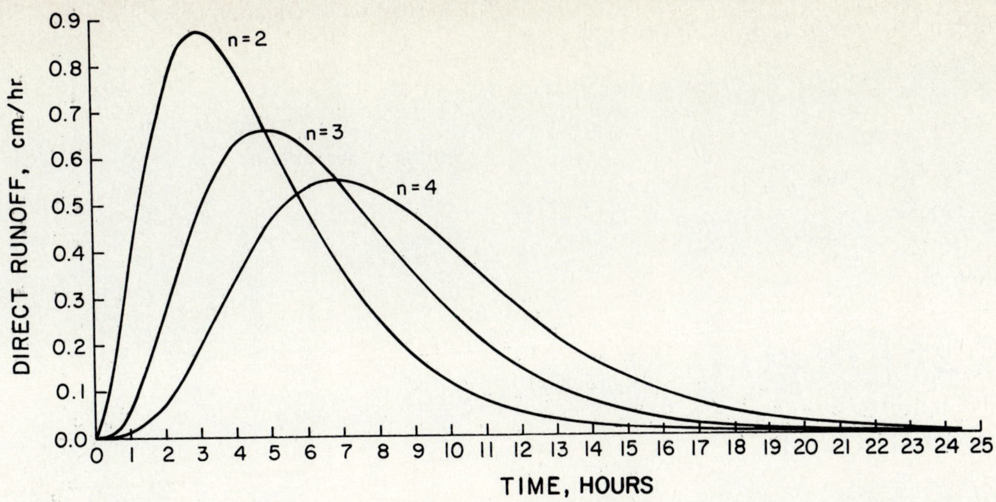

FIGURE 13.34 Example 13.16: The direct runoff hydrograph by the Nash model for the ERH in Example 13.2. k = 2 hrs.

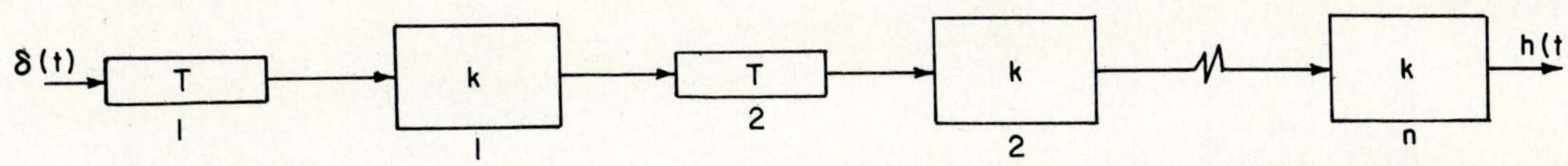

FIGURE 13.35 A cascade of equal linear reservoirs and channels arranged in series.

13.9 THE CASCADE OF UNEQUAL LINEAR RESERVOIRS

This model, as shown in Fig. 13.36, consists of a number of unequal linear reservoirs in series, where the outflow from one reservoir forms inflow to the next. We can derive its IUH as follows:

$$S_i = k_i Q_i, \qquad i = 1, 2, \ldots, n \tag{13.116}$$

For $n = 1$ we can write

$$Q_1 + k_1 \frac{dQ_1}{dt} = I_1$$

Then

$$(1 + k_1 D)Q_1 = I_1$$

When $n = 2$,

$$Q_2 + k_2 \frac{dQ_2}{dt} = I_2$$

$$I_2 = \frac{I_1}{1 + k_1 D} = Q_1$$

Thus

$$(1 + k_1\text{D})\,(1 + k_2 D)Q_2 = I_1$$

Similarly,

$$Q_n = \frac{I_1}{\prod_{i=1}^{n} (1 + k_i D)} = \frac{I_1}{\prod_{i=1}^{n} k_i(1/k_i + D)} \tag{13.117}$$

$$h_n(t) = h(t) = \frac{\delta(t)}{\prod_{i=1}^{n} (1 + k_i D)}, \qquad t > 0 \tag{13.118}$$

This can be easily solved since the inverse operator is already divided into factors. The solution of equation (13.118) can be written as

$$h(t) = \sum_{i=1}^{n} C_i e^{-t/k_i} \tag{13.119}$$

The problem then reduces to determining the coefficients C_i's. To that end, we can utilize the volume continuity to generate n equations, whose solution gives C_i's. Since the flow from any one of the reservoirs decreases to zero after an infinite time, we can write for each reservoir,

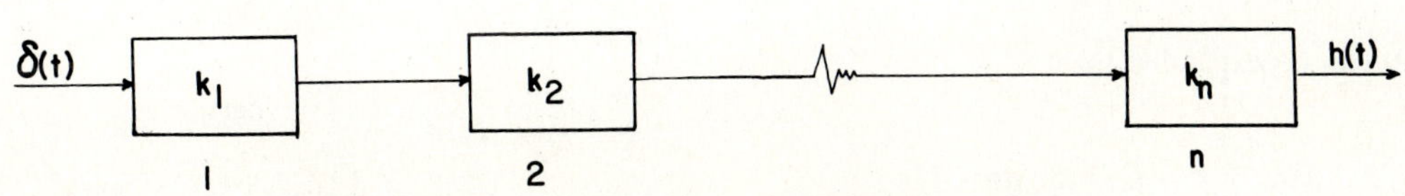

FIGURE 13.36 A cascade of unequal linear reservoirs.

$$\int_0^\infty h_i \, dt = 1, \qquad i = 1, 2, 3, \ldots, n \tag{13.120}$$

That is,

$$\int_0^\infty \sum_{i=1}^n C_i e^{-t/k_i} \, dt = 1 \tag{13.121}$$

which can be written as

$$\sum_{i=1}^n C_i \int_0^\infty e^{-t/k_i} \, dt = 1 \tag{13.122}$$

The solution of equation (13.122) gives

$$\sum_{i=1}^n C_i k_i = 1 \tag{13.123}$$

Similarly, we can write

$$\int_0^\infty h_{n-1} \, dt = 1 \tag{13.124}$$

Equation (13.118) can be written in alternative form

$$\int_0^\infty (1 + k_n D) h_n \, dt = 1 \tag{13.125}$$

Thus

$$\int_0^\infty h_n \, dt + \int_0^\infty k_n \frac{d}{dt} h_n \, dt = 1 \tag{13.126}$$

Then

$$\sum_{i=1}^n C_i k_i + k_n [h_n] \Big|_0^\infty = 1$$

This reduces to

$$\sum_{i=1}^n C_i = 0 \tag{13.127}$$

Similarly,

$$\int_0^\infty h_{n-2} \, dt = 1 \tag{13.128}$$

$$\int_0^\infty (1 + k_{n-1} D)(1 + k_n D) \, h_n \, dt = 1 \tag{13.129}$$

Equation (13.129) can be expanded to

$$\int_0^\infty (h_n + k_{n-1} D h_n + k_n D h_n + k_{n-1} k_n D^2 h_n) \, dt = 1 \tag{13.130}$$

The solution of equation (13.130) gives

$$\sum_{i=1}^n \frac{C_i}{k_i} = 0 \tag{13.131}$$

Similarly, we obtain

$$\sum_{i=1}^n \frac{C_i}{k_i^2} = 0 \tag{13.132}$$

$$\sum_{i=1}^n \frac{C_i}{k_i^3} = 0 \tag{13.133}$$

$$\vdots$$

$$\sum_{i=1}^n \frac{C_i}{k_i^{n-2}} = 0 \tag{13.134}$$

Equations (13.127) and (13.131)–(13.134) can be written as

$$\sum_{i=1}^n C_i / k_i^j = 0, \qquad j = 0, 1, \ldots, n-2 \tag{13.135}$$

On solving these equations, the general equation can be written as

$$h(t) = \frac{k_1^{n-2} e^{-t/k_1}}{\prod_{i=2}^{n} (k_1 - k_i)} + \frac{k_1^{n-2} e^{-t/k_2}}{\prod_{\substack{i=1 \\ i \neq 2}}^{n} (k_2 - k_i)} + \cdots + \frac{k_1^{n-2} e^{-t/k_n}}{\prod_{i=1}^{n-1} (k_n - k_i)} \tag{13.136}$$

We can write equation (13.136) in a more compact form,

$$h(t) = \sum_{j=1}^n \frac{k_j^{n-2} e^{-t/k_j}}{\prod_{\substack{i=1 \\ i \neq j}}^{n} (k_j - k_i)} \qquad \text{for } n \geq 2 \tag{13.137}$$

Equation (13.137) shows that the outflow from a cascade of n unequal linear reservoirs can be expressed as a sum of n terms, each of which is a simple exponential decay curve. This equation, however, is difficult to use. If we were to assume that some of the n linear reservoirs are equal, repeated roots would be obtained. This complicates some of the individual terms without reducing the number of terms in the series.

13.9.1 A Special Case: $k_j = r^{j-1} k, j = 1, 2, \ldots, n$

If the storage is constant for the series of reservoirs and decreases in geometric proportion, that is,

$$k_j = k \, r^{j-1}, \qquad j = 1, 2, \ldots, n \tag{13.138}$$

Then,

$$h_n(t) = \frac{\delta(t)}{\prod_{i=1}^n (1 + k \, r^{i-1} D)}, \qquad t > 0 \tag{13.139}$$

This can be solved as before. From equation (13.137) it can be obtained as

$$h(t) = \frac{1}{k^{n-3}} \sum_{j=1}^{n} \frac{(r^{j-1})^{(n-2)}}{\sum\limits_{\substack{i=1\\ j\neq 1}}^{n} (r^{j-1} - r^{i-1})}, \qquad \text{for } n \geq 2 \tag{13.140}$$

Now the DRH can be derived for any $I(t)$. Let us use the Laplace transform. Since

$$h(s) = \frac{1}{\prod\limits_{i=1}^{n} (1 + kr^{i-1}s)}$$

we obtain

$$\begin{aligned} Q(s) &= \frac{1}{\prod\limits_{i=1}^{n} (1 + kr^{i-1}s)} \sum_{j=0}^{m} w_j \exp(-jDs)\frac{1}{s} \\ &= \sum_{j=0}^{m} w_j \frac{1}{\prod\limits_{i=1}^{n} (1 + kr^{i-1}s)} \exp(-jDs)\frac{1}{s} \end{aligned} \tag{13.141}$$

For purposes of analysis, let us suppose that $m = 1$. Then the first pulse leads to

$$\frac{w_0}{s\prod\limits_{i=1}^{n}(1 + kr^{i-1}s)} = w_0\left[\frac{c_0}{s} + \frac{c_1}{1 + ks} + \frac{c_2}{1 + krs} + \frac{c_3}{1 + kr^2s} + \cdots\right]$$

$$c_0 = \left.\frac{1}{s\prod\limits_{i=1}^{n}(1 + kr^{i-1}s)} s\right|_{s=0} = 1$$

$$\begin{aligned} c_1 &= \left.\frac{1}{s\prod\limits_{i=1}^{n}(1 + kr^{i-1}s)}(1 + ks)\right|_{s=-1/k} \\ &= -\frac{k}{\prod\limits_{i=1}^{n}(1 - r^{i-1})}, \qquad i \neq 1 \end{aligned}$$

$$\begin{aligned} c_2 &= \left.\frac{1}{s\prod\limits_{i=1}^{n}(1 + kr^{i-1}s)}(1 + krs)\right|_{s=-1/(kr)} \\ &= -\frac{kr}{\prod\limits_{i=1}^{n}(1 - r^{i-1})}, \qquad i \neq 2 \end{aligned}$$

$$\vdots$$

$$\begin{aligned} c_n &= \left.\frac{1}{s\prod\limits_{i=1}^{n}(1 + kr^{i-1}s)}(1 + kr^{n-1}s)\right|_{s=-1/(kr^{n-1})} \\ &= -\frac{kr^{n-1}}{\prod\limits_{i=1}^{n}(1 - r^{i-1})}, \qquad i \neq 2 \end{aligned} \tag{13.142}$$

Therefore,

$$Q(s) = \frac{w_0}{s\prod\limits_{i=1}^{n}(1 + kr^{i-1}s)} = 1 - \sum_{j=1}^{n} \frac{kr^{j-1}}{\prod\limits_{\substack{i=1\\ i\neq j}}^{n}(1 - r^{i-1})} \frac{1}{(1 + kr^{j-1}s)} \tag{13.143a}$$

Taking the inverse Laplace transform,

$$Q(t) = w_0u_1(t) - \sum_{j=1}^{n}\left[\frac{w_0u_1(t)}{\sum\limits_{\substack{i=1\\ i\neq j}}^{n}(1 - r^{i-1})} \exp\left(-\frac{t}{kr^{j-1}}\right)\right]$$

If $m > 1$, then equation (13.143a) can be generalized to

$$Q(t) = \sum_{x=0}^{m} w_x u_1(t - xD) - \sum_{j=1}^{n}\left[\frac{1}{\prod\limits_{\substack{i=1\\ i\neq j}}^{n}(1 - r^{i-j})} \sum_{x=0}^{m} w_x u_1(t - xD) \exp\left(-\frac{(t - xD)}{kr^{j-1}}\right)\right] \tag{13.143b}$$

EXAMPLE 13.17

Determine the DRH for the cascade of n unequal reservoirs due to the ERH of Example 13.2. Take $n = 3$, $r = .5$, and $k = 2$ h.

$$\begin{aligned} I(t) &= 5[u_1(t) - u_1(t - .5)] + 2[u_1(t - .5) - u_1(t - 1)] \\ &\quad + [u_1(t - 1) - u_1(t - 1.5)] + 2[u_1(t - 1.5) - u_1(t - 2)] \\ &= 5u_1(t) - 3u_1(t - .5) - u_1(t - 1) + u_1(t - 1.5) - 2u_1(t - 2) \end{aligned}$$

$$h(t) = \frac{4}{3}\exp\left(\frac{-t}{2}\right) - 2\exp(-t) + \frac{2}{3}\exp(-2t)$$

On performing convolution,

$$\begin{aligned} Q(t) &= 5u_1(t)\int_0^t h(t - s)\,ds - 3u_1(t - 0.5)\int_{0.5}^t h(t - s)\,ds \\ &\quad - u_1(t - 1)\int_1^t h(t - s)\,ds + u_1(t - 1.5)\int_{1.5}^t h(t - s)\,ds \\ &\quad - 2u_1(t - 2)\int_2^t h(t - s)\,ds \end{aligned}$$

On inserting $h(t)$ and simplifying,

$$\begin{aligned} Q(t) &= 5u_1(t)\left\{\frac{8}{3}\left[1 - \exp\left(\frac{-t}{2}\right)\right] - 2[1 - \exp(-t)] + \frac{1}{3}[1 - \exp(-2t)]\right\} \\ &\quad - 3u_1(t - 0.5)\left(\frac{8}{3}\left\{1 - \exp\left[\frac{-(t - 0.5)}{2}\right]\right\}\right. \\ &\quad \left. - 2\{1 - \exp[-(t - 0.5)]\} + \frac{1}{3}\{1 - \exp[-2(t - 0.5)]\}\right) \\ &\quad - u_1(t - 1)\left(\frac{8}{3}\left\{1 - \exp\left[\frac{-(t - 1}{2}\right]\right\}\right. \\ &\quad - 2\{1 - \exp[-(t - 1)]\} \end{aligned}$$

$$+\frac{1}{3}\{1-\exp[-2(t-1)]\}\Big)$$

$$+u_1(t-1.5)\left(\frac{8}{3}\left\{1-\exp\frac{[-(t-1.5]}{2}\right)\right\}$$

$$-2\{1.0-\exp[-(t-1.5)]\}$$

$$+\frac{1}{3}\{1.0-\exp[-2(t-1.5)]\}\Big)$$

$$-u_1(t-2)\left(\frac{8}{3}\left\{1-\exp\left[-\frac{(t-2)}{2}\right]\right\}\right.$$

$$-2\{1.0-\exp[-(t-2)]\}$$

$$+\frac{1}{3}\{1-\exp[-2(t-2)]\}\Big)$$

For $t \geq 2$, this simplies to

$$Q(t) = 10.187\exp\left(\frac{-t}{2}\right) - 25.94\exp(-t) + 33.22\exp(-2t)$$

Fig. 13.37 shows the IUH and the resulting DRH.

13.10 THE DOOGE MODEL

Dooge (1959) developed a general unit hydrograph theory, which embraced all previous models as its special cases. The three elements TAC curve, linear channel, and linear reservoir were included in the theory. The basic premise of the Dooge model is that a watershed can be represented by some combination of linear channels and reservoirs. The watershed is drained by a network of channels composed of a complex network of linear channels and linear reservoirs placed in series. Following Dooge (1959) we derive his model as follows.

Consider an elemental area in the watershed. Since the watershed is assumed to be linear, the distribution of any one portion can be considered independently. To reach the outlet the effective rainfall falling on the elemental area must pass through a linear reservoir representing either overland flow or interflow and then pass through a chain of linear channels and linear reservoirs representing the watershed system between the elemental area under consideration and the outlet. Since the order of operation is immaterial, it can be considered that the flow passes through a succession of linear channels and then through the series of linear reservoirs appropriate to the portion of the watershed under consideration. Thus, the outflow ΔQ due to the effective rainfall $I(t)$ on an element of area ΔA will be given by

$$\Delta Q(t) = \frac{1}{\prod_{j=1}^{n}(1+k_jD)}[I(t-\tau)]\,\Delta A \tag{13.142}$$

where τ denotes the total translation time between the element and the outlet and k_j is the storage lag time of the jth reservoir between the element and the outlet. Then the contribution of ΔA to the IUH is given by

$$\Delta h(0, t) = 0, \qquad t < 0$$

$$\Delta h(0, t) = \frac{1}{\prod_{j=1}^{n}(1+k_jD)}\delta(t-\tau)\frac{\Delta A}{A} \tag{13.143}$$

For specified τ and k_j's, it can compute $\Delta h(0, t)$.

The total runoff from the watershed after a time t will clearly be the sum of the contributions of the individual elements of the area. It goes without saying that these areas, when the total translation time T is greater than the elapsed time τ since the start of the effective rainfall, will contribute

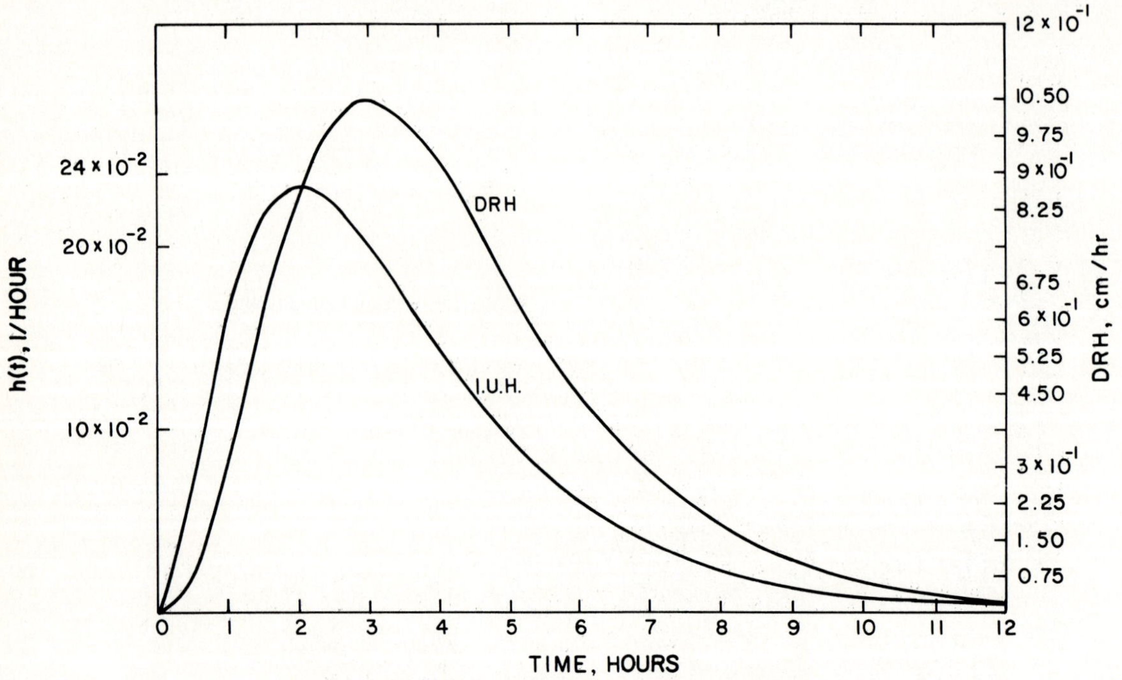

FIGURE 13.37 Example 13.17: IUH and DRH.

nothing to the runoff at the outlet, since the initial effect of the effective rainfall will still be in transit. Summation over the area that is contributing gives the following expression for IUH:

$$h(0, t) = \sum_{A(0)}^{A(t)} \frac{1}{\prod_{j=1}^{n(A)} (1 + k_j D)} \delta(t - \tau)\, I(A) \frac{\Delta A}{A} \tag{13.144}$$

where $I(A)$ is the ratio of local rainfall intensity to the average of the watershed. If the integral is taken in the Stieltjes sense to allow for discontinuities, we can write

$$h(0, t) = \frac{1}{A} \int_0^{A(t)} \frac{\delta(t - \tau)}{\prod_{j=1}^{n(A)} (1 + k_j D)} I(A)\, \mathrm{d}A \tag{13.145}$$

The above equation is the general equation derived by Dooge (1959) for the UH of an ideal linear watershed in which translatory action and reservoir action are separated. It is reasonable to suppose that any watershed suitable for UH analysis can be represented by an equivalent ideal linear watershed consisting of linear channels and linear reservoirs arranged in order to give the same UH as the natural watershed to the required degree of accuracy.

The exact solution of the general equation requires the evaluation for each element of the watershed area of the n-term series (generated by the n reservoirs between the element and the outlet) for all the values of $(t - \tau)$ and then the combination of the contributions from each element to obtain the complete UH. From a practical point of view, the accuracy of the basic data renders the tedium of such a computation pointless, and in any case the accuracy required in applied hydrology makes this tedium unnecessary. Since the IUH, itself an integral, is integrated to obtain the actual hydrograph, small differences in the watershed characteristics are not reflected in the hydrograph of direct runoff due to the double smoothing involved. This suggests that the general equation of the UH can be made tractable by carefully making physical assumptions without undue loss of accuracy.

It is obviously desirable to transform the general equation from a surface integral to a simple integral. This can be done as follows. For purposes of solving the general equation, any point in the watershed can be characterized by its translation time T from the outlet and the set of reservoirs with lag time $k_1, k_2, \ldots, k_n$ lying between the point and the outlet. Imagine a contour joining all points in the watershed that are separated from the outlet by the same translation time. In general, it is possible for the points on an individual isochrone to have different characteristic reservoir chains lying between the point and outlet. This is illustrated for a simple case in Fig. 13.38. Natural watersheds conform to certain laws of equilibrium (Horton, 1945). Thus, the assumption that all points on the same isochrone have the same reservoir chain would not be unduly restrictive. This assumption is shown in Fig. 13.39. It is clear that all the tributaries can be folded onto the main stream to give a single chain of reservoirs, as shown in Fig. 13.40, the inflow at any point being proportional to the length of the isochrone ($dA/d\tau$) cutting the main stream at that point. In order to take advantage of this assumption for spatial variability of rainfall, it is necessary to average the rainfall intensity along each isochrone so that the intensity varies only with distance from the outlet.

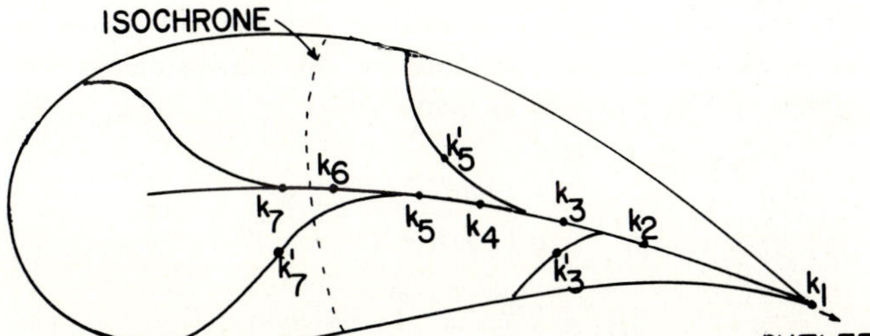

FIGURE 13.38 General distribution of reservoirs (after Dooge, 1959).

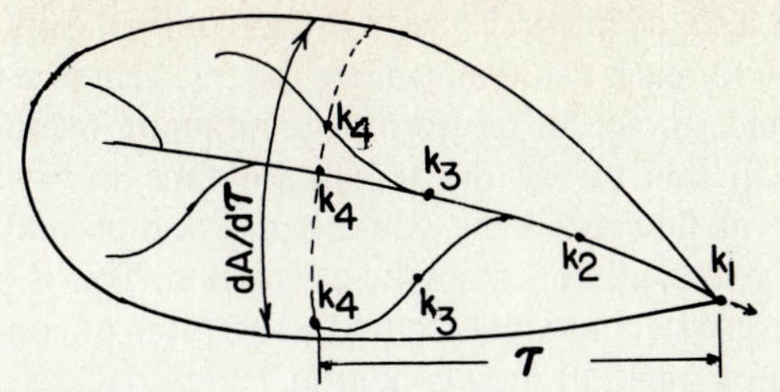

FIGURE 13.39 Uniform distribution of reservoirs (after Dooge, 1959).

The general equation can then be modified as

$$h(0, t) = \frac{1}{A} \int_0^{t \le T} \frac{\delta(t - \tau)}{\prod_{j=1}^{n(\tau)} (1 + k_j D)} I(A) \frac{dA}{d\tau}\, d\tau \tag{13.146}$$

where $dA/d\tau$ denotes the length of the isochrone (that is, the ordinate of the TA curve) and T is the time of concentration (that is, the maximum translation time in the watershed). This can be written further as

$$h(0, t) = \frac{1}{T} \int_0^{t \le T} \frac{\delta(t - \tau)}{\prod_{j=1}^{n(\tau)} (1 + k_j D)} w\left(\frac{\tau}{T}\right) d\tau \tag{13.147}$$

where $w(\tau/T)$ denotes the ordinate of the dimensionless TAC curve, adjusted for the variation in rainfall intensity. In compact dimensionless form,

$$h(0, t) = \frac{1}{T} \int_0^{t' \le 1} \frac{\delta(t' - \tau')}{\prod_{j=1}^{n(\tau')} (1 + k_j D)} w(\tau')\, d\tau' \tag{13.148}$$

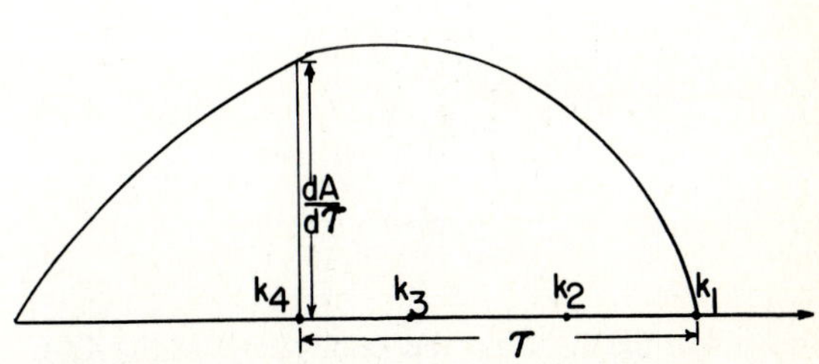

FIGURE 13.40 Folding of tributaries onto main river (after Dooge, 1959).

which is a single integral since the chain of reservoirs is defined uniquely for each value of τ. Here the translation may include distributed storage in the form of continuing overbank flow. Hence, in this theory the translation time is based on the velocity of flow from the whole cross section and does not exclude overbank flow from the calculation. This is in contrast with the classical rational method where time of concentration is based on bankfull flow condition.

The integral given above can be further simplified by assumptions in relation to the TAC curve. A complex $w(t)$-curve can be broken down into simpler elements for which solutions are easily found, and these solutions can be recombined to give the complex solution for the complex curve. Or more generally, a TAC curve can be expressed in terms of step functions, as done in case of the ERH, that is,

$$w(\tau) = \sum_{j=0}^{m} z_j u_1(\tau - j\Delta\tau) \tag{13.149}$$

where z_is have the same interpretation as w_j's have for the ERH. Then the IUH can be given as

$$h(t') = \frac{1}{T}\sum_{j=0}^{m} z_i u_1(\tau' - j\Delta\tau) \int_0^{t'} \frac{\delta(t' - \tau')}{\prod_{i=1}^{n(\tau')}(1 + k_i D)}\, d\tau' \tag{13.150}$$

In order to compute an actual UH from the general equation, the size and distribution of the linear reservoirs must be known or assumed. In general, these reservoirs can be placed at any given translation time from the outlet. The Zoch and Clark models assume, in effect, that there is a single reservoir in the watershed. With this assumption the UH can be easily computed. If more reservoirs are inserted in the idealized watershed, the computational work is increased and would become prohibitively large for a large number of unequal reservoirs. If, however, the reservoirs are all taken as equal, there is a great simplification and the number of reservoirs can be increased without much extra work. The various models discussed previously can easily be derived as special cases of the Dooge model.

13.10.1 The Rational Method

This method treats the problem of direct runoff as one of translation only. This assumption leads to

$$k_j = 0, \qquad j = 1, 2, \ldots, n \tag{13.151}$$

which gives

$$\begin{aligned} h(0, t) &= \frac{1}{T}\int_0^t \delta(t - \tau)\, w(\tau')\, d\tau \\ &= \frac{1}{T} w(t') \end{aligned} \tag{13.152}$$

This says that the IUH has the same shape as the TAC curve. In the original rational method, the critical storm duration was taken as equal to T, the time of concentration. In the modified rational method, the critical storm is taken to maximize the product of the rainfall intensity and the peak of the finite period unit hydrograph.

13.10.2 The Zoch, Clark, and O'Kelly Models

Zoch (1936) assumed the storage to occur in the soil, whereas Clark (1945) more correctly regarded it as channel storage. This assumption corresponds to

$$k_1 = k$$
$$k_j = 0, \qquad j = 2, 3, \ldots, n$$

The general equation then becomes

$$\begin{aligned} h(0, t) &= \frac{1}{T}\int_0^t \frac{\delta(t - \tau)}{1 + kD} w(\tau')\, d\tau \\ &= \frac{1}{T}\int_0^t \frac{1}{k} e^{-(t-\tau)/k} w(\tau')\, d\tau \\ &= \frac{1}{T}\frac{e^{-t/k}}{k}\int_0^t e^{\tau/k} w(\tau')\, d\tau \end{aligned} \tag{13.153}$$

The latter equation is identical to the solution obtained when a TAC curve $w(\tau')$ is routed through a reservoir of delay time k in accordance with the Clark model. In practice such a routing can be carried out very rapidly by using the coefficient form of the Muskingum method to obtain the IUH.

13.10.3 The Nash Model

For this model

$$k_j = k, \qquad j = 1, 2, \ldots, n$$

This assumption transforms the general equation of the IUH as

$$\begin{aligned} h(t) &= \frac{1}{T}\int_0^t \frac{\delta(t - \tau)}{(1 + kD)^n} w(\tau')\, d\tau \\ &= \frac{1}{T}\int_0^t \frac{1}{k}\frac{m^{n-1} e^{-m}}{(n-1)!} w(\tau')\, d\tau \end{aligned} \tag{13.154}$$

where $m = (t - \tau)/k$.

Since Nash (1957) ignored variation in translation time over the watershed, he assumed in fact that all points had the same translation time. That is, the dimensionless TAC curve is a δ-function. This gives

$$h(t) = \frac{1}{T}\int_0^t \frac{m^{n-1} e^{-m}}{k(n-1)!}\delta(0)\, d\tau$$

$$h(t) = \frac{1}{k}\frac{m^{n-1} e^{-m}}{(n-1)!}, \qquad m = \frac{t}{k}$$

which is the Nash model.

TABLE 13.4

Characteristic parameters of some linear conceptual models (after Dooge, 1959)

Method	T	w(τ′)	k	n	n(τ′)	Degree of Freedom
Rational	Variable	Variable	0	0	—	2
Zoch-Clark	Variable	Variable	Variable	1	At outlet	3
O'Kelly	Variable	Triangle	Variable	1	At outlet	2
Nash	—	—	Variable	Variable	—	2
Using equal reservoirs	Variable	Variable	Variable	Variable	Variable	5

In all these models, we notice that there are three numerical parameters, T, k, and n, and two functional parameters, $w(\tau')$ and $n(\tau')$, defining the adjusted TAC curve and the reservoir distribution curve, respectively. The relationship is given in Table 13.4.

The assumption made in the Zoch-Clark method, that the reservoir storage in all parts of the watershed can be represented by a single linear reservoir at the outlet, seems unduly restrictive. Similarly, the neglect of translation in the Nash model makes it impossible to analyze the effect on the UH of such factors as the shape of the watershed and the ratio of overbank storage to channel storage. The equal linear reservoir model has five degrees of freedom, as compared with two for the Nash and Zoch-Clark models. The flexibility is valuable in both direct analysis of the effect of watershed characteristics on the UH and improving statistical analysis of derived UHs.

13.10.4 Equal Linear Reservoirs

For equal linear reservoirs the general equation is

$$\begin{aligned} h(t) &= \frac{1}{T}\int_0^t \frac{\delta(t-\tau)}{(1+kD)^n} w(\tau')\,d\tau \\ &= \frac{1}{T}\int_0^t \frac{m^{n-1}e^{-m}}{(n-1)!} w(\tau')\,dm, \qquad m = \frac{t-\tau}{k} \\ &= \frac{1}{T}\int_0^{t/k} p(m, n-1)w(\tau')\,dm \end{aligned} \tag{13.155}$$

where $p(m, n-1)$ is the Poisson probability function. Since this function is extensively tabulated, computation of any ordinate of the IUH by numerical integration is comparatively easy. For any ordinate (that is, for fixed t) the variable under the integral sign can be evaluated for each value of τ and the products integrated using Simpson's rule or some other method. For any value of τ, $w(\tau)$ is given by the TAC curve, n is given by the distribution $n(\tau)$ of the n reservoirs each of delay time k, and m is given by the difference $t - \tau$. It is important to remember that each reservoir introduces a change in the value of n and hence a discontinuity in the value $p(m, n-1)$. For this reason, it is necessary to evaluate the integral immediately upstream and downstream of each reservoir.

13.11 THE CASCADE OF UNEQUAL LINEAR RESERVOIRS WITH DISTRIBUTED EFFECTIVE RAINFALL

The cascade model of unequal linear reservoirs discussed by Maddaus and Eagleson (1969), O'Meara (1969), Bravo-S. et al. (1970), and Singh and McCann (1980a), is shown in Fig. 13.41. Each element in the model may or may not correspond to a particular portion of the watershed. For generality, it will be assumed that each element represents a particular portion of the watershed area. The system of equations representing this cascade can then be written as:

$$\dot{S}_1 = \frac{c_1}{A} I_1 - a_1 S_1 \tag{13.156}$$

$$\dot{S}_i = \frac{c_i}{A} I_i + a_{i-1} S_{i-1} - a_i S, \qquad i = 2, 3, \ldots, n \tag{13.157}$$

where S is storage; $\dot{S}$ is the time derivative of S; I is the intensity of effective rainfall, $a = 1/k$; k is the reservoir delay parameter; c is the area represented by reservoir; and A is the watershed area. The subscript, i, $i = 1, 2, \ldots, n$, denotes the reservoir number. We assume that $k_i \neq k_j$ for $i \neq j$. The watershed area A equals

$$A = \sum_{i=1}^{n} c_i \tag{13.158}$$

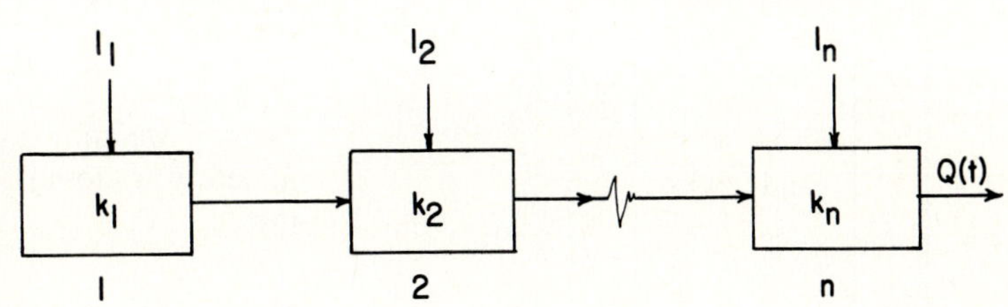

FIGURE 13.41 Cascade of unequal linear reservoirs with distributed effective rainfall.

The linear reservoir is subject to

$$S_i = k_i Q_i \tag{13.159}$$

where Q is the discharge from the ith reservoir having the parameter k_i. The effective rainfall falling on a reservoir must be multiplied by the ratio of its area to the watershed area. Writing equations (13.156)–(13.157) in a compact form using matrix notation, [ ],

$$[\dot{S}] = [a]\,[S] + [I] \tag{13.160}$$

where

$$[\dot{S}] = \begin{bmatrix} \dot{S}_1 \\ \dot{S}_2 \\ \cdot \\ \cdot \\ \cdot \\ \dot{S}_n \end{bmatrix}; \qquad [S] = \begin{bmatrix} S_1 \\ S_2 \\ \cdot \\ \cdot \\ \cdot \\ S_n \end{bmatrix}; \qquad [I] = \frac{1}{A}\begin{bmatrix} c_1 I_1 \\ c_2 I_2 \\ \cdot \\ \cdot \\ \cdot \\ c_n I_n \end{bmatrix}; \quad \text{and}$$

$$[a] = \begin{bmatrix} -a_1 & 0 & 0 & & & 0 \\ a_1 & -a_2 & 0 & & & 0 \\ 0 & a_2 & -a_3 & & & 0 \\ 0 & 0 & 0 & & & 0 \\ \cdot & \cdot & \cdot & \cdot & & \cdot \\ \cdot & \cdot & \cdot & & \cdot & \cdot \\ \cdot & \cdot & \cdot & & \cdot & \cdot \\ 0 & 0 & a_{n-1} & & & -a_n \end{bmatrix}$$

The initial condition for solution of equation (13.160) can be assumed to be

$$[S(0)] = [0] \tag{13.161}$$

The solution of equation (13.160) subject to equation (13.161) can be shown to be (Coddington and Levinson, 1955)

$$[S(t)] = \int_0^t e^{(t-\tau)\,[a]}\,[I(\tau)]\,d\tau \tag{13.162}$$

Therefore,

$$[Q] = \frac{[S]}{[k]} \tag{13.163}$$

where

$$[Q] = \begin{bmatrix} Q_1 \\ Q_2 \\ \cdot \\ \cdot \\ \cdot \\ Q_n \end{bmatrix} \quad \text{and} \quad [k] = \begin{bmatrix} k_1 \\ k_2 \\ \cdot \\ \cdot \\ \cdot \\ k_n \end{bmatrix}$$

Hence,

$$[Q(t)] = \frac{1}{[k]}\int_0^t e^{(t-\tau)\,[a]}\,[I(\tau)]\,d\tau \tag{13.164}$$

Thus, the problem reduces to evaluating $e^{t[a]}$, where $[a]$ is specified before. Since $[a]$ is a triangular matrix, its eigenvalues are the diagonal entries $-a_1, -a_2, \ldots, -a_n$. A direct computation shows that an eigenvector corresponding to the eigenvalue $-a_i$ is given by

$$[x^i] = \begin{bmatrix} x_1^i \\ x_2^i \\ \cdot \\ \cdot \\ \cdot \\ x_n^i \end{bmatrix}$$

where

$$x_j^i = \begin{cases} 0, & \text{if } j < i \\ (a_i - a_{i+1}) \cdots (a_i - a_n), & \text{if } j = i \\ (-1)^{i+j} a_i a_{i+1} \cdots a_{j-1} (a_i - a_{j+1}) \cdots (a_j - a_n), & \text{if } j > k \end{cases}$$

if $i < n$ and

$$[x^n] = \begin{bmatrix} 0 \\ 0 \\ \cdot \\ \cdot \\ \cdot \\ 0 \\ \cdot \\ \cdot \\ \cdot \\ 1 \end{bmatrix}$$

Set $[D] = \text{diag}\,[-a_1, a_2, \ldots, -a_n]$ and $[x] = [x^1, x^2, \ldots, x^n]$. Then it is well known that

$$[a] = [x]\,[D]\,[x]^{-1}$$
$$e^{t[a]} = [x] e^{t[D]} [x]^{-1} \tag{13.165}$$

where

$$e^{t[D]} = \begin{bmatrix} e^{-a_1 t} & 0 & \cdot & \cdot & \cdot & 0 \\ 0 & e^{-a_2 t} & \cdot & \cdot & \cdot & 0 \\ \cdot & \cdot & \cdot & & & \cdot \\ \cdot & \cdot & & \cdot & & \cdot \\ \cdot & \cdot & & & \cdot & \cdot \\ 0 & 0 & \cdot & \cdot & \cdot & e^{-a_n t} \end{bmatrix}$$

Thus, if we know $a_1, \ldots, a_n$, the matrix $e^{t[a]}$ can be determined in a closed form (Lancaster, 1969). It can be shown that

$$e^{t[a]} = [b_{ij}(t)] \tag{13.166}$$

where

$$b_{ij}(t) = \begin{cases} 0, & i < j \\ e^{-a_i t}, & i = j \\ a_j \cdots a_{i-1} \sum_{m=j}^{i} \dfrac{e^{-a_m t}}{\prod_{\substack{r=j \\ r \neq m}}^{i} (a_r - a_m)}, & i > j \end{cases}$$

Since the system is assumed to be linear, it is sufficient to determine its IUH. To that end we suppose that

$$[I(t)] = \frac{\delta(t)}{A} \begin{bmatrix} c_1 \\ c_2 \\ \cdot \\ \cdot \\ \cdot \\ c_n \end{bmatrix} = \delta(t)[C] \tag{13.167}$$

where

$$[C] = \begin{bmatrix} \dfrac{c_1}{A} \\ \dfrac{c_2}{A} \\ \cdot \\ \cdot \\ \cdot \\ \dfrac{c_n}{A} \end{bmatrix}$$

Then equation (13.162) becomes

$$[h(t)] = \frac{e^{t[a]}}{[k]} \left[\int_0^t \delta(t)\, e^{-\tau[a]} [C]\, d\tau \right] = \frac{1}{[k]} e^{t[a]} [C] \tag{13.168}$$

where $[h(t)]$ is the IUH vector. Equation (13.168) is a general solution. To illustrate, consider the case when $n = 3$. Then

$$[a] = \begin{bmatrix} -a_1 & 0 & 0 \\ a_1 & -a_2 & 0 \\ 0 & a_2 & -a_3 \end{bmatrix},$$

$$[x] = \begin{bmatrix} (a_1 - a_2(a_1 a_3) & 0 & 0 \\ -a_1(a_1 - a_3) & (a_2 - a_3) & 0 \\ a_1 a_2 & -a_2 & 0 \end{bmatrix}$$

and

$$[D] = \begin{bmatrix} -a_1 & 0 & 0 \\ 0 & -a_2 & 0 \\ 0 & 0 & -a_3 \end{bmatrix}$$

A straightforward calculation shows that

$$[x^{-1}] = \begin{bmatrix} \dfrac{1}{(a_1 - a_2)(a_1 - a_3)} & 0 & 0 \\ \dfrac{a_1}{(a_1 - a_2)(a_2 - a_3)} & \dfrac{1}{a_2 - a_3} & 0 \\ \dfrac{-a_1 a_2}{(a_1 - a_3)(a_2 - a_3)} & \dfrac{a_2}{a_2 - a_3} & 1 \end{bmatrix}$$

$$e^{t[a]} = \begin{bmatrix} e^{-a_1 t} & 0 & 0 \\ \dfrac{a_1}{a_1 - a_2}(e^{-a_2 t} - e^{-a_1 t}) & e^{-a_2 t} & 0 \\ \dfrac{a_1 a_2[(a_2 - a_3)e^{-a_1 t} - (a_1 - a_3)e^{-a_2 t} + (a_1 - a_2)e^{-a_3 t}]}{(a_1 - a_2)(a_1 - a_3)(a_2 - a_3)} & \dfrac{a_2(e^{-a_3 t} - e^{-a_2 t})}{a_2 - a_3} & e^{-a_3 t} \end{bmatrix}$$

Then $[h(t)]$ can be computed by using equation (13.168).

The IUH of this model can also be derived in an alternative manner by recalling that an equivalent representation of the model consists of a parallel arrangement of cascades of unequal linear reservoirs, as shown in Fig. 13.42. For purposes of notational convenience, the reservoir numbering is reversed. Note that the IUH of each cascade is already derived in Section 13.2.7. Therefore, the IUH of this model is the weighted sum of the IUHs of the individual cascades.

$$h(t) = h_n(t) + h_{n-1}(t) + \cdots + h_1(t) = \sum_{j=1}^{n} h_j(t)$$
$$= \sum_{j=1}^{n} \frac{c_j}{A \prod_{i=1}^{j} (1 + k_i D)} \tag{13.169}$$

By employing equation (13.137),

$$h(t) = \frac{c_1}{A\, k_1} e^{-t/k_1} + \sum_{m=2}^{n} \frac{c_m}{A} \sum_{j=1}^{m} \frac{k_j^{m-2}\, e_j^{-t/k}}{\sum_{\substack{i=1 \\ i \neq j}}^{m} (k_j - k_i)}, \quad n \geq 2 \tag{13.170}$$

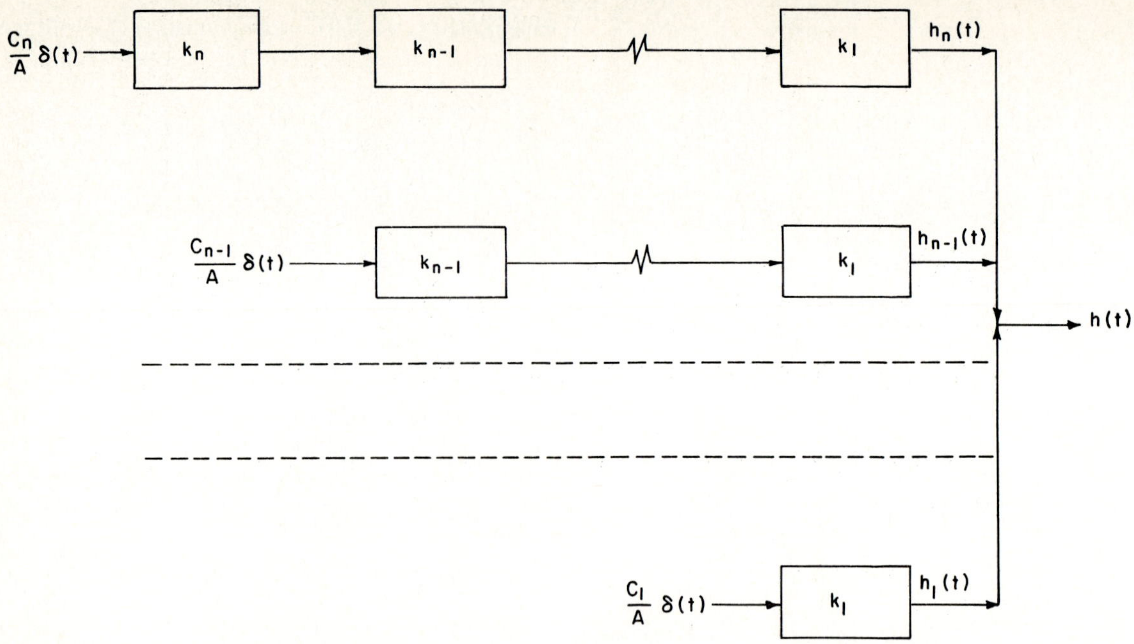

FIGURE 13.42 An equivalent representation of the model of Fig. 13.41.

This model can be extended to include translation. If there are as many linear channels each having translation T_i, $i = 1, 2, \ldots, n$, as there are linear reservoirs, then the IUH can be expressed as

$$h(t) = \frac{c_1}{Ak_1} e^{-(t-T_1)} + \sum_{m=2}^{n} \frac{c_m}{A} \sum_{j=1}^{m} \frac{k_j^{m-2} \exp\left[-\left(t - \sum_{i=1}^{j} T_i/k_j\right)\right]}{\prod_{\substack{i=1\\ i\neq j}}^{m} (k_j - k_i)}, \quad t > T \tag{13.171}$$

where $T = \sum_{i=1}^{n} T_i$.

However, from the application standpoint it may not be appropriate to use equations (13.170) and (13.171) to account fully for spatial variability of the ER. Instead, each individual h_i, $i = 1, 2, \ldots, n$, must be convoluted with its corresponding areal ER to generate Q_i, $i = 1, 2, \ldots, n$, which must then be summed up to produce the DR. Symbolically,

$$\begin{aligned} Q(t) &= \sum_{j=1}^{n} Q_j(t) \\ &= \sum_{j=1}^{n} \int_0^t h_j(t-\tau) I_j(\tau)\, d\tau \\ &= \frac{c_1}{AK_1} \int_0^t I_1(\tau)\, e^{-(t-\tau)/k_1}\, d\tau \\ &\quad + \sum_{m=2}^{n} \frac{c_m}{A} \sum_{j=1}^{m} \int_0^t k_j^{m-2} e^{-(t-\tau)/k_j)} \frac{I_m(\tau)}{\prod_{\substack{i=1\\ i\neq j}}^{n} (k_j = k_i)}\, d\tau \end{aligned} \tag{13.172}$$

Following equation (13.172) we can easily extend equation (13.142) to include translation.

In a similar manner we can get solutions for several special cases that are of hydrological significance. One case of particular interest is when $I_2 = I_3 = \cdots = I_n = 0$. Then $e^{t[a]}\,[C]$ is the first column of $e^{t[a]}$ in equation (13.168). This is the Dooge model (Dooge, 1959). Another case of importance is $c_1 = c_2 \cdots = c_n$.

13.12 THE CASCADE OF EQUAL LINEAR RESERVOIRS WITH DISTRIBUTED EFFECTIVE RAINFALL

The cascade model of equal linear reservoirs is shown in Fig. 13.43. This is a special case of the previous model but requires an independent treatment. The system of equations can be written as

$$\dot{S}_i = \frac{c_1}{A} I_1 - a s_1 \tag{13.173a}$$

$$\dot{S}_i = \frac{c_i}{A} I_i + a s_{i-1} - a s_i, \qquad i = 2, 3, \ldots, n \tag{13.173b}$$

This system of equations can be represented by equation (13.160), where all the forms remain unaltered except for $[a]$, which can be defined as

$$[a] = \begin{bmatrix} -a & 0 & 0 & \cdots & 0 \\ a & -a & 0 & \cdots & 0 \\ 0 & a & -a & \cdots & 0 \\ \vdots & \vdots & \vdots & & \vdots \\ 0 & 0 & 0 & a & -a \end{bmatrix}$$

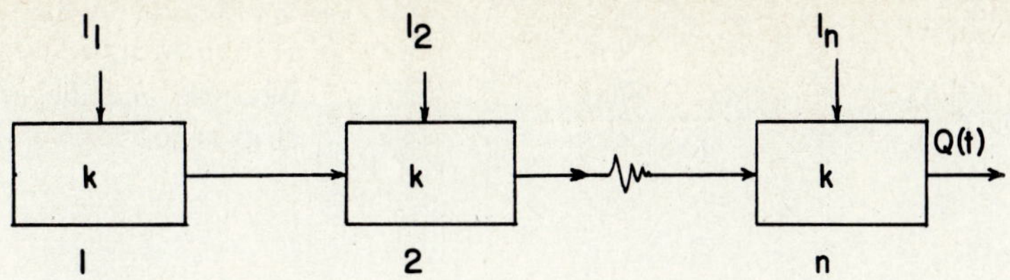

FIGURE 13.43 Cascade of equal linear reservoirs with distributed effective rainfall.

The solution of equation (13.173) subject to equation (13.161) is given by equation (13.162) or equation (13.164). Its IUH is given by equation (13.158). The terms of $e^{t[a]}$ can be obtained in a closed form.

$$e^{t[a]} = e^{-at}\begin{bmatrix} 1 & 0 & \cdots & 0 \\ at & 1 & \cdots & \\ \dfrac{(at)^2}{2!} & at & \cdots & \\ \cdot & \cdot & \cdot & \cdot \\ \cdot & \cdot & \cdot & \cdot \\ \cdot & \cdot & \cdot & \cdot \\ \dfrac{(at)^{n-1}}{(n-1)!} & \dfrac{(at)^{n-1}}{(n-2)!} \cdots & \cdots at & 1 \end{bmatrix}$$

Several special cases of hydrological importance can be obtained from the general solution in equation (13.173). If $I_2 = I_3 = \cdots = I_n = 0$, then this case is equivalent to the Nash model (Nash, 1960). Here $e^{t[a]}$ is the first column in equation (13.168).

Alternatively, the cascade of Fig. 13.43 can be represented by a number of cascades arranged in parallel, as discussed in the previous section as well as in Chapter 12. Consequently, the system of equations (13.173) will have the solution

$$h(t) = \sum_{j=1}^{n} \frac{c_j}{A}\left(\frac{t}{k}\right)^{j-1} \frac{e^{-t/k}}{(n-1)!k} \qquad (13.174)$$

This model can be easily extended to include translation. Let there be as many linear channels each of translation T_i, $i = 1, 2, \ldots, n$, as there are linear reservoirs such that these elements are arranged in series. Then the IUH can be written as

$$h(t) = \sum_{j=1}^{n} \frac{c_j}{A}\left(\frac{t - \sum_{i=1}^{j} T_i}{k}\right)^{n-1} \frac{\exp\left[-\left(t - \sum_{i=1}^{j} T_i\right)/k\right]}{k\Gamma(n)}, \qquad t > \sum_{i=1}^{j} T_i \qquad (13.175)$$

If all linear channels have equal translation T,

$$h(t) = \sum_{j=1}^{n} \frac{c_j}{Ak\Gamma(n)}\left(\frac{t - jT}{k}\right)^{n-1} e^{-(t-jT)/k} \qquad (13.176)$$

The moments and cumulants of equation (13.175) with $c_1 = c_2 = \cdots = c_n = c$ can be written as

$$M_1 = \frac{k}{2}(n+1) \qquad (13.177a)$$

$$M_2 = \frac{k^2}{3}(n+1)(n+2) \qquad (13.177b)$$

and

$$K_1 = \frac{k}{2}(n+1) \qquad (13.178a)$$

$$K_2 = \frac{k^2}{12}(n+1)(n+5) \qquad (13.178b)$$

$$K_3 = \frac{k^3}{4}(n+1)(n+3) \qquad (13.178c)$$

$$K_4 = \frac{k^4}{120}(n+1)(251 + 109n + n^2 - n^3) \qquad (13.178d)$$

where $c/A = 1/n$.

13.13 MODELS BASED ON DIFFERENTIAL EQUATIONS

A general differential equation for linear time-invariant systems is given by equation (8.11). Writing it in terms of operators (with $a_0 = b_0 = 1$),

$$\left(\sum_{i=0}^{N} a_i D^i\right) Q(t) = \left(\sum_{j=0}^{M} b_j D^j\right) I(t), \qquad N > M \qquad (13.179)$$

It may be useful to write equation (13.179) in terms of the Laplace transform.

$$L\left[a_N \frac{d^N Q}{dt^N}\right] = a_N s^N Q(s) - a_N[s^{N-1}Q(0) + s^{N-2}Q'(0) + \cdots + Q^{(N-1)}(0)]$$

$$L\left[b_M \frac{d^M I}{dt^M}\right] = b_M s^M I(s) - b_M[s^{M-1}I(0) + s^{M-2}I'(0) + \cdots$$

Thus taking Laplace transform of equation (13.179),

$$D(s)Q(s) - Y_Q(s) = X(s)I(s) - Y_I(s)$$

resulting in

$$Q(s) = \frac{X(s)}{D(s)} I(s) + \frac{Y_Q(s) - Y_I(s)}{D(s)} \quad (13.180)$$

in which

$$D(s) = \sum_{j=0}^{N} a_j s^j$$

$$X(s) = \sum_{j=0}^{M} b_j s^j$$

$$Y_Q(s) = a_N Q(0) s^{N-1} + [a_N Q'(0) + a_{N-1} Q(0)] s^{N-2} + \cdots + [a_N Q^{(N-1)}(0) + a_{N-1} Q^{(N-2)}(0) + \cdots + a_2 Q'(0) + a_1 Q(0)]$$

$$Y_I(s) = b_M I(0) s^{M-1} + [b_M I'(0) + b_{M-1} I(0)] s^{N-2} + \cdots + [b_M I^{(M-1)}(0) + b_{M-1} I^{(M-2)}(0) + \cdots + b_2 I'(0) + b_1 I(0)]$$

If we let $h(s) = X(s)/D(s)$ and $h_0(s)[Y_Q(s) - Y_I(s)]/D(s)$, then equation (13.180) can be written as

$$Q(s) = h(s)I(s) + h_0(s) \quad (13.181)$$

where

$$h(s) = \frac{\sum_{j=0}^{M} b_j s^j}{\sum_{j=0}^{N} a_j s^j} \quad (13.182)$$

Equation (13.182) can be considered as the transfer function of the system or the system IUH in the Laplacian domain. It does not depend on the initial condition or the input. However, $h_0(s)$ depends on the initial state of the system and the input. If $h_0(s) = 0$, then

$$Q(s) = h(s)I(s) \quad \text{or} \quad h(s) = \frac{Q(s)}{I(s)} \quad (13.183)$$

which is also obtained directly from the convolution integral as discussed in Chapter 4. Several of the models discussed earlier can be derived from the general differential equation (13.179).

13.13.1 TA Models

Kulandaiswamy (1964) showed that if

$$b_j = (-1)^j \frac{(T)^j}{j!} \quad (13.184a)$$

where T is some time, say translation time, then the effect of the operator on the right side of the differential equation is to lag $I(t)$ by the time equal to T. Therefore, if the watershed is divided into R segments by drawing isochrones at Δt hour intervals and the effective rainfall moving down from each strip is successively operated by the right-side operator, then

$$b_j = (-1)^j \frac{(\Delta T)^j}{j!} \quad (13.184b)$$

and the TAC curve will follow

$$w(t) = \sum_{j=0}^{R-1} [H(D)]^j \frac{\Delta A_j}{A} I_j \delta(t, \Delta t) \quad (13.185)$$

where

$$H(D) = \sum_{j=0}^{M} b_j D^j$$

with b_j defined as before.

Substituting this into the differential equation (13.179)

$$h(t) = \frac{1}{\sum_{i=0}^{N} a_i D^i} w(t) \quad (13.186)$$

If $N = 1$, then

$$h(t) = \frac{1}{1 + a_1 D} w(t) \quad (13.187)$$

It is readily seen that equation (13.187) corresponds to the models of Zoch (1934, 1936, 1937), Clark (1945), and O'Kelly (1965).

13.13.2 The Nash Model

If we let $b_j = 0$, $j = 1, 2, \ldots, M$, and the roots of the polynomial in the denominator be real and equal, say k, then

$$h(t) = \frac{1}{(1 + kD)^N} \delta(t)$$

This corresponds to the Nash model (Nash, 1957).

13.13.3 The Singh Model

If we let $N = 2$ in equation (13.186) and the roots of the polynomial in the denominator have real and unequal roots, say k_1 and k_2, then

$$h(t) = \frac{1}{(1 + k_1 \mathrm{D})(1 + k_2 \mathrm{D})} w(t) \quad (13.188)$$

This corresponds to the Singh model (K. P. Singh, 1964) if k_1 and k_2 are considered constant for a given watershed.

Kulandaiswamy (1964) argued that the roots of the characteristic polynomial in the denominator can be real or complex in a hydrologic system. The complex roots are plausible under certain conditions. For example, when a system is disturbed from the rest or equilibrium condition, some oscillation may be expected. The conceptual models, based on linear elements, exclude the occurrence of complex roots.

13.13.4 The Chow-Kulandaiswamy (CK) Models

13.13.4.1 Determination of the Order of Derivatives *N* and *M*

An application of the general linear differential equation in practice would require determination of the order of derivatives N and M of Q and I, respectively. Since Q and I are normally available in discrete-time form, there will obviously be errors in computation of their derivatives. Therefore, beyond certain values of M and N, the advantage that might be gained by the use of higher-order derivatives will be offset by these errors. It has been shown previously that the effect of the operator in the numerator is one of translation. The effect of the operator in the denominator is both lag and change in shape or attenuation. Therefore, dropping off higher-order derivatives of Q and I will mainly reduce the translatory effects. Further, an analysis of the effective rainfall–direct runoff data can indicate the values of N and M that should be used for a particular watershed. Kulandaiswamy (1964) and Chow and Kulandaiswamy (1971, 1982) analyzed six cases, including (1) $N = 2$, $M = 2$, (2) $N = 3$, $M = 2$, (3) $N = 3$, $M = 1$, (4) $N = 3$, $M = 0$, (5) $N = 2$, $M = 1$, and (6) $N = 2$, $M = 0$, some of which have been investigated by Chaudhry (1976), Chaudhry, Simoes, and Ferreira Filho (1976), McCann and V. P. Singh (1980, 1981), and V. P. Singh and McCann (1979).

13.13.4.2 Third-Order Differential Equation

The differential equation can be expressed as

$$Q(t) = \frac{b_2D^2 + b_1D + 1}{a_3D^3 + a_2D^2 + a_1D + 1} I(t) \qquad (13.189)$$

The characteristic polynomial in equation (13.189) is of third degree and has three roots. Let these be m, n, and p. If the system represented by equation (13.189) is stable, then four possibilities may arise:

1. All the roots are real and unequal, that is, $m \neq n \neq p$.
2. All the roots are real with two being equal, that is, $n = p$.
3. All the roots are real and equal, that is, $m = n = p$.
4. One root is real and two are complex conjugates, that is, $n = \gamma + jw$, $p = \gamma - jw$.

***Case 1*: m ≠ n ≠ p.** Equation (13.189) can be written as

$$Q(t) = \left[\frac{c_1}{1 + k_1D} + \frac{c_2}{1 + k_2D} + \frac{c_3}{1 + k_3D}\right] I(t), \qquad Q(0) = 0 \qquad (13.190)$$

The IUH and S-hydrograph can be written as

$$h(t) = Ae^{mt} + Be^{nt} + Ce^{pt} \qquad (13.191)$$

$$U(t) = 1 + \frac{A}{m}e^{mt} + \frac{B}{n}e^{nt} + \frac{C}{p}e^{pt} \qquad (13.192)$$

where

$$A = -\,mnp\frac{(1 + b_1m + b_2m^2)}{m^2 - m(n + p) + np}$$

$$B = -\,mnp\frac{(1 + b_1n + b_2n^2)}{n^2 - n(m + p) + pm}$$

$$C = -\,mnp\frac{(1 + b_1p + b_2p^2)}{p^2 - p(m + n) + mn}$$

$$\frac{A}{m} + \frac{B}{n} + \frac{C}{p} = -1$$

The system represented by this case is shown in Fig. 13.44.

***Case 2*: m, n = p.** Equation (13.189) can be written as

$$Q(t) = \left[\frac{c_1}{1 + k_1D} + \frac{c_2}{1 + k_2D} + \frac{c_3}{(1 + k_2D)^2}\right] I(t), \qquad Q(0) = 0 \qquad (13.193)$$

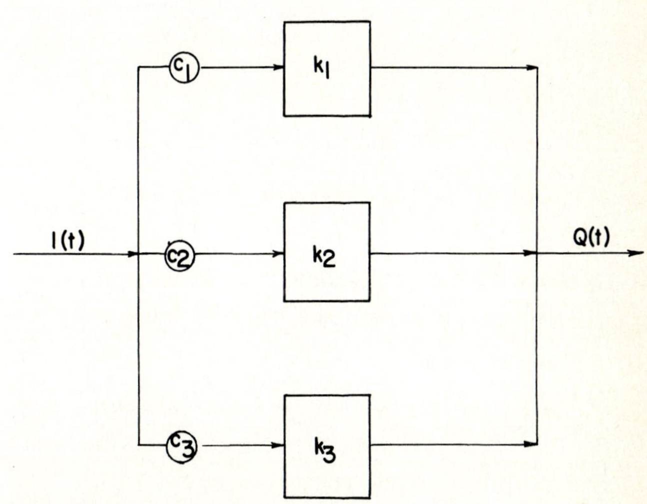

FIGURE 13.44 System representation of equation (13.190).

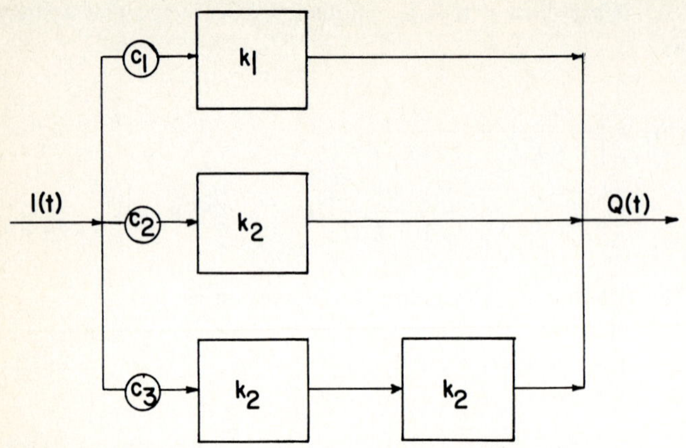

FIGURE 13.45 System representation of equation (13.193).

The IUH and S-hydrograph can be written as

$$h(t) = Ae^{mt} + Be^{nt} + Cte^{nt} \tag{13.194}$$

$$U(t) = -\frac{A}{m}(1 - e^{pt}) - \frac{B}{n}(1 - e^{nt}) - \frac{C}{n^2}[1 - e^{nt}(1 - nt)] \tag{13.195}$$

where

$$A = \frac{-mn^2(1 + b_1m + b_2m^2)}{(m-n)^2}$$

$$B = \frac{mn^2[1 + b_1m + b_2(2m - n)n]}{(m-n)^2}$$

$$C = \frac{(1 + b_1n + b_2n)}{(n - m)}$$

$$\frac{A}{m} + \frac{B}{n} = 0$$

Fig. 13.45 shows the system represented by this case.

***Case 3*: m = n = p.** Equation (13.189) can be expressed as

$$Q(t) = \left[\frac{c_1}{1 + kD} + \frac{c_2}{(1 + kD)^2} + \frac{c_3}{(1 + kD)^2}\right] I(t) \qquad Q(0) = 0 \tag{13.196}$$

The IUH and S-hydrograph follow

$$h(t) = A\,[A + Bt + Ct^2]e^{mt} \tag{13.197}$$

$$U(t) = -\frac{A}{m}(1 - e^{mt}) - \frac{B}{m^2}[1 - e^{mt}(1 - mt)] + \frac{C}{m^3}[1 - {}^{mt}(1 - mt + \frac{1}{2}m^2t^2)] \tag{13.198}$$

where

$$A = -\,b_2m^3$$

$$B = -\,b_1m^3 - 2b_2m^4$$

$$C = -\,m^3 + b_1m^4 + b_2m^5$$

$$\frac{A}{m} = 0$$

The system represented by this case is shown in Fig. 13.46.

***Case 4*: m *real*, n = z = γ + jw, *and* p = z̄ = γ − jw.** Equation (13.189) for this case can be written as

$$Q(t) = \left[\frac{c_1}{1 + D} + \frac{c_2}{1 - (\gamma - jw)D} + \frac{c_3}{1 + (\gamma - jw)D}\right] I(t), \qquad Q(0) = 0 \tag{13.199}$$

The IUH and S-hydrograph are

$$h(t) = A\,e^{mt} + e^{\gamma t}\left[B\cos wt + \frac{C + B\gamma}{w}\sin wt\right] \tag{13.200}$$

$$U(t) = \frac{A}{m}(e^{mt} - 1) + \frac{e^{\gamma t}}{\gamma^2 + w^2}\left[B(\gamma\cos wt + w\sin wt) + \frac{C + B\gamma}{w}(\gamma\sin wt - w\cos wt)\right] + \frac{C}{\gamma^2 + w^2} \tag{13.201}$$

where

$$A = \frac{-\,m(\gamma^2 + w^2)(1 + b_1m + b_2m^2)}{m^2 - 2\gamma m + \gamma^2 + w^2}$$

$$B = -\,m(\gamma^2 + m^2)b_2 - A$$

$$C = (\gamma^2 + w^2)\left(1 + \frac{A}{m}\right)$$

$$-\frac{A}{m} + \frac{c}{\gamma^2 + w^2} = -\frac{1}{m(\gamma^2 + w^2)}$$

The system representation of this case is shown in Fig. 13.47.

13.14.4.3 Second-Order Differential Equation

The second-order differential equation can be expressed as

$$Q(t) = \frac{b_2D^2 + b_1D + 1}{a_2D^2 + a_1D + 1} I(t) \tag{13.202}$$

The characteristic polynomial is of second order and has two roots, m and n. If the system represented by this equation is stable, three possibilities may be distinguished:

1. Both roots are real and unequal, $m \neq n$.
2. Both roots are real and equal, $m = n$.
3. Both roots are complex conjugates, $m = z = \gamma + jw$, $n = \bar{z} = \gamma - jw$.

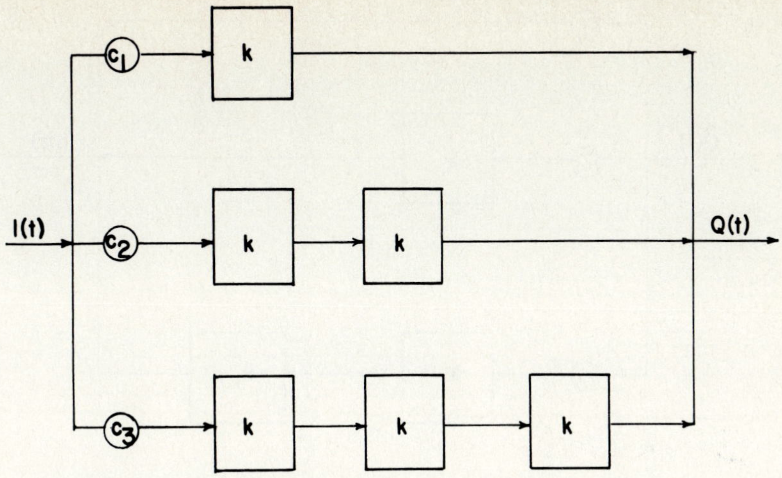

FIGURE 13.46 System representation of equation (13.196).

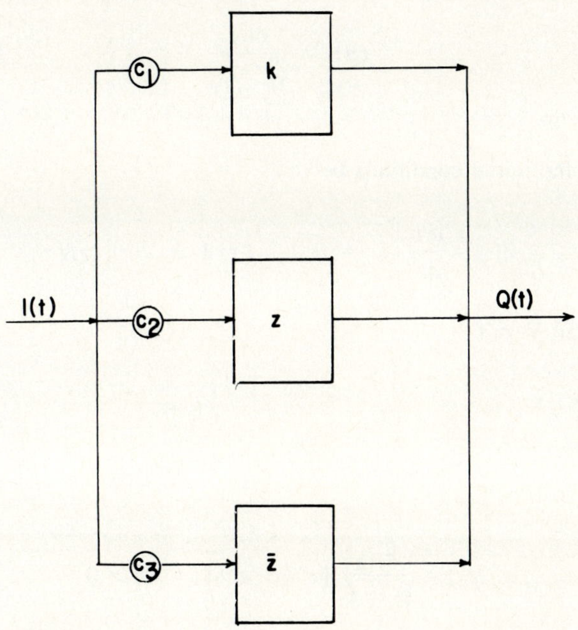

FIGURE 13.47 System representation of equation (13.199).

***Case 1*: m ≠ n.** Equation (13.202) can be written in the form

$$Q(t) = \left[C_3' + \frac{C_1'}{1 + k_1 D} + \frac{C_2'}{1 + k_2 D}\right] I(t), \qquad Q(0) = 0 \quad (13.203)$$

The IUH and S-hydrograph of this system can be written as

$$h(t) = \frac{1}{a_2(m - 1)}\left[(C_1 m + C_2)\, e^{mt} - (C_1 n + C_2)\, e^{nt}\right] \quad (13.204)$$

$$U(t) = \frac{1}{a_2(m - 1)}\left[\frac{1}{m}(C_1 m + C_2)\,(e^{mt} - 1) - \frac{1}{n}(C_1 n + C_2)\,(e^{nt} - 1)\right] \quad (13.205)$$

where

$$C_1 = b_1 - \frac{b_2}{a_2} a_1$$

$$C_2 = 1 - \frac{b_2}{a_2}$$

$$C_3 = \frac{b_2}{a_2}$$

The system representation is shown in Fig. 13.48.

***Case 2*: m = n.** Equation (13.202) can be expressed in the form

$$Q(t) = \left[C_3' + \frac{C_1'}{1 + k_1 D} + \frac{C_2'}{(1 + k_1 D)^2}\right] I(t), \qquad Q(0) = 0 \quad (13.206)$$

The IUH and S-hydrograph can be expressed as

$$h(t) = \frac{1}{a_2}[C_1 e^{nt} + (C_1 n + C_2)\, t e^{nt} \quad (13.207)$$

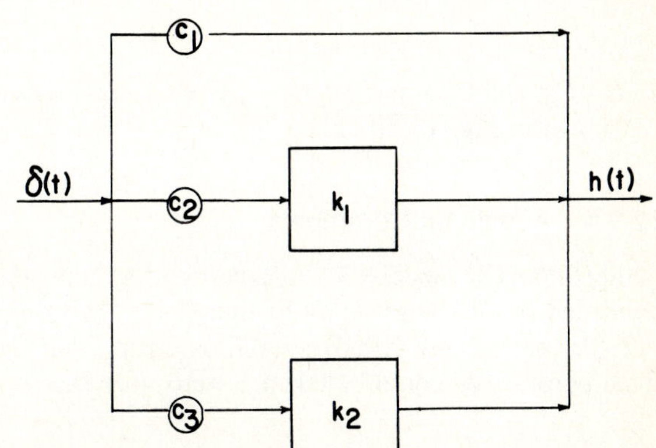

FIGURE 13.48 System representation of equation (13.203).

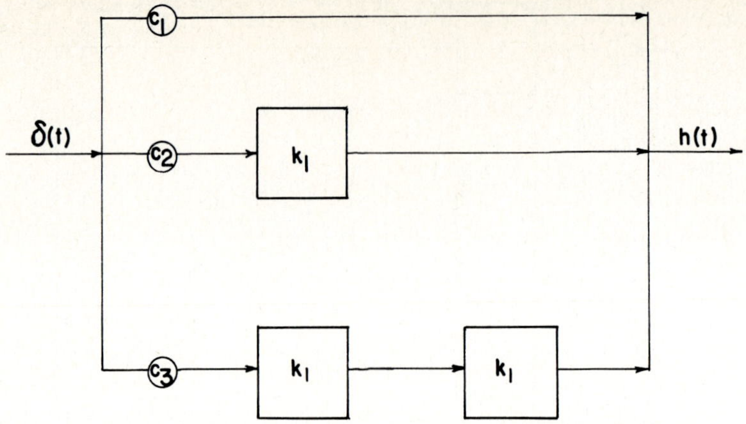

FIGURE 13.49 System representation of equation (13.206).

$$U(t) = \frac{1}{a_2}\left[\frac{C_1}{n}(e^{nt} - 1) + \frac{1}{n}(C_1 n + C_2)\left\{te^{nt} - \frac{e^{nt}}{n} + \frac{1}{n}\right\}\right] \quad (13.208)$$

where C_1, C_2, and C_3 are as in case 1. The system representation is shown in Fig. 13.49.

***Case 3*: m = z = γ + jw, n = z̄ = γ − jw.** Equation (13.202) can be written in the form

$$Q(t) = \left[C_3' + \frac{C_1'}{1 + zd} + \frac{C_2'}{1 + \bar{z}D}\right] I(t),$$

$$Q(0) = 0 \quad (13.209)$$

The IUH and S-hydrograph follow:

$$h(t) = \frac{1}{a_2}\left[e^{\gamma t}\left(C_1 \cos wt + \frac{C_1\gamma + C_2}{w}\sin wt\right)\right] \quad (13.210)$$

$$U(t) = \frac{e^{\gamma t}}{a_2(\gamma^2 + w^2)}\left[C_1(\gamma \cos wt + w \sin wt) + \frac{C_1\gamma + C_2}{2}(\gamma \sin wt - w \cos wt)\right] + \frac{C_1}{a_2(\gamma^2 + w^2)} \quad (13.211)$$

where C_1, C_2, and C_3 are as in case 1. The system representation is shown in Fig. 13.50.

13.13.4.4 Additional Comments

In a study of the CK models, V. P. Singh and McCann (1979), McCann and V. P. Singh (1980b) found, by analyzing the data of Kulandaiswamy (1964) as well as others, that for all practical purposes M could be taken as zero, that is,

$$Q(t) = \frac{I(t)}{\sum_{j=0}^{N} a_j D^i} \quad (13.212)$$

Therefore,

$$Q(t) = \frac{\delta(t)}{\sum_{j=0}^{N} a_j D^i} \quad (13.213)$$

Let the initial conditions be

$$h(t) = \left.\frac{d^j h}{dt^j}\right|_{t=0} = 0, \qquad j = 1, 2, \ldots, N$$

When $N = 1$

$$h(t) = \frac{e^{-t/a_1}}{a_1}, \qquad t > 0$$

When $N = 2$,

$$h(t) = \frac{k_1 k_2}{k_2 - k_1}\left(e^{k_2 t} - e^{k_1 t}\right), \qquad t > 0$$

where k_1 and k_2 are the roots of

$$f(x) = a_2 x^2 + a_1 x + 1$$

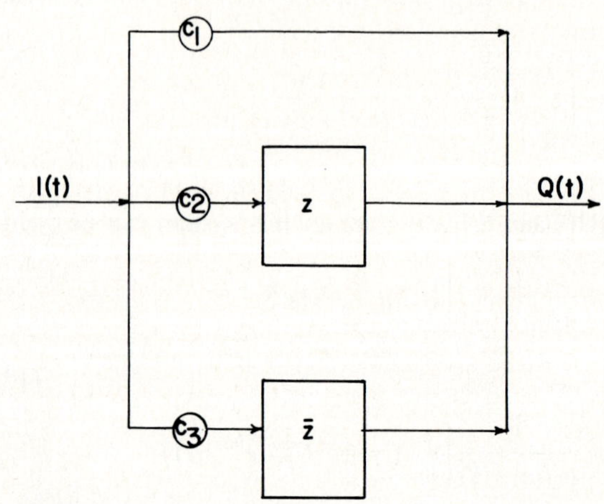

FIGURE 13.50 System representation of equation (13.209).

When $N = 3$,

$$h(t) = -k_1k_2k_3\left[\frac{e^{k_1t}}{(k_1-k_2)(k_1-k_3)} + \frac{e^{k_2t}}{(k_2-k_1)(k_2-k_3)} + \frac{e^{k_3t}}{(k_3-k_1)(k_3-k_2)}\right]$$

where k_1, k_2, and k_3 are the roots of

$$g(x) = a_3x^3 + a_2x^2 + a_1x + 1$$

Similarly, we can obtain $h(t)$ for higher values of N. Singh and McCann (1980b) found that in most cases $N \leq 3$ suffice.

Chow and Kulandaiswamy (1971) used the least squares method to estimate the coefficients a_i and b_i in their models, which implies that these coefficients depend on how accurately the derivatives of I and Q can be obtained. Given I, usually in histogram form, it is preferable to use a method to estimate a_i and b_i that is independent of the derivatives of I and Q. V. P. Singh and McCann (1979, 1980) showed the method of moments or cumulants would be preferable for this purpose. They derived a criterion to determine M and N. The method of moments for equation (13.213) yields

$$m_1 = a_1$$

$$m_2 = -2a_2 + 2a_1^2$$

$$m_3 = 6a_3 - 12a_1a_2 + 6a_1^3$$

and so on. These moments can be used to determine the coefficients a_i. Note that since moments, especially third and higher, are sensitive to small changes in the function, it is expected that the coefficients will be sensitive to small errors in the ERH-DRH data. This is one of the reasons that choosing $N > 3$ would not normally increase the accuracy of the model.

13.14 STRUCTURAL RELATION BETWEEN DETERMINISTIC AND STOCHASTIC MODELS

Close analogies have been shown to exist between stochastic and deterministic linear models (Dooge, 1972; Quimpo, 1971a, 1971b, 1973; Selvalingam, 1979; Salas et al., 1980). A short discussion is provided here. A key characteristic in stochastic modeling of hydrologic time series is preserving its persistence characteristics. The persistence is usually measured by the covariance function $C(t)$, which—when divided by the variance $C(0)$—yields the autocorrelation function $\rho(t)$. Quimpo (1971a, 1971b, 1973) suggested deriving a relation between the IUH and $\rho(t)$, which can be obtained by determining the autocorrelation function of $Q(t)$. Recall the convolution integral,

$$Q(t) = \int_0^\infty h(u)\, I(t-u)\, du \tag{13.214}$$

For mathematical convenience $I(t)$ and $Q(t)$ are assumed to have zero means. The autocovariance function of $Q(t)$ at lag τ can be defined as

$$C_{QQ}(\tau) = E[Q(t), Q(t+\tau)] \tag{13.215}$$

in which $E[\cdot]$ denotes mathematical expectation. Inserting equation (13.214) in equation (13.215),

$$C_{QQ}(\tau) = E\left[\int_0^\infty h(u)\, I(t-u)d\tau \int_0^\infty h(v)\, I(t+\tau-v)\, dv\right]$$
$$= \int_0^\infty\int_0^\infty h(u)\, h(v)\, E[I(t-u)I(t+\tau-v)]\, du\, dv \tag{13.216}$$

If $I(t)$ is assumed stationary, equation (13.216) becomes

$$C_{QQ}(\tau) = \int_0^\infty\int_0^\infty h(u)h(v)C_{II}(\tau+u-v)\, du\, dv \tag{13.217}$$

where $C_{II}(\tau + u - v)$ is the autocovariance function of $I(t)$. This relates autocovariance function of the effective rainfall. By choosing an appropriate model of $I(t)$, equation (13.217) can be integrated. For example, Hino (1970) and Hino, Sukigara, and Kikkawa (1971) suggested that $C_{II}(\tau)$ could be represented by a negative exponential,

$$C_{II}(\tau) = \alpha \exp[-\beta\tau] \tag{13.218}$$

On the other hand, if $I(t)$ is assumed to be a white noise, then its $C_{II}(\tau)$ is a Dirac delta function,

$$C_{II}(\tau) = C_{II}(0)\delta(\tau)$$

in which $C_{II}(0)$ denotes the variance of $I(t)$.

Substituting equation (13.218) in equation (13.217),

$$C_{QQ}(\tau) = \int_0^\infty\int_0^\infty h(u)\, h(v)\, \alpha\exp[-\beta(u-v+\tau)]\, du\, dv \tag{13.220}$$

Equation (13.220) can be solved for various models yielding $h(t)$.

Substituting equation (13.219) in equation (13.217),

$$C_{QQ}(\tau) = \int_0^\infty\int_0^\infty h(u)h(v)C_{II}(0)\delta(u+\tau-v)\, du\, dv$$
$$= C_{II}(0)\int_0^\infty h(u)h(u+\tau)\, du \tag{13.221}$$

Dividing both sides by $C_{QQ}(0)$,

$$\rho_{QQ}(\tau) = \frac{\int_0^\infty h(u)\, h(u+\tau)\, du}{\int_0^\infty h^2(u)\, du}$$
$$= \frac{C_{QQ}(\tau)}{C_{QQ}(0)} \tag{13.222}$$

Since the autocorrelation function $\rho(\tau)$ defines the parameters of the autoregression models, equation (13.222) can be used to determine the relation between stochastic and deterministic models.

13.14.1 Linear Reservoir

First, consider the autocovariance function of $I(t)$ as given by equation (13.218). Substituting equations (13.23) and (13.218) in equation (13.217),

$$C_{QQ}(\tau) = \frac{1}{k^2}\int_0^\infty\int_0^\infty \exp\left(\frac{u}{k}\right)\exp\left(\frac{v}{k}\right) \alpha\exp[-\beta(\tau + u - v)]\,du\,dv$$

$$= \frac{\alpha\exp(-\beta\tau)}{k^2}\int_0^\infty \exp\left[-u\left(\beta - \frac{1}{k}\right)\right]du \int_0^\infty \exp\left[-v\left(\beta - \frac{1}{k}\right)\right]dv$$

$$= \alpha\frac{\exp(-\beta\tau)}{(\beta k - 1)^2} \tag{13.223}$$

and

$$\rho_{QQ}(\tau) = \exp(-\beta\tau) \tag{13.224}$$

Thus it is seen that the autocovariance of Q depends on characteristics of both the reservoir and $I(t)$, whereas its autocorrelation function depends only on the characteristics of $I(t)$.

Now consider the autocovariance function of $I(t)$, as given by equation (13.219). Substituting equations (13.23) and (13.219) in equation (13.217),

$$C_{QQ}(\tau) = C_{II}(0)\int^\infty \frac{1}{k^2}\exp\left(\frac{-u}{k}\right)\exp\left(\frac{-u+\tau}{k}\right)du$$

$$= \frac{\sigma_I^2}{2k}\exp\left(\frac{-\tau}{k}\right) \tag{13.225}$$

and

$$\rho_{QQ}(\tau) = \exp\left(\frac{-\tau}{k}\right) = kh(\tau) \tag{13.226}$$

Equation (13.225) may be recognized as the autocorrelation function, which is expressed as

$$\rho(\tau) = \rho^\tau(1) \tag{13.227}$$

the simple Markov model used extensively in stochastic hydrology (Fiering, 1967; Matalas, 1966; Yevjevich, 1963; Quimpo, 1973),

$$Q(t) = \rho(1)\,Q(t-1) + [1-\rho(1)]^{0.5}\sigma_I I(t) \tag{13.228}$$

Equation (13.226) yields

$$k = \frac{-\tau}{[\ln\rho(\tau)]} \tag{13.229}$$

This implies that the simple Markov model and the single linear reservoir represent the same system (Quimpo, 1971b).

13.14.2 The Nash Model

First, we represent $C_{II}(\tau)$ by equation (13.218). Inserting equations (13.76) and (13.218) in equation (13.217),

$$C_{QQ}(\tau) = \frac{\alpha\exp(-\beta\tau)}{k^{2n}[\Gamma(n)]^2}\int_0^\infty \exp\left[-u\left(\beta+\frac{1}{k}\right)\right]u^{n-1}du \cdot \int_0^\infty \exp\left[-v\left(\beta+\frac{1}{k}\right)\right]v^{n-1}dv \tag{13.230}$$

The last integral can be evaluated by letting $w = v[\beta + (1/k)]$,

$$\int_0^\infty \exp\left[-v\left(\beta+\frac{1}{k}\right)\right]v^{n-1}dv = \left(\frac{k}{1+k\beta}\right)^n\Gamma(n)$$

By doing a similar calculation, equation (13.230) results in

$$C_{QQ}(\tau) = \frac{\alpha\exp(-\beta\tau)}{(1+\beta K)^{2n}} \tag{13.231}$$

and

$$\rho_{QQ}(\tau) = \exp(-\beta\tau) \tag{13.232}$$

Now consider $C_{II}(\tau)$, as represented by equation (13.218). Following the same procedure,

$$C_{QQ}(\tau) = \frac{\exp(-\tau/k)}{k^{2n}[\Gamma(n)]^2}\int_0^\infty \exp\left[\frac{-2u}{k}\right]u^{n-1}(u+\tau)^{n-1}du \tag{13.233}$$

The integral can be evaluated as

$$\int_0^\infty \exp\left(\frac{-2u}{k}\right)u^{n-1}(u+\tau)^{n-1}du = \left(\frac{k\tau}{2}\right)^{n-1}\frac{\exp(\tau/k)}{\sqrt{2\pi\tau/k}}\Gamma(n)J_{n-(1/2)}\left(\frac{\tau}{k}\right)$$

in which $J_{n-(1/2)}(\cdot)$ is the modified Bessel function of the second kind of order $n - (\frac{1}{2})$. Therefore,

$$C_{QQ}(\tau) = \left(\frac{\tau}{2}\right)^{n-1}\frac{1}{\sqrt{2\pi\tau/k}\,k^{n+1}\,\Gamma(n)}J_{n-(1/2)}\left(\frac{\tau}{k}\right) \tag{13.234}$$

A similar calculation yields

$$C_{QQ}(0) = \frac{1}{k^{2n}[\Gamma(n)]^2}\int_0^\infty \exp(-2u/k)\,u^{2(n-1)}du$$

$$= \frac{\Gamma(2n-1)}{k^{2n}[\Gamma(n)]^2}\left(\frac{k}{2}\right)^{2n-1} \tag{13.235}$$

Therefore,

$$\rho_{QQ}(\tau) = \left(\frac{2\tau}{k}\right)^n \left(\frac{k}{2\pi\tau}\right)^{1/2} \frac{\Gamma(n)}{\Gamma(2n-1)} J_{n-(1/2)}\left(\frac{\tau}{k}\right) \quad (13.236)$$

Note that

$$J_{n-(1/2)}(s) = \sqrt{\frac{\pi}{2s}} \exp(-s) \sum_{i=0}^{n} \frac{(n+i)!}{i!(n-i)!(2s)^i} \quad (13.237)$$

Specifically, for integral values of n,

For $n = 1$, $\rho_{QQ}(\tau) = \exp\left(\frac{-\tau}{k}\right)$

For $n = 2$, $\rho_{QQ}(\tau) = \left(1 + \frac{\tau}{k}\right) \exp\left(\frac{-\tau}{k}\right)$

For $n = 3$, $\rho_{QQ}(\tau) = \left(1 + \frac{\tau}{k} + \frac{\tau^2}{3k^2}\right) \exp\left(\frac{-\tau}{k}\right)$

For $n = 4$, $\rho_{QQ}(\tau) = \left[1 + \frac{\tau}{k} + \frac{2}{5}\left(\frac{\tau}{k}\right)^2 + \frac{1}{15}\left(\frac{\tau}{k}\right)^3\right] \exp\left(\frac{-\tau}{k}\right)$

EXERCISES

13.1 A watershed has an area of 1 km^2. Compute the direct runoff hydrograph due to an effective rainfall occurring at a uniform rate of 2 cm/h for a period of 4 h. The TAC diagram is represented by a triangle that peaks at the origin and has a base of 6 h. The linear reservoir has k of 3 h. Repeat the calculations for the effective rainfall durations of 6 and 8 h.

13.2 Assume that the TAC diagram is represented by a triangle that peaks at the midpoint of its base of 6 h. Solve the problem in Question 13.1.

13.3 Solve Question 13.1 if the TAC curve is represented by a rectangle of base 6 h and height of $\frac{1}{6}$. Compare the results obtained for the three problems in Questions 13.1–13.3. Discuss the effect of TAC curve shape and the duration of the ERH.

13.4 The ERH-DRH data for a laboratory watershed are given in Table 13.5. It is desired to fit the Nash model to the DRH. Estimate the parameters n and k of this model in at least three different ways. Which method yields a better fit? Which method of parameter estimation would you recommend?

TABLE 13.5

Rainfall-runoff data from an experimental facility located at Colorado State University (area = 3.193.90 square feet, surface plastic flaps, September 19, 1969)

Time (s)	*Rainfall Rate (cm/min)*	*Measured Outflow (cm/min)*
.0	.019	.0006
12.1	.019	.0006
25.5	.019	.0006
44.6	.019	.0006
55.9	.019	.0007
72.1	.019	.0011
84.9	.019	.0014
101.0	.00 at 111.32 s	.0023
118.0		.0032
139.7	.00	.0047
158.2	.00	.0057
179.2	.00	.0070
205.3	.00	.0070
223.4	.00	.0070
242.9	.00	.0066
265.3	.00	.0062

TABLE 13.5 *(continued)*

Time (s)	*Rainfall Rate (cm/min)*	*Measured Outflow (cm/min)*
291.9	.00	.0055
317.0	.00	.0048
349.1	.00	.0039
373.9	.00	.0035
393.7	.00	.0031
417.5	.00	.0028
446.8	.00	.0022
480.1	.00	.0020
508.5	.00	.0017
541.2	.00	.0014
562.3	.00	.0012
589.0	.00	.0011
614.6	.00	.0010
635.1	.00	.0009
660.5	.00	.0008
685.6	.00	.0008
710.9	.00	.0008
731.1	.00	.0007
765.6	.00	.0006
795.8	.00	.0005

13.5 Take $n = 3$ in a cascade of unequal reservoirs. Determine for the data of Question 13.4 the values of k_1, k_2 and k_3 by using an appropriate method of parameter estimation. Compare the fit obtained here with that obtained in Question 13.4. Determine if there is any advantage of using this cascade over the Nash model?

13.6 Consider the case $k_j = kr^{j-1}$, $j = 1, 2, 3$. Solve Question 13.5 with this stipulation for $r = .5$ as well as for $r = .75$.

13.7 A watershed has an area of 10 km^2. It receives a uniform effective rainfall of 1 cm/h for 5 h. Represent the watershed by a cascade of three unequal linear reservoirs ($k_1 = 5$ h, $k_2 = 4$ h, and $k_3 = 3$ h). Compute its IUH and then its DRH. Each reservoir represents the same drainage area.

13.8 Let $k_1 = k_2 = k_3 = 4$ h. Solve Question 13.7 and compare the results.

13.9 Solve Question 13.7 if k_1 receives $I = 1.5$ cm/h, k_2 receives $I = 0.5$ cm/h and k_3 receives $I = 1$ cm/h.

13.10 Solve Question 13.8 for the effective rainfall data of Question 13.9.

13.11 Compare the results of Questions 13.7–13.10.

13.12 Consider the following form of the Chow-Kulandaiswamy model:

$$[a_1 D^2 + a_0 D + 1]\, Q(t) = [1 - b_0 D]\, I(t)$$

Determine the IUH of this model.

13.13 Determine the DRH due to any $I(t)$ for the model of Question 13.12.

13.14 Derive the DRH due to an arbitrary $I(t)$ using equation (13.191).

13.15 Derive the DRH due to an arbitrary $I(t)$ using equation (13.194).

13.16 Derive the DRH due to an arbitrary $I(t)$ using equation (13.197).

13.17 Derive the DRH due to an arbitrary $I(t)$ using equation (13.204).

13.18 Derive the DRH due to an arbitrary $I(t)$ using equation (13.207).

CHAPTER FOURTEEN

Linear Conceptual Models for Discrete-Time Systems

Hydrologic processes and their models are usually continuous. However, the data, experimental or field, taken on these processes and used for calibrating, verifying, and applying their models are discrete. Therefore, for practical reasons we have to use discrete-time models. O'Connor (1982) has discussed three discrete forms (or difference equations) of some well-known continuous conceptual models: (1) finite difference form (FDF), (2) discretely coincident form (DCF), and (3) discretely analogous form (DAF). As the discretization interval Δt (or data or sampling interval) tends to zero, all three forms of the difference equations tend to the corresponding differential equations from which they were derived. A major disadvantage of the FDF is that its accuracy is closely tied with the size of Δt. As Δt becomes larger and larger, this form becomes less and less meaningful. Frequently inflow, outflow, and storage are interpolated linearly over Δt, which may or may not be adequate.

The term *discretely coincident* corresponds to the relationship existing between a discrete time series and a continuous time function that coincide at discrete intervals. A discrete system and a continuous system can be called discretely coincident when their outputs corresponding to discretely coincident inputs are themselves coincident. A major advantage of the DCF is that it is exact for any chosen value of Δt (large or small) for the chosen order of the input polynomial over intervals of Δt for which it is derived. Furthermore, once a model is calibrated for one value of Δt, it is calibrated for all values of Δt and can be adjusted for different data intervals simply by adjusting the value of Δt in the difference equation. The DAF does not easily lend itself to changing the data interval. The FDF performs even more poorly in this respect because its accuracy decreases with increasing value of Δt.

The three forms of discrete-time conceptual models have been discussed by several investigators but perhaps the best and most comprehensive account of some of these models is given by O'Connor (1976, 1982). In the ensuing discussion we follow his work closely.

14.1 DISCRETE-TIME LINEAR RESERVOIR

The discrete form of the storage-discharge relation of equation (13.20) can be written as

$$S_j = kQ_j \tag{14.1}$$

where S_j is the storage corresponding to Q_j, the jth direct runoff ordinate, and k is the reservoir coefficient or average residence time as before. The time interval of discretization Δt has been taken as 1. If Δt is retained, then

$$S_{j\Delta t} = kQ_{j\Delta t}$$

Similarly, the discrete form of equation (13.19) can be written as

$$\sum_{r=0}^{j} I_r - \sum_{r=0}^{j} Q_r = S_j \tag{14.2}$$

where I_r is the rth ordinate of the effective rainfall. Also,

$$\Delta t \sum_{r=0}^{j} I_{r\Delta t} - \Delta t \sum_{r=0}^{j} Q_{r\Delta t} = S_{j\Delta t}$$

It is assumed that the system is initially relaxed. From equations (14.1) and (14.2),

$$S_0 = I_0 - Q_0 = kQ_0$$
$$S_1 = S_0 + I_1 - Q_1 = kQ_1$$
$$S_2 = S_1 + I_2 - Q_2 = kQ_2$$
$$\vdots$$
$$S_j = S_{j-1} + I_j - Q_j = kQ_j$$

This system of equations can simply be written as

$$kQ_{j-1} + I_j - Q_j = kQ_j, \qquad j = 1, 2, 3, \ldots$$

Therefore,

$$(1 + k)Q_j - kQ_{j-1} = I_j \tag{14.3}$$

or

$$(\Delta t + k)Q_{j\Delta t} - kQ_{(j-1)\Delta t} = \Delta t I_{j\Delta t}$$

which is a simple first-order difference equation, a discrete-time representation of equation (13.21). If we use the backward difference operator ∇ defined by

$$\nabla Q_j = Q_j - Q_{j-1}$$

then equation (14.3) becomes

$$(1 + k\nabla)Q_j = I_j \tag{14.4}$$

or

$$\left(1 + \frac{k}{\Delta t}\nabla\right)Q_{j\Delta t} = I_{j\Delta t}$$

Equation (14.4) is the discrete form of equation (13.21). We can derive the IUH of discrete linear reservoir using the Z-transform. Note that the IUH of the discrete-time system will be interpreted as the sampled pulse response or the UH (of the sample time interval) of the continuous time system. Taking the Z-transform of equation (14.3),

$$(1 + k)Q(z) - \frac{k}{z}Q(z) = I(z)$$

$$Z[Q_j] = Q(z) = \sum_{j=0}^{\infty} Q_j z^{-j} \tag{14.5}$$

Recall the convolution summation from Chapter 11,

$$Q_j = \sum_{i=0}^{i} I_i h_{j-i} = \sum_{i=0}^{j} h_i I_{j-i}, \qquad j = 0, 1, 2, \ldots \tag{14.6}$$

The Z-transform of equation (14.6) yields

$$Q(z) = I(z)\cdot h(z) = h(z)\cdot I(z) \tag{14.7}$$

where $Q(z)$, $I(z)$ and $h(z)$ are the Z-transforms of the direct runoff Q_j, the effective rainfall I_j, and the IUH h_j, respectively. Evidently,

$$h(z) = \frac{Q(z)}{I(z)} \tag{14.8}$$

Thus, if the Z-transforms of Q and I are available, then the Z-transform of h can be obtained. By inverting $h(z)$ we obtain the IUH. Equation (14.5) yields

$$h(z) = \frac{Q(z)}{Iz} = \left[\frac{z}{(1 + k)z - k}\right]$$
$$= \left(\frac{1}{1 + k}\right)\left(\frac{z}{z - \dfrac{k}{1 + k}}\right) = A\frac{c}{z - C}$$

where $A = 1/(1 + k)$ and $C = k/(1 + k)$. The inverse of $h(z)$, namely, the IUH of the discrete reservoir, is

$$h_j = AC^j = \left(\frac{1}{1 + k}\right)\left(\frac{k}{1 + k}\right)^j, \qquad j = 0, 1, 2, \ldots \tag{14.9}$$

or

$$h_{j\Delta t} = \left(\frac{\Delta t}{\Delta t + k}\right)\left(\frac{k}{\Delta t + k}\right)^j, \qquad j = 0, 1, 2, \ldots$$

$$h_0 = \frac{1}{1 + k}$$

$$\sum_{j=0}^{\infty} h_j = 1$$

Equation (14.9) is a simple geometric series.

We can also derive the S-hydrograph or U_j as well as the IUH or h_j either by solving equation (14.4) directly using Z-transform or from equation (14.9). If $I_j = 1$, $j \geq 1$, then Q_j reduces to U_j, that is,

$$U_j - \frac{k}{1 + k}U_{j-1} = \frac{1}{1 + k}$$

Let $U_j = \beta^j$. Then the characteristic equation corresponding to its homoegeneous difference equation is

$$\beta^j - \frac{k}{1 + k}\beta^{j-1} = 0$$

Its solution is

$$\beta = \frac{k}{1 + k}$$

To determine the particular solution let $U_j = \gamma$. Then

$$\gamma\left(1 - \frac{k}{1+k}\right) = \frac{1}{1+k}$$

yielding

$$\gamma = 1$$

Therefore, the general solution is

$$U_j = a\beta^j + \gamma = a\left(\frac{k}{1+k}\right)^j + 1$$

If the initial condition $U_0 = 0$ is used, then $a = -1$. This then leads to

$$U_j = 1 - \left(\frac{k}{1+k}\right)^j \tag{14.10}$$

Alternatively,

$$U_j = \sum_{i-0}^{j} h_i = \sum_{i-0}^{j}\left(\frac{1}{1+k}\right)\left(\frac{k}{1+k}\right)^i$$
$$= 1 - \left(\frac{k}{1+k}\right)^j$$

which is the same as equation (14.10).

Likewise, when $I_j = 1, j = 0, I_j = 0, j > 0$, then Q_j reduces to h_j. From the preceding derivation, it is clear that

$$h_j = h_0\left(\frac{k}{1+k}\right)^j$$

where h_0 is constant to be determined by knowing the initial condition. Note that

$$\sum_{j=0}^{\infty} h_j = 1$$

Therefore,

$$1 = h_0 \sum_{j=0}^{\infty}\left(\frac{k}{1+k}\right)^j$$
$$= \frac{h_0}{1 - \dfrac{k}{1+k}} = (1+k)h_0$$

This yields $h_0 = 1/(1 + k)$. Therefore,

$$h_j = \left(\frac{1}{1+k}\right)\left(\frac{k}{1+k}\right)^j, \qquad j = 0, 1, 2, \ldots$$

which is the same as equation (14.9).

14.1.1. Characteristics and Comparison with Continuous-Time Counterpart

For purposes of comparison, let the IUH of equation (13.23) be written as

$$h(0, t) = \frac{1}{k_l}\exp\left[\frac{-t}{k_1}\right] \tag{14.11}$$

Thus, the UH of unit interval duration $h(1, t)$ is

$$h(1, t) = U(t) - U(t-1)$$
$$= \exp\left[-\frac{(t-1)}{k_1}\right] - \exp\left[\frac{-t}{k_1}\right] \tag{14.12a}$$

From equation (14.11), the UH at unit time intervals can be expressed as

$$h(1, j) = \left[1 - \exp\left(\frac{-1}{k_1}\right)\right]\left[\exp\left(\frac{-1}{k_1}\right)\right]^j, \qquad j > 0$$
$$= 0, \qquad j = 0 \tag{14.12b}$$

Let us compare equation (14.12b) with equation (14.9). The shape of the UH given by equation (14.12b) is more realistic than the geometric series IUH of equation (14.9). The peak value of $h(1, j)$ always occurs at $t = j = 1$. Furthermore, its IUH does not have any rising limb; hence neither of these models is satisfactory as a watershed model. However, if a linear reservoir is used as a component of an overall watershed model, then one would seem to prefer the continuous linear reservoir to its discrete counterpart.

O'Connor (1976) showed an equivalence between h_j and $h(1, j)$. If a one-time interval lag is introduced into equation (14.4) as

$$(1 + k\nabla)Q_j = I_{j-1} \tag{14.13}$$

for which

$$h_j = \begin{cases} 0, & j = 0 \\ \left(\dfrac{1}{1+k}\right)\left(\dfrac{k}{1+k}\right)^{j-1}, & j > 0 \end{cases} \tag{14.14}$$

then the substitution

$$\frac{k}{1+k} = \exp(-1/k)$$

transforms h_j exactly into $h(1, j)$ given by equation (14.12b). Thus, the two models are *discretely coincident*; that is, the continuous output exactly equals the discrete output at unit time intervals. This result follows from a more general relationship between discrete and continuous systems for the case of pulsed input derived by Box and Jenkins (1970). They show that if a continuous linear system is represented by an *m*-

TABLE 14.1

The IUH of a linear reservoir

j	*Sampled Continuous* h(t)	*Continuous* h(*1*, t)	*Discrete*, h_j *Equation (14.9)*	*Discrete*, h_j *Equation(14.21)*
0	.333	.0	.25	.286
1	.239	.283	.188	.204
2	.171	.203	.141	.146
3	.123	.146	.105	.104
4	.088	.104	.079	.074
5	.063	.075	.059	.053
6	.045	.054	.044	.038
7	.032	.038	.033	.027
8	.023	.027	.025	.020
9	.017	.020	.019	.013
10	.012	.014	.014	.010
11	.009	.010	.011	.007
12	.006	.007	.008	.005
13	.004	.005	.006	.004
14	.003	.004	.004	.003
15	.002	.003	.003	.001

order differential equation with $P(D)$ as polynomial of differential operator,

$$P(D)Q(t) = I(t) \tag{14.15}$$

then there exists a difference equation of the form

$$B(\nabla)Q_j = E(\nabla)I_{j-1} \tag{14.16}$$

of the order $(m, m-1)$ such that the output from the continuous system is identical to that of its discrete counterpart at the discrete times provided that the input is pulsed. $B(\nabla)$ and $E(\nabla)$ are polynomials of difference operators.

EXAMPLE 14.1

Compute the IUH of the discrete linear reservoir with $k = 3$ h, and compare it with that of the continuous linear reservoir sampled at one time unit interval.

The IUH of the discrete linear reservoir is given by equation (14.9), and the sampled IUH of the continuous linear regression is given by $h(t) = \exp(-t/k)/k$. These are tabulated in Table 14.1. Also computed is the UH, $h(1, t)$, by using equation (14.12b). It is clear that the sampled $h(t)$ differs from the discrete h_j in the beginning. However, if $h(t)$ is converted to $h(1, t)$ by using equation (14.12b), then the two UHs are in closer agreement.

EXAMPLE 14.2

Compute the SH of the discrete linear reservoir with $k = 3$ h, and compare it with that of its continuous counterpart sampled at one time unit interval.

The SH of discrete linear reservoir is given by equation (14.10), and that of continuous linear reservoir is given by $U(t) = 1 - \exp(-t/k)$. For $k = 3$ these are tabulated in Table 14.2. It is clear that the continuous linear reservoir responds more quickly than the discrete linear reservoir given by equation (14.10) but the difference between the two S-hydrographs is marginal.

TABLE 14.2

The S-hydrograph of a linear reservoir

Discrete Equation (14.10)		*Discrete Equation (14.21)*	*Continuous*	
j	U_j		t	U(t)
0	.0	.0	0	.0
1	.250	.286	1	.284
2	.438	.490	2	.487
3	.578	.636	3	.632
4	.684	.740	4	.736
5	.763	.814	5	.811
6	.822	.867	6	.865
7	.867	.905	7	.903
8	.900	.932	8	.931
9	.925	.952	9	.950
10	.944	.965	10	.964
11	.958	.975	11	.974
12	.968	.982	12	.982
13	.976	.987	13	.987
14	.982	.991	14	.991
15	.987	.994	15	.993
16	.990	.995	16	.995
17	.993	.997	17	.997
18	.994	.998	18	.998
19	.996	.998	19	.998
20	.997	.999	20	.999
21	.998	.999	21	.999
22	.998	.999	22	.999
23	.999	1.000	23	1.000
24	.999	1.000	24	1.000
25	.999	1.000	25	1.000
26	.999	1.000	26	1.000
27	1.000	1.000	27	1.000
28	1.000	1.000	28	1.000
29	1.000	1.000	28	1.000
30	1.000	1.000	28	1.000

14.1.2 Moments and Cumulants

The moments of h_j about its origin can be obtained directly from their definition or by using $h(z)$ as a moment-generating function. The moments of h_j about its origin are:

$$M_0 = \sum_{j=0}^{\infty} h_j = h(z)\,|_{z=1} = 1 \quad (14.17a)$$

$$M_1 = \left[-z\frac{d}{dz}h(z) \right] |_{z=1} = k \quad (14.17b)$$

$$M_2 = \left[-z\frac{d}{dz}\left\{ -z\frac{d}{dz}h(z) \right\} \right] |_{z=1}$$

$$= k(1 + 2k) \quad (14.17c)$$

$$M_3 = k(1 + 6k + 6k^2) \quad (14.17d)$$

$$M_4 = k(1 + 2k)\,(1 + 12k + 12k^2) \quad (14.17e)$$

Likewise the cumulants are:

$$K_1 = k \quad (14.18a)$$

$$K_2 = k(1 + k) \quad (14.18b)$$

$$K_3 = k(1 + k)(1 + 2k) \quad (14.18c)$$

$$K_4 = k(1 + k)(1 + 6k + 6k^2) \quad (14.18d)$$

It is seen that the parameter k is equal to the lag time, which can be computed variously as discussed in Chapter 9. Some workers (Golding, 1980; Stubchner, 1981) take k equal to the time of concentration, as done in the rational method and the Santa Barbara urban hydrograph (SBUH) method.

14.1.3 Finite-Difference Approximation

Some investigators use a finite difference approximation (in the sense of the Muskingum method of flood routing) to describe a continuous-time linear reservoir (Pederson, Peters, and Helweg, 1980; Ponce, 1980). If we choose a sufficiently small time interval Δt such that a reasonable approximation of equation (13.19) can be obtained,

$$\frac{I_j + I_{j+1}}{2} - \frac{Q_j + Q_{j+1}}{2} = \frac{S_{j+1} - S_j}{\Delta_t}, \quad j = 1, 2, \ldots \quad (14.19)$$

where $\Delta_t = t_{j+1} - t_j$. Combining equations (14.19) and (14.1),

$$\frac{I_j + I_{j+1}}{2} - \frac{Q_j + Q_{j+1}}{2} = \frac{k}{\Delta t}(Q_{j+1} - Q_j)$$

Solving for Q_{j+1},

$$Q_{j+1} = 2c_1 \frac{I_j + I_{j+1}}{2} - c_2 Q_j \quad (14.20)$$

in which

$$c_1 = \frac{\Delta t}{2k + \Delta t}$$

$$c_2 = \frac{2k - \Delta t}{2k + \Delta t}$$

Note that $(I_j + I_{j+1})/2$ is the average effective rainfall for the time interval Δt. Since the ERH is normally given as a histogram, Δt is chosen to coincide with the duration of ERH pulses. This implies $I_j = I_{j+1}$. Then equation (14.20) becomes

$$Q_{j+1} = 2c_1 I_{j+1} - c_2 Q_j \quad (14.21)$$

If I is reduced to take place with a unit volume over the duration Δt, then equation (14.21) prescribes the UH of duration Δt. This would practically be identical to the UH obtained from equation (14.14).

EXAMPLE 14.3

Compute the IUH by using equation (14.21) with $k = 3$ h and $\Delta t = 1$ h, and compare the answer with that obtained in Example 14.1 and of the continuous counterpart.

The computations of the IUH are shown in Table 14.1. The IUH by equation (14.21) is closer to $h(1, t)$ than the IUH $h(t)$ sampled at 1 = h intervals.

EXAMPLE 14.4

Compute the S-hydrograph by employing equation (14.21) with $k = 3$ h and $\Delta t = 1$ h. Compare it with the S-hydrograph obtained in Example 14.2 and of the continuous counterpart.

Table 14.2 lists the S-hydrograph given by equation (14.21). Clearly, this S-hydrograph is almost the same as that of the continuous linear reservoir.

14.2 DISCRETE-TIME LAG AND ROUTE MODEL

The discrete form of the equation for a simple lag and route model can be expressed as

$$(1 + k\nabla)Q_j = I_{j-b} \quad (14.22a)$$

which can also be expressed as

$$(k + 1)Q_j - kQ_{j-1} = I_{j-b} \quad (14.22b)$$

where b is the lag parameter assuming integer values. If we take the Z-transform of equation (14.22b),

$$(k + 1)Q(z) - \frac{k}{z}Q(z) = z^{-b}I(z) \quad (14.23)$$

Therefore,

$$h(z) = \frac{Q(z)}{I(z)} = z^{-b}\left(\frac{1}{1 + k}\right)\left\{\frac{z}{z - [k/(1 + k]}\right\} = z^{-b}A\left(\frac{z}{z - C}\right)$$

in which $A = 1/(1 + k)$ and $C = k/(1 + k)$. Taking its inverse,

$$h_j = \begin{cases} 0, & \text{for } j < b \\ AC^{j-b}, & \text{for } j \geq b \end{cases} \tag{14.24}$$

Equation (14.24) shows that h_j is the IUH of a linear reservoir delayed by the lag b.

The S-hydrograph of the lag and route model is

$$\begin{aligned} U_j &= \sum_{i=0}^{j} h_i \\ &= AC^{-b}\left[\sum_{i=0}^{j} C^i - \sum_{i=0}^{b} C^i\right] \\ &= 1 - \left(\frac{k}{1+k}\right)^{j-b}, \quad j \geq b \\ &= 0, \quad j \leq b \end{aligned} \tag{14.25}$$

14.2.1 Moments and Cumulants

The first four moments about origin of equation (14.24) are

$$M_1 = k + b \tag{14.26a}$$

$$M_2 = k(1 + 2k) \tag{14.26b}$$

$$M_3 = k(1 + 6k + 6k^2) \tag{14.26c}$$

$$M_4 = k(1 + 2k)(1 + 12k + 12k^2) \tag{14.26d}$$

Likewise, the first four cumulants of equation (14.24) are

$$K_1 = k + b \tag{14.27a}$$

$$K_2 = k(1 + k) \tag{14.27b}$$

$$K_3 = k(1 + k)(1 + 2k) \tag{14.27c}$$

$$K_4 = k(1 + k)(1 + 6k + 6k^2) \tag{14.27d}$$

14.3 DISCRETE-TIME LINEAR CHANNEL

The IUH of the discrete linear channel subject to the unit function of the ER at its upstream can be written as

$$h_j = \begin{cases} 0, & j < T \\ 1, & j = T \text{ or } \delta(j - T) \\ 0, & j > T \end{cases} \tag{14.28}$$

where T is the channel translation time. The S-hydrograph can therefore be written as

$$U_j = u_1(j - T)$$

or

$$U_j = \begin{cases} 0, & j < \mathrm{T} \\ 1, & j \geq T \end{cases} \tag{14.29}$$

If the ER is introduced to the channel laterally along its side, then

$$h_j = \begin{cases} \dfrac{1}{T}, & 0 \leq j \leq T \\ 0, & j \geq T \end{cases} \tag{14.30}$$

Its S-hydrograph can be expressed as

$$U_j = \sum_{i=0}^{j} h_i = \sum_{i=0}^{j} \frac{1}{T}$$

This leads to

$$U_j = \begin{cases} \dfrac{j}{T}, & 0 \leq j \leq \mathrm{T} \\ 1, & j \geq T \end{cases} \tag{14.31}$$

14.3.1 Moments and Cumulants

The first four moments of equation (14.28) can be derived by employing their definition,

$$\begin{aligned} M_r &= \sum_{j=0}^{\infty} j^r h_j = \sum_{j=0}^{\infty} j^r \delta(j - T) \\ &= T^r \end{aligned}$$

Therefore,

$$M_1 = T \tag{14.32a}$$

$$M_2 = T^2 \tag{14.32b}$$

$$M_3 = T^3 \tag{14.32c}$$

$$M_4 = T^4 \tag{14.32d}$$

Likewise, the first four cumulants can be expressed as

$$K_1 = T \tag{14.33a}$$

$$K_2 = 0 \tag{14.33b}$$

$$K_3 = 0 \tag{14.33c}$$

$$K_4 = 0 \tag{14.33d}$$

The moments and cumulants of equation (14.31) are derived similarly:

$$M_r = \sum_{j=0}^{\infty} j^r h_j = \sum_{j=0}^{T} \frac{j^r}{T}$$

$$M_1 = \frac{1}{2}(1 + T) \tag{14.34a}$$

$$M_2 = \frac{1}{6}(T + 1)(2T + 1) \tag{14.34b}$$

$$M_3 = \frac{T}{4}(T+1)^2 \quad (14.34c)$$

$$M_4 = \frac{1}{T}(T+1)(2T+1)(3T^2+3T-1) \quad (14.34d)$$

and

$$K_1 = \frac{1}{2}(1+T) \quad (14.35a)$$

$$K_2 = \frac{1}{10}(T^2-1) \quad (14.35b)$$

$$K_3 = 0 \quad (14.35c)$$

14.4 THE DISCRETE-TIME NASH MODEL

For the Nash model the discrete-time form of the governing equation (13.71) can be expressed as

$$(1 + k\nabla)^n Q_j = I_j \quad (14.36)$$

Taking the Z-transform,

$$\left[\frac{(1+k)z - k}{z}\right]^n Q(z) = I(z)$$

Therefore,

$$h(z) = \frac{Q(z)}{I(z)} = \left[\frac{z}{(1+k)z - k}\right]^n = A^n\left(\frac{z}{z - C}\right)^n \quad (14.37)$$

in which A and C are as defined before. Taking its inverse,

$$h_j = \frac{1}{2\pi i}\int h(z)z^{j-1}\,dz$$

$$= \text{residue of } h(z)z^{j-1} \text{ at the multiple pole } z = \frac{k}{1+k} = C$$

Therefore,

$$h_j = \frac{1}{(n-1)!}\frac{d^{n-1}}{dz^{n-1}}[(z-C)^n h(z)\,z^{j-1}]\,|_{z=C}$$

$$= \frac{(j+n-1)!}{(n-1)!j!}A^nC^j$$

$$= \binom{j+n-1}{j}A^nC^j \quad (14.38)$$

Equation (14.38), which is the negative binomial or Pascal distribution, was derived by O'Connor (1976) and Bree (1977). To allow for noninteger values of n, this may be written as

$$h_j = \frac{\Gamma(j+n)}{\Gamma(n)\,\Gamma(j+1)}A^nC^j, \qquad j \geq 0 \quad (14.39)$$

Using $\Gamma(j+1) = j\Gamma(j)$ and the definition of the beta function,

$$\beta(j, n) = \frac{\Gamma(j)\,\Gamma(n)}{\Gamma(j+n)}$$

equation (14.39) can be written as

$$h_j = \frac{1}{j\beta(j, n)}A^nC^j \quad (14.40)$$

Bree (1977) argued that the time interval Δt, taken here as equal to 1, should be retained in the derivations. If so, equation (14.40) becomes

$$h_{j,\Delta t} = \frac{1}{j\beta(k, n)}\left[\frac{\Delta t}{\Delta t + k}\right]^n\left[\frac{k}{\Delta t + k}\right]^j \quad (14.41)$$

He showed that as $\Delta t \to 0$, $h_{j,\Delta t} \to h(t)$, derived in Chapter 11. O'Connor (1977), however, showed that taking $\Delta t = 1$ was a matter of convenience and not a restriction on the model.

EXAMPLE 14.5

Compute the IUH of the discrete Nash model for $n = 4$ and $k = 5$ hours. Compare it with the sampled IUH of its continuous counterpart.

For the discrete Nash model,

$$h_j = \frac{(j+3)!}{6j!}\left(\frac{1}{6}\right)^4\left(\frac{5}{6}\right)^j$$

$$= 1.286 \times 10^{-4}\frac{(j+3)!}{j!}(.833)^j$$

For the continuous Nash model,

$$h(t) = \frac{1}{5}\frac{1}{6}\frac{t^3}{125}\exp\left(\frac{-t}{5}\right)$$

$$= \frac{1}{3750}t^3\exp(-t/5)$$

h_j and $h(t)$ are computed in Table 14.3. The differences between the two IUHs are very small except at the beginning.

EXAMPLE 14.6

Compute the S-hydrograph of the discrete Nash model for $n = 4$ and $k = 5$ h. Compare it with that of its continuous counterpart.

The S-hydrograph of the continuous Nash model is

$$U(t) = \int_0^t h(t)\,dt = \int_0^t \frac{1}{3750}t^3\exp\left(\frac{-t}{5}\right)dt$$

$$= 1 - \frac{1}{6}\exp\left(\frac{-t}{5}\right)\left[\frac{t^3}{125} + \frac{3}{25}t^2 + \frac{6}{5}t + 6\right]$$

The S-hydrograph of the discrete Nash model is

$$U_m = \sum_{j=0}^{m}\frac{(j+n-1)!}{(n-1)!j!}\left(\frac{1}{1+k}\right)\left(\frac{k}{1+k}\right)^j$$

$$= \sum_{j=0}^{m}\frac{(j+3)!}{j!}1.286 \times 10^{-4}\left(\frac{5}{6}\right)^j$$

The S-hydrograph for both cases is computed in Table 14.4.

TABLE 14.3

The IUH of the Nash model

j	h_j	$h(t)$
0	.001	.0
1	.003	.000
2	.005	.001
3	.009	.004
4	.013	.008
5	.017	.012
6	.022	.017
7	.026	.023
8	.030	.028
9	.033	.032
10	.036	.036
11	.038	.039
12	.039	.042
13	.040	.044
14	.041	.045
15	.041	.045
16	.040	.045
17	.040	.044
18	.039	.043
19	.037	.040
20	.036	.039
21	.034	.037
22	.032	.035
23	.030	.033
24	.028	.030
25	.027	.028
26	.026	.026
27	.023	.024
28	.021	.022
29	.019	.020
30	.018	.018
31	.016	.016
32	.015	.015
33	.013	.013
34	.012	.012
35	.011	.010
36	.010	.009
37	.010	.008

TABLE 14.4

The S-hydrograph of the Nash model

m	U_m	$U(t)$
1	.003	.000
2	.008	.001
3	.017	.003
4	.030	.009
5	.047	.019
6	.069	.034
7	.095	.054
8	.124	.079
9	.157	.109
10	.193	.143
11	.231	.181
12	.270	.221
13	.311	.264
14	.351	.308
15	.392	.353
16	.433	.398
17	.472	.442
18	.511	.485
19	.548	.527
20	.584	.567
21	.618	.605
22	.650	.641
23	.680	.674
24	.708	.706
25	.735	.735
26	.759	.762
27	.782	.788
28	.803	.809
29	.823	.830
30	.840	.849
31	.856	.866
32	.871	.881
33	.885	.895
34	.897	.907
35	.908	.918
36	.918	.928
37	.927	.937
∞	1	1

14.4.1 Moments and Cumulants

The moments about origin of equation (14.40) can be derived by using the Z-transform,

$$M_1 = nk \tag{14.42a}$$

$$M_2 = nk + n(n+1)k^2 \tag{14.42b}$$

$$M_3 = nk + 3n(n+1)k^2 + n(n+1)(n+2)k^3 \tag{14.42c}$$

$$M_4 = nk + 7n(n+1)k^2 + 6n(n+1)(n+2)k^3 + n(n+1)(n+2)(n+3)k^4 \tag{14.42d}$$

Similarly the cumulants are

$$K_1 = nk \tag{14.43a}$$

$$K_2 = nk(1+k) \tag{14.43b}$$

$$K_3 = nk(1+k)(1+2k) \tag{14.43c}$$

$$K_4 = nk(1+k)(1+6k+6k^2) \tag{14.43d}$$

Bree (1977) derived equations (14.42) and (14.43) by using factorial moments and factorial cumulants. He showed that for discrete functions, these lead to a more straightforward method of estimating their parameters and may therefore be preferable to ordinary moments and cumulants. He further noted that the theorems of ordinary moments and cumulants also applied to their factorial counterparts.

14.5 GENERAL LINEAR DIFFERENCE EQUATION

Equation (8.11) can be expressed in discrete form by substituting the finite difference approximations for the derivatives. Let the central difference approximations of the derivatives be defined as

$$\frac{dQ}{dt} \cong \frac{Q_{j+1} - Q_{j-1}}{2\Delta t}$$

$$\frac{d^2Q}{dt^2} \cong \frac{Q_{j+1} - 2Q_j + Q_{j-1}}{\Delta t^2}$$

and so on. Therefore,

$$Q_j + \gamma_1 Q_{j-1} + \gamma_2 Q_{j-2} + \ldots = \beta_0 I_j + \beta_1 I_{j-1} + \ldots$$

This can be written in compact form as

$$\Phi(B)Q_j = \theta(B)I_j \qquad (14.44)$$

where B is a backward shift operator such that

$$BQ_j = Q_{j-1}$$
$$B^r Q_j = Q_{j-r}$$

and where B is related to the difference operator ∇,

$$\nabla Q_j = Q_j - Q_{j-1} = (1 - B)Q_j$$

$\Phi(B)$ and $\theta(B)$ are the polynomials $1 + \gamma_1 B + \gamma_2 B^2 + \ldots$, and $\beta_0 + \beta_1 B + \beta_2 B^2 + \ldots$, respectively. Equation (14.44) can be considered as a particular case of the stochastic difference equation examined by Box and Jenkins (1970),

$$\Phi(B)Q_j = \theta(B)I_j + \psi(B)\epsilon_j \qquad (14.45)$$

where ϵ_j is a random variable with the properties of *white noise*, and $\psi(B)$ a polynomial in operator B. At this point we are not concerned with stochastic models.

From equation (14.44),

$$Q_j = \frac{\theta(B)}{\Phi(B)} I_j = \psi(B)I_j \qquad (14.46)$$

The operator $\psi(B)$ is the transfer function of the system or simply the system function; it is just another form of $h(z)$. For stability the roots of the characteristic polynomial

$$\Phi(B) = 0$$

with B considered as a variable, lie outside the unit circle. It can be shown that the discrete models discussed previously are special cases of equation (14.46).

14.5.1 Linear Reservoir

Consider equation (14.4) describing the discrete-time linear reservoir, which can be written in B notation as

$$(1 - CB)Q_j = AI_j$$

or

$$Q_j = \left(\frac{A}{1 - CB}\right) I_j$$

with $A = 1/(1 + k)$ and $C = k/(1 + k)$. Therefore, $\psi(B) = A/(1 - CB)$ for a discrete linear reservoir. If we expand it as an infinite series,

$$\frac{A}{1 - CB} = A(1 + CB + C^2B^2 + C^3B^3 + \cdots + C^iB^i + \cdots)$$

Hence,

$$\begin{aligned} Q_j &= A(1 + CB + C^2B^2 + C^3B^3 + \cdots + C^iB^i + \cdots)I_j \\ &= A(I_j + CI_{j-1} + C^2I_{j-2} + \cdots + C^4I_{j-4} + \cdots) \\ &= \sum_{r=0}^{\infty} AC^r I_{j-r} \end{aligned} \qquad (14.47)$$

which is simply the discrete convolution of I with $h_j = AB^j$, $j = 0, 1, 2, \ldots$. The operator $\psi(B)$ is the system function, where the coefficient of B^r is the rth ordinate of the IUH. In general, we can write

$$\psi(B) = \sum_{j=0}^{\infty} h_j B^j \qquad (14.48)$$

which is analogous to the Z-transform of h_j,

$$h(z) = \sum_{j=0}^{\infty} h_j z^{-j}$$

If $I_j = \delta(j)$, then $Q_j = h_j$. Therefore,

$$h_j = \psi(B)\,\delta(j)$$

Taking the Z-transform,

$$h(z) = \psi(B) \qquad (14.49)$$

It is then clear that $h(z)$ can be directly obtained from $\psi(B)$ by replacing B by z^{-1} such that $h(z) = \psi(z^{-1})$. This explains the advantage of writing the difference equation in operator B notation.

14.5.2 The Nash Model

Equation (14.36) can be written in operator B notation as

$$(1 - CB)^n Q_j = A^n I_j \qquad (14.50)$$

This can be considered, following the analogy with the continuous case, as a special case of

$$(1 - a_1B)(1 - a_2B) \cdots (1 - a_nB)Q_j = I_j$$

or

$$Q_j = \frac{1}{\prod_{i=1}^{n} (1 - a_iB)} I_j \qquad (14.51)$$

for which all the roots are equal.

14.5.3 The Chow-Kulandaiswamy Model

Consider a special case of the Chow-Kulandaiswamy model (Chow and Kulandaiswamy, 1971, 1982) given by equation (14.44) such that both left and right sides represent equal root cascades, that is,

$$(1 + k\nabla)^n Q_j = (1 + a\nabla)^m A_j, \qquad n \geq m \qquad (14.52)$$

This model has four parameters k, n, a, and m, and is flexible and parsimonious. Writing equation (14.52) in operator B notation as

$$x^m(1 - CB)^n Q_j = A^n(1 - yB)^m I_j, \qquad n \geq m$$

where $C = k/(1 + k)$, $A = 1/(1 + k)$, $x = 1/(1 + a)$, and $y = a/(1 + a)$. From the Z-transform of equation (14.52)

$$h(z) = \frac{A^n}{x^m}\left[\frac{(z - y)^m z^{n-m}}{(z - C)^n}\right] = \sum_{i=0}^{m} \frac{(-1)^i A^n \binom{m}{i} y^i z^{n-i}}{x^m (z - C)^n}$$

On inverting,

$$h_j = \sum_{i=0}^{m} \binom{m}{i} (-1)^i \frac{A^n y^i}{X^m (n - 1)!} \frac{d^{n-1}}{dz^{n-1}} [z^{j+n-i-1}]_{z=C}$$
$$= \sum_{i=0}^{m} \binom{m}{i} (-1)^i \frac{A^n y^i (j + n - r + 1)!}{x^m (n - 1)!\,(j - i)!} C^{j-i}$$
$$= \sum_{i=0}^{m} \binom{m}{i} (-1)^i \frac{A^n y^i \,\Gamma(j + n - i)}{x^m\,\Gamma(n)\,\Gamma(j - i + 1)} C^{j-i} \qquad (14.53)$$

where the gamma function allows for nonintegral values of n but not of m, j, or i. Note that if $m = 0$, equation (14.53) reduces to equations (14.39) and (14.40). If $m = 1$,

$$h_j = \frac{A^n}{x(n - 1)!}\left[\frac{(j + n - 1)!}{j!} C^j - \frac{(j + n - 2)!}{(j - 1)!} y C^{j-1}\right] \qquad (14.54)$$

Similar derivations can be obtained for higher values of m.

The parameters of the model in equation (14.52) can be estimated more conveniently by the method of cumulants. Since there are four parameters, it is necessary to derive the first four cumulants of h_j. An application of the theorem of moments or cumulants makes this task straightforward. Let

$$W_j = (1 + k\nabla)^n Q_j = (1 + a\nabla)^m I_j$$

The theorem of cumulants for $W_j = (1 + k\nabla)^n Q_j$ yields

$$K_r[Q] = K_r[h_1(n, k)] + K_r[W], \qquad r = 1, 2, \ldots \qquad (14.55a)$$

where h_1 (n, k) is the IUH of the cascade having the parameter n and k. Likewise, for $W_j = (1 + a\nabla)^m I_j$,

$$K_r[I] = K_r[h_2(m, a)] + K_r[W], \qquad r = 1, 2, \ldots \qquad (14.55b)$$

where h_2 (m, a) denotes the IUH of the cascade having the parameters m and a.

Substracting equation (14.55b) from equation (14.55a),

$$K_r[Q] - K_r[I] = K_r[h_1(n, k)] - K_r[h_2(m, a)] \qquad (14.56)$$

If h_j is the IUH of the model, then it is known

$$K_r[Q] - K_r[I] = K_r[h]$$

Therefore,

$$K_r[h] = K_r[h_1(n, k)] - K_r[h_2(m, a)] = K_r[Q] - K_r\,[I] \qquad (14.57)$$

Thus, given I_j and Q_j the rth cumulant of h_j can be evaluated. The first four cumulants of equation (14.52) can be obtained from equations (14.43) and (14.57) as

$$K_1 = nk - ma \qquad (14.58a)$$

$$K_2 = nk(1 + k) - ma(1 + a) \qquad (14.58b)$$

$$K_3 = nk(1 + k)(1 + 2k) - ma(1 + a)(1 + 2a) \qquad (14.58c)$$

$$K_4 = nk(1 + k)(1 + 6k + 6k^2) - ma(1 + a)(1 + 6a + 6a^2) \qquad (14.58d)$$

14.6 GENERAL DISCRETE-TIME LINEAR RESERVOIR

Dingman (1985) expressed a general discrete-time linear reservoir as

$$V_i = \alpha[(1 - a)S_i + a\,S_{i-1}] \qquad (14.59)$$

in which V_i is the volume of outflow during the time period i, S_i is the storage at the end of time period i, S_{i-1} is the storage at the end of time period $(i - 1)$ or at the beginning of time period i, and α and a are dimensionless constants. Equation (14.59) expresses outflow volume as a weighted sum of storages. The constant a is an arbitrary weighting constant, and α is related to the reservoir constant k. Physical considerations require that $0 \leq \alpha a \leq 1$ and $0 \leq \alpha(1 - a) \leq 1$, which are satisfied if and only if

$$0 \leq \alpha \leq 1 \qquad (14.60)$$

and

$$0 \leq a \leq 1 \qquad (14.61)$$

Equation (14.1) can be related to equation (14.59). The volume of outflow during time interval i, that is, between t_{i-1} and t_i, is

$$V_i = \int_{t_{i-1}}^{t_i} Q(w)\,dw = \frac{1}{k}\int_{t_{i-1}}^{t_i} S(w)\,dw \tag{14.62}$$

If the storage-time relation is assumed linear, then equation (14.62) can be approximated as

$$\begin{aligned} V_i &= \frac{1}{k}\int_{t_{i-1}}^{t_i}\left[S_{i-1} + \frac{S_i - S_{i-1}}{t_i - t_{i-1}}(w - t_{i-1})\right] dw \\ &= \frac{1}{k}\left[\frac{\Delta t}{2} S_i + \frac{\Delta t}{2} S_{i-1}\right], \qquad \Delta t = t_i - t_{i-1} \end{aligned} \tag{14.63}$$

On comparing equation (14.63) with equation (14.59),

$$\frac{\Delta t}{2k} = \alpha(1 - a) = \alpha a \tag{14.64}$$

which can be true only if $a = .5$. Then equation (14.64) yields

$$\frac{\Delta t}{k} = \alpha \tag{14.65a}$$

or

$$\frac{\Delta t}{\alpha} = k \tag{14.65b}$$

Thus, α is the product of the inverse of the storage constant k and the discretization interval. It is minimum when Δt is minimum, meaning that the continuous reservoir with linear variation in storage has the least value of α. From equations (14.60) and (14.65) it is clear that

$$\Delta t \le k < \infty \tag{14.66}$$

Equation (14.66) states that the discretization time interval should be less than the storage lag or reservoir residence time.

The requirement $a = .5$ assigns equal weight to S_i and S_{i-1} in computation of V_i. Most studies in hydrologic modeling have used $a = 0$ and some have $a = 1$. Thus, three cases can be distinguished: (1) $a = 0$, a backward case as given by equation (14.1); (2) $a = .5$, an average case; and (c) $a = 1$, forward case. The choice of a affects the relations among statistics of I, Q, and S (Gould, 1959; Melentijevich, 1965).

This formulation is analogous to the Muskingum method of flood routing. Combining equations (14.59) and (14.2),

$$S_i = ar_i + \left(\frac{1}{\alpha} - a\right) V_i \tag{14.67}$$

in which r_i is the volume of inflow during the interval i. Appropriate substitutions of equations (14.2) and (14.59) lead to the Muskingum type formulation,

$$V_i = C_0 r_i + C_1 r_{i-1} + C_2 V_i \tag{14.68}$$

where

$$C_0 = \frac{\alpha - a\alpha}{1 + \alpha - a\alpha}$$

$$C_1 = \frac{a\alpha}{1 + \alpha - a\alpha}$$

and

$$C_2 = \frac{1 - a\alpha}{1 + \alpha - a\alpha}$$

These expressions show that C_i, $i = 0, 1, 2$, depend on the value of a for a given α.

14.7 STRUCTURAL RELATION BETWEEN DETERMINISTIC AND STOCHASTIC MODELS

The linear deterministic models are structurally analogous to linear stochastic models (Dooge, 1972; Quimpo, 1971). This analogy has been exploited to estimate parameters of one type of models from those of the other type (Quimpo, 1971, 1973; Moss and Dawdy, 1973; Klemes, 1974; Klemes and Boruvka, 1975; Spolia and Chander, 1974; Strupczewski et al., 1975; Tao and Delleur, 1976; Selvalingam, 1979). The structural relations between these two types of models can be visualized in two ways: (1) convolution summation, and (2) difference equations. In either case, the models are of autoregressive moving average (ARMA) type. Let us recall the discrete-time convolution,

$$Q_i = \sum_{j=0}^{M} h_j I_{i-j} \tag{14.69}$$

where M represents the finite system memory. In matrix notation

$$[Q] = [I][h] \tag{14.70}$$

This equation expresses the direct runoff as a weighted moving average of the effective rainfall, that is, this is a moving average (MA) model of order M.

If the direct runoffs due to separate effective rainfall events overlap, then the linear form cannot be reduced to the convenient form of equation (14.59). However, it can be shown that the cross-correlation function ρ_{IQ} of the effective rainfall and direct runoff is related to the autocorrelation function ρ_{II} of the effective rainfall by the convolution-type relation,

$$\rho_{IQ}(i) = \sum_{j=0}^{M} h_j \rho_{II}(i - j) \tag{14.71}$$

Equation (14.71) holds for both deterministic and stochastic effective rainfall (Lee, 1960). Note that ρ_{II} is symmetrical and decays as the value of its argument moves away from zero in both directions; upon suitable truncation it represents an input of finite length. If M is unusually long, then equation (14.71) may be replaced by a more convenient relation (Lee, 1960),

$$\rho_{QQ}(i) = \sum_{j=0}^{N} \rho_{hh}(j)\rho_{II}(i-j) \tag{14.72}$$

in which N represents the length of ρ_{II} and is less than M. Equations (14.71) and (14.72) are equivalent to equation (14.69) if ρ_{II} and ρ_{QQ} are considered as a normal input-output pair.

Consider the Markov model frequently used in streamflow generation (Kottegoda, 1979; Salas et al., 1980; Yevjevich, 1972),

$$Q_i = \rho_1 Q_{i-1} + I_i \tag{14.73}$$

where ρ_1 is the first autocorrelation coefficient of Q_i. A repeated application of equation (14.73), with $Q_0 = I_0$, leads to

$$Q_i = \sum_{u=0}^{i} \rho_1^u I_{i-u} \tag{14.74}$$

On comparing equation (14.74) with equation (14.69),

$$h_u = \rho_1^u \tag{14.75}$$

Quimpo (1973) reasoned that the use of the Markov model for streamflow generation has been motivated by persistence or carryover effects of storage in a basin (Dawdy and Matalas, 1964). The IUH, on the other hand, reflects the storage characteristics. The parameter ρ_1 of the Markov model must be related to the IUH parameters. Quimpo (1971a) pointed out that fitting a Markov model to a hydrologic time series was equivalent to saying that the system generating the series was a simple continuous-time linear reservoir whose storage constant is k having the relation

$$k = -(\ln \rho_1)^{-1} \tag{14.76}$$

We now extend equation (14.73):

$$Q_i = \sum_{j=1}^{M} a_j Q_{i-j} + I_i \tag{14.77}$$

in which the a_j terms are constants. Recall that the Q_i and I_i sequences possess the following properties: $E[Q_i] = E[I_i] = 0$; $\mathrm{var}[Q_i] = E[Q_i^2] = \sigma_Q^2 = C_{QQ}(0)$; $\mathrm{var}[I_i] = E[I_i^2] = C_{II}(0) = \sigma_I^2$; $\rho_j = E[Q_i Q_{i-j}]/\sigma_Q^2$; and $E[I_i I_{i-j}] = E[I_i Q_{i-j}] = 0$ for $j = 1, 2, \ldots$. The last equality asserts that I_t is independent of the past values of I and Q. Let it be assumed that σ_Q^2 is equal to 1.

If $M = 1$, equation (14.73) is obtained and $a_1 = \rho_1$. Wold (1954) shows that these constants are determined by the Yule-Walker equations (Yule, 1921, 1927; Walker, 1931), which can be derived as follows. Multiply equation (14.77) by Q_{i-1} and take the expectation of both sides;

$$\rho_1 = a_1 + a_2\rho_1 + a_3\rho_2 + \cdots + a_M\rho_{M-1}$$

Likewise, multiply equation (14.77) by Q_{i-2} and take the expectation on both sides:

$$\rho_2 = a_1\rho_1 + a_2 + a_3\rho_1 + \cdots + a_M\rho_{M-2}$$

Similarly, equation (14.77) is multiplied by $Q_{i-3}, Q_{i-4}, \ldots, Q_{i-M}$, and expectations are taken on both sides each time. This leads to M equations, which can be written in matrix form as

$$\begin{bmatrix} 1 & \rho_1 & \rho_2 & \cdots & \rho_{M-1} \\ \rho_1 & 1 & \rho_1 & \cdots & \rho_{M-2} \\ \rho_2 & \rho_1 & 1 & \cdots & \rho_{M-3} \\ \vdots & \vdots & \vdots & & \vdots \\ \rho_{M-1} & \rho_{M-2} & \rho_{M-3} & \cdots & 1 \end{bmatrix} \begin{bmatrix} a_1 \\ a_2 \\ a_3 \\ \vdots \\ a_M \end{bmatrix} = \begin{bmatrix} \rho_1 \\ \rho_2 \\ \rho_2 \\ \vdots \\ \rho_M \end{bmatrix} \tag{14.78}$$

where ρ_i, $i = 1, 2, \ldots, M$, are discrete-time equivalents of $\rho(\tau)$ with τ being the lag. Equations (14.74) and (14.75) provide the key to the link between the IUH and stochastic streamflow models, which we discuss for some simple IUH models. In compact matrix notation

$$[\rho] = [W]\,[a] \tag{14.79}$$

where $[W]$ is the autocorrelation matrix. A necessary condition for stationarity is that $[W]$ is positive definite.

The connection between various formulations of the linear model can also be seen by considering the Z-transform of h. Let $y(z) = 1/h(z)$. Equation (14.69) can be expressed as

$$I(z) = y(z)\,Q(z) \tag{14.80}$$

On inversion,

$$I_i = \sum_{j=0}^{\infty} y_j Q_{i-j} \tag{14.81}$$

Equation (14.81) is the reciprocal version of equation (14.59); it expresses that the convolution of the reciprocal of h and the direct runoff gives the effective rainfall of a linear time-invariant system. Rewriting,

$$y_0\, Q_i = -\sum_{j=1}^{\infty} y_j\, Q_{i-j} + I_i \tag{14.82}$$

Equation (14.82) expresses the DR at any instant in terms of the ER at that instant and the DR at preceding instants. If equation (14.82) is divided by y_0, it reduces to an autoregressive (AR) model of infinite order. When I and Q are stochastic, then equation (14.82) expresses the relation between ρ_{IQ} and ρ_{II}. In practice the order of an AR model will be finite.

The connection between deterministic and stochastic linear models can also be derived by considering equation (14.44), which is analogous to an ARMA model. Consequently, the parameters in this equation can be related to those of the stochastic ARMA model (Kottegoda, 1979; Salas et al., 1980). The ensuing discussion further clarifies this linkage.

14.7.1 Linear Reservoir

The discrete-time linear reservoir may be compared directly with the first order AR model obtained from equation (14.69),

$$Q_j = \rho Q_{j-1} + I_j$$

where ρ is the first autocorrelation coefficient of Q_j. Expressing in B operator notation,

$$Q_j = \frac{1}{1 - \rho B} I_j = \sum_{i=0}^{\infty} \rho^i B^i I_j \qquad (14.83)$$

On comparing equation (14.83) with equation (14.69),

$$h_j = \rho^j \qquad (14.84)$$

which is identical to equation (14.75).

14.7.2 The Nash Model

The two-parameter discrete-time Nash model given by equation (14.36),

$$(1 + k\nabla)^n Q_j = I_j$$

can be expressed in B notation as in equation (14.50),

$$(1 - CB)^n Q_j = A^n I_j$$

This can be considered as a special case of the n-order AR model. Thus, the parameters of the AR model can be obtained from cumulants or moments derived previously.

14.7.3 General Linear Reservoir

The general linear reservoir, discussed in the preceding section, can be expressed as an MA formulation. Combining equations (14.59) and (14.2).

$$S_i = \left[\frac{1}{1 + \alpha - a\alpha}\right] r_i + \left[\frac{1 - a\alpha}{1 + \alpha - a\alpha}\right] S_{i-1} \qquad (14.85)$$

in which r_i is the volume of I during time interval i. Successive substitution yields

$$S_i = \frac{1}{(1 + \alpha - a\alpha)} \sum_{j=0}^{\infty} \left[\frac{1 - a\alpha}{1 + \alpha - a\alpha}\right]^j r_{i-j} \qquad (14.86)$$

Likewise, the outflow volume can be derived as

$$V_i = \left[1 - \frac{1}{1 + \alpha - a\alpha}\right] r_i + \left[1 - \frac{1 - a\alpha}{1 + \alpha - a\alpha}\right] \left[\frac{1}{1 + \alpha - a\alpha}\right] \sum_{j=1}^{\infty} \left(\frac{1 - a\alpha}{1 + \alpha - a\alpha}\right)^{j-1} r_{i-j} \qquad (14.87)$$

The coefficients of r_k in equations (14.86) and (14.87) are the weights w_j and can be viewed as the ordinates of the IUH for S and V. That is, for storage,

$$w_j = \left[\frac{1}{1 + \alpha - a\alpha}\right] \left[\frac{1 - a\alpha}{1 + \alpha - a\alpha}\right]^j, \quad j = 0, 1, 2, \ldots \qquad (14.88)$$

and for outflow volume

$$w_0 = 1 - \frac{1}{1 + \alpha - a\alpha} \qquad (14.89a)$$

$$w_j = \left[1 - \frac{1 - a\alpha}{1 + \alpha - a\alpha}\right] \left[\frac{1}{1 + \alpha - a\alpha}\right] \left[\frac{1 - a\alpha}{1 + \alpha - a\alpha}\right]^{j-1}, \quad j = 1, 2, \ldots \qquad (14.89b)$$

The parameter a has a greater influence on w_j when α is large than when it is small.

EXERCISES

14.1. Compute the IUHs of the discrete linear reservoir for k = 1, 2, 3, 4, and 5 h. Compare them with those of the continuous counterpart. Discuss the effect of varying k.

14.2. Compute the IUHs of the discrete linear reservoir for k = 3 h and Δt = .25, .5, .75, 1.0, 1.5, and 2.0 h. Compare them with those of the continuous counterpart. Discuss the effect of varying Δt.

14.3. Compute the S-hydrographs of the discrete linear reservoir for k = 1, 2, 3, 4, and 5 h. Compare them with those of the continuous counterpart. Discuss the effect of varying k.

14.4. Compute the S-hydrographs of the discrete linear reservoir for k = 3 h and Δt = .25, .5, .75, 1.0, 1.5, and 2.0 h. Compare them with those of the continuous counterpart. Discuss the effect of varying Δt.

14.5. Compute the IUH of the discrete linear reservoir by using equation (14.21) with k = 1, 2, 3, 4, and 5 h, and Δt = 1 h. Compare the results with those of Question 14.1. Discuss the effect of varying k.

14.6. Compute the IUHs of the discrete linear reservoir by using equation (14.21) with k = 3 h and Δt = .25, .5, .75, 1.0, 1.5, and 2.0 h. Compare the results with those of Question 14.2. Discuss the effect of varying Δt.

TABLE 14.5

Effective rainfall–direct runoff hydrograph data for the watershed Tevere St. Lucia at Umberia (934 km²), Central Italy

Rainfall-Runoff Event of December 12–14, 1981

Effective Rainfall Hydrograph

Time (min)	Intensity (cm/h)	Time (min)	Intensity (cm/h)	Time (min)	Intensity (cm/h)
.0	.023	180.0	.105	360.0	.122
60.0	.025	240.0	.098	420.0	.034
120.0	.055	300.0	.082	480.0	.0000

Direct Runoff Hydrograph

Time	Discharge	Time	Discharge	Time	Discharge
.0	.004	660.0	.0429	1320.0	.0042
60.0	.0054	720.0	.0318	1380.0	.0040
120.0	.0118	780.0	.0321	1440.0	.0031
180.0	.0245	840.0	.0175	1500.0	.0026
240.0	.0364	900.0	.0178	1560.0	.0023
300.0	.0360	960.0	.0104	1620.0	.0022
360.0	.0356	1020.0	.0090	1680.0	.0021
420.0	.0352	1080.0	.0075	1740.0	.0018
480.0	.0466	1140.0	.0066	1800.0	.0000
540.0	.0524	1200.0	.0056		
600.0	.0528	1260.0	.0049		

14.7. Compute the IUHs of the discrete Nash model for $\Delta t = 1$ h and $k = 1, 2, 3, 4$, and 5 h. Compare them with those of the continuous counterpart. Discuss the effect of varying k.

14.8. Compute the IUHs of the discrete Nash model for $n = 4$, $k = 5$ h, and $\Delta t = .5, 1.0, 1.5$, and 2.0 h. Compare them with those of the continuous counterpart. Discuss the effect of varying Δt.

14.9. Compute the S-hydrographs of the discrete Nash model for $\Delta t = 1$ h, $n = 4$, and $k = 1, 2, 3, 4$, and 5 h. Compare them with those of the continuous counterpart. Discuss the effect of varying k.

14.10. Compute the S-hydrographs of the discrete Nash model for $n = 4$, $k = 5$ h, and $\Delta t = .5, 1.0, 1.5$, and 2.0 h. Compare them with those of the continuous counterpart. Discuss the effect of varying Δt.

14.11. The effective rainfall–direct runoff hydrograph data are given in Table 14.5. Compute the parameters of the model in equation (14.52) with $n = 3$ and $m = 1$ by the method of moments or cumulants.

14.12. Compute the IUH with the data and parameters of Question 14.11.

14.13. Compute the first autocorrelation coefficient of direct runoff series in Question 14.11. Then compute the IUH of the single linear reservoir using this coefficient.

14.14. Compute the first four moments and first four cumulants for the ER-DR data in Question 14.11.

14.15. Compute the parameters of the single linear reservoir and the Nash model for the ER-DR data in Question 14.11.

CHAPTER FIFTEEN

Linear Time-Variant Models

Most hydrologic systems are time-variant. They are time-invariant only by assumption. O'Donnell (1966) was perhaps the first to propose time-variant linear models for hydrograph synthesis. Several investigators have since undertaken studies on various aspects of linear time-variant modeling of rainfall-runoff relationship (Carrasquilla and Perkins, 1967; Chiu and Bittler, 1969; Hino, 1970; Snyder, Mills and Stephens, 1970; Mandeville and O'Donnell, 1973; Young and Whitehead, 1977; Whitehead, 1979). In this chapter we draw heavily from these studies, particularly from the one by Mandeville and O'Donnell. Although the assumption of linearity is retained, considerable complexity is introduced into modeling by mere relaxation of time-invariance. This is partly the reason that the bulk of the research effort on rainfall-runoff modeling has been directed at linear time-invariant models.

Two directions of inquiry have been pursued: (1) black-box modeling or analysis and (2) conceptual modeling or synthesis. We discuss both these modes of approaches in this chapter. The second approach employs either linear elements or a differential equation. We consider time-variant forms of these element in this chapter.

15.1 TIME-VARIANT LINEAR RESERVOIR

A time-variant linear reservoir can be defined by equation (13.20) with the reservoir storage coefficient $k(t)$ varying in time, that is,

$$S = k(t)Q \tag{15.1}$$

The response of this reservoir due to given effective rainfall can be obtained by solving the spatially lumped form of continuity equation given by equation (13.19) in conjunction with equation (15.1),

$$k(t)\frac{dQ(t)}{dt} + \left(1 + \frac{dk}{dt}\right)Q(t) = I \tag{15.2}$$

Its integrating factor F is

$$F(t) = k(t)\exp\left[\int \frac{dt}{k(t)}\right] \tag{15.3}$$

Therefore, the general solution of equation (15.2) is

$$Q(t) = \frac{1}{F(t)}\int I\frac{F(w)}{k(w)}dw + \frac{C}{F(t)} \tag{15.4}$$

where C is a constant of integration. If the initial condition is $Q(0) = 0$, then $C = 0$ and equation (15.2) has the solution

$$Q(t) = \frac{1}{F(t)}\int I\frac{F(w)}{k(w)}dw \tag{15.5}$$

On the other hand, if $Q(0) = Q_0$, then $C = Q_0F(0) = Q_0k(0)$ and equation (15.2) has the solution

$$Q(t) = \frac{1}{F(t)}\int I\frac{F(w)}{k(w)}dw + \frac{Q_0F(0)}{F(t)} \tag{15.6}$$

Equation (15.4) can be utilized to yield response of a linear time-variant reservoir due to a D-hour unit pulse. The rising hydrograph is given (with $Q(0) = 0$) by

$$Q(t) = h(D, t)$$
$$= \frac{1}{DF(t)} \int_0^D \exp\left[\int_0^w ds/k(s)\right] dw, \qquad 0 \le t \le D \quad (15.7)$$

and the recession hydrograph, by

$$Q(t) = h(D, t) = \frac{1}{DF(t)} \int_0^D \exp\left[\int_0^w ds/k(s)\right] dw$$
$$= \frac{Q_p F(D)}{F(t)}, \qquad t \ge D \quad (15.8)$$

where $F(D)$ is given by equation (15.3) and $Q_p = Q(D)$ is from equation (15.7). Equations (15.7) and (15.8) also define the D-hour unit hydrograph.

In a like manner we can determine the IUH of this reservoir that is of particular interest. If the impulse effective rainfall occurs at time τ, then $I(t) = \delta(t - \tau)$. Denoting the IUH by $h(t, \tau)$ we obtain from equation (15.5)

$$h(t,\tau) = \frac{1}{F(t)} \int_0^t \delta(s - \tau) \frac{F(s)}{k(s)} ds$$
$$= \frac{1}{F(t)} \frac{F(\tau)}{k(\tau)}$$

Substituting equation (15.3),

$$h(t, \tau) = \frac{1}{k(t)\exp\left[\int_0^t dw/k(w)\right]} \; \frac{k(\tau)\exp\left[\int_0^\tau dw/k(w)\right]}{k(\tau)}$$
$$= \frac{1}{k(t)}\left\{\exp\left[-\int_0^t \frac{dw}{k(w)}\right] + \left[\int_0^\tau \frac{dw}{k(w)}\right]\right\}$$
$$= \frac{1}{k(t)} \exp\left[-\int_\tau^t \frac{dw}{k(w)}\right], \qquad t > \tau \quad (15.9)$$

Because equation (15.9) is valid for $t > \tau$, we simply write

$$h(t, \tau) = u_1(t - \tau) \frac{1}{k(t)} \exp\left[-\int_\tau^t \frac{dw}{k(w)}\right] \quad (15.10)$$

where u_1 is the unit step function. This equation was derived by Mandeville and O'Donnell (1973). Chiu and Bittler (1969) found an expression for $h(t, \tau)$, which can be written as

$$h(t, \tau) = \frac{1}{k(\tau)} \exp\{-[y(t) - y(\tau)]\} \quad (15.11)$$

where

$$y(s) = \int_0^s \frac{[1 + (dk(w)/dw)]}{k(w)} dw$$

Equation (15.11) looks somewhat different from equation (15.10). It can be shown that these two equations are essentially the same. Writing

$$y(t) - y(\tau) = \int_0^t \frac{1 + k'}{k} dw - \int_0^\tau \frac{1 + k'}{k} dw; \qquad k' = \frac{dk}{dw}$$
$$= \int_\tau^t \frac{1 + k'}{k} dw$$
$$= [\ln k(t) - \ln k(\tau)] + \int_\tau^t \frac{1}{k(w)} dw$$

Therefore,

$$\exp\{-[y(t) - y(\tau)]\} = \exp\left[-\frac{\ln k(t)}{\ln k(\tau)} - \int_\tau^t \frac{1}{k(w)} dw\right]$$
$$= \frac{k(\tau)}{k(t)} \exp\left[-\int_\tau^t \frac{1}{k(w)} dw\right]$$

Hence,

$$h(t, \tau) = \frac{1}{k(t)} \exp\left[-\int_\tau^t \frac{dw}{k(w)}\right]$$

which is the same as equation (15.9). We can determine the UH as well as the S-hydrograph for the reservoir by utilizing equation (15.10). The IUH, $h(t, \tau)$, is a function of two variables t and τ, where τ denotes the time of application of the effective rainfall and t denotes the time of observation. Both t and τ are measured from a common fixed origin. It may be instructive to consider an example of a time-variant reservoir and compare its response with a time-invariant one.

EXAMPLE 15.1

Let $k(t) = 10 + t$. Compute $h(t, \tau)$ and compare it with $h(t - \tau)$ of the time-invariant reservoir. The effective rainfall is applied at $\tau = 0$, 2, and 4 h.

From equation (15.10)

$$h(t, \tau) = 0, \qquad t < \tau$$
$$h(t, \tau) = \frac{10 + \tau}{(10 + t)^2} = \frac{10 + \tau}{[k(t)]^2}$$

We can graph $h(t, \tau)$ for the different values of τ, as shown in Fig. 15.1.

For time-invariant reservoirs, $k(t) = k$, a constant. Substituting k into equation (15.10) and noting $b = t - \tau$,

$$h(b) = u_1(b) \frac{e^{-b/k}}{k}$$

It is readily seen that $h(b)$ becomes $1/k$ immediately after the time of application of the effective rainfall and decays exponentially thereafter, and its appearance remains independent of the time of application. We also graph $h(b)$ in Fig. 15.1.

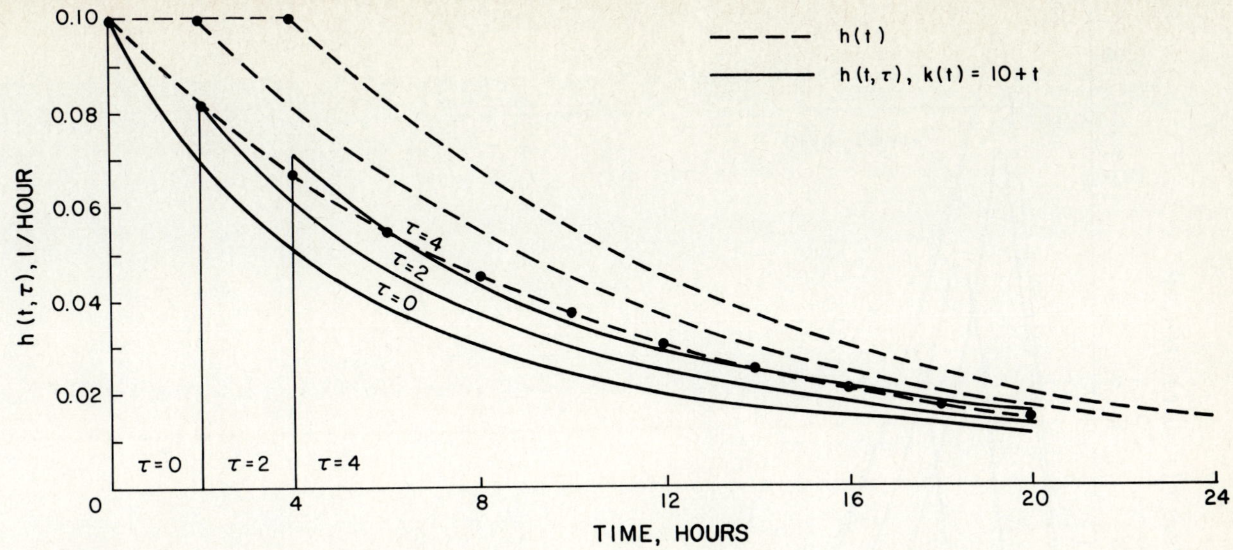

FIGURE 15.1 Example 15.1: The IUH of a time-variant linear reservoir for different times of application of the effective rainfall, $\tau = 0$, 2, and 4 hours.

It may be instructive to compare $h(t, \tau)$ with $h(t - \tau)$. There are certain common features. Both of these functions tend to be 0 for increasing t. This fact is ensured by the presence of negative exponential term in their respective equations. Further, for a fixed t and therefore fixed $k(t)$, the value of $h(t, \tau)$ is always greater for later values of τ.

On the other hand, $h(t, \tau)$ does not normally take the same value immediately after different input times, as shown in Fig. 15.1. Also, $h(t, \tau)$ may possess stationary points with regard to t. If these exist, they will occur at the same t for any $h(t, \tau)$ regardless of the time of input and these coincide with the stationary points of the function

$$F(t) = -k(t) - t$$

To emphasize again, $h(t, \tau)$ depends on the individual values of both t and τ, whereas $h(t - \tau)$ depends only on their difference, that is, $t - \tau$.

EXAMPLE 15.2

Compare the S-hydrograph, $U(t, \tau)$, if $k = 10 + t$. Compare it with the S-hydrograph of the time-invariant linear reservoir. The effective rainfall is applied at $\tau = 0$, 2, and 4 h.

The S-hydrograph can be written as

$$U(t, \tau) = \int_{\tau}^{t} h(w, \tau)\, dw$$

$$= \int_{\tau}^{t} \frac{10 + \tau}{(10 + w)^2}\, dw$$

$$= 1 - \frac{10 + \tau}{10 + t}, \qquad t \geq \tau$$

For the time-invariant case $k = 10$, the S-hydrograph is

$$U(t) = 1 - \exp\left(-\frac{t}{10}\right)$$

which is the independent of the time of application of I. The S-hydrographs for both bases are given as follows:

Time (h)	U(t, τ) τ = 0	τ = 2	τ = 4	Time (h)	U(t)
0	.000	.000	.000	0	.00
10	.500	.400	.300	1	.10
20	.667	.600	.533	2	.18
30	.750	.700	.650	3	.26
40	.800	.760	.720	4	.33
50	.833	.800	.767	5	.39
100	.909	.891	.873	10	.63
150	.938	.925	.913	15	.78
200	.952	.943	.933	20	.86
250	.962	.954	.946	30	.95
300	.967	.961	.955	40	.98
350	.972	.967	.961	50	.99
400	.977	.970	.966		
450	.978	.974	.970		
500	.980	.976	.973		

EXAMPLE 15.3

Compute the 4-h UH, $h(t, D; \tau)$, if $k = 10 + t$. Compare it with the UH of the time-invariant linear reservoir. The effective rainfall is applied at $\tau = 0$, 2, and 4 h.

The 4-h UH is

$$h(t,4;\tau) = \tfrac{1}{4}\left[1 - \frac{10 + \tau}{10 + t}\right], \qquad 0 \leq t \leq t$$

$$= \frac{10 + \tau}{4}\left[\frac{1}{6 + t} - \frac{1}{10 + t}\right], \qquad t \geq 4$$

For the time-invariant case,

$$h(t, 4) = \tfrac{1}{4}\left[1 - \exp\left(\frac{-t}{10}\right)\right], \qquad 0 \leq t \leq 4$$

$$= \tfrac{1}{4}\left[1 - \exp\left(\frac{-2}{5}\right)\right] \exp\left[\frac{-(t - 4)}{10}\right], \qquad t \geq 4$$

The UH for both cases is graphed in Fig. 15.2.

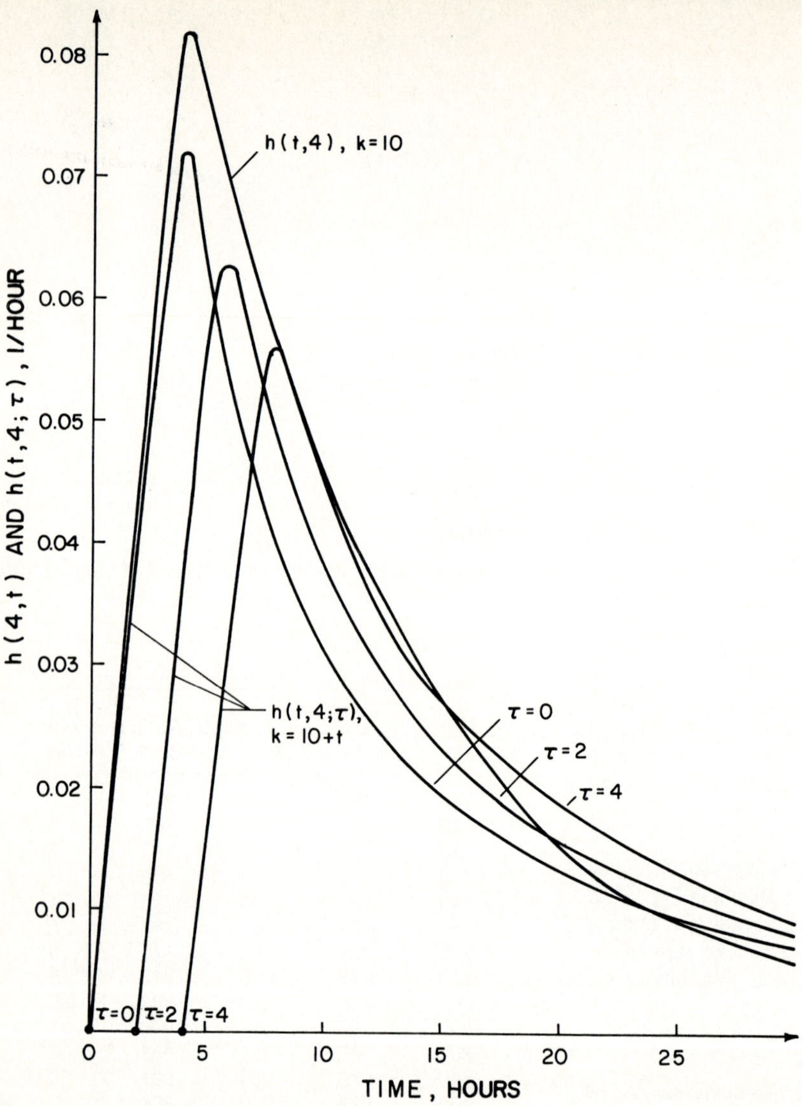

FIGURE 15.2 Example 15.3: The 4-hour UH of a time-variant linear reservoir for different times of application of the effective rainfall, τ = 0, 2, and 4 hours.

15.2 TIME-VARIANT LINEAR CHANNEL

The continuity equation for open channel flow with no lateral inflow is given by equation (13.1). Rewriting it,

$$\frac{\partial A}{\partial t} + \frac{\partial Q}{\partial x} = 0 \tag{15.12}$$

where the discharge Q and flow cross-sectional area are functions of space x and time t. The distance x is measured along the channel and the time t measured from a fixed origin. Mandeville and O'Donnell (1973) defined a particular class of channels for which Q and A are explicitly related by

$$A(x, t) = \left[\frac{a'(x)}{B'(t)}\right] Q(x, t) \tag{15.13}$$

where $a'(x) = da/dx$ and $B'(t) = dB/dt$; here $a(x)$ is a function of x alone and $B(t)$ is a function of t alone. Equation (15.13) specifies a rating curve for such channels. These channels are both linear and time-variant. It may be noted that the proportional change in the rating curve with the progress of time remains the same at every point. If $B'(t)$ is constant, these channels assume the form of those defined by Dooge (1959).

Substituting equation (15.13) into equation (15.12),

$$\frac{\partial Q}{\partial x} + \frac{\partial}{\partial t}\left[\frac{a'(x)}{B'(t)} Q\right] = 0 \tag{15.14}$$

whose solution is

$$Q(x, t) = B'(t)\, G[B(t) - a(x)] \tag{15.15}$$

where G is any differentiable function of x. Equation (15.15) is the general solution for flow in channels defined by equation (15.13). The form of the function G can be determined by using initial conditions. For example, if the inflow hydrograph at an upstream point x_0 is known, $Q(x_0, s) = Q_0(s)$, then we can write

$$Q_0(s) = B'(s)G[B(s) - a(x_0)] \tag{15.16}$$

Since $B(s)$ and $a(x_0)$ are known, equation (15.16) defines the form of the function $G(w)$ for certain values of w. The discharge $Q(x_1, t) = Q_1(t)$ at time t at a downstream point x_1 is given by

$$Q_1(t) = B'(t)G[B(t) - a(x_1)] \tag{15.17}$$

However, we need to determine $Q_1(t)$ in terms of $Q_0(s)$ and other known functions. For a specified t we can select s such that s and t are related by

$$B(s) - a(x_0) = B(t) - a(x_1) \tag{15.18}$$

then

$$Q_1(t) = \left[\frac{B'(t)}{B'(s)}\right] Q_0(s) \tag{15.19}$$

Equation (15.19) implies that the channels defined by equation (15.13) are linear. This can be seen by noting that for a matched pair of s and t defined by equation (15.18), the ratio term in equation (15.19) is constant, and hence proportionality and superposition hold between $Q_1(t)$ and $Q_0(s)$. Because the value of the ratio term varies with the matched pair of s and t, such channels are time-variant.

The behavior of a time-variant linear channel obviously depends on the particular forms of $a(x)$ and $B(t)$ and on the values of x_0 and x_1 in equations (15.18) and (15.19). Furthermore, a channel having $B(t)$ and $B'(t)$ fixed can have different reaches, for each of which $a(x_1) - a(x_0)$ is the same. For such reaches equations (15.18) and (15.19) are the same and consequently have indistinguishable behavior. Similarly, distinct reaches in different channels for which $B(t)$ functions are proportional have indistinguishable behavior as long as the differences $a(x_1) - a(x_0)$ are proportional by the same factor. Such indistinguishable channels can be described by a function,

$$J(t) = \frac{a(x_1) - a(x_0)}{B'(t)} \tag{15.20}$$

Integrating equation (15.20) between the limits s and t and then inserting equation (15.18),

$$\int_s^t \frac{dw}{J(w)} = \frac{B(t) - B(s)}{a(x_1) - a(x_0)} = 1 \tag{15.21}$$

Implicit here is the assumption that $J(t)$ is physically realistic; that is, $J(t)$ is positive and finite. From equation (15.21) it is possible to determine t for specified s, or vice versa. Using equation (15.20) we can express equation (15.19) as

$$Q_1(t) = \left[\frac{J(s)}{J(t)}\right] Q_0(s) \tag{15.22}$$

Equation (15.22) uniquely defines the behavior of a linear time-variant channel.

The IUH of a linear time-variant channel can now be specified. If the instantaneous unit inflow is given at the upstream end x_0 at time $s = \tau$, then the IUH at the downstream end x_1 can be determined by inserting $Q_0(s) = \delta(s - \tau)$ in equation (15.22),

$$h(t, \tau) = \delta(t - t_1) \tag{15.23}$$

where

$$\int_\tau^{t_1} \frac{dw}{J(w)} = 1 \tag{15.24}$$

This can be seen to be true as follows. When $s \neq \tau$, $Q_0(s) = 0$ and therefore $h(t, \tau) = 0$. When $s = \tau$, both $Q_0(s)$ and $Q_1(t)$ are infinite. Since the total volume of outflow must equal that of inflow, which is unity, the IUH must be a delta function at some time t_1, as defined earlier.

EXAMPLE 15.4

Let the function $J(t) = 1 - e^{-t}$. Determine the response of a linear time-variant channel and compare it with that of its time-invariant counterpart.

The function $J(t)$ determines not only the varying response interval $(t_1 - \tau)$ but also the ratio of matching inflow and outflow values through equation (15.22). From equation (15.24),

$$\int_\tau^{t_1} \frac{dw}{(1 - e^{-w})} = \ln\,[e^w - 1]\Big|_\tau^{t_1} = 1$$

Therefore,

$$t_1 = \ln[1 + e(e^\tau - 1)]$$

The IUH is

$$h(t, \tau) = \delta\{t - \ln[1 + e(e^\tau - 1)]\}$$

and the outflow hydrograph is

$$Q_1(t) = \frac{1 - e^{-s}}{1 - e^{-t}} Q_0(s)$$

Before proceeding further, let us contrast it with the corresponding time-invariant channel. Here $J(t) = 1$, a constant. Therefore, $t_1 - \tau = 1$. This says that the time interval between the inflow and outflow impulses always remains a fixed value. This value is what is called as the translation time. Equations (15.23) and (15.24) reduce to

$$h(t, \tau) = \delta(t - t_1)$$

$$t_1 - \tau = 1$$

which can be written simply as

$$h(t - \tau) = \delta(t - \tau - 1)$$

The response of such a channel is illustrated in Fig. 15.3.

On the other hand, for a time-variant linear channel, the response interval varies for different inflow times. This variation alters the characteristics of more complicated inflow functions. Further, for a specified $J(t)$ the inflow time s corresponding to a given outflow time t is uniquely determined by equation (15.21). Equation (15.22) shows that the particular outflow value depends only on the inflow at this corresponding time and is independent of all other inflow values. As shown in Fig. 15.4, it says that if the outflow is composed of a number of blocks, then each block will have the same corresponding response whether or not the other blocks are present.

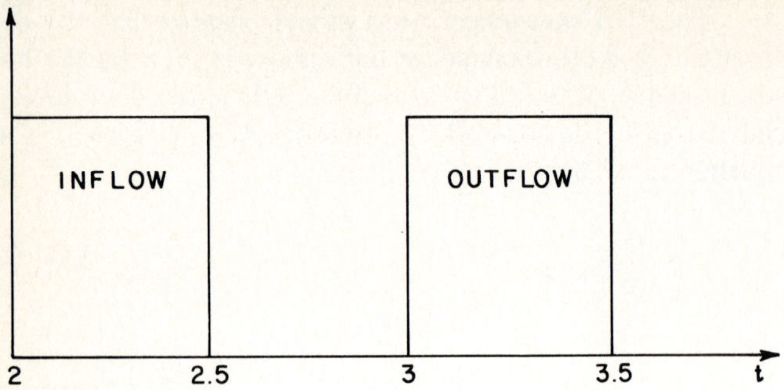

FIGURE 15.3 Example 15.4: Response of a time-invariant linear channel, with translation time = 1 unit, to a uniform pulse of the inflow.

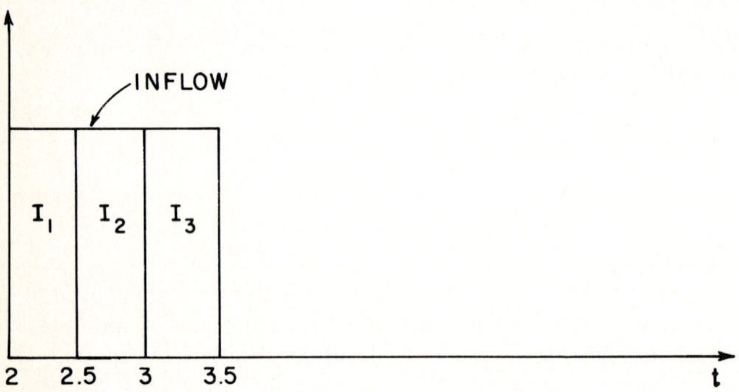

FIGURE 15.4 Example 15.4: Response of a time-variant linear channel due to a uniform pulse of inflow.

EXAMPLE 15.5

Derive the IUH of the linear time-variant channel when the inflow takes place laterally all along its length. Take $J(t)$ as in Example 15.4.

Since the channel is linear, the inflow taking place along the channel simply gets translated. Therefore, its IUH will be a rectangular pulse of duration equal to the time of translation t_*,

$$h(t, \tau) = \frac{1}{t_*}[u_1(t, \tau) - u_1(t - t_*, \tau)]$$

t_* is as before, that is,

$$t_* = \ln\{1 + e[\exp(\tau) - 1]\}$$

Thus the IUH depends upon the time of application of $I(t)$, as shown in Fig. 15.5.

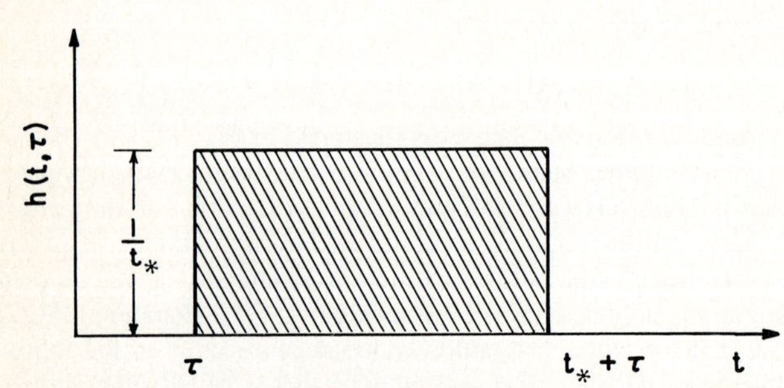

FIGURE 15.5 Example 15.5: The IUH of a linear time-variant channel with lateral inflow.

EXAMPLE 15.6

Compute the S-hydrograph of the linear time-variant channel when the inflow occurs at the upstream. The function $J(t)$ is as in Example 15.4.

The IUH of this channel is dervied in the previous example. Its S-hydrograph is given as

$$U(t, \tau) = 0, \quad t \le t_*$$
$$= 1, \quad t \ge t_*$$

where $t_* = \ln\{1 + \exp[\exp(\tau) - 1]\}$. It is graphed in Fig. 15.6.

15.3 COMBINATIONS OF RESERVOIRS AND CHANNELS

The reservoirs and channels can be variously combined to yield different conceptual models. To gain further insight we consider several combinations. Let us rewrite the convolution integral for the time-variant linear systems.

15.3.1 Convolution Integral

If $h(t, \tau)$ is the IUH of a system, then the response $Q(t)$ to any input $I(t)$ can be found by using the integral

$$Q(t) = \int_{-\infty}^{\infty} I(\tau)\, h(t, \tau)\, d\tau \tag{15.25}$$

If two systems are placed in series whose IUHs are $h_1(t, \tau)$ and $h_2(t, \tau)$, respectively, then the IUH of the combined system can be found from equation (15.25), wherein $I(\tau)$ can be replaced by $h_1(t, \tau)$,

$$h(t, \tau) = \int_{-\infty}^{\infty} h_1(s, \tau) h_2(t, s)\, ds \tag{15.26}$$

When the order of the systems is reversed, the IUHs h_1 and h_2 will be interchanged. This change usually alters the IUH $h(t, \tau)$, an important characteristic that distinguishes the time-variant systems from time-invariant systems.

15.3.2 Time-Variant Linear Channels in Series

Let us first consider two channels in series with time functions $J_1(t)$ and $J_2(t)$. Subscript 1 denotes the upstream channel and subscript 2 denotes the downstream channel. Let $h_1(t, \tau)$ and $h_2(t, \tau)$ be their respective IUHs defined by equation (15.23) with a different t_1 for each channel. Then the IUH of the combined system, using equation (15.26), would be

$$h(t, \tau) = \delta(t - t_2) \tag{15.27}$$

where

$$\int_{\tau}^{t_1} \frac{dw}{J_1(w)} = 1 \tag{15.28}$$

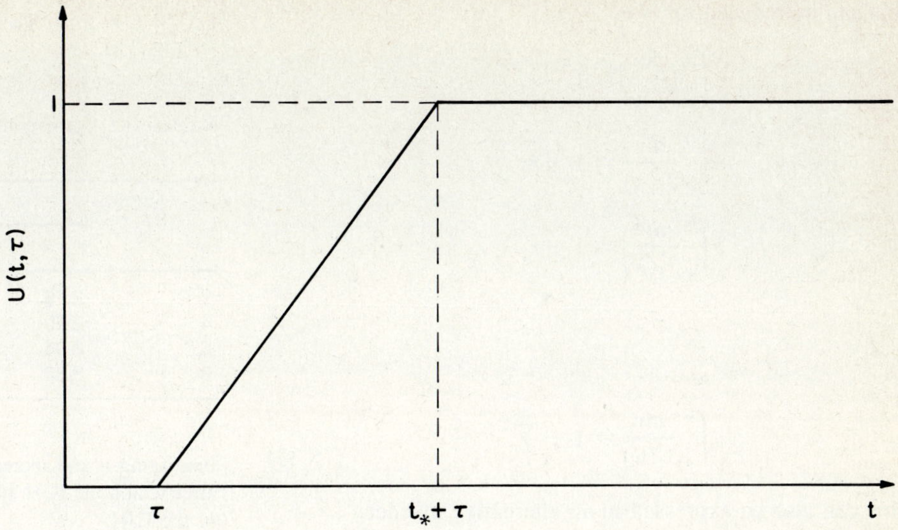

FIGURE 15.6 Example 15.6: The SH of a linear time-variant channel with lateral inflow.

and

$$\int_{t_1}^{t_2} \frac{dw}{J_2(w)} = 1 \tag{15.29}$$

Equation (15.27) shows that the IUH is a δ-function at t_2 defined through equations (15.28) and (15.29).

If the order of channels is reversed, the IUH will still be a δ-function with a different value of t_2. Considerable simplification results if it can be assumed that the time functions of the channels are proportional; that is, $J_2(t) = rJ_1(t)$, where r is factor of proportionality. Then equations (15.28) and (15.29) can be reduced to

$$\int_{\tau}^{t_2} \frac{dw}{J_1(w)} = 1 + r \tag{15.30}$$

Alternatively,

$$\int_{\tau}^{t_2} \frac{dw}{J_1(w) + rJ_1(w)} = 1 \tag{15.31}$$

Equations (15.30) and (15.31) show that the order of the channels is immaterial in this case because t_2 is independent of this order. Further, the combined system can be thought of as a single channel with time function $J(t) = J_1(t) + J_2(t)$. This property can be extended to any number of channels in a series with proportional time functions.

To illustrate, consider a cascade of channels, as shown in Fig. 15.7, where the time functions are proportional; that is,

$$J_2(t) = r_2 J_1(t)$$
$$J_3(t) = r_3 J_1(t)$$
$$\vdots$$
$$J_i(t) = r_i J_1(t)$$
$$\vdots$$
$$J_n(t) = r_n J_1(t)$$

The IUH of this cascade is

$$h(t, \tau) = \delta(t - t_n) \tag{15.32}$$

where

$$\int_{\tau}^{t_1} \frac{dw}{J_1(w)} = 1$$
$$\int_{t_1}^{t_2} \frac{dw}{J_2(w)} = 1$$
$$\vdots$$
$$\int_{t_{n-1}}^{t_n} \frac{dw}{J_n(w)} = 1$$

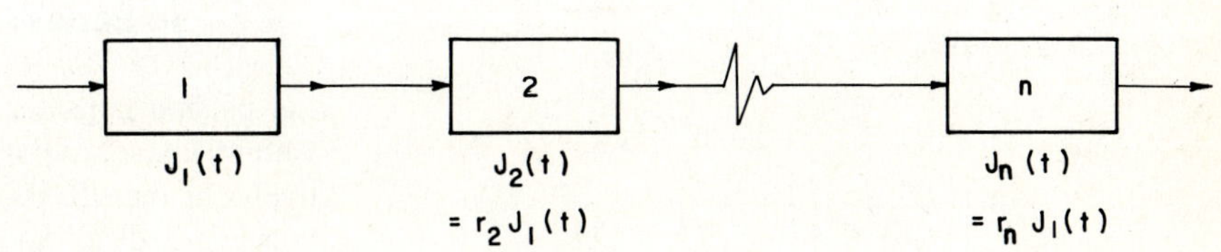

FIGURE 15.7 A cascade of time-variant linear channels with proportional time functions; that is, $J_i(t) = r_i J_i(t)$ with $r_i = 1$.

This can be reduced to

$$\int_{\tau}^{t_1} \frac{dw}{J_1(w)} = 1$$

$$\int_{\tau}^{t_2} \frac{dw}{J_1(w)} = 1 + r_2$$

$$\int_{\tau}^{t_3} \frac{dw}{J_1(w)} = 1 + r_2 + r_3$$

$$\vdots$$

$$\int_{\tau}^{t_n} \frac{dw}{J_1(w)} = 1 + \sum_{i=2}^{n} r_i \tag{15.33}$$

This can also be expressed in an alternative manner,

$$\int_{\tau}^{t_1} \frac{dw}{J_1(w)} = 1$$

$$\int_{\tau}^{t_2} \frac{dw}{J_1(w) + J_2(w)} = 1$$

$$\int_{\tau}^{t_3} \frac{dw}{J_1(w) + J_2(w) + J_3(w)} = 1$$

$$\vdots$$

$$\int_{\tau}^{t_n} \frac{dw}{\sum_{i=1}^{n} J_1(w)} = 1 \tag{15.34}$$

If it is assumed that $r_i = 1$, $i = 2, 3, \ldots, n$, then the behavior of this cascade will simply be equivalent to that of a single channel with time function $nJ(t)$; that is,

$$\int_{\tau}^{t_n} \frac{dw}{nJ_1(w)} = \int_{\tau}^{t_n} \frac{dw}{nJ(w)} = 1 \tag{15.35}$$

EXAMPLE 15.7

Compute the IUH of the cascade of n linear time-variant channels. Take $J_i(t) = r_i[1 - \exp(-t)]$ with $r_i = 1/i$, $i = 1, 2, \ldots, n$.

The IUH will simply be a δ-function translated by time t_n,

$$h(t, \tau) = \delta(t - t_n)$$

$$\int_{\tau}^{t_n} \frac{dw}{J(w)} = 1 + \sum_{i=2}^{n} r_i$$

We are given $r_i = 1/i$ and $J_i(t) = r_i[1 - \exp(-t)]$. Therefore,

$$\int_{\tau}^{t_n} \frac{dw}{J(w)} = \int_{\tau}^{t_n} \frac{dw}{[1 - \exp(-w)]} = \ln\left[\frac{\exp(t_n) - 1}{\exp(\tau) - 1}\right]$$

So

$$\ln\left[\frac{\exp(t_n) - 1}{\exp(\tau) - 1}\right] = \sum_{i=1}^{n} r_i$$

Let $\sum_{i=1}^{n} r_i = \ln A$. Then

$$t_n = \ln\{A[\exp(\tau) - 1) + 1\}$$

Values of t_n corresponding to values of τ are:

	Values of t_n			
n	τ = 1	τ = 2	τ = 3	τ = 4
2	2.16	3.39	4.46	5.48
4	2.69	3.96	5.03	6.06
6	3.04	4.32	5.40	6.43
8	3.30	4.58	5.67	6.70

Increasing n and increasing τ lead to increasing t_n, that is, the time after which the IUH appears at the outlet. Fig. 15.8 shows the graph of the IUH.

EXAMPLE 15.8

Compute the IUH of the cascade of n linear time-variant channels. Take $j_i = r_i[1 - \exp(-t)]$, with $r_i = 1/n$, $i = 1, 2, \ldots, n$. Compare this IUH with the one obtained in Example 15.7.

Since

$$\int_{\tau}^{t_n} \frac{dw}{J(w)} = n$$

we obtain

$$\int_{\tau}^{t_n} \frac{dw}{1 - \exp(-w)} = 1$$

Therefore,

$$t_n = \ln\{1 + e[\exp(\tau) - 1]\}$$

The IUH of this cascade is

$$h(t, \tau) = \delta(t - t_n) = \delta(t - \ln\{1 + e[\exp(\tau) - 1]\})$$

The cascade is hydrologically equivalent to a single channel with $J(t) = nJ_1(t)$. For various values of τ, we compute t_n. The IUH is graphed in Fig. 15.9.

τ	t_n	τ	t_n
.0	.00	6.0	7.00
1.0	1.73	7.0	8.00
2.0	2.91	8.0	9.00
3.0	3.96	9.0	10.00
4.0	4.99	10.0	11.00
5.0	5.99		

15.3.3 Time-Variant Linear Reservoirs in Series

Let us first consider a simple cascade of two unequal time-variant linear reservoirs having $h_1(t, \tau)$ and $h_2(t, \tau)$, respectively, as their IUHs, subscript 1 denoting the upstream reservoir and 2 denoting the downstream reservoir. Let their time

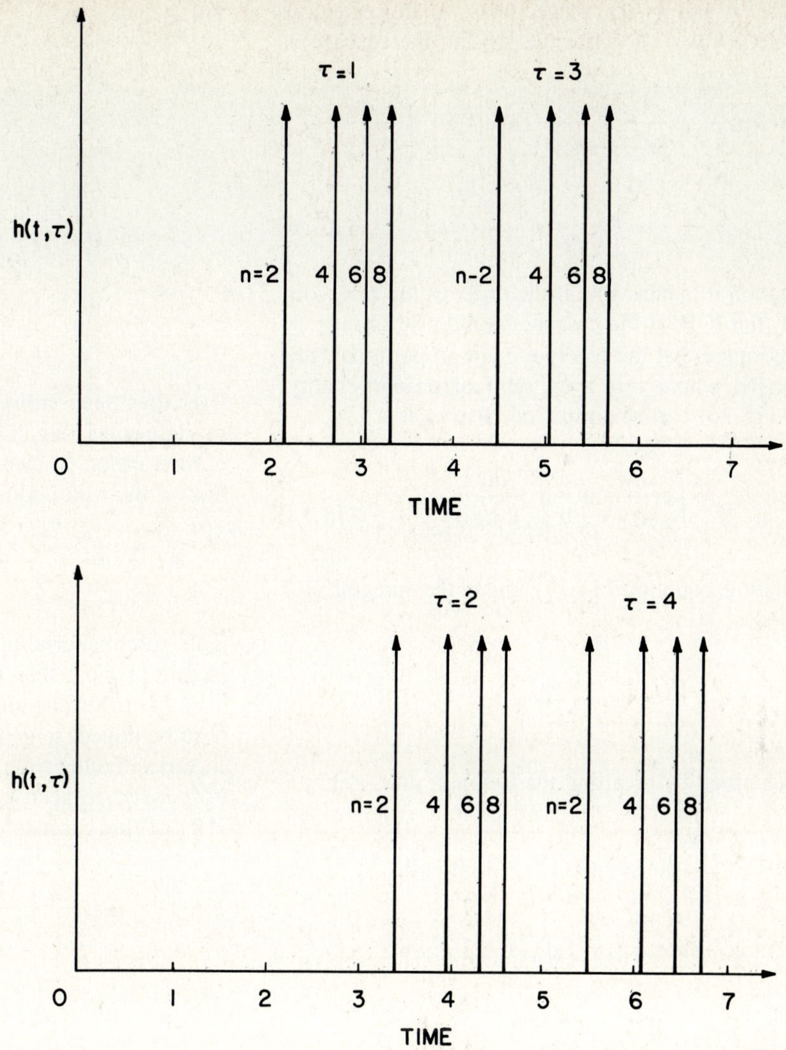

FIGURE 15.8 Example 15.7: The IUH of a cascade of linear time-variant proportional channels.

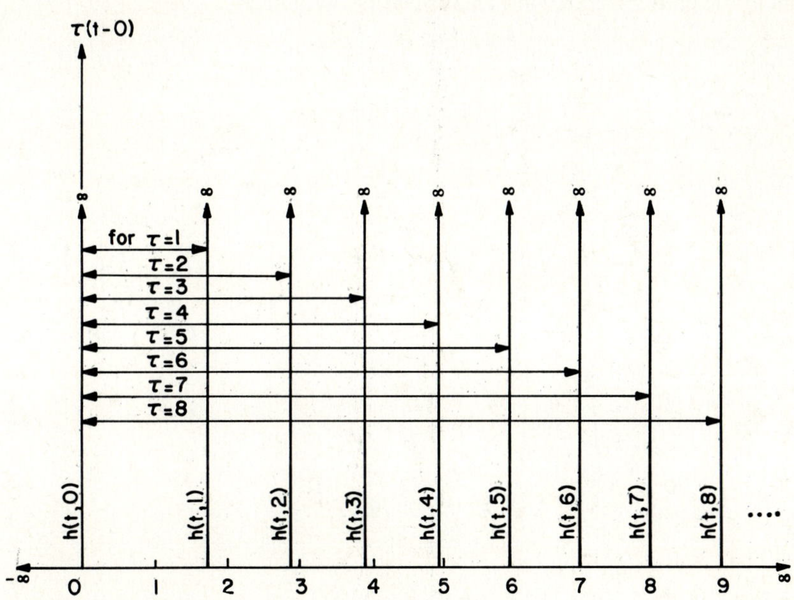

FIGURE 15.9 Example 15.8: The IUH of a cascade of equal linear time-variant channels.

functions be $k_1(t)$ and $k_2(t)$, respectively. Using equations (15.10) and (15.26) we can write the IUH of the cascade as

$$h(t, \tau) = u_1(t-\tau)\frac{1}{k_2(t)}\int_\tau^t \frac{1}{k_1(s)}\exp\left\{-\left[\int_\tau^s \frac{dw}{k_1(w)} + \int_s^t \frac{dw}{k_2(w)}\right]\right\} ds \quad (15.36)$$

From this equation it is clear that if the order of the reservoirs were changed, the IUH of the cascade would also change.

If it is assumed that the reservoirs are proportional, that is, $k_2(t) = rk_1(t)$, where r is a constant of proportionality, then equation (15.36) can be simplified. If we set

$$z = \int_\tau^s \frac{dw}{k_1(w)} + \frac{1}{r}\int_s^t \frac{dw}{k_1(w)} \quad (15.37)$$

then differentiating equation (15.37) under the integral sign gives

$$\frac{dz}{ds} = \frac{r-1}{r}\frac{1}{k_1(s)}$$

This may be obtained by recalling the Leibnitz rule. Let

$$\phi(\alpha) = \int_{u_1}^{u_2} f(x, \alpha)\, dx, \qquad a \le \alpha \le b$$

where u_1 and u_2 may depend on the parameter α. Then applying the Leibnitz rule gives

$$\frac{d\phi}{d\alpha} = \int_{u_1}^{u_2} \frac{\partial f}{\partial \alpha} dx + f(u_2, \alpha)\frac{du_2}{d\alpha} - f(u_1, \alpha)\frac{du_1}{d\alpha}$$

for $a \le \alpha \le b$. Therefore, equation (15.36) can be written as

$$h(t, \tau) = u_1(t-\tau)\frac{1}{(r-1)k_1(t)}\int_{z_2}^{z_1} \exp(-z)\, dz$$

where

$$z_2 = \frac{1}{r}\int_\tau^t \frac{dw}{k_1(w)}$$

and

$$z_1 = \int_\tau^t \frac{dw}{k_1(w)}$$

Therefore,

$$h(t, \tau) = u_1(t-\tau)\frac{1}{(r-1)k_1(t)}[\exp(-z_1) - \exp(-z_2)]$$

This yields

$$\begin{aligned} h(t, \tau) &= u_1(t-\tau)\frac{1}{(r-1)\,k_1(t)}\left\{\exp\left[-\int_\tau^t \frac{dw}{k_1(w)}\right] - \exp\left[-\frac{1}{r}\int_\tau^t \frac{dw}{k_1(w)}\right]\right\} \\ &= u_1(t-\tau)\frac{1}{k_2(t) - k_1(t)}\left\{\exp\left[-\int_\tau^t \frac{dw}{k_1(w)}\right] - \exp\left[-\int_\tau^t \frac{dw}{k_2(w)}\right]\right\} \end{aligned} \quad (15.38a)$$

It is thus seen from equation (15.38a) that the IUH of the cascade consisting of the proportional reservoirs is independent of their order. In fact, it can be expressed as a linear combination of their individual IUHs.

$$h(t, \tau) = -\frac{r}{(r-1)}h_2(t, \tau) + \frac{1}{(r-1)}h_1(t, \tau) \quad (15.38b)$$

This combinatorial property can be extended to a cascade consisting of more than two reservoirs.

To illustrate further this combinatorial property and, it is to be hoped, to generalize it, let us consider three reservoirs in series having time functions $k_1(t)$, $k_2(t) = r_2k_1(t)$, and $k_3(t) = r_3k_1(t)$, respectively. Following the same procedure, the IUH of this cascade can be derived as

$$h(t, \tau) = u_1(t-\tau)\frac{1}{k_3(t)}\int_\tau^t \frac{1}{k_2(s) - k_1(s)}\left\{\exp\left(-\int_\tau^s \frac{dw}{k_1(w)}\right) - \exp\left(-\int_\tau^s \frac{dw}{k_2(w)}\right)\right\}\exp\left(-\int_s^t \frac{dw}{k_3(w)}\right) ds$$

Upon simplification,

$$h(t, \tau) = u_1(t-\tau)\left[\frac{r_3^2}{(r_3 - r_2)(r_3 - 1)}h_3(t, \tau) + \frac{r_2^2}{(r_2 - 1)(r_2 - r_3)}h_2(t, \tau) + \frac{1}{(1 - r_2)(1 - r_3)}h_1(t, \tau)\right] \quad (15.39a)$$

Again, it is seen that the IUH of the cascade is a linear combination of the IUHs of its individual reservoirs. It is plausible to generalize the relationship in equation (15.39a) for any arbitrary number of reservoirs, n greater than or equal to two,

$$h(t, \tau) = u_1(t-\tau)\left[\frac{r_n^{n-1}}{\prod_{\substack{j=1\\ j\ne n}}^{n}(r_n - r_j)}h_n(t, \tau) + \frac{r_{n-1}^{n-1}}{\prod_{\substack{j=1\\ j\ne n-1}}^{n}(r_{n-1} - r_j)}h_{n-1}(t, \tau) + \cdots\right.$$

which can be written in a compact form as

$$h(t, \tau) = u_1(t - \tau) \sum_{i=1}^{n} \frac{r_i^{n-1}}{\prod_{\substack{j=1 \\ j \neq i}}^{n} (r_i - r_j)} h_i(t, \tau) \tag{15.39b}$$

where $r_1 = 1$.

We can now specialize in derivation of the IUH by assuming that the linear reservoirs arranged in series are all equal, that is, $k_1(t) = k_2(t) = \cdots = k_n(t)$. Its time-invariant counterpart is the Nash model (Nash, 1960). To derive the IUH for such a cascade, let us first consider two equal time-variant linear reservoirs arranged in series. By inserting $k_1(t) = k_2(t) = k(t)$ in equation (15.36), we obtain

$$h(t, \tau) = u_1(t - \tau) \frac{1}{k(t)} \exp\left[- \int_\tau^t \frac{dw}{k(w)} \right] \left[\int_\tau^t \frac{dw}{k(w)} \right] \tag{15.40}$$

Equation (15.40) can also be obtained from equation (15.38) by putting $r = 1 + \epsilon$ and letting ϵ tend to 0. The IUH given by equation (15.40) is simply the IUH of a single reservoir modified by a factor that increases from 0 as t goes beyond τ. If there are three equal reservoirs in the cascade, then

$$h(t, \tau) = u_1(t - \tau) \int_\tau^t \frac{1}{k(s)} \exp\left[- \int_\tau^s \frac{dw}{k(w)} \right] \left[\int_\tau^s \frac{dw}{k(w)} \right] \frac{1}{k(t)} \cdot \exp\left[- \int_s^t \frac{dw}{k(w)} \right] ds$$

Let $z = \int_\tau^s dw/k(w)$;
then using the Leibnitz rule for differentiating under the integral sign,

$$\frac{dz}{ds} = \frac{1}{k(s)}$$

Therefore,

$$h(t, \tau) = u_1(t - \tau) \frac{1}{k(t)} \exp\left[- \int_\tau^t \frac{dw}{k(w)} \right] \int_{z_1}^{z_2} z \, dz$$

where $z_1 = 0$ and $z_2 = \int_\tau^t dw/k(w)$
This can be simplified to

$$h(t, \tau) = u_1(t - \tau) \frac{1}{2k(t)} \exp\left[- \int_\tau^t \frac{dw}{k(w)} \right] \left[\int_\tau^t \frac{dw}{k(w)} \right]^2 \tag{15.41}$$

Equation (15.41) can be extended to the case where there are n reservoirs,

$$h(t, \tau) = u_1(t - \tau) \frac{1}{(n-1)!} \frac{1}{k(t)} \exp\left[- \int_\tau^t \frac{dw}{k(w)} \right] \left[\int_\tau^t \frac{dw}{k(w)} \right]^{n-1} \tag{15.42}$$

As Nash (1960) demonstrated, if n is nonintegral, the term $(n - 1)!$ can be replaced by $\Gamma(n)$.

Equation (15.42) can be specialized in the case when $k(t) = k$ as

$$h(t - \tau) = u_1(t - \tau) \frac{1}{\Gamma(n)} \frac{1}{k} \left(\frac{t - \tau}{k} \right)^{n-1} \exp\left[\frac{-(t - \tau)}{k} \right] \tag{15.43}$$

Note that here the IUH depends on $(t - \tau)$, not on t and τ separately.

Equation (15.42) can be utilized to obtain specific forms of the IUH depending upon the form of the function $k(t)$. O'Donnell (1966) used the following relationship for k(t) and derived the *IUH*:

$$k(t) = k_0 + \alpha t \tag{15.44}$$

where k_0 and α are constants. Note that

$$\int_\tau^t \frac{dw}{k_0 + \alpha w} = \ln \left[\frac{k_0 + \alpha t}{k_0 + \alpha \tau} \right]^{1/\alpha} \tag{15.45}$$

Substituting in equation (15.42),

$$h(t, \tau) = u_1(t - \tau) \frac{1}{\Gamma(n)} \frac{1}{(k_0 + \alpha t)} \left[\frac{k_0 + \alpha \tau}{k_0 + \alpha t} \right]^{1/\alpha} \left[\ln \left(\frac{k_0 + \alpha t}{k_0 + \alpha \tau} \right)^{1/\alpha} \right]^{n-1} \tag{15.46}$$

On the other hand, Chiu and Bittler (1969), based on an analysis of 28 storms that occurred during the 17-year period from 1940 through 1957 on the Chartiers Creek basin near Pittsburgh, Pennsylvania, proposed $k(t)$ as

$$k(t) = \alpha t^\beta \tag{15.47}$$

where α and β are constants. Inserting this value in equation (15.42),

$$h(t, \tau) = u_1(t - \tau) \frac{1}{\Gamma(n)} \frac{1}{\alpha t^\beta} \exp\left[- \frac{1}{\alpha(1 - \beta)} (t^{1-\beta} - \tau^{1-\beta}) \right] \left[\frac{t^{1-\beta} - \tau^{1-\beta}}{\alpha(1 - \beta)} \right]^{n-1} \tag{15.48}$$

Similarly, the IUH can be derived for any $k(t)$. If $k(t)$ is given in a form not amenable to simple integration, then equation (15.41) can best be solved by numerical integration.

Before closing this section, it may be appropriate to examine the peak characteristics of equation (15.42). From Chapter 13 it is known that the peak of the IUH in equation (15.43) occurs when the time $(t - \tau)$ between the centers of areas under the effective rainfall and direct runoff is $(n - 1)k$ and is specified by

$$h_{\max} = \frac{1}{k\Gamma(n)}(n-1)^{n-1}\exp[-(n-1)] \tag{15.49}$$

To derive this result for equation (15.42), we observe that (1) $n \geq 2$, (2) $h(t, \tau)$ is a function of both t and τ, and (3) $t > \tau$. Let there be a point (t_0, τ_0), called a relative maximum point of $h(t, \tau)$. Then a necessary condition for $h(t, \tau)$ to have a relative maximum is

$$\frac{\partial h(t,\tau)}{\partial t} = 0, \quad \frac{\partial h(t,\tau)}{\partial \tau} = 0 \tag{15.50}$$

Differentiating equation (15.42) partially with respect to t,

$$\int_{\tau_0}^{t_0} \frac{dw}{k(w)} - \frac{(n-1)}{1 + k'(t_0)} = 0 \tag{15.51}$$

where $k'(t_0) = dk/dt$ at $t = t_0$. Similarly, differentiating equation (15.42) partially with respect to τ,

$$\int_{\tau_0}^{t_0} \frac{dw}{k(w)} - (n-1) = 0 \tag{51.52}$$

These partial derivatives are obtained by noting the fact that if

$$z = \int_{\tau}^{t} \frac{dw}{k(w)}$$

then

$$\frac{\partial z}{\partial t} = \frac{1}{k(t)}$$

and

$$\frac{\partial z}{\partial \tau} = -\frac{1}{k(\tau)}$$

Combining equations (15.51) and (15.52), we found that $k'(t_0) = 0$, and t_0 and τ_0 are related by

$$\int_{\tau_0}^{t_0} \frac{dw}{k(w)} = n - 1 \tag{15.53}$$

This means that $k(t)$ has a stationary value at t_0. To ensure that $h(t_0, \tau_0)$ is a maximum value, $d^2k/dt^2 > 0$ at $t = t_0$; this ensures $k(t_0)$ to be minimum.

It is then clear that there may be many sets of (t_0, τ_0) leading to local or relative maxima of $h(t, \tau)$ provided that $k(t_0)$ is minimum and t_0 and τ_0 are related by equation (15.53). These maxima can be expressed by substituting equation (15.53) into equation (15.42),

$$h_{max}(t_0) = \frac{1}{k(t_0)\Gamma(n)}(n-1)^{n-1}\exp[-(n-1)] \tag{15.54}$$

This can be simply written as

$$h_{\max}(t_0) = \frac{E(n)}{k(t_0)}$$

where

$$E(n) = \frac{1}{\Gamma(n)}(n-1)^{n-1}\exp[-(n-1)]$$

Therefore, the global maximum of $h(t, \tau)$ occurs when $k(t_0)$ is minimum. This result is also valid for the case $n = 1$. Further, for a specified t_0, $h_{\max}$ decreases with increasing n. To illustrate this point, let us consider the cascade of identical reservoirs where as n increases, each reservoir becomes proportionately smaller. If the individual time functions are defined by $k(t)/n$, the IUH of equation (15.42) becomes

$$h(t, \tau) = u_1(t-\tau)\frac{1}{k(t)}\frac{1}{(n-1)!}z^{n-1}\, n^n \exp[-nz] \tag{15.55}$$

where

$$z = \int_{\tau}^{t} \frac{dw}{k(w)}$$

For large n, we can use Stirling's formula to approximate $n!$,

$$n! \cong (2\pi)^{.5}\exp(-n)\, n^{n+.5} \tag{15.56}$$

Inserting equation (15.56) in equation (15.55) and allowing $n \to \infty$,

$$h(t,\tau) \to \infty \quad \text{if } z = 1$$
$$h(t,\tau) \to 0 \quad \text{otherwise}$$

However, it can be seen that z increases monotonically from 0 as t exceeds τ. Therefore, a unique value of t^* exists for which $z = 1$,

$$\int_{\tau}^{t^*} \frac{dw}{k(w)} = 1 \tag{15.57}$$

Therefore, $h(t, \tau)$ becomes $\delta(t - t^*)$ for an infinite cascade of infinitesimal time-variant linear reservoirs. This is the IUH of a time-variant linear channel whose time function is $k(t)$, as seen from equation (15.23). This result is an extension of Dooge's result (Dooge, 1959), that is, a time-variant linear channel may also be considered to be an infinite cascade of infinitesimal time-variant linear reservoirs.

EXAMPLE 15.9

Compute the IUH of the cascade of n linear time-variant equal reservoirs. Take $k(t) = \alpha t^{\beta}$, where $\beta = -2.0$, $\alpha = 100$, and t is in hours.

$$k_1(t) = k_2(t) = \cdots = k_n(t) = k(t) = \frac{100}{t^{-2}}$$

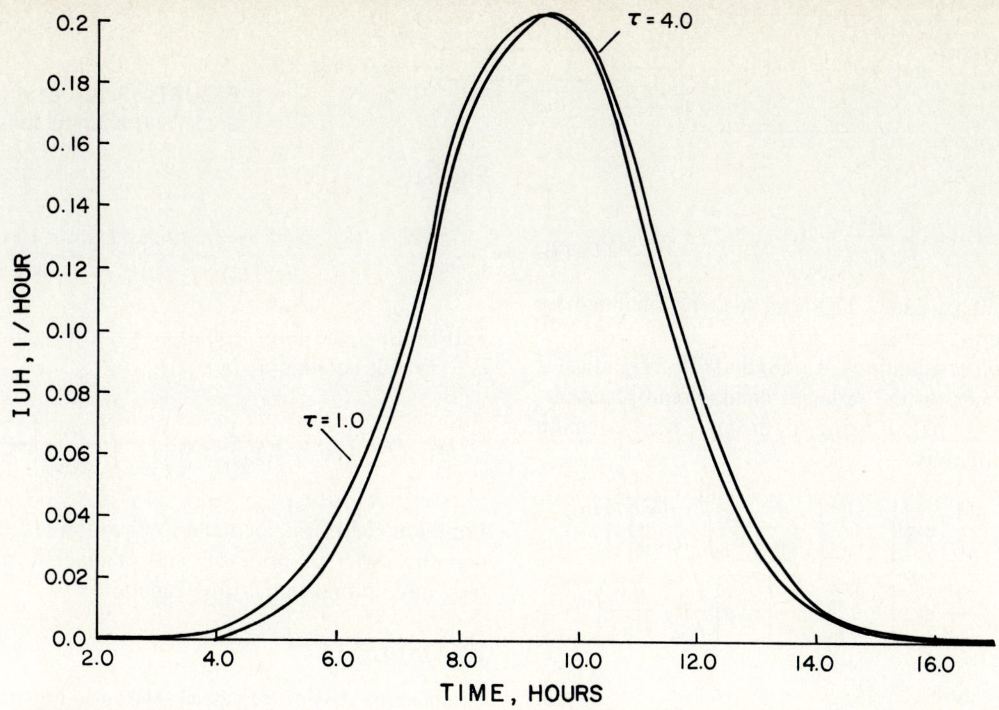

FIGURE 15.10 Example 15.9: The IUH of a cascade of equal linear time-variant reservoirs.

The IUH of the cascade is

$$h(t, \tau) = u_1(t - \tau)\frac{1}{(n-1)!}\left(\frac{t^2}{100}\right)\left(\frac{t^3 - \tau^3}{300}\right)^{n-1} \exp\left(\frac{\tau^3 - t^3}{300}\right), \qquad t \geq \tau$$

For various values of n, $h(t, \tau)$ can be computed for different times of application $\tau = 0, 1, 2, \ldots$ For $n = 3$ and $\tau = 1$ and $\tau = 4$, the IUH is graphed in Fig. 15.10.

EXAMPLE 15.10

Compute the peak and time to peak of the IUH in Example 15.9.

Consider $n = 3$ and $\tau_0 = 1$. We want to find h_{max} and its time t_0. From Equation (15.53)

$$\int_{\tau_0}^{t_0} \frac{w^2}{100}\, dw = \frac{t_0^3 - \tau_0^3}{300} = 2$$

Thus

$$t_0^3 = (\tau_0^3 + 600)^{1/3} = (1 + 600)^{1/3} = 8.44 \text{ h}$$

Substituting in equation (15.54),

$$h_{max}(8.44) = \frac{t_0^2}{100}\frac{1}{(3-1)!}(3-1)^{3-1}\exp[-(3-1)]$$

$$= .192$$

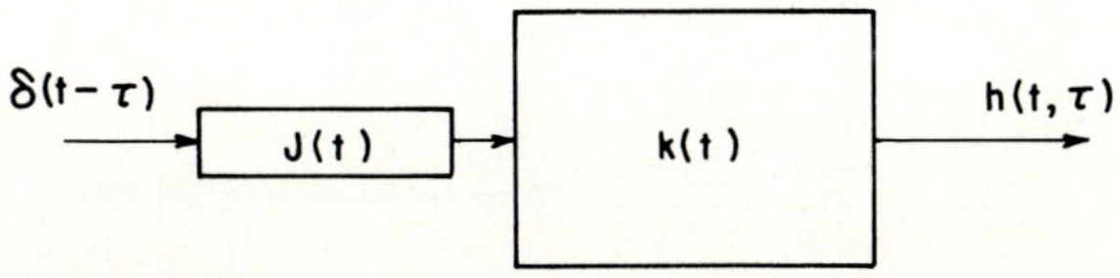

FIGURE 15.11 Linear time-variant channel and reservoir in series.

If $\tau_0 = 2$, following the same procedure gives $t_0 = 8.47$ h and $h_{max} = .194$. In this way h_{max} and t_0 can be determined for any value of τ_0. Doing this we obtain

$$t_0 = 9 \text{ h}$$
$$h_{max} = .219/\text{h}$$

15.3.4 Linear Time-Variant Channels and Reservoirs

The channels and reservoirs can be variously combined. Let us first consider a channel and a reservoir in series with time functions as $J(t)$ and $k(t)$, respectively, as shown in Fig. 15.11. The IUH of this cascade can be obtained simply as follows.

The IUH of the channel due to $\delta(t - \tau)$ is given by equations (15.23) and (15.24). This constitutes input to the reservoir. Therefore, replacing τ by t_1 in equation (15.10),

$$h(t, \tau) = u_1(t - t_1)\frac{1}{k(t)}\exp\left[-\int_{t_1}^{t} \frac{dw}{k(w)}\right] \tag{15.58}$$

where t_1 is given by equation (15.24). This is the IUH of the cascade. Equation (15.58) can also be obtained by employing equation (15.26).

If the order of these elements is reversed, that is, if a reservoir is followed by a channel as shown in Fig. 15.12, the IUH can be obtained as follows. The IUH of a reservoir is given by equation (15.10). This constitutes input to the channel whose response is given by equations (15.21) and (15.22). Therefore,

$$h(t, \tau) = u_1(s - \tau)\frac{J(s)}{J(t)}\frac{1}{k(s)}\exp\left[-\int_{\tau}^{s} \frac{dw}{k(w)}\right] \tag{15.59}$$

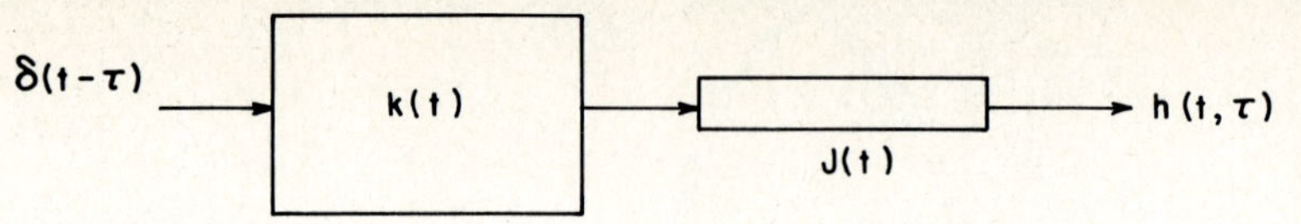

FIGURE 15.12 Line time-variant reservoir and channel in series.

where

$$\int_s^t \frac{dw}{J(w)} = 1 \tag{15.60}$$

This is the IUH of cascade. This can also be obtained by employing equation (15.26).

A comparison of equations (15.58) and (15.59) indicates that the IUH depends on the order of channel and reservoir. If, however, $k(t) = rJ(t)$, r being a constant, then equation (15.58) can be written as

$$h(t, \tau) = u_1(t - t_1)\frac{1}{k(t)}\exp\left[-\left\{\int_\tau^t \frac{dw}{k(w)} + \int_{t_1}^\tau \frac{dw}{k(w)}\right\}\right]$$

$$= u_1(t - t_1)\frac{1}{k(t)}\exp\left[-\int_\tau^t \frac{dw}{k(s)}\right]\exp\left[\frac{1}{r}\int_\tau^{t_1} \frac{dw}{J(w)}\right]$$

Making use of equation (15.24),

$$h(t, \tau) = u_1(t - t_1)\frac{1}{k(t)}\exp\left[-\int_\tau^t \frac{dw}{k(w)}\right]\exp(1/r) \tag{15.61}$$

Let us now use equation (15.59). The exponential terms can be modified as

$$\exp\left[-\int_\tau^s \frac{dw}{k(w)}\right] = \exp\left[-\int_\tau^t \frac{dw}{k(w)}\right]\exp\left[-\int_t^s \frac{dw}{k(w)}\right]$$

$$= \exp\left[-\int_\tau^t \frac{dw}{k(w)}\right]\exp\left(\frac{1}{r}\right)$$

Similarly,

$$\frac{J(s)}{J(t)}\frac{1}{k(s)} = \frac{rJ(s)}{rJ(t)}\frac{1}{k(s)} = \frac{1}{k(t)}$$

Substituting these in equation (15.59) and noting that u_1 $(s - \tau)$ will become u_1 $(t - t_1)$,

$$h(t, \tau) = u_1(t - t_1)\frac{1}{k(t)}\exp\left[-\int_\tau^t \frac{dw}{k(w)}\right]\exp\left(\frac{1}{r}\right) \tag{15.62}$$

Equation (15.62) is identical to equation (15.61). It is evident that the order of reservoir and channel is immaterial when they have proportional time functions.

EXAMPLE 15.11

Compute the IUH of the cascade shown in Fig. 15.11. Take $J(t) = 1 - e^{-t}$ and $k(t) = 10/(1 + t)$.

The IUH of the channel h_1 (s, t_1) is $\delta(s - t_1)$, where

$$\int_\tau^{t_1} \frac{dw}{J(w)} = 1$$

By substituting $J(t) = 1 - \exp(-t)$,

$$t_1 = \ln\{1 + e[\exp(\tau) - 1]\}$$

The IUH of the channel is the input to the reservoir. Therefore, the IUH of the cascade is

$$h(t, \tau) = u_1(t - t_1)\frac{1}{k(t)}\exp\left[-\int_{t_1}^t \frac{dw}{k(w)}\right]$$

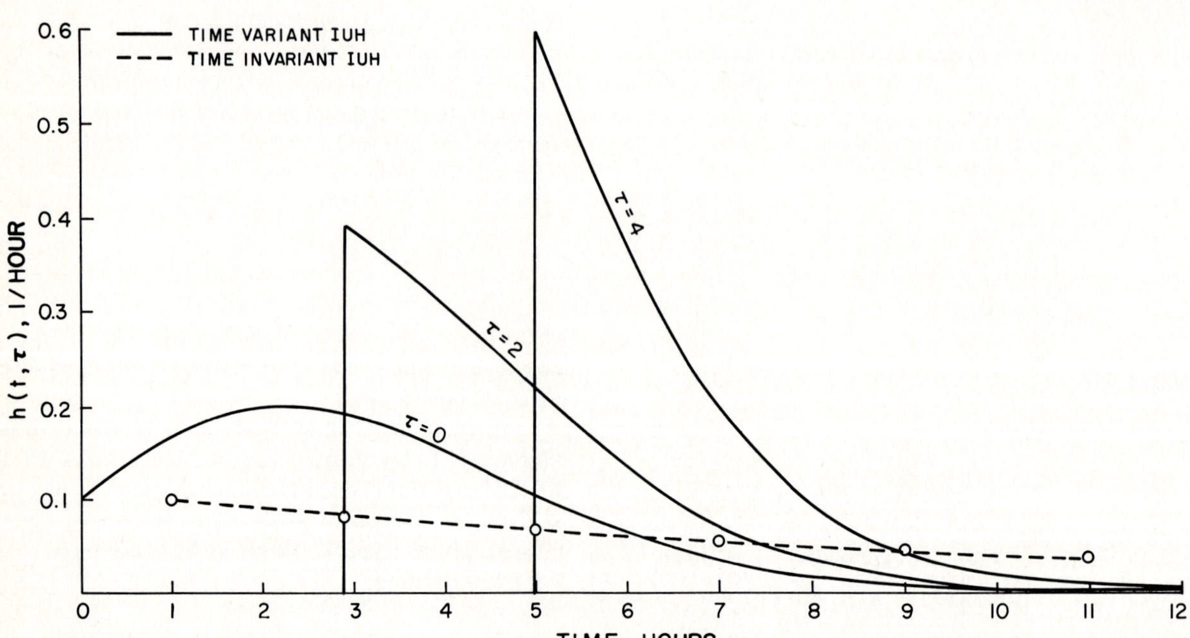

FIGURE 15.13 Example 15.11: The IUH of a linear cascade of a time-variant channel and time-variant reservoir.

Inserting $k(t) = 10/(1 + t)$,

$$h(t, \tau) = 0, \; t \le t_1$$

$$= \frac{(1+t)}{10} \exp\left\{ -\frac{1}{10}(t - t_1)\left[1 + \frac{1}{2}(t + t_1)\right]\right\}, \qquad t \ge t_1$$

where

$$t_1 = \ln\{1 + e[\exp(\tau) - 1]\}$$

The IUH of the cascade can be computed for $\tau = 0$, 2, and 4 h, as shown in Fig. 15.13.

EXAMPLE 15.12

Compute the IUH of the cascade shown in Fig. 15.12. Take $J(t) = 1 - e^{-t}$ and $k(t) = 10/(1 + t)$.

The IUH of the reservoir $h(s, \tau)$ is

$$h(s, \tau) = u_1(s - \tau)\frac{1}{k(s)}\exp\left[-\int_\tau^s \frac{dw}{k(w)}\right]$$

The argument of the exponential function can be written as

$$-\int_\tau^s \frac{(1+w)}{10}\,dw = -\frac{1}{10}(s - \tau)\left[1 + \frac{1}{2}(s + \tau)\right]$$

Therefore, the IUH of the reservoir is

$$h(s, \tau) = \frac{1+s}{10}\exp\left\{-\frac{1}{10}(s - \tau)\left[1 + \frac{1}{2}(s + \tau)\right]\right\}$$

This constitutes input to the channel. Thus the IUH of the cascade is

$$h(t, \tau) = \frac{J(s)}{J(t)} h(s, \tau), \qquad \int_s^t \frac{dw}{J(w)} = 1$$

$$\int_s^t \frac{dw}{1 - \exp(-w)} = 1$$

This yields

$$t = \ln\{1 + e[\exp(s) - 1]\}$$

Therefore,

$$h(t, \tau) = \frac{1 - \exp(-s)}{1 - \exp\{-\ln[1 + e(\exp(s) - 1)]\}}\,\frac{1+s}{10}$$

$$\exp\left\{-\frac{1}{10}(s - \tau)\left[1 + \frac{1}{2}(s + \tau)\right]\right\}$$

$$= \frac{1 - \exp(-s)}{1 - \dfrac{1}{1 + e(\exp(s) - 1)}}\,\frac{1 + s)}{10}$$

$$\exp\left\{-\frac{1}{10}(s - \tau)\left[1 + \frac{1}{2}(s + \tau)\right]\right\}, \qquad s \ge \tau$$

The IUH can now be computed for various values of τ, as graphed in Fig. 15.14.

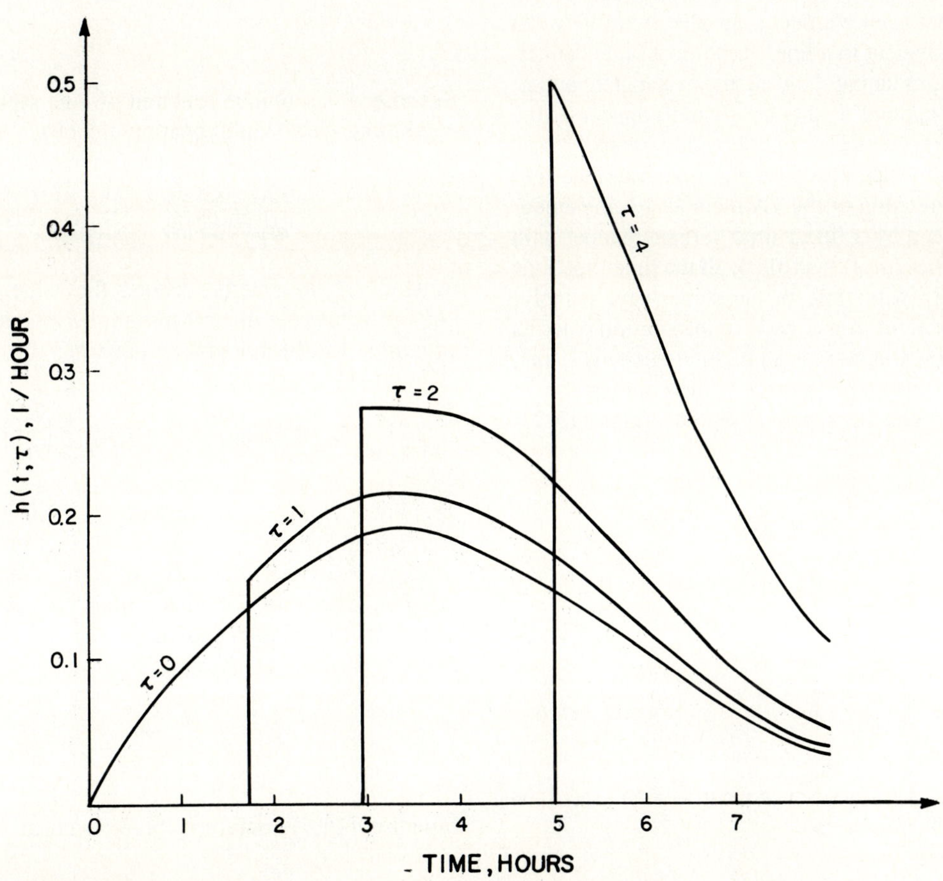

FIGURE 15.14 Example 15.12: The IUH of a linear cascade of a time-variant reservoir and a time-variant channel.

15.4 TA METHOD

Let the TA diagram be represented by Fig. 15.15, where a is the area contributing to runoff at time c after the start of the effective rainfall. A is the total watershed area and T_c is the time of concentration. The IUH due to $\delta(t - \tau)$ is given by equation (10.11). Rewriting it,

$$h(t - \tau) = \frac{1}{A}\left(\frac{da}{dc}\right)\bigg|_{c=t-\tau} = \frac{1}{A}\frac{da(t - \tau)}{dc} \tag{15.63}$$

The derivative, which is 0 for c outside the range $0 < c < T_c$, is the TAC curve. The factor $1/A$ insures the correct dimension of $h(t - \tau)$, that is,

$$\int_{-\infty}^{+\infty} h(t - \tau)\, dt = \int_0^{T_c} \frac{1}{A}\left(\frac{da}{dc}\right) dc = \int_0^A \frac{da}{A} = 1.0 \tag{15.64}$$

The IUH in equation (15.63) may be interpreted to include the effect of the watershed shape transforming $\delta(t - \tau)$, which occurs uniformly over the entire area. The mean lag time is the distance between the start and the center of the TAC curve. This time-invariant version assumes that the time of travel for each element of area is independent of the rate and time of input as well as the amount of water flowing in the watershed. According to Mandeville and O'Donnell (1973) and the discussion in Chapter 11, an alternative interpretation of the TA method is that each elemental area ∂a lying between two isochrones is equivalent to a linear time-invariant channel, with constant $c(a)$, producing flow at the watershed outlet. Therefore, $\delta(t - \tau)$ applied to this area would appear at the outlet at $t = \tau + c(a)$ without a modification in shape.

It is now plausible to introduce the time variance into this alternative interpretation of the TA method. Each elemental area can be replaced by a linear time-variant channel with time function defined as $J(a, t) = c(a)f(t)$; all the time functions in the watershed vary with time in the same way, namely, $f(t)$; but at any instant of time, each is proportional to its corresponding time-invariant translation time, namely, $c(a)$. For $\delta(t - \tau)$ falling over the watershed, the contribution of an elemental area ∂a can be specified by equations (15.23) and (15.24) as

$$\partial h(t, \tau) = \frac{\partial a}{A}\delta(t - t_1) \tag{15.65}$$

with

$$\int_\tau^{t_1} \frac{dw}{J(a, w)} = 1 \tag{15.66}$$

Integrating equation (15.65) over the entire area, the IUH for the time variant TA model is

$$h(t, \tau) = \frac{1}{A}\int_0^A \delta(t - t_1)\, da \tag{15.67}$$

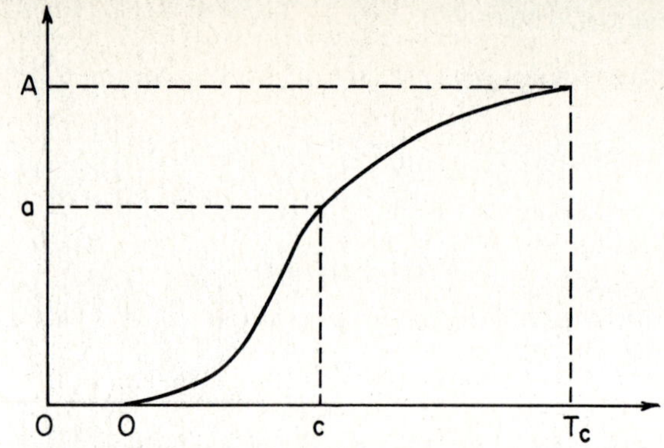

FIGURE 15.15 Cumulative watershed area a contributing to direct runoff at time c after the onset of the effective rainfall. This is TA diagram.

For evaluation of equation (15.67) it is necessary to know the relation between t_1 and a; note that t and τ remain fixed during integration. Replacing $J(a, t)$ by $c(a)f(t)$ in equation (15.66),

$$\int_\tau^{t_1} \frac{dw}{f(w)} = c(a) \tag{15.68}$$

For a fixed τ, equation (15.68) relates t_1 to c. Therefore, its derivative with respect to t_1 is

$$\frac{dc}{dt_1} = \frac{1}{f(t_1)} \tag{15.69}$$

Because a is a unique function of c as seen from Fig. 15.15, combining da/dc with equation (15.69),

$$\frac{da}{dt_1} = \frac{da}{dc}\frac{dc}{dt_1} = \frac{1}{f(t_1)}\left(\frac{da}{dc}\right)\bigg|_{c=c_{t_1}} \tag{15.70}$$

in which c_{t_1}, in general, denotes the value of $c(a)$ in equation (15.68) where the upper limit on the integrand is t_1. Inserting equation (15.70) in equation (15.67),

$$h(t, \tau) = \frac{1}{A}\int_\tau^d \delta(t - t_1)\frac{1}{f(t_1)}\left(\frac{da}{dc}\right)\bigg|_{c=c_{t_1}} dt_1 \tag{15.71}$$

The lower limit of integration of $a = 0$ in equation (15.67) corresponds to $c(a) = 0$ and, hence, from equation (15.68) for $t_1 = \tau$ we get the lower limit of integration in equation (15.71). Similarly the upper limit of integration $a = A$ in equation (15.67) corresponds to $c(a) = T_c$; hence, from equation (15.68) we get an upper limit of $t_1 = d$, where

$$\int_\tau^d \frac{dw}{f(w)} = T_c \tag{15.72}$$

Equation (15.71) may now be reduced to

$$h(t, \tau) = \frac{1}{A}\frac{1}{f(t)}\left(\frac{da}{dc}\right)\bigg|_{c=c_t} \tag{15.73}$$

where, from equation (15.68), c_t is

$$c_t = \int_{\tau}^{t} \frac{dw}{f(w)} \tag{15.74}$$

Equation (15.73) is valid only within the range $\tau < t < d$; otherwise $h(t, \tau)$ is 0. Also, this range corresponds to $0 < c_t < T_c$, since $f(t) > 0$. da/dc has been assumed 0 outside this range. Equations (15.73) and (15.74) specify the time-variant TA model. Note that if $f(t)$ is assumed to be unity, equations (15.73) and (15.74) reduce to equation (15.63).

15.5 LINEAR DIFFERENTIAL EQUATION

Linear time-varying models for rainfall-runoff relationships can be formulated in terms of linear differential equations whose coefficients are time-varying. These coefficients can, in turn, be estimated from the effective rainfall–direct runoff data using appropriate optimization procedures, as was done by Carrasquilla and Perkins (1967). We briefly outline this approach here.

Let a linear differential equation relate direct runoff to the effective rainfall as

$$\frac{dQ^i(t)}{dt} + A(t)Q^i(t) = B(t)I^i(t), \qquad i = 1, 2, \ldots, N \tag{15.75}$$

where $Q^i(t)$, $I^i(t)$, $A(t)$, and $B(t)$ are matrices of real functions of the real variable t. The problem reduces to the determination of the operator H such that

$$HI^i(t) = Q^i(t)$$

over the set $= \{I^i(t), Q^i(t)\}$, $i = 1, 2, \ldots, M$. From the theory of linear differential equations (Coddington and Levinson, 1955), it is known that if $A(t)$ can be an (n, n) continuous matrix of real functions of real variable t, then the linear differential equation

$$\frac{dQ(t)}{dt} + A(t)Q(t) = B(t)I(t) \tag{15.76}$$

where $Q(t)$ is an n-dimensional vector, $I(t)$ is an m-dimensional vector, and $B(t)$ is an (n, m) matrix of real functions of t, will have its solution expressed as

$$Q(t) = h(t, 0)\, Q(0) + \int_0^t h(t, s)B(s)I(s)\, ds \tag{15.77}$$

where $h(t, s)$ is the matrix-kernel function satisfying

$$\frac{dh(t, s)}{dt} + A(t)h(t, s) = 0 \tag{15.78}$$

$$h(t, t) = I \qquad \text{(identity matrix)} \tag{15.79}$$

The matrix $h(t, s)$ is unique and characterizes the system completely. Here the initial condition is $Q(0)$. This means that given the input $B(t)I(t)$ and the initial condition $Q(0)$, the knowledge of $h(t, s)$ implies that of $Q(t, s)$. The principal property of such systems is that $h(t, s)$ depends only on $A(t)$ and not on $I(t)$ or $Q(t)$.

Furthermore, the homogeneous system

$$\frac{dQ(t)}{dt} + A(t)\, Q(t) = 0 \tag{15.80}$$

has n linearly independent solutions. A matrix $\Phi(t)$, whose n columns are n linearly independent solutions of equation (15.80), is called a fundamental matrix. The kernel function can then be expressed in terms of $\Phi(t)$ as

$$k(t, s) = \Phi(t)\Phi^{-1}(s) \tag{15.81}$$

EXERCISES

15.1 Consider $k = 5 + [1/(1 + t)]$. Determine and graph the IUHs of the linear reservoir if the effective rainfall is applied at 0, 2, and 4 h measured from the origin. Compare these with those of the time-invariant linear reservoir.

15.2 Determine and graph the 2-h, 4-h, and 6-h UHs for the data in Question 15.1. Compare them with those of the time-invariant linear reservoir.

15.3 Determine and graph the S-hydrographs for the data in Question 15.1. Compare them with those of the time-invariant linear reservoir.

15.4 Compute and graph the S-hydrographs of the linear time-variant channel when the inflow occurs at the upstream. Take $J(t) = 1 - [1/(1 + t)]$.

15.5 Compute and graph the IUH of the linear time-variant channel when the inflow occurs laterally all along its length. Take $J(t) = 1 - [1/(1 + t)]$.

15.6 Compute and graph the S-hydrograph of the linear time-variant channel subject to lateral inflow along its length. Take $J(t) = 1 - [1/(1 + t)]$.

15.7 Compute and graph the IUH of the cascade of n linear time-variant channels with $J_i(t) = r_i[1 - (1/(1 + t))]$, $r_i = 1/i$, $i = 1, 2, \ldots, n$.

15.8 Compute and graph the IUH of the cascade in Question 15.7 with $r_i = 1/n$, and compare it with that in that question.

15.9 Compute and graph the IUH of the cascade of n equal linear time-variant reservoirs for $k(t) = 10 + [1/(1 + t)]$. Take $n = 2, 4$, and 6.

15.10 Determine the peak and time to peak of the IUH in Question 15.9. Graph them against n.

15.11 Compute the IUH of the cascade of n unequal time-variant reservoirs. Take $k_i(t) = r_i[10 + (1/1 + t))]$, where $r_i = 1/i$, $i = 1, 2, \ldots, n$. Consider $n = 2$, 4, and 6.

15.12 Determine the peak and time to peak of the IUH in Question 15.11. Graph them against n.

15.13 Compute the IUH of the cascade of Fig. 15.11. Take $J(t) = 1 - [1/(1 + t)]$ and $k(t) = 10 + [1/(1 + t)]$.

15.14 Compute the IUH of the cascade of Fig. 15.12. Take $J(t)$ and $k(t)$ as the same as in Question 15.13.

15.15 Compute the IUH of the cascade of Fig. 15.11. Take $J(t) = 1 + [1/(1 + t)]$ and $k(t) = 10 + [10/(1 + t)]$.

CHAPTER SIXTEEN

Nonlinear Conceptual Models

There exist a multitude of nonlinear models of rainfall-runoff relationships. Indeed the past two decades have witnessed a phenomenal increase in research on nonlinear analyses of hydrologic systems. In this chapter we discuss time-invariant nonlinear conceptual models at length. A short discussion of time-variant nonlinear reservoir is also given.

16.1 EVIDENCE FOR NONLINEARITY OF DIRECT RUNOFF

Rainfall-runoff systems are nonlinear. Their linear representation is only an assumption. As early as 1952, Paynter indicated the possible existence of nonlinear effects in rainfall-runoff relationship and suggested that those might be due to seasonal variations and relative magnitude of floods. Minshall (1960) showed that on small watersheds (up to about 500 acres) one UH could not adequately define the shape of the hydrograph derived from a storm of unit duration and that the UH peak and its time varied with rainfall intensity and storm pattern. In a discussion of Minshall's paper, Amorocho (1961) found that the UHs obtained by Minshall could be represented by the Nash model with parameters n and k and derived relationships between those parameters and rainfall intensity. Overton (1967) reached the same conclusion. Later, in a series of papers, Amorocho and his associates (Amorocho, 1963a, 1967; Amorocho and Orlob, 1961; Amorocho and Hart, 1965; Amorocho and Brandstetter, 1971) as well as others (Jacoby, 1966; Hino, Sukigara and Kikkawa, 1971; Hino and Nadaoka, 1979; Diskin and Boneh, 1972, 1973; Diskin, 1982; Diskin, Boneh and Golan, 1984; A. R. Rao and R. G. S. Rao, 1974; Liu and Brutsaert, 1975, 1978; Napiorkowski and Strupczewski, 1979, 1981) investigated a procedure for nonlinear analysis of the rainfall-runoff relationship by employing a generalized functional series introduced originally by Wiener (1942). In 1962, K. P. Singh undertook a systematic study to establish variability of the IUH with storm and watershed characteristics. Diskin (1964) also found the IUH to vary systematically with rainfall intensity. Kulandaiswamy (1964) found that some of the coefficients of his linearized model decreased exponentially with a decrease in the direct runoff peak. In a series of papers, Pilgrim and his associates (Pilgrim, 1966a, 1966b; 1976, 1977, 1978, 1980; Pilgrim and Huff, 1978; Pilgrim, Huff, and Steele, 1978) used radioactive tracers to show that the watershed runoff was highly nonlinear, particularly at low discharges. This was inferred from variation of travel time and average velocity with discharge, as well as from spatial variations of travel times and hydrograph shape. Clearly there is an accumulated evidence to show that all rainfall-runoff systems are, to varying degrees, nonlinear.

16.1.1 Variation of Storage Effects with Flood Magnitude

A number of workers have noted variations in storage characteristics of watersheds with flood magnitude (Minshall, 1960; Amorocho, 1961; K. P. Singh, 1962; Kulandaiswamy, 1964; Pilgrim, 1977, 1980; Boyd, 1976). As the flood magnitude increases, the storage effects decrease. This causes the lag time to decrease as the event magnitude increases and causes an increase in discharge magnitude that is proportionately greater than the increase in the effective rainfall magnitude. This may partly explain why larger UHs are derived from larger floods rather than from small floods. This is also seen by the decrease in magnitude of the storage parameters of linear models with increase in event magnitude.

Pilgrim (1976), using radioactive tracing on the Research Creek watershed (.39 square-kilometer area) located near Sydney, New South Wales, Australia, found that the travel time from the point of injection to the centroid of the tracer outflow at the watershed outlet decreased as the hydrograph peak increased. A linear regression of the data (20 events) yielded (with a correlation coefficient of −.935),

$$t_{cm} = 49.5\, Q_p{}^{-.492} \tag{16.1a}$$

in which t_{cm} is the time of travel in minutes and Q_p is the peak discharge in m^3/s. The average velocity $v_a(m/s)$ of the centroid of molecules from the site to the outlet was

$$v_a = .255\, Q_p{}^{.492} \tag{16.1b}$$

Laurenson (1962) measured the variation of lag time t_L between the centroids of the effective rainfall and direct runoff with the mean discharge for 23 events on the South Creek (89.6 square kilometers) watershed located near Sydney, New South Wales, Australia, and showed

$$t_L = 64 Q_m{}^{-.27} \tag{16.2}$$

where t_L is in hours and Q_m in cubic feet per second. Askew (1968) extended this analysis to 201 events on five watersheds (varying in area from .39 to 89.6 square kilometers) also located near Sydney, New South Wales, and found

$$t_L = a Q_{wm}{}^{-b} \tag{16.3}$$

where t_L is in hours, and Q_{wm} is a weighted mean discharge in cubic feet per second, a is a parameter that varied from 3.61 to 72.1, and b is an exponent that varied from +.328 to +.190. These analyses support the hypothesis that storage effects decrease as the flood magnitude increases. Askew took b as +.23.

16.1.2 Variation of Peak Discharge with Runoff Volume

In a series of papers, Rogers and co-workers (Rogers, 1980, 1982; Rogers and Zia, 1982) investigated the relation between peak discharge Q_p and runoff volume V for 42 watersheds (4.5 to 701 square kilometers), which was found to be

$$\log\left(\frac{Q_P}{V^2}\right) = a + m \log V \tag{16.4}$$

where a is the intercept and m is the slope of the line. For hydrologically linear watersheds, $|m| = 1$, and for nonlinear watersheds, $|m| < 1$. This criterion was used to assess the degree of nonlinearity. Most of the watersheds were nonlinear. Mimikou (1983) employed this same criterion for investigating 8 watersheds (200 to 6000 square kilometers) in Greece, and V. P. Singh and Aminian (1976) investigated 139 watersheds in the United States, Italy, and Greece.

16.1.3 Variation of Travel Time with Stream Discharge

From the equations of open channel hydraulics (such as Manning's or Chezy's equation) it is observed that the velocity of flow increases as the depth of flow (or hydraulic radius) increases. This corresponds to a decrease in travel time as the flood magnitude increases. This provides evidence of nonlinear behavior of stream hydraulics. Note that here the travel time refers to water rather than to the flood wave. Expressions for travel time can be derived by employing either Manning's or Chezy's equation. Consider a small channel segment of length ΔL. If it is of wide rectangular cross section, then

$$t = \frac{\Delta L}{V} = \frac{\Delta L\, n_m^{0.6}}{S_0^{.3}} Q^{-.4} = a\, Q^{-.4} \tag{16.5a}$$

where n_m is Manning's coefficient, S_0 is bed slope, and Q is discharge. The quantities ΔL, n_m, and S_0 are lumped together because they are assumed to be constant. If the cross-sectional area of flow A_x is arbitrary, then

$$t = \frac{\Delta L A_x}{Q} \tag{16.5b}$$

Leopold et al. (1964) reported a discharge-area relation of wide applicability,

$$A_x \propto Q^{.57} \tag{16.6}$$

where Q is bankfull discharge. Inserting equation (16.6) in equation (16.5),

$$t \propto Q^{-.43} \tag{16.7}$$

The exponent for nonlinearity is −.43. Note that the exponents of equations (16.4) and (16.7) are closer to the exponent of equation (16.1), which also refers to the travel of water rather than to the exponents for the lag time of the flood wave given by equations (16.2) and (16.3).

16.1.4 Representation of Nonlinearity

The nonlinearity in rainfall-runoff relationship can be represented in two ways. First, the lag time of each storage element k can be expressed as a nonlinear function of discharge,

$$k = aQ^b \tag{16.8}$$

where a and b are parameters. Second, the storage S can be expressed as a power function of discharge

$$S = aQ^c \tag{16.9a}$$

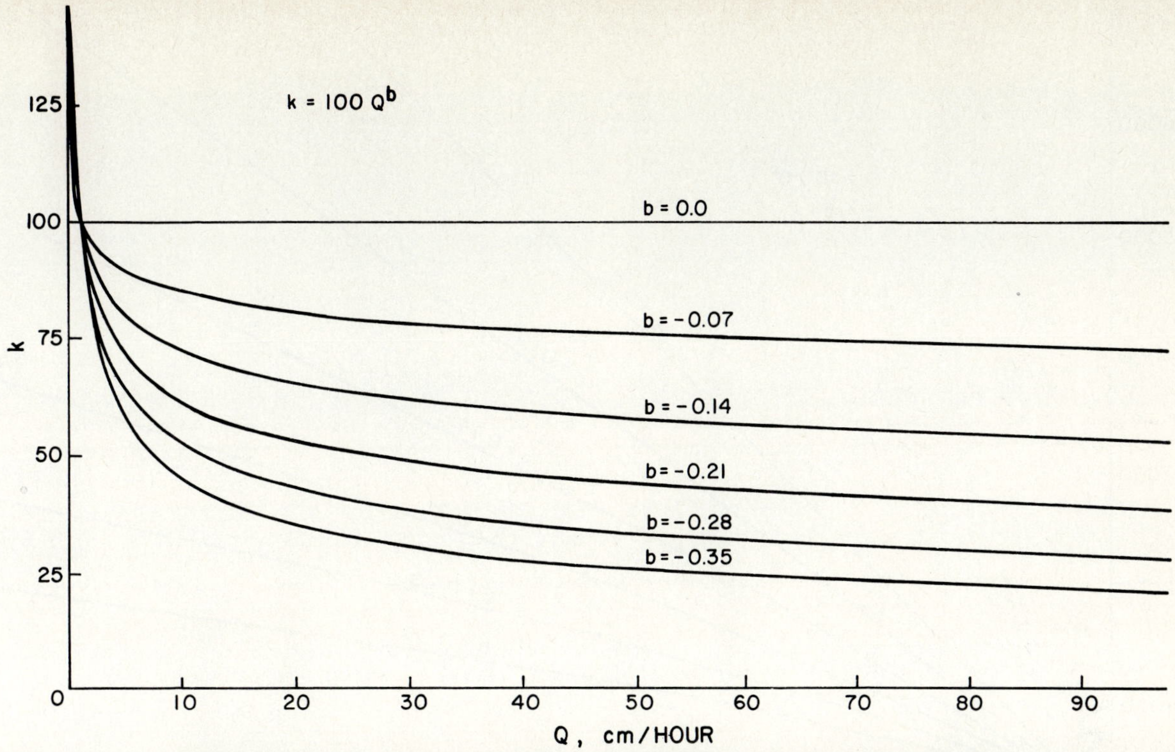

FIGURE 16.1 Variation of lag time k with discharge Q for various values of b.

Equations (16.8) and (16.9a) are alternative expressions of nonlinearity and are connected by $S = kQ$, discussed extensively in Chapters 8 and 13. If k is constant, the system is linear; if it is a function of Q, the system is nonlinear. Thus it is seen that $c = b + 1$.

The parameter a controls the magnitude of k, whereas b controls the degree of nonlinearity. The variation of k with Q becomes more pronounced as b becomes progressively less than zero, as shown in Fig. 16.1; this implies that the degree of nonlinearity increases with increasing departure of b from zero. If b is fixed, k decreases with increasing discharge, indicating diminishing storage effects. Furthermore, the variation of k with Q is considerably larger for lower discharges than for higher discharges.

The variation of S with Q for various values of b is shown in Fig. 16.2. Clearly, S is less sensitive to Q as b departs from zero. The smaller the value of b, the more nonlinear is the relation between S and Q. The variation of S with Q is considerably larger for smaller discharges than for larger discharges.

Equation (16.9a) is a special case of the more general storage-discharge relation proposed by Kulandaiswamy (1964) and discussed in Chapter 8. Simple forms of this equation have been used by Kulandaiswamy and co-workers (Kulandaiswamy and Subramanian, 1967; Kulandaiswamy and Babu Rao, 1970; Babu Rao, 1973) such as

$$S = a_0(Q)Q + a_1 \frac{dQ}{dt} + b_0 I \qquad (16.9b)$$

in which the parameters a_1 and b_0 have been related to the characteristics of ER, I. Also the quantity defined as dS/dQ, representing the nonlinearity, has been determined.

16.2 TIME-INVARIANT MODELS

The nonlinear conceptual models reported in the literature have employed some arrangement of nonlinear elements. The nonlinearity of a conceptual element like reservoir is defined by its storage equation. There are several storage equations that have been proposed and applied in hydrologic modeling. Many of these are special cases of the general storage equation derived in Chapter 8. We consider some of these equations here.

16.2.1 Nonlinear Reservoir: The Case $S = kQ^n$

The simplest nonlinear reservoir is perhaps the one defined by equation (16.9a). Recasting it in a form more directly comparable to its linear counterpart,

$$S = kQ^n \qquad (16.10)$$

where S is storage, Q is discharge, and k and n are parameters. In addition, n is dimensionless and k has the dimensions as $L^{1-n}T^n$ if S and Q are in L and L/T dimensions, respectively. Evidently, the physical significance of k is different here from that in the case of linear reservoirs. When n is 1, the reservoir becomes linear. Prasad (1966, 1967) argued that although these parameters are empirically determined, they may represent

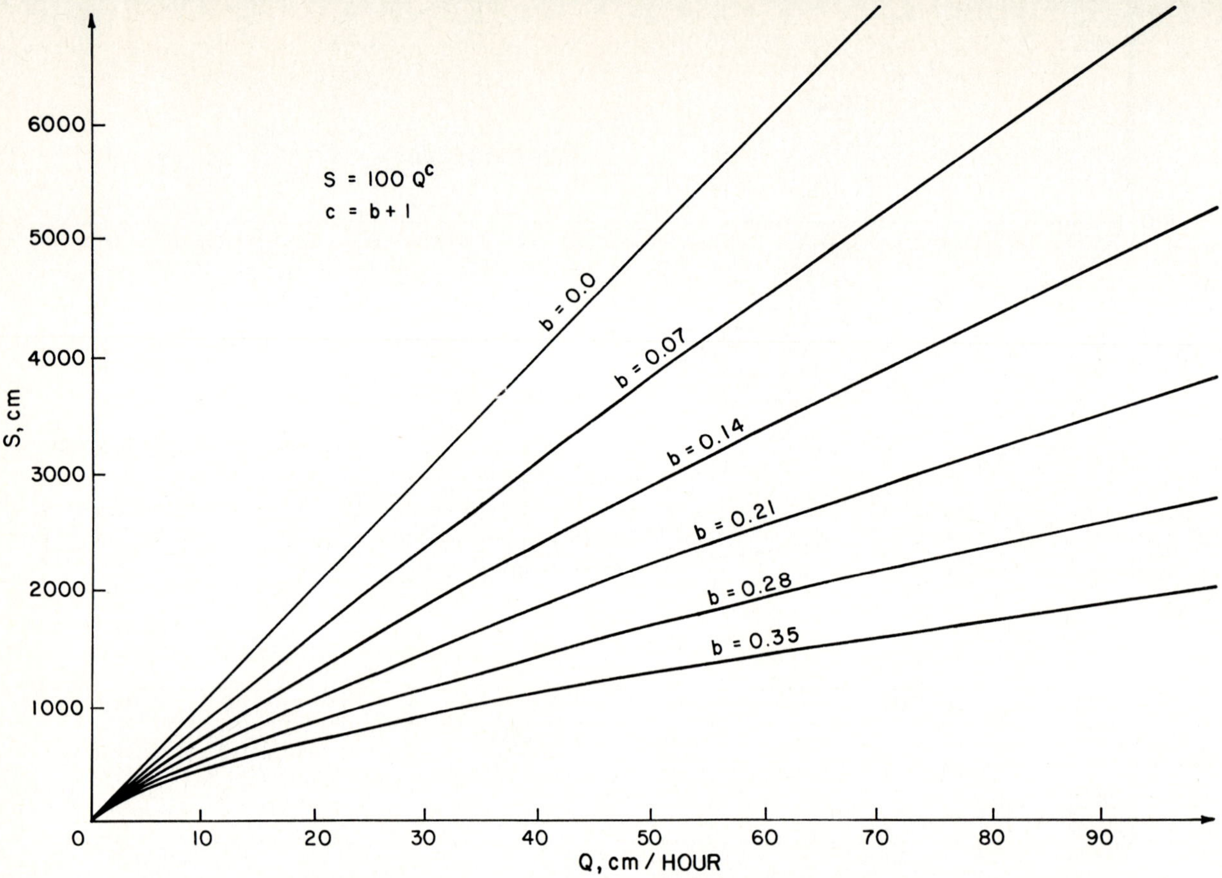

FIGURE 16.2 Variation of S with Q for various values of c.

physical characteristics of the watershed system. To illustrate, the storage-stage relationship representing the characteristics of some reservoirs can be expressed as

$$S = ay^b \tag{16.11}$$

where y is storage or water level and a and b are constants. It is believed that the value of b for most reservoirs varies between 1 and 5. Fig. 16.3(a) illustrates equation (16.11) if it is assumed that the lumped storage system has an outlet control for which the stage-discharge relation is given by

$$Q = cy^d \tag{16.12}$$

where c and d are constants. This may be adequate for most rivers except under very low flow conditions. The values of d for various geometric outlet controls are 1.5 for rectangular sections, 2 for hyperbolic sections, and 2.5 for triangular sections (Chow, 1959). This range should encompass the characteristics for the gage control of most natural channels. Fig. 16.3(b) illustrates the outlet control of equation (16.12). The storage-discharge relationship of a lumped system with an outlet control, as shown in Fig. 16.3(c), can be obtained by eliminating y from equations (16.11) and (16.12):

$$S = \frac{a}{(c)^{b/d}}(Q)^{b/d} \tag{16.13}$$

Comparing equation (16.13) with equation (16.10),

$$k = \frac{a}{c^n} \tag{16.14}$$

$$n = \frac{b}{d} \tag{16.15}$$

Therefore, n will normally be between .4 and 3.33. A commonly used value is $n = .75$. Table 16.1 presents interpretations of the storage element for various values of $m(= 1/n)$ (Ding, 1967a). Prasad (1966, 1967) estimated k and n of the nonlinear reservoir model by correlating them with the watershed and effective rainfall characteristics using the least squares method. He applied the method to seven east central Illinois drainage basins ranging in area from 28.8 square miles to 1280 square miles. As a result, k was found to vary between 3.13 and 514.22 and n, between .804 and 1.597, with discharge and storage measured in inches per hour and inches, respectively.

16.2.1.1 Governing Equation

Equation (16.10) can be used in a spatially lumped form of the continuity equation

$$\frac{dS}{dt} = I - Q \tag{16.16}$$

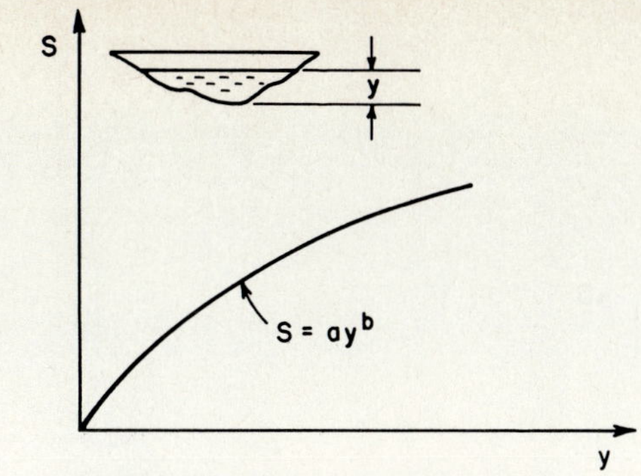

FIGURE 16.3(a) Discharge-stage curve.

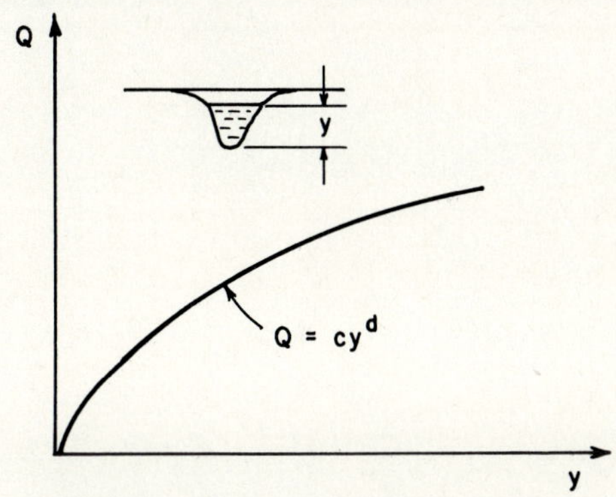

FIGURE 16.3(b) Storage-stage curve.

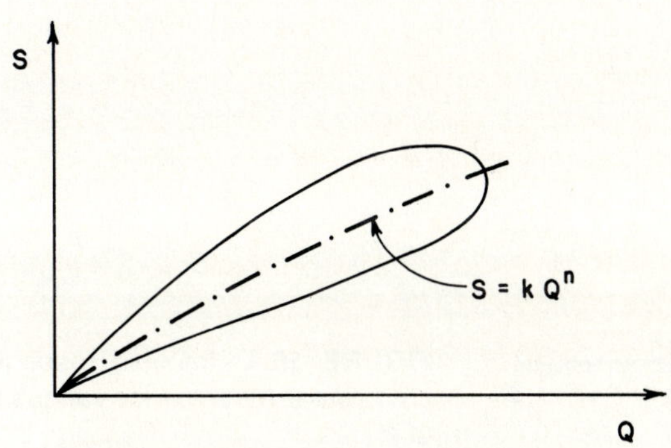

FIGURE 16.3(c) Storage-discharge curve.

to obtain the DRH for a given ERH. When equations (16.10) and (16.16) are combined, the resulting equation is the governing differential equation of the nonlinear reservoir,

$$knQ^{n-1}\frac{dQ}{dt} + Q = I \tag{16.17}$$

The solution of equation (16.17) depends upon n and the effective rainfall I. In general, the closed form solution will not be tractable, but solutions can be found for some particular values of n.

16.2.1.2 The Impulse Response

The instantaneous effective rainfall occurs during a very short duration dt. When the reservoir is subjected to this impulse, it reaches an initial state Q_0 during dt. Thereafter, $I = 0$ for all t, and its response decays. Thus, for $t > 0$, equation (16.17) reduces to

$$knQ^{n-1}\frac{dQ}{dt} + Q = 0 \tag{16.18}$$

Integrating,

$$kn\int_{Q_0}^{Q} Q^{n-2}\,dQ = -\int_0^t dt$$

$$\frac{kn}{n-1}(Q^{n-1} - Q_0^{n-1}) = -t$$

This yields

$$Q(t) = \begin{cases} \left[Q_0^{n-1} - \dfrac{(n-1)t}{kn}\right]^{1/(n-1)}, & t > 0 \\ Q_0, & t = 0 \end{cases} \tag{16.19}$$

Note that $Q(t)$ reduces to the IUH, $h(t)$, when the volume V of the impulse reduces to unity. Fig. 16.4 plots $Q(t)$ for various values of V.

TABLE 16.1

Storage elements (Ding, 1967a)

Exponent m	*Storage Element*	*Source*
0	Infinite storage	—
.5	Orifice	King and Brater (1963)
1	Confined or artesian aquifers	Werner and Sandquist (1951)
1.5	Rectangular weir Wide channel (Chezy formula)	King and Brater (1963)
$\frac{5}{3}$	Turbulent overland flow Wide channel (Manning formula)	Chow (1959)
2	Unconfined or nonartesian aquifer	Werner and Sandquist (1951)
2.5	Triangular weir	King and Brater (1963)
3	Laminar overland flow	King and Brater (1963)
∞	Transient storage	—

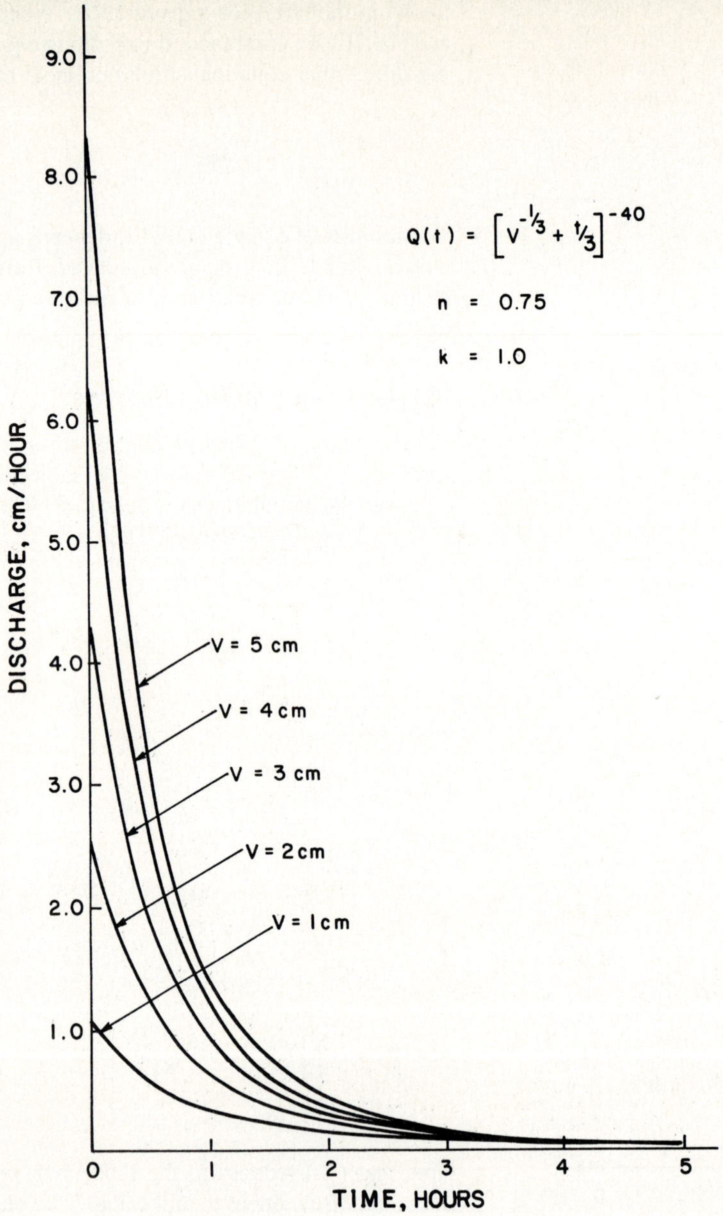

FIGURE 16.4 Impulse responses of a nonlinear reservoir for various values of *V*.

16.2.1.3 Properties of the Impulse Response

Usually $n = .75$ for watershed runoff. This is close to the value observed by Laurenson (1962), Askew (1968), and Boyd (1976). If $n < 1$, equation (16.19) can be integrated to yield the volume of the impulse, V.

$$V = \int_{0+}^{\infty} Q(t)\,dt = \int_{0+}^{\infty} \left[Q_0^{n-1} - \frac{n-1}{kn} t \right]^{1/(n-1)} dt$$

$$= -k \left[Q_0^{n-1} - \frac{n-1}{kn} t \right]^{n/(n-1)} \Bigg|_{0+}^{\infty}$$

$$= kQ_0^n \tag{16.20}$$

Equation (16.20) can be used to define the initial state Q_0,

$$Q_0 = \left(\frac{V}{k}\right)^{1/n} \tag{16.21}$$

More specifically, if $n = .75$,

$$V = kQ_0^{.75}$$

and

$$Q_0 = \left(\frac{V}{k}\right)^{4/3}$$

We can derive the first moment of $Q(t)$ about the Q axis,

$$M_1^Q = \int_{0+}^{\infty} Q(t)t\,dt = \int_{0+}^{\infty} t \left[Q_0^{n-1} - \frac{n-1}{kn} t \right]^{1/(n-1)} dt$$

Integrating by parts,

$$M_1^Q = -kt\left(Q_0^{n-1} - \frac{n-1}{kn}t\right]^{n/(n-1)} - \frac{k^2 n}{(2n-1)}\left[Q_0^{n-1} - \frac{n-1}{kn}t\right]^{(2n-1)/(n-1)}\Bigg|_0^{\infty}$$

$$= \frac{k^2 n}{2n-1} Q_0^{2n-1} \qquad (16.22)$$

If $n = .75$, then

$$M_1^Q - 1.5k^2 Q_0^{2n-1}$$

We can now determine the lag time t_L of the nonlinear reservoir,

$$t_L = \frac{M_1}{V} = \frac{k^2\, n\, Q_0^{2n-1}}{kQ_0^n\,(2n-1)} = \frac{nk}{2n-1} Q_0^{n-1} \qquad (16.23a)$$

For $n = .75$,

$$t_L = 1.5kQ_0^{-0.25} \qquad (16.23b)$$

We can also derive the first moment of $Q(t)$ about the t axis,

$$M_1^t = \int_{0+}^{\infty} \frac{Q(t)^2}{2}\,dt = \frac{1}{2}\int_{0+}^{\infty}\left(Q_0^{n-1} - \frac{n-1}{kn}t\right)^{2/(n-1)} dt$$

$$= -\frac{kn}{2(n+1)}\left(Q_0^{n-1} - \frac{n-1}{kn}t\right)^{(n+1)/(n-1)}\Bigg|_{0+}^{\infty}$$

$$= \frac{kn}{2(n+1)} Q_0^{n+1} \qquad (16.24)$$

For $n = .75$,

$$M_1^t = \frac{3}{14} Q_0^{1.75}$$

The mean discharge Q_m can thus be computed:

$$Q_m = \frac{M_1^t}{V} = \frac{knQ_0^{n+1}}{2(n+1)\,kQ_0^n} = \frac{n}{2(n+1)} Q_0 \qquad (16.25a)$$

For $n = .75$

$$Q_m = \tfrac{3}{14} Q_0 \qquad (16.25b)$$

The quantities Q_0, t_L, and Q_m can be construed as hydrograph measures. These are plotted for various values of V in Fig. 16.5. It is clear from equation (16.23) that as Q_0 increases, t_L decreases. Further, Q_0 and Q_m experience proportionately higher increases than does V. These measures therefore reflect nonlinearity of the reservoir.

16.2.1.4 The Response to Steady ER

If $I(t) = I_0$ for all t, then equation (16.17) can be expressed as

$$dt = \frac{knQ^{n-1}}{(I_0 - Q)} dQ$$

Let $y = Q/I_0$. Then $dy = dQ/I_0$. Therefore,

$$dt = knI_0^{n-1} \frac{y^{n-1}}{1-y} dy$$

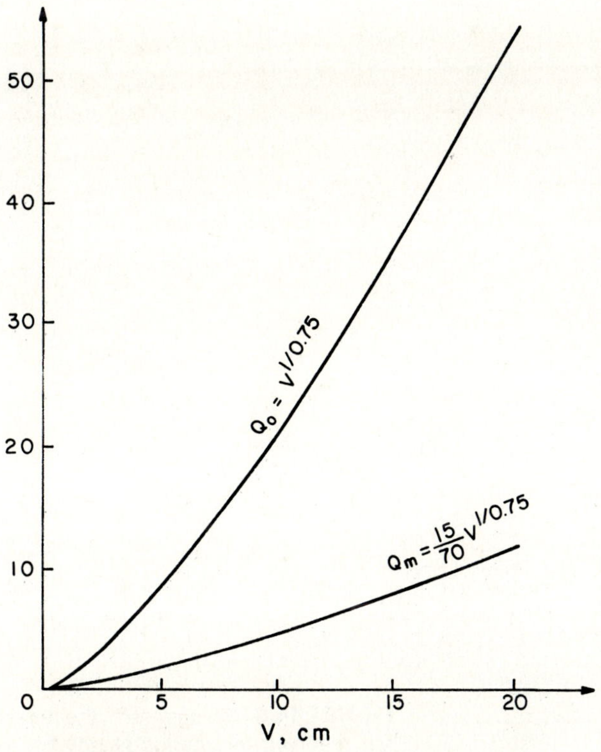

FIGURE 16.5(a) Variation of Q_0 and Q_m with V.

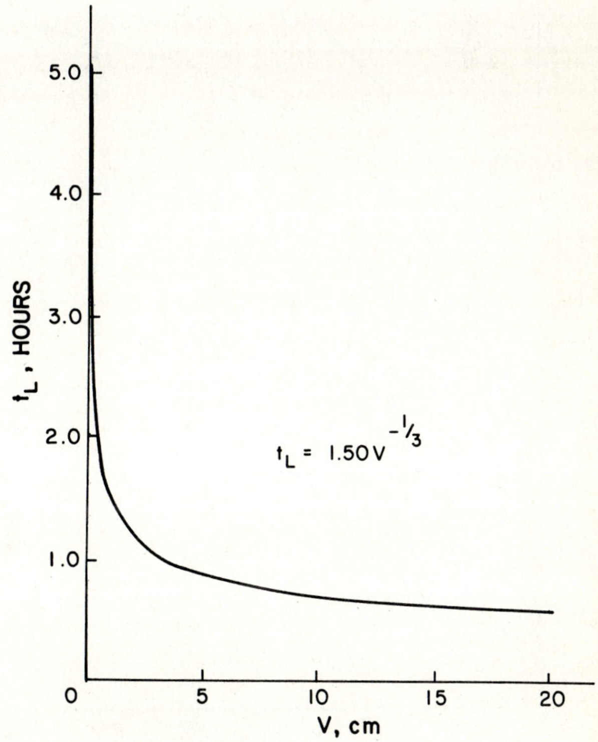

FIGURE 16.5(b) Variation of t_L with V.

Upon integration,

$$t = knI_0^{n-1} \int \frac{y^{n-1}}{1-y} dy$$

If the initial condition is assumed to be $Q(0) = 0$,

$$t = knI_0^{n-1} \int_0^y \frac{y^{n-1}}{1-y} dy \tag{16.26}$$

We can obtain analytical solutions for some particular values of n. When $n = 2$,

$$t = knI_0 \left[-\frac{Q}{I_0} + \ln\left(1 - \frac{Q}{I_0}\right) \right] \tag{16.27a}$$

When $n = 3$,

$$t = -knI_0^2 \left[\frac{1}{2}\left(1 - \frac{Q}{I_0}\right)^2 - \left(1 - \frac{Q}{I_0}\right) + \ln\left(1 - \frac{Q}{I_0}\right) \right] \tag{16.27b}$$

When $n = .5$,

$$Q = I_0 \tan h^2\left(\frac{t}{K}\right), \qquad K = \frac{I_0^{n-1}}{k} \tag{16.27c}$$

When $n = .75$,

$$t = \frac{3}{4} kI_0^{-.25} \left[-2\tan^{-1}\left(\frac{Q}{I_0}\right)^{.25} + \ln(I_0^{.25} + Q^{0.25}) - \ln(I_0^{.25} - Q^{.25}) \right) \tag{16.28}$$

For other values of n equation (16.26) can be solved numerically. Note that $Q < I_0$. Therefore, we can obtain a series approximation,

$$t = knI_0^{n-1} \sum_{j=0}^{\infty} \frac{(Q/I_0)^{j+n}}{(n+j)} \tag{16.29}$$

Furthermore, we can express (16.26) alternatively in terms of S,

$$t = kI_0^{n-1} \int_0^w \frac{1}{1-w^m} dw \tag{16.30}$$

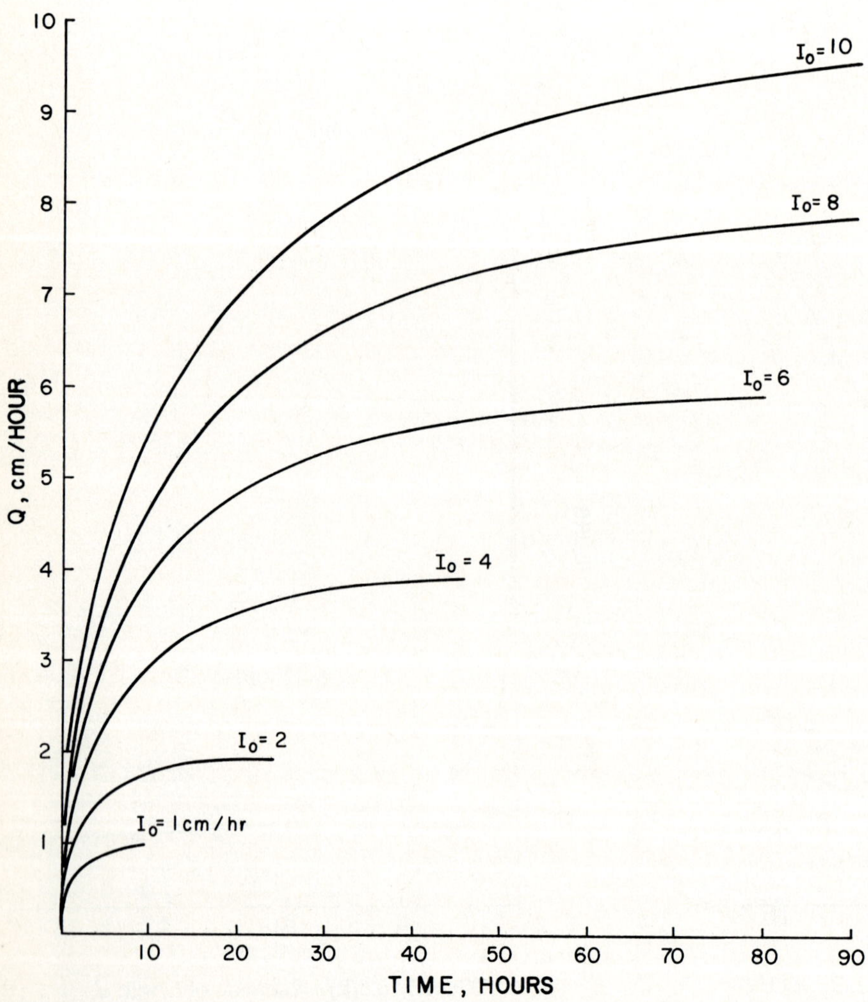

FIGURE 16.6 Response of the nonlinear reservoir for various steady ER intensities.

where $m = 1/n$, and $w = [(1/I_0)^n](S/k)$. As before, for some particular values of n, equation (16.30) can be solved analytically. It may then be noted that equations (16.26) and (16.30) are of the same form as the Bakhmeteff varied flow function (Bakhmeteff, 1932), whose values are commonly tabulated in books on hydraulics.

Equation (13.60) can be written as

$$F(u, m) - F(0, m) = \frac{t}{k} I_0^{1-n}$$

where

$$F(u, m) = \int_0^u \frac{ds}{1 - s^m}$$

If $F(0, m) = 0$, then

$$F(u, m) = \frac{t}{k} I_0^{1-n}$$

It may be interesting to note that for $I > 0$, $Q(0) = 0$, $F(u, m) \to \infty$ as $t \to \infty$. Therefore,

$$u = \left(\frac{Q}{I}\right)^n \to 1 \quad \text{and} \quad Q \to 1 \quad \text{as} \quad t \to \infty$$

In other words, the equilibrium condition is approached asymptotically.

Fig. 16.6 plots the responses of the nonlinear reservoir to various rates of steady effective rainfall. It is clear that—due to its larger k—the nonlinear reservoir responds less quickly than its linear counterpart. Initially, dQ/dt is zero and then increases to a maximum positive value before declining to zero as Q approaches I_0; it depends upon $(I_0 - Q)$ and Q^{1-n}. In the case of a linear reservoir, dQ/dt is initially positive, decreases to 0 as Q approaches I_0, and depends upon the difference $(I_0 - Q)$.

16.2.1.5 The Response to Pulsed ER

Consider the case where $I(t)$ is defined by a rectangular pulse of duration D and intensity I_0 and, therefore, volume $I_0 D$. Equation (16.17) can be solved subject to specified initial conditions. The solutions, equations (16.27)–(16.30), are valid for $t \leq D$, that is, the rising part of the hydrograph. When I ceases, that is, $t \geq D$, then equation (16.17) has the solution

$$Q = \left[Q_p^{n-1} - \frac{n-1}{nk}(t - D)\right]^{1/(n-1)} \tag{16.31}$$

where $Q_p = Q(D)$ to be obtained from equation (16.26). This gives the recession part of the hydrograph. Fig. 16.7 shows the rising and recession hydrographs for various values of I_0 and Fig. 16.8 for various values of D. It is clear that for a nonlinear reservoir as in the case of a linear reservoir, the duration of the rising hydrograph is equal to that of the rectangular pulse. Moreover, the time base T of the entire hydrograph will be

$$T = D + \frac{nk}{n-1} Q_p^{n-1} \tag{16.32}$$

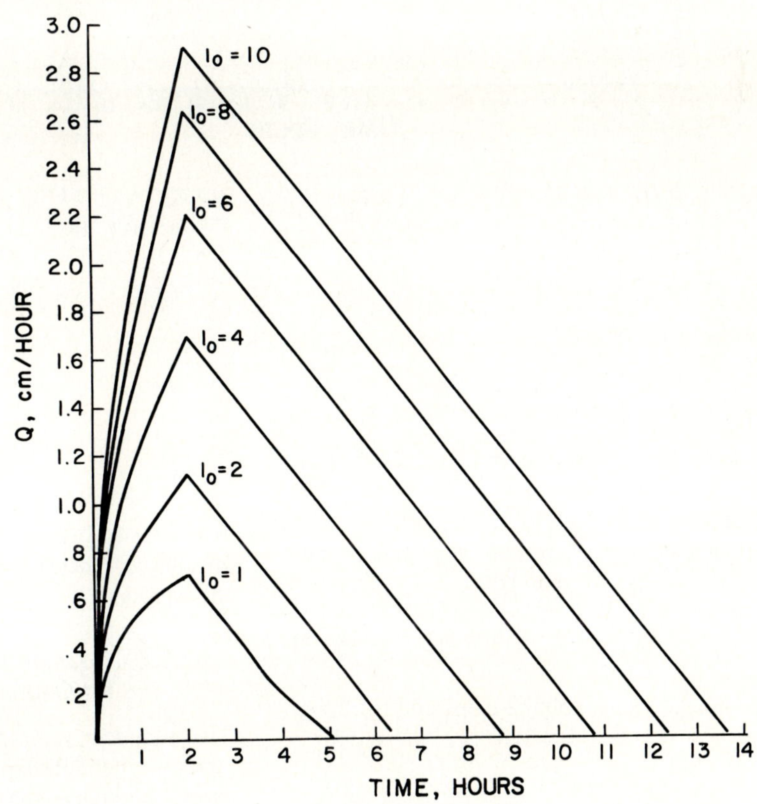

FIGURE 16.7 Responses of the nonlinear reservoir for pulses of various magnitudes (duration = 2 hours, n = 2 and k = 2).

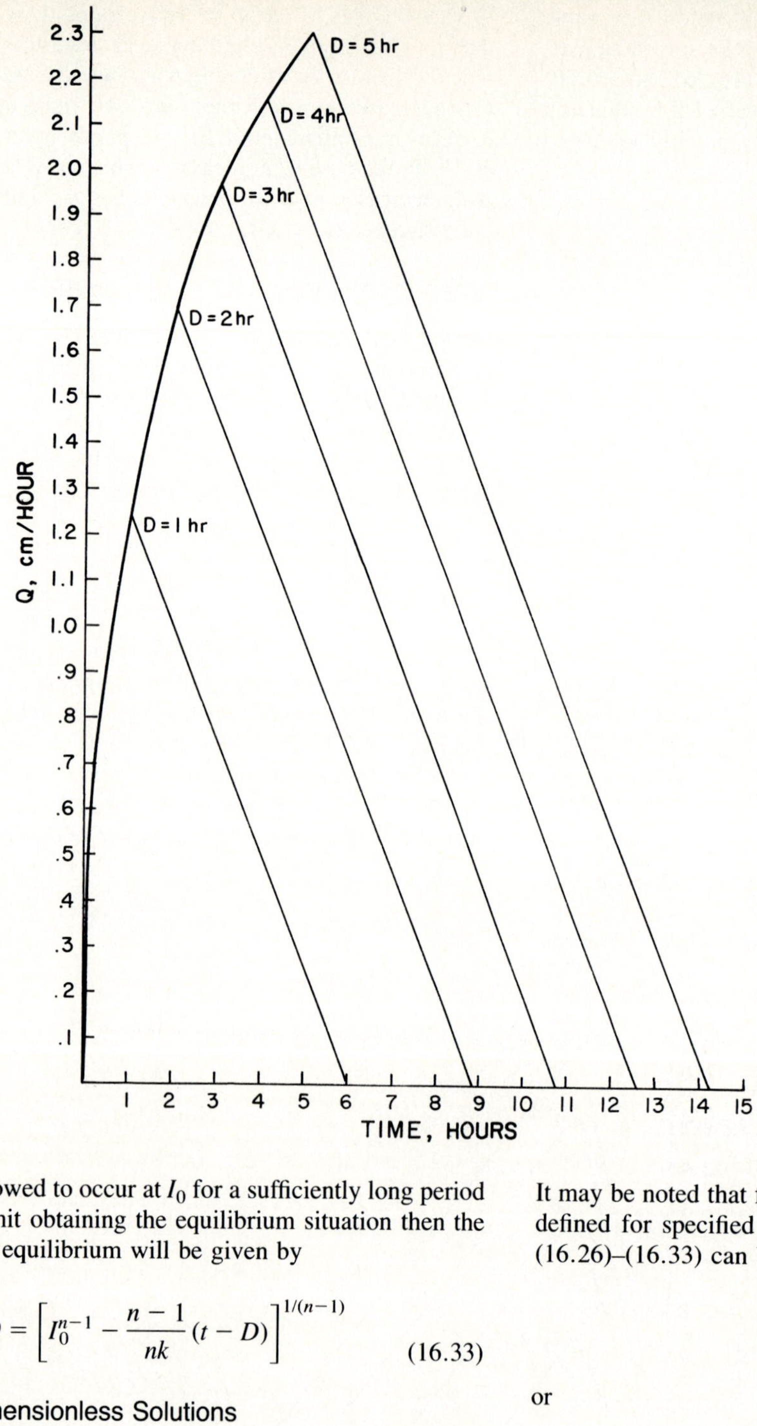

FIGURE 16.8 Response of the nonlinear reservoir for pulses of various durations ($I_0 = 4$ cm/h, $n = 2$, and $k = 2$).

If I is allowed to occur at I_0 for a sufficiently long period of time to permit obtaining the equilibrium situation then the recession from equilibrium will be given by

$$Q = \left[I_0^{n-1} - \frac{n-1}{nk}(t - D) \right]^{1/(n-1)} \tag{16.33}$$

16.2.1.6 Dimensionless Solutions

The solutions of the nonlinear reservoir model given by equations (16.26)–(16.33) can be recast in a general form by employing some characteristic variables (Dooge, 1967a, 1967b, 1973). Let I_0, as defined earlier, be the characteristic rate. The characteristic storage S_0 will then be

$$S_0 = kI_0^n \tag{16.34}$$

The characteristic time t_0 can simply be expressed by

$$t_0 = \frac{S_0}{I_0} = kI_0^{n-1} \tag{16.35}$$

It may be noted that for a choice of I_0, S_0, and t_0 are uniquely defined for specified k and n. Then it is seen that equations (16.26)–(16.33) can be expressed as

$$\frac{t}{t_0} = \Phi\left(\frac{Q}{I_0}\right) \tag{16.36}$$

or

$$\frac{Q}{I_0} = \Psi\left(\frac{t}{t_0}\right) \tag{16.37}$$

where Φ and Ψ are appropriate functions such that $\Phi^{-1} = \Psi$. These functions can be calculated for various values of the exponent n.

16.2.1.7 Relation to Hydraulic Approaches for Overland Flow

The nonlinear reservoir model can be related to at least two types of hydraulic approaches to overland flow: (1) Horton-Izzard approach and (2) kinematic wave approach. Dooge

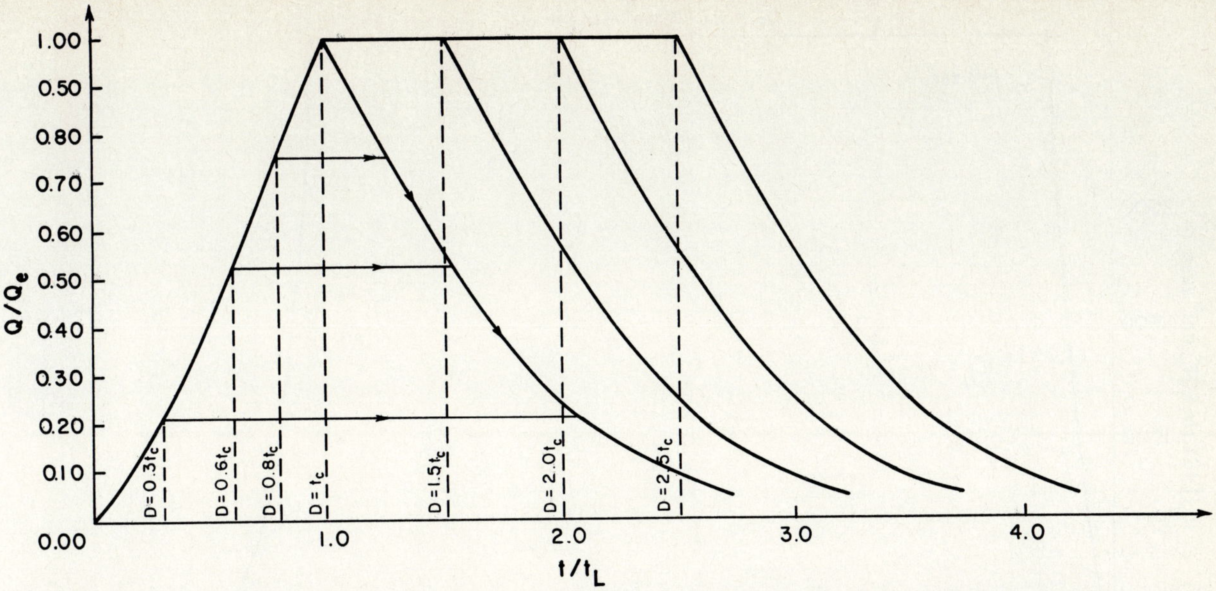

FIGURE 16.11 Dimensionless hydrographs by kinematic wave solutions for various durations of I_0 (m = 1.25).

which is less than D = .5 h. This shows that the equilibrium is reached. Thus for $t \le t_c$,

$$\left(\frac{Q}{2777.0}\right) = \left(\frac{t}{148.66}\right)^{5/3}$$

yields

t (s)	Q(cm^3/s/m)
0	0
50	451.72
100	1434.0
120	1934.0
148.66	2777.0

For $t_c \le t \le D$, $Q/Q_c = 1$. For $t \ge D_e$, $t = t_c + 90\ (Q/2777)^{0.6}\ (2777/Q - 1)$ yields

t (s)	Q(cm^3/s/m)
1809	2500
1829	2000
1853	1500
1897	1000
1946	500
1967	400
1996	300
2039	200
2127	100
2241	50
2443	20

The DRH is plotted in Fig. 16.12.

16.2.1.8 Lag Time–Discharge Relations

It has been pointed out that equation (16.8), relating lag time to discharge, expresses nonlinearity in the watershed response. It involves instantaneous discharge. However, equations (16.6) and (16.7) relate lag time to mean discharge; the lag time is averaged over time and area, and discharge is averaged over the duration of the hydrograph. Laurenson (1962) employed the average form in place of the instantaneous relation for each element in his model. Equations (16.23a) and (16.25a) can be combined to yield the lag time in terms of the mean discharge as

$$t_L = \frac{nk}{2n-1}\left(\frac{2n+2}{n}\right)^{n-1} Q_m^{n-1} \tag{16.52}$$

For various values of n, t_L can be determined as graphed in Fig. 16.13. If n = .75,

$$t_L = \frac{3}{2}k\left(\frac{14}{3}\right)^{-.25} Q_m^{-.25}$$

16.2.2 Nonlinear Reservoir: The Case $S = k_1Q^n + k_2(dQ/dt)$

Equation (16.10) does not account for unsteady effects, as evident from Fig. 16.3(c). To allow for these effects, the storage-discharge relationship can be hypothesized (Prasad, 1966, 1967) as

$$S = k_1Q^n + k_2\frac{dQ}{dt} \tag{16.54}$$

where k_1 and k_2 are parameters, and n is an exponent as defined previously. In addition, k_2 may depend upon the factors affecting the wedge storage as well as the storage-discharge relationship. It primarily causes the lag between the effective rainfall and hydrograph peak. For practical purposes, however, k_2 may be assumed constant for a particular hydrograph.

Combining equation (16.54) with equation (16.16),

$$k_2\frac{d^2Q}{dt^2} + k_1nQ^{n-1}\frac{dQ}{dt} + Q = I \tag{16.55}$$

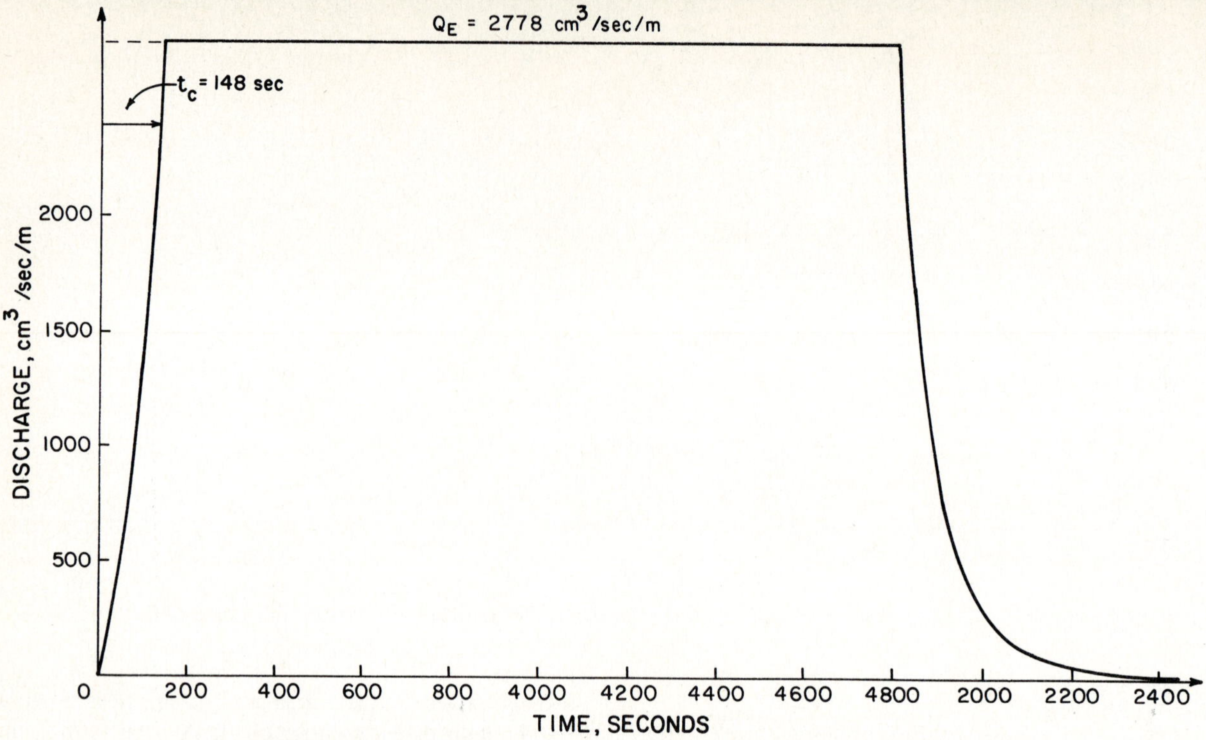

FIGURE 16.12 Example 16.1: Runoff hydrograph due to I = 10 cm/h.

Due to nonlinearity, equation (16.55) cannot, in general, be solved analytically. However, for some particular values of n analytical solutions are tractable. If $n = 1$, equation (16.55) becomes linear and its solution providing the IUH, subject to $Q(0) = dQ(0)/dt = 0$, for $I(t) = \delta(t)$ is

$$h(t) = \frac{ab}{b - a}(e^{bt} - e^{at}) \tag{16.56}$$

where a and b are the roots of

$$f(x) = k_2x^2 + k_1x + 1 \tag{16.57}$$

When $n \neq 1$, equation (16.55) can best be solved numerically. This was done by Prasad (1966, 1967).

16.2.2.1 Parameter Estimation

Prasad (1966) discussed the significance of parameters n, k_1, and k_2 and related them to the physical characteristics of seven watersheds in the east-central Illinois region. He enumerated their influence on hydrograph peak characteristics: (1) The peak discharge Q_p increases with increasing n; the rate of increase in Q_p is higher than that of n. (2) Q_p and its time t_p both decrease with increase in k_1. (3) Q_p decreases only moderately with increase in k_2 but t_p increases at a greater rate than Q_p.

We can get a first approximation of the parameters from observed rainfall-runoff records. Note that at $t = t_p$, $dQ/dt = 0$. This implies that $S = k_1Q^n$ at $t = t_p$. Therefore, equation (16.10) can be integrated from $t = 0$ to $t = t_p$ in conjunction with equation (16.54). Since there are two unknowns, n and k_1, at least two pairs of $S(Q_p)$ and Q_p are needed. For more data on Q_p and $S(Q_p)$, a least squares method can be utilized to obtain n and k_1. Since n would probably be more independent of storm characteristics, Prasad (1966) held n constant for each watershed.

Prasad (1966) correlated n, k_1, and k_2 with watershed and effective rainfall characteristics and found n related to channel slope, circularity ratio, and the parameters c and d in equation (16.12); k_1 was related to n, drainage area, circularity ratio, and d; and k_2 was related to channel slope, duration of the effective rainfall, drainage area, n, d, and circularity ratio. In each case the relationship was of the form (with coefficient of determination > 0.96)

$$y = a_0 \prod_{i=1}^{N} x_i a^i \tag{16.58}$$

in which y is the dependent variable, x_i, $i = 1, 2, \ldots, N$, are independent variables, and a_i, $i = 1, 2, \ldots, N$, are parameters.

EXAMPLE 16.2

Determine the direct runoff hydrograph by using equation (16.55). Take I = 5 cm/h for a period of 1 h, n = 1.25, k_1 = 100, and k_2 225.

Inserting values of given parameters in equation (16.55),

$$225\frac{d^2Q}{dt^2} + 125\,Q^{0.25}\frac{dQ}{dt} + Q = \begin{cases} 5, & t \le 1 \\ 0, & t \ge 1 \end{cases}$$

It is easier to solve this equation numerically, say, using a fourth-order Runga-Kutta method. The results are plotted in Fig. 16.14.

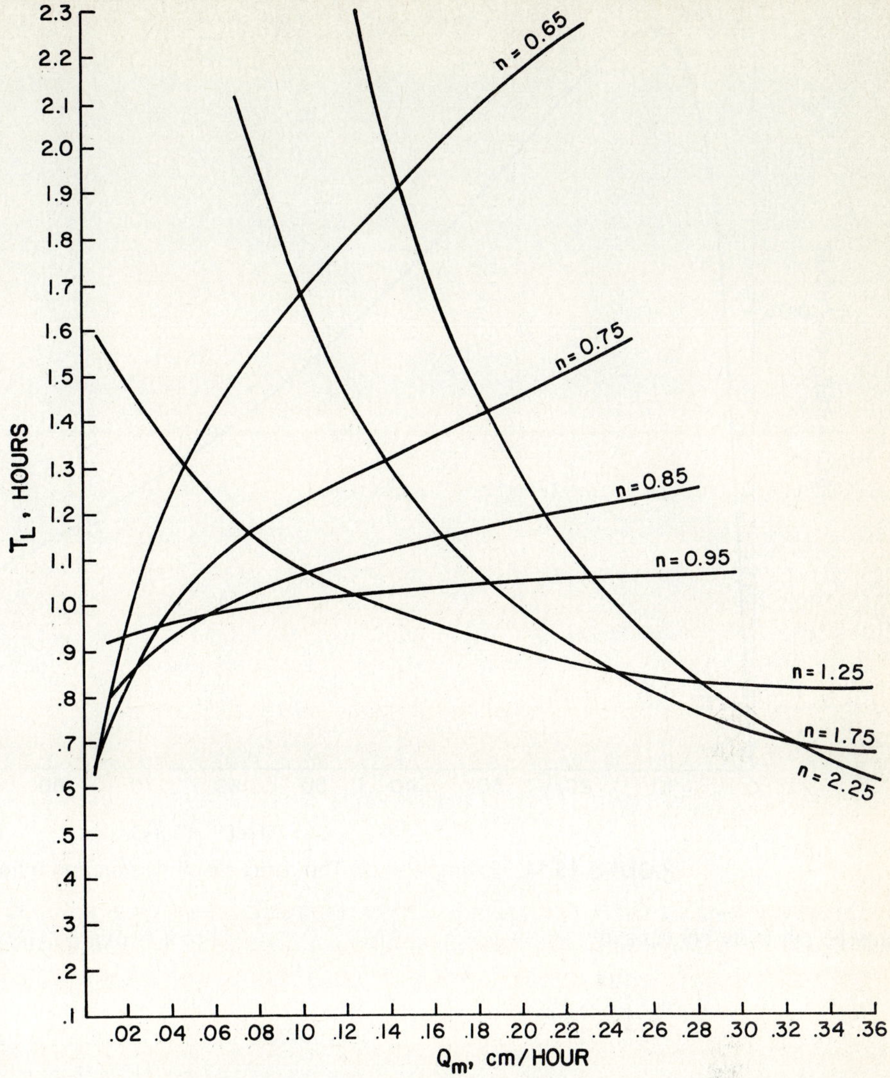

FIGURE 16.13 Variation of lag with mean discharge for various values of n ($k = 2$).

The peak of the direct runoff hydrograph is very small. This is possibly due to particular values of k_1 and k_2.

16.2.3 Nonlinear Reservoir: The Case $S = k \ln Q$

Based on an analysis of streamflow recession data from the River Ceiriog (drainage area 114 square kilometers), a tributary of the River Dee in Great Britain, Lambert (1969) proposed a storage-discharge relationship of the form

$$S = k \ln Q \tag{16.59}$$

He argued that hydrograph separation into surface flow, interflow, and baseflow was unnecessary. The streamflow can be uniquely related to the amount of water stored within the watershed area (in aquifers, soil strata, ponds, lakes, and so on). Therefore, S implies the collective storage and Q implies total runoff. He noted that a difference of 10 mm in S changed the runoff twofold, that is, $S_2 - S_1 = \Delta S = 10 = k \ln(Q_2/Q_1) = k \ln 2$, yielded a value of k as 14.5 mm. Note that in the limit, when $S = 0$, $Q = 1$, which is a contradiction.

Combining equation (16.59) with equation (16.16),

$$\frac{k}{Q}\frac{dQ}{dt} + Q = I \tag{16.60}$$

It may be interesting to compare characteristics of this reservoir with the one described by equation (16.10).

16.2.3.1 The Impulse Response

Let the initial state be defined by Q_0. Since $I = 0$ for all $t > 0$, equation (16.60) reduces to

$$\frac{k}{Q}\frac{dQ}{dt} + Q = 0 \tag{16.61}$$

whose solution is

$$Q(t) = \begin{cases} \dfrac{kQ_0}{[k + tQ_0]}, & t > 0 \\ Q_0, & t = 0 \end{cases}$$

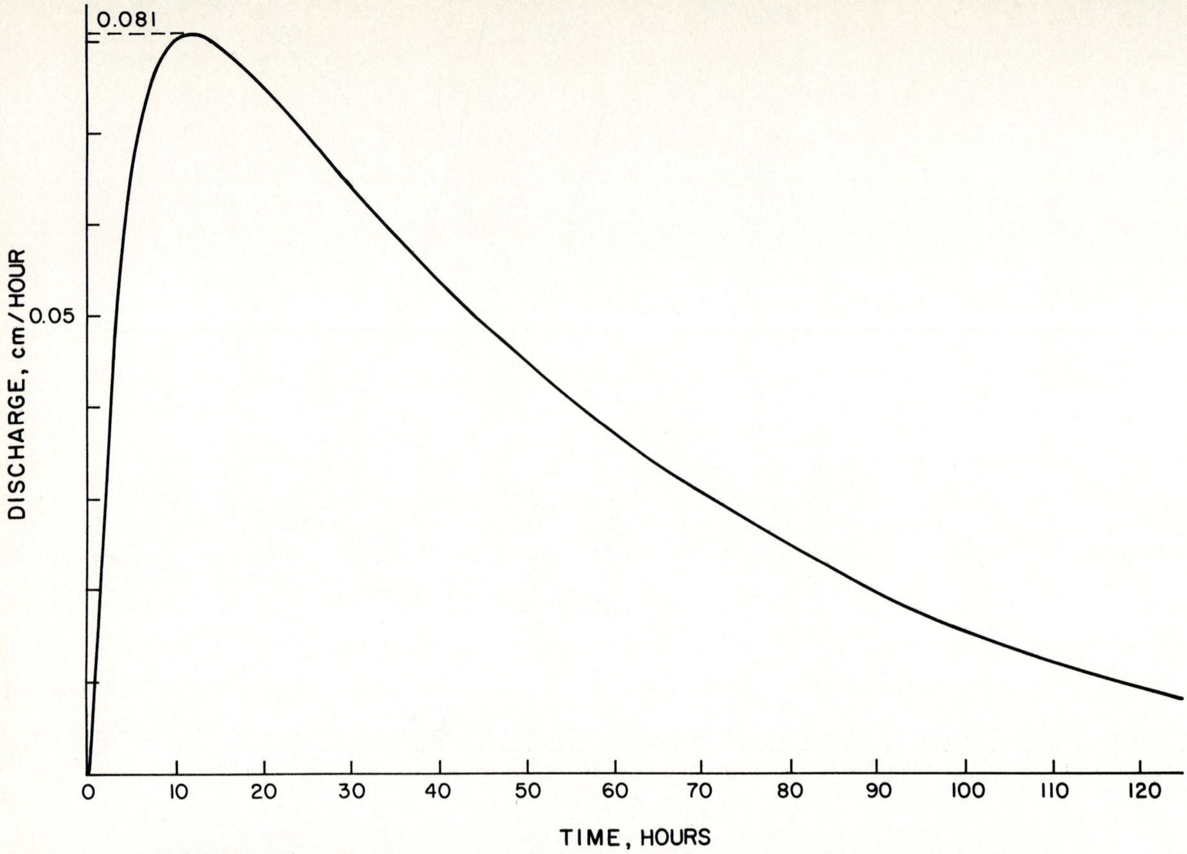

FIGURE 16.14 Example 16.2: The direct runoff hydrograph from the Prasad model.

It can be put more conveniently as

$$\frac{Q}{Q_0} = \frac{1}{1+z} \tag{16.62}$$

where

$$z = \frac{Q_0 t}{k} \tag{16.63}$$

Note that $Q(t) = h(t)$, the IUH, when the volume of the impulse V reduces to unity. Fig. 16.15 shows graphs of $Q(t)$ for various values of Q_0. The parameter k has the dimension of L in this model. Lambert (1969) found that $k = 14$ mm was satisfactory for flood prediction on the River Ceiriog.

16.2.3.2 The Response to Steady Rainfall

If I is assumed to be constant I_0 for all $t > 0$, then subject to $Q(0) = Q_0$,

$$\int_0^t \frac{dt}{k} = -\int_{Q_0}^{Q} \frac{ds}{s(s-I_0)} = -\int_{Q_0}^{Q} \frac{ds}{I_0}\left(\frac{1}{s-I_0} - \frac{1}{s}\right) \tag{16.64}$$

Therefore,

$$-\frac{I_0 t}{k} = \ln\left[\frac{Q_0(Q-I_0)}{Q(Q_0-I_0)}\right] \tag{16.65}$$

Let $x = \exp(-I_0 t/k)$ and $y = I_0/Q_0$. Then

$$Q = \frac{Q_0}{x + [(1-x)/y]}, \qquad 0 \le t \tag{16.66}$$

Fig. 16.16 depicts the dimensionless response to steady rainfall occurring for an indefinite period of time.

16.2.3.3 The Response to Pulsed Rainfall

We consider the case when $I(t)$ is defined by a rectangle of duration D and the intensity I_0 and, therefore, by the volume $I_0 D$. Equation (16.60) can be solved subject to specified initial conditions. The solution given by equation (16.66) is valid for $0 \le t \le D$. For $t \ge D$, $I(t) = 0$, and equation (16.60) will have the solution

$$Q = \frac{Q_p}{1+w}, \qquad t \ge D \tag{16.67}$$

where $w = Q_p(t-D)/k$ and Q_p is specified by equation (16.66) for $t = D$. Equation (16.60) specifies the rising part of the hydrograph and equation (16.67) the recession part.

EXAMPLE 16.3

Compute the streamflow hydrograph by the Lambert model. The beginning streamflow Q_0 at 09:00 hours on October 27, 1967, was 126 cusec (2.72 mm/day). The value of k is 14 mm. Compare with

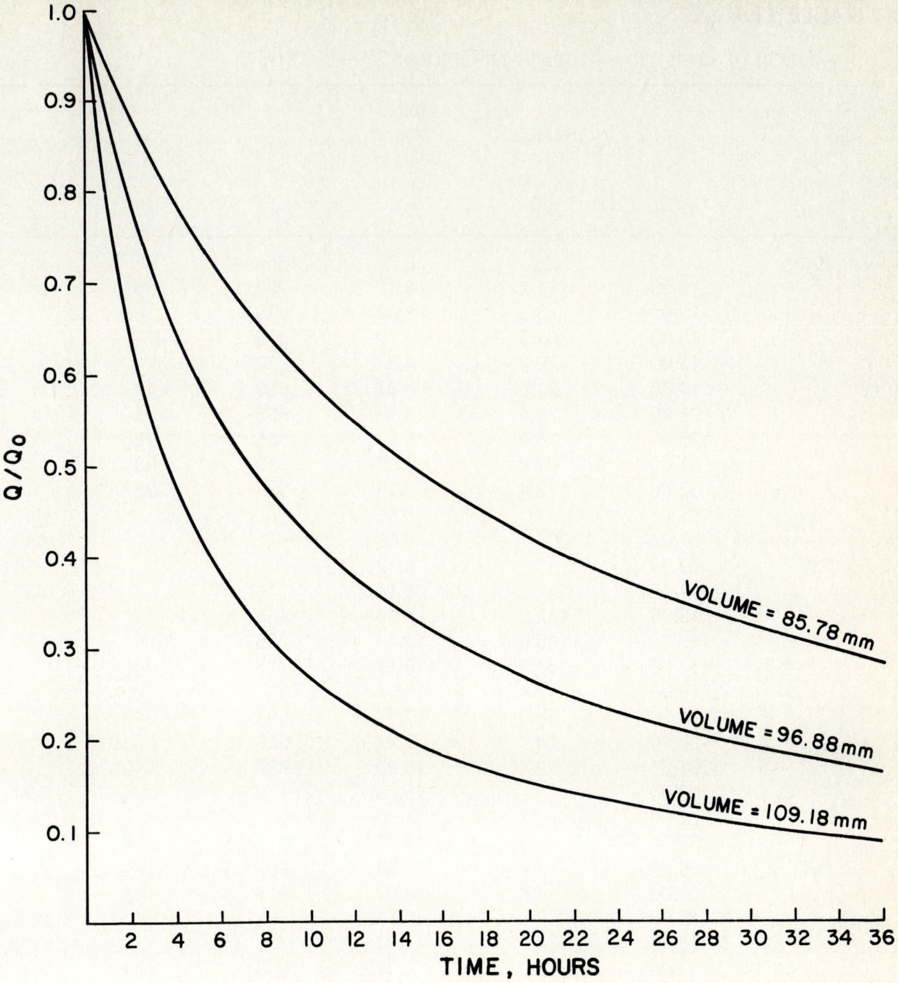

FIGURE 16.15 Dimensionless impulse response for various values of the initial discharge Q_0.

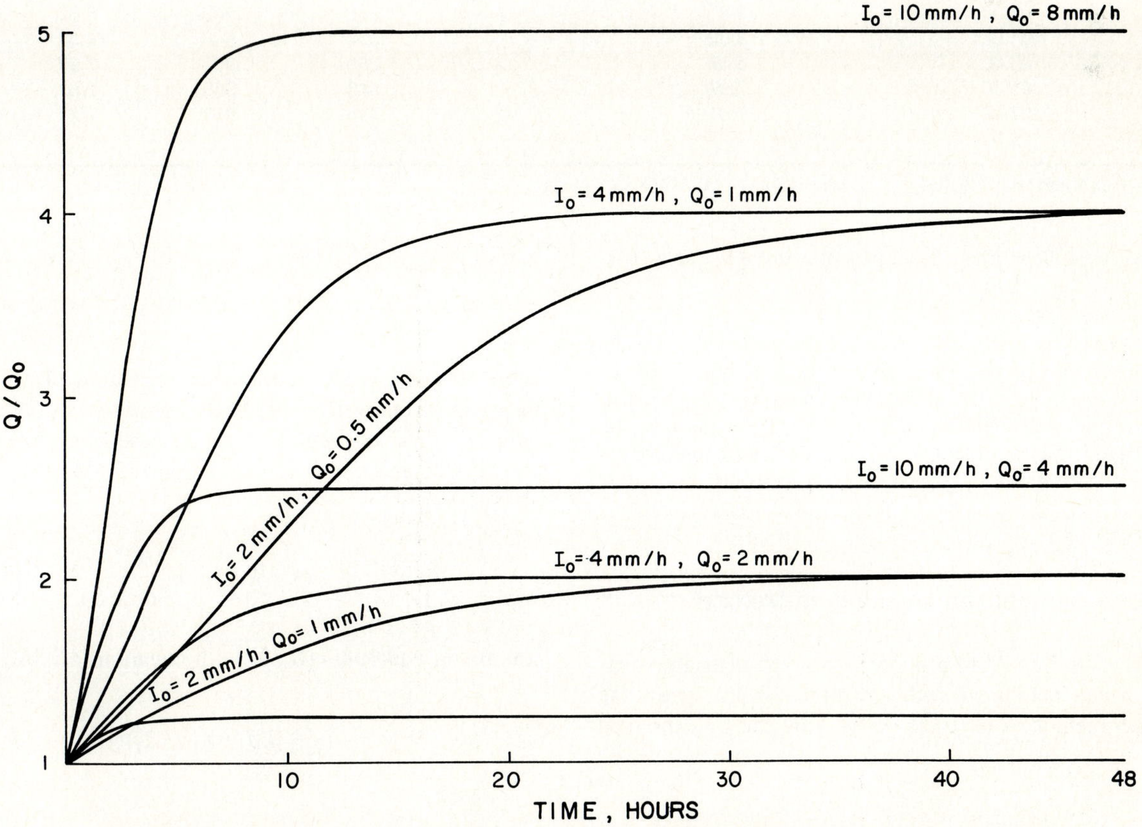

FIGURE 16.16 Dimensionless response to steady ER for various values of ER intensities.

TABLE 16.1

Prediction of streamflow hydrograph October 27–29, 1967

Date October, 1967	*Time*	*Rainfall Rate R mm Per Day*	*Initial Runoff Rate q_0 mm Per Day*	x	y	z	*Q_0/Q_0 Eq. (16.12) or (16.66)*	*Final Runoff Rate q_0 mm Per Day, Day*	
27th	09.00	26.4	2.72	.855	9.71		1.150	3.13	145
	11.00	1.2	3.13	.993	.383		.989	3.09	143
	13.00	3.6	3.09	.979	1.165		1.003	3.10	143
	15.00	61.2	2.10	.695	19.75		1.407	4.37	202
	17.00	22.8	4.37	.873	5.22		1.115	4.87	226
	19.00	21.6	4.87	.879	4.43		1.104	5.37	249
	21.00	26.4	5.37	.855	4.91		1.131	6.07	281
	23.00	6.0	6.07	.965	.988		1.000	6.07	281
	01.00	33.6	6.07	.819	5.53		1.174	7.13	330
	03.00	2.4	7.13	.987	.337		.975	6.95	322
	05.00		6.95			.041	.960	6.68	309
	07.00		6.68			.040	.962	6.42	297
28th	09.00		6.42			.038	.963	6.18	286
	11.00		6.18			.037	.965	5.96	276
	13.00	132.0	5.96	.455	22.15		2.085	12.43	575
	15.00	46.8	12.43	.756	3.76		1.220	15.17	703
	17.00	51.6	15.17	.735	3.40		1.230	18.67	865
	19.00	2.4	18.67	.987	.129		.920	17.17	795
	21.00	18.0	17.17	.899	1.048		1.004	17.23	797
	23.00	14.4	17.23	.918	.835		.985	16.96	785
	01.00	2.4	16.96	.987	.142		.928	15.73	728
	03.00		15.73			.094	.914	14.39	667
	05.00		14.39			.086	.921	13.25	613
	07.00	14.4	13.25	.918	1.088		1.007	13.34	618
29th	09.00	2.4	13.34	.987	.179		.943	12.60	583
	11.00	14.4	12.60	.918	1.143		1.012	12.73	590
	13.00		12.73			.076	.930	11.84	548
	15.00		11.84			.071	.934	11.07	512
	17.00	1.2	11.07	.993	.108		.946	10.47	485
	19.00		10.47			.062	.942	9.86	456
	21.00		9.86			.059	.944	9.31	431
	23.00		9.31			.055	.948	8.82	408
	01.00		8.82			.053	.948	8.37	388
	03.00		8.37			.050	.953	7.97	369
	05.00		7.97			.048	.955	7.61	353
	07.00		7.61			.045	.957	7.27	337
	09.00		7.27						

k = 14 mm; 1 mm/day = 46.3 cusec; $x = \exp(-It/k)$; $y = I/Q_0$; Q_0t/k; $\Delta t = \frac{1}{12}$ day

the observed streamflow. The rainfall pattern is given in Table 16.1. This example is due to Lambert (1969).

***Solution*:** To solve this problem, we choose a time of computation Δt = 2 h. During the period of rainfall, equation (16.66) is utilized. First, the factors x and y are computed. Then Q/Q_0 is obtained, which yields $Q(t = 2)$. This value is used as Q_0 for the next 2-h period and $Q(t = 4)$ is obtained. The calculation is repeated. If there is no rainfall in the 2-h period, then equation (16.69) is used to compute the factor z and Q/Q_0. Results of the calculations are shown in Table 16.1.

16.2.4 The Generalized Muskingum Method

Ding (1967a, 1967b, 1974) proposed a storage-discharge relation that, when combined with equation (16.16), gives the Muskingum method as a special case. This can be expressed as

$$Q = k\,S^n - \sum_{i=1}^{M} k_i \frac{d^iS}{dt^i}, \quad n > 0 \tag{16.68}$$

where k and k_i's are coefficients. Equation (16.68) encompasses linear as well as nonlinear reservoirs as special cases. Ding (1974) investigated a special case of equation (16.68) with M = 1 and called it the generalized Muskingum method which is

$$Q = k\,S^n - k_1 \frac{dS}{dt} \tag{16.69}$$

Combining equation (16.69) with equation (16.16),

$$S = \frac{1}{k^{1/n}} [k_1 I + (1 - k_1) Q]^{1/n} \tag{16.70}$$

Equations similar to equation (16.70) have recently been used by Gill (1978) and V. P. Singh (1983b). If $n = 1$, equation (16.70) becomes the linear Muskingum method. We can obtain solutions for the generalized Muskingum method if $I(t)$ is constant or zero. For simplicity, let us write $c^n = k$ and $c_1 = k_1$. Then

$$Q = c^n S^n - c_1 \frac{dS}{dt}, \qquad n > 0, c > 0, c_1 \geq 0 \tag{16.71}$$

16.2.4.1 The Response to Steady ER

When the effective rainfall is constant for all $t \geq 0$, that is, $I(t) = I(0) = I_0$, equation (16.70) can be written as

$$\frac{dS(t)}{dt} = \left[\frac{I_0}{c}\right]^{1/n} \frac{dw(t)}{dt} \tag{16.72}$$

where

$$w(t) = \left[c_1 + (1 - c_1)\frac{Q(t)}{I_0}\right]^{1/n} \tag{16.73}$$

Rearranging equation (16.16) by inserting equation (16.73),

$$\frac{dS}{dt} = \frac{I_0}{1 - c_1}[1 - w^n] \tag{16.74}$$

Eliminating dS/dt in equations (16.72) and (16.74),

$$\frac{dw}{dt} = \frac{c}{1 - c_1} I_0^{(n-1)/n} [1 - w^n] \tag{16.75}$$

On integration,

$$F[w(t), n] - F[w(0), n] = \frac{c}{1 - c_1} I_0^{(n-1)/n}\, t \tag{16.76}$$

where $F(w, n)$ is the Bakhmeteff varied flow function (Bakhmeteff, 1932) as defined previously. Equation (16.76) defines the rising hydrograph. It can be shown that

$$Q(0) = 0, \qquad Q(t) \rightarrow I_0 \quad \text{as} \quad t \rightarrow \infty, \quad \text{and} \quad \frac{dQ(t)}{dt} \geq 0$$

16.2.4.2 The Response to Pulsed ER

When $I(t)$ is defined by a rectangular pulse whose duration is D and intensity I_0 then equation (16.76) defines the rising hydrograph for $0 \leq t \leq D$. For $t \geq D$, $I(t) = 0$, the solution of equations (16.16) and (16.71) will yield the hydrograph for $t \geq D$. Writing equation (16.16) as

$$\frac{dS}{dt} = -Q \tag{16.77}$$

Inserting equation (16.71),

$$\frac{dS}{S^n} = \frac{c^n}{1 - c_1} dt$$

Upon integration with the condition $S = S(D)$ at $t = D$,

$$S^{1-n}(t) = S^{1-n}(D) + \frac{c^n(1 - n)}{1 - c_1}(t - D) \tag{16.78}$$

Note that equations (16.16) and (16.77) yield

$$Q(t) = \frac{c^n}{1 - c} S^n(t) \tag{16.79}$$

Substituting in equation (16.78),

$$Q^{-(n-1)/n} = Q_D^{-(n-1)/n} + \frac{c}{(1 - c_1)^{1/n}}(t - D) \tag{16.80}$$

where $Q_D = Q(D)$ to be obtained from equation (16.76). Equation (16.80) gives the hydrograph for $t \geq D$.

Ding (1974) noted further that

$$Q = c^n(S + A)^n - c_1\frac{dS}{dt} + B \tag{16.81}$$

can also be used in equation (16.16), and the above procedure can be employed to obtain explicit solutions. A and B are arbitrary constants. From equation (16.76) it can be shown that the hydrograph peak occurs at the end of rainfall or shortly thereafter.

16.2.4.3 The Impulse Response

Ding (1974) utilized the generalized Muskingum method to derive the IUH. We follow his procedure here. Writing the convolution integral,

$$Q(t) = \int_0^t I(t - \tau)h[I(t - \tau), \tau]\, d\tau \tag{16.82}$$

where the IUH, $h[I(t - \tau), \tau]$, depends on the intensity of the effective rainfall $I(t - \tau)$ as well as its time of application. To simplify notations $h(t) = h[I(t - \tau), \tau]$ with the implication that h is I-dependent. For the derivation of the IUH we differentiate equation (16.82),

$$h(t) = \frac{dQ(t)}{I_0 dt} \tag{16.83}$$

This is evident from the fact that the IUH is the first derivative of the S-hydrograph. The IUH can, therefore, be derived from the definition of equation (16.83). Differentiating equation (16.73) with respect to t,

$$\frac{dw}{dt} = \frac{1}{nw^{n-1}}(1 - c_1)h(t) \tag{16.84}$$

Using equation (16.83) and equating to equation (16.75),

$$\frac{c}{1 - c_1} I_0^{(n-1)/n} [1 - w^n] = \frac{1 - c_1}{nw^{n-1}} h(t)$$

Therefore,

$$h(t) = \frac{cnI_0^{(n-1)/n}}{(1 - c_1)^2} w^{n-1} [1 - w^n] \tag{16.85}$$

and from equation (16.76)

$$t = \frac{F[w(t), n] - F[w(0), n]}{c\, I_0^{(n-1)/n}} (1 - c_1) \tag{16.86}$$

Equations (16.85) and (16.86) are a parametric representation of the IUH with w as an independent variable. If the IUH, as noted in the previous chapters, is assumed to satisfy $h(0) = 0$, then it is seen from equation (16.85),

$$h(0) = \frac{cnI_0^{(n-1)/n}}{(1 - c_1)^2} c_1^{(n-1)/n} \tag{16.87}$$

Evidently, c_1 must vanish for $h(0) = 0$. Hence, equations (16.85) and (16.86) reduce to

$$h(t) = ncI_0^{(n-1)/n} w^{n-1} (1 - w^n) \tag{16.88a}$$

or

$$h(t) = ncQ^{(n-1)/n} (I_0 - Q)/I_0 \tag{16.88b}$$

$$t = \frac{F[w(t), n]}{c\, I_0^{(n-1)/n}} \tag{16.88c}$$

with $w(t) = [Q(t)/I_0]^{1/n}$. It must be noted that the IUH depends upon the intensity of the effective rainfall; its ordinates vary directly, whereas its time element varies inversely, with $I_0^{(n-1)/n}$. The IUH given by equation (16.88) is a two-parameter model.

Chen (1984) and Chen and V. P. Singh (1986) noted that equation (16.88b) can be obtained directly by using equations (16.16), (16.71), and (16.83). Note that $c_1 = 0$. Therefore, $Q = c^n S^n$. This yields

$$\frac{dS}{dt} = \frac{1}{nc} Q^{(1-n)/n}$$

Inserting this in equation (16.16) and considering equation (16.83),

$$h(t) = ncQ^{(n-1)/n} (I_0 - Q)/I_0$$

which is the same as equation (16.88b).

The impulse response can also be derived by noting that $I(t)$ occurs during a very small time dt and is zero thereafter. Therefore,

$$Q(0) = Q_0, \qquad t = 0$$

and from equation (16.80)

$$Q^{-(n-1)/n}(t) = Q_0^{-(n-1)/n} + \frac{c}{(1 - c_1)^{1/n}} t \tag{16.89}$$

The initial condition Q_0 can be determined by invoking continuity of volume,

$$V = \int_0^\infty Q(t)\, dt = \int_0^\infty \left[Q_0^{-(n-1)/n} + \frac{c}{(1 - c_1)^{1/n}} t \right]^{-n/(n-1)}$$

$$= - \frac{\left[Q_0^{-(n-1)/n} - \dfrac{c}{(c - c_1)^{1/n}} \right]^{-1/(n-1)}}{\left(\dfrac{1}{n-1} \right) \left[\dfrac{c}{(1 - c_1)^{1/n}} \right]} \Bigg|_0^\infty$$

$$= \frac{(n-1)(1 - c_1)^{1/n}}{c} Q_0^{1/n}$$

Therefore,

$$Q_0 = \frac{1}{1 - c_1} \left(\frac{cV}{n-1} \right) n \tag{16.90}$$

where $V = 1$, $Q(t) = h(t)$

16.2.4.4 Parameter Estimation

We can establish the limits of the parameter c and n. For the IUH to have a positive peak, it must satisfy

$$h(t) > 0, \qquad \frac{dh(t)}{dt} = 0, \qquad \frac{d^2h(t)}{dt^2} < 0 \text{ at } t = t_p \tag{16.91a}$$

where t_p is the time to the IUH peak. Equation (16.91a) yields

$$w(t_p) = \left(\frac{n-1}{2n-1} \right)^{1/n} \tag{16.91b}$$

Therefore, $n > 1$ and $c > 0$. Substituting equation (16.91b) into equation (16.88),

$$h(t_p) = \frac{n^2(n-1)}{(2n-1)^{(2n-1)/n}} cI_0^{(n-1)/n} \tag{16.92}$$

The parameters c and n can be obtained using the method of moments (Nash, 1959). The ith moment M_i of the IUH about its origin can be defined as

$$M_i = \frac{\int_0^\infty h(t)\, t^i\, dt}{\int_0^\infty h(t)\, dt}$$

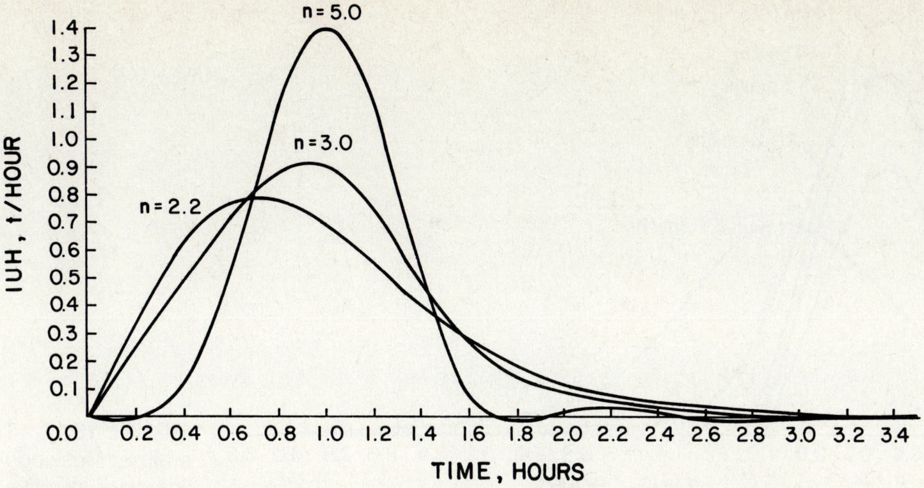

FIGURE 16.17 The IUH from the generalized Muskingum method for various values of *n*.

Using equation (16.88),

$$M_i = \frac{n}{c_i I_0^{i(n-1)/n}} \int_0^1 w^{n-1} [F(w, n)]^i \, dw \tag{16.93a}$$

When $i = 0$,

$$M_0 = n \int_0^1 w^{n-1} \, dw = 1$$

When $i \geq 1$, integrating equation (16.93a) by parts,

$$M_1 = t_L = \frac{1}{c\, I_0^{(n-1)/n}} \tag{16.93b}$$

$$M_i = [M_1]^i \, G_i(n) \tag{16.93c}$$

where

$$G_i(n) = i \int_0^1 [F(w, n)]^{i-1} \, dw$$

and t_L is the time of centroid or lag of the IUH. It may be added that $G_i(n)$ is a dimensionless function of the parameter n. When $i > 1$, $G_i(n)$ can be characterized as the shape factor because it reflects the IUH shape (Nash, 1960). It is apparent that $G_i(n)$ is always unity.

16.2.4.5 Properties of the IUH

It may now be interesting to examine the properties of the IUH. Fig. 16.17 and 16.18 show the IUH for various values of n and c, respectively. It is observed that the higher the value of n, the more symmetrical the IUH. The higher the value of c, the greater the peak and shorter the time to peak. The effect of variation of I on $h(t)$ is shown in Fig. 16.19. Here the variation has a striking similarity to the UHs of Minshall (1960).

It may be advantageous to express the IUH in a dimensionless form and then examine its sensitivity to changes in its parameters. Dividing equation (16.88) by equation (16.92).

$$\frac{h(t)}{h(t_p)} = h_* = \frac{(2n-1)^{(2n-1)/n}}{n(n-1)^{(n-1)/n}} w^{n-1} (1 - w^n) \tag{16.94}$$

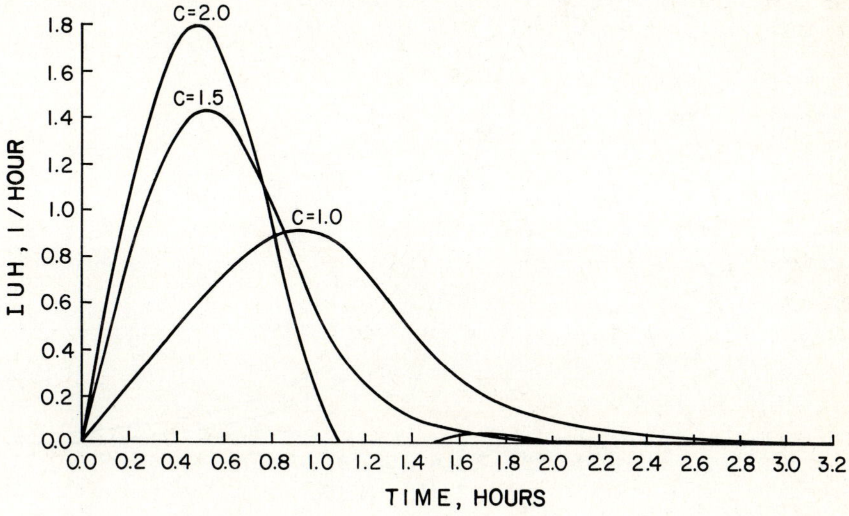

FIGURE 16.18 The IUH from the generalized Muskingum method for various values of *c*.

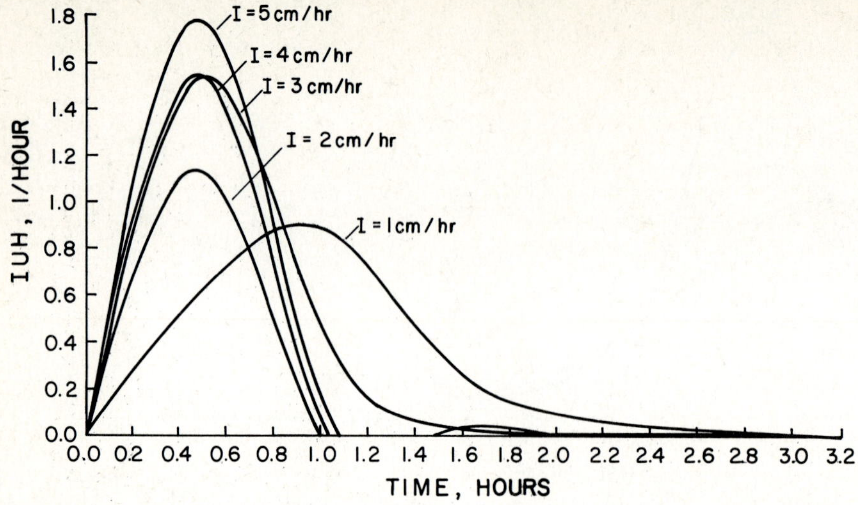

FIGURE 16.19 The IUH from the generalized Muskingum method for various values of effective rainfall itensity.

Dividing equation (16.88c) by equation (16.93b),

$$\frac{t}{t_L} = T_* = F(w, n) \tag{16.95}$$

Fig. 16.20 shows variation of the dimensionless IUH with n. The variation of $h(t_p)$ with t_L can be obtained by multiplying equations (16.92) and (16.93b),

$$h(t_p)t_L = \frac{n^2(n-1)^{(n-1)/n}}{(2n-1)^{(2n-1)/n}} \tag{16.96a}$$

In a similar vein, from equations (16.91b), (16.95), and (16.96a),

$$h(t_p)t_p = h(t_p)t_L\, F[w(t_p), n] \tag{16.96b}$$

Fig. 16.21 graphs equations (16.96a) and (16.96b).

16.2.4.6 Comparison with the Nash Model

It may be appropriate to compare the variable IUH model with the Nash model (Nash, 1957). The direct comparison may be somewhat difficult but can be facilitated by comparing their moments. For the Nash model, with N and K as parameters,

$$M_1 = NK$$

and

$$M_2 = N(N+1)K^2$$

Equating these moments with those of the variable IUH model,

$$G_2(n) = 1 + \frac{1}{N} \tag{16.97a}$$

$$c = \frac{1}{N} K I_0^{(n-1)/n} \tag{16.97b}$$

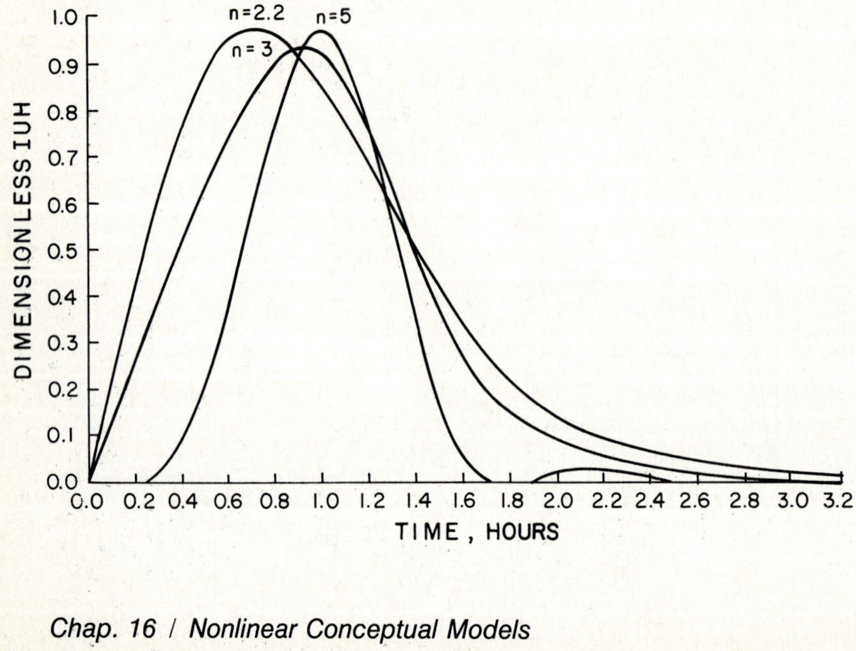

FIGURE 16.20 The dimensionless IUH from the generalized Muskingum method for various values of n.

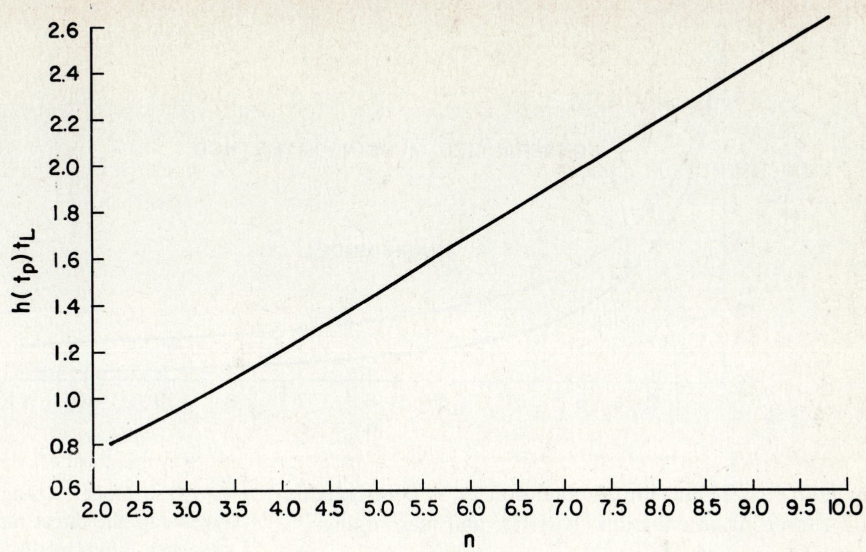

FIGURE 16.21(a) $h(t_p)t_L$ versus n for the generalized Muskingum method.

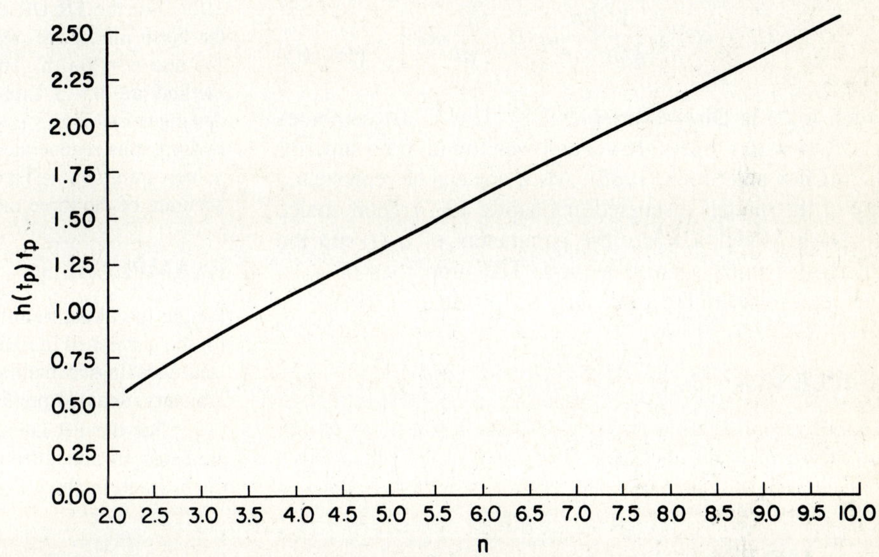

FIGURE 16.21(b) $h(t_p)t_p$ versus n for the generalized Muskingum method.

The function G_2 for both models is shown in Fig. 16.22. For a given IUH the value of n is always less than N.

16.2.5 Chen-Singh Method

Chen and Singh (1976) proposed a method for deriving a variable IUH using the second-order convolution integral, which can be written as

$$Q(t) = \int_0^t I(t-\tau)u[I(t-\tau);\tau]\,d\tau + \int_0^t\int_0^t g[I(t-\tau_1), I(t-\tau_2); \tau_1, \tau_2]\, I(t-\tau_1)\,d\tau_1\,d\tau_2 \tag{16.98}$$

Differentiating equation (16.98) and observing that the nonlinear kernel $u_N(t)$, which is to be obtained, equals the sum of the second-order kernel $g(t, t)$ and the linear kernel $u(t)$,

$$\frac{dQ(t)}{dt} = I(0)u(t) + [I(0)]^2 \int_0^t [u_N(t) - u(t)]\,dt \tag{16.99}$$

Differentiating equation (16.73) and obtaining

$$\frac{dQ(t)}{dt} = \frac{I(0)}{1-c_1} nw^{n-1}\frac{dw}{dt} \tag{16.100}$$

Equating equations (16.99) and (16.100) and using the initial condition $c_1 = 0$,

$$I(0)nw^{n-1}(t)\frac{dw(t)}{dt} = u(t)I(0) + I^2(0)\int_0^t [u_N(t) - u(t)]\,dt \tag{16.101}$$

The term $dw(t)/dt$ can be eliminated by equating equation (16.72) and (16.74),

$$\frac{dw(t)}{dt} = [I(0)]^{(n-1)/n}\frac{c}{1-c_1}[1 - w^n(t)] \tag{16.102}$$

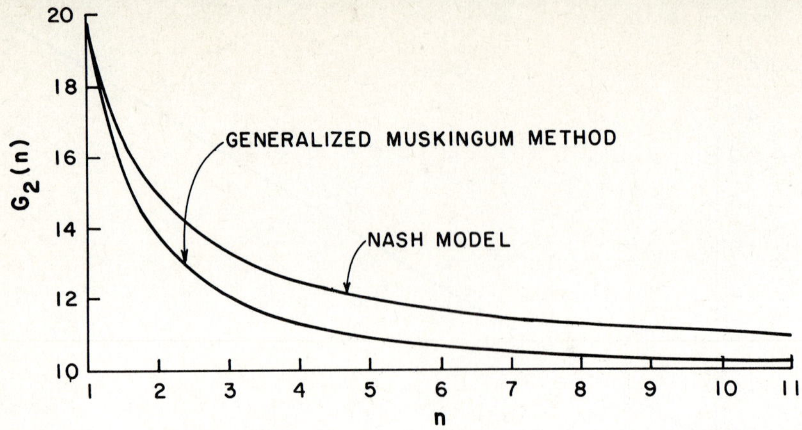

FIGURE 16.22 The shape factors for the Nash model and the variable IUH.

Taking into account the initial condition $c_1 = 0$ in equation (16.102), inserting in equation (16.101), and rearranging,

$$u_N(t) = u(t) + \frac{1}{I(0)}\left[\frac{d}{dt}u_D(t) - \frac{d}{dt}u(t)\right] \tag{16.103}$$

in which $u_D(t)$ is Ding's variable IUH. Thus $u_N(t)$ combines Ding's IUH with a linear kernel and was found to be superior to $u_D(t)$ (Chen and Singh, 1985). Also, $u(t)$ can be represented by any of the models discussed in Chapter 13. A good choice is the Nash Model. Clearly the parameters of $u_N(t)$ are the parameters of $u_D(t)$ as well as $u(t)$. The properties of $u_N(t)$ can be determined in the same way as for $u_D(t)$.

EXAMPLE 16.4

An ER-DR record is given in Table 16.2 for a watershed of 656 km^2 in the People's Republic of China. Compute the parameters n and c of the generalized Muskingum method as well as N and k of the Nash model. Using these parameters, determine the IUHs and regenerate the direct runoff hydrograph by these methods as well as the Chen-Singh method with the Nash model for a linear kernel.

For the ER-DR data in Table 16.2, the parameters N and k of the Nash model are obtained using the method of moments as $N = 1.6$ and $k = 6.0$ h. The parameters of the generalized Muskingum method are $n = 3.4$ and $c = 0.005$. The IUHs by the three methods designated as u_{Nash}, u_{Ding}, and u_N are given in Table 16.3. The DR hydrographs regenerated by the three methods using these parameter values are given in Table 16.4. The Chen-Singh method is the most accurate of the three methods.

EXAMPLE 16.5

Predict the direct runoff for the effective rainfall given in Table 16.5 for the watershed in Example 16.4 using the Nash model, the generalized Muskingum method, and the Chen-Singh method. Use the parameters obtained previously.

For the ER-DR data in Table 16.5, the DR hydrographs are predicted by the three methods and are given in Table 16.6. The Chen-Singh method is the most accurate of the three methods.

TABLE 16.2

Measured ER and DR data for a watershed in China

Time Interval $\Delta t = 6\,h$	*0*	*1*	*2*	*3*	*4*	*5*	*6*	*7*	*8*	*9*	*10*	*11*
Q (m^3/s)	0	562	1412	1147	900	703	600	450	386	332	288	0
I (mm)		28.0	58.1	7.4								

TABLE 16.3

u_{Nash}, u_{Ding} and u_N for varying ER

Time Interval $\Delta t = 6\,h$	*0*	*1*	*2*	*3*	*4*	*5*	*6*	*7*	*8*	*9*	*10*	*11*
u_{Nash}	0	.071	.074	.056	.041	.028	.020	.012	.009	.005	.003	
u_{Ding} for												
$I(0) = 28$ mm	0	.025	.078	.055	.033	.020	.011	.005	.001			
$I(0) = 58.1$ mm	0	.040	.084	.060	.024	.010	.006	.002				
$I(0) = 7.4$ mm	0	.015	.061	.057	.045	.030	.020	.015	.010	.009	.005	.003
u_N for												
$I(0) = 28$ mm	0	.086	.074	.050	.038	.029	.020	.014	.008	.004	.001	
$I(0) = 58.1$ mm	0	.096	.074	.046	.029	.026	.019	.013	.008	.004	.001	
$I(0) = 7.4$ mm	0	.078	.074	.054	.039	.027	.020	.014	.008	.004	.001	

TABLE 16.4

Observed and regenerated direct runoff hydrographs

Time Interval $\Delta t = 6\ h$	0	1	2	3	4	5	6	7	8	9	10	Error in Peak
$Q_{Observed}$ (m^3/s)	0	562	1412	1147	900	703	600	450	386	332	288	
Q_{Nash} (m^3/s)	0	360	1127	1165	892	655	455	309	200	137	80	−17.5%
Q_{Ding} (m^3/s)	0	127	819	1188	880	431	222	129	100	18	1	−15.9%
Q_N (m^3/s)	0	439	1392	1142	779	517	430	308	206	122	58	−1.4%

TABLE 16.5

Measured ER and DR data for the watershed in China in Example 16.4

Time Interval $\Delta t = 6\ h$	0	1	2	3	4	5	6	7	8	9	10	11	12
Q (m^3/s)	0	400	1350	1126	836	602	441	320	220	143	81	52	31
I (mm)		25.4	55.6	9.2									

TABLE 16.6

Observed and predicted direct runoff hydrographs

Time Interval														Error in Peak	
$\Delta t = 6\ h$	0	1	2	3	4	5	6	7	8	9	10	11	12	Discharge	Time
$Q_{Observed}$ (m^3/s)	0	400	1350	1126	836	602	441	320	220	143	81	52	31		
Q_{Nash} (m^3/s)	0	328	1056	1128	864	637	455	291	200	127	91	36	12	−16.4%	6 h
Q_{Ding} (m^3/s)	0	116	764	1130	863	430	228	135	58	26	17	8	2	−16.3%	6 h
Q_N (m^3/s)	0	397	1314	1123	806	593	430	313	211	122	58	17	2	−2.7%	0

16.2.6 Cascade of Uniformly Nonlinear Reservoirs

Dooge (1967a, 1967b) proposed a conceptual model composed of equal nonlinear reservoirs. The underlying concept was referred to as uniform nonlinearity. This concept has been used by Singh (1976d, 1979a, 1979b) for modeling direct runoff from small agricultural watersheds. The model consists of an arrangement of equal nonlinear reservoirs in series, parallel or a combination of both. Fig. 16.23 shows a series arrangement of N uniformly nonlinear reservoirs with a lateral inflow. Each reservoir may or may not correspond to a particular portion of the watershed in question. We can formulate this model mathematically as follows.

Let us write equation (16.10) as

$$Q = \alpha S^x \tag{16.104}$$

where α and x are parameters defined in equation (16.13). Substituting equation (16.104) into equation (16.16),

$$\frac{dS}{dt} = I - \alpha S^x \tag{16.105}$$

Then for the cascade of Fig. 16.23, we get a system of equations of the type:

$$\begin{aligned}
\frac{dS_1}{dt} &= I_1 - \alpha S_1^x \\
\frac{dS_2}{dt} &= I_2 + \alpha S_1^x - \alpha S_1^x \\
&\vdots \\
\frac{dS_{N-1}}{dt} &= I_{N-1} + \alpha S_{N-2}^x - \alpha S_{N-1}^x \\
\frac{dS_N}{dt} &= I_N + \alpha S_{N-1}^x - \alpha S_n^x
\end{aligned} \tag{16.106}$$

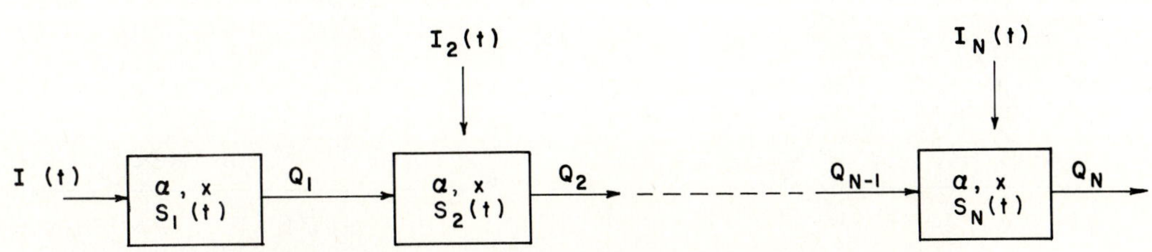

FIGURE 16.23 A cascade of uniformly nonlinear reservoirs.

It is easier to write the system of equations (16.106) in matrix form,

$$[\dot{S}] = [I] + \alpha[B][S] \tag{16.107}$$

where

$$[\dot{S}] = \begin{bmatrix} \dot{S}_1 \\ \dot{S}_2 \\ \cdot \\ \cdot \\ \dot{S}_N \end{bmatrix}; \quad I = \begin{bmatrix} I_1 \\ I_2 \\ \cdot \\ \cdot \\ I_N \end{bmatrix}; \quad [S] = \begin{bmatrix} S_1 \\ S_2 \\ \cdot \\ \cdot \\ S_N \end{bmatrix}; \text{ and}$$

$$B = \begin{bmatrix} S_1^{x-1} & 0 & 0 & \cdots & 0 \\ S_1^{x-1} & -S_2^{x-1} & 0 & \cdots & 0 \\ 0 & S_2^{x-1} & -S_3^{x-1} & \cdots & 0 \\ \cdot & \cdot & & \cdot & \cdot \\ \cdot & \cdot & & \cdot & \cdot \\ \cdot & \cdot & & \cdot & \cdot \\ 0 & 0 & 0 & S_{N-1}^{x-1} & -S_N^{x-1} \end{bmatrix}$$

Our interest is in the relationship between Q_N and the set of lateral inflows $I_1, I_2, \ldots, I_N$, which can be expressed as

$$[Q] = \alpha[C][S] \tag{16.108}$$

where

$$[C] = \begin{bmatrix} S_1^{x-1} & \cdots & 0 & \cdots & 0 \\ 0 & \cdots & S_2^{x-1} & \cdots & 0 \\ \cdot & \cdot & \cdot & \cdot & \cdot \\ \cdot & \cdot & \cdot & \cdot & \cdot \\ \cdot & \cdot & \cdot & \cdot & \cdot \\ 0 & \cdots & \cdot & \cdots & S_N^{x-1} \end{bmatrix}$$

Equations (16.107) and (16.108) constitute what may be called *state variable representation*. Given initial conditions, the calculations can be carried out explicitly with the progress of time. If $j = 0, 1, 2, \ldots, M$ are the points in time at which the computation is to be made then

$$[S_j] = [S_{j-1}] + [\dot{S}_{j-1}] \tag{16.109}$$

This specifies storage for each reservoir at the jth point in time. Using equation (16.108) we can compute discharge at that point in time.

16.2.6.1 Mathematical Properties

In general, nonlinear models do not obey the principles of proportionality and superposition, unlike their linear counterparts. However, Dooge (1967a) showed that the principle of proportionality would hold for uniformly nonlinear models provided the time scale was properly transformed in accordance with the input intensity. To prove it, we define equation (16.107) in a dimensionless form by using the following normalizing quantities (V. P. Singh, 1976d) in the same manner as used previously,

P_0 = normalizing rate of inflow. This can be taken as average input rate, or any other input rate that may be suitable.

S_0 = normalizing storage. This can be expressed in terms of P_0 as

$$S_0 = \left(\frac{P_0}{\alpha}\right)^{1/x}$$

T_0 = normalizing time. This can be expressed in terms of P_0 and S_0 as

$$T_0 = \frac{S_0}{P_0} = \frac{1}{(\alpha)^{1/x}\,(P_0)^{(x^2-1)/x}}$$

Once the normalizing rate P_0 has been defined, the normalizing storage and the normalizing time can be calculated from it using the parameters α and x. Thus the normalized quantities, denoted by an asterisk, are

$$I_* = \frac{I(t/T_0)}{P_0}; \quad t_* = \frac{t}{T_0}; \quad S_* = \frac{S}{S_0}; \quad \frac{dS_*}{dt_*} = \frac{d(S/S_0)}{d(t/T_0)};$$

and $Q_* = \dfrac{Q}{P_0}$

Equations (16.107) and (16.108) can be written as

$$[\dot{S}_x] = [I_*] + \alpha[B_*]\,[S_*] \tag{16.110a}$$

$$[Q_*] = \alpha[C_*]\,[S_*] \tag{16.110b}$$

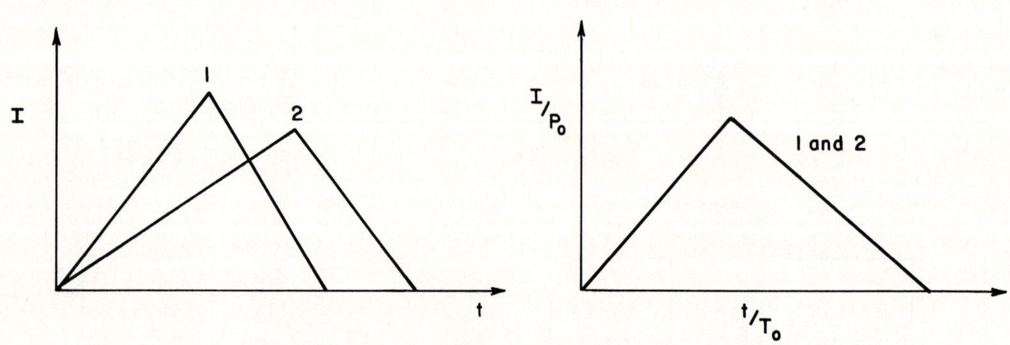

FIGURE 16.24 Similar effective rainfall patterns.

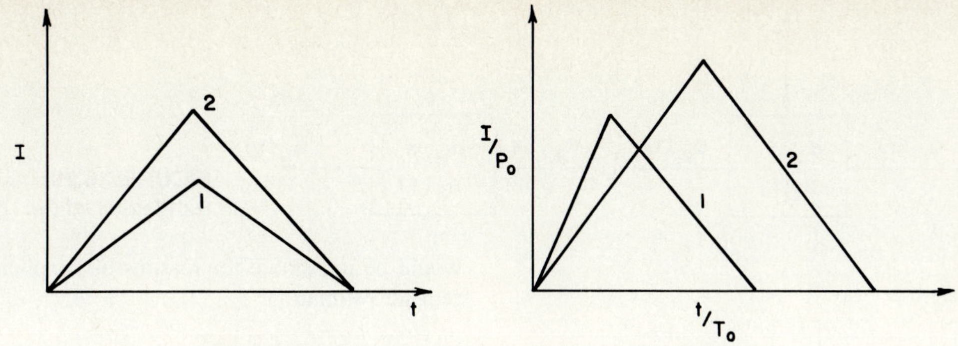

FIGURE 16.25 Dissimilar effective rainfall patterns.

Let us now suppose that two distinctly different patterns of the effective rainfall are given, as shown in Fig. 16.24, for which direct runoff hydrograph is desired. If these patterns are such that they lead to one and the same I_*, then Q_* will also be the same in both cases, as is evident from equations (16.110a) and (16.110b). Such inputs are defined by Dooge (1967a) as similar inputs, and for these inputs the principle of proportionality holds for the uniformly nonlinear models. On the other hand, if the given input patterns do not lead to the same I_*, as shown in Fig. 16.25, then the principle of proportionality will no longer hold even though the inputs may be proportional in an absolute sense.

An examination of equations (16.110a) and (16.110b) reveals that

$$\frac{Q_N}{P_0} = \phi\left(\frac{t}{T_0}\right) \tag{16.111}$$

which is of the same form as for the single nonlinear reservoir, the Horton-Izzard approach, and the kinematic wave approach discussed previously. As Dooge (1967a) has pointed out, these approaches may be considered as particular cases of uniform nonlinearity.

16.2.6.2 Parameter Estimation

The uniformly nonlinear cascade has three parameters: N, α, and x. N signifies the number of equal nonlinear reservoirs. The value of N must be greater than 1 to obtain the proper hydrograph shape and will depend on watershed topography. This can essentially be called a shape parameter. A natural watershed entails a network of channels and overland flow planes. The combined action of a channel and a plane is accounted for by a nonlinear reservoir. V. P. Singh (1976d) applied this model to 21 small watersheds (ranging in area from 112 hectares to 1773 hectares) in Texas and Nebraska and found that although N would vary from one watershed to another, it could be fixed for watersheds in a certain range of areas having the same order of drainage evolution. For these small watersheds, $N = 3$ was found satisfactory.

The parameter x signifies the degree of nonlinearity of the direct runoff. Although the value of x may change through the development of a runoff hydrograph, it seems plausible that it can be fixed for watersheds in certain range of areas and that this fixed value would provide a good approximation to the degree of nonlinearity. For the preceding small watersheds, V. P. Singh (1976d) found $x = 1.4$ yielding satisfactory results.

The precise physical significance of the parameter α is not clear. It appears that it accounts for both translatory and attenuation effects; consequently, it may change considerably from one watershed to another. The topographic factors would seem to be dominant factors affecting it. V. P. Singh (1976) related α to watershed area A, shape, width, channel length L, and slope S_c. He found three regression equations for estimating α from these characteristics. One of them is

$$\alpha = \frac{S_c^{.10079}}{A^{.22889}\,L^{.26395}} \tag{16.112}$$

with a correlation coefficient of .9880 and a standard error of estimate of .1834.

16.2.7 Cascade of Nonlinear Reservoirs

The assumption of uniform nonlinearity can be relaxed by considering unequal nonlinear reservoirs in the model. As before the reservoirs can be arranged in series, parallel, or a combination of both. A simple model may be a cascade of nonlinear reservoirs arranged in series. This model has been used by Buapeng and V. P. Singh (1977) and V. P. Singh and Buapeng (1979). Mathematically this model can be formulated in a manner similar to that of the cascade of uniformly nonlinear reservoirs. Referring to Fig. 16.26, the nonlinear cascade can be represented by a system of equations,

$$\begin{aligned} \dot{S}_1 &= I_1 - \alpha_1 S_1^x \\ \dot{S}_2 &= I_2 + \alpha_1 S_1^x - \alpha_2 S_2^x \\ &\;\;\vdots \\ \dot{S}_N &= I_N + \alpha_{N-1} S_{N-1}^x - \alpha_N S_N^x \end{aligned} \tag{16.113}$$

Note that the parameter x does not change from one reservoir to another, whereas α does. Writing in compact matrix form,

$$[\dot{S}] = [I] + [\alpha][B][S] \tag{16.114}$$

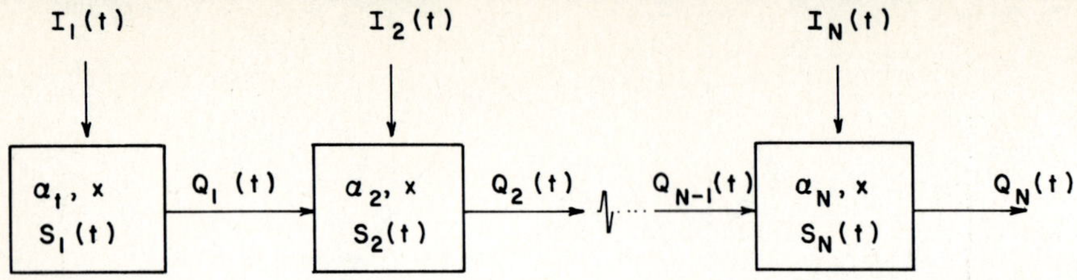

FIGURE 16.26 A nonlinear cascade with lateral inflow.

where

$$[\dot{S}] = \begin{bmatrix} \dot{S} \\ \dot{S} \\ \cdot \\ \cdot \\ \cdot \\ \dot{S}_N \end{bmatrix}; \quad [I] = \begin{bmatrix} I_1 \\ I_2 \\ \cdot \\ \cdot \\ \cdot \\ I_N \end{bmatrix};$$

$$[\alpha] = \begin{bmatrix} \alpha_1 \\ \alpha_2 \\ \cdot \\ \cdot \\ \cdot \\ \alpha_N \end{bmatrix}; \quad [S] = \begin{bmatrix} S_1 \\ S_2 \\ \cdot \\ \cdot \\ \cdot \\ S_N \end{bmatrix}; \quad \text{and}$$

$$[B] = \begin{bmatrix} S_1^{x-1} & 0 & \cdots & 0 \\ S_1^{x-1} & -S^{x-1} & \cdots & 0 \\ \cdot & \cdot & \cdot & \cdot \\ \cdot & \cdot & \cdot & \cdot \\ \cdot & \cdot & \cdot & \cdot \\ 0 \;\cdots & 0 & \cdots\; S_{N-1}^{x-1} & -S_N^{x-1} \end{bmatrix}$$

The relationship between Q_N and I_j, $j = 1, 2, \ldots, N$, is what is required. Given initial conditions,

$$[S_j] = [S_{j-1}] + [\dot{S}_{j-1}], \qquad j = 1, 2, \ldots, M \tag{16.115}$$

where the subscript j denotes the jth point in time at which computation is to be performed. Therefore,

$$[Q] = [\alpha][C][S] \tag{16.116}$$

where

$$[C] = \begin{bmatrix} S_1^{x-1} & 0 & \cdots & 0 \\ 0 & S_x^{x-1} & \cdots & 0 \\ \cdot & \cdot & \cdot & \cdot \\ \cdot & \cdot & \cdot & \cdot \\ \cdot & \cdot & \cdot & \cdot \\ 0 & 0 & \cdots & S_N^{x-1} \end{bmatrix}$$

The computation time interval is very important here. Guapeng and V. P. Singh (1977) took it as 1 minute when applying it to 37 small agricultural watersheds (ranging in area from 0.59 hectares to 3,055 hectares) located in various parts of the United States. They found $N = 3$ and $x = 1.4$ gave satisfactory results but could not establish significant relationships between the parameters α_i, $i = 1, 2, 3, \ldots$ and watershed characteristics. They concluded that the uniformly nonlinear cascade would be preferable for reasons of simplicity and ease of parameter estimation.

16.2.8 Variable Lag Model

Reed, Johnson, and Firth (1975) developed a variable lag model that is nonlinear and time invariant. The model hypothesizes a simple power law for lag time $t_L(\tau)$,

$$t_L(\tau) = a[S(\tau)]^b \tag{16.117}$$

where $t_L(\tau)$ is the lag time applicable to an input introduced at time τ, $S(\tau)$ is the volume of potential runoff or storage at τ, and a and b are constants. The exponent b will, in general, be negative. For the River Ray catchment, b was found to be $-.867$. The volume of potential runoff can be expressed as

$$S(t) = \int_t^\infty Q(t)\, dt \tag{16.118}$$

If equation (16.117) is substituted into a procedure of determining the IUH or UH, the resulting model is a variable lag model. Reed, Johnson, and Firth (1975) utilized the Nash model (Nash, 1957, 1960) with n and k are parameters for the IUH,

$$h(t) = \frac{1}{k\Gamma(n)} \left(\frac{t}{k}\right)^{n-1} \exp\left(-\frac{t}{k}\right) \tag{16.119}$$

The lag time of this model is given by

$$t_L = nk \tag{16.120}$$

Substituting equation (16.120) into equation (16.119) to eliminate k,

$$h(t) = \frac{1}{t_L\Gamma(n)} \left(\frac{nt}{t_L}\right)^{n-1} \exp\left(-\frac{tn}{t_L}\right) \tag{16.121}$$

The finite period UH is given by

$$h(t, D) = \frac{1}{D}\left[g\left(\frac{nt}{t_L}, n\right) - g\left(\frac{n(t-D)}{t_L}, n\right)\right] \tag{16.122}$$

where

$$g(x, n) = \frac{1}{\Gamma(n)} \int_0^x \exp(-w) w^{n-1} dw \tag{16.123}$$

is the incomplete gamma function.

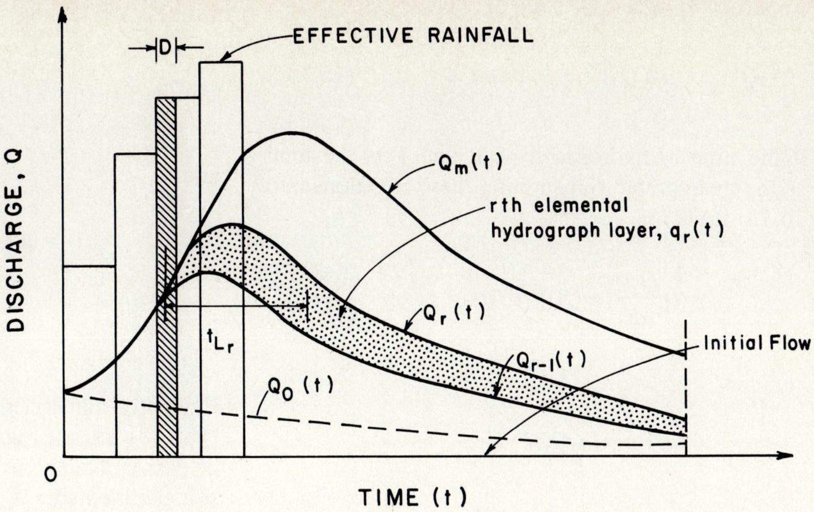

FIGURE 16.27 Hydrograph computation from variable lag model.

Let us visualize a pattern of effective rainfall for which the runoff hydrograph is to be computed. Referring to Fig. 16.27, this pattern can be regarded as a sequence of successive elements of the effective rainfall of equal duration D. Each element is associated with a D-unit hydrograph, $h(t, D)$, to produce an elemental layer of runoff. Equation (16.122) can be used to yield $h(D, t)$. It must be noted that t_L will vary with the time of application of the effective rainfall. Conceptually, t_L applied to a given element will depend on the cumulative effect of the hydrograph layers due to all previous elements.

The contribution $q(t)$ of the rth element of the effective rainfall to the runoff hydrograph, designed as the elemental hydrograph layer, can be expressed as

$$q_r(t) = u_1[t - (r-1)D]\, I_r h[t - (r-1)D, D], \qquad r = 1, 2, \ldots, m \tag{16.124}$$

The cumulative hydrograph $Q_r(t)$, after the elemental layer has been superposed on the layers due to previous elements, can be written as

$$Q_r(t) = Q_{r-1}(t) + q_r(t), \qquad r = 1, 2, \ldots, m \tag{16.125}$$

with $Q_0(t)$ defining the initial flow.

For purposes of computation we can write equation (16.117) as

$$t_L = aS_r^b \tag{16.126}$$

S_r represents the average value of the storage in the system during the time of input,

$$S_r = \frac{1}{2}\left\{\int_{(r-1)D}^{\infty} Q_{r-1}(t)\, dt + \int_{rD}^{\infty} Q_r(t)\, dt\right\} \tag{16.127}$$

To start the computation $Q_0(t)$ must be estimated from the runoff hydrograph observed prior to the effective rainfall under consideration. Therefore,

$$S_0 = \int_0^{\infty} Q_0(t)\, dt \tag{16.128}$$

If $Q_0(t)$ is expressed by

$$Q_0(t) = Q_0(0)\exp(-\alpha t) \tag{16.128}$$

then

$$S_0 = \frac{Q_0(0)}{\alpha} \tag{16.130}$$

with $Q_0(0)$ signifying the observed flow at the commencement of the effective rainfall and α recession constant. The computation can be carried out for each elemental input. An initial estimate of S_r can be obtained from

$$S_r = S_{r-1} + I_r D \tag{16.131}$$

t_L can then be obtained from equation (16.126). In addition, q_r can be computed from equations (16.122) and (16.124). Using equation (16.124), Q_r can be obtained. Then an improved estimate of S_r is obtained from equation (16.127). We can go back and compute t_{Lr}, q_r, and then Q_r until the value of t_{Lr} remains fixed. Finally, $Q_m(t)$, the runoff hydrograph, can be obtained.

16.3 NONLINEAR TIME-VARIANT CONCEPTUAL MODELS

16.3.1 Nonlinear Reservoir

Based on an analysis of rainfall-runoff data from Chartiers Creek basin near Pittsburgh, Pennsylvania, Chiu and Huang (1970) proposed a time-varying nonlinear reservoir model. They hypothesized that

$$S(t) = k_1 Q^{n_1}, \qquad t_p < t \leq T \tag{16.132}$$

and

$$S(t) = k_1 Q^{n_1} + k_2^*(t)\, Q^{n_2^*(t)}, \qquad 0 \le t \le t_p \tag{16.133}$$

where t_p is the time to hydrograph peak and T is the total base time of the hydrograph. Substituting these equations into equation (16.16) leads to

$$f_1(t)\frac{Q(t)}{dt} + Q(t) = I(t) \tag{16.134}$$

where

$$f_1(t) = k_1 n_1 Q(t)^{n_1 - 1}$$

and

$$k_2^*(t) n_2^*(t) \left(\frac{dQ}{dt}\right)^{n_2^*(t)-1} \frac{d^2Q}{dt^2} + k_1 n_1 Q^{n_1 - 1}\frac{dQ}{dt} + k_2^*(t)\left(\frac{dQ}{dt}\right)^{n_2^*(t)}\left(\ln\frac{dQ}{dt}\right) n_2^*(t) + \left(\frac{dk_2^*(t)}{dt}\right)\left(\frac{dQ}{dt}\right)^{n_2^*(t)} + Q = I \tag{16.135}$$

Considering the difficulty of estimating the functions $n_2^*(t)$ and $k_2^*(t)$, Chiu and Huang (1970) simplified equation (16.133) by assuming $n_2 = n_2^*(t)$. Therefore, equation (16.133) becomes, with $k_2(t) = k_2^*(t)$,

$$S(t) = k_1 Q(t)^{n_1} + k_2(t)\left[\frac{dQ(t)}{dt}\right]^{n_2} \tag{16.136}$$

The time function $k_2(t)$ differs from $k_2^*(t)$ in equation (16.133) such that it accounts for the time variability of $n_2^*(t)$ in the following sense:

$$k_2(t) = \frac{S(t) - k_1 Q(t)^{n_1}}{\left[\dfrac{dQ(t)}{dt}\right]^{n_2}} \tag{16.137}$$

Then equation (16.135) becomes

$$f_2(t)\frac{d^2Q(t)}{dt^2} + f_3(t)\frac{dQ(t)}{dt} + Q(t) = I(t), \qquad 0 \le t \le t_p \tag{16.138}$$

where

$$f_2(t) = n_2 k_2(t)\left(\frac{dQ(t)}{dt}\right)^{n_2 - 1}$$

and

$$f_3(t) = n_1 k_1 Q(t)^{n_1 - 1} + \frac{dk_2(t)}{dt}\left(\frac{dQ(t)}{dt}\right)^{n_2 - 1} + n_2 k_2(t) \ln\left(\frac{dQ}{dt}\right)\left(\frac{dQ}{dt}\right)^{n_2 - 1}$$

Evidently, analtyical solutions of equations (16.134) and (16.138) are not tractable. However, numerical solutions can be easily obtained using the Runga-Kutta method or any other suitable method.

By analyzing the rainfall-runoff data, Chiu and Huang (1970) obtained the following empirical expressions for the model parameters. The parameter n_2 was found to be a constant.

$$n_1 = c_1 + c_2 \sin\left(\frac{2\pi}{12}(M - M_1)\right) \tag{16.139}$$

where c_1 and c_2 are constants. M is the month number and M_1 is another constant.

$$k_1 = c_3 V_I^{c_4} \tag{16.140}$$

where c_3 and c_4 are empirical constants and

$$V_I = \int_0^T I(w)\, dw \tag{16.141}$$

T is the total runoff duration.

$$k_2(t) = \alpha \exp(\beta t) \int_0^t I(w)\, dw \tag{16.142}$$

where

$$\alpha = c_5 V_I^{c_6} \tag{16.143}$$

in which c_5 and c_6 are constants.

$$\beta = c_7 + c_8 \sin\left[\frac{2}{12}(M - M_2)\right] \tag{16.144}$$

where c_7, c_8, and M_2 are constants.

EXERCISES

16.1. A residential lot 100 m deep and 50 m wide is inclined with a slope of .005 toward the street. The street is 20 m wide and has a 15-cm crown. Determine the peak flow into the gutter due to a rainfall intensity of 5 cm/h and 1-h duration. Assume Izzard's roughness factor of .05 for the lot and .007 for the street. Compute the peak flow if the rainfall duration is 15 min, 30 min, and 120 min.

16.2. Consider a parking lot 200 m long in the direction of slope and 100 m wide. It has a tar and gravel pavement on a slope of .0025. Construct the runoff hydrograph by Izzard's method due to a rainfall intensity of 10 cm/h for 1 h.

16.3. Use the kinematic wave approach to compute the runoff hydrograph for the data in Question 16.2.

16.4. Compute the direct runoff hydrograph by using equation (16.55). Take I = 10 cm/h for 30 min, $n = 0.8$, $k_1 = 50$, and $k_2 = 250$.

16.5. Compute the direct runoff hydrograph by using equation (16.55) with I = 10 cm/h for 30 min, $n = 0.8$, $k_1 = 50$ and $k_2 = 0$. Compare the results of Question 16.4.

16.6. Compute the direct runoff by the uniformly nonlinear cascade model for the following rainfall event on watershed SW-17 near Riesel (Waco), Texas. Take $x = 1.4$ and α estimated from equation (16.112).

Time (min)	Effective Rainfall (cm/h)
0	0.2
8	2.83
12	1.48
16	5.08
20	2.8
24	3.79
27	3.29
31	2.09
39	1.31
44	0.97
50	0.29
56	0.04
78	0.0

16.7. Determine the effect of computation interval Δt for Question 16.6.

16.8. Compute the parameters n and c of the generalized Muskingum method for an effective rainfall-direct runoff record of a watershed in your area.

16.9. Compute the direct runoff due to another rainfall event on the watershed in Question 16.8 by using the generalized Muskingum method with parameters estimated previously.

16.10. Compute the dimensionless and streamflow hydrograph by the model $S = k \ln Q$. Take k = 14 mm and I = 10 cm/h for 1 h. Do the calculations for varying I, I = 5 cm/h and 15 cm/h.

16.11. Solve Question 16.8 by varying the duration of I = 10 cm/h as 30 min, 1 h, 1.5 h, and 2 h.

16.12. Compute and graph the Bakhmeteff varied flow function $F(u, m)$ for various values of m.

CHAPTER SEVENTEEN

Nonlinear Black-Box Models

The nonlinear black-box models of the ER-DR relationship have principally been based on the Volterra integral series (VIS) (Volterra, 1913, 1930). The first application of this series to the study of nonlinear systems in communications theory is, however, due to Wiener (1942). A number of workers, following the work of Wiener, have used and further investigated the VIS (Bose, 1956; Brilliant, 1958; Barrett, 1957, 1964, 1965; George, 1959; Schetzen, 1965a, 1965b, 1965c; Bush, 1966; Bruce, 1966). Amorocho and Orlob (1961) and Jacoby (1962) were probably the first to have applied the VIS for analysis of hydrologic systems. It has since been the basis of several studies (Amorocho, 1963, 1967a, 1967b, 1973; Amorocho and Brandstetter, 1971; Brandstetter and Amorocho, 1970; Amorocho and Hart, 1964, 1965; Jacoby, 1965, 1966; Chiu, 1967; Harder and Zand, 1969; Bidwell, 1971; Zand and Harder, 1973; Boneh and Diskin, 1973; Diskin and Boneh, 1972, 1973, 1974, 1975; Boneh and Golan, 1978, 1979; Diskin, Boneh and Golan, 1975, 1984; Kuchment and Borshchevsky, 1971; Hino, 1972; Hino, Sukigara and Kikkawa, 1971; Hino and Nadaoka, 1979; A. R. Rao and R. G. S. Rao, 1974; Liu and Brutsaert, 1975, 1978; Papazafiriou, 1976; Napiorkowski and Strupczewski, 1979, 1981; Norton and Watson, 1979; Diskin, 1982; Helweg, Amorocho and Finch, 1982; Patry and Marino, 1983a, 1983b; Muftuoglu, 1984) on various aspects of the ER-DR relationship. In this chapter we discuss the models based on this series as well as some others.

17.1 VIS MODELS

A large class of nonlinear black box models of the ER-DR relationship are based on a functional series,

$$
\begin{aligned}
Q(t) &= Q_0 + \int_{-\infty}^{\infty} h_1(t, s)I(s)\,ds \\
&\quad + \int_{-\infty}^{\infty}\int_{-\infty}^{\infty} h_2(t; s_1, s_2)I(s_1)I(s_2)\,ds_1\,ds_2 \\
&\quad + \int_{-\infty}^{\infty}\int_{-\infty}^{\infty}\int_{-\infty}^{\infty} h_3(t; s_1, s_2, s_3)I(s_1)I(s_2) \\
&\quad\quad I(s_3)\,ds_1\,ds_2\,ds_3 + \cdots \\
&= Q_0 + \sum_{n=1}^{\infty} {}_{(n)}\int_{-\infty}^{\infty} \\
&\quad\quad h_n(t; s_1, s_2, \ldots, s_n) \prod_{i=1}^{n} I(s_i)\,ds_i
\end{aligned}
\tag{17.1}
$$

where ${}_{(n)}\int_{-\infty}^{\infty}$ is the n-tuple integral; the symbol $\prod_{i=1}^{n}$ denotes the product of n values of the ER at times $s = -\infty, \ldots t, \ldots, \infty$; s is the variable of integration (a time variable); and h_n $(t; s_1, s_2, \ldots, s_n)$ is a generalized kernel function. The term Q_0 represents an internal sink or source and is usually deleted. Equation (17.1) is the VIS and is also referred to as generalized convolution or nonlinear generalization of convolution integral. If the upper limit of the summation is finite, say p, then equation (17.1) reduces to a VIS of the pth order; p also signifies the highest order of nonlinearity. The preceding functional series expansion may be shown to be unique if all the kernels are completely symmetric in the variables, that is,

$$
\begin{aligned}
h_n(t; s_1, s_2, \ldots, s_n) &= h_n(t; s_2, s_3, \ldots, s_n, s_1) \\
&= h_n(t; s_3, s_4, \ldots, s_n, s_1, s_2) = \cdots
\end{aligned}
$$

for any arrangement of the variables, $s_1, s_2, \ldots, s_n$. An unsymmetric kernel may always be changed to symmetrical kernel (Barrett, 1964). Equation (17.1) represents a time-variant relationship, since the kernels depend on the time of application of the effective rainfall. The kernels satisfy

$$h_n(t; s_1, s_2, \ldots, s_n) = 0 \quad \text{for any} \quad s_1, s_2, \ldots, s_n > t$$

Each term in the functional series of equation (17.1) represents a functional of certain order. The first term is the operation of the zero-order functional, which is usually a constant; it may be different from zero in the presence of sink or source. The second term is the first order functional or one-dimensional familiar convolution integral and represents a linear element in the functional series. This is analogous to representation of a linear system operation wherein $h_1(t)$ constitutes the IUH, or first-order kernel. The third term is the second-order functional, or two-dimensional, convolution integral and represents a nonlinear element. It has a second-order kernel $h_2(t_1, t_2)$, which operates on the squares and cross-products of the ER $I(t)$, and so on. Thus the $(n+1)$st term is the n-dimensional generalization of the convolution integral wherein $h_n(t_1, t_2, \ldots, t_n)$ can be thought of as a generalized IUH.

Further, equation (17.1) can be conveniently written as a sum of time functions,

$$Q(t) = Q_0 + \sum_{n=1}^{\infty} g_n(t) \tag{17.2}$$

where

$$g_n(t) = {}_{(n)}\int_{-\infty}^{\infty} h_n(t; s_1, s_2, \ldots, s_n) \prod_{i=1}^{n} I(s_i)\, ds_i \tag{17.3}$$

In particular, the second-order term is

$$g_2(t) = \int_{-\infty}^{\infty}\int_{-\infty}^{\infty} h_2(t; s_1, s_2) I(s_1) I(s_2)\, ds_1\, ds_2 \tag{17.4}$$

As n increases, so does the order of the function represented by $g_n(t)$, which is a result of the convolution operation on the ER $I(t)$. If, as before, we let $g(t) = H[I(t)]$, then equation (17.2) can simply be written as

$$Q(t) = H[I(t)] = g(t) = Q_0 + \sum_{n=1}^{\infty} H_n[I(t)] \tag{17.5}$$

where the subscript n indicates the degree. In this representation the symbol H_n, representing the integral, is called an nth order Volterra operator and $h_n(t; s_1, s_2, \ldots, s_n)$ is called an nth order Volterra kernel because if $I(t)$ is replaced by $aI(t)$ with a being a constant greater than 0, then $g_n(t)$ is replaced by $a^n g_n(t)$. Thus the VIS is a power series with memory because

$$Q(t) = Q_0 + \sum_{n=1}^{\infty} a^n g_n(t)$$

which is a power series in a and has memory, since the integrals for $g_n(t)$ are convolutions.

Equation (17.5) can be interpreted as a summation of the results of a linear operation and a series of nonlinear operations; each nonlinear operation is associated with a nonlinear system having a generalized IUH. This is equivalent to parallel combination of systems denoted by H_n, $n = 1, 2, \ldots$, as shown in Fig. 17.1. A useful conclusion of this interpretation is that if the contributions of the higher-order systems to the total discharge become progressively smaller, the discharge of a nonlinear watershed can be approximated to any desired accuracy by combining an appropriate finite number of systems. Perhaps that is why the second-order systems have been more frequently employed in modeling the ER-DR relationship (Diskin and Boneh, 1972, 1973; A. R. Rao and R. G. S. Rao, 1974).

Equation (17.1) represents a time variant relationship. In this section we are concerned with time-variant systems only. Therefore, we can write

$$Q(t) = Q_0 + \sum_{n=1}^{\infty} {}_{(n)}\int_{-\infty}^{\infty} h_n(t - s_1, t - s_1, \ldots, t - s_n) \prod_{i=1}^{n} I(s_i)\, ds_i \tag{17.6a}$$

where

$$h_n(t - s_1, t - s_2, \ldots, t - s_n) = 0 \quad \text{for any } s_1, s_2, \ldots, s_n > t$$

or

$$Q(t) = Q_0 + \sum_{n=1}^{\infty} {}_{(n)}\int_{-\infty}^{\infty} h_n(t; s_1, s_2, \ldots, s_n) \prod_{i=1}^{n} I(t - s_i)\, ds_i \tag{17.6b}$$

in which

$$h_n(t; s_1, s_2, \ldots, s_n) = 0 \quad \text{for any } s_j < 0, j = 1, 2, \ldots, n$$

Invoking the property of hydrologic systems to be initially relaxed, finite memory and nonanticipatory, we can write

$$Q(t) = \sum_{n=1}^{\infty} {}_{(n)}\int_{0}^{t_m} h_n(t - s_1, t - s_2, \ldots, t - s_n) \prod_{i=1}^{n} I(s_i)\, ds_i \tag{17.7}$$

in which t_m represents the system memory. Alternatively,

$$Q(t) = \sum_{n=1}^{\infty} {}_{(n)}\int_{0}^{t_m} h_n(s_1, s_2, \ldots, s_n) \prod_{i=1}^{n} I(t - s_i)\, ds_i \tag{17.8}$$

Equation (17.8) is found to be more suitable in the analysis and synthesis of nonlinear hydrologic systems.

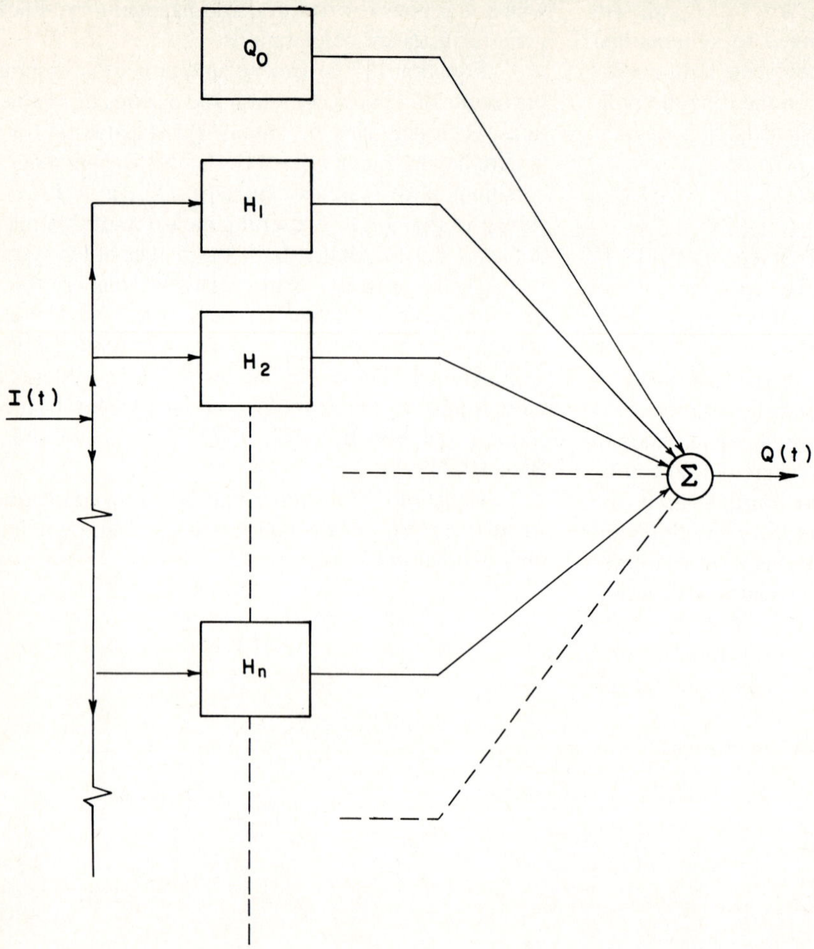

FIGURE 17.1 Schematic of the functional series representation of a nonlinear system.

The notation employed in equation (17.5) is convenient to express many functional operations. To illustrate, we consider the Volterra representation of a second-order system, that is, $n = 2$. Therefore,

$$Q(t) = H_1[I(t)] + H_2[I(t)]$$

Let $I = y + z$. Then $g_2(t) = H_2[y + z]$, which can be expressed as

$$g_2(t) = \int\int h_2(s_1, s_2)\,\{y(t - s_1) + z(t - s_1)\}\,\{y(t - s_2) + z(t - s_2)\}\,ds_1\,ds_2$$

This can be written as

$$g_2(t) = \int\int h_2(s_1, s_2)\,\{y(t - s_1)\,y(t - s_2) + y(t - s_1)\,z(t - s_2) + y(t - s_2)\,z(t - s_1) + z(t - s_1)\,z(t - s_2)\}\,ds_1\,ds_2$$

Using the symmetric property of the kernel (see the next section for discussion),

$$g_2(t) = \int\int h_2(s_1,s_2)\,\{y(t - s_1)\,y(t - s_2) + 2y(t - s_1)z(t - s_2) + z(t - s_1)z(t - s_2)\}\,ds_1\,ds_2$$

Therefore,

$$g_2(t) = H_2(y^2) + 2H_2(yz) + H_2(z^2)$$

This can be further written as

$$g_2(t) = H_2[I(t)] = H_2[y + z] = H_2((y + z)^2)$$

Note that the parentheses indicate a mathematical operation on a pair of functions, whereas the brackets denote an actual system operation.

EXAMPLE 17.1

Consider a simple second-order system, as shown in Fig. 17.2. This represents a linear system in series with an imperfect square-law system having the characteristic

$$Q(t) = az(t) + bz^2(t)$$

Let the IUH of the linear system be denoted by $h(t) = \exp(-t/k)/k$, $t \geq 0$, the ER be denoted by $I(t)$, and the *DR* be denoted by $z(t)$. Therefore,

$$Q(t) = a\int_0^t h(s_1)\,I(t - s_1)\,ds_1 + b\int_0^t h(s_1)I(t - s_1)\,ds_1 \int_0^t h(s_2)\,I(t - s_2)\,ds_2$$

$$= a\int_0^t h(s_1)\,I(t - s_1)\,ds_1 + \int_0^t\int_0^t bh(s_1)h(s_2)\,I(t - s_1)\,I(t - s_2)\,ds_1\,ds_2$$

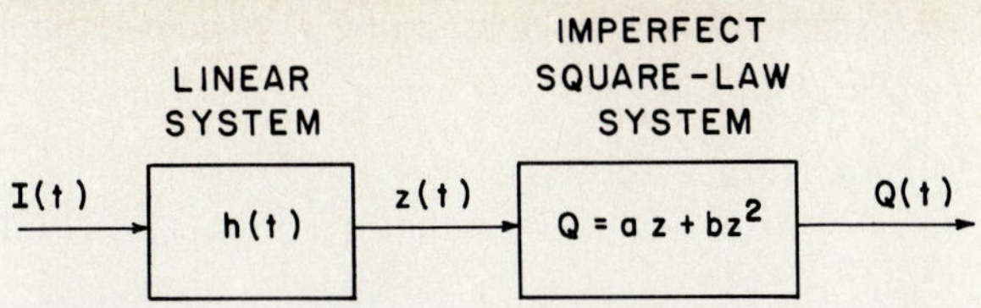

FIGURE 17.2 Example 17.1: A second-order system consisting of a linear system in series with an imperfect square-law system (after Schetzen, 1965a).

This is the same as equation (17.8) with order 2 in which

$$h_1(s_1) = ah(s_1)$$
$$h_2(s_1, s_2) = bh(s_1)h(s_2)$$

Therefore,

$$h_1(s_1) = \begin{cases} \frac{a}{k} \exp\left(\frac{-s_1}{k}\right), & s_1 \geq 0 \\ 0, & s_1 < 0 \end{cases}$$

and

$$h_2(s_1, s_2) = \begin{cases} \frac{b_2}{k^2} \exp\left[\frac{-(s_1 + s_2)}{k}\right], & s_1 \geq 0 \text{ and } s_2 \geq 0 \\ 0, & \text{otherwise} \end{cases}$$

17.1.1 Some Properties of the ER, DR, and Kernels

Amorocho (1973) discussed properties of the kernels of the functional series in equation (17.7) in general, whereas Diskin and Boneh (1972) discussed properties of the kernels of second-order systems in particular. The properties of the kernels are affected by the properties of the ER and DR, which belong to special classes of functions. Let us first discuss properties of the effective rainfall and direct runoff.

17.1.1.1 Properties of the ER

The ER has the following properties,

$$\begin{aligned} I(t) &= 0 && \text{for } t < 0 \\ I(t) &> 0 && \text{for } t = 0 \\ I(t) &\geq 0 && \text{for } t > 0 \end{aligned} \tag{17.9}$$

Note that the system is assumed to be initially relaxed. It is therefore convenient to measure the time from the beginning of the occurrence of the ER. At $t = 0$ the system is considered to be subjected to a sudden change in the function $I(t)$.

17.1.1.2 Properties of the DR

Because the watershed is initially relaxed for $t < 0$ by assumption, $Q(t)$ and its derivatives must be zero at $t = 0$. However, it is difficult to determine from streamflow records which derivative(s) will be significantly different from zero. $Q(t)$ and its derivatives are assumed to be all continuous at $t = 0$. Further,

$$\begin{aligned} Q(t) &= 0 && \text{for } t \leq 0 \\ \frac{d^iQ}{dt^i} &= 0 && \text{for } t = 0, \text{i} = 1, 2, \ldots, M - 1 \\ \frac{d^MQ}{dt^M} &> 0 && \text{for } t = 0, M \text{ some finite positive number} \end{aligned} \tag{17.10}$$

Similarly, we need to impose some conditions on $Q(t)$ at some time T_b when the watershed, having been subjected to an isolated rainfall storm, returns to its orginally relaxed position. These can be expressed as

$$\begin{aligned} I(t) &= 0 && \text{for } t = T_b \\ Q(t) &= 0 && \text{for } t = T_b \\ \frac{d^iQ}{dt^i} &= 0 && \text{for } t = T_b, i = 1, 2, \ldots, N - 1 \\ \frac{d^NQ}{dt^N} &= 0 && \text{for } t = T, N \text{ some finite positive number} \end{aligned} \tag{17.11}$$

Note that the value of N here is not necessarily equal to that of M above. Furthermore,

$$0 \leq Q(t) \leq Q_P \tag{17.12}$$

where Q_P is some upper bound.

17.1.1.3 Volume Continuity

The law of continuity requires that

$$\int_0^t Q(t)\, dt \leq \int_0^t I(t)\, dt \qquad \text{for all } t \geq 0 \tag{17.13}$$

Further, for an isolated rainfall-runoff event where $Q(t) = 0$ after T_b,

$$\int_0^{T_b} Q(t)\, dt = \int_0^{T_b} I(t)\, dt \tag{17.14}$$

17.1.1.4 Properties of Kernels

Although the kernels of the functional series are not known algebraically, certain conditions are imposed on them in order to make the functionals mathematically wieldy. These are given next. (Amorocho, 1973; Diskin and Boneh, 1972).

1. To satisfy the initial condition $Q(0) = 0$ in equation (17.10),

$$\begin{aligned} h_1(0) &= 0 \\ h_2(0, 0) &= 0 \\ h_3(0, 0, 0) &= 0 \end{aligned} \tag{17.15}$$

and so on. The derivation of these can be illustrated by considering the second-order kernel in equation

(17.8). Using Leibnitz rule for differentiating an integral, the first derivative of equation (17.8) with respect to t at $t = 0$ is

$$\frac{dQ(0)}{dt} = h_1(0)I(0) \tag{17.16a}$$

Using equation (17.9) it is seen that

$$h_1(0) = 0 \tag{17.16b}$$

Similarly, by considering the second derivative of equation (17.8) at $t = 0$,

$$\frac{dh_1(0)}{dt} = 0 \tag{17.16c}$$

$$h_2(0, 0) = 0 \tag{17.16d}$$

Additional conditions on derivatives of the kernels can be derived in an analogous manner.

2. To satisfy the condition in equation (17.11),

$$\begin{aligned} &h_1(s) \to 0 && \text{for } s \to \infty \\ &h_2(s_1, s_2) \to 0 && \text{for } s_1 \to \infty \text{ or } s_2 \to \infty \\ &h_3(s_1, s_2, s_3) \to 0 && \text{for } s_1 \to \infty \text{ or } s_2 \to \infty \\ & && \text{or } s_3 \to \infty \end{aligned} \tag{17.17}$$

and so on. This can be easily illustrated for the second-order kernel. If equation (17.11) is inserted in the first few derivatives of equation (17.9), then, in a manner analogous to equation (17.16), for the first derivative equal to 0,

$$h_1(s) = 0 \quad \text{for } s \geq T_b \tag{17.18a}$$

and for the second derivative equal to zero,

$$\frac{dh_1}{ds} = 0 \quad \text{for } s \geq T_b \tag{17.18b}$$

and

$$h_2(s_1, s_2) = 0 \quad \text{for } s_1 \geq T_b \text{ or } s_2 \geq T_b \tag{17.18c}$$

3. To satisfy the condition in equation (17.12), the kernels must be bounded:

$$|h_1(s)| < B_1 \quad \text{for all } s \tag{17.19a}$$

$$|h_2(s_1, s_2)| < B_2 \quad \text{for all } s_1 \text{ and } s_2 \tag{17.19b}$$

$$|h_3(s_1, s_2, s_3)| < B_3 \quad \text{for all } s_1,\ s_2 \text{ and } s_3 \tag{17.19c}$$

and so on. The quantities B_1, B_2, and B_3 are some finite constants and are related to the constant Q_p and the upper bound of the depth of the effective rainfall.

4. All kernels of any arbitrary order n can be defined as the sums of products of linear kernel functions $h_{ij}(s_j)$,

$$h_n(s_1, s_2, \ldots, s_n) = \sum_{i=1}^{r} \sum_{j=1}^{n} h_{ij}(s_j) \tag{17.20a}$$

where r is a finite integer.

5. A kernel is said to be *separable* if it can be expressed as

$$h_n(s_1, s_2, \ldots, s_n) = k_1(s_1)k_2(s_2) \cdots k_n(s_n) \tag{17.20b}$$

where $k_i(s_i)$ is some function of s_i. In general, a kernel is not separable. To this end, Wiener (1942) suggested expanding the nonseparable kernels in terms of orthogonal functions that are realizable as kernels of linear systems. For example, Schetzen (1965b) used Laguerre functions $L_i(t)$ for such an expansion as

$$h_1(s_1) = \sum_{i_1=0}^{\infty} c_{i_1} L_{i_1}(s_1)$$

$$h_2(s_1, s_2) = \sum_{i_1=0}^{\infty} \sum_{i_2=0}^{\infty} c_{i_1 i_2} L_{i_1}(s_1)\, L_{i_2}(s_2)$$

$$\vdots$$

$$h_n(s_1, s_2, \ldots, s_n) = \sum_{i=0}^{\infty} \cdots \sum_{i_n=0}^{\infty} c_{i_1, \ldots, i_n} L_{i_1}(s_1)\, L_{i_2}(s_2) \cdots L_{i_n}(s_n)$$

in which

$$c_{i_1, i_2, \ldots, i_n} = \int_0^{\infty} \cdots \int_0^{\infty} h_n(s_1, s_2, \ldots, s_n)\, L_{i_1}(s_1) L_{i_2}(s_2) \cdots L_{i_n}(s_n)\, ds_1\, ds_2 \cdots ds_n \tag{17.20c}$$

Papazafiriou (1976) used a form of orthogonal polynomials analyzed by Forsythe (1957).

6. The kernels of all orders except the first are symmetric with respect to the arguments s_j,

$$\begin{aligned} h_n(s_1, s_2, \ldots, s_n) &= h_n(s_j, \ldots, s_1, \ldots, s_n) \\ &= h_n(s_n, \ldots, s_j, \ldots, s_1) \end{aligned} \tag{17.21}$$

This property of symmetricity implies that a kernel remains independent of the order of its argument s_j.

If a kernel is not symmetric with respect to its arguments, it can be replaced by a symmetric kernel without changing the value of the functional. Let $h_n^*(s_1, s_2, \ldots, s_n)$ be an asymmetric kernel. The symmetric kernel $h_n(s_1, s_2, \ldots, s_n)$ can then be obtained from

$$h_n(s_1, s_2, \ldots, s_n) = \sum_{j=1}^{n_*} \frac{h_n^*(s_1, s_2, \ldots, s_n)^{(j)}}{n_*}$$

where $n_* = n!$, interpreted as equal to the number of permutations of n values $(s_1, s_2, \ldots, s_n)$, takes

n at a time and $h_n^*(s_1, s_2, \ldots, s_n)^{(j)}$ the jth permutation of the arguments s_j of $h_n^*(s_1, s_2, \ldots, s_n)$.

To illustrate, let us consider the second-order kernel $h_2(s_1, s_2)$. This can be taken, without any loss of generality, to be symmetrical with respect to the principal diagonal $s_1 = s_2$, that is,

$$h_2(s_1, s_2) = h_2(s_2, s_1) \qquad \text{for all } s_1 \text{ and } s_2$$

If $h_2^*(s_1, s_2)$ is not symmetric, then it can be replaced by the symmetric counterpart

$$h_2(s_1, s_2) = \frac{h_2^*(s_1, s_2) + h_2^*(s_2, s_1)}{2}$$

Thus, $h_2(s_1, s_2)$ can be used in place of $h_2^*(s_1, s_2)$ without altering the value of the functional.

7. The kernels are subject to

$$h_n(s_1, s_2, \ldots, s_n) = 0 \quad \text{for all } s_1, s_2, \ldots, s_n < 0 \tag{17.22}$$

This is known as the Volterra condition and is essential to render the system as nonanticipatory.

8. To satisfy the condition in equation (17.14), the kernels of order higher than 1 may assume negitive as well as positive values. To illustrate, let us consider the case $n = 2$. Integrating equation (17.8), with appropriate changes of variables,

$$\int_0^\infty Q\,dt = \int_0^\infty I\,dt \int_0^\infty h(s_1)\,ds_1 + \int_0^\infty \int_0^\infty h_2(s_1, s_2)\,\rho(s_1 - s_2)\,ds_1\,ds_2 \tag{17.23}$$

where $\rho(s_1 - s_2)$ is the autocorrelation function of $I(t)$,

$$\rho(w) = \int_{-\infty}^\infty I(t)\,I(t + w)\,dt \tag{17.24}$$

The only way equation (17.14) can be satisfied is if

$$\int_0^\infty h(s)\,ds = 1.0 \tag{17.25a}$$

and

$$\int_0^\infty \int_0^\infty h_2(s_1, s_2)\rho(s_1 - s_2)\,ds_1\,ds_2 = 0 \tag{17.25b}$$

Equation (17.25b) will be satisfied for any arbitrary $I(t)$ if and only if

$$\int_0^\infty h_2(s_1, s_1 + w)\,ds_1 = 0 \qquad \text{for all } w \geq 0 \tag{17.25c}$$

This condition states that the kernel $h_2(s_1, s_2)$ is such that its line integrals along any straight line, parallel to the principal diagonal $s_1 = s_2$, always vanish. This means that $h_2(s_1, s_2)$ takes negative values in some portion of the (s_1, s_2) plane and positive values elsewhere. Similarly, we can derive required conditions for the kernels of higher order.

EXAMPLE 17.2

Prove whether or not the second-order kernel $h(t; s_1, s_2)$ can be a separable function. We can write $h(s_1, s_2)$ as a product of two functions—for instance, f_1 and f_2—each a function of one variable,

$$h(s_1, s_2) = f_1(s_1)f_2(s_2)$$

By virtue of symmetry the functions, $f_1(s_1)$ and $f_2(s_2)$ can be considered to be identical, that is,

$$h(s_1, s_2) = f_1(s)f_1(s)$$

Inserting in the expression for $g_2(t)$,

$$g_2(t) = \int_0^t \int_0^t f_1(s)f_1(s)I(t - s_1)\,I(t - s_2)\,ds_1\,ds_2 = \left[\int_0^t f_1(s)I(t - s_1)\,ds_1\right]^2$$

Obviously $g_2(t) \geq 0$ for all t and $I(t)$. This results in a contradiction with the requirement of equation (17.25c). Indeed, equation (17.25c) cannot be satisfied if $h(s_1, s_2)$ is considered separable. Thus we conclude that $h(s_1, s_2)$ cannot be a separable function.

EXAMPLE 17.3

Diskin and Boneh (1972) proposed the following second-order kernel:

$$h(s_1, s_2) = \begin{cases} a\left[\dfrac{s_1 + s_2}{b} - \dfrac{2s_1s_2}{b^2}\right]\exp\left[-\dfrac{s_1 + s_2}{b}\right], & s_1 \geq 0, s_2 \geq 0 \\ 0, & s_1 < 0 \text{ or } s_2 < 0 \end{cases}$$

where a is a scale parameter affecting the magnitude and sign of the function at any point and b is a time constant affecting the time scale and the recession rate of the function for large values of time.

Graph and compute to show that this function satisfies all the conditions imposed on the second-order kernel. Fig. 17.3 shows a contour diagram for the surface $h(s_1, s_2)$.

It can be seen that $h(s_1, s_2)$ is zero at the origin and is also zero if either s_1 or s_2 is too large because of the exponential function containing them. Also, the integral of $h(s_1, s_2)$ over the positive quadrant is zero. The intersection of the surface represented by $h(s_1, s_2)$ with a perpendicular plane along the principal diagonal $s_1 = s_2$ is a curve given by

$$h(t, t) = \frac{2at}{b}\left(1 - \frac{t}{b}\right)\exp\left(\frac{-2t}{b}\right)$$

The curve obtained by the intersection of the surface with a perpendicular plane on a line $s_2 = s_1 - t_1$ parallel to the principal diagonal is given by

$$h(s_1, s_1 - t_1) = a\left[\frac{2s_1 - t_1}{b} - \frac{2s_1(s_1 - t_1)}{b^2}\right]\exp\left[\frac{-(2s_1 - t_1)}{b}\right]$$

The curves of intersection are shown in Fig. 17.4. Thus, the line integrals along any of the diagonal lines $s_2 = s_1 + c$ as just given are zero.

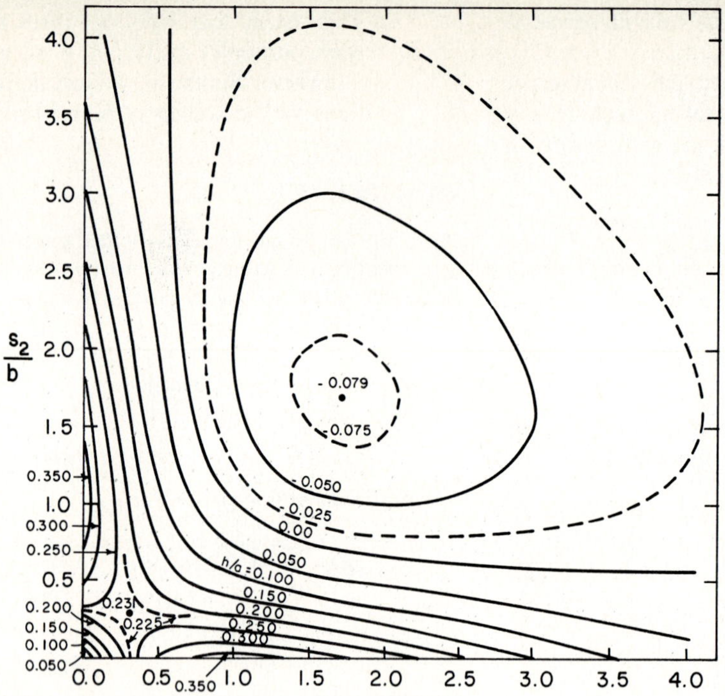

FIGURE 17.3 Example 17.3: A contour diagram representing a second-order kernel function (after Diskin and Boneh, 1972).

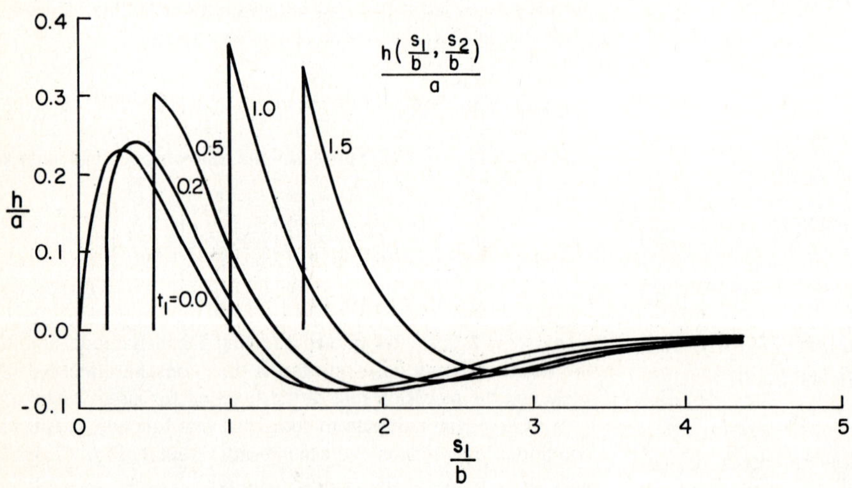

FIGURE 17.4 Example 17.3: Lines of intersection of second-order kernel with perpendicular planes parallel to principal diagonal (after Diskin and Boneh, 1972).

17.1.1.5 Dimensions of Kernels

It is easy to obtain the dimensions of the kernels h_n, $n = 1, 2, \ldots$, from equation (17.8). They are given by

$$[h_1] = \frac{[g_1]}{[I]\,[t]}$$

$$[h_2] = \frac{[g_2]}{[I]^2\,[t]^2}$$

$$[h_3] = \frac{[g_3]}{[I]^3\,[t]^n}$$

$$\vdots$$

$$[h_n] = \frac{[g_n]}{[I]^n\,[t]^n}$$

where the square brackets stand for the dimensions of the quantity enclosed. If the ER and DR are assumed to have the same dimensions, then

$$[h_1] = \frac{1}{T}$$

$$[h_2] = \frac{1}{LT}$$

$$[h_3] = \frac{1}{L^2T}$$

$$\vdots$$

$$[h_n] = \frac{1}{L^{n-1}T} \qquad (17.26)$$

17.1.1.6 System Characteristics

The ER and DR constitute a passive pair (Saeks, 1972; Boneh and Golan, 1978) because

$$\int_0^t I(s)\,ds \geq \int_0^t Q(s)\,ds \qquad \text{for all } t$$

These also form a lossless pair because

$$\int_0^t I(s)\,ds = \int_0^t Q(s)\,ds \qquad \text{for all } t \geq T$$

where T is at least equal to the time base of the ER-DR graph. This pair is conservative, since it is both passive and lossless (Diskin, Boneh and Golan, 1975). Furthermore, ER and DR make a copositive pair because they are nonegative for all times (Hall and Newman, 1963). The rainfall-runoff system is a passive, lossless, conservative or copositive system with respect to ER because ER and DR constitute a passive, lossless, conservative or copositive pair, respectively.

17.1.2 Functional Response and Kernel Properties

Greater familiarity with equation (17.8) can be developed by considering its response to certain types of $I(t)$. From this response the properties of kernels derived previously can be further elucidated. As before, we restrict our discussion to the second-order case. We follow the work of Diskin and Boneh (1972).

17.1.2.1 Response to a Step Input

Let $I(t) = Au_1(t)$ with $A > 0$ being a constant. Then

$$Q(t) = A\int_0^t h_1(s)\,ds + A^2\int_0^t\int_0^t h_2(s_1, s_2)\,ds_1\,ds_2 \tag{17.27}$$

For $t \to \infty$, $Q(t) \to A$. Obviously this will be possible if

$$\int_0^\infty h_1(s)\,ds = 1.0 \tag{17.28a}$$

$$\int_0^\infty\int_0^\infty h_2(s_1, s_2)\,ds_1\,ds_2 = 0 \tag{17.28b}$$

This first condition is identical to equation (17.25a). The second condition is a corollary of equation (17.25c).

17.1.2.2 Response to a Finite Pulse

Let $I(t) = A > 0$ for $0 \leq t \leq D$ and $I(t) = 0$ for $t < 0$ or $t > D$, where D is the duration of $I(t)$.
Then

$$Q(t) = A\int_{t_0}^{t_1} h_1(s)\,ds + A^2\int_{t_0}^{t_2}\int_{t_0}^{t_2} h_2(s_1, s_2)\,ds_1\,ds_2 \tag{17.29a}$$

The limits of integration are specified by

$$t_0 = \max\,[0, t - D] \tag{17.29b}$$

$$t_1 = \min\,[t, t_h] \tag{17.29c}$$

$$t_2 = \min\,[t, t_g] \tag{17.29d}$$

where t_h and t_g represent the times beyond which the kernels h_1 and h_2, respectively, vanish. Obviously each of these times will be smaller than or at most equal to the T_b needed for the system to relax.

The contribution of the linear part due to a nonuniform effective rainfall of duration D is determined by $h_1(s)$ in the interval $t - D \leq s \leq t$, as shown in Fig. 17.5. Similarly, the contribution of the second-order nonlinear part is determined by $h_2(s_1, s_2)$ in the square of size D, centered on the principal diagonal in the geometric plane, and defined by $t - D \leq s_1 \leq t$ and $t - D \leq s_2 \leq t$. This implies that if the longest available duration is D_m, then $h_2(s_1, s_2)$ outside the band defined by $s_1 - D_m \leq s_2 \leq s_1 + D_m$ is immaterial. If the response to a high-intensity, short-duration rainfall is of principal interest, then $h_2(s_1, s_2)$ should be accurately known on the principal diagonal and within a relatively narrow band centered around it.

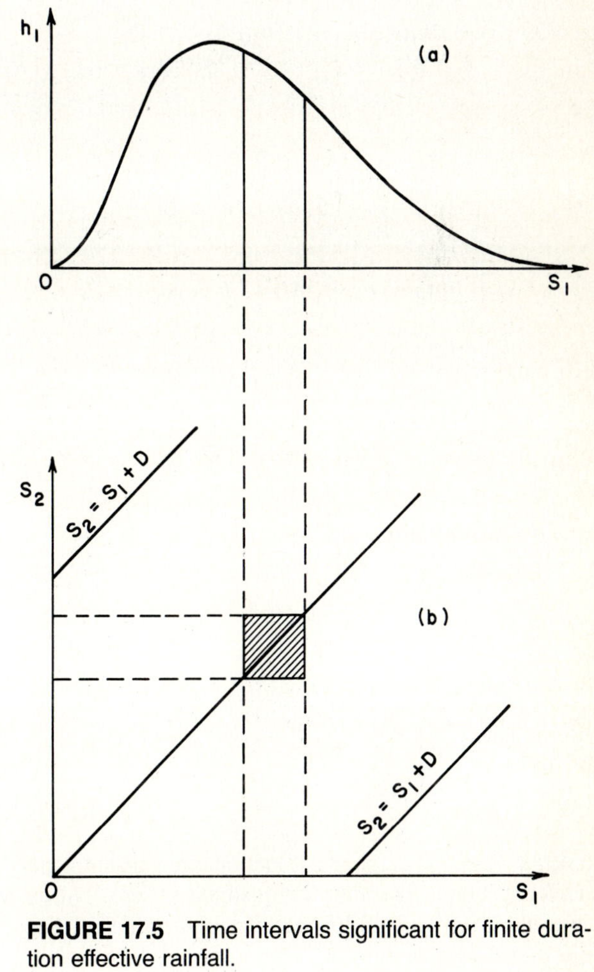

FIGURE 17.5 Time intervals significant for finite duration effective rainfall.

17.1.2.3 Response to a Finite Impulse

The response of the second-order system due to $I(t)$ represented by an isolated impulse of magnitude $A > 0$ applied at $t = 0$ is

$$Q(t) = Ah_1(t) + A^2h_2(t, t)$$

Integrating,

$$\int_0^\infty Q\,dt = A\int_0^\infty h_1(s)\,ds + A^2\int_0^\infty h_2(s, s)\,ds$$

Applying the law of continuity,

$$\int_0^\infty h_1(s)\,ds = 1.0 \quad (17.30a)$$

$$\int_0^\infty h_2(s, s)\,ds = 0.0 \quad (17.30b)$$

which is the same as before.

17.1.2.4 Response to a Pair of Impulses

Let $I(t)$ be presented by the impulses of magnitudes A and B applied at $t = 0$ and $t = t_1$, respectively, where $0 < t_1 < t_b$. Then, using the property of symmetry,

$$Q(t) = Ah_1(t) + Bh_1(t - t_1) + A^2h_2(t, t) + B^2h(t - t_1, t - t_1) + 2ABh(t, t - t_1) \quad (17.31)$$

On integrating,

$$\int_0^\infty Q(t)\,dt = (A + B)\int_0^\infty h_1(s)\,ds + (A^2 + B^2)\int_0^\infty h_2(s, s)\,ds + 2AB\int_0^\infty h_2(s, s - t_1)\,ds \quad (17.32)$$

which must equal

$$\int_0^\infty I(t)\,dt = A + B \quad (17.33)$$

To maintain continuity,

$$\int_0^\infty h_1(s)\,ds = 1.0 \quad (17.34a)$$

$$\int_0^\infty h_2(s, s)\,ds = 0.0 \quad (17.34b)$$

$$\int_0^\infty h_2(s, s - t_1)\,ds = 0.0 \quad (17.34c)$$

It follows that $h_2(s_1, s_2)$ is a function such that not only is its surface integral over the first quadrant is zero, but its line integral along any straight line parallel to the principal diagonal $s_1 = s_2$ must also be zero.

EXAMPLE 17.4

Compute the response of the second-order system due to a unit impulse at $t = 0$, where the first-order kernel is given as

$$h(s) = s\exp[-s]$$

and the second-order kernel is as in Example 17.3.

The response $Q(t)$ can be expressed as

$$Q(t) = t\exp(-t) + \frac{2at}{b}\left(1 - \frac{t}{b}\right)\exp\left(-\frac{2t}{b}\right)$$

For the case $a = b = 1$, values of $Q(t)$ are:

Time (h)	Q(t) (cm/h)	*Time* (h)	Q(t) (cm/h)	*Time* (h)	Q(t) (cm/h)
.0	.0	.5	.49	3.0	.12
.1	.24	.7	.45	4.0	.07
.2	.38	1.0	.37	5.0	.03
.3	.45	1.5	.26	6.0	.02
.4	.48	2.0	.20	7.0	.01

EXAMPLE 17.5

Compute the response of the second-order system of Example 17.4 due to a uniform pulse of D-hour duration.

The uniform pulse is $I(t) + A$, $0 \le t < D$, $I(t) = 0$, $t \ge D$, where D is duration and A is magnitude. Equation (17.29) specifies response in integral form of the system due to such a pulse and can be solved using the given kernels.

$$\int_{t_0}^{t_1} h_1(s)\,ds = \int_{t_0}^{t_1} s\exp(-s)\,ds$$
$$= -(1 + t_1)\exp(-t_1) + (1 + t_0)\exp(-t_0)$$

Using the second-order kernel,

$$\int_{t_0}^{t_2}\int_{t_0}^{t_2} h_2(s_1, s_2)\,ds_1\,ds_2 = \frac{a}{b}\int\int\left(s_1 + s_2 - \frac{2s_1s_2}{b}\right)\exp\left(-\frac{s_1 + s_2}{b}\right)ds_1\,ds_2$$
$$= \frac{a}{b}\int_{t_0}^{t_2}\exp\left(-\frac{s_2}{b}\right)\left[\int_{t_0}^{t}\left(s_1 + s_2 - \frac{2s_2}{b}s_1\right)\exp\left(-\frac{s_1}{b}\right)ds_1\,ds_2\right.$$

Let $A = \int_{t_0}^{t_2} s_1\exp(-s_1/b)\,ds_1$. Then

$$A = -(bt_2 + b^2)\exp\left(-\frac{t_2}{b}\right) + (bt_0 + b^2)\exp\left(-\frac{t_0}{b}\right)$$

Therefore,

$$\int_{t_0}^{t_2}\int_{t_0}^{t_2} h_2(s_1, s_2)\,ds_1\,ds_2 = \frac{a}{b}\int_{t_0}^{t_2}\exp\left(-\frac{s_2}{b}\right)\left[\left\{1 - \left(\frac{2s_2}{b}\right)\right\}A + s_2\left\{-b\exp\left(-\frac{t_2}{b}\right) + b\exp\left(-\frac{t_0}{b}\right)\right\}\right]ds_2$$
$$= a\left[(t_0 + b)\exp\left(-\frac{t_0}{b}\right) - (t_2 + b)\exp\right.$$

$$\left(-\frac{t_2}{b}\right)\Big]\int_{t_0}^{t_2}\left(1-\frac{2s_2}{b}\right)\exp\left(-\frac{s_2}{b}\right)$$

$$ds_2 - a\Big[\exp\left(-\frac{t_2}{b}\right)$$

$$-\exp\left(-\frac{t_0}{b}\right)\Big]\int_{t_0}^{t_2} s_2\exp\left(-\frac{s_2}{b}\right)ds_2$$

$$= a\Big[(t_0+b)\exp\left(-\frac{t_0}{b}\right)-(t_2+b)$$

$$\exp\left(-\frac{t_2}{b}\right)\Big]$$

$$\Big[2t_2\exp\left(-\frac{t_2}{b}\right)-2t_0\exp\left(-\frac{t_0}{b}\right)\Big]$$

Therefore,

$$Q(t) = A\{-\exp(t_1)(1+t_1)+\exp(-t_0)(1+t_0)\}$$
$$+A^2\Big\{a\Big[(t_0+b)\exp\left(-\frac{t_0}{b}\right)-(t_2+b)\exp\left(-\frac{t_2}{b}\right)\Big]$$
$$\cdot\Big[2t_2\exp\left(-\frac{t_2}{b}\right)-2t_0\exp\left(-\frac{t_0}{b}\right)\Big]\Big\}$$

in which $t_0 = \max[0, t - D]$, $t_1 = \min[t, t_h]$, $t_2 = \min[t, t_g]$, t_h is the time when h_1 vanishes, and t_g is the time when h_2 vanishes.

Let $A = 0.5$ cm/h, $D = 2$ h, and $a = b = 1$. Then

$$Q(t) = 0.5\{-\exp(-t_1)(1+t_1)+\exp(-t_0)(1+t_0)\}$$
$$+0.5\{[(1+t_0)\exp(-t_0)-(1+t_2)\exp(-t_2)]$$
$$\cdot[2t_2\exp(-t_2)-2t_0\exp(-t_0)]\}$$

For $Q(t)$,

$0 \le t \le 2,$	$t_0 = 0,$	$t_1 = t_2 = t$
$10 \ge t \ge 2,$	$t_0 = t - D = t - 2,$	$t_1 = t_2 = t$
$10 < t \le 15,$	$t_0 = t - 2, t_1 = t,$	$t_2 = 10$
$t_g = t = 10$ h,	$h_2(t, t) = -3.71E-07$	
$t_h = t = 15$ h,	$h_1(t) = 4.6E - 0.6$	

Values of $Q(t)$ computed are:

Time (h)	Q(t) (cm/h)	*Time* (h)	Q(t) (cm/h)	*Time* (h)	Q(t) (cm/h)
.0	.0	1.2	.23	4.0	.13
.2	.01	1.6	.31	5.0	.07
.4	.04	2.0	.38	6.0	.04
.6	.08	2.4	.30	7.0	.02
.8	.13	2.8	.23	8.0	.01
1.0	.18	3.0	.21	9.0	.00

17.1.3 Requirements of Q(t) ≥ 0

Here, again, we consider the second-order case. $I(t)$ is, of course, always nonnegative. Physically, $Q(t)$ must also be nonnegative. However, to obtain nonnegative $Q(t)$ from the second-order nonlinear system, some constraints must be imposed on the volume of $I(t)$. This can perhaps be best illustrated by an example. Consider the case where the effective rainfall is represented by a finite impulse of magnitude A as considered in the preceding section. Then we can write

$$Q(t) = A[h_1(t) + Ah_2(t, t)] \tag{17.35}$$

For satisfaction of equation (17.28b), some of the ordinates of $h_2(t_1, t_2)$ must be negative and some, of course, must be positive. In particular, equation (17.30b) requires that some of the ordinates must be negative on the principal diagonal. This means that there must always be at least one time interval, say T_1 and T_2, in which $h_2(t, t)$ is negative. Further, this time interval will have some overlap with the duration D of specified effective rainfall. Therefore, from equation (17.35), it is clear that $Q(t)$ will be negative for those times when the quantity within the bracket is negative. The occurrence of negative ordinates can be avoided in this case by specifying the upper bound of A as

$$A^* = \min_{t\in\theta}\left[\frac{h_1(t)}{|h_2(t,t)|}\right] \tag{17.36}$$

where θ represents the interval $(T_2 - T_1)$ over which $h_2(t, t) < 0$.

The concept just illustrated can be extended to an arbitrarily defined $I(t)$ (Boneh, 1971). In other words, for known $h_1(t)$ and $h_2(t_1, t_2)$, the minimum value of the upper bound, say I^*, can be determined for $I(t)$ to insure nonnegative $Q(t)$, that is,

$$0 \le I(t) \le I^* \qquad \text{for all } t \ge 0 \tag{17.37}$$

If $I(t)$ is represented by a finite pulse A of duration $D(0 < D < \infty)$ as considered previously then the upper bound can be obtained from equation (17.29a) as

$$I^* = \min_{0\le D}\min_{t\in\theta}\left[\frac{\int_{t_0}^{t_1} h_1(s)\,ds}{\left|\int_{t_0}^{t_2}\int_{t_0}^{t_2} h_2(s_1, s_2)\,ds_1\,ds_2\right|}\right] \tag{17.38}$$

where t_0, t_1, and t_2 are defined by equation (17.29) and θ represents the time interval over which the denominator is negative. In like manner, results on the upper bound can be obtained for arbitrarily defined $I(t)$.

It may be noted that the emergence of negative $Q(t)$ for the second-order system leads to two interpretations. First, the ER must have an upper bound to insure positivity of $Q(t)$. This bound will, in general, vary with the order of the VIS model. Placing an upper bound on the magnitude of the ER is not necessarily a contradiction, since there will probably be an upper limit to the intensity and depth of short-duration rainfalls that could be obtained in nature. Second, if the preceding interpretation is unacceptable, then the second-order VIS model is unsuitable for modeling the ER-DR relationship where the continuity and nonnegative requirements must be satisfied.

17.1.4 Estimation of Kernels

Several methods of estimation of nonlinear kernels are available and will be discussed here. First, we briefly discuss the method amenable to laboratory experimentation (Amorocho and Orlob, 1961). This method also helps augment understanding of nonlinear analysis.

17.1.4.1 Measurement Using Step Functions

Let $I(t)$ be multiplied by a scale factor a. Then from equation (17.8),

$$g_n(t) = H_n[aI(t)] = \int\int \cdots \int h_n(s_1, s_2, \ldots, s_n)\, a^n \prod_{i=1}^{n} I(t - s_i)\, ds_i$$
$$= a^n H_n[I] \tag{17.39}$$

Therefore, if $g(t)$ is the response of equation (17.8) for $I(t)$ then for $aI(t)$ the response $g_a(t)$ follows

$$g_a(t) = aH_1[I] + a^2H_2[I] + a^3H_3[I] + \cdots + a^nH_n[I] + \cdots$$
$$= \sum_{i=1}^{\infty} a^iH_i[I] \tag{17.40}$$

If $I(t)$ is constant, say at $t = t_1$, $I(t) = I(t_1)$ and $g_a(t)$ becomes a function of a only, that is,

$$g_a(t_1) = g(a) \tag{17.41}$$

Expanding $g(a)$ in the Maclaurin series,

$$g(a) = g(0) + ag'(0) + \frac{a^2}{2!}g''(0) + \cdots + \frac{a^n}{n!}g^{(n)}(0) + \frac{a^{n+1}}{(n+1)!}g^{(n+1)}(\theta) \tag{17.42}$$

where the last term is the error term containing some number θ between 0 and a. It is thus seen that a correspondence exists between the functional series representation of the discharge and equation (17.42) for $t = t_1$. This analogy can be utilized in the employment of step functions for the measurement of a system.

Let $I(t) = Au_1(t)$. Using equation (17.40)

$$g(t) = A\int_0^t h_1(s)\, ds + A^2\int_0^t\int_0^t h_2(s_1, s_2)\, ds_1\, ds_1 + \cdots + A^n\int_0^t\int_0^t \cdots \int_0^t h_n(s_1, s_2, \ldots, s_n)\, ds_1\, ds_2 \cdots ds_n$$

For $t = t_1$, a particular instant,

$$g(t_1) = \sum_{i=1}^{\infty} A^i m_i(t_1) \tag{17.43}$$

where

$$m_1(t_1) = \int_0^{t_1} h_1(s)\, ds$$

$$m_2(t_1) = \int_0^{t_1}\int_0^{t_1} h_2(s_1, s_2)\, ds_1\, ds_2$$

$$\vdots$$

$$m_n(t_1) = \int_0^{t_1}\int_0^{t_1} \cdots \int_0^{t_1} h_n(s_1, s_s, \ldots, s_n)\, ds_1\, ds_2 \cdots ds_n$$

Equation (17.43) is equivalent to a Taylor series in A. This means that if $g(t_1)$ is determined experimentally or otherwise for different values of A and if its Taylor series expansion can be effected, then the coefficients of this expansion corresponds to the function $g_n(t_1)$ of equation (17.43). Thus $g_n(t)$ for different times $t_1, t_2, \ldots, t_n$ can be obtained. From equations (17.42) and (17.43), we can see that the IUH of H_1 is the derivative of $g_1(t)$,

$$h_1(t) = \frac{d}{dt}m_1(t) \tag{17.44}$$

To determine $h_2(t_1, t_2)$ in this manner is somewhat more complicated. Consider $I(t) = y(t) + z(t)$, where $y(t) = Au_1(t)$ and $z(t) = Au_1(t + T)$ with T being some positive number. Then

$$g_2(t) = H_2[y + z] = H_2(y^2 + 2yz + z^2)$$
$$= m_2(t) + 2\int_0^t\int_0^{t+T} h_2(s_1, s_2)ds_1ds_2 + m_2(t + T)$$

We can determine $m_2(t)$ from the Taylor series expansion of $g(t)$. Therefore,

$$f_2(t, t + T) = \int_0^t\int_0^{t+T} h_2(s_1, s_2)\, ds_1\, ds_2$$
$$= \frac{g_2(t) - m_2(t) - m_2(t + T)}{2} \tag{17.45}$$

Therefore,

$$h_2(t_1, t_2) = \frac{\partial^2}{\partial t_1 \partial t_2} f_2(t_1, t_2)$$

This procedure can be repeated for higher-order kernels, with increasing complexity. Let us consider an example to illustrate this procedure.

EXAMPLE 17.6

Let $g(t)$ be represented by

$$g(t) = a(t)\,[e^{A\alpha(t)} - 1]$$

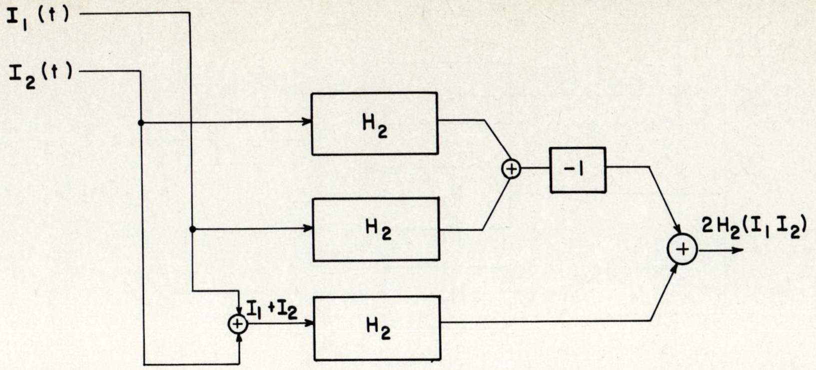

FIGURE 17.6 A second-order system with only a second-order kernel for the output $2H_2(I_1 + I_2)$.

where α is some function of time, $a(t)$ can be determined graphically or by trials, and $g(t)$ is instantaneous response at t. Expanding it,

$$g(t) = a(t)[\alpha(t)A + \frac{\alpha^2(t)}{2!}A^2 + \frac{\alpha^3(t)}{3!}A^3 + \cdots] \quad (17.46)$$

Equation (17.46) is equivalent to equation (17.43). Therefore,

$$m_1(t) = a(t)\,\alpha(t)$$

$$m_2(t) = a(t)\frac{\alpha^2(t)}{2!}$$

$$\vdots$$

$$m_n(t) = a(t)\frac{\alpha^n(t)}{n!} = \frac{m_1^n(t)}{n!}a^{n-1}(t)$$

Therefore, $m_i(t)$, $i = 1, 2, \ldots, n$, can be determined, as can the kernels.

17.1.4.2 Measurement of Volterra Kernels

Schetzen (1965a) developed a method of measuring the Volterra kernels of a finite-order nonlinear system. The kernels are obtained individually as a multidimensional impulse response. We discuss the method for determining the kernels of second-order systems, which can be extended to higher-order systems.

The Volterra representation of a second-order system is

$$Q(t) = N_2[I(t)] = H_1[I(t)] + H_2[I(t)]$$

in which N_2 is used to denote the operator of the second-order system. Let us consider the special case of this system composed of only a second-order kernel, that is, $Q(t) = N_2[I(t)] = H_2[I(t)] = g_2(t)$. George (1959) has presented a method for isolating such a second-order kernel by observing that

$$2H_2(I_1I_2) = H_2[I_1 + I_2] - H_2[I_1] - H_2[I_2] \quad (17.47)$$

in which the argument t is dropped for convenience, and

$$H_2(I_1I_2) = \int_0^\infty\int_0^\infty h_2(t; s_1, s_2)\,x_1(t - s_1)\,x_2(t - s_2)\,ds_1\,ds_2$$

Equation (17.47) can be verified by direct substitution in equation (17.6b) and assuming $h_2(s_1, s_2)$ to be symmetrical. Fig. 17.6 shows a system with $Q(t) = 2H_2(I_1I_2)$.

Consider $I(t)$ as a unit impulse $\delta(t - T)$ occurring at $t = T$. Also, $I_1(t) = \delta(t - T_1)$ and $I_2(t) = \delta(t - T_2)$. Then

$$g_2(t) = 2H_2(I_1I_2) = 2h_2(t - T_1, t - T_2) \quad (17.48)$$

which is the second-order impulse of the kernel $h_2(s_1, s_2)$ along a 45° line in the s_1s_2 plane, as shown in Fig. 17.7.

Since the input is applied at t equal to the smaller of T_1 or T_2, there cannot be any output before then. Also, the system is initially at rest, $g_2(t) = 0$ for t less than the smaller of T_1 or T_2. It is therefore concluded from equation (17.48) that $h_2(s_1, s_2) = 0$ for s_1 or s_2 less than 0. This implies that $h_2(s_1, s_2)$ can be nonzero only in the first quadrant of the s_1s_2 plane.

If the second-order system also contains a first-order kernel, then the output will still be $2H_2(I_1, I_2)$ because

$$H_1[I_1 + I_2] - H_1[I_1] - H_1[I_2] = 0$$

from equation (17.8). Fig. 17.8 shows the second-order system containing both first-order and second-order kernels. The first-order kernel can be obtained as follows. From equation (17.6b) the unit impulse response (UIR) of the second-order system is $h_1(t) + h_2(t, t)$. Thus, $h_1(s_1)$ can be determined from this UIR once $h_2(s_1, s_2)$ along the line $s_1 = s_2$ is known.

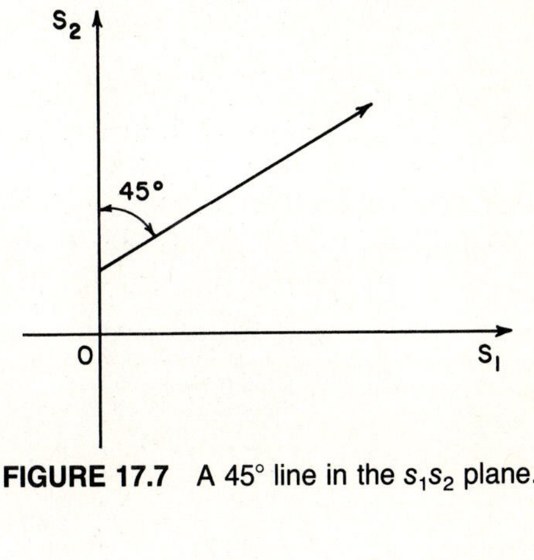

FIGURE 17.7 A 45° line in the s_1s_2 plane.

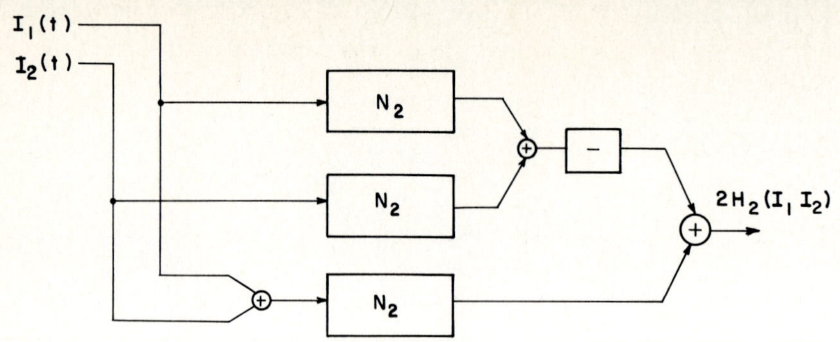

FIGURE 17.8 A general second-order system for the output $2H_2(I_1I_2)$.

EXAMPLE 17.7

Compute $g_2(t)$ for the Example 17.1. Since $g_2(t) = 2h_2(t - T_1; t - T_2)$,

$$g_2(t) = \frac{b}{k^2}\exp\left[-\left(\frac{t-T_1}{k}+\frac{t-T_2}{k}\right)\right]$$

$$= \frac{b}{k^2}\exp\left(\frac{T_1+T_2}{k}\right)\exp\left(\frac{-2t}{k}\right), \quad \text{for } T \geq \text{larger of } T_1 \text{ or } T_2$$

$$g_2(t) = 0, \quad \text{otherwise}$$

17.1.4.2 Discrete-Time Representation

The methods of estimation of kernels to be discussed in the ensuing sections have been developed using discrete-time representation. Even hydrologic data are normally given in discrete-time form rather than as continuous functions. Therefore, we must express equation (17.8) in discrete-time form,

$$Q(t) = \sum_{s=0}^{M} h_1(s)I(t-s) + \sum_{s_1=0}^{M}\sum_{s_2=0}^{M} h_2(s_1, s_2)I(t-s_1)I(t-s_2) \cdots + \sum_{s_1=0}^{M}\sum_{s_2=0}^{M}\cdots\sum_{s_r=0}^{M} h_r(s_1, s_2, \ldots, s_r) I(t-s_1)I(t-s_2)\cdots I(t-s_r) \tag{17.49}$$

where it is understood that t is a discrete-time variable defined by

$$t = \{0, \pm 1, \pm 2, \pm 3, \ldots\}$$

Similarly,

$$s = \{s_1, s_2, \ldots, s_n = 0, \pm 1, \pm 2, \pm 3, \ldots\}$$

Equation (17.49) can be written as

$$Q(t) = \sum_{n=0}^{r} Q_n(t) \tag{17.50}$$

where

$$Q_n(t) = \sum_{s_1=0}^{M}\cdots\sum_{s_n=0}^{M} h_n(s_1, s_2, \ldots, s_n) I(t-s_1)I(t-s_2)\cdots I(t-s_n)$$

For a given sequence of N values of $I(t)$ and $Q(t)$, equation (17.49) can be written in matrix notation as

$$[Q] = [I]\,[h] \tag{17.51}$$

where $[Q]$ is an $(N \times 1)$ vector of observed discharge values; $[h]$ is an $(R \times 1)$ vector of the R unknown ordinates of the kernel functions $h_1(s)$, $h_2(s_1, s_2)$, . . . , $h_4(s_1, s_2, \ldots, s_4)$, $s_1, s_2, \ldots, s_r = 0, 1, 2, \ldots, M$; and $[I]$ is an $(N \times R)$ matrix of $I(t-s)$ and its products $I(t-s_1)$, $I(t-s_2)$, . . . , $I(t-s_r)$; $N \geq R$. The matrix $[I]$ and the vectors $[Q]$ and $[h]$ can be expressed (for $n = 2$) as follows:

$$[I] = (a_1, a_2, \ldots, a_i, \ldots, a_N)^T$$

$$a_i = [I(i), I(i-1), \ldots, I(i-M), \ldots, I^2(i), \ldots, I(i)I(i-1), \ldots, I(i-M)I(i-M)]$$

$$[Q] = [Q(1), Q(2), \ldots, Q(N)]^T$$

$$[h] = [h_1(0), h_1(1), \ldots, h_1(M), h_2(0, 1), h_2(0, 2), \ldots, h_2(M, M)]^T$$

Equation (17.51) is linear in $[h]$; therefore, for $N = R$,

$$[h] = [I]^{-1}[Q] \tag{17.52}$$

If $N > R$, equation (17.51) is overdetermined and its solution is inconsistent. In addition, further difficulties may arise. The matrix $[I]$ can be very large for large rainfall-runoff values. Therefore, its inversion may require large computer storage. If some of the elements of $[I]$ are zero, its inversion may not be accurate. These difficulties can be circumvented by choosing a method that minimizes the error between the observed Q and the Q computed by equation (17.52). The least squares method is an example of such a method. These difficulties prevent the matrix inversion method from being a viable tool for estimation of kernels.

17.1.4.3 The Brandstetter-Amorocho (BA) Method

Brandstetter and Amorocho (1970) and Amorocho and Brandstetter (1971) expanded the kernel functions in orthogonal series by using Meixner polynomials. These polynomials are discussed in Chapter 11. It may suffice to recall that the Meixner functions $P_i(t)$ can be generated by using the recurrence relationship

$$2(i+1)P_{i+1}(t) = \sqrt{2}(3i + 1 - t)P_i(t) - 2iP_{i-1}(t) \tag{17.53}$$

The Meixner functions are orthogonal in the interval $0 \leq t < \infty$. They are symmetric, $P_i(t) = P_t(i)$, and are equal to zero for even values of i. Further details on these functions and their polynomials in the estimation of kernels can be found in Brandstetter and Amorocho (1970) and Amorocho (1973).

Orthogonal expansion of first-order kernel: The first-order kernel $h_1(s)$ can be approximated by

$$h_1(s) = \sum_{i=0}^{M_1} \alpha_i P_i(s) \qquad (17.54)$$

where M_1 is the number of terms in the truncated series, $M_1 \leq M$, and α_i are the coefficients to be determined. Substituting equation (17.54) into the first-order response in equation (17.49),

$$Q_1(t) = \sum_{s=0}^{M} \sum_{i=0}^{M_1} \alpha_i P_i(s)\, I(t-s) \qquad (17.55)$$

Let

$$a_i(t) = \sum_{s=0}^{M} P_i(s)\, I(t-s) \qquad (17.56)$$

Then

$$Q_1(t) = \sum_{i=0}^{M_1} \alpha_i a_i(t) \qquad (17.57)$$

Therefore, the linear response $Q_1(t)$, the contribution to $Q(t)$ by the first-order term, can be computed by knowing the coefficients α_i and the functions $P_i(t)$. This procedure has been discussed by Dooge (1973) and by Singh and McCann (1983) for linear system identification. Note that equation (17.57) is linear in α_i, so these coefficients can also be computed by using the least squares procedure.

Orthogonal expansion of the second-order kernel: The second-order kernel $h_2(s_1, s_2)$ can be written as

$$h_2(s_1, s_2) = \sum_{k=1}^{M_2^*} h_{k1}(s_1)\, h_{k2}(s_2) \qquad (17.58)$$

where M_2^* is some finite integer less than M and $h_{k1}(s_1)$ and $h_{k2}(s_2)$ are linear kernel functions, and can be expanded as in equation (17.54),

$$h_{k1}(s_1) = \sum_{i=1}^{M_2} \beta_{ki} P_i(s_1), \qquad M_2 \leq M \qquad (17.59)$$

$$h_{k2}(s_2) = \sum_{j=0}^{M_2} \gamma_{kj} P_j(s_2), \qquad M_2 \leq M \qquad (17.60)$$

Therefore,

$$h_2(s_1, s_2) = \sum_{k=1}^{M_2^*} \left[\sum_{i=0}^{M_2} \beta_{ki} P_i(s_1)\right] \left[\sum_{j=0}^{M_2} \gamma_{jk} P_j(s_2)\right] \qquad (17.61)$$

In equation (17.61) β_{ki} and γ_{kj} are orthogonal coefficients and $P_i(s)$ are the Meixner functions. Let

$$\alpha_{ij} = \sum_{r=1}^{M_2^*} \beta_{ri} \gamma_{rj} \qquad (17.62)$$

Then

$$h_2(s_1, s_2) = \sum_{i=0}^{M_2} \sum_{j=0}^{M_2} \alpha_{ij} P_i(s_1)\, P_j(s_2) \qquad (17.63)$$

Substituting equation (17.63) into the second-order term in equation (17.49),

$$Q_2(t) = \sum_{s_1=0}^{M} \sum_{s_2=0}^{M} \left[\sum_{i=0}^{M_2} \sum_{j=0}^{M_2} \alpha_{ij} P_i(s_1) P_j(s_2)\right] I(t-s_1) I(t-s_2) \qquad (17.64)$$

Interchanging the order of summation and separating the sums with respect to the arguments s_1 and s_2,

$$Q_2(t) = \sum_{i=0}^{M_2} \sum_{j=0}^{M_2} \alpha_{ij} \left[\sum_{s_1=0}^{m} p_i(s_1) I(t-s_1)\right] \left[\sum_{s_2=0}^{M} p_j(s_2) I(t-s_2)\right] \qquad (17.65)$$

The terms inside the brackets are analogous to equations (17.56). Likewise,

$$a_i(t) = \sum_{s_1=0}^{M} P_i(t-s) I(t-s);$$

$$a_j(t) = \sum_{s_2=0}^{M} P_j(t-s) I(t-s) \qquad (17.66)$$

Hence,

$$Q_2(t) = \sum_{i=0}^{M_2} \sum_{j=0}^{M_2} \alpha_{ij} a_i(t) a_j(t) \qquad (17.67)$$

Separating the diagonal terms from the double summation,

$$Q_2(t) = \sum_{j=0}^{M_2} \alpha_{ji} a_j^2(t) + \sum_{i=1}^{M_2} \sum_{j=1}^{i-1} (\alpha_{ij} + \alpha_{ji}) a_i(t) a_j(t) \qquad (17.68)$$

Recalling that the second-order kernel is symmetric, $h_2(s_1, s_2) = h_2(s_2, s_1)$. Brandstetter and Amorocho (1970) have shown that the coefficients of the expansion of a symmetric kernel are also symmetric, that is, $\alpha_{ij} = \alpha_{ji}$. This means that

$$\alpha_{ij} + \alpha_{ji} = 2\alpha_{ij} = 2\alpha_{ji} \qquad (17.69)$$

Therefore, equation (17.68) becomes

$$Q_2(t) = \sum_{j=0}^{M_2} \alpha_{ij} a_j^2(t) + 2 \sum_{i=1}^{M_2} \sum_{j=0}^{i-1} \alpha_{ij} a_i(t)\, a_j(t) \qquad (17.70)$$

This is linear in the coefficients α_{ij}. Therefore, linear regression analysis can be used to compute them.

Orthogonal expansion of higher-order kernels: This method of derivation can be extended to higher-order kernels, although the manipulations become increasingly complex. For example, for the third-order kernel,

$$h_3(s_1, s_2, s_3) = \sum_{i=0}^{M_3} \sum_{j=0}^{M_3} \sum_{k=0}^{M_3} \alpha_{ijk} P_i(s_1)P_j(s_2)P_k(s_3) \tag{17.71}$$

and

$$Q_3(t) = \sum_{k=0}^{M_3} \alpha_{kkk} a_k^3(t) + 3 \sum_{j=1}^{M_3} \sum_{k=0}^{j-1} \alpha_{jkk} a_j(t) a_k^2(t) + 6 \sum_{i=2}^{M_3} \sum_{j=1}^{i-1} \sum_{k=0}^{j-1} \alpha_{ijk} a_i(t) a_j(t) a_k(t) \tag{17.72}$$

Likewise, for the nth-order kernel,

$$h_n(s_1, s_2, \ldots, s_n) = \sum_{m_1=0}^{M_n} \sum_{m_2=0}^{M_n} \cdots \sum_{m_n=0}^{M_n} \alpha_{m_1 \ldots m_n} P_{m_1}(t) \cdots P_{m_n}(t) \tag{17.73}$$

and

$$Q_n(t) = \sum_{m_1=0}^{M_n} \sum_{m_2=0}^{M_n} \cdots \sum_{m_n=0}^{M_n} \alpha_{m_1 \ldots m_n} a_{m_1}(t) \cdots a_{m_n}(t) \tag{17.74}$$

where M_n is a finite integer. In a manner similar to equation (17.56),

$$a_{m_n}(t) = \sum_{s=0}^{M} P_{m_n}(s) I(t-s) \tag{17.75}$$

Using equation (17.75) in equation (17.74), we can obtain, after some manipulation, $Q_n(t)$ in terms of $\alpha_{m_1 \ldots m_n}$ and $a_{m_n}(t)$.

It is now possible to express $Q(t)$ of equation (17.49) in terms of the coefficients α and a,

$$Q(t) = \sum_{i=1}^{M_1} \alpha_i a_i(t) + \sum_{j=0}^{M_1} \alpha_{jj} a_j^2(t) + 2 \sum_{i=1}^{M_2} \sum_{j=0}^{i-1} \alpha_{ij} a_i(t) a_j(t) + \cdots \tag{17.76}$$

Equation (17.76) can be written as

$$Q(t) = B_0 A_0(t) + B_1 A_1(t) = B_2 A_2(t) + \cdots \tag{17.77}$$

or, in matrix form,

$$[Q] = [A]\,[B] \tag{17.78}$$

where $A_i(t)$, $i = 1, 2, \ldots, R$, represents the ordered set of functions formed by $a(t)$ and their powers and products; B_i, $i = 1, 2, \ldots, R$, represents the ordered set of coefficients α; and R is the total number of coefficients in equation (17.77).

Equation (17.78) can be compared with equation (17.51); it is evident that the former is a transformed version of the latter. The kernel functions of all orders up to r can then be determined from their expansions in terms of the coefficients α, which are obtained from the ordered set B_i.

The total number of coefficients α that need to be evaluated for the nth order kernel is

$$C_{M_n+n,n} = \binom{M_n + n}{n} \tag{17.79}$$

and for the nth order system,

$$M_\alpha = \sum_{n=0}^{n=r} \binom{M_n + n}{n} \tag{17.80}$$

where M_n is the order of truncation of the series expansion of the nth-order kernel. Similarly, the total number r_h of ordinates of all kernels up to order r, counting symmetric ordinates only once, is

$$R_h = \sum_{n=0}^{r} \binom{R + n}{n} \tag{17.81}$$

The number M_α of the coefficients α is many times smaller than R_h to be determined through the inversion of the matrix in equation (17.51).

The BA model has been applied to model runoff from rural watersheds (Norton and Watson, 1979; Helweg, Amorocho and Finch, 1982) as well as from urban watersheds (Patry and Marino, 1982, 1983a, 1983b). From a practical viewpoint, the system memory t_m and the order of truncation M_1 and M_2 should be determined. Helweg, Amorocho, and Finch found that there were no optimum values for these parameters. Hence, they have to be determined by trial and error.

17.1.4.4 The Harder-Zand (HZ) Method

Harder and Zand (1969) and Zand and Harder (1973) used orthogonal functions presented by Forsythe (1957), which are different from those of Brandstetter and Amorocho (1970). The functions in equations (17.54) and (17.61) have the particular form

$$P_i(s) = F_i(s) E^{-s} \tag{17.82}$$

where the functions $F_i(s)$ are orthonormal with respect to the weighting function E^{-s} in the interval $[0, R]$,

$$\sum_{s=0}^{R} F_i(s) F_j(s) E^{-2s} = \delta_{ij}, \qquad \delta_{ij} = \begin{cases} 1, & i = j \\ 0, & \text{otherwise} \end{cases} \tag{17.83}$$

where R is the length of the system memory as before, E is a coefficient, and the functions $F_i(s)$ can be generated by the recursive relation

$$F_{i+1}(s) = [sF_i(s) - \eta_{i+1} F_i(s) - \xi_i F_{i-1}(s)] E^{-s} \tag{17.84}$$

The parameters η_i and ξ_i are obtained from

$$\eta_i = \frac{\sum_{s=1}^{R} s[F_i(s)]^2 E^{-2s}}{\sum_{s=1}^{R} [F_i(s)]^2 E^{-2s}} \tag{17.85}$$

and

$$\xi_i = \frac{\sum_{s=1}^{R} [sF_i(s)F_{i-1}(s)]E^{-2s}}{\sum_{s=1}^{R} [F_{i-1}(s)]^2 E^{-2s}} \tag{17.86}$$

The decay factor E is determined such that for $t = s$, the quantity E^{-2s} is of the order of the expected error level of the effective rainfall data. These functions have two intuitive advantages: (1) The interval of orthonormality is finite and can be used to the length of the system memory. (2) The weighting function E^{-s} decreases as s increases, thus associating gradually decreasing weight to the values of $I(t)$ that are remote in the past.

In applying these functions we can obtain equations equivalent to equations (17.76)–(17.78) following the same reasoning. Harder and Zand (1969) utilized stepwise regression to evaluate the coefficients α. They used this procedure and a third-order model to predict the gage height of the Mekong River in South Vietnam with the discharge measured at two upstream locations taken as inputs.

17.1.4.5 The Papazafiriou (PA) Method

Papazafiriou (1973, 1976) employed a form of the orthogonal polynomials analyzed by Forsythe (1957) for analyzing second-order nonlinear systems. These polynomials are orthogonal over a finite interval and can be generated by the following recurrence relation:

$$P_{j+1}(t) = [(t - a_{j+1})\, P_j(t) - b_j P_{j-1}(t)]\, w^2(t)$$
$$P_o(t) = w^2(t) \tag{17.87a}$$

in which $w^2(t)$ is an arbitrary, positive weight function. Their orthonormal form with weight function unit can be obtained as

$$F_j(t_i) = \frac{P_j(t_i)}{\left[\sum_{i=1}^{N} P_j^2(t_i)\, w^2(t_i)\right]^{0.5}} \tag{17.87b}$$

Following the procedure discussed by Papazafiriou (1975), which parallels the reasoning of Amorocho and Brandstetter (1970), the first-order and second-order kernels in equation (17.49) can be expressed in terms of these polynomials.

$$h_1(s) = \sum_{j=0}^{N_1} a_j F_j(s) \tag{17.88a}$$

$$h_2(s_1, s_2) = \sum_{i=1}^{N_2} \sum_{j=0}^{N_2} a_{ij} F_i(s_1) F_j(s_2) \tag{17.88b}$$

where N_i, $i = 1, 2, \ldots$, specifies the number of terms in the polynomial expansion. This yields

$$Q(t) = \sum_{j=0}^{N_1} a_j \sum_{s=0}^{M} F_j(s) I(t-s) + \sum_{i=0}^{N_2} \sum_{j=0}^{N_2} a_{ij} \sum_{s_1=0}^{M} F_i(s_1) I(t-s_1) \sum_{s_2=1}^{M} F_j(s_2) I(t-s_2) \tag{17.89}$$

For brevity let

$$b_i(t) = \sum_{s_1=0}^{M} F_i(s_1) I(t-s_1) \tag{17.90a}$$

$$b_j(t) = \sum_{j=0}^{M} F_j(s_2) I(t-s_2) \tag{17.90b}$$

Therefore,

$$Q(t) = \sum_{j=0}^{N_1} a_j b_j(t) + \sum_{i=0}^{N_2} \sum_{j=0}^{N_2} a_{ij} b_i(t) b_j(t) \tag{17.91}$$

This can be further simplified by using the property of symmetry of the second-order kernel,

$$Q(t) = \sum_{j=0}^{N_1} a_j b_j(t) + \sum_{i=0}^{N_2} a_{ii} b_i^2(t) + 2\sum_{i=1}^{N_2} \sum_{j=0}^{i-1} a_{ij} b_i(t) b_j(t) \tag{17.92}$$

This significantly minimizes the number of operations involved in the second-order terms. Equation (17.92) is similar to equation (17.70) and can be written in matrix notation

$$[Q_i] = [b_{i,j}]\, [a_j], \qquad i = 0, 1, 2, \ldots, L;\ j = 0, 1, 2, \ldots, N \tag{17.93}$$

in which L is the length of the DR record and

$$N = N_1 + \sum_{i=1}^{N_2+1} i \tag{17.94}$$

Papazafiriou (1976) applied the second-order model to the Hopland watershed (.85 km^2) draining into the Russian River in California. The optimum memory length was 96 hours, the order of polynomial expansion of the first-order kernel N_1 was 7 and that of the second-order kernels N_2 was 4. The second-order model improved the test and prediction results obtained by the linear model by more than 69 percent. The second-order kernel could be considered as more correct than the first-order kernel.

17.1.4.6 The Liu-Brutsaert (LB) Method

Liu and Brutsaert (1975, 1978) employed Chebyshev polynomials for analysis of a second-order nonlinear system. These polynomials can be defined as (M. A. Snyder, 1966),

$$P_j(s) = \cos(j \cos^{-1} s), \qquad j = 0, 1, 2, \cdots \qquad (17.95)$$

Clearly the polynomials of degree 0 and 1 equal unity. The polynomials of higher degrees can be generated from the recurrence relation

$$P_{j+1}(s) - 2sP_j(s) + P_{j-1}(s) = 0 \qquad (17.96)$$

Chebyshev polynomials are orthogonal over the interval (−1, 1) with respect to the weighting function $(1 - s)^{-.5}$, that is,

$$\int_{-1}^{1} P_i(s)P_j(s)(1 - s)^{-.5}\, ds = c_j\delta_{ij} \qquad (17.97)$$

where $c_0 = \pi$ and $c_i = \pi/2$ for $j \neq 0$ and δ_{ij} is Kronecker's delta. The first- and second-order kernels in equation (17.49) can be expressed in terms of these polynomials as

$$h_1(s) = \sum_{j=0}^{N_1} c_j P_j(s) \qquad (17.98a)$$

$$h_2(s_1, s_2) = \sum_{i=0}^{N_2}\sum_{j=0}^{N_2} c_{ij}P_i(s_1)P_j(s_2) \qquad (17.98b)$$

in which N_i, $i = 1, 2$, specifies the number of terms in polynomial expansion, and c_j and c_{ij} are coefficients of the series expansion. Liu and Brutsaert took $N_1 = N_2 = 8$. Inserting these in the expression for $Q_2(t)$ results in the same equation as equation (17.70). Liu and Brutsaert applied the second-order model to the Cowanesque watershed (764 square kilometers) in the Chumung River basin.

17.1.4.7 The Kuchment-Borshchevsky (KB) Method

Kuchment and Borshchevsky (1971) developed a procedure similar to the Brandstetter-Amorocho method using a continuous-time representation. An analog network was used to compute the coefficients α and the function $a(t)$ in equation (17.76). They noted that the function $a(t)$, given by equations (17.56), (17.66), and (17.75), can be responses of linear systems with $P(s)$ representing the kernels operating on $I(t - s)$. The continuous-time representation of $a(t)$ can be expressed as

$$a(t) = \int_0^M P(s)\, I(t - s)\, ds \qquad (17.99)$$

where $P(s)$ was represented by the normalized Laguerre functions,

$$P(s) = f_r(\beta s) = \begin{cases} \dfrac{1}{r!}\exp\left(\dfrac{\beta s}{2}\right) d^r \dfrac{s^r \exp(\beta)}{ds^r}, & s \geq 0 \\ 0, & \text{otherwise} \end{cases} \qquad (17.100)$$

where r is an integer defining the order of the functions and β is a time scale. Equation (17.76) can be written for the second-order case as

$$\begin{aligned} Q(t) &= \alpha_0 a_0(t) + \alpha_1 a_1(t) + \alpha_2 a_2(t) + \alpha_{00}a_0^2(t) \\ &\quad + \alpha_{11}a_1^2(t) + \alpha_{22}a_2^2(t) + 2\alpha_{01}a_0(t)a_1(t) \\ &\quad + 2\alpha_{02}a_0(t)a_2(t) + 2\alpha_{12}a_1(t)a_2(t) \end{aligned} \qquad (17.101)$$

or

$$\begin{aligned} Q(t) &= k_0a_0 + k_1a_1 + k_2a_2 + k_3a_0^2 + k_4a_1^2 \\ &\quad + k_5a_2^2 + k_6a_0a_1 + k_7a_0a_2 + k_8a_1a_2 \end{aligned} \qquad (17.102)$$

where k's represent α's and their combinations and a's are the responses of the linear Laguerre systems:

$$a_0 = a_0(t) = \int_0^t \beta \exp(-\beta s/2) I(t - s)\, ds \qquad (17.103a)$$

$$a_1 = a_1(t) = \int_0^t \beta\,(1 - \beta s)\exp\left(\frac{-\beta s}{2}\right) I(t - s)\, ds \qquad (17.103b)$$

$$a_2 = a_2(t) = \int_0^t \beta\left(1 - 2\beta s + \frac{1}{2}\beta s^2\right) \exp\left(\frac{-\beta s}{2}\right) I(t - s)\, ds \qquad (17.103c)$$

It is thus seen that two sequential operations are involved in representation of a nonlinear system: (1) a transformation of $I(t)$ by finite memory, linear Laguerre systems and (2) a nonlinear no-memory combination of the responses of these linear systems. This characteristic of nonlinear systems was first postulated by Wiener (1942).

Taking the Laplace transform of equation (17.100),

$$L[f_n(\beta t)] = f_r(w) = \left[\frac{(2\beta)^{1/2}}{w + \beta}\right]\left[\frac{w - \beta}{w + \beta}\right]^r \qquad (17.104)$$

where w is the variable in frequency domain. The operator $(2\beta)^{1/2}/(w + \beta)$ is equivalent to the operator of the equation,

$$\frac{dy}{dt} + \beta y = \beta(2\beta)^{1/2}x \qquad (17.105)$$

Therefore, the transformation of input by this operator results in flattening of the input hydrograph without translation. The operation $[(w - \beta)/(w + \beta)]^r$ produces an r-fold shift or translation. Each one of these shifts can be expressed by

$$\frac{dy}{dt} + \frac{dx}{dt} = \beta(y - x) \qquad (17.106)$$

Equation (17.106) is of the same form as the Muskingum equation with weighting factor of .5. The reciprocal of β signifies the lag between the centers of areas of $t(t)$ and $y(t)$.

Kuchment and Borshchevsky (1971) represented equation (17.103) by an analog network and gave its hydrologic interpretation. However, the convolutions represented by these equations can be easily undertaken by means of Laguerre networks (Lee, 1932).

17.1.4.8 The Bidwell (BI) Method

Bidwell (1971) developed a method that replaces equation (17.49) by a regression equation, which contains as independent variables a set of arbitrary groupings of the values of $I(t)$ defined by

$$y_i(t) = \sum_{m=1}^{2i-1} I(t - m + 1),$$
$$i = 1, 2, \ldots, N_u; t = 1, 2, \ldots, N \quad (17.107)$$

where N_u is the number of the grouped variables y_i to be selected by considering the history of rainfall-runoff sequences. The purpose of grouping is to reduce the number of unknown coefficients in the prediction equation. The original kernel functions $h_1(s)$, $h_2(s_1, s_2)$, . . . , $h_n(s_1, s_2, \ldots, s_n)$ are not explicitly evaluated in the method.

Equation (17.49) can be written in terms of the grouped variables as

$$Q(t) = \sum_{i=0}^{N_u} b_i y_i + \sum_{i=1}^{N_u} \sum_{j=1}^{N_u} b_{ij} y_i y_j + \cdots \quad (17.108)$$

where b's are coefficients that are obviously not the same as the kernels in equation (17.49). We can write equation (17.108) in matrix form,

$$[Q] = [y][B] \quad (17.109)$$

where $[Q]$ is a vector of direct runoff values, $[Q_1(t), Q_2(t), \ldots, Q_N(t)]^T$; $[y]$ is a matrix of grouped variables, and $[B]$ is a vector of unknown coefficients. This equation is similar to equation (17.51). The unknown vector $[B]$ can be estimated by using the regression techniques. However, the kernel function cannot be obtained by this method.

Bidwell applied this method to the analysis of rainfall and streamflow data from two small watersheds—one was 27 acres and the other 17 acres. It stimulated streamflow with multiple correlation coefficient of approximately .92. It was, however, not used for prediction. A somewhat similar method has been used by Muftuoglu (1984).

17.1.4.9 The Diskin-Boneh (DB) Method

Boneh and Diskin (1972) and Diskin and Boneh (1973) developed a procedure for determination of kernel functions in which several storms can be utilized simultaneously, unlike most other methods, which handle rainfall-runoff values from a single storm or a sequence of storms. This procedure was used for a second-order model. Let i denote the serial number of the rainfall-runoff event. Then

$$Q_{i,j} = \sum_{k=w}^{L} h_{1,k} I_{i,i+1-k} + \sum_{k=w}^{L} \sum_{r=w}^{L} h_{2,k} I_{i,j+1-k} I_{i,j+1-r} \quad (17.110)$$

where the limits of the summations in equation (17.110) have been modified, in comparison with equation (17.49), to take into account the finite durations of $I(t)$ and the kernels. These new limits are

$$w = \max[1, j - NI_i + 1] \quad (17.111)$$

$$L = \min[j, N_k] \quad (17.112)$$

where NI_i is the number of rainfall values for the ith event and $N_k = \max[NI_i]$.

Since the number of runoff values in different storms are usually different, the time base of runoff values for all storms can be made equal by using a common memory length M, which can be obtained from

$$M = \max_{1 \le i \le N} [NQ_i - NI_i] \quad (17.113)$$

where NQ_i is the number of runoff values in the ith storm and N is the number of storms. The time bases of runoff for all storms are given by

$$NQ_i^* = NI_i + M \quad (17.114)$$

If the actual storm duration is less than the extended time base, the runoff values for the time period beyond the actual storm duration are set to zero.

The kernels satisfy the properties discussed previously, namely, equations (17.19) and (17.28). Casting equation (17.110) in a matrix form,

$$[Q] = [I][h] \quad (17.115)$$

where $[I]$ is an $(N_e \times N_u)$ matrix containing effective rainfall values, their squares and products; $[h]$ is an $(N_u \times 1)$ column vector of unknown kernels; $[Q]$ is an $(N_e \times 1)$ column vector of runoff values; $N_e = \sum_{i=1}^{N} NQ_i^*$ is the number of equations; N_u is the number of unknown kernels $= N_k + .5\, NI^*[2N_k - N_m + 1]$, where $N_m = \min\,[NI^*, N_k]$.

Equation (17.115) was solved by minimizing the sum of squares of deviations E_{ij} between the observed and computed $Q(t)$, that is,

$$\min F = \sum_{i=1}^{N} \sum_{j=1}^{NQ_i} E_{ij}^2 \quad (17.116a)$$

and subject to

$$\sum_{k=1}^{N_k} h_{1,k} = 1.0 \quad (17.116b)$$

$$\sum_{k=1}^{N_k+J} h_{2,k,k+j} = 0.0, J = 0, 1, 2, \ldots \quad (17.116c)$$

$$h_{1,k} \geq 0 \quad \text{for all } k \quad (17.116d)$$

To obtain the vector $[h]$ according to the preceding optimization problem, Diskin and Boneh (1973) used the Fletcher-

Powell algorithm (Fletcher and Powell, 1963). This is accomplished in two steps. First, min F is obtained subject only to the inequality constraint with the initial value of $[h]$ set equal to zero. The resulting solution of $[h]$ is then used as the initial value to obtain once again min F, subject now to both equality and inequality constraints. The equality constraints are accommodated in the system of equations by using the penalty function method (Osborne and Ryan, 1970). The final solution satisfies both types of constraints.

Diskin and Boneh successfully applied the second-order model to the Cache River watershed (630 km^2) in southern Illinois. The number of equations and the number of unknowns depend upon the number of storm events available and the basic time interval used for optimization. If storm events are many and discretization interval is small, there may be more equations than unknowns. The least squares criterion may be used for optimization. The solution yields values of the first-order kernel over its entire range. The values of the second-order kernel are, however, obtainable only over a restricted zone parallel to the main diagonal in the plane of its two independent variables. The width of this zone depends on the longest duration of the effective rainfall available in the storm records.

17.1.4.10 The Roy-Sherman (RS) Method

Roy and Sherman (1967) proposed a recursive method for estimation of kernel functions, which can handle a large number of storms. The kernel functions are determined by an adaptive or learning algorithm. In this method various products of the past effective rainfall are formed and are weighted by error correction weights represented by weight vector. The weights are continuously adjusted by an error-correcting algorithm minimizing the norm of the error between the observed and computed discharges. The resulting weight vector is equivalent to the kernels.

Equation (17.49) represents a pth-order polynomial-type surface; for $r = 1$, the transformation surface is a hyperplane, and for $r = 2$, it is a quadratic surface. This method first transforms the $[I]$ vector into a vector $[F]$ in the Φ-space. The vector $[F]$ is a set of linearly independent functions $f_i[I]$ which are made up of products of the past $I(t)$ values. The coordinates in Φ-space are a set of functions that span the space. These functions can be rth-order polynomials. For example, if

$$I(t) = [I(t), I(t-1), \ldots, I(t-M)]^T \tag{17.117}$$

The vector $[I]$ has dimensions of $(M^*, 1)$, where $M^* = M + 1$. If $r = 2$, various linear and quadratic functions of $I(t)$ are formed and are denoted by the vector $[F]$, expressed as

$$[F] = [I(t), I(t-1), \ldots, I(t-M), I^2(t), I(t)$$
$$I(t-1), \ldots, I(t)\,I(t-m), \ldots, I^2(t-m)]^T \tag{17.118}$$

Thus

$$[F] = [f_1(t), f_2(t), \ldots, f_L(t)]^T \tag{17.119}$$

is defined in an M-dimensional space, where

$$L = \binom{M^* + r}{r} - 1 \tag{17.120}$$

where r is the order of nonlinearity considered in the analysis. For illustration, let $[I] = [I_1, I_2]$ and $r = 2$; then

$$f_1 = I_1, \quad f_2 = I_2, \quad f_3 = I_1^2, \quad f_4 = I_2^2, \quad \text{and} \quad f_5 = I_1 I_2$$

Therefore, the vector $[F]$, in general, contains all the possible products of the components of I, from linear terms up to rth-order terms.

The second operation of the Φ-machine is a linear summation of the functions $f_1[I]$. Consequently,

$$Q_M[I] = \sum_{i=1}^{L} w_i f_i[I] = [w]^T[F] \tag{17.121}$$

where Q_M represents a hyperplane in Φ-space and an rth-order polynomial surface in the original $[I]$ space and $[w]$ is the weight vector.

In the transformed space, or the Φ-space, the surface $\Phi[I]$ is adjusted by an iterative error-correcting algorithm. The method is similar to the multiple regression technique, where a least squares fit to a given surface is achieved.

Equation (17.121) is equivalent to equation (17.49). The surface, implemented by an rth-order Φ-machine, is given by

$$Q_M(I) = \sum_{i=1}^{L} w_i I_i + \sum_{i=1}^{L}\sum_{j=1}^{L} w_{ij} I_i I_j + \cdots$$
$$+ \sum_{i=1}^{L} \cdots \sum_{r=1}^{L} w_i \cdots I_i \cdots I_r \tag{17.122}$$

Therefore, the determination of weight vector $[w]$ is equivalent to the determination of kernel vector $[h]$. The vector $[w]$ is determined by an error-correcting gradient algorithm subject to minimizing the norm between the observed and computed $[Q]$.

17.1.5 Comparison of Different Methods

A comparative analysis of the BA, BI, DB, and RS methods was made by A. R. Rao and R. G. S. Rao (1974). We follow their work here. The BA method does not require satisfaction of the continuity equation. Only the DB method incorporates the continuity equation explicitly. However, the methods—not including continuity equation—explicitly reflect such factors as infiltration rate and watershed storage through their kernels. Since the BA and BI methods are based on regression analysis techniques, a large number of observations would be needed for reliable estimation of kernels. The data requirements of the DB and RS methods are also comparable. The ER-DR data are required by the DB method, whereas other methods can use rainfall-runoff data directly. In the BI method, the rainfall values are grouped. The DB method is computationally the most complex, followed by the BA, BI, and RS methods, in order of decreasing complexity.

For a number of rainfall-runoff events on an urban watershed (Ross Ade) in West Lafayette, Indiana, A. R. Rao and R. G. S. Rao found that the BA and DB methods were similar in their regeneration and prediction performance. The BA method is easier to use than the DB method. The regeneration results of the BI method were excellent, but the predictions were less satisfactory, perhaps due to grouping of rainfall data. The regeneration performance of the RS method was satisfactory, but the prediction performance was not. The BA and DB methods were superior to the linear model. The other two methods were comparable to the linear model. They concluded that the BA method was preferable because of its flexibility and accuracy.

17.2 RELATION BETWEEN VOLTERRA KERNELS AND CASCADE OF NONLINEAR RESERVOIRS

Napiorkowski and Strupczewski (1979, 1981) have, using a novel method of analysis, shown that the VIS can be used to represent a cascade of nonlinear reservoirs. By deriving this equivalence they were able to suggest a set of equations for the first- and second-order kernels in the truncated VIS. For the cascade of equal N nonlinear reservoirs with the storage (S)–discharge (Q) relation specified as

$$Q = a_1 S + a_2 S^2 \tag{17.123}$$

they specified the first two kernels $h_1(s_1)$ and $h_2(s_1, s_2)$. This cascade has three parameters: N, a_1, and a_2. Letting

$$F_N(t) = \frac{(a_1 t)^{N-1} \exp(-a_1 t)}{(N-1)!}$$

the expressions for the two kernels were expressed as

$$h_1(s_1) = a_1 F_N(s_1) \tag{17.124}$$

and

$$h_2(s_1, s_2) = a_2 \left[F_N(s_1) \sum_{j=1}^{N} F_j(s_2) + F_N(s_2) \sum_{j=1}^{N} F_j(s_1) - F_N\{\max(s_1, s_2)\} \right] \tag{17.125}$$

Equations (17.124) and (17.125) are linked through the parameters a_1 and N. To ensure convergence of the truncated VIS, certain restrictions were imposed on the parameters.

These two kernels satisfy all the conditions discussed in Part 17.1.1.4. In addition, for any set of N, a_1, and a_2, the model satisfies two other optional conditions proposed by Boneh and Diskin (1973) to explain the type of nonlinearity observed in small watersheds: (1) The second-order kernel $h_2(s_1, s_2)$ has positive ordinates near the origin. (2) The peak of $h_2(t, t)$ precedes the first peak of $h_1(t)$. It is also noted that the zero crossing of $h_2(t, t) = 0$ occurs after the peak of $h_1(t)$, irrespective of the values of a_1, a_2, and N.

Equation (17.124) is the IUH of the cascade of N equal linear reservoirs with a storage constant of $1/a_1$. The parameter a_2 measures the degree of nonlinearity of the cascade. With a_2 tending to zero, the VIS as well as the nonlinear cascade reduce to a cascade of linear reservoirs.

17.3 TWO-STAGE CONVOLUTION

W. M. Snyder, Mills, and Stephens (1970) proposed a two-stage convolution for storm hydrograph analysis and applied it to coastal plain watersheds in Georgia. It allows variation of unit response functions during multiple periods of rainfall and produces nonlinear watershed response. As the name suggests, this consists of two stages of convolution, as shown in Fig. 17.9. The first stage is used to construct variable unit response functions. This is accomplished by convoluting a characteristic function $C(t)$, which assumes the role of input with a variable state function $g(t)$, which assumes the role of kernel; obviously the watershed unit response function $h(t)$ is the output. In the second stage the effective rainfall is convolved with the variable unit response functions to produce the DR hydrograph.

The characteristic function $C(t)$, also referred to as core function, is a time-transformation of a watershed potential runoff map and can be considered analogous to the TAC curve. It is constructed as a histogram of runoff from unit areas versus time of travel classes. W. M. Snyder, Mills, and Stephens considered two forms of $C(t)$: (1) $C(t) = a$, constant for $0 \le t \le T$, and $C(t) = 0$, $t > T$, where T is the base of a rectangular pulse; (2) $C(t)$ is an isosceles triangle with time base T, whose sides can be expressed as $C(t) = a + bt$, in which a and b are constants.

It is difficult to establish a specific form of the state function $g(t)$, also referred to as routing function. It should reflect certain watershed characteristics. As a first approximation it can be taken as a simple descending exponential or a gamma function. Thus, the first-stage convolution can be expressed for an ith effective rainfall pulse as

$$h(t) = \int_0^t C(\tau)\, g_i(t - \tau)\, d\tau \tag{17.126}$$

If $C(t)$ is a rectangular pulse as defined above and $g_i(t) = A_i \exp(-A_i t)$, then

$$h_i(t) = \begin{cases} a[1 - \exp(-A_i t)], & 0 \le t \le T \quad (17.127a) \\ a[\exp(A_i T) - 1] \exp(A_i t), & t \ge T \quad (17.127b) \end{cases}$$

where A is a parameter reflecting the wetness of the watershed. If $C(t)$ is an isosceles triangle, then

$$h_i(t) = A_i \int_0^t (a + b\tau) \exp[-A_i(t - \tau)\, d\tau$$

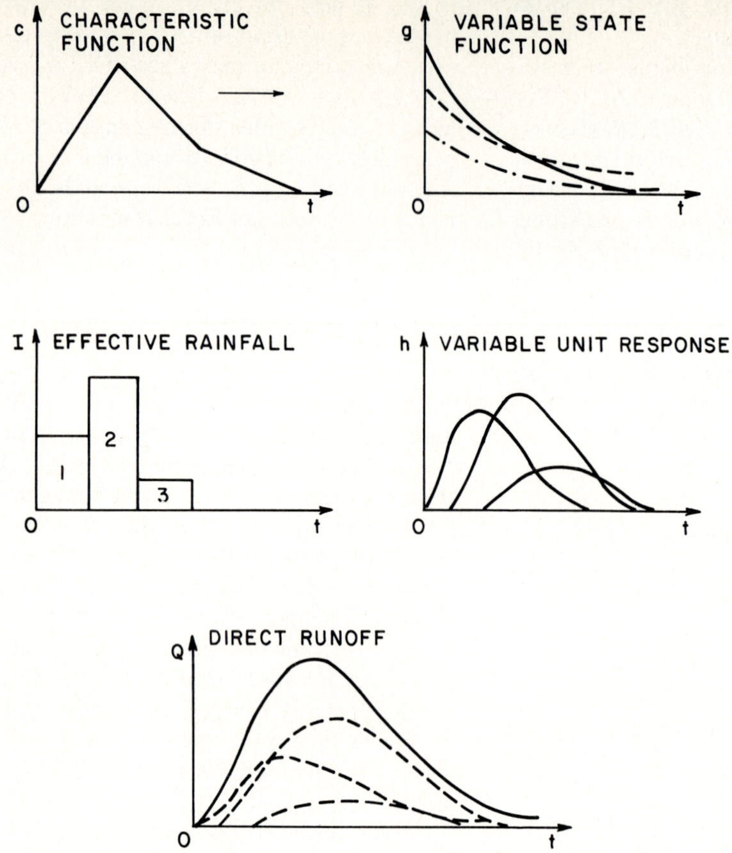

FIGURE 17.9 A schematic of a two-stage convolution (after Snyder, Mills and Stephens, 1970).

which yields

$$h_i(t) = (a + bt)[1 - \exp(-A_i)] + \frac{b}{A_i}[(A_i t + 1)\exp(-A_i t) - 1], \quad 0 \le t \le T \qquad (17.128a)$$

$$h_i(t) = (a + bt)\,[\exp(TA_i) - 1]\exp(-A_i t) - \frac{b}{A_i}[\{A_i(t - T) + 1\}\exp\{-A_i(t - T)\} - (A_i t + 1)\exp(-A_i t)] \qquad (17.128b)$$

Each effective rainfall pulse is distributed by $h_i(t)$ to produce the direct runoff. Let us illustrate this procedure by an example.

EXAMPLE 17.8

The DR hydrograph is desired by the two-stage convolution for the ER of three pulses shown in Fig. 17.9. Let $C(t)$ and $g(t)$ be defined as before. The parameter A is defined as

$$A(t) = w_1 + w_2[Q(t - 1) + B(t - 1)] \qquad (17.129)$$

where w_1 and w_2 are empirical constants, $B(t - 1)$ is antecedent baseflow, and $Q(t - 1)$ is discharge at the beginning of each discrete-time interval $\Delta t(= 1)$. The following steps are involved in carrying out the calculations.

1. To start the computation, we obtain, for the first time interval selected as being less than the duration of the shortest pulse,

$$A(1) = w_1 + w_2 B(0)$$

2. Get $g_1(t)$. W. M. Snyder and Asmussen (1973) have shown that to change $g(t)$ for discrete convolution, the area under the curve (exponential) must be computed. Thus

$$g_1(1) = 1 - \exp(-A(1))$$

$$g_1(2) = \exp\{-A(1)\} - \exp\{-2A(1)\}$$

$$g_1(j) = \exp\{-(j - 1)A(1)\} - \exp\{-jA(1)\}$$

3. Obtain $h_1(t)$ by convolution

$$h_1(1) = C(1)g_1(1)$$

$$h_1(2) = C(1)g_1(2) + C(2)g_1(1)$$

$$\vdots$$

$$h_i(j) = C(1)g_1(j) + C(2)g_1(j - 1) + \cdots + C(5)g_1(1)$$

4. Compute the DR $Q_1(t)$ due to the effective rainfall during the first time interval,

$$Q_1(1) = h_1(1)I(1)$$

$$Q_2(2) = h_1(2)I(1)$$

$$\vdots$$

$$Q_1(j) = h_1(j)I(1)$$

5. Proceed to the second time interval. Compute $A(2)$ as

$$A(2) = w_1 + w_2[Q(1) + B(1)]$$

6. Compute $g_2(j) = \exp\{-(j-1)A(2)\} - \exp\{-jA(2)\}$.
7. Compute $h_2(t)$ by convolution,

$$h_2(1) = g_2(1)C(1)$$

$$h_2(2) = g_2(2)C(1) + g_2(1)C(2)$$

$$\vdots$$

$$h_2(j) = g_2(j)C(1) + g_2(j-1)C(2) + \cdots + g_2(1)C(j)$$

8. Compute the DR $Q_2(t)$ due to the effective rainfall during the second time interval,

$$Q_2(j) = h_2(j)I(2),\ j = 1, 2, \ . \ . \ .$$

9. Repeat the steps for each subsequent time interval.
10. Add the discharge contributions computed for each time interval to obtain total discharge $Q(t)$. For example,

$$Q(1) = Q_1(1) = h_1(1)I(1)$$

$$Q(2) = Q_1(2) + Q_2(1) = h_1(2)I(1) + h_2(1)I(2)$$

$$Q(3) = Q_1(3) + Q_2(2) + Q_3(1) = h_1(3)I(1) + h_2(2)I(2) + h_3(1)I(3)$$

$$\vdots$$

$$Q(j) = Q_1(j) + Q_2(j-1) + \cdots + Q_j(1) = \sum_{i=1}^{j} Q_i(j-i+1)$$

$$= h_1(j)I(1) + h_2(j-1)I(2) + \cdots + h_j(1)I(j)$$

$$= \sum_{i=1}^{k} h_i(j-i+1)I(i)$$

Although two-stage convolution is an empirical procedure, it has some attractive features. First, through a characteristic function, it attempts to represent spatial characteristics (rainfall as well as soil moisture profile). Second, the state function accounts explicitly for antecedent wetness for each increment of the effective rainfall. Thus the variable unit response function is produced. In this manner the time-varying characteristics of a watershed are accommodated.

EXERCISES

17.1. Enumerate the properties of the second-order kernel.

17.2. Show that for $I(t) = x(t) + y(t) + z(t)$, $g_2(t) = H_2[I(t)] = H_2[x(t) + y(t) + z(t)] = H_2[(x + y + z)^2]$.

17.3. Derive the method of measuring the Volterra kernels of a third-order system following the procedure of Schetzen (1965a).

17.4. Consider the second-order system of Example 17.4. The input is composed of two pulses of durations D_1 and D_2, where $D_1 > D_2$ and their heights are $1/D_1$ and $1/D_2$, respectively. Compute the system response.

17.5. Solve Question 17.4, where the order of pulses is reversed.

17.6. Catalog different methods of estimating nonlinear kernels along with their authors and list their advantages and disadvantages. Which method is preferable for rainfall-runoff modeling?

17.7. Consider a second-order system consisting of a first-order kernal (1/h) as

$$h(t) = t\exp(-t)$$

and a second-order kernel (1/(mm/h)) as

$$h(t, s) = \frac{1}{10}(t + s - 2ts)\exp[-(t + s)]$$

Determine the impulse response of this system for various magnitudes of impulses. Assume the impulse input at $t = 0$.

17.8. The second-order system of Question 17.7 is subjected to a uniformly steady input of 10 mm/h for indefinite duration. Determine the system response for various input intensities.

17.9. The second-order system of Question 17.7 is subjected to a uniformly steady input of 10 mm/h for 2 h. Determine the system response.

17.10. The second-order system of Question 17.7 is subjected to a uniformly steady input of 5 mm/h for 1 h, followed by a uniformly steady input of 10 mm/h for another hour. Determine the system response.

17.11. The second-order system of Question 17.7 is subjected to a uniformly steady input of 10 mm/h for 1 h, followed by a uniformly steady input of 5 mm/h for another hour. Determine the system response.

CHAPTER EIGHTEEN

Models for Ungaged Watersheds

Chapters 10 through 17 presented various kinds of rainfall-runoff models. Clearly the black-box models are entirely based on analysis of rainfall-runoff data. Consequently, these models are not applicable to those watersheds that lack such data. On the other hand, conceptual models contain parameters that may be determined from either rainfall-runoff data or physically measurable watershed characteristics. Frequently the model parameters are optimized for some selected rainfall-runoff events over a given watershed, using a suitable optimization algorithm subject to an objective function. The optimized parameter values are then utilized in the model to predict runoff for the rainfall events of interest not used in the optimization. This approach is obviously not applicable to ungaged watersheds. Further, it has other shortcomings. For example, the optimized parameters can best represent the watershed only for the events used in the optimization. No sooner does the optimization set of rainfall-runoff events change than the optimum parameter values change. The extensive amount of data normally required for reliable optimization is often lacking and may prove prohibitive in the widespread model applicability. Equally important, optimization is often expensive.

The other approach attempts to establish relationships between model parameters and physically measurable watershed characteristics. These relationships are then assumed to hold for ungaged watersheds having similar hydrologic characteristics. Thus, the models can also be utilized in evaluating the effect of land-use practices on watershed response. In this chapter, the various methods applicable to ungaged watersheds are discussed in detail.

Rainfall-runoff relationships for ungaged watersheds have been developed along two complementary lines: (1) Empirical equations have been developed to relate some individual runoff hydrograph characteristics to watershed characteristics. These characteristics can be partitioned into (a) time characteristics and (b) discharge characteristics. The time characteristics were discussed in Chapter 9. (2) Procedures have been developed to synthesize the entire runoff hydrograph from watershed characteristics. These procedures or models are either linear or nonlinear in character.

18.1 DISCHARGE CHARACTERISTICS

Many empirical relationships have been developed for estimating discharge characteristics from rainfall and watershed characteristics (Kinnison and Colby, 1945; Chow, 1962; Thomas and Benson, 1970; Duru, 1976; Chang and Boyer, 1977; Aron and A. C. Miller, 1978; Dingman, 1978; Crippen, 1982; Aron, Kibler and Tagliati, 1981; Mosley, 1981; Adejuwon, Jeje and Ogunkoya, 1983; Mimikou, 1983; Harlin, 1984). Of these characteristics the peak discharge has been the most popular; it has been frequently utilized in hydrograph synthesis (Snyder, 1938; Carter, 1961; Wu, 1963; Rao, Assenzo and Harp, 1966; Bell, 1967; Larson and Machmeier, 1968; Bell and Omkar, 1969; Cordery, 1971) as well as frequency analysis (Durant and Blackwell, 1959; Gray, 1970; Alexander, 1972). An excellent discussion of peak discharge formulas is presented by Chow (1962) and Gray (1970). These relationships should, however, be used with caution, and they should be used in the regions for which they were developed. A major drawback of these equations arises not so much from their empirical basis but more so from the lack of knowledge of exact conditions of their applicability.

The simplest of the formulas used in frequency analysis

are those based on drainage area, which are of the following forms:

$$Q_m = aA^b \tag{18.1a}$$

$$Q_m = aA^{cA^{-b}} \tag{18.1b}$$

$$Q_m = \frac{aA}{(a_1 + a_2A)^c} + dA \tag{18.2}$$

in which Q_m is maximum flood flow rate, A is drainage area, c and b are exponents, and a, a_1, a_2 and d are coefficients that depend upon the area, climatic characteristics, and frequency. A major drawback of these formulas is that they do not contain an explicit frequency term. Equations similar to equation (18.1) have also been proposed for estimating hydrograph peak (Black, 1972; Black and Cronn, 1975), as will be clear in the following sections.

In a similar vein, many workers (Getty and McHughs, 1962; Morgan and S. M. Johnson, 1962; Rao, Assenzo and Harp, 1966; Reich, 1965, 1968; Stephens and Mills, 1965; Harp and Hiemstra, 1968, 1966; Colombi, 1978; McCuen and Bondelid, 1982) have proposed empirical formulas for computing peak of the UH of a given duration. Many of them are of the form proposed by Getty and McHughs (1962):

$$h_p = \left[\frac{110{,}860}{A^{0.45}(LL_{ca}/S^{0.5})^{0.32}}\right]^{0.5} \tag{18.3}$$

where h_p is the 2-hour UH peak in square feet per second per square mile, A is watershed area in square miles, L is the length of the main channel in miles, L_{ca} is the length to the center of gravity of the watershed in miles, and S is the stream slope in feet per mile.

In a series of papers, Rogers (1980, 1982) and Rogers and Zia (1982) investigated the relation between peak discharge Q_p and volume of direct runoff V for rainfall and runoff events of any duration, spatial distribution, or uniformity.

$$\log\left(\frac{Q_p}{V^2}\right) = b + m_*\log V, \qquad m_* = m - 2 \tag{18.4}$$

where b is intercept and (m_*) is slope of the line in logarithmic domain. This is referred to as the *second-order standardized peak discharge distribution* and is used to interpret the degree of hydrologic nonlinearity of the watershed. If $m_* = -1$ then the watershed is hydrologically linear; if $m < -1$, the watershed is nonlinear. This relation was tested on 42 watersheds ranging in size from 2.4 km^2 to 701.9 km^2. The watersheds were from Nebraska, Pennsylvania, Maryland, and New York. This relation accounted for an average of 86 percent of variation in the dependent variable. Mimikou (1983) applied equation (18.4) to 8 watersheds between 200 and 6000 square kilometers in northern and western Greece. She supported Rogers' finding that the slope has no geographic, climatic, or morphologic influence. Furthermore, she found that the coefficient of determination of this relation was linearly related to the degree of hydrologic nonlinearity. The intercept b was found to be significantly related to the logarithm of any of the two watershed morphological indices AS/L and A/L, where S and L are the main channel slope and length. Singh and Aminian (1986) applied equation (18.4) to effective rainfall-direct runoff records from 134 watersheds located in Australia, Greece, Italy, and the United States and obtained similar results.

18.2 LINEAR BLACK-BOX MODELS

These models do not account for variations in watershed characteristics or their effect on the UH. An empirical curve giving a unique shape of the UH is proposed. Several different models have been proposed, as discussed by Benson (1962), Chow (1962), Allison (1967), and Cordery and Pilgrim (1970).

18.2.1 The Bernard Model

The model by Bernard (1935) appears to be perhaps the first attempt to synthesize the UH from watershed characteristics. He studied six watersheds in Ohio varying in area from 500 to 6000 square miles and in channel slope from 3.5 to 12 feet per mile. He assumed that the peak of the UH should be inversely proportional to the time of concentration, which, in turn, was assumed to be proportional to a so-called watershed factor u. Using a 24-hour duration, he developed a diagram relating the daily ordinates of various DGs to the watershed factor u. In the diagram, u was plotted on a logarithmic paper against the ordinates of the 1-day DG at 1 day, 2 days, 3 days, . . . , 10 days, after the rainfall. Straight parallel lines were drawn, one to fit all the points representing the ordinates of the 1-day DG on the first day after the rainfall, and another to fit the points representing the ordinates on the second day, and so on, as shown in Fig. 18.1.

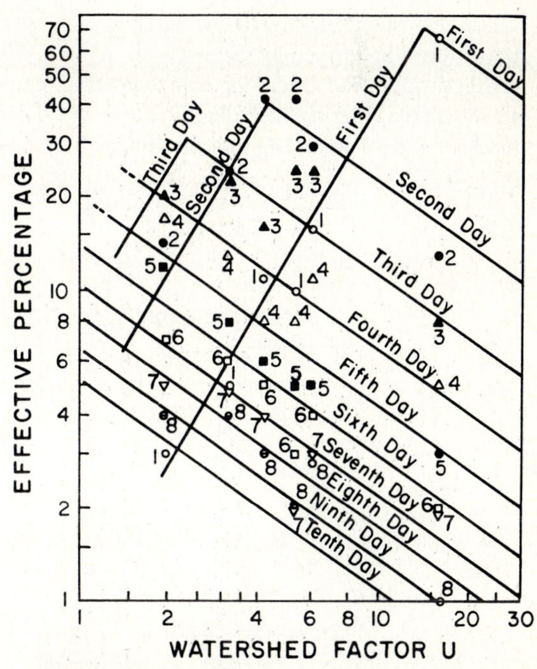

FIGURE 18.1 Relation of watershed factor u to daily increments of the various distribution graphs (after Bernard, 1935).

The watershed factor was expressed as

$$u = \left(\frac{60P}{L}\right)^{4eg} F^{8eg} \frac{S^{1.5eg}}{1000^{2eg}} \tag{18.5}$$

where P is a constant depending upon the shape and area of the watershed, L is the length of the mainstream in feet from the most remote portion of the watershed to the outlet, F is a constant depending upon the shape and condition of the main channel, S is the slope in feet per 1000 feet of the main channel, e is an exponent in the rainfall-intensity equation

$$I = \frac{KF^n}{D^e} \tag{18.6}$$

in which I is the intensity of rainfall in inches per hour for a storm of duration of D in hours, and F, n, and e are constants. The term g is

$$g = \frac{1}{4 - e} \tag{18.7}$$

If we look at equation (18.5), we see immediately that there are only two variables on the right side. Therefore, u can be more simply expressed as a function of S and L, lumping all the constants into one. Jarvis (1935), in a discussion of the Bernard method, suggested we express u simply as

$$u = 20\frac{S^{.5}}{L} \tag{18.8}$$

In his closure, Bernard (1935) agreed with the usefulness of this equation. The factor u is supposed to embody the watershed characteristics that govern the UH. It was obtained from a runoff formula derived by Gregory and Arnold (1932). For an ungaged watershed, u can be computed from equation (18.5) or (18.8). By entering the value of u into the diagram, the ordinates of the 24-hour DG can be obtained by reading the ordinates on the first day, the second day, the third day, and so on. Bernard found that for any particular day within the period of rainfall, the daily value of the DG increased directly as the five-thirds power of u to a maximum daily value and then receded inversely as the two-thirds power of u. Furthermore, for a particular value of u, the daily value of DG increased directly as the square of the rainfall duration in days to a maximum value and then receded inversely as the square of the storm duration in days.

The Bernard model is a one-parameter model since it involves only one measure of watershed characteristics. Therefore, this measure must determine both the scale and the shape of the UH. An implication of this hypothesis is, as Nash (1958) emphasized, that all 1-day UHs having the same peak will be identical. Further, as the duration changes, a separate diagram of the above type will be required. This diagram may also change with the location. The concept of a DG is nevertheless appealing and has been frequently used in hydrology (Laden, Reilly, and Minnotte, 1940; Jetter, 1944).

18.2.2 The Snyder Model

Snyder (1938) was perhaps the first to have established a set of formulas relating the physical geometry of the watershed to three basic parameters of the unit hydrograph. These formulas were based on a study of 20 watersheds located mainly in the Appalachian Highlands, which varied in size from 10 to 10,000 square miles.

The basic parameter that Snyder defined is t_p, the time of lag to peak in hours taken as the time from the center of mass of the effective rainfall of unit duration to the peak of the UH or simply the watershed lag (see Fig. 18.2). He derived all other characteristic parameters of the UH in terms of t_p. According to the Snyder model,

$$t_p = C_t (LL_c)^{.3} \tag{18.9}$$

where L is the length of main stream in miles from the outlet to divide, L_c is the distance in miles from the outlet to a

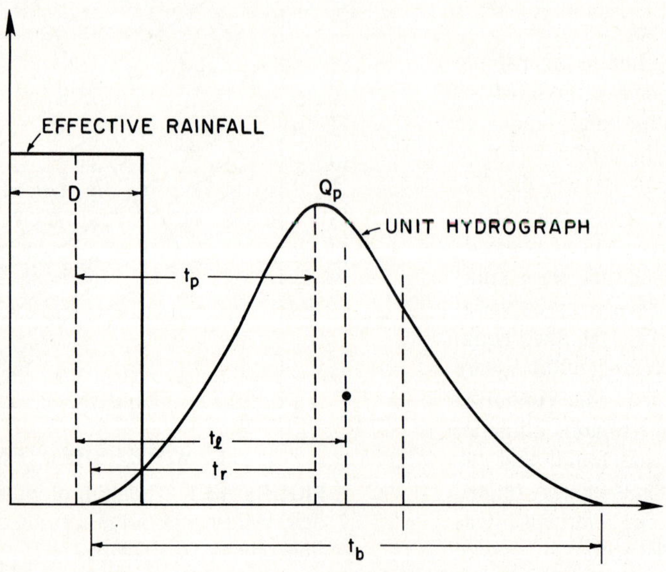

FIGURE 18.2 Schematic for the Snyder model.

point on the stream nearest the center of area of the watershed, and C_t is a constant varying normally in the range from 1.8 to 2.2, with some indication of lower values for watersheds with steeper slopes. In a study of 27 watersheds in Pennsylvania, A. C. Miller, Kerr, and Spaeder (1983) found C_t to vary from 1.01 to 4.33. This much variation was also observed by Bull (1968) for 25 watersheds in the lower Missouri River basin in Kansas and Missouri. Hudlow and R. A. Clark (1969) found its range from .40 to 2.26 for 13 watersheds (.5 to 75 square miles) in central Texas. Linsley (1943) noted its range from .3 to .7 for northwestern states in the United States. Cordery (1968) found C_t to vary from .4 to 2.4 for 12 watersheds (.021 to 248 square miles) in eastern New South Wales, Australia. The quantity LL_c is a measure of the size and shape of the watershed. We thus see that watershed slopes were not considered.

The unit duration of the effective rainfall was defined by

$$D = \frac{t_p}{5.5} \tag{18.10}$$

Due to this effective rainfall, the peak h_p of the resulting UH was given by

$$h_p = \frac{C_p}{t_p} \tag{18.11}$$

where C_p is a constant ranging from .56 to .69 and h_p has the dimensions L/T. If h_p is desired in the dimensions of L^3/T, then equation (18.11) must be multiplied by watershed area A and an appropriate conversion factor. A. C. Miller, Kerr, and Spaeder (1983) found C_p to vary from .23 to .67 for 27 watersheds in Pennsylvania. Cordery found its range as .4 to 1.1 for 12 watersheds in eastern New South Wales, Australia. Hudlow and R. A. Clark (1968) found its variability from .31 to 1.22 for watersheds in central Texas. For Sacramento and lower San Joaquin Rivers in California, Linsley (1943) noted its variability from .35 to .59. For example, the time base t_b in days of the UH was defined as

$$t_b = 3 + 3\left(\frac{t_p}{24}\right) \tag{18.12}$$

The constants in equation (18.12) are fixed by the procedure of hydrograph separation. Having determined t_p, h_p, and t_b, the UH can be constructed such that its volume is equal to 1 unit. Obviously, many UHs can be sketched that will satisfy this requirement. As an aid to sketching a reasonable UH, the U.S. Army Corps of Engineers (1940) developed a relation between h_p and the width of the UH at values of 50 percent (W_{50}) and 75 percent (W_{75}) of h_p, as shown in Fig. 18.3.

These time widths can be evaluated from the following equations:

$$W_{50} = \frac{830}{q_p^{1.1}} \tag{18.13}$$

$$W_{75} = \frac{470}{q_p^{1.1}} \tag{18.14}$$

where q_p is the UH peak h_p in square feet per second per square mile. The Corps of Engineers have suggested, as a guide for shaping the UH, that these widths should be positioned such that one-third of the width is placed to the left of the peak and two-thirds is to the right. However, Hudlow and R. A. Clark (1969) found in their study that an allocation of four-tenths of the width to the left and six-tenths of the width to the right produced optimum results. On the other hand, Viessman, et al. (1977) stated that as a general rule of thumb, the time widths W_{50} and W_{75} should be proportioned each side of peak in a ratio of 1:2, with the larger proportion on the right of the peak. Equation (18.12) is reasonable for large watersheds but is known to produce exceptionally large t_b for smaller watersheds. As a general rule of thumb, t_b can be taken as three to five times the time to peak for sketching

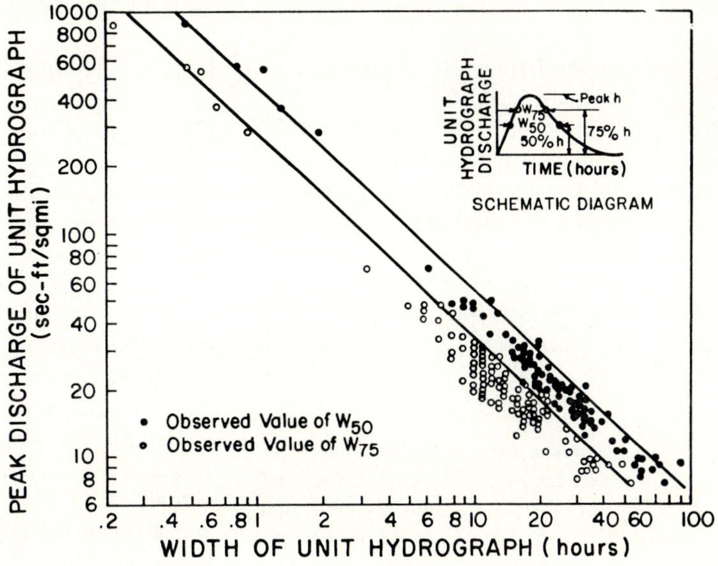

FIGURE 18.3 Unit hydrographs width at 50 and 75 percent of peak value (after U.S. Army Corps of Engineers, 1940).

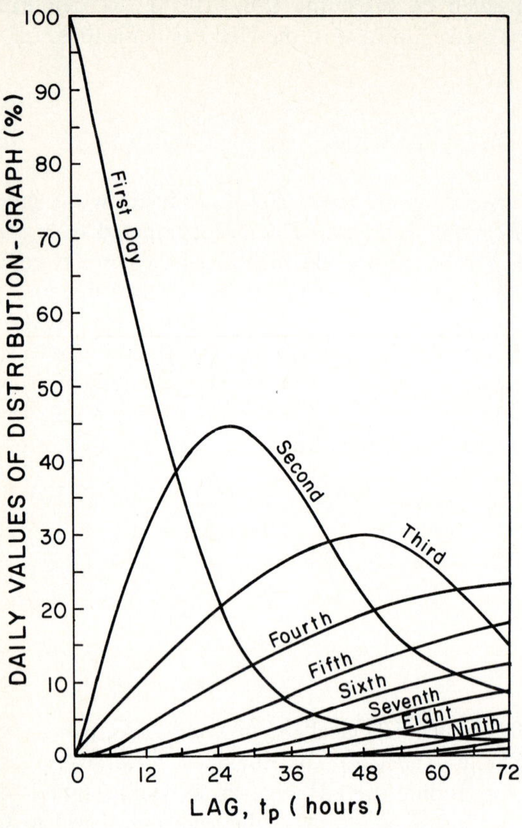

FIGURE 18.4 Relation between basin lag and distribution graph (after Snyder, 1938).

a UH. Some workers (Hudlow and R. A. Clark, 1969) have attempted to fit a mathematical function to the UH.

If the duration of the effective rainfall, say t_R, is different from D as previously defined, a modified lag time t_p^* can be obtained from

$$t_p^* = t_p + \frac{t_R - D}{4} \tag{18.15}$$

Using this t_p^*, h_p and t_b can be determined from equations (18.11) and (18.12), and then the UH can be constructed.

The Snyder model has been widely used in applied hydrology (Laden, Reilly, and Minnotte, 1940; Linsley, 1943; Bull, 1968; Cordery, 1968; Hudlow and R. A. Clark, 1969; A. C. Miller, Kerr, and Spaeder, 1983), and is a one-parameter method. Snyder also developed a relationship between t_p and the ordinates of the distribution graph, as shown in Fig. 18.4. From a hydrographic standpoint, equations (18.9)–(18.11) appear to have a real physical significance, but equations (18.12) and (18.13) and Fig. 18.4 lack generality.

EXAMPLE 18.1

Derive a 3-h UH by the Snyder model for a watershed of area 54 km^2. It has a main stream that is 10 km long. The distance measured from the watershed outlet to a point on the stream nearest to the centroid of the watershed is 3.75 km. Take C_t as 2.0 and C_p as .65.

Solution: The watershed area is $A = 54$ km^2 $= 20.86$ mi^2. The length of main stream is

$$L = 10 \text{ km} = 6.215 \text{ mi}$$

$$L_c = 3.75 \text{ km} = 2.331 \text{ mi}$$

Therefore,

$$t_p = 2(6.215 \times 2.331)^{0.3} = 4.46 \text{ h}$$

$$D = \frac{t_p}{5.5} = 0.81$$

Since the desired duration t_R is 3 h, we need to modify the value of t_p,

$$t_p = 4.46 + \frac{(3 - 0.81)}{24} = 5 \text{ h}$$

Therefore,

$$h_p = \frac{0.65}{5} = 0.13/\text{h}$$

The peak ordinate in terms of discharge is

$$q_p = 640 \frac{c_p}{t_p}$$

$$= 0.65 \frac{640}{5} = 83.075 \text{ ft}^3/\text{s/mi}^2$$

$$Q_p = q_p A = 83.075 \times 20.86 = 1732.8 \text{ ft}^3/\text{s} = 49.09 \text{ m}^3/\text{s}$$

The widths of the UH at 50 percent and 75 percent of peak discharge, respectively, are

$$W_{50} = \frac{830}{(83.075)^{1.1}} = 6.42 \text{ h}$$

$$W_{75} = \frac{470}{(83.075)^{1.1}} = 3.64 \text{ h}$$

W_{50} and W_{75} are partitioned about $t = t_p$ such that one-third is toward the rising limb and two-thirds is toward the recession limb. The 3-h UH is plotted in Fig. 18.5. The UH is drawn such that the area under the curve is 1 in.

18.2.3 The McCarthy Model

In 1938 McCarthy analyzed data from 22 watersheds ranging in area from 74 to 716 square miles located in the Connecticut River basin. He related three parameters of 6-hour UH, including the time of rise, the peak discharge, and the base length, to watershed characteristics such as area, overland slope expressed as the average slope of the hypsometric curve (and hence in feet per square miles), and stream pattern. This work was not published by McCarthy, but its account is available in the U.S. Army Corps of Engineers (1940).

The area is considered by converting the UHs and the watershed characteristics to the corresponding quantities for model watersheds of 10 square miles by employing the Froude model law. The overland slope is obtained by constructing a graph of percentage of watershed area greater than a given altitude above the gage versus elevation (hypsometric curve) and dividing the area under this curve by half the square of the watershed area. This quantity, being area divided by

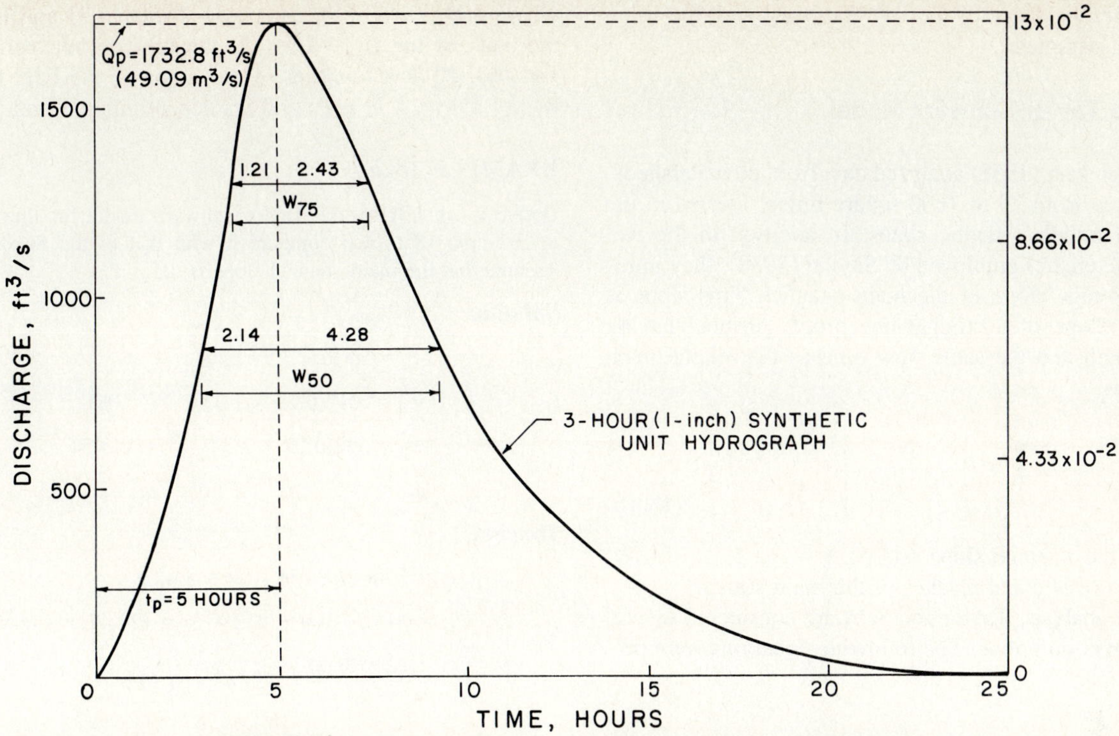

FIGURE 18.5 Example 18.1: Three-hour UH from the Snyder model.

length, has the dimension of length and has, therefore, to be multiplied by the length scale in converting to the model watershed.

The stream pattern number can be obtained by an inspection of the watershed map. The number is defined to have a value of 1 if no stream has a tributary draining more than 25 percent of the total watershed area, a value of 2 if there are two tributaries of approximately the same size draining at least 50 percent of the total watershed area, and a value of 3 if three tributaries drain 75 percent of the watershed area.

McCarthy developed a diagram on a logarithmic paper by plotting the peaks of the UHs against the slope of the respective watersheds for various values of the stream pattern number. He essentially constructed a surface whose coordinates are these quantities. Simply put, for each of the three stream pattern numbers, a curve was drawn relating peaks to slopes. Except for this McCarthy did not correlate any other parameter of the UH with watershed characteristics. Instead, he separately correlated the lag time and the base time with the UH peak; the lag time was defined as the time from the beginning of the effective rainfall to the peak.

This method has several drawbacks: (1) The slope is tedious to determine. (2) No information is obtained on the UH shape. (3) The diagram developed in the method applies only to one rainfall duration and possibly only to one region. Further, as Nash (1958) pointed out, an implicit assumption in expressing the lag time and base time of the UH as functions of the peak is that all UHs having the same peak are identical. This model is a one-parameter method. The use of the model law implies that small watersheds are similar to large ones. This is hardly a tenable assumption.

18.2.4 The Commons Model

Commons (1942) hypothesized that the dimensionless hydrographs, the so-called basic hydrographs, derived from all short-period UHs would be identical. The dimensionless hydrograph represents, in a sense, the shape of the UH. He suggested that, as a first approximation, all short-period UHs were of the same basic shape and differed only in scale. Some may be high and of short duration, whereas others may be low and of long duration. This means that a dimensionless hydrograph is all that is needed to be synthesized for a watershed. The dimensionless hydrograph can be obtained by dividing all the ordinates of the UH by the value of the peak ordinate and multiplying all the abscissas by the peak divided by the volume of the UH. Commons divided the base time into 100 units and the peak discharge into 60 units. The area under the graph would then be 1196.5 square units (time multiplied by discharge). The choice of these numbers of units was dictated by the units that he used. For example, if the total flood discharge in acre-feet is divided by 1196.5, it gives the value of 1 square unit in acre-feet. Dividing the peak flow in cubic feet per second by 60 gives the value of 1 unit of flow in cubic feet per second. Since 1 cubic foot per second is approximately 1 acre-foot in 12 hours, the value of 1 square unit multiplied by 12 and divided by the value of 1 unit of flow yields the value of 1 unit of time in hours. In fact, this rather awkward choice of units can be avoided by dealing with discharge in the dimensions of *L*/*T*.

The Commons method is also a one-parameter method. The method assumes that the knowledge of one parameter alone, say the peak of the UH, implies knowledge of the

UH. However, the shape of the UH may vary with the value of the chosen parameter.

18.2.5 The Taylor-Schwarz Model

Taylor and Schwarz (1952) analyzed data from 20 watersheds, ranging in area from 20 to 1600 square miles, located in the northern and middle Atlantic states. In addition to the watershed characteristics employed by Snyder (1938), they introduced the average slope of the main channel. This slope is equal to the slope of a straight-line profile channel having the same length and the same flow time as that of the main stream; that is,

$$S_c = \left[\frac{N}{\sum_{i=1} 1/S_i^{.5}} \right]^2 \tag{18.16}$$

where S_c is the average slope and S_i, $i = 1, 2, \ldots, N$, are the slopes of N equal reaches of the main stream.

In their analysis, Taylor and Schwarz considered several UHs of different durations. The following equations were presented:

$$t_p = C_1 \exp(m_1 D) \tag{18.17}$$

where

$$m_1 = .212(LL_c)^{-.36} \tag{18.18}$$

$$C_1 = \frac{.6}{S_c^{.5}} \tag{18.19}$$

and

$$h_p = C_2 \exp(m_2 D) \tag{18.20}$$

where

$$m_2 = .121 S_c^{.142} - .05 - m_1 \tag{18.21}$$

$$C_2 = 382(LL_c)^{-.36} \tag{18.22}$$

Here h_p is in square feet per second per square mile. Finally,

$$t_b = 5\left(t_p + \frac{D}{2}\right) \tag{18.23}$$

The remainder of the symbols carry the meanings defined in the Snyder model.

Taylor and Schwarz presented a nomograph for solving these equations. They also recommended the use of the Corps of Engineers' width graphs for drawing the UH. This model is a two-parameter model. As Nash (1958) pointed out, equation (18.20) shows that they correlated, for each watershed, the peaks of the UHs of different periods with the periods of the UHs and a function of the form

$$h(D, P) = h(0, P)\exp(m_2 D) \tag{18.24}$$

where $h(D, P)$ is the peak of the D-hour UH and $h(0, P)$ is the peak of the IUH. Clearly, the relation between $h(D, P)$ and $h(0, P)$ is a function of the shape of the IUH measured by m_2. Further, $h(0, P)$ and m_2 depend upon S and (LL_c).

EXAMPLE 18.2

Derive a 3-h UH by the Taylor-Schwarz model for the watershed in Example 18.1, and compare it with that of the Snyder model. Assume that the main channel slope is .05.

Solution

$$c_1 = .6(.05)^{-.5} = 2.68$$

$$m_1 = .212(6.22 \times 2.33)^{-.36} = .081$$

$$c_2 = 382(6.22 \times 2.33)^{-.36} = 145.34$$

$$m_2 = .121(.05)^{.142} - .05 - m_1 = -.052$$

Therefore,

$$t_p = 2.68 \exp(.081 \times 3) = 3.42 \text{ h}$$

$$h_p = 145.34 \exp(-.052 \times 3) = 124.47 \text{ ft}^2\text{/s/mi}^2$$

$$= \frac{12 \times 3600}{(5280)^2} \times 124.47 = 0.19\text{/h}$$

$$t_b = 5\left(3.42 + \frac{3}{2}\right) = 24.6 \text{ h}$$

We can compute W_{50} and W_{75} as in the case of the Snyder model,

$$W_{50} = \frac{830}{(124.47)^{1.1}} = 4.116$$

$$W_{75} = \frac{470}{(124.47)^{1.1}} = 2.33$$

The UH is plotted in Fig. 18.6. A comparison with the Snyder model results shows the following:

	h_p (1/h)	t_p (h)	t_b (h)
Taylor-Schwarz	.19	3.42	24.6
Snyder	.13	5.00	25.0

18.2.6 The SCS Model

The method of hydrograph synthesis employed by the Soil Conservation Service (SCS) (1955, 1971), U.S. Department of Agriculture, uses an average dimensionless hydrograph derived from an analysis of a large number of natural UHs for watersheds varying widely in size and geographical locations. The ordinates of this dimensionless hydrograph, as shown in Fig. 18.7, are expressed as h/h_p, and its abscissas are expressed as t/t_p, in which h is the discharge at any time t, h_p is the peak discharge, and t_p is the time from the beginning of rise to the peak. Its ordinates can also be expressed as V_a/V, where V_a is the accumulated value of h at time t and V the total volume. This UH has a point of inflection approximately 1.7 times t_p, and t_p is approximately $.2t_b$, the base time.

The dimensionless UH (see Table 18.1) has approximately 37.5 percent of the total volume in the rising side, which can be represented by 1 unit of time and 1 unit of discharge. The SCS method suggests that the dimensionless

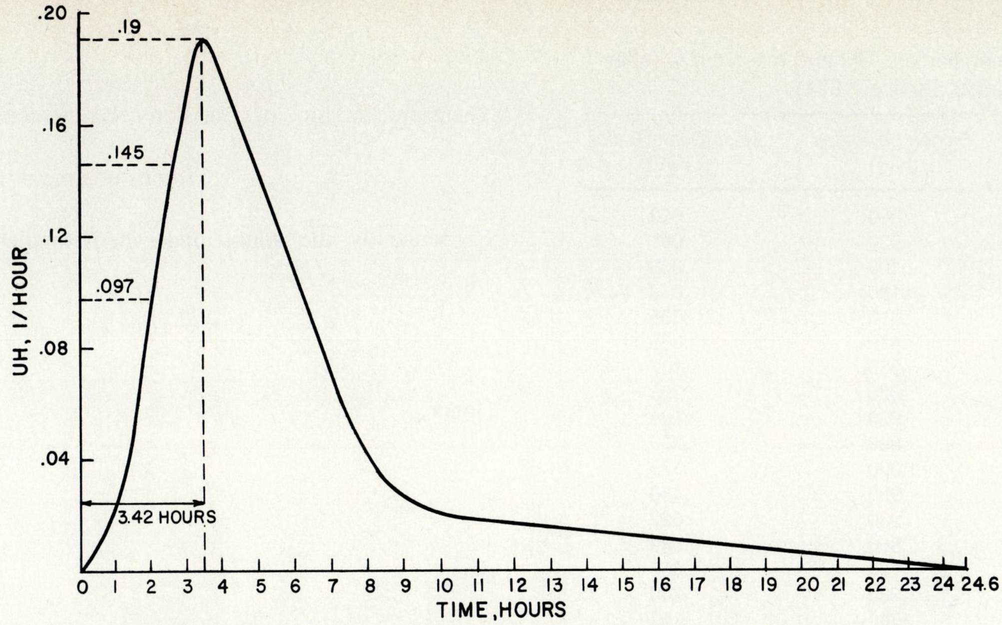

FIGURE 18.6 Example 18.2: Three-hour UH from the Taylor-Schwarz model.

UH can also be represented by an equivalent triangular hydrograph, as shown in Fig. 18.8, having the same units of time and discharge, having in turn the same percent of volume on the rising side of the triangle. This enables defining t_b, the base of the triangle, in relation to t_p. If 1 unit of time t_p equals .375 of volume, then

$$t_b = \frac{1}{.375} = 2.67 \text{ units of time}$$

Chu and Lytle (1972) found for 23 watersheds (.18 to 750 square miles) in South Dakota that

$$t_b = .1021 \left(\frac{L}{S^{.5}}\right)^{.968} \tag{18.25a}$$

in which L is the main channel length in miles and S is the main channel slope in feet per foot. The correlation coefficient was approximately .982. For simplicity we may use

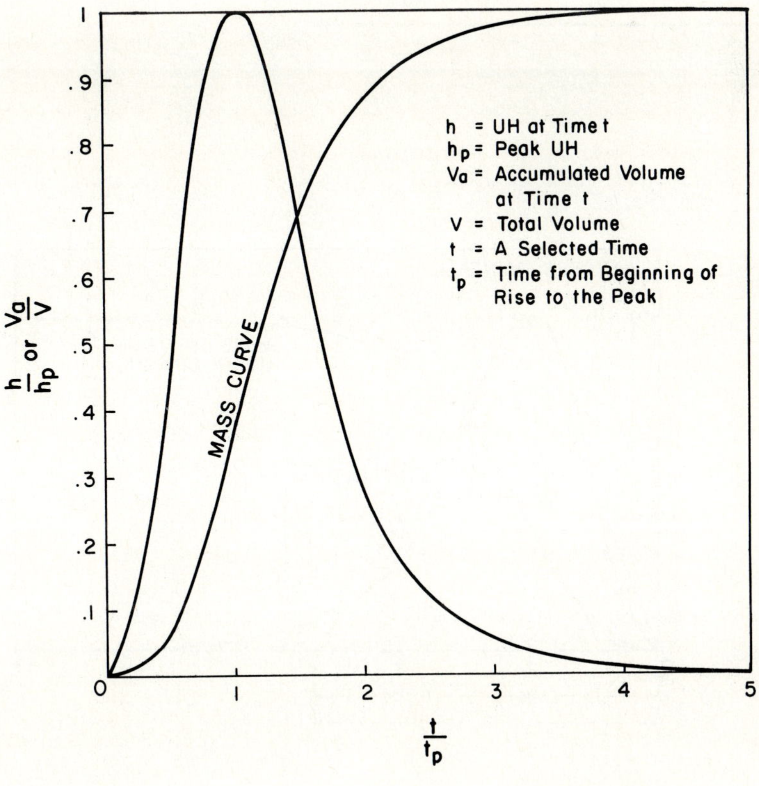

FIGURE 18.7 Dimensionless UH and mass curve (after Soil Conservation Service, 1971).

TABLE 18.1

Ratios for dimensionless UH and mass curve (after Soil Conservation Service, 1971)

Time Ratios (t/t_p)	*Discharge Ratios* (h/h_p)	*Mass Curve Ratios* (V_a/V)
0	.000	.000
.1	.030	.001
.2	.100	.006
.3	.190	.012
.4	.310	.035
.5	.470	.065
.6	.660	.107
.7	.820	.163
.8	.930	.228
.9	.990	.300
1.0	1.000	.375
1.1	.990	.450
1.2	.930	.522
1.3	.860	.589
1.4	.780	.650
1.5	.680	.700
1.6	.560	.751
1.7	.460	.790
1.8	.390	.822
1.9	.330	.849
2.0	.280	.871
2.2	.207	.908
2.4	.147	.934
2.6	.107	.953
2.8	.077	.967
3.0	.055	.977
3.2	.040	.984
3.4	.029	.989
3.6	.021	.993
3.8	.015	.995
4.0	.011	.997
4.5	.005	.999
5.0	.000	1.000

$$t_b = .1 \frac{L}{S^{.5}} \tag{18.25b}$$

Therefore, the time of recession t_r can be computed as

$$t_r = t_b - t_p = 1.67 \text{ units of time} = 1.67t_p$$

Consequently, the volume under the triangular UH V is

$$V = \frac{h_p}{2}(t_p + t_r) \tag{18.25c}$$

Hence,

$$h_p = \frac{KV}{t_p} \tag{18.26}$$

where

$$K = \frac{2}{1 + (t_r/t_p)} = 0.75$$

Therefore,

$$h_p = .75 \frac{V}{t_p} \tag{18.27}$$

If h_p is to be expressed in volumetric units (L^2/T), then the watershed area and appropriate conversion factor can be multiplied with h_p.

From Fig. 18.6 it is clear that

$$t_p = \frac{D}{2} + t_L \tag{18.28}$$

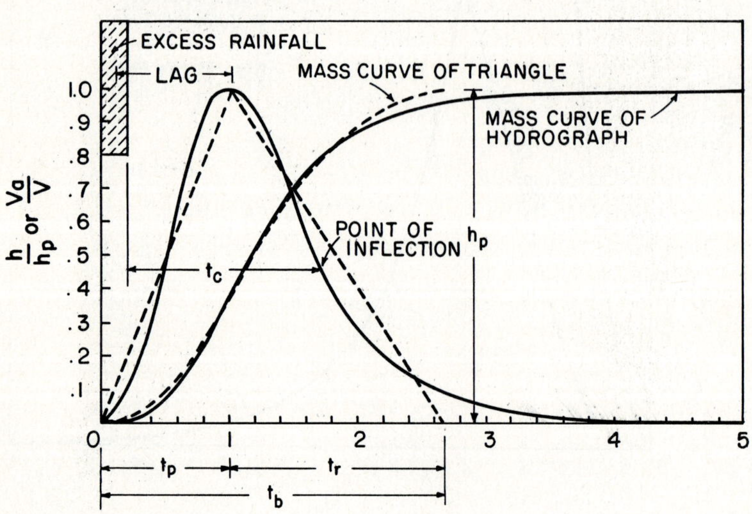

FIGURE 18.8 Dimensionless curvilinear UH and equivalent triangular hydrograph (after Soil Conservation Service, 1971).

where t_L is the lag time between the center of the effective rainfall and peak discharge and D is the duration of the effective rainfall. Further,

$$t_L = .6t_c \tag{18.29}$$

where t_c is the time of concentration. Substituting equations (18.28) and (18.29) into equation (18.27),

$$h_p = \frac{1.5V}{D + 1.2t_c} \tag{18.30}$$

Thus, the SCS model reduces to a one-parameter model. The parameter t_c can be computed from watershed characteristics, as discussed in Chapter 9.

The Soil Conservation Service defined t_c in two ways: (1) the time of travel from the divide to the outlet and (2) the time from the end of the effective rainfall to the point of inflection on the recession limb of the UH. The first definition was used to compute t_c in the SCS model. However, the point of inflection, or t_c as just mentioned, is 1.7 times t_p,

$$t_c = 1.7t_p \tag{18.31}$$

Using equations (18.28), (18.29), and (18.31),

$$t_c + D = 1.7t_p$$
$$1.2t_c + 2D = 2t_p$$

Solving these equations gives

$$D = .133t_c \tag{18.32}$$

Thus the UH is completely specified by one parameter. The preceding relationships can be used to construct either the triangular UH or the curvilinear UH. The SCS model is one of the popular models for synthesizing the UH for small watersheds of less than 500 square miles (Hanson and H. P. Johnson, 1964; Reich, 1965, 1968; Hiemstra, 1968; Wu, 1969a; R. Y. Wang and Wu, 1971; Chu and Lytle, 1972; Reich and Wolf, 1973; Mostaghimi and Mitchell, 1982; Wheater, Shaw, and Rutherford, 1982; McCuen and Bondelid, 1983; Sangal, 1983). Morgan and S. M. Johnson (1962) compared the SCS, Snyder, Commons, and Mitchell models on 12 watersheds (10.1 to 101 square miles) in Illinois and found that errors in prediction of peak discharge ranged up to nearly 200 percent. The four methods were comparable. Recently, Mostaghimi and Mitchell (1982) undertook a comparison on 15 small watersheds (18.4 to 558.1 hectares) in central Illinois of four models: the Cypress Creek, Rational, Chow, and SCS. They found that the SCS model was the best for predicting peak runoff rates.

EXAMPLE 18.3

Derive a 3-h UH for the watershed of Example 18.1, using the SCS model and compare it with that of the Snyder model.

Solution

$$t_p = \frac{D}{2} + t_L = \frac{3}{2} + 4.46 = 5.96 \text{ h}$$

$$t_r = 1.67 \times 5.96 = 9.95 \text{ h}$$

$$t_p = 2.67 \times 5.96 = 15.91 \text{ h}$$

$$h_p = \frac{.75}{5.96} = .125 \text{ h}$$

The UH is sketched in Fig. 18.9. A comparison with the results of the Snyder model shows the following:

	h_p (1/h)	t_p (h)	t_b (h)
SCS	.125	5.96	15.91
Snyder	.13	5.0	25.0

18.2.7 The Hickok-Keppel-Rafferty (HKR) Model

The HKR model (Hickok, Keppel, and Rafferty, 1959) is similar to the SCS method and was developed entirely for small watersheds. The basic data involved the runoff characteristics of 14 watersheds varying in size from 11 to 790 acres located in semiarid regions (Arizona, Colorado, and New Mexico). An average dimensionless hydrograph was developed by plotting h/h_p versus t/t_L. However, t_L was defined as the time difference between the centroid of a limited block of intense rainfall and the resulting peak discharge.

For reasonably homogeneous semiarid range lands up to about 1000 acres in area,

$$t_L = K_1 \left[\frac{A^{.3}}{S_a D_w^{.5}} \right]^{.61} \tag{18.33}$$

where t_L is the lag time in minutes, A is the watershed area in acres, S_a is the average land slope in percent, D_w is drainage density in feet per acre, and K_1 is a constant equal to 106.

For watersheds with widely different physiographic characteristics,

$$t_L = K_2 \left[\frac{(L_{sa} W_{sa})^{.5}}{S_{sa} D_w^{.5}} \right] \tag{18.34}$$

where L_{sa} is the length from the outlet of watershed to the center of gravity of source area in feet, W_{sa} is the average width of the source area in feet, S_{sa} is the average land slope of the source area in percent, D_w is the drainage density of the entire watershed in feet per acre, and K_2 is a constant equal to 23.

Lag time was found to be a major determinant of the hydrograph shape. Therefore, the ratio of the peak rate of runoff h_p to the total runoff volume V was expressed as a function of lag time,

$$\frac{h_p}{V} = \frac{K_3}{t_L} \tag{18.35}$$

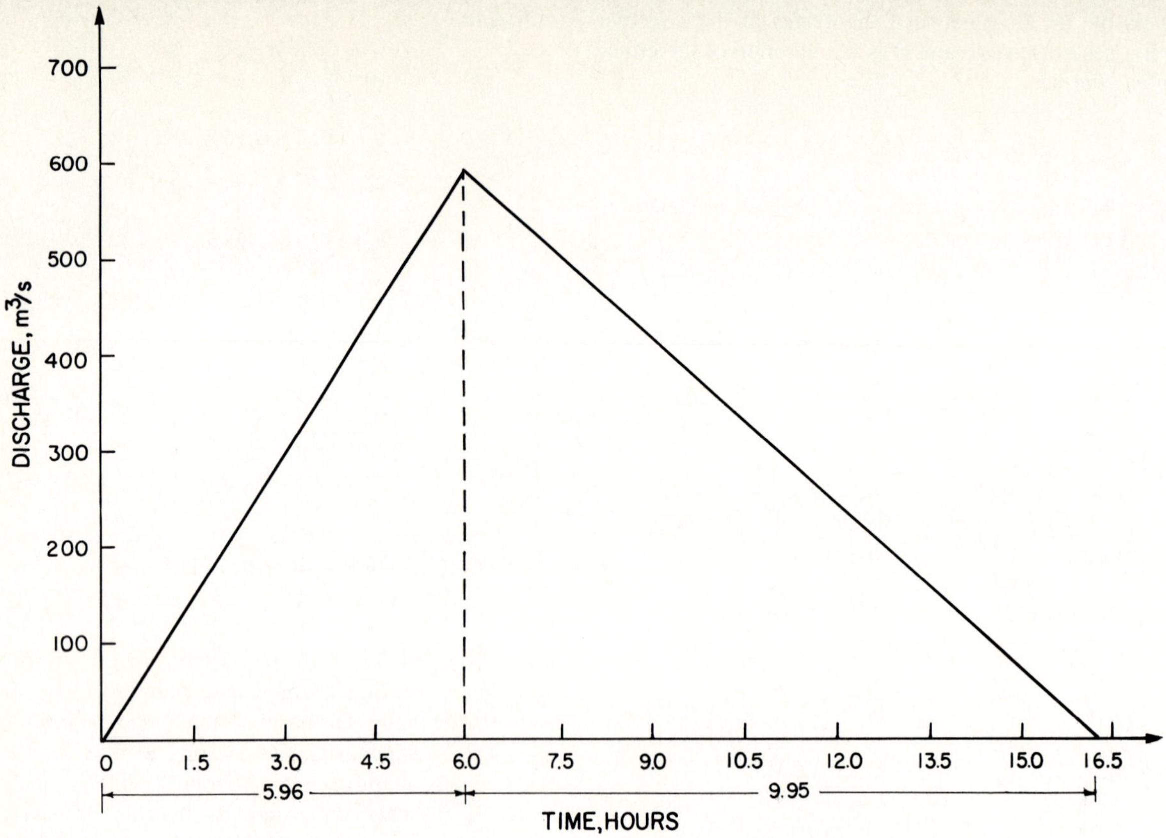

FIGURE 18.9 Example 18.3: Three-hour UH from the SCS model.

For h_p in cubic feet per second, V in acre-feet and t_L in minutes, K_3 equals 545.

Using these equations, a generalized dimensionless hydrograph and mass curve were prepared, as shown in Fig. 18.10. Data for these curves are shown in Table 18.2. This model is, again, a one-parameter model and therefore has the same drawbacks as other one-parameter models.

18.2.8 The Double-Triangle Model

The double-triangle method for hydrograph synthesis on small watersheds was developed by Ardis (1972, 1973) and the Tennessee Valley Authority (1972) and has been used by Reich and Wolf (1972), Diskin and Lane (1976), and Tay and Singh (1978). This model is essentially an extension of the SCS model using a simple triangular UH. The double-triangle UH is obtained by assuming the watershed response to be composed of two parts, each represented by a triangular shape. The first part represents a fast response with a high peak and short time base, whereas the second represents a slow response

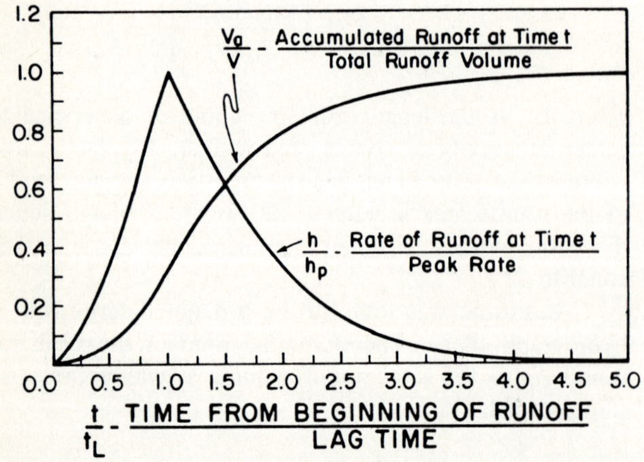

FIGURE 18.10 General dimensionless hydrograph and mass curve (after Hickok, Keppel, and Rafferty, 1959).

TABLE 18.2

Dimensionless hydrograph and mass curve for the HKR model

t/t_L	h/h_p	V_a/V	t/t_L	h/h_p	V_a/V
0	0	0	1.6	.545	.671
.1	.025	.002	1.7	.482	.707
.2	.087	.007	1.8	.424	.742
.3	.160	.020	1.9	.372	.773
.4	.243	.036	2.0	.323	.799
.5	.346	.063	2.2	.241	.841
.6	.451	.096	2.4	.179	.875
.7	.576	.136	2.6	.136	.900
.8	.738	.180	2.8	.102	.917
.9	.887	.253	3.0	.078	.932
1.0	1.000	.325	3.4	.049	.953
1.1	.924	.400	3.8	.030	.965
1.2	.839	.464	4.2	.020	.973
1.3	.756	.523	4.6	.012	.979
1.4	.678	.578	5.0	.008	.983
1.5	.604	.627	7.0	0	0

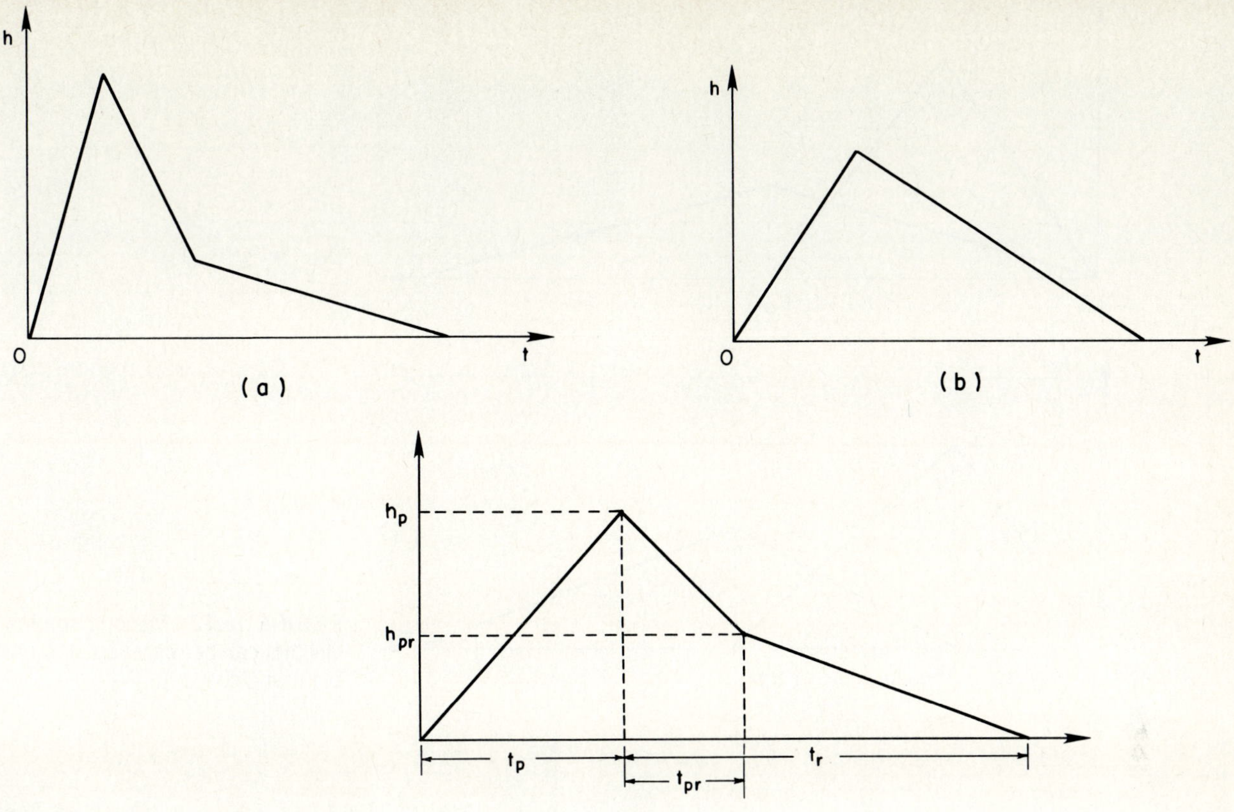

FIGURE 18.11 The double-triangle UH.

having a low peak and longer time base. The latter may start either at the same time as the former or at the time corresponding to the peak of the fast response. In either case, the superimposition of the ordinates of the two individual responses results in a three-line polygon, a double triangle. Evidently the shape of this polygon depends on the relative proportions of the two responses. Fig. 18.11 illustrates the double triangle UH. However, it is tacitly assumed that the delayed response peaks where the initial response ends.

The shape of the double triangle UH can be specified by four parameters. A fifth parameter required for complete specification can be obtained from the condition that UH encloses a unit area. As shown in Fig. 18.11, the five parameters are: t_p, time to peak; t_r, time of recession; h_p, peak ordinate; h_{pr}, ordinate at the break point (or peak ordinate of delayed response), which can be expressed as $h_{pr} = ah_p$; and t_{pr}, time from the peak ordinate to the break point, which can be written as $t_{pr} = bt_r$. Therefore, the parameter h_p can be written as

$$h_p = \frac{2}{t_p + (a + b)t_r} \tag{18.36}$$

For the recession limb to be concave,

$$0 < a + b < 1.0 \tag{18.37}$$

Any four parameters can be used to define the double-triangle UH. For example, if h_p, t_p, t_{pr}, and t_r are taken as parameters, then

$$h_{pr} = \frac{[2 - h_p(t_p + t_{pr})]}{t_{pr} + t_r} \tag{18.38}$$

Each parameter measures a specific attribute of the UH.

As Ardis (1972) has shown, the double triangle has considerable flexibility to generate UHs of various shapes. This is illustrated in Fig. 18.12 for a constant time base ($t_p + t_{pr} + t_r$). For convenience, let $t_1 = t_p$, $t_2 = t_p + t_{pr}$, and $t_3 = t_2 + t_r$. Therefore, the basic four parameters defining the double-triangle UH can be h_p, t_1, t_2, and t_3. Ardis selected the following watershed characteristics, which were related to these parameters using multiple linear regression analysis, with the duration of the UH as 1 hour. Eleven watersheds (7.04 to 16.8 square miles) lying in the Tennessee River basin were used.

1. Drainage area A in square miles.
2. Watershed shape, a dimensionless measure defined as

$$S_p = \frac{L_c^2}{A} \tag{18.39}$$

where S_p is the shape factor measuring watershed elongation and L_c is the length of the mainstream channel measured to the watershed boundary.

3. Mainstream sinuosity, S_u, a measure of sinuosity defined as

$$S_u = \left(\frac{L_c}{L_1}\right) - 1 \tag{18.40}$$

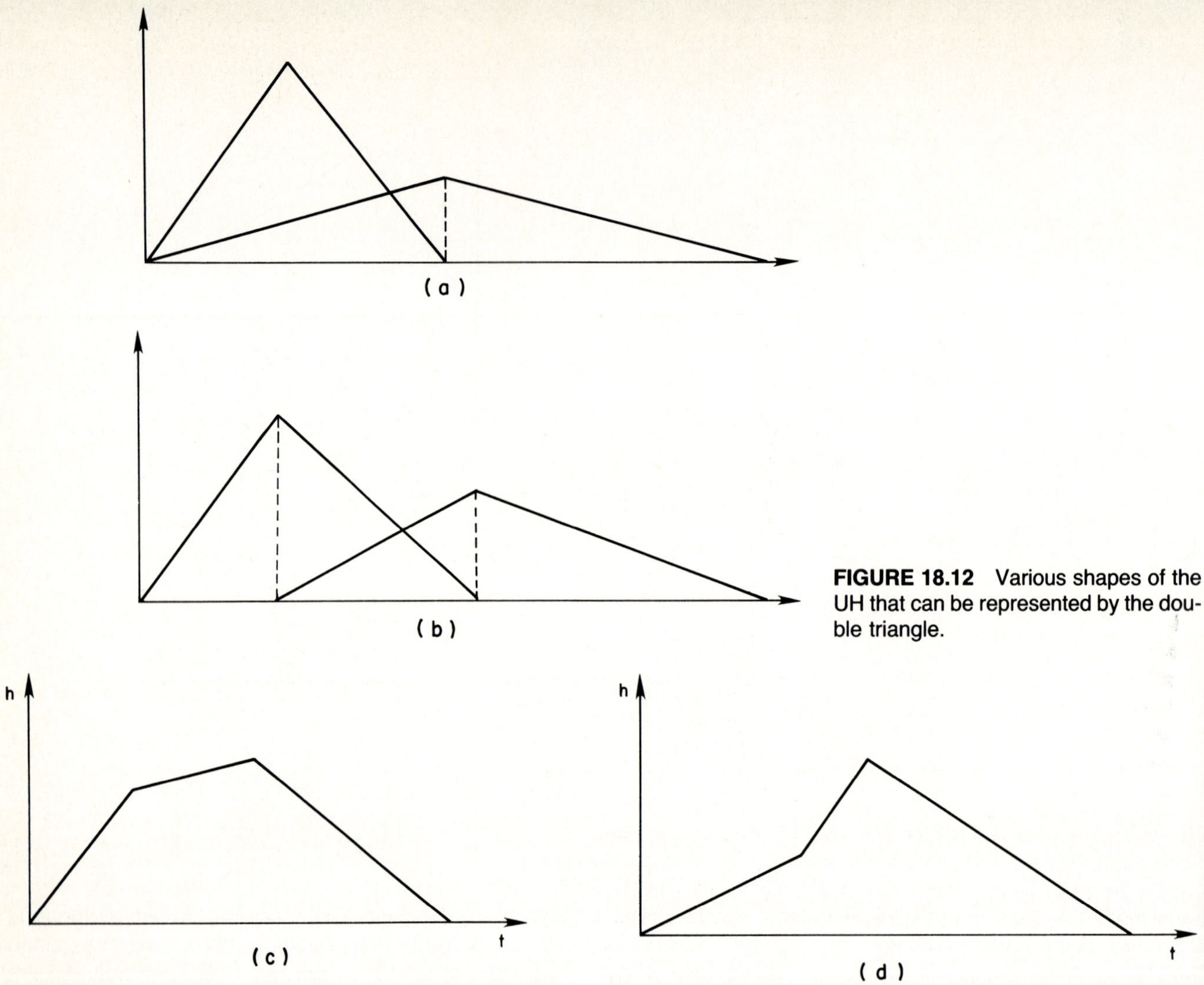

FIGURE 18.12 Various shapes of the UH that can be represented by the double triangle.

where L_1 is the length of the mainstream channel measured in 1-mile chords.

4. Drainage density D_w, a measure of the extent of channelization in a watershed and hence the runoff-carrying capacity, defined as

$$D_w = \frac{1.571\, \Sigma N}{\Sigma L_i} \tag{18.41}$$

where ΣN denotes the total number of intersections all streams and extended channels make with the grid system and ΣL_i is the total length of all grid lines within the watershed in miles.

5. A measure of soils from low-flow data, S_0, defined as

$$S_0 = Q_{70} \tag{18.42}$$

where Q_{70} is the discharge equaled or exceeded 70 percent of the time.

6. Percent of watershed in forest, W_f.
7. Slope and time measures. The slope of a watershed or channel is a characteristic related to the corresponding velocity of flow. The corresponding time of travel for a specific reach is then its length divided by its slope. Therefore,

$$S_c = \left(\frac{L_c}{T_M}\right)^2 \tag{18.43}$$

where T_M is the time factor and S_c is the weighted mean channel slope in feet per mile.

8. Total direct runoff, V.
9. Duration of the effective rainfall in multiples of D, equal to 1 hour in this case.

The coefficients of correlation between watershed characteristics and the UH parameters varied from .79 to .95. The double-triangle model was found to be a significant improvement over the SCS model (Ardis, 1973; Diskin and Lane, 1976).

EXAMPLE 18.4

Derive a 3-h UH by the double-triangle method for the watershed in Example 18.1, and compare it with that of the Snyder model.

Solution: As a first approximation we assume $a = .5$ and $b = .4$. Let $t_r = 1.67t_p$. From the Snyder model $t_L = 4.47$ h. Therefore,

$$t_p = \frac{3}{2} + 4.47 = 5.97 \text{ h}$$

$$h_p = \frac{0.8}{5.97} = 0.134/\text{h}$$

The other parameters are computed as follows:

$$t_r = 1.67 \times 5.97 = 9.97 \text{ h}$$

$$h_{pr} = ah_p = 0.5 \times 0.134 = 0.067/\text{h}$$

$$t_{pr} = at_r = 0.5 \times 9.97 = 3.99 \text{ h}$$

h_p and h_{pr} can be expressed in cubic meters per second, taking into account that $A = 54 \text{ km}^2$,

$$h_p = 22.10 \text{ m}^3/\text{s}$$
$$h_{pr} = 10.05 \text{ m}^3/\text{s}$$

A comparison with the Snyder model results might suggest that b should be around .3 and t_r/t_p, around 2.0. With these modifications we obtain

$$t_p = 5.97 \text{ h}$$

$$h_p = \frac{.77}{t_p} = .129/\text{h}$$

$$t_r = 11.94 \text{ h}$$

$$t_{pr} = 3.58 \text{ h}$$

$$t_{pr} = 0.0645/\text{h}$$

The UH is sketched in Fig. 18.13.

18.3 LINEAR CONCEPTUAL MODELS

18.3.1 The Clark Model

This model has been discussed in Chapter 13. It involves determination of the TAC diagram and the storage coefficient k. C. O. Clark (1943) suggested that k can be approximated by the relation

$$k = \frac{CL}{S_c^{.5}} \tag{18.44}$$

where L is the channel length in miles, S_c is the mean channel slope, and C is a constant varying from about .8 to 2.2.

Linsley (1945), in a discussion of Clark's paper, suggested a modification of equation (18.44) by including a square root of area term,

$$k = bL\left(\frac{A}{S_c}\right)^{.5} \tag{18.45}$$

where b is a coefficient found to vary from .04 to .075 for 11 rivers in California and Virginia. This better represents the effect of storage in tributaries on flood flows.

The time base of the TAC diagram t_c was considered to equal the time interval between the end of rain and the point of contraflexure on the hydrograph recession limb. This time base was measured from the recorded floods, and not related to watershed characteristics. It can, however, be determined by using any of the methods discussed in Chapter 9.

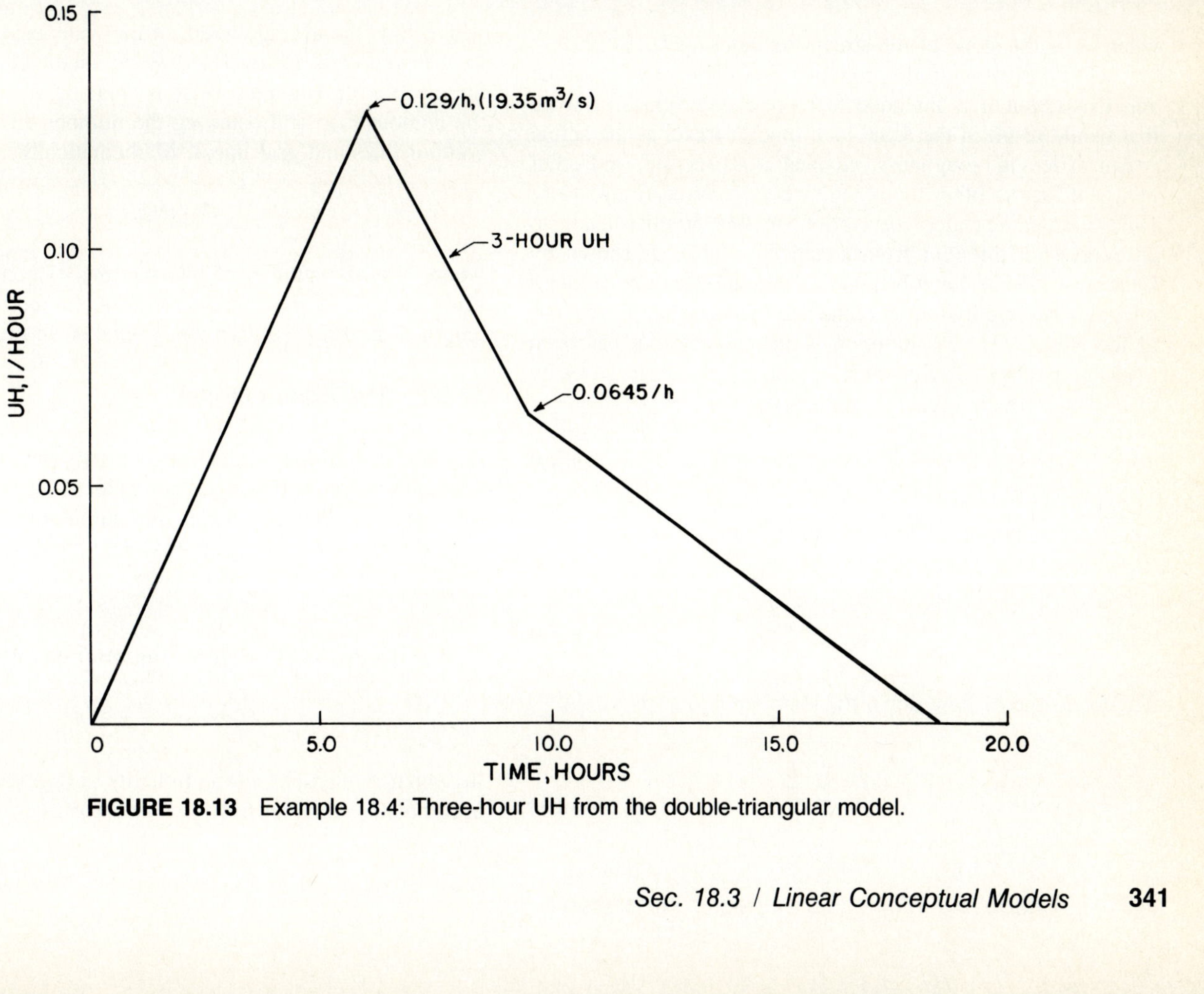

FIGURE 18.13 Example 18.4: Three-hour UH from the double-triangular model.

18.3.2 The Johnstone-Cross (JC) Model

Johnstone and Cross (1949) correlated the parameters t_c and k of the Clark model for 19 watersheds, varying from 25 to 1624 square miles in the Scotie and Sandusky River basins. They found, using correlation analysis on 15 of their 19 watersheds, that

$$t_c = \frac{4.7}{2}\left(\frac{L}{S_c}\right)^{1/2} \tag{18.46}$$

where t_c is the time base of the TAC diagram in hours, L is the length of the main stream in miles, S_c is the average slope of the main stream in feet per mile, and r is a branching factor based on the stream pattern.

The factor r is found by dividing the area under a graph showing the total area tributary to the main stream above any point by the area under a similar curve constructed, assuming the watershed has a single branch and is of uniform width. Mathematically,

$$r = \frac{\int_0^L \int_y^L w\,dx\,dy}{.5AL} \tag{18.47}$$

where w is the effective width of the watershed.

The average slope S_c was computed as

$$S_c = \left[\frac{\Sigma L_i S_i^{.5}}{L}\right]^2 \tag{18.48}$$

where S_i is the slope of the stream segment L_i.

The derivation of this equation is not given by Johnstone and Cross, but it is intended to represent the uniform slope that would result in the same total time of travel as the actual stream if length, roughness, channel cross-section, and other pertinent factors other than slope were unaltered (Laurenson, 1962). An independent derivation on this premise leads to an expression different from equation (18.48), as shown by Laurenson (1962). Let it be assumed that the effects on velocity of roughness and hydraulic radius are the same for all reaches of the stream. This assumption is questionable but has been made previously (Taylor and Schwarz, 1952). The velocity of flow v_i through a reach i can be expressed as

$$v_i = BS_i^{1/2} \tag{18.49}$$

where B is a constant. The time of flow t_i then is

$$t_i = \frac{L_i}{v_i} = \frac{L_i}{BS_i^{1/2}} \tag{18.50}$$

The total time of flow down the main stream, t_c, is

$$t_c = \frac{1}{B}\sum_{i=1}^{M}\frac{L_i}{S_i^{1/2}} \tag{18.51}$$

where M is the number of reaches. The mean velocity of flow v_m can be expressed as

$$v_m = \sum_{i=1}^{M}\frac{L_i}{t_c} = \frac{B\sum_{i=1}^{M} L_i}{\sum_{i=1}^{M} L_i/S_i^{1/2}} \tag{18.52}$$

Since

$$v_m = BS_c^{1/2} \tag{18.53}$$

we can write

$$S_c = \left(\frac{\Sigma L_i}{\Sigma L_i/S_i^{1/2}}\right)^2 \tag{18.54}$$

Equation (18.54) is different from equation (18.48) and should be preferred.

Johnstone and Cross noted that equation (18.46) could be modified, without appreciable loss of accuracy, to

$$t_c = 5\left(\frac{L}{S_c}\right)^{1/2} \tag{18.55}$$

From a correlation analysis they found that

$$k = 1.5 + 90\left(\frac{W}{R}\right)$$

where W is the average width of the watershed in miles equal to A/L and R is an overland slope factor computed by the intersection grid method (that is, placing a square grid over the contour map and counting the number of intersections of contour lines and grid lines). Mathematically,

$$R = NE/g \tag{18.56}$$

where N is the number of intersections of contour lines with grid lines, E is the contour interval in feet and g is total length of grid lines within the watershed in miles.

18.3.3 The Edson Model

Edson (1951) proposed that if isochrones were drawn to represent the time required for each element of the effective rainfall to reach the watershed outlet, the cumulation of area A with time t would result in an approximate parabola,

$$A \propto t^x, \qquad x > 1 \tag{18.57}$$

so that the discharge of direct runoff Q might become

$$Q \propto t^x, \qquad x > 1 \tag{18.58}$$

in which $\propto$ signifies proportionality. However, when runoff accumulates simultaneously over all parts of a watershed, the time of travel required for each component is so affected by the presence of other components that the hypothetical isochrones no longer remain valid. It is reasonable to regard the consequent delay in discharge as a result of valley storage. In other words, a portion of the total runoff must be stored in the valley upstream of the watershed outlet to produce discharge. The valley of the watershed acts as a reservoir; the discharge from its storage decreases exponentially with time:

$$Q \propto \exp(-yt), \qquad y > 0 \tag{18.59}$$

where y is recession constant greater than 0. Thus, the reservoir action of the valley storage exerts a damping effect on the flow given by equation (18.59), which is valid for an indefinite period of time. Further, the valley storage must exist for even the smallest Q, thereby implying that equation (18.59) is valid from $t = 0$.

The combined effect of translation and storage effects can be expressed by coupling equations (18.58) and (18.59),

$$Q \propto t^x \exp(-yt) \tag{18.60}$$

or

$$Q = B\, t^x \exp(-yt) \tag{18.61}$$

where B is a proportionality constant. It may be noted that equation (18.59) is dominant during the recession, whereas equation (18.58) prevails during the rise.

If V is the total volume of discharge, then

$$V = \int_0^\infty Q\, dt \tag{18.62}$$

where Q is specified by equation (18.61). To facilitate integration of equation (18.62), let $m = x + 1$ and $w = yt$. Then

$$V = \int_0^\infty B\left(\frac{w}{y}\right)^{m-1} \exp(-w)\frac{dw}{y} \tag{18.63}$$

Recalling that

$$\Gamma(m) = \int_0^\infty w^{m-1}\exp(-w)\, dw$$

equation (18.63) becomes

$$V = By^{-m}\,\Gamma(m)$$

Therefore,

$$Q = \frac{V y^m}{\Gamma(m)} t^{m-1}\exp(-yt) \tag{18.64a}$$

Edson argued by taking $dQ/dt = 0$ that $m > 2$ and that $Q = 0$ when $t = 0$ and $t = \infty$. Furthermore, the peak discharge and its time would be given as

$$Q_p = \frac{Vy[(m-1)e]^{m-1}}{\Gamma(m)} \tag{18.64b}$$

$$t_p = \frac{m-1}{y} \tag{18.64c}$$

Similarly, by taking $d^2Q/dt^2 = 0 = [m - 1) - yt]^2 - (m - 1)$, it follows that $t = [(m - 1) \pm (m - 1)^{0.5}]/y$. It then follows that

$$t_r = \frac{(m-1) - (m-1)^{.5}}{y}$$

$$t_f = \frac{(m-1) + (m-1)^{.5}}{y}$$

where t_r is the time of maximum dQ/dt and t_f is the time of minimum dQ/dt.

Equation (18.64) can be taken to represent the IUH or UH with $V = 1$. Therefore,

$$h(t, 0) = \frac{y^m}{\Gamma(m)} t^{m-1}\exp(-yt) \tag{18.65}$$

which is identical to the Nash model. Equation (18.65) has two parameters m and y, which can be estimated from watershed characteristics. Edson, except for alluding to it, did not attempt to determine them in this manner. Instead, he computed values from flow records and found m to vary from 3.0 to 7.2 and y to vary from .55 to 1.28 per day.

18.3.4 The Eaton Model

Eaton (1954) also used the Clark method and correlated its parameters t_c and k for seven Tasmanian rivers in Australia with watershed areas ranging from 48 to 322 square miles. His analysis yielded

$$t_c = 1.35\left(\frac{AL}{r}\right)^{.37} \tag{18.66}$$

and

$$k = 1.2\left(\frac{WA}{Lr}\right)^{1/3} \tag{18.67}$$

where the symbols are as defined previously. The branching factor r varies between 1 and 2. Use of slope factor would be preferable to r, but a lack of contour maps for these watersheds prevented Eaton from using it.

18.3.5 The O'Kelly Model

The O'Kelly model (O'Kelly, 1955) like the Clark model, involves determination of the TAC diagram and the storage coefficient k. His basic data included 10 flood-free watersheds in Ireland, varying in area from 43 to 366 square miles and in ground slope from 20 to 330 feet per mile. O'Kelly, as discussed in Chapter 13, proposed to represent, without loss of accuracy, the TAC diagram by an isosceles triangle. Therefore, his method is a two-parameter method containing the base of the isosceles triangle t_c and k as parameters.

O'Kelly modified the values of t_c and k to correspond with a watershed of 100 squares miles in area, using a hydrologic time-scale factor based on a fourth root of the area. The modified values of t_c and k were then plotted against the overland slope S_0, defined as the median value of the maximum slope occurring at the intersections of a grid of square mesh imposed on the watershed map. Thus he concluded that

$$t_c = a\,S_0^b \tag{18.68}$$

and

$$k = c\,S_0^d \tag{18.69}$$

where a, b, c, and d are empirically derived constants. It may be noted that if $b = d$, $t_c/k = a/c$. This implies that the shape of the IUH would be fixed, as suggested by Commons (1942). However, O'Kelly used slightly different values of b and d and thus obtained a basic shape that varied slightly with S_0 and, therefore, with k.

Nash (1958) used O'Kelly's data and found that the evidence for varying the basic shape of the IUH with S_0 was inadequate; practically equally good results could be obtained by using Commons' basic shape and varying a single parameter with S_0. In either case, since t_c and k are both correlated with S_0, a functional relationship between t_c and k is implied. Therefore, the IUH can be completely characterized by a single parameter S_0.

In a discussion of O'Kelly's work, Dooge (1955) proposed to express t_c in hours as

$$t_c = a\frac{A^{1/4}}{S_0^{1/2}} \tag{18.70}$$

where A is the watershed area in square miles, S is the slope in parts per 10,000, and a is an empirical constant. For the values of t_c obtained by O'Kelly, a varied from 10 for a slope of 10 in 10,000 to 14 for a slope of 500 in 10,000. Similarly, he expressed k in hours as

$$k = b\frac{A^{1/4}}{S_0^{1/2}} \tag{18.71}$$

where b is an empirical constant. For the values of k obtained by O'Kelly, b varied from 13 for a slope of 10 in 10,000 to 10 for a slope of 500 in 10,000.

Based on a least squares analysis of O'Kelly's data, Dooge (1956) derived

$$t_c = 2.58\frac{A^{.41}}{S_0^{.17}} \tag{18.72}$$

$$k = 100.5\frac{A^{.28}}{S_0^{.7}} \tag{18.73}$$

The tedium of evaluating average ground slope might be avoided by finding the average channel slope and using the approximate formula

$$S_0 = 4(S_c)^{.8} \tag{18.74}$$

where S_0 is overland slope in degrees and S_c is channel slope in degrees. If the slopes are expressed in feet per mile, equation (18.74) becomes

$$S_0 = 10(S_c)^{0.8} \tag{18.75}$$

EXAMPLE 18.5

Derive a 3-h UH by the O'Kelly model for the watershed in Example 18.1, and compare it with that of the Snyder model.

Solution: We use equations (18.72) and (18.73) to compute t_c and k. S_0 = 865.53 ft/mi.

$$t_c = \frac{2.58(54 \times .3861)^{.41}}{(865.53)^{.17}} = 2.84 \text{ h}$$

$$k = \frac{100.5(54 \times .3861)^{0.28}}{865.53^{.7}} = 2.07 \text{ h}$$

The TAC diagram is an isosceles triangle with peak = 2/2.84. Its base is $t_c = t_1 + t_2$, $t_1 = t_2$. The TAC diagram is

$$w(t) = \frac{2t}{t_c t_1}, \qquad 0 \le t \le t_1$$

$$= \frac{2}{t_c - t_1} - \frac{2t}{t_c(t_c - t_1)}, \qquad t_1 \le t \le t_1 + t_2 = t_c$$

The IUH of the O'Kelly model is

$$h(t) = \frac{4}{t_c^2}\left[t - k + k\exp\left(-\frac{t}{k}\right)\right], \qquad 0 \le t \le t_1$$

$$= -1.012 + .496t + 1.012\exp(-0.49\,t), \qquad 0 \le t \le 1.42$$

$$h(t) = \frac{4}{t_c^2}\exp\left[-\frac{(t-t_1)}{k}\right]\left[t_1 - k + k\exp\left(\frac{-t_1}{k}\right)\right]$$
$$+ \frac{2}{(t_c - t_1)}\left(1 - \frac{t-k}{t_c}\right)$$
$$+ \frac{2}{t_c - t_1}\exp\left[-\frac{(t-t_1)}{k}\right]\left[\frac{t_1 - k}{t_c} - 1\right], \qquad t_1 \le t \le t_c$$

$$= 2.42 - 0.496\,t - 3.047\exp[-0.49\,\mathrm{t}], \qquad t_1 \le t \le t_c$$

$$h(t) = \frac{4}{t_c^2}\exp\left[-\frac{(t-t_1)}{k}\right]\left[t_1 - k + k\exp\left(-\frac{t_1}{k}\right)\right]$$
$$+ \frac{2}{t_c - t_1}\exp\left[-\frac{(t-t_c)}{k}\right]$$
$$\left[1 - \frac{t_c - k}{t_c}\right] + \frac{2}{t_c - t_1}\exp\left[-\frac{(t-t_1)}{k}\right]$$
$$\left[\frac{t_1 - k}{t_c} - 1\right], \qquad t \ge t_c$$

$$= 1.021\exp(-49t), \qquad t \ge t_c$$

The 3-h UH, $h(3, t)$ is given as

$$h(3, t) = \frac{1}{3}\int_{t-3}^{t} h(0, t)\,dt$$

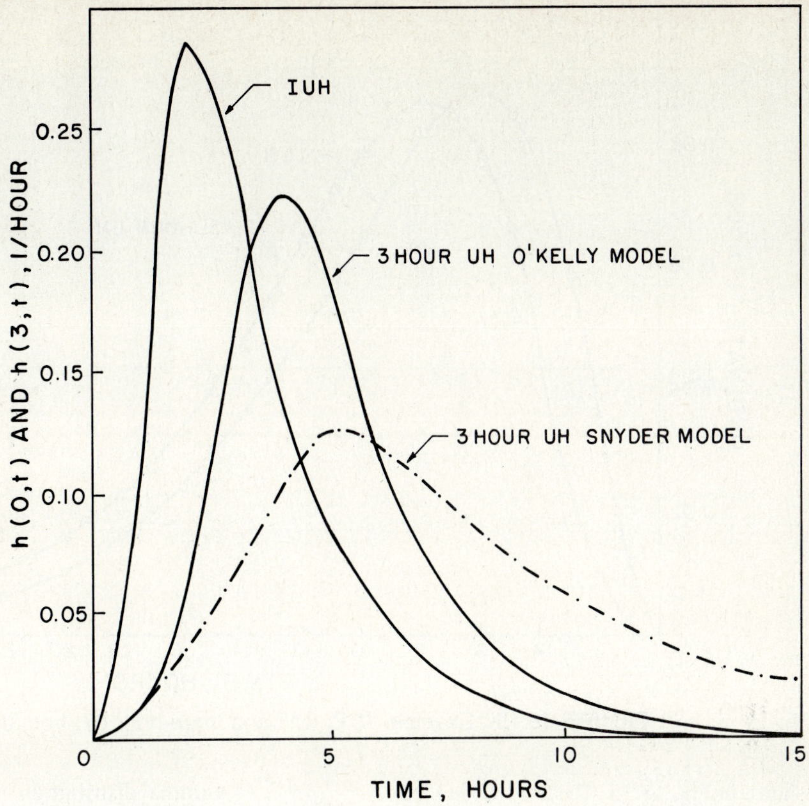

FIGURE 18.14 Example 18.5: Three-hour UH from the O'Kelly model.

Fig. 18.14 shows the graph of the IUH and the 3-h UH of the O'Kelly model. A comparison with the Snyder model results shows the following:

	h_p (1/h)	t_p (h)	t_b (h)
O'Kelly Model	.223	4	15
Snyder Model	.13	5	25

18.3.6 The Nash Model

Nash (1962) applied his model to British catchments. As in Chapter 13, this model has two parameters n and k. Nash showed that these parameters were related to the first and second moments of the IUH about the origin as

$$m_1 = nk \tag{18.76}$$

$$m_2 = \frac{1}{n} \tag{18.77}$$

These moments were then correlated empirically with watershed characteristics as

$$m_1 = 27.6\, A^{.3}\, S_0^{-.3} \tag{18.78}$$

and

$$m_2 = .41\, L^{-.1} \tag{18.79}$$

where S_0 is the overland slope in parts per 10,000, calculated as the mean of a grid sample of slopes, A is the watershed area in square miles, and L is the channel length in miles.

EXAMPLE 18.6

Compute the IUH by the Nash model for the watershed in Example 18.1. Compute the 3-h UH and compare it with that of the Snyder model.

Solution: In order to use equation (18.78), we must compute S_0. From equation (18.75),

$$S_0 = 10(.05 \times 5{,}280)^{.8} = 865.53 \text{ ft/mi}$$

$$A = \frac{54}{2.589} = 20.86 \text{ mi}^2$$

$$L = 10 \text{ km} = 6.215 \text{ mi}$$

Therefore,

$$m_1 = 27.6(20.86)^{.3}(865.53)^{-.3} = 9.026$$

$$m_2 = .41(6.215)^{-.1} = .342$$

This leads to

$$n = \frac{1}{m_2} = \frac{1}{0.342} = 2.92 \cong 3$$

$$k = \frac{m_1}{n} = \frac{9.026}{2.92} = 3.08 \text{ h}$$

Thus

$$h(t) = \frac{1}{3.08(3-1)!}\left(\frac{t}{3.08}\right)^2 \exp\left(-\frac{t}{3.08}\right)$$

$$= 0.0171t^2 \exp\left(\frac{-t}{3.08}\right)$$

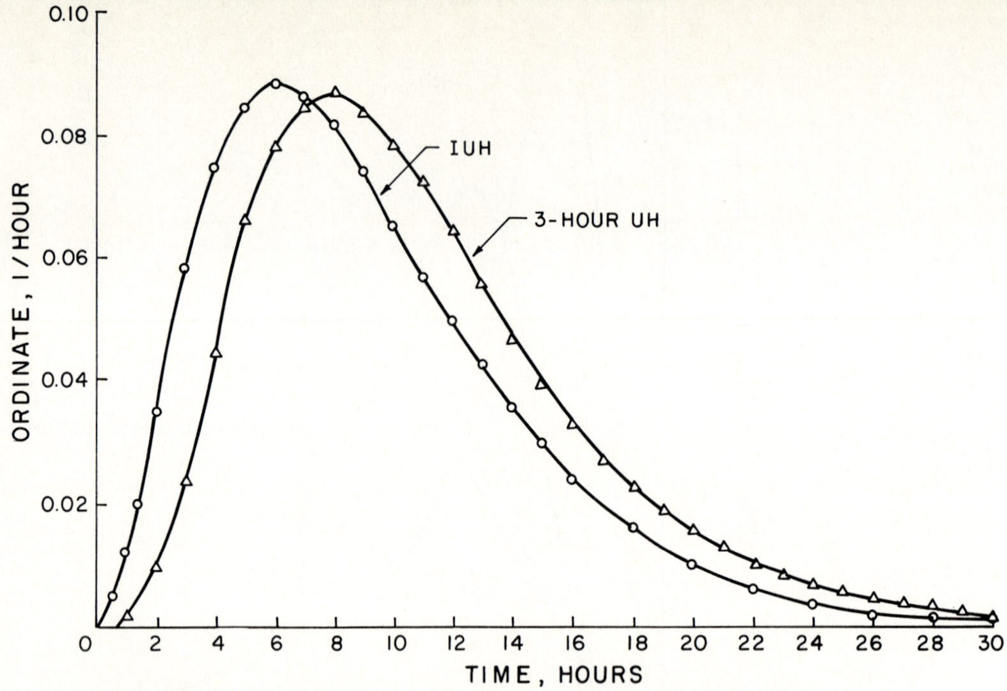

FIGURE 18.15 Example 18.6: IUH and three-hour UH from the Nash model.

The IUH is shown in Fig. 18.15. The 3-h UH, $h(3, t)$, is

$$h(3,t) = \frac{1}{3}\int_{t-3}^{t} 0.0171\, t^2 \exp(-t/3.08)dt$$

$$= -0.1662\exp(-t/3.08)[0.1056\, t^2 + 0.65\, t + 2]\Big|_{t-3}^{t}$$

$$= \exp(-t/3.08)[-0.029\, t^2 + 0.099\, t - 0.108]$$

Fig. 18.15 shows the graph of $h(3, t)$. The following compares the Nash model with the Snyder model:

	h_p (1/h)	t_p (h)	t_b (h)
Nash Model	.0873	8	30
Snyder Model	.13	5	25

18.3.7 The Gray Model

Gray (1961, 1962), using the Edson model, developed a procedure for derivation of the UH from physically measurable watershed characteristics. He analyzed topographic and hydrologic characteristics of 42 small watersheds, varying in size from .23 to 33.00 square miles, located in the states of Illinois, Iowa, Missouri, Nebraska, Ohio, and Wisconsin. Gray derived the UHs for these small watersheds and reduced them to a dimensionless form by expressing the ordinate in percent flow based on a time increment equal to one-quarter of the period of rise t_r (percent flow/.25t_r) and the abscissa as the ratio of any time t divided by the period of rise (t/t_r). Although each ordinate is expressed as percent flow/.25t_r, the connotation simply infers that it is the percentage of the total volume of flow based on a time increment duration of .25 t_r.

The dimensionless graph represents a modified form of the UH and can be derived as follows. The normalizing time is taken as t_r, which is $t_r = k(n - 1)$ for the 2-parameter gamma distribution (Nash model). The Nash model can be reduced to a dimensionless form by defining

$$t_* = \frac{t}{t_r}, \qquad h_*(t) = \frac{h(t)}{h_0}; \qquad k_* = \frac{k}{t_r}$$

in which h_0 is the normalizing value of h and is defined later. Using these quantities in the Nash model,

$$h_*(t_*)h_0 = \frac{1}{\Gamma(n)}\left(\frac{1}{k_*t_r}\right)^n (t_*t_r)^{n-1}\exp\left[-\frac{t_*t_r}{k_*t_r}\right]$$

$$= \frac{1}{\Gamma(n)}(t_*)^{n-1}\left(\frac{1}{k_*}\right)^n \frac{1}{t_r}\exp\left[-\frac{t_*}{k_*}\right] \qquad (18.80)$$

Let $h_0 = 1/(.25t_r)$. Then

$$h_*(t_*) = \frac{0.25}{\Gamma(n)}\left(\frac{1}{k_*}\right)^n (t_*)^{n-1}\exp\left[-\frac{t_*}{k_*}\right]$$

To express $h_*(t_*)$ as a percent,

$$h_*(t_*) = \frac{25}{\Gamma(n)}(\gamma^*)^n\left(\frac{t}{t_r}\right)^{n-1}\exp\left[-\frac{\gamma^* t}{t_r}\right] \qquad (18.81)$$

in which $\gamma^* = 1/k_* = n - 1$. If $h_*(t_*)$ is expressed in discharge units, then $Q_*(t/t_r) = h_*(t/t_r)V_*$, where V_* is the volume in percent, t_r is the period of rise, $Q(t/t_r)$ the percent flow/0.25 t_r for any value of t/t_r, and k and n are parameters. The time increment of .25t_r was chosen for the following reasons: (1) The period of rise is an important time characteristic. (2) The use of .25t_r enables definition of the rising limb at four points.

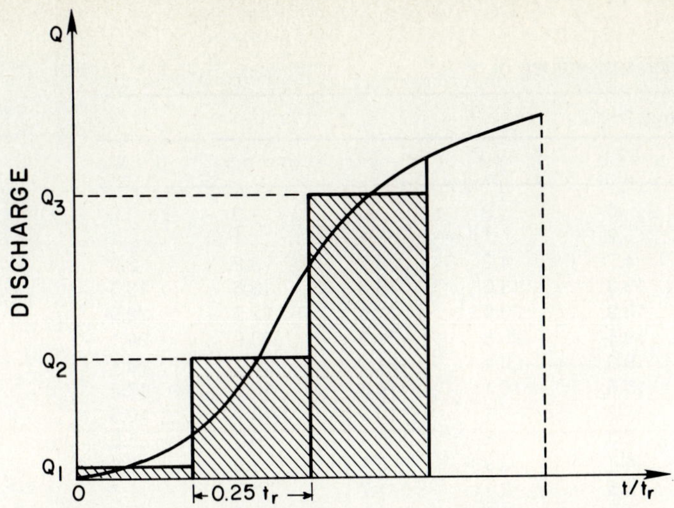

FIGURE 18.16 A dimensionless plot showing percentage of flow occurring during $.25t_r$ for the Gray model.

(3) The shape of the hydrograph is retained by using this size of increment.

Here, $Q(t/t_r)$ represents the percent flow/$.25t_r$ or the percent of the total volume of flow that occurs during a time increment of $.25t_r$ at a specific value of t/t_r. Referring to Fig. 18.16,

$$\frac{\text{Percent flow}}{.25t_r} = \frac{Q_1}{Q_1 + Q_2 + \cdots + Q_N} \times 100$$

where N is the number of increments in t. Therefore,

$$Q_1 = \left(\frac{\text{percent flow}}{.25\ t_r}\right) \times \Sigma \frac{Q_i}{100}$$

which can be expressed as

$$Q_1 = \frac{\text{percent flow}/.25t_r}{100} \frac{AP}{0.25t_r} \qquad (18.82)$$

where A is area and P is the depth of effective rainfall.

Equation (18.81) has two parameters, γ^* and n, which can be evaluated from the following relationships:

$$\frac{t_r}{\gamma^*} = a\left(\frac{L}{S_c^{.5}}\right)^b \qquad (18.83)$$

where L is the length of the mainstream in miles and S_c is the slope of the main channel in percent. The values of the coefficient a and exponent b were found to be 11.4 and 0.531, respectively, for Ohio watersheds, 7.4 and .498 for Nebraska and western Iowa watersheds, and 9.27 and .562 for Illinois, Missouri, Wisconsin, and central Iowa watersheds.

$$n = \gamma^* + 1 \qquad (18.84)$$

$$\frac{t_r}{\gamma^*} = \frac{1}{.262/t_r + .0139} \qquad (18.85)$$

in which t_r/γ^* and t_r are in minutes. Gray prepared a table giving percent flow/$.25t_r$ versus t/t_r for different values of n (see Table 18.3).

EXAMPLE 18.7

Compute the IUH by the Gray model for the watershed in Example 18.1. Then compute the 3-h UH and compare it with that of the Snyder model.

Solution: To use equation (18.83), we take a = 9.77 and b = 0.475, which represent values for all the watersheds used by Gray.

$$\frac{t_r}{\gamma_*} = 9.77\left[\frac{2.331}{5}\right]^{0.475} = 15.87 \text{ min}$$

Substituting it in equation (18.85),

$$15.87 = \frac{1}{\frac{.262}{t_r} + .0139}$$

This gives

$$t_r = 50 \text{ min}$$

$$\gamma_* = \frac{50}{15.87} = 3.1506$$

$$n = 3.1506 + 1 = 4.1506$$

The IUH $h(t/t_r)$ is

$$h\left(\frac{t}{t_r}\right) = \left[\frac{3.1506}{\Gamma(4.1506)}\right]\left[\left(\frac{t}{50}\right)3.1506\right]^{3.1506}$$
$$\exp\left[-3.1506\left(\frac{t}{50}\right)\right]V^*$$

$$h_{\text{peak}}(t = t_r) = \frac{25 \times (3.1506)^{4.1506}}{3!} \exp[-3.1506]$$
$$= 20.899 \text{ percent}$$

The volume of the IUH (1 cm) is

$$V = \frac{1}{10^2} \times 54 \times 10^6 = 5.4 \times 10^5 \text{ m}^3$$

The normalizing IUH ordinate is $.25 \times 50 \times 60 = 750$ m^3/s. Therefore, the IUH in cubic meters per second is equal to dimensionless IUH $\times$ 720A/100. This IUH is equivalent to the UH of duration $.25t_r$. The IUH is shown in Fig. 18.17. The 3-h UH is obtained by performing convolution; it is graphed in Fig. 18.18.

18.3.8 The Cordery Model

Cordery (1968, 1971) used the Clark model and estimated its parameters t_c and k using data from a number of watersheds in eastern New South Wales, Australia. His equations are:

$$t_c = 56\left[\frac{(LL_{ca})^{.8}n}{S_c}\right]^{.8} \qquad (18.86a)$$

and

$$k = 11\left[\frac{W}{S_0^{.5}} + \frac{Ln}{2S_c^{.5}}\right]^{0.8} \qquad (18.86b)$$

TABLE 18.3

Dimensionless graph coordinates for the Gray model for different values of *n*

	Percent Flow/.25 t_r^*								
t/t_r	n = 2.0	n = 2.5	n = 3.0	n = 3.5	n = 4.0	n = 4.5	n = 5.0	n = 5.5	n = 6.0
.000	.0	.0	.0	.0	.0	.0	.0	.0	.0
.125	1.2	.8	.5	.3	.2	.1	.1	.1	.1
.375	6.6	6.3	5.8	5.2	4.7	4.2	3.7	3.2	2.8
.625	11.2	12.0	12.6	13.0	13.3	13.6	13.6	13.6	13.5
.875	13.3	14.9	16.4	17.7	18.9	20.0	21.0	22.0	22.9
1.000	13.5	15.3	16.8	18.2	19.5	20.8	21.9	23.0	24.1
1.125	13.4	15.0	16.5	17.8	19.0	20.1	21.1	22.1	23.1
1.375	12.1	13.3	14.2	14.9	15.6	16.1	16.6	16.9	17.2
1.625	10.3	10.7	11.1	11.2	11.2	11.1	10.9	10.7	10.5
1.875	8.3	8.2	8.0	7.7	7.3	6.8	6.4	6.0	5.5
2.125	6.5	6.0	5.5	5.0	4.4	3.9	3.4	3.0	2.6
2.375	4.9	4.3	3.6	3.1	2.5	2.1	1.7	1.4	1.1
2.625	3.6	2.9	2.3	1.8	1.4	1.0	.8	.6	.5
2.875	2.6	2.0	1.4	1.0	.8	.6	.4	.3	.2
3.125	1.9	1.3	.9	.6	.4	.3	.2	.1	
3.375	1.3	.9	.5	.3	.2	.1	.1		
3.625	1.0	.6	.3	.2	.1				
3.875	.6	.3	.2	.1					
4.125	.4	.2	.1	.1					
4.375	.3	.1	.1						
4.625	.2	.1							
4.875	.1	.1							
5.125	.1								
5.375	.1								
Sum**	100.0	100.0	100.0	100.0	100.0	100.0	100.0	100.0	100.0

*Rounded to the nearest .10 percent.
**Sums do not include peak percentages.

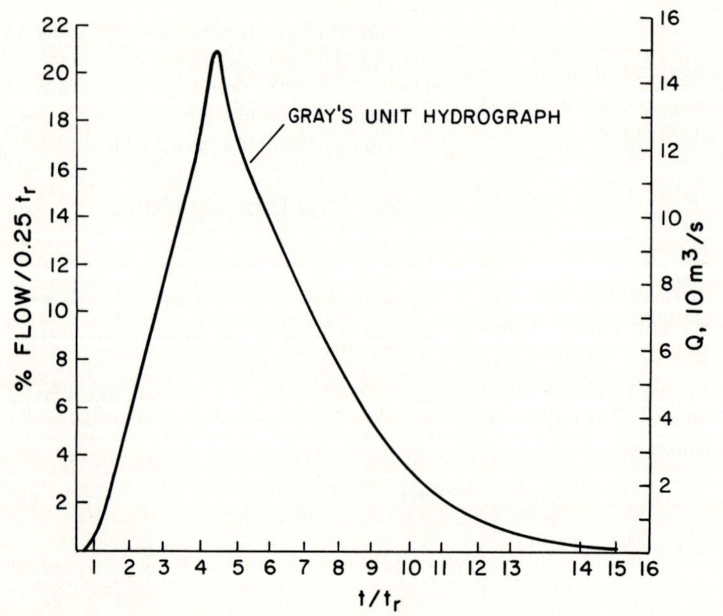

FIGURE 18.17 Example 18.7: IUH of the Gray model.

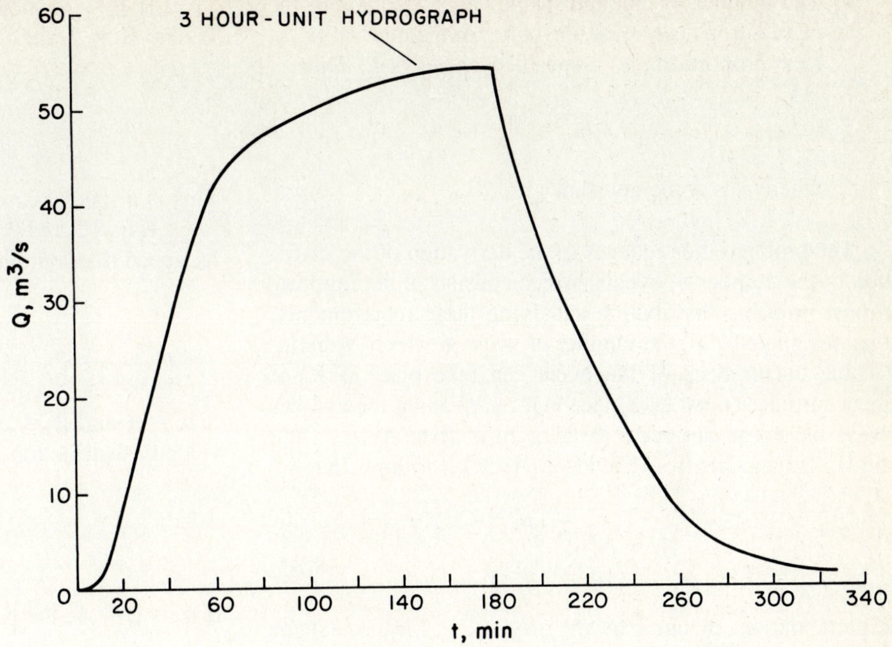

FIGURE 18.18 Example 18.7: Three-hour UH from the Gray Model.

with

$$W = \frac{A}{L}$$

where A is the watershed area in square miles, L is the length of the main stream measured to watershed divide in miles, L_{ca} is the length in miles measured along the path water would follow from the watershed center of area to the outlet, S_c is the slope of the main stream in feet per mile, S_0 is the overland slope in feet per mile, and n is Manning's roughness coefficient estimated for the highest-order stream in the watershed.

18.3.9 The Lienhard Model

In a series of papers (Lienhard, 1964; Lienhard and Meyer, 1967; Lienhard and Davis, 1971; Lienhard, 1972) Lienhard, using the Boltzman statistics, derived a generalized gamma distribution for representation of the IUH. His analysis can be briefly summarized as follows.

Let $(N + M)$ raindrops constitute an instantaneous burst of rainfall. Of these raindrops, M disappear into the ground or evaporate into the air; that is, these represent abstractions. The remaining N raindrops constitute the effective rainfall and eventually find their way to the watershed outlet. The direct runoff hydrograph can be interpreted in terms of raindrops N_i in each of a series of time increments of duration Δt. The result is a histogram, as shown in Fig. 18.19.

It is assumed that each raindrop is distinguishable from and independent of the other raindrops. Each raindrop has the same a priori probability of reaching the gaging station during the ith time interval. Whether or not it does reach the station depends upon where it lands and what obstacles it encounters during its journey. Let us consider the slug of water composed of N_i raindrops that reach the watershed outlet during the time t_{i-1} to t_i. There can be g_i ways in which these raindrops can find their way to the outlet. It is reasonable to assume that there are, on the average, a constant number of ways by which a raindrop must leave each unit of watershed area. Furthermore, the time required for a raindrop to reach the outlet after traveling a distance L_i will be approximately proportional to L_i raised to some power.

The following requirements can now be imposed on N_i's:

1. The total number of occurrences of the event is fixed,

$$\sum_{i=1}^{\infty} N_i = N$$

The N_i's and N are assumed to be large numbers.

2. For each choice of β the following sum is a positive constant,

$$\sum_{i=1}^{\infty} \left(\frac{N_i}{N}\right) t_i^{\beta} = k, \qquad \beta > 0, k > 0$$

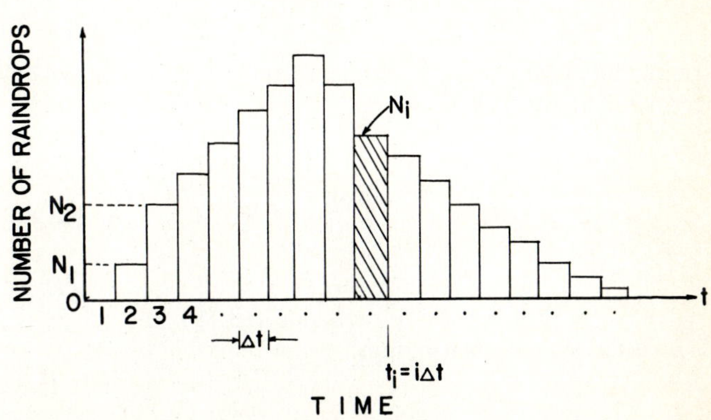

FIGURE 18.19 The runoff hydrograph interpreted as a raindrop histogram.

3. The number of distinguishable ways g_i in which the N_i raindrops can reach the outlet in the interval $[t_{i-1}, t_i)$ is proportional to a specified power of t_i. Thus

$$g_i = c\, t_i^{\alpha-1}, \qquad \alpha > 0$$

where c is some constant.

The problem thus reduces to the derivation of the distribution of the number of events, which consists of determining the most probable distribution satisfying these requirements. Let us designate by W the number of ways in which N distinguishable occurrences of the event can take place if N_i of these occurrences must take place in $[t_{i-1}, t_i)$ and if the number of ways the event can occur in $[t_{i-1}, t_i)$ is given by g_i. Then using Boltzman statistics (Davidson, 1952) it follows that

$$W = N! \prod_{i=1}^{\infty} \frac{g_i^{N_i}}{N_i!} \tag{18.87}$$

The particular set of numbers $(N_1, N_2, \ldots, N_i, \ldots)$ for which W is maximum will be the most probable distribution. Since the numbers involved are very large, the most probable distribution will be certain for all practical purposes. Let $\tilde{N}_i$ denote those values of N_i that give maximum of W. Then, subject to the foregoing three requirements, it can be shown that

$$\frac{\tilde{N}_i}{N} = \frac{\Delta t[\beta(\beta k/\alpha)^{-\alpha/\beta}]}{\Gamma(\alpha/\beta)} t_i^{\alpha-1} \exp\left[-\frac{\alpha}{\beta}\frac{t_i^{\beta}}{k}\right] \tag{18.88}$$

We may suppose that the $\tilde{N}_i/N$'s represent the discrete probability distribution associated with the random variable T, where T is the time at which the first occurrence of the event under consideration takes place. Therefore,

$$P(t_{i-1} \le T < t_i) = \frac{\tilde{N}_i}{N}, \qquad i = 1, 2, \ldots$$

Approximating this discrete distribution with a continuous probability density function $f(t)$,

$$\frac{\tilde{N}_i}{N} = \int_{t_{i-1}}^{t_i} f(t)\, dt$$

Using the mean value theorem and letting $\Delta t \to 0$, we obtain, from equation (18.88),

$$f(t) = \left[\frac{\beta}{\Gamma(\alpha/\beta)}\left(\frac{\alpha}{\beta k}\right)^{\alpha/\beta}\right] t^{\alpha-1} \exp\left[-\frac{\alpha}{\beta}\frac{t^{\beta}}{k}\right], \qquad t \ge 0 \tag{18.89}$$

It may be of interest to note that several distributions can be obtained as special cases of equation (18.89). If $\alpha = \beta$, we obtain the Weibul distribution,

$$f(t) = \frac{\alpha}{k} t^{\alpha-1} \exp\left[-\frac{t^{\alpha}}{k}\right] \tag{18.90}$$

If $\beta = 1$, it leads to the familiar gamma distribution. If $\alpha = \beta = 2$, then we obtain the Rayleigh distribution,

$$f(t) = \frac{2}{k} t \exp\left[-\frac{t^2}{k}\right] \tag{18.91}$$

This is a special case of the Weibull distribution.

If $\alpha = 3$ and $\beta = 2$, then we obtain the Maxwell molecular speed distribution,

$$f(t) = \left[\frac{(54/\pi)^{.5}}{k^{1.5}}\right] t^2 \exp\left[-\frac{3}{2}\frac{t^2}{k}\right] \tag{18.92}$$

If $\alpha = 1$ and $\beta = 2$, then we obtain the Maxwell molecular velocity distribution,

$$f(t) = \left(\frac{2}{k\pi}\right)^{.5} \exp\left[-\frac{t^2}{2k}\right] \tag{18.93}$$

If $\alpha = \beta = 1$, the resulting distribution is exponential,

$$f(t) = \frac{1}{k}\, exp\left[-\frac{t}{k}\right] \tag{18.94}$$

The function $f(t)$ can be interpreted as the IUH. Equation (18.89) is a generalized three-parameter gamma distribution from which several distributions can be obtained as special cases (Lienhard and Meyer, 1967). Lienhard (1964) applied his model to two natural watersheds in Illinois. He observed that for representation of the IUH, β can be taken as 2 and α can be taken as 2 and 3, for long slender and fan shaped watersheds, respectively. The parameter k establishes the scale of the distribution and will be the second moment of the IUH about its origin. Betson and Green (1968) used an empirical expression similar to equation (18.89) for analysis of streamflow of the Upper Bear Creek in the Tennessee Valley region.

18.3.10 The Linear Watershed Bounded Network (LWBN) Model

Boyd (1978, 1982) developed the linear watershed bounded network (LWBN) model for synthesis of the IUH employing geomorphologic and hydrologic properties of the watershed. The conceptual framework embodied in this model parallels the one developed independently by Singh and McCann (1980a). The model divides a watershed into subareas bounded by watershed lines using large-scale topographic maps. Following Leopold, Wolman, and J. P. Miller (1964), two types of subareas having different collection and drainage of water are distinguished: (1) ordered basin and (2) interbasin area. An ordered basin is a complete subwatershed and does not allow for flow across its boundaries. Its operation is independent of other subwatersheds. These basins have been shown to be geomorphologically similar to the complete watersheds (Strahler, 1964). An interbasin area is a subarea with a stream flowing through it. The direct runoff at its outlet consists of both runoff from upstream areas and that generated within it. Each ordered basin and interbasin area is represented by a

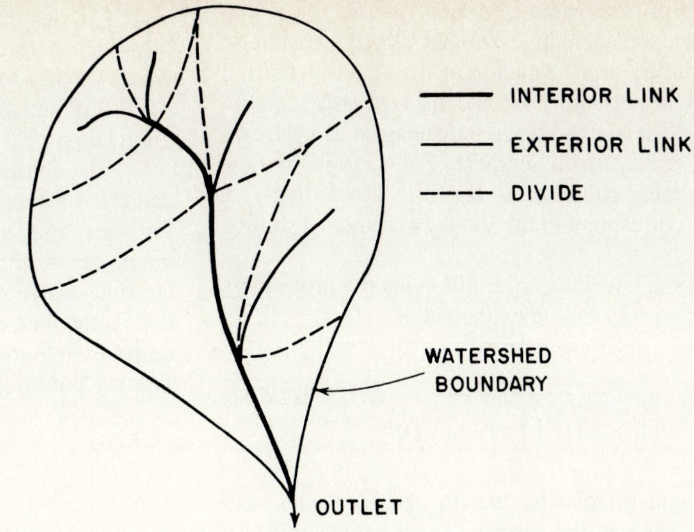

(a)

FIGURE 18.20(a) A hypothetical watershed and its stream network.

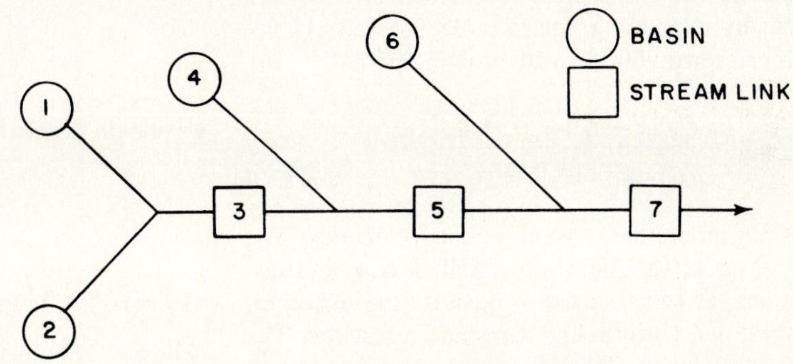

(b)

FIGURE 18.20(b) A branched network model structure for watershed in (a).

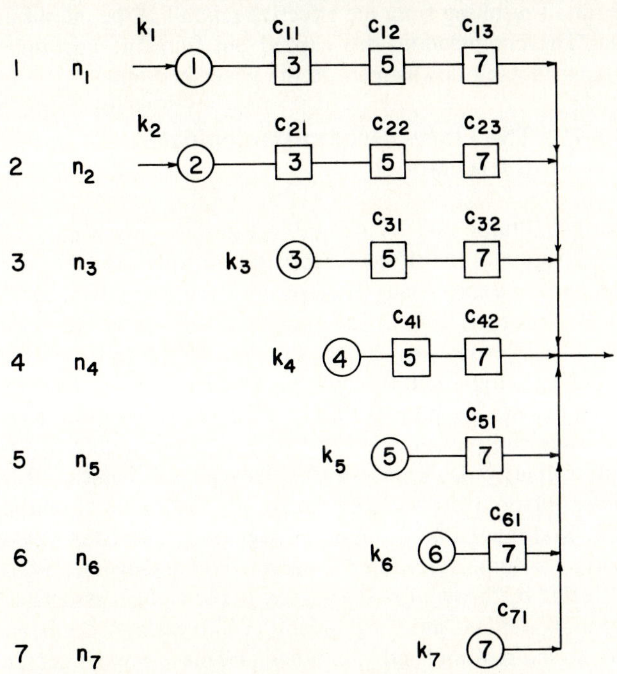

FIGURE 18.20(c) An equivalent model structure.

lumped storage element. These elements are connected in the topology as the watershed stream network. Thus, the watershed is represented by a branched network of storage elements. For linear elements, this can be represented by an arrangement of parallel cascades wherein the number of cascades is equal to the number of links (interior as well as exterior) plus the number of subareas. To be more precise, if there are N sources, then $(2N - 1)$ parallel cascades of storage elements will result. The number of elements in a cascade is specified by the position of the link or subarea. This suggests that a knowledge of stream network implies a knowledge of the arrangement of cascades. Fig. 18.20 illustrates the model structure (a) showing a second-order hypothetical watershed, (b) outlining a branched network model structure, and (c) giving an equivalent model structure. The linear storage element can be described by

$$\frac{dS}{dt} = I - Q \tag{18.95}$$

$$S = kQ \tag{18.96}$$

in which S is the volume of water stored in the area, I is the inflow to the area, Q is the outflow from the area, k is the storage coefficient or time parameter, and t is time. Note that the LWBN model considers two storage effects separately: (1) those acting during transformation of the effective rainfall to direct runoff via overland flow and flow in small stream channels and (2) those acting during transmission of upstream runoff through a main stream segment. For a single storage element, k is precisely equal to the lag time. Accordingly, k and C are used to correspond to the above two types of storage effects separately.

The IUH of the LWBN model, following the procedures discussed in Chapter 13, can be expressed as

$$h(t) = \frac{1}{A} \sum_{i=1}^{2N-1} \frac{1}{(1 + k_i D) \prod_{j=1}^{n_i} (1 + C_{ij} D)} \delta(t)\, a_i \quad (18.97)$$

where a_i is the area drained by the ith cascade and n_i the number of stream links in the cascade. Equation (18.97) is a special case of the Dooge model with pure translation deleted. Its solutions have been presented in Chapter 13. In practice, the model equations can easily be solved numerically for each storage element by expressing equation (18.95) and (18.96) in finite-difference form with Δt as the routing period,

$$Q_{t+\Delta t} = \frac{Q_t(2k - \Delta t) + (I_t + I_{t+\Delta t})\Delta t}{2k + \Delta t} \quad (18.98)$$

The LWBN model has several attractive features. The model allows for areal variations of rainfall, losses, and watershed conditions. It can be used to quantify the effects of land use changes, and constructing dams and reservoirs. The model structure is uniquely defined by watershed stream network. In other words, components of the model can be identified with specific parts of the watershed. The model has a large number of parameters. These include the number n of storage elements, the n values of k values for these elements, and the n_i values of C for the n_i interbasin stream segments. However, most of these parameters are deduced from geomorphologic properties, as will be clear from the ensuing discussion. A drawback of this model is that there may be hundreds of links in the network representation requiring larger computation than actually needed. If some of the links, exterior or interior, are lumped or suppressed, then an objective way of accomplishing this is lacking.

18.3.10.1 Parameter Estimation

The parameters to be estimated are k_i; C_{ij}, $j = 1, 2, \ldots,$ n_i; and n_i, $i = 1, 2, \ldots, (2N - 1)$. n_i is specified by the stream network. For each storage element i representing the area A, the parameter k is estimated as

$$k = aA^b \quad (18.99)$$

where a and b are constants. For 10 watersheds (drainage area .39 to 251 km^2), Boyd, Pilgrim and Cordery (1979a, 1979b) reported $a = 2.51$ and $b = .38$ with A in square kilometers and k in hours. Similarly, for each element, C is computed as

$$C = a_1 A^{b_1} \quad (18.100)$$

where a_1 and b_1 are constants. For the same 10 watersheds, $a_1 = 1.5$ and $b_1 = .38$, with C in hours and A in square kilometers.

The constants, especially a and a_1, in equations (18.99) and (18.100) may vary from one region to another. It is therefore desirable to discuss the method by which to estimate them. This can be accomplished by computing the watershed lag time which can be obtained from inflow-outflow records, that is, the time difference between the centroid of the effective rainfall hyetogrpah and that of the direct runoff hydrograph. The lag time t_L of the LWBN model is

$$t_L = \frac{1}{A} \sum_{i=1}^{2N-1} a_i g_i \quad (18.101)$$

where A is the total watershed area, a_i is the area corresponding to the cascade i, and g_i is the total lag along the cascade i, defined as

$$g_i = k_i + \sum_j C_{ij} \quad (18.102)$$

Equation (18.101) can be written alternatively as

$$t_L = \frac{1}{A}\left(\sum_{i=1}^{2N-1} a_i k_i + \sum_j C_{ij} \sum_{m=1}^{j-1} a_m \right) \quad (18.103)$$

18.3.10.2 Model Operation

For any effective rainfall pattern, inflows are routed through each storage element using equations (18.95) and (18.96). Computed outflows at the confluence of basins are added together. For interbasin areas, the runoff from upstream is routed through the main channel segment, and to this outflow is added the runoff resulting from the effective rainfall of the interbasin area. The computations are carried out from the uppermost element moving downstream to the watershed outlet.

18.3.11 The Linear Geomorphological (LGM) Model

Rodriguez-Iturbe and Valdes (1979) developed an approach for derivation of the IUH by explicitly incorporating the characteristics of drainage basin composition (Horton, 1945; Strahler, 1964; Smart, 1972). The approach coupled the empirical laws of geomorphology with the principles of linear hydrologic systems. Rodriguez-Iturbe and his associates have since extended this approach by explicitly incorporating climatic characteristics and have studied several aspects including hydrologic similarity (Rodriguez-Iturbe, Devoto and Valdes, 1979; Valdes, Fiallo and Rodriguez-Iturbe, 1979; Rodriguez-Iturbe, 1982; Rodriguez-Iturbe, Gonzalez-Sanabria, and Bras, 1982; Rodriguez-Iturbe, Gonzalez-Sanabria, and Caamano, 1982). Motivated by the work of Rodriguez-Iturbe and his associates, Gupta, Waymire, and C. T. Wang (1980) examined this approach, and reformulated, simplified and made it more general and even more elegant. Its assumptions, limitations, and potential for application to synthesis of direct runoff from ungaged watersheds became more apparent. Others (C. T. Wang, Gupta, and Waymire, 1981; Hebson and Wood, 1982; Kirshen and Bras, 1983; Singh, 1983b; Hill, Singh, and Aminian, 1984) have also applied this approach to hydrologic analysis.

The geomorphologic approach is probabilistic in character. However, we develop here a deterministic interpretation that is much simpler, easier for hydrologists to understand, equally versatile, and retains the essential import of the original approach (Singh and Aminian, 1984, 1985). This interpretation can be explained as follows: A watershed is of a fixed order W. It contains channels $c_1, c_2, \ldots, c_W$, where c_i denotes the ith order stream. The watershed network is ordered according to the Strahler-Horton ordering scheme (Strahler, 1964, Smart, 1972). When it rains, runoff is generated. The time taken by a body of water to travel to the watershed outlet depends upon the position where the travel is initiated and the path it follows. In a basin there can be an infinite number of positions where raindrops will land and initiate their travel in association with physiographic characteristics. Likewise, there can be an infinite number of paths of travel. These paths are, however, carved by topographic slope configuration and channel network.

A watershed surface is comprised of overland regions and channels. The surface area occupied by the overland regions is many times greater than that occupied by the channels. It can therefore be assumed without undue loss of accuracy that the raindrops fall and start their journey on the overland regions and travel through the channel network to their destination at the watershed outlet. Let the overland regions be assigned the same order as the channels to which they directly contribute. For notational convenience, let an overland region of order i be denoted by r_i. To recount, rain water, depending upon the position of its landing, starts its travel in the overland region and then goes through a string of channels. Logically its travel follows the following rules:

Rule 1. The only possible route from the overland region r_i is to the channel c_i, $1 \le i \le W$.

Rule 2. The only possible route from the channel c_i is to the channel c_j for some $j > i$, $1 \le i \le W$.

Rule 3. The only possible route from the channel c_W is the watershed outlet.

These rules define a collection of paths that water can take from the start of its journey to the watershed outlet. The number of these paths N equals $N = 2^{W-1}$. To illustrate, consider a third-order basin as shown in Fig. 18.21. Then paths R_i, $i = 1, 2, \ldots, N$, can be demarcated as follows:

Path R_1: $r_1 \rightarrow c_1 \rightarrow c_2 \rightarrow c_3$
Path R_2: $r_1 \rightarrow c_1 \rightarrow c_3$
Path R_3: $r_2 \rightarrow c_2 \rightarrow c_3$
Path R_4: $r_3 \rightarrow c_3$

Every third-order basin has four or fewer paths.

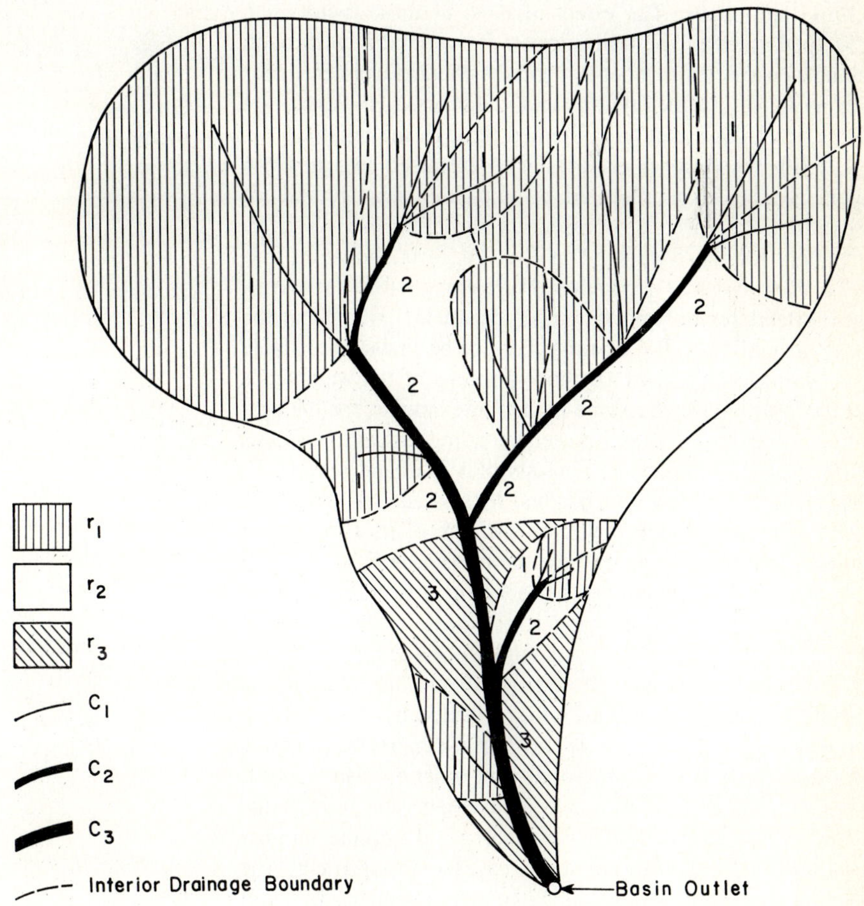

FIGURE 18.21 A hypothetical third-order watershed with Strahler-Horton order scheme.

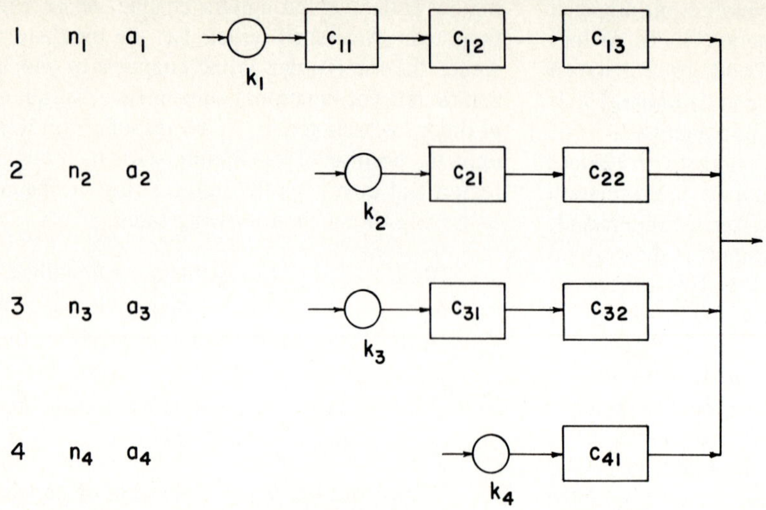

FIGURE 18.22 An equivalent representation of the linear geomorphological model.

These paths specify the spatial evolution of rainwater through a geomorphologic network of channels and overland regions. It is clear that the time of travel that water spends in following a given path depends upon its composition. Obviously, different paths may have different travel times. A path is composed of one overland region and one or more channels. This means that the travel time of a path can be specified by computing travel times of its overland region and channel(s) and then summing them up. The transformation of rainfall to runoff from the overland region is determined by its characteristics, such as area, roughness, vegetal cover, slope, length of flow, and shape. The effect of most of these factors can be lumped as area. The transmission of overland flow through channels downstream depends upon the channel characteristics such as length, slope, width, and roughness. However, the effects of all these can be combined through the length.

With these considerations it is surmised that a given watershed can be represented by a parallel arrangement of cascades. Each cascade represents a particular path. The arrangement for the watershed in Fig. 18.20 is shown in Fig. 18.22. Each cascade is composed of elements (one representing overland region and the other channels), which can be linear or nonlinear. We assume them to be linear here. Note that a cascade represents all those portions of the watersheds that drain following the same path. For example, for the preceding watershed, R_1 (or its corresponding cascade) consists of 10 overland regions, R_2 consists of 1 overland region, R_3 consists of 2 overland regions, and R_4 consists of 1 overland region. The area drained by R_1 is the sum of its 10 overland regions, and the same is true for the remaining R_i's.

An overland region can be represented by a linear reservoir of lag time k and a channel by a linear reservoir of lag time C. Since a cascade corresponds to a specific path, the lag time of an element is an average of the lag times of the portions of the actual watershed represented by this element. For example, the lag time of the first-order overland region is the average of the lag times of all first-order overland regions in the watershed. Therefore, k is an average value. This same is true of the channels. This averaging reduces the number of parameters needed to define the cascade arrangement. The total number M of parameters to be determined is $2W$, half of which are for overland regions and half for channels. Referring to Fig. 18.21, the unknown parameters are $k_1 = k_2$, k_3, k_4; $C_{11} = C_{21}$, $C_{12} = C_{31}$, $C_{13} = C_{22} = C_{32} = C_{41}$. These parameters can be determined following the procedure employed in the LWNB model:

$$k_i = a(A_i)^{.3}, \qquad 1 < i < W \tag{18.104a}$$

and

$$C_i = a(\bar{L}_i)^{0.3}, \qquad 1 < i < W \tag{18.104b}$$

where a is an empirical constant, $\bar{L}_i$ is the average channel length of order i, and A_i is the drainage area of order i. The only unknown is a, which is computed by equating the first moment of the IUH to watershed lag K_b, which was found to be

$$K_b = .875A^{.3} \tag{18.104c}$$

in which A is in square kilometers and K_b is in hours.

The IUH of this model can be expressed as

$$h(t) = \frac{1}{A}\sum_{i=1}^{N} \frac{1}{(1 + k_i D)\prod_{j=1}^{n_i}(1 + C_{ij}D)}\,\delta(t)\,a_i \tag{18.105a}$$

Since each cascade can be considered as a cascade of linear unequal elements, the solution of equation (18.104) is as given in Chapter 13. Since the cascades are connected in a parallel arrangement, they are summed to obtain the watershed IUH. For any cascade with n elements or path, the IUH is

$$h_m(t) = \frac{1}{\prod_{i=1}^{n}(1 + K_i D)}\,\delta(t) = \sum_{j=1}^{n} \frac{K_j^{n-2}\exp(-t/K_j)}{\prod_{\substack{i=1\\ i\neq j}}^{n}(K_j - K_i)}, \quad n \geq 2 \tag{18.105b}$$

in which K_i, $i = 1, 2, \ldots, n$, are the parameters of the elements, with 1 denoting the most upstream element and n

denoting the most downstream elements. Evidently, k's and C's define K's.

18.3.11.1 Contrast with the Probabilistic Approach

The probabilistic approach leading to the IUH is analogous to its deterministic counterpart. Both modes are comparable and equally flexible in virtually all respects. The hydrology of an element in a path (or cascade) is assumed in terms of its IUH in the deterministic mode and in terms of probability density function (pdf) in the stochastic mode. For example, assuming an element to be a linear reservoir with lag time k yields a negative exponential IUH. This is equivalent to saying that the pdf of the lag time of the element is negative exponential. The notion of probability arises from the observation that this element represents an average behavior of an ensemble of portions of the watershed. Likewise, assuming an element to behave like the rational formula is equivalent to stating that its IUH is uniform over a finite period. Stochastically, this implies that the pdf of its lag time is a uniform distribution.

18.3.11.2 A Comparison with the LWBN Model

The LGM and LWBN models share several common features. Both treat the watershed as a quasi-distributed system. The degree of lumping is considerably greater with the LGM model than the LWBN model. Both contain parameters that can be estimated from geomorphological attributes in essentially the same fashion. Both models are linear and time-invariant. Although the concepts in representing a watershed by a network of planes and channels are different for the two models, their operation is identical. The LGM model prescribes a unique network representation of a watershed, whereas the LWBN model does not. Of course, depending upon the watershed size, the representations can be made less detailed. However, the degree of lumping can be achieved more objectively with the LGM than with the LWBN model.

18.4 NONLINEAR CONCEPTUAL MODELS

18.4.1 Nonlinear Reservoir

The nonlinear storage-discharge relation

$$S = k\,Q^m \qquad (18.106)$$

has been used in a number of nonlinear models (Laurenson, 1965; Laurenson, Mein and McMahon, 1975; Boyd, Pilgrim, and Cordery, 1979a, 1979b). Alternatively,

$$S = k\,Q \qquad (18.107)$$

$$k = k_1\,Q^n \qquad (18.108)$$

Therefore, $m = n + 1$.

Equation (18.106) has two parameters, m and k. Laurenson, Mein, and McMahon suggested a value of .75 for m, unless evidence supported otherwise. They argued that k can be conveniently expressed as a product of two factors:

$$k = k_1 k_2 \qquad (18.109)$$

where k_1 is a constant for a watershed and k_2 is proportional to the delay time of the individual model storage. Thus k_1 and k_2 depend upon the size and physical characteristics of the watershed.

Both k_1 and k_2 are dimensional quantities. Laurenson, Mein, and McMahon (1975) made k_2 dimensionless by expressing it as

$$k_r = \frac{d_i - d_j}{d_c} \qquad (18.110)$$

where k_r is dimensionless k_2, d_c is the mean travel distance for the watershed, and $(d_i - d_j)$ is the travel distance of the ith subwatershed having area a_i in a watershed composed of N subwatersheds marked by equal increments of travel distance (i denoting upstream and j denoting downstream). Therefore,

$$d_c = \sum_{i=1}^{N} \frac{a_i d_i}{A} \qquad (18.111)$$

where A is the area of the watershed.

With k_2 replaced by k_r, k_1 can be supplanted by k_c, which can be determined for ungaged watersheds as

$$k_c = 2.12\,A^{.57} \qquad (18.112)$$

where A is in square kilometers. For gaged watersheds an optimization is suggested. Thus

$$k = k_c k_r \qquad (18.113)$$

18.4.2 Uniformly Nonlinear Cascade

The uniformly nonlinear model has three parameters; N, k, and m. Based on a study of 39 small agricultural watersheds, Singh (1976d, 1977, 1979a, 1976b) suggested m to be 1.4, N to be 3, and k to be estimated from watershed characteristics. He correlated k to watershed area A, stream length L_c, watershed width W, channel slope S_c, and watershed shape S_h, and developed three alternative relationships for k:

$$k = A^{-.3087} L_c^{-.2161}\, S_c^{-.0833}\, S_h^{.3038} \qquad (18.114)$$

$$k = A^{-.0283} W^{-.2431} \qquad (18.115)$$

and

$$k = A^{-.2289} L_c^{-.264} S_c^{.1008} \qquad (18.116)$$

where A is in hectares, L_c is in meters, W is in meters, and S is a percent. The shape factor was computed as proposed by Chorley, Malm, and Pagorzelski (1957).

18.4.3 The Laurenson Model

This model was developed by Laurenson (1962). It divides the watershed into 10 subareas by drawing isochrones at equal time intervals. These subareas can be delineated differently if one chooses to do so. A node is placed at the center of each subarea, where the inflow is assumed to be concentrated. Between each pair of subareas is placed a nonlinear storage element (reservoir) defined by equation (18.107), with k also being a funciton of Q. In this way a watershed is represented by 10 nonlinear storage elements connected in series. Each element receives the effective rainfall occurring over its corresponding subarea of the watershed as well as the direct runoff from all upstream elements, as shown in Fig. 18.23.

The total watershed travel time is divided evenly between 10 isochrones. The travel time between each pair of nodes is one-tenth of the total travel time. This is true for all storage elements except the lowest, which has one-twentieth of the total travel time. Storage routing is performed between each pair of adjacent nodes, with storage allocated to each element based on the storage properties between the nodes. Since the upstream runoff is routed between nodes, allocation of one-tenth of the total travel time to the upper nine elements and one-twentieth to the lowest element is reasonable. For the effective rainfall occurring on each subarea, the travel time ranges from one-twentieth from the downstream isochrone to one-fifteenth from the upstream isochrone; this also yields an average of one-tenth of the travel time for the upper nine elements. For the lowest subarea it ranges from 0 to one-tenth of the travel time, giving an average of one-twentieth of the travel time.

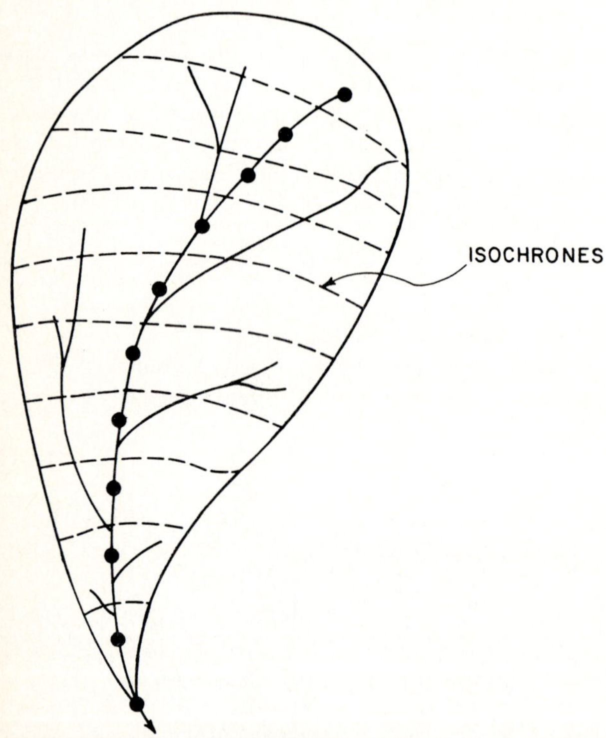

FIGURE 18.23(a) A hypothetical watershed with isochrones drawn at equal interval.

The time of travel is analogous to k in equation (18.107) and is related to the physical geometry of the watershed by the assumption that it is proportional to the distance and inversely proportional to the square root of slope. This can be estimated for any point in the watershed from its topographic map as discussed in Chapter 11. The centroid of the TA diagram is the mean time of travel τ_m for the watershed and can be expressed as

$$\tau_m = \frac{\int_A \tau\, da}{\int_A da} = \frac{1}{A}\int_A \tau\, da \tag{18.117}$$

where τ is the dimensionless time of travel and A is the watershed area. Because of the analogy between time of travel and storage delay time, τ_m is equivalent in dimensional terms to k. Thus, using equation (18.108) for any value of Q, the dimensionless τ can be made dimensional for any point in the watershed as

$$k_* = \frac{\tau\, k_1 Q^n}{\tau_m} \tag{18.118}$$

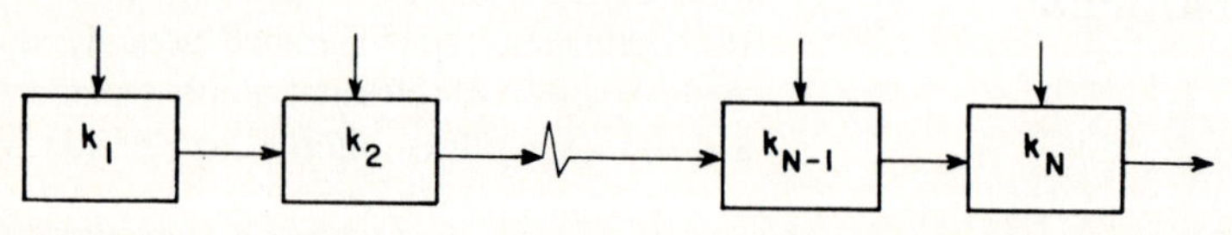

FIGURE 18.23(b) The Laurenson model for the watershed in (a).

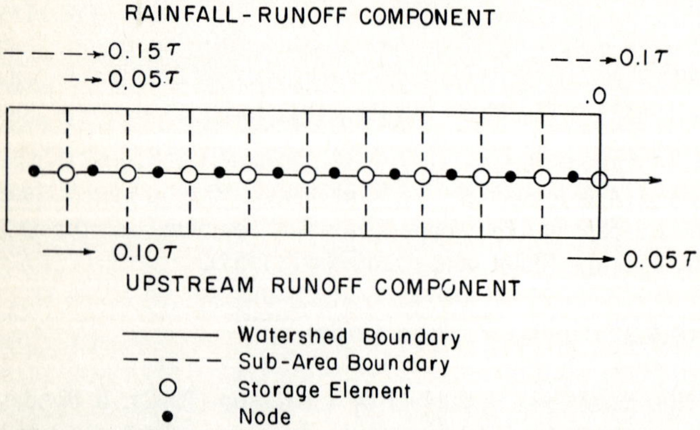

FIGURE 18.23(c) Allocation of travel time (or storage) for ten series elements in the Laurenson model.

Note that this does not correspond to the mean delay time. If $\Delta\tau$ is the interval of time of travel, then each storage element has the time of travel as

$$k_i = \frac{\Delta\tau\, k_1 Q^n}{\tau_m} \tag{18.119}$$

The effective rainfall occurring on the topmost subarea is first routed through the topmost storage element. The direct runoff from this element is combined with the effective rainfall to the second subarea, and the combined flow is routed through the second element. This operation is repeated all the way up to the watershed outlet. Since each element is nonlinear and its effect depends on the magnitude of its inflow, it is not permissible to route the inflow of each subarea individually to the outlet and then to combine the 10 outflows there. Note that the watershed storage is distributed rather than concentrated because each inflow is routed through a series of concentrated storages, thus providing for both translation and attenuation effects.

This model may be considered to have 3 parameters: number of storage elements N, coefficient k, and exponent m of the storage discharge relation of equation (18.108). N is fixed at 10. Based on a regression of lag time versus mean discharge using a number of recorded rainfall and runoff hydrographs of the South Creek watershed in New South Wales, Australia, Laurenson obtained

$$t_L = 64\, Q_m^{-.27} \tag{18.120}$$

where t_L is the lag time in hours and Q_m is the mean discharge in cubic feet per second. He adopted m as $-.27$. He computed k by assuming that the coefficient 64 was applicable to the points on the watershed corresponding to the centroid of the TA diagram. For the South Creek watershed, the relative time to the centroid was 0.66. The value of k for the entire watershed was $64.0/.66 = 97.0$. For each of the 10 equal storage elements, the coefficient k was $97.0/10 = 9.7$.

The Laurenson model considers the internal properties of the watershed. Specifically, it accounts for (1) areal variation of effective rainfall, (2) areal variation of watershed characteristics, (3) nonlinear storage effects, and (4) effective rainfall subjected to different storage amounts, depending upon the point of its occurrence. Thus the model is physically realistic. However, there are some weaknesses of the model that should be recognized. Since the method of constructing isochrones is uncertain, the representation of a watershed into subareas is less than objective. Storage properties between each pair of nodes cannot be measured, so the storage allocation is difficult. The assumption of the watershed lag applying to the centroid of the TA diagram holds only for a linear model and for spatially uniform effective rainfall. Equation (18.120) refers to average values in time and magnitude but is applied to instantaneous values in equation (18.119). The time parameter k is assumed to be proportional to travel times, which, in turn, are assumed to be proportional to $L/S_0^{.5}$, with L being length of travel and S_0 slope. Boyd (1976) has shown that this is a poor parameter for measuring travel time. These limitations introduce uncertainty into model results and the parameters as previously estimated. When subareas are defined by isochrones, flows in more than one well-defined channel may be lumped together for determination of lag time; this superimposition of flows is not valid because of the nonlinearity.

18.4.4 The Porter Model

The Porter model is a variation of the Laurenson model. It divides the watershed into a number of subareas defined by internal watershed boundaries or stream network so that each subarea contains only one well-defined channel. Streams of lower order may be lumped, but major tributaries are not. Each subarea is represented by a nonlinear storage element defined by equation (18.106). Subareas are determined by higher-order streams such that the maximum number of storage elements is about five. The resulting arrangement of elements is in parallel and series, as shown in Fig. 18.24. It is a more versatile arrangement and may be of value in modeling large watersheds.

The storage elements represent specific subareas rather than storage routing between subareas. The storage is allocated to each element by assuming that it is proportional to travel time. Since travel time is not known, it is assumed to be a function of the properties of the stream segment in each subarea, that is, it is proportional to the ratio $L/S_c^{.5}$, which was

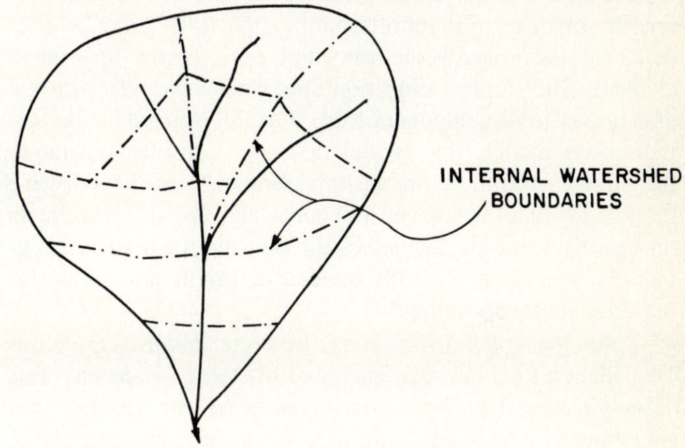

FIGURE 18.24(a) A hypothetical watershed with sub-areas demarcated by internal watershed boundaries.

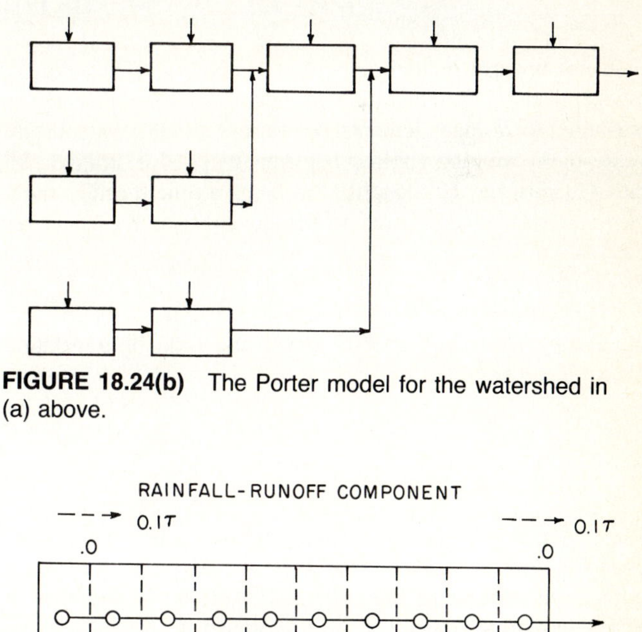

FIGURE 18.24(b) The Porter model for the watershed in (a) above.

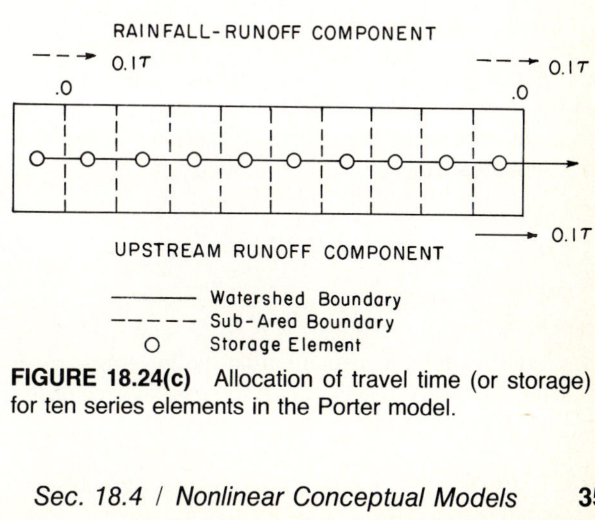

FIGURE 18.24(c) Allocation of travel time (or storage) for ten series elements in the Porter model.

shown by Boyd (1976) to be a poor indicator. The storage allocation is essentially the same as in the Laurenson model. The storage routing, however, occurs in each subarea.

Porter (1975) criticized the Laurenson model for halving the lag time for the lowest subarea on the grounds that the runoff from all upstream subareas passes through this subarea and is hence subject to modification by its storage. In addition, it transforms its share of rainfall to runoff. Porter used the full lag time for the lowest subarea but only one-half for the uppermost subarea. His argument was that this subarea had no upstream runoff. For example, assume that there are 10 storage elements in series. The travel time for the effective rainfall ranges from zero for the lower isochrone to one-tenth for the upper, giving an average travel time of one-twentieth for all subareas. For runoff routing, the travel time is one-tenth for the lower 9 elements but zero for the uppermost element. The storage allocation appears inconsistent with the effective rainfall component from 9 of the elements. Like the Laurenson model, this model does not consider separately the effects of storage on the transformation of the effective rainfall to direct runoff in a subarea and the storage effects on upstream runoff that passes through the main stream segment of the subarea. Both sources of runoff are subject to the same storage routing.

Another modification made by Porter involves applying a translation time between each pair of storage elements. The observed watershed lag is considered to be comprised of two components, a storage delay time k for each element and a translation time T between the elements:

$$t_L = k + T \tag{18.121}$$

t_L is not linearly related to Q. For simplicity, T is assumed independent of Q. Thus all nonlinearity of the watershed is attributed to storage elements. To further simplify, a constant value of T_* may be applied between successive storage elements. (This may be thought of as being a time for movement of water from one subarea to the next.) T_* is the same for each subarea, even though its size varies from one element to the next.

The translation time T for the watershed is obtained by integrating over the watershed area the time of translation incurred to the outlet as

$$T = \frac{1}{A} \int_A t \, da \tag{18.122}$$

where A is watershed area, t is translation time incurred between the point in question and the watershed outlet, and a is the area variable of integration. If there are N subareas in series with areas $a_1, a_2, \ldots, a_N$, then

$$\int_A t \, da \cong T_* \sum_{j=1}^{N-1} (N-j) a_j \tag{18.123}$$

in which the number of subareas is in the downstream direction. Note that the direct runoff from the lowermost element (N) incurs no translation, that from the next subarea upstream ($N-1$) incurs one translation T_*, and so on. Equation (18.122) can be expressed differently:

$$\frac{1}{A} \int_A t \, da \cong \frac{T_*}{A} \sum_{j=1}^{N-1} \alpha_j \tag{18.124}$$

where α_j is the area drained by the jth subarea. Therefore,

$$T \cong \frac{T_*}{A} \sum_{j=1}^{N-1} \alpha_j \tag{18.125}$$

If T is known, then k can be determined from equation (18.121) with t_L obtained from observations.

From a physical standpoint, introduction of travel time may be difficult to justify. The elements in the model are storage elements. The storage allocation is based on the travel time between the upper and lower boundaries of the subarea. Hence, there is no room to place this additional storage between the subareas. Delay time and translation are considered separately to account independently for the effects of hydrograph attenuation and translation; since they both result from storage routing, this consideration is inconsistent.

18.4.5 The Monash Model

The Monash model is also a variation of the Laurenson model and has been presented by Mein, Laurenson, and McMahon (1974). The watershed is divided into several subwatersheds such that subwatersheds are based on the major tributaries, gaging stations and the proposed dam site fall on subwatershed boundaries, and all subwatersheds are of the same order of area. Nodes are placed so that there is a node on the main stream in each subwatershed at the point nearest its centroid, at each confluence where subwatershed flows combine, at each gaging station, and at the proposed dam site. If large storage, natural or artificial, exists in the watershed, then nodes should be placed at each subwatershed entry to the storage and at a point immediately downstream of the storage. This allows the flows to be routed through the reservoir or storage, using, of course, an appropriate storage-discharge relation.

Storage effects are represented by nonlinear storage elements. A storage element is placed between each pair of adjacent nodes. Thus storage routing occurs between the nodes. However, if there is a reservoir, it is represented by a single storage element even if it has more than one upstream node. The resulting model structure therefore consists of a network of nonlinear storage elements having a shape similar to the watershed stream network. Storage is allocated to each element based on the properties of the stream segment between the nodes. Three relations are suggested:

$$k = \frac{L}{S_c^{.5}} \tag{18.126}$$

$$k = L \tag{18.127}$$

$$k = \left[\frac{L^2}{S_c^{.5}}\right]^{.71} \tag{18.128}$$

The allocation of storage elements is basically the same as in the Laurenson model, assuming that storage effects are proportional to travel time between each pair of nodes.

18.4.6 The Nonlinear Watershed Bounded Network (NWBN) Model

The nonlinear watershed bounded network (NWBN) model was developed by Boyd (1978, 1982) and is conceptually analogous to its linear counterpart, the LWBN model. The lag parameter k in hours in equation (18.96) is replaced by equation (18.108), with k_1 varying with the size of the subarea as

$$k = \alpha A^{.57} Q^{-.23} \tag{18.129}$$

and

$$C = .6\alpha A^{.57} Q^{-.23} \tag{18.130}$$

where α is a parameter evaluated by optimization; an average value of 1.68 was reported. Note that the ratio $C/k = .6$ is the same as maintained in the LWBN model. The units of A and Q are square kilometers and cubic meters per second. Computations can be carried out numerically as in equation (18.98). Boyd, Pilgrim, and Cordery (1979) found the NWBN model to be physically realistic and sufficiently accurate on 10 small watersheds in New South Wales, Australia. The model is of general applicability because it can represent a range of nonlinear responses exhibited by most watersheds.

18.4.7 The Nonlinear Geomorphological (NGM) Model

The nonlinear geomorphological (NGM) model is analogous to the LGM model, where equation (18.104) is replaced by equation (18.108) or any other appropriate nonlinear relation. Calculations can then be carried out numerically as for the NWBN model. Thus the NGM and NWBN models are comparable in the same way as their linear versions are.

EXERCISES

18.1. Develop a 6-h UH using the Snyder model for a watershed of area 60 ha. The main stream length is 2 km, and the distance from the point on the stream nearest to the watershed centroid to the outlet is 1 km. (Note that the average of C_t and C_p in the Snyder model is generally available for FPS system of units.) The average slope of the main channel is .05.

18.2. Derive a 6-h UH for the watershed in Question 18.1 using the SCS model and compare it with that of the Snyder model.

18.3. Derive a 6-h UH for the watershed in Question 18.1 using the Taylor-Schwarz model and compare it with that of the Snyder model as well as the SCS model.

18.4. Use the O'Kelly model to derive a 6-h UH for the watershed in Question 18.1.

18.5. Derive the instantaneous unit hydrograph by the Nash model for the watershed in Question 18.1. Then derive the 6-h UH and compare with that of the Snyder model.

18.6. Derive a 6-h UH by the Gray model for the watershed in Question 18.1.

18.7. Select a watershed in your area for which a topographic map is available. Determine its lag time using different definitions.

18.8. Establish a linear watershed bounded network model representation for the watershed in Question 18.7.

18.9. Compute the IUH using the LWBN model for the watershed in Question 18.7.

18.10. Establish the linear geomorphological model representation for the watershed in Question 18.7, and compare with that in Question 18.8.

18.11. Compute the IUH using the LGM model for the watershed in Question 18.7.

PART III
FLOOD ROUTING

CHAPTER NINETEEN
Reservoir Flood Routing

19.1 FLOOD ROUTING

The term *flood routing* is used in a much broader sense in this and the following chapter than suggested by its wording. Flood routing is variously defined in hydrologic literature. It usually represents a procedure by which to determine the outflow hydrograph at a downstream point of a river (or reservoir), given the inflow hydrograph at an upstream point thereof. To illustrate, consider a river reach as shown in Fig. 19.1 for which the streamflow hydrograph $Q(t)$ is known at location A. The flood routing can be utilized to determine the streamflow hydrograph at the location B. We can do likewise for location C, given the inflow hydrograph at location B. In this manner, the movement and shape of a flood wave in a river can be predicted. Fig. 19.1 shows that the streamflow hydrograph experiences a decrease in its peak and an increase in its time base, while preserving its volume as the flood wave moves downstream. This is a manifestation of the effect of channel storage on a flood hydrograph.

Although flood routing is normally discussed in relation to routing a flood wave from upstream to downstream in a stream or a reservoir or routing lateral inflow in a channel, its scope is much broader indeed. The routing can be reversed; that is, given the flood hydrograph at a downstream location, say location C in Fig. 19.1, the routing can be performed to determine the flood hydrograph upstream at the location B. This operation can be called *upstream routing*. Also, the routing can be done with mass curves or merely peak rates or peak stages of runoff. It may include determining inflow or outflow hydrographs, mass curves, or peak rates of runoff in reservoirs, farm ponds, lakes, swamps, and hydraulic structures. On the other hand, the routing can be extended to include low flows.

The objective of flood routing is manifold. It is used in the design of detention storage, channels, and many types of hydraulic structures. It is also used in the operation of control structures. The effect of a water-control structure on flood flows can be determined using routing procedures. It is used in (1) short-term forecasting of floods, (2) determining the UHs at various points, and (3) deriving synthetic UHs.

19.2 METHODS OF FLOOD ROUTING

Several methods have been developed for routing floods in reservoirs and rivers. Excellent reviews of these methods have been presented by Weinmann and Laurenson (1979), Dooge (1980), and Fread (1982). Broadly speaking, these methods can be classified as (1) hydraulic and (2) hydrologic. The hydraulic methods employ partial differential equations (usually the St. Venant equations or their approximations) for unsteady flow in open channels (Hayami, 1951; Chow, 1959; Brakensiek and Comer, 1965; Becker and Yeh, 1972; Yeh and Becker, 1973; Quick and Pipes, 1975; Huang, 1978; Weinmann and Laurenson, 1979; Smith, 1980; Dooge, 1980; Dooge, Kundzewicz, and Napiorkowski, 1983; Peters, 1983; Stepien, 1984). The hydrologic methods use a conceptual or systems approach (Wilson, 1941; Paynter, 1952; Chow, 1959; Mitchell, 1962; Dooge, 1973; Sauer, 1971; Strelkoff, 1980; Butler, 1982). As pointed out by Weinmann and Laurenson (1979), this classification is pertinent for well-defined regular channels but may be fallacious for natural channels, whose

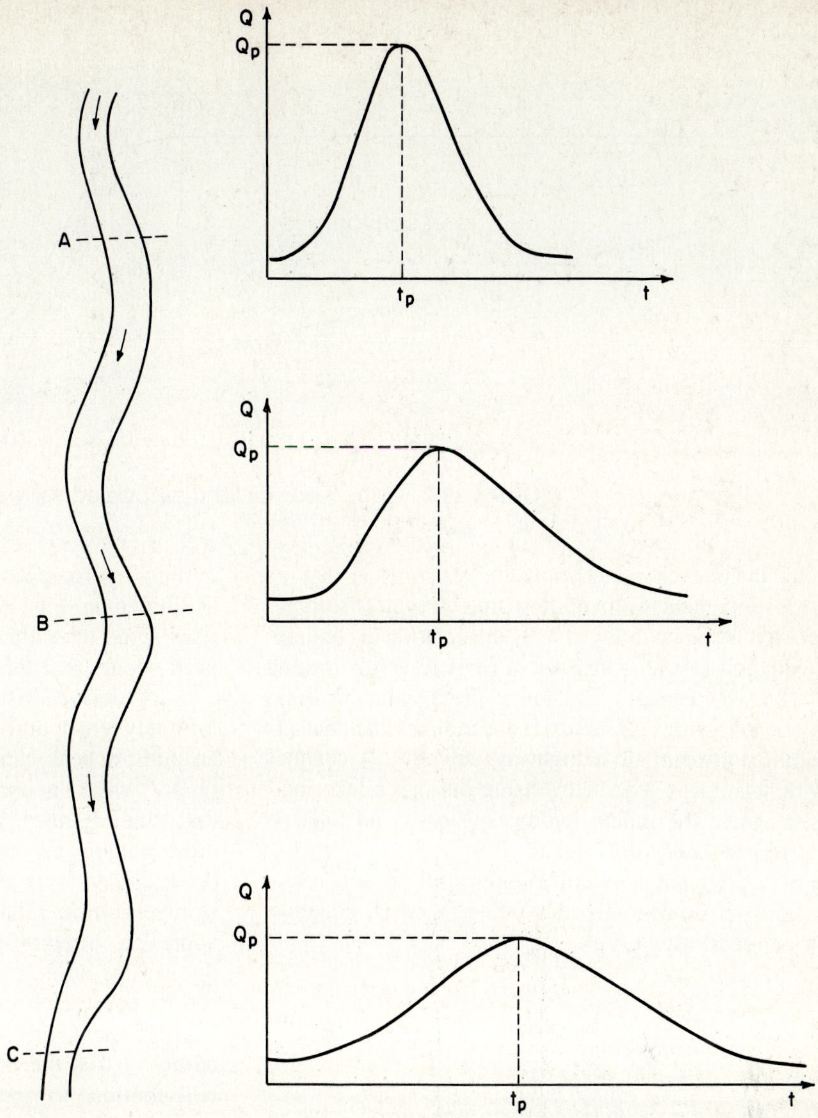

FIGURE 19.1 Steam flow hydrographs at locations A, B, and C along a river representing the movement of a flood wave. Q_p represents hydrograph peak and t_p represents its time.

complex physical properties may defy exact mathematical representation of their hydraulics and necessarily lead to some degree of conceptualization.

Each method possesses distinctive features that make it suitable for particular applications. The hydrologic methods are, in general, simpler but may not be satisfactory in problems other than those determining the progress of a flood wave down a long river. For example, when a flood passes through a junction, it produces backwater; when it is regulated by a dam, it produces surges. The backwater effect and the effect of surges cannot be accurately evaluated by a hydrologic method. The hydraulic methods are generally more accurate and versatile. However, their high demands on computing technology as well as on quantity and quality of input data may restrict their efficiency in practical applications.

Only the hydrologic methods are considered in this chapter. These methods (Laden, Reilly, and Minnotte, 1940; Kohler, 1944; Nemec, 1964; Gray, 1970; Hjelmfelt and Cassidy, 1975; Haan and Barfield, 1979; Wanielista, 1978) are based on the spatially lumped form of continuity equation,

$$\frac{dS}{dt} = I - Q \tag{19.1}$$

where S is the storage or volume of water, I is the inflow rate, Q is the outflow rate, t is time, and dS/dt is the rate of change of storage. Another equation relating S to I and Q is required to obtain a unique solution of equation (19.1).

19.3 RESERVOIR FLOOD ROUTING

When a flood wave passes through a full reservoir, it experiences attenuation in its peak and lengthening in its time base. This is the effect of a reservoir on the flood hydrograph. The water stored in the reservoir may discharge in a controlled or an uncontrolled manner. The water released as pipe flow through turbines or outlet works, called *prinicpal spillways*, represents an example of controlled conditions, whereas the flow over an emergency spillway represents uncontrolled conditions (see Fig. 19.2).

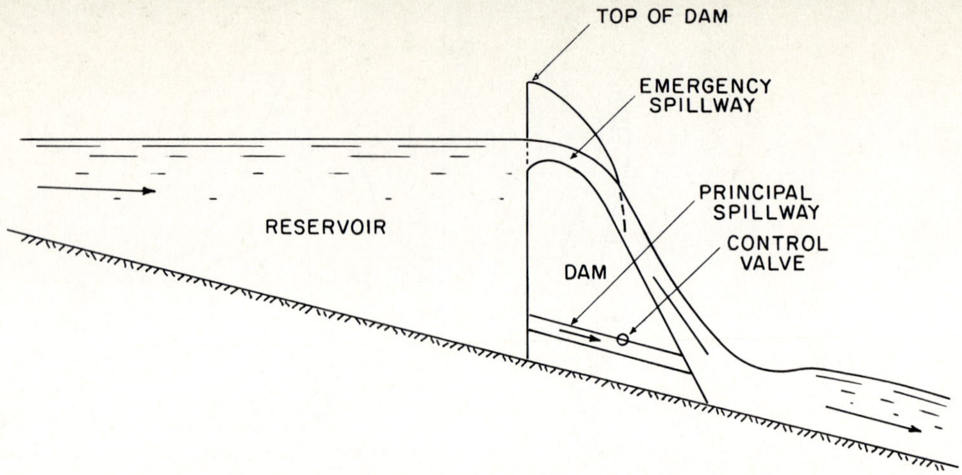

FIGURE 19.2 Dam, reservoir, and associated spillway works.

A fundamental characteristic of a reservoir is that its storage is closely related to its outflow rate. A typical storage-outflow relation is shown in Fig. 19.3. This relation in conjunction with equation (19.1) is utilized in most reservoir routing methods. This characteristic is seen to hold in many other situations not involving a dam. A few examples include: (1) water ponding upstream of a highway culvert, (2) channel reaches with swamps, (3) a river passing through a lake, and (4) channels where the inflow hydrograph rises and falls so slowly that nearly steady flow occurs.

It may be pertinent to examine implications of this reservoir characteristic. Since S depends uniquely on Q, equation (19.1) can be recast as

$$\frac{dS}{dQ}\frac{dQ}{dt} + Q = I \tag{19.2}$$

The term dS/dQ is the theoretical travel time of the wave in the reach. With $I(t)$ and dS/dQ known, $Q(t)$ can be computed. Rewriting this equation as

$$\frac{dQ}{dt} = \frac{(I - Q)}{dS/dQ} \tag{19.3}$$

For $0 < dS/dQ < \infty$, as in Fig. 19.3, it is clear that the outflow hydrograph must begin to rise immediately, without a lag, upon a rise in inflow. This is contrary to the notion of a flood wave, where a flood occurs at a point and is perceived only some time later at a point downstream. Furthermore, the outflow and storage peak concurrently. The peak outflow intersects the inflow hydrograph on its falling limb. This is easily seen from equation (19.1). For maximum S we can set dS/dt to zero. Therefore,

$$I = Q$$

where the maximum of Q also occurs. The storage volume required in the reservoir is the area between the inflow and outflow hydrographs from the beginning of outflow to its peak as shown in Fig. 19.4. In reality, the outflow hydrograph rises some time after the inflow hydrograph, and the outflow peak occurs past the time when $I = Q$.

A delayed rise in outflow can occur only if dS/dQ is infinitely great during the early stages of inflow. Likewise, an outflow peak can occur when $Q \neq I$—for instance, when $Q > I$, which is normally the case—provided $dS/dQ \to \infty$ at that time. Further, the inflow and outflow hydrographs can cross without an outflow peak, that is, $Q = I$ and $dQ/dt \neq 0$, if $dS/dQ = 0$ at that time. These characteristics of the storage-outflow relation are usually mimicked by a hydraulic approach. Strelkoff (1980) has shown that by manipulating the storage-outflow relation, any desired outflow hydrograph can be constructed for a given inflow hydrograph.

If the storage-discharge relation is a loop, then with $dS/dQ = 0$ at the time when $I = Q$, the outflow hydrograph will continue to rise beyond that time. At the time when the outflow ultimately peaks, $dS/dQ \to \infty$. The subsequent decline of the outflow hydrograph corresponds to that branch of the loop that descends monotonically to the starting value of S.

Flood routing through a reservoir may represent the simplest of all routing problems. The methods commonly used for reservoir flood routing are graphical, semigraphical, and analytical. We present some of these methods here.

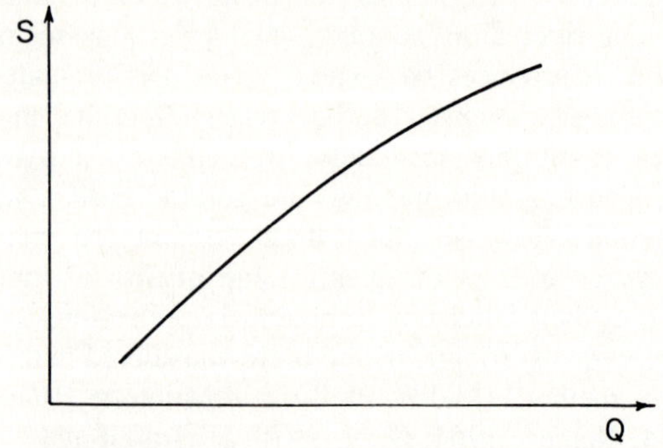

FIGURE 19.3 A typical storage-outflow discharge relation.

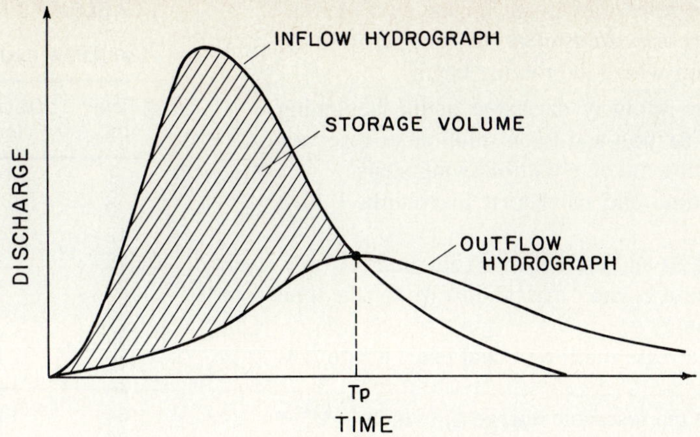

FIGURE 19.4 Inflow and outflow hydrographs.

19.4 THE MASS CURVE METHOD

This is one of the most versatile methods of reservoir routing (Soil Conservation Service, 1964). The various versions of this method available in hydrologic literature include: (1) direct, (2) trial and error, and (3) graphical. The first two are described next.

19.4.1 Direct Solution

In the direct solution method we write equation (19.1) as

$$\int_0^S dS = \int_0^t I\,dt - \int_0^t Q\,dt \tag{19.4}$$

If we select Δt as the routing time interval $(t_2 - t_1)$, with t_1 denoting the beginning and t_2 denoting the end of the time interval, and initial storage as zero, then

$$\int_0^{S_{t+\Delta t}} dS = \int_0^{t+\Delta t} I\,dt - \int_0^t Q\,dt - \int_t^{t+\Delta t} Q\,dt$$

$$S_{t+\Delta t} = M_{t+\Delta t} - V_t - \left(\frac{Q_t + Q_{t+\Delta t}}{2}\right)\Delta t \tag{19.5}$$

where $S_{t+\Delta t}$ is storage at time $t + \Delta t$, $M_{t+\Delta t}$ is mass (accumulated) inflow at time $t + \Delta t$, V_t is mass (accumulated) outflow at time t, and Q_t and $Q_{t+\Delta t}$ are outflows at the beginning and the end of the routing time interval Δt, respectively. The unknowns are $Q_{t+\Delta t}$ and $S_{t+\Delta t}$ in equation (19.5).

From equation (19.4) it is clear that mass outflow at any time equals the difference between mass inflow and storage,

$$V_t = M_t - S_t \tag{19.6}$$

Substituting equation (19.6) into equation (19.5) and rearranging gives

$$M_{t+\Delta t} - M_t + \left(S_t - \frac{\Delta t}{2} Q_t\right) = S_{t+\Delta t} + \frac{\Delta t}{2} Q_{t+\Delta t} \tag{19.7}$$

Equation (19.7) is the working equation for direct version of the mass curve method and can be solved numerically. This is the mass-curve version of the storage-indication method to be discussed later. The steps involved in these two methods are analogous and are, therefore, not repeated. It may suffice to point out that in equation (19.7), the left side is completely known and the right side is unknown. Therefore, to solve this equation requires a knowledge of the storage-outflow relation.

19.4.2 Trial-and-Error Solution

The trial-and-error method seeks solution to equation (19.1) written as

$$M_{t+\Delta t} - (V_t + \bar{Q}\,\Delta t) = S_{t+\Delta t} \tag{19.8}$$

where $\bar{Q}$ represents the average discharge during Δt. The solution to equation (19.8) is a trial-and-error one and involves the use of elevation-storage and elevation-discharge curves, either separately or in combination. The routing operation is simple, easily done, and efficiently adapted to complex routing problems.

As evident from equation (19.8), the input includes the mass inflow, whereas the output includes mass outflow, outflow hydrograph, and reservoir storage. The routing interval Δt can be either variable or constant. However, it is more convenient as well as realistic to use a variable Δt, keeping it small for a large change in mass inflow and large for a small change therein.

Before obtaining the trial-and-error solution, the following steps must be undertaken.

1. Develop an elevation-discharge curve for the spillway or whatever structure is under consideration.
2. Develop an elevation-storage curve for the structure. Steps 1 and 2 can be combined, if desired, into a single storage-discharge relationship.
3. Plot the mass curve of inflow.
4. Develop a working table for routing, as shown in the ensuing example.

5. Determine the reservoir storage for the start of routing. Make necessary adjustments to show zero storage at the elevation where the routing begins.
6. Determine the spillway discharge at the beginning of routing. The trial-and-error solution to reservoir routing may now involve the following steps.
   - **(a)** Choose time and tabulate it in column 1 of a table.
   - **(b)** Compute Δt and enter it into column 2.
   - **(c)** Into column 3, enter mass inflow M_t for the time in column 1.
   - **(d)** Assume a mass outflow V_t and enter it into column 4.
   - **(e)** Compute the reservoir storage S_t, which is $M_t - V_t$ or column 3 − column 4, and enter it into column 5.
   - **(f)** Determine the instantaneous discharge of the spillway. This can be obtained as follows. Determine the elevation for S_t in column 5 from the elevation-storage curve. Then determine discharge for this elevation from the elevation-discharge curve. On the other hand, if the storage-discharge curve is available, use it directly to obtain discharge for specified S_t. Tabulate it in column 6.
   - **(g)** Compute the average discharge for Δt. This is taken to be the arithmetic mean of the discharge determined in Step 6 (f) and the one for the previous time. Tabulate it in column 7.
   - **(h)** Compute the outflow Q_t for Δt. This is obtained by multiplying Δt from column 2 by the average discharge in Step 6 (g). Tabulate it in column 8.
   - **(i)** Determine the mass outflow V_t by successively accumulating Q_t of column 8, and tabulate it in column 9.
   - **(j)** Compute V_t of column 9 with the assumed V_t of column 4. If the two entries agree within an acceptable degree of accuracy, then the routing is complete. If their agreement is not acceptable, assume another V_t in column 4 and repeat the Steps 3–6 (j).

EXAMPLE 19.1

Using the trial-and-error version of the mass-curve method, route the flood given in Table 19.1 through a reservoir having a spillway crest at the elevation 130 m. Assume that the pool level prior to the flood is also 130 m. The spillway discharging capacity and reservoir storage at various elevations are given in Table 19.2 and are shown in Fig. 19.5 and 19.6. Compute the peak of the outflow hydrograph and the maximum water level. Draw the routed outflow hydrograph and indicate the reduction in flood peak.

***Solution*:** The procedure just discussed is followed step by step. The governing equation (19.8) is used for computations with Δt = 6 h. A storage-discharge curve is constructed as shown in Fig. 19.7. Five trial-and-error iterations conclude the routing. Calculations for the five trials are given in Tables 19.3–19.5. Mass inflow and mass outflow curves are plotted as shown in Fig. 19.8. This gives the maximum storage during the flood or minimum required reservoir storage, for which the maximum water level in the reservoir is gotten from Fig. 19.5. Finally, the inflow and outflow hydrographs are plotted in Fig. 19.9. Thus

$$\text{Maximum water level} = 138.4 \text{ m}$$

$$\text{Maximum storage during flood} = 90 \times 10^6 \text{ m}^3$$

$$\text{Peak of the outflow hydrograph} = 955 \text{ m}^3/\text{s}$$

$$\text{Reduction in flood peak} = 845 \text{ m}^3/\text{s}$$

TABLE 19.1

An inflow hydrograph

Time (h)	*Discharge (m^3/s)*	*Time (h)*	*Discharge (m^3/s)*
0	0	78	620
6	50	84	510
12	130	90	420
18	250	96	320
24	350	102	270
30	540	108	200
36	735	114	150
42	1215	120	100
48	1800	126	72
54	1400	132	45
60	1050	138	25
66	900	144	10
72	740	150	0

TABLE 19.2

Spillway discharge and reservoir storage at various elevations

Elevation (m)	*Spillway Discharge (m^3/s)*	*Reservoir Storage (10^6 m^3)*
128	0	0
130	0	0
131	35	7
132	90	16
133	165	25
134	270	33
136	540	51
137	710	70
138	890	85
139	1080	100
140	1310	120

EXAMPLE 19.2

Route the flood in Example 19.1 by the direct numerical version of the mass-curve method.

***Solution*:** The routing is performed numerically on computer. By regression analysis, a storage discharge is obtained as

$$S = 5.98 + .088Q$$

with a correlation coefficient of .997, as shown in Fig. 19.10.

The time interval used in computations is allowed to vary. This allows for better accounting of changes in flow. The criterion used for varying the time interval is: if the change in inflow during a 6-h period is larger than 100 m^3/s, then the time interval is reduced

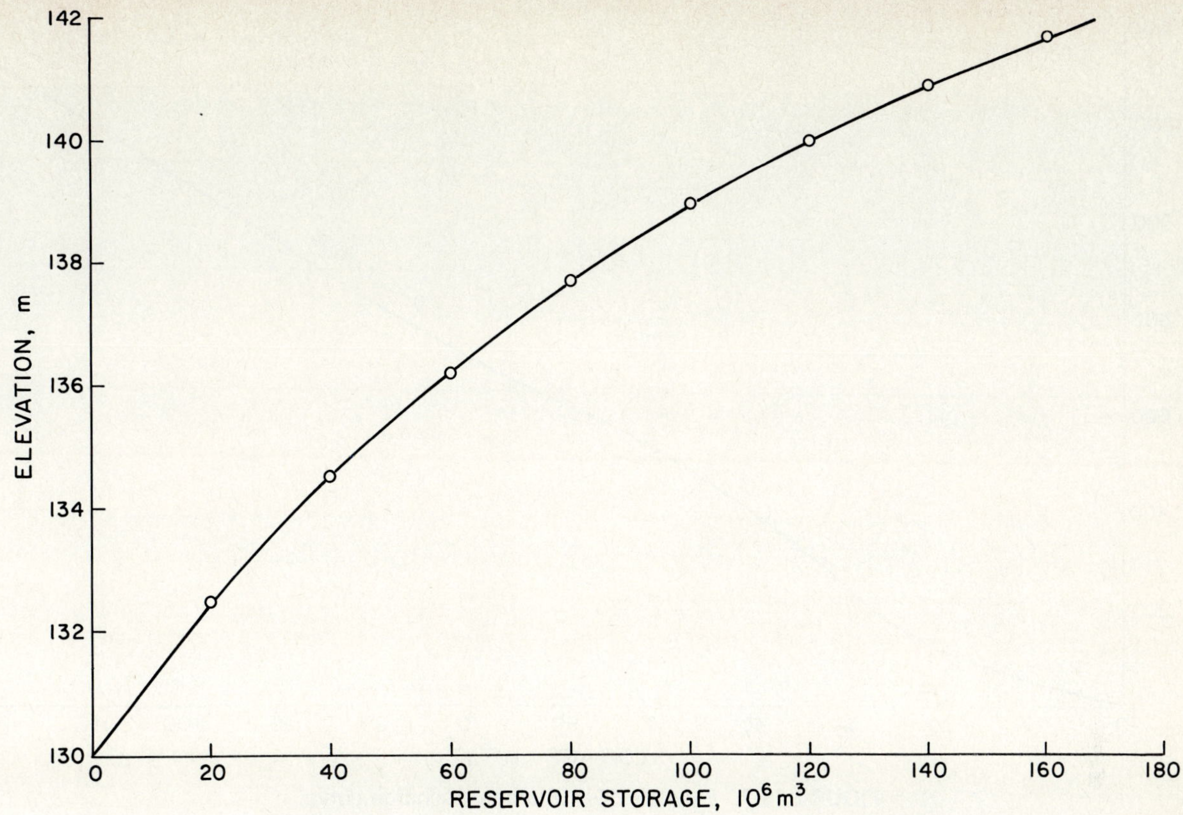

FIGURE 19.5 Example 19.1: Elevation versus reservoir storage.

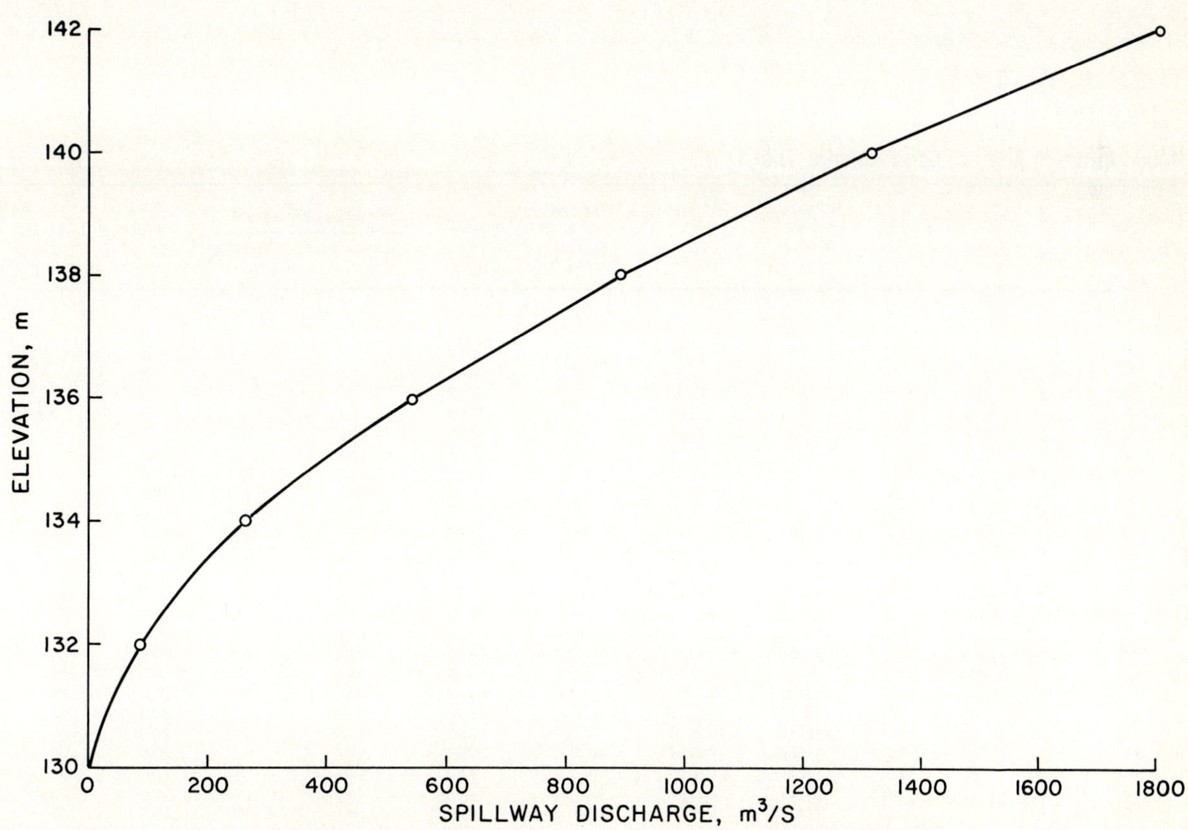

FIGURE 19.6 Example 19.1: Elevation versus spillway discharge.

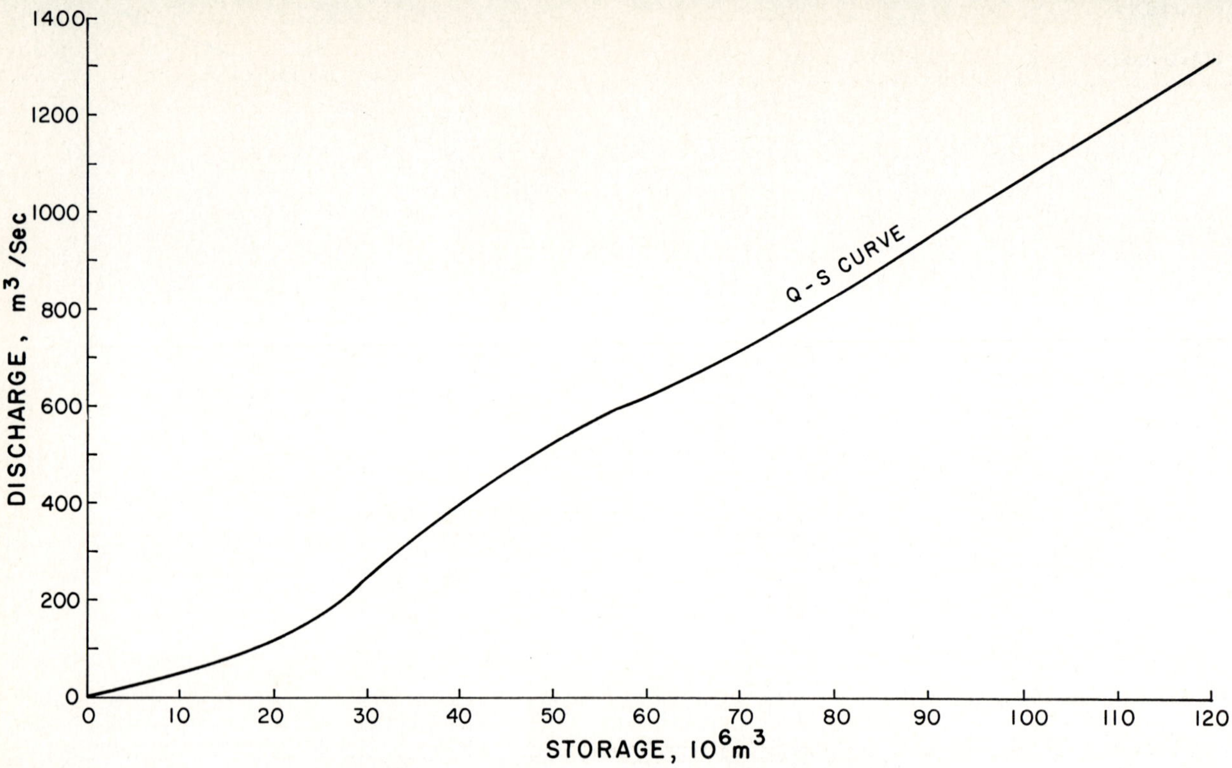

FIGURE 19.7 Example 19.1: Storage-discharge curve.

TABLE 19.3

Mass curve method of flood routing: Trial 1

Time Accumulated (h)	*Δt (h)*	*Inflow (m^3/s)*	*Mass Inflow (m^3/s h)*	*Cumulative Mass Inflow (m^3/s h)*	*Cumulative Mass Inflow (10^6 m^3)*	*Assumed Accumulated Outflow (10^6 m^3)*	*Reservoir Volume (10^6 m^3)*	*Spillway Discharge: Instantaneous (m^3/s)*	*Spillway Discharge: Average (m^3/s)*	*Outflow for Δt (10^6 m^3)*	*Cumulative Outflow (10^6 m^3)*
0	0	0	0	0	0	0	0	0	0	0	0
6	6	50	150	150	.54	.10	.44	5	3	.06	.06
12	6	130	540	690	2.48	.80	1.68	10	8	.2	.26
18	6	250	1140	1830	6.59	2.50	4.09	20	15	.3	.56
24	6	350	1800	3630	13.07	6.50	6.57	32	26	.6	1.20
30	6	540	2675	6300	22.68	11.00	11.68	60	46	1.0	2.2
36	6	735	3825	10125	36.45	18.00	18.45	110	85	2.0	3.2
42	6	1215	5850	15975	57.51	28.00	29.51	240	175	4.0	7.2
48	6	1800	9045	25020	90.07	45.00	45.07	465	353	8.0	15
54	6	1400	9600	34620	124.63	60.00	64.63	665	565	10.0	25
60	6	1060	7350	41970	151.09	80.00	71.09	730	698	20.0	45
66	6	900	5850	47820	172.15	85.00	77.15	800	765	20.0	65
72	6	740	4920	52740	189.86	110.00	79.86	825	813	20.0	85
78	6	620	4080	56820	204.55	120.00	84.55	880	853	20.0	105
84	6	510	3390	60210	216.76	130.00	86.76	910	895	20.0	125
90	6	420	2790	63000	226.80	140.00	86.80	915	913	20.0	145
96	6	320	2220	65220	234.79	150.00	84.79	890	903	20.0	165
102	6	270	1770	66990	241.16	155.00	86.16	910	900	20.0	185
108	6	200	1410	68400	246.24	160.00	86.24	910	910	20.0	205
114	6	150	1050	69450	250.02	165.00	85.02	890	900	20.0	225
120	6	100	750	70200	252.72	170.00	82.72	860	875	20.0	245
126	6	72	516	70716	254.58	174.00	80.58	840	850	20.0	265
132	6	45	351	71067	255.84	178.00	77.84	800	820	20.0	285
138	6	25	210	71277	256.60	182.00	74.60	770	785	20.0	305
144	6	10	105	71382	256.97	185.00	71.97	740	755	20.0	325
150	6	0	30	71412	257.08	190.00	67.08	690	715	20.0	345

TABLE 19.4

Mass curve method of flood routing: Trials 2 and 3

Time Accumulated (h)	Δt (h)	Cumulative Mass Inflow ($10^6 m^3$)	Assumed Accumulated Outflow ($10^6 m^3$)	Reservoir Volume ($10^6 m^3$)	Spillway Discharge Instantaneous (m^3/s)	Spillway Discharge Average (m^3/s)	Outflow for Δt ($10^6 m^3$)	Cumulative Outflow ($10^6 m^3$)	Assumed Accumulated Outflow ($10^6 m^3$)	Reservoir Volume ($10^6 m^3$)	Spillway Discharge Instantaneous (m^3/s)	Spillway Discharge Average (m^3/s)	Outflow for Δt ($10^6 m^3$)	Cumulative Outflow ($10^6 m^3$)
0	0	0	0	0	0	0	0	0	0	0	0	0	0	0
6	6	.54	.08	.46	5	3	.06	.06	6	1	5	3	.06	.06
12	6	2.48	.40	2.08	10	8	.20	.26	.2	1.8	10	8	.20	.26
18	6	6.59	1.00	5.59	28	19	.40	.66	.7	6.3	30	20	.40	.7
24	6	13.07	2.50	10.57	55	42	.90	1.56	1.8	11.2	60	45	.97	1.7
30	6	22.68	3.00	19.68	116	86	2.00	3.56	4	19	112	96	2.00	4
36	6	36.45	7.00	29.45	240	178	4.00	7.6	8	28	210	161	3.50	8
42	6	57.51	12.00	45.51	470	355	8.00	15.6	15	43	435	323	6.98	15
48	6	90.07	25.00	65.07	710	590	10.00	25.6	25	65	665	550	11.88	27
54	6	124.83	35.00	89.63	350	830	20.00	45.6	45	80	830	748	16.15	43
60	6	151.09	60.00	91.09	965	958	20.00	65.6	64	87	915	873	18.85	62
66	6	172.15	75.00	97.15	1040	1003	20.00	85.6	80	92	975	945	20.41	82
72	6	189.86	95.00	94.86	1015	1028	20.00	105.6	105	85	890	933	20.14	102
78	6	204.55	110.00	94.55	1012	1014	20.00	125.6	122	83	865	878	18.95	121
84	6	216.76	127.00	89.76	952	982	20.00	145.6	140	77	795	830	17.93	139
90	6	226.80	142.00	84.80	890	921	20.00	165.6	158	69	710	753	16.25	155
96	6	234.79	160.00	74.79	770	830	20.00	185.6	176	59	612	691	14.91	170
102	6	241.16	170.00	71.16	725	748	20.00	205.6	185	56	585	599	12.93	183
108	6	246.24	180.00	66.24	680	703	20.00	225.6	195	51	535	560	12.09	195
114	6	250.02	190.00	60.02	625	653	10.00	236	206	44	450	493	10.64	206
120	6	252.72	200.00	52.72	555	590	10.00	246	217	36	340	395	8.53	215
126	6	254.58	210.00	44.58	455	505	10.00	256	222	33	295	317	6.86	222
132	6	255.84	220.00	35.84	335	395	9.00	265	227	29	225	260	5.62	228
138	6	256.60	230.00	26.60	190	263	6.00	271	233	24	155	190	4.10	232
144	6	256.97	240.00	16.97	95	143	3.00	274	235	22	140	148	3.19	235
150	6	257.08	250.00	7.08	40	68	1.00	275	237	20	120	130	2.81	238

TABLE 19.5

Mass curve method of flood routing: Trials 4 and 5

Time Accumulated (h)	Δt (h)	Cumulative Mass Inflow ($10^6 m^3$)	Assumed Accumulated Outflow ($10^6 m^3$)	Reservoir Volume ($10^6 m^3$)	Spillway Discharge Instantaneous (m^3/s)	Spillway Discharge Average (m^3/s)	Outflow for Δt ($10^6 m^3$)	Cumulative Outflow ($10^6 m^3$)	Assumed Accumulated Outflow ($10^6 m^3$)	Reservoir Volume ($10^6 m^3$)	Spillway Discharge Instantaneous (m^3/s)	Spillway Discharge Average (m^3/s)	Outflow for Δt ($10^6 m^3$)	Cumulative Outflow ($10^6 m^3$)	Absolute Error in Outflow Determined ($10^6 m^3$)
0	0	0	0	0	0	0	0	0	0	0	0	0	0	0	0
6	6	1	0	1	5	3	.06	.06	0	1	5	3	.06	.06	.06
12	6	2	.2	1.8	10	8	.20	.26	.2	1.8	10	8	.20	.26	.06
18	6	7	.7	6.3	30	20	.40	.66	.7	6.3	30	20	.40	.66	.04
24	6	13	1.8	11.2	60	45	.97	1.63	1.8	11.2	60	45	.97	1.6	.20
30	6	23	4	19	112	86	2.00	3.63	4	19	112	86	2.00	3.63	.37
36	6	36	8	28	210	161	3.50	7.13	8	28	210	161	3.50	7.13	.87
42	6	58	15	43	435	323	6.98	14.11	15	43	435	323	6.98	14.11	.89
48	6	90	27	63	650	543	11.72	25.83	26	64	660	548	11.83	25.94	.16
54	6	125	43	82	855	753	16.25	42.08	42	83	865	763	16.47	42.41	.41
60	6	151	62	89	940	898	19.35	61.47	62	89	940	903	19.49	61.90	.10
66	6	172	82	90	955	948	20.47	81.94	82	90	955	948	20.47	82.37	.37
72	6	190	102	88	930	943	20.36	102.30	102	88	930	943	20.36	102.27	.27
78	6	205	121	84	880	905	19.55	121.85	122	83	865	898	19.38	121.66	.34
84	6	219	139	78	807	844	18.22	140.07	140	77	795	830	17.93	139.59	.41
90	6	227	156	71	730	769	16.60	156.67	156	71	730	763	16.47	156.06	.06
96	6	235	172	63	650	690	14.90	171.57	171	64	660	695	15.01	171.07	.07
102	6	241	184	57	595	622	13.45	185.02	185	56	585	623	13.45	184.52	.48
108	6	246	197	49	510	553	11.93	196.95	196	50	525	555	11.99	196.51	.51
114	6	250	208	42	425	468	10.10	207.10	207	43	435	480	10.37	206.88	.12
120	6	253	215	38	370	398	8.59	215.69	215	38	370	403	8.70	215.58	.58
126	6	255	223	32	280	325	7.02	222.71	223	32	280	325	7.02	222.60	.40
132	6	256	228	28	210	245	5.29	228.00	228	28	210	245	5.29	227.89	.11
138	6	257	232	25	165	188	4.05	232.00	232	25	165	188	4.05	231.94	.06
144	6	257	235	22	140	153	3.25	235.29	235	22	140	153	3.29	235.23	.23
150	6	257	238	19	112	126	2.72	238.00	238	19	112	126	2.72	237.95	.05

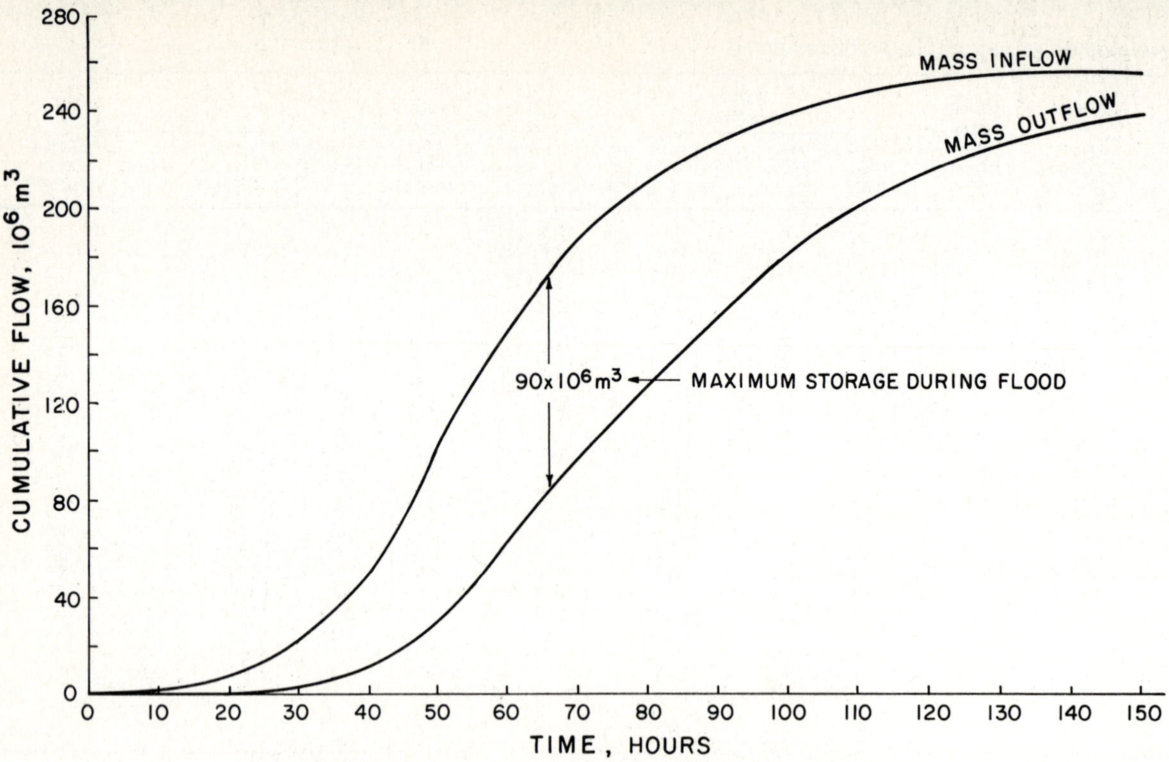

FIGURE 19.8 Example 19.1: Cumulative inflow and outflow curves.

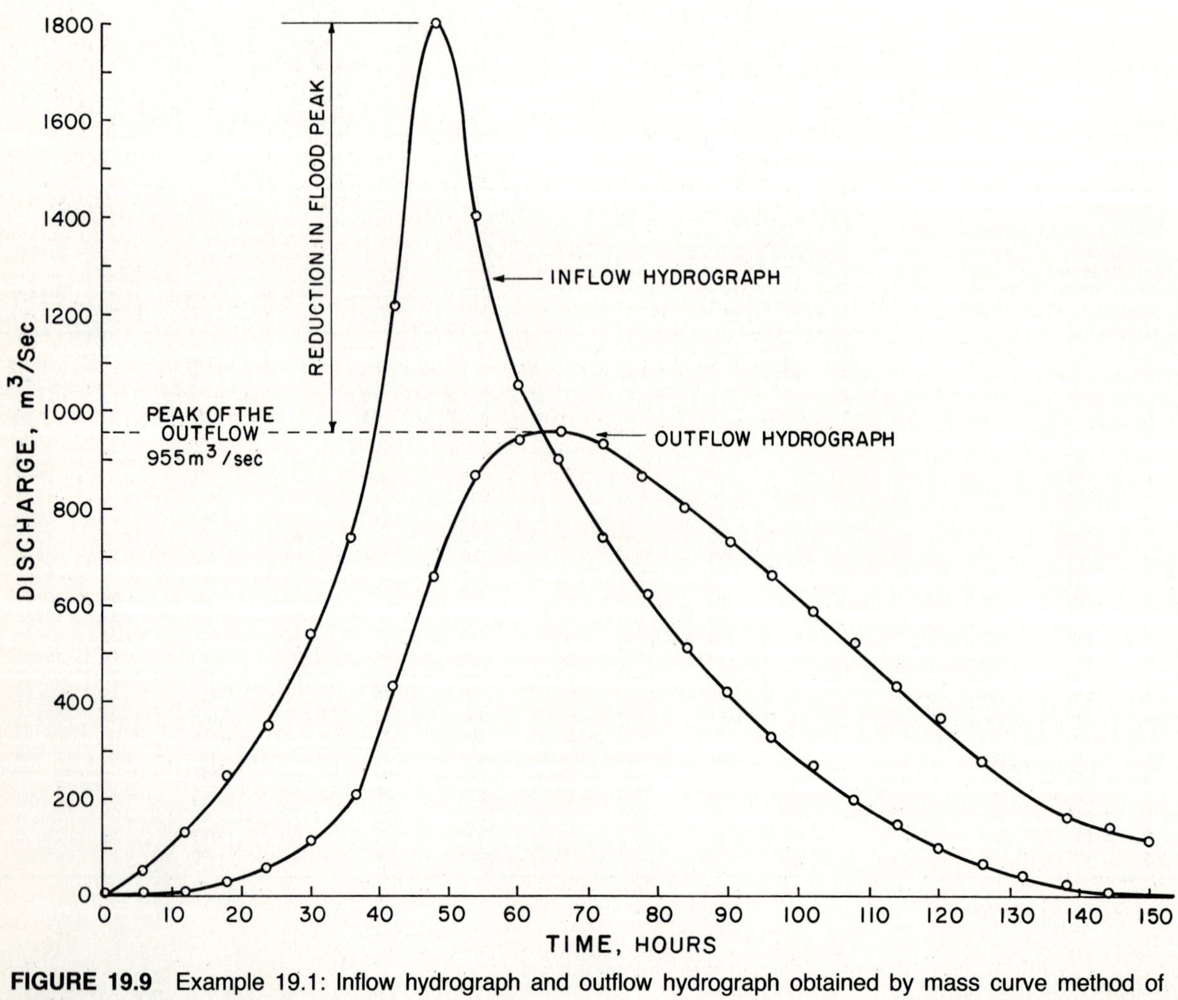

FIGURE 19.9 Example 19.1: Inflow hydrograph and outflow hydrograph obtained by mass curve method of routing.

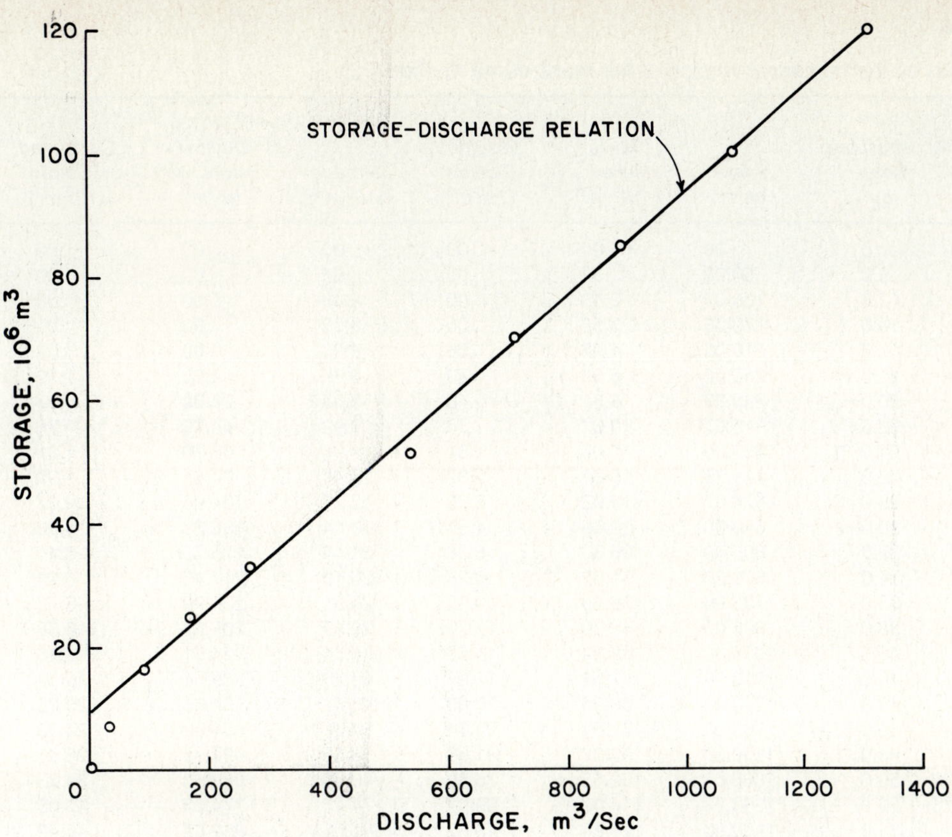

FIGURE 19.10 Example 19.2: A line fit to storage-discharge data.

to 2 h. The procedure for iterative numerical computations is as discussed previously and is illustrated in Table 19.6. To summarize,

$$\text{Maximum water level} = 138.5\text{m}$$

$$\text{Peak of the outflow hydrograph} = 988.43 \text{ m}^3/\text{s}$$

$$\text{Reduction in flood peak} = 811.57 \text{ m}^3/\text{s}$$

$$\text{Maximum storage during flood} = 92.96 \times 10^6 \text{ m}^3$$

A comparison of results obtained here with those obtained in Example 19.1 indicates that both graphical and numerical versions of the mass curve perform satisfactorily.

19.5 THE STORAGE-INDICATION METHOD

The storage-indication method of reservoir routing is also referred to as the modified Puls method. This method represents equation (19.1) in finite difference form,

$$S_{t+\Delta t} - S_t = (I_{t+\Delta t} + I_t)\frac{\Delta t}{2} - (Q_{t+\Delta t} + Q_t)\frac{\Delta t}{2} \quad (19.9)$$

in which Q_t may incorporate controlled Q_c as well as uncontrolled discharge Q_s,

$$Q_t = C_{c_t} + Q_{s_t}$$

Equation (19.9) assumes that the inflow during a time interval Δt is equal to the average of the inflows at the beginning and the end of the time interval; the same is assumed for the outflow. In addition, the change in storage is equal to the storage at the end of Δt minus the storage at the beginning of Δt.

Separating the known quantities from the unknown ones and rearranging,

$$(I_t + I_{t+\Delta t}) - (Q_{c_t} + Q_{c_{t+\Delta t}})\left(\frac{2S_t}{\Delta t} - Q_{s_t}\right) = \frac{2S_{t+\Delta t}}{\Delta t} + Q_{s_{t+\Delta t}} \quad (19.10)$$

The left side is known and the right side is unknown. The inflow hydrograph is known. Q_{c_t}, which is known, is the controlled discharge, which may pass through the turbines, outlet works, or over the spillway. The remaining quantity on the left side is assumed known at the start of the routing operation, that is, S_1 and Q_1 are initially known; they are determined for each subsequent step.

The uncontrolled discharge Q_{s_t} goes freely over the spillway. It depends upon the depth of flow over the spillway and the spillway geometry. Further, the depth of flow over the spillway depends upon the depth of water in the reservoir or storage. Therefore,

$$S = S(y)$$

$$Q_s = Q_s(y)$$

TABLE 19.6

Flood routing by the numerical version of the mass-curve method

Time Interval	Accumulated Time (h)	Inflow (m^3/s)	Mass Inflow ($10^6\,m^3$)	Assumed Outflow ($10^6\,m^3$)	Storage ($10^6\,m^3$)	Outflow Discharge (m^3/s)	Computed Outflow ($10^6\,m^3$)	Error in Computed Discharge (%)
0	.0	.00	.00	.00	.00	.00	.00	.00
6	6.0	50.00	.54	.00	.54	.00	.00	.00
6	12.0	130.00	2.48	.00	2.48	.00	.00	.00
2	14.0	170.00	3.56	.00	3.56	.00	.00	.00
2	16.0	210.00	4.93	.00	4.93	.00	.02	.00
2	18.0	250.00	6.59	.20	6.39	4.66	.12	.39
2	20.0	283.33	8.51	.43	8.08	23.89	.37	.14
2	22.0	316.67	10.67	.75	9.92	44.79	.76	.02
2	24.0	350.00	13.07	1.31	11.76	65.70	1.31	.00
2	26.0	415.33	15.82	2.37	13.44	84.81	1.99	.16
2	28.0	476.67	19.02	3.80	15.22	104.95	2.87	.25
2	30.0	540.00	22.68	4.54	18.14	138.23	4.00	.12
2	32.0	605.00	26.80	5.36	21.44	175.70	5.41	.01
2	34.0	670.00	31.39	6.28	25.11	217.43	6.99	.11
2	36.0	735.00	36.45	10.93	25.51	221.99	8.76	.20
2	38.0	895.00	42.32	12.70	29.62	268.67	10.89	.14
2	40.0	1055.00	49.34	14.80	34.54	324.51	13.46	.09
2	42.0	1215.00	57.51	17.25	40.26	389.51	16.54	.04
2	44.0	1410.00	66.96	20.09	46.87	464.68	20.20	.01
2	46.0	1605.00	77.81	23.34	54.47	551.02	24.33	.04
2	48.0	1800.00	90.07	31.53	58.55	597.35	28.96	.08
2	50.0	1666.67	102.55	35.89	66.66	689.63	34.23	.05
2	52.0	1533.33	114.07	39.93	74.15	774.62	40.09	.00
2	54.0	1400.00	124.63	43.62	81.01	852.62	46.49	.07
2	56.0	1283.33	134.29	47.00	87.29	923.97	52.73	.12
2	58.0	1166.67	143.11	65.83	77.28	810.23	58.74	.11
2	60.0	1050.00	151.09	69.50	81.59	859.20	65.09	.06
2	62.0	1000.00	158.47	72.90	85.57	904.49	71.76	.02
2	64.0	950.00	165.49	76.13	89.37	947.56	78.73	.03
2	66.0	900.00	172.15	79.19	92.96	988.43	85.33	.08
2	68.0	846.67	178.44	98.14	80.30	844.52	91.52	.07
2	70.0	793.33	184.34	101.39	82.95	874.71	97.91	.03
2	72.0	740.00	189.86	104.43	85.44	902.94	104.51	.00
2	74.0	700.00	195.05	107.28	87.77	929.45	111.29	.04
2	76.0	660.00	199.94	109.97	89.97	954.49	117.00	.06
2	78.0	620.00	204.55	143.19	61.39	629.38	121.58	.15
2	80.0	583.33	208.88	146.22	62.67	644.15	126.27	.14
2	82.0	546.67	212.95	149.07	63.89	658.02	131.05	.12
2	84.0	510.00	216.76	151.73	65.03	670.99	145.92	.04
6	90.0	420.00	226.80	158.76	68.04	705.23	150.09	.05
2	92.0	386.67	229.70	183.76	45.94	454.10	153.38	.17
2	94.0	353.33	232.37	185.89	46.47	460.15	156.71	.16
2	96.0	320.00	234.79	187.83	46.96	465.66	166.93	.11
6	102.0	270.00	241.16	192.93	48.23	480.15	177.42	.08
6	108.0	200.00	246.24	196.99	49.25	491.68	187.22	.05
6	114.0	150.00	250.02	207.52	42.50	415.04	196.24	.05
6	120.0	100.00	252.72	209.76	42.96	420.25	205.35	.02
6	126.0	72.00	254.58	211.30	43.28	423.84	214.54	.02
6	132.0	45.00	255.84	212.35	43.49	426.28	223.76	.05
6	138.0	25.00	256.60	212.98	43.62	427.74	232.38	.09
6	144.0	10.00	256.98	218.43	38.55	370.07	240.37	.10

where y represents the water surface elevation. The right side of equation (19.10) can be written as

$$\frac{2S}{\Delta t} + Q = f(y)$$

In order to utilize equation (19.10), elevation-storage and elevation-discharge relationships must be specified. For simplicity let us assume that Q_c is negligible. Therefore, Q can be taken to imply Q_s.

The elevation-discharge relationship is utilized to determine the outflow, given the depth of water over the spillway, and can be constructed using the general weir formula

$$Q = CLh^{3/2} \tag{19.11}$$

where C is the weir coefficient, L is the length of spillway crest, and h is the depth of the water above the spillway crest or stage. From topographic information, y corresponding to h can be obtained. Then the curve for either y versus Q or h versus Q can be constructed, as shown in Fig. 19.11.

To construct the elevation-storage curve, we again utilize the topographic map. The depth of water above the spillway h multiplied by the reservoir surface area can yield the (temporary) storage of the reservoir. The reservoir surface area is known from the topographic map. Therefore, the temporary storage S for any h is approximately

$$S = \frac{h}{2}(A_1 + A_2) \tag{19.12}$$

where A_1 denotes the surface area of the reservoir when $h = 0$, and A_2 denotes the surface area of the reservoir when the depth of flow is h. Then curves for either y versus S or h versus S can be constructed, as shown in Fig. 19.12. It may be noted that elevation-discharge and elevation-storage curves of Fig. 19.11 and 19.12 can be easily combined to form a storage-discharge curve. Thus the storage in a reservoir depends only on the outflow.

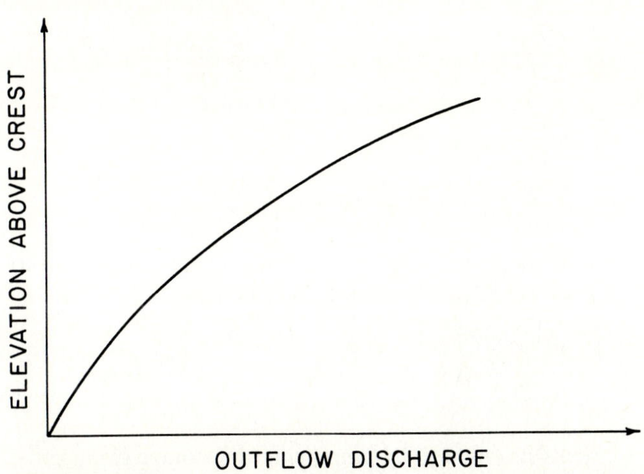

FIGURE 19.11 Elevation above spillway crest versus reservoir outflow.

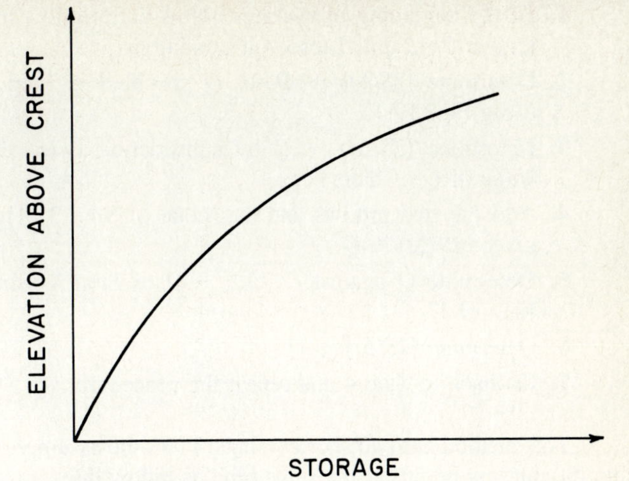

FIGURE 19.12 Elevation above spillway crest versus storage in the reservoir.

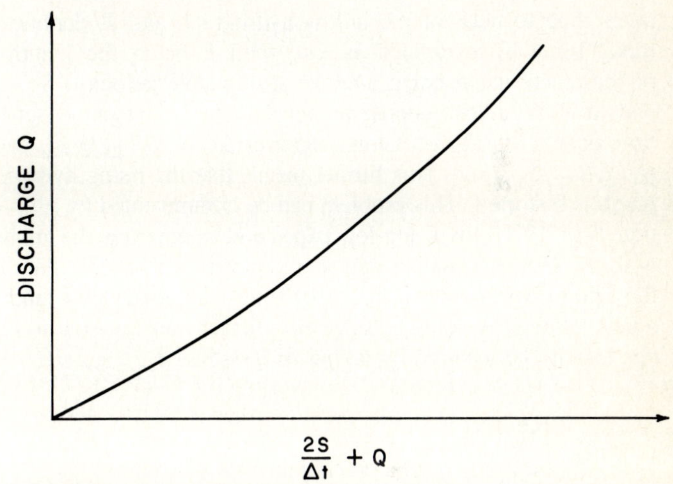

FIGURE 19.13 Discharge versus $(2S/\Delta t) + Q$ for reservoir routing.

The next step in routing by equation (19.10) is the selection of the routing time interval Δt. Its value should be such that it is neither too long nor too short. If it is too long and exceeds the travel time through the reservoir, then the crest segment of outflow containing the peak discharge could pass through the reservoir between time intervals and could therefore not be computed. On the other hand, if Δt is too short, then it takes longer to perform flood routing. Further, Δt is assumed so short that I is approximately linear during it. Linsley, Kohler, and Paulhus (1975) recommended Δt to be one-third ot one-half the travel time through the reservoir.

Before routing can be performed, the curves of $(2S/\Delta t) \pm Q$ versus Q should be constructed. This can be accomplished as follows. Select an initial outflow discharge Q. Estimate the corresponding elevation using the elevation-discharge curve of Fig. 19.11. Determine the corresponding storage from the elevation-storage curve of Fig. 19.12. Then determine $(2S/\Delta t) + Q$ and plot against Q, as shown in Fig. 19.13.

The routing by equation (19.10) can now be performed as follows:

1. At the beginning of routing, inflow and outflow are known from initial reservoir conditions.
2. Determine $(2S/\Delta t) + Q$ for Q specified in Step 1 using Fig. 19.13.
3. Determine $(2S/\Delta t) - Q$ by substracting twice the value of Q of Step 1.
4. Add the two inflows and the value of Step 3. This gives $(2S/\Delta t) + Q$.
5. Determine Q against $(2S/\Delta t) + Q$ of Step 4 using Fig. 19.13.
6. Determine $(2S/\Delta t) - Q$.
7. Go back to Step 4 and repeat the procedure.

This method has two shortcomings. First, the assumption that the outflow begins at the same time as inflow does imply that the inflow passes through the reservoir instantaneously regardless of its length. However, this weakness is not serious if the ratio T_t/T_m is less than or equal to $\frac{1}{2}$, where T_m denotes time to peak of the inflow hydrograph and T_t denotes travel time. T_t is defined as L/u, with L being the length of the reach and u being average steady-state velocity. Second, it is difficult to choose an appropriate Δt. Negative outflow occurs during recession whenever $\Delta t > 2S_{t+\Delta t}/Q_{t+\Delta t}$ or $Q_{t+\Delta t}/2 > S_{t+\Delta t}/\Delta t$. This further means that the rising hydrograph is distorted. This problem can be circumvented by plotting Fig. 19.13 on a log-log paper and comparing the plot with the line of equal values. If the plotted values lie above the line of equal values, Fig. 19.13 must be abandoned, and a new value at Δt must be selected. Further, negative outflow can usually be avoided by taking Δt less than T_c.

TABLE 19.7

Storage and discharge characteristics of a reservoir at the dam

Elevation (m)	*Discharge (10^8 m^3)*	*Storage (m^3/s)*
125	.02	.210
130	.13	.900
140	.45	1.59
160	2.70	2.40
180	7.50	3.03
200	15.50	3.54
220	30.00	3.97
240	50.0	4.54
260	75.0	4.75
280	105.0	5.10
300	142.0	5.40
320	185.0	5.70
340	230.0	6.00
350	256.0	6.15
355	268.0	6.19
360	280.0	6.30
365	293.0	6.39
370	306.0	6.43
375	321.0	6.49
390	360.0	6.70
400	392.0	6.80
420	456.0	7.15

TABLE 19.8

An inflow hydrograph

Time (h)	*Discharge (m^3/s)*	*Time (h)*	*Discharge (m^3/s)*
00	20	114	360
06	60	120	330
12	220	126	280
18	350	132	200
24	380	138	160
30	430	144	130
36	480	150	110
42	530	156	90
48	610	162	60
54	670	168	50
60	730	174	30
66	780	180	22
72	870	186	21
78	770	192	20
84	700	198	19
90	670	204	18
96	630	210	17
102	550	216	17
108	490	222	17

EXAMPLE 19.3

The storage and discharge characteristics for a reservoir are given in Table 19.7. Consider the routing period as 6 h. The dam has an unvalved orifice 5 m high, with its elevation at 125 m, and a spillway crest elevation of 355 m. Discharges for pool elevations above 355 m include both orifice and spillway discharge. Route the inflow hydrograph given in Table 19.8 by the storage indication method. Determine the attenuation and lag.

***Solution*:** The reservoir elevation-discharge-storage characteristics are plotted in Fig. 19.14. The quantities $(2S/\Delta t) - Q$ and $(2S/\Delta t) + Q$ are computed and plotted in Fig. 19.15. Clearly, the two curves are very close to one another because Q is much smaller than $(2S/\Delta t)$. The inflow hydrograph is routed following the steps previously outlined, and calculations are given in Table 19.9. The inflow and outflow hydrographs are shown in Fig. 19.16. Thus

$$\text{Reduction in peak flow} = 440 \text{ m}^3/\text{s}$$

$$\text{Lag between inflow and outflow peaks} = 39.5 \text{ h}$$

$$\text{Outflow peak} = 430 \text{ m}^3/\text{s}$$

19.6 THE PULS METHOD

The Puls method is similar to the storage-indication method. It also expresses the continuity equation as

$$\frac{I_t + I_{t+\Delta t}}{2}\Delta t - S_t - \frac{Q_t}{2}\Delta t = S_{t+\Delta t} + \frac{Q_{t+\Delta t}}{2}\Delta t \qquad (19.13)$$

and seeks its solution by employing the storage characteristic curves in a like manner. In order to use equation (19.13), curves for $S \pm (Q/2)\Delta t$ versus stage h or outflow Q must be

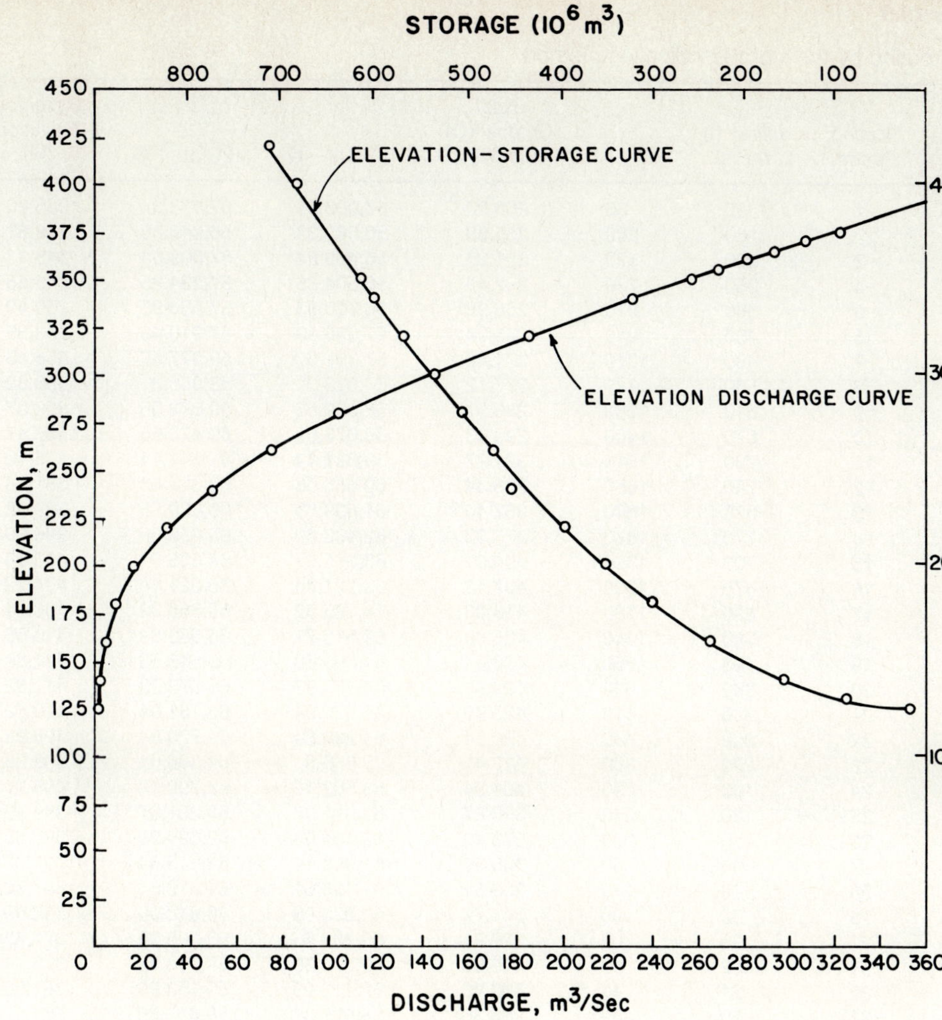

FIGURE 19.14 Example 9.3: Reservoir elevation-discharge-storage characteristics.

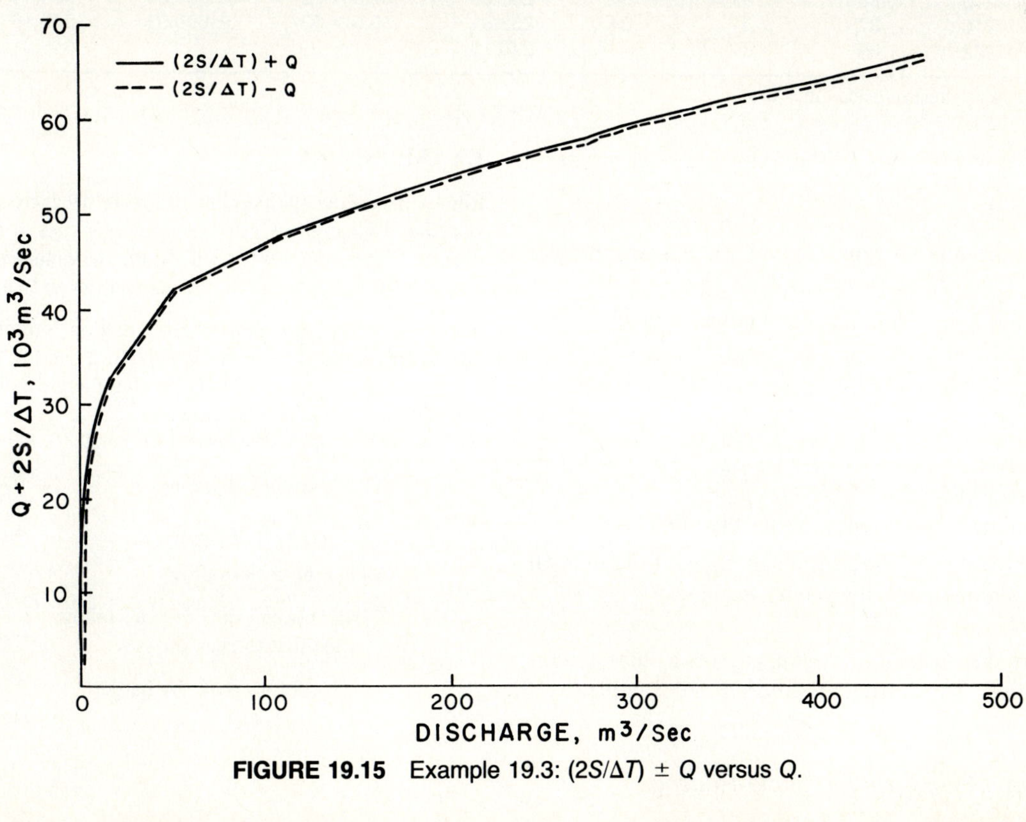

FIGURE 19.15 Example 19.3: $(2S/\Delta T) \pm Q$ versus Q.

TABLE 19.9

Flood routing by the storage indication method

Time Hours	*Routing Period*	*Inflow* (I) (m^3/s)	$I_1 + I_2$	*Total Outflow* (Q) (m^3/s)	$(2S/\Delta t) - Q$	$(2S/\Delta t) + Q$	*Reservoir Elevation* (M)
0	1	20	80	268.00	57,093.34	57,173.34	355.00
6	2	60	280	255.50	56,662.35	56,942.35	349.81
12	3	220	570	251.25	56,439.84	57,009.84	348.17
18	4	350	730	252.49	56,504.85	57,234.85	348.65
24	5	380	810	256.96	56,720.93	57,530.93	350.40
30	6	430	910	265.24	57,000.43	57,910.43	353.85
36	7	480	1010	271.43	57,367.57	58,377.57	356.43
42	8	530	1140	277.12	57,823.31	58,963.31	358.80
48	9	610	1280	285.37	58,392.55	59,672.55	362.07
54	10	670	1400	299.43	59,073.68	60,473.68	367.47
60	11	730	1510	321.27	59,831.14	61,341.14	375.10
66	12	780	1650	338.32	60,665.58	62,314.48	381.66
72	13	870	1640	357.46	61,599.55	63,239.55	389.02
78	14	770	1470	377.93	62,483.69	63,953.69	395.60
84	15	700	1370	394.01	63,165.67	64,535.67	400.63
90	16	670	1300	407.12	63,721.44	65,021.44	404.72
96	17	630	1180	418.06	64,185.32	65,365.32	408.14
102	18	550	1040	425.80	64,513.71	65,553.69	410.56
108	19	490	850	430.04	64,693.60	65,543.56	411.89
114	20	360	690	429.81	64,683.93	65,373.93	411.82
120	21	330	610	425.99	64,521.94	65,131.94	410.62
126	22	280	480	420.54	64,290.84	64,770.84	408.92
132	23	200	360	412.41	63,946.01	64,306.01	406.38
138	24	160	290	401.94	63,502.12	63,792.12	403.11
144	25	130	240	390.37	63,011.37	63,251.37	399.49
150	26	110	200	378.19	62,494.98	62,694.98	395.68
156	27	90	150	365.66	61,963.64	62,113.64	391.77
162	28	60	110	353.51	61,406.61	61,516.61	387.50
168	29	50	80	341.77	60,833.06	60,913.06	382.99
174	30	30	52	329.91	60,253.25	60,305.25	378.43
180	31	22	43	316.93	59,671.39	59,714.39	373.64
186	32	21	41	300.70	59,112.99	59,153.99	367.96
192	33	20	39	288.30	58,577.38	58,616.38	363.19
198	34	19	37	280.04	58,056.28	58,093.28	360.02
204	35	18	35	273.66	57,545.95	57,580.95	357.36
210	36	17	34	266.65	57,047.66	57,081.66	354.44
216	37	17	34	253.81	56,574.03	56,608.03	349.16
222	38	17		245.11			345.81

Note: Routing Period (Δt) = 6 h.

plotted as shown in Fig. 19.17. The following steps are used in this method.

1. Select the routing time interval Δt. For any Δt, we know I_t and $I_{t+\Delta t}$ as well as h_t or Q_t.
2. Determine $S_t - (Q_t/2)\Delta t$ for known h_t or Q_t, using Fig. 19.8.
3. Using equation (19.13) determine $S_{t+\Delta t} + (Q_{t+\Delta t}/2)\Delta t$.
4. Determine $h_{t+\Delta t}$ or $Q_{t+\Delta t}$ for $S_{t+\Delta t} + (Q_{t+\Delta t}/2)\Delta t$ in Step 3, using Fig. 19.17.
5. Determine $Q_{t+\Delta t}$ or $h_{t+\Delta t}$ using Fig. 19.11.
6. Repeat this procedure for a subsequent time interval and continue until the routing is complete.

This procedure has the same weaknesses as the storage-indication method. To illustrate it, consider an example.

EXAMPLE 19.4

Route the inflow hydrograph of Table 19.1 by the Puls method. Use the discharge and storage characteristics for the reservoir as given in Table 19.2. What will be the maximum reservoir elevation and at what time? Compute the attenuation and lag.

Solution: The storage-discharge relation is plotted in Fig. 19.7. Using this curve we derive $S + (Q\Delta t/2)$ and $S - (Q\Delta t/2)$ versus Q, as graphed in Fig. 19.18. Following the procedure just discussed, calculations are made as shown in Table 19.10. The inflow and outflow hydrographs are shown in Fig. 19.19. Thus

Peak outflow discharge	= 970 m^3/s
Peak elevation reached	= 138.4 m
Time to peak outflow	= 62.4 h
Lag between time to peak inflow and that to peak outflow	= 14.4 h
Reduction in peak discharge	= 830 m^3/s

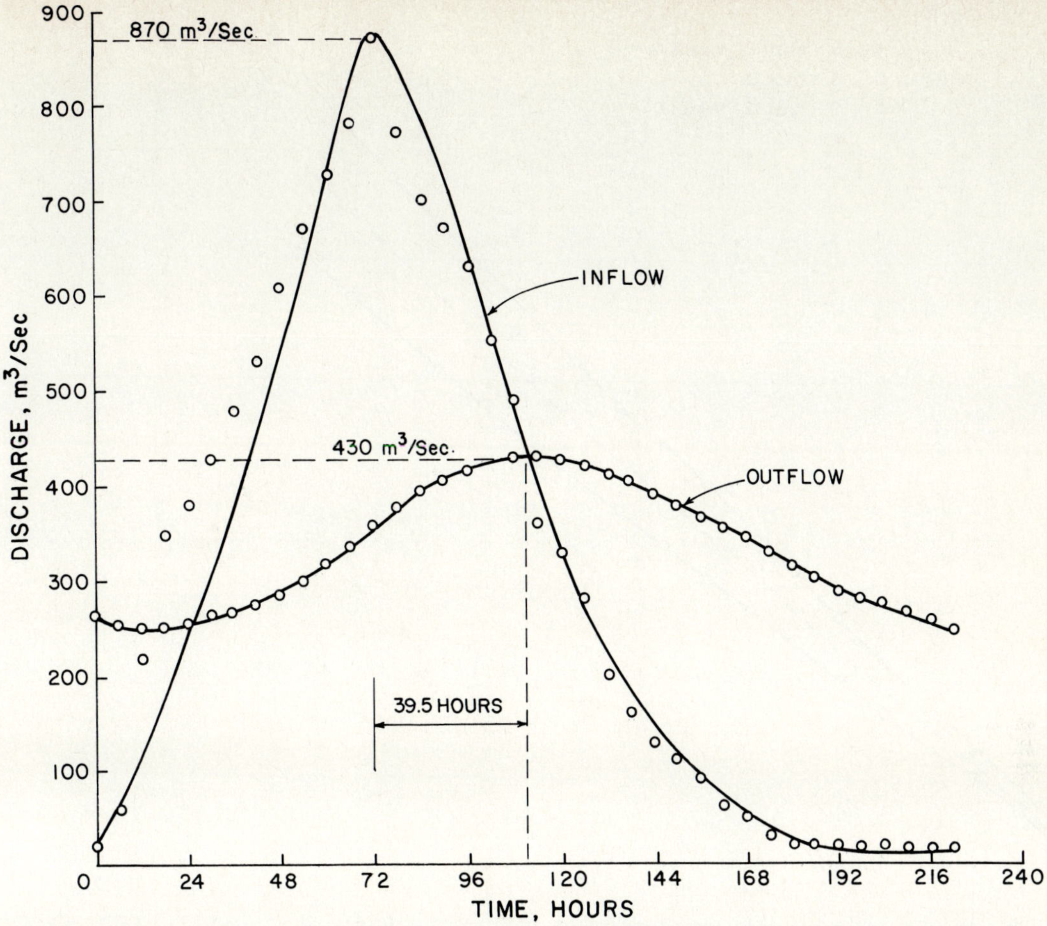

FIGURE 19.16 Example 19.3: Inflow hydrograph and outflow hydrograph obtained by storage indication method of routing.

The results obtained here and by the mass-curve method indicate that these two methods are comparable.

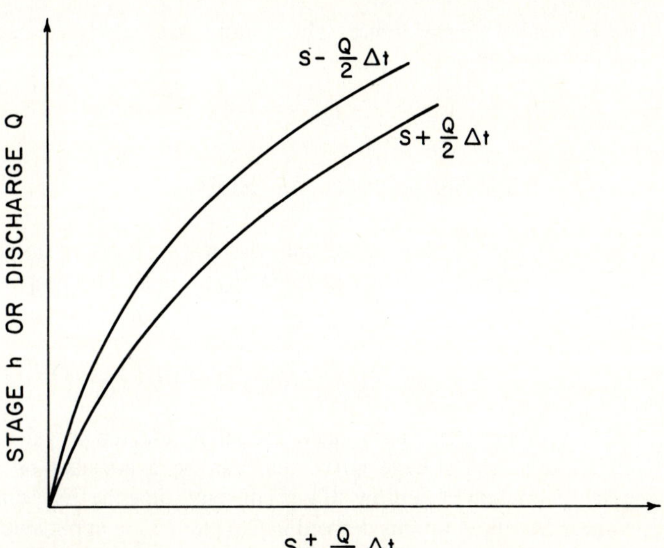

FIGURE 19.17 Curves showing stage versus $S \pm (Q/2)\,\Delta t$.

19.7 THE WISLER-BRATER (WB) METHOD

Wisler and Brater (1959) presented a method that expresses storage as a function of the sum of inflow and outflow and constructed storage curves for $I + Q$ versus $[(2S/\Delta t) + I + Q]$. Equation (19.1) is written as

$$\frac{2S_t}{\Delta t} + I_t + 2I_{t+\Delta t} - Q_t = \frac{2S_{t+\Delta t}}{\Delta t} + I_{t+\Delta t} + Q_{t+\Delta t} \qquad (19.14)$$

Since every term on the left side of equation (19.14) is known for a given routing period, the right side can be computed. The value of $I_{t+\Delta t} + Q_{t+\Delta t}$ can then be read from the storage curves. Since $I_{t+\Delta t}$ is known, $Q_{t+\Delta t}$ is immediately obtained. This procedure can be repeated for subsequent routing periods. It can easily be extended to the case where storage is a function of the weighted sum of inflow and outflow.

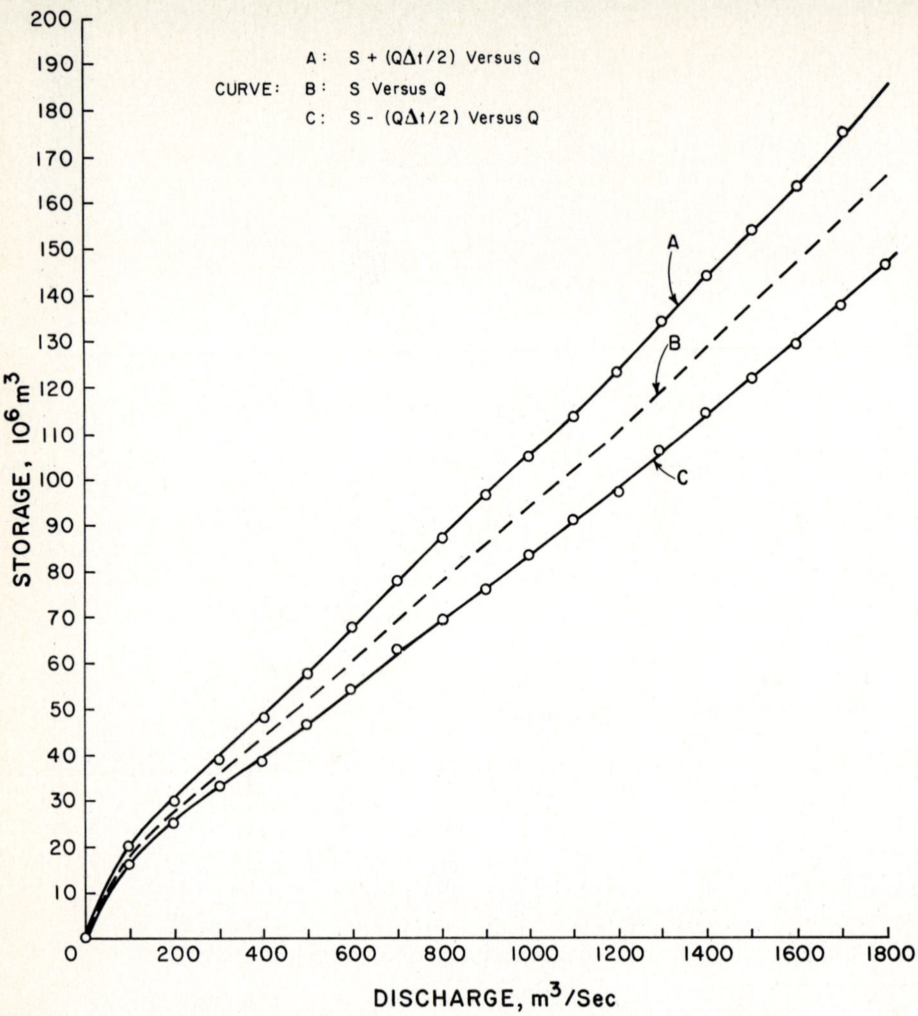

FIGURE 19.18 Example 19.4: $S \pm (Q\Delta t/2)$ versus Q.

19.8 THE GOODRICH METHOD

Goodrich (1931) proposed a method that involves construction of a family of routing curves for $[(2S/\Delta_t) + Q]$ against Q for various values of I, as shown in Fig. 19.20. Equation (19.1) is written as

$$I_t + I_{t+\Delta t} + \frac{2S_t}{\Delta t} - Q_t - \frac{2S_{t+\Delta t}}{\Delta t} = Q_{t+\Delta t} \tag{19.15}$$

Since all terms on the left side of equation (19.15) are known, the right side can readily be obtained for a routine period Δt. The value of $Q_{t+\Delta t}$ can be read from the routine curves against $[(2S_{t+\Delta t}/\Delta t]$ and then $2S_{i+\Delta t}/\Delta t$, can be computed. The routing can be carried out for subsequent time periods. Rutter, Greaves, and Snyder (1939) used this method.

19.9 THE STEINBERG METHOD

The Steinberg method (Steinberg, 1947) expresses equation (19.1) as

$$(I_t + I_{t+\Delta t} - Q_t)\frac{\Delta t}{2} + S_t = \frac{\Delta t}{2} Q_{t+\Delta t} + S_{t+\Delta t} + K \tag{19.16}$$

The term $K = S(Q\ \Delta t/2)$ was called the *storage factor*. The storage curves, showing storage as a function of inflow and outflow, are superimposed over K-curves, as shown in Fig. 19.21. All terms on the left side of equation (19.16) are known, so the right side is readily obtained. Then $Q_{t+\Delta t}$ and $S_{t+\Delta t}$ can be read from the figure. The routing is similarly carried out for subsequent time periods.

19.10 THE COEFFICIENT METHOD

The coefficient method represents the reservoir by a single conceptual storage element, where storage S is directly proportional to outflow Q,

$$S = kQ \tag{19.17}$$

and k is a proportionality factor equal to the reciprocal of the slope of the storage curve that can be a constant or a variable function of outflow. If k is constant, then the reservoir is linear and the solutions derived in Chapter 13 are applicable. Otherwise, the reservoir is nonlinear, as used by Chander (1970), and the solutions derived in Chapter 17 are applicable.

TABLE 19.10

Flood routing by the Puls method

Time Hours	*Inflow* (I) (m^3/s)	$(I_1 + I_2)0.5\ \Delta t$ (10^6 m)	*Discharge* (Q) (m^3/s)	$S - (\frac{1}{2})Q\ \Delta t$ ($10^6\ m^3$)	$S + (\frac{1}{2})Q\Delta t$ ($10^6\ m^3$)	*Elevation* (m)
0	0	.54	0	0	.54	
6	50	1.94	0	.54	2.48	130.00
12	130	4.10	5	2.40	6.50	130.20
18	250	6.48	20	4.00	10.48	130.60
24	350	9.61	38	7.50	17.11	131.05
30	540	13.77	70	12.5	27.27	131.70
36	735	21.06	165	23.5	44.56	133.00
42	1215	32.56	365	39.5	72.06	134.70
48	1800	34.56	650	58.5	93.06	136.70
54	1400	26.46	870	74.0	100.46	137.90
60	1050	21.06	950	79.5	100.56	138.30
66	900	17.71	950	79.5	97.21	138.30
72	740	14.69	915	77.0	91.69	138.15
78	620	12.20	850	73.0	85.20	137.75
84	510	10.04	780	68.0	78.04	137.40
90	420	7.99	707	62.51	70.49	137.00
96	320	6.37	635	57.5	63.87	136.60
102	270	5.08	560	51.5	56.68	136.10
108	200	3.78	490	46.0	49.78	135.60
114	150	2.70	420	41.0	43.70	135.15
120	100	1.86	350	36.5	38.36	134.60
126	72	1.26	295	33.0	34.26	134.20
132	45	.76	250	29.0	29.76	133.80
138	25	.38	195	25.5	25.88	133.30
144	10	.11	150	22.0	22.11	132.80
150	0	.00	120	19.5	19.50	132.40
156	0	.00	90	16.0	16.00	132.00
162	0	.00	65	12.0	12.00	131.60
168	0	.00	42.5	8.0	8.00	131.20
174	0	.00	22	4.5	4.50	130.80
180	0	.00	5	2.0	2.00	130.20

The solutions are in both dimensional and dimensionless forms. For flood routing, a finite difference approximation is normally employed. Equations (19.1) and (19.17) can be combined and written as

$$k(Q_{t+\Delta t} - Q_t) = \frac{\Delta t}{2}(I_t + I_{t+\Delta t}) - \frac{\Delta t}{2}(Q_t + Q_{t+\Delta t})$$

Rearranging the terms,

$$Q_{t+\Delta t} = Q_t + C(I_t - Q_t) + .5C(I_{t+\Delta t} - I_t) \quad (19.18)$$

in which

$$C = \frac{\Delta t}{k + .5\Delta t} \quad (19.19)$$

If k is variable, then C can be derived and plotted as a function of Q. For each routing period, the appropriate value of C must be obtained corresponding to the outflow under consideration. By using equation (19.18), the flood routing can then be performed.

EXAMPLE 19.5

Use the coefficient method (numerical version) to route a triangular inflow hydrograph, which has a base of 24 h and peak of 500 m^3/s located at 8 h. Assume the value of $k = 2$ h. Determine the attenuation and lag.

Solution: The value of Δt is chosen as 1 h. Then $C = .4$. Equation (19.18) is used to perform flood routing, and calculations are given in Table 19.11. The inflow hydrograph and the outflow hydrograph obtained by routing are shown in Fig. 19.22. Therefore,

Peak of outflow hydrograph	= 434 m^3/s
Peak of inflow hydrograph	= 500 m^3/s
Reduction in peak discharge	= 66 m^3/s
Time to peak outflow	= 10.1 h
Time to peak inflow	= 8 h
Difference in peak times	= 2.1 h
Lag defined in terms of the time difference between the centroids of inflow and outflow hydrographs	= 2.0 h

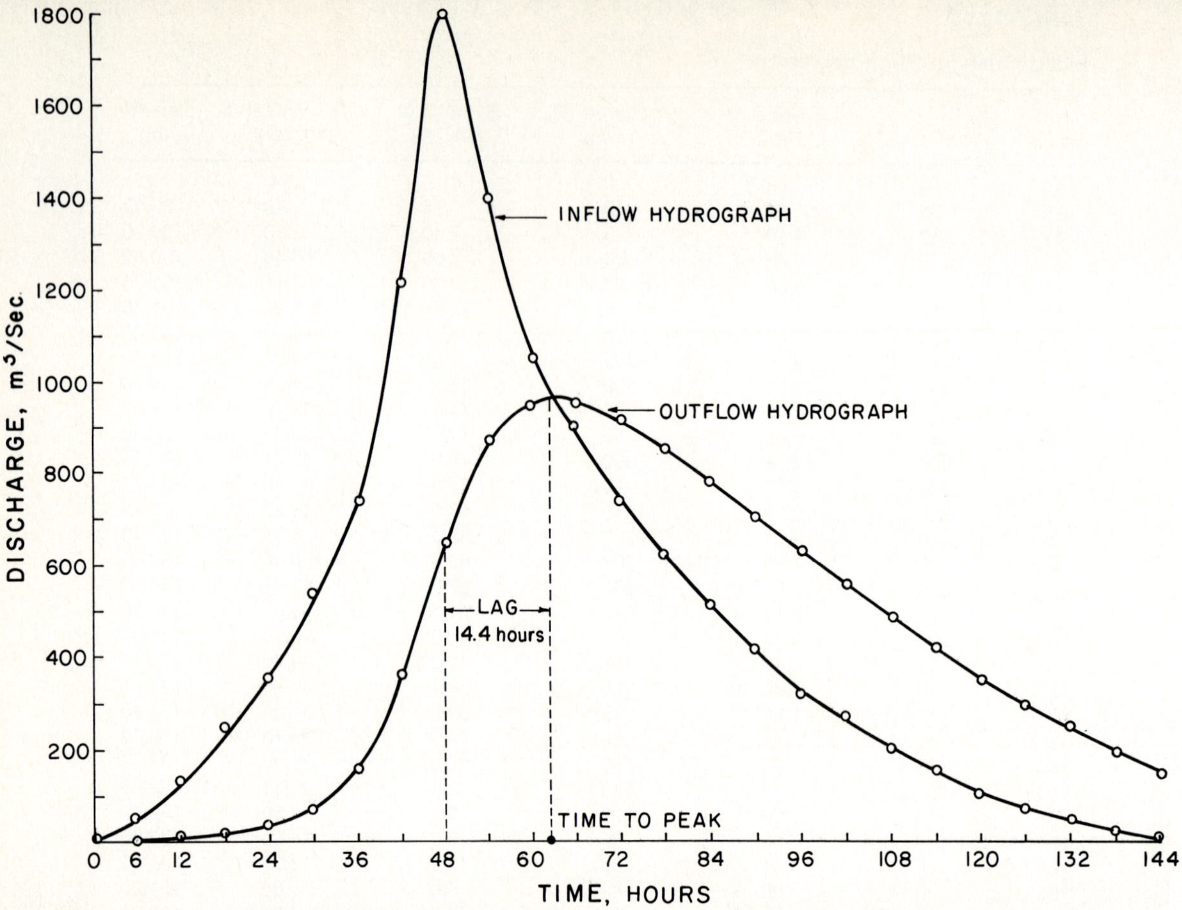

FIGURE 19.19 Example 19.4: Inflow hydrograph and outflow hydrograph obtained by the Puls method of routing.

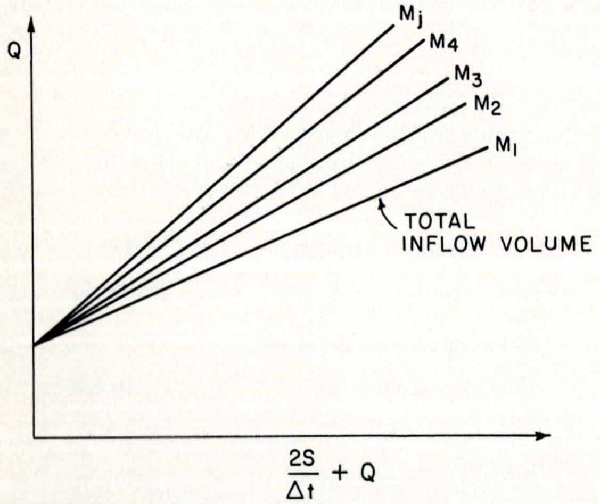

FIGURE 19.20 Typical routing curves for the Goodrich method.

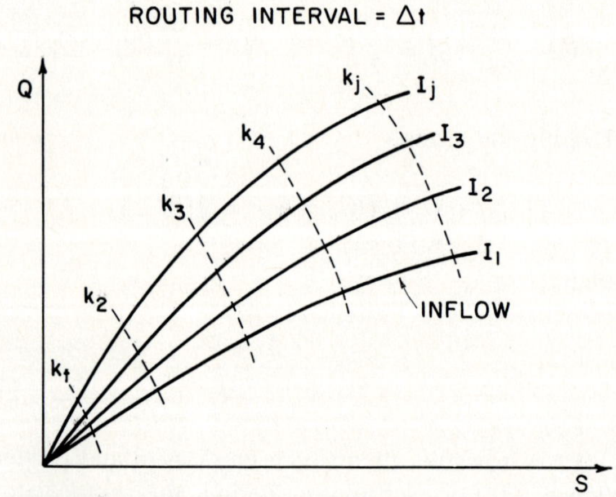

FIGURE 19.21 Typical routing curves for the Steinberg method.

EXAMPLE 19.6

Route the flood of Example 19.5 by the analytical version of the coefficient method.

Solution: The coefficient method involves a single linear reservoir whose IUH is $h(t) = \exp(-t/k)/k$. The inflow hydrograph can be expressed as

$$I(t) = \begin{cases} 62.5t, & 0 \le t \le 8 \\ 31.25(24 - t), & 8 \le t \le 24 \end{cases}$$

The outflow hydrograph can be expressed by convolution as

$$Q(t) = \frac{62.5}{2}\int_0^t \tau \exp\left[-\frac{(t-\tau)}{2}\right] d\tau$$

$$= 62.5\left[t - 2 + 2\exp\left(-\frac{t}{2}\right)\right], \qquad 0 \le t \le 8$$

$$Q(t) = \frac{31.25}{2}\int_8^t (24-\tau)\exp\left[-\frac{(t-\tau)}{2}\right] d\tau$$

$$= 31.25\left\{26 - t - 14\exp\left[-\frac{(t-8)}{2}\right]\right\}, \qquad 8 \le t \le 24$$

$$Q(t) = 62.31\exp[-24 - t)/2], \qquad t \ge 24$$

Therefore, $Q(t)$ is computed for various values of t and is virtually the same as yielded by the numerical version reported in Table 19.11. To summarize,

Peak of outflow hydrograph	$= 431.87\ m^3/s$
Reduction in peak discharge	$= 68.13\ m^3/s$
Time difference between peak times of inflow and outflow	$= 2$ h
Lag between centroids of inflow and outflow	$= 2$ h

19.11 THE WOODWARD METHOD

Woodward (1920) presented a method for making a preliminary appraisal of reservoir behavior required for flood regulation. It considers the inflow hydrograph as a rectangular pulse of duration D, the area-depth curve as a semicubical parabola ($a = ch^{3/2}$), and the outflow-depth curve as an ordinary parabola ($Q = bh^{1/2}$). Here, a is surface reservoir area, b and c are constants, and h is depth of water in the reservoir or effective head on the outlet. This method does not account for flow over the spillway. If the outlets are relatively small, such that the maximum outflow is small in comparison with the average inflow, then this method yields satisfactory results.

Writing equation (19.1) as

$$a\frac{dh}{dt} = I - Q \tag{19.20}$$

Inserting the expression for a and Q in terms of h,

$$ch^{1.5}\frac{dh}{dt} + bh^{.5} = I \tag{19.21}$$

TABLE 19.11

Flood routing by the coefficient method

Time t (h)	t + Δt (h)	*Inflow* I_t (m^3/s)	$I_t + \Delta t$ (m^3/s)	*Outflow* Q_t (m^3/s)	$Q_t + \Delta t$ (m^3/s)
0	1	0	62.5	.0	12.5
1	2	62.5	125	12.5	45.0
2	3	125	187.5	45.0	89.5
3	4	187.5	250	89.5	151.2
4	5	250	312.5	151.2	197.22
5	6	312.5	375	197.22	255.83
6	7	375	437.5	255.83	316.00
7	8	437.5	500	316.00	377.10
8	9	500	468.75	377.10	420.01
9	10	468.75	437.5	420.01	433.26
10	11	437.5	406.25	433.25	428.70
11	12	406.25	375	428.70	413.47
12	13	375	343.75	413.47	391.83
13	14	343.75	312.5	391.83	366.35
14	15	312.5	281.25	366.35	338.56
15	16	281.25	250	338.56	309.39
16	17	250	218.75	309.39	279.39
17	18	218.75	187.5	279.38	248.88
18	19	187.5	156.25	248.88	218.08
19	20	156.25	125	218.08	187.10
20	21	125	93.75	187.10	156.01
21	22	93.75	62.5	156.01	124.86
22	23	62.5	31.25	124.86	93.66
23	24	31.25	0	93.66	62.45
24	25	0		62.45	37.47
25	26			37.47	22.48
26	57			22.48	13.49
27	28			13.49	8.09
28	29			8.09	4.86
29	30			4.86	2.91
30	31			2.91	1.75
31	32			1.75	1.05
32	33			1.05	.63
33	34			.38	.23
34	35			.38	.23
35	36			.23	.14
36				.	.
				.	.
				.	.

Since I is constant for $0 \le t < D$ and 0 for $t \ge D$, a series solution can be derived. Let $w = h^{.5}$. Then $h = w^2$, and $dh = 2w\,dw$. Equation (19.21) can now be written as

$$\frac{2c}{b}\frac{w^4\,dw}{(I/b) - w} = dt \tag{19.22}$$

Dividing the numerator by denominator,

$$dt = \frac{2c}{b}\left[\left(\frac{I}{b}\right)^4 \frac{1}{(I/b) - w} - \left(\frac{I}{b}\right)^3 - \left(\frac{I}{b}\right)^2 w - \left(\frac{I}{b}\right) w^2 - w^3\right] dw$$

Upon integrating with the initial condition when $t = t_0$, $w = w_0$,

$$t - t_0 = \frac{2c}{b}\left[-\left(\frac{I}{b}\right)^4 \ln\left(1 - \frac{b}{I}w\right) - \left(\frac{I}{b}\right)^3 w - \frac{1}{2}\left(\frac{I}{b}\right)^2 w^2 - \frac{1}{3}\left(\frac{I}{b}\right)^3 w^3 - \frac{1}{4}w^4\right]\Bigg|_{w_0}^{w}$$

If the initial condition is $t = 0$, $w = 0$, then

$$t = \frac{2c}{b}\left[-\left(\frac{I}{b}\right)^4 \ln\left(1 - \frac{b}{I}w\right) - \left(\frac{I}{b}\right)^3 w - \frac{1}{2}\left(\frac{I}{b}\right)^2 w^2 - \frac{1}{3}\left(\frac{I}{b}\right)^3 w^3 - \frac{1}{4}w^4\right] \quad (19.23)$$

This is the working equation for the Woodward method. It leads to a simple working rule: The average rate of outflow during a uniform flood is approximately five-sixths of the maximum rate of outflow during this period. We derive this rule shortly. Further, it can be shown during a sharp sudden flood of much grater intensity than the maximum spillway discharge that the average spillway discharge is about two-fifths of the maximum. This is left for the reader to work out.

The logarithmic term in equation (19.23) can be expanded by Maclaurin's series:

$$\ln(1 - x) = -\left(x + \frac{x^2}{2} + \frac{x^3}{3} + \frac{x^4}{4} + \cdots\right)$$

Therefore,

$$-\left(\frac{I}{b}\right)^4 \ln\left(1 - \frac{b}{I}w\right) = \left(\frac{I}{b}\right)^3 w + \frac{1}{2}\left(\frac{I}{b}\right)^2 w^2 + \frac{1}{3}\left(\frac{I}{b}\right) w^3 + \frac{1}{4}w^4 + \frac{1}{5}\left(\frac{I}{b}\right)^{-1} w^5 + \cdots \quad (19.24)$$

Inserting equation (19.24) in equation (19.23),

$$t = \frac{2c}{b}\left[\frac{1}{5}\left(\frac{b}{I}\right)w^5 + \frac{1}{6}\left(\frac{b}{I}\right)^2 w^6 + \frac{1}{7}\left(\frac{b}{I}\right)^3 w^7 + \cdots\right]$$

or

$$t = \frac{2c}{5I}w^5\left[1 + \frac{5}{6}\left(\frac{b}{I}w\right) + \frac{5}{7}\left(\frac{b}{I}w\right)^2 + \frac{5}{8}\left(\frac{b}{I}w\right)^3 + \cdots\right]$$

Reverting back to the original variables, $S = \frac{2}{5}cw^5$ and $Q = bw$, so

$$t = \frac{S}{I}\left[1 + \frac{5}{6}\frac{Q}{I} + \frac{5}{7}\left(\frac{Q}{I}\right)^2 + \frac{5}{8}\left(\frac{Q}{T}\right)^3 + \cdots\right] \quad (19.25)$$

This solution is valid during the period $0 \le t \le D$. At $t = D$, the storage S and the outflow Q achieve their maximum values. At this instant,

$$DI = S(D)\left[1 + \frac{5}{6}y + \frac{5}{7}y^2 + \frac{5}{8}y^3 + \frac{5}{9}y^4 + \cdots\right]$$

in which $y = Q(D)/I$ = (maximum outflow)/(average inflow).

The volume of outflow during the period of inflow can be expressed as

$$\int_0^D Q\,dt = ID - S(D)$$

$$= S(D)\left[\frac{5}{6}y + \frac{5}{7}y^2 + \frac{5}{8}y^3 + \cdots\right]$$

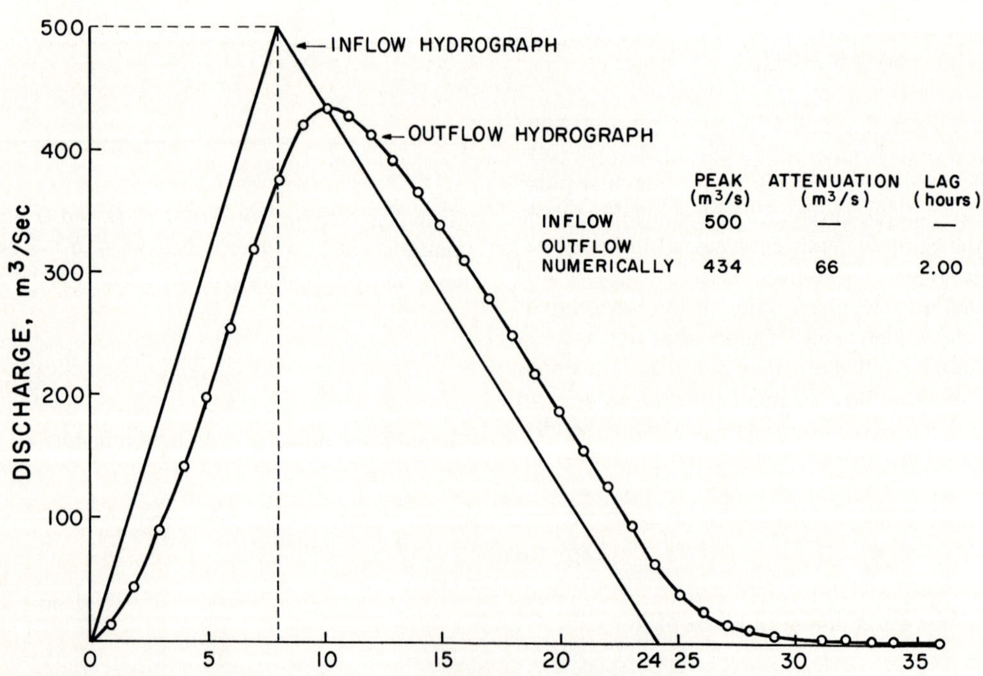

FIGURE 19.22 Example 19.5: Inflow hydrograph and outflow hydrograph obtained by the coefficient method of routing.

The average rate of outflow during D is

$$\frac{\int_0^D Q\,dt}{D} = \frac{S(D)\left[\frac{5}{6}y + \frac{5}{7}y^2 + \frac{5}{8}y^3 + \cdots\right]}{\frac{S(D)}{I}\left[1 + \frac{5}{6}y + \frac{5}{7}y^2 + \frac{5}{8}y^3 + \cdots\right]}$$

$$= \frac{Iy\left[\frac{5}{6} + \frac{5}{7}y + \frac{5}{8}y^2 + \cdots\right]}{\left[1 + \frac{5}{6}y + \frac{5}{7}y^2 + \frac{5}{8}y^3 + \cdots\right]}$$

The ratio r of the average rate to the maximum rate of outflow can be expressed as

$$r = \frac{\int_0^D Q\,dt}{DQ(D)} = \frac{\frac{5}{6} + \frac{5}{7}y + \frac{5}{8}y^2 + \frac{5}{9}y^3 + \cdots}{1 + \frac{5}{6}y + \frac{5}{7}y^2 + \frac{5}{8}y^3 + \cdots}$$

This can be shown to be

$$r = \frac{5}{6} + \frac{5}{6}\frac{1}{6 \cdot 7}y + \frac{5 \cdot 32}{6 \cdot 6 \cdot 6 \cdot 7 \cdot 8}y^3 + \frac{5 \cdot 8712}{6 \cdot 6 \cdot 6 \cdot 6 \cdot 7 \cdot 7 \cdot 8 \cdot 9}y^4 + \cdots$$

$$= \frac{5}{6}\left(1 + \frac{1}{42}y + \frac{1}{63}y^3 + \frac{121}{10584}y^4 + \cdots\right)$$

$$= \frac{5}{6}(1 + .02381y + .01587y^3 + .01143y^4 + \cdots) \tag{19.26}$$

If y is small then $r \cong \frac{5}{6}$.

When $t \geq D$, then $I = 0$. The solution can be obtained directly from equation (19.21),

$$bh^{.5} + ch^{1.5}\frac{dh}{dt} = 0$$

This can be integrated with the condition $t = D$, $h(D) = H$:

$$t - D = \frac{c}{2b}(H^2 - h^2) \tag{19.27a}$$

Inserting $Q = bh^{.5}$ and $Q(D) = bH^{.5}$ and recalling $S(D) = \frac{2}{5}cH^{5/2}$,

$$t - D = \frac{5}{2}\frac{S(D)H^{.5}}{H^{5/2}\,2Q}(H^2 - h^2)$$

$$= \frac{5}{4}\frac{S(D)}{Q(D)}\left\{1 - \frac{Q^4}{[Q(D)]^4}\right\} \tag{19.27b}$$

Equation (19.27b) can be employed to determine the time required to empty the reservoir after cessation of inflow. This is obtained by equating Q to zero:

$$T = \frac{5}{4}\frac{S(D)}{Q(D)}$$

Rewriting it as

$$\frac{S(D)}{T}\frac{1}{Q(D)} = \frac{4}{5}$$

But $S(D)/T = Q$. Therefore,

$$\frac{Q(t)}{Q(D)} = r = \frac{4}{5}$$

This shows that during the period of depletion, the average rate of outflow is four-fifths of the maximum outflow.

EXAMPLE 19.7

The average rate of inflow is 2 million cubic meters per hour for a period of 30 h. The maximum outflow is found to be 0.5 million cubic meters per hour. Compute the storage capacity required for the reservoir.

Average outflow during the period of inflow	= (5/6) × .5 million = .416 million m³/h
Volume of outflow during this period	= .416 million × 30 = 12.48 million m³
Storage capacity required	= the volume of inflow − the volume of outflow = 30 × 2 million − 12.48 million = 47.52 million m³

EXAMPLE 19.8

The average rate of inflow is 2 million cubic meters per hour for a period of 30 h. The maximum storage capacity of the reservoir is 47.52 million cubic meters. Determine the maximum rate of outflow from the reservoir and the time needed to empty the reservoir after the cessation of flood.

Volume of outflow during the period of inflow	= 2 million × 30 − 47.52 million = 12.48 million m³
Average outflow rate during this period	$= \frac{12.48 \text{ million}}{30}$ = .416 million m³/h
The maximum rate of outflow	$= .416 \text{ million} \times \frac{6}{5}$ = 0.5 million m³/h
The average outflow during the period of depletion	$= \frac{4}{5} \times .5$ million = 0.4 million m³/h
The time to empty the reservoir	$= \frac{\text{maximum storage}}{\text{average outflow}}$ $= \frac{47.52}{.4} = 11.88$ h

19.12 THE POINT-SLOPE METHOD

The point-slope method was developed by Butler (1982) and is particularly applicable for routing design floods determined from statistically derived design storms. It determines the hydrograph slope at various points and the inflection point of the rising limb of the outflow hydrograph by using the inflow hydrograph, the topography of the reservoir, and the hydraulic characteristics of the outflow structures. Because it accommodates simple, single-peaked inflow hydrographs, it is less versatile than the traditional methods. However, within this limitation it is fast and adaptable for investigating alternate designs.

It is assumed that in the beginning inflow and outflow hydrographs are horizontal and identical. The outflow hydrograph rises simultaneously with the inflow hydrograph but at a slower rate because part of the inflow is retained as storage. If A_s is water surface area and h is elevation of the surface, then equation (19.1) can also be written as

$$\frac{dS}{dt} = A_s \frac{dh}{dt} = \beta \qquad (19.28)$$

If the reservoir has vertical walls, then A_s is constant. If the walls slope outward, A_s increases with elevation of the water surface and can be expressed as a function of h, the head relative to the crest or zero-flow water surface elevation:

$$A_s = K_1(h + z)^m \qquad (19.29)$$

in which K_1, m, and z are constants that can be determined by plotting A_s versus h obtained from topographic data.

The reservoir outflow from spillway, conduit, or orifice can be expressed as

$$Q = Kh^n \qquad (19.30)$$

The slope of the outflow hydrograph dQ/dt is

$$\alpha = \frac{dQ}{dt} = nKh^{n-1}\frac{dh}{dt} \qquad (19.31)$$

Replacing h by $(Q/K)^{1/n}$ and dh/dt by $\beta/\{K_1[(Q/K)^{1/n} + z]^m\}$ from equations (19.28)–(19.30),

$$\alpha = \frac{K_2 Q^{[1-(1/n)]}\beta}{[(Q/K)^{1/n} + z]^m} \qquad (19.32)$$

where $K_2 = nK^{1/n}/K_1$. Equation (19.32) is the general equation for hydrograph slope α, which can be computed for specific values of K, K_2, m, n, and z. This equation can be solved numerically. If the reservoir walls are vertical or if the percent change in A_s is negligible within the range of h under consideration, then $m = 0$ and $K_1 = A_s$. Equation (19.32) then reduces to

$$\alpha = K_2\beta Q^{(n-1)/n} \qquad (19.33)$$

where $K_2 = nK^{1/n}/A_s$.

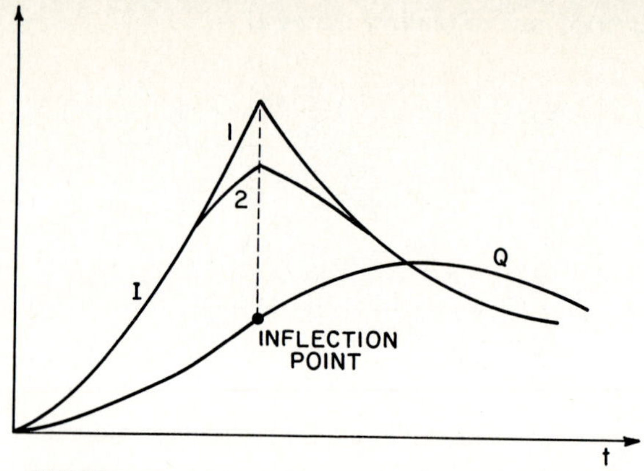

FIGURE 19.23 Inflection point on the rising part of the outflow hydrograph resulting from a point-peaked inflow hydrograph.

A standard outflow hydrograph for a usual inflow hydrograph is shown in Fig. 19.23. A knowledge of the inflection point is helpful for derivation of the outflow hydrograph. From equation (19.30),

$$\frac{dh}{dt} = \left(\frac{1}{K}\right)^{1/n}\frac{d}{dt}Q^{1/n} \qquad (19.34)$$

Replacing dh/dt by β/A_s,

$$\beta = A_s\left(\frac{1}{K}\right)^{1/n}\frac{d}{dt}Q^{1/n} \qquad (19.35)$$

Equation (19.34) indicates that α is maximum, that is, the inflection point occurs, when dh/dt is maximum. Equation (19.35) indicates that for A_s constant, α is maximum (at the inflection point) when β is maximum. At this time the inflow and outflow hydrographs are found to have the same slope. As A_s is not constant, the inflection point occurs where the quantity β/A_s achieves its maximum; both β and A_s vary with time. Theoretically, the inflection point will occur no later than the time of maximum β. Butler (1982) has reasoned that in all probability the inflection point occurs at the time of maximum β where the two hydrographs have the same slope.

There are other possibilities for the inflection point to occur, depending upon the shapes of inflow and outflow hydrographs. If the rising limb of a single-peaked inflow hydrograph is continuously steeper than the outflow hydrograph, the maximum β and, therefore, the outflow inflection point occur at the time of peak inflow, as shown in Fig. 19.23. This always happens when the inflow hydrograph continuously maintains or increases its slope until its peak time. On the other hand, if the outflow hydrograph is steeper than the inflow hydrograph for a period of time prior to the time of peak inflow, then the maximum β and, therefore, the inflection point of the outflow hydrograph occur prior to the time of peak inflow, as shown in Fig. 19.23. If the inflow hydrograph has a round peak, as shown in Fig. 19.24, then the inflection point must occur within a range of time. As noted already, the inflection point must occur when the inflow and outflow hydrographs have the same slope. The latest time for this is the time of

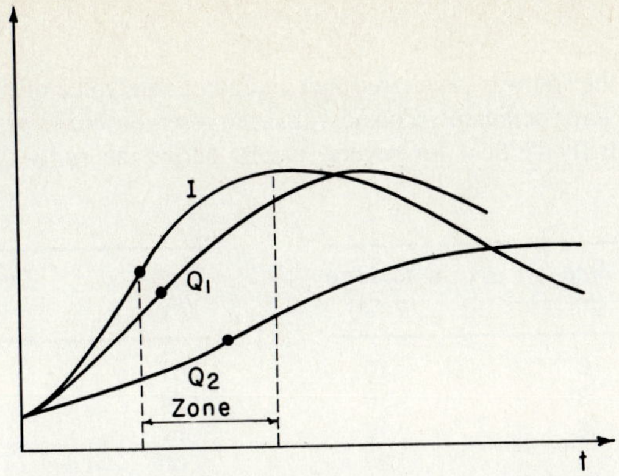

FIGURE 19.24 Time zone for outflow inflection point. Case 1 shows an early inflection point, and case 2 shows a late inflection point.

peak inflow and the earliest time is the time of the inflection point of the inflow hydrograph.

For deriving the outflow hydrograph, some checks and adjustments must be made. If A_s is constant, then storage S equals $A_s h$. Using equation (19.30), $S = (A_s/K^{1/n})Q^{1/n}$. S can be converted from cubic meters to cubic meters per second per day (1 m^3/s/da is a volume equal to 1 m^3/s for a duration of 1 da):

$$S = K_3 Q^{1/n} \tag{19.36}$$

in which $K_3 = A_s/(86{,}400\ K^{1/n})$, S is storage in cubic meters per second per day, A_s is in square meters, and Q in cubic meters per second. The number 86,400 specifies the number of seconds in a day. If A_s is variable, then equation (19.29) can be recast as

$$A_s = K_1 \left[\left(\frac{Q}{K} \right)^{1/n} + z \right]^m \tag{19.37}$$

and equation (19.36) is written as

$$S = \frac{K_1}{86{,}400 K^{1/n}} \left[\left(\frac{Q}{K} \right)^{1/n} + z \right]^m \tag{19.38}$$

The net increase in the live storage during the period $t = t_0$ to $t = t_p$ is

$$\Delta S = \int_{t_0}^{t_p} S(t)\, dt \tag{19.39}$$

If A_s is constant, then $\Delta S = K_3(\Delta Q^{1/n})$ from equation (19.36). Dividing ΔS by equation (19.39),

$$\frac{K_3(Q_0^{1/n} - Q_0^{1/n})}{\int_{t_0}^{t_p} S(t)\, dt} = 1.0 \tag{19.40}$$

where $Q_P = Q$ at $t = t_p$ and $Q_0 = Q$ at $t = t_0$. Equation (19.40) provides a check on the solution and the basis for adjustments only. For example, if the left side of equation is less than unity, then either the numerator is too small or the denominator is erroneously large or both. The outflow hydrograph must be moved upward slightly. As a first approximation, the adjustment can be made by assuming a linear variation in Q.

EXAMPLE 19.9

Route the inflow hydrograph given in Table 19.12 by the point-slope method. The reservoir water surface area is 5 million square meters with topography such that the percentage change in A_s is negligible in the range of water surface elevation under consideration. The outflow is over a rectangular spillway such that $Q = 25h^{1.5}$ with Q in cubic meters per second and h in meters. Assume that $m = 0$ and $z = 1.0$ m in equation (19.29).

Solution: The point-slope method can be used either graphically or numerically for routing floods. The numerical version is used here. The slope of the outflow hydrograph is computed from equation (19.33) as

$$\alpha(\text{m}^3/\text{s/da}) = .22161288\beta Q^{1/3}$$

Assume that during the rise, outflow discharge is about 60 percent of the inflow and that during the recession, it is about 30 percent more than the inflow. This is a good initial guess for $Q(t)$. To illustrate, select a point in the vicinity of the outflow hydrograph. Let that point be $t = 3$ da. Then $Q = 21.6$ m^3/s. The slope is 8.862 m^3/s/da. A line can be drawn through this point having that slope. Likewise, other point-slope lines can be drawn. The rising outflow hydrograph can then be drawn, guided by the trend of the point-slope lines. The peak of outflow is found to be 25.5 m^3/s at t = 4.1 da. To check the continuity,

$$K_3 = \frac{5 \times 10^6}{(86{,}400)(25)^{2/3}} = 6.8$$

$$\int_{t_0}^{t_p} S(t)\, dt = 78.7 - 33.93 = 44.77 \text{ m}^3/\text{s/da}$$

Then

$$\frac{6.8(25.5^{2/3} - 3^{2/3})}{44.77} = 0.9999$$

This ensures continuity up to time to peak outflow. This same procedure is followed for the entire hydrograph. The routing computations are given in Table 19.12.

TABLE 19.12

Inflow hydrograph and routing by the point-slope method

Time (da)	Inflow Discharge (m^3/s)	Outflow Discharge Assumed (m^3/s)	Outflow Discharge Final (m^3/s)
0	3	3.0	3.0
1	8	4.8	5.2
2	19	11.4	11.6
3	36	21.6	21.1
4	27	16.2	25.5
5	15	19.0	23.5
6	10	13.0	16.5
7	7	9.1	13.5
8	5	6.5	7.6
9	4	5.2	6.5
10	3	3.9	5.0

EXERCISES

19.1 A survey of a potential reservoir site provided the following data:

Elevation (m)	Area (ha)
510	0
511	1
512	5
513	8
514	10
515	12
516	16
517	18
518	21
519	25
520	30

Calculate the stage-storage curves. Assume that storage below inlet is zero. The inlet elevation is 511 m and there is no permanent pool.

19.2 A reservoir has the following characteristics:

$$S = .01h^2$$

$$Q = 60h^{.5}$$

where h is the depth of water, Q is outflow in cubic meters per second, and S is the storage in hectare-meters. Determine the maximum depth to which water will rise as a result of the following inflow hydrograph:

Time	Discharge (m^3/s)
6:00 A.M.	30
7:00	30
8:00	60
9:00	120
10:00	240
11:00	420
12:00 noon	450
1:00 P.M.	430
2:00	400
3:00	300
4:00	240
5:00	180
6:00	30

The depth h at 6:00 A.M. is 0.1 m.

19.3 A rectangular reservoir equipped with an outflow-controlling weir has the following characteristics:

$$S = 5h$$

$$Q = 2h$$

with S in cubic meters per second-day, Q is cubic meters per second, and h is in meters. Route the inflow hydrograph of Question 19.2, and determine the head on the weir as a function of time. The initial elevation h is .25 m.

19.4 A reservoir has an area of 500 ha at spillway level. The spillway is 10 m long. Compute the maximum pool level and maximum discharge to be expected as a result of the following inflow hydrograph. The reservoir is initially at the spillway level. Assume an appropriate value of the spillway coefficient. Consider that the reservoir banks are essentially vertical for several meters above the spillway level.

Time (h)	Discharge (m^3/s)	Time (h)	Discharge (m^3/s)
0	10	51	107
3	10	54	100
6	12	57	95
9	20	60	88
12	30	63	80
15	45	66	75
18	65	69	70
21	90	72	65
24	110	75	60
27	130	78	55
30	150	81	50
33	155	84	42
36	150	87	35
39	140	90	28
42	130	93	20
45	122	96	15
48	115	99	10
		102	10

19.5 Compute the maximum pool elevation and maximum discharge as a result of the inflow hydrograph of Question 19.2. Assume the initial pool elevation to be 500. The reservoir has the following elevation-storage and elevation-discharge data.

Elevation (m)	Storage (ha−m)	Discharge (m^3/s)
480	0	0
485	2.5	0
490	15	0
495	35	0
500	60	0
501	70	10
502	80	30
503	100	60
504	110	80
505	135	120
506	150	140
507	200	180
508	230	200
509	280	270
510	300	300

19.6 A reservoir has an area of 5000 ha at the spillway level. The spillway is 50 m long and has a coefficient of 3.5. The reservoir banks are essentially vertical for several meters above the spillway level. Compute the maximum pool level and maximum discharge to be expected as a result of the inflow hydrograph of Question 19.4. The reservoir is initially at the spillway level.

19.7 Consider the data of Question 19.5 as elevation in meters, storage in 10^5 m^3 and discharge in cubic meters per second. Determine the maximum pool level and maximum outflow rate.

Time (*h*)	*Inflow* (m^3/s)
0	40
2	65
4	80
6	110
8	120
10	170
12	200
14	190
16	160
18	150
20	120
22	110
24	100
26	80
28	60
30	40

19.9 Determine the maximum outflow rate from the reservoir for the inflow hydrograph of Question 19.4. The outflow rate and storage for an emergency spillway are found to be related by $S = 3Q$, where S is in cubic meters per second-hour and Q is in cubic meters per second. The number 3 is in hours.

19.10 A linear storage indication curve for a simple reservoir can be defined as

$$Q = \frac{Q}{2} + \frac{S}{\Delta t}$$

Assume $\Delta t = 3$ h. Route the inflow hydrograph of Question 19.8. Assume the storage to be zero at the beginning.

19.11 A vertical walled reservoir has a surface area of 1000 ha. The spillway is 50 m long. The initial inflow and outflow are both taken as 100 m^3/s. Determine reservoir storage for the following water surface elevations above the spillway crest: $h(m) = 0, .2, .4, .6, .8, 1.0, 1.2, 1.4, 1.6, 1.8, 2.0$. Determine the outflow rates for these values of h. Plot the storage-discharge curve as well as the storage-indication curve. Determine the outflow rates over the spillway at the end of each 6-h interval for the inflow hydrograph: 100, 200, 500, 1000, 1500, 1200, 8000, 400, 300, 2000, 100, 100 m^3/s.

CHAPTER TWENTY

River Flood Routing

Flood routing in a stream is somewhat more complicated than in a reservoir. In the case of a reservoir, the water level is assumed to be level at all times. Storage in the reservoir and outflow therefrom are assumed to be unique functions of the depth of water behind the dam. In other words, storage and outflow are uniquely related. However, in a stream the water surface is not always parallel to its bed. In fact, the average free surface slope is higher than the bed slope during the stream rise but lower during the stream recession. Therefore, the relation between storage and outflow is more complicated in a channel than in a reservoir. Different methods developed for streamflow routing differ in the mechanics of accounting for this relationship. Excellent discussions on many of these methods are available in Gilcrest (1950), Yevjevich (1964), Lawler (1964), Dooge (1980), and Fread (1982). In this chapter we discuss some of these methods as well as some other methods developed more recently. In addition, the Puls method, the storage-indication method, and other methods of reservoir flood routing (Strelkoff, 1980) can also be used for channel-flow routing but are not repeated here.

It may be noted that local inflows and seepage may occur during the passage of a flood wave in a channel reach. In discussion of the routing methods, these will not be explicitly considered. There are, however, ways to incorporate their effect. For example, one common practice to include seepage is to subtract it from the routed outflow hydrograph. Similarly, local inflows can be considered by adding them to the outflow hydrograph.

20.1 CONVEX METHOD

This method is based on the assumption that the outflow at any time t depends upon the outflow Q and inflow I one routing time interval Δt earlier. Let c be some constant, $0 \leq c \leq 1$. We can write

$$Q_{t+\Delta t} = cI_t + (1 - c)Q_t \tag{20.1}$$

This assumption requires that for the rising hydrograph,

$$\text{If} \quad I_t \geq Q_t, \quad \text{then} \quad I_t \geq Q_{t+\Delta t} \geq Q_t$$

and for the recession hydrograph,

$$\text{If} \quad I_t \leq Q_t, \quad \text{then} \quad I_t \leq Q_{t+\Delta t} \leq Q_t$$

Therefore, if Δt is properly chosen, $Q_{t+\Delta t}$ will fall somewhere on or between I_t and Q_t in magnitude but not above or below either one. It can be shown that the quantities I_t, Q_t, and $Q_{t+\Delta t}$ are members of a convex set, whence the name of this method.

Equation (20.1) can also be expressed as

$$c = \frac{Q_{t+\Delta t} - Q_t}{I_t - Q_t} \tag{20.2}$$

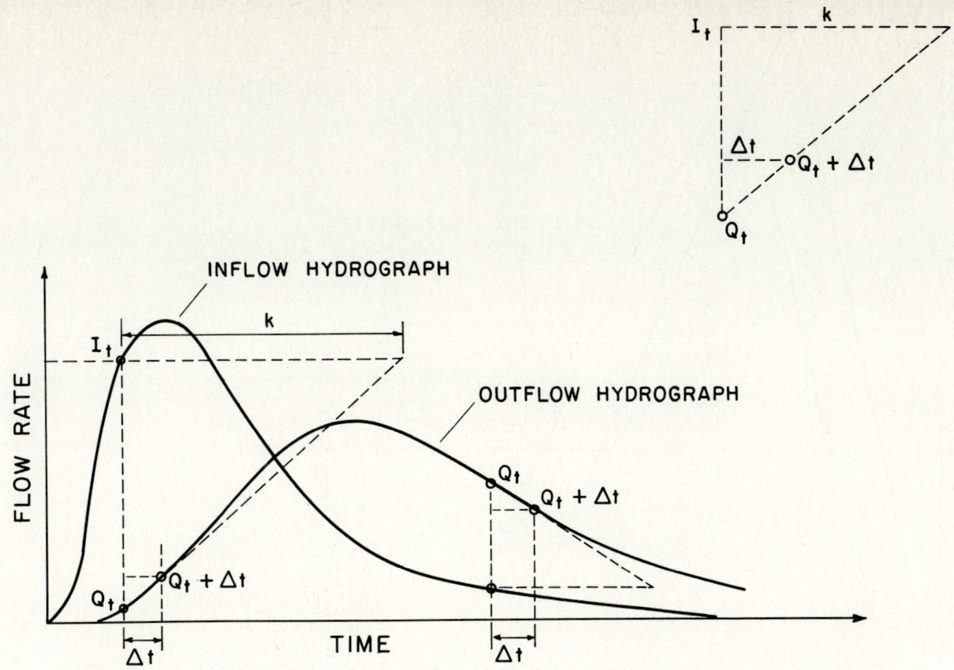

FIGURE 20.1 Construction of triangles for the convex method of flood routing.

Referring to Fig. 20.1, we see that

$$\frac{Q_{t+\Delta t} - Q_t}{\Delta t} = \frac{I_t - Q_t}{k}$$

where k is some time parameter yet to be determined. Therefore,

$$\frac{\Delta t}{k} = \frac{Q_{t+\Delta t} - Q_t}{I_t - Q_t} \tag{20.3}$$

Comparing equations (20.2) and (20.3),

$$c = \frac{\Delta t}{k} \quad \text{or} \quad \Delta t = ck \tag{20.4}$$

The convex method is, therefore, a two-parameter method of flood routing in a channel. This method has the advantage that $I_{t+\Delta t}$ does not enter into computation of $Q_{t+\Delta t}$ and can, therefore, be used for forecasting. To illustrate, if the routing interval is 1 day and if the inflow and outflow are known today, then the outflow can be predicted for tomorrow without a knowledge of tomorrow's inflow.

This method differs from other methods of streamflow routing in that the outflow begins one routing time interval after inflow begins, which is more realistic. In this method, the peak outflow does not intersect with the inflow hydrograph, as in the case of reservoir routing methods. However, the maximum storage in the stream, as in most routing methods, occurs at the time when inflow equals outflow.

20.1.1 Estimation of Parameters

The parameter k has the same dimension as time. It has the same physical connotation as the time of travel and can be computed by

$$k = \frac{LA}{Q} = \frac{L}{u} \tag{20.5}$$

where L is the reach length, A is the average cross-sectional area, Q is the steady-state discharge and u is the average steady-state velocity of Q. (This k is approximately equal to k of the Muskingum method to be discussed later.)

The parameter c is the ratio of two velocities,

$$c = \frac{u}{u_a} \tag{20.6}$$

where u_a is a special wave function. Soil Conservation Service (1964) has derived an empirical expression for c,

$$c = \frac{u}{1.7 + u} \tag{20.7}$$

The parameter c is approximately related to the weighting factor x of the Muskingum method,

$$c = 2x \tag{20.8}$$

A question arises: How can one select the steady-state velocity? In a natural channel, u varies with stage and from

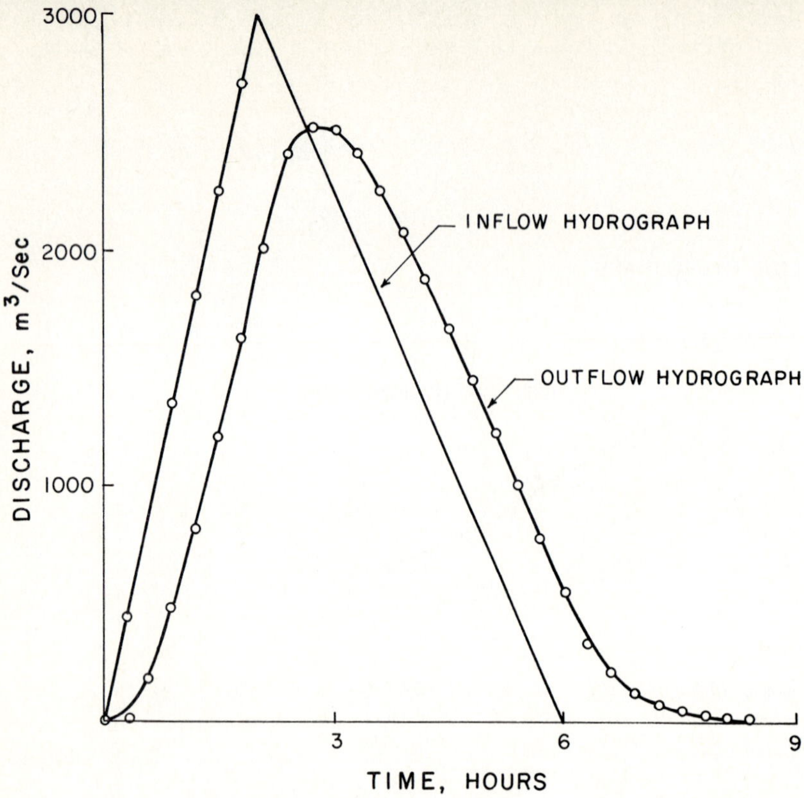

FIGURE 20.2 Example 20.1: Inflow hydrograph and outflow hydrograph obtained by convex method of routing.

one season to another. This variation is not consistent with increase or decrease in stage either. A flood wave is an unsteady flow phenomenon. Its velocity continuously varies in time. Although it may therefore appear that the parameters c and k should change from one step of computation to another, mass conservation requires them to be constant. Furthermore, since one inflow hydrograph will always be different from another, these parameters will change likewise. One way to select u is to compute the velocity corresponding to a discharge equal to one-half the peak inflow. Another way is to determine u for a discharge equal to three-fourths of the peak inflow. If the inflow hydrograph has two or more peaks, then u should correspond to the discharge with the largest value TQ defined as TQ is equal to three-fourths of discharge multiplied by duration of three-fourths discharge. Whichever peak yields the highest TQ should be selected.

From equation (20.4) it is clear that Δt can be determined from known values of c and k. The value of Δt computed in this manner may not be a convenient one for routing calculations. Soil Conservation Service (1964) recommends selecting a desired value Δt^* and then modifying c by

$$c^* = 1 - (1 - c)^{\Delta t^*/\Delta t} \tag{20.9}$$

where c^* is the modified c required with the use of the desired time interval Δt^*. A choice of Δt^* can be such that

$$\Delta t^* \leq \frac{t_p}{5} \tag{20.10}$$

where t_p is the time from the beginning of rise to the peak of the inflow hydrograph. If the inflow hydrograph has more than one peak, then Δt^* should be chosen using t_p for the shortest of the rise periods of the peaks. Let us consider an example to illustrate this method.

EXAMPLE 20.1

Consider a channel reach receiving an inflow hydrograph defined by a triangle such that its base is 6 h and its peak of 3000 m^3/s occurs at t = 2 h. There is no tributary inflow. Assume that c = 0.4 and Δt = 0.3 hours. Route this inflow by the convex method.

Solution: Equation (20.1) can be written as

$$Q_{t+0.3} = 0.4\, I_t + 0.6\, Q_t$$

which is used for routing the inflow hydrograph. The inflow hydrograph and the outflow hydrograph obtained by routing are graphed in Fig. 20.2.

EXAMPLE 20.2

Consider a rectangular channel which is 5 m wide and 1000 m long, with a bed slope of .001 and the Manning roughness coefficient of .03. The discharge occurring concurrently at the upstream and downstream ends of the channel respectively is 10 m^3/s and 6 m^3/s. What is the discharge downstream for the upstream flow of 10 m^3/s? Neglect evaporation, infiltration, seepage, and other losses.

Solution: We need to determine c and Δt. Equation (20.7) can be used to determine c by knowing u. By using the Manning equation, the depth of flow y can be obtained for given inflow and outflow. Assume the friction slope to be equal to the bed slope:

$$10 = \frac{(.001)^{1/2}}{0.03} \frac{(5y)^{5/3}}{(2y + 5)^{2/3}}$$

This yields y = 1.83 m and wetted area A = 9.15 m^2. Therefore, u = 10/9.15 = 1.093 m/s. Also

$$6 = \frac{(.001)^{1/2}}{.03} \frac{(5y)^{5/3}}{(2y + 5)^{2/3}}$$

This yields $y = 1.275$ m, $A = 6.375$ m^2, and $u = 0.941$ m/s. The average velocity $u = (1.093 + .941)/2 = 1.017$ m/s. Therefore, equation (20.7) yields

$$c = \frac{1.017}{1.7 + 1.017} = .374$$

Equation (20.5) yields

$$k = \frac{1000}{1.017} = 983.28 \text{ s}$$

Now equation (20.4) gives

$$\Delta t = ck = 0.374 \times 983.28 = 367.8 \text{ s} = 0.102 \text{ h}$$

Therefore,

$$\begin{aligned} Q_{t+\Delta t} &= .374 I_t + (1 - .374) Q_t \\ &= .374 \times 10 + (1 - 0.374) \times 6 \\ &= 7.5 \text{ m}^3/\text{s} \end{aligned}$$

This is the answer.

20.2 LINEAR MUSKINGUM METHOD

Since its development around 1934 (Gilcrest, 1950; Chow, 1959; Linsley, Kohler, and Paulhus, 1975) by McCarthy (1938), the Muskingum method has been one of the most popular methods for routing flood waves in rivers and channels. Therefore, it is not surprising that many investigators have studied the method and discussed the implications involved in its use (Gilcrest, 1950; Nash and Farell, 1955; Linsley, Kohler and Paulus, 1958; Chow, 1959; Nash, 1959a; Carter and Godfrey, 1960; Kulandaiswamy, 1966; Overton, 1966; Diskin, 1967; Harley, 1967; O'Meara, 1968; Venetis, 1969; Cunge, 1969; O'Connor, 1976; Gill, 1977b, 1978, 1979a, 1979b, 1979c, 1980, 1984; Koussis, 1978, 1980; Ponce and Yevjevich, 1978a, 1978b; Ponce, 1979; Meehan, 1979; Singh and Chowdhury, 1980; Singh and McCann, 1979, 1980a, 1980b; Stephenson, 1979; Cameron, 1980; Jones and moore, 1980; Dooge, Strupczewski, and Napiorkowski, 1982; Peterson and Verhoff, 1982; Chang, Singer, and Koussis, 1983; Hjelmfelt, 1984; Singh, 1983a; Heggen, 1984). The Muskingum method is a two-parameter lumped linear model, although its nonlinear versions have been proposed and used (Chow, 1959; Gill, 1978; Singh, 1983a).

This method consists of a spatially lumped form of continuity equation

$$\frac{dS}{dt} = I - Q \tag{20.11}$$

and a linear storage-discharge relationship,

$$S = k[xI + (1 - x)Q] \tag{20.12}$$

where k and x are routing parameters. It is assumed that at $t = 0$, $I = Q$. Physically, k is considered as the average reach travel time, and x the coefficient used to weigh the relative effects of inflow and outflow on reach storage. Theoretically, x can vary from 0 to 1.

Expressing equation (20.11) in finite-difference form and then substituting equation (20.12),

$$\left(\frac{I_t + I_{t+\Delta t}}{2}\right)\Delta t - \left(\frac{Q_t + Q_{t+\Delta t}}{2}\right)\Delta t = k[x(I_{t+\Delta t} - I_t) + (1 - x)(Q_{t+\Delta t} - Q_t)]$$

in which Δt is routing period or discretization time interval. Rearranging and solving for $Q_{t+\Delta t}$,

$$Q_{t+\Delta t} = C_0 I_{t+\Delta t} + C_1 I_t + C_2 Q_t \tag{20.13}$$

where

$$C_0 = \frac{-kx + 0.5\,\Delta t}{C_3} \tag{20.14a}$$

$$C_1 = \frac{kx + 0.5\,\Delta t}{C_3} \tag{20.14b}$$

$$C_2 = \frac{k - kx - .5\,\Delta t}{C_3} \tag{20.14c}$$

and

$$C_3 = k - kx + 0.5\,\Delta t \tag{20.14d}$$

The coefficients C_0, C_1, and C_2 are such that

$$C_0 + C_1 + C_2 = 1 \tag{20.15}$$

Clearly the coefficients C_0, C_1, and C_2 are appropriate only for a specified Δt. To transform these coefficients to a different time step, x and k can be computed as

$$x = \frac{C_1 - C_0}{2(1 - C_0)} \tag{20.16a}$$

$$k = \frac{\Delta t(1 - C_0)}{C_1 + C_2} \tag{20.16b}$$

Equation (20.13) is known as the Muskingum routing equation. If x and k are known then streamflow routing of any inflow hydrograph can be performed. This equation can be recast such that the right side contains only inflow terms. This suggests that the Muskingum method can be used to route only a portion of the inflow hydrograph. To this end we rewrite equation (20.13) as

$$Q_j = C_0 I_j + C_1 I_{j-1} + C_2 Q_{j-1} \tag{20.17a}$$

in which Δt is replaced by 1 and t by $j - 1$. The outflow Q_{j-1} can be replaced by the substitution,

$$Q_{j-1} = C_0 I_{j-1} + C_1 I_{j-2} + C_2 Q_{j-2} \tag{20.17b}$$

resulting in

$$Q_j = C_0 I_j + (C_0 C_2 + C_1) I_{j-1} + C_1 C_2 I_{j-2} + C_2^2 Q_{j-2}$$

By repeated substitutions for the outflow term, Q_{j-2}, Q_{j-3}, Q_{j-4}, . . . , on the right side, Q_j can be expressed solely as a function of inflow as

$$Q_j = \sum_{i=1}^{j} k_i I_{j-i+1} \tag{20.18}$$

where

$$k_1 = C_0$$
$$k_2 = C_0 C_2 + C_1$$
$$k_i = C_{i-1} C_2, \qquad i > 2 \tag{20.19}$$

Equation (20.18) can be viewed as a discrete-time convolution of I and k with k being considered as the Muskingum IUH or UH. Thus k_1, k_2, . . . , represent ordinates of the IUH.

If $x = .5$ and $k = \Delta t$, then from equations (20.13) and (20.14) it follows that $C_0 = 0$, $C_1 = 1$, and $C_2 = 0$. Therefore,

$$Q_{t+\Delta t} = I_t \tag{20.20}$$

This says that the inflow hydrograph is translated with a time lag of k. The validity of this conclusion has considerably been debated. Some workers consider it to be erroneous, whereas some state that it happens because a large value of routing time Δt is chosen and is equated with k, which roughly equals the travel time of the flood wave in the reach under consideration. This explanation is based on the observation that when Δt and k are not the same, C_0 and C_2 are not zero even for $x = .5$. Thus

$$Q_{t+\Delta t} \neq I_t \tag{20.21}$$

The validity of equations (20.20) and (20.21) is discussed further in the next section. The choice of Δt is more or less arbitrary but must be such that it is much smaller than the travel time of the flood wave. Furthermore, Δt should be such that the linear variation of I and Q is valid. In general, the smaller Δt, the more accurate should be the result. A heuristic argument is developed next for a reasonable Δt in relation to k and x.

If $\Delta t = 2kx$, then $C_0 = 0$; $C_1 = 2x$ and $C_2 = 1 - 2x$.
Therefore, $Q_{t+\Delta t} = 2xI_t + (1 - 2x)Q_t$

This equation parallels equation (20.1) for the convex method. In order for $Q_{t+\Delta t}$ to be positive at all times, x must be less than or equal to .5. At the same time, we see that

$$\Delta t \geq 2kx$$

These observations are useful in estimating k and x. Later in the chapter, we discuss various methods of estimating these parameters.

Although equation (20.13) can be used to perform flood routing numerically, considerable insight is obtained by solving equations (20.11) and (20.12) analytically subject to specified initial conditions.

20.2.1 Analytical Solutions

Analytical solutions of the Muskingum equation have been obtained for various initial conditions. We derive these solutions here. Combining equations (20.11) and (20.12),

$$k(1 - x)\frac{dQ}{dt} + kx\frac{dI}{dt} = I - Q \tag{20.22}$$

Let us define the Laplace transform of I and Q:

$$L[I] = P = \int_0^{\infty} e^{-st} I\, dt$$

$$L[Q] = q = \int_0^{\infty} e^{-st} Q\, dt$$

Taking the Laplace transform of equation (20.22),

$$a[sq - Q(0)] + b[sP - I(0)] = P - q$$

where $a = k(1 - x)$ and $b = kx$. $Q(0)$ and $I(0)$ are the initial conditions. Rearranging the terms,

$$q = \frac{aQ(0) + bI(0)}{1 + as} + \frac{P}{1 + as} - \frac{bs\,P}{1 + as}$$

The last term in this equation can be arranged as

$$\frac{bsP}{1 + as} = P\frac{b}{a}\left[-\frac{1}{(1 + as)} + 1\right]$$

Inserting this and rearranging gives

$$q = \frac{qQ(0) + bI(0)}{1 + as} - \frac{b}{a}P + \left(a + \frac{b}{a}\right)\frac{P}{1 + as}$$

Taking the inverse Laplace transform,

$$Q(t) = \frac{qQ(0) + bI(0)}{a} e^{-t/a} - \frac{b}{a}I(t)$$
$$+ \left(1 + \frac{b}{a}\right)\frac{1}{a}\int_0^{\infty} I(u)e^{-(t-u)/a}\, du$$
$$= \frac{k(1 - x)q(0) + kxI(0)}{k(1 - x)} e^{-t/(k(1-x))} - \frac{x}{1 - x}I(t)$$
$$+ \frac{1}{k(1 - x)^2}\int_0^{t} e^{-(t-u)/(k(1-x))} I(u)\, du \tag{20.23}$$

Equation (20.23) is the solution of equation (20.22) subject to the given initial conditions. This equation can be utilized to study the effect of initial conditions on the Muskingum solution as well as to determine the IUH, UH, or SH. If initially $Q(0) = I(0)$,

$$Q(t) = \frac{I(0)}{1-x} e^{-t/[k(1-x)]} - \frac{x}{1-x} I(t) + \frac{1}{k(1-x)^2} \int_0^t e^{-(t-u)/(k(1-x))} I(u)\, du \qquad (20.24)$$

This equation was also obtained by Kulandaiswamy (1966) and Diskin (1967). If, on the other hand, $I(0) = Q(0) = 0$,

$$Q(t) = \frac{1}{k(1-x)^2} \int_0^t e^{-(t-u)/[k(1-x)]} I(u)\, du - \frac{x}{1-x} I(t) \qquad (20.25)$$

If we use the initial condition proposed by Gill (1979a, 1979b),

$$Q(\tau) = I(0), \qquad \tau > 0 \qquad (20.26)$$

then we cannot use the preceding method unless $Q(0)$ is specified. But we can still solve equation (20.22) in a standard manner. Set $y = 1 - x$; then $x = 1 - y$ and equation (20.22) becomes

$$ky \frac{dQ}{dt} + k(1-y) \frac{dI}{dt} = I - Q, \qquad t \geq \tau$$

or, equivalently,

$$ky \left(\frac{dQ}{dt} - \frac{dI}{dt} \right) + (Q - I) = -k \frac{dI}{dt}$$

Thus

$$\frac{d}{dt}(Q - I) + \frac{1}{ky}(Q - I) = -\frac{1}{y} \frac{dI}{dt}$$

which, if dI/dt is known, is a first-order linear differential equation in $Q - I$. Hence, its integrating factor is $e^{t/(kg)}$. Then

$$Q(t) - I(t) = -\frac{e^{-t/(ky)}}{y} \left[\int_\tau^t e^{s/(ky)} \frac{dI}{ds}\, ds \right]$$

Integrating by parts,

$$\begin{aligned} Q(t) &= I(t) + [Q(\tau) - I(\tau)] e^{-(t-\tau)/(ky)} \\ &\quad - \frac{e^{-t/(ky)}}{y} \left[I(s) e^{s/(ky)} \Big|_\tau^t - \frac{1}{ky} \int_\tau^t I(s) e^{s/(ky)}\, ds \right] \\ &= I(t) + [Q(\tau) - I(\tau)] e^{-(t-\tau)/ky} \\ &\quad - \frac{e^{-t/(ky)}}{y} \left[I(t) e^{t/(ky)} - I(\tau) e^{t/(ky)} - \frac{1}{ky} \int_\tau^t I(s) e^{s)/(ky)}\, ds \right] \end{aligned}$$

$$\begin{aligned} Q(t) &= -\frac{x}{1-x} I(t) + Q(\tau) e^{-(t-\tau)/[k(1-x)]} \\ &\quad + \frac{x}{1-x} I(\tau) e^{-(t-\tau)/[k(1-x)]} \\ &\quad + \frac{1}{k(1-x)^2} \int_\tau^t e^{-(t-s)/[k(1-x)]} I(s)\, ds, \qquad t \geq \tau \end{aligned} \qquad (20.27)$$

20.2.2 Continuous-Time Response Functions

The Muskingum model represents a linear system, as shown in Fig. 20.3 it may, therefore, suffice to determine its IUH (see Fig. 20.4). If $I(0)$ is interpreted as $\lim_{t \to 0} + I(t)$, then in the case $I(t) = \delta(t)$, the IUH can be determined. Because $\delta(t)$ is not defined at $t = 0$, we can determine the IUH meaningfully only for initial condition used in equation (20.25). Therefore, replacing $I(t)$ there by $\delta(t)$,

$$h(0, t) = -\frac{x}{1-x} \delta(t) + \frac{1}{k(1-x)^2} \int_0^t e^{-(t-s)/[k(1-x)]} \delta(s)\, ds$$

$$h(0, t) = -\frac{x}{1-x} \delta(t) + \frac{1}{k(1-x)^2} e^{-t/[k(1-x)]} \qquad (20.28)$$

where $h(0, t)$ is the IUH. This equation was obtained by Venetis (1969). This shows a negative impulse at the origin plus a positive logarithmic recession.

However, Morel-Seytoux (1979) has argued that it is preferable to obtain the SH because it is better behaved; then obtain $h(0, t)$ and $h(D, t)$ from the SH. To obtain the SH, we combine equations (20.11) and (20.12) with S as dependent variable and solve:

$$k(1-x) \frac{dS}{dt} + S = kI \qquad (20.29)$$

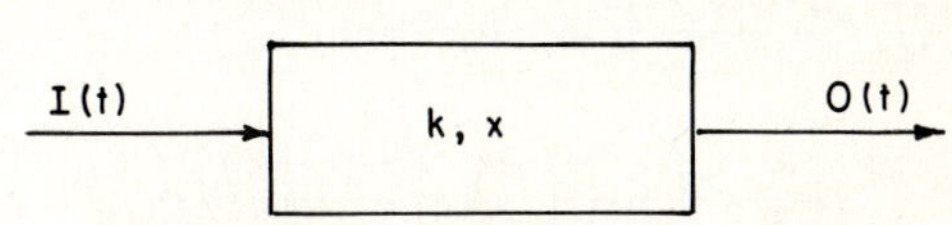

FIGURE 20.3 The Muskingum model.

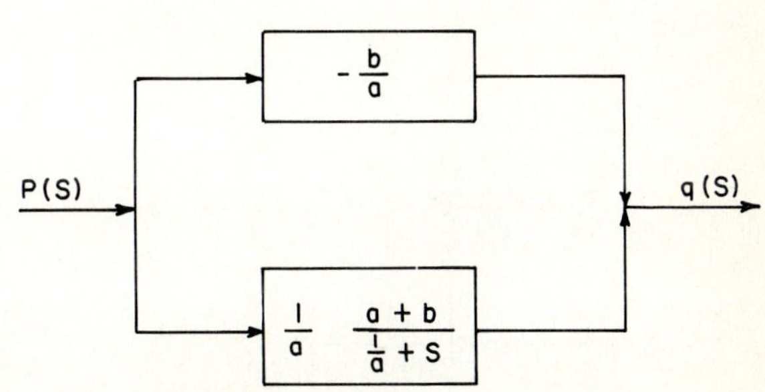

FIGURE 20.4 The response function representation of the Muskingum method.

The solution of equation (20.29) with right side equal to 1, subject to $S(0) = 0$, is

$$U_s(t) = 1 - \exp\{-t/[k(1-x)]\} \tag{20.30}$$

where $U_s(t)$ is the Muskingum storage step response function. The Muskingum storage IUH is obtained by simply differentiating $U_s(t)$,

$$h_s(0, t) = \frac{1}{k(1-x)} \exp\{-t/[k(1-x)]\} \tag{20.31}$$

We can express $S(t)$ as

$$S(t) = \int_0^t kI(\tau) \frac{1}{k(1-x)} \exp\left[\frac{-t+\tau}{k(1-x)}\right] d\tau$$

Substituting $S(t)$ into equation (20.12), we obtain equation (20.25),

$$Q(t) = \frac{1}{k(1-x)^2} \int_0^t \exp\left[\frac{-(t-\tau)}{k(1-x)}\right] I(\tau)\, d\tau - \frac{x}{1-x} I(t)$$

The UH of the Muskingum model can be derived from equation (20.28). If the period of the UH is 1, then

$$h(1, t) = \int_0^t h(0, t-\tau)\, d\tau, \qquad t \le 1$$

and

$$h(1, t) = \int_0^1 h(0, t-\tau)\, d\tau, \qquad t > 1$$

Therefore, using equation (20.28),

$$h(1, t) = -\frac{x}{1-x} \int_0^t \delta(t-\tau)\, d\tau + \frac{1}{k(1-x)^2} \int_0^t \exp\left[\frac{-(t-\tau)}{k(1-x)}\right] d\tau, \qquad t \le 1$$

which yields

$$h(1, t) = \begin{cases} -\dfrac{x}{1-x} + \dfrac{\exp(-t/(k(1-x)))}{(1-x)} \left\{\exp\left[\dfrac{t}{k(1-x)}\right] - 1\right\}, & t \le 1 \\ 1 - \dfrac{1}{1-x} \exp\left[\dfrac{-t}{k(1-x)}\right], & t \le 1 \end{cases} \tag{20.32}$$

Similarly for $t > 1$,

$$\begin{aligned} h(1, t) &= -\frac{x}{1-x} \int_0^1 \delta(t-\tau)\, d\tau + \frac{1}{k(1-x)^2} \int_0^1 \exp\left[\frac{-(t-\tau)}{k(1-x)}\right] d\tau \\ &= 0 + \frac{\exp\{-t/[k(1-x)]\}}{1-x} \left\{\exp\left[\frac{1}{k(1-x)}\right] - 1\right\} \\ &= \left\{\frac{\exp[1/(k(1-x))] - 1}{1-x}\right\} \exp\left[\frac{-t}{k(1-x)}\right] \end{aligned} \tag{20.33}$$

The ordinates of the UH, $h(1, t)$, can be computed by replacing t by j, $j = 1, 2, \ldots$, in equations (20.32) and (20.33),

$$h(1) = 1 - \frac{1}{1-x}\left[\frac{-1}{k(1-x)}\right]$$

$$h(j) = \left\{\frac{\exp[1/(k(1-x))] - 1}{1-x}\right\} \exp\left[-\frac{j}{k(1-x)}\right], \qquad j = 2, 3, \ldots$$

20.2.3 Discrete-Time Response Functions

Equations (20.11) and (20.12) can be written in discrete form as

$$S_j = \sum_{i=0}^{j} I_i - \sum_{i=0}^{j} Q_i, \qquad j = 0, 1, 2, \ldots \tag{20.34a}$$

or

$$S_j = S_{j-1} + I_j - Q_j \tag{20.34b}$$

and

$$S_j = k[xI_j + (1-x)Q_j], \qquad j = 0, 1, 2, \ldots \tag{20.35}$$

Combining equations (20.34b) and (20.35),

$$[1 + k(1-x)]Q_j = [k(1-x)]Q_{j-1} + (1-kx)I_j + kx\, I_{j-1} \tag{20.36}$$

Using the backward difference operator ∇ defined as, for example, $\nabla Q_j = Q_j - Q_{j-1}$, equation (20.36) can be written as

$$[1 + k(1-x)\nabla]Q_j = (1 - kx\nabla)I_j \tag{20.37}$$

This is the governing difference equation of the Muskingum reach. It can be solved either by Z-transform (O'Connor, 1976) or directly. Taking Z-transforms of both sides,

$$H(z) = \frac{Q(z)}{I(z)} = \frac{(1-kx)z + kx}{[1 + k(1-x)]z - k(1-x)}$$

where $H(z) = Z[h_j]$, $Q(z) = Z[Q_j]$, and $I(z) = Z[I_j]$. Following O'Connor (1976), divide the numerator and denominator by $1 + k(1 - x)$ and simplify:

$$H(z) = \frac{\dfrac{(1 - kx)z}{1 + k(1 - x)} + \dfrac{kx}{1 + k(1 - x)}}{z - \dfrac{k(1 - x)}{1 + k(1 - x)}}$$

$$= \frac{1 - kx}{1 + k(1 - x)} + \frac{k}{[1 + k(1 - x)]^2}\left[\frac{1}{z - \dfrac{k(1 - x)}{1 + k(1 - x)}}\right]$$

Upon inversion,

$$h_j = \begin{cases} \dfrac{1 - kx}{1 + k(1 - x)}, & j = 0 \\ \left[\dfrac{k}{[1 + k(1 - x)]^2}\right]\left[\dfrac{k(1 - x)}{1 + k(1 - x)}\right]^{j-1}, & j > 0 \end{cases} \tag{20.38}$$

This is the unit pulse response of unit interval duration.

It may be useful to compare equation (20.38) with equations (20.32) and (20.33). To this end, let $C = k(1 - x)$. Equation (20.38) then becomes

$$h_j = \begin{cases} \left(\dfrac{1}{1 - x}\right)\left(\dfrac{1}{1 + C}\right) - \left(\dfrac{x}{1 - x}\right), & j = 0 \\ \left(\dfrac{1}{1 - x}\right)\left(\dfrac{1}{1 + C}\right)\left[\dfrac{C}{1 + C}\right]^j, & j > 0 \end{cases} \tag{20.39}$$

Again, there is a negative term at $j = 0$ and geometric recession thereafter. The shape of the UH derived from the continuous-time representation appears more realistic than the geometric series pulse response of the corresponding discrete representation given by equation (20.39)

If one time-interval lag is introduced into the Muskingum model,

$$[1 + k(1 - x)\nabla]Q_j = [1 - kx\nabla]I_{j-1} \tag{20.40}$$

The discrete pulse response (UH) is

$$h_j = \begin{cases} 0, & j = 0 \\ \left(\dfrac{1}{1 - x}\right)\left(\dfrac{1}{1 + C}\right) - \left(\dfrac{x}{1 - x}\right), & j = 1 \qquad (20.41) \\ \left(\dfrac{1}{1 - x}\right)\left(\dfrac{1}{1 + C}\right)\left(\dfrac{C}{1 + C}\right)^{j-1}, & j > 1 \end{cases} \tag{20.39}$$

If we let $C/(1 + C) = \exp[- 1/C]$, then $1/(1 + C) = 1 - \exp(- 1/C)$. Now equations (20.41), (20.32), and (20.33) are discretely coincident; that is, the continuous UH is exactly equal to the discrete UH at unit intervals.

20.2.4 A Hydraulic Analogy

Various attempts have been made to establish a hydraulic analog of the Muskingum method and then estimate its parameters (Cunge, 1969; Dooge, 1973; Weinmann and Laurenson, 1979; Koussis, 1978, 1980, 1983; Ponce, 1979; Ponce and Theurer, 1982; Peterson and Verhoff, 1982; Dooge, Strupczewski, and Napiorkowski, 1982; Kundzewicz and Dooge, 1985). Fundamental to these attempts is the assumption that the Muskingum method has a hydraulic basis, which may be questionable, however, on both physical and mathematical grounds. This method is purely a black-box conceptual, lumped model. It says nothing about the spatial variability of inflow, outflow, or storage within the specified river reach. Therefore, it is physically unrealistic to transform a black-box system into a distributed one. Let us, nevertheless, consider a river reach of length ΔX. Then, following Koussis (1978), equation (20.11) and (20.12) can be written as

$$\frac{dS}{dt} = Q(X, t) - Q(X + \Delta X, t) \tag{20.42}$$

$$S = k[x\,Q(X, t) + (1 - x)Q(X + \Delta X, t)] \tag{20.43}$$

in which X denotes distance measured from the upstream end. Expanding $Q(X + \Delta X, t)$ about X by a Taylor series,

$$Q(X + \Delta X, t) = Q(X, t) + \Delta X\frac{\partial Q}{\partial X} + \frac{(\Delta X)^2}{2}\frac{\partial^2 Q}{\partial X^2} + \text{HOT} \tag{20.44}$$

where HOT denotes higher-order terms and is henceforth neglected. Substituting equation (20.44) into equations (20.42) and (20.43),

$$\frac{dS}{dt} = -\Delta X\frac{\partial Q}{\partial X} - \frac{(\Delta X)^2}{2}\frac{\partial^2 Q}{\partial X^2} \tag{20.45a}$$

$$S = k\left[xQ + (1 - x)Q + (1 - x)\,\Delta X\frac{\partial Q}{\partial X} + (1 - x)\frac{(\Delta X)^2}{2}\frac{\partial^2 Q}{\partial X^2}\right] \tag{20.45b}$$

Differentiating equation (20.45b) with respect to time and neglecting the derivative terms higher than second order,

$$\frac{\partial S}{\partial t} = k\left[\frac{\partial Q}{\partial t} + (1 - x)\,\Delta X\frac{\partial}{\partial X}\left(\frac{\partial Q}{\partial t}\right)\right] \tag{20.46}$$

Equating equation (20.46) with equation (20.45),

$$k\left[\text{B}\frac{\partial Q}{\partial t} + (1 - x)\,\Delta X\frac{\partial}{\partial X}\left(\frac{\partial Q}{\partial t}\right)\right] = -\Delta X\frac{\partial Q}{\partial X} - \frac{(\Delta X)^2}{2}\frac{\partial^2 Q}{\partial X^2} \tag{20.47}$$

Writing the continuity equation for the case of no lateral inflow or outflow,

$$\frac{\partial Q}{\partial X} + \frac{\partial A}{\partial t} = 0 \tag{20.48a}$$

or

$$\frac{\partial Q}{\partial X} + \left(\frac{dA}{dQ}\right)_{X=\text{constant}} \frac{\partial Q}{\partial t} = 0 \tag{20.48b}$$

where A is the wetted cross-sectional area. This yields

$$\frac{\partial Q}{\partial t} = -\left(\frac{dQ}{dA}\right)_{X=\text{constant}} \frac{\partial Q}{\partial X} \tag{20.48c}$$

From equation (20.48b), the celerity of the discharge is

$$\frac{dx}{dt} = \left(\frac{dQ}{dA}\right)_{X=\text{constant}}$$

If Q is uniquely related to A, then the velocity of propagation of kinematic waves c is

$$c = \left(\frac{dQ}{dA}\right)_{X=\text{constant}} \tag{20.49}$$

Inserting equations (20.48c) and (20.49) in equation (20.47) and considering c as constant,

$$\frac{\partial Q}{\partial t} + \frac{\Delta X}{k}\frac{\partial Q}{\partial X} = \left[c(1-x)\,\Delta X - \frac{(\Delta X)^2}{2k}\right]\frac{\partial^2 Q}{\partial X^2} \tag{29.50}$$

Equation (20.50) is of the same form as the convection-diffusion equation (Lighthill and Whitham, 1955; Cunge, 1969), which is simply written as

$$\frac{\partial Q}{\partial t} + \theta_1\frac{\partial Q}{\partial X} = \theta_2\frac{\partial^2 Q}{\partial X^2} \tag{20.51}$$

where θ_1 is analogous to the celerity of discharge Q and θ_2, to the diffusion coefficient. If the assumption of a single-valued discharge-area relationship holds, then

$$\theta_2 = \frac{Q_0}{2B_0S_0} \tag{20.52}$$

$$\theta_1 = \text{kinematic celerity} = c \tag{20.53}$$

where B_0 is surface width of the wetted channel cross section, S_0 bed slope and Q_0 reference discharge for linearization.

20.2.5 Parameter Estimation

Determining the Muskingum parameters depends upon their interpretation. For example, if the Muskingum method is related to the convection-diffusion model, then the parameters can be expressed in terms of hydraulic and morphological characteristics of the channel reach. On the other hand, if the method is considered to be a black-box type, then its parameters can be estimated using standard techniques applicable to linear systems. It is generally recognized that the effectiveness of this method depends upon the accuracy with which we determine its parameters. We discuss some methods of estimating its parameters.

20.2.5.1 Least Squares Method

The least squares method is based on minimizing the sum of squares of deviations between observed storage and estimated storage for a given inflow and outflow sequence. That is,

$$E = \sum_{j=1}^{N} [S_0(j) - S_e(j)]^2 \rightarrow \min \tag{20.54}$$

where E is the error function to be minimized, $S_0(j)$ is the observed storage for the jth time interval, $S_e(j)$ is the estimated storage for the jth time interval, and N is the number of observations or data.

The estimated storage should be given by equation (20.12). In actual practice the storage that is given is a relative storage, not an absolute one. Therefore, equation (20.12) can be modified to incorporate the difference C between relative and absolute storages:

$$S = k[xI + (1-x)Q] + C \tag{20.55}$$

If the flood wave is starting on a dry river bed, the parameter C will vanish. First, consider the case $C \neq 0$. Substituting equation (20.55) into equation (20.54) and dropping the index j for brevity,

$$E = \sum_{1}^{N} [S_0 - kxI - k(1-x)Q - C]^2 \rightarrow \min \tag{20,56}$$

Let $A = kx$ and $B = k(1-x)$. Differentiating equation (20.56) with respect to A, B, and C, equating each time to zero, and rearranging,

$$\Sigma S_0 I = A\Sigma I^2 + B\Sigma QI + C\Sigma I$$
$$\Sigma S_0 Q = A\Sigma IQ + B\Sigma Q^2 + C\Sigma Q$$
$$\Sigma S_0 = A\Sigma I + B\Sigma Q + CN$$

Solving for A, B, and C,

$$A = \frac{y_1}{y_2} - \frac{y_3}{y_2}$$

$$B = \frac{y_1z_2 - z_1y_2}{y_3z_2 - y_2z_3}$$

$$C = \frac{\Sigma S_0 - A\Sigma I - B\Sigma Q}{N} \tag{20.57}$$

where

$$y_1 = \Sigma\, S_0 I - \frac{\Sigma\, S_0 \Sigma\, I}{N}$$

$$y_2 = \Sigma\, I^2 - \frac{(\Sigma\, I)^2}{N}$$

$$y_3 = \Sigma\, IQ - \frac{\Sigma\, Q \,\Sigma\, I}{N}$$

$$z_1 = \Sigma\, S_0 Q - \frac{\Sigma\, S_0 \Sigma\, Q}{N}$$

$$z_2 = \Sigma\, IQ - \left(\frac{\Sigma\, I \,\Sigma\, Q}{N}\right)$$

$$z_c = \Sigma\, Q^2 - \frac{(\Sigma\, Q)^2}{N}$$

Therefore,

$$k = A + B \tag{20.58a}$$

$$x = \frac{A}{k} \tag{20.58b}$$

Equations (20.57) and (20.58) give the parameters k, x, and C for a known inflow-outflow sequence.

If $C = 0$, then the error function E follows:

$$E = \sum_{1}^{N} [S_0 - kxI - (1 - x)kQ]^2 \rightarrow \min \tag{20.59}$$

Using A and B as defined before, differentiating E with respect to A and B and equating each time to zero,

$$\Sigma\, S_0 I = A\, \Sigma\, I^2 + B\, \Sigma\, IQ$$

$$\Sigma\, S_0 Q = A\, \Sigma\, IQ + B\, \Sigma\, Q^2$$

Solving for A and B,

$$A = \frac{S_0 I\, \Sigma\, Q^2 - \Sigma\, S_0 Q\, \Sigma\, IQ}{\Sigma\, I^2 \Sigma\, Q^2 - (\Sigma\, IQ)^2} \tag{20.60a}$$

$$B = \frac{\Sigma\, S_0 Q\, \Sigma\, I^2 - \Sigma\, S_0 I\, \Sigma\, IQ}{\Sigma\, I^2 \Sigma\, Q^2 - (\Sigma\, IQ)^2} \tag{20.60b}$$

Therefore,

$$k = A + B \tag{20.61a}$$

$$x = \frac{A}{k} \tag{20.61b}$$

Thus k and x can be obtained objectively and conveniently. The computation involved in the method can be easily carried out on a desk calculator. This method was used by Gill (1978), Singh and McCann (1980c), Heggen (1984), and Wu, King, and Wang (1985).

20.2.5.2 Graphical Method

This method consists in choosing x such that the loop resulting from the plot of S versus $xI + (1 - x)Q$ becomes as close to a straight line as possible. The slope of the straight line fitted through the loop gives k. Obviously, this method, as practiced, results in a trial-and-error procedure. However, if we closely follow the steps involved in the graphical method and the result obtained therefrom, it becomes immediately clear that the least squares method is a numerical expression of the graphical method; hence, the two methods are equivalent. Therefore, the least squares method should constitute a natural replacement for the graphical method.

In graphical method we try to choose x such that the correlation coefficient r between S and $xI + (1 - x)Q$ is maximum. It is well known that

$$r = \frac{\text{cov}(S, Y)}{\sigma_S \sigma_Y} \tag{20.62}$$

where cov(S, Y) is covariance of S and Y, σ_S is the standard deviation of S, σ_Y is the standard deviation of Y, and $Y = xI + (1 - x)Q$. For computation,

$$r = \frac{\Sigma\, SY}{(\Sigma\, S^2 \Sigma\, Y^2)^{.5}} \tag{20.63a}$$

$$k = \frac{r\sigma_s}{\sigma_Y} \tag{20.63b}$$

$$C = \bar{S} - k\bar{Y} \tag{20.63c}$$

where $\bar{S}$ is the mean of S and $\bar{Y}$ is the mean of Y. Thus, the problem of determining k, x, and C is one of maximizing r as a function of x. This is the same as in the least squares method.

20.2.5.3 Method of Moments

This method is based on first determining the moments of the IUH, $h(t)$, of the Muskingum reach and then expressing these moments in terms of the moments of inflow and outflow. If we define

$$M_r^0(h) = \int_0^\infty t^r h(t)\, dt$$

$$M_r(h) = \int_0^\infty (t - \bar{t}\,)^r h(t)\, dt$$

where $M_r^0(h)$ is the rth moment of the function $h(t)$ about the origin, and $M_r(h)$ the rth moment of $h(t)$ about the centroid, $\bar{t}$. Note $\bar{t} = M_1^0(h)$. Using equation (20.28), it is easy to show that

$$M_1^0(h) = k \tag{20.64a}$$

$$M_2(h) = (1 - 2x)k^2 \tag{20.64b}$$

These moments of $h(t)$ can be determined from the moment of I and Q using the Nash theorem of moments (Nash, 1959b),

$$M_r^0(Q) = \sum_{j=0}^{r} \binom{r}{j} M_j^0(I) M_{r-j}^0(h)$$

$$M_r(Q) = \sum_{j=0}^{r} \binom{r}{j} M_j(I) M_{r-j}(h)$$

Thus we obtain

$$M_1^0(Q) = M_1^1(I) + M_1^0(h)$$

$$M_2(Q) = M_2(I) + M_2(h)$$

Therefore,

$$k = M_1^0(Q) - M_1^0(I) \tag{20.65a}$$

$$x = \frac{1}{2}\left\{1 - \frac{M_2(Q) - M_2(I)}{[M_1^0(Q) - M_1^0(I)]^2}\right\} \tag{20.65b}$$

If the discrete representation of equation (20.38) is employed, then

$$M_1^0(h) = k$$

$$M_2(h) = k[1 + 2k(1 - x)]$$

By using the theorem of moments,

$$k = M_1^0(Q) - M_1^0(I) \tag{20.66a}$$

$$x = 1 + \frac{1}{2k} - \frac{M_2(Q) - M_2(I)}{2k^2} \tag{20.66b}$$

Note that these parameter estimates differ from those given by equation (20.65).

Thus, k and x can be obtained objectively and conveniently by computing the first two moments of inflow and outflow, one about the origin and the other about the centroid. The calculation involved in the method can easily be carried out on a desk calculator.

20.2.5.4 Method of Cumulants

Let the Laplace transform of $h(t)$ be defined as

$$y(s) = \int_0^\infty e^{-st} h(t)\, dt \tag{20.67}$$

Then the rth cumulant of the IUH, $h(t)$, can be defined as

$$K_r = (-1)^r \frac{d^r}{ds^r} [\log y(s)]|_{s=0} \tag{20.68}$$

Substituting equation (20.67) into equation (20.68),

$$K_r = (-1) \frac{d^r}{ds^r} \log \left[\int_0^\infty e^{-st} h(t)\, dt\right]\Big|_{s=0} \tag{20.69}$$

The notation $|_{s=0}$ means that the associated quantity is to be evaluated at $s = 0$. Equation (20.69) yields

$$K_1 = k \tag{20.70a}$$

$$K_2 = (1 - 2x)k^2 \tag{20.70b}$$

Hence,

$$k = K_1 \tag{20.71a}$$

$$x = \frac{1}{2K_1^2}(K_1^2 - K_2) \tag{20.71b}$$

If the discrete representation of equation (20.38) is used, then

$$K_1 = k \tag{20.72a}$$

$$K_2 = k[1 + k(1 - 2x)] \tag{20.72b}$$

which are not the same as equation (20.70) for continuous representation. The cumulants K_1 and K_2 of $h(t)$ can be determined from inflow-outflow data using the theorem of cumulants for linear, time-invariant hydrologic systems (Dooge, 1973),

$$K_r(Q) = K_r(I) + K_r(h)$$

Thus, k and x can be computed objectively and conveniently by computing the first two cumulants of inflow and outflow. A distinct advantage of cumulants is that they, except the first, remain unaffected by the change of origin (Dooge, 1973).

20.2.5.5 Direct Optimization

The direct optimization method (Stephenson, 1979; O'Donnell, 1985) is based on directly determining the coefficients C_0, C_1, and C_2 in equation (20.13) without first estimating k and x in equation (20.14). This would involve minimizing the error function in comparing observed hydrograph with computed hydrograph. Again, the error function can be defined in a least squares sense or differently. There are, in fact, only two unknowns, since the third can be obtained from equation (20.15). Let us consider C_1 and C_2 to be the unknowns. Then equation (20.13) reduces to

$$Q_{t+\Delta t} - I_{t+\Delta t} = C_1(I_{t+\Delta t} - I_t) + C_2(I_{t+\Delta t} - Q_t) \tag{20.73}$$

The coefficients C_1 and C_2 can be derived by comparing flows of a known inflow hydrograph with flows of the corresponding outflow hydrograph. This comparison can be done in a least squares sense or differently. If we employ the least squares–type error function, then C_1 and C_2 can be estimated in exactly the same manner as k and x evaluated using equation (20.59). Let us define

$$R = Q_{t+\Delta t} - I_{t+\Delta t}$$

$$F = I_{t+\Delta t} - I_t$$

$$G = I_{t+\Delta t} - Q_t$$

Equation (20.73) can be rewritten as

$$R = C_1F + C_2G$$

which is identical in form to equation (20.13) with $C_0 = 0$. Therefore,

$$C_1 = \left(\frac{\Sigma R_0F \Sigma G^2 - \Sigma R_0G \Sigma FG}{DET}\right)$$

$$C_2 = \left(\frac{\Sigma R_0G \Sigma F^2 - \Sigma R_0F \Sigma FG}{DET}\right)$$

where

$$DET = \Sigma F^2 \Sigma G^2 - (\Sigma FG)^2$$

and R_0 is observed R. Then C_0 can be computed from equation (20.15). Further, a little algebraic manipulation shows that

$$x = \frac{C_1 + .5C_2 - .5}{C_1 + C_2} \tag{20.74a}$$

$$k = \frac{\Delta t(C_1 + C_2)}{1 - C_2} \tag{20.74b}$$

Thus, k and x can be objectively determined. The amount of computation involved is quite small and can be undertaken on a desk calculator.

20.2.5.6 Hydraulic Method

A hydraulic analog of the Muskingum method was discussed previously. If we compare the coefficients of equations (20.50) and (20.51) and employ equations (20.52) and (20.53), we see that

$$k = \frac{\Delta X}{c} = \frac{\Delta X}{\left(\frac{dQ}{dA}\right)_{X=\text{constant}}} \tag{20.75a}$$

$$x = \frac{1}{2} - \frac{kQ_0}{2B_0S_0(\Delta X)^2} \tag{20.75b}$$

Equation (20.75a) indicates that k can be interpreted as the propagation time of discharge. Equation (20.75b) relates the weighting factor x to hydraulic and morphological characteristics of the channel reach. For attenuation of waves during propagation, the diffusion coefficient must be positive. This implies that $x < .5$. Since $x > 0$ by definition, it is concluded that $0 < x < .5$. This was described by Cunge (1969) and Koussis (1978). It may be noted that if k and x are specified, then ΔX and Δt for numerical computation can be determined in an objective manner as was done for a special case of $\Delta t = k$ and $x = 0$ by Ponce (1979).

On the other hand, Dooge (1973) derived the Muskingum parameters by matching the moments or cumulants of the Muskingum IUH with the corresponding moments or cumulants of the IUH of linear hydrodynamic model. By so doing,

$$k = \frac{\Delta X}{1.5V_0} \tag{20.76a}$$

$$x = \frac{1}{2} - \frac{1}{3}\left(1 - \frac{F^2}{4}\right)\left(\frac{y_0}{S_0\Delta X}\right) \tag{20.76b}$$

where V_0 is reference velocity, y_0 is reference depth, and F is the Froude number equal to $V_0/(gy_0)^{1/2}$, with g being acceleration due to gravity. In another study Dooge, Strupczewski, and Napiorkowski (1982) derived the Muskingum parameters hydrodynamically.

The hydraulic method assumes small variations in Q, which may be unrealistic in case of floods. Koussis (1978) suggested using discharge-dependent parameters.

20.2.5.7 Comparison of Parameter Estimation Methods

Three example problems were worked out to estimate k and x by the foregoing methods except for the hydraulic method (Singh and McCann, 1980b). The first example problem is from Linsley, Kohler, and Paulhus (1975), the second is from Lawler (1964), and the third is from Viessman, Knapp, Lewis, and Harbough (1977).

The parameters k and x were determined for each inflow-outflow data set, as shown in Table 20.1. It is clear that these parameters change significantly as their method of computation changes. It is particularly true between the values given by the optimization method and the method of moments or cumulants as well as between the least squares method or the graphical method and the method of moments or cumulants. However, the values given by the least squares method or the graphical method for the two cases with parameter $C \neq 0$ and $C = 0$ are relatively close. The inclusion of parameter C appears to influence only the parameter k and not x. The value of x takes on a negative value for the direct optimization method for the second example problem. From a physical standpoint, it may look absurd, but for mathematical matching, it is the correct value to use (Dooge, 1973). It can therefore be argued that x does not have a strict physical meaning.

Table 20.2 lists the values of the routing coefficients C_0, C_1, and C_2 for the four methods. It is clear that the variability in the coefficients for different methods is far greater than that in their respective parameters. The effect of inclusion of the parameter C seems to be great on the coefficient C_0 and quite small on C_1 and C_2.

A plot of observed storage versus weighted flow $xI + (1 - x)Q$ showed that loops of the four methods were not as far apart from one another as their corresponding values of k and x might suggest and were indeed quite close to each other (see Fig. 20.5 as an example). This confirms that more than one set of the parameters k and x may exist for the Muskingum method of flood routing. A comparison of observed and computed outflows showed that all four methods were comparable on the whole (see Fig. 20.6 as an example).

TABLE 20.1

Values of k, x, and C for three example data sets

Example Problems	Least Squares + Graphical, $C \neq 0$	Graphical + Least Squares $C = 0$	Moments	Cumulants	Optimization
	Method				
First					
k (da)	.78	.636	.61	.61	.731
x	.272	.25	.446	.446	.16
C (m^3)	−162,391.8	—	—	—	—
Second					
k (da)	1.78	1.401	1.4	1.4	1.839
x	.21	.25	.138	.138	−.296
C (m^3)	−754,381.3	—	—	—	—
Third					
k (da)	1.84	1.416	1.29	1.29	2.367
x	.245	.25	.256	.245	.005
C (m^3)	−36,358,899	—	—	—	—

TABLE 20.2

Values of C_0, C_1, and C_2 for three example data sets

Example Problems	Least Squares + Graphical, $C \neq 0$	Graphical + Least Squares $C = 0$	Moments	Cumulants	Optimization
	Method				
First					
C_0	.0473	.1251	−.0395	−.0395	.1542
C_1	.5447	.5626	.8886	.8886	.4241
C_2	.3881	.3124	.1509	.1509	.4217
Second					
C_0	.0411	.0966	.1801	.1801	.3623
C_1	.4429	.5483	.4064	.4064	−.0155
C_2	.5160	.3551	.4135	.4135	.6532
Third					
C_0	.0265	.0934	.1227	.1227	.1707
C_1	.5032	.5467	.5555	.5555	.1796
C_2	.4703	.3598	.3218	.3218	.6496

20.2.6 Interpretation of the Muskingum Parameters

The interpretation of the Muskingum parameters has been a subject of considerable discussion in hydrologic literature. A consensus, however, appears to have emerged over the interpretation of k but not x. The parameter k is a storage constant and has the dimension of time. From the method of moments it turns out to be the first moment of the Muskingum IUH, which measures the time difference between the centroids of inflow and outflow, as shown by equation (20.65a). On the other hand, using the hydraulic methods, k can be interpreted as the propagation time of discharge. Its value depends upon the length of the channel reach and wave velocity, as shown by equations (20.75a) and (20.76a).

The parameter x has been variously interpreted, as has its range. In addition, the interpretation and range of x are interdependent. From equation (20.12) we see that x is a weighting factor and measures the relative effects of inflow and outflow on storage. It would then appear that x should have a positive value from the range [0, 1]. This appears to be a more popular opinion, although some argue for a narrow range [0, .5], as seen from equations (20.75b) and (20.76b). On the other hand, Yevjevich and Barnes (1970) argue that x relates to the difference between flow rates in the cross sections terminating the channel reach. Therefore, it measures the shape of the wave passing through the reach and differs with the type of wave.

A somewhat different approach was presented by Dooge (1973), in which even negative values of x are admitted. He stated:

It will be noted that for certain short distance the value of x will be negative. From the point of view of classical hydrology which views x as a measure of wedge storage present, this appears physically unreasonable. From the point of view of mathematical modeling, the negative value of x is the correct value.

Along similar lines Strupczewski and Kundzewicz (1980a, 1980b) state that the preceding conclusion relates to highly simplified hydrodynamic considerations, which constitute a basis of traditional development of the Muskingum model. Furthermore, they present arguments justifying the assumption of x values from the range $(-\infty, .5]$. In particular, they show that if hysteresis is assumed for a rating curve in derivation of equation (20.12), the negative values of x may be consistent with the physical interpretation of wedge storage. The mathematical argument is as follows.

The IUH of the Muskingum method is given by equation (20.28), whose representation in the Laplacian domain is

$$H(s) = \frac{q}{P} = \frac{(1 - bs)}{(1 + as)} = \frac{a + b}{a(1 + as)} - \frac{b}{a}$$

Its graphical representation is shown in Fig. 20.4. Here $a = k(1 - x)$ and $b = kx$. It is clear that the Muskingum method is a parallel combination of two systems. One of them is a δ-function with the coefficient $-b/a$. This term provides a direct transformation of the damped inflow for $x > 0$ or the inverted inflow (damped for $x \in (0, .5)$ and amplified for $x > 0.5$) to the model output. The remaining term modulates the system performance. Its response function is equivalent to that of a linear reservoir amplified by $(a + b)/a$. For $x = 0$, the δ-function terms goes to zero and the Muskingum method reduces to a linear reservoir.

From physical considerations it is evident that the flood wave is not amplified as it progresses down the channel. In frequency domain, inflow is not amplified at any frequency. In this regard, if we take the Fourier transfer function of the Muskingum method,

$$H(j\sigma) = \frac{1 - j\sigma b}{1 + j\sigma a}$$

Taking its absolute value,

$$|H(j\sigma)| = \left|\frac{1 - j\sigma b}{1 + j\sigma a}\right| = \left[1 + \frac{\sigma^2(b^2 - a^2)}{1 + \sigma^2 a^2}\right]^{1/2} \leq 1$$

Therefore,

$$b^2 - a^2 = k^2(2x - 1) \leq 0 \quad \text{and} \quad x \leq .5$$

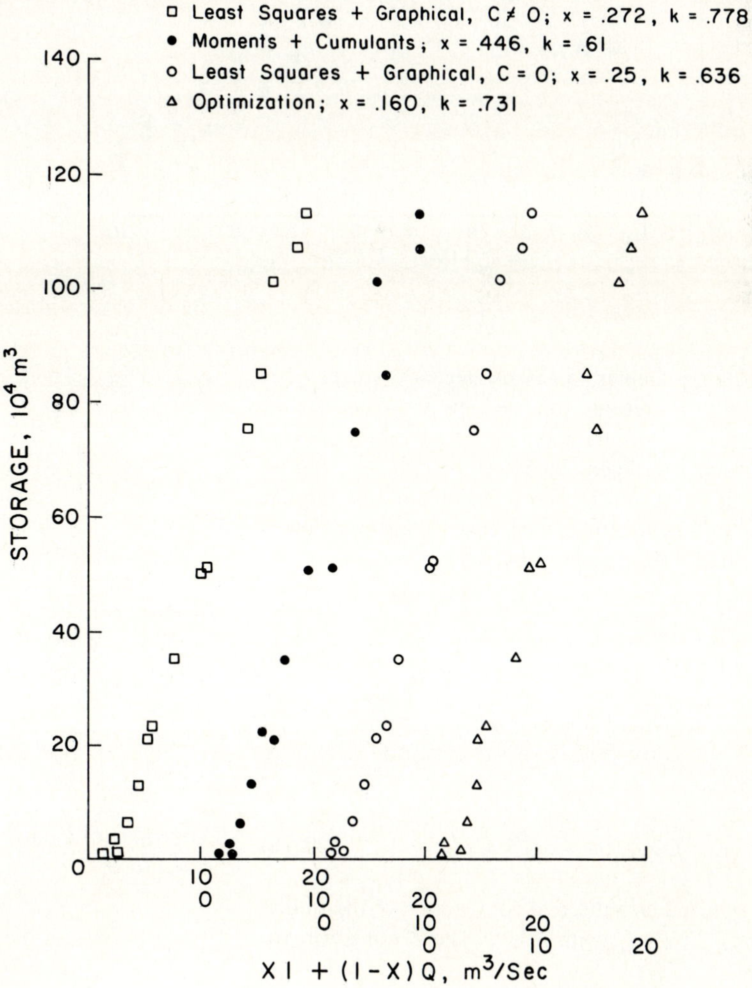

FIGURE 20.5 Storage versus weighted flow for different methods of parameter estimation for data of Linsley, Kohler, and Paulhus (1975).

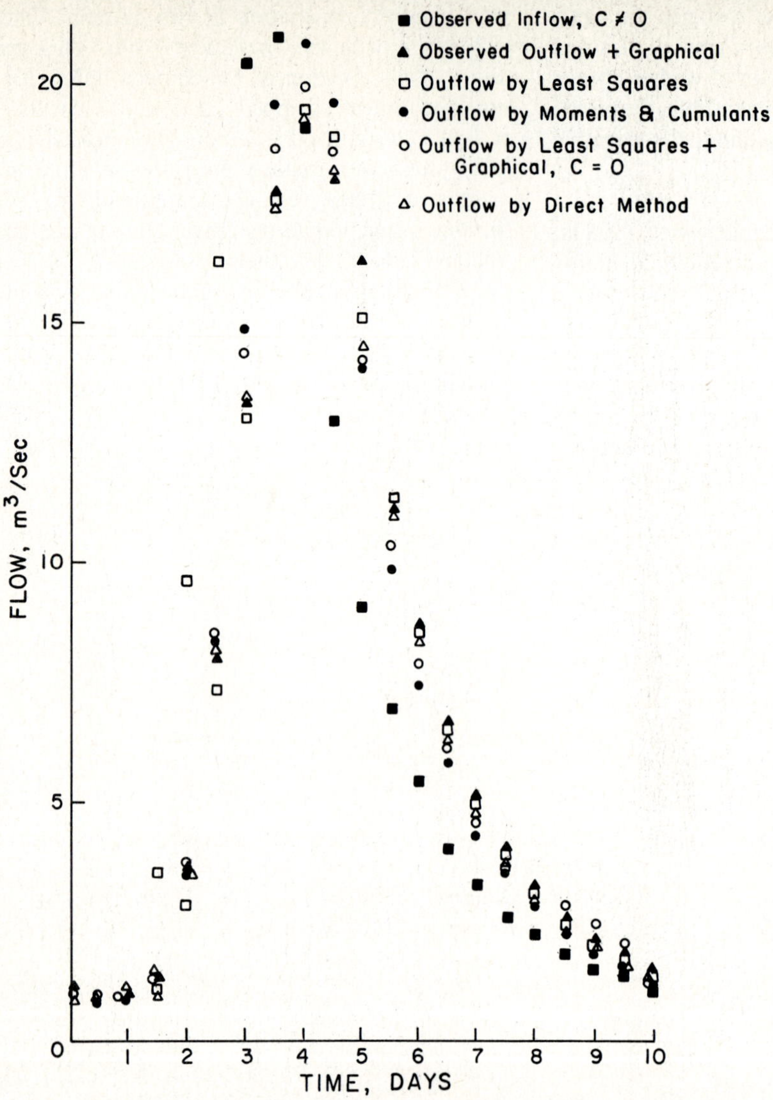

FIGURE 20.6 Flood routing by the Muskingum method for different methods of parameter estimation for data of Linsley, Kohler, and Paulhus (1975).

The upper limit for x then is .5. Since negative values of x do not violate any conditions imposed on the Muskingum IUH and are consistent with the moments and their properties, the admissible range of values of x is $(-\infty, .5]$.

By using the finite difference form of the Muskingum method, Hjelmfelt (1984) derived a criterion to insure positive ordinates of the unit pulse response as

$$x < .5\frac{\Delta t}{k} < (1 - x), \qquad x < 1 \tag{20.77}$$

and argued that this would ensure positive outflow for all positive inflows. Its derivation is straightforward. Rewriting equation (20.16) as

$$Q_j - C_2Q_{j-1} = C_0I_j + C_1I_{j-1} \tag{20.78}$$

This can be recognized as a discrete convolution of the inflow sequence $(I_0, I_1, \ldots, I_j)$ with a UH. The Z-transform for that UH is

$$H(z) = \frac{C_0 + C_1z}{1 - C_2z} = C_0 + \left(\frac{C_1}{C_2} + C_0\right)C_2z + \left(\frac{C_1}{C_2} + C_0\right)C_2^2z^2 + \cdots + \left(\frac{C_1}{C_2} + C_0\right)C_2^j z^j + \cdots \tag{20.79}$$

The coefficients in the series are the ordinates of the UH. The exponent j coincides with the number of Δt intervals between the start of the UH and the jth ordinate. Since positive ordinates of the UH ensure positive outflow, it is sufficient to consider positivity of the UH. The first ordinate is

$$C_0 = \frac{kx - .5\,\Delta t}{k(1 - x) + .5\,\Delta t} > 0 \tag{20.80}$$

The value of k must be positive by virtue of equation (20.75a). Therefore,

$$x < .5\frac{\Delta t}{k}, \qquad x < 1 \tag{20.81a}$$

$$x > .5\frac{\Delta t}{k}, \qquad x > 1 \tag{20.81b}$$

The second condition is of little practical interest.

The other ordinates of the UH are of the form

$$\left(\frac{C_1}{C_2} + C_0\right)C_2^j, \qquad j = 1, 2, \ldots \tag{20.82}$$

For each of these ordinates to be positive, two conditions must be satisfied: (1) $C_2 > 0$, and (2) $((C_1/C_2) + C_0) \geq 0$. Consider the first condition first.

$$C_2 = \frac{k(1-x) - .5\,\Delta t}{k(1-x) + .5\,\Delta t} > 0 \tag{20.83}$$

This will be true if

$$.5\frac{\Delta t}{k} < 1 - x, \qquad x < 1 \tag{20.84a}$$

or

$$.5\frac{\Delta t}{k} > 1 - x, \qquad x > 1 \tag{20.84b}$$

Equation (20.84b) is of little practical significance.

Now consider the second condition. For its satisfaction, $C_1 > 0$,

$$C_1 = \frac{kx + .5\,\Delta t}{k(1-x) + .5\,\Delta t} \geq 0 \tag{20.85}$$

This yields

$$.5\frac{\Delta t}{k} > x, \qquad x < 1 \tag{20.86}$$

The criterion for $x > 1$ is more complex but can be derived and is of little practical value. Furthermore, the second condition will hold even for negative C_1 provided $|C_1| < C_0C_2$.

The criteria to ensure positive UH ordinates can be summarized as

$$x < .5\frac{\Delta t}{k} < (1 - x), \qquad x < 1$$

This criterion has also been suggested by Koussis (1976). Note that even negative x meets this criterion, as noted empirically by Ponce and Theurer (1982). If, however, $\Delta t/k > 1$, then this criterion will be violated, but this is seldom the case.

20.2.7 Translatory Characteristics

Recently there has been a great deal of discussion on translatory characteristics of the Muskingum method (Kulandaiswamy, 1967; Cunge, 1969; Gill, 1979a, 1979c, 1980; Singh and McCann, 1980a, 1980b; Strupczewski and Kundzewicz, 1980a, 1980b). In particular, attention has been focused on determining conditions under which this method yields a translatory solution. The motivation for this discussion has primarily arisen from equation (20.20)

If a routing system is purely translatory, it transforms the inflow into outflow without change of shape, that is,

$$Q(t) = I(t - k) \tag{20.87}$$

It is evident that the Muskingum method does not obey equation (20.87) and is, therefore, not translatory regardless of the values of x and k. However, it may exhibit some weak quasi-translatory properties (Strupczewski and Kundzewicz, 1980b). These can be examined by considering the characteristics of the Muskingum IUH and inflow.

20.2.7.1 The IUH Characteristics

Following Strupczewski and Kundzewicz (1980b), the Laplace transform of the Muskingum UH is

$$H(s) = \frac{q}{P} = \frac{(1 - kxs)}{1 + k(1-x)s} \tag{20.88}$$

The response function for a system (linear channel) exhibiting equation (20.87),

$$H(s) = \exp(-ks) \tag{20.89}$$

Let us write

$$\exp(-kx) = \frac{\exp(-ks/2)}{\exp(ks/2)} \cong \frac{1 - (ks/2)}{1 + (ks/2)} \tag{20.90}$$

Equation (20.89) is the response function of the so-called phase shift system. Fig. 20.7 shows that unit step responses for equations (20.89) and (20.90).

If $x = .5$, equation (20.88) is identical to equation (20.90). This shows that the Muskingum model behaves as a phase-shift system regardless of the values of k or Δt. The parameter k is not a translation time but is equivalent to the first moment of the Muskingum IUH. It is, therefore, clear that the Muskingum method exhibits weak quasi-translatory properties at best. Increased insight is obtained by expressing equation (20.89) as

$$\exp(-ks) = \exp\left[\frac{-xks}{\exp(1-x)kx}\right] \cong \frac{1 - xks}{1 + k(1-x)s} \tag{20.91a}$$

If we note

$$\exp(w) = \sum_{j=0}^{\infty}\frac{w}{j!}, \qquad |w| < \infty \tag{20.91b}$$

then the linear approximation of equation (20.91b)—and, consequently, of equation (20.91a)—improves with decreasing value of the argument w. Therefore, under the assumption x

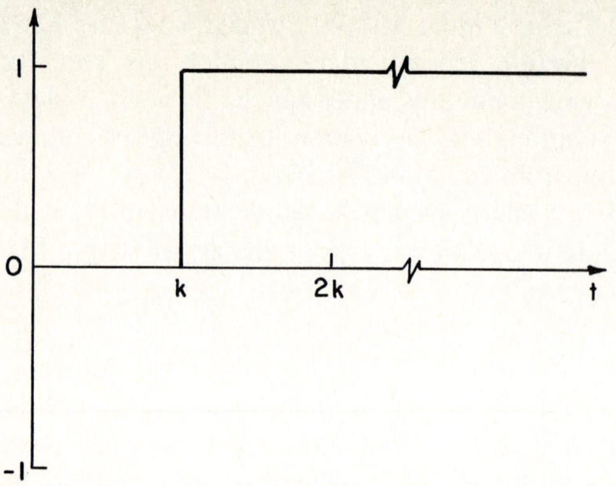

FIGURE 20.7(a) Unit step response of a translatory system.

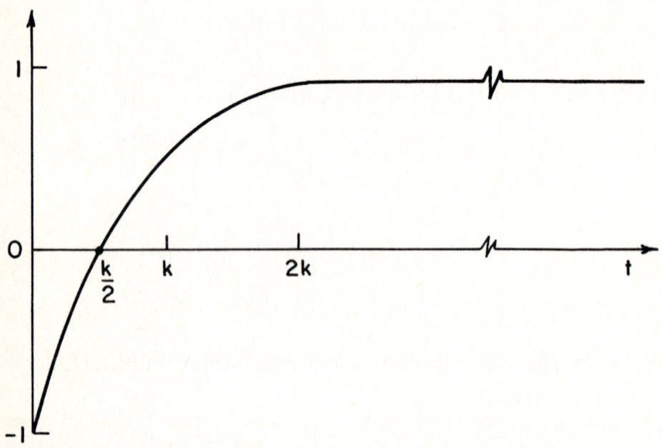

FIGURE 20.7(b) Unit step response of a phase-shift system.

is a member of [0, .5], equation (20.90) is the best approximation to equation (20.91) corresponding to $x = .5$. As x decreases, the Muskingum method departs farther and farther from even quasi-translatory behavior. A special case is when $x = 0$, which corresponds to a linear reservoir. For this case,

$$\exp(-ks) \cong (1 + ks)^{-1}$$

20.2.7.2 Inflow Characteristics

Following Kulandaiswamy (1966), we can write equation (20.24) in operator form,

$$Q(t) = \frac{1 - xkD}{1 + k(1 - x)D} I(t) \qquad D \equiv \frac{d}{dt}$$

For $x = .5$,

$$Q(t) = \frac{1 - .5kD}{1 + .5kD} I(t)$$

Expanding $1/(1 + .5kD)$ in series and writing

$$Q(t) = \left[1 - \frac{1}{2}kD + \frac{1}{4}k^2D^2 - \frac{1}{8}k^3D^3 + \cdots\right][1 - .5kD]\, I(t)$$

Since we can treat the operator as an algebraic quantity,

$$Q(t) = (1 - kD + \frac{1}{2}k^2D^2 - \frac{1}{4}k^3D^3 + \cdots)\, I(t) \tag{20.92}$$

This equation is a particular solution of equation (20.24).

If we consider the case where $I(t)$ is merely translated, equation (20.87) must hold. Expanding $I(t - k)$ in a Taylor series,

$$I(t - k) = I(t) - kI'(t) + \frac{k^2}{2!}I''(t) - \frac{k^3}{3!}I'''(t) + \cdots$$

$$= (1 - kD + \frac{k^2}{2!}D^2 - \frac{k^3}{3!}D^3 + \cdots)\, I(t) \tag{20.93}$$

Comparing equation (20.93) with equation (20.92), we note that the first three terms are identical; the difference between the two equations arises beginning with the fourth term. If $I(t)$ is such that its third and higher-order derivatives are vanishingly small, then equation (20.87) will approximately hold.

However, hidden in the assumptions used by Kulandaiswamy (1966) to derive the above result is the fact that inflow I can be written in the form

$$I(t) = \sum_{n=0}^{\infty} a_n t^n \tag{20.94}$$

and that the series converges for all t. Hence, the values of $I(t)$ on an interval $[0, \epsilon]$ uniquely defines $I(t)$ for any $t \geq 0$. This clearly is an unrealistic hypothesis. Even if I satisfies this hypothesis, there need not be a translatory solution.

Gill (1979a) discussed translatory characteristics of the Muskingum method. He attempted to generalize the previously discussed results obtained by Kulandaiswamy (1966). A careful study of Gill's paper by Singh and McCann (1980a, 1980b) shows that he has made mathematically inconsistent assumptions that would seldom be true in nature. Therefore, his analysis is incorrect.

EXAMPLE 20.3

Consider

$$I(t) = e^{-t/[k(1-x)]}$$

and examine translatory characteristics of the Muskingum method. Then $I_0 = 1$ and, using equation (20.24),

$$Q(t) = \left(1 + \frac{t}{k(1 - x)^2}\right) e^{-t/[k(1-x)]}$$

$$\neq I(t - \tau) \qquad \text{for any } \tau$$

Furthermore,

$$Q(t) - I(t - k) = \left[1 + \frac{t}{k(1 - x)^2} - e^{1/(1-x)}\right] e^{-t/[k(1-x)]}$$

When $x = 0.5$, $\tau = k$,

$$Q(t) - I(t - k) = (1 + \frac{4t}{k} - e^2)\, e^{-2t/k}$$

When $t = k$,

$$Q(k) - I(0) = (1 + 4 - e^2)\, e^{-2} = -.3233$$

Hence, $Q(k)$ is approximately 32 percent too small. On the other hand, third- and higher-order derivatives of $I(t)$ are far from zero. Further, the appearance of $I(t)$, frequently encountered in nature, resembles a gamma or lognormal distribution. Obviously their third- and higher-order derivatives do not vanish. Therefore, the assumptions for approximate translatory behavior are not valid.

EXAMPLE 20.4

Let the inflow be expressed by

$$I(t) = C, \qquad t \geq 0$$

Route this inflow by the Muskingum method and examine the translatory characteristics. Substituting into equation (20.25),

$$\begin{aligned} Q(t) &= \frac{C}{k(1-x)^2}\int_0^t e^{-(t-u)/(k(1-x))}\,du - \frac{Cx}{1-x} \\ &= \frac{C}{1-x}\{1 - e^{-t/[k(1-x)]}\} - \frac{Cx}{1-x} \\ &= C\left\{1 - \frac{e^{-t/(k(1-x))}}{1-x}\right\} \end{aligned}$$

It is thus seen that as $t \to \infty$, $Q(t) \to C$, which is a case of pure translation. This result was also given by Paynter (1952). However, for small times initially, the outflow becomes negative. This was noted by Nash (1959a). When $t = 0$, $Q(t) = -Cx/(1 - x)$. The period for which the outflow would be negative is

$$t = -k(1 - x)\ln(1 - x)$$

If $x = .5$, then $t = .3465k$. These observations indicate that the $Q(t)$ is not a translation of $I(t)$. Now consider the case associated with equation (20.25). Therefore,

$$\begin{aligned} Q(t) &= \frac{C}{1-x}e^{-t/[k(1-x)]} - \frac{xC}{1-x} + \frac{C}{k(1-x)^2}\int_0^t e^{-(t-u)/[k(1-x)]}\,du \\ &= \frac{C}{1-x}e^{-t/[k(1-x)]} - \frac{xC}{1-x} + \frac{C}{1-x}\{1 - e^{-t/[k(1-x)]}\} \\ &= c, \qquad t \geq 0 \end{aligned}$$

This says that the outflow equals inflow and the flood wave is translated. This is easily understood from the steady-state condition of the channel flow.

If, however, $0 < I(0) < C$, then

$$\begin{aligned} Q(t) &= \frac{I(0)}{1-x}e^{-t/[k(1-x)]} + C\left\{1 - \frac{e^{-t/[k(1-x)]}}{1-x}\right\} \\ &= \frac{(I(0) - C)}{1-x}e^{-t/[k(1-x)]} + C \end{aligned}$$

Let $I(0) = xC$. Then

$$Q(t) = C[1 - e^{-t/(k(1-x))}]$$

Thus we notice that $Q(t) > 0$ for $t > 0$ if $I(0) > xC$, unlike the previous case. An important implication of this is that negative outflow ordinates obtained in the Muskingum method depend upon the initial conditions to be used.

EXAMPLE 20.5

Let the inflow be expressed by

$$I(t) = \begin{cases} C, & 0 \leq t \leq T \\ 0, & \text{otherwise} \end{cases}$$

Route the inflow by the Muskingum method and examine the translatory characteristics.

$$Q(t) = C\left\{1 - \frac{e^{-t/[k(1-x)]}}{1-x}\right\}, \qquad 0 \leq t < T$$

$$Q(t) = Q_p e^{-(t-T)/[k(1-x)]}, \qquad t > T$$

where

$$Q_p = C\left\{1 - \frac{e^{-T/[k(1-x)]}}{1-x}\right\}$$

It is clear that $I(t)$ is not translated. Here again, negative outflow is produced for some time in the beginning.

If we consider the case associated with equation (20.25),

$$Q(t) = \begin{cases} C, & 0 \leq t < T \\ 0, & t > T \end{cases}$$

Again the flood wave is simply translated. If $0 < I(0) < C$, we obtain

$$Q(t) = \begin{cases} \dfrac{(I(0) - C)}{1-x}e^{-t/[k(1-x)]} + C, & 0 < t < T \\[2ex] \dfrac{Q_p}{k(1-x)}e^{-(t-T)/[k(1-x)]}, & t \geq T \end{cases}$$

where

$$Q_p = \frac{(I(0) - C)}{1-x}e^{-T/[k(1-x)]} + C$$

If $I(0) = xC$, then

$$\begin{aligned} Q(t) &= \begin{cases} C\{1 - e^{-t/[k(1-x)]}\}, & 0 < t < T \\[1ex] \dfrac{Q_p}{k(1-x)}e^{-(t-T)/[k(1-x)]}, & t \geq T \end{cases} \\ Q_p &= C\{1 - e^{-T/[k(1-x)]}\} \end{aligned}$$

It is immediately clear that the flood wave is not translated, unlike the previous case. This further illustrates the effect of initial conditions on Muskingum solutions.

EXAMPLE 20.6

Let the inflow be expressed by

$$I(t) = \begin{cases} 2\dfrac{h}{T}t, & 0 \leq t \leq \dfrac{T}{2} \\[2ex] 2\dfrac{h}{T}(T - t), & \dfrac{T}{2} \leq t \leq T \end{cases}$$

Route the inflow by the Muskingum method and examine the translatory characteristics.

If this value of $I(t)$ is substituted into equation (20.25),

$$Q(t) = \frac{1}{k(1-x)^2}\int_0^t e^{-(t-u)/[k(1-x)]}\frac{2h}{T}u\,du$$
$$-\frac{2hx}{T(1-x)}t, \qquad 0 \le t < \frac{T}{2}$$
$$= \frac{2he^{-t/[k(1-x)]}}{Tk(1-x)^2}\left\{k(1-x)^2 e^{u/[k(1-x)]}\left[\frac{u}{k(1-x)} - 1\right]\Big|_0^t\right\}$$
$$-\frac{2hx}{T(1-x)}t, \qquad 0 \le t < \frac{T}{2}$$
$$Q(t) = \frac{2hk}{T}\left\{\frac{t}{k(1-x)} + e^{-t/[k(1-x)]} - 1\right\}$$
$$-\frac{2hxt}{T(1-x)}, \qquad 0 \le t \le \frac{T}{2}$$

For $T/2 \le t \le T$,

$$Q(t) = \frac{2h}{T}(T-t) + Q(T/2)\,e^{-[t-(T/2)]/[k(1-x)]}$$
$$-he^{-[t-(T/2)]/(k(1-x))}\left(1 + \frac{2k}{T}\right) + \frac{2hk}{T}, \qquad \frac{T}{2} \le t \le T$$

where

$$Q\left(\frac{T}{2}\right) = h - \frac{2hk}{T}\{1 - e^{-T/[2k(1-x)]}\}$$

For $t \ge T$,

$$Q(t) = Q_* e^{-(t-T)/[k(1-x)]}$$

where

$$Q_* = Q\left(\frac{T}{2}\right)e^{-T/[2k(1-x)]} - he^{-T/[2k(1-x)]}\left(1 + \frac{2k}{T}\right) + \frac{2hk}{T}$$

These solutions are not translatory. As before, negative outflow results for some time in the beginning.

If $I(t)$ is assumed to be

$$I(t) = \begin{cases} I(0) + \dfrac{2h}{T}t, & 0 \le t \le \dfrac{T}{2} \\ I(0) + 2h - \dfrac{2ht}{T}, & \dfrac{T}{2} \le t < \dfrac{[I(0) + 2h]T}{2h} \end{cases}$$

When we substitute $I(t)$ into equation (20.25), we obtain

$$Q(t) = \frac{I(0)}{1-x}e^{-t/[k(1-x)]} - \frac{x}{1-x}\left[I(0) + \frac{2h}{T}t\right]$$
$$+\frac{1}{k(1-x)^2}\int_0^t e^{-(t-u)/[k(1-x)]}(I(0)$$
$$+\frac{2h}{T}u)\,du, \qquad 0 \le t \le \frac{T}{2}$$
$$= \frac{I(0)}{1-x}e^{-t/[k(1-x)]}$$
$$-\frac{x}{1-x}\left[I(0) + \frac{2h}{T}t\right]$$
$$+\frac{I(0)}{1-x}\{1 - e^{-t/[k(1-x)]}\} + \frac{2hk}{T}\left\{\frac{t}{k(1-x)}\right.$$
$$\left. + e^{-t/[k(1-x)]} - 1\right\}, 0 \le t \le \frac{T}{2}$$

On simplifying,

$$Q(t) = I(0) + \frac{2h}{T}(t-k) + \frac{2hk}{T}e^{-t/[k(1-x)]}$$

For $T/2 \le t < [I(0) + 2h]T/2h$,

$$Q(t) = t(0) + 2h - \frac{2ht}{T} + Q\left(\frac{T}{2}\right)e^{-[t-(T/2)]/[k(1-x)]}$$
$$-e^{-[t-(T/2)]/[k(1-x)]}\left[I(0) + h + \frac{2hk}{T}\right] + \frac{2hk}{T}$$

where

$$Q\left(\frac{T}{2}\right) = I(0) + h - \frac{2hk}{T}\{1 - e^{-T/[2k(1-x)]}\}$$

For $t \ge [I(0) + 2h]T/2h$,

$$Q(t) = Q_* e^{-(t-T)/[k(1-x)]}$$

where Q_* is obtained by substituting $t = [I(0) + 2h]T/2h$ into $Q(t)$.

$$Q_* = Q\left(\frac{T}{2}\right)e^{-[I(0)+h]T/[2hk(1-x)]}$$
$$-e^{-[I(0)+h]T/[2hk(1-x)]}\left[I(0) + h + \frac{2hk}{T}\right] + \frac{2hk}{T}$$

Thus, we see that the flood wave is not translated. Further, $Q(t) > 0$ for $t \ge 0$, unlike the previous cases.

20.2.8 Spatial and Temporal Resolution

The preceding discussion on interpretation of the Muskingum parameters throws some light on temporal resolution but provides little information on spatial resolution to be used in flow routing. Koussis (1976) was among the first to propose a criterion linking spatial and temporal resolutions (or grid sizes in numerical computations),

$$\Delta t \ge \frac{2x\Delta X}{c} \quad \text{or} \quad \Delta X \le \frac{c\Delta t}{2x} \tag{20.95}$$

in which ΔX is the space step or reach length. This follows from the condition $C_0 \ge 0$.

Weinmann and Laurenson (1979) suggested

$$\frac{\Delta X}{c} \le \frac{t_r}{2x} \tag{20.96}$$

where t_r denotes the period of rise of inflow hydrograph. Equations (20.95) and (20.96) are similar, as pointed out by Ponce and Theurer (1982). This can be seen by expressing t_r in terms of Δt as $t_r/\Delta t = m$, where m is an integer greater than 5. Replacing t_r in equation (20.96),

$$\frac{\Delta X}{c} \le \frac{m\Delta t}{2x} \tag{20.97}$$

This shows that equation (20.95) is more conservative than equation (20.96). It appears that equations (20.95) and (20.96) would not apply if $x = 0$. However, in that case Koussis (1983) showed that $\Delta X < \infty$, implying no restriction.

Another criterion was proposed by Ponce and Theurer (1980, 1982):

$$CD \geq a \tag{20.98}$$

in which C and D are Courant and cell Reynolds numbers defined, respectively, as

$$C = c\frac{\Delta t}{\Delta x} \tag{20.99}$$

$$D = \frac{Q_0}{S_0 B_0 c \Delta X} \tag{20.100}$$

A value of $a = .25$ provided satisfactory results. Note that equation (20.98) is in terms of D rather than x. The upper limit on ΔX based on this expression is

$$\Delta X \leq \left[\frac{Q_0 \Delta t}{B_0 S_0 a}\right]^{.5} \tag{20.101}$$

A lower limit on ΔX can also be derived, especially from equation (20.75b), by setting $x \geq 0$:

$$\Delta X \geq \frac{Q_0}{B_0 S_0 c} \tag{20.102}$$

This specifies the *characteristic reach length* of Kalinin and Milyukov (1957) and supports the arguments of Weinmann and Laurenson (1979). Ponce and Theurer (1980), Kundzewicz (1980), and Weinmann (1977) have, on the other hand, reasoned that there is no theoretical lower limit on Δx.

On the other hand, there is clearly a need to define the range of Δt. Usual practice is to define

$$\Delta t \leq \frac{t_r}{m} \tag{20.103}$$

where m is an integer having a minimum value of 5. Ponce and Theurer (1982) stated a lower bound based on Courant numerical stability criterion,

$$\Delta t \leq \frac{\Delta X}{c} \tag{20.103}$$

which follows from

$$C \leq 1 \tag{20.104}$$

However, they confirmed that violating equations (20.103) and (20.104) did not result in numerical instability. This implies that there is no theoretical justification for a lower limit on Δt. This same is true for ΔX.

Jones (1981, 1983) recommended

$$\frac{T_r}{5} > \Delta t > 1.43\left(\frac{Q_0}{2B_0 S_0 c}\right) \tag{20.105a}$$

for accuracy to be maintained. This is possible when

$$N = \frac{S_0 c B_0 t_r}{Q_0} > 3.58 \tag{20.105b}$$

with low baseflow $N = .288$.

EXAMPLE 20.7

For a river reach, the inflow-outflow hydrographs are given in Table 20.3. Determine the Muskingum parameters x and k.

Solution: The parameters x and k are determined graphically. To this end, the weighted discharge $(xI + (1 - x)Q]$ is computed corresponding to $x = 0, .1, .2, .3, .4$, and $.5$, as shown in Table 20.4. The weighted discharge is then plotted against storage for various values of x. Two plots showing the narrower loops are graphed in Fig. 20.8. The narrowest loop is obtained for $x = .2$. The value of $k = 2.33$ days is thus found. To summarize,

$$x = .2$$

$$k = 2.33 \text{ da}$$

EXAMPLE 20.8

Using the values of x and k estimated in the preceding example, route the inflow hydrograph of Example 20.7.

Solution: The value of $\Delta t = 1$ da is selected, which is greater than $2kx$. From equation (20.14),

$$C_3 = 2.3 - (2.33)(0.2) + (0.5)1 = 2.36$$

$$C_0 = \frac{-(2.33)(0.2) + (0.5)1}{2.36} = 0.013$$

$$C_1 = \frac{(2.33)(0.2) + (0.5)1}{2.36} = 0.411$$

$$C_2 = \frac{2.33 - (2.33)(0.2) - (0.5)1}{2.36} = 0.576$$

Equation (20.13) is used to perform routing. Fig. 20.9 shows the inflow and outflow hydrographs.

TABLE 20.3

An inflow-outflow record for a river reach

Time (da)	*Inflow (m^3/s)*	*Outflow (m^3/s)*	*Time (da)*	*Inflow (m^3/s)*	*Outflow (m^3/s)*
1	180	160	16	1160	1220
2	270	200	17	1100	1190
3	420	280	18	1000	1150
4	650	415	19	950	1100
5	890	590	20	900	1040
6	1100	770	21	790	980
7	1270	950	22	710	920
8	1360	1090	23	650	860
9	1380	1180	24	590	790
10	1390	1250	25	510	710
11	1370	1280	26	450	650
12	1350	1290	27	380	590
13	1310	1300	28	300	510
14	1260	1280	29	230	460
15	1210	1250	30	180	410

TABLE 20.4

Computations for weighted discharge

Time (da)	*Inflow* (m^3/s)	*Outflow* (m^3/s)	*Storage* (m^3/s-*da*)	$xI + (1 - x)Q$					
				x = .0	x = .1	x = .2	x = .3	x = .4	x = .5
1	180	160	20	160	180	164	166	168	170
2	270	200	90	200	207	214	221	228	235
3	420	280	230	280	294	308	322	336	350
4	650	415	465	415	438	462	485	509	532
5	890	590	765	590	620	650	680	710	740
6	1100	770	1095	770	803	836	869	902	935
7	1270	950	1415	950	982	1014	1046	1078	1110
8	1360	1090	1685	1090	1117	1144	1171	1198	1225
9	1380	1180	1885	1180	1200	1220	1240	1260	1280
10	1390	1250	2025	1250	1264	1278	1292	1306	1320
11	1370	1280	2115	1280	1289	1298	1307	1316	1325
12	1350	1290	2175	1290	1296	1302	1308	1314	1320
13	1310	1300	2185	1300	1301	1302	1303	1304	1305
14	1260	1280	2165	1280	1278	1276	1274	1272	1270
15	1210	1250	2125	1250	1246	1242	1238	1234	1230
16	1160	1220	2065	1220	1214	1208	1202	1196	1190
17	1100	1190	1975	1190	1082	1172	1163	1154	1145
18	1000	1150	1825	1150	1135	1120	1105	1090	1075
19	950	1100	1675	1100	1085	1070	1055	1040	1025
20	900	1040	1535	1040	1026	1012	996	984	970
21	790	980	1345	980	961	942	923	904	885
22	710	920	1135	920	895	878	857	836	815
23	650	860	925	860	839	818	797	776	755
24	690	790	825	790	780	770	760	750	740
25	510	710	625	710	690	670	650	630	610
26	450	650	425	650	630	610	590	570	550
27	380	590	215	590	569	548	527	506	485
28	300	510	5	510	489	468	447	426	405
29	230	460	−225	460	437	414	391	368	345
30	180	410	−455	410	387	364	341	318	295

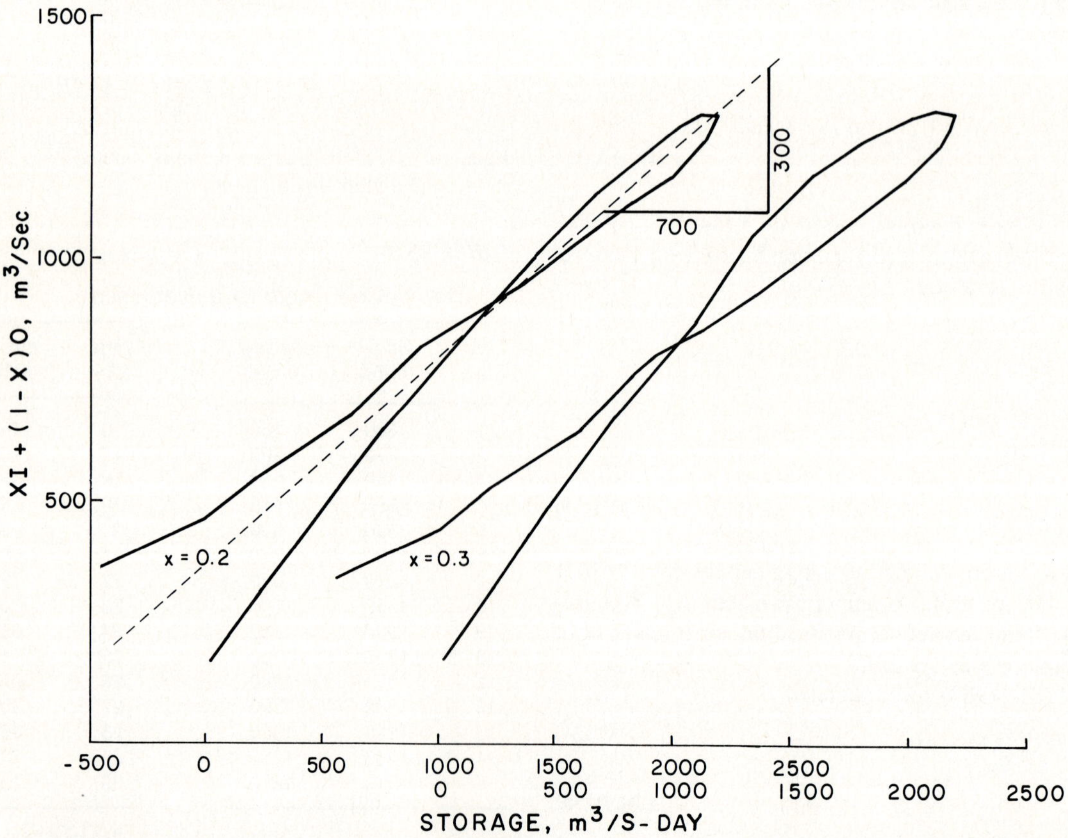

FIGURE 20.8 Example 20.7: Weighted discharge versus storage for determining the Muskingum parameters.

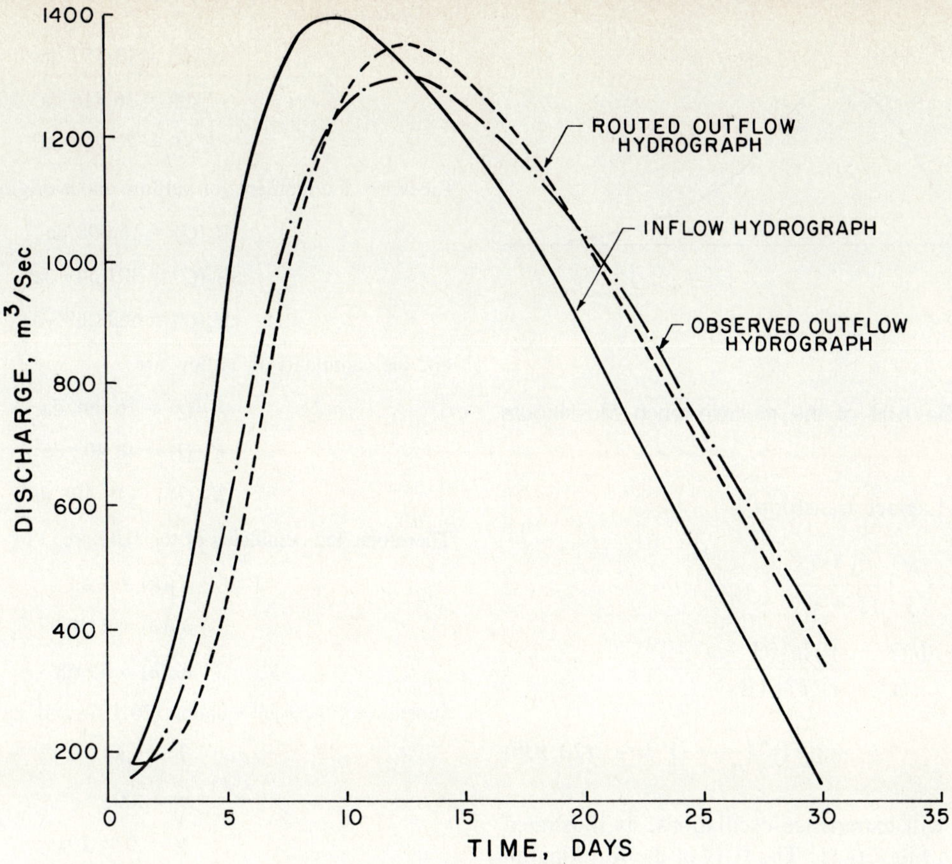

FIGURE 20.9 Example 20.8: Observed inflow and outflow hydrographs and outflow hydrograph obtained by the Muskingum method of routing.

20.3 MULTIPLE-REACH MUSKINGUM METHOD

In the preceding discussion, the Muskingum model was considered for a single channel reach. This model can be extended (Harley, 1967; O'Meara, 1968) to a cascade of N Muskingum reaches, as shown in Fig. 20.10 or to a network of reaches. We consider the case where all Muskingum reaches are equal. We can derive the IUH in this case in a standard manner. For jth reach,

$$\frac{dS_j}{dt} = I_j - Q_j, \qquad j = 1, 2, \ldots, N$$

$$S_j = k[xI_j + (1-x)Q_j]$$

For a single reach, with vanishing initial conditions, the IUH is given by equation (20.28), that is,

$$h(0, t) = \frac{1}{k(1-x)^2}\exp\left[-\frac{t}{k(1-x)}\right] - \frac{x}{1-x}\delta(t)$$

The Laplace transform of the IUH of a cascade of N equal reaches is equal to the Laplace transform of the IUH of a single reach raised to the power N. Therefore, taking Laplace transform of this IUH,

$$L[h(0, t)] = L\left[\frac{1}{k(1-x)^2}\exp\left[-\frac{t}{k(1-x)}\right] - \frac{x}{1-x}\delta(t)\right]$$

$$= \frac{1}{(1-x)}\left[\frac{1}{k(1-x)s+1}\right] - \frac{x}{1-x}$$

Thus

$$L[h_N(0, t)] = \left\{\frac{1}{(1-x)}\left[\frac{1}{k(1-x)s+1}\right] - \frac{x}{1-x}\right\}^N$$

Expanding it by the binomial series,

$$L[h_N(0, t)] = \sum_{j=0}^{N} (-1)^j \binom{N}{j} \frac{1}{(1-x)^j} \left[\frac{1}{k(1-x)s+1}\right]^j \left(\frac{x}{1-x}\right)^{N-j}$$

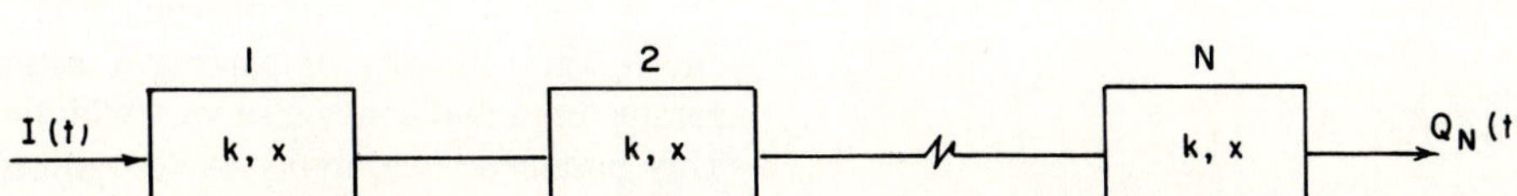

FIGURE 20.10 A cascade of N equal Muskingum reaches.

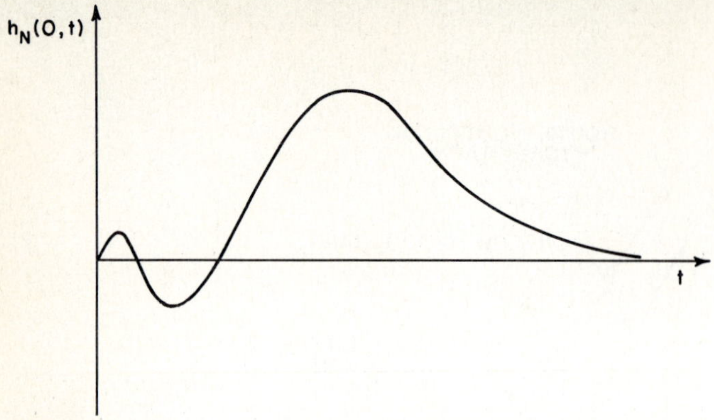

FIGURE 20.11 The IUH of the multiple-reach Muskingum method.

Taking the inverse Laplace transform,

$$h_N(0, t) = \sum_{j=0}^{N} (-1)^j \binom{N}{j} \frac{x^{N-j}}{(1-x)^N} \left(\frac{\exp\{-t/[k(1-x)]\}\{t/[k(1-x)]\}^{j-1}}{k(1-x)(j-1)!}\right) + (-1)^N \left(\frac{x}{1-x}\right)^N \delta(t) \quad (20.106)$$

Equation (20.106) will experience oscillations, as illustrated for a typical case in Fig. 20.11. The IUH of the Muskingum cascade can also be derived by using either the Muskingum storage IUH as shown for the case of a single reach or the matrix algebra as shown in Chapter 13. The same approach can be extended to the case where the Muskingum elements are unequal.

The parameters of the Muskingum cascade can be obtained using some of the methods discussed previously. If the method of cumulants is used, then it turns out that the cumulants for this case are simply N times those for the single reach,

$$K_1 = Nk \quad (20.107a)$$

$$K_2 = (1 - 2x)Nk^2 \quad (20.107b)$$

$$K_3 = (6x^2 - 6x + 2)Nk^3 \quad (20.107c)$$

$$K_4 = 6(1 - 4x + 6x^2 - 4x^3)Nk^4 \quad (20.107d)$$

Using equation (20.107), we can estimate the parameters x, k, and N.

EXAMPLE 20.9

Determine the parameters N, x, and k of the multiple-reach Muskingum method for the data in Table 20.5. Graph the IUH. Route the inflow hydrograph using these parameters.

Solution: For the data of Table 20.5, we obtain the moments of inflow about origin as

$$M_1(I) = 13.557 \text{ da}$$

$$M_2(I) = 220.208 \text{ da}^2$$

$$M_3(I) = 4050.445 \text{ da}^3$$

and cumulants as

$$K_1(I) = 13.557 \text{ da}$$

$$K_2(I) = 36.416 \text{ da}^2$$

$$K_3(I) = 77.719 \text{ da}^3$$

Likewise, the moments of outflow about origin are

$$M_1(Q) = 16.098 \text{ da}$$

$$M_2(Q) = 307.549 \text{ da}^2$$

$$M_3(Q) = 6628.695 \text{ da}^3$$

and the cumulants of outflow are

$$K_1(Q) = 16.098 \text{ da}$$

$$K_2(Q) = 48.402 \text{ da}^2$$

$$K_3(Q) = 119.406 \text{ da}^3$$

Therefore, the cumulants of the IUH are:

$$K_1(h) = 2.54$$

$$K_2(h) = 11.98$$

$$K_3(h) = 41.68$$

Substituting these in equation (20.107) and solving for N, x, and k,

$$k = 5.83$$

$$x = .095$$

$$N = .9 \cong 1.0$$

Thus the IUH in this case is the same as equation (20.28). It is graphed in Fig. 20.12. Using these parameter values, the inflow hydrograph is then routed. Fig. 20.13 shows observed and routed outflow hydrographs.

20.4 NONLINEAR MUSKINGUM METHOD

Two principal types of nonlinear forms of the Muskingum method have been proposed and used (Singh and Scarlatos, 1986): (1) hydrologic, and (2) hydraulic. We first discuss the nonlinear hydrologic forms.

20.4.1 Nonlinear Case 1

The storage-discharge relationship is nonlinear. It is linear only by assumption. If the degree of nonlinearity it pronounced, equation (20.12) is modified (Chow, 1959) to

$$S = k[x\, I^n + (1 - xQ^m)] \quad (20.108)$$

where m and n are exponents and k is a parameter. Chow, based on the Manning formula, suggested that $m = n = .6$ for prismatic channels and that m is higher for natural channels. This relationship has seldom been used for flood routing largely due to the difficulty of parameter estimation. In nature the parameters x and k may also vary with the flow characteristics. This parameter variability is recognized but there are very few, if any, methods developed to account for the variation of these parameters.

TABLE 20.5

An inflow-outflow record for a channel reach

Time *(da)*	*Inflow* (m^3/s)	*Outflow* (m^3/s)
1	0	0.00
2	90	36.83
3	240	110.48
4	470	234.78
5	710	395.90
6	920	561.62
7	1090	727.35
8	1180	856.24
9	1200	939.11
10	1210	1003.56
11	1190	1031.18
12	1170	1040.38
13	1130	1049.59
14	1080	1031.18
15	1030	1003.56
16	980	975.94
17	920	948.31
18	820	911.49
19	770	865.45
20	720	810.21
21	610	754.97
22	530	699.73
23	470	644.49
24	410	580.04
25	330	506.38
26	270	451.14
27	200	395.90
28	120	322.24
29	50	276.21
30	0	230.17
31	0	184.14
32		138.10
33		101.28
34		64.45
35		27.62
36		

Approximate integral solutions have been proposed by Singh (1983a). Coupling equations (20.12) and (20.108),

$$k\frac{d}{dt}[xI^n + (1-x)Q^m] = I - Q \tag{20.109}$$

Integrating and simplifying,

$$k(1-x)Q^m = -\int_0^t Q\,dt + \int_0^t I\,dt - kxI^n + S_0 \tag{20.110}$$

$$S_0 = k[xI_0^n + (1-x)Q_0^m] \tag{20.111}$$

The first and second terms on the right side of equation (20.110) denote the volumes of inflow and outflow at time t, respectively. Equation (20.110) is integral solution, in implicit form, of equation (20.109). If we choose a small Δt, equation (20.110) can be written as

$$k(1-x)Q^m = -\int_0^{t-\Delta t} Q\,dt - \int_{t-\Delta t}^{t} Q\,dt + \int_0^t I\,dt - kxI^n + S_0 \tag{20.112}$$

The second term on the right side can be simplified by assuming Q to vary linearly during Δt. This assumption may not be unduly restrictive for very small values of Δt. Thus this term can be replaced by $[Q(t) + Q(t - \Delta t)]\Delta t/2$. Equation (20.112) can be written as

$$k(1-x)Q^m + \frac{1}{2}Q\Delta t = G \tag{20.113}$$

where

$$G = -Q(t-\Delta t)\frac{\Delta t}{2} - \int_0^{t-\Delta t} Q\,dt + \int_0^t I\,dt - kxI^n + S_0$$

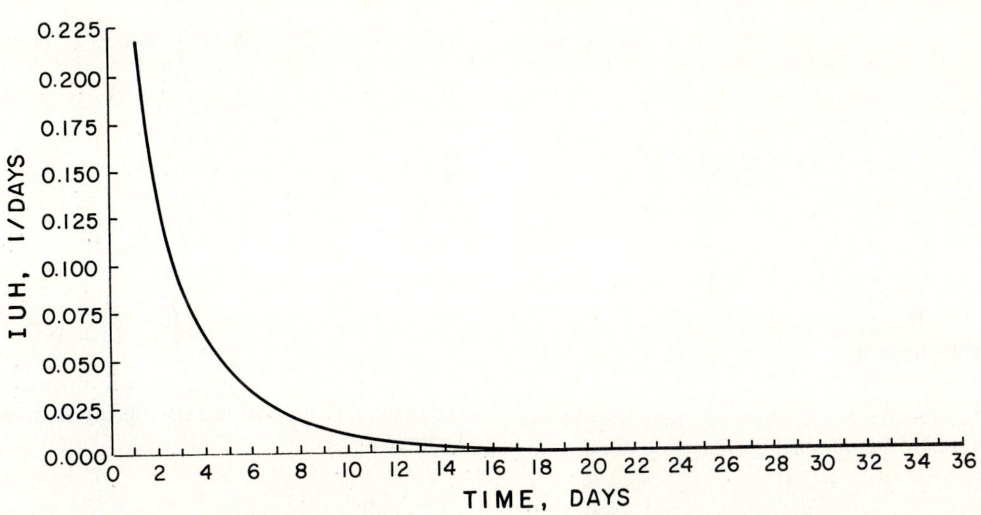

FIGURE 20.12 Example 20.9: The instantaneous UH.

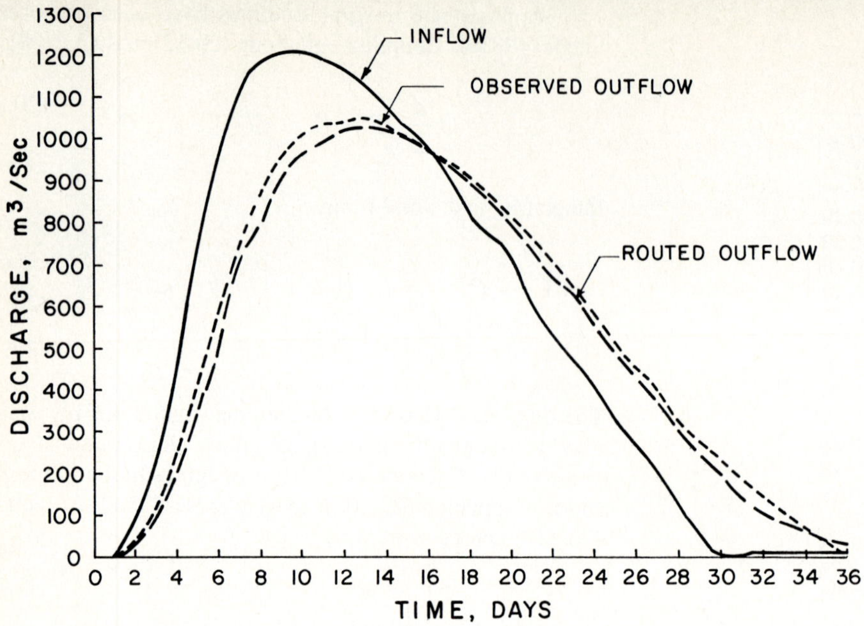

FIGURE 20.13 Example 20.9: Observed inflow and outflow hydrographs and outflow hydrograph obtained by routing by a single Muskingum reach model.

At time t, G is known provided the parameters k, m, n, and x are known. Therefore, Q can be obtained by solving equation (20.113) in a standard manner applicable to nonlinear equations.

If $m = 1$, then

$$Q = \frac{2G}{2k(1-x) + \Delta t} \tag{20.114}$$

If n also equals 1, then equation (20.114) becomes a solution of the linear Muskingum method, which can be carried out conveniently in a tabular form.

For purposes of computation, equation (20.113) can be expressed in a finite difference form.

$$.5k(1-x)[Q_j^m + Q_{j-1}^m] + .5Q_j\Delta t = -.5\Delta t\, Q_{j-1} - V_{j-1} + W_j - .5kx[I_j^n + I_{j-1}^n] + S_0, \quad j = 1, 2, 3, \ldots \tag{20.115}$$

where

$$V_{j-1} = \left(0.5Q_0 + \sum_{i=1}^{j-2} Q_i + 0.5\, Q_{j-1}\right)\Delta t$$

$$W_j = \left(0.5I_0 + \sum_{i=1}^{j-1} I_i + 0.5I_j\right)\Delta t$$

20.4.2 Nonlinear Case 2

Gill (1978) proposed an alternative nonlinear storage equation and showed that it would be satisfactory for some flood situation,

$$S = k[x\,I + (1-x)Q]^p + C \tag{20.116}$$

where p is an exponent. This form has been used by Tung (1985). Gill (1978) proposed a segmented curve method to estimate k, x, p, and C. This method is a trial-and-error procedure and accounts for variation of k with flow characteristics. The results obtained by Gill (1978) indicate that a nonlinear storage equation does not appear to have significant advantage to offset its complexity. This is clearly seen to be the case from the table and figure reported by Gill (1978).

Approximate integral solutions can also be developed following the above procedure. Coupling equation (20.166) with equation (20.12),

$$k\frac{d}{dt}[xI + (1-x)Q]^p = I - Q \tag{20.117}$$

Integrating and simplifying,

$$k[xI + (1-x)Q]^p = S_0 + \int_0^t I\,dt - \int_0^t Q\,dt$$

$$S_0 = k[xI_0 + (1-x)Q_0]^p \tag{20.118}$$

Choosing a small Δt and writing equation (20.118) as

$$k[xI + (1-x)Q]^p = S_0 + \int_0^t I\,dt - \int_0^{t-\Delta t} Q\,dt - \int_{t-\Delta t}^t Q\,dt \tag{20.119}$$

We assume Q to vary linearly during Δt in equation (20.119). Therefore,

$$k[xI + (1-x)Q]^p + \frac{1}{2}\Delta t Q = G \tag{20.120}$$

in which

$$G = S_0 + \int_0^t I\,dt - \int_0^{t-\Delta t} Q\,dt - \frac{\Delta t}{2} Q(t - \Delta t)$$

The right side of equation (20.120) is known at anytime t if the parameters k, p, and x are known. Thus, Q can be obtained by solving it. If $p = 1$,

$$Q = \frac{G - kxI}{k(1 - x) + 0.5\,\Delta t} \tag{20.121}$$

Equation (20.121) is a linear solution that can be calculated in tabular form.

For computing discharge, equation (20.120) can be expressed in finite different form as

$$0.5k[xI_j + (1 - x)Q_j] + 0.5k[xI_{j-1} + (1 - x)Q_{j-1}]^p = S_0 + W_j - V_{j-1} - 0.5\Delta t Q_{j-1}, \quad j = 1, 2, 3, \ldots \tag{20.122}$$

where W_j and V_{j-1} are as before.

20.4.3 Hydraulic Forms

The hydraulic forms of the nonlinear Muskingum method hypothesize that the parameters x and k vary in space and time according to flow variability (Ponce and Yevjevich, 1978; Koussis, 1978, 1980). The Muskingum method can be expressed, referring to Fig. 20.14 as

$$Q_{j+1}^{i+1} = C_1 Q_j^i + C_2 Q_j^{i+1} + C_3 Q_{j+1}^i \tag{20.123}$$

where C_1, C_2, and C_3 are as defined previously in equation (20.14). The parameters k and x are defined by equation (20.70), which when substituted into equation (20.14), gives

$$C_1 = \frac{1 + C - C}{1 + C + D} \tag{20.124}$$

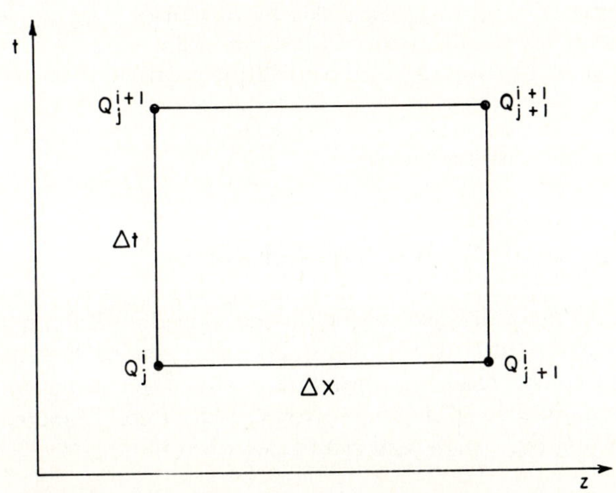

FIGURE 20.14 Space-time representation of the Muskingum method.

$$C_2 = \frac{-1 + C + D}{1 + C + D} \tag{20.125}$$

$$C_3 = \frac{1 - C + D}{1 + C + D}$$

where $C = c\Delta t/\Delta x$ is the Courant number, and $D = (q/S_0)c\Delta x$ is a type of cell Reynolds number, C is a ratio of celerities, and D is a ratio of diffusivities, and q is the unit width discharge.

Usually Δt is fixed, and Δx and S_0 are specified for each grid spacing. Therefore, for each grid spacing it is necessary to determine the flood wave celerity c and the unit width discharge q. These may be defined at each grid point (j, i) by

$$c = \left.\frac{dQ}{dA}\right|_{j,i} \tag{20.126}$$

$$q = \left.\frac{Q}{B}\right|_{j,i} \tag{20.127}$$

in which Q is discharge, A is flow area, and B is top width. Thus the parameters k and x vary from one grid spacing to another and depend on discharge. This makes the Muskingum method nonlinear.

20.5 LAG AND ROUTE MODEL

The lag and route model represents streamflow routing by a linear channel and a linear reservoir arranged in series as shown in Fig. 20.15. Mathematically, it is based on equation (20.11) and a storage-discharge relationship,

$$S(t) = kQ(t + \tau) \tag{20.128}$$

Equation (20.128) assumes that the storage at any time t is directly proportional to the outflow occurring after the lapse of lag time τ and k is a proportionality factor associated with a linear storage. Thus the lag and route model is a two-parameter model.

This is a linear model and is one of the simplest models. Therefore, it will suffice to determine its IUH, $h(0, t)$ or its T-hour UH, $h(T, t)$.

Substituting equation (20.128) into equation (20.11),

$$Q(t) + k\frac{d}{dt}Q(t + \tau) + I(t) \tag{20.129}$$

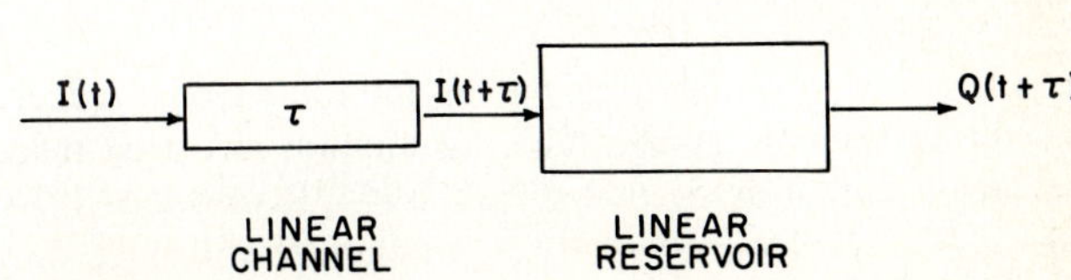

FIGURE 20.15 A block diagram representation of lag and route model.

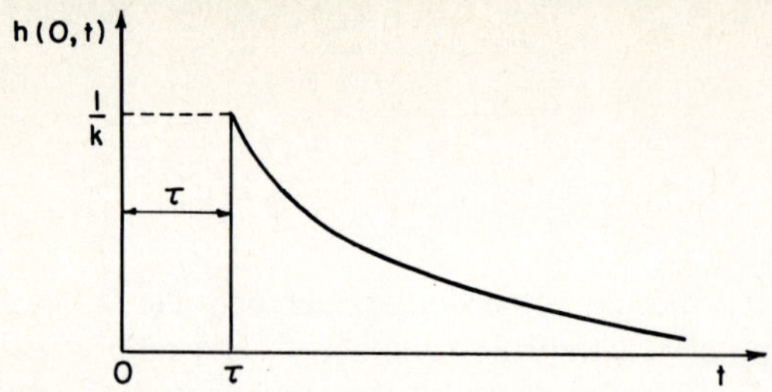

FIGURE 20.16 The IUH of the lag and route model.

Denoting $I(t)$ by $\delta(t)$ and solving equation (20.129),

$$h(0, t) = \begin{cases} 0, & t < \tau \quad (20.130a) \\ \frac{1}{k}\exp\left[-\left(\frac{t-\tau}{k}\right)\right], & t \geq \tau \quad (20.130b) \end{cases}$$

The IUH is shown in Fig. 20.16.

On the other hand, the UH of duration T will be

$$h(D, t) = \begin{cases} \frac{1}{T}\left[\exp\left(\frac{T}{k}\right) - 1\right]\exp\left[-\frac{(t-\tau)}{k}\right], & t \geq \tau \quad (20.131a) \\ 0, & t \leq \tau \quad (20.131b) \end{cases}$$

20.5.1 Parameter Estimation

The parameters τ and k can be estimated using standard techniques. If the method of cumulants is used, then the first four cumulants are

$$K_1 = \tau + k \quad (20.132a)$$
$$K_2 = k^2 \quad (20.132b)$$
$$K_3 = 2k^3 \quad (20.132c)$$
$$K_4 = 6k^4 \quad (20.132d)$$

and in general

$$K_j = (j-1)!k^j, \qquad k > 1 \quad (20.132e)$$

On the other hand, k and τ can also be estimated by equating the first two moments of equation (20.130) with those of the complete linear dynamic equation (Dooge, 1973):

$$k = \frac{x}{1.5u_0}\left[\frac{2}{3}\left(1 - \frac{F^2}{4}\right)\left(\frac{h_0}{S_0 x}\right)\right] \quad (20.133a)$$

$$\tau = \frac{x}{1.5u_0}\left[1 - \frac{2}{3}\left(1 - \frac{F^2}{4}\right)\left(\frac{h_0}{S_0 x}\right)\right] \quad (20.133b)$$

20.5.2 Single Reservoir

If τ is assumed to vanish, then the lag and route model reduces to a single linear reservoir. This assumption has been made in channel routing in the past (Clark, 1944; Rockwood, 1958; Sauer, 1973; Ponce, 1979). Thus, the model will have only one parameter k. This can be estimated using the method of moments or cumulants, as done previously. However, Rockwood (1958) estimated k from storage-elevation and discharge-elevation curves as

$$k = \frac{dS/dy}{dQ/dy}$$

where dS/dy represents the slope of the storage-elevation curve and dQ/dy represents the slope of the discharge-elevation curve at elevation y. In applying this model to the Columbia River, he found that

$$k = aQ^{-.2}$$

where a is the basin storage coefficient.

On the other hand, Sauer (1973) estimated k from recession of the outflow hydrograph as

$$k = \frac{\Delta t \log e}{\log(1/r)}$$

where r is the recession coefficient computed for the time interval Δt. The coefficient r is computed from the recession hydrograph beyond the point of inflection. Several values of discharge are computed along the recession of intervals at Δt, baseflow is subtracted from each, and r is computed for each pair of discharges by dividing the second by the first. An average of these individual values of r is taken as the best estimate of r.

EXAMPLE 20.10

Using the data of Table 20.5, determine the parameters of the lag and route model. Using these parameters, route the inflow hydrograph.

Solution: To determine the parameters k and τ, we need only the first two cumulants, K_1 and K_2, which are estimated in Example 20.9. From equation (20.132),

$$k = 3.46$$

$$\tau = .0$$

Actually, τ turns out to be negative from computations and is therefore taken to be zero. This implies that for these data, a single linear reservoir would suffice. Recalculating the value of k, we obtain k = 2.54 da. The IUH is a negative exponential with an initial value of 1/2.54. Its appearance will be similar to Fig. 20.12. Using these parameters values, the inflow hydrograph is routed, and the results are graphed in Fig. 20.17.

20.6 KALININ—MILYUKOV MODEL

The Kalinin-Milyukov model was developed by Kalinin and Milyukov (1957) and involves successive routing through a characteristic reach. Its operation is equivalent to that of a cascade of n equal linear reservoirs, each of delay time k, as shown in Fig. 20.18, and can be described by

$$(1 + kD)^n Q(t) = I(t), \quad D \equiv \frac{d}{dt} \quad (20.134)$$

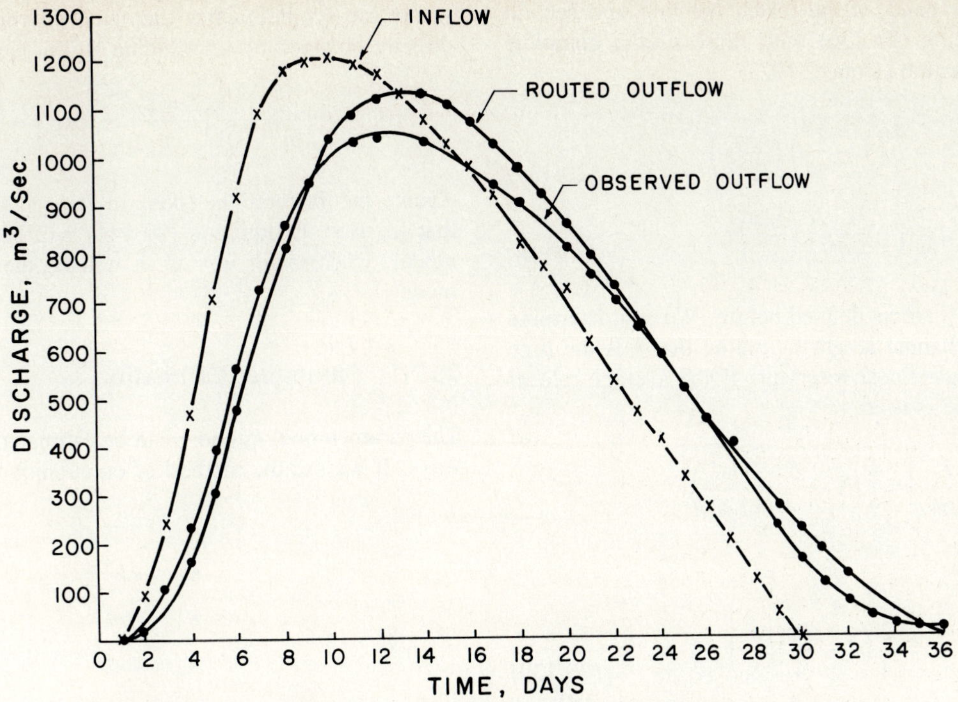

FIGURE 20.17 Example 20.10: Observed inflow and outflow hydrographs and outflow hydrograph obtained by lag and route model.

where D is an operator. Because equation (20.134) is linear, it may suffice to determine its IUH, which is

$$h(0, t) = \frac{1}{k(n-1)!}\left(\frac{t}{k}\right)^{n-1} \exp\left(-\frac{t}{k}\right) \tag{20.135}$$

which is the well-known gamma distribution used extensively in watershed hydrology (see Chapter 13). The factorial in equation (20.135) should be interpreted as a generalized factorial, including positive and negative integer or noninteger values. The $h(0, t)$ will have the typical appearance shown in Fig. 20.19. Using a population balance concept, Himmelblau and Yates (1968) derived a similar model.

20.6.1 Parameter Estimation

The parameters k and n can be estimated using standard techniques. If the method of cumulants is used, then the first four cumulants are

$$K_1 = nk$$
$$K_2 = nk^2$$
$$K_3 = 2nk^3$$
$$K_4 = 6nk^4 \tag{20.136}$$

FIGURE 20.18 A block-diagram representation of the Kalinin-Milyukov model.

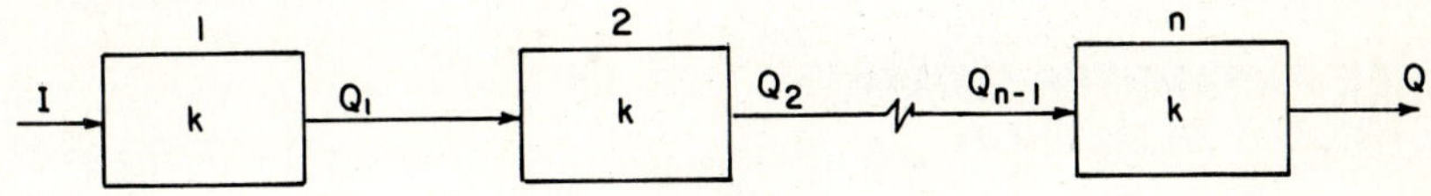

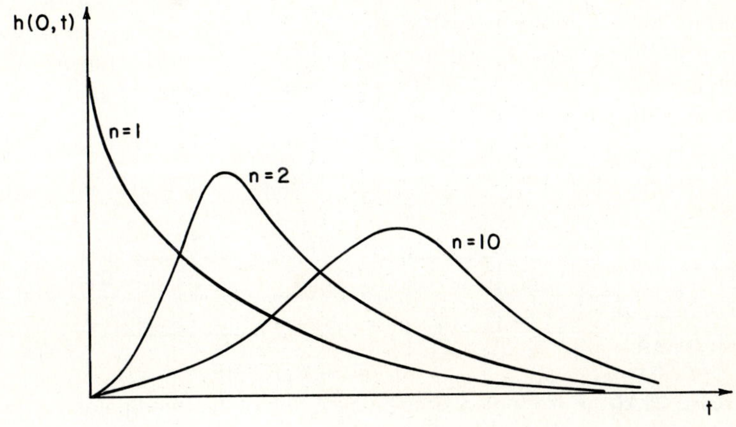

FIGURE 20.19 The IUH of the Kalinin-Milyukov model.

On the other hand, if we match the first and second moments of equation (20.136) with those of the complete linear dynamic equation (Dooge, 1973).

$$k = \frac{4}{9}\left(1 - \frac{F^2}{4}\right)\left(\frac{y_0}{S_0 V_0}\right) \tag{20.137}$$

$$n = \left(\frac{6}{4 - F^2}\right)\left(\frac{S_0 x}{y_0}\right) \tag{20.138}$$

where y_0, F, and V_0 are as defined before. We can determine the characteristic channel length by noting that k is the time parameter of a single linear reservoir. If the average rate of travel of flood wave is used,

$$\frac{L}{1.5V_0} = \frac{4}{9}\left(1 - \frac{F^2}{4}\right)\left(\frac{h_0}{S_0 V_0}\right)$$

Therefore,

$$L = \frac{2}{3}\left(1 - \frac{F^2}{4}\right)\left(\frac{h_0}{S_0}\right) \tag{20.139}$$

We consider an example to illustrate this method.

EXAMPLE 20.11

Determine the parameters of the Kalinin-Milyukov model for the data of Table 20.5. Graph its IUH. Using these parameter values, route the inflow hydrograph.

Solution: The cumulants of the IUH are already computed from the data in Example 20.9. Therefore, from equation (20.136),

$$k = 4.717$$

$$n = 0.539$$

For purposes of computation, n was taken to be 1. Then k is recommended with this value of n and found to be 2.54 da, as in the previous example. Consequently, the IUH and outflow hydrographs would be the same as before.

20.7 THREE-PARAMETER GAMMA DISTRIBUTION METHOD

The three-parameter gamma distribution method consists of a series combination of linear channel and a cascade of n equal linear reservoirs as shown in Fig. 20.20. Therefore, the governing differential equation is

$$(1 + kD)^n Q(t + \tau) = I(t) \tag{20.140}$$

where the symbols carry the same interpretation. Its IUH is directly obtained by introducing lag in equation (20.135),

$$h(0, t) = \frac{1}{k(n-1)!}\left(\frac{t-\tau}{k}\right)^{n-1} \exp\left[-\frac{(t-\tau)}{k}\right] \tag{20.141}$$

Again, the factorial is taken in its generalized form. Note that for $n = 1$, equation (20.141) will be the lag and route model, whereas for $\tau = 0$, it will be the Kalinin-Milyukov model.

20.7.1 Parameter Estimation

The parameters n, k, and τ can be estimated in many different ways. If we use the method of cumulants, then

$$K_1 = \tau + nk$$
$$K_2 = nk^2$$
$$K_3 = 2nk^3$$
$$K_4 = 6nk^4 \tag{20.142}$$

We consider an example to illustrate this method.

EXAMPLE 20.12

Using the data of Example 20.7, estimate parameters of the three-parameter gamma distribution method. Then graph its IUH.

Solution: Using the cumulants computed in Example 20.9 in equation (20.142),

$$k = 1.739$$

$$n = 3.964$$

$$\tau = -4.352$$

Since τ is negative, it loses its physical significance and is therefore considered to be zero. Then the three-parameter model reduces to the two-parameter model of the preceding example, for which calculations have already been reported.

20.8 VARIABLE TRAVEL TIME (VTT) METHOD

The variable travel time (VTT) method of flood routing, developed by Williams (1969), takes into account the change in travel time with the change in stage of flow. Its mathematical derivation is as follows.

The finite difference form of equation (20.11) can be rearranged as

$$\frac{I_t + I_{t+\Delta t}}{2} + \left[\frac{S_t}{\Delta t} - \frac{Q_t}{2}\right] = \left[\frac{S_{t+\Delta t}}{\Delta t} + \frac{Q_{t+\Delta t}}{2}\right] \tag{20.143}$$

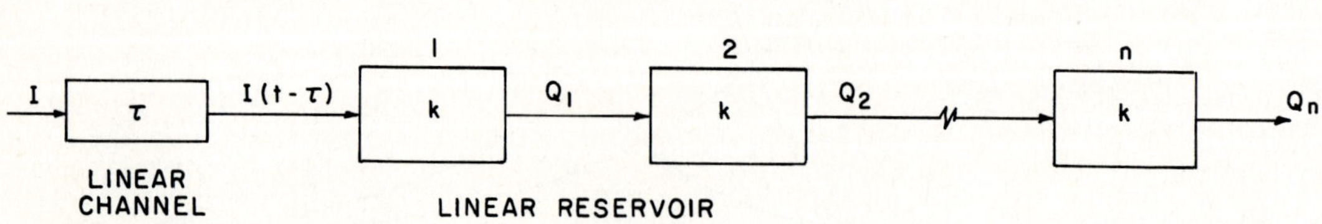

FIGURE 20.20 A block-diagram representation of the three-parameter gamma distribution model.

Consider an extremely short channel reach with small travel time and with approximately equal inflow and outflow rates. The travel time T for a given discharge can be obtained from

$$T = \frac{S}{Q} \tag{20.144}$$

Since the inflow and outflow rates are approximately equal, equation (20.144) can be written as

$$T = \frac{S_t}{Q_t} = \frac{S_{t+\Delta t}}{Q_{t+\Delta t}} \tag{20.145}$$

Substituting this in place of Δt into equation (20.143),

$$Q_{t+\Delta t} = \frac{2}{3}\left[\frac{I_t + I_{t+\Delta t}}{2}\right] + \frac{1}{3}Q_t \tag{20.146}$$

Equation (20.146) is similar to the coefficient method,

$$Q_{t+\Delta t} = C\left[\frac{I_t + I_{t+\Delta t}}{2}\right] + (1 - C)\,Q_t \tag{20.147}$$

in which C is storage coefficient. Equation (20.147) forms the basis of the convex method or the Muskingum method, neither of which allows the travel time to vary within the reach. It is thus seen that if $\Delta t = T$, the storage coefficient is $\frac{2}{3}$.

It can be shown, using equations (20.146) and (20.147), that

$$(1 - C)\,Q_t = \frac{CS_t}{T} \tag{20.148}$$

Then equation (20.146) can be written using equation (20.148) as

$$Q_{t+\Delta t} = \frac{2}{3}\left[\frac{I_t + I_{t+\Delta t}}{2} + \frac{S_t}{T}\right] \tag{20.149}$$

which can be written as

$$TQ_{t+\Delta t} = \frac{2}{3}\left[\frac{I_t + I_{t+\Delta t}}{2} \cdot T + S_t\right] \tag{20.150}$$

It is preferable to use equation (20.150) because this eliminates the problem of gain or loss of volume that is experienced in using equation (20.149). Equation (20.150) is the governing equation for the VTT method.

It should be noted that the fundamental assumption underlying equation (20.150) pertains to an extremely short reach with equal inflow and outflow rates. Hence, the travel time cannot be computed correctly. The error introduced by this assumption can, however, be minimized by a careful selection of the reach lengths. A reasonable guide is to keep $T < t_p/5$, where t_p is the time to peak inflow. Furthermore, if a hydrograph with a short time to peak must be routed through a reach with a long travel time, several routings through partial storage must be carried out using the relation.

$$T_t = \frac{T}{n}$$

where T_t is partial travel time and n the number of routings. Perhaps the best way to illustrate the VTT method is by considering an example.

EXAMPLE 20.13

A flood event, as a result of heavy rainfall storm of March 29, 1965, occurred on the Brushy Creek watershed near Riesel, Texas. For purposes of routing the valley section of channel reach is measured at four points (two end points and two in between). The end areas of valley sections for selected discharges are given in Table 20.6. The inflow hydrograph at the upstream, called valley section VS-3, is given in Table 20.7. Route the flood from the valley section 3 to the gaging station. This example is due to Williams (1966).

Solution: The travel time of the flood wave for the reach must be computed. The length L of the reach is reasured from the watershed map and then the travel time T is computed as

$$T = \frac{\bar{A}L}{Q}$$

in which $\bar{A}$ is average measured cross-sectional area and Q is the discharge. The computation is shown in Table 20.6. In Fig. 20.21 T is plotted as a function of Q. The procedure for routing is described as follows:

1. Compute the time increment Δt from Table 20.7 and compute the volume of inflow during Δt as $(I_t + I_{t+\Delta t}) \times \Delta t/2$.
2. Calculate inflow volume plus storage as $(I_t + I_{t+\Delta t}) \times 0.5 \times \Delta t + S_t$. Take storage at $t = 0$ (the beginning of flood) as zero.
3. Compute outflow volume ($\Delta t \times Q_{t+\Delta t}$) by multiplying inflow volume by the constant storage coefficient of $\frac{2}{3}$.
4. Determine the storage for the next time interval as $S_{t+\Delta t}$ = outflow volume at t − inflow volume at t.
5. At each step calculate the outflow by dividing the outflow volume by time increment of traveling. Table 20.8 shows the computations. The resulting outflow hydrograph is plotted in Fig. 20.22.

Because the channel reach length is relatively small and the channel is narrow, the flood peak is reduced by about 3 m^3/s and the outflow hydrograph is only slightly translated.

20.9 VARIABLE STORAGE COEFFICIENT (VSC) METHOD

The variable storage coefficient (VSC) method, developed by Williams (1969), considers variation in storage coefficient C. From the VTT method we see that the storage coefficient is $\frac{2}{3}$ for $\Delta t = T$. In actuality, C must deviate from $\frac{2}{3}$. Toward

TABLE 20.6

Valley cross sections and travel time computations

Q (m^3/s)	VS-3 (m^2)	VS-2 (m^2)	VS-1 (m^2)	VS-G (m^2)	Average Ā (m^2)	S_t (m^3)	Time of Travel T (s)	T (h)
0	0	0	0	0	0	0		
.85	2.79	2.79	1.86	2.79	2.56	2,578	3,033	.84
1.13	3.25	3.25	2.32	3.25	3.02	3,042	2,692	.75
1.42	3.72	3.72	2.79	3.72	3.49	3,515	2,476	.69
2.12	4.65	5.57	3.72	4.65	4.65	4,685	2,210	.61
2.83	6.04	9.29	6.04	6.50	6.97	7,023	2,482	.69
4.25	10.22	13.01	9.29	9.29	10.45	10,536	2,479	.69
5.66	15.79	17.65	13.94	11.15	14.63	14,750	2,606	.72
8.50	24.15	20.90	21.37	15.79	20.55	20,717	2,437	.68
11.33	30.19	25.08	26.94	20.44	25.66	25,868	2,283	.63
14.16	36.23	28.80	30.66	25.55	30.31	30,552	2,158	.60
16.93	40.88	32.52	35.30	29.73	34.61	34,884	2,053	.57
19.82	45.52	36.23	39.02	33.45	38.56	38,863	1,961	.55
22.65	49.24	39.65	43.66	37.16	42.43	42,767	1,888	.52
25.49	54.81	43.66	47.85	40.41	46.68	47,056	1,846	.51
28.32	58.53	47.38	52.49	45.52	50.98	51,388	1,815	.50
35.40	69.68	56.67	65.03	54.81	61.55	62,040	1,752	.49
42.48	77.57	65.03	77.11	65.03	71.19	71,754	1,689	.47
49.55	85.94	74.32	89.19	78.97	82.10	82,762	1,670	.46
56.63	96.62	82.68	100.34	92.90	93.13	93,880	1,658	.46
63.71	107.77	92.90	109.63	902.19	103.12	103,997	1,632	.45
70.79	120.77	103.12	119.84	113.81	114.38	115,300	1,629	.45
84.95	147.72	120.77	138.43	132.39	134.83	135,906	1,600	.44
98.11	178.84	137.03	153.29	149.57	154.68	155,920	1,573	.44
113.27	218.32	150.97	169.55	163.51	175.59	176,992	1,563	.43
127.43	264.77	164.90	183.48	177.44	197.65	199,229	1,563	.43
141.58	315.87	178.37	197.42	190.45	220.53	222,292	1,570	.43

TABLE 20.7

An inflow hydrograph

Time (h)	Inflow (m^3/s)	Time (h)	Inflow (m^3/s)
13.33	.00	18.62	38.23
14.03	28.32	19.08	31.43
14.54	56.63	19.59	25.20
15.00	71.64	20.10	19.26
15.45	76.74	20.65	14.16
15.90	75.32	21.25	9.63
16.35	70.79	21.91	5.66
16.80	65.13	22.64	3.11
17.25	58.90	23.30	2.55
17.70	52.39	23.93	1.93
18.16	46.72		

derivation of this method, we substitute equation (20.144) into equation (20.143) and assume various ratios of $\Delta t/T$,

$$\left[\frac{I_t + I_{t+\Delta t}}{2}\right] + \frac{S_t}{(\Delta t/T)\,(S_t/Q_t)} - \frac{Q_t}{2} = \frac{S_{t+\Delta t}}{(\Delta t/T)\,(S_{t+\Delta t}/Q_{t+\Delta t})} + \frac{Q_{t+\Delta t}}{2}$$

Simplifying and rearranging,

$$Q_{t+\Delta t} = \left(\frac{2\Delta t}{2T + \Delta t}\right)\left[\frac{I_t + I_{t+\Delta t}}{2}\right] + \left[1 - \frac{2\Delta t}{2T + \Delta t}\right] Q_t \tag{20.151}$$

Equation (20.151) is similar to the coefficient method given by equation (20.147). Since equation (20.148) still holds,

$$Q_{t+\Delta t} = \left(\frac{2\Delta t}{2T + \Delta t}\right)\left[\frac{I_t + I_{t+\Delta t}}{2} + \frac{S_t}{T}\right] \tag{20.152}$$

Therefore,

$$C = \frac{2\Delta t}{2T + \Delta t} \tag{20.153}$$

Equation (20.153) is shown in Fig. 20.23. Equation (20.152) is the governing equation for the VSC method. This method is more suitable than the VTT method. Only one Δt can be used for the entire watershed. Short-duration hydrographs can be routed through long reaches. Therefore, less judgment is required in selection of routing reaches.

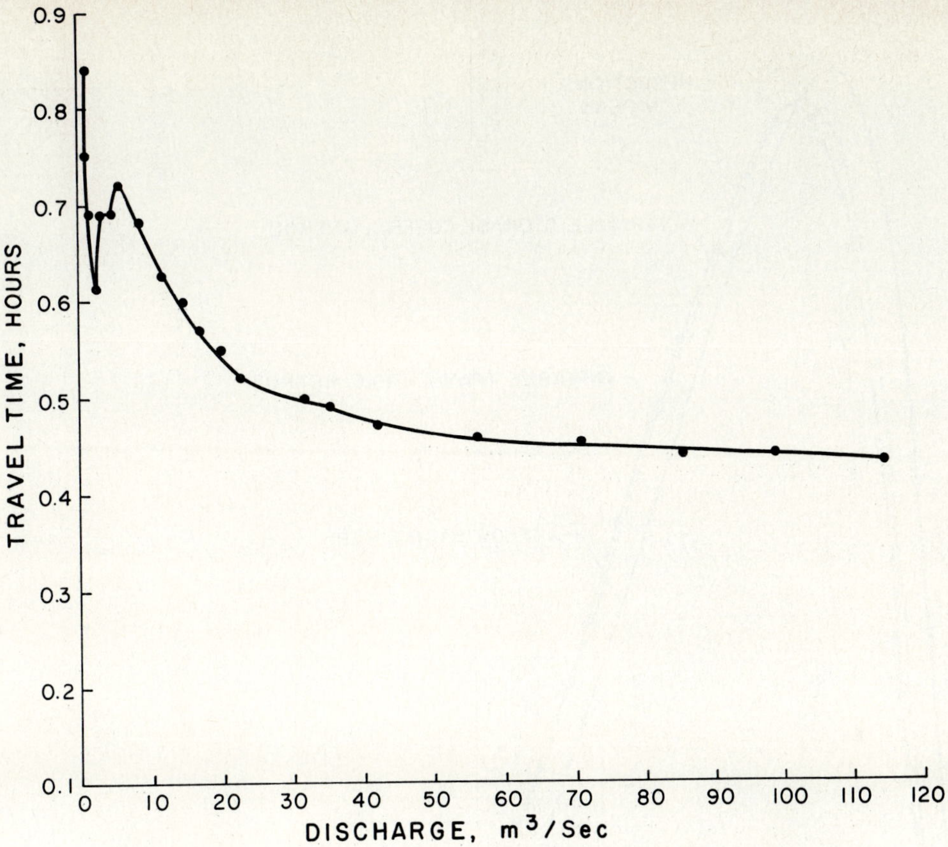

FIGURE 20.21 Example 20.13: Travel time as a function of discharge.

TABLE 20.8

Computations for variable travel time routing procedure

Time (*h*)	*Time Increment*	*Inflow* I (m^3/s)	*Inflow* $\bar{I} = (I_t + I_{t+\Delta t})/2$	*Inflow Volume* $\bar{I}T$ (m^3)	*Storage* S (m^3)	*Volume* $S_t + \bar{I}T$ (m^3)	*Outflow Volume* (m^3) $\frac{2}{3}(S_t + \bar{I}T)$	*Outflow* $Q = (TQ_{t+\Delta t})/2$
13.33	0	0	0	0	0	0	0	0
14.05	2,592	30	15.00	38,880	0	38,880	25,920	10
14.55	1,800	56.70	43.35	78,030	12,960	90,990	60,660	33.7
15.01	1,638	72.00	64.35	105,405	30,330	135,735	90,490	55.2
15.46	1,620	76.70	74.35	120,447	45,245	165,692	110,462	68.2
15.90	1,602	75.30	76.00	121,752	55,230	176,982	117,988	73.65
16.35	1,602	70.80	73.05	117,026	58,994	176,020	117.347	73.25
16.80	1,620	65.00	67.90	109,998	58,673	168,671	112,447	69.41
17.25	1,620	58.90	61.95	100,359	56,223	156,582	104,388	64.44
17.72	1,674	52.33	55.62	93,108	52,193	145,302	96,868	57.87
18.18	1,674	46.30	49.32	82,562	48,434	130,936	87,331	52.17
18.65	1,692	38.66	42.48	71,876	43,665	115,541	77,027	45.52
19.13	1,728	31.80	35.23	60,877	38,514	99,391	66,261	38.35
19.62	1,782	25.00	28.40	50,608	33,130	83,738	55,826	31.33
30.13	1,854	19.00	22.00	40,788	27,912	68,700	45,800	24.70
20.62	1,962	14.20	16.60	32,569	22,900	55,469	36,979	18.85
21.28	2,160	9.70	11.95	25,812	18,490	44,302	29,535	13.67
21.94	2,376	5.70	7.70	18,295	14,767	33,062	22,041	9.28
22.66	2,592	3.00	4.35	11,275	11,021	22,296	14,864	5.73
23.35	2,484	2.30	2.65	6,583	7,432	14,015	9,343	3.76
23.96	2,196	2.00	2.15	4,721	4,672	9,393	6,262	2.85

Volume check: $\Sigma\, \bar{I}T = 1{,}290{,}977\ m^3$

$S + \Sigma$ outflow volume $\cong 1{,}287{,}839 + 4{,}672 \cong 129 \times 10^4\ m^3$

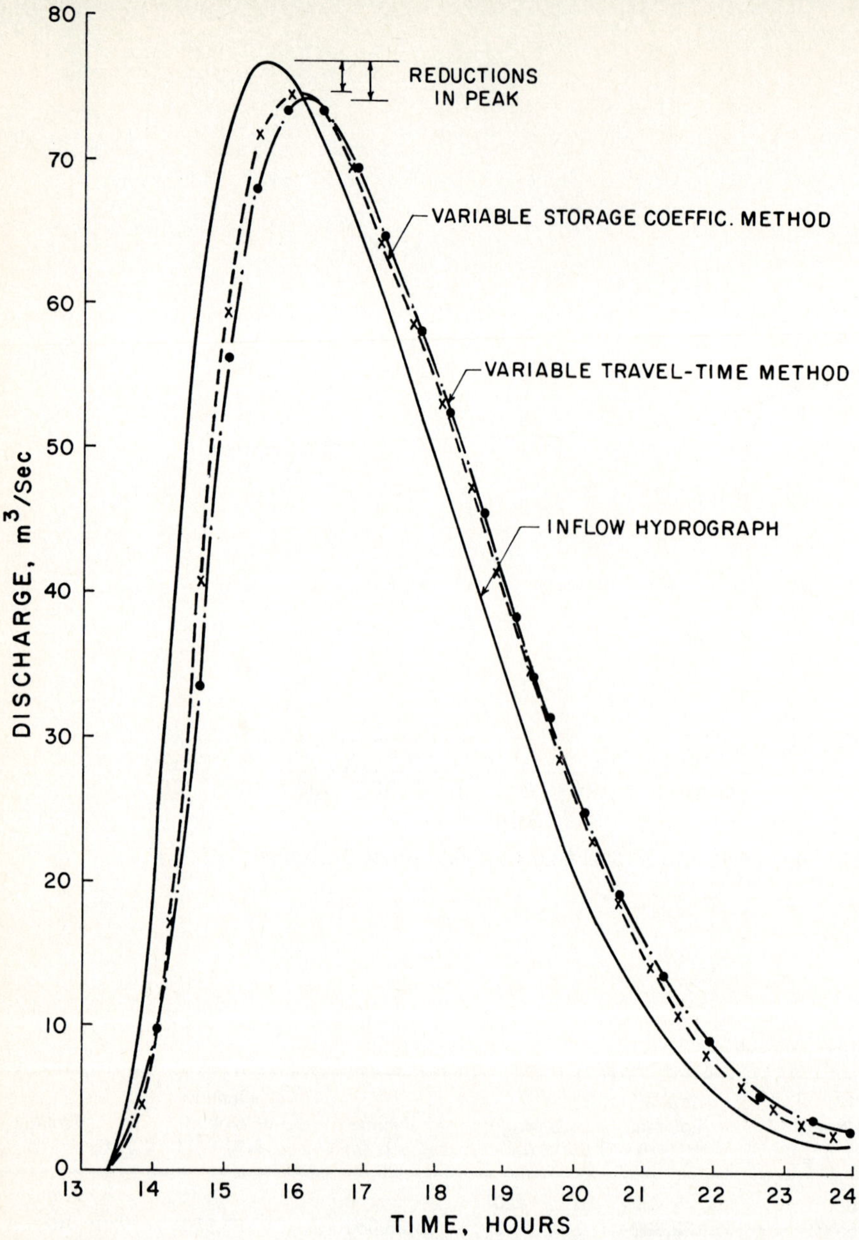

FIGURE 20.22 Example 20.13: Flood routing by VTT and VSC methods.

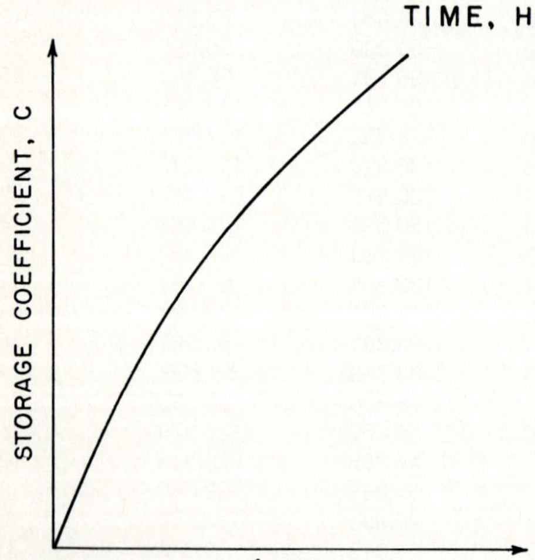

FIGURE 20.23 Relationship among storage coefficient, routing interval, and travel time.

However, the VSC method has a serious shortcoming, that is, the peak rate may increase as the flood travels downstream. This defect becomes more pronounced when routing through reaches with little storage, such as channels without floodplain flow. We illustrate this procedure with an example.

EXAMPLE 20.14

Perform flood routing by the VSC method for the data in Example 20.13.

***Solution*:** The routing by the VSC method is similar to that by the VTT method. The routing interval Δt is taken as an average value of .43 h from the previous method. The storage coefficient is computed from equation (20.153), as shown in Fig. 20.24 and Table 20.9, and routing is performed by equation (20.152). To start with, $S_t = 0$.

The quantity $(I_t + I_{t+\Delta t}).5 + (S_t/\Delta t)$ is considered as the sum of the average inflow and storage discharge. For this value, C is

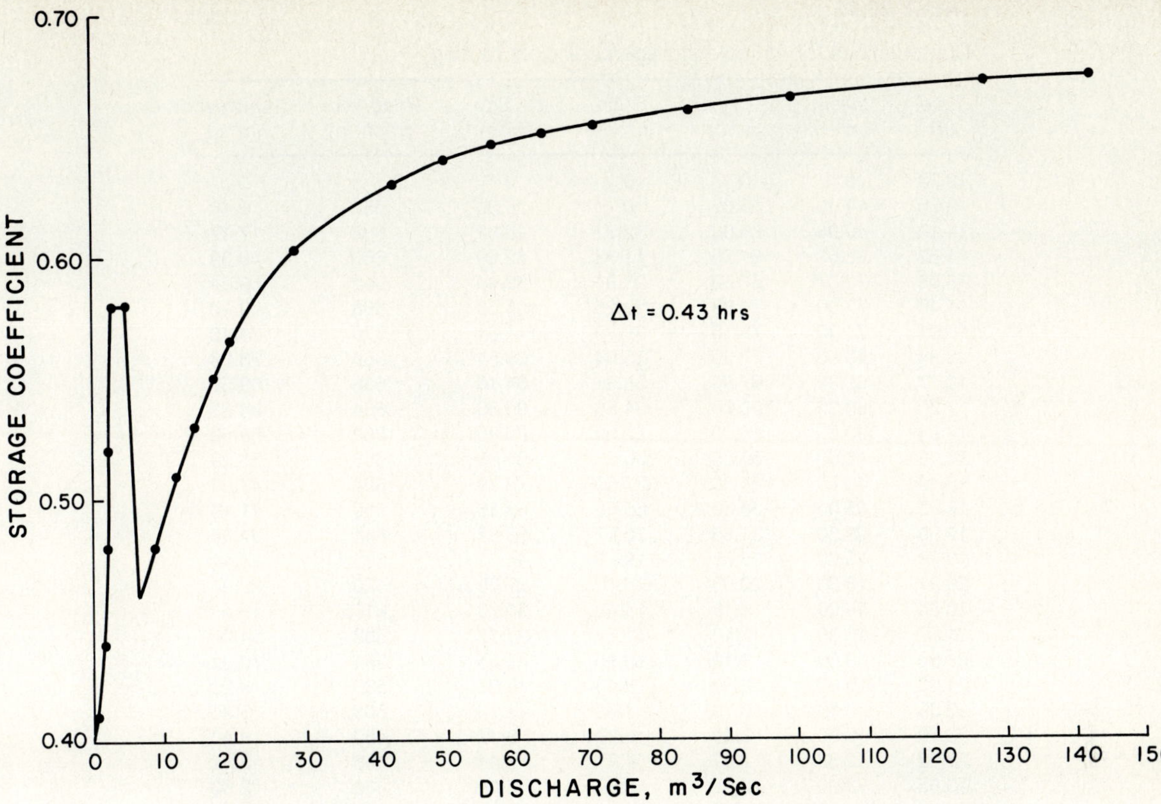

FIGURE 20.24 Example 20.14: Storage coefficient as a function of discharge.

TABLE 20.9

Computation of storage coefficient

T (*h*)	$c = \frac{2\Delta t}{2T + .43}$	Q (m^3/s)
.84	.41	.85
.75	.44	1.13
.69	.48	1.42
.61	.52	2.12
.69	.48	2.83
.69	.48	4.25
.72	.46	5.66
.68	.48	8.50
.63	.51	11.33
.60	.53	14.16
.57	.55	16.99
.55	.56	19.82
.52	.59	22.65
.51	.59	25.49
.50	.60	28.32
.49	.61	35.40
.47	.63	42.48
.46	.64	49.55
.46	.64	56.63
.45	.65	63.71
.45	.65	70.79
.44	.66	84.95
.44	.66	99.11
.43	.67	113.27
.43	.67	127.43

read from Fig. 20.24, and then outflow Q is computed. The storage discharge for the next time increment is computed as

$$\frac{S_{t+\Delta t}}{\Delta t} = \frac{I_t + I_{t+\Delta t}}{2} + \frac{S_t}{\Delta t} - Q_t$$

That is the remaining storage. Calculations are now repeated. Table 20.10 shows routing computations. At the end of calculations, a volume check is made by equating $\bar{I} + (S/\Delta t)$ for all Δt to Q for all $\Delta t + (S/\Delta t)$ for the last time increment. The outflow hydrograph is plotted in Fig. 20.22. Clearly, both the VIT and VSC methods yield comparable results.

20.10 THE HYMO METHOD

The HYMO method, developed by Williams and Hann (1972) and Williams (1975), is a storage routing method but includes the effect of a changing water surface slope using the VTT method. Its mathematical derivation is similar to the VTT method. for a specified reach, the travel time T can be obtained from

$$T = \frac{L}{u}$$

where L is the reach length and u is the average velocity within the reach. u can be computed by dividing the average

TABLE 20.10

Computations for variable storage coefficient routing

Time (h)	Inflow (m^3/s)	$\bar{I}$ (m^3/s)	$S/\Delta t$ (m^3/s)	$\bar{I} + S/\Delta t$ (m^3/s)	Read from graph (c)	Outflow (m^3/s)
13.33	0	0	0	0	0	0
13.76	12.00	6.00	0	6.00	.58	3.48
14.19	40.00	26.00	2.52	28.52	.602	17.17
14.62	62.50	51.25	11.35	62.60	.650	40.69
15.05	72.67	67.58	21.91	89.49	.662	59.24
15.48	77.00	74.83	30.25	105.08	.668	70.19
15.91	75.30	76.15	34.89	111.04	.670	74.40
16.34	70.50	72.90	36.64	109.54	.668	73.28
16.77	65.30	67.90	36.26	104.16	.668	69.57
17.20	59.30	62.30	34.58	96.88	.664	64.33
17.63	53.30	56.30	32.55	88.85	.662	58.82
18.06	48.00	50.65	30.03	80.68	.658	53.09
18.49	41.60	44.80	27.59	72.39	.654	47.34
18.92	35.00	38.30	25.05	63.35	.650	41.18
19.35	28.30	31.65	22.17	53.82	.644	34.66
19.78	23.00	25.65	119.16	44.81	.634	28.41
20.21	18.30	20.65	16.40	37.05	.620	22.97
20.64	14.00	16.15	14.08	30.23	.6105	18.45
21.07	10.80	12.40	11.77	24.17	.588	14.22
21.50	8.00	9.40	9.95	19.35	.564	10.92
21.93	5.60	6.80	8.43	15.23	.538	8.20
22.36	3.80	4.70	7.03	11.73	.508	5.96
22.79	3.00	3.40	5.77	9.17	.480	4.40
23.22	2.33	2.67	4.77	7.44	.466	3.47
23.65	2.00	2.17	3.97	6.14	.460	2.83

flow by the average end area. Since storage is the product of the average end area and the reach length, T can be expressed as

$$T = \frac{2S}{I + Q} \tag{20.154}$$

If we express Δt as $T_{t+\Delta t} - T_t$, then equation (20.143) can be used to compute it, that is,

$$T_t = \frac{2S_t}{I_t + Q_t}$$

and

$$T_{t+\Delta t} = \frac{2S_{t+\Delta t}}{I_{t+\Delta t} + Q_{t+\Delta t}}$$

Rearranging these expressions for S_t and $S_{t+\Delta t}$ and then substituting into equation (20.143),

$$Q_{t+\Delta t} = \frac{I_t(\Delta t + T_t) - Q_t(\Delta t - T_t) + I_{t+\Delta t}(\Delta t - T_{t+\Delta t})}{\Delta t - T_{t+\Delta t}} \tag{20.155}$$

This is the VTT routing equation. All the terms on the right side are known except $T_{t+\Delta t}$, which must be computed. To this end we compute u as

$$u = u_n \left(\frac{S_w}{S_c}\right)^{.5} \tag{20.156}$$

where u_n is the normal velocity, S_w is the water surface slope, and S_c is the average channel slope. Also, S_w can be computed as

$$S_w = S_c + \frac{y_I - y_0}{L} \tag{20.157}$$

where y_I and y_0 are the inflow and outflow depths, respectively. Substituting equation (20.157) into equation (20.156),

$$T = \frac{L}{u_n[1 + (y_I - y_0)/(S_c L)]^{.5}} \tag{20.158}$$

Equation (20.158) can be used to compute travel time by taking into account the slope variation. However, to ensure continuity, equation (20.158) must be expressed in terms of storage and flow terms:

$$T = \frac{2S}{(I_n + Q_n)[1 + (y_I - y_0)/(S_c L)]^{.5}} \tag{20.159}$$

where I_n and Q_n are the normal flow rates at depths y_I and y_0. The values of $T_{t+\Delta t}$ can be obtained from equation (20.159) and substituted into equation (20.155). Note that T and y_0 are interrelated. Therefore, computation of T is not straightforward. However, an iterative solution is simple to get. To achieve this, a first approximation can be obtained by assuming that $T_{t+\Delta t} = T_t$ and solving equation (20.155). Next, we determine $T_{t+\Delta t}$ from equation (20.159). Here we use the first approximation of $Q_{t+\Delta t}$ to determine y_0 and Q_n,

$$Q_n = \frac{Q_{t+\Delta t}}{[1 + (y_I - y_0)/(LS_c)]^{.5}} \qquad (20.160)$$

where y_I and y_0 are the normal flow depths for $I_{t+\Delta t}$ and $Q_{t+\Delta t}$, respectively. With this value of Q_n, y_0 can be computed. If desired, y_I and y_0 can be used again in equation (20.160) to obtain a more refined estimate of Q_n, and this process can be repeated.

The values of y_0 and Q_n can now be used in equation (20.159) to obtain a better estimate of $T_{t+\Delta t}$. If this value of Q does not agree with its previous estimate, the iterative procedure is repeated until desired accuracy is obtained. The convergence is expected to be rapid.

EXAMPLE 20.15

Consider a rectangular channel 10 m wide at the bottom with n = .03 and bed slope of .001. The length is 10,000 m. Route the flood of Example 20.12 by the HYMO method.

Solution: The time of computation is chosen as Δt = .5. Consequently, the inflow is interpolated at the spacing of .5 h, as given in Table 20.11. The computations are carried out as follows:

At $t = 0$, $I_0 = 0$, $Q_0 = 0$, and $T_0 = 0$. $I_{0.5} = 18$ m³/s.

Therefore, from equation (20.155),

$$Q_{.5} = \frac{18(.5 - 0)}{.5 + 0} = 18 \text{ m}^3/\text{s}$$

From Manning's equation, for Q = 18 m³/s,

$$18 = \frac{1}{.03} \frac{(10y)^{5/3}}{(10 + 2y)^{5/3}} (.001)^{1/2}$$

This yields y = 1.536 m. Using equation (20.159),

$$T_{.5} = \frac{27.7778(1.536 + 1.536)}{(18 + 18)[1 + (1.536 - 1.536)/10]^{.5}}$$

Recomputing $Q_{.5}$ from equation (20.155),

$$Q_{.5} = \frac{18(.5 - 2.3673)}{.5 + 2.3673} = -11.72 \text{ m}^3/\text{s}$$

Since this is less than 0, $Q_{.5} = 0$. Therefore, $y_{.5} = 0$. $I_{.5} = 18$ m³/s and $yI_{.5} = 1.536$ m. Hence, from equation (20.160),

$$I_n = \frac{18}{[1 + (1.534 - 0)/10]^{.5}} = 16.76 \text{ m}^3/\text{s}$$

For $I_n = 16.76$ m³/s, $y_I = 1.464$ m and $I_n = 18/[1 + (1.464/10)]^{.5}$ = 16.81 m³/s. Recomputing,

$$T_{.5} = \frac{27.7778(1.464 + 0)}{(16.81)(1.464)^{.5}} = 2.259 \text{ h}$$

$$Q_{.5} = \frac{18(.5 - 2.259)}{.5 + 2.259} = -11.48 \text{ m}^3/\text{s}$$

TABLE 20.11

An inflow hydrograph

Time (h)	Inflow (m³/s)	Time (h)	Inflow (m³/s)
0	.0	6.0	28.0
.5	18.0	6.5	22.0
1.0	47.0	7.0	17.0
1.5	67.0	7.5	12.5
2.0	76.5	8.0	9.0
2.5	76.0	8.5	6.0
3.0	71.0	9.0	4.0
3.5	65.0	9.5	3.0
4.0	61.5	10.0	2.5
4.5	51.0	10.5	2.0
5.0	44.0	11.0	1.5
5.5	35.0		

Thus, $Q_{.5} = 0$.

At t = .5 h, $I_{.5}$ = 18 m³/s, $Q_{.5}$ = 0, $T_{.5}$ = 2.259 h, and I_1 = 47 m³/s. From equation (20.155),

$$Q_1 = \frac{18(.5 + 2.259) + 47(.5 - 2.259)}{.5 + 2.259} = 11.96$$

$$= .0$$

For I_1 = 47 m³/s, yI_1 = 2.952 m, and

$$I_n = \frac{47}{[1 + (2.952 - 0)/10]^{.5}} = 41.30 \text{ m}^3/\text{s}$$

For I_n = 41.30 m³s, yI_1 = 2.696 and $I_n = 47/[1.2696]^{.5}$ = 41.71 m³/s. Then

$$T_1 = \frac{27.778(2.696 + 0)}{(41.71)[1 + (2.696 - 0)/10]^{.5}} = 1.593 \text{ h}$$

$$Q_1 = \frac{18(.5 + 2.259) + 47(.5 - 1.593)}{.5 + 1.593} = -.817 \text{ m}^3/\text{s}$$

$$= .0$$

At t = 1.0 hour, I_1 = 47 m³/s, Q_1 = 0.9, T_1 = 1.593 h, and $I_{1.5}$ = 67 m³/s. Therefore,

$$Q_{1.5} = \frac{47(.5 + 1.593) + 67(.5 - 1.593)}{.5 + 1.593} = 12.01 \text{ m}^3/\text{s}$$

For $Q_{1.5}$ = 12.01 m³/s, $y_{1.5}$ = 1.177 m, $I_{1.5}$ = 67 m³/s, $y_{1.5}$ = 3.803. Hence, $Q_n = 12/[1 + ((3.803 - 1.177)/10)]^{.5}$ = 10.68 m³/s. $I_n = 67/[1 + ((3.803 - 1.177)/10)]^{.5}$ = 59.63 m³/s. For Q_n = 10.68, y = 1.091 m and for I_n = 59.63, y = 3.496 m. These yield Q_n = 10.77 m³/s and I_n = 60.15 m³/s. Then

$$T_{1.5} = \frac{27.778(1.091 - 3.496)}{(10.77 + 60.15)[1 + (3.496 - 1.091/10]^{.5}} = 1.613 \text{ h}$$

$$Q_{1.5} = \frac{47(.5 + 1.593) + 67(.5 - 1.613)}{0.5 + 1.613} = 11.26 \text{ m}^3\text{s}$$

Repeating the procedure, $T_{1.5}$ = 1.623 h and $Q_{1.5}$ = 10.89 m³/s. Again, $T_{1.5}$ = 1.622 h and $Q_{1.5}$ = 10.93 m³/s.

The procedure is repeated to compute the entire outflow hydrograph given in Table 20.12.

TABLE 20.12

Routing computation by the HYMO method

t (h)	t + Δt (h)	T_t (m^3/s)	T_t (h)	Q_t (m^3/s)	$T_{t+\Delta t}$ (m^3/s)	$T_{t+\Delta t}$ (h)	$Q_{t+\Delta t}$ (m^3/s)
.0	.5	0	0	0	18	2.26	0
.5	1.0	18	2.26	0	47	1.59	0
1.0	1.5	47	1.59	0	67	1.62	10.93
1.5	2.0	67	1.62	10.93	76.5	1.60	35.65
2.0	2.5	76.5	1.60	35.65	76	1.54	56.91
2.5	3.0	76	1.54	56.91	71	1.56	67.87
3.0	3.5	71	1.56	67.87	65	1.58	70.64
3.5	4.0	65	1.58	70.64	57.5	1.63	69.23
4.0	4.5	57.5	1.63	69.23	51	1.67	64.60
4.5	5.0	51	1.67	64.60	44	1.73	59.36
5.0	5.5	44	1.73	59.36	35	1.82	53.95
5.5	6.0	35	1.82	53.95	28	1.92	46.60
6.0	6.5	28	1.92	46.60	22	2.03	39.51
6.5	6.5	22	2.03	39.51	17	2.16	33.00
7.0	7.5	17	2.16	33.00	12.5	2.32	27.42
7.5	8.0	12.5	2.32	27.42	9	2.50	22.36
8.0	8.5	9	2.50	22.36	6	2.71	18.22
8.5	9.0	6	2.71	18.22	4	2.95	14.39
9.0	9.5	4	2.95	14.39	3	3.25	10.90
9.5	10.0	3	3.25	10.90	2.5	3.57	8.24
10.0	10.5	2.5	3.57	8.24	2	3.87	6.57
10.5	11.0	2.0	3.87	6.57	1.5	4.16	5.45

20.11 STORAGE ROUTING METHOD A

Storage routing method A, developed by Rice and Larson (1972, 1974), assumes that the flood wave movement can be adequately represented by instantaneous nonuniform steady flow. It expresses equation (20.1) in finite difference form as

$$Q_{t+\Delta t} = I_t + I_{t+\Delta t} - Q_t + \frac{2}{\Delta t}(S_t - S_{t+\Delta t}) \qquad (20.161)$$

and employs a relation between S and Q, following the notation in Fig. 20.25, in dimensionless form as

$$\frac{S}{S_n} = f\left(\frac{Q}{I}\right) \qquad (20.162)$$

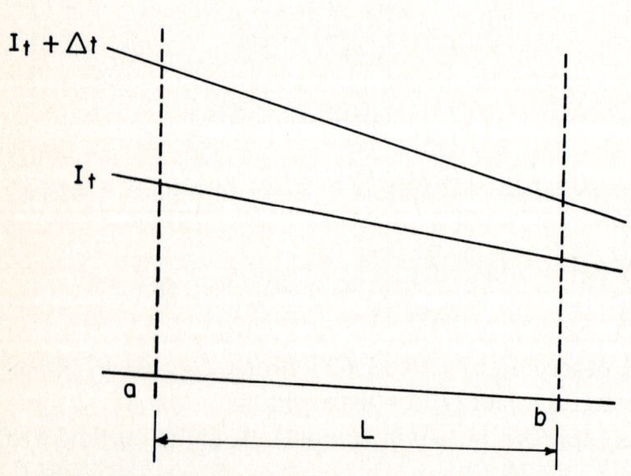

FIGURE 20.25 Channel schematic for storage routing methods A and B.

where S, the storage volume, is computed by the average end areas and S_n is the storage corresponding to a uniform flow equal to I. We can write, from the definition of S,

$$\frac{S}{S_n} = \frac{(A_a + A_b)L/2}{A_a L} = \frac{1}{2}\left(1 + \frac{A_b}{A_a}\right)$$

Therefore, equation (20.162) has two features. First, S/S_n is independent of L. Second, it appears to be independent of channel roughness and friction slope but may be indirectly dependent. Therefore, S/S_n depends theoretically on channel depth and geometry. Its likely appearance is shown in Fig. 20.26. To take advantage of this relation, for Q/I less than 1, we can write

$$\frac{S}{S_n} = a + b\frac{Q}{I} \qquad (20.163)$$

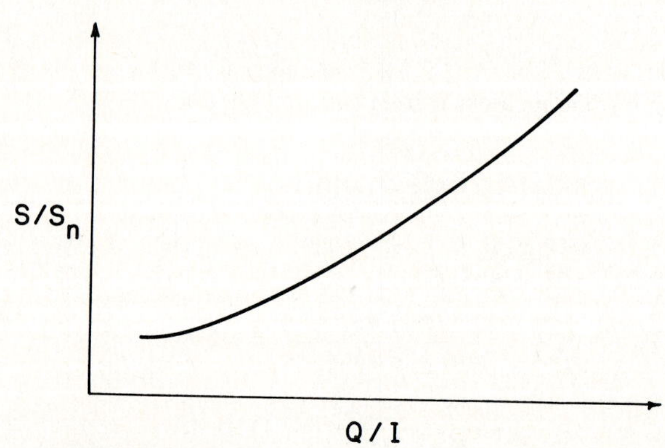

FIGURE 20.26 Storage-outflow relation for nonuniform steady flow.

TABLE 20.13

Routing computations for storage routing method A

	Discharge		*Areas (m^2)*		*Storage (m^3)*					
Time (min)	I (m^3/s)	Q (m^3/s)	A_a	A_b	$S_t = (A_a + A_b)L/2$	$S_n = (A_a)_{t+\Delta t}L$	P_1	Q_1	R_1	$\frac{P_1 + R_1}{Q_1} = Q$
0	0	0	0	0	0	3,512	.3	10.463	− 3.22	− .28 → 0
20	.3	0	.878	0	1,756	27,240	7.3	4.146	−22.04	−3.56 → 0
40	7.0	0	6.81	0	13,620	35,040	17	3.832	− 9.42	1.98
60	10	1.98	8.76	2.92	23,360	24,480	14.02	4.298	16.49	7.10
80	6	7.10	6.12	6.87	25,980	11,760	.9	5.753	32.52	5.81
100	2	5.81	2.94	5.98	17,840	7,520	−2.81	7.079	22.84	2.83
120	1	2.83	1.88	3.69	11,140	3,512	−1.53	10.463	15.35	1.32
140	.3	1.32	.878	2.25	6,256	2,320	− .87	13.502	8.30	.55
160	.15	.55	.58	1.30	3,760	1,200	− .35	20.400	5.16	.24
180	.05	.236	.30	.77	2,140	0	− .186	∞	3.56	0
200	.00	0	0	0	0	0				

Note: L = reach length = 4000 m

$$P_1 = (I_t + I_{t+\Delta t}) - Q_t$$

$$Q_1 = 1 + 3.2333 \frac{A_{a_{t+\Delta t}}}{I_{t+\Delta t}}$$

$$R_1 = (S_t - 0.55\, S_n) \frac{1}{600}$$

This corresponds to the case of rising hydrograph. For data used by Rice and Larson (1972), a was .55 and b was .485. Equation (20.163) can be used in equation (20.161) by expressing

$$\frac{S_{t+\Delta t}}{(S_n)_{t+\Delta t}} = a + b \frac{Q_{t+\Delta t}}{I_{t+\Delta t}}$$

For this procedure, $Q_{t+\Delta t}$ may be less than the baseflow for some time in the beginning. In that case, it can arbitrarily set equal to the baseflow.

EXAMPLE 20.16

A channel has the following characteristics: n = .04, L = 4000 m, and slope = .02. Route the following flood through the storage routing method A.

Time (min)	*Inflow (m^3/s)*	*Time (min)*	*Inflow (m^3/s)*	*Time (min)*	*Inflow (m^3/s)*
0	0	80	6	160	.15
20	.3	100	2	180	.05
40	7.0	120	1	200	.00
60	10	140	.3		

Solution: Substituting equation (20.163) into equation (20.161),

$$Q_{t+\Delta t}\left[1 + \frac{2b}{\Delta t}\frac{(S_n)_{t+\Delta t}}{I_{t+\Delta t}}\right] = I_t + I_{t+\Delta t} - Q_t + \frac{2}{\Delta t}[S_t - a(S_n)_{t+\Delta t}]$$

The parameter a is taken as .55 and b is .485. For flow depths at the spacing of .005 m, pertinent geometric and hydraulic characteristics, including A, P, R, and Q, are computed. These data are then used to compute end areas A_a and A_b. It is assumed that equation (20.163) is also valid for the recession hydrograph. Table 20.13 shows the computations. The notations used there are:

$$Q_1 = 1 + \frac{2b}{\Delta t}\frac{(S_n)_{t+\Delta t}}{I_{t+\Delta t}}$$

$$P_1 = I_t + I_{t+\Delta t} - Q_t$$

$$R_1 = \frac{2}{\Delta t}[S_t - a(S_n)_{t+\Delta t}]$$

This yields

$$Q_{t+\Delta t} = \frac{P_1 + R_1}{Q_1}$$

20.12 STORAGE ROUTING METHOD B

Storage routing method B was also developed by Rice and Larson (1972, 1974). It employs data on inflow, channel slope, roughness, and geometry for flood routing by equation (20.161). Combining known quantities,

$$G = \left[(I_t + I_{t+\Delta t} - Q_t)\frac{\Delta t}{2} + S_t\right] \quad (20.164)$$

and so equation (20.161) is

$$\frac{\Delta t}{2} Q_{t+\Delta t} + S_{t+\Delta t} - G = 0 \quad (20.165)$$

Equation (20.165) contains two unknowns $Q_{t+\Delta t}$ and $S_{t+\Delta t}$ which in turn, need to be expressed in terms of one unknown.

If the reach length ΔX is not too large, $S_{t+\Delta t}$ can be approximated by average end areas,

$$S_{t+\Delta t} = [A_{a_{t+\Delta t}} + A_{b_{t+\Delta t}}]\frac{\Delta x}{2}$$

TABLE 20.14

Routing computations for storage routing method B

Time (*min*)	*Inflow* (m^3/s)	*Outflow* (m^3/s)	A_a (m^2)	A_b (m^2)	G	$G - \frac{\Delta x}{2} A_{a_{t+\Delta t}}$
0	0	0	0	0	180	− 1,576
20	.3	0	.878	0	6,136	− 7,484
40	7.0	0	6.81	0	23,820	6,300
60	10	1.70	8.76	2.64	31,380	19,140
80	6	7.68	6.12	7.266	26,964	21,084
100	2	8.69	2.94	7.93	18,326	14,566
120	1	5.37	1.88	5.67	12,658	10,902
140	.3	3.64	.878	4.36	8,562	7,402
160	.15	2.13	.58	3.06	6,122	5,522
180	.05	1.40	.30	2.34	4,470	4,470
200	.00	1.04	.00	1.92	3,216	3,216
220	.00	.63	.00	1.42	2,462	2,462
240	.00	.42	.00	1.10	1,948	1,948

Note: Δx = reach length (L) = 4000 m

$$G = (I_t + I_{t+\Delta t} - Q_t)\frac{\Delta t}{2} + S_t$$

$$= (I_t + I_{t+\Delta t} - T_t)\frac{\Delta t}{2} + \frac{A_{a_t} + A_{b_t}}{2}\Delta x$$

where $A_{a_{t+\Delta t}}$ can be computed from given inflow data. Substituting into equation (20.165),

$$\frac{\Delta t}{2} Q_{t+\Delta t} + \frac{\Delta x}{2} A_{a_{t+\Delta t}} + \frac{\Delta x}{2} A_{b_{t+\Delta t}} - G = 0 \qquad (20.166)$$

Equation (20.166) has two unknowns, $Q_{t+\Delta t}$ and $A_{b_{t+\Delta t}}$. However, $Q_{t+\Delta t}$ can be defined by Manning's equation (with n as roughness). Therefore,

$$\frac{\Delta t}{n} \frac{A_{b_{t+\Delta t}}^{5/3}}{P_{b_{t+\Delta t}}^{2/3}} S_{b_{t+\Delta t}}^{1/2} + \Delta x\, A_{b_{t+\Delta t}} = -2\left(\frac{\Delta x}{2} A_{a_{t+\Delta t}} - G\right) \qquad (20.167)$$

Both $A_{b_{t+\Delta t}}$ and $P_{b_{t+\Delta t}}$ can be expressed in terms of $y_{b_{t+\Delta t}}$ as

$$A = By + y^2 \qquad (20.168a)$$

$$P = B + 2y(1 + z^2)^{1/2} \qquad (20.168b)$$

where z is the side slope and B is the bottom width of the trapezoidal section. Substitution of equation (20.168) into equation (20.167) will leave only one unknown, $Y_{bt+\Delta t}$. The term $S_{bt+\Delta t}$, the friction slope at section b, can be computed by using $S_{at+\Delta t}$:

$$S_{b_{t+\Delta t}} \cong S_{a_{t+\Delta t}} = \left[I_{a_{t+\Delta t}} \frac{n P_{a_{t+\Delta t}}^{2/3}}{A_{a_{t+\Delta t}}^{5/3}} \right]^2 \qquad (20.169)$$

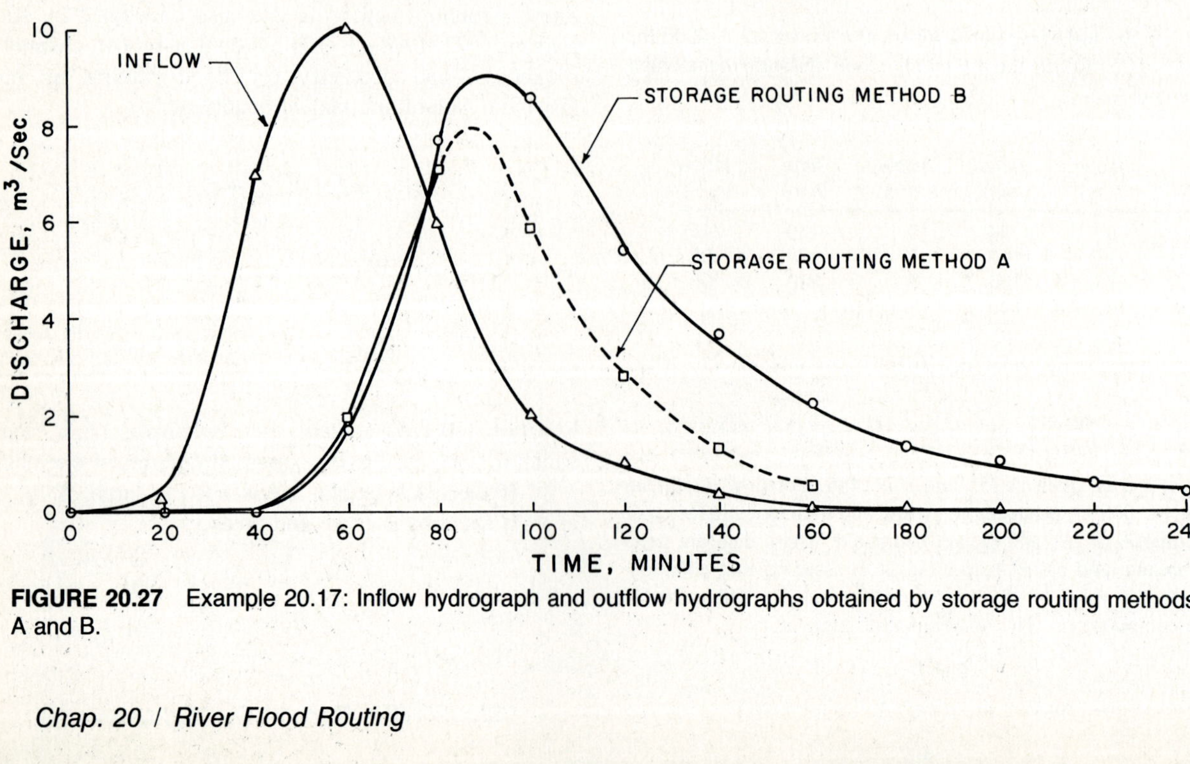

FIGURE 20.27 Example 20.17: Inflow hydrograph and outflow hydrographs obtained by storage routing methods A and B.

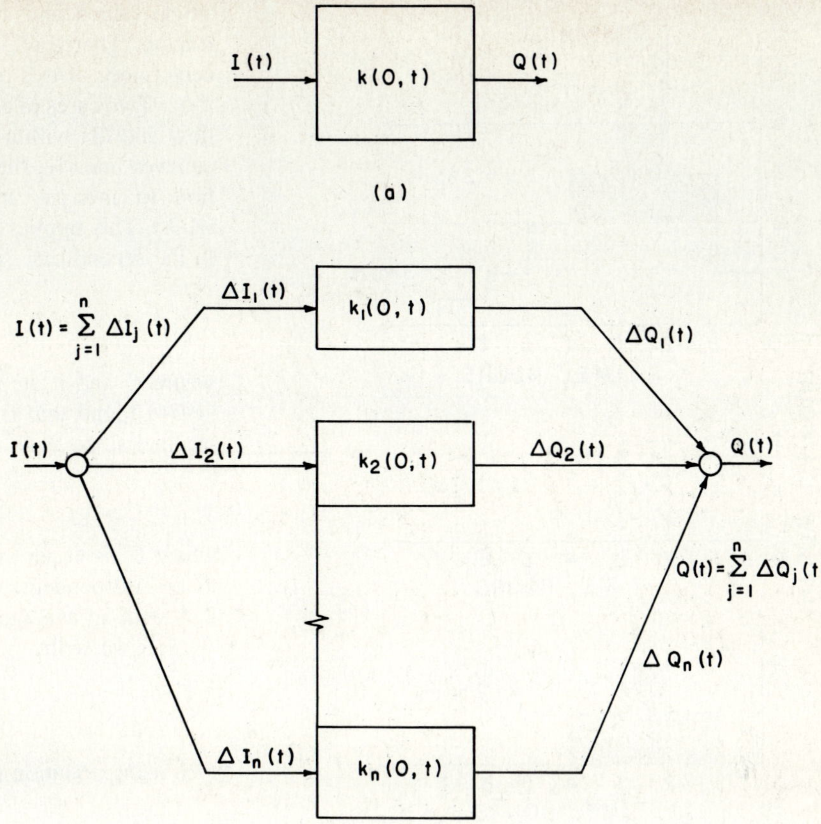

FIGURE 20.28 Linearization models: (a) single, (b) multiple.

Using $S_{bt+\Delta t}$ from equation (20.169), we can compute $y_{bt+\Delta t}$ from equation (20.167) using an iterative numerical procedure. To illustrate it, we consider an example.

EXAMPLE 20.17

Solve Example 20.16 by storage routing method B.

***Solution*:** The slope in equation (20.167) is replaced by the channel slope. The left side of equation (20.167), cross-sectional area, and discharge are calculated for water depths at the spacing of .005 m. These are used to compute $Q_{t+\Delta t}$, with the right side of the equation known at various iterations. Table 20.14 shows the computations. Fig. 20.27 shows an inflow hydrograph and outflow hydrographs obtained by routing methods A and B. Clearly, the two methods yield comparable rising hydrographs. The recession hydrographs differ because equation (20.163) is not valid in method A.

20.13 MULTIPLE LINEARIZATION MODEL

The multiple linearization model was developed by Keefer and McQuivey (1974). The fundamental assumption is that most channels behave very nearly as linear systems (Harley, 1967; Sauer, 1973) over small ranges of discharges. Therefore, it is plausible to divide the channel up into a set of linear subsystems, which when summed constitute the entire flow. This concept is illustrated in Fig. 20.28.

Operationally, we first define the range of flows over which to apply the model. This range is then partitioned horizontally in such a way that nonlinearities are minimized and a family of response functions is generated; one response function corresponds to one part of the range. These response functions are then convoluted with appropriate portions of inflow to perform flood routing.

The response functions can be derived using any of the linear models discussed previously. Keefer and McQuivey (1974) used the diffusion analogy (Hyami, 1951). In this model, better accounting of nonlinear effects can be made. It appears that a vertical partitioning of the inflow should be attempted because this is more natural and will accomplish just as well the purpose of multiple linearization.

EXAMPLE 20.18

An inflow hydrograph is described by a histogram comprised of seven pulses each of duration 1 h, which have the intensities 1, 3, 6, 3, 2, 1, and .5 cm/h, respectively. The watershed area drained by this channel is 1 km^2. Route the inflow by the multiple linearization model. Assume that three response functions will suffice. Each response function is obtained from a linear reservoir having a value of k = 3 h for flows up to 2 cm/h, k = 2 h for flows between 2 and 4 cm/h, and k = 1 h for flows between 4 and 6 cm/h. Determine the attentuation and lag due to routing.

***Solution*:** The inflow hydrograph is divided horizontally into three flow ranges, as shown in Fig. 20.29. For each flow range the linear reservoir with appropriate value of k is applied, and the outflow hydrograph is computed. The three outflow hydrographs are superposed to produce the total outflow hydrograph. The routing computations are given in Table 20.15.

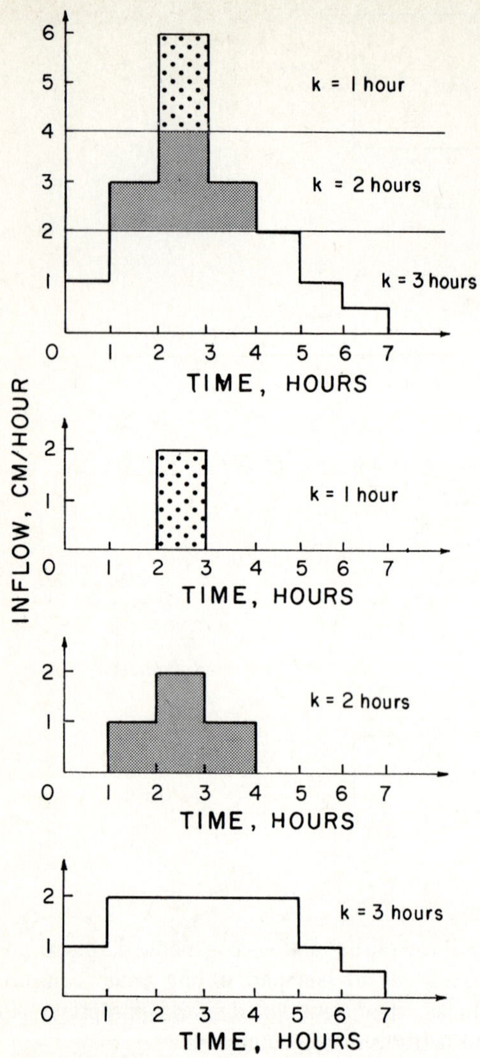

FIGURE 20.29 Example 20.18: Partitioning of an inflow hydrograph for the multiple linearization model.

The lag time is computed as the time difference between the centroid of inflow and outflow. The centroid of inflow is 2.92 h, and that of outflow is 5.55. Therefore, the lag time is 2.63 h. The attentuation is 6.0 − 5.95 = .05 cm/h.

EXAMPLE 20.19

Solve Example 20.18 by partitioning the inflow vertically.

Solution: The inflow hydrograph is divided vertically into three flow ranges as shown in Fig. 20.30. For each flow range, the linear reservoir with appropriate value of k is applied to produce the outflow hydrograph. The outflow hydrographs of all the flows are then combined to produce the total outflow hydrograph, which is given in the last column of Table 20.15. This outflow hydrograph compares well with that in the previous example.

20.14 NONLINEAR STORAGE MODEL

The nonlinear storage model, developed by Ardis (1974), is based on equation (20.1) and a rating function as suggested by Brakensiek (1966). The rating function is assumed to be single-valued and supposedly approximates the equation of motion. Therefore, characterization of this rating function is centerpiece of this method.

Two cases of channel flow can be identified: (1) overbank flow and (2) within-bank flow. In the first case, the channel behaves linearly, that is, discharge increases in direct proportion to increase in channel cross-section area ($\partial Q/\partial t \propto \partial A/\partial t$). This implies that equation (20.12) with $x = 0$ is valid. In the second case, however,

$$Q = Cu^n \tag{20.170}$$

where C and n are constants and u is mean velocity. Ardis (1974) found that C per unit drainage area can be related to channel slope S_c as

$$C_a = 0.132 S_c^{1.16}$$

where C_a is C per unit drainage area. The constant n appears to be independent of slope and is close to the range (1.6, 2.7) with an average value of 2.

If we write

$$Q = Au$$

then using equation (20.170) it follows that

$$A = Cu^{n-1}$$

If equation (20.170) is assumed to apply over a reach of channel length, then water stored, S, in this reach length is

$$S = AL$$

Therefore,

$$S = CLu^{n-1}$$

Using equation (20.170)

$$S = \left[\frac{LC^{1/n}}{Q^{1/n}}\right] Q \tag{20.171}$$

This can be written as

$$S = kQ \tag{20.172}$$

where

$$k = \frac{LC^{1/n}}{Q^{1/n}}$$

It is thus seen that k is not constant. It depends on channel flow and rating function coefficient. As Q increases, k—the routing time characteristic—decreases, since the channel is more efficient with higher velocities. For a specified Q, k increases with L. On the other hand, when overbank flow occurs, k is essentially constant because u is constant and

TABLE 20.15

Routing computations for the multiple linearization model

	K = 1	K = 2	K = 3		Total Outflow Produced by Vertical Partitioning of Inflow (cm/h)
Time (h)	*Discharge (cm/h)*	*Discharge (cm/h)*	*Discharge (cm/h)*	*Total Outflow (cm/h)*	
0	0.00	0.00	0.00	0.00	0.00
1	.00	.00	.28	.28	.28
2	.00	.63	1.17	1.80	2.10
3	1.90	1.94	2.11	5.95	5.93
4	.70	2.04	2.98	5.72	5.49
5	.26	1.24	3.76	5.26	4.47
6	.10	.75	3.56	4.41	3.58
7	.04	.46	3.00	3.50	2.78
8	.01	.28	2.15	2.44	1.88
9	.01	.17	1.54	1.72	1.28
10	.0	.10	1.10	1.20	.88
11		.06	.79	.85	.61
12		.04	.57	.61	.43
13		.02	.41	.43	.30
14		.01	.29	.30	.21
15		.01	.21	.22	.15
16		.01	.16	.16	.11
17		.00	.11	.11	.07
18		.00	.08	.08	.05
19		.00	.06	.06	.04
20		.00	.04	.04	.03
21		.00	.03	.03	.02
22			.02	.02	.01
23			.01	.01	.01
24			.00	.00	.01
25			.00	.00	.01

independent of Q. This implies channel linearity, and $k = L/b$, $b =$ constant u.

Differentiating equation (20.171),

$$\frac{dS}{dt} = LC^{1/n}\frac{(n-1)}{n}Q^{-1/n}\frac{dQ}{dt} \tag{20.173}$$

$$\frac{dS}{dt} = k\frac{(n-1)}{n}\frac{dQ}{dt} \tag{20.174}$$

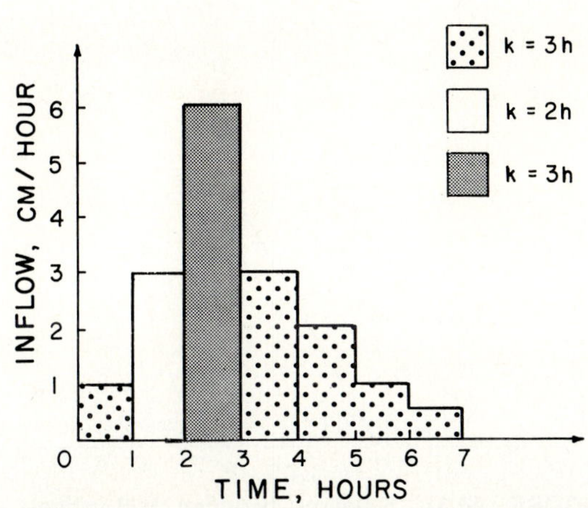

FIGURE 20.30 Example 20.19: Vertical partitioning of an inflow hydrograph for the multiple linearization model.

Let $m = k(n-1)/n$. Writing equation (20.174) in finite different form,

$$\frac{\Delta S}{\Delta t} = m\frac{\Delta Q}{\Delta t}$$

Using double routing (Holtan and Overton, 1965; Overton, 1967, 1970),

$$\frac{\Delta S}{\Delta t} = \frac{m}{2}\frac{\Delta Q}{\Delta t} \tag{20.175}$$

Combining equation (20.175) with equation (20.1),

$$\frac{I_t + I_{t+\Delta t}}{2} - \frac{Q_t + Q_{t+\Delta t}}{2} = \frac{m}{2}\frac{Q_{t+\Delta t} - Q_t}{\Delta t}$$

Separating out $Q_{t+\Delta t}$,

$$Q_{t+\Delta t} = \left(\frac{\Delta t}{m+\Delta t}\right)(I_{t+\Delta t} + I_t) + \left(\frac{m-\Delta t}{m+\Delta t}\right)Q_t \tag{20.176}$$

where $Q_{t+\Delta t}$ is the outflow from the first routing through the first half of storage at $t + \Delta t$. This becomes inflow at $t + \Delta t$ because routing is analogous to routing through a series of two reservoirs. Equation (20.176) is nonlinear with m as

$$m = \left(\frac{n-1}{n}\right)\frac{LC^{1/n}}{Q_t^{1/n}}, \qquad m_k < m < \infty \tag{20.177}$$

when $m = m_k$, the minimum value of m at which time flow becomes overbank flow. This limit for m can be evaluated from equation (20.170) by performing its plot, whose appearance is shown in Fig. 20.31. We consider an example to illustrate this nonlinear model.

EXAMPLE 20.20

Consider a channel reach having the following characteristics: $n = 2.0$, $S_c = .002$, drainage area $= 10\text{ km}^2$, and $L = 1000$ m. Route a triangular inflow whose base is 12 h and peak is 50 cm occurring at 4 h by the nonlinear storage method.

Solution: First the value of c is computed by

$$\frac{C}{A_d} = .132S_c^{1.16}$$

in which S_c is channel slope expressed in feet per mile and A_d is drainage area in square miles. From equation (20.170) it is clear that the dimensions of C depend on n:

$$[c] = \frac{L^3T^{-1}}{L^nT^{-n}} = L^{3-n}T^{n-1}$$

Since $n = 2$, $[c] = LT$ (in foot-seconds). In order to use the mks system of units, the preceding equation has to be modified accordingly:

$$\frac{C}{A_d} = K_1S_c^{1.16}, \qquad K_1 = .132\frac{s}{\text{mi}^{.84}\,\text{ft}^{.16}}$$

Then

$$K_1 = \frac{1}{3081.66}\frac{s}{m}$$

and

$$\frac{C}{A_d} = \frac{1}{3081.66}S_c^{1.16}$$

where C is expressed in meter-seconds, A_d in m^2, and S_c is in meters per meter. For $A_d = 107\text{ m}^2$, $S_c = 0.002$, $C = 2.4011$ meter-seconds. From equation (20.177),

$$m = \frac{2-1}{2}\frac{100(2.4011)^{0.5}}{Q_t^{0.5}} = \frac{774.77}{Q_t^{0.5}}$$

Peak inflow is 50 cm/h. In terms of cubic meters per second,

$$I_{\text{peak}} = .50\frac{107}{3600} = 1388.89\text{ m}^3/\text{s}$$

The time of computation is taken as $\Delta t = 0.25$ h. The computations are carried out as follows.

At $t = 0$, it is necessary to assume $Q_0 \neq 0$; otherwise $m \to \infty$. Take $Q_0 = 1.0\text{ m}^3/\text{s}$. From equation (20.177),

$$m = \frac{774.77}{1^{.5}} = 774.77$$

From equation (20.176), for the first storage routing,

$$Q_{.5} = \left(\frac{.25 \times 3600}{774.77 + 900}\right)(0 + 86.81) + \left(\frac{774.77 - 900}{774.77 + 900}\right)(1.0) = 46.57\text{ m}^3/\text{s}$$

Now for the second storage routing, $I_{.5} = 46.57\text{ m}^3/\text{s}$ and $I_0 = 1.0\text{ m}^3/\text{s}$. Therefore,

$$Q_{.5} = \frac{900}{774.77 + 900}(1.0 + 46.57) + \frac{774.77 - 900}{774.77 + 900}(1.0) = 25.49\text{ m}^3\text{s}$$

At $t = .5$,

$$m = \frac{776.77}{(25.49)^{.5}} = 163.46$$

For the first storage routing,

$$Q_1 = \frac{900}{153.46 + 900}(86.81 + 173.61) + \frac{153.46 - 900}{153.46 + 900}(46.57) = 189.48\text{ m}^3/\text{s}$$

For the second storage routing,

$$Q_1 = \frac{900}{153.46 + 900}(46.57 + 189.48) + \frac{153.46 - 900}{153.46 + 900}(25.49) = 183.6\text{ m}^3/\text{s}$$

The computations are carried out for subsequent time intervals and are given in Table 20.16.

A comment is in order. The nonlinear method is very sensitive to Δt. A very high value of Δt may lead to instability. In this problem there is practically no attentuation or lag because the reach is short and inflow is relatively high. Consider the peak inflow $I_p = 1388.89\text{ m}^3/\text{s}$. From equation (20.170), for $Q = 1388.89\text{ m}^3/\text{s}$ and $C = 2.4011$ meter-seconds, $u = 24.05$ m/s. Time of travel for this velocity is $1000/24.05 = 42$ s, assuming constant conditions within the reach.

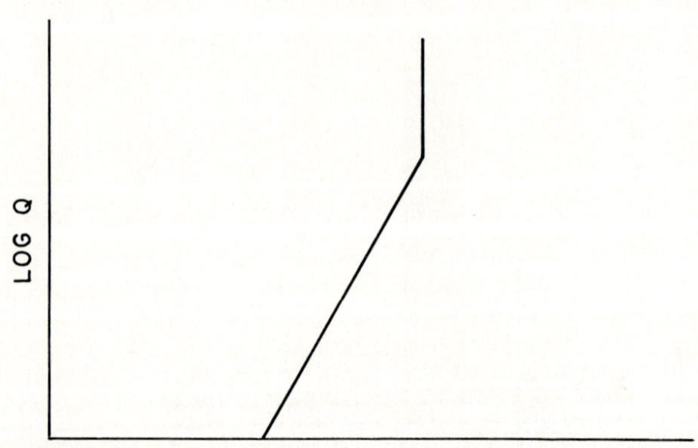

FIGURE 20.31 Relation between within-bank discharge and mean velocity u.

TABLE 20.16

Routing computations for the nonlinear storage routing method

t (h)	I_t (m^3/s)	$I_t+\Delta_t$ (m^3/s)	I_t (m^3/s) / Q_{1_t} (m^3/s)	$I_t+\Delta_t$ (m^3/s) / $Q_{2_{t+\Delta t}}$ (m^3/s)	Q_t (m^3/s)	$Q_{t+\Delta t}$ (m^3/s)
(1)	(2)	(3)	(4)	(5)	(6)	(7)
0	0	86.81	1.0	46.57	1.0	25.49
.25	86.81	173.61	46.57	189.48	25.49	183.60
.50	173.61	260.42	189.48	241.26	183.60	243.34
.75	260.42	347.22		359.83		351.77
1	347.22	434.03		418.71		423.49
1.25	434.03	520.83		531.43		522.52
1.50	520.83	607.64		594.66		600.63
1.75	607.64	694.45		703.60		694.33
2	694.45	781.25		769.93		776.51
2.25	781.25	868.06		876.10		866.73
2.50	868.06	954.86		944.81		951.69
2.75	954.86	1,041.67		1,048.81		1,039.48
3	1,041.67	1,128.47		119.44		1,126.45
3.25	1,128.47	1,215.28		1,221.69		1,212.47
3.50	1,215.28	1,302.08		1,293.89		1,300.92
3.75	1,302.08	1,388.89		1,394.68		1,385.63
4	1,388.89	1,345.49		1,340.94		1,350.80
4.25	1,345.49	1,302.08		1,307.42		1,298.78
4.50	1,302.08	1,258.68		1,254.61		1,264.08
4.75	1,258.68	1,215.28		1,220.18		1,211.98
5	1,215.28	1,171.88		1,168.25		1,177.32
5.25	1,171.88	1,128.47		1,132.98		1,125.22
5.50	1,128.47	1,085.07		1,081.88		1,090.52
5.75	1,085.07	1,041.67		1,045.80		1,038.51
6	1,041.67	998.26		995.47		1,003.69
6.25	998.26	954.86		958.65		951.84
6.50	954.86	991.46		909.05		916.84
6.75	991.46	868.06		871.53		865.21
7	868.06	824.65		822.61		829.96
7.25	824.65	781.25		784.43		778.61
7.50	781.25	737.85		736.15		743.07
7.75	737.85	694.44		697.36		692.06
8	694.44	651.04		649.68		656.16
8.25	651.04	607.64		610.32		605.54
8.50	607.64	564.24		563.20		569.25
8.75	564.24	520.83		523.31		519.07
9	520.83	477.43		476.72		482.34
9.25	477.43	434.03		436.32		432.65
9.50	434.03	390.63		390.24		395.45
9.75	390.63	347.22		349.38		346.29
10	347.22	303.82		303.77		308.60
10.25	303.12	260.42		262.49		260.04
10.50	260.42	217.01		217.35		221.84
10.75	217.01	173.61		175.68		173.96
11	173.61	130.21		313.05		135.30
11.25	130.21	86.81		89.07		88.30
11.50	86.81	43.40		45.16		49.49
11.75	43.40	0		3.36		4.53
12	0	0		0		0

Note: Columns (1) through (5) are used for first storage routing, and (4) through (7) are used for second storage routing.

20.15 GENERAL STORAGE MODEL

The general storage model was developed by Kulandaiswamy, Krishnaswami, and Ramalingam (1967), employing the storage equation derived by Kulandaiswamy (1964). The general storage equation can be written as

$$S = \sum_{m=1}^{M} a_{m-1} \frac{d^{m-1}Q}{dt^{m-1}} + \sum_{n=1}^{N} b_{n-1} \frac{d^{n-1}I}{dt^{n-1}}$$

where the coefficients a_i and b_i are either constants or functions of variables involving inflow I and outflow Q. In this model, they are treated as constants. Substituting into equation (20.1),

$$I - Q = \sum_{m=1}^{M} a_{m-1} \frac{d^{m}Q}{dt^{m}} + \sum_{n=1}^{N} b_{n-1} \frac{d^{n}I}{dt^{n}} \quad (20.178)$$

Let the backward difference be defined as $\Delta S = S_j - S_{j-1}$. Then equation (20.178) can be written as

$$Q_j = \sum_{i=1}^{M} (-1)^{i+1} A_{i-1} Q_{j-i} + \sum_{i=1}^{N} (-1)^{i+1} B_{i-1} I_{j-i} + \frac{1}{2k}(CI_j + I_{j-1} - Q_{j-1}) \quad (20.179)$$

where

$$A_i = \frac{1}{k} \sum_{r=0}^{M-i-1} \frac{a_{r+i}}{(\Delta t)^{t+i-1}} \binom{r+i+1}{r}$$

$$B_i = \frac{1}{k} \sum_{r=0}^{N-i-1} \frac{b_{r+i}}{(\Delta t)^{t+i-1}} \binom{r+i+1}{r}$$

$$k = \frac{1}{2} \sum_{r=0}^{M-1} \frac{a_r}{(\Delta t)^{r+1}}$$

$$C = 1 - 2 \sum_{r=0}^{N-1} \frac{b_r}{(\Delta t)^{r+1}}$$

It is clear from equation (20.179) that the outflow at the end of the jth interval can be determined if the outflow values for M intervals of time prior to the jth interval and the inflow values for $N + 1$ intervals of time prior to and including the jth interval are known.

Kulandaiswamy, Krishnaswami, and Ramalingam (1967) investigated the following special cases:

$$S = a_0 Q \quad (20.180a)$$

$$S = a_0 Q + a_1 \frac{dQ}{dt} + a_2 \frac{d^2Q}{dt^2} \quad (20.180b)$$

$$S = a_0 Q + a_1 \frac{dQ}{dt} + b_0 I \quad (20.180c)$$

$$S = a_0 Q + a_1 \frac{dQ}{dt} + b_0 I + b_1 \frac{dI}{dt} \quad (20.180d)$$

Their study showed that equation (20.180c) was satisfactory for channel routing. Considering higher-order derivatives beyond $M = 2$ and $N = 1$ did not improve the results significantly. For this case, equation (20.179) reduces to

$$Q_j = AQ_{j-1} + BQ_{j-1} + CI_j + DI_{j-1} \quad (20.181)$$

where

$$A = \frac{2a_1 + a_0\Delta t - \frac{1}{2}(\Delta t)^2}{E}$$

$$B = \frac{a_1}{E}$$

$$C = \frac{-b_0\,\Delta t + \frac{1}{2}(\Delta t)^2}{E}$$

$$D = \frac{b_0\,\Delta t + \frac{1}{2}(\Delta t)^2}{E}$$

$$E = a_1 + a_0\Delta t \frac{1}{2}(\Delta t)^2$$

The parameters a_0, a_1, and b_0 can be estimated using any one of the standard techniques, including least squares method and method of moments.

20.16 THE WORKING VALUE METHOD

The working value method assumes a discharge, called a working value discharge W, representing a steady state, which would produce the same storage as that produced by actually existing inflow I and outflow Q. On referring to Fig. 20.32, the wedge storage given by $kx(I - Q)$ equals $k(W - Q)$. Therefore,

$$W = xI + (1 - x)Q \quad (20.182)$$

Equation (20.182) can be recast as

$$Q = W - \frac{x}{1-x}(I - W)$$

Inserting it in equation (20.11),

$$\frac{dS}{dt} = \frac{1}{1-x}(I - W) \quad (20.183)$$

Expressing it in finite difference form,

$$S_{t+\Delta t} - S_t = \frac{\Delta t}{2(1-x)}(I_{t+\Delta t} + I_t) - \frac{\Delta t}{2(1-x)}(W_{t+\Delta t} + W_t)$$

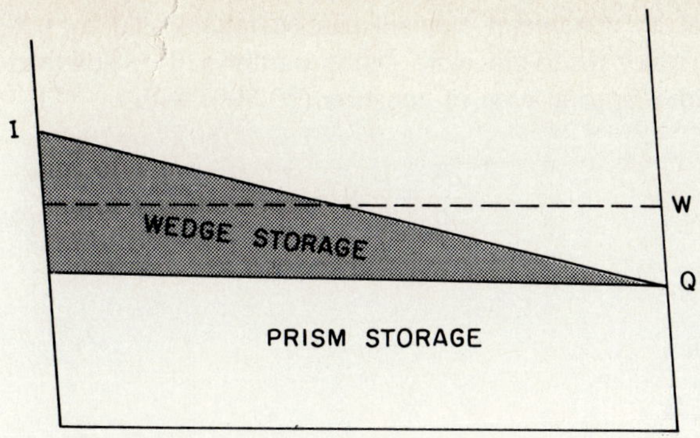

FIGURE 20.32 Prism storage, wedge storage, and working value discharge.

or

$$(1 - x) S_t + \frac{\Delta t}{2}(I_t + I_{t+\Delta t}) - \frac{\Delta t}{2} W_t = (1 - x) S_{t+\Delta t} + \frac{\Delta t}{2} W_{t+\Delta t} \quad (20.184)$$

Let

$$R_1 = (1 - x)S_t + \frac{\Delta t}{2} W_t \quad (20.185)$$

$$R_2 = (1 - x)S_{t+\Delta t} + \frac{\Delta t}{2} W_{t+\Delta t} \quad (20.186)$$

The terms R_1 and R_2 are working indices of storage. Equation (20.184) can be written as

$$R_2 = R_1 + \frac{\Delta t}{2}(I_t + I_{t+\Delta t}) - \Delta t\, W_t \quad (20.187)$$

From equations (20.12) and (20.182),

$$S = kW \quad (20.188)$$

Equations (20.187) and (20.188) can be used to perform flow routing. The calculations can be carried out as follows. Compute W_t from equation (20.182) for known Q_t, I_t, and x. Compute S_t from equation (20.188) for known k_t (if k is variable). Compute R_1 and R_2 from equations (20.185) and (20.187). Then calculate $Q_{t+\Delta t}$ from equations (20.186) and (20.188) using $k_{t+\Delta t}$ for an assumed $Q_{t+\Delta t}$ if k is variable. If necessary, an iteration can be carried out by assuming a new value of $Q_{t+\Delta t}$. This method has an advantage over the Muskingum method, especially where tributary inflow or controlled discharge through a dam is concerned.

20.17 ROUTING IN ABSTRACTING CHANNELS

In the preceding discussion it was assumed that reservoirs or channels were not subject to seepage or infiltration. However, in ephemeral streams this assumption is far from valid. In the southwestern United States such streams are commonly found. These streams may abstract large quantities of runoff (Babcock and Coshing, 1942; Keppel, 1960; Keppel and Renard, 1962; Allis, Dragoun, and Sharp, 1964; Renard and Keppel, 1966; Quashu and Buol, 1967; Wallace and Renard, 1967; Burkham, 1979a, 1979b; Lane, Diskin, and Renard, 1971; Lane, 1972; Peebles, 1975). These abstractions or transmission losses are important in diminishing streamflow, supporting riparian vegetation, and recharging local aquifers (Renard, 1970).

Flood routine in ephemeral streams in arid or semiarid areas is somewhat more complicated due to transmission losses (Keppel and Renard, 1962). There is a reduction in both peak discharge and volume of flow during the progression of a flood wave downstream, as shown in Fig. 20.33. Therefore, we need to modify equation (20.1) for flood routing in ephemeral streams:

$$\frac{ds}{dt} = I - Q - W \quad (20.189)$$

where W represents the transmission loss rate. Lane (1972) expressed Q and W as functions of S,

$$W = a_1 S^{b_1}$$

$$Q = a_2 S^{b_2}$$

Substituting into equation (20.189),

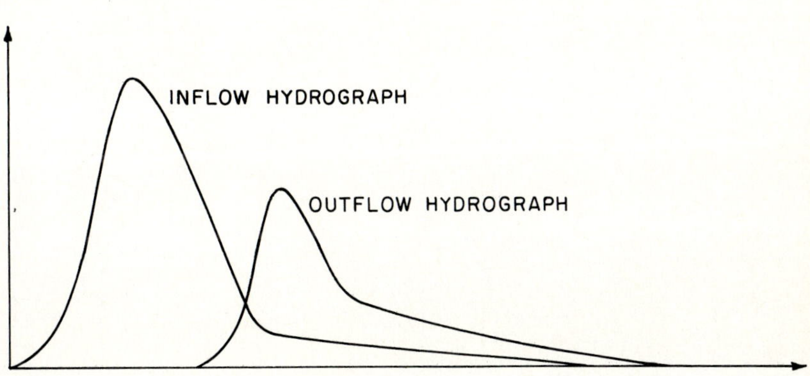

FIGURE 20.33 Flood routing in an abstracting ephemeral channel.

$$\frac{dS}{dt} + a_1 S^{b_1} + a_2 S^{b_2} = I \tag{20.190}$$

Equation (20.190) serves as a governing differential equation for an abstracting channel system or a component thereof. This is equivalent to a leaky reservoir. Lane (1972) suggested that an ephemeral channel can be represented by a cascade of leaky reservoirs with piston outflows. He, however, dealt with a special case of equation (20.190) with $b_1 = 1$,

$$\frac{dS}{dt} + a_1 S + a_2 S^{b_2} = I \tag{20.191}$$

EXERCISES

20.1. Consider a rectangular channel reach 2000 m long. An inflow of 10 m^3/s is introduced at the upstream of the reach. At the same time an outflow of 6 m/s is observed at the downstream end. Let the channel be 2 m wide with a side slope of .001 and the roughness coefficient of 0.025. Compute the outflow rate for this inflow using the convex method.

20.2. An inflow-outflow record is available for a river reach. Compute the storage in the reach and graph storage as a function of outflow.

Time (h)	Inflow (m^3/s)	Outflow (m^3/s)	Time (h)	Inflow (m^3/s)	Outflow (m^3/s)
0	10	10	66	72	110
6	12	10	72	60	92
12	15	12	78	43	66
18	37	16	84	34	54
24	105	40	90	28	42
30	175	85	96	24	32
36	220	140	102	20	27
42	210	160	108	16	22
48	190	180	114	14	18
54	135	190	120	12	16
60	100	172	126	10	14
			132	10	12

20.3. Determine the Muskingum parameters x and k for the data in Question 20.2 by using different methods and compare them.

20.4. Using the values of x and k obtained in Question 20.3 route the inflow hydrograph of Question 20.2. Determine the attenuation and lag.

20.5. Compute appropriate values of the Muskingum coefficients C_0, C_1, and C_2 using the data of Question 20.2. Determine the outflow at 54 h when the inflow is known between 24 and 54 h. The outflow at 24 h is also known.

20.6. Compute the 6-h UH for the Muskingum method using equation (20.18). Compare it with the one derived by using equation (20.28), as well as with that obtained from equation (20.38). Using the data of Question 20.2.

20.7. Compute the parameters of the multiple-reach Muskingum method for the data of Question 20.2. Then route its inflow hydrograph.

20.8. Determine the parameters of the lag and route model for the data of Question 20.2.

20.9. Determine the parameters of the Kalinin-Milyukov model for the data of Question 20.2.

20.10. Determine the parameters of the three-parameter gamma distribution method for the data of Question 20.2.

APPENDIX A

Hydrologic Books and Periodicals

A.1 TEXTBOOKS ON SURFACE WATER HYDROLOGY

Agricultural Research Service. 1962. *Field Manual for Research in Agricultural Hydrology*. USDA-ARS-SWCRD-Agricultural Handbook 224, U.S. Department of Agriculture, Washington, D.C.

ALBERTSON, M. L., L. S. TUCKER, AND D. C. TAYLOR, eds. 1971. *Treatise on Urban Water Systems*. Colorado State University, Fort Collins.

American Society of Civil Engineers. 1949. *Hydrology Handbook*. Manual of Engineering Practice no. 28, ASCE Headquarters, New York.

ANDERSON, M. G. AND T. P. BURT, eds. 1985. *Hydrological Forecasting*. New York: John Wiley.

APOLLOV, B. A., G. P. KALININ, AND V. D. KOVAROV. 1964. *Hydrological Forecasting*. Translated from Russian, published for the U.S. Department of Commerce, Washington, D.C.

BACK, W., AND D. A. STEPHENSON, eds. 1979. *Contemporary Hydrology*. Amsterdam: Elsevier.

BALEK, J. 1977. *Hydrology and Water Resources in Tropical Africa*. Amsterdam: Elsevier.

BARROWS, R. K. 1948. *Floods, Their Hydrology and Control*. New York: McGraw-Hill.

BENNETT, D., AND J. F. THOMAS, eds. 1982. *On Rational Grounds: Systems Analysis in Carchment Land Use Planning*. Amsterdam: Elsevier.

BISWAS, A. K. 1972. *History of Hydrology*. New York: American Elsevier.

______, ed. 1976. *Systems Approach to Water Management*. New York: McGraw-Hill.

BOWEN, R. 1982. *Surface Water*. New York: John Wiley.

BRAS, R., AND I. RODRIGUEZ-ITURBE. 1984. *Random Functions and Hydrology*. Reading, Mass.: Addison-Wesley.

BRUTSAERT, W. H. 1982. *Evaporation into the Atmosphere*. Boston: D. Reidel.

Bureau of Reclamation. 1951. "Flood Hydrology." Pt. 6 of Vol. 4 on Water Studies, U.S. Department of Interior, Washington, D.C.

BUTLER, S. S. 1957. *Engineering Hydrology*. Englewood Cliffs, N.J.: Prentice-Hall, Inc.

CHAPMAN, T. G., AND F. X. DUNIN, eds. 1975. *Prediction in Catchment Hydrology*. Canberra, Australia: Australian Academy of Science.

CHEBOTAREV, N. P. 1966. *The Theory of Stream Runoff*. Translated from Russian, published for the U.S. Department of Commerce, Washington, D.C.

CHORLEY, R. J. 1969. *Water, Earth, and Man*. England: Methuen.

CHOW, V. T., ed. 1964. *Handbook of Applied Hydrology: A Compendium of Water Resources Technology*. New York: McGraw-Hill.

CIRIANI, T. A., V. MAIONE, AND J. R. WALLIS, eds. 1971. *Mathematical Models for Surface Water Hydrology*. New York: John Wiley.

DEBSKI, K. 1965. *Continental Hydrology*. Published for the U.S. Department of the Interior and the National Science Foundation, Washington, D.C., by the Scientific Publication Foreign Cooperation Center of the Central Institute for Scientific, Technical, and Economic Information, Springfield, Va.

DEJU, R. A. 1971. *Regional Hydrology Fundamentals*. New York: Gordon and Breach.

DEMING, H. G. 1975. *Water: The Fountain of Opportunity*. Oxford: Oxford Univ. Press.

DEUTSCH, M., D. R. WIESNET, AND A. RANGO, eds. 1981. *Satellite Hydrology*. Bethesda, Md.: American Water Resources Association.

DINGMAN, S. L. 1984. *Fluvial Hydrology*. New York: W. H. Freeman and Company Publishers.

Dooge, J. C. I. 1973. *Linear Theory of Hydrologic Systems*. Agricultural Research Service, U.S. Department of Agriculture, Technical Bulletin No. 1468. Washington, D.C.: U.S. Government Printing Office.

Eagleson, P. S. 1970. *Dynamic Hydrology*. New York: McGraw-Hill.

Featherstone, R. E. and A. James, eds. 1982. *Urban Drainage Systems*. Marshfield, Mass.: Pitman.

Fleming, G. 1972. *Computer Simulation Techniques in Hydrology*. New York: Elsevier.

Forest Service. 1959. *Forest and Range Hydrology Handbook*. USDA-FS Category 2, Handbook 2518, U.S. Department of Agriculture, Forest Service, Washington, D.C.

Foster, E. E. 1948. *Rainfall and Runoff*. New York: Macmillan.

Fox, C. S. 1972. *Water*. Westport, Conn.: Greenwood Press Publishers.

Furon, R. 1967. *The Problem of Water: A World Study*. Amsterdam: Elsevier.

Garg, S. K. 1982. *Water Resources and Hydrology*. Delhi, India: Khanna.

Garstka, W. U. 1978. *Water Resources and the National Welfare*. Fort Collins, Colo.: Water Resources Publications.

Gottschalk, L., G. Lindh, and L. de Mare. 1977. *Stochastic Processes in Water Resources Engineering*. Fort Collins, Colo.: Water Resources Publications.

Gray, D. M., ed. 1970. *Handbook on the Principles of Hydrology*. National Research Council of Canada, Ottawa, Canada.

Gregory, K. J., and D. E. Walling. 1973. *Drainage Basin Form and Process: A Geomorphological Approach*. London: Edward Arnold.

Haan, C. T., and B. J. Barfield. 1979. *Hydrology and Sedimentology of Surface Mined Lands*. Office of Continuing Education and Extension, College of Engineering, University of Kentucky, Lexington.

Haan, C. T., H. P. Johnson, and D. L. Brakensiek, eds. 1982. *Hydrologic Modeling of Small Watersheds*. ASAE Nomograph no. 5, American Society of Agricultural Engineers, St. Joseph, Mich.

Hall, M. J. 1984. *Urban Hydrology*. London: Elsevier.

Halldin, S., ed. 1979. *Comparison of Forest Water and Energy Exchange Models*. Amsterdam: Elsevier.

Hammer, M. J., and K. A. MacKichan. 1981. *Hydrology and Quality of Water Resources*. New York: John Wiley.

Hazen, A. 1930. *Flood Flows*. New York: John Wiley.

Hewlett, J. D. 1982. *Principles of Forest Hydrology*. Athens, Ga.: Univ. of Georgia Press.

Hjelmfelt, A. T., and J. J. Cassidy. 1975. *Hydrology for Engineers and Planners*. Ames, Ia.: Iowa State Univ. Press.

Horton, R. E. 1935. *Surface Runoff Phenomena*. Ann Arbor, Mich.: Edwards Brothers.

Hromadka, T. V. 1983. *Computer Methods in Urban Hydrology*. Mission Viejo, Calif.: Lighthouse Publications.

Hromadka, T. V., J. Clements, and H. Saluja. 1984. *Computer Methods in Watershed Hydraulics*. Mission Viejo, Calif.: Lighthouse Publications.

Hromadka, T. V., T. J. Durbin, and J. J. DeVries. 1984. *Computer Methods in Water Resources*. Mission Viejo, Calif.: Lighthouse Publications.

Hudson, N. 1981. *Soil Conservation*. Ithaca, N.Y.: Cornell Univ. Press.

Hydrologic Engineering Center. 1970. *Hydrologic Engineering Methods for Water Resources Development*. Vol. 1, *Requirements and General Procedures*. U.S. Army Corps of Engineers, U.S. Department of Army, Davis, Calif.

______. 1970. *Hydrologic Engineering Methods for Water Resources Development*. Vol. 2, *Hydrologic Data Management*. Army Corps of Engineers, U.S. Department of Army, Davis, Calif.

______. 1970. *Hydrologic Engineering Methods for Water Resources Development*. Vol. 3, *Hydrologic Frequency Analysis*. U.S. Army Corps of Engineers, U.S. Department of Army, Davis, Calif.

______. 1970. *Hydrologic Engineering Methods for Water Resources Development*. Vol. 4, *Hydrograph Analysis*. U.S. Army Corps of Engineers, U.S. Department of Army, Davis, Calif.

______. 1970. *Hydrologic Engineering Methods for Water Resources Development*. Vol. 5, *Hypothetical Floods*. U.S. Army Corps of Engineers, U.S. Department of Army, Davis, Calif.

______. 1970. *Hydrologic Engineering Methods for Water Resources Development*. Vol. 6, *Water Surface Profiles*. U.S. Army Corps of Engineers, U.S. Department of Army, Davis, Calif.

______. 1970. *Hydrologic Engineering Methods for Water Resources Development*. Vol. 7, *Flood Control by Reservoirs*. U.S. Army Corps of Engineers, U.S. Department of Army, Davis, Calif.

______. 1970. *Hydrologic Engineering Methods for Water Resources Development*. Vol. 8, *Reservoir Yield*. U.S. Army Corps of Engineers, U.S. Department of Army, Davis, Calif.

______. 1970. *Hydrologic Engineering Methods for Water Resources Development*. Vol. 9, *Reservoir System Analysis for Conservation*. U.S. Army Corps of Engineers, U.S. Department of Army, Davis, Calif.

______. 1970. *Hydrologic Engineering Methods for Water Resources Development*. Vol. 10, *Principles of Ground-Water Hydrology*. U.S. Army Corps of Engineers, U.S. Department of Army, Davis, Calif.

______. 1970. *Hydrologic Engineering Methods for Water Resources Development*. Vol. 11, *Water Quality Determinations*. U.S. Army Corps of Engineers, U.S. Department of Army, Davis, Calif.

______. 1970. *Hydrologic Engineering Methods for Water Resources Development*. Vol. 12, *Sediment Transport*. U.S. Army Corps of Engineers, U.S. Department of Army, Davis, Calif.

Johnstone, D., and W. P. Cross. 1949. *Elements of Applied Hydrology*. New York: Ronald Press Company.

Kaczmarek, Z. 1977. *Statistical Methods in Hydrology and Meteorology*. Translated from Polish. National Center, Scientific, Technical and Economic Information, Warsaw, Poland.

Kalinin, G. P. 1971. *Global Hydrology*. Translated from Russian. Published for NOAA, U.S. Department of Commerce, Washington, D.C.

Kartvelishvili, N. A. 1969. *Theory of Stochastic Processes in Hydrology and River Runoff Regulation*. Translated from Russian. Published for the U.S. Department of Commerce, Washington, D.C.

Kashyap, R. L., and A. R. Rao. 1976. *Dynamic Stochastic Models from Empirical Data*. New York: Academic Press.

Kazmann, R. G. 1965. *Modern Hydrology*. New York: Harper and Row.

Kinori, B. Z. 1970. *Manual of Surface Drainage Engineering*, vol. 1. Amsterdam: Elsevier.

Kinori, B. Z., and J. Mevorach. 1984. *Manual of Surface Drainage Engineering*. Vol. 2, *Stream Flow Engineering and Flood Protection*. Amsterdam: Elsevier.

Kirby, M. J., ed. 1978. *Hillslope Hydrology*. New York: John Wiley.

KITE, G. W. 1977. *Frequency and Risk Analysis in Hydrology*. Fort Collins, Colo.: Water Resources Publications.

KNAPP, B. J. 1979. *Elements of Geographical Hydrology*. Boston: George Allen and Unwin.

KOTTEGODA, N. T. 1980. *Stochastic Water Resources Technology*. New York: Halsted Press.

LAZARO, T. R. 1979. *Urban Hydrology: A Multidisciplinary Perspective*. Ann Arbor, Mich.: Ann Arbor Science Pub.

LEE, R. 1978. *Forest Climatology*. New York: Columbia Univ. Press.

______. 1980. *Forest Hydrology*. New York: Columbia Univ. Press.

LI, W. H. 1983. *Fluid Mechanics in Water Resources Engineering*. Boston: Allyn and Bacon.

LINSLEY, R. K., AND J. B. FRANZINI. 1972. *Water Resources Engineering*. New York: McGraw-Hill.

LINSLEY, R. K., M. A. KOHLER, AND J. L. H. PAULHUS. 1949. *Applied Hydrology*. New York: McGraw-Hill.

______. 1975. *Hydrology for Engineers*. New York: McGraw-Hill.

L'VOVICH, M. I. 1979. *World Water Resources and Their Future*. English translation from Russian, edited by R. L. Nace. American Geophysical Union, Washington, D.C.

MCCUEN, R. H., AND W. M. SNYDER. 1986. *Hydrologic Modeling: Statistical Methods and Applications*. Englewood Cliffs, N.J.: Prentice-Hall Inc.

MCMAHON, T. A., AND R. G. MEIN. 1978. *Reservoir Capacity and Yield*. Amsterdam: Elsevier.

MEAD, D. W. 1904. *Notes on Hydrology*. Chicago: Press of Smith and Company.

______. 1919. *Hydrology*. 1st ed. New York: McGraw-Hill; 2d edition. Revised by H. W. Mead and H. J. Hunt. Madison, Wis.: Mead and Hunt, 1950.

MEINZER, O. E., ed. 1942. *Hydrology*. New York: McGraw-Hill. Reprinted 1949. New York: Dover.

MEYER, A. F. 1928. *The Elements of Hydrology*. New York: John Wiley.

MOREL-SEYTOUX, H. J., J. D. SALAS, T. G. SANDERS, AND R. E. SMITH. eds. 1979. *Modeling Hydrologic Processes*. Fort Collins, Colo.: Water Resources Publications.

MURRAY, D. L. AND P. ACKROYD, eds. 1979. *Physical Hydrology: New Zealand Experience*. Wellington, New Zealand: The New Zealand Hydrological Society.

National Research Council. 1977. *Climate, Climatic Change, and Water Supply*. Washington, D.C.: National Academy of Sciences.

______. 1978. *Geophysical Predictions*. Washington, D.C.: National Academy of Sciences.

National Resources Council. 1982. *Scientific Basis of Water-Resource Management*. Washington, D.C.: National Academy Press.

Natural Environment Research Council. 1975. *Flood Studies Report*. Vol. 1, *Hydrological Studies*; Vol. 2, *Meteorological Studies*; Vol. 3, *Flood Routing Studies*. London.

NEMEC, J. 1972. *Engineering Hydrology*. London: McGraw-Hill.

OVERTON, D. E., AND M. E. MEADOWS. 1976. *Stormwater Modeling*. New York: Academic Press.

POWLEDGE, F. 1982. *Water: The Nature, Uses, and Future of Our Most Precious and Abused Resource*. New York: Farrar, Strauss and Giroux.

RAUDKIVI, A. J. 1979. *Hydrology: An Advanced Introduction to Hydrological Processes and Modeling*. New York: Pergamon.

RODDA, J. C., ed. 1976. *Facets of Hydrology*. New York: John Wiley.

RODDA, J. C., R. A. DOWNING, AND F. M. LAW. 1976. *Systematic Hydrology*. London: Butterworths.

SALAS, J. D., J. W. DELLEUR, V. YEVJEVICH, AND W. L. LANE. 1980. *Applied Modeling of Hydrologic Time Series*. Littleton, Colo.: Water Resources Publications.

SANDERS, T. G., ed. 1980. *Hydrology for Transportation Engineers*. Federal Highway Administration, U.S. Department of Transportation, Washington, D.C.

SCHULZ, E. F. 1974. *Problems in Applied Hydrology*. Fort Collins, Colo.: Water Resources Publications.

SCHULZ, E. F., V. A. KOELZER, AND K. MAHMOOD, eds. 1973. *Floods and Droughts*. Fort Collins, Colo.: Water Resources Publications.

SHARMA, R. K. 1983. *Hydrology and Water Resources*. Delhi, India: Dhanpat Rai.

SHAW, E. M. 1983. *Hydrology in Practice*. Berkshire, England: Van Nostrand Reinhold (UK).

SHEAFER, J. R., K. R. WRIGHT, W. C. TAGGART, AND R. M. WRIGHT. 1982. *Urban Storm Drainage Management*. New York: Marcel Dekker.

SHEN, H. W., ed. 1976. *Stochastic Approaches to Water Resources*. Fort Collins, Colo.: Water Resources Publications.

SIMONS, LI AND ASSOCIATES, INC. 1982. *Engineering Analysis of Fluvial Systems*. Fort Collins, Colo.: Simons, Li and Associates.

SINGH, V. P., ed. 1982. *Applied Modeling in Catchment Hydrology*. Littleton, Colo.: Water Resources Publications.

______. 1982. *Modeling Components of Hydrologic Cycle*. Littleton, Colo.: Water Resources Publications.

______. 1982. *Rainfall-Runoff Relationship*. Littleton, Colo.: Water Resources Publications.

______. 1982. *Statistical Analysis of Rainfall and Runoff*. Littleton, Colo.: Water Resources Publications.

SMEDEMA, L. K., AND D. W. RYCROFT. 1983. *Land Drainage*. Ithaca, N.Y.: Cornell Univ. Press.

SMITH, D. J., AND P. STOPP. 1979. *The River Basin*. Cambridge: Cambridge Univ. Press.

SOKOLOVSKII, D. C. 1971. *River Runoff Theory and Analysis*. Translated from Russian. Published for the U.S. Department of Commerce, Washington, D.C.

Soil Conservation Service. 1958. *Hydrology*. Section 4, *SCS National Engineering Handbook*, U.S. Department of Agriculture, Washington, D.C.

STEPHENSON, D. 1981. *Stormwater Hydrology and Drainage*. Amsterdam: Elsevier.

STEPHENSON, D., AND M. E. MEADOWS. 1986. *Kinematic Hydrology and Modeling*. Amsterdam: Elsevier.

SUBRAMANYA, K. 1984. *Engineering Hydrology*. New Delhi, India: Tata McGraw-Hill.

The Institution of Engineers. 1966. *River Flood Hydrology*. London.

______. 1977. *Australian Rainfall and Runoff: Flood Analysis and Design*. Australia.

TOEBES, C. 1963. *Applied Hydrology*, vols. 1 and 2. Technical Correspondence School, New Zealand Department of Education, Wellington, New Zealand.

UNNY, T. E., AND E. A. MCBEAN, eds. 1982. *Decision Making for Hydrosystems: Forecasting and Operation*. Littleton, Colo.: Water Resources Publications.

______. 1982. *Experience in Operation of Hydrosystems*. Littleton, Colo.: Water Resources Publications.

U.S. Army Corps of Engineers, 1956. *Snow Hydrology*. U.S. Army Corps of Engineers, North-Pacific Division, Portland, Oregon.

VARSHNEY, D. V., AND L. NARAIN. 1983. *Water Resources Engineering*. Delhi, India: Khanna.

VARSHNEY, R. S. 1979. *Engineering Hydrology*. Roorkee, India: Nem Chand.

VIESSMAN, W., J. W. KNAPP, G. L. LEWIS, AND T. E. HARBAUGH. 1977. *Introduction to Hydrology*. New York: Harper and Row.

WALTON, W. C. 1970. *The World of Water*. New York: Tapliger.

WANIELISTA, M. P. 1978. *Stormwater Management: Quantity and Quality*. Ann Arbor, Mich.: Ann Arbor Science Publ.

WARD, R. C. 1968. *Principles of Hydrology*. New York: McGraw-Hill.

WHITTINGTON, D. 1983. *Water Management Models in Practice: A Case Study of the Aswan High Dam*. Amsterdam: Elsevier.

WILSON, E. M. 1969. *Engineering Hydrology*. London: Macmillan.

WISLER, C. O., AND E. F. BRATER. 1949. *Hydrology*. New York: John Wiley.

WOOD, E. F., ed. 1980. *Real-Time Forecasting/Control of Water Resources Systems*. New York: Pergamon.

WOOLHISER, D. A., ed. 1973. *Decisions with Inadequate Hydrologic Data*. Fort Collins, Colo.: Water Resources Publications.

WRIGHT, C. E., ed. 1980. *Surface Water and Groundwater Interaction*. Paris: UNESCO Press.

YEN, B. C., ed. 1982. *Urban Stormwater Hydraulics and Hydrology*. Littleton, Colo.: Water Resources Publications.

______. 1982. *Urban Stormwater Quality, Management, and Planning*. Littleton, Colo.: Water Resources Publications.

YEVJEVICH, V. 1971. *Systems Approach to Hydrology*. Fort Collins, Colo.: Water Resources Publications.

______. 1972. *Probability and Statistics in Hydrology*. Fort Collins, Colo.: Water Resources Publications.

______. 1972. *Stochastic Processes in Hydrology*. Fort Collins, Colo.: Water Resources Publications.

______. 1976. *Karst Hydrology and Water Resources*, vols. 1 and 2. Fort Collins, Colo.: Water Resources Publications.

A.2 TEXTBOOKS ON SUBSURFACE HYDROLOGY

BACHMAT, Y., J. BREDEHOEFT, B. ANDREWS, D. HOLTZ, AND S. SEBASTIAN. 1980. *Groundwater Management: The Use of Numerical Models*. Water Resources Monograph 5, American Geophysical Union, Washington, D.C.

BEAR, J., D. ZASLAVSKY, AND S. IRMAY, eds. 1968. *Physical Principles of Water Percolation and Seepage*. Paris: UNESCO Press.

BEAR, J. 1972. *Dynamics of Fluids in Porous Media*. New York: American Elsevier.

______. 1976. *Hydraulics of Groundwater*. New York: McGraw-Hill.

BOUWER, H. 1978. *Groundwater Hydrology*. New York: McGraw-Hill.

BOWEN, R. 1980. *Groundwater*. New York: Halsted.

BREDEHOEFT, J. D. 1982. *Groundwater Models*. Paris: UNESCO Press.

CEDERGREN, H. R. 1967. *Seepage, Drainage, and Flow Nets*. New York: John Wiley.

CHILDS, E. C. 1947. *An Introduction to the Physical Basis of Soil-Water Phenomena*. New York: John Wiley.

COLLINS, R. E., ed. 1976. *Flow of Fluids Through Porous Materials*. Tulsa, Okla.: Petroleum Publishing Co.

COREY, A. T. 1977. *Mechanics of Heterogeneous Fluids in Porous Media*. Littleton, Colo.: Water Resources Publications.

DAVIS, S. N., AND R. J. M. DEWIEST. 1966. *Hydrogeology*. New York: John Wiley.

DEWIEST, R. J. M. 1965. *Geohydrology*. New York: John Wiley.

DEWIEST, R. J. M., ed. 1969. *Flow Through Porous Media*. New York: Academic Press.

DOMENICO, P. A. 1972. *Concepts and Models in Groundwater Hydrology*. New York: McGraw-Hill.

E. E. JOHNSON, INC. 1966. *Ground Water and Wells*. St. Paul, Minn.: Edward E. Johnson.

FETTER, C. W., JR. 1980. *Applied Hydrogeology*. Columbus, Ohio: C. E. Merrill.

FREEZE, R. A., AND CHERRY, J. A. 1979. *Groundwater*. Englewood Cliffs, N.J.: Prentice-Hall, Inc.

FYFE, W. S. 1978. *Fluids in the Earth's Crust*. New York: Elsevier.

GLOVER, R. E. 1977. *Transient Ground Water Hydraulics*. Littleton, Colo.: Water Resources Publications.

GREENKORN, R. 1983. *Flow Phenomena in Porous Media*. New York: Marcel Dekker.

HALEK, V., AND J. SVEC. 1979. *Groundwater Hydraulics*. New York: Elsevier.

HARR, M. E. 1962. *Groundwater and Seepage*. New York: McGraw-Hill.

HEATH, R. C., AND F. W. TRAINER. 1981. *Introduction to Ground Water Hydrology*. Worthington, Ohio: Water Well Journal Publishing Co.

HUNT, B. 1983. *Mathematical Analysis of Groundwater Resources*. London: Butterworths.

International Association for Hydraulic Research. 1972. *Fundamentals of Transport Phenomena in Porous Media. Developments in Soil Science* 2. New York: Elsevier.

JACOB, C. E. 1950. *Flow of Groundwater*. New York: John Wiley.

JAVANDEL, I., C. DOUGHTY, AND C. F. TSANG. 1984. *Groundwater Transport: Handbook of Mathematical Models*. Water Resources Monograph Series 10, American Geophysical Union, Washington, D.C.

KIRKHAM, D., AND W. L. POWERS. 1972. *Advanced Soil Physics*. New York: Wiley-Interscience.

KOVACS, G. 1981. *Seepage Hydraulics*. Amsterdam: Elsevier.

LIGGETT, J. A., AND P. L. F. LIU. 1983. *The Boundary Integral Equation Method for Porous Media Flow*. Boston: George Allen and Unwin.

LUTHIN, J. N., ed. 1967. *Drainage of Agricultural Lands*. Madison, Wis.: American Society of Agronomy Publishers.

MANDEL, S., AND Z. L. SHIFTEN. 1981. *Groundwater Resources: Investigation and Development*. New York: Academic Press.

MARINO, M. A. 1982. *Seepage and Ground Water*. Amsterdam: Elsevier.

MARSHALL, T. J., AND J. W. HOLMES. 1979. *Soil Physics*. Cambridge: Cambridge Univ. Press.

MCWHORTER, D. B., AND D. K. SUNADA. 1977. *Ground-Water Hydrology and Hydraulics*. Littleton, Colo.: Water Resources Publications.

MOREL-SEYTOUX, H. J., J. D. SALAS, T. G. SANDERS, AND R. E. SMITH, eds. 1977. *Surface and Subsurface Hydrology*. Colorado State University, Fort Collins: Water Resources Publications.

MORISON, R. D. 1983. *Groundwater Monitoring Technology: Procedures, Equipment, and Applications*. Timco, Mfg., Prairie du Sac, Wisconsin.

MUSKAT, M. 1907. *The Flow of Homogeneous Fluid through Porous Media*. Boston: International Human Resources.

NARASIMHAN, T. N., ed. 1982. *Recent Trends in Hydrogeology*. Special Paper 189, The Geological Society of America, Boulder, Colo.

National Research Council. 1984. *Groundwater Contamination*. Washington, D.C.: National Academy Press.

NYE, P. H., AND P. B. TINKER. 1977. *Solute Movement in the Soil-Root System*. Oxford, England: Blackwells.

PINDER, G. F., ed. 1983. *Flow through Porous Media*. Southampton, England: CML Publications.

PINDER, G. F., AND W. G. GRAY. 1977. *Finite Element Simulation in Surface and Subsurface Hydrology*. New York: Academic Press.

POLUBARINOVA-KOCHINA, P. Y. 1962. *The Theory of Ground Water Movement*. Translated from the Russian by Roger J. M. DeWiest. Princeton, N.J.: Princeton Univ. Press.

PYE, V. I., R. PATRIC, AND J. QUARTES. 1983. *Groundwater Contamination in the United States*. Philadelphia: Univ. of Pennsylvania Press.

RAGHUNATH, H. M. 1982. *Ground Water*. New Delhi, India: Wiley Eastern.

RAUDKIVI, A. J. AND R. A. CALLANDER. 1976. *Analysis of Groundwater Flow*. London: Edward Arnold.

REMSON, I., G. M. HORNBURGER, AND F. J. MOLZ. 1971. *Numerical Methods in Subsurface Hydrology*. New York: Wiley-Interscience.

ROSE, C. W. 1969. *Agricultural Physics*. New York: Pergamon.

ROSENSHEIM, J. AND G. D. BENNETT, eds. 1984. *Groundwater Hydraulics*. Water Resources Monograph Series 9, American Geophysical Union, Washington, D.C.

SCHEIDEGGER, A. E. 1974. *The Physics of Flow through Porous Media*. Toronto, Canada: Univ. of Toronto Press.

TODD, D. K. 1980. *Ground Water Hydrology*. 2d ed. New York: John Wiley.

VERRUIJT, A. 1970. *Theory of Groundwater Flow*. New York: Gordon Breach.

WALTON, W. C. 1970. *Ground Water Resource Evaluation*. New York: McGraw-Hill.

WANG, H. F. 1982. *Introduction to Groundwater Modeling: Finite Differences and Finite Element Methods*. San Francisco: W. H. Freeman.

Water and Power Resources Service. 1981. *Ground Water Manual*. A Water Resources Technical Publication, U.S. Department of Interior, Washington, D.C.

WOOD, E. F., R. A. FERRARA, W. G. GRAY, AND G. F. PINDER. 1984. *Groundwater Contamination from Hazardous Wastes*. Englewood Cliffs, N.J.: Prentice-Hall, Inc.

A.3 TEXTBOOKS ON WATER RESOURCES: PLANNING, OPERATION, AND MANAGEMENT

BURAS, N. 1972. *Scientific Allocation of Water Resources*. New York: American Elsevier.

ECKSTEIN, O. 1958. *Water Resource Development: The Economics of Project Evaluation*. Cambridge: Harvard Univ. Press.

GOODMAN, A. S. 1984. *Principles of Water Resources Planning*. Englewood Cliffs, N.J.: Prentice-Hall, Inc.

HAIMES, Y. Y. 1977. *Hierarchical Analyses of Water Resources Systems*. New York: McGraw-Hill.

HALL, W. A. AND J. A. DRACUP. 1970. *Water Resources Systems Engineering*. New York: McGraw-Hill.

JAMES, L. D. AND R. R. LEE. 1971. *Economics of Water Resources Planning*. New York: McGraw-Hill.

LOUCKS, D. P., J. R. STEDINGER, AND D. A. HAITH. 1981. *Water Resource Systems Planning and Analysis*. Englewood Cliffs, N.J.: Prentice-Hall, Inc.

MAJOR, D. C., AND R. L. LENTON, eds. 1979. *Applied Water Resource Systems Planning*. Englewood Cliffs, N.J.: Prentice-Hall, Inc.

MASS, A., M. M. HUFSCHMIDT, R. DORFMAN, H. A. THOMAS, S. A. MARGLIN, AND G. M. FAIR. 1962. *Design of Water-Resource Systems*. Cambridge: Harvard Univ. Press.

Meta Systems, Inc. 1975. *Systems Analysis in Water Resources Planning*. Water Information Center, Inc., Port Washington, N.Y.

NAPIER, T. L., D. SCOTT, K. W. EASTER, AND R. SUPALLA. 1983. *Water Resources Research*. Soil Conservation Society of America, Ankeny, Ia.

PETERSON, M. S. 1984. *Water Resources Planning and Development*. Englewood Cliffs, N.J.: Prentice-Hall, Inc.

VIESSMAN, W. AND C. WELTY. 1984. *Water Management Technology and Institutions*. New York: Harper and Row.

A.4 PERIODICALS IN AND PERTINENT TO HYDROLOGY

Advances in Water Resources. Southampton, England: CML.

Agricultural Water Management. Amsterdam: Elsevier.

Canadian Agricultural Engineering. Ottawa, Canada: Canadian Society of Agricultural Engineers.

Canadian Journal of Civil Engineering. Ottawa, Canada: Canadian Society of Civil Engineers.

Canadian Journal of Earth Sciences. Ottawa, Canada: National Research Council.

Canadian Journal of Soil Science. Ottawa, Canada: Agricultural Institute of Canada.

Canadian Water Resources Journal. Ottawa, Canada: Canadian Water Resources Association.

CATENA. Leuz-Verlay, Germany: Giesseu.

Earth Surface Processes. New York: John Wiley.

Environmental Geology and Water Science. Berlin: Springer-Verlag.

Ground Water. Worthington, Ohio: National Water Well Association.

Hydrological Processes. New York: John Wiley.

Hydrological Sciences Journal. International Association of Hydrological Sciences.

Hydrological Science and Technology: Short Papers. American Institute of Hydrology.

Hydrology, Journal of the Indian Association of Hydrologists. Roorkee, India: Indian Association of Hydrologists.

Journal of the Institution of Engineers. Canberra, Australia: Institution of Engineers.

Journal of Contaminant Hydrology. Amsterdam: Elsevier.

Journal of Engineering Mechanics. American Society of Civil Engineers.

Journal of Environmental Engineering. American Society of Civil Engineers.

Journal of Environmental Quality. American Society of Agronomy, Crop Science Society of America, and the Soil Science Society of America.

Journal of Geophysical Research. American Geophysical Union.

Journal of Harbors and Coastal Engineering. American Society of Civil Engineers.

Journal of Hydraulic Engineering. American Society of Civil Engineers.

Journal of Hydraulic Research. International Association for Hydraulic Research.

Journal of Hydrological Sciences. Warsaw, Poland: Polish Academy of Sciences.

Journal of Hydrology (New Zealand). Wellington, New Zealand: New Zealand Hydrological Society.

Journal of Hydrology. Amsterdam: North-Holland.

Journal of Indian Water Resources Society. Roorkee, India: Indian Water Resources Society.

Journal of Irrigation and Drainage Engineering. American Society of Civil Engineers.

Journal of Soil and Water Conservation. Soil Conservation Society of America.

Journal of Water Resources Planning and Management. American Society of Civil Engineers.

Journal of the Institution of Water Engineers. Institution of Water Engineers, Great Britain.

Nordic Hydrology. Copenhagen: Munksgaard.

Proceedings of the Institution of Civil Engineers. London: Institution of Civil Engineers.

Soil Science. Baltimore, Md.: Williams and Wilkins.

Soil Science Society of America Journal. Soil Science Society of America.

Soviet Hydrology. Selected papers translated from current hydrologic literature appearing in the Soviet Serial publications. American Geophysical Union.

Stochastic Hydrology and Hydraulics. Berlin: Springer Produktions-Gesellschaft.

Transactions of the American Geophysical Union. American Geophysical Union.

Transactions of the American Society of Agricultural Engineers. American Society of Agricultural Engineers.

Transport in Porous Media. D. Reidel.

Water International. International Water Resources Association.

Water Resources Bulletin. American Water Resources Association.

Water Resources Management. Athens, Greece: European Water Resources Association.

Water Resources Research. American Geophysical Union.

Water Supply and Management. London: Pergamon.

APPENDIX B

Conversion from One System of Units to Another

B.1 COMMONLY USED ABBREVIATIONS

Unit	*Abbreviation*
Acre	ac
Acre-foot	acre-ft or ac-ft
British thermal unit	BTU
Cubic feet per second	ft^3/s or cfs
Cubic meters per second	m^3/s or cms
Centimeter	cm
degrees Celsius	°C
degrees Fahrenheit	°F
Foot or feet	ft
Gallon per minute	gal/min or gpm
Gram	g
Hectare	ha
Hour	h
Inch	in.
Joule	J
Kilogram	kg
Kilometer	km
Liter	l
Langley	ly
Log to the base *e*	ln
Log to the base 10	log
Meter	m
Mile	mi
Millibar	mb
Minute	min
Millimeter	mm
Newton	N
Parts per million by weight	ppm
Pound	lb
Second	s
Second-foot-day	sfd
Yard	yd
Watt	W

B.2 COMMONLY USED UNITS OF MEASUREMENT

B.2.1 Length

1 meter (or metre) = 39.37 in. = 3.2808 ft

1 inch = 2.54 cm

1 foot = 12 in.

1 mile = 5280 ft

1 kilometer = 1000 m

B.2.2 Area

Square meter

Square foot

1 acre = 43,560 ft^2 = 0.4047 ha

1 hectare = 0.01 sq km = 2.471 acres

1 square kilometer = 0.3861 sq mi = 247.1 acres = 100 ha = 10^6 m^2

1 square yard = 9 ft^2

B.2.3 Volume

1 cubic meter = 1000 l = 1.308 yd^3 = 35.31 ft^3

1 liter (or litre) = 0.001 m^3

Gallon

1 acre-foot = 1233 m^3

Acre-inch

Cubic foot

Cubic yard

B.2.4 Weight

1 kilogram = 2.205 lb
Pounds
1 long ton = 2240 lb
1 short ton = 2000 lb
1 metric ton (tonne) = 1000 kg = 2204.62 lb

B.2.5 Pressure

Bar
Pounds per square inch
Newtons per square meter

B.2.6 Force

Newton, N (kg·m/s)
Dyne
1 pound = 4.448 N
1 kilogram = 9.807 N

B.2.7 Power

1 watt = 1 J/s
1 horsepower = 745.7 W
1 kilowatt = 1.341 horsepower = 737 ft-lb/s
= 102 kg·m/s

B.2.8 Energy

Joule (newton × meter)
1 erg = 10^{-7} J
1 calorie = 4.184 J = .00397 BTU
1 BTU = 1055 J
1 kilowatt-hour = 3.6×10^6 J

B.2.9 Temperature

Degrees Celsius (°C)
Degrees Fahrenheit (°F)
Degrees Kelvin (K)

B.3 COMMONLY USED CONVERSION FACTORS

$e = 2.71828$
$\text{Log}_{10}e = 0.43429$
$\ln 10 = \log_e 10 = 2.30259$
Acceleration due to gravity (g) = 32.17 ft/s^2
= 980.6 cm/s^2
1 ft^3/s = .9917 acre-in./h
= .000214 mi^3/year
= 1 s-ft
= 1 s^3
1 horsepower = .746 kw
= 550 ft-lb/s
1 in. of runoff per square mile = 26.9 sfd
= 53.3 acre-ft
= 2,323,200 ft^3
1 sfd/mi^2 = .03719 in.
1 in.^{3} of water = .0361 lb
Heat of fusion = 79.7 cal/g = 144 BTU/lb
Heat of vaporization of water = 540 cal/g = 970 BTU/lb
1 cal/cm^2 = 3.69 BTU/ft^2

B.4 CONVERSION OF LENGTH UNITS (after Schulz, 1973)

1 unit	Equivalent of Unit Value, that is, 1 m = 3.28 ft									
	mm	*cm*	*dm*	*m*	*km*	*in.*	*ft*	*yd*	*mi*	*naut mi*
millimeters (mm)	1	0.1	.01	.001	10^{-6}	.03937	.00328	.00109	6.21×10^{-7}	5.39×10^{-7}
centimeters (cm)	10	1	0.1	.01	.00001	.3937	.0328	.0109	6.21×10^{-6}	5.39×10^{-6}
decimeters (dm)	100	10	1	0.1	.0001	3.937	.3281	.1094	6.21×10^{-5}	5.39×10^{-5}
meters (m)	1000	100	10	1	.001	39.37	3.281	1.094	6.21×10^{-4}	5.39×10^{-4}
kilometers (km)	10^6	10^5	10^4	10^3	1	39370	3281	1093.6	0.621	.5396
inches (in.)	25.4	2.54	.254	.0254	2.54×10^{-5}	1	.0833	.0278	1.58×10^{-5}	1.37×10^{-5}
feet (ft)	304.8	30.48	3.048	.3048	3.05×10^{-4}	12	1	.333	1.89×10^{-4}	1.64×10^{-4}
yards (yd)	914.4	91.44	9.144	.9144	9.14×10^{-4}	36	3	1	5.68×10^{-4}	4.93×10^{-4}
miles (mi)	1.61×10^6	1.61×10^5	1.61×10^4	1.61×10^3	1.6093	63360	5280	1760	1	.8684
nautical mile (naut mi)	1.85×10^6	1.85×10^5	1.85×10^4	1.853	1.853	72963	6076	2025	1.151	1

B.5 CONVERSION OF AREA UNITS (after Schulz, 1973)

1 unit	Equivalent of Unit Value, that is, 1 mi² = 640 acres										
	cm²	*m²*	*km²*	*ha*	*in.²*	*ft²*	*yd²*	*mi²*	*ac*	*morgen*	*rai*
square centimeters	1	.0001	10^{-10}	10^{-8}	.155	1.08×10^{-3}	1.2×10^{-4}	3.861×10^{-11}	2.471×10^{-8}	—	—
square meters (m²)	10^4	1	10^{-6}	10^{-4}	1,550	10.76	1.196	3.861×10^{-7}	2.47×10^{-4}	1.17×10^{-4}	6.25×10^{-4}
square km (km²)	10^{10}	10^6	1	100	1.55×10^9	10.76×10^6	1.196×10^6	.3861	247.1	117	625
hectares (ha)	10^8	10^4	.01	1	1.55×10^7	10.76×10^4	1.196×10^4	3.861×10^{-3}	2.471	1.17	6.25
square inch (in.²)	6.452	6.4×10^{-4}	6.45×10^{-10}	6.45×10^{-8}	1	6.94×10^{-3}	7.7×10^{-4}	2.49×10^{-10}	1.574×10^{-7}	7.55×10^{-8}	4.03×10^{-7}
square feet (ft²)	929	.0929	9.29×10^{-8}	9.29×10^{-6}	144	1	.1111	3.587×10^{-8}	2.3×10^{-5}	1.09×10^{-5}	5.80×10^{-5}
square yard (yd²)	8361	.8361	8.36×10^{-7}	8.36×10^{-5}	1296	9	1	3.23×10^{-7}	2.07×10^{-4}	9.82×10^{-5}	5.22×10^{-4}
square mile (mi²)	—	2,589,998	2.59	259	—	27.87×10^6	3.098×10^6	1	640	302.5	1,619
acre (ac)	40.4×10^6	4,047	4.047×10^{-3}	.4047	6.27×10^6	43,560	4,840	1.562×10^{-3}	1	.473	2.53
morgen	85.5×10^6	8,550	8.55×10^{-3}	.855	1.322×10^7	92,000	10,220	3.3×10^{-3}	2.111	1	5.34
rai	16×10^6	1,600	1.600	.1600	2.48×10^6	17,250	1,917	6.18×10^{-4}	0.395	.187	1

B.6 CONVERSION OF VOLUME UNITS (after Schulz, 1973).

1 unit	Equivalent of Unit Value, that is, 1 acre-ft = 43,560 cu ft									
	cm^3	l	cu meters	cu in.	cu ft	U.S. gal	Imp. gal	acre-ft	sfd	mil gal
cubic centimeters (cm^3)	1	.001	10^{-6}	.06102	3.53×10^{-5}	2.64×10^{-4}	2.2×10^{-4}	8.1×10^{-10}	4.09×10^{-10}	2.64×10^{-10}
liters (l)	1,000	1	.001	61.023	.0353	.26417	.22008	8.1×10^{-7}	4.09×10^{-7}	2.64×10^{-7}
cubic meter (cm)	10^6	1,000	1	61,023	35.314	264.17	220.08	8.107×10^{-4}	4.09×10^{-4}	2.642×10^{-4}
cubic inches ($in.^3$)	16.39	1.64×10^{-2}	1.64×10^{-4}	1	5.79×10^{-4}	4.33×10^{-3}	3.61×10^{-3}	1.218×10^{-8}	6.698×10^{-9}	4.329×10^{-9}
cubic feet (ft^3)	28,317	28.317	.02832	1728	1	7.481	6.232	2.296×10^{-5}	1.157×10^{-5}	7.48×10^{-6}
U.S. gallon	3,785.4	3.785	3.785×10^{-3}	231	.13368	1	.83311	3.069×10^{-6}	1.547×10^{-6}	10^{-6}
Imperial gallon (Imp. gal)	4,542.5	4.542	4.54×10^{-3}	277.3	.16046	1.200	1	3.684×10^{-6}	1.86×10^{-6}	1.2×10^{-6}
acre feet (acre-ft)	1.233×10^9	1.233×10^6	1,233.5	75.27×10^6	43,560	3.26×10^5	2.75×10^5	1	.5042	.3260
second-foot-day (sfd)	2.45×10^9	2.45×10^6	2,451.3	1.493×10^8	86,400	6.46×10^5	5.38×10^5	1.9835	1	.6463
million gallons (mil gal)	3.785×10^9	3.785×10^6	3,785	2.31×10^8	1.338×10^5	10^6	8.33×10^5	3.0684	1.5472	1

B.7 CONVERSION OF FLOW UNITS (after Schulz, 1973).

1 Unit	Equivalent of Unit Value, that is, 1 ft^3/s = 28.317 l/s										
	m^3/c	m^3/da	l/s	ft^3/s	ft^3/da	ac-ft/da	gal/min	gal/da	million gal/da	Imp. gal/min	Imp gal/da
cubic meters per second (m^3/s)	1	86,400	1,000	35.314	3.051×10^6	70.045	15,850	22.82×10^6	22.824	13,208	19.02×10^6
cubic meters per day (m^3/da)	1.157×10^{-5}	1	.116	4.09×10^{-4}	35.314	8.1×10^{-4}	.1835	264.17	2.64×10^{-4}	.1528	220.14
liters/second (l/s)	.001	86.4	1	.0353	3,051.2	.070	15.85	22,824	2.28×10^{-2}	13.208	19,020
cubic feet per sec (ft^3/s)	.0283	2,446.6	28.317	1	86,400	1.9835	448.83	646,323	.6463	374.03	538.603
cubic feet per day (ft^3/da)	3.28×10^{-7}	.02832	3.28×10^{-4}	1.16×10^{-5}	1	2.3×10^{-5}	5.19×10^{-3}	7.48	7.48×10^{-6}	4.33×10^{-3}	6.233
acre-foot per day (ac-ft/da)	.0143	1,233.5	14.276	.5042	43,560	1	226.28	325,850	.3258	188.57	271,542
U.S. gallons/minute (gal/min)	6.309×10^{-5}	5.451	.0631	2.23×10^{-3}	192.5	4.42×10^{-3}	1	1,440	1.44×10^{-3}	.8333	1,200
U.S. gallons/day (gal/da)	4.38×10^{-8}	3.785×10^{-3}	4.382×10^{-5}	1.55×10^{-6}	.1337	3.07×10^{-6}	6.94×10^{-4}	1	10^{-6}	5.79×10^{-4}	.833
million gallons per day (mill gal/da)	4.38×10^{-2}	3,785	43.82	1.55	1.337×10^5	3.07	694	10^6	1	579	8.33×10^5
Imperial gallons per minute (Imp gal/min)	7.47×10^{-5}	6.541	.0757	2.67×10^{-3}	231.12	5.3×10^{-3}	1.2	1,728	1.7×10^{-3}	1	1,440
Imperial gallons per day (Imp gal/da)	5.26×10^{-8}	4.54×10^{-3}	5.26×10^{-5}	1.86×10^{-6}	.160	3.68×10^{-6}	8.33×10^{-4}	1.2	1.2×10^{-6}	6.94×10^{-4}	1

Reference: Schulz, E. F. 1973. *Problems in Applied Hydrology*. Fort Collins, Colo.: Water Resources Publications.

APPENDIX C

Derivation of the Continuity Equation

Consider an elemental area in a watershed. Let Δx and Δy be the two sides of this elemental area. It has a depth of water d, which varies with x and y. We can write the rate of change of mass for the entire watershed as

$$\frac{\partial}{\partial t}\int\int_A \rho d \, dx \, dy \tag{C.1}$$

where ρ is the density of water, a constant. The water enters through part of the surface and leaves through part of the surface. Therefore, the outward flux is

$$\int\int_A \rho \mathbf{v} \cdot \mathbf{m} \, dA \tag{C.2}$$

where dA is an elemental surface area, $\mathbf{m}$ is the unit outward normal vector, and $\mathbf{v}$ the velocity vector. The principle of conservation of mass requires that

$$\int\int_A \rho \mathbf{v} \cdot \mathbf{m} \, dA + \frac{d}{dt}\int\int_A \rho d \, dx \, dy = 0 \tag{C.3}$$

We cannot integrate equation (C.3) because we do not know the variation of d and $\mathbf{v}$ with x and y. If the watershed is assumed to be a lumped system, then the space coordinates can be eliminated. Therefore, equation (C.3) becomes

$$I = Q + \frac{dS}{dt} \tag{C.4}$$

where I is the rate of inflow, Q is the rate of outflow, and S is the storage. This is one of the basic equations used in systems hydrology.

APPENDIX D

Derivation of a General Storage Equation

We assume that the watershed behavior can be described by analytic functions. This assumption may not be unduly restrictive and the deviation from reality may not be large. It is acknowledged that storage is governed by the characteristics of inflow and outflow for a given watershed. It must be noted that the geometry is fixed for that watershed. Its influence on storage will be reflected through coefficients in the functional relationship. In the derivation of a general storage equation, we follow the work of Kulandaiswamy (1964) and Chow and Kulandaiswamy (1971).

In its most general form, the watershed storage may be expressed as

$$S = F\left(Q, \frac{dQ}{dt}, \frac{d^2Q}{dt^2}, \frac{d^3Q}{dt^3}, \ldots; I, \frac{dI}{dt}, \frac{d^2I}{dt^2}, \ldots\right) \quad \text{(D.1)}$$

If I continues at a constant rate for an indefinite period of time, Q will reach a maximum value after some time and will remain so from then on. This is the equilibrium state, or steady state, of the system. Denoting storage, inflow, and outflow under equilibrium conditions by $\bar{S}$, $\bar{I}$, $\bar{Q}$, we can express equation (D.1) as

$$\bar{S} = F(\bar{Q}, 0, 0, 0, \ldots; \bar{I}, 0, 0, 0, \ldots)$$
$$\bar{S} = F(\bar{Q}, \bar{I}) \quad \text{(D.2)}$$

Equation (D.2) is the equilibrium equation of the watershed system. By the assumption of S being analytic, it can be expanded in Taylor series about equilibrium. Let

$$y_i = \frac{d^iQ}{dt^i}, \qquad i = 0, 1, 2, \ldots, M \quad \text{(D.3a)}$$

and

$$x_j = \frac{d^jI}{dt^j}, \qquad j = 0, 1, 2, \ldots, N \quad \text{(D.3b)}$$

where M and N represent the order of derivatives to be chosen in equation (D.1). Expressing equation (D.1) as

$$S = F(y_0, y_1, y_2, \ldots, y_M; x_0, x_1, x_2, \ldots, x_N) \quad \text{(D.4)}$$

let us define

$$w = \left[\sum_{i=1}^{M} (y_i - \bar{y}_i)\frac{\partial}{\partial y_i} + \sum_{j=0}^{N} (x_j - \bar{x}_j)\frac{\partial}{\partial x_j}\right] F \quad \text{(D.5)}$$

Performing the Taylor series expansion of equation (D.4) and using the notation of equation (D.5) we obtain

$$S = F(\bar{y}_0, \bar{x}_0) + \sum_{j=1}^{\infty} \frac{w^j}{j!} F \quad \text{(D.6)}$$

where the exponent of w is to be interpreted in accord with the binomial expansion:

$$\left(\delta\frac{\partial}{\partial x} + \gamma\frac{\partial}{\partial y}\right)^k Z(x, y) = \sum_{j=0}^{k} \binom{k}{j} \delta^{k-j}\gamma^j Z_x\, k - j_{y^j} \quad \text{(D.7)}$$

Consider an example when $j = 2$. We can write

$$\left[(y_0 - \bar{y}_0)\frac{\partial}{\partial y_0} + (x_0 - \bar{x}_0)\frac{\partial}{\partial x_0}\right]^2 F = (y_0 - \bar{y}_0)^2\frac{\partial^2 F}{\partial y_0^2} + 2(y_0 - \bar{y}_0)(x_0 - \bar{x}_0)\frac{\partial^2 F}{\partial y_0 \partial x_0} + (x_0 - \bar{x}_0)^2\frac{\partial^2 F}{\partial x_0^2} \quad \text{(D.8)}$$

All the derivatives of F in equation (D.6) must be evaluated at $y_0 = \bar{y}_0$ and $x_0 = \bar{x}_0$. If we can assume that the quantities $(y_0 - \bar{y}_0)$, $(y_1 - \bar{y}_1)$, . . . , $(y_M - \bar{y}_M)$; $(x_0 - \bar{x}_0)$, $(x_1 - \bar{x}_1)$, $(x_2 - \bar{x}_2)$, . . . , $(x_N - \bar{x}_N)$ are small relative to $\bar{y}_0$ and $\bar{x}_0$, then the powers and products of these terms can be dropped. That is,

$$w^j = 0, \qquad j \geq 2 \quad \text{(D.9)}$$

Equation (D.6) reduces to

$$S = F(\bar{y}_0, \bar{x}_0) + wF \quad \text{(D.10)}$$

Recalling the equilibrium condition,

$$\frac{d^iQ}{dt^i} = 0, \qquad i \geq 1 \text{ for } Q = \bar{Q}$$

$$\frac{d^jI}{dt^j} = 0, \qquad j \geq 1 \text{ for } I = \bar{I} \quad \text{(D.11)}$$

Hence

$$\bar{y}_i = 0, \qquad i \geq 1$$

$$\bar{x}_j = 0, \qquad j \geq 1$$

Then equation (D.10) reduces to:

$$S = F(\bar{y}_0, \bar{x}_0) + (y_0 - \bar{y}_0)\frac{\partial F}{\partial y_0} + \sum_{i=2}^{M} y_i \frac{\partial F}{\partial y_i} + (x_0 - \bar{x}_0)\frac{\partial F}{\partial x_0} + \sum_{j=2}^{N} x_j \frac{\partial F}{\partial x_j} \quad \text{(D.12)}$$

Let us define

$$a_i = \frac{\partial F}{\partial y_i}, \qquad i \geq 1, \quad \text{at } y_0 = \bar{y}_0, x_0 = \bar{x}_0 \quad \text{(D.13)}$$

$$b_j = \frac{\partial F}{\partial x_j}, \qquad i \geq 1, \quad \text{at } y_0 = \bar{y}_0, x_0 = \bar{x}_0 \quad \text{(D.13)}$$

Equation (D.12) reduces to:

$$S = [F(\bar{y}_0, \bar{x}_0) - (a_0\bar{y}_0 + b_0\bar{x}_0)] + \sum_{i=0}^{M} a_i y_i + \sum_{j=0}^{N} b_j x_j \quad \text{(D.14)}$$

If the function $F(\bar{y}_0, \bar{x}_0)$ is linear, it can be shown that

$$F(\bar{y}_0, \bar{x}_0) = a_0\bar{y}_0 + b_0\bar{x}_0 \quad \text{(D.15)}$$

If $F(\bar{y}_0, \bar{x}_0)$ is not linear, we can express

$$\epsilon = F(\bar{y}_0, \bar{x}_0) - a_0\bar{y}_0 - b_0\bar{x}_0 \quad \text{(D.16)}$$

where ϵ depends on $\bar{x}_0$ and $\bar{y}_0$. If $F(\bar{y}_0, \bar{x}_0)$ is not highly nonlinear, ϵ may be small in comparison with the remaining terms in equation (D.14), that is,

$$\epsilon <<< \sum_{i=0}^{M} a_i y_i + \sum_{j=0}^{N} b_j x_j \quad \text{(D.17)}$$

If equation (D.17) is valid, that is, the nonlinear effects in the storage equation under equilibrium conditions are vanishingly small, then equation (D.14) can be written as

$$S = \sum_{i=0}^{M} a_i y_i + \sum_{j=0}^{N} b_j x_j \quad \text{(D.18)}$$

Here a_i's and b_j's are either constants or functions of $\bar{y}_0$ and/or $\bar{x}_0$. This, of course, depends on whether the system is linear or not.

Consider that the runoff hydrograph is divided into R intervals of duration Δt each. The storage equation for the jth interval can be written as

$$S_j = \sum_{i=0}^{M} a_{ij} y_{ij} + \sum_{i=0}^{N} b_{ij} x_{ij}, \qquad j = 1, 2, \ldots, R \quad \text{(D.19)}$$

where S_j, y_{ij}, and x_{ij} are the values of S, y, and x restricted to the jth interval. In that case, coefficients a_i and b are precisely

$$a_{ij} = \frac{\partial F}{\partial y_i}, \qquad y_0 = \bar{y}_{0j}, \qquad x_0 = \bar{x}_{0j} \quad \text{(D.20)}$$

$$b_{ij} = \frac{\partial F}{\partial x_i}, \qquad y_0 = \bar{y}_{0j}, \qquad x_0 = \bar{x}_{0j} \quad \text{(D.21)}$$

where $\bar{y}_{0j}$ and $\bar{x}_{0j}$ are the equilibrium values of y_0 and x_0 for the jth interval. The values of these coefficients may change when R is changed. We can also write them as $a_{ij}(R)$ and $b_{ij}(R)$, just to emphasize their dependence on R. When Δt becomes small, the values of $\bar{y}_{0j}(R)$ and $\bar{x}_{0j}(R)$ become more and more representative of the values of y_0 and x_0 in that interval. It may therefore be reasonable to conclude that the values of $a_{ij}(R)$ and $b_{ij}(R)$ should ultimately approach the functional values of $\partial F/\partial y_i$ and $\partial F/\partial x_i$ as $R \to \infty$ or $\Delta t \to 0$. The storage equation then becomes

$$S = \sum_{i=0}^{M} a_i(x_0, y_0) y_i + \sum_{i=0}^{N} b_i(x_0, y_0) x_i \quad \text{(D.22)}$$

Replacing x_0 and y_0 by I and Q, equation (D.22) can be replaced by

$$S = \sum_{i=0}^{M} a_i(I, Q)\frac{d^iQ}{dt^i} + \sum_{j=0}^{N} b_j(I, Q)\frac{d^jI}{dt^j} \quad \text{(D.23)}$$

If the a_i's and b_i's are constant, that is, independent of I and Q, S is obtained as a weighted linear sum of I, Q, and their time derivatives, where the coefficients are the weights.

A Comment

The general storage equation can be derived more simply as follows. Let equation (D.1) be expressed as

$$S = F(Q, Q', Q'', \ldots ; I, I', I'', \ldots) \quad \text{(D.24)}$$

Differentiating on both sides

$$dS = \frac{\partial F}{\partial Q} dQ + \frac{\partial F}{\partial Q'} dq' + \frac{\partial F}{\partial Q''} dQ'' + \cdots + \frac{\partial F}{\partial I} dI + \frac{\partial F}{\partial I'} dI' + \frac{\partial F}{\partial I''} dI'' + \cdots \quad \text{(D.25)}$$

Integrating equation (D.25) over the interval $[t_0, t]$, with t_0 denoting the time when the system is in steady state,

$$S - S_0 = \frac{\partial F}{\partial Q}(Q - Q_0) + \frac{\partial F}{\partial Q'}(Q' - Q_0') + \frac{\partial F}{\partial Q''}(Q'' - Q_0'') + \cdots \frac{\partial F}{\partial I}(I - I') + \frac{\partial F}{\partial I}(I' - I_0') + \frac{\partial F}{\partial I}(I'' - I_0'') + \cdots \quad \text{(D.26)}$$

where $S_0 = S(t_0)$, $I_0^{(i)} = I^{(i)}(t_0)$, and $Q_0^{(i)} = Q^{(i)}(t_0)$, $i = 1, 2, \ldots$. At $t = t_0$, all the derivatives of I and Q vanish. Therefore,

$$S = S_0 - \frac{\partial F}{\partial Q} Q_0 - \frac{\partial F}{\partial I} I_0 + \sum_{i=0} \frac{\partial^i F}{\partial Q^i} Q^{(i)} + \sum_{j=0} \frac{\partial^j F}{\partial I^j} I^{(j)} \quad \text{(D.27)}$$

If the system is linear, then

$$S_0 - \frac{\partial F}{\partial Q} Q_0 - \frac{\partial F}{\partial I} I_0 = 0 \quad \text{(D.28)}$$

and

$$S = \sum_{i=0} \frac{\partial^i F}{\partial Q^i} Q^{(i)} + \sum_{j=0} \frac{\partial^j F}{\partial I^j} I^{(j)} \quad \text{(D.29)}$$

which is similar to equation (D.22).

APPENDIX E

Derivation of Overland Flow Travel Time

Let there be a plane of unit width and length L, subject to a uniform effective rainfall intensity I. The plane is inclined at a slope of S. If v is the velocity of flow, then the time of concentration of the plane T_c can be expressed as

$$T_c = \int_0^L \frac{1}{v}\,dx \tag{E.1}$$

Here, $v = q/d$, where d is the depth of flow and q is the discharge unit width. On a unit-width basis, d is identical to cross-sectional area. The discharge in cubic feet per second is equal to 1.008 times I in inches per hour multiplied by the drainage area in acres. Therefore,

$$q = \frac{1.008IL}{43{,}560}$$

or

$$q = \frac{IL}{(12 \times 3600)} = \frac{IL}{43{,}200} \tag{E.2}$$

Let us assume that

$$q = Kd^aS^b \tag{E.3}$$

By using equation (E.3) we can express v as

$$\frac{1}{v} = \frac{d}{q} = \frac{1}{q}\left[\frac{q}{KS^b}\right]^{1/a} \tag{E.4}$$

In equation (E.4) q can be eliminated by inserting equation (E.2):

$$\frac{1}{v} = \frac{d}{q} = \left[\frac{IL}{43{,}200}\right]^{-1}\left[\frac{IL}{43{,}200 \times K \times S^b}\right]^{1/a} \tag{E.5}$$

Substituting equation (E.5) in equation (E.1) and simplifying,

$$\begin{aligned} T_c &= \left(\frac{1}{KS^b}\right)^{1/a}\left(\frac{I}{43{,}200}\right)^{(1-a)/a}\int_0^L x^{(1-a)/a}\,dx \\ &= a\left(\frac{1}{KS^b}\right)^{1/a}\left(\frac{I}{43{,}200}\right)^{(1-a)/a} L^{1/a} \end{aligned} \tag{E.6}$$

The values of a and b for laminar flow are equal to 3 and 1, respectively, in equation (E.6). Therefore, Izzard's K may be used:

$$K = \frac{1}{(.007I + c)^3} \tag{E.7}$$

in which I is the actual (gross) rate of rainfall in inches per hour and c is a friction factor (but not equal) to Manning's n_m. Replacing K in equation (E.6) by equation (E.7),

$$T_c = 3(43{,}200)^{2/3}\left(\frac{1}{S}\right)^{1/3}\left[\frac{(.0007I + c)(L^{1/3})}{I^{2/3}}\right] \tag{E.8}$$

where T_c is in seconds. If T_c is in minutes, then

$$T_c = \frac{3}{60}(43{,}200)^{2/3}\left(\frac{1}{S}\right)^{1/3}\left[\frac{(.0007I + c)(L^{1/3})}{I^{2/3}}\right]$$

Butler (1977a, 1977b) extended equation (E.6) to incorporate both laminar and turbulent flow regimes. The Reynolds number can be used to distinguish between laminar and turbulent flows. For overland flow this can be defined as

$$NR = \frac{vd\rho}{\mu} \tag{E.9}$$

where ρ is fluid density and μ is dynamic viscosity. It can be assumed that laminar flow occurs at NR up to about 1000, after which the flow becomes turbulent with transition around $NR = 1000$. For planar flow (with unit width), $q = vd$. Therefore, $NR = q/\upsilon$, in which υ is kinematic viscosity and is equal to μ/ρ. Clearly, NR will be affected by the temperature under consideration and, hence, the quality of flow. For $NR = 1000$ and a temperature of 60°F with $\upsilon = 1.23 \times 10^{-5}$,

$$\frac{q}{\upsilon} = \frac{(IL/43{,}200)}{1.23 \times 10^{-5}} = 1000$$

This yields $IL = 530$. This says that at 60°F, the flow is laminar for values of IL below 530 and turbulent for values above 530, with a transition region in the neighborhood of $IL = 530$. Equation (E.8) will be valid if NR is smaller than 1,000 for the entire length of planar flow. Let us assume that laminar flow occurs over only a portion of the flow length from $L = 0$ to $L = L_1$. Then equation (E.8) should be modified to

$$T_c = 3(43{,}200)^{2/3}\left(\frac{1}{S}\right)^{1/3}\left[\frac{(.0007I + c)L_1^{1/3}}{I^{2/3}}\right] \tag{E.10}$$

Turbulent flow occurs over the distance L_1 to L. To compute the travel time for this flow, equation (E.1) must be integrated from L_1 to L. Therefore, equation (E.6) becomes

$$T_c = a\left(\frac{1}{KS^b}\right)^{1/a}\left(\frac{I}{43{,}200}\right)^{(1-a)/a}[L^{1/a} - L_1^{1/a}] \tag{E.11}$$

For turbulent flow, $a = \frac{5}{3}$, $b = 0.5$, and $K = 1.49/n_m$, where n_m is Manning's roughness coefficient. Equation (E.11) becomes

$$T_c = \frac{5}{3}\left(\frac{n_m}{1.49S^{0.5}}\right)^{3/5}\left(\frac{I}{43{,}200}\right)^{-2/5}[L^{3/5} - L_1^{3/5}] \tag{E.12}$$

Thus, the time of concentration for the planar can be computed as

$$T_c = T_c(\text{laminar portion}) + T_c(\text{turbulent portion}) \tag{E.13}$$

Equations (E.11) and (E.12) can be modified by replacing L_1 by $530/I$ in accord with $NR = 1000$ and temperature of 60°. If the temperature is different from 60°, then an adjustment will have to be made. Turbulent flow time is essentially independent of temperature. If the case of laminar flow, K varies inversely with μ. T_c then varies inversely with $K^{1/a}$ and directly with $\mu^{1/a}$, where $a = 3$. Thus, an adjustment to a temperature other than 60° can be made by multiplying equations representing laminar flow of the quantity $(\mu_T/\mu_{60})^{1/3}$.

APPENDIX F

Derivation of Linkage Equations Using Meixner Functions

For purpose of clarity the derivation of linkage equations is divided into three parts: (1) transform of Meixner functions, (2) convolution summation, and (3) linkage equations.

F.1 TRANSFORM OF MEIXNER FUNCTIONS

One convenient way to derive the aforementioned linkage equations is by knowing the Z-transform of Meixner functions. The Z-transform is frequently employed for the analysis of discrete-time systems. Thus, we want to determine the Z-transform $F_n(z)$ of Meixner functions $f_n(s)$, that is,

$$z[f_n(s)] = F_n(z)$$

Therefore,

$$F_n(z) = \sum_{s=0}^{\infty} \sum_{k=0}^{n} (-1)^k \binom{n}{k} \binom{s}{k} \left(\frac{1}{2}\right)^{(s+n+1)/2} z^{-s}$$

$$= \left(\frac{1}{2}\right)^{(n+1)/2} \sum_{k=0}^{n} (-1)^k \binom{n}{k} \sum_{s=0}^{\infty} \binom{s}{k} (2^{1/2}z)^{-s}$$

We now simplify the second summation:

$$\sum_{s=0}^{\infty} \binom{s}{k} (2^{1/2}z)^{-s} = \sum_{s=k}^{\infty} \binom{s}{k} (2^{1/2}z)^{-s}$$

$$= \sum_{t=0}^{\infty} \binom{t+k}{k} (2^{1/2}z)^{-t-k}$$

$$= (2^{1/2}z)^{-k} \sum_{t=0}^{\infty} \frac{1}{t!} \frac{(t+k)!}{k!} (2^{1/2}z)^{-t} \quad \text{(F.1)}$$

Now consider the function $G(z) = (1 - z)^{-(k+1)}$. Its tth derivative can be written as

$$G^{(t)}(z) = (k+1)(k+2) \ . \ . \ . \ (k+t)(1-z)^{-(k+1+t)}$$

$$= \frac{(k+t)!}{k!}(1-z)^{-(k+1+t)} \quad \text{(F.2)}$$

Since

$$G(z) = \sum_{t=0}^{\infty} \frac{G^{(t)}(0)}{t!} z^t$$

$$= \sum_{t=0}^{\infty} \frac{1}{t!} \frac{(t+k)!}{k!} z^t$$

for $|z| < 1$, we have from equations (F.1) and (F.2)

$$\sum_{s=0}^{\infty} \binom{s}{k} (2^{1/2}z)^{-s} = (2^{1/2}z)^{-k} G\left(\frac{1}{2^{1/2}z}\right)$$

$$= (2^{1/2}z)(2^{1/2}z - 1)^{-(k+1)}$$

Now

$$F_n(z) = \left(\frac{1}{2}\right)^{(n+1)/2} \sum_{k+0}^{N} (-1)^k \binom{n}{k} (2^{1/2}z)(2^{1/2}z - 1)^{-(k+1)}$$

$$= \frac{2^{1/2}z}{(2^{1/2}z - 1)^{n+1}} \left(\frac{1}{2}\right)^{(n+1)/2} \sum_{k=0}^{n} \binom{n}{k} (-1)^k (2^{1/2}z - 1)^{n-k}$$

The second summation is the binomial expansion of $((2^{1/2}z - 1) - 1)^n$. Therefore,

$$F_n(z) = \frac{2^{1/2}z}{(2^{1/2}z - 1)^{n+1}} \left(\frac{1}{2}\right)^{(n+1)/2} (2^{1/2}z - 2)^n$$
$$= \frac{z(z - 2^{1/2})n}{(2^{1/2}z - 1)^{n+1}} \tag{F.3}$$

F.2 CONVOLUTION SUMMATION

Equation (11.1) expresses the relation between the effective rainfall, the UH, and the direct runoff. To utilize this relation we must determine the convolution of two Meixner functions,

$$g(s) = \sum_{w=0}^{w=s} f_m(w)f_n(s - w) = f_n*f_m$$

This can be obtained by taking the Z-transform,

$$G(z) = f_m(z)f_n(z)$$

Substituting equation (F.3),

$$G(z) = \frac{z(z - 2^{1/2})^m}{(2^{1/2}z - 1)^{m+1}} \cdot \frac{z(z - 2^{1/2})^n}{(2^{1/2}z - 1)^{n+1}}$$
$$= \frac{z^2(z - 2^{1/2})^{m+n}}{(2^{1/2}z - 1)^{m+n+2}}$$

We can simplify $G(z)$ such that its inverse is easily obtained. Thus,

$$G(z) = \frac{z(z - 2^{1/2})^{m+n}}{(2^{1/2}z - 1)^{m+n+2}} [2^{1/2}(2^{1/2}z - 1) - (z - 2^{1/2})]$$
$$= \frac{2^{1/2}z(z - 2^{1/2})^{m+n}}{(2^{1/2}z - 1)^{m+n+1}} - \frac{z(z - 2^{1/2})^{m+n+1}}{(2^{1/2}z - 1)^{m+n+2}}$$

Taking the inverse Z-transform,

$$f_n(s) * f_m(s) = g(s) = 2^{1/2}f_{m+n}(s) - f_{m+n+1}(s) \tag{F.4}$$

F.3 LINKAGE EQUATIONS

To derive the linkage equations for Meixner coefficients a_m, α_m, and A_m, we substitute equations (11.50) through (11.52) into equation (11.2),

$$\sum_{p=0}^{\infty} A_p f_p(s) = \sum_{t=0}^{s} \sum_{m=0}^{\infty} A_m f_m(s) \sum_{n=0}^{\infty} \alpha_n f_n(s - t)$$
$$= \sum_{m=0}^{\infty} \sum_{n=0}^{\infty} A_m \alpha_n \sum_{t=0}^{\infty} f_m(s) f_n(s - t)$$
$$= \sum_{m=0}^{\infty} \sum_{n=0}^{\infty} A_m \alpha_n [f_n * f_m(s)]$$

Substituting equation (F.4) gives

$$\sum_{p=0}^{\infty} A_p f_p(s) = \sum_{m=0}^{\infty} \sum_{n=0}^{\infty} A_m \alpha_n [2^{1/2} f_{n+m}(s) - f_{n+m+1}(s)]$$

Since the functions $\{f_q\}$ are orthogonal, we can equate the coefficients of like functions f_q to obtain

$$A_p = \sum_{k=0}^{p} 2^{1/2} \alpha_k a_{p-k} - \sum_{k=0}^{p-1} \alpha_k a_{p-k-1} \tag{F.5}$$

Using equation (F.5) we have

$$\sum_{k=0}^{p} \left(\frac{1}{2}\right)^{(p-k+1)/2} A_k = \sum_{k=0}^{p} \left(\frac{1}{2}\right)^{(p-k+1)/2} \left(\sum_{j=0}^{k} 2^{1/2} \alpha_j a_{k-j} - \sum_{j=0}^{k-1} \alpha_j a_{p-1-j}\right)$$
$$= \sum_{k=0}^{p} \left(\sum_{j=0}^{k} \left(\frac{1}{2}\right)^{(p-k)/2} \alpha_j a_{k-j} - \sum_{j=0}^{k-1} \left(\frac{1}{2}\right)^{(p-k+1)/2} \alpha_j a_{k-1-j}\right)$$

If we set

$$B_k = \sum_{j=0}^{k} \left(\frac{1}{2}\right)^{(p-k)/2} \alpha_j a_{k-j}$$

then we have, where $B_{-1} = 0$,

$$\sum_{k=0}^{p} \left(\frac{1}{2}\right)^{(p-k+1)/2} A_k = \sum_{k=0}^{\infty} (B_k - B_{k-1})$$
$$= B_p$$
$$= \sum_{j=0}^{p} \alpha_j a_{p-j}$$

Therefore,

$$\alpha_p a_0 = \sum_{k=0}^{p} \left(\frac{1}{2}\right)^{(p-k+1)/2} A_k - \sum_{k=0}^{p-1} \alpha_k a_{p-k} \tag{F.6}$$

This is the desired linkage equation.

References

Abramowitz, M., and I. A. Stegun. 1965. *Handbook of Mathematical Functions*. U.S. Department of Commerce, National Bureau of Standards, Washington, D.C.

Ackerman, W. D., E. A. Coleman, and H. O. Ogrosky. 1955. "From ocean to sky to land to ocean." In *Water, The Yearbook of Agriculture,* 41–51. U.S. Department of Agriculture, Washington, D.C.

Ackoff, R. L. 1962. *Scientific Methods*. New York: John Wiley.

Adejuwan, J. O., L. K. Jeje, and O. O. Ogunkoya. 1983. Hydrological response patterns of some third order streams on the basement complex of southwestern Nigeria. *Hydrological Science Journal* 28(3):377–89.

Agiralioglu, N., and V. P. Singh. 1981. Kinematic wave nomograph for time of concentration and lag time. *Irrigation and Power Journal* 38(4):351–58.

Aitken, A. P. 1973. Assessing systematic errors in rainfall-runoff models. *Journal of Hydrology* 20:121–36.

Alexander, G. N. 1972. Effect of catchment area on flood estimation. *Journal of Hydrology* 16:225–40.

Allis, J. A., F. J. Dragoun, and A. L. Sharp. 1964. Transmission losses in Loessial watersheds. *Transactions of the American Society of Agricultural Engineers* 7(3):209–12.

Allison, W. V. 1967. Review of small basin runoff prediction methods. *Journal of the Irrigation and Drainage Division, Proceedings of the American Society of Civil Engineers* 93(IR1):1–9.

American Concrete Pipe Association. 1958. *Concrete Pipe Handbook*. Arlington, Va.

______. 1970. *Concrete Pipe Design Manual*. Vienna, Va.

American Society of Civil Engineers. 1969. *Basic information needs in urban hydrology*. ASCE Urban Hydrology Research Council, New York.

______. 1976. *Design and Construction of Sanitary Storm Sewers*. Manual and Reports on Engineering Practice No. 37, New York.

Amerman, C. R. 1965. The use of unit source watershed data for runoff prediction. *Water Resources Research* 1(4):499–508.

Amorocho, J. 1961. Discussion of "Predicting storm runoff on small experimental watersheds by N. E. Minshall." *Journal of the Hydraulics Division, Proceedings of the American Society of Civil Engineers* 87(2):185–91.

______. 1963. Measures of the linearity of the hydrologic systems. *Journal of Geophysical Research* 68(8):2237–49.

______. 1967a. The nonlinear prediction problem in the study of the runoff cycle. *Water Resources Research* 3(3):861–80.

______. 1967b. Role of infiltration in nonlinear watershed analysis processes. *Transactions of the American Society of Agricultural Engineers* 10:396–99 and 404.

______. 1973. "Nonlinear hydrologic analysis." In *Advances in Hydroscience*, edited by V. T. Chow, vol. 9, 203–51. New York: Academic Press.

Amorocho, J., and A. Brandstetter. 1971. Determination of nonlinear functional response functions in rainfall-runoff processes. *Water Resources Research* 7(5):1087–1101.

Amorocho, J., and W. E. Hart. 1964. A critique of current methods of hydrologic systems investigation. *Transactions of the American Geophysical Union* 45:307–21.

______. 1965. The use of laboratory catchments in the study of hydrologic systems. *Journal of Hydrology* 3:106–23.

Amorocho, J., and G. T. Orlob. 1961. *Nonlinear analysis of hydrologic systems*. Water Resources Center Contribution 40, University of California, Berkeley.

Anderson, M. G., and T. P. Burt. 1980. Interpretation of recession flow. *Journal of Hydrology* 46:89–101.

Ardis, C. V., Jr. 1972. A storm-hydrograph model for the response and variation of small rural watersheds. Ph.D. diss. University of Wisconsin, Madison.

———. 1973. A double triangular model for generating storm hydrographs. *Proceedings of the First World Congress on Water Resources* vol. 4, 350–60, Chicago.

———. 1974. A nonlinear channel routing model. *Proceedings of the First Symposium on Flow, Its Measurement, and Control in Science and Industry*, part I, 99–106. Pittsburgh, Penn: Instrument Society of America.

Aron, G., D. F. Kibler, and C. J. Tagliati. 1981. Use of flood potential indices for flood peak estimation on ungaged watersheds. *Water Resources Bulletin* 17(4):691–98.

Aron, G., and A. C. Miller. 1978. Adaptation of flood peaks and design hydrographs from gaged to ungaged watersheds. *Water Resources Bulletin* 14(2):313–21.

Aron, G., and E. L. White. 1982. Fitting a gamma distribution over a synthetic unit hydrograph. *Water Resources Bulletin* 18(1):95–98.

Arteaga, F. E., and S. E. Rantz. 1973. Application of the source-area concept of storm runoff to a small Arizona watershed. *Journal of Research, U.S. Geological Survey*, 1(4):493–98.

Askew, A. J. 1968. Lag time of natural catchments. Technical Report No. 107, Water Research Laboratory, University of New South Wales, Kensington, New South Wales, Australia.

———. 1970a. Derivation of formulae for variable lag time. *Journal of Hydrology* 10:225–42.

———. 1970b. Variation in lag time for natural catchments. *Journal of the Hydraulics Division, Proceedings of the American Society of Civil Engineers* 96(HY2):317–30.

Ayer, A. J. 1952. Language, Truth and Logic. New York: Dover.

Babcock, H. M., and E. M. Coshing. 1942. Recharge to ground water from floods in a typical desert wash, Pinal County, Arizona. *Transactions of the American Geophysical Union* 33(pt. 1):49.

Babu Rao, T. 1973. A mathematical model for basin runoff. Ph.D. diss., University of Madras, Madras, India.

Bakhmeteff, B. A. 1932. *Hydraulics of Open Channels*. New York: McGraw-Hill.

Barnes, B. S. 1939. The structure of discharge recession curves. *Transactions of the American Geophysical Union* 20(pt. 4):721–25.

———. 1940. Discussion of analysis of runoff characteristics. *Transactions of the American Society of Civil Engineers* 105:106.

———. 1959. Consistency in unitgraphs. *Journal of the Hydraulics Division, Proceedings of the American Society of Civil Engineers* 85(HY8):39–61.

Barrett, J. F. 1957. The use of functionals in the analysis of nonlinear physical systems. *Statistical Advisory Unit Report* 1/57, Ministry of Supply, Great Britain.

———. 1964. The use of functionals in the analysis of nonlinear physical systems. *Journal of Electronics Control*. 15(6):567–615.

———. 1965. Hermite functional expansions and the calculation of output autocorrelation and spectrum for any time-invariant nonlinear system with noise input. *Journal of Electronics Control* 16(1):107.

Bartlett, M. S. 1966. *An Introduction to Stochastic Processes with Special Reference to Methods and Applications*. Cambridge: Cambridge Univ. Press.

Bayazit, M. 1966. Instantaneous unit hydrograph derivation by spectral analysis and its numerical application. *Proceedings of the Central Treaty Organization Symposium on Hydrology and Water Resources Development*, 127–49, Ankara, Turkey.

Becker, A. 1973. Applied principles of catchment simulation. *Proceedings, IASH/UNESCO Symposium on Mathematical Models in Hydrology*, Warsaw, Poland, International Association of Hydrological Sciences Publication 101,762–74.

Becker, L., and W. W. G. Yeh. 1972. Identification of parameters in unsteady open channel flows. *Water Resources Research* 8(4):956–65.

Beer, S. 1959. *Cybernetics and Management*. New York: John Wiley.

Bell, F. C. 1967. Estimating design floods from extreme rainfall (with special reference to small watersheds in western U.S.A.). *Technical Report* CER67-68FCB5, Civil Engineering Department, Colorado State University, Fort Collins.

Bell, F. C., and S. Omkar. 1969. Characteristic response times in design flood estimation. *Journal of Hydrology*, 8:173–96.

Bellman, R. E., R. E. Kalaba, and J. Lockett. 1966. *Numerical Inversion of the Laplace Transform*. New York: American Elsevier Publishing Co.

Benson, M. A. 1962. Evolution of methods for evaluating the occurrence of floods. U.S. Geological Survey Water Supply Paper no. 180-A, Washington, D.C.

Ben-zvi, A. 1984. Runoff peaks from two-dimensional laboratory watersheds. *Journal of Hydrology* 68:115–39.

Bernard, M. 1935. An approach to determinate stream flow. *Transactions of the American Society of Civil Engineers* 100:347–62.

———. 1938. Modified rational method of estimating flood flows. Appendix A in *Low Dams*, National Resources Committee, Washington, D.C.

Betson, R. P. 1964. What is watershed runoff? *Journal of Geophysical Research* 69(8):1541–51.

Betson, R. P., and R. F. Green. 1968. Analytically derived unit graph and runoff. *Journal of the Hydraulics Division, Proceedings of the American Society of Civil Engineers* 94(HY6):1489–1505.

Betson, R. P., and J. B. Marius. 1969. Source areas of storm runoff. *Water Resources Research* 5(3):574–82.

Beven, K., K. Gilman, and M. Newson. 1979. Flow and flow routing in upland channel networks. *Hydrological Sciences Bulletin* 24(3):303–25.

Beven, K. J., and J. J. Kirkby. 1979. A physically based, variable contributing area model of basin hydrology. *Hydrological Sciences Bulletin* 24(1):43–49.

Beveridge, W. I. B. 1957. *The Art of Scientific Investigation*. New York: Random House, Vintage Books.

Bidwell, V. J. 1971. Regression analysis of nonlinear catchment systems. *Water Resources Research* 7:1118–26.

Bissell, V. C., and R. M. Ragan. 1969. The determination of zones of intense contribution to streamflow as related to the concept of partial area contributions. *Technical Report* no. 21, Water Resources Research Center, University of Maryland, College Park.

Black, P. E. 1972. Hydrograph responses to geomorphic model watershed characteristics and precipitation variables. *Journal of Hydrology* 17:309–27.

Black, P. E., and J. W. Cronn, Jr. 1975. Hydrograph responses to watershed model size and similitude relations. *Journal of Hydrology* 26:255–66.

Blaney, H. F., and K. V. Morin. 1942. Evaporation and consumptive use of water-empirical formulas. *Transactions of the American Geophysical Union* 23(1):76–83.

Blank, D., and J. W. Delleur. 1968. A program for estimating runoff from Indiana watersheds: 1. Linear system analysis in surface hydrology and its application to Indiana watersheds. *Technical Report* 4, Water Resources Research Center, Purdue University, West Lafayette, Ind.

Blank, D., J. W. Delleur, and A. Giorgini. 1971. Oscillatory kernel functions in linear hydrologic models. *Water Resources Research* 7(5):1102–17.

Bloomsburg, G. L. 1960. A hydrograph. *Research Report no.* 325, Agricultural Research Service, U.S. Department of Agriculture, Washington, D.C.

Bohm, D. 1980. *Causality and Chance in Modern Physics*. 2d ed. Philadelphia: Univ. of Pennsylvania Press.

______. 1982. *Wholeness and the Implicate Order*. London: Routledge and Kegan Paul.

Bolin, B., and H. Rodhe. 1973. A note on the concepts of age distribution and transit time in natural reservoirs. *Tellus* 25:58–62.

Bondelid, T. R., and R. H. McCuen. 1979. Critical storm duration for the rational method. *Journal of Civil Engineering Design* 1(3):273–85.

Boneh, A. 1971. Identification of the kernel functions of hydrologic systems from independent multi-storm records. D.Sc. thesis, Technion-Israel Institute of Technology, Haifa.

Boneh, A., and M. H. Diskin. 1973. "Demonstration of nonlinear effects of a second order runoff model by field data." In *Floods and Droughts*, edited by E. F. Schulz et al., 157–68. Fort Collins, Colo: Water Resources Publications.

Boneh, A., and A. Golan. 1978. Optimal identification of nonlinear surface runoff systems with copositivity threshold constraints. *Advances in Water Resources* 1(3):121–29.

______. 1979. Application of generalized network concepts to water resources systems. *Proceedings of the International Conference on Operations Research in Agriculture and Water Resources*, edited by D. Yaron and C. S. Tapiero, 25–29, Jerusalem, Israel.

Bose, A. G. 1956. A theory of nonlinear systems. Research Laboratory of Electronics, *Technical Report no.* 309, Massachusetts Institute of Technology, Cambridge.

Box, G. E. P., and G. M. Jenkins. 1970. *Time Series Analysis Forecasting and Control*. San Francisco: Holden-Day.

Boyd, M. J. 1976. A storage routing model for flood hydrograph synthesis based on geomorphologic relations. Ph.D. diss., University of New South Wales, Kensington, New South Wales, Australia.

______. 1978. A storage-routing model relating drainage basin hydrology and geomorphology. *Water Resources Research* 14(2):921–28.

______. 1982. "A linear branched network model for storm rainfall and runoff." In *Rainfall-Runoff Relationship*, edited by V. P. Singh, 111–25. Littleton, Colo. Water Resources Publications.

Boyd, M. J., D. H. Pilgrim, and I. Cordery. 1979a. A storage routing model based on catchment geomorphology. *Journal of Hydrology* 42:209–30.

______. 1979b. An improved runoff modeling based on geomorphological relations. Hydrology Symposium 1979, Perth, The Institution of Engineers, Australia, *National Conference Publication* no. 79/10, 229–233.

Brakensiek, D. L. 1966. Storage flood routing without coefficients. ARS 41–122, U.S. Department of Agriculture, Agricultural Research Service. Washington, D.C.

Brakensiek, D. L., and G. H. Comer. 1965. A re-examination of a flood routing method comparison. *Journal of Hydrology* 3:225–30.

Brandstetter, A., and J. Amorocho. 1970. Generalized analysis of small watershed responses. *Water Science and Engineering Papers* 1035, Department of Water Science and Engineering, University of California, Davis.

Bravo, S. C. A., B. M. Harley, F. E. Perkins, and P. S. Eagleson. 1970. A linear distributed model of catchment model. *Hydrodynamics Laboratory Report no.* 123, Massachusetts Institute of Technology, Cambridge.

Bree, T. 1977. A comment on "A discrete linear cascade model for hydrology by K. M. O'Connor." *Journal of Hydrology* 33:403–6.

______. 1978a. The stability of parameter estimation in the general linear model. *Journal of Hydrology* 37:47–66.

______. 1978b. The general linear model with prior information. *Journal of Hydrology* 39:113–27.

Brilliant, M. B. 1958. Theory of the analysis of nonlinear systems. *Technical Report no.* 345, Research Laboratory of Electronics, Massachusetts Institute of Technology, Cambridge.

Brillouin, L. 1962. *Scientific Uncertainty and Information*. 2d ed. New York: Academic Press.

Bronowski, J. 1970. *Science and Human Values*. New York: Harper and Row.

Bruce, J. P., and R. H. Clark. 1966. *Introduction to Hydrometeorology*. Elmsford, N.Y.: Pergamon Press.

Bruce, R. 1966. Functional analysis of systems characterized by nonlinear differential equations. Technical Report No. 444, Research Laboratory of Electronics, Massachusetts Institute of Technology, Cambridge.

Brutsaert, W. 1982. *Evaporation into the Atmosphere*. Boston: D. Reidel Publishing Company.

Brutsaert, W., and J. A. Mawdsley. 1976. The applicability of planetary boundary layer theory to calculate regional evapotranspiration. *Water Resources Research* 12:852–58.

Buapeng, S., and V. P. Singh. 1977. Studies on rainfall-runoff modeling: 7. A nonlinear hydrologic cascade. *WRRI Report no.* 087, New Mexico Water Resources Research Institute, New Mexico State University, Las Cruces.

Bull, J. A. 1968. Unit graphs for nonuniform rainfall distribution. *Journal of the Hydraulics Division, Proceedings of the American Society of Civil Engineers* 94(HY1):235–57.

Burke, C. B., and D. D. Gray. 1979. A comparative application of several methods for the design of storm sewers. *Technical Report* 118, Water Resources Research Center, Purdue University, West Lafayette, Ind.

Burkham, D. E. 1970a. A method of relating infiltration rates to streamflow rates in perched streams. Geological Survey Research 1970: U.S. Geological Survey Professional Paper 700-D. Washington, D.C.

______. 1970b. Depletion of streamflow by infiltration in the main channels of the Tucson basin, southeastern Arizona. U.S. Geological Survey Water Supply Paper 1939-B. Washington, D.C.

Bush, A. M. 1966. Some techniques for the synthesis of nonlinear systems. *Technical Report no.* 441, Research Laboratory of Electronics, Massachusetts Institute of Technology, Cambridge.

Butler, S. S. 1977a. Overland-flow travel time versus Reynolds number. *Journal of Hydrology* 32:175–82.

______. 1977b. Closing discussion of paper "Overland-flow travel time versus Reynolds number." *Journal of Hydrology* 35:205–8.

______. 1979. Deterministic-systemic relationships in overland flow analysis. *Water Resources Bulletin* 15(2):519–29.

______. 1982. Point-slope approach for reservoir flood routing. *Journal of the Hydraulics Division, Proceedings of the American Society of Civil Engineers* 108(HY10):1102–13.

CAMERON, R. J. 1980. An updating version of the Muskingum-Cunge flow routing technique. *Proceedings of the Oxford Symposium on Hydrological Forecasting*, International Association of Hydrological Sciences Publication no. 129, 381–87.

CARRASQUILLA, A. B., AND F. E. PERKINS. 1967. An extension of the role of linear systems analysis in hydrograph theory. Report no. 106, Hydrodynamics Laboratory, Massachusetts Institute of Technology, Cambridge.

CARTER, R. W. 1961. Magnitude and frequency of floods in suburban areas. U.S. Geological Survey Professional Paper no. 424-B, U.S. Dept. of Interior, Washington, D.C.

CARTER, R. W., AND R. G. GODFREY. 1960. "Storage and flood routing." In *Manual of Hydrology*: Pt. 3, *Flood Flow Techniques*, 80–104. U.S. Geological Survey, U.S. Dept. of Interior, Washington, D.C.

CAUCHY, A. L. 1847. Méthode *générale pour la résolution de systèmes d'equations simultanées*, vol. 25, 536–38. Paris: C. R. Read Science.

CHAMBERS, J. B. 1965. A study of Laplace transformation techniques on an analog computer. M.S. thesis, Pennsylvania State University, University Park.

CHANDER, S. 1970. Flood routing using dimensionless parameters. *Journal of Irrigation and Power* 27(4):295–420.

CHANG M., AND D. G. BOYER. 1977. Estimates of low flows using watershed and climatic characteristics. *Water Resources Research* 13(6):997–1001.

CHANG, C. N., E. D. M. SINGER, AND A. D. KOUSSIS. 1983. On the mathematics of storage routing. *Journal of Hydrology* 61:357–70.

CHAPMAN, T. G. 1985. Convolution with hydrologic data. *Water Resources Research* 21(6):847–52.

CHAUDHRY, F. H. 1976. General linear hydrologic response models. *Proceedings of the National Symposium of Waterways, Harbors, and Coastal Engineering Division, American Society of Civil Engineers*, vol. 1, 367–76.

CHAUDHRY, F. H., M. A. SIMOES, AND W. M. FERREIRA FILHO. 1976. Chow's general hydrologic model. *Journal of the Hydraulics Division, Proceedings of the American Society of Civil Engineers* 102(HY9):1387–90.

CHEN, C. N., AND R. R. EVANS. 1977. Application of kinematic wave method to predict overland-peak flows. *Proceedings of the International Symposium on Urban Hydrology, Hydraulics, and Sediment Control*, University of Kentucky, Lexington, 113–18.

CHEN, S. J. 1984. The kernel function for watershed runoff modeling. *Journal of Hydrology* 68:29–37.

CHEN, S. J., AND V. P. SINGH. 1986. Derivation of a new variable instantaneous unit hydrograph. *Journal of Hydrology*, 86:25–42.

CHERY, D. L. 1966. Design and tests of a physical watershed model. *Journal of Hydrology* 4:224–35.

CHIU, C. L. 1967. Analysis of "black box" relationship between rainfall and runoff. *Proceedings of the International Hydrology Symposium*, Fort Collins, Colo. Vol. 1, 608–15.

______. 1968. Stochastic open channel flow. *Journal of the Engineering Mechanics Division, Proceedings of the American Society of Civil Engineers* 94(EM3):811–22.

CHIU, C. L., AND R. P. BITTLER. 1969. Linear time-varying model of rainfall-runoff relation. *Water Resources Research* 5(2):426–37.

CHIU, C. L., AND J. T. HUANG. 1970. Nonlinear time-varying model of rainfall-runoff relation. *Water Resources Research* 6(1):1277–86.

CHORLEY, R. J., D. E. G. MALM, AND H. A. PAGORZELSKI. 1957. A new standard for estimating drainage shape. *American Journal of Science* 255:138–41.

CHOW, V. T. 1959. *Open Channel Hydraulics*. New York: McGraw-Hill.

______. 1962. Hydrologic determination of waterway areas for the design of drainage structures in small drainage basins. *Engineering Experiment Station Bulletin*, no. 462, University of Illinois, Urbana.

CHOW, V. T. 1964. "Runoff." Section 14 in *Handbook of Applied Hydrology*, edited by V. T. Chow. New York: McGraw-Hill.

______, ed. 1964. *Handbook of Applied Hydrology*. New York: McGraw-Hill.

______. 1967. Laboratory study of watershed hydrology. *Proceedings of the International Hydrology Symposium*, September 6–8, Fort Collins, Colo., vol. 1, 194–202.

______. 1970. Systems approaches in hydrology and water resources. *Proceedings of the Second International Seminar for Hydrology Professors*, August 2–14, Utah State University, Logan, 490–509.

CHOW, V. T., AND V. C. KULANDAISWAMY. 1971. General hydrologic system model. *Journal of the Hydraulics Division, Proceedings of the American Society of Civil Engineers* 97(HY6):791–804.

______. 1982. The IUH of the general hydrologic system model. *Journal of the Hydraulics Division, Proceedings of the American Society of Civil Engineers* 108(HY7):830–44.

CHU, S. T., AND W. F. LYTLE. 1972. Time base of watersheds in South Dakota. *Transactions of the American Society of Agricultural Engineers* 15:276–79.

CLARK, C. O. 1944. Flood-storage accounting. *Transactions of the American Geophysical Union* 25, 1014–39.

______. 1945. Storage and the unit hydrograph. *Transactions of the American Society of Civil Engineers* 110:1419–46.

CLARKE, K. D. 1968. Applications of Stanford watershed model concepts to predict flood peaks for small drainage areas. Kentucky Department of Highways, Division of Research, Lexington.

CLARKE, R. T. 1971. The use of the term "linearity" as applied to hydrological models. *Journal of Hydrology* 13:91–95.

______. 1973. A review of some mathematical models used in hydrology, with observations on their calibration and use. *Journal of Hydrology* 19:1–19.

CODDINGTON, E. A., AND N. LEVINSON. 1955. *The Theory of Ordinary Differential Equations*. New York: McGraw-Hill.

COLLINS, M. A. 1983. Discussion of "Fitting a gamma distribution over a synthetic unit hydrograph by G. Aron and E. L. White." *Water Resources Bulletin* 19(2):303–4.

COLLINS, W. T. 1939. Runoff distribution graphs from precipitation occurring in more than one time unit. *Civil Engineering* 9(9):559–61.

COLOMBI, J. S. 1978. The modified Richards method of flood estimation. *Proceedings of the Institution of Civil Engineers*, pt. 2, vol. 65, 299–306.

COMMONS, G. G. 1942. Flood hydrographs. *Civil Engineering* 12(10):571–72.

CONTE, S. D. 1965. *Elementary Numerical Analysis*. New York: McGraw-Hill.

CORDERY, I. 1968. Synthetic unit graphs for small catchments in Eastern New South Wales. *Civil Engineering Transactions of the Institution of Engineers*, Australia, vol. 10, 47–58.

______. 1971. Estimation of design hydrographs for small rural catchments. *Journal of Hydrology* 13:263–77.

CORDERY, I., AND D. H. PILGRIM. 1970. Design hydrograph methods of flood estimation for small rural catchments. *Civil Engineering Transactions of the Institution of Engineers*, Australia, vol. 12, 175–180.

CRAWFORD, N. H., AND R. K. LINSLEY. 1962. The synthesis of continuous streamflow hydrograph on a digital computer. Technical Report no. 12, 121 p., Department of Civil Engineering, Stanford University, Palo Alto, Calif.

______. 1964. A conceptual model of the hydrologic cycle. *International Association of Scientific Hydrology Publication* 63, 573–87.

______. 1966. Digital simulation of hydrology: The Stanford Watershed Model IV. *Technical Report* 39, Department of Civil Engineering, Stanford University, Palo Alto, Calif.

CRIPPEN, J. R. 1982. Envelope curves for extreme flood events. *Journal of the Hydraulics Division, Proceedings of the American Society of Civil Engineers* 108(HY10):1208–12.

CROLEY II, T. E. 1980. Gamma synthetic hydrographs. *Journal of Hydrology* 47:41–52.

CUNGE, J. A. 1969. On the subject of a flood propagation computation method (Muskingum method). *Journal of Hydraulic Research* 7(2):205–30.

DANTZIG, G. B. 1963. *Linear Programming and Extensions*. Princeton, N.J.: Princeton Univ. Press.

DAVIDSON, N. 1962. *Statistical Mechanics*. New York: McGraw-Hill.

DAWDY, D. R. 1969a. Considerations involved in evaluating mathematical modeling of urban hydrologic systems. *U.S. Geological Survey Water Supply Paper* 1591-D, U.S. Dept. of Interior, Washington, D. C., D1–D18.

______. 1969b. Mathematical modeling in hydrology. *Proceedings of the First International Seminar for Hydrology Professors*, University of Illinois, Urbana, vol. 1, 346–61.

DAWDY, D. R., AND G. P. KALININ. 1969. Mathematical modeling in hydrology. *International Association of Scientific Hydrology Report, Mid-Decade Conference of the International Hydrological Decade*, held in August in Surány, Hungary.

DAWDY, D. R., R. W. LITCHY, AND J. R. BERGMANN. 1970. A rainfall-runoff simulation model for estimation of flood peaks for small drainage basins: a progress report. U.S. Geological Survey, U.S. Dept. of Interior, Washington, D.C., Open File Report.

DAWDY, D. R., AND N. C. MATALAS. 1964. "Analysis of variance, covariance and time series." Section 8-III in *Handbook of Applied Hydrology*, edited by V. T. Chow. New York: McGraw-Hill.

DAWDY, D. R., AND T. O'DONNELL. 1965. Mathematical model of catchment behavior. *Journal of the Hydraulics Division, Proceedings of the American Society of Civil Engineers* 91(HY4):123–37.

———. 1966. A runoff hydrograph equation. ARS-41-116, Agricultural Research Service, U.S. Department of Agriculture, Washington, D.C.

DECOURSEY, D. 1971. The stochastic approach to watershed modeling. *Nordic Hydrology* 11:186–216.

DECOURSEY, D. G., AND SNYDER, W. M. 1969. Computer-oriented method of optimizing hydrologic model parameters. *Journal of Hydrology* 9:34–56.

DEININGER, R. A. 1969. Linear programming for hydrologic analyses. *Water Resources Research* (5):1105–9.

DELAINE, R. J. 1970. Deriving the unitgraph without using rainfall data. *Journal of Hydrology* 10:379–90.

DELLEUR, J. W., AND A. R. RAO. 1971a. "Linear systems analysis hydrology, the transform approach, the kernel oscillations and the effect of noise." In *Systems Approach to Hydrology*, edited by V. Yevjevich, 116–130. Fort Collins, Colo.: Water Resources Publications.

———. 1971b. Characteristics and filtering of noise in linear hydrologic systems. *Proceedings of the Warsaw Symposium on Mathematical Models in Hydrology*, Warsaw, Poland, 570–79.

DE NEUFVILLE, R., AND J. H. STAFFORD. 1971. *System Analysis for Engineers and Managers*. New York: McGraw-Hill.

Denver Regional Council of Governments. 1969. *Urban storm drainage criteria manual, vol. 1, Denver Regional Council of Governments, Denver, Colo.*

DICKINSON, W. T., AND H. D. AYERS. 1965. The effect of storm characteristics on the unit hydrograph. *Transactions of the Engineering Institute of Canada* 8 (A-15):3–7.

DING, J. W. 1967a. Flow routing by direct integration method. *Proceedings of the International Hydrology Symposium*, September 6–8, Fort Collins, Colo., vol. 1, 113–20.

———. 1967b. Comments on mathematical models for hydrologic systems. *Proceedings of the International Hydrology Symposium*, September 6–8, Fort Collins, Colo., vol. 2, 421–22.

———. 1974. Variable unit hydrograph. *Journal of Hydrology* 22:53–69.

DINGMAN, S. L. 1978. Drainage density and streamflow: a closer look. *Water Resources Research* 14(6):1183–87.

———. 1985. Effects of routing function form on outflow and storage statistics for discrete linear reservoir. Paper submitted to *Water Resources Research*.

DISKIN, M. H. 1964. A basic study of the linearity of the rainfall-runoff process in watersheds. Ph.D. diss. University of Illinois, Urbana.

———. 1967a. A Laplace transform proof of the theorem of moments for the instantaneous unit hydrograph. *Water Resources Research* 3(2):385–88.

———. 1967b. On the solution of the Muskingum method of flood routing equation. *Journal of Hydrology* 5:286–89.

———. 1970. Definition and uses of the linear regression model. *Water Resources Research* 6:1668–73.

———. 1977. On the derivation of linkage equations for Laguerre function coefficients. *Journal of Hydrology* 32:321–27.

———. 1979. Some dimensional considerations in the unit hydrograph theory. *Journal of Hydrology* 42:199–208.

———. 1982. "Nonlinear hydrologic models." In *Rainfall-Runoff Relationship*, edited by V. P. Singh, 127–46. Littleton, Colo.: Water Resources Publications.

DISKIN, M. H., AND A. BONEH. 1968. Moments of the input, output and impulse response functions of linear systems about arbitrary points. *Water Resources Research* 4(4):727–36.

———. 1972. Properties of the kernels for time-invariant, initially relaxed, second order surface runoff systems. *Journal of Hydrology* 17:115–41.

———. 1973. Determination of optimal kernels for second order surface runoff systems. *Water Resources Research* 9(2):311–25.

———. 1974. The kernel function of linear, nonstationary surface runoff systems. *Water Resources Research* 10(4):753–61.

———. 1975. Determination of an optimal IUH for linear, time-invariant systems from multistorm records. *Journal of Hydrology* 24:57–76.

———. 1980. Discussion of "Optimization of unit hydrograph determination by L. W. Mays and L. Coles." *Journal of the Hydraulics*

Division, Proceedings of the American Society of Civil Engineers 106(HY11):1951–54.

Diskin, M. H., A. Boneh, and A. Golan. 1975. On the impossibility of a partial mass violation in surface runoff systems. *Water Resources Research* 11(2):236–44.

———. 1984. Identification of a Volterra series conceptual model based on a cascade of nonlinear reservoirs. *Journal of Hydrology* 68:231–45.

Diskin, M. H., and L. J. Lane. 1976. Application of double triangle unit hydrograph to a small semiarid watershed. Proceedings of the 1976 Meetings of the Arizona–American Water Resources Association and the Hydrology Section, Arizona Academy of Science, vol. 6, Hydrology and Water Resources in Arizona and the Southwest, 125–35.

Diskin, M. H., and E. Simon. 1977. A procedure for the selection of objective functions for hydrologic simulation models. *Journal of Hydrology* 34(1/2):129–49.

Dooge, J. C. I. 1955. Discussion, "The employment of unit hydrograph to determine the flows of Irish arterial drainage channels" by J. J. O'Kelly. *Proceedings of the Institution of Civil Engineers* (Ireland), vol. 4, 436–442.

———. 1956. Synthetic unit hydrographs based on triangular inflow. M.S. thesis, University of Iowa, Iowa City.

———. 1957. The rational method for estimating flood peaks. *Engineering* 184:311–74, London.

———. 1959. A general theory of the unit hydrograph. *Journal of Geophysical Research* 64(2):241–56.

———. 1965. Analysis of linear systems by means of Laguerre functions. *Society of Industrial and Applied Mathematics (SIAM) Journal of Control* 2(3):396–408.

———. 1966. Response of heavily damped discrete systems. Unpublished manuscript, Department of Civil Engineering, University College, Cork, Ireland.

———. 1967a. A new approach to nonlinear problems in surface water hydrology: hydrologic systems with uniform nonlinearity. *International Association of Scientific Hydrology Publication*, 76, 409–413.

———. 1967b. Conceptual models of surface runoff. *Proceedings of the International Symposium on Floods, Their Prediction and the Defense of the Soil*, Rome, Italy, 179–207.

Dooge, J. C. I. 1967. The hydrologic cycle as a closed system. *International Association of Scientific Hydrology Bulletin* 13(1):58–68.

Dooge, J. C. 1972. Mathematical models of hydrologic systems. *Proceedings of the International Symposium on Modeling Techniques in Water Resources Systems*, Ottawa, Canada, vol. 1, 171–89.

———. 1973. Linear theory of hydrologic systems. *Technical Bulletin No. 1468*, U.S. Department of Agriculture, Agricultural Research Service, Washington, D.C.

———. 1977. "Problems and methods of rainfall-runoff modeling." In *Mathematical Models for Surface Water Hydrology*, edited by T. A. Ciriani, U. Maione, and J. R. Wallis, 71–108. New York: John Wiley.

———. 1980. Flood routing in channels. Unpublished lecture notes, Department of Civil Engineering, University of College, Dublin, Ireland.

Dooge, J. C. I., and Garvey, B. J. 1970. Discussion of "Data error effects in unit hydrograph derivation by E. M. Laurenson and T. O'Donnell." *Journal of the Hydraulics Division, Proceedings of the American Society of Civil Engineers* 96(HY9):1888–94.

Dooge, J. C. I., Z. W. Kundzewicz, and J. J. Napiorkowski. 1983. On backwater effects in linear diffusion flood routing. *Hydrological Sciences Journal* 28(3):391–402.

Dooge, J. C. I., W. G. Strupczewski, and J. J. Napiorkowski. 1982. Hydrodynamic derivation of storage parameters of the Muskingum model. *Journal of Hydrology* 54:371–87.

Douglas, J. R., R. T. Clarke, and S. G. Newton. 1976. The use of likelihood functions to fit conceptual models with more than one dependent variable. *Journal of Hydrology* 29:181–98.

Dreiss, S. J. 1982. Linear kernels for karst aquifers. *Water Resources Research* 18(4):865–76.

Dunne, T. 1970. Runoff production in humid areas. ARS 41-160, Agriculture Research Service, U.S. Department of Agriculture, Washington, D.C.

Dunne, T., and R. D. Black. 1970a. An experimental investigation of runoff prediction in permeable soils. *Water Resources Research* 6(2):478–90.

———. 1970b. Partial area contributions to storm runoff in a small New England watershed. *Water Resources Research* 6(5):129–138.

Dunne, T., T. R. Moore, and C. H. Taylor. 1975. Recognition and prediction of runoff producing zones in humid regions. *Hydrological Sciences Bulletin* 20(3):305–27.

Durant, E. F., and S. R. Blackwell. 1959. The magnitude and frequency of floods on the Canadian prairies. *Proceedings of Symposium No. 1: Spillway Design Floods*, 101–54. Ottawa, Canada: National Research Council of Canada.

Duru, G. H. 1976. Discharge prediction, present and former, from channel dimensions. *Journal of Hydrology* 30:219–45.

Dwight, H. B. 1966. *Tables of Integrals and other Mathematical Data*. 4th ed. New York: Macmillan.

Eagleson, P. S. 1962. Unit hydrograph characteristics for sewered areas. *Journal of the Hydraulics Division, Proceedings of the American Society of Civil Engineers* 88(HY2):1–25.

———. 1967. A linear distributed model for peak catchment discharge. *Proceedings of the International Hydrology Symposium*, Fort Collins, Colo., vol. 1, 1–8.

———. 1970. *Dynamic Hydrology*. New York: McGraw-Hill.

———. 1972. Dynamics of flood frequency. *Water Resources Research* 8(4):878–98.

Eagelson, P. S., R. Mejia-R., and F. March. 1966. Computation of optimum realizable unit hydrographs. *Water Resources Research* 2(4):755–64.

Eaton, T. D. 1954. The derivation and synthesis of unit hydrographs when rainfall records are inadequate. *Institution of Engineers Journal* (Australia) 26:239–46.

Edson, C. G. 1951. Parameters for relating unit hydrographs to watershed characteristics. *Transactions of the American Geophysical Union* 32(4):591–96.

Engman, E. T., and A. S. Rogowski. 1974. A partial area model for storm flow synthesis. *Water Resources Research* 10(3):464–72.

Espey, W. H., Jr., C. W. Morgan, and F. D. Masch. 1965. A study of some effects of urbanization on storm runoff from a small watershed. Technical Report, Center for Research in Water Resources, University of Texas, Austin.

Espey, W. H., and D. E. Winslow. 1968. The effects of urbanization on unit hydrographs for small watersheds, Houston, Texas, 1964–67. Tracor Doc. No. 68-975-U, TRACOR, Inc., Austin, Texas.

Eykhoff, P. 1974. *System Identification: Parameter and State Estimation*. New York: John Wiley.

Federal Aviation Agency. 1970. Advisory circular on airport drainage. Report A/C 150-5320-58, U.S. Department of Transportation, Washington, D.C.

Fiering, M. B. 1967. *Streamflow Synthesis*. Cambridge: Harvard University Press.

Fletcher, R., and M. J. D. Powell. 1963. A rapidly convergent descent method for minimization. *Computer Journal* 6(2):163–68.

Forsythe, G. E. 1957. Generation and use of orthogonal polynomials for data fitting with a digital computer. *Journal of the Society of Industrial and Applied Mathematics* 5(2):74–88.

Fread, D. L. 1982. "Flood routing: a synopsis of past, present, and future capability." In *Rainfall-Runoff Relationship*, edited by V. P. Singh, 521–42. Littleton, Colo.: Water Resources Publications.

Freeze, R. A. 1971. Three-dimensional, transient, saturated-unsaturated flow in a groundwater basin. *Water Resources Research* 7(2):347–66.

———. 1972a. Role of subsurface flow in generating surface runoff. 1. Base flow contributions to channel flow. *Water Resources Research* 8(3):609–23.

———. 1972b. Role of subsurface flow in generating surface runoff: 2. Upstream source areas. *Water Resources Research* 8(5):1272–83.

———. 1974. Streamflow generation. *Reviews of Geophysics and Space Physics* 12(4):627–47.

George, D. A. 1959. Continuous nonlinear systems. Technical Report no. 355, Research Laboratory of Electronics, Massachusetts Institute of Technology, Cambridge.

Getty, H. C., and J. H. McHughs. 1962. Synthetic peak discharges for design criteria. *Journal of the Hydraulics Division, Proceedings of the American Society of Civil Engineers* 88(HY5):1–12.

Gilcrest, B. R. 1950. "Flood routing." Chapter 10 in *Engineering Hydraulics*, edited by H. Rouse, 635–76. New York: John Wiley.

Gill, M. A. 1977a. Overland-flow travel time versus Reynolds number—a comment. *Journal of Hydrology* 35:203–4.

———. 1977b. Routing of floods in river channels. *Nordic Hydrology* 8:163–70.

———. 1978. Flood routing by the Muskingum method. *Journal of Hydrology* 36:353–63.

———. 1979a. Translatory characteristics of the Muskingum method of flood routing. *Journal of Hydrology* 40:17–29.

———. 1979b. Flood routing by Muskingum method—reply. *Journal of Hydrology* 41:169–70.

———. 1979c. Critical examination of the Muskingum method. *Nordic Hydrology* 10:261–70.

———. 1980. Translatory characteristics of the Muskingum method of flood routing—reply. *Journal of Hydrology* 48:369–72.

———. 1984. Time lag solution of the Muskingum flood routing equation. *Nordic Hydrology* 15:145–54.

Gilman, K. 1977a. Dilution gauging on the recession limb: 1. Constant rate injection method. *Hydrological Sciences Bulletin* 22(3):353–69.

———. 1977b. Dilution gauging on the recession limb: 2. The integration method. *Hydrologic Sciences Bulletin* 22(4):469–81.

Golding, B. L. 1980. Hydrograph synthesis by the HNV-SBUH method. Paper presented at the UWMM Users Group Meeting, June 19–20, Toronto, Canada.

Goodrich, R. D. 1931. Rapid calculation of reservoir storage. *Civil Engineering* 1:417–18.

Gould, B. W. 1959. Queing theory and water storage. *Journal of the Hydraulics Division, Proceedings of the American Society of Civil Engineers* 85(HY4):117–19.

Grace, R. A., and P. S. Eagleson. 1966. The use of scale models in rainfall-runoff studies. *Water Resources Research* 2(3):393–403.

Gray, D. M. 1961. Synthetic unit hydrographs for small watersheds. *Journal of the Hydraulics Division, Proceedings of the American Society of Civil Engineers* 87(HY4):33–54.

———. 1962. Derivation of hydrographs for small watersheds from measurable physical characteristics. Research Bulletin 506, 514–79, Agricultural and Home Economics Experiment Station, Iowa State University, Ames.

———. ed. 1970. *Handbook of the Principles of Hydrology*. The Secretariat, Canadian National Committee for the International Hydrological Decade, Ottawa, Canada.

Green, R. F. 1970. Optimization by the pattern search method. Tennessee Valley Authority Research Paper 7, Knoxville, Tenn.

Greenwood, J. A., J. M. Landwehr, N. C. Matalas, and J. R. Wallis. 1979. Probability weighted moments: definition and relation to parameters of several distributions expressible in inverse form. *Water Resources Research* 15(5):1049–54.

Gregory, J. M. 1982. Time of concentration determination for small watersheds. *Transactions of the American Society of Agricultural Engineers* 25:686–88 and 694.

Gregory, R. L., and C. E. Arnold. 1932. Runoff-rational runoff formulas. *Transactions of the American Society of Civil Engineers* 96:1038–1177.

Guillemin, E. A. 1954. What is nature's criterion? *Transactions of the Institute of Electronic and Electrical Engineers*, PGCT CT-1:76.

Gumbel, E. J. 1941. The return period of flood flows. *Annals of Mathematical Statistics* 12(4):163–90.

Gupta, S. C. 1971. *Transform and State Variable Methods in Linear Systems*. Huntington, N.Y.: Robert E. Krieger.

Gupta, V. K., and S. Sorooshian. 1983. Uniqueness and observability of conceptual rainfall-runoff model parameters: the percolation process examined. *Water Resources Research* 19(1):269–76.

Gupta, V. K., and E. Waymire. 1983. On the formulation of an analytical approach to hydrologic response and similarity at the basin scale. *Journal of Hydrology* 65:95–123.

Gupta, V. K., E. Waymire, and C. T. Wang. 1980. A representation of an instantaneous unit hydrograph from geomorphology. *Water Resources Research* 16(6):855–62.

Gupta, V. L., and S. A. Moin. 1974. Surface runoff hydrograph equation. *Journal of the Hydraulics Division, Proceedings of the American Society of Civil Engineers* 100(HY10):1353–68.

Gupta, V. L., V. Thongchareon, and S. A. Moin. 1974. Analytical modeling of surface runoff hydrographs for major streams in northeast Thailand. *Hydrological Sciences Bulletin* 19(4/12):523–40.

Haan, C. T. 1970. A dimensionless hydrograph equation. File report, Agricultural Engineering Department, University of Kentucky, Lexington.

Haan, C. T., and B. J. Barfield. 1978. *Hydrology and Sedimentology of Surface Mined Lands*, 85–88, Office of Continuing Education and Extension, College of Engineering, University of Kentucky, Lexington.

Hall, F. R. 1968. Base flow recessions: a review. *Water Resources Research* 4(5):973–83.

———. 1982. "Subsurface water contribution to stream flow." In *Rainfall-Runoff Relationship*, edited by V. P. Singh. Littleton, Colo.: Water Resources Publications.

Hall, M., Jr., and M. Newman. 1963. Copositive and completely positive quadratic forms. *Proceedings of the Cambridge Philosophical Society*, vol. 59, 329–39.

HALL, M. J. 1977. On the smoothing of oscillations in finite period unit hydrographs derived by the harmonic method. *Hydrological Sciences Bulletin* 127(2/6):313–24.

HANSON, T. L., AND H. P. JOHNSON. 1964. Unit hydrograph methods compared. *Transactions of the ASAE* 7:448–51.

HARDER, J. A., AND S. M. ZAND. 1969. The identification of nonlinear hydrologic systems. *Technical Report HELS-2*, Hydraulic Engineering Lab, University of California, Berkeley.

HARLEY, B. M. 1967. Linear routing in uniform open channels. M. Eng. Sc. thesis, National University of Ireland, University College, Cork, Ireland.

HARLIN, J. M. 1984. Watershed morphometry and time to hydrograph peak. *Journal of Hydrology* 67:141–54.

HATHAWAY, G. A. 1945. Design of drainage facilities. *Transactions of the American Society of Civil Engineers* 110:697–730.

HAYAMI, S. 1951. On the propagation of flood waves. *Bulletin no. 1*, Disaster Prevention Research Institute, Kyoto University, Kyoto, Japan.

HAZEN, A. 1930. *Flood Flows*. New York: John Wiley.

HEBSON, E., AND E. F. WOOD. 1982. A design flood frequency distribution using Horton order ratios. *Water Resources Research* 18(6):1509–18.

HEGGEN, R. J. 1984. Univariate least squares muskingum flood routing. *Water Resources Bulletin* 20(1):103–7.

HELWEG, O. J., J. AMOROCHO, AND R. H. FINCH. 1982. Improvement of nonlinear rainfall-runoff model. *Journal of the Hydraulics Division, Proceedings of the American Society of Civil Engineers* 108(HY7):813–22.

HEMPEL, C. G. 1966. *Philosophy of Natural Science*. Englewood Cliffs, N.J.: Prentice-Hall, Inc.

HENDERSON, F. M., AND R. A. WOODING. 1964. Overland flow and groundwater flow from a steady rain of finite duration. *Journal of Geophysical Research* 69(8):1531–40.

HERSHFIELD, D. 1961. Rainfall frequency atlas of the United States. U.S. Weather Bureau Technical Paper 40, Washington, D.C.

HERTZ, H. 1956. *The Principles of Mechanics*. New York: Dover Publications.

HEWLETT, J. D. 1961. Soil moisture as a source of base flow from steep mountain watersheds. Southeast Forest Experiment Station Research Paper SE 132, U.S. Forest Service.

HEWLETT, J. D., AND A. R. HIBBERTT. 1967. Factors affecting the response of small watersheds to precipitation in humid areas. *Proceedings of the International Symposium on Forest Hydrology*, Pennsylvania State University, University Park, 275–90.

HEWLETT, J. D., AND W. L. NUTTER 1970. The varying source area of streamflow from upland basins. Paper presented at the Symposium on Interdisciplinary Aspects of Watershed Management, Montana State University, Bozeman.

HEWLETT, J. D., AND C. A. TROENDLE. 1975. Nonpoint and diffused water sources: a variable source area problem. *Proceedings of the American Society of Civil Engineers Symposium on Watershed Management*, Utah State University, Logan, 21–45.

HICKOK, R. B., R. V. KEPPEL, AND B. R. RAFFERTY. 1959. Hydrograph synthesis. *Agricultural Engineering* 40:608–11, 615.

HIEMSTRA, L. A. V. 1968. Modification on the "Soil Conservation Service Method" for the estimation of design floods on very small catchments of the arid western part of the United States of America. *Technical Report*, Department of Civil Engineering, Colorado State University, Fort Collins.

HILL, J. M., V. P. SINGH, AND H. AMINIAN. 1984. A computerized data base for hydrologic modeling of the Amite River basin, Louisiana. *Final Report*, Remote Sensing and Image Processing Laboratory, Louisiana State University, Baton Rouge.

HIMMELBLAU, D. M. 1972. *Applied Nonlinear Programming*. New York: McGraw-Hill.

HIMMELBLAU, D. M., AND R. V. YATES. 1968. A method of flow routing. *Water Resources Research* 4(6):1193–99.

HINO, M. 1970. Runoff forecasts by predictive filter. *Journal of the Hydraulics Division, Proceedings of the American Society of Civil Engineers* 96(HY3):681–702.

———. 1970. Runoff forecasting by variable transformation. *Journal of the Hydraulics Division, Proceedings of the American Society of Civil Engineers* 96(HY4):871–78.

———. 1972. Stochastic approach to linear and nonlinear runoff analysis. *Technical Report no.* 12, Department of Civil Engineering, Tokyo Institute of Technology, Tokyo, Japan.

HINO, M., AND K. NADAOKA. 1979. Mathematical derivation of linear and nonlinear runoff kernels. *Water Resources Research* 15(4):918–28.

HINO, M., T. SUKIGARA, AND H. KIKKAWA. 1971. "Nonlinear runoff kernels of hydrologic systems." In *Systems Approach to Hydrology*, edited by V. Yevjevich, 102–12. Fort Collins, Colo.: Water Resources Publications.

HIRANO, M. 1980. On distribution of concentration time in runoff process. *Proceedings of the Third International Symposium on Stochastic Hydraulics*, August 5–7, Tokyo, Japan, 253–61.

HJELMFELT, JR., A. T. 1984. Negative outflows from Muskingum flood routing. ASCE *Journal of Hydraulic Engineering* 111(6):1010–14.

HJELMFELT, A. T., JR., AND J. J. CASSIDY. 1975. *Hydrology for Engineers and Planners*. Ames, Iowa: Iowa State Univ. Press.

HOLLAND, M. E. 1969. Design and testing of rainfall-runoff system. *Report CER 69-70 ME H*, Department of Civil Engineering, Colorado State University, Fort Collins.

HOLTAN, H. N. 1970. Representative and experimental basins as dispersed systems. *Proceedings of the International Symposium on Results of Research on Representative and Experimental Basins*, Wellington, New Zealand, International Association of Scientific Hydrology Publication 96, 113–126.

HOLTAN, H. N., AND N. C. LOPEZ. 1973. USDAHL–73 revised model of watershed hydrology. *Plant Physiology Institute Report no.* 1, U.S. Department of Agriculture, Agricultural Research Service, Beltsville, Md.

HOLTAN, H. N., AND D. E. OVERTON. 1965. Storage-flow hysteresis in hydrograph synthesis. *Journal of Hydrology* 2:309–23.

HORN, D. L., AND G. O. SCHWAB. 1963. Evaluation of rational runoff coefficients for small agricultural watersheds. *Transactions of the American Society of Agricultural Engineers* 6:105–201, 223.

HORNER, W. W. 1910. Modern procedure in district sewer design. *Engineering News*, 64.

HORNER, W. W., AND F. L. FLYNT. 1936. Relation between rainfall and runoff from small urban areas. *Transactions of the American Society of Civil Engineers* 101:140–83.

HORNER, W. W., AND S. W. JENS. 1941. Surface runoff determination from rainfall without using coefficients. *Transactions of the American Society of Civil Engineers* 107:1039–75.

HORTON, R. E. 1931. The field, scope and status of the science of hydrology. *Transactions of the American Geophysical Union* 12:189–202.

———. 1933. The role of infiltration in the hydrologic cycle. *Transactions of the American Geophysical Union* 14:446–60.

———. 1938. The interpretation and application of runoff plot experiments with reference to soil erosion problems. *Soil Science Society of America Proceedings* 3:340–49.

———. 1939. Analysis of runoff-plot experiments with varying infiltration capacity. *Transactions of the American Geophysical Union* 20:693–711.

———. 1940. An approach toward a physical interpretation of infiltration capacity. *Soil Science Society of America Proceedings* 5:399–417.

———. 1945. Erosional development of streams and their drainage basins: hydrophysical approach to quantitative morphology. *Geological Society of America Bulletin* 56:275–370.

Houghton, J. C. 1978. The incomplete means estimation procedure applied to flood frequency analysis. *Water Resources Research* 14(6):1111–15.

Hoyt, W. G. 1942a. An outline of the runoff cycle. *Technical Bulletin* 27, 57–69, Pennsylvania State College, School of Engineering, State College.

———. 1942b. "The runoff cycle." Chapter 11D in *Physics of the Earth*, Vol 9: Hydrology, edited by O. E. Meinzer. New York: McGraw-Hill.

Hoyt, W. G., and W. B. Langbein. 1955. *Floods*. Princeton, N.J.: Princeton Univ. Press.

Huang, Y. H. 1978. Channel routing by finite difference method. *Journal of the Hydraulics Division, Proceedings of the American Society of Civil Engineers* 104(HY10):1379–93.

Hudlow, M. D., and R. A. Clark. 1969. Hydrological synthesis by digital computer. *Journal of the Hydraulics Division, Proceedings of the American Society of Civil Engineers* 95(HY3):839–60.

Hursh, C. R., and E. F. Brater. 1941. Separating storm hydrographs from small drainage areas in surface and subsurface flow. *Transactions of the American Geophysical Union* 22(pt. 3):863–71.

Hyami, S. 1951. On the propagation of flood waves. *Disaster Prevention Research Institute Bulletin* 1, Kyoto University Kyoto, Japan.

Ibbit, R. P., and T. O'Donnell. 1971. Fitting methods for conceptual catchment models. *Journal of the Hydraulics Division, Proceedings of the American Society of Civil Engineers* 100(HY7):995–1109.

Institution of Engineers, Australia, 1977. *Australian Rainfall and Runoff. Cunberra, A.C.T., Australia.*

Izzard, C. F. 1946. Hydraulics of runoff from developed surfaces. *Proceedings of the Highway Research Board* 26:129–50.

Jackson, D. R. 1968. A dissipative river flow model. *Journal of Hydrology* 6:33–44.

Jacoby, S. L. S. 1965. An output prediction model for nonlinear time lag systems. Document D6-17443, Airplane Division, The Boeing Company. Seattle, Washington.

———. 1966. A mathematical model for nonlinear hydrologic systems. *Journal of Geophysical Research* 71(20):4811–4824.

James, I. C., II, B. T. Bower, and N. C. Matalas. 1969. Relative importance of variables in water resources planning. *Water Resources Research* 5(6):1165–73.

James, L. D., and W. O. Thompson. 1970. Least squares estimation of constants in a linear recession model. *Water Resources Research* 6(4):1062–69.

Jarvis, C. S. 1935. Discussion, "An approach to determinate stream flow" by M. Bernard. *Transactions of the American Society of Civil Engineers* 100:363–65.

Jaynes, E. T. 1969. *Probability Theory in Science and Engineering*. New York: McGraw-Hill.

———. 1982. On the rationale of entropy methods. *Proceedings of the IEEE* 70(9):939–52.

Jenkins, G. M., and D. G. Watts. 1968. *Spectral Analysis and Its Applications*. San Francisco: Holden-Day.

Jetter, K. 1944. Evaluation of runoff-distribution values from basic data and study of related drainage-area characteristics. *Transactions of the American Geophysical Union* 25:990–1004.

Johnstone, D., and W. P. Cross. 1949. *Elements of Applied Hydrology*. New York: Ronald Press.

Jones, D. A., and R. J. Moore. 1980. A simple channel flow routing model for real time use. *Proceedings of the Oxford Symposium on Hydrological Forecasting*, IAHS Publication no. 129, 397–408.

Jones, L. E. 1971. Linearizing weight factors for least squares fitting. *Journal of the Hydraulics Division, Proceedings of the American Society of Civil Engineers* 97(HY5):665–75.

Jones, P. N., and C. A. McGilchrist. 1978. Analysis of hydrological recession curves. *Journal of Hydrology* 36:365–74.

Jones, S. B. 1981. Choice of space and time steps in the Muskingum-Cunge flood routing method. *Proceedings of the Institution of Civil Engineers* 55(pt. 1):913–30.

———. 1983. Discussion of "Accuracy criteria in diffusion routing by V. M. Ponce and F. D. Theurer." ASCE *Journal of Hydraulic Engineering* 109(5):601–3.

Kalinin, G. P., and P. I. Milyukov. 1957. On the computation of unsteady flow in open channels. *Meteorologiya Gidrologiya Zhuzurnal* 10:10–18, Leningrad, USSR.

Kavvas, M. L. 1972. Derivation of stable non-oscillating unit hydrographs. M.Sc. Thesis, Department of Civil Engineering, Colorado State University, Fort Collins.

Kavvas, M. L., and E. F. Schulz. 1972. "Removal of unit hydrograph oscillations by filtering." In: *Floods and Droughts*, edited by E. F. Schulz, V. A. Koelzer, and K. Mahmood, 167–74. Fort Collins, Colo.: Water Resources Publications.

Keefer, T. N., and R. S. McQuivey. 1974. Multiple linearization flow routing model. *Journal of the Hydraulics Division, Proceedings of the American Society of Civil Engineers* 100(HY7):1031–46.

Kendall, M. G., and A. Stuart. 1958. "Distribution theory." Vol. 1 of *The Advanced Theory of Statistics*. London: Charles Griffin Co.

Kennedy, R. J., and W. E. Watt. 1967. The relationship between lag time and the physical characteristics of drainage basins in Southern Ontario. *International Association of Scientific Hydrology Publication* no. 85, 866–74.

Keppel, R. V. 1960. Transmission losses on Walnut Gulch watershed. *Proceedings of the Joint ARS-SWC Hydrology Workshop*, New Orleans, Louisiana.

Keppel, R. V., and K. G. Renard. 1962. Transmission losses in ephemeral stream beds. *Journal of the Hydraulics Division, Proceedings of the American Society of Civil Engineers* 88(HY3):59–68.

Kerby, W. S. 1959. Time of concentration of overland flow. *Civil Engineering* 60:174.

King, H. W., and E. F. Brater. 1963. *Handbook of Hydraulics*. New York: McGraw-Hill.

Kinnison, H. B., and B. R. Colby. 1945. Flood formulas based on drainage basin characteristics. *Transactions of the American Society of Civil Engineers* 110:849–76.

Kirpich, T. P. 1940. Time of concentration of small agricultural watersheds. *Civil Engineering* (New York) 10(6):362.

Kirshen, D. M., and R. L. Bras. 1983. The linear channel and its effect on the geomorphologic IUH. *Journal of Hydrology* 65:175–208.

Kisiel, C. C. 1969a. "Time series analysis of hydrologic data." In *Advances in Hydroscience*, edited by V. T. Chow, vol. 5, 1–119. New York: Academic Press.

———. 1969b. Mathematical methodology in hydrology. *Proceedings of the First International Seminar for Hydrology Professors*, July, University of Illinois, Urbana, vol. 1, 362–99.

Kitanidis, P. K., and R. L. Bras. 1979. Collinearity and stability in the estimation of rainfall-runoff model parameters. *Journal of Hydrology* 42:91–108.

Klassen, B., and D. H. Pilgrim. 1975. Hydrograph recession constants for New South Wales Streams. *Civil Engineering Transactions* (Australia) CE17(1):43–49.

Klemes, V. 1972. Comment on "Mathematical models of hydrologic systems by J. C. I. Dooge." *Proceedings of the International Symposium on Modeling Techniques in Water Resources Systems*, May 9–12, 1972, Ottawa, Canada, vol. 3, 701–5.

———. 1974. Probability distribution of outflow from a linear reservoir. *Journal of Hydrology* 21:305–14.

———. 1978. "Physically based stochastic hydrologic analysis." In *Advances in Hydroscience*, edited by V. T. Chow, vol. 2, 285–356. New York: Academic Press.

———. 1982. "Empirical and causal models in hydrology." Chapter 8 in *Scientific Basis of Water-Resource Management*, 95–104. Washington, D.C.: National Academy Press.

———. 1983. Conceptualization and scale in hydrology. *Journal of Hydrology* 65:1–23.

Klemes, V., and L. Boruvka. 1975. Output from a cascade of discrete linear reservoirs with stochastic input. *Journal of Hydrology* 27:1–13.

Knisel, W. G., Jr. 1963. Baseflow recession analysis for comparison of drainage basins and geology. *Journal of Geophysical Research* 68(12):3649–53.

Kohler, M. A. 1944. A forecasting technique of routing and combining flow in terms of stage. *Transactions of the American Geophysical Union* 25(pt. 6):1030–35.

Kostyakov, A. M. 1932. On the dynamics of the coefficient of water percolation in soils and of the necessity of studying it from a dynamic point of view for purposes of amelioration (in Russian). *The Sixth International Congress of Soil Science Transactions*, 17–21

Kottegoda, N. T. 1979. *Stochastic Water Resources Technology*. London: John Wiley.

Koussis, A. D. 1976. An approximate dynamic flood routing method. *Proceedings of the International Symposium on Unsteady Flow in Open Channels*, Paper L1, Newcastle-Upon-Tyne, England.

———. 1978. Theoretical estimation of flood routing parameters. *Journal of the Hydraulics Division, Proceedings of the American Society of Civil Engineers* 104(HY1):109–15.

———. 1980. Comparison of Muskingum method difference schemes. *Journal of the Hydraulics Division, Proceedings of the American Society of Civil Engineers* 106(HY5):925–29.

———. 1983. Unified theory for flood and pollution routing. ASCE Journal of Hydraulic Engineering 109(12):1652–84.

Kuczera, G. 1982. On the relationship between the reliability of parameter estimates and hydrologic time series data used in calibration. *Water Resources Research* 18(1):146–54.

———. 1983a. Improved parameter inference in catchment models: 1. Evaluating parameter uncertainty. *Water Resources Research* 19(5):1151–62.

———. 1983b. Improved parameter inference in catchment models: 2. Combining different kinds of hydrologic data and testing their compatibility. *Water Resources Research* 19(5):1163–72.

Kuchment, L. S., and E. N. Borshchevsky. 1971. Identification of nonlinear hydrologic systems. *Meteorology, Gidrology*, no. 1, 42–47.

Kuhn, T. S. 1973. *The Structure of Scientific Revolutions*. Chicago: The Univ. of Chicago Press.

Kuichling, E. 1889. The relation between the rainfall and the discharge of sewers in populous districts. *Transactions of the American Society of Civil Engineers* 20:37–40.

Kulandaiswamy, V. C. 1964. A basic study of the rainfall excess-surface runoff relationship in a basin system. Ph.D. diss., University Illinois, Urbana.

———. 1966. A note on the Muskingum method of flood routing. *Journal of Hydrology* 4:273–76.

Kulandaiswamy, V. C., and T. Babu Rao. 1970. An investigation of nonlinearity in storage-discharge relationship for watersheds. Report 4, C.S.I.R. Project, Hydraulics and Water Resources Department, College of Engineering, Madras, India.

Kulandaiswamy, V. C., M. Krishnaswami, and T. N. Ramalingam. 1967. Flood routing through linear channels. *Journal of Hydrology* 5:279–85.

Kulandaiswamy, V. C., and S. Seetharaman. 1969. A note on Barnes' method of hydrograph separation. *Journal of Hydrology* 9:222–29.

Kulandaiswamy, V. C., and C. V. Subramanian. 1967. A nonlinear approach to runoff studies. *Proceedings of the International Hydrology Symposium*, Fort Collins, Colo., vol. 1, 72–79.

Kundzewicz, Z. W. 1980. Discussion of "Approximate flood routing methods: a review by P. E. Weinmann and E. M. Laurenson." *Journal of the Hydraulics Division, Proceedings of the American Society of Civil Engineers* 106(HY12):2072–75.

Kundzewicz, Z. W., and J. C. I. Dooge. 1985. Unified structural approach to linear flood routing. *Advances in Water Resources* 8:37–43.

Kunkle, G. R. 1962. The baseflow duration curve, a technique for the study of groundwater discharge from a drainage basin. *Journal of Geophysical Research* 67(4):1543–54.

Laden, N. R., T. L. Reilly, and J. S. Minnotte. 1940. Synthetic unit-hydrographs, distribution-graphs and flood routing in the upper Ohio River basin. *Transactions of the American Geophysical Union* 21:649–59.

Lambert, A. O. 1969. A comprehensive rainfall-runoff model for an upland catchment area. *Journal of Institution of Water Engineers* 23(4):231–38.

Lancaster, P. 1969. *Theory of Matrices*. New York: Academic Press.

Lanczos, C. 1956. *Applied Analysis*. Englewood Cliffs, N.J.: Prentice Hall, Inc.

Lane, L. J. 1972. A proposed model for flood routing in abstracting ephemeral channels. *Proceedings of the Arizona Section of the American Water Resources Association*, Prescott, Ariz., 439–53.

Lane, L. J., M. H. Diskin, and K. G. Renard. 1971. Input-output relationships for an ephemeral channel system. *Journal of Hydrology* 13:22–44.

LANE, L. J., M. H. DISKIN, D. E. WALLACE, AND R. M. DIXON. 1978. Partial area response on small semiarid watersheds. *Water Resources Bulletin* 14(5):1143–58.

LANE, L. J., AND D. E. WALLACE. 1976. Simulation of partial area response from a small semiarid watershed. *Proceedings of the 1976 Joint Meetings of Arizona Section of AWRA and Hydrology Section*, Arizona Academy of Science, April, Tucson, Ariz., 137–47.

LARSON, C. L., AND R. E. MACHMEIER. 1968. Peak flow and critical duration for small watersheds. *Transactions of the American Society of Agricultural Engineers* 11:208–13.

LAURENSON, E. M. 1961. A study of hydrograph recession curves of an experimental catchment. *The Journal of the Institution of Engineers* (Australia) 33:253–58.

LAURENSON, E. M. 1962. Hydrograph synthesis by runoff routing. *Technical Report No.* 66, Water Research Laboratory, University of New South Wales, Kensington, New South Wales, Australia.

———. 1964. A catchment storage model for runoff routing. *Journal of Hydrology* 2:141–63.

———. 1965. Storage routing methods of flood estimation. *The Civil Engineering Transactions of the Institution of Engineers* (Australia) CE7:39–47.

LAURENSON, E. M., R. G. MEIN, AND T. A. MCMAHON. 1975. Flood estimating by runoff routing. Lecture notes, Department of Civil Engineering, Monash University, Australia.

LAURENSON, E. M., AND T. O'DONNELL. 1969. Data error effects in unit hydrograph derivation. *Journal of the Hydraulics Division, Proceedings of the American Society of Civil Engineers* 95(HY6):1899–1917.

LAWLER, E. A. 1964. "Hydrology of flow control." 2. "Flow control." In *Handbook of Applied Hydrology*, edited by V. T. Chow, 25–34 to 25–59, New York: McGraw-Hill.

LEE, M. T., AND J. W. DELLEUR. 1976. A variable source area model of rainfall-runoff process based on the watershed stream network. *Water Resources Research* 12(5):1029–36.

LEE, P. A. 1966. Properties of the Meixner polynomials. Report, Department of Electrical Engineering, Monash University, Monash, Australia.

LEE, R. C. K. 1964. *Optimal Estimation, Identification, and Control*. Cambridge: MIT Press.

LEE, T. H., G. E. ADAMS, AND W. H. GAINES. 1968. *Computer Process Control: Modeling and Optimization*. New York: John Wiley.

LEE, Y. W. 1932. Synthesis of electric networks by means of the Fourier transforms of Laguerre's functions. *Journal of Mathematical Physics* (Cambridge, Mass.) 11:83–113.

———. 1960. *Statistical Theory of Communication*. New York: John Wiley & Sons, Inc.

LEOPOLD, L. B., M. G. WOLMAN, AND J. P. MILLER. 1964. *Fluvial Processes in Geomorphology*. San Francisco: W. H. Freeman.

LETTAU, H. 1954. A study of mass, momentum, and energy budget of the atmosphere. Archives, Meteorology, Geophysics, and Bioclimatology, Vienna 7(ser. A):131–53.

LEVI, E., AND E. VALDES. 1964. A method for direct analysis of hydrographs. *Journal of Hydrology* 2:182–90.

LIEBELT, P. B. 1967. *An Introduction to Optimal Estimation*. Reading, Mass.: Addison-Wesley.

LIENHARD, J. H. 1964. A statistical mechanical prediction of the dimensionless unit hydrograph. *Journal of Geophysical Research* 69:5231–38.

———. 1972. Prediction of the dimensionless unit hydrograph. *Nordic Hydrology* 3:107–9.

LIENHARD, J. H., AND L. B. DAVIS. 1971. An extension of statistical mechanics to the description of a broad class of macroscopic systems. *Journal of Applied Mathematics and Mechanics* 22:85–96.

LIENHARD, J. H., AND P. L. MEYER. 1967. A physical basis for the generalized gamma distribution. *Quarterly of Applied Mathematics* 25(3):330–34.

LIGHTHILL, M. J., AND G. B. WHITHAM. 1955. On kinematic waves: flood movement in long rivers. *Proceedings of the Royal Society* 22(ser. A):281–316.

LINSLEY, R. K. 1943. Application of the synthetic unitgraph in the Western Mountain States. *Transactions of the American Geophysical Union* 24:581–87.

———. 1945. Discussion of "storage and the unit hydrograph" by C. O. Clark. *Transactions of the American Society of Civil Engineers* 110:1452–55.

LINSLEY, R. K., M. A. KOHLER, AND J. L. H. PAULHUS. 1958. *Hydrology for Engineers*. New York: McGraw-Hill.

LIU, C. K., AND W. BRUTSAERT. 1975. Optimal identification of a watershed rainfall-runoff system. *International Association of Hydrological Sciences Publication No.* 117, 333–41.

———. 1978. A nonlinear analysis of the relationship between rainfall and runoff for extreme floods. *Water Resources Research* 14(1):75–83.

LLOYD-DAVIES, D. E. 1906. The elimination of storm water from sewerage systems. Minutes of Proceedings, Institution of Civil Engineers, 164:41, London, England.

LOPEZ, O. G. 1973. Response time in urban Colorado basins. M.S. thesis, Department of Civil Engineering, Colorado State University, Fort Collins.

LOWING, M. J., AND R. G. MEIN. 1981. Flood event modeling: a study of two methods. *Water Resources Bulletin* 17(4):599–606.

MADDAUS, W. O., AND P. S. EAGLESON. 1969. A distributed linear representation of surface runoff. *Hydrodynamics Laboratory Report No.* 115, Massachusetts Institute of Technology, Cambridge.

MANDELBROT, B. B. 1970. Comment on "Stochastic models in hydrology by A. E. Scheidegger." *Water Resources Research* 6:1791.

MANDELBROT, B. B., AND J. R. WALLIS. 1969a. Computer experiments with fractional gaussian noises: 1. Averages and variances. *Water Resources Research* 5:228–241.

———. 1969b. Computer experiments with fractional gaussian noises: 2. Rescaled ranges and spectra. *Water Resources Research* 5:242–59.

———. 1969c. Computer experiments with fractional gaussian noises: 3. Mathematical appendix. *Water Resources Research* 5:260–67.

MANDEVILLE, A. N., AND T. O'DONNELL. 1973. Introduction of time variance to linear conceptual catchment models. *Water Resources Research* 9(2):298–310.

MATALAS, N. C. 1966. Some aspects of time series analysis in hydrologic studies. *Proceedings of Hydrology Symposium No.* 5, McGill University, Canada.

MATERN, B. 1960. "Spatial variations." In *Stochastic Models and Their Application to Some Problems in Forest Surveys, and other Sampling Investigations*, 5(5):144. Stockholm, Sweden: Maddelander Ser. Skogsforsknints.

MATHUR, B. S. 1974a. Natural catchment representation by a series of linear channels, 1. *Proceedings of the Warsaw Symposium on Mathematical Models in Hydrology*, vol. 2, 634–42.

———. 1974b. Natural catchment representation by a series of linear channels, 2. *Proceedings of the Warsaw Symposium on Mathematical Models in Hydrology*, vol. 2, 643–52.

Mays, L. W., and L. Coles. 1980. Optimization of unit hydrograph determination. *Journal of the Hydraulics Division, Proceedings of the American Society of Civil Engineers* 106(HY1):85–97.

Mays, L. W., and C. K. Taur. 1982. Unit hydrographs via nonlinear programming. *Water Resources Research* 18(4):744–52.

Mawdsley, J. A., and A. F. Tagg. 1981. Identification of unit hydrographs from multi-event analysis. *Journal of Hydrology* 49:315–27.

McCabe, J. A. 1962. Floods in Kentucky: magnitude and frequency. *Information Circular* 9, Kentucky Geological Survey, Lexington, Ky.

McCann, R. C., and V. P. Singh. 1980. An analysis of the Chow-Kulandaiswamy GHS model. *Advances in Water Resources* 3:173–80.

———. 1981. The general hydrologic system model. *Journal of the Hydraulics Division, Proceedings of the American Society of Civil Engineers* 107(HY12):1581–82.

McCarthy, G. T. 1938. The unit hydrograph and flood routing. Unpublished manuscript presented at a conference of the North Atlantic Division, June 24, U.S. Army Corps of Engineers.

McCuen, R. H. 1973a. The role of sensitivity analysis in hydrologic modeling. *Journal of Hydrology* 18:37–53.

———. 1973b. Component sensitivity: a tool for the analysis of complex water resource systems. *Water Resources Research* 9(1):243–46.

McCuen, R. H., and T. R. Bondelid. 1983. Estimating unit hydrograph peak rate factors. *ASCE Journal of Irrigation and Drainage Engineering* 109(2):238–50.

McCuen, R. H., S. L. Wong, and W. J. Rawls. 1984. Estimating urban time of concentration. *ASCE Journal of Hydraulic Engineering* 110(7):887–904.

McPherson, M. B. 1969. Some notes on the rational method of storm drain design. *Technical Memorandum no.* 6, American Society of Civil Engineers Water Resources Research Program, Harvard University, Cambridge, Mass.

Meehan, P. J. 1979. Flood routing by the Muskingum method: comments. *Journal of Hydrology* 41:167–68.

Mein, R. G., E. M. Laurenson, and T. A. McMahon. 1974. Simple nonlinear model for flood estimation. *Journal of the Hydraulics Division, Proceedings of the American Society of Civil Engineers* 100(HY11):1507–18.

Meinzer, O. E., ed. 1942. *Hydrology*. Chapter 1, 1–31, New York: McGraw-Hill.

Melentijevich, M. 1965. The analysis of range with output linearly dependent upon storage. Hydrology Paper 11, Colorado State University, Fort Collins.

Meredith, D. D., and T. E. Byrones. 1981. Discussion of "Optimization of unit hydrograph determination by L. W. Mays and L. Coles." *Journal of the Hydraulics Division, Proceedings of the American Society of Civil Engineers* 107(HY1):146–49.

Meynink, W. J. C. 1978. Catchment time response. Project 44/1977, Darling Downs Institute of Advanced Education, Australia.

Meynink, W. J. C., and I. Cordery. 1979. Critical duration of rainfall for flood estimation. *Water Resources Research* 12(6):1209–14.

Miller, A. C., S. N. Kerr, and D. J. Spaeder. 1983. Calibration of Snyder coefficients for Pennsylvania. *Water Resources Bulletin* 19(4):625–30.

Miller, C. F. 1968. Evaluation of runoff coefficients from small natural drainage areas. *Research Report no.* 14, Water Resources Institute, University of Kentucky, Lexington, Ky.

Miller, W. A., and D. A. Woolhiser. 1975. "Choice of models." Chapter 9 in *Unsteady Flow in Open Channels*, edited by K. Mahmood and V. Yevjevich, 367–78. Littleton, Colo.: Water Resources Publications.

Mimikou, M. 1983. A study of drainage basin linearity and nonlinearity. *Journal of Hydrology* 64:113–34.

———. 1984. Envelope curves for extreme flood events in northwestern and western Greece. *Journal of Hydrology* 67:55–67.

Minshall, N. E. 1960. Predicting storm runoff on small experimental watersheds. *Journal of the Hydraulics Division, Proceedings of the American Society of Civil Engineers* 86(HY8):17–38.

Mitchell, W. D. 1948. Unit hydrographs in Illinois. Department of Public Works and Buildings, Division of Waterways, Ill.

———. 1962. Effect of reservoir storage on peak flow. Geological Survey Water Supply Paper 1580-C, U.S. Department of Interior, Washington, D.C.

———. 1972. Model hydrographs. U.S. Geological Survey Water Supply Paper 2005, U.S. Dept. of Interior, Washington, D.C.

Mockus, V. 1957. Use of storm and watershed characteristics in synthetic hydrograph analysis and application. Paper presented at the Annual Meeting of the American Geophysical Union, Pacific Southwest Region, held in Sacramento, California, February.

Morel-Seytoux, H. J. 1973. "Two-phase flows in porous media." In *Advances in Hydroscience*, edited by V. T. Chow, vol. 9, 119–202. New York: Academic Press.

———. 1978. Derivation of equations for variable rainfall infiltration. *Water Resources Research* 14(4):561–68.

———. 1979. "Flow forecasting based on preseason conditions." In *Modeling of Rivers*, edited by H. W. Shen, 4-1–4-46. New York: John Wiley.

———. 1982. "Optimization methods in rainfall-runoff modeling." In *Rainfall-Runoff Relationship*, edited by V. P. Singh, 487–506. Littleton, Colo.: Water Resources Publications.

Morgan, P. E., and S. M. Johnson. 1962. Analysis of synthetic unit graph methods. *Proceedings of the American Society of Civil Engineers, Journal of the Hydraulics Division* 88(HY5):199–220.

Moss, M. E., and D. R. Dawdy. 1973. Stochastic simulation for basins with short or no records of streamflow. *Proceedings of the International Symposium on the Design of Water Resources Projects with Inadequate Data*, Madrid, 365–75.

Mosley, M. P. 1981. Delimitation of New Zealand hydrologic regions. *Journal of Hydrology* 49:173–92.

Mostaghimi, S., and J. K. Mitchell. 1982. Peak runoff model comparison on central Illinois. *Water Resources Bulletin* 18(1):9–13.

Muftuoglu, R. F. 1984. New models for nonlinear catchment analysis. *Journal of Hydrology* 73:335–57.

Murray, D. L. 1970. Boughton's daily rainfall-runoff model modified for the Brenig catchment. Proceedings of the International Symposium on Results of Research on Representative and Experimental Basins, Wellington, New Zealand, *International Association of Scientific Hydrology Publication* 96, 112–26.

Mulvany, T. J. 1850. On the use of self-registering rain and flood gauges. *Proceedings of the Institution of Civil Engineers* 4(2):1–8, Dublin, Ireland.

Napiorkowski, J. J., and W. G. Strupczewski. 1979. The analytical determination of the kernels of the Volterra series describing the cascade of nonlinear reservoirs. *Journal of Hydrological Sciences* 6(3-4):121–42.

———. 1981. The properties of the kernels of the Volterra series describing flow deviations from a steady state in an open channel. *Journal of Hydrology* 52:185–98.

NASH, J. E. 1957. The form of the instantaneous unit hydrograph. *International Association of Scientific Hydrology Publication* 45(3):114–21.

———. 1958. Determining runoff from rainfall. *Proceedings of the Institution of Civil Engineers* 10:163–84, Dublin, Ireland.

———. 1959a. A note on the Muskingum method of flood routing. *Journal of Geophysical Research* 64:1953–56.

———. 1959b. Systematic determination of unit hydrograph parameters. *Journal of Geophysical Research* 64(1):111–15.

———. 1960. A unit hydrograph study, with particular reference to British catchments. *Proceedings of the Institution of Civil Engineers* 17:249–82, London.

———. 1966. "Applied Flood Hydrology." Chapter 6 in *River Engineering and Water Conservation Works*, ed. R. B. Thorn, 63–110, Butterworths, London.

———. The role of parametric hydrology. *Journal of the Institution of Water Engineers* 21: 1–23, Dublin, Ireland.

———. 1969. A course of lectures in parametric or analytical hydrology. Great Lakes Institute, University of Toronto, Ontario, Canada.

NASH, J. E., AND J. P. FARELL. 1955. A graphical solution of the flood routing equation for linear storage discharge relation. *Transactions of the American Geophysical Union* 36:319–20.

NASH, J. E., AND J. V. SUTCLIFFE. 1969. Flood-wave formation. *World Meteorological Organization Publication* 238, TP 122, 147–55.

NATALE, L., AND TODINI, E. 1974. A constrained parameter estimation technique for linear models in hydrology. *Proceedings of the Workshop on Mathematical Models in Hydrology*, Pisa, Italy.

———. 1974. A constrained parameter estimation technique for linear models in hydrology. *Publication no.* 13, Institute of Hydraulics, University of Pavia, Italy.

———. 1976a. A stable estimator for linear models: 1. theoretical development and Monte Carlo experiments. *Water Resources Research* 12(4):667–76.

———. 1976b. A stable estimator for linear models: 2. real world hydrologic applications. *Water Resources Research* 12(4):67.

National Research Council. 1982. *Scientific Basis of Water Resource Management*, 3–13. Washington, D.C.: National Academy Press.

National Weather Service. 1972. National Weather Service river forecast system: forecast procedures. *NOAA Technical Memorandum NWS HYDRO*-14, U.S. Department of Commerce, Silver Springs, Md.

NEELY, B. L. 1976. Floods in Louisiana: magnitude and frequency. Report, Department of Highways, State of Louisiana, Baton Rouge.

NELDER, J. A. 1972. "Summary and assessment: a statistician's point of view." In *Mathematical Models in Ecology*, edited by J. N. R. Jeffers, Oxford: 367–74. Blackwell.

NEMEC, J. 1964. *Engineering Hydrology*. London: McGraw-Hill.

NEWMAN, S. P., AND G. DEMARSILY. 1976. Identification of linear systems response by parametric programming. *Water Resources Research* 12(4):253–62.

NEWTON, D. W., AND J. W. VINYARD. 1967. Computer-determined unit hydrograph from floods. *Journal of the Hydraulics Division, Proceedings of the American Society of Civil Engineers* 93(HY5):219–35.

NORRIS, W. H. 1946. Estimation of runoff from impervious surfaces. *Proceedings of the Institution of Municipal and County Engineers* 72:425.

NORTON, J. P., AND B. WATSON. 1979. On-line hourly river flow prediction under widely varying catchment conditions. *Proceedings of the Hydrology and Water Resources Symposium*, Perth, Australia, 239–43.

O'CONNOR, K. M. 1976. A discrete linear cascade model for hydrology. *Journal of Hydrology* 29:213–42.

———. 1977. A discrete linear cascade model for hydrology: a reply. *Journal of Hydrology* 33:407–8.

———. 1982. Derivation of discretely coincident forms of continuous linear time-invariant models using the transfer function approach. *Journal of Hydrology* 59:1–48.

O'CONNOR, K. M., AND J. E. NASH. 1968. Comment on "A Laplace transform proof of the theorem of moments by M. H. Diskin." *Water Resources Research* 4(3):675–77.

O'DONNELL, T. 1960. Instantaneous unit hydrograph derivation by harmonic analysis. *International Association of Scientific Hydrology Publication no.* 51, 546–57.

———. 1966. Methods of computation in hydrograph analysis and synthesis. P. 3 in Recent Trends in Hydrograph Synthesis, Proceedings, Technical Meeting 21, Central Organization for Applied Scientific Research, T.N.O., The Hague, The Netherlands.

———. 1985. A direct three-parameter Muskingum procedure incorporating lateral inflow. *Hydrological Sciences Journal* 30(4):479–96.

O'KELLY, J. J. 1955. The employment of unit hydrograph to determine the flows of Irish arterial drainage channels. *Proceedings of the Institution of Civil Engineers* 4(3):365–412, Dublin, Ireland.

O'MEARA, W. A. 1968. Linear routing of lateral inflow in uniform open channels. M.E.Sc. thesis, The National University of Ireland, University College, Cork.

OSBORNE, M. R., AND D. M. RYAN. 1970. On penalty function methods for nonlinear programming problems. *Journal of Mathematical Analysis and Applications* 31:559–78.

OVERTON, D. E. 1966. Muskingum flood routing of upland streamflow. *Journal of Hydrology* 4:185–200.

———. 1967. Analytical simulation of watershed hydrographs from rainfall. *Proceedings of the International Symposium on Hydrology*, Fort Collins, Colorado, 1:9–17.

———. 1970. Route or convolute? *Water Resources Research* 6(1):43–52.

———. 1971. Estimation of surface water lag time from kinematic wave equations. *Water Resources Bulletin* 7(1):428–40.

OVERTON, D. E., AND M. E. MEADOWS. 1976. *Stormwater Modeling*. New York: Academic Press.

PALMER, D. B., AND H. P. JOHNSON. 1964. Runoff prediction from point rainfall data by application of the digital computer. *Transactions of the American Society of Agricultural Engineers* 7:424–26.

PALMER, J. R. 1969. An improved procedure for orthogonalizing the search vectors in Rosenbrock's and Swann's direct search optimization methods. *Computer Journal* 12:69–71.

PAPADAKIS, C., AND H. C. PREUL. 1972. University of Cincinnati urban runoff model. *Journal of the Hydraulics Division, Proceedings of the American Society of Civil Engineers* 98(HY10):789–1804.

PAPAZAFIRIOUS, Z. G. 1973. Hydrologic analysis of small watersheds using a systems analysis approach. Ph.D. diss. University of California, Davis.

———. 1975. Polynomial approximation of the kernels of closed linear hydrologic systems. *Journal of Hydrology* 27(3/4):319–29.

———. 1976. Linear and nonlinear approaches for short-term runoff estimations in time-invariant open hydrologic systems. *Journal of Hydrology* 30:63–80.

PATRY, G. G., AND MARINO, M. A. 1983a. Nonlinear runoff modeling: parameter identification. *ASCE Journal of Hydraulic Engineering* 109(6):865–80.

______. 1983b. Time-variant nonlinear functional runoff model for real-time forecasting. *Journal of Hydrology* 66:227–44.

PATTERN, R. T. 1975. A watershed volume response model considering contributing area. M.S. thesis, University of Arizona, Tucson.

PAYNTER, H. M. 1952. Methods and results from MIT studies in unsteady flow. *Journal of the Boston Society of Civil Engineers* 39(2):120–65.

PEARCE, A. J., AND A. I. MCKERCHAR. 1979. "Upstream generation of storm runoff." In *Physical Hydrology*: New Zealand Experience, edited by D. L. Murray and P. Ackroyd. New Zealand Hydrological Society, New Zealand.

PEDERSEN, J. T., J. C. PETERS, AND O. J. HELWEG. 1980. Hydrographs by single linear reservoir model. *Journal of the Hydraulics Division, Proceedings of the American Society of Civil Engineers* 106(HY5):837–52.

PEEBLES, R. W. 1975. Flow recession in the ephemeral stream. Ph.D. diss. University of Arizona, Tucson.

PENMAN, H. L. 1948. Natural evaporation from open water, bare soil, and grass. *Proceedings of the Royal Society* (London), ser. A, vol. 193, 120–45.

PETERS, J. C. 1983. Flood routing through a flat, complex floodplain using a one-dimensional unsteady flow computer program. *Water Resources Bulletin* 19(6):913–20.

PETERSON, W. C., AND F. H. VERHOFF. 1982. Muskingum-like approximations for water routing. *Journal of the Hydraulics Division Proceedings of the American Society of Civil Engineers* 108(HY11):1387–93.

PHIEN, H. N., AND T. JIVAJIRAJAH. 1984. Fitting the S_B curve by the method of maximum likelihood. *Journal of Hydrology*, 67:67–75.

PHILLIPPE, J. T., AND J. M. WIGGERT. 1969. Instantaneous unit hydrograph response by harmonic analysis. *Water Resources Research Center Bulletin* 15, Virginia Polytechnic Institute, Blacksburg.

PILGRIM, D. H. 1966a. Radioactive tracing of storm runoff on a small catchment: 1. experimental technique. *Journal of Hydrology* 4:289–305.

______. 1966b. Radioactive tracing of storm runoff on a small catchment: 2. discussion of results. *Journal of Hydrology* 4:306–26.

______. 1976. Travel times and nonlinearity of flood runoff from tracer measurements on a small watershed. *Water Resources Research* 12(3):487–96.

______. 1977. Isochrones of travel time and distribution of flood storage from a tracer study on a small watershed. *Water Resources Research* 13(3):587–95.

______. 1978. Adsorption losses and performance ^{51}Cr-EDTA and ^{198}Au in tracing of flood runoff. *Journal of Hydrology* 36:47–64.

______. 1980. Characteristics of flood nonlinearity from two tracing studies. *Proceedings of the Hydrology and Water Resources Symposium*, Institute of Engineers, Australia, held November 4–6 in Adelaide, Australia, 17–22.

PILGRIM, D. H., AND D. D. HUFF. 1978. A field evaluation of subsurface and surface runoff: 1. Tracer studies. *Journal of Hydrology* 38:299–318.

PILGRIM, D. H., D. D. HUFF, AND T. D. STEELE. 1978. A field evaluation of subsurface and surface runoff: Runoff processes. *Journal of Hydrology* 38:319–41.

PONCE, V. M. 1979. Simplified Muskingum routing equation. *Journal of the Hydraulics Division, Proceedings of the American Society of Civil Engineers* 195(HY1):85–91.

______. 1980. Linear reservoirs and numerical diffusion. *Journal of the Hydraulics Division, Proceedings of the American Society of Civil Engineers* 106(HY5):691–99.

PONCE, V. M., AND THEURER, F. D. 1980. Discussion of "Approximate flood routing methods: a review, by P. E. Weinmann and E. M. Laurenson." *Journal of the Hydraulics Division, Proceedings of the American Society of Civil Engineers* 106(HY11):1945–47.

______. 1982. Accuracy criteria in diffusion routing. *Journal of the Hydraulics Division, Proceedings of the American Society of Civil Engineers* 108(HY6):747–57.

PONCE, V. M., AND V. YEVJEVICH. 1978a. Muskingum-Cunge method with variable parameters. *Journal of the Hydraulics Division, Proceedings of the American Society of Civil Engineers* 104(HY12):1663–67.

______. 1978b. The numerical dispersion of the Muskingum method. Proceedings of 26th Hydraulics Division Specialty Conference, American Society of Civil Engineers, College Park, Md, August.

PORTER, J. W. 1975. A comparison of hydrologic and hydraulic catchment routing procedures. *Journal of Hydrology* 24:333–49.

PRASAD, R. 1966. Nonlinear simulation of a regional hydrologic system. Ph.D. diss. University of Illinois, Urbana.

______. 1967. A nonlinear hydrologic system response model. *Journal of the Hydraulics Division, Proceedings of the American Society of Civil Engineers* 93(HY4):201–21.

PRASAD, S. N., AND M. J. M. ROMKENS. 1983. An approximate integral solution of vertical infiltration under changing boundary conditions. *Water Resources Research* 18(4):1022–28.

QASHU, H. K., AND S. W. BUOL. 1967. Hydraulic and micromorphological properties of stream channel sediments. *Water Resources Research* 3(2):465–69.

QUICK, M. C., AND A. PIPES. 1975. Nonlinear channel routing by computer. *Journal of the Hydraulics Division, Proceedings of the American Society of Civil Engineers* 101(HY6):651–54.

QUIMPO, R. G. 1971a. Structural relation between parametric and stochastic hydrology models. Proceedings of the International Symposium on Mathematical Models in Hydrology held in Warsaw, Poland, 301–12.

______. 1971b. "Kernels of stochastic linear hydrologic systems." In *Systems Approach to Hydrology*, edited by Yevjevich, 163–77. Fort Collins, Colo. Water Resources Publications.

______. 1973. Link between stochastic and parametric hydrology. *Journal of the Hydraulics Division, Proceedings of the American Society of Civil Engineers* 99(HY3):461–70.

RAGAN, R. M. 1968. An experimental investigation of partial-area contributions. *International Association of Hydrology Publication* 76,251–49.

RAGAN, R. M., AND J. O. DURU. 1972. Kinematic wave nomograph for times of concentration. *Journal of the Hydraulics Division, Proceedings of the American Society of Civil Engineers* 98(HY10):1765–71.

RAGHAVENDRAN, R., AND REDDY, P. J. 1975. Synthesis of basin response with inadequate data. *Nordic Hydrology* 6:14–27.

RAMASESHAN, S., U. C. CHAUBE, AND A. K. SHARMA. 1973. A discrete linear model for the hydrologic basin. *Proceedings of the First World Congress on Water Resources*, vol. 4, 367–76.

RAMSER, C. E. 1927. Runoff from small agricultural areas. *Journal of Agricultural Research* 34:797–823.

Ramser, C. E., R. A. Norton, and W. D. Ellison. 1929. The rational method of estimating runoff from small agricultural areas. *Agricultural Engineering* 10:351–53.

Rao, A. R., and J. W. Delleur. 1971. The instantaneous unit hydrograph: its calculation by the transform method and noise control by digital filtering. *Technical Report* no. 20, Water Resources Research Center, Purdue University, West Lafayette, Ind.

______. 1974. Instantaneous unit hydrographs, peak discharges, and time lags in urban areas. *Hydrological Sciences Bulletin* 19(2):185–98.

Rao, A. R., J. W. Delleur, and P. B. S. Sarma. 1972. Conceptual hydrologic models for unbanizing basins. *Journal of the Hydraulics Division, Proceedings of the American Society of Civil Engineers* 92(HY7):1205–20.

Rao, A. R., and R. G. S. Rao. 1974. Comparative analysis of estimation methods in nonlinear functional models of the rainfall-runoff process. *Technical Report no.* 56, Water Resources Research Center, Purdue University, West Lafayette, Ind.

Rao, D. V. 1980. Log Pearson type 3 distribution: method of mixed moments. *Journal of the Hydraulics Division, Proceedings of the American Society of Civil Engineers* 106(HY6):999–1019.

______. 1983. Estimating log Pearson parameters by mixed moments. *ASCE Journal of Hydraulic Engineering* 109(8):1118–1032.

Rao, G. N. S., J. R. Assenzo, and J. F. Harp. 1966. Estimating peak rates of runoff and design hydrographs from rainfall in the state of Oklahoma. *Journal of Hydrology* 4:141–70.

Reed, D. W., P. Johnson, and J. M. Firth. 1975. A nonlinear rainfall-runoff model, providing for variable lag time. *Journal of Hydrology* 25:295–305.

Reich, B. M. 1965. Estimating flood peaks from small South African catchments. *Journal of Hydrology* 3:231–53.

______. 1968. Rapid flood-peak determination of small watersheds. *Transactions of the ASAE* 11:448–51.

Reich, B. M., and D. A. Wolf. 1973. The triangle as a tentative unit hydrograph. *Proceedings of First World Congress of Water Resources*, Chicago, Ill., vol. 4, 337–49.

Reid, J. 1927. The estimation of storm water discharge. *Proceedings of the Institution of Municipal and County Engineers*, vol. 53, 997.

Reid, J. K. 1967. A note on the least squares solution of a band system of linear equations by Householder reductions. *Computer Journal* 10:188–89.

Renard, K. G. 1970. The hydrology of semiarid rangeland watershed. U.S. Department of Agriculture Publication ARS 41–162.

Renard, K. G., and R. V. Keppel. 1966. Hydrographs of ephemeral streams in the southwest. *Journal of the Hydraulics Division, Proceedings of the American Society of Civil Engineers* 92(HY2):33–52.

Rendon-Herrero, O. 1978. Unit sediment graph. *Water Resources Research* 14(5):889–901.

Rice, C. E., and C. L. Larson. 1972. Methods for routing hydrographs through open channels. University of Minnesota Water Resources Research Center, Bulletin 51, Minneapolis.

______. 1974. Storage routing methods without fitted coefficients. *Transactions of the American Society of Agricultural Engineers* 17:152–56.

Rockwood, D. M. 1958. Columbia basin streamflow routing by computer. *Journal of the Waterways and Harbors Division, Proceedings of the American Society of Civil Engineers* 94(WW5):1874-1–1874-15.

Rodriguez-Iturbe, I. 1982. "The coupling of climate and geomorphology in rainfall-runoff analysis." In *Rainfall-Runoff Relationship*, edited by V. P. Singh, 431–48. Littleton, Colo.: Water Resources Publications.

Rodriguez-Iturbe, I., G. Devoto, and J. B. Valdes. 1979. Discharge response analysis and hydrologic similarity: the interrelation between the geomorphologic IUH and the storm characteristics. *Water Resources Research* 15(6):1435–44.

Rodriguez-Iturbe, I., M. Gonzalez-Sanabria, and R. L. Bras. 1982. A geomorphoclimatic theory of the instantaneous unit hydrograph. *Water Resources Research* 18(4):877–86.

Rodriguez-Iturbe, I., M. Gonzalez-Sanabria, and G. Caamano. 1982. On the climatic dependence of the IUH: a rainfall-runoff analysis of the Nash model and the geomorphoclimatic theory. *Water Resources Research* 18(4):887–903.

Rodriguez-Iturbe, I., and J. B. Valdes. 1979. The geomorphologic structure of hydrologic response. *Water Resources Research* 15(6):1409–20.

Roe, H. B., and C. E. Snyder. 1943. A logical modification of the rational runoff formula for runoff from small agricultural areas. *Agricultural Engineering* 24:423–27.

Rogers, W. F. 1971. "Hydrograph analysis and some related geomorphic variables." In *Quantitative Geomorphology: Some Aspects and Applications*, edited by M. Morisawa, 245–57. Publications in Geomorphology, State University of New York, Binghamton.

______. 1972. New concept in hydrograph analysis. *Water Resources Research* 8:973–81.

______. 1980. A practical model for linear and nonlinear runoff. *Journal of Hydrology* 46:51–78.

______. 1982. Some characteristics and implications of drainage basin linearity and nonlinearity. *Journal of Hydrology* 55:247–65.

Rogers, W. F., and H. A. Zia. 1982. Linear and nonlinear runoff from large drainage basins. *Journal of Hydrology* 55:267–78.

Rosenblueth, A., and N. Wiener. 1945. The role of models in science. *Philosophy of Science* 12(4):316–21.

Rosenbrock, H. H. 1960. An automatic method of finding the greatest or least value of a function. *Computer Journal* 3:175–84.

Ross, C. N. 1921. The calculation of flood discharges by a time contour plan. *Transactions of the Institution of Engineers* (Australia), 2:85.

Ross, G. J. S. 1970. The efficient use of function minimization in nonlinear maximum likelihood estimation. *Applied Statistics* 19:205–21.

______. 1972. "Stochastic model fitting by evolutionary operation." In *Mathematical Models in Ecology*, edited by J. N. R. Jeffers, Oxford: Blackwell. 29–308.

Rossmiller, R. L. 1980. The rational formula revised. *Proceedings of the International Symposium on Urban Storm Runoff* held July 28–31, 1980, at University of Kentucky, Lexington, 1–9.

Roy, R. J., and J. Sherman. 1967. A learning technique for Volterra series representation. *Institute of Electronics and Electrical Engineers Transactions, Automatic Control* 12:761–64.

Rutter, E. J., Q. B. Greaves, and F. F. Snyder. 1939. Flood routing. *Transactions of the American Society of Civil Engineers* 104:276–373.

Saeks, R. 1972. *Generalized Networks*. New York: Holt, Rinehart & Winston.

Salas, J. D., J. W. Delleur, V. Yevjevich, and W. L. Lane. 1980. *Applied Modeling of Hydrologic Time Series*. Littleton, Colo. Water Resources Publications.

Sangal, B. P. 1983. Practical method of estimating peak flow. *ASCE Journal of Hydraulic Engineering* 109(4):549–63.

Sarma, P. B. S., J. W. Delleur, and A. R. Rao. 1969. A program in urban hydrology. P. 2, an evaluation of rainfall-runoff models for small watersheds and the effects of urbanization on runoff. *Technical Report no.* 9, Water Resources Research Institute, Purdue University, West Lafayette, Ind.

Sauer, V. B. 1971. The unit hydrograph: a technique for routing reservoir releases. U.S. Geological Survey Professional Paper 750-B, B259–B264.

______. 1973. Unit-response method of open-channel flow routing. *Journal of the Hydraulics Division, Proceedings of the American Society of Civil Engineers* 99(HY1):1979–93.

Schaake, J. C., J. C. Geyer, and J. W. Knapp. 1967. Experimental examination of the rational method. *Journal of the Hydraulics Division, Proceedings of the American Society of Civil Engineers* 93(HY6):353–70.

Schetzen, M. 1965a. Synthesis of a class of nonlinear systems. *International Journal of Control* 1(5):401–14.

______. 1965b. Determination of optimum nonlinear systems for generalized error criteria based on the use of gate functions. *IEEE Transactions on Information Theory* IT-11(1):117–25.

______. 1965c. Measurement of the kernels of a nonlinear system of finite order. *International Journal of Control* 1(3):251–63.

Schulz, E. F. 1969. Dimensionless unit hydrographs from tropical watersheds. Paper presented at the National Fall Meeting of the American Geophysical Union, December 15–16, San Francisco, Calif.

______. 1973. *Problems in Applied Hydrology*. Fort Collins, CO: Water Resources Publications.

Schulz, E. F., and O. G. Lopez. 1974. Determination of urban response time. *Contract Report* DACW05-73-C-0029, Hydrologic Engineering Center, U.S. Army Corps of Engineers, Davis, Calif.

______. 1974. Determination of urban watershed response time. *Hydrology Paper no.* 71, Colorado State University, Fort Collins.

Schulz, E. F., S. Pinkayan, and C. Komsartra, C. 1971. Comparison of dimensionless unit hydrographs in Thailand and Taiwan. *Nordic Hydrology* 2:23–46.

Schwab, G. O., R. K. Frevert, T. W. Edminster, and K. K. Barnes. 1971. *Soil and Water Conservation Engineering*. New York: John Wiley.

Schwarz, R. J., and B. Friedland. 1965. *Linear Systems*. New York: McGraw-Hill.

Seelye, E. E. 1968. *Design-Data Book for Civil Engineers*, vol. 1, 18–101. New York: John Wiley.

Sehitoglu, H. 1983. Some algorithms for parameter estimation in water resources systems. *Water Resources Research* 19(5): 1193–1202.

Selvalingam, S. 1979. "ARMA and linear tank models." In *Modeling Hydrologic Processes*, edited by H. J. Morel-Seytoux et al., 297–325. Fort Collins, Colo.: Water Resources Publications.

Serban, P. 1976. A mathematical model of the "reservoir" type designed for flood-wave modeling and forecasting. *Hydrological Sciences Bulletin* 21(1/3):139–47.

Shannon, C. E. 1948. The mathematical theory of communication, I and II. *Bell System Technical Journal* 27:379–423.

Shelburne, K. L., and V. P. Singh. 1976. Studies on rainfall-runoff modeling: 4. Estimation of parameters of two mathematical models of surface runoff. *WRRI Report no.* 076, New Mexico Water Resources Research Institute, New Mexico State University, Las Cruces.

Shen, J. W. 1965. Use of analog models in the analysis of flood runoff. Geological Survey Professional Paper 506-A, U.S. Department of Interior, Washington, D.C.

Sherman, L. K. 1932. Stream flow from rainfall by the unit graph method. *Engineering News-Record* 108:501–5.

Shrader, M. L., W. J. Rawls, W. M. Snyder, and R. H. McCuen. 1981. Flood peak regionalization using mixed-mode estimation of the parameters of the lognormal distribution. *Journal of Hydrology* 52:229–37.

Singh, K. P. 1962. A nonlinear approach to the instantaneous unit hydrograph theory. Ph.D. diss. University of Illinois, Urbana.

______. 1964. Nonlinear instantaneous unit hydrograph theory. *Journal of the Hydraulics Division, Proceedings of the American Society of Civil Engineers* 90(HY2):313–47.

______. 1976. Unit hydrographs: a comparative study. *Water Resources Bulletin* 12(2):381–392.

Singh, K. P., and J. B. Stall. 1971. Derivation of base flow recession curves and parameters. *Water Resources Research* 7(2):292–303.

Singh, V. P. 1975. Derivation of surface water lag time for converging overland flow. *Water Resources Bulletin* 11(3):505–13.

______. 1976a. Comparison of two mathematical models of surface runoff. *International Association of Hydrological Sciences Bulletin* 21(2):285–99.

______. 1976b. Derivation of time of concentration. *Journal of Hydrology* 30:147–65.

______. 1976c. Mathematical aspects of surface runoff. Proceedings of the ASCE Symposium on Inland Waterways for Navigation, Flood Control, and Water Diversions held at Colorado State University, Fort Collins, vol. 1, 773–92.

______. 1976d. Studies on rainfall-runoff modeling: 5. A uniformly nonlinear hydrologic cascade model. *WRRI Report no.* 078, New Mexico State University, New Mexico Water Resources Research Institute, Las Cruces.

______. 1977. Estimation of parameters of a uniformly nonlinear surface runoff model. *Nordic Hydrology* 8:33–45.

______. 1979a. A systematic evaluation of urban runoff models. *Engineering and Industrial Research Station Report MSSU-EIRS-CE-79-3*, Mississippi State University, Mississippi State, Mississippi.

______. 1979b. A uniformly nonlinear hydrologic cascade model. *Irrigation and Power* 36(3):301–17.

______. 1979c. "A uniformly nonlinear model for surface runoff prediction." In *Modeling Hydrologic Processes*, ed. H. J. Morel-Seytoux et al., 385–404. Fort Collins, Colo. Water Resources Publications.

______. 1983a. Approximate integral solutions for flood routing by the Muskingum method. Proceedings of the 20th Congress of the International Association for Hydraulic Research, vol. 6, 480–86, Moscow, USSR.

______. 1983b. A geomorphic approach to hydrograph synthesis, with potential for application to ungaged watersheds. Technical Report, Louisiana Water Resources Research Institute, Louisiana State University, Baton Rouge.

Singh, V. P., and N. Agiralioglu. 1980. A mathematical investigation of diverging flow: 1. Analytical solutions. *Technical Report* MSSU-EIRS-CE-80-3, *Engineering and Industrial Research Station*, Mississippi State University, Mississippi State, Mississippi.

______. 1982. Lag time for diverging overland flow. *Nordic Hydrology* 13:39–48.

Singh, V. P., and H. Aminian. 1984. A rainfall-runoff model for ungaged basins. *Proceedings of the IV Congress of APD-IAHR on Water Resources Development*, vol. 2, 1095–1109, Chiang Mai, Thailand.

______. 1985. The watershed hydrology simulation (WAHS) model. *Proceedings of the International Workshop on Operational Applications of Mathematical Models in Developing Countries*, vol. 1, 243–69, New Delhi, India.

______. 1986. An empirical relation between volume and peak or direct runoff. *Water Resources Bulletin*. In press.

Singh, V. P., and A. Baniukiewicz. 1980. A study of some empirical methods of determining the unit hydrograph. *Interim Report* no. 1, Water Resources Research Institute, Mississippi State University, Mississippi State.

Singh, V. P., A. Baniukiewicz, and V. J. Chen. 1982. "An instantaneous unit sediment graph study for small upland watersheds." In *Modeling Components of Hydrologic Cycle*, edited by V. P. Singh, 539–54. Littleton, Colo.: Water Resources Publications.

Singh, V. P., A. Baniukiewicz, and R. S. Ram. 1982. "Some empirical methods of determining the unit hydrograph." In *Rainfall-Runoff Relationship*, edited by V. P. Singh, 67–90. Littleton, Colo. Water Resources Publications.

Singh, V. P., and S. Buapeng. 1977. Effect of rainfall excess determination on runoff computation. *Water Resources Bulletin* 13(3):499–514.

______. 1979. A nonlinear hydrologic cascade. *Proceedings of the Third World Congress on Water Resources*, Mexico City, Mexico, vol. 6, 2633–47.

Singh, V. P., and V. J. Chen. 1982. "On the relation between sediment yield and runoff volume." In *Modeling Components of Hydrologic Cycle*, edited by V. P. Singh, 555–70. Littleton, Colo. Water Resources Publications.

Singh, V. P., and P. K. Chowdhury. 1980. Determining parameters of Muskingum method of flood routing. *Journal of the Institution of Engineers* (India), Civil Engineering Division, vol. 60, pt. CL.6, 319–25.

______. 1985. On fitting gamma distribution to synthetic runoff hydrographs. *Nordic Hydrology* 16:177–92.

Singh, V. P., and W. T. Dickinson. 1975. An analytical method to determine daily soil moisture. *Proceedings of the Second World Congress on Water Resources*, Delhi, India, vol. 4, 355–65.

Singh, V. P., and R. C. McCann. 1979. A mathematical study of the general hydrologic system model. *Technical Report* MSSU-EIRS-CE-80-1, Engineering and Industrial Research Station, Mississippi State University, Mississippi State.

______. 1979. Quick estimation of parameters of Muskingum method of flood routing. *Proceedings of the 14th Annual Mississippi Water Resources Conference*, September 24–25, Jackson, Miss.

______. 1980a. Mathematical modeling of catchment response. *Proceedings of the International Conference on Water Resources Development*, Taipei, Taiwan, vol. 2, 587–97.

______. 1980b. Some notes on Muskingum method of flood routing. *Journal of Hydrology* 48:343–61.

______. 1980c. A study of the Muskingum method of flood routing. *Technical Report* MSSU-EIRS-CE-80-2, Engineering and Industrial Research Station, Mississippi State University, Mississippi State.

______. 1980d. Use of cumulants to estimate coefficients in Chow-Kulandaiswamy's GHS model. *Nordic Hydrology* 11:83–92.

______. 1983. Meixner functions for derivation of the unit hydrograph. *Advances in Water Resources* 6:157–64.

Singh, V. P., S. N. Prasad, and L. Ubertini. 1983. A continuum mechanics approach to streamflow modeling. *Technical Completion Report*, Louisiana Water Resources Research Institute, Louisiana State University, Baton Rouge.

Singh, V. P., A. K. Rajagopal, and K. Singh. 1986. Derivation of some frequency distributions using the principle of maximum entropy (POME). *Advances in Water Resources* 9:91–106.

Singh, V. P., and R. S. Ram. 1983. A kinematic model for surface irrigation: verification by experimental data. *Water Resources Research*. In press.

Singh, V. P., and R. Regl. 1983. Analytical solutions of kinematic equations for erosion on a plane: 1. Rainfall of infinite duration. *Advances in Water Resources* 6:2–10.

Singh, V. P., and P. D. Scarlatos. 1987. Analysis of the nonlinear Muskingum flood routing. *Journal of Hydraulic Engineering*, ASCE. 113(1):61–79.

Singh, V. P., and B. Sherman. 1983. A kinematic model of surface irrigation: mathematical solutions. *Technical Report* WRR4, Water Resources Program, Department of Civil Engineering, Louisiana State University, Baton Rouge.

Singh, V. P., and D. A. Woolhiser. 1976. Sensitivity of linear and nonlinear surface runoff models to input errors. *Journal of Hydrology* 29:243–49.

Smart, J. S. 1972. Channel networks. *Advances in Hydroscience*, ed. V. T. Chow, vol. 8, 305–46.

______. 1979. Comment on "a storage routing model relating drainage basin hydrology and geomorphology by M. J. Boyd." *Water Resources Research* 15(4):985–86.

Smith, A. A. 1980. A generalized approach to kinematic flood routing. *Journal of Hydrology* 45:71–89.

Smith, R. E. 1972. Border irrigation advance and ephemeral flood waves. *Journal of the Irrigation and Drainage Division, Proceedings of the American Society of Civil Engineers* 98(IR2):289–307.

Snyder, F. F. 1938. Synthetic unit graphs. *Transactions of the American Geophysical Union* 19:447–54.

Snyder, M. A. 1966. *Chebyshev Methods in Numerical Approximation*. Englewood Cliffs, N.J.: Prentice-Hall, Inc.

Snyder, W. M. 1955. Hydrograph analysis by the method of least squares. *Journal of the Hydraulics Division, Proceedings of the American Society of Civil Engineers*, vol. 81, Paper No. 793, 25 pp.

______. 1961a. Continuous parabolic interpolation. *Journal of the Hydraulics Division, Proceedings of the American Society of Civil Engineers* 87(HY4):99–111.

______. 1961b. Matrix operations in hydrograph computations. *Research Paper* no. 1, Tennessee Valley Authority, Knoxville, Tenn.

______. 1971. The parametric approach to watershed modeling. *Nordic Hydrology* 11:167–85.

______. 1972. Fitting of distribution functions by nonlinear least squares. *Water Resources Research* 8(6):1423–32.

______. 1976. Interpolation and smoothing of experimental data with sliding polynomials. Report ARS-S-83, U.S. Department of Agriculture, Agricultural Research Service, Washington, D.C.

______. 1980. Statistical Hydrologic Modeling Methods and Case Studies. Unpublished manuscript, Athens, Ga.

Snyder, W. M., and L. E. Asmussen. 1973. Subsurface hydrograph analysis by convolution. *Journal of the Irrigation and Drainage Division, Proceedings of the American Society of Civil Engineers* 98(IR3):405–18.

SNYDER, W. M., W. C. MILLS, AND J. C. STEPHENS. 1970. A method of derivation of nonconstant watershed response function. *Water Resources Research* 6(1)261–74.

SNYDER, W. M., AND J. B. STALL. 1965. Men, models, methods, and machines in hydrologic analysis. *Journal of the Hydraulics Division*, Proceedings of the American Society of Civil Engineers 91(HY2):85–99.

Soil Conservation Service. 1964. Hydrology. *SCS National Engineering Handbook*, sec. 4, chaps. 17 and 21, U.S. Department of Agriculture, soil Conservation Service, Washington, D.C.

______. 1969. Computer program for project formulation: hydrology. Technical Release no. 2, U.S. Department of Agriculture, Washington, D.C.

______. 1971. Hydrology. SCS National Engineering Handbook, sec. 4. U.S. Department of Agriculture, Washington, D.C.

______. 1972. Hydrology. SCS National Engineering Handbook, sec. 4. U.S. Department of Agriculture, Washington, D.C.

______. 1972. Travel time, time of concentration, and lag. National Engineering Handbook, chap. 15, 15-1–15-16, U.S. Department of Agriculture, Washington, D.C.

______. 1975. Urban hydrology for small watersheds. Technical Release no. 55, U.S. Department of Agriculture, Washington, D.C.

SOLODOVNIKOV, V. V. 1969. *Statistical Dynamics of Automatic Systems*. New York: Dover.

SONUGA, J. O. 1972. Principle of maximum entropy in hydrologic frequency analysis. *Journal of Hydrology* 17:177–91.

SOROOSHIAN, S. 1980. "Comparison of two direct search algorithms used in calibration of rainfall-runoff models." In *Water Related Land Resource Systems*, edited by Y. Y. Haimes and J. Kindler, 477–85, New York: Pergamon.

SOROOSHIAN, S., AND V. K. GUPTA. 1983. Automatic calibration of conceptual rainfall-runoff models: the question of parameter observability and uniqueness. *Water Resources Research* 19(1):260–68.

SOROOSHIAN, S., V. K. GUPTA, AND J. L. FULTON. 1983. Evaluation of maximum likelihood parameter estimation techniques for conceptual rainfall-runoff models: influence of calibration data variability and length on model credibility. *Water Resources Research* 19(1):251–59.

SPOLIA, S. K., AND S. CHANDER. 1974. Modeling of surface runoff systems by an ARMA model. *Journal of Hydrology* 22:317–32.

STEINBERG, I. H. 1947. A flood routing device. *Transactions of the American Geophysical Union* 28(2):247–54.

STEPHENS, J. C., AND W. C. MILLS. 1965. Using the Cypress Creek formula to estimate runoff rates in the southern coastal plain and adjacent flatland land resource areas. ARS-41-95, U.S. Department of Agriculture, Washington, D.C.

STEPHENSON, D. 1979. Direct optimization of Muskingum routing coefficients. *Journal of Hydrology* 41:161–65.

STEPIEN, I. 1984. On the numerical solution of the Saint Venant equations. *Journal of Hydrology* 67:1–11.

STIEFEL, E. L. 1955. Kernel polynomials in linear algebra and their numerical applications. U.S. National Bureau of Standards, Washington, D.C.

STRAHLER, A. N. 1964. "Quantitative geomorphology of drainage basins and channel networks." Section 4-2 in *Handbook of Applied Hydrology*, edited by V. T. Chow. New York: McGraw-Hill.

STRELKOFF, T. 1980. Modified-Puls routing in Chuquatonchee Creek. Report, The Hydrologic Engineering Center, U.S. Army Corps of Engineers, Davis California.

STRUPCZEWSKI, W. G., R. J. KICZKO, Z. W. KUNDZEWICZ, J. J. NAPIORKOWSKI, AND H. T. MITOSEK. 1975. Stochastic properties of the processes transformed in linear hydrological systems. Proceedings of the Symposium on the Application of Mathematical Models in Hydrology and Water Resources Systems, *IAHS Publication* 115, 231–37.

STRUPCZEWSKI, W., AND Z. KUNDZEWICZ. 1980a. Muskingum method revisited. *Journal of Hydrology* 48:327–42.

______. 1980b. Translatory characteristics of the Muskingum method of flood routing: a comment. *Journal of Hydrology* 48:363–68.

STUBCHNER, J. M. 1981. Discussion of "Hydrographs by single linear reservoir model by Pederson et al., 1980." *Journal of the Hydraulics Division, Proceedings of the American Society of Civil Engineers* 107(HY2):254–55.

TAO, P, C., AND J. W. DELLEUR. 1976. Seasonal and nonseasonal ARMA models in hydrology. *Journal of the Hydraulics Division, Proceedings of the American Society of Civil Engineers* 102,(HY10):1541–59.

TAY, A. J. H., AND V. P. SINGH. 1978. A coupled runoff-chloride model. *Proceedings of the International Symposium on Stormwater Management*, July, 1978, University of Kentucky, Lexington.

TAYLOR, A. B., AND H. E. SCHWARZ. 1952. United hydrograph lag and peak flow related to basin characteristics. *Transactions of the American Geophysical Union* 33:235–46.

TE, W. G., AND W. KAY. 1983. The unit-step response functions for several hydrological conceptual models. *Journal of Hydrology* 62:119–28.

Tennessee Valley Authority. 1972. Storm hydrographs using a double triangle model. *Research Paper* no. 9, Division of Water Control Planning, Hydrologic Research and Analysis Staff, Hydraulic Date Branch, Tennessee Valley Authority, Knoxville, Tenn.

Texas Water Development Board. 1971. Stochastic optimization and simulation techniques for management of regional water resource systems. Report 131, Austin, Tex.

THOMAS, D. M., AND M. A. BENSON. 1970. Generalization of streamflow characteristics from drainage-basin characteristics. *Geological Survey Water Supply Paper 1975*, U.S. Department of the Interior, Washington, D.C.

THORNTHWAITE, C. W. 1931. The climates of North America according to a new classification. *Geographical Review* 21:633–55.

TJOMSLAND, T., E. RUUD, AND K. NORDSETH. 1978. The physiographic influence on recession runoff in small Norwegian rivers. *Nordic Hydrology* 9:17–30.

TODINI, E., AND J. R. WALLIS. 1977. "Using CLS for daily or longer period of rainfall-runoff modeling." In *Mathematical Models for Surface Water Hydrology*, edited by T. A. Ciriani, U. Miaone, and J. R. Wallis, 149–68. New York: John Wiley.

TOEBES, C., W. B. MORRISSEY, R. SHORTER, AND M. HENDY. 1969. "Baseflow-recession curves." In *Handbook of Hydrological Procedures*: Procedure No. 8, Ministry of Works, Wellington, New Zealand.

TOEBES, C., AND D. D. STRANG. 1964. On recession curves. 1. Recession equations. *Journal of Hydrology* (New Zealand) 3(2):2–15.

TRIBUS, M. 1969. *Rational Descriptors, Decisions, and Designs*. New York: Pergamon.

TROENDLE, C. A. 1979. A variable source area model for streamflow prediction on first order forested watersheds. Ph.D. diss. University of Georgia, Athens.

TRUXAL, J. G. 1955. *Automatic Feedback Control System Synthesis*. New York: McGraw-Hill.

Tung, Y. K. 1985. River flood routing by nonlinear Muskingum method. *Journal of Hydraulic Engineering*, ASCE 111(12):1447–1460.

UNESCO. 1977. *Research on Urban Hydrology*, vol. 1 and 2, edited by M. B. McPherson, United National Educational, Scientific, and Cultural Organization, Paris, France.

Unver, O., and L. W. Mays. 1984. Optimal determination of loss rate functions and unit hydrographs. *Water Resources Research* 20(2):203–14.

U.S. Army Corps of Engineers. 1940. Engineering construction—flood control. Engineering School, Fort Belvoir, Va.

______. 1949. Unit hydrograph computations, pt. 2. Office, District Engineer, Washington, D.C.

U.S. Bureau of Reclamation. 1965. Unit hydrograph procedures. Supplement to Chaps. 6 and 7 in vol. 4, Reclamation Manual, Denver, Colorado.

Valdes, J. B., Y. Fiallo, and I. Rodriguez-Iturbe. 1979. A rainfall- runoff analysis of the geomorphologic IUH. *Water Resources Research* 15(6):1421–34.

van de Nes, Th. J., and M. H. Hendriks. 1971. Analysis of a linear distributed model of surface runoff. Rapport 1, Laboratory of Hydraulics and Catchment Hydrology, Agricultural University, Wageningen, The Netherlands.

Vemuri, V., J. A. Dracup, R. C. Erdmann, and N. Vemuri. 1969. Sensitivity analysis method of system identification and its potential in hydrologic research. *Water Resources Research* 5(2):341–49.

Vemuri, V., and N. Vemuri. 1970. On the systems approach in hydrology. *International Association of Scientific Hydrology Bulletin* 15(2):17–38.

von Neumann, J., and J. J. Goldstine. 1947. Numerical inverting of matrices of high order. *Bulletin of the American Mathematical Society* 53:1021–99.

Venetis, C. 1969. The IUH of the Muskingum channel reach. *Journal of Hydrology* 4:185–200.

Viessman, W., J. W. Knapp, G. L. Lewis, and T. E. Harbaugh. 1977. Introduction to Hydrology. 2d ed. New York: Harper and Row.

Volterra, V. 1913. Leçons sur les Équations Intégrales et les Equations Integro-Différentielles. Paris: Gauthier-Villars.

Volterra, V. 1930. *Theory of Functionals and Integral and Integro-Differential Equations*. London: Blackie & Sons.

Walker, G. 1931. On periodicity in series of related terms. Proceedings of the Royal Society, London 131(ser. A.):518–32.

Wallace, D. E., and K. G. Renard. 1967. Contribution to regional water table from transmission losses of ephemeral streambeds. *Transactions of the American Society of Agricultural Engineers* 10(6):786–89.

Wang, C. T., V. K. Gupta, and E. Waymire. 1981. A geomorphologic synthesis of nonlinearity in surface runoff. *Water Resources Research* 17(3):545–54.

Wang, R. Y., and I. P. Wu. 1972. Characteristics of short-duration unit hydrograph. Transactions of the American Society of Agricultural Engineers 15:452–56.

Wanielista, M. P. 1978. *Stormwater Management: Quantity and Quality*, 163–86, Ann Arbor, Mich.: Ann Arbor Science.

Watt, W. E. 1971. A relation between peak discharge and maximum twenty-four hour flow for rainfall records. *Journal of Hydrology* 14:285–92.

Wei, T. C., and C. L. Larson. 1971. Effects of areal and time distribution of rainfall on small watershed runoff hydrographs. WRRC Bulletin no. 30, Water Resources Research Center, University of Minnesota, Minneapolis.

Weinmann, P. E. 1977. Comparison of flood routing methods for natural rivers. Report no. 2/1977, Department of Civil Engineering, Monash University, Clayton, Australia.

Weinmann, P. E., and E. M. Laurenson. 1979. Approximate flood routing methods: a review. *Journal of the Hydraulics Division, Proceedings of the American Society of Civil Engineers* 105(HY12):1521–36.

Weisman, R. N. 1978. Some comments on source-area catchment modeling. American Society of Civil Engineers, Hydraulics Division Specialty Conference, University of Maryland, College Park.

Werner, P. W., and K. J. Sundquist. 1951. On the groundwater recession curve for large watersheds. *International Association of Hydrological Sciences Publication* no. 33, 202–12.

Wheater, H. S., T. L. Shaw, and J. C. Rutherford. 1982. Storm runoff from small lowland catchments in southwest England. *Journal of Hydrology* 55:321–37.

Whipkey, R. Z. 1965. Subsurface stormflow from forested slopes. *International Association of Scientific Hydrology Bulletin* 10(2):74–85.

Whitehead, P. G. 1979. Applications of recursive estimation techniques to time variable hydrological systems. *Journal of Hydrology* 40:1–16.

Whittle, P. 1962. Topographic correlation, power-law covariance function, and diffusion. *Biometrika* 49:305–14.

______. 1964. On stationary processes in the plane. *Biometrika* 41:434–49.

Wiener, N. 1942. Response of a nonlinear device to noise. *Radiation Laboratory Report* no. 129, Massachusetts Institute of Technology, Cambridge.

______. 1949. *The Interpolation, Extrapolation, and Smoothing of Stationary Time Series*. New York: John Wiley.

______. 1958. *Nonlinear Problems in Random Theory*. New York: Wiley, Technology Press.

Wilkinson, J. H. 1965. *The Algebraic Eigenvalue Problem*. Oxford: Oxford Univ. Press.

Williams, G. B. 1922. Flood discharges and the dimensions of spillways in India. *Engineering* (London) 134:321.

______. 1952. Flood intensities and frequencies on British catchments. *Engineer* 194:247.

Williams, J. R. 1969. Flood routing with variable travel time for variable storage coefficients. *Transactions of the American Society of Agricultural Engineers* 12(1):100–3.

______. 1975. HYMO flood routing. *Journal of Hydrology* 26:17–27.

______. 1978. A sediment graph model based on an instantaneous unit sediment graph. *Water Resources Research* 14(4):659–64.

Williams, J. R., and R. W. Hann. 1972. HYMO, a problem-oriented computer language for building hydrologic models. *Water Resources Research* 8:79–86.

Wilson, W. A. 1972. Unit hydrograph response times. M.S. thesis, Department of Civil Engineering, Colorado State University, Fort Collins.

Wilson, W. T. 1941. A graphical flood-routing method. *Transactions of the American Geophysical Union* 22:893–97.

Wischmeir, W. H., and D. D. Smith. 1960. A universal soil loss equation to guide conservation farm planning. The 7th International Congress of Soil Science Transactions, vol 1, 418–25.

WISLER, C. O., AND E. F. BRATER. 1959. *Hydrology. New York: John Wiley.*

WOLD, H. 1954. *A Study in the Analysis of Stationary Time Series.* Upsala, Sweden: Almquist and Wiksell.

WOODING, R. A. 1965. A hydraulic modeling of the catchment-stream problem. 1. Kinematic wave theory. *Journal of Hydrology* 3:254–67.

WOODWARD, S. M. 1920. Hydraulics of Miami flood control project. Pt. 7, routing floods through retarding basins. Miami Conservancy District, Ohio, Technical Reports, pt. 7, chap. 7, 156–200.

WOOLHISER, D. A. 1971. Deterministic approach to watershed modeling. *Nordic Hydrology* 11:146–66.

______. 1973. Hydrologic and watershed modeling: state of the art. *Transactions of the American Society of Agricultural Engineers* 16(3):553–59.

______. 1975. The watershed approach to understanding our environment. *Journal of Environmental Quality* 4(1);17–21.

______. 1982a. ''Hydrologic modeling of small watersheds.'' Chapter 1 in *Hydrologic Modeling of Small Watersheds*, edited by C. T. Haan, H. P. Johnson, and D. L. Brakensiek. St. Joseph, Mich.: American Society of Agricultural Engineers.

______. 1982b. ''Physically based models of watershed runoff.'' In *Rainfall-Runoff Relationship*, edited by V. P. Singh. Littleton, Colo. Water Resources Publications.

WOOLHISER, D. A., AND J. A. LIGGETT. 1967. Unsteady, one dimensional flow over a plane: the rising hydrograph. *Water Resources Research* 3(3):753–71.

WOOLHISER, D. A., AND E. F. SCHULZ. 1973. Large material models in watershed hydrology research. International Symposium on River Mechanics, January 9–12, Asian Institute of Technology, Bangkok, Thailand, 63–74.

WORMLEATON, P. R. 1980. Discussion of ''Optimization of unit hydrograph determination by L. W. Mays and L. Coles.'' *Journal of the Hydraulics Division, Proceedings of the American Society of Civil Engineers* 106(HY12):2076–78.

WU, I. P. 1963. Design hydrographs for small watersheds in Indiana. *Journal of the Hydraulics Division, Proceedings of the American Society of Civil Engineers* 89(HY6):35–66.

______. 1969a. Flood hydrology of small watersheds: evaluation of time parameters and determination of peak discharge. *Transactions of the American Society of Agricultural Engineers* 12:655–60 and 663.

______. 1969b. Hydrograph study and peak discharge determination of Hawaiian small watersheds: Island of Oahu. *Technical Report* no. 30, Water Resources Research Center, University of Hawaii, Honolulu.

WU, J. S., E. L. KING, AND M. WANG. 1985. Optimal identification of Muskingum routing coefficients. *Water Resources Bulletin* 21(3):417–21.

WYCOFF, R. L., AND U. P. SINGH. 1976. Preliminary hydrologic design of small flood detention reservoirs. *Water Resources Bulletin* 12(2):337–49.

YAIR, A., AND H. LAVEE. 1974. Areal contribution to runoff on scree slopes in an extreme arid environment—a simulated rainstorm experiment. *Zeitschrift fur Geomorphologie*, 21:106–21.

______. 1976. Runoff generative process and runoff yield from arid talus mantled slopes. *Earth Surface Processes* 1:235–47.

YANO, K., AND Y. ISHIHARA. 1968. Experimental facility for research of river disaster. *Journal of Disaster Prevention Research Institute* no. 128, 237–46.

YEH, W. W. G., AND L. BECKER. 1973. Linear programming and channel flow identification. *Journal of the Hydraulics Division, Proceedings of the American Society of Civil Engineers* 99(HY11):2013–21.

YEVJEVICH, V. M. 1963. Fluctuations of wet and dry years. Pt. 1, research data assembly and mathematical models. *Hydrology Paper no.* 1, Colorado State University, Fort Collins.

______. 1964. Bibliography and discussion of flood routing methods and unsteady flow in channels. U.S. Geological Survey Water Supply Paper 1690.

______. 1972a. *Probability and Statistics in Hydrology*. Fort Collins, Colo: Water Resources Publications.

______. 1972b. *Stochastic Processes in Hydrology*. Fort Collins, Colo: Water Resources Publications.

YEVJEVICH, V., AND A. H. BARNES. 1970. Flood routing through storm drains. Pt. 1, solution of problems of unsteady free surface flow in storm drains. *Hydrology Paper no.* 43, Colorado State University, Fort Collins.

YOUNG, P., AND P. G. WHITEHEAD. 1977. A recursive approach to time-series analysis for multivariable systems. *International Journal of Control* 25(3):457–82.

YULE, G. U. 1921. On the time correlation problem with special reference to the variate-difference correlation method. *Journal of the Royal Statistical Society* 84:497–537.

______. 1927. On a method of investigating periodicities in disturbed series, with special reference to Wolfer's sunspot numbers. *Philosophical Transactions*, 226(ser. A):267–98.

ZADEH, L. A., AND C. A. DESOER. 1963. *Linear System Theory*. New York: McGraw-Hill.

ZAND, S. M., AND J. A. HARDER. 1973. Application of nonlinear system identification to the lower Mekong River, Southeast Asia. *Water Resources Research* 9(2):290–97.

ZOCH, R. T. 1934. On the relation between rainfall and stream flow, pt. 1. *Monthly Weather Review* 62(9):315–22.

______. 1936. On the relation between rainfall and stream flow, pt. 2. *Monthly Weather Review* 65(4):105–21.

______. 1937. On the relation between rainfall and stream flow, pt. 2. *Monthly Weather Review* 65(4):135–47.

Author Index

Subject Index